Clays in Natural and Engineered Barriers for Radioactive Waste Confinement

Geological Society books refereeing procedures

The Society makes every effort to ensure that the scientific and production quality of its books matches that of its journals. Since 1997, all book proposals have been refereed by specialist reviewers as well as by the Society's Books Editorial Committee. If the referees identify weaknesses in the proposal, these must be addressed before the proposal is accepted.

Once the book is accepted, the Society Book Editors ensure that the volume editors follow strict guidelines on refereeing and quality control. We insist that individual papers can only be accepted after satisfactory review by two independent referees. The questions on the review forms are similar to those for *Journal of the Geological Society*. The referees' forms and comments must be available to the Society's Book Editors on request.

Although many of the books result from meetings, the editors are expected to commission papers that were not presented at the meeting to ensure that the book provides a balanced coverage of the subject. Being accepted for presentation at the meeting does not guarantee inclusion in the book.

More information about submitting a proposal and producing a book for the Society can be found on its website: www.geolsoc.org.uk.

It is recommended that reference to all or part of this book should be made in one of the following ways:

NORRIS, S., BRUNO, J., CATHELINEAU, M., DELAGE, P., FAIRHURST, C., GAUCHER, E. C., HÖHN, E. H., KALINICHEV, A., LALIEUX, P. & SELLIN, P. (eds) 2014. *Clays in Natural and Engineered Barriers for Radioactive Waste Confinement*. Geological Society, London, Special Publications, **400**.

WEETJENS, E., MAES, N. & VAN RAVESTYN, L. 2014. Model validation based on *in situ* radionuclide migration tests in Boom Clay: status of a large-scale migration experiment, 24 years after injection. *In:* NORRIS, S., BRUNO, J., CATHELINEAU, M., DELAGE, P., FAIRHURST, C., GAUCHER, E. C., HÖHN, E. H., KALINICHEV, A., LALIEUX, P. & SELLIN, P. (eds) *Clays in Natural and Engineered Barriers for Radioactive Waste Confinement*. Geological Society, London, Special Publications, **400**, 613–623. First published online May 15, 2014, http://dx.doi.org/10.1144/SP400.39

GEOLOGICAL SOCIETY SPECIAL PUBLICATION NO. 400

Clays in Natural and Engineered Barriers for Radioactive Waste Confinement

EDITED BY

S. NORRIS
Radioactive Waste Management Limited, UK

J. BRUNO
Amphos 21, Spain

M. CATHELINEAU
UMR Georessources, Université de Lorraine, CNRS, France

P. DELAGE
Ecole des ponts ParisTech (Navier/CERMES), France

C. FAIRHURST
Itasca Consulting Group, Inc. & University of Minnesota, USA

E. C. GAUCHER
Total, CSTJF, France

E. H. HÖHN
Eawag, Swiss Federal Institute of Aquatic Science and Technology, Switzerland

A. KALINICHEV
Ecole des Mines de Nantes, France

P. LALIEUX
ONDRAF/NIRAS, Belgium

and

P. SELLIN
SKB, Sweden

2014
Published by
The Geological Society
London

THE GEOLOGICAL SOCIETY

Published by The Geological Society from:
The Geological Society Publishing House, Unit 7, Brassmill Enterprise Centre, Brassmill Lane, Bath BA1 3JN, UK

The Lyell Collection: www.lyellcollection.org
Online bookshop: www.geolsoc.org.uk/bookshop
Orders: Tel. +44 (0)1225 445046, Fax +44 (0)1225 442836

British Library Cataloguing in Publication Data

A catalogue record for this book is available from the British Library.
ISBN 978-1-86239-654-8
ISSN 0305-8719

Distributors
For details of international agents and distributors see:
www.geolsoc.org.uk/agentsdistributors

Typeset by Techset Composition India (P) Ltd., Bangalore and Chennai, India.
Printed by Berforts Information Press Ltd, Oxford, UK

Contents

Geomechanics

Mass transfer/Gas transfer

Mass transfer mechanisms

Clays in Natural and Engineered Barriers for Radioactive Waste Confinement: an introduction

SIMON NORRIS

Radioactive Waste Management Limited, Building 587, Curie Avenue, Harwell Campus, Didcot, Oxon, OX11 0RH, UK

Corresponding Editor, Scientific Committee, Montpellier 2012

(e-mail: simon.norris@nda.gov.uk)

The concept of engineered geological disposal has been developed for the safe long-term management of long-lived radioactive waste. This involves emplacement of radioactive waste in deep geological repositories that contain and isolate the waste and, consequently, protect humans and the environment.

The 'multiple barrier concept' is the cornerstone of all proposed schemes for the geological disposal of radioactive waste. Based on the principle that uncertainties in performance can be minimized by conservatism in design, the concept invokes a series of complementary barriers, both engineered and natural (geological), between the waste and the surface environment. Each successive barrier represents an additional impediment to the movement of radionuclides.

Regarding engineered barriers, for low- and intermediate-level waste the waste may be incorporated in a relatively stable and inert matrix such as cement, bitumen, lead-alloy or polymer resin (the choice varying depending on the waste management organization); glass may be used in the case of certain high-level reprocessing wastes. Owing to the very low leach-rate of glass in groundwater, vitrification is widely accepted to be one of the best methods of immobilizing the aqueous products from the reprocessing of spent fuel. Many waste containers will provide some form of physical barrier to groundwater. However, because of the relatively small volumes of waste involved, spent fuel, vitrified waste and other highly active wastes will be totally encapsulated in corrosion-resistant metal canisters that are designed to prevent groundwater entry for extended time periods in excess of 100 000 years.

Depending on the disposal concept, engineered barriers may comprise the buffer/backfill medium enclosing the waste containers, the tunnel/borehole liner, and the backfill and high-integrity seals placed in the repository access ways or emplacement boreholes. The buffer/backfill medium enclosing the waste will often also provide both a physical and a chemical barrier to radionuclide migration. The functions of the engineered/chemical barriers are:

- to reduce the rate of corrosion of the waste containers and thus extend their life;
- to limit the rate of hydraulic transport;
- to limit the release of radionuclides from the waste-form to the far-field (geosphere) after container failure;
- to limit the migration of radionuclides along the pathway provided by the access tunnels and shafts of a repository or the boreholes in the case of a deep borehole emplacement.

For low- and intermediate-level waste disposal in vaults, the backfill may be a porous, cementitious grout, which is intended to pH-buffer the pore water for an extended time period. Typically the buffer/backfill for high-level waste/spent fuel might comprise compacted bentonite or other clay-based material, providing a low-permeability, alkaline pH-buffered pore water to limit solubility and mobility of certain radionuclides (e.g. actinides), plus good retention/retardation properties including high sorption and a capacity to filter colloids.

The geological barrier is the final impediment to radionuclide migration. Depending on details of the local geology, this may be considered to constitute the host formation itself, extending above, below and laterally away from the repository. Alternatively, the entire sequence of low-permeability rocks, which may separate the repository from the surface and/or more permeable, water-bearing strata, may be included. The practical realization of the multiple barrier concept is the primary objective of all stages of a disposal programme, from site appraisal and characterization through to design and construction. The general performance of the repository as a whole (waste, buffer, engineering disturbed zone and host rock) is being intensively studied in many national programmes.

In that context of geological repositories, argillaceous media (ranging from soft clays to indurated shales) are being considered in numerous countries as potential host rocks for the geological disposal

From: Norris, S., Bruno, J., Cathelineau, M., Delage, P., Fairhurst, C., Gaucher, E. C., Höhn, E. H., Kalinichev, A., Lalieux, P. & Sellin, P. (eds) 2014. *Clays in Natural and Engineered Barriers for Radioactive Waste Confinement*. Geological Society, London, Special Publications, **400**, 1–5.
First published online July 16, 2014, http://dx.doi.org/10.1144/SP400.43

of radioactive waste. Compacted bentonite or other clay-based material may additionally form part of the engineered barrier. Argillaceous media – present as the geological barrier and/or as part of the engineered barrier – have a number of favourable generic properties, for example, homogeneity, low groundwater flow, chemical buffering, propensity for plastic deformation and self-sealing of fractures by swelling, and marked capacity to chemically and physically retard the migration of radionuclides.

Organizations responsible for implementing geological repositories are undertaking significant programmes of research – nationally and internationally – to advance understanding of argillaceous media and their contribution to ensuring safe long-term management of long-lived radioactive waste. Multidisciplinary approaches including geology, mineralogy, geochemistry, rheology, physics and chemistry of clay minerals and assemblages are required in order to provide a detailed characterization of the geological host formations considered for the disposal of radioactive waste and to assess the behaviour of engineered and natural barriers when submitted to various types of perturbations induced by such facilities. The evaluation of the performances of the natural barrier as well as of the impact of repository-induced disturbances upon the confinement properties of clay-rich geological formations constitutes a major objective for the experimental programmes being conducted and/or to be conducted in underground research laboratories, for interpreting the subsequent scientific results, for modelling the long-term behaviour of radioactive waste repositories and for carrying out safety assessment exercises.

This Geological Society Special Publication contains 43 papers of scientific studies presented at the 5th conference on *Clays in Natural and Engineered Barriers for Radioactive Waste Confinement* held in Montpellier, France, in 2012. Papers presented in this Special Publication cover a range of outputs from the Montpellier meeting, and provide insight for the reader into the range of clay-related work currently being undertaken internationally in relation to natural and engineered barriers for radioactive waste confinement.

Since 2002, ANDRA, the French National Radioactive Waste Management Agency, has developed this conference into the most important event for all kinds of scientists from all over the world dealing with the disposal of radioactive waste. These conferences, each attracting a large number of papers and attendees from around the world, have contributed significantly to the outstanding scientific level of research relevant to argillaceous media in the context of the disposal of radioactive waste.

The 2012 meeting covered all the aspects of clay characterization and behaviour considered at various temporal and spatial scales relevant to the confinement of radionuclides in clay, from the description of basic phenomenological processes to the global understanding of the performance and safety at repository and geological scales. Special emphasis was put on the modelling of processes occurring at the mineralogical level within the clay barriers.

The 43 papers published in this Special Publication have been classified according to different areas within the field of disposal of radioactive waste research. The assignment of a study to one of the topic area was not always easy because many studies provide information on, and for, different aspects of disposal research. The papers in this Special Publication consider research into argillaceous media under the following topic areas:

- Keynote: **Delay *et al.* (2014)**.
- Large scale geological characterization: **Beerten *et al.* (2014)**; **Delcourt-Honorez & Scholz (2014)**.
- Clay based concept/Large scale experiments: **Dixon *et al.* (2014)**; **Gaus *et al.* (2014)**; **Lanyon *et al.* (2014)**; **Van Marcke *et al.* (2014)**; **Mokni *et al.* (2014)**; **Yamada *et al.* (2014)**.
- Hydrodynamical modelling: **Bénet *et al.* (2014*a*)**; **Vandersteen *et al.* (2014)**.
- Geochemistry: **Meleshyn (2014)**; **Dymitrowska *et al.* 2014**; **Vinsot *et al.* (2014*a*); Takayama *et al.* (2014)**; **Wersin & Birgersson (2014)**.
- Geomechanics: **Malmberg & Kristensson (2014)**; **Yildizdag *et al.* (2014)**; **de La Vaissière *et al.* (2014*a*)**; **Hausmannova & Vasicek (2014)**; **Graham *et al.* (2014)**; **Kobayashi *et al.* (2014)**; **Zhang (2014)**; **Ababou *et al.* (2014)**; **Xu *et al.* (2014)**; **Deleruyelle *et al.* (2014)**; **Czaikowski *et al.* (2014)**; **Priyanto *et al.* (2014)**; **Bénet *et al.* (2014*b*)**.
- Mass transfer/gas transfer: **de La Vaissière *et al.* (2014*b*)**; **Brommundt *et al.* (2014)**; **Enssle *et al.* (2014)**; **Bénet *et al.* (2014*c*)**; **Bennett *et al.* (2014)**; **Cuss *et al.* (2014)**; **Namiki *et al.* (2014)**; **Senger *et al.* (2014)**; **Tawara *et al.* (2014)**; **Vinsot *et al.* (2014*b*).**
- Mass transfer mechanisms: **Savoye *et al.* (2014)**; **Harrington *et al.* (2014)**; **Gondolli & Večerník (2014)**; **Weetjens *et al.* (2014)**.

The collection of different topics presented in this Special Publication demonstrates the diversity of geological repository research. Of course, the understanding of all topics presented here and investigated elsewhere (such as corrosion or detachment of colloidal particles) still requires more work. At the same time, more studies focusing on the complex connection of different processes are expected to be required (e.g. change in pH, change in adsorption properties, change in micro-structural arrangement, change in swelling pressure and change in hydraulic conductivity). This will be a challenging and engaging future task for scientists working on the implementation of geological repositories for the safe long-term management of long-lived radioactive waste.

The 6th conference on *Clays in Natural and Engineered Barriers for Radioactive Waste Confinement* will be held in Brussels, Belgium, in March 2015, hosted by the Belgian National Radioactive Waste Management Agency, ONDRAF-NIRAS.

References

Ababou, R., Cañamón, I. & Poutrel, A. 2014. Equivalent upscaled hydro-mechanical properties of a damaged and fractured claystone around a gallery (Meuse/Haute-Marne Underground Research Laboratory). *In*: Norris, S., Bruno, J. *et al.* (eds) *Clays in Natural and Engineered Barriers for Radioactive Waste Confinement*. Geological Society, London, Special Publications, **400**. First published online July 16, 2014, http://dx.doi.org/10.1144/SP400.44

Beerten, K., De Craen, M. & Leterme, B. 2014. Long-term evolution of the surface environment of the Campine area, northeastern Belgium: first assessment. *In*: Norris, S., Bruno, J. *et al.* (eds) *Clays in Natural and Engineered Barriers for Radioactive Waste Confinement*. Geological Society, London, Special Publications, **400**. First published online April 30, 2014, http://dx.doi.org/10.1144/SP400.23

Bénet, L.-V., Bouillet, C. & Wendling, J. 2014*a*. Analysis of the ambient condition in an IL-LLW storage cell in a deep clay repository during the waiting closure period. *In*: Norris, S., Bruno, J. *et al.* (eds) *Clays in Natural and Engineered Barriers for Radioactive Waste Confinement*. Geological Society, London, Special Publications, **400**. First published online April 9, 2014, http://dx.doi.org/10.1144/SP400.18

Bénet, L.-V., Tulita, C., Calsyn, L. & Wendling, J. 2014*b*. Evolution of temperature and humidity in an underground repository over the operation period. *In*: Norris, S., Bruno, J. *et al.* (eds) *Clays in Natural and Engineered Barriers for Radioactive Waste Confinement*. Geological Society, London, Special Publications, **400**. First published online June 16, 2014, http://dx.doi.org/10.1144/SP400.41

Bénet, L.-V., Tulita, C., Pasteau, A. & Wendling, J. 2014*c*. Analysis of the long-term hydraulic-gas transient in the central zone of a deep clay repository. *In*: Norris, S., Bruno, J. *et al.* (eds) *Clays in Natural and Engineered Barriers for Radioactive Waste Confinement*. Geological Society, London, Special Publications, **400**. First published online April 9, 2014, http://dx.doi.org/10.1144/SP400.19

Bennett, D. P., Cuss, R. J., Vardon, P. J., Harrington, J. F. & Thomas, H. R. 2014. Phenomena exposure from the large scale gas injection test (Lasgit) dataset using a bespoke data analysis toolkit. *In*: Norris, S., Bruno, J. *et al.* (eds) *Clays in Natural and Engineered Barriers for Radioactive Waste Confinement*. Geological Society, London, Special Publications, **400**. First published online March 5, 2014, http://dx.doi.org/10.1144/SP400.5

Brommundt, J., Kaempfer, Th. U., Enssle, C. P., Mayer, G. & Wendling, J. 2014. Full-scale 3D modelling of a nuclear waste repository in the Callovo-Oxfordian clay. Part 1: thermo-hydraulic two-phase transport of water and hydrogen. *In*: Norris, S., Bruno, J. *et al.* (eds) *Clays in Natural and Engineered Barriers for Radioactive Waste Confinement*. Geological Society, London, Special Publications, **400**. First published online May 7, 2014, http://dx.doi.org/10.1144/SP400.34

Cuss, R., Harrington, J., Giot, R. & Auvray, C. 2014. Experimental observations of mechanical dilation at the onset of gas flow in Callovo-Oxfordian claystone. *In*: Norris, S., Bruno, J. *et al.* (eds) *Clays in Natural and Engineered Barriers for Radioactive Waste Confinement*. Geological Society, London, Special Publications, **400**. First published online May 8, 2014, http://dx.doi.org/10.1144/SP400.26

Czaikowski, O., Miehe, R. & Rothfuchs, T. 2014. Self-sealing barriers of sand/clay-mixtures – lessons learnt from *in-situ* experiment and retrospective modelling. *In*: Norris, S., Bruno, J. *et al.* (eds) *Clays in Natural and Engineered Barriers for Radioactive Waste Confinement*. Geological Society, London, Special Publications, **400**. First published online April 7, 2014, http://dx.doi.org/10.1144/SP400.16

de La Vaissière, R., Morel, J. *et al.* 2014*a*. Excavation-induced fractures network surrounding tunnel: properties and evolution under loading. *In*: Norris, S., Bruno, J. *et al.* (eds) *Clays in Natural and Engineered Barriers for Radioactive Waste Confinement*. Geological Society, London, Special Publications, **400**. First published online May 6, 2014, http://dx.doi.org/10.1144/SP400.30

de La Vaissière, R., Gerard, P. *et al.* 2014*b*. Gas injection test in the Callovo-Oxfordian claystone: data analysis and numerical modelling. *In*: Norris, S., Bruno, J. *et al.* (eds) *Clays in Natural and Engineered Barriers for Radioactive Waste Confinement*. Geological Society, London, Special Publications, **400**. First published online March 7, 2014, http://dx.doi.org/10.1144/SP400.10

Delay, J., Bossart, P. *et al.* 2014. Three decades of underground research laboratories: what have we learned? *In*: Norris, S., Bruno, J. *et al.* (eds) *Clays in Natural and Engineered Barriers for Radioactive Waste Confinement*. Geological Society, London, Special Publications, **400**. First published online March 5, 2014, http://dx.doi.org/10.1144/SP400.1

Delcourt-Honorez, M. & Scholz, E. 2014. Earth tidal and barometric responses observed in the Callovo-Oxfordian clay Formation at Andra Meuse/Haute-Marne Underground Research Laboratory. *In*: Norris, S., Bruno, J. *et al.* (eds) *Clays in Natural and Engineered Barriers for Radioactive Waste Confinement*. Geological Society, London, Special Publications, **400**. First published online April 7, 2014, http://dx.doi.org/10.1144/SP400.17

DELERUYELLE, F., BUI, T. A., WONG, H. & DUFOUR, N. 2014. Analytical modelling of a deep tunnel in a visco-plastic rock mass accounting for a simplified life cycle and extension to a particular case of porous media. *In*: NORRIS, S., BRUNO, J. ET AL. (eds) *Clays in Natural and Engineered Barriers for Radioactive Waste Confinement*. Geological Society, London, Special Publications, **400**. First published online April 2, 2014, http://dx.doi.org/10.1144/SP400.14

DIXON, D. A., PRIYANTO, D. G. ET AL. 2014. Enhanced Sealing Project (ESP): evolution of a full-sized bentonite and concrete shaft seal. *In*: NORRIS, S., BRUNO, J. ET AL. (eds) *Clays in Natural and Engineered Barriers for Radioactive Waste Confinement*. Geological Society, London, Special Publications, **400**. First published online May 1, 2014, http://dx.doi.org/10.1144/SP400.33

DYMITROWSKA, M., PAZDNIAKOU, A. & ADLER, P. M. 2014. Two-phase-flow pore-size simulations in Opalinus clay by the Lattice Boltzmann Method. *In*: NORRIS, S., BRUNO, J. ET AL. (eds) *Clays in Natural and Engineered Barriers for Radioactive Waste Confinement*. Geological Society, London, Special Publications, **400**. First published online April 7, 2014, http://dx.doi.org/10.1144/SP400.20

ENSSLE, C. P., BROMMUNDT, J., KAEMPFER, TH. U., MAYER, G. & WENDLING, J. 2014. Full-scale 3D modelling of a nuclear waste repository in the Callovo-Oxfordian clay. Part 2: thermo-hydraulic two-phase transport of water, hydrogen, ^{14}C and ^{129}I. *In*: NORRIS, S., BRUNO, J. ET AL. (eds) *Clays in Natural and Engineered Barriers for Radioactive Waste Confinement*. Geological Society, London, Special Publications, **400**. First published online May 7, 2014, http://dx.doi.org/10.1144/SP400.35

GAUS, I., WIECZOREK, K. ET AL. 2014. EBS behaviour immediately after repository closure in a clay host rock: HE-E experiment (Mont Terri URL). *In*: NORRIS, S., BRUNO, J. ET AL. (eds) *Clays in Natural and Engineered Barriers for Radioactive Waste Confinement*. Geological Society, London, Special Publications, **400**. First published online March 7, 2014, http://dx.doi.org/10.1144/SP400.11

GONDOLLI, J. & VEČERNÍK, P. 2014. The uncertainties associated with the application of through-diffusion, the steady-state method: a case study of strontium diffusion. *In*: NORRIS, S., BRUNO, J. ET AL. (eds) *Clays in Natural and Engineered Barriers for Radioactive Waste Confinement*. Geological Society, London, Special Publications, **400**. First published online March 5, 2014, http://dx.doi.org/10.1144/SP400.3

GRAHAM, C. C., HARRINGTON, J. F., CUSS, R. J. & SELLIN, P. 2014. Pore-pressure cycling experiments on Mx80 Bentonite. *In*: NORRIS, S., BRUNO, J. ET AL. (eds) *Clays in Natural and Engineered Barriers for Radioactive Waste Confinement*. Geological Society, London, Special Publications, **400**. First published online May 7, 2014, http://dx.doi.org/10.1144/SP400.32

HARRINGTON, J. F., VOLCKAERT, G. & NOY, D. J. 2014. Long-term impact of temperature on the hydraulic permeability of bentonite. *In*: NORRIS, S., BRUNO, J. ET AL. (eds) *Clays in Natural and Engineered Barriers for Radioactive Waste Confinement*. Geological Society, London, Special Publications, **400**. First published online May 12, 2014, http://dx.doi.org/10.1144/SP400.31

HAUSMANNOVA, L. & VASICEK, R. 2014. Measuring hydraulic conductivity and swelling pressure under high hydraulic gradients. *In*: NORRIS, S., BRUNO, J. ET AL. (eds) *Clays in Natural and Engineered Barriers for Radioactive Waste Confinement*. Geological Society, London, Special Publications, **400**. First published online May 15, 2014, http://dx.doi.org/10.1144/SP400.36

KOBAYASHI, I., SUZUKI, K., ASANO, H., SELLIN, P., SVEMAR, C. & HOLMQVIST, M. 2014. Mechanical interpretations of the homogeneous nature of bentonite due to swelling. *In*: NORRIS, S., BRUNO, J. ET AL. (eds) *Clays in Natural and Engineered Barriers for Radioactive Waste Confinement*. Geological Society, London, Special Publications, **400**. First published online April 2, 2014, http://dx.doi.org/10.1144/SP400.21

LANYON, G. W., MARSCHALL, P., TRICK, T., DE LA VAISSIÈRE, R., SHAO, H. & LEUNG, H. 2014. Self-sealing experiments and gas injection tests in a backfilled microtunnel of the Mont Terri URL. *In*: NORRIS, S., BRUNO, J. ET AL. (eds) *Clays in Natural and Engineered Barriers for Radioactive Waste Confinement*. Geological Society, London, Special Publications, **400**. First published online March 5, 2014, http://dx.doi.org/10.1144/SP400.8

MALMBERG, D. & KRISTENSSON, O. 2014. Thermo-hydraulic modelling of the bentonite buffer in deposition hole 6 of the Prototype Repository. *In*: NORRIS, S., BRUNO, J. ET AL. (eds) *Clays in Natural and Engineered Barriers for Radioactive Waste Confinement*. Geological Society, London, Special Publications, **400**. First published online May 1, 2014, http://dx.doi.org/10.1144/SP400.25

MELESHYN, A. 2014. Microbial processes relevant for the long-term performance of high-level radioactive waste repositories in clays. *In*: NORRIS, S., BRUNO, J. ET AL. (eds) *Clays in Natural and Engineered Barriers for Radioactive Waste Confinement*. Geological Society, London, Special Publications, **400**. First published online March 5, 2014, http://dx.doi.org/10.1144/SP400.6

MOKNI, N., OLIVELLA, S. ET AL. 2014. Hydro-chemical modelling of *in situ* behaviour of bituminized radioactive waste in Boom Clay. *In*: NORRIS, S., BRUNO, J. ET AL. (eds) *Clays in Natural and Engineered Barriers for Radioactive Waste Confinement*. Geological Society, London, Special Publications, **400**. First published online June 16, 2014, http://dx.doi.org/10.1144/SP400.40

NAMIKI, K., ASANO, H., TAKAHASHI, S., SHIMURA, T. & HIROTA, K. 2014. Laboratory gas injection tests of compacted bentonite buffer material for TRU waste disposal. *In*: NORRIS, S., BRUNO, J. ET AL. (eds) *Clays in Natural and Engineered Barriers for Radioactive Waste Confinement*. Geological Society, London, Special Publications, **400**. First published online May 12, 2014, http://dx.doi.org/10.1144/SP400.27

PRIYANTO, D. G., DIXON, D. A., KIM, C.-S., KORKEAKOSKI, P. & VILLAGRAN, J. E. 2014. Preliminary modelling of the saturation of a full-sized clay and concrete shaft seal. *In*: Norris, S., Bruno, J. et al. (eds) *Clays in Natural and Engineered Barriers for Radioactive Waste Confinement*. Geological Society, London, Special Publications, **400**. First published online May 15, 2014, http://dx.doi.org/10.1144/SP400.38

Savoye, S., Imbert, C., Fayette, A. & Coelho, D. 2014. Experimental study on diffusion of tritiated water and anions under variable water-saturation and clay mineral content: comparison with the Callovo-Oxfordian claystones. *In*: Norris, S., Bruno, J. et al. (eds) *Clays in Natural and Engineered Barriers for Radioactive Waste Confinement*. Geological Society, London, Special Publications, **400**. First published online March 5, 2014, http://dx.doi.org/10.1144/SP400.9

Senger, R., Romero, E., Ferrari, A. & Marschall, P. 2014. Characterization of gas flow through low-permeability claystone: laboratory experiments and two-phase flow analyses. *In*: Norris, S., Bruno, J. et al. (eds) *Clays in Natural and Engineered Barriers for Radioactive Waste Confinement*. Geological Society, London, Special Publications, **400**. First published online April 9, 2014, http://dx.doi.org/10.1144/SP400.15

Takayama, Y., Tsurumi, S. et al. 2014. Effect of montmorillonite content on mechanical and hydraulic properties of bentonite and its numerical modelling. *In*: Norris, S., Bruno, J. et al. (eds) *Clays in Natural and Engineered Barriers for Radioactive Waste Confinement*. Geological Society, London, Special Publications, **400**. First published online April 7, 2014, http://dx.doi.org/10.1144/SP400.13

Tawara, Y., Hazart, A. et al. 2014. Extended two-phase flow model with mechanical capability to simulate gas migration in bentonite. *In*: Norris, S., Bruno, J. et al. (eds) *Clays in Natural and Engineered Barriers for Radioactive Waste Confinement*. Geological Society, London, Special Publications, **400**. First published online March 11, 2014, http://dx.doi.org/10.1144/SP400.7

Vandersteen, K., Gedeon, M., Marivoet, J. & Wouters, L. 2014. Regional groundwater flow modelling of the confined aquifers below the Boom Clay in NE Belgium. *In*: Norris, S., Bruno, J. et al. (eds) *Clays in Natural and Engineered Barriers for Radioactive Waste Confinement*. Geological Society, London, Special Publications, **400**. First published online April 30, 2014, http://dx.doi.org/10.1144/SP400.22

Van Marcke, P., Li, X. L., Chen, G. J., Verstricht, J., Bastiaens, W. & Sillen, X. 2014. Installation of the PRACLAY Seal and Heater. *In*: Norris, S., Bruno, J. et al. (eds) *Clays in Natural and Engineered Barriers for Radioactive Waste Confinement*. Geological Society, London, Special Publications, **400**. First published online March 5, 2014, http://dx.doi.org/10.1144/SP400.4

Vinsot, A., Leveau, F., Bouchet, A. & Arnould, A. 2014*a*. Oxidation front and oxygen transfer in the fractured zone surrounding the Meuse/Haute-Marne URL drifts in the Callovian–Oxfordian argillaceous rock. *In*: Norris, S., Bruno, J. et al. (eds) *Clays in Natural and Engineered Barriers for Radioactive Waste Confinement*. Geological Society, London, Special Publications, **400**. First published online June 2, 2014, http://dx.doi.org/10.1144/SP400.37

Vinsot, A., Appelo, C. A. J. et al. 2014*b*. *In situ* diffusion test of hydrogen gas in the Opalinus Clay. *In*: Norris, S., Bruno, J. et al. (eds) *Clays in Natural and Engineered Barriers for Radioactive Waste Confinement*. Geological Society, London, Special Publications, **400**. First published online April 2, 2014, http://dx.doi.org/10.1144/SP400.12

Weetjens, E., Maes, N. & van Ravestyn, L. 2014. Model validation based on *in situ* radionuclide migration tests in Boom Clay: status of a large-scale migration experiment, 24 years after injection. *In*: Norris, S., Bruno, J. et al. (eds) *Clays in Natural and Engineered Barriers for Radioactive Waste Confinement*. Geological Society, London, Special Publications, **400**. First published online May 15, 2014, http://dx.doi.org/10.1144/SP400.39

Wersin, P. & Birgersson, M. 2014. Reactive transport modelling of iron-bentonite interaction within the KBS-3H disposal concept: the Olkiluoto site as case study. *In*: Norris, S., Bruno, J. et al. (eds) *Clays in Natural and Engineered Barriers for Radioactive Waste Confinement*. Geological Society, London, Special Publications, **400**. First published online May 1, 2014, http://dx.doi.org/10.1144/SP400.24

Xu, W. J., Shao, H., Hesser, J. & Kolditz, O. 2014. Numerical modelling of moisture controlled laboratory swelling/shrinkage experiments on argillaceous rocks. *In*: Norris, S., Bruno, J. et al. (eds) *Clays in Natural and Engineered Barriers for Radioactive Waste Confinement*. Geological Society, London, Special Publications, **400**. First published online May 8, 2014, http://dx.doi.org/10.1144/SP400.29

Yamada, A., Akiyama, Y., Nakajima, M., Yada, T., Chijimatsu, M. & Nakajima, T. 2014. Studies of construction methods for Bentonite engineered barrier systems for sub-surface disposal: vibratory compaction. *In*: Norris, S., Bruno, J. et al. (eds) *Clays in Natural and Engineered Barriers for Radioactive Waste Confinement*. Geological Society, London, Special Publications, **400**. First published online March 5, 2014, http://dx.doi.org/10.1144/SP400.2

Yildizdag, K., Shao, H., Hesser, J., Noiret, A. & Soennke, J. 2014. Coupled hydromechanical modelling of the mine-by experiment at Meuse-Haute-Marne underground rock laboratory France. *In*: Norris, S., Bruno, J. et al. (eds) *Clays in Natural and Engineered Barriers for Radioactive Waste Confinement*. Geological Society, London, Special Publications, **400**. First published online June 16, 2014, http://dx.doi.org/10.1144/SP400.42

Zhang, C.-L. 2014. Characterization of excavated claystone and claystone-bentonite mixtures as backfill/seal material. *In*: Norris, S., Bruno, J. et al. (eds) *Clays in Natural and Engineered Barriers for Radioactive Waste Confinement*. Geological Society, London, Special Publications, **400**. First published online April 30, 2014, http://dx.doi.org/10.1144/SP400.28

Three decades of underground research laboratories: what have we learned?

JACQUES DELAY[1]*, PAUL BOSSART[2], LI XIANG LING[3], INGO BLECHSCHMIDT[4], MATS OHLSSON[5], AGNÈS VINSOT[1], CHRISTOPHE NUSSBAUM[6] & NORBERT MAES[7]

[1]*Andra – Centre de Meuse–Haute-Marne, Route Départementale 960, F-55290 Bure, France*

[2]*Federal Office of Topography Swisstopo, Seftigenstrasse 264, CH-3084 Wabern, Switzerland*

[3]*ESV EURIDICE, Boeretang 200, B-2400, Mol, Belgium*

[4]*Nagra, National Cooperative for the Disposal of Radioactive Waste, Hardstrasse 73, CH-430 Wettingen, Switzerland*

[5]*SKB, Swedish Nuclear Fuel and Waste Management Co., Box 929, SE-572 29 Oskarshamn, Sweden*

[6]*Federal Office of Topography Swisstopo, Fabrique de Chaux, CH-2882 St-Ursanne, Switzerland*

[7]*SCK.CEN, Belgian Nuclear Research Centre, Boeretang 200, B-2400 Mol, Belgium*

**Corresponding author (e-mail: jacques.delay@andra.fr)*

Abstract: This paper describes how four scientific and safety relevant issues have been addressed in special-purpose research laboratories focusing on the geological disposal of high level and long-lived radioactive waste. These are: (a) the effects of heat on the engineered barriers and the geological environment; (b) the geochemical characterization of pore-water in argillaceous rocks; (c) the diffusion and retention of radionuclides; and (d) the full-size sealing of a waste emplacement. They are illustrated by experiments conducted in five underground research laboratories (URLs), three of which are in clay formations (Mol in Belgium, Centre de Meuse–Haute-Marne in France, and Mont Terri Rock Laboratory in Switzerland) and two in granite (Aspö Hard Rock Laboratory in Sweden and Grimsel Test Site in Switzerland).

This paper highlights how the various types of experiments are related and how their results have been applied to foster progress. The most complex experiments have revealed artefacts and technical or methodological difficulties associated with interactions among multiple phenomena, the occurrence or intensity of which cannot be analysed by simple models. In turn, these difficulties have prompted experiments targeted at elementary phenomena, thereby encouraging the development of new investigation protocols and monitoring tools.

More than 30 years of investigations in special-purpose URLs show the benefits of *in-situ* experimental programmes in the context of radioactive waste management. The laboratories have opened up avenues for research and advanced knowledge and technology. Thanks to a large component of international cooperation, they have made it possible to mobilize the financial and human resources required for this type of research. They have, above all, shared thoughts and promoted interdisciplinary studies around the same subject. They make common strategies possible at international level.

In the context of radioactive waste disposal, a URL is a facility in which experiments are conducted so as to establish and to be able to demonstrate the feasibility of constructing and operating a radioactive waste disposal facility within a geological formation (NEA 2001*a*, *b*). Twenty-six URLs were set up in 10 countries between 1965 and 2006. They are located in a range of geological formations: argillaceous sedimentary rocks, magmatic rocks, evaporites and volcanic tuff. Some laboratories have been installed in existing facilities; others have been purpose built.

Experiments in URLs meet two sets of needs: (a) characterization, that is, acquiring knowledge of the geological, hydro-geological, geochemical, structural and mechanical properties of the host rock and of its response to perturbations; and (b) construction and operation, that is, developing equipment to acquire know-how about the construction of all the components of a disposal facility up to

From: Norris, S., Bruno, J., Cathelineau, M., Delage, P., Fairhurst, C., Gaucher, E. C., Höhn, E. H., Kalinichev, A., Lalieux, P. & Sellin, P. (eds) 2014. *Clays in Natural and Engineered Barriers for Radioactive Waste Confinement*. Geological Society, London, Special Publications, **400**, 7–32.
First published online March 5, 2014, http://dx.doi.org/10.1144/SP400.1

its closure, and the emplacement and/or retrieval of the waste. The examples discussed in this article fall into the two categories. Thus, the diffusion experiments and pore-water composition assessment are related to the need for 'characterization'. The sealing experiments illustrate the construction and operation needs. The thermal experiments were initially set up for characterization of the rock properties. However, the most complicated ones foreshadow the behaviour of disposal cells under thermal load.

The experiments conducted over more than 30 years have been of very variable scope in terms of objective, volume, duration, sophistication of installation, budget, and so on. The simplest have been elementary measurements of temperature or pressure. The most complex have replicated full-scale or scaled-down disposal components under actual conditions and/or simulating conditions that might be found in a disposal facility.

Beginning with an overview of the different URLs and a description of five of them, this paper presents the experimental developments that they have made possible. Rather than attempting to be all-inclusive, four key issues common to deep geological repositories have been selected to exemplify the developments: (a) characterization of the rock's response to thermal perturbation; (b) characterization of pore-water composition; (c) characterization of the diffusive properties of the rock; and (d) seal construction. The paper traces the stages in the development of characterization and construction methods suitable for the specific features of impermeable clay rocks used as geological barriers or engineered barriers. Finally, it presents considerations for the construction of disposal facilities in geological formations.

Underground research laboratories

Although there is a continuum of possibilities in the design of URLs, two main categories stand out (Blechschmidt & Vomvoris 2010): 'methodological laboratories' and 'site-specific laboratories'.

Methodological laboratories

Methodological URLs (Table 1) have been designed for the purposes of (IAEA 2001): (a) developing technologies for the construction of disposal facilities, emplacing packages and constructing engineered barriers; and (b) developing techniques for *characterizing* the geological environment to provide an understanding of the hydraulic, chemical, geotechnical and geological parameters that contribute to the confinement properties. For financial and technical (maintenance) reasons, most methodological research laboratories have been set up within or as extensions of existing underground facilities in mines or tunnels. Table 1 presents the main features of methodological laboratories.

Site-specific laboratories

The underground research laboratories specific to a site (Table 2) are used to confirm the suitability of the selected host rock, to guide the specific design and architecture of the disposal facility and to validate the various technological operations in the conditions that are particular to the site. They are constructed close to or within the footprint of the potential future disposal facility. There are limits to the engineering work that can be done on the potential disposal site insofar as the activities in a site-specific underground laboratory must not jeopardize the subsequent safety and security of the disposal facility.

Presentation of five laboratories

High-activity disposal experiment site – HADES (Belgium). The HADES at Mol is the oldest methodological URL in clay. The first construction phase started in 1980 and since then HADES has been extended several times. In 1995, ONDRAF/NIRAS, the Belgian Agency for Radioactive Waste and Fissile Materials and the Belgian Nuclear Research Centre SCK•CEN launched the 'Preliminary demonstration test for CLAY disposal of highly radioactive waste' (PRACLAY experiment), which is managed by EIG EURIDICE, the economic interest grouping between ONDRAF/NIRAS and the SCK•CEN. The PRACLAY experiment is part of the Belgian R&D programme to assess the safety and feasibility of disposal of radioactive waste in a deep clay layer.

The HADES is situated at a depth of 225 m. The PRACLAY gallery (Fig. 1) can be reached via the second shaft. The first shaft is used only for emergencies or during test operations.

Geological setting. The HADES Laboratory is situated in the Boom Clay Formation (Rupelian, Oligocene). This formation outcrops in the central part of Flanders and extends under cover beneath large areas of the North Sea sedimentary basin. The Boom Clay displays a 1–2% dip towards the NNE and thickens in this direction. In the Mol–Dessel area where the HADES is located, the Boom Clay is present at a depth of approximately 190–290 m, attaining a maximum thickness of about 100 m.

The mineralogical composition is qualitatively very homogeneous in the vertical profile of the

Table 1. *List of methodological laboratories (modified from IAEA 2001; NEA 2001*a, b*; Blechschmidt & Vomvoris 2010)*

Generic URL *Operation*	Country *Organization*	Host rock, depth	Comments
Whiteshell Underground Research Laboratory (URL) *1984–2003*	Canada *AECL*	Granite, 240–420 m	Purpose built; generic Shaft sealed
Olkiluoto Research Tunnel *1992–*	Finland *Posiva*	Granite (tonalite), 60–100 m	Purpose built; parallel to repository facilities
Amelie *1986–1992*	France *Andra*	Bedded salt	Pre-existing tunnels
Fanay-Augères *1980–1990*	*IRSN*	Granite Uranium mine	Pre-existing tunnels
Tournemire (Tournemire Research Tunnel) *1990–*	*IRSN*	Sediments (shale), 250 m	Pre-existing tunnels
Asse Mine *1965–1997*	Germany *GSF*	Permian rock salt anticline, mining levels 490–800 m, cavern 950 m	Pre-existing tunnels
Tono *1986–2006*	Japan *JNC/JAEA*	Sediments, uranium mine, 130 m	Pre-existing tunnels
Kamaishi *1988–1998*		Granite, Fe mine	Pre-existing tunnels
Mizunami Underground Research Laboratory (MIU) *2004– (Shaft sinking initiation currently at 500 m)*		Granite, 1000 m (shaft)	Purpose built; generic Surface investigations since 1996
Horonobe Underground Research Laboratory *2005– (Shaft sinking initiation)*		Sedimentary rock, 500 m (shaft)	Purpose built; generic Shafts under construction
KURT – Korean Underground Research Tunnel *2006–*	South Korea *KAERI*	Granite, 90 m	Purpose built; generic
Stripa Mine *1976–1992*	Sweden *SKB*	Granite, Fe mine, 360–410 m	Pre-existing tunnels
Aspö Hard Rock Laboratory (HRL) *1995–*		Granite, 200–460 m (ramp/spiral)	Purpose built; generic
High-Activity Disposal Experiment Site URL, Mol (HADES) *1984–*	Belgium *GIE EURDICE*	Boom Clay (plastic clay), 230 m	Purpose built; generic
Grimsel Test Site (GTS) 1984–	Switzerland *Nagra*	Granite, 450 m	Purpose built; parallel to existing tunnels
Mt Terri *1995–*	*SWISSTOPO*	Opalinus Clay, 400 m	Purpose built; parallel to existing tunnels
Climax, Nevada *1978–1983*	USA *US-DOE*	Granite, mine, 420 m	Pre-existing tunnels
G-Tunnel, Nevada *1979–1990*		Tuff, 300 m	Pre-existing tunnels
Busted Butte, Yucca Mountain, Nevada *1998–*		Bedded tuff, 100 m	Purpose built; generic

sedimentary bed. However, significant quantitative variations are found. They are associated with the grain-size distributions in the beds with variable silt and clay contents. In the studied profile at Mol site, the proportion of clay minerals ranges from 23 to 59%: by dry weight.

Mont Terri Rock Laboratory (Switzerland). The Mont Terri Rock Laboratory, directed and operated by SWISSTOPO – Federal Office of Topography, is located close to the town of St-Ursanne in Canton Jura. The research facilities are located at a depth of around 300 m below ground level and

Table 2. *Site-specific laboratories (modified from IAEA 2001; NEA 2001*a, b*; Blechschmidt & Vomvoris 2010)*

Site-specific URL *Operation*	Country *Organization*	Host rock, depth	Comments
ONKALO, Olkiluoto *2003–*	Finland *Posiva*	Granite (tonalite), 500 m (ramp)	Purpose built; site-specific
Meuse/Haute-Marne (Bure URL) *2000–*	France *ANDRA*	Shale (indurated clays), 450–500 m	Purpose built; site-specific
Gorleben *1985–1990 and 2010*	Germany *BfS, DBE*	Salt dome, below 900 m	Purpose built; site-specific, moratorium cancelled in 2010
Konrad *1980–*		Limestone, Fe mine, 800–1300 m	Facility in former iron mine
Morsleben (ERAM) *1981–1998*		Salt dome, K/salt mine, below 525 m	Facility decommissioned
Waste Isolation Pilot Plant (WIPP) *1982–(1999)*	USA *US-DOE*	Salt (bedded), 655 m	Operating repository since 1999
Exploratory Studies Facility (ESF) Yucca Mountain, Nevada *1996–2010*		Welded tuff, 300 m (ramp)	Purpose built; site-specific; activities stopped in 2010

accessed via the security gallery of the 4 km-long Mont Terri motorway tunnel, which passes through the Jura Mountains.

The research galleries in the Opalinus Clay Formation have a total length of around 500 m (Fig. 2). They serve for investigation and analyses of the hydrogeological, geochemical and rock mechanical properties of the Opalinus Clay (Bossart & Thury 2008). Currently 15 Project Partners from eight countries participate in the Mont Terri Project. The experiments are funded by the Project Partners. The Federal Government (SWISSTOPO) finances the operation and the maintenance of the rock laboratory. The current experimental research programme encompasses about 30 *in-situ* experiments. Some of these are long-term experiments, extending over several yearly research phases.

Geological setting. The Mont Terri rock laboratory is located in the Opalinus Clay, an argillaceous formation consisting mainly of incompetent, silty and sandy shales, deposited in the Aalenian around 180 Ma ago. The Opalinus Clay Formation can be characterized as an over-consolidated shale formation (present overburden 300 m, past overburden estimated to at least 1350 m by Mazurek *et al.* (2006)). It has an apparent thickness of 160 m and a true thickness of 90 m, and can be divided into five lithostratigraphic sub-units (Schaeren & Norbert 1989; Blaesi *et al.* 1991; Bossart & Thury 2008), which are grouped into two main facies: a shaly facies in the lower part of the formation, and a thin carbonate-rich, sandy facies in the middle of the formation (calcareous sandstones intercalated with bioturbated limestone beds, the latter with a high detrital quartz content). The two facies can be explained by the occurrence of different sedimentary environments in a shallow coastal basin at the time of deposition.

The total dry weight percentage of clay minerals varies between 28 and 93%, with an average of 66%.

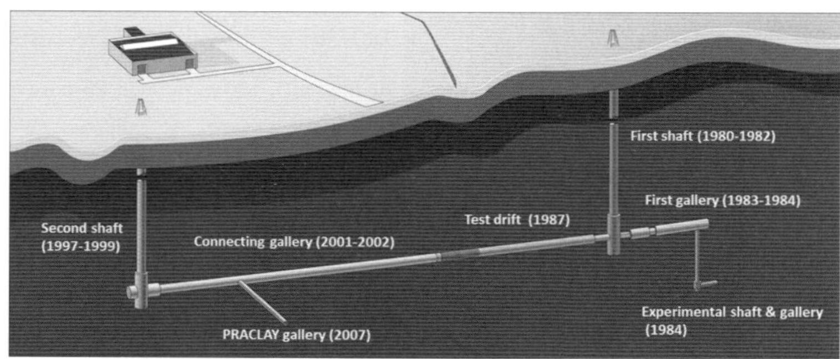

Fig. 1. Architecture of the HADES laboratory.

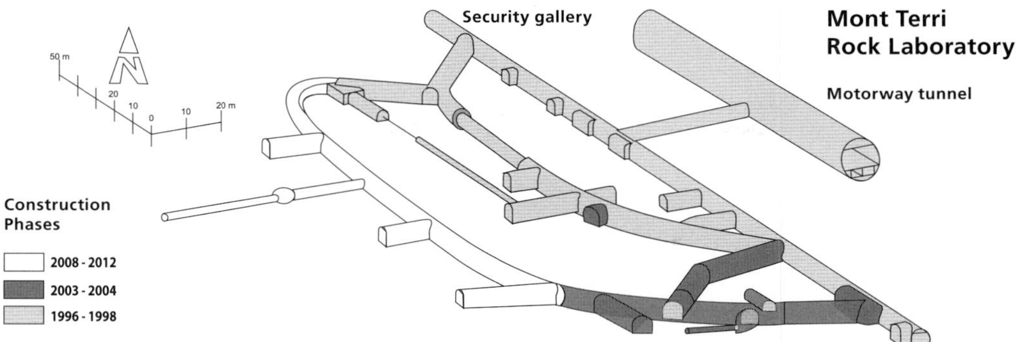

Fig. 2. Galleries of the Mont Terri rock laboratory.

Quartz varies between 10 and 32%, with an average of 14%, and calcite varies between 4 and 22% with an average of 13%. The average for pyrite is 1.1% and that for organic carbon is 0.8%.

The Mont Terri underground laboratory was excavated in the back limb of the Mont Terri anticline, a NNW-vergent imbricate fault-bend fold. In the rock laboratory the bedding planes dip moderately towards the SSE. Between the lower and upper boundaries of the Opalinus Clay Formation, dip angles steepen from 30 to 50° (Fig. 3). The Jura belt was folded during the Late Miocene to Pliocene (from 10.5 to 3 Ma).

The tectonic faults in the rock laboratory can be separated into three different fault systems (Nussbaum *et al.* 2011): (a) moderately SSE-dipping reverse faults; (b) low-angle SW-dipping fault planes and flat-lying (sub-horizontal) faults; and (c) moderately to steeply inclined north- to NNE-striking sinistral strike-slip faults. Sealing of the tectonic fractures with calcite slickenfibres and fault gouge minerals (i.e. illite) is most probably the reason why there is no advective flow in these discontinuities. Tectonic fractures are distinguished from artificial fractures, which are primarily a consequence of tunnel excavation and the associated stress redistribution. The zone around the tunnel where fractures are induced on the micro and macro-scale is called the excavation damaged zone (EDZ) (Bossart *et al.* 2002, 2004).

Meuse–Haute-Marne Research Centre (France). The French National Radioactive Waste Management Agency (Andra) has built and operates the Underground Research Laboratory in the Meuse–Haute-Marne Research Centre (CMHM) on the border of the Meuse and Haute-Marne districts on the eastern boundary of the Paris Basin (Delay *et al.* 2007). On the site, the Callovian–Oxfordian argillaceous rock layer is about 130 m thick and lies at a depth of 422–552 m. Construction began in August 2000. Two shafts provide access to two levels of drifts at depths of 445 and 490 m. In 2012, 1200 m of scientific and technical drifts were excavated and are used for the experimental programme (Fig. 4; Andra 2005).

More than 530 boreholes have been drilled and more than 9000 sensors are monitored in real time on an integrated data acquisition and management system. The system displays the operation of the experiments and holds records of all of the data acquired since the experimental drifts were put into service. It is also used for monitoring the environmental parameters that may influence the phenomena observed (temperature, hygrometry, effects of excavation in adjacent drifts, etc.).

Geological setting. The north of the Haute-Marne district and south of the Meuse district form a simple geological domain of the Paris Basin, comprising a sequence of near-horizontal limestone layers, marls and argillaceous rocks that were deposited at the bottom of former oceans (Fig. 5). The detailed study of seismic geophysical profiles of the sector shows that the tectonic deformations affecting the region over the past 150 Ma have been mild and essentially limited by the Gondrecourt fault system and the River Marne fault system at the boundaries of the study sector. Between these faults, the Callovian–Oxfordian layer is even and planar (Bergerat *et al.* 2007).

Clay minerals consist of illite and mica, interstratified illite/smectite, kaolinite and chlorite (Pellenard & Deconinck 2006). The comparative properties of the Meuse–Haute-Marne and Mont Terri laboratories are set out in Mazurek *et al.* (2008). Table 3 lists the main mineralogical characteristics of the Callovian–Oxfordian argillaceous rock.

Grimsel Test Site (Switzerland). The Grimsel Test Site (GTS) is an underground research laboratory

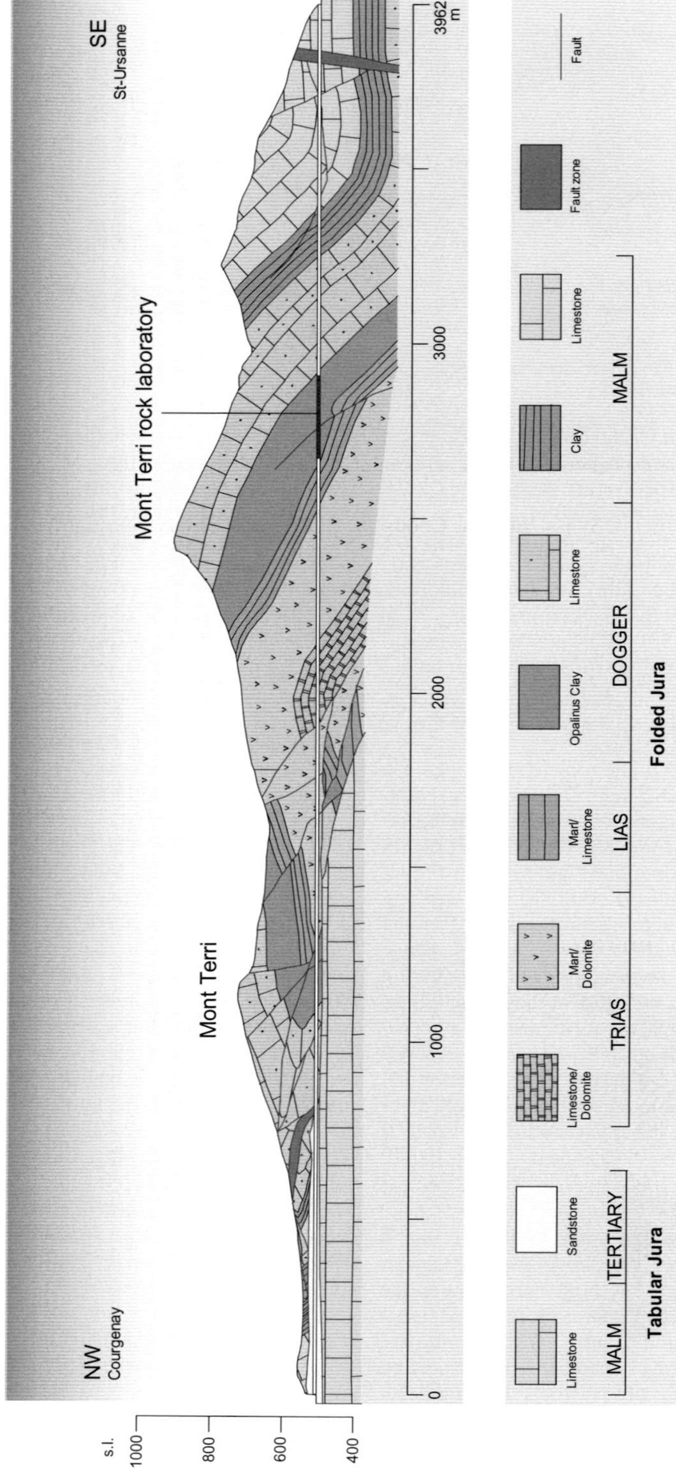

Fig. 3. Geological cross section of the Mont Terri anticline (after Freivogel & Huggenberger 2003).

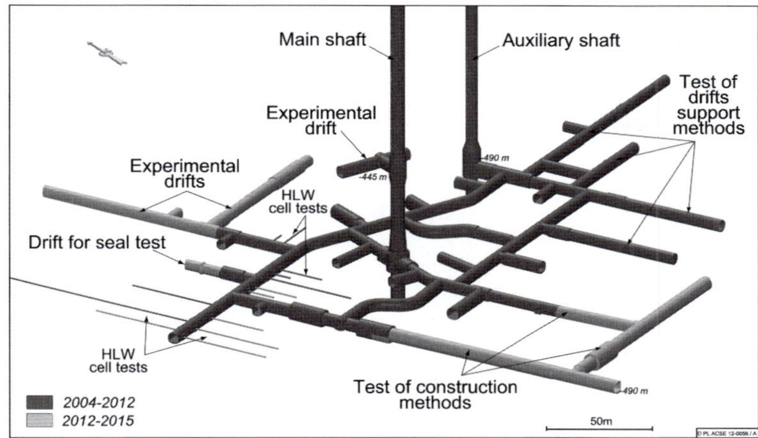

Fig. 4. Overall view of the underground technical and experimental drifts of the Meuse–Haute-Marne Research Centre.

owned and operated by the Swiss National Cooperative for the Disposal of Radioactive Waste (Nagra). It is situated in the Bernese Oberland in the Swiss Alps. The GTS has been in operation since 1984 and for each decade of operation, the main focus has mirrored the evolution of the Swiss national radioactive waste management programme in both scientific and strategic perspectives (Blechschmidt & Vomvoris 2010). It was set up to serve as a methodological laboratory to study both the rock and engineered barrier systems. It can under no circumstances be converted into a disposal facility.

After preliminary investigations, the tunnels and caverns of the GTS were excavated in 1983/1984. The tunnel system about 1.1 km long is located near an access gallery to an underground electrical power plant at an elevation of approximately 1730 m above sea-level, about 450 m beneath the Juchlistock mountain top, in the crystalline rock formations of the Aar Massif (Fig. 6). Additional galleries and caverns were excavated in 1995 and 1998 for two large-scale demonstration tests.

Most of the GTS projects are organized as stand-alone international cooperation projects. The participating organizations form an Experiment

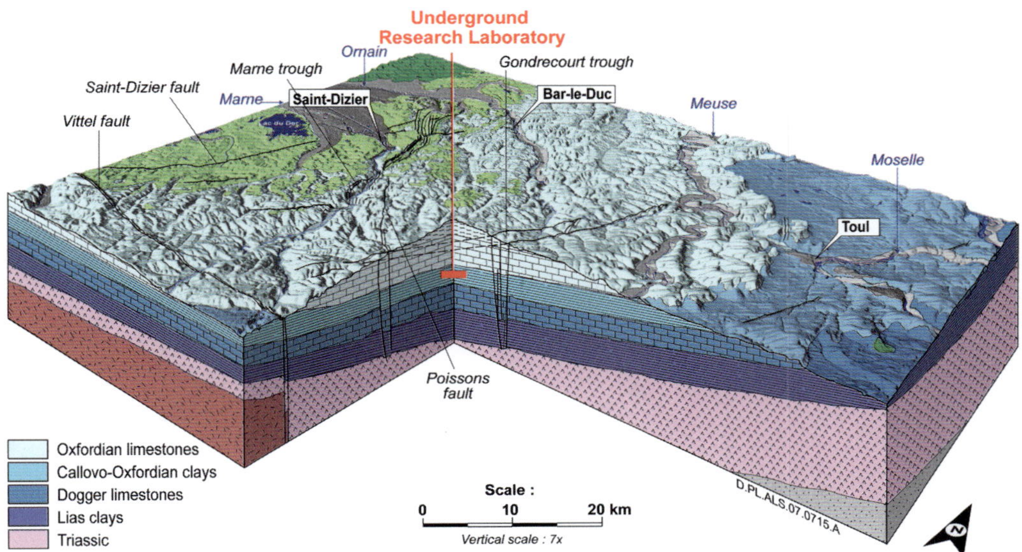

Fig. 5. Geological 3D block diagram of the Meuse–Haute-Marne site (Andra 2005).

Table 3. *Callovian–Oxfordian formation – major physical and chemical parameters*

Property/parameter	Callovian–Oxfordian at Bure
Age	Middle Callovian–Lower Oxfordian, 158–152 Ma
Thickness (m) (at the URL location)	130 m
Clay minerals (weight-%)	40 ± 12
Clay-mineral species (in order of decreasing abundance)	Mixed illite/smectite layers, illite (chlorite, kaolinite)
Other mineral species ordered by decreasing abundance	Calcite, dolomite and other minor carbonates, quartz, feldspars
Pyrite (weight-%)	1
Organic carbon (weight-%)	1
Pore-water type	$Na–Cl–SO_4$
Mineralization ($g\ l^{-1}$)	4–6
Eh (mV SHE); SHE = Standard Hydrogen Electrode	Eh: < -150
Bulk wet density ($g\ cm^{-3}$)	2.46 ± 0.05
Water content (weight-% relative to dry weight)	6–8
Physical porosity (—)	0.16 ± 0.04
Anion accessible porosity (—)	0.06–0.09
Effective diffusion coefficient De (HTO) ($m^2\ s^{-1}$)	$2 \times 10^{-11} \pm 0.1 \times 10^{-11}$
Hydraulic conductivity K normal to bedding ($m\ s^{-1}$), anisotropy factor	1×10^{-14} to 2×10^{-12}, low anisotropy (within the range of variation)
Thermal conductivity normal to bedding ($W\ m^{-1}\ K^{-1}$)	1.5 ± 0.3 (⊥)
Thermal conductivity parallel to bedding ($W\ m^{-1}\ K^{-1}$)	2.0 ± 0.1 (//)
Uniaxial compressive strength normal to bedding (MPa)	21–29

Team which has the overall responsibility for the planning, implementation and evaluation of the Experiment. A typical project has anywhere from four to 10 partner organizations. A total of 17 partner organizations and research institutes from 11 countries were involved in one or more projects as of September 2012. The subsurface 'climatic' conditions are reasonably stable with the temperature at approximately 13°C and the humidity varying between 60 and 80%. Unique to the GTS is an IAEA level B/C radiation-controlled zone, which allows *in-situ* experiments with radioactive tracers in the geosphere under realistic conditions, that is, in a natural groundwater flow field.

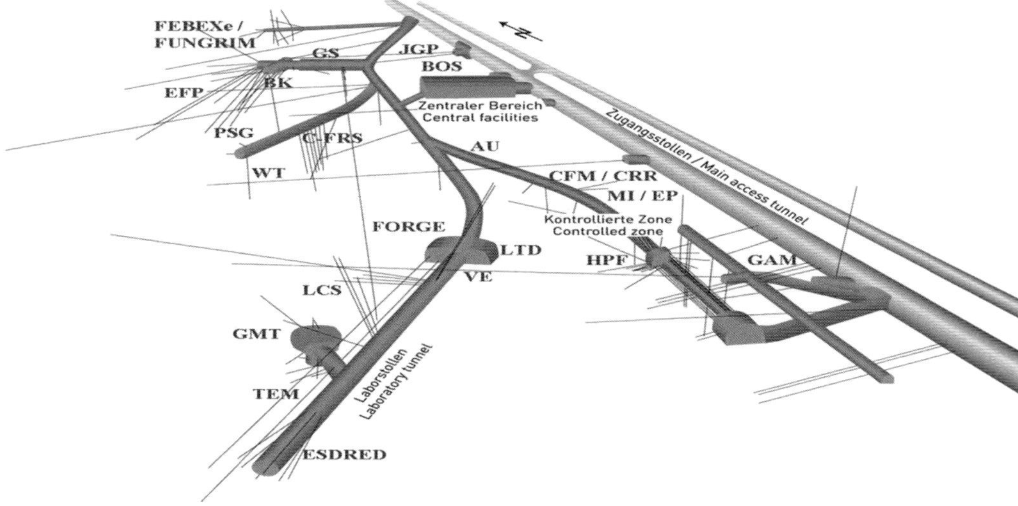

Fig. 6. The system of galleries at the Grimsel Test Site.

Currently, one key area of activities is the study and demonstration of a bentonite-based engineered barrier system (EBS) part of many geological disposal concepts. Of particular interest are the thermal, hydraulic, mechanical, chemical and gas migration properties of these materials under static and dynamic boundary conditions.

The acronyms refer to a small selection of test sites: FEBEX, Full-scale High-level Waste Engineered Barriers; PSG, Pore Space Geometry; GMT, Gas Migration in EBS and geosphere; CRR, Colloid and Radionuclide Retardation; HPF, Hyperalkaline Plume in Fractured rock; CFM, Colloid Formation and Migration; LTD, Long Term Diffusion, GAST, Gas Permeable Seal Test.

Geological setting. The main rock type at the GTS is a foliated crystalline rock with leucocratic Central Aar Granite in the northern part of the laboratory and a Grimsel Granodiorite in the southern part. These plutonic rock formations were formed during the Variscian Orogenesis (290 Ma) and intruded into a Palaeozoic framework of metamorphic sediments. During the cooling phase, the intense extensional stresses caused regional fracturing and faulting. These features served as pathways for the intrusion of granodioritic and aplitic dykes. Later lamprophyres intrude the younger fracture and fault systems.

At the end of the Variscian Orogenesis, the area became part of the Tethys and was buried beneath several kilometres of sediments. After 200 Ma of tectonic rest, the area was rejuvenated by the Alpidian Orogeny (40 Ma). During this latest phase of tectonic compression, the rock mass was affected by regional shear displacement and weak to intermediate metamorphosis. The shear displacement caused cataclastic and mylonitic shear zones, while the metamorphosis caused a significant foliation. The rise of the pluton within the granite is evidenced by brittle deformation, whereas the lamprophyres were affected by ductile deformation. The combination of brittle and ductile deformation leads to the present system of fault and fracture zones (Fig. 7). The uplift of about 0.5–0.8 mm per year and coinciding erosion of the Aar Massif still continue today. The uplift is responsible for the reactivation of alpine structural elements.

Aspö Hard Rock Laboratory (Sweden). The Aspö HRL, located in the Simpevarp area in the municipality of Oskarshamn, was constructed between 1990 and 1995. The underground facility is set directly below Aspö Island, north of the Oskarshamn nuclear power plant. The objectives of the laboratory are to study how the components of a disposal facility (packages, engineered barrier, backfill, sealing and properties of the rock environment) are liable to confine radioactivity. A summary of the work performed at Aspö HRL is published annually (SKB 2012). The Aspö laboratory continues the research and development action begun by the Swedish Fuel and Waste Management Company (SKB) in the Stripa mine (Fairshurst *et al.* 1993; Gnirk 1993; Gray 1993).

Many currently on-going activities are focused on demonstrations that address the performance of the engineered barriers, and practical means of constructing a repository and emplacement of spent fuel canisters. It is also expected that the laboratory will be a training site for staff who will work in the disposal facility. Future plans for the laboratory also include using this unique facility as a national/international research infrastructure.

The underground part of the laboratory consists of a tunnel from the Simpevarp peninsula to the

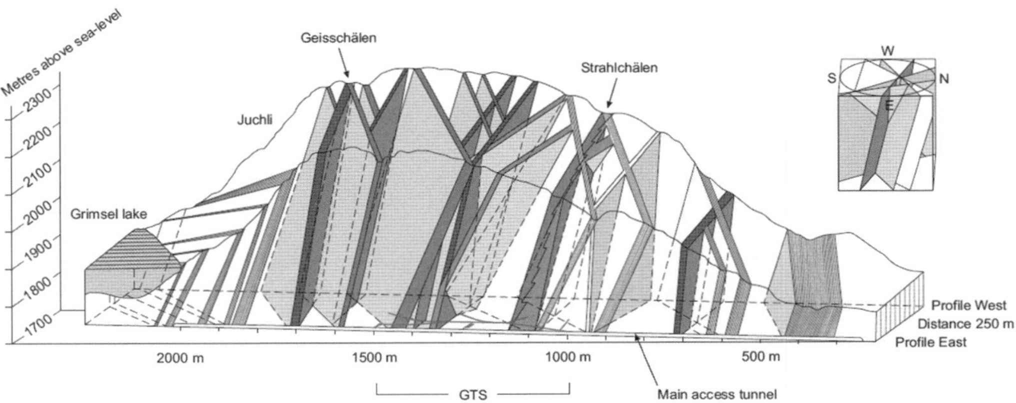

Fig. 7. Structural geological model of Grimsel Test Site (indicated GTS and marked in red in online version) along the main access tunnel (large, steeply dipping shear zones, interpretation from surface and tunnel geological mapping).

southern part of the Aspö Island. The tunnel spirals down to a depth of 460 m into the granite rock (Fig. 8). The total length of the tunnel is 3600 m. The main part of the tunnel was excavated by conventional drill and blast technique and the last 400 m by a tunnel boring machine with a diameter of 5 m. The lower parts of the tunnel are connected to the ground surface via a hoist shaft and two ventilation shafts.

An aspect of the Swedish approach is to have identified 'production lines' of study (copper canister, engineered barriers and closure) which each require development and investigation on various scales for industrialization. Plant or pilot workshops have been developed to produce canisters and to check that they are up to standard. The same goes for the bentonite that is to act as an engineered barrier. The 'bentonite' pilot workshop installed in 2007 on the laboratory site is used for studying the behaviour of bentonite rings and monitoring their responses to various re-saturation kinetics and different physical/chemical water conditions.

During the construction, the experiments were primarily aimed at characterizing the rock properties and improving the technical performances of excavation.

Different types of bentonite have been tested (MX80 – North American Wyoming bentonite, a natural sodium bentonite – and clays from India and Greece). To test the concept in full scale, a prototype repository has been constructed. This is a disposal gallery with six vertical drifts with canisters (without fuel), divided into two sections separated by a plug. The heat produced by the waste is simulated by heater probes. The gallery is sealed in the manner planned for the disposal facility.

Geological setting. The major parts of the Precambrian bedrock of southeastern Sweden are dominated by intrusive rocks of the so-called Trans-Scandinavian Igneous Belt, which was formed during repeated periods of intense calc-alkaline magmatism between 1850 and 1650 Ma, that is, during the waning stages of the Svecokarelian orogeny. At the Aspö Island they intruded at *c.* 1810–1760 Ma (Kornfält *et al.* 1997). The dominating rocks comprise granitoids to dioritoids and gabbroids, and related rocks, of possible volcanic origin, although not positively identified in the area (Fig. 9). The dominating felsic portions of the granitoids to dioritoids are by tradition collectively referred to as 'Småland granites', although the latter comprise a variety of rock types regarding texture, mineralogy and chemical composition. Magma-mingling and mixing processes, exemplified by the occurrence of enclaves, hybridization

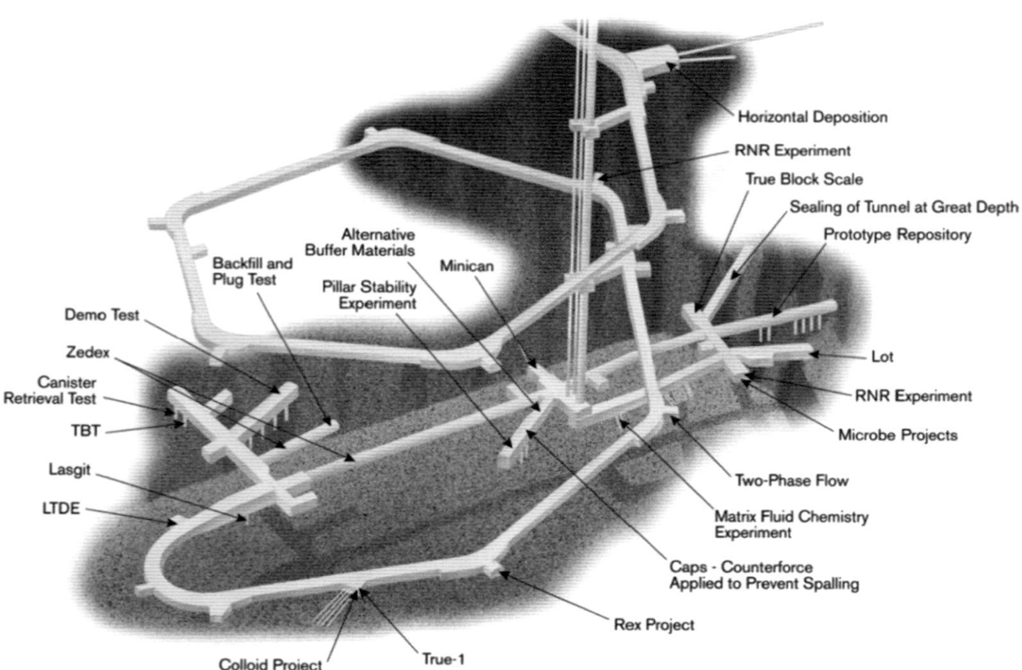

Fig. 8. View of the underground galleries of the Aspö HRL.

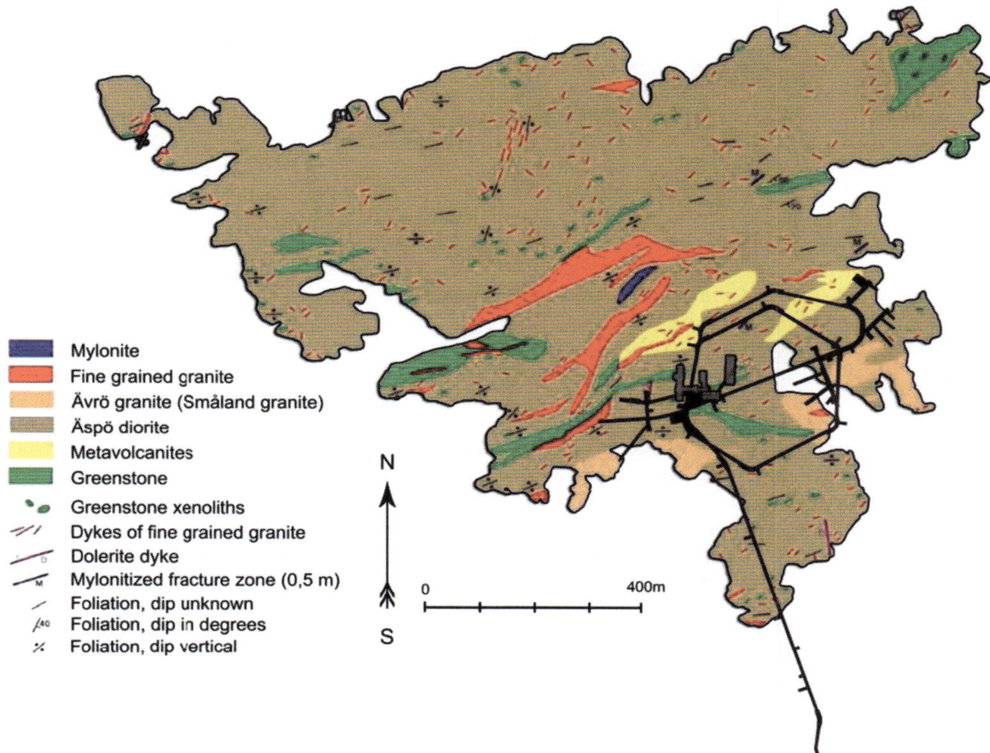

Fig. 9. Bedrock map of the Aspö Island (Rhén *et al.* 1997).

and diffuse transitions between different lithologies, are typical for Trans-Scandinavian Igneous Belt rocks. In mesoscopic scale, these processes often result in a more or less inhomogeneous rock mass. However, if larger rock volumes are considered, these may be regarded as being more or less homogeneous, despite some internal variations.

Examples of experimental topics

This section describes the main characteristics of selected experiments that have been conducted in the various laboratories on four issues: (a) thermal effects; (b) pore-water geochemistry; (c) diffusion, migration and retention; and (d) the sealing of the drift and vaults.

Thermal experiments

In the course of the last 30 years, some 20 *in-situ* experiments have been conducted in URLs for the purpose of studying the consequences on clay rock and/or the barrier elements of heat given off by high-level radioactive waste. The main technical differences between the experiments are related to their geometry (drift length and diameter, sensor network design), the drift heater equipment (sealed or unsealed tubing, bentonite), the mode of heating (energy source within the borehole or outside, electrical resistor) and the control mode (programmable controller, inverter).

The earliest thermal experiments in clay formations were conducted at HADES, and the Mont Terri laboratory. In HADES, these were the CACTUS thermal-hydro-mechanical tests conducted from 1989 and at Mont Terri a series of 'Heater Experiments' launched in 1997. Since 2005 at HADES, Mont Terri and at the Meuse–Haute-Marne Laboratory, experiments have focused on understanding coupled thermal-hydro-mechanical processes in different conditions of power, geometry and thermal loading. In URLs hosted in crystalline rock, the thermal expressions have been investigated in the clay elements of the engineered barrier systems.

FEBEX at Grimsel Test Site. At the Grimsel Test Site, the experimental setup of FEBEX (Full-scale Engineered Barrier system Experiment) was installed in the mid-1990s (Blechschmidt & Vomvoris 2010; Vomvoris *et al.* 2011). The FEBEX experiment consists of an *in-situ* full-scale EBS test performed under natural conditions (Fig. 10). A mock-up test, at almost full scale, runs in parallel in the laboratories at Ciemat in Madrid, Spain (a comprehensive description of the *in-situ* and the mock-up tests for the period 1994–2004 is included in Huertas *et al.* (2006)).

The experiment follows the Spanish reference concept for the disposal of high-level radioactive waste in crystalline rock, in which the canisters enclosing the conditioned waste are emplaced horizontally in drifts and surrounded by a clay barrier consisting of highly compacted bentonite blocks. The aims of FEBEX have been to demonstrate the engineering feasibility of the disposal concept (handling and constructing an EBS) and to study the behaviour of the disposal near-field in terms of thermo-hydro-mechanical processes and thermo-hydro-geochemical processes.

In a tunnel excavated with a tunnel boring machine, two heaters were emplaced surrounded by compacted bentonite blocks. The heaters were turned on in February 1997, maintaining a constant temperature of 100°C at the heater/bentonite contact, while the bentonite buffer slowly hydrated with the naturally infiltrating formation water. In total, 632 sensors were installed in the clay barrier, the rock formation, the heaters and the service zone to monitor a range of parameters including temperature, humidity, total pressure, displacement and pore pressure.

Heater no. 1 was removed in 2002 after five years of heating. The recovered material samples and sensors were investigated for traces of processes and conditions that they had been exposed to (Huertas *et al.* 2006). Heater no. 2 stayed in place and has continued to generate a temperature of 100°C at the interface with the bentonite. Results from almost 15 years of continuous heating (Vomvoris *et al.* 2011) show that, after a rapid increase during the first year, the temperatures in the bentonite buffer are currently fairly stable (Fig. 11). A radial pattern of decreasing temperatures has been established away from the heaters within the bentonite and the adjacent granite. The hydration pattern in the bentonite buffer is relatively symmetrical, with no major differences along the experiment axis. As the granite is characterized by significant heterogeneities, it was concluded that the re-saturation process is controlled mainly by the suction behaviour of the bentonite rather than the water availability from the adjacent rock formation.

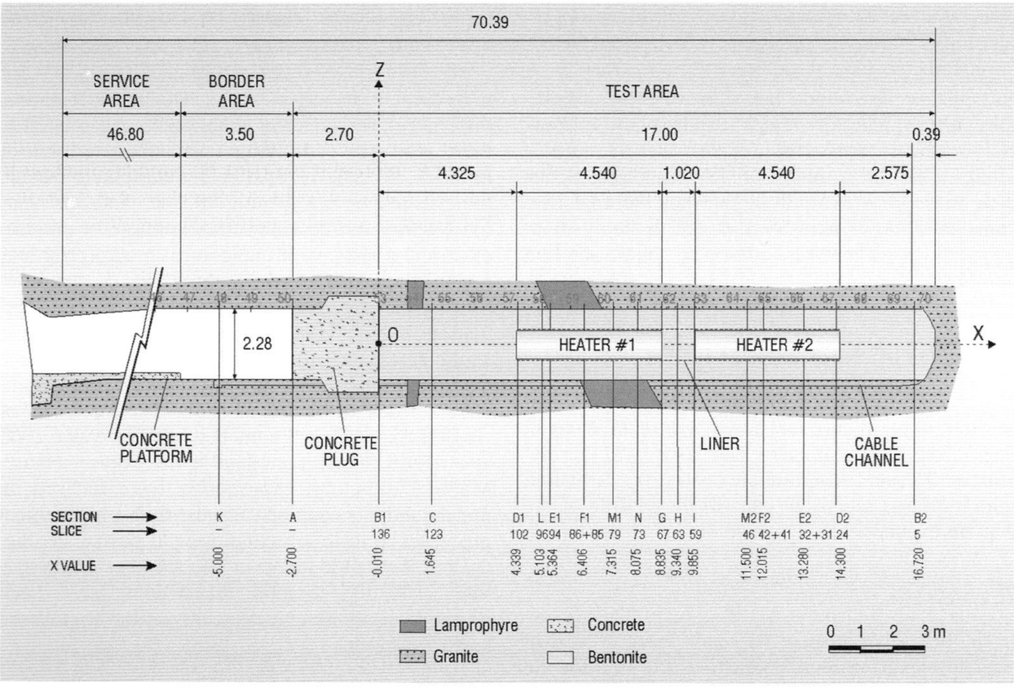

Fig. 10. FEBEX experiment; schematic of the installations during the heating period 1997–2002.

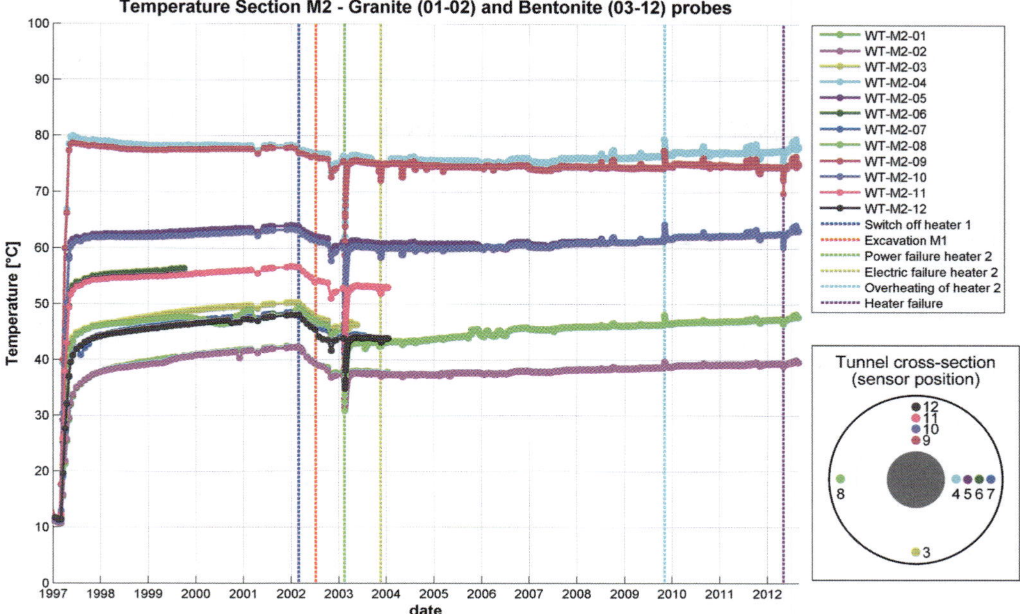

Fig. 11. Temperature measurements in the bentonite of the FEBEX *in-situ* test at different distances from the heater (mid section).

With continuous heating to 100°C for about 15 years and natural saturation, FEBEX is the longest running experiment of this type. The FEBEX *in-situ* test has been the subject of benchmark modelling exercises (Alonso & Hoffmann 2007). Analysis of the modelled distributions of the radial displacements show that the temperature-induced drying processes causes shrinkage close to the heater, while the hydration of the outer part produces swelling. In addition, it is shown that the bentonite is prone to plastic processes during heating and/or wetting cycles (Dupray *et al.* 2013), although there is no indication that these processes influence the anticipated long-term behaviour of the engineered barrier system. The current plans are to continue with constant heating and monitoring at least until 2014, followed by the excavation of the remaining heater provisionally planned in 2015 (Vomvoris *et al.* 2011). The final natural saturation of the inner bentonite will provide insight into how the swelling pressures might be affected by vapour at temperatures around 100°C.

PRACLAY at HADES. The main objectives of the PRACLAY experiment are to (a) demonstrate the feasibility of constructing a crossing between access and disposal galleries from both a technical and an economic point of view and (b) carry out the heater test at a large scale and for a period of at least 10 years (PRACLAY heater test) to study the effects of the heat emitted by the high-level waste on the deep clay layer. The experimental programme is conducted in a 30 m-long tunnel (PRACLAY gallery).

The thermal and hydraulic boundary conditions to be applied in the PRACLAY heater test is supposed to represent the least favourable situation in terms of thermo-hydro-mechanical responses that the disposal repository might encounter:

- The temperature at the interface between the clay and the gallery will be set at 80 °C, which is 5 °C higher for spent fuel and 15 °C higher for vitrified waste than expected in future repository.
- An undrained hydraulic boundary condition will be applied for the heated section, allowing pore-water pressure build-up in the heated clay. This undrained boundary condition will be achieved by backfilling the heated section of the PRACLAY Gallery with water-saturated sand and by installing a bentonite hydraulic seal at the intersection between the heated and the non-heated parts of the gallery (PRACLAY seal test).

The heater test and the seal test make up what is termed the PRACLAY *in-situ* experiment, shown in Figure 12.

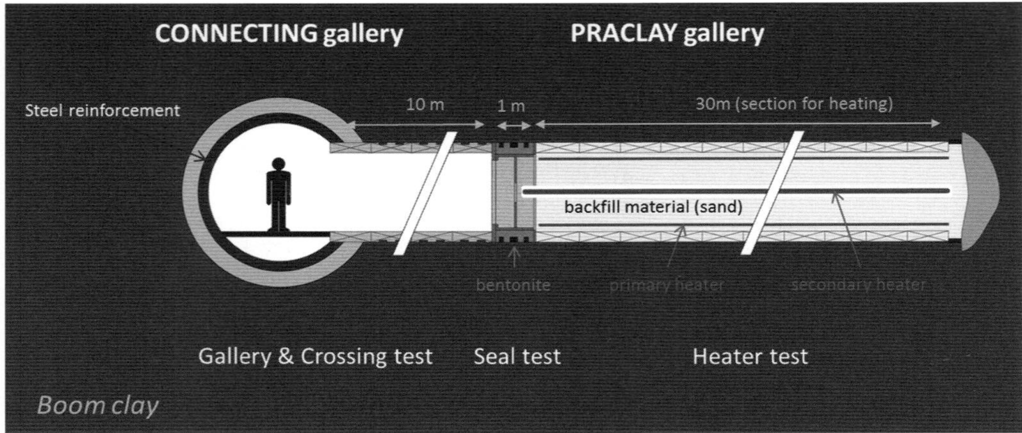

Fig. 12. General feature of the PRACLAY experiment at HADES.

The construction of the PRACLAY gallery was completed in 2007. It was excavated with an open-face tunnelling machine mounted in the connecting gallery. It took a month to excavate the 45 m of gallery with a nominal progress rate about 2 m/day. The support was designed to ensure that the gallery was stable under mechanical and thermal load (Li *et al.* 2010). The design allows two types of loading: geotechnical loading by pressure from the ground on the lining and thermal loading induced by the heater test. Ultimately, a support formed by 81 cement rings (Wedge Block System) was selected. Each ring is made up of eight blocks and a tapered keystone.

The heating system comprises two sub-systems: a primary heating system close to the gallery intrados (Fig. 13) and a second back-up system inside the central tube. Both systems are electrical resistances.

An extensive measuring device network was installed inside gallery, lining, seal and peripheral boreholes to monitor temperatures, deformation, stresses and pore pressure and thus investigate the interactions between the host formation, the seal and the lining.

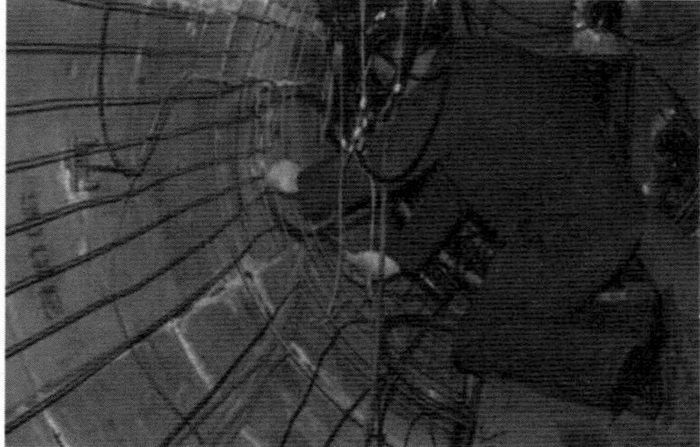

Fig. 13. Installation of the heating systems.

Heater experiment (HE) at Mont Terri. Several experimental programmes (HE, HE-C, SE-H, HE-D) have been conducted at Mont Terri to investigate the thermo-hydro-mechanical and chemical behaviour of the Opalinus Clay Formation (Wileveau 2005*a*, *b*; Göbel *et al.* 2007). The response to thermal loading was studied by two approaches, one on the sample scale (Schnier & Stührenberg 2007) and the other on the experimental scale (several metres).

The first programme, initially HE then HE-C, began in 1997. It was a complex experiment simulating a reduced-scale disposal concept by means of a heater probe surrounded by bentonite rings. The heater element was 2 m long. A total of 19 observation boreholes were drilled to observe the different physical changes in the environment. Electrical power was controlled so as to maintain a constant temperature of 100 °C. The heating phase with electrical power of 200–230 W lasted 18 months in both experiments. Measurements were made of changes induced in the bentonite and the adjacent clay formation. These HE and HE-C experiments highlighted the temperature and pressure behaviour of the rock and engineered barriers as a result of saturation and heating.

The HE-D experimental programme was conducted in the argillaceous facies of the Opalinus Clay Formation without any bentonite. A location 30 m from the other excavations was chosen to limit interferences and artefacts (Wileveau 2006; Fig. 14). A 30 cm-diameter borehole was made

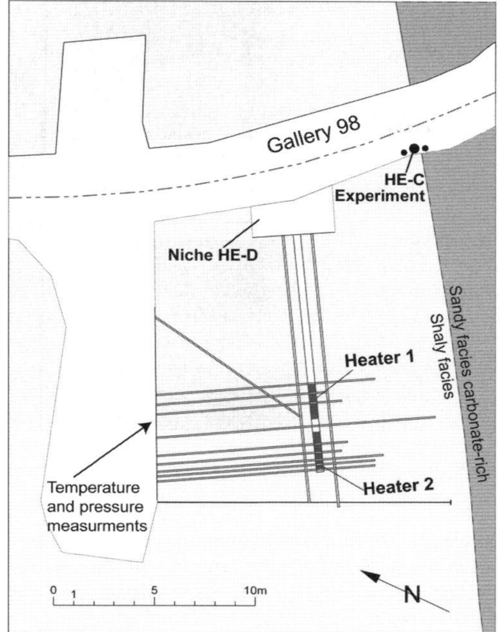

Heater 1

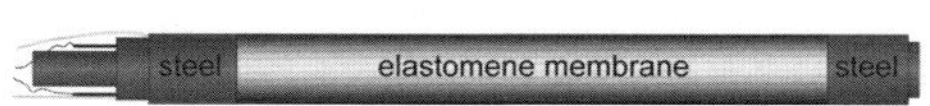

Fig. 14. Layout of the HE-D experimental programme and view of the heater probe at Mont Terri.

over a length of 14 m starting from the HE-D recess; this borehole was horizontal. The heater probe was 5.4 m long with two 2 m-long heater elements. The reaction of the rock was monitored beyond the EDZ. A series of 24 peripheral boreholes with 80 measurement sensors were fitted with instruments and the temperature monitored using fibre optics.

The heating phase began in April 2004 with a power of 650 W for three months and power was then stepped up to 1950 W and maintained at that level until March 2005. The HE-D experimental programme was accompanied by a substantial modelling effort involving four different teams (Wileveau 2006). It was possible (a) to develop the modelling tools for optimizing the position and number of sensors, (b) to test fibre-optic temperature sensors, and (c) to develop numerical interpretation programmes testing several coupled and non-coupled numerical codes in parallel.

Conclusions. Between CACTUS 1, the first thermal experiment at HADES in 1989 (Bernier & Neerdael 1996), and the TED experiment at CMHM, started in 2010 (Conil *et al.* 2012), no less than seven experimental setups were designed with progressive improvements. The first experimental programmes often fell foul of heating system or sensor failures, but each time the regulation systems and the equipment resilience were improved. The main improvements are related to the sealing of the probe connections, the emplacement of sensors and the use of heating power regulation. Concerning the monitoring devices, their number was reduced because they created measurement artefacts.

All of these experiments have helped to understand the processes involved in the THM behaviour of natural argillaceous rocks or bentonite of the engineered barrier. The results from these experiments were crucial to develop predictive models to optimize the number and the location of sensors.

Geochemical experiments to characterize pore-waters in clay media

Knowledge of pore-water in clays was limited in the early 1990s because of the difficulty (a) in extracting pore-water from clay without perturbing it, and (b) in describing the water–rock interactions at the micro- or nanometre scale of the pore size in argillaceous rocks. Studies of radioactive waste disposal have led the organizations involved to develop (a) theoretical research into the state of the water in the rock matrix, (b) experiments aimed at *in-situ* measurements of the physical and chemical properties of water, and (c) speciation modelling taking into account interactions at the surface of clay minerals. Knowledge of pore-water composition and of the mechanisms regulating it is required to evaluate the deterioration of disposal materials and radionuclide mobility.

The initial hydraulic load at the depth of underground laboratories is of several tens of bars. Underground drifts are suitable settings to drill boreholes that reach the hydraulically unperturbed zone where a pressure of 1 bar can easily be imposed. With a pressure difference of several tens of bars between the rock and a borehole, even a very low-permeability formation allows water to flow through. Thus, for example by applying Darcy's law, the production of water per square metre of rock of 10^{-20} m^2 permeability is of several tens of millilitres per day for a pressure difference of the order of 30 bars between a borehole and the adjacent rock formation (Vinsot *et al.* 2011).

The Archimede project conducted at Mol from 1991 and then the Water Sampling experimental programme at Mont Terri from 1996 onwards were among the first tests in underground laboratories aimed at such studies.

Archimede at HADES. The objectives of the Archimede–Argile project were: (a) to understand the mechanisms of acquisition and regulation of the chemistry of interstitial water in a clay sedimentary medium; (b) to test and validate the physical and chemical parameters measured in clays and on which geochemical models of the behaviour of radio-elements are based; (c) to predict the response of the fluid–rock system to perturbation of its geochemical equilibrium; and (d) to model the fluid–rock system, on the scale of local transfers (equilibrium models) and on the scale of global transfers (global models).

The experiment's ambitious objectives were reached in experimental terms, in part thanks to the high water content of the rock (22–27% mass). The water could be sampled directly with relatively simple equipment. The technical and analytical developments (Griffault *et al.* 1996; Sacchi *et al.* 2000) paved the way to directly sample water in less porous and permeable formations at Mont Terri and CMHM.

Water Sampling at Mont Terri. The Water Sampling experimental programme was conducted as part of a project to determine the chemical composition of pore-water *in-situ*. Two experimental principles were implemented at Mont Terri: (a) direct sampling by gravitational flow; and (b) the equilibration of synthetic water. Alongside these, a series of analyses and tests on rock cores were conducted to collate data to reconstruct the composition of water by a speciation model (Bradbury & Baeyens 1998). The results obtained with these two approaches were brought together by the Mont Terri Geochemical Modelling Group and integrated into a synthesis (Pearson *et al.* 2003). That document gathers the knowledge acquired and methods applied for pore–water studies at Mont Terri. It forms the basis for a pore-water conceptual model (Gaucher *et al.* 2009) that provides the framework to interpret experiments on diffusion, retention and the interaction with air and various materials such as concrete, steel or glass.

In addition, recommendations have been made for characterizing pore-water in clay formations: (a) drill boreholes with an anoxic gas to avoid oxidation of the rock, particularly pyrite; (b) clean tools in a suitable fashion and apply rules of cleanliness when handling core samples or installing the equipment to limit bacterial contamination of the borehole test interval; (c) do not use metal parts where water passes so as to prevent oxidation–reduction reactions caused by the presence of such materials; (d) do not use organic materials in the equipment as they could constitute a nutrient for bacteria; and (e) limit the volume of porous materials which represent large surfaces that could facilitate the formation of bacterial colonies.

Pore-water Chemistry (PC and PC-C) at Mont Terri and Prélèvement et Analyse Chimique (PAC) at CMHM. In the Pore-water Chemistry (PC and PC-C) experiments at Mont Terri (Vinsot *et al.* 2008*a*; Wersin *et al.* 2011) and the Prélèvement et Analyse Chimique (Sampling and Chemical Analysis, PAC) experiments at CMHM (Appelo *et al.* 2008; Vinsot *et al.* 2008*b*) the boreholes were designed following the recommendations of the Geochemical Modelling Group (Pearson *et al.* 2003). Moreover, naturally dissolved gases were characterized by circulating initially pure argon in some of the boreholes and monitoring the change in the gas by infrared spectrometry (Cailteau *et al.* 2011) and by sampling.

Data obtained from these experiments contributed to improve the pore-water conceptual models (Gaucher *et al.* 2009). These *in-situ* experiments highlighted the role of microorganisms in the kinetics of chemical processes.

Conclusion. The current development in the URLs is to design sophisticated experiments to analyse and measure on-line the water composition in boreholes submitted to oxidizing, thermal or other kinds of perturbations (e.g. hydrogen injection in a borehole at Mont Terri in the HT experiment). The advances achieved in the understanding of geochemistry of waters in clays are the outcome of transmission of knowledge and know-how among the different sites. This was made possible by (a) the close association of the specialists from different laboratories in all the experimental programmes, (b) the involvement of the scientists in the smallest technological details of the equipment, and (c) the existence of working groups and think-tanks to compare and confront the ideas, and produce documents of high scientific standard.

Diffusion migration and retention experiments

The purpose of the diffusion and retention experiments is to characterize the conditions under which radionuclides diffuse in the low-permeability clay-rich natural formations. An understanding of the transfer mechanisms of radionuclides in the Boom Clay, Opalinus Clay and Callovian–Oxfordian formations was achieved initially by diffusion experiments, with tracers like HTO and I^-, in surface laboratories on centimetre-sized samples.

Subsequently, to analyse the predictive capacity of diffusion/retention models used for the safety analyses, *in-situ* tracer injection experiments were conducted at HADES (CP1, Tribicarb-3D, TD41H/V, etc.), Mont Terri (DI, DI-B, FM-C, DI-A1, DI-A3 and DR) and CMHM (DIR, Diffusion of Inert and Reactive tracers).

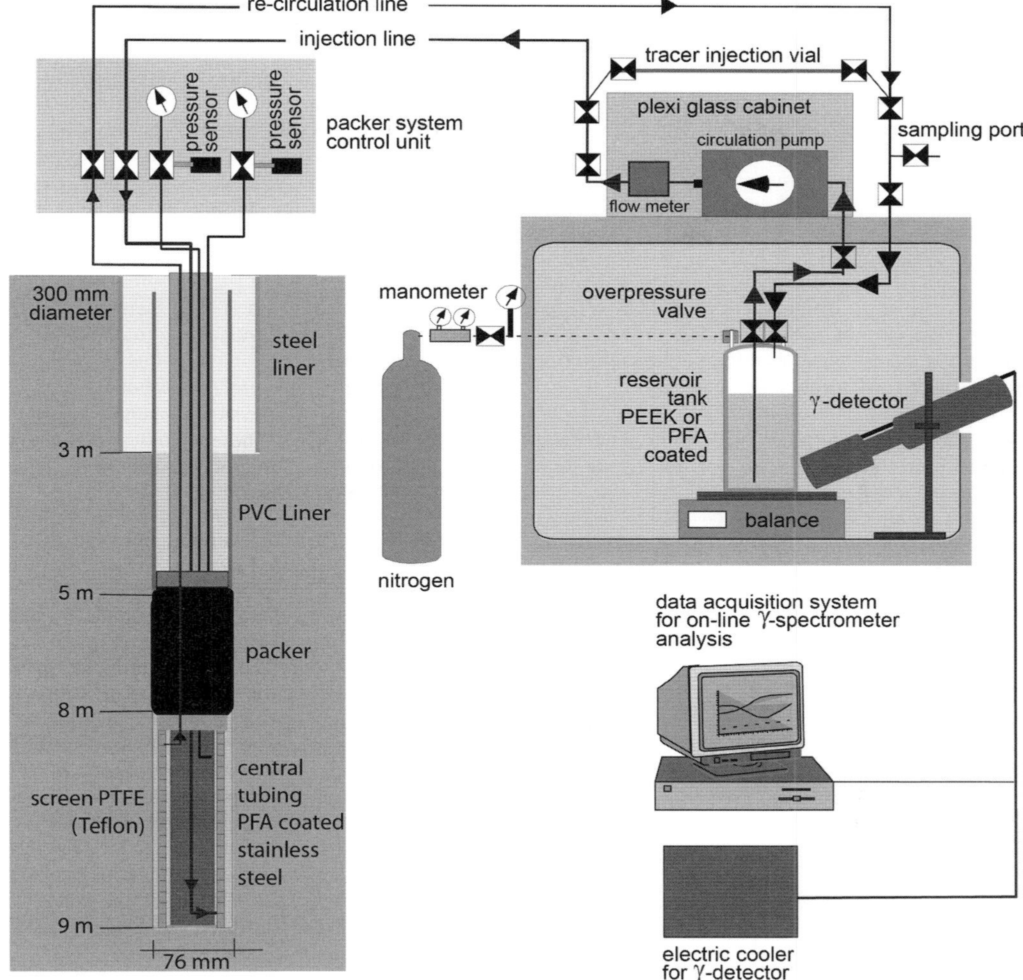

Fig. 15 Setup for an *in-situ* diffusion experiment; DI experimental setup.

These experiments are designed to identify over a diffusion distance of several centimetres up to metres (Boom Clay) the four typical behaviours of tracers highlighted in the sample tests, namely, (a) tracers representing the neutral behaviour of the water molecule (HTO), (b) tracers experiencing anionic exclusion ($^{36}Cl^-$, $^{125}I^-$), (c) tracers experiencing sorption ($^{22}Na^+$, $^{85}Sr^{++}$, $^{134}Cs^+$) and (d) colloidal transport behaviour of humic substances (^{14}C-labelled natural organic matter).

Principle of diffusion tests. An *in-situ* diffusion test (Palut *et al.* 2003) involves putting tracers in solution into contact with the rock in a test chamber (Fig. 15). The test chamber (or diffusion chamber) is located well away from the gallery floor so that the test is conducted in a saturated zone that is little affected by the drainage induced by the gallery. The tracer concentration in the test chamber (injection signal) is monitored over at least a year. Monitoring makes it possible to estimate initial diffusion parameters and to refine the overcoring data predictive calculations providing the migration distance of the tracer with time.

After this monitoring phase, the zone where the tracers have diffused is sampled by large-diameter overcoring (about 300 mm). The core itself is sampled and analysed so that the tracer concentration profiles can be plotted from the wall of the test chamber. The diffusion parameters are inferred from the interpretation of such profiles (Palut *et al.* 2003).

DI at Mont Terri and DIR at CMHM. The methodology presented in the previous section and developed as part of the DI experimental programme proved to be highly relevant and was further enhanced over the course of subsequent experiments at Mont Terri and CMHM. The results from the changing tracer concentrations and the rock profiles were very consistent with the diffusion experiments conducted on small samples in surface laboratories. The different simulation approaches were also fine-tuned to provide a consistent set of values for effective diffusion, porosity and retention (Wersin *et al.* 2008; Naves *et al.* 2010). Technically, the use of the gamma spectroscopy on circulating water helped to improve the monitoring of the tracer concentration decrease.

Moreover, injections have been performed with fluids whose composition is close to the pore-water composition and microbiological conditions have been increasingly well controlled. Lastly the large-diameter overcoring technique has been constantly improved, allowing good-quality samples to be recovered at ever-greater depths, reaching 17 m at the CMHM.

CP1 and Tribicarb-3D at HADES. Two *in-situ* HTO migration experiments, CP1 and Tribicarb-3D, were installed in the HADES laboratory to verify radionuclide migration behaviour in Boom Clay at a large scale and over the long term. Both are still going on today. CP1 stands for Concrete Plug 1 and Tribicarb-3D stands for TRItium and BICARBonate migration experiment in three dimensions. The principle of both migration tests consists of installing multifilter piezometers, into which at a certain location a known quantity of radioactive tracer is injected and breakthroughs are recorded at other filters on the same piezometer or on neighbouring piezometers (Weetjens *et al.* 2011). Tritiated water (HTO) is used to study migration in Boom Clay in both *in-situ* tests.

The CP1 experiment, or Concrete Plug 1 experiment, was installed in 1986. One multiple-screen piezometer, consisting of a 4.6 cm-diameter pipe, penetrates 10 m into the Boom Clay and contains ninr filters with a pitch of 1 m. In 1988, 1.25 GBq of HTO was injected in central filter (Fig. 16) and samples were taken from all filter screens at regular time intervals.

This test also aimed at validating the migration behaviour of unretarded species with emphasis on the effect of anisotropy. A cocktail of HTO and $H^{14}CO_3^-$ was injected in October 1995 in a central filter screen of piezometer R32-3. Because of difficulties in ^{14}C measurements owing to degassing of $^{14}CO_2$, the experimental data points of the test with injected bicarbonate are less reliable.

Excellent predictions of the tracer's breakthrough curves at a few metres from the source have been obtained using the conventional advection–diffusion–retardation equation to describe solute transport and parameters obtained from small-scale migration experiments in the laboratory (Maes *et al.* 2011).

Other large-scale migration experiments in Boom Clay were conducted with $^{125}I^-$ and ^{14}C-labelled Natural Organic Matter (TD41H/V). Details on all *in-situ* migration experiments conducted in Boom Clay are summarized in Aertsens *et al.* (2013).

Conclusions. The technology developed in the URLs has allowed substantial progress in the study of radionuclide transport mechanisms in argillaceous rock. Currently, the focus is being shifted towards the diffusion/retention of more strongly sorbing radionuclides, the role of organic molecules and the diffusion behaviour of redox-sensitive radionuclides. The technical challenges involved in investigating more strongly sorbing nuclides in the field are significant, in particular because transport distances into the rock within a few years are

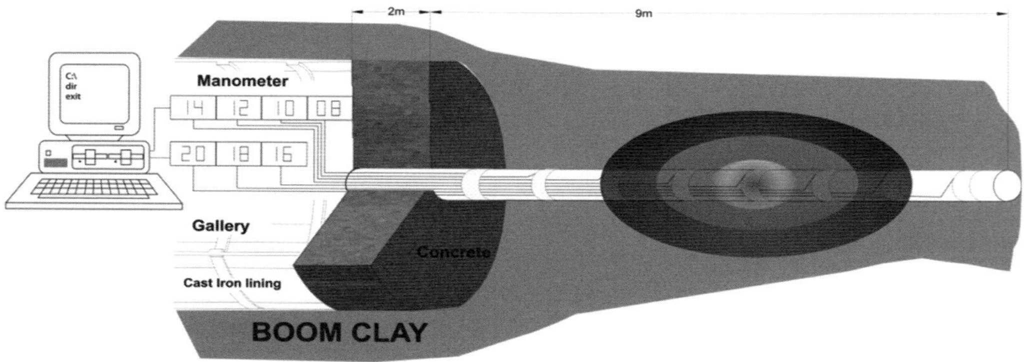

Fig. 16. Set-up of the horizontal piezometer nest for the *in-situ* HTO injection experiment CP1 at HADES.

likely to be shorter than the thickness of disturbed rock created during the installation of the experiments. This means that prolonged experimental durations must be envisaged with the corresponding technical challenges that need to be resolved beforehand. At Mont Terri, the technology used is based on X-ray fluorescence, which detects non-active isotopes in a similar concentration range to scintillation counting. A version with a solid-passive source and micro-observation boreholes has been designed for the DR-B experiment. At the CMHM, *in-situ* beta-ray and gamma-ray detectors have been designed and will be used for more than 10 years (DRN). For the sorbing tracers, such as $^{137}Cs^{+}$ and $^{60}Co^{2+}$, the geometry and features in the borehole and the transport limitations in the circulation system need to be considered.

Full-scale sealing experiments

Principles of sealing demonstration tests. The concepts for the sealing of shafts and drifts have a special place in all repository projects as they are intended to restore as far as possible the confinement properties of the geological medium and to delay the migration of radionuclides by stopping anything that might form preferential channels for water flow.

Three types of test are involved: (a) tests of general scope (permeability, rock damage, self-sealing of cracks), some aspects of which are of interest for seals (EH, Selfrac – Mont Terri); (b) tests on seal components (EZ-A – Mont Terri; KEY, TSS – CMHM); and, (c) tests representing full-size seals (TSX – URL Canada; RESEAL – Mol; Prototype Repository – Aspö; GAST – GTS; NSC – CMHM; DOPAS – Onkalo/Aspö/CMHM/Czech Republic). Backfill testing is also part of the sealing tests. Only the final point is illustrated below for current or future demonstration experiments.

Aspö Prototype Repository. The Swedish KBS-3V disposal concept is based on excavating vertical drifts in granite into which containers of waste or spent fuels are placed (SKB 2009*a*, *b*). The impermeable copper canisters are placed in crystalline basement rock at a depth of about 500 m, embedded in bentonite clay. After waste disposal, the tunnels and rock caverns are sealed. The features of the KBS-3 V concept are (a) a copper canister with a steel insert, (b) a bentonite engineered barrier surrounding the canisters, (c) a disposal depth of 400–700 m, and (d) backfilling of the repository galleries with a bentonite-based material.

The main objectives for the Prototype Repository are to: (a) test and demonstrate the integrated function of the final repository components under realistic conditions in full scale and to compare results with model predictions and assumptions; (b) develop, test and demonstrate appropriate engineering standards and quality assurance methods; and (c) simulate appropriate parts of the repository design and construction processes (Johannesson *et al.* 2007).

The project began in 1998, installation was completed in 2003 and the dismantling of the outer section was carried out during 2011 after approximately eight years of water uptake of the buffer and backfill. Tests will continue in the inner section over a period of at least 10 years.

Located at the −450 level in the laboratory, the project comprises two sections, one with four canister emplacements and one with two emplacements. The two sections are separated by a self-emplacing poured concrete seal. A second concrete seal separates the test section from the remainder of the laboratory.

The main objective of the instrumentation is to provide information on the behaviour of the near-field and far-field under thermal and hydraulic loading during the construction and heating phases. The equipment was fitted both in unfractured blocks and in fractures (Fig. 17).

Most of the drifts are outfitted with thermal probes, water saturation probes and stress gauges. Instruments for measuring canister displacement are also installed in the two drifts.

The six emplacement holes are 8 m deep and 1.75 m in diameter. The engineered barrier is made up of MX80 bentonite blocks 1.65 m in diameter and 50 cm high. Bentonite pellets are used as backfill between the blocks and the rock. The blocks were prepared by compacting bentonite granulates at a pressure of 100 MPa, resulting in a mean dry density of 1.67 g cm^{-3} and a block weight of 2 tonnes.

This global and large-scale experimental programme led to the conclusion that the actual geometry of the massif should be precisely modelled and that the most difficult point is the process of re-saturation of the engineered barrier and the backfilling. In addition, this experiment gave rise to many innovations, for example, (a) the manufacturing of the canister and large bentonite blocks, (b) the design of tools for handling the canister and bentonite blocks, and (c) backfilling methods.

EGTS at GTS. A large-scale permeability test on a seal is underway at the GTS. It involves testing the concept of a special-purpose seal plug that should be gas-permeable but nonetheless have low hydraulic permeability and good radionuclide retention. The Swiss term is Engineered Gas Transport System (EGTS; Fig. 18). It comprises a specially designed backfill such as highly porous grouts for

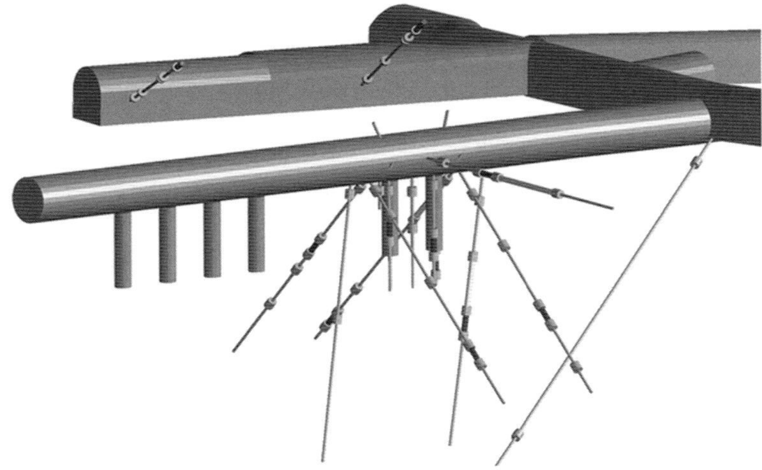

Fig. 17. View of the instrumentation of the #2-drift zone of the Prototype Repository at Aspö.

the disposal drifts, and sand/bentonite mixes for the backfilling of the other underground structures and for the seals. Preliminary experiments have shown the high transport capacity of gases in sand/bentonite mixtures.

The experiment aims to understand at actual scale the behaviour of a seal in the re-saturation phase and when subsequently invaded by gas. This test is conducted in the granitic formation at GTS so as to eliminate the complexity represented by the rheological behaviour of the clay formation.

As part of the clay medium concepts, the feedback elements will be: (a) to demonstrate the global behaviour of a gas-permeable seal on a realistic scale and under realistic hydraulic boundary conditions; (b) to validate conceptual models of gas migration in a seal plug during re-saturation; (c) to evaluate the technology of fitting a seal plug and study the procedures to be developed; (d) to develop non-destructive sounding and monitoring techniques; (e) to develop rules for transposing laboratory measurements to a relevant scale with respect to an actual size project; and (f) to provide information on the geochemical evolution of water and seal materials.

Conclusions. Experiments on seals are now the focus of a significant part of the URL work. They

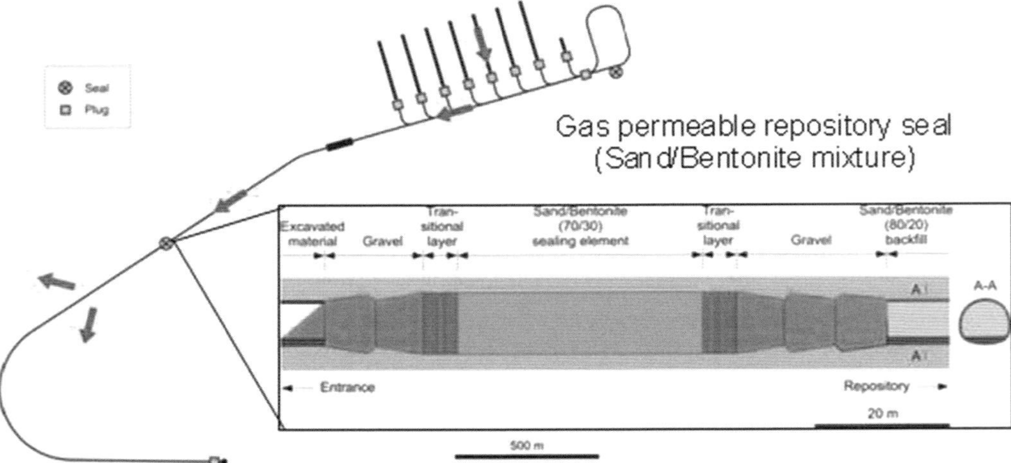

Fig. 18. Design of a gas permeable seal (Engineered Gas Transport System, EGTS) tested at GTS.

make use of techniques and characterization methods acquired over the last 30 years in specialized domains. They benefit both from scientific findings in rock mechanics and geochemistry and from advanced data acquisition and processing systems. The demonstration experiments share the common features of implementing large quantities of materials and requiring heavy-duty technological means for excavating and supporting (and materials handling) similar to those used in mining and tunnelling.

The experiments presented above will be pursued by the Large Scale Sealing Experiment (NSC experiment) at the CMHM where the objective is to put in place a seal with mechanical and confinement components. However, re-saturation of bentonite remains a complex phenomenon to control and to model even over the time scale of an experiment.

Discussion

Present and future benefits from underground rock laboratories

The exploration of intrinsic properties of the geological environment. Over time, the work done in URLs has evolved both in type and in importance (Read 2004). In the early days, some 25–30 years ago, the objective was to identify the main scientific and technical issues and to develop methods to cater for them. The priorities then were (a) to define programmes relating to the study of confinement by the medium or by engineered barriers, (b) to develop experimental equipment and methods, (c) to conduct elementary technical feasibility studies, and (d) to collate fundamental geological data.

For geomechanics, the first stage of study was to observe the effects of excavation of underground galleries: that is essentially their convergence and the characterization of the EDZ. In this domain, many observation and experimental techniques have been developed, not only in road tunnels but also in URLs (HADES, GTS, Mont Terri, CMHM; Tsang *et al.* 2005; Bernier *et al.* 2006, 2003; Armand *et al.* 2014).

Verification of the characteristic parameters of processes expected in the lifetime of a repository. The second evolutionary phase made it possible to verify *in-situ* and under actual conditions the basic understanding of the elementary processes associated with various phenomena. This phase included: (a) tests on phenomena induced by the operation of a repository and its consequences for confinement properties; (b) tests on interactions between disposal materials and the geological environment; and (c) the study of migration and the fate of gases.

For the concepts where the performance of the engineered barriers is essential, this stage offered bentonite emplacement and behaviour tests. The task was then to verify the re-saturation and swelling properties of the material, and its physical and chemical stabilities.

From demonstration to implementation. Since the early 2000s and more recently since 2005, all of the URLs have turned towards the creation of demonstrators. Demonstration experiments are supposed to represent, test and optimize actual potential disposal systems and repository components. Among the more advanced demonstrators, the PRACLAY demonstrator at the HADES laboratory prefigures the Belgian concept (Li *et al.* 2010). In Sweden, the Prototype Repository project (Johannesson *et al.* 2007) is a life-size test of the KBS-3 concept. At the CMHM, the initial tests of *in-situ* demonstrators were of horizontal drifts that were first lightly instrumented, and then progressively outfitted (Morel *et al.* 2010). At Mont Terri, the Full Scale Emplacement experiment (Vomvoris *et al.* 2012) intends to demonstrate a new concept for high-level waste emplacement in horizontal drifts in clay formations. At GTS, the GAST project developed for the Swiss concept tests gas migration through a sand-bentonite plug.

Future perspectives

The avenues for the development of the experiments for the coming decades include: (a) the improvement of excavation and sealing techniques; (b) hydro-geochemical and gas experiments focused on the global observation of disturbances of confinement properties in connection with repository materials; (c) the development of robust monitoring devices for long-term surveillance; (d) experiments dedicated to international training, education and cooperation; and (e) opinion building for the public and stakeholder involvement.

Excavation and sealing. The improvement of excavation techniques focuses on both increasing excavation progress and limiting the damage to the rock. For disposal emplacements containing long-lived, high-level radioactive waste, the initial concepts have been tested in Sweden, Belgium, France and Switzerland.

At CMHM just like at HADES or Mont Terri, the applied excavation methods (drill and blast, road header, pneumatic hammer) have not caused any particular problems, whereas tunnel lining, support, evacuation of dust and emplacement of engineered barriers require further testing. In addition to the technique applied at CMHM and Mont Terri, we might cite the horizontal steel auger and the horizontal raise-boring machine. For the time being, the most advanced experience is the successful excavation of the connecting gallery and PRACLAY gallery using the tunnel machine in an industrial way at HADES.

In granite, a complete gallery system with drifts and materials handling equipment for emplacing packages has been developed at Aspö. This experience is informative on the actual complexity of all of the handling operations and of the design of the necessary radioprotection devices.

For all the geological disposal concepts, the design, construction and performance of seals are the subjects of special-purpose in-depth studies. The sealing materials (swelling bentonite clays in compact blocks or granular backfills) are artificially saturated because of the very low production of water in argillaceous formations. However, sealing experiments with natural saturation and with measurement devices to monitor over several decades are also planned or have already started, for example, *in-situ* heater experiment, HE-E at Mont Terri (Gauss *et al.* 2012).

Hydro-geochemical experiments and experiments involving gases. Hydro-geochemical and gas experiments cover a large range of topics, such as: (a) characterization of pore-water and its thermal, oxidizing, bacterial perturbations; (b) characterization of interactions between water/rock and repository materials; (c) continued study of the radionuclide diffusion mechanisms; (d) study programmes into the consequences of rock saturation and de-saturation; and (e) study programmes into the generation of gases from packages and their migration through the rock and sealing devices. These experiments have been designed on two separate scales: the scale of the borehole or sample and the scale of a repository component. Experiments on borehole scale are underway in several URL programmes. Testing on the larger scale is in preparation in conjunction with demonstration experiments.

The strategy is to weight for each issue the relative importance of the size of the experiment against its duration. In some cases, it will be better to give precedence to the implementation of large quantities of material to the detriment of the accuracy or precision of the results. In other cases, small volumes will be preferred so as to obtain a better understanding of the phenomena. Once a higher degree of maturity has been achieved in the understanding of elementary processes, it will be possible to address the study of the relative rates of competing processes, such as corrosion, gas production and re-saturation.

In clay environments the study of actual-size gas transfer in the formation raises specific problems. The gas migration mechanisms in bentonite or in the natural medium may be studied in small-diameter boreholes. However, to analyse how gases are likely to migrate through a seal plug and above all in the damaged zone between the engineered barrier materials and surrounding rock, a life-size device must be constructed. It is only at this scale that the mechanical perturbations related to the de-stressing of the ground following excavation can be examined. The final verification of the performance of seals with respect to gases can only be achieved when the first test seals are constructed in an environment already influenced by repository access engineering work.

Development of long-term monitoring devices. The development of strategies and techniques for repository surveillance, called monitoring, meets the need to ensure that the engineering works do have the expected quality and long-term performance. In addition, these strategies must provide input for decision-making about the operation of a repository, that is, for deciding on the transition from one operating phase to the next and to the final closure, or a potential decision to retrieve waste packages.

For example, the substantial EDZ observation work done especially in clay laboratories has made it possible to propose phenomenological models that restore observations at distances ranging from one metre to several metres. Moreover, they make it possible to extend interpretations of phenomena the occurrence or amplitude of which had not initially been envisaged and to maintain a watch on the 'unexpected' deviations relative to the processes studied.

Training and education and international cooperation. For 30 years, various countries have worked together through multilateral agreements to pool their research potential and their facilities. Scientists, engineers, technicians and operators from many countries are involved in the experiments. Under the aegis of international instances (European Union (EU), Nuclear Energy Agency (NEA), International Atomic Energy Agency (IAEA)), extensive knowledge and experience have been accumulated and quality and controlling standards have been developed. In this spirit, the laboratories are a privileged setting for learning for countries that do not have such facilities and will be used in the future to train teams involved with repositories.

The transfer of knowledge among scientists from three decades of experiments in URLs can be fostered through direct interaction among teams from different origins over long periods of time and by allowing the teams to develop their own expertise (Delay *et al.* 2007). URLs are important for filling this gap and gaining tacit knowledge. All of the laboratories have thus felt the need to create groups for knowledge exchange and for international cooperation in terms of both programme strategy and the evaluation of the results.

Opinion building for the public. Rock laboratories are also essential elements for the information and communication policy of implementers, safety authorities and regulators. During open days, citizens can come into contact with scientists. They can ask detailed questions about repository concepts and technical issues, but also about phenomena and time scales that cannot easily be understood. During such visits, people build their opinion about geological disposal in a neutral and informative manner.

The way forward for the two types of URLs. URLs have by now accumulated substantial knowledge on the use of technologies and the development of research programmes for the geological disposal of nuclear waste. They have fulfilled their initial purpose by showing that the idea of geological disposal is feasible in principle and by opening up pathways for the development of concepts.

Based on the strength of this experience, non-site-specific laboratories provide an opportunity for comparing methods and concepts for countries that are at a preliminary stage in defining solutions for their disposal. Their position outside of any specific disposal programme provides flexibility in the types of experiments that they can conduct.

Site-specific rock laboratories are dedicated to the development of studies and concepts directly related to a national programme. They allow technical developments but are not necessarily adapted for that by reasons of cost, accessibility or operational programming. They make it possible to have partial or scaled-down demonstration experiments. They ensure continuity in the approach to repository design between scientific R&D, the initial technological developments and the verification of performance in the early pilot stages of the repository. The value of maintaining a site-specific rock laboratory in operation after construction of a repository will depend essentially on its accessibility insofar as it does not support the constraints of a nuclear installation.

Summary and conclusion

The five URLs presented in this paper are now in a mature stage and the issues addressed are ever more sophisticated. As a result of a large component of international cooperation, the URLs have made it possible to mobilize the financial and human resources required for this type of research.

The URLs make it possible to accompany the detailed design stages of a disposal facility, allowing the various constructors to test industrial solutions for construction or operation on prototypes. They also make it possible to envisage scenarios of normal or deteriorated operation of equipment, to test incident and accident situations, and to validate remediation arrangements.

The experience gained from investigations in existing rock laboratories is noteworthy for the development of site investigations and planning of programmes as well as for developing new URLs in Eastern Europe and Asia. This is important both in terms of experimental arrangements and model studies, as well as in adapting and further developing investigation methods.

Beyond the scientific communities which may take advantage of the research centres created by such facilities, URLs offer unique possibilities to demonstrate to the interested general public the scientific and technical work that is carried out to assess the safety of a repository and therefore significantly contribute to building public confidence in radioactive waste disposal in geological formations.

References

Aertsens, M., Maes, N., van Ravestyn, L. & Brassinnes, S. 2013. Overview of radionuclide migration experiments in the HADES Underground Research Facility at Mol (Belgium). *Clay Minerals*, **48**, 153–166.

Alonso, E. & Hoffmann, C. 2007. Modelling the field behaviour of a granular expansive barrier. *Physics and Chemistry of the Earth, Parts A/B/C*, **32**, 850–865.

ANDRA. 2005. *Dossier 2005 Argile: Synthesis. Evaluation of the Feasibility of a Geological Repository in An Argillaceous Formation. Meuse/Haute-Marne site*. Andra Report Series **266 VA**.

Appelo, C. A. J., Vinsot, A., Mettler, S. & Wechner, S. 2008. Obtaining the porewater composition of a clay rock by modeling the in- and out-diffusion of anions and cations from an *in-situ* experiment. *Journal of Contaminant Hydrology*, **101**, 67–76.

Armand, G., Leveau, F. *et al.* 2014. Geometry and properties of the excavation-induced fractures at the Meuse/Haute-Marne URL drifts. *Rock Mechanics and Rock Engineering*, **47**, 21–41.

Bergerat, F., Elion, P. *et al.* 2007. 3D multiscale structural analysis of the eastern Paris basin: the Andra contribution. *Mémoires de la Société géologique de France*, **178**, 15–35.

Bernier, F. & Neerdael, B. 1996. Overview of *in-situ* thermomechanical experiments in clay: concept, results and interpretation. *Engineering Geology*, **41**, 51–64.

BERNIER, F., LI, X. L. ET AL. 2003. *Clipex – Clay Instrumentation Programme for the Extension of an Underground Research Laboratory*. Final Report. Nuclear Science and Technology EUR 20619 EN.

BERNIER, F., LI, X. L. ET AL. 2006. *Fractures and Self-Healing within the Excavation Disturbed Zone in Clays (SELFRAC)*. Final Report. Nuclear Science and Technology EUR 22585.

BLAESI, H. R., PETERS, T. J. & MAZUREK, M. 1991. *Der Opalinus-Ton des Mt. Terri (Kanton Jura): Lithologie, Mineralogie und physiko-chemische Gesteinsparameter*. Nagra Interner Bericht NAGRA NIB 90 60.

BLECHSCHMIDT, I. & VOMVORIS, S. 2010. Underground research facilities and rock laboratories for the development of geological disposal concepts and repository systems. *In*: AHN, J. & APTED, M. J. (eds) *Geological Repository Systems for Safe Disposal of Spent Nuclear Fuels and Radioactive Waste*. Woodhead Publishing, Cambridge, 82–118, Chapter 4.

BOSSART, P. & THURY, M. (eds) 2008. *Mont Terri Rock Laboratory Project: Programme 1996 to 2007 and Results*. Swiss Geological Survey, Wabern.

BOSSART, P., MEIER, P., MOERI, A., TRICK, T. & MAYOR, J. C. 2002. Geological and hydraulic characterisation of the excavation disturbed zone in the Opalinus Clay of the Mont Terri Rock Laboratory. *Engineering Geology*, **66**, 19–38.

BOSSART, P., TRICK, T., MEIER, P. M. & MAYOR, J. C. 2004. Structural and hydrogeological characterisation of the excavation-disturbed zone in the Opalinus Clay (Mont Terri Project, Switzerland). *Applied Clay Science*, **26**, 429–448.

BRADBURY, M. H. & BAEYENS, B. 1998. A physicochemical characterisation and geochemical modelling approach for determining porewater Chemistries in Argillaceous rocks. *Geochimica et Cosmochimica Acta*, **62**, 783–795.

CAILTEAU, C., PIRONON, J., DE DONATO, P., VINSOT, A., FIERZ, T., GARNIER, C. & BARRES, O. 2011. FT-IR metrology aspects for on-line monitoring of CO_2 and CH_4 in underground laboratory conditions. *Analytical Methods*, **3**, 877–887.

CONIL, N., ARMAND, G. ET AL. 2012. In situ heating test in Callovo-Oxfordian claystone, Measurement and Interpretation. *5th International Meeting on Clays and Engineered Barriers for Radioactive Waste Confinement*, Montpellier, 22–25 October 2012, 68–69, Abstract, http://www.montpellier2012.com/doc/abstracts-montpellier-2012/index.html

DELAY, J., VINSOT, A., KRIEGUER, J.-M., REBOURS, H. & ARMAND, G. 2007. Making of the underground scientific experimental programme at the Meuse/Haute-Marne underground research laboratory, North Eastern France. *Physics and Chemistry of the Earth, Parts A/B/C*, **32**, 2–18.

DUPRAY, F., FRANÇOIS, B. & LALOUI, L. 2013. Analysis of the FEBEX multi-barrier system including thermoplasticity of unsaturated bentonite. *International Journal for Numerical and Analytical Methods in Geomechanics*, **37**, 399–422.

FAIRSHURST, C., GERA, F., GNIRK, P., GRAY, M. & STILLBORG, B. 1993. *OECD/NEA International Stripa Project. Overview. Volume 1: Executive Summary*. SKB, Oskarshamn.

FREIVOGEL, M. & HUGGENBERGER, P. 2003. Modellierung bilanzierter Profile im Gebiet Mont Terri La Croix (Kanton Jura). *In*: HEITZMANN, P. & TRIPET, J. P. (eds) *Mont Terri Project Geology, Paleohydrology and Stress Field of the Mont Terri Region*. Reports of the FOWG. Geology Series 4. Federal Office for Water and Geology, Bern, 7–43.

GAUCHER, E. C., TOURNASSAT, C., PEARSON, F. J., BLANC, P., CROUZET, C., LEROUGE, C. & ALTMANN, S. 2009. A robust model for pore-water chemistry of clayrock. *Geochimica et Cosmochimica Acta*, **73**, 6470–6487.

GAUSS, I., WIECZOREK, K. ET AL. 2012. EBS Behavior immediately after repository closure in a clay host rock: the HE-E Experiment (Mont Terri). *5th International Meeting on Clays and Engineered Barriers for Radioactive Waste Confinement*, Montpellier, 22–25 October 2012, 58–59, http://www.montpellier2012.com/doc/abstracts-montpellier-2012/index.html

GNIRK, P. 1993. *OECD/NEA International STRIPA Project. Overview. Volume 2: Natural Barriers*. SKB, Oskarshamn.

GÖBEL, I., ALHEID, H. J. ET AL. 2007. Heater experiment: rock and bentonite thermohydro-mechanical (THM) processes in the near field of a thermal source for development of deep underground high level radioactive waste repositories. *In*: BOSSART, P. & NUSSBAUM, C. (eds) *Mont Terri Project: Heater Experiment, Engineered Barriers Emplacement, and Ventilation Tests*. Berichte der Landesgeologie, Swiss Geological Survey, Wabern, **1**, 7–114.

GRAY, M. 1993. *OECD/NEA International Stripa Project Overview. Volume 3: Engineered Barriers*. SKB, Oskarshamn.

GRIFFAULT, L., MERCERON, T. ET AL. 1996. *Acquisition et régulation de la chimie des eaux en milieu argileux pourle projet de stockage de déchets radioactifs en formation géologique Projet 'Archimède argile'*. Rapport final, Sciences et techniques nucléaires EUR 17454 FR.

HUERTAS, F., FARIÑA, P. & FARIAS, J. 2006. *FEBEX Full-scale Engineered Barriers Experiment*. Updated Final Report 1994–2004. ENRESA Publicación técnica ENRESA PT 05-0/2006.

IAEA. 2001. *The use of scientific and technical results from underground research laboratory investigations for the geological disposal of radioactive waste*. TECDOC IAEA-TECDOC-1243. IAEA, Vienna.

JOHANNESSON, L.-E., BÖRGESSON, L., GOUDARZI, R., SANDÉN, T., GUNNARSSON, D. & SVEMAR, C. 2007. Prototype repository: a full scale experiment at Äspö HRL. *Physics and Chemistry of the Earth, Parts A/B/C*, **32**, 58–76.

KORNFÄLT, K. A., PERSSON, P. O. & WIKMAN, H. 1997. Granitoids from the Äspö area, southeastern Sweden geochemical and geochronological data. *GFF*, **119**, 109–114.

LI, X. L., BASTIAENS, W., VAN MARCKE, P., VERSTRICHT, J., CHEN, G., WEETJENS, E. & SILLEN, X. 2010. Design and development of large-scale *in-situ* PRACLAY heater test and horizontal high-level radioactive waste disposal gallery seal test in Belgian HADES. *Journal of Rock Mechanics and Geotechnical Engineering*, **2**, 103–110.

Maes, N., Weetjens, E., Aertsens, M., Govaerts, J. & Van Ravestyn, L. 2011. Added value & lessons learned from *in-situ* experiments: radionuclide migration. *In*: *HADES, 30 Years of Underground Research Laboratory. The Changing Role of URLs in Radioactive Waste Disposal Research*, Antwerp, 23–25 May 2011, SCK CEN.

Mazurek, M., Hurford, A. J. & Leu, W. 2006. Unravelling the multi-stage burial history of the Swiss Molasse Basin: integration of apatite fission track, vitrinite reflectance and biomarker isomerisation analysis. *Basin Research*, **18**, 27–50.

Mazurek, M., Gautschi, A., Marschall, P., Vigneron, G., Lebon, P. & Delay, J. 2008. Transferability of geoscientific information from various sources (study sites, underground rock laboratories, natural analogues) to support safety cases for radioactive waste repositories in argillaceous formations. *Physics and Chemistry of the Earth, Parts A/B/C*, **33**, S95–S105.

Morel, J., Armand, G. & Renaud, V. 2010. Feasibility of excavation of disposal cells in 500 meter deep clay formation. *In*: Vrkljan, I. (ed.) *Rock Engineering in Difficult Ground Conditions. Soft Rocks and Karst.* Taylor & Francis, London, 561–566.

Naves, A., Dewonck, S. & Samper, J. 2010. In situ diffusion experiments: effect of water sampling on tracer concentrations and parameters. *Physics and Chemistry of the Earth, Parts A/B/C*, **35**, 242–247.

NEA. 2001*a*. *Going underground for testing, characterisation and demonstration (A Technical Position Paper).* NEA/RWM(2001)6/rev1.

NEA. 2001*b*. *The role of underground laboratories in nuclear waste disposal programmes.* Radioactive Waste Management, NEA 3142.

Nussbaum, C., Bossart, P., Amann, F. & Aubourg, C. 2011. Analysis of tectonic structures and excavation induced fractures in the Opalinus Clay, Mont Terri underground rock laboratory (Switzerland). *Swiss Journal of Geosciences*, **104**, 187–210.

Palut, J. M., Montarnal, P., Gautschi, A., Tevissen, E. & Mouche, E. 2003. Characterisation of HTO diffusion properties by an in situ tracer experiment in Opalinus clay at Mont Terri. *Journal of Contaminant Hydrology*, **61**, 203–218.

Pearson, F. J., Arcos, D. et al. 2003. *Mont Terri Project – Geochemistry of Water in the Opalinus Clay Formation at the Mont Terri Rock Laboratory Rapport de l'OFEG*, Série Géologie, **5**.

Pellenard, P. & Deconinck, J.-F. 2006. Mineralogical variability of Callovo-Oxfordian clays from the Paris Basin and the Subalpine Basin. *Comptes Rendus Geoscience*, **338**, 854–866.

Read, R. S. 2004. 20 years of excavation response studies at AECL's Underground Research Laboratory. *International Journal of Rock Mechanics and Mining Sciences*, **41**, 1251–1275.

Rhén, I., Gustafson, G., Stanfors, R. & Wikberg, P. 1997. *Äspö HRL – Geoscientific Evaluation 1997/5. Models based on site Characterisation 1986–1995.* SKB Technical report SKB TR 97 06.

Sacchi, E., Pitsch, H. & Michelot, J. L. 2000. *Porewater Extraction from Argillaceous Rocks for Geochemical Characterisation.* OECD, Paris.

Schaeren, G. & Norbert, J. 1989. *Tunnels du Mont Terri et du Mont Russelin. La traversée des 'roches à risques': marnes et marnes à anhydrite.* La traversée du Jura – les projets de nouveaux tunnels. Journée d'étude, 6 et 7 avril 1989, Delémont. SIA.

Schnier, H. & Stührenberg, D. 2007. *LT Experiment: Strength Tests on Cylindrical Specimens, Documentation and Evaluation (phases 8 & 9).* Mont Terri Project Technical Report **2003-04**.

SKB. 2009. *Design Premises for a KBS-3 V Repository based on Results from the Safety Assessment SR-Can and some Subsequent Analyses.* SKB Technical Report SKB TR **09 22**.

SKB. 2010*a*. *Design and Production of the KBS-3 Repository.* SKB Technical Report SKB TR **10 12**.

SKB. 2010*b*. *RD&D Programme 2010. Programme for Research, Development and Demonstration of Methods for the Management and Disposal of Nuclear Waste.* SKB Technical report SKB TR **10 63**.

SKB. 2012. *Äspö Hard Rock Laboratory.* Annual Report 2011. SKB Technical Report SKB TR **12 03**.

Tsang, C.-F., Bernier, F. & Davies, C. 2005. Geohydromechanical processes in the Excavation Damaged Zone in crystalline rock, rock salt, and indurated and plastic clays – in the context of radioactive waste disposal. *International Journal of Rock Mechanics and Mining Sciences*, **42**, 109–125.

Vinsot, A., Appelo, C. A. J. et al. 2008*a*. CO_2 data on gas and pore water sampled in situ in the Opalinus Clay at the Mont Terri rock laboratory. *Physics and Chemistry of the Earth, Parts A/B/C*, **33**, S54–S60.

Vinsot, A., Mettler, S. & Wechner, S. 2008*b*. In situ characterisation of the Callovo-Oxfordian pore water composition. *Physics and Chemistry of the Earth, Parts A/B/C*, **33**, S75–S86.

Vinsot, A., Delay, J., de La Vaissière, R. & Cruchaudet, M. 2011. Pumping tests in a low permeability rock: results and interpretation of a four-year long monitoring of water production flow rates in the Callovo-Oxfordian argillaceous rock. *Physics and Chemistry of the Earth, Parts A/B/C*, **36**, 1679–1687.

Vomvoris, S., Gaus, I., Rueedi, J. & Marschall, P. 2011. Large-scale tests of the behaviour of bentonite-based barriers: FEBEX and GAST. *13th International High-Level Radioactive Waste Management Conference 2011 (IHLRWMC 2011)*, 10–14 April 2011. American Nuclear Society, Albuquerque, NM.

Vomvoris, S., Claudel, A., Blechschmidt, I. & Müller, H. R. 2013. The Swiss radioactive waste management program – brief history, status and outlook. *Journal of Nuclear Fuel Cycle and Waste Technology*, **1**, 9–27.

Weetjens, E., Govaerts, J. & Aertsens, M. 2011. *Model and Parameter Validation based on in Situ Experiments in Boom Clay.* External Report of the Belgian Nuclear Research Centre ER-171.

Wersin, P., Soler, J. M. et al. 2008. Diffusion of HTO, Br^-, I^-, Cs^+, $^{85}Sr^{2+}$ and $^{60}Co^{2+}$ in a clay formation: results and modelling from an in situ experiment in Opalinus Clay. *Applied Geochemistry*, **23**, 678–691.

Wersin, P., Leupin, O. X. et al. 2011. Biogeochemical processes in a clay formation in situ experiment: part a – overview, experimental design and water data of an experiment in the Opalinus Clay at the Mont Terri Underground Research Laboratory, Switzerland. *Applied Geochemistry*, **26**, 931–953.

Wileveau, Y. 2005*a*. *Summary of the HE-C Experiment at Mont Terri: Meuse/Haute-Marne Underground Research Laboratory*. Mont Terri Project Technical Report, Mont Terri TR **2005 02**.

Wileveau, Y. 2005*b*. *THM Behaviour of host rock: (HE-D experiment): Progress Report September 2003–October 2004. Part 1*. Mont Terri Project Technical Report, Mont Terri TR **2005 03**.

Wileveau, Y. 2006. *HE-D Experiment: Synthesis*. Mont Terri Project Technical Report. Mont Terri TR **2006 01**.

Long-term evolution of the surface environment of the Campine area, northeastern Belgium: first assessment

K. BEERTEN*, M. DE CRAEN & B. LETERME

SCK•CEN, Belgian Nuclear Research Centre, Institute Environment-Health-Safety, Boeretang 200, 2400 Mol, Belgium

**Corresponding author (e-mail: kbeerten@sckcen.be)*

Abstract: In this paper we describe the characteristics of the surface environment of the Boom Clay in the Campine area (outcrop and subcrop zone), which is regarded as a potential host formation for disposal of radioactive waste. A good description and understanding of the relationship between surface variables – the geomorphology, hydrography, vegetation, soils, land use and hydrology – is needed to evaluate the past evolution and assess the future evolution of the surface environment. Changing climatic conditions (glacials and interglacials), global sea-level variations and tectonic movements (uplift and subsidence) may cause the surface environment to change profoundly over long timescales. Starting from the present status, the palaeogeographical and palaeohydrological evolution of the Campine area is described in the framework of the Quaternary geological history of the area. Finally, a first assessment of possible future conditions of the surface environment is given, based on the integration of the palaeorecord and several published modelling studies.

High-level and long-lived radioactive waste is a major radiological hazard to man and the environment. ONDRAF/NIRAS, the Belgian agency for radwaste management and dedicated to finding a solution for this problem, is currently investigating the safety and feasibility of geological disposal in poorly indurated plastic clays such as the Boom Clay and Ypresian clays. The Boom Clay is a thick and relatively homogeneous clay layer that is found in the outcrop and subcrop of a large part of northeastern Belgium. Its thickness varies between several tens of metres (outcrop) to more than 100 m (subcrop), while its top reaches depths of up to 200 m and more in the Campine area. The Geosynthesis project launched by ONDRAF/NIRAS in 2004 aims to bring together all 'geological' data that demonstrate and underpin the safety and feasibility of geological disposal of high-level and long-lived radioactive waste in Boom Clay in the Campine area (northeastern Belgium). Geological data include the geology of northeastern Belgium, the geology of the Boom Clay, its relevant properties, the regional hydrogeological system, and the geography of northeastern Belgium. The scope of the Geosynthesis report covers the entire Boom Clay outcrop and subcrop zone, which, roughly speaking, coincides with an area tentatively denoted northeastern Belgium, and/or the Campine region.

The Earth's surface is a complex interface where atmosphere, hydrosphere, geosphere, pedosphere and biosphere processes interfere. Before it infiltrates, precipitation interacts with the topography, vegetation, land use and soil such that recharge to the aquifers is the direct result of processes acting between the various spheres. Similarly, discharge of groundwater eventually proceeds through soil horizons and surficial geology into the surface environment, including lakes, rivers and seepage zones. The Earth's surface is also the place where biosphere receptors live and interfere with the hydrosphere and geosphere. It is clear that a solid and reliable safety analysis needs to consider aspects of the surface environment, here grouped under the term 'physical geography'. In the long term, the surface environment will undergo important changes, so the potential impact of the changing environment on the safety of a geological waste repository should be considered. During the timeframe considered in safety cases for geological disposal of nuclear waste (up to one million years), the climate will change (global warming, glacial–interglacial cycles), as will the vegetation, soils, topography and hydrology. In this paper, the palaeogeographical and palaeohydrological development of the Boom Clay outcrop and subcrop zone in northeastern Belgium is described in light of the climatic and tectonic evolution. The focus will be on the Quaternary period, given the relatively large climatic variability observed for this particular period. Subsequently, this knowledge is combined with results from modelling studies (climate, recharge, groundwater flow etc.) to give a first assessment of the future evolution of the Earth surface

From: Norris, S., Bruno, J., Cathelineau, M., Delage, P., Fairhurst, C., Gaucher, E. C., Höhn, E. H., Kalinichev, A., Lalieux, P. & Sellin, P. (eds) 2014. *Clays in Natural and Engineered Barriers for Radioactive Waste Confinement*. Geological Society, London, Special Publications, **400**, 33–51.
First published online April 30, 2014, http://dx.doi.org/10.1144/SP400.23

environment of the Campine and surrounding areas, including the possible impact on the future burial depth and hydrogeological system. The burial depth is crucial to ensure sufficient isolation of the waste, while groundwater flow in the aquifers above the Boom Clay is responsible for the transport and dilution of potentially released radionuclides. Whereas the impact from the changing surface environment on the thickness of the Boom Clay overburden and hydrogeological system will briefly be evaluated qualitatively, the present work does not include a real safety/performance assessment.

Characteristics of the present-day surface environment

Geology and hydrogeology

The location of the Campine area and the Campine Basin with regard to the overall tectonic setting is given in Figure 1. The basin is characterized by consistent burial with some isolated erosional events, especially during the Quaternary (Beerten *et al.* 2013). It is situated north of the Caledonian and Variscan massifs, between the uplifting Ardennes and the rapidly subsiding Roer Valley Graben. The pre-Quaternary outcrop map of the Campine area is shown in Figure 2. The geometry of the map basically shows the progressive infill of the Campine Basin during the Neogene. This results in an off-lap pattern, as shown in the cross-section in Figure 3. The Boom Formation, considered a potential host formation for radioactive waste disposal, is clearly visible in the profile, and constitutes an important aquitard in the Campine Basin. The Miocene Diest Formation is part of the so-called 'Neogene' aquifer that is overlying the Boom Clay.

Orohydrography

Figure 4 shows the orohydrography of the Campine area. The relief in the outcrop and subcrop zone stretches from *c.* 100 m TAW (above sea-level according to the 'Tweede Algemene Waterpassing') in the SE along the Meuse-Demer water divide, to *c.* 0 m in the NW along the River Scheldt in the west. Two major rivers and their tributaries traverse the Boom Clay zone: the Meuse in the east and the Scheldt in the west. Relative uplift since the Pliocene has forced these rivers to incise, which resulted in significant fluvial erosion and denudation of the landscape. The most important morphological and structural entities are the partially dissected (e.g. by the river Scheldt) Boom Clay cuesta ('1' in Fig. 4), the Campine Clay cuesta ('2'), the Nete basin ('3'), the Hageland area (with the so-called 'Diest Hills', '4'), the Campine Plateau ('5') and the Roer Valley Graben ('6'). More details on the palaeohydrological and palaeogeographical

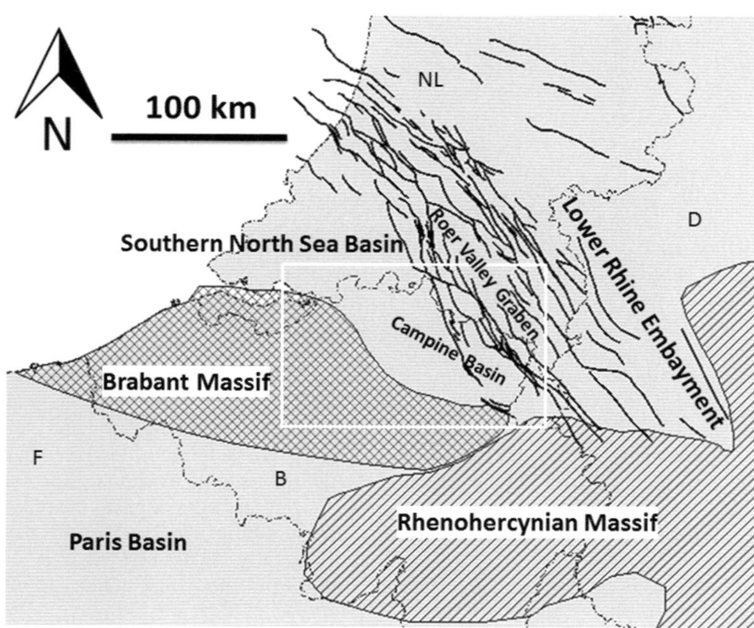

Fig. 1. Geological and tectonic setting of the Campine Basin and the Campine area (Beerten *et al.* 2013).

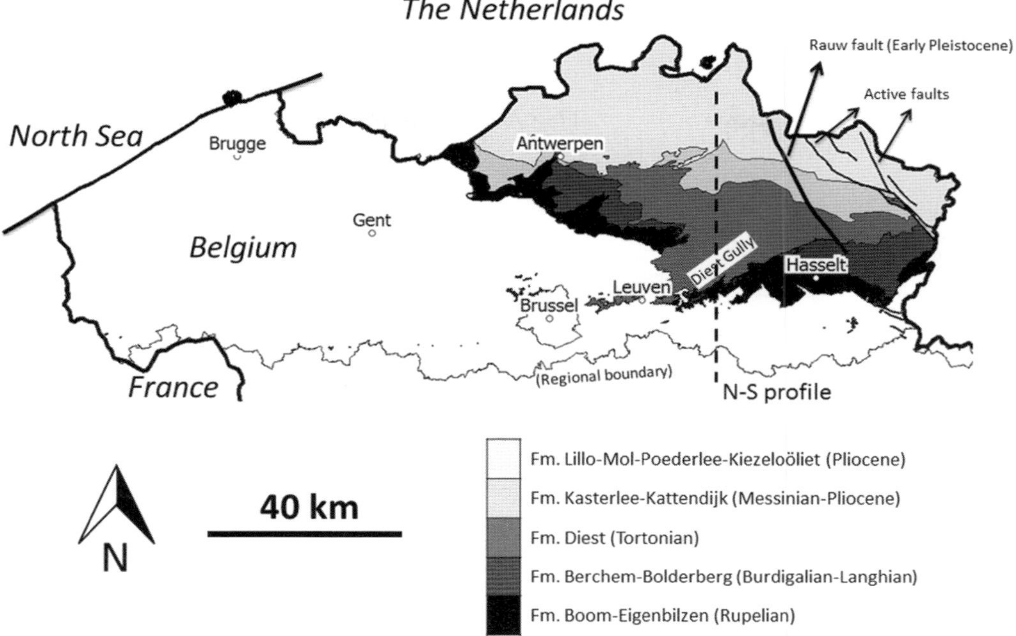

Fig. 2. Simplified geological map of the Campine area, northern Belgium, which more or less coincides with the Boom Clay outcrop and subcrop zone (Geologisch 3D Model Vlaanderen (v1.2011)). Geological cross-section along the indicated profile line is given in Figure 3. Note the extent of the Diest Formation towards the SW in a gully where it reshaped the Boom Clay outcrop.

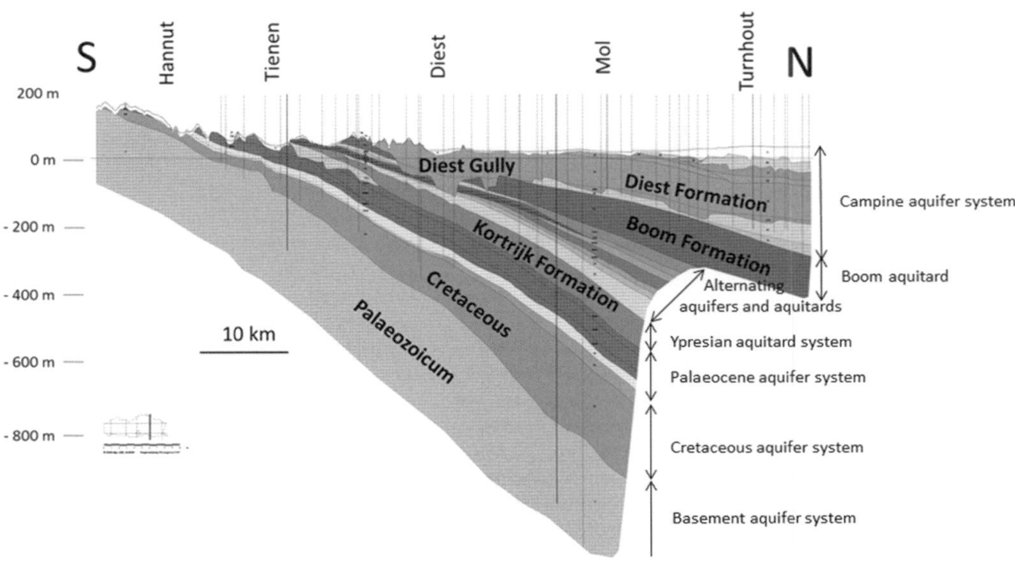

Fig. 3. Geological cross-section according to the profile line in Figure 2, showing the geological architecture of the Campine Basin, and the most important aquifers and aquitards (from Matthijs *et al.* 2003).

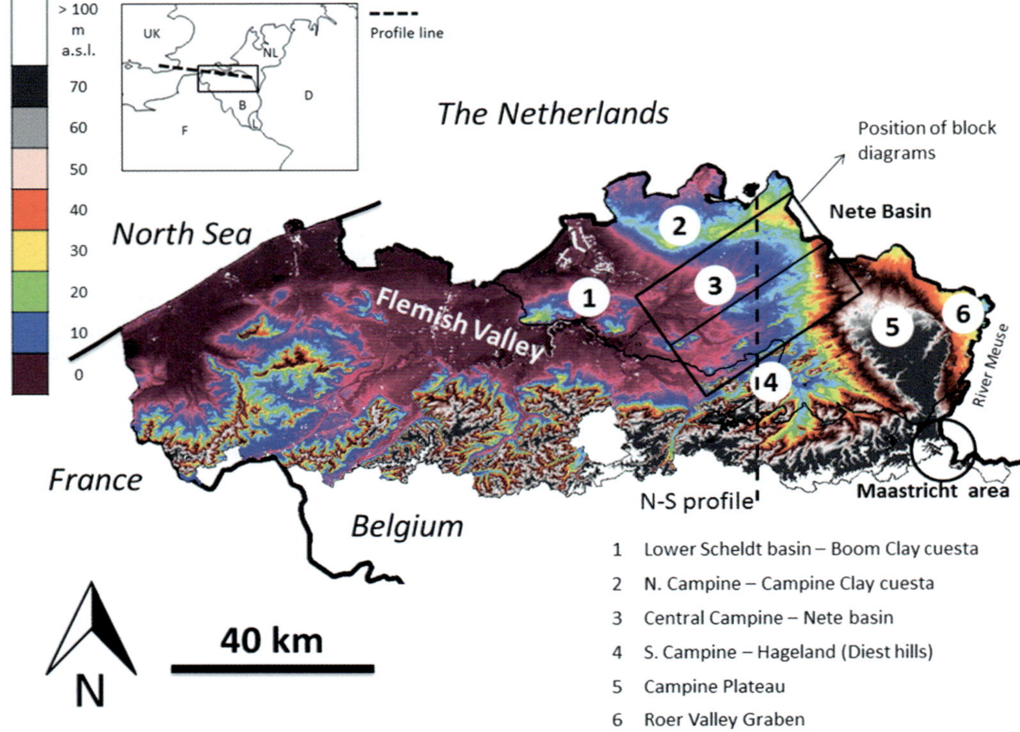

Fig. 4. Digital terrain model of Flanders (AGIV 2006), with major morphological entities (1–6) marked, as well as the (buried) Flemish Valley, the Nete basin, the Meuse Valley and the Maastricht area. Positions of block diagrams (Figs 10–13) are indicated by rectangles.

evolution of the Campine area during the Quaternary period will be given in the section 'Quaternary geological and geomorphological development'.

Climate, soils, vegetation and (surface) hydrology

The present-day climate in Belgium is a temperate oceanic climate (RMI 2011). Its middle latitude position (around 50°50′N) is characterized by the convergence of cold polar air masses with warm tropical air masses, forming the polar front zone. In general, summers are relatively cool and rainy, and winters are relatively mild and wet. Compared to the average climate described for Belgium, the Campine has a more continental character than one would expect at this altitude and distance from the sea. It is characterized by relatively warmer summers and colder winters, hotter days and colder nights compared to the national reference station at Uccle (Brussels, central Belgium). This difference with the rest of the country can be explained by the coarse-grained sandy soils. The sand absorbs heat very quickly during the daytime, but at night this warmth is also radiated quickly.

Soils in the region of interest are predominantly sandy and loamy sandy, with various degrees of wetness (wet in alluvial plains and dry on interfluves) (AGIV 1998). These soils are characterized by the presence of a well-developed humus-B or iron-B horizon (podzol soil) or by the presence of a thick anthropogenic humus-A horizon (plaggen soils), especially in the northern and central Campine area (van Mourik *et al.* 2010, 2011; Beerten *et al.* 2012). They developed on sandy parent material consisting of wind-blown quartz sands.

The landscape in the western Campine area is basically treeless as a result of large-scale forest-clearing and is dominated by meadows and pastures (Van Landuyt *et al.* 2006). In the valleys there is more grassland, most of which is in intensive agricultural use. On the drier sandy areas there are some forests and some rare relics of heathland. In the northern, central and eastern Campine area, much of the former heathlands are urbanized or converted to arable land or commercial forest, with non-native species such as Scots pine (*Pinus sylvestris*) and red oak (*Quercus rubra*).

The Lower Scheldt together with two of its tributaries (the Nete and Demer) and the River Meuse in the east are the most important rivers in the Boom Clay outcrop and subcrop zone (Fig. 4). They can all be classified as meandering rivers (Strahler & Strahler 1992). A large part of the Scheldt basin is influenced by tides. A tidal wave penetrates the Lower Scheldt and Nete basins twice a day. The tidal discharges in the Scheldt river are enormous in comparison with the basal river discharges and mount up to 2500 $m^3 s^{-1}$ and more (De Smedt 1992). Basal discharges range from *c.* 100 $m^3 s^{-1}$ for the downstream part of the Lower Scheldt to *c.* 5 $m^3 s^{-1}$ for the Nete river. Peak discharges are several times higher than these values and may be associated with overbank flow. Heavy winter storms, in combination with high tides and strong westerlies that increase the water level in the Scheldt estuarium, may cause inundations in the Lower Scheldt basin and the Nete basin. The Meuse is a typical rain-fed river with an average discharge of *c.* 170 $m^3 s^{-1}$ (Maaseik), and extreme values as low as *c.* 5 $m^3 s^{-1}$ and as high as 2000 $m^3 s^{-1}$ (values for 2011; Waterbouwkundig Laboratorium 2012). Aquifer recharge in the investigation area between the Meuse/Scheldt watershed on the Campine Plateau in the east, the Dijle–Demer axis in the south, and the River Scheldt in the west ranges between *c.* 0 mm a^{-1} (on urban impervious areas) and >400 mm a^{-1}, with an average value of *c.* 250–300 mm a^{-1} (Meyus *et al.* 2004; Batelaan & De Smedt 2007). Leterme & Mallants (2012) calculated the average groundwater recharge for the Nete catchment to be *c.* 390 mm a^{-1}, which is *c.* 100 mm a^{-1} larger than that calculated by Batelaan & De Smedt (2007). The difference is due to the models used in the studies (HYDRUS 1-D v. WetSpass, respectively), and the degree of soil and land-use variability that was introduced into the model (stylized v. detailed, respectively). Around 80–90% of the groundwater recharge is drained by rivers, while the rest of the infiltrating water is removed by pumping wells (Vandersteen *et al.* 2013).

Quaternary geological and geomorphological development

Climatic evolution

Pre-Weichselian (c. 2.6 Ma–110 ka). The Neogene climate is characterized by a gradual cooling that is clearly expressed in eustatic sea-level lowering (Miller *et al.* 2005). The permanent development of the East Antarctic ice sheet at the end of the middle Miocene resulted in further lowering of the sea-level (see Miller *et al.* 2005). Permanent and large Northern Hemisphere ice sheets developed during the Quaternary (*c.* 2.6 Ma), and their growth and decay were controlled by 10^4–10^5-yr scale Milankovitch changes (Miller *et al.* 2005). During the earlier part of the Quaternary (from 2.6 Ma up to *c.* 400–800 ka ago), the warm to cool alternations presented a dominant periodicity of *c.* 41 ka. Possibly since 800 ka, and certainly during the last 400 ka, the situation became more complex. In general, the glacial periods during the last 800 ka occurred with a mean periodicity of *c.* 100 ka (ranging between 50 and 130 ka), and displayed a slow (70–90 ka) and uneven cooling followed by a rapid deglaciation (Hays *et al.* 1976; Berger 1977; Imbrie *et al.* 1984). Sea-level changes during the Quaternary (last 2.6 Ma) are typically between +25 and −125 m (Miller *et al.* 2005; Fig. 5). During the Elsterian, Saalian and Weichselian glaciations, a large ice sheet originating in Scandinavia and Great Britain spread across northwestern Europe (Ehlers *et al.* 2011; Fig. 6). During the Saalian glaciation, the ice sheet reached the Hoge Veluwe in the Netherlands, which is *c.* 100 km north of the Belgian–Dutch border (Zagwijn 1974; Lambeck *et al.* 1998). The glaciation resulted in a sea-level drop of *c.* 125 m, and the rivers Meuse and Rhine were forced in a western course parallel to the southern limit of the ice sheet at that time. Glaciotectonic deformation resulted in the formation of ice-pushed ridges in the Netherlands (100 km north of the Campine area) up to 100 m high, and glacial basins up to 150 m deep (Zagwijn 1974). South of the glaciated area, tundra-like conditions existed, resulting in permafrost.

Weichselian (c. 110–11.7 ka). The Last Glacial Maximum (LGM, *c.* 20 ka) is the last ice volume

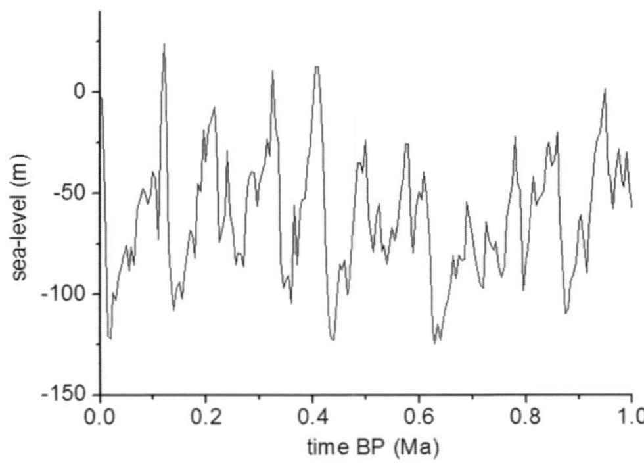

Fig. 5. Global sea-level change during the last 1 Ma, based on data published by Miller *et al.* (2005).

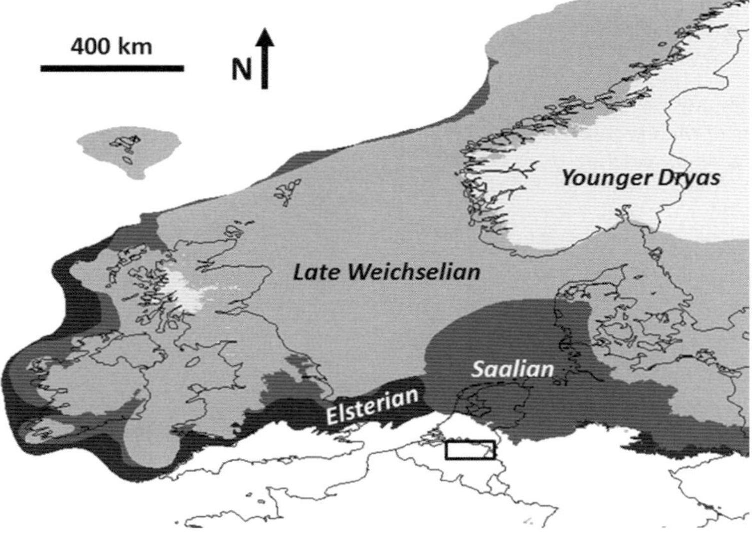

Fig. 6. Extent of the Elsterian (*c.* 450 ka), Saalian (*c.* 150 ka), Weichselian LGM (*c.* 20 ka) and Younger Dryas (*c.* 12 ka) ice sheets in northern Europe (Ehlers *et al.* 2011; high-resolution map datasets retrieved from http://www.dov.vlaanderen.be in September 2013). The location of the Campine area is indicated by the rectangle.

culmination that took place in the Late Weichselian, when *c.* 50 × 10^6 km^3 of ice spread over a huge territory in the Northern Hemisphere. Mean global air temperature was *c.* 4–6 °C lower than at present, and the global sea-level was lower by almost 120 m (Clark *et al.* 2001), considerably modifying the landscape. In northern Europe, the most important feature was undoubtedly the development of the Fennoscandian ice sheet (Svendsen *et al.* 2004; Ehlers *et al.* 2011). It is important to mention that, during the Weichselian glacial stage, the extent of the Fennoscandian ice sheet was more restricted than during the Saalian glacial stage: the ice sheet never extended further south than southern Denmark and northern Germany, that is, some 400 km north of the Belgian–Dutch border. Like most of northern Europe, the northeastern part of Belgium experienced dry, tundra-like surface conditions during the LGM, together with permafrost, causing periglacial soil deformation (Gullentops *et al.* 1981; Huijzer & Vandenberghe 1998). Vegetation was sparse and dominated by steppic plants. The LGM was overall characterized by a strong decrease in global forest cover, and a concomitant increase in non-forest environments (mainly steppe, tundra, desert and ice caps). Our present-day climate started developing *c.* 11.7 ka BP, when temperatures and sea-level rose again to values comparable to the present ones. Northwestern Europe became reforested, first with birch and pine, and later on with a mixed deciduous forest. At the same time, soils started developing and further contributed to the stabilization of the landscape, together with a dense vegetation cover (Hoek 2001).

Uplift and subsidence during the Quaternary

Maximum subsidence rates can be inferred from the Quaternary record in the Roer Valley Graben (Figs 1 & 2) where *c.* 60 m of sediment was deposited during the last 1 Ma (van Balen *et al.* 2000; Gullentops *et al.* 2001; Beerten 2010; Beerten *et al.* 2013). Maximum uplift rates in the Boom Clay outcrop and subcrop zone are recorded near the southeastern outcrop area (Maastricht area; Fig. 4) where the Meuse Main terrace is *c.* 50 m above the current floodplain, yielding an average uplift rate of *c.* 70 m myr^{-1} during the last 0.7 Ma (see also van Balen *et al.* 2000; van den Berg & van Hoof 2001; Westaway 2001; Demoulin & Hallot 2009). The uplift is probably the result of a combination of forces: an isostatic response to the subsiding raben, intraplate tectonic stresses in the Alpine foreland, and mantle plume dynamics in the Ardennes–Eifel area (van Balen *et al.* 2000). Furthermore, glacio-isostatic modelling suggests that the northern part of Belgium may have experienced isostatic uplift and subsidence of *c.* 25 m as a result of the waxing and waning of the Weichselian ice sheet (Busschers *et al.* 2007). The uplift and subsidence values given here are considered to be extreme values, and the real values for the next 1 Ma will probably be less in the northern part of the Boom Clay subcrop zone, outside the Roer Valley Graben.

Geological and geomorphological consequences of the Quaternary climatic and tectonic evolution

Palaeogeography and palaeohydrology. The post-marine hydrographical evolution of the Campine area started with the final retreat of the sea during the Neogene, as a result of systematic sea-level lowering and overall uplift of the bordering areas around the southern North Sea (Miller *et al.* 2005; Cloething *et al.* 2007). Marine conditions in the area, which is now covered by the Nete basin, are known to have occurred during deposition of the Diest Formation (glauconite-rich sands, deposited 11.7–7.5 Ma ago), the Kasterlee Formation (glauconite-rich sands, deposited 7.5–5.3 Ma ago), the Mol (quartz-rich sands), Poederlee (glauconite-rich sands) and Lillo formations (glauconite-rich sands) (3.6–2.8 Ma), and the Kempen Group (clays and sands, deposited 2.2–1.7 Ma ago) (Kasse & Bohncke 2001; Vandenberghe *et al.* 2004; Louwye *et al.* 2007). The outcrop areas of these deposits are relatively well known, but reconstruction of former shorelines remains somewhat hypothetical (Fig. 2). Around 2.5 Ma ago (Plio-Pleistocene boundary), the North Sea coastline was located in the northern Campine area where mainly peri-marine sediments were deposited (clays from the Kempen Group; Gullentops *et al.* 2001). The hydrography was relatively simple, with rivers consequently flowing from south to north (Fig. 7). Most likely, these rivers were not incised deeply into the substrate. The situation may have been somewhat different further to the south in the upstream regions of these consequent rivers where continental conditions had already been prevailing for a much longer time.

From the Middle Pleistocene onwards, the hydrographical network was gradually changing. The coastline took a NNE–SSW direction (as is the case nowadays) due to the 'opening' of the English Channel (Vandenberghe & De Smedt 1979). Recent studies (Gibbard 2007; Gupta *et al.* 2007; Toucanne *et al.* 2009) link the opening of the English Channel to the catastrophic drainage of a large proglacial lake during marine isotope stage 12 (MIS 12), *c.* 450 ka ago (Elsterian) (Fig. 7). As a consequence, a westward component was superimposed onto the hydrographical network. In those areas where the river network was incising into the Diest Formation, this east–west component became accentuated because of the gradually exposed Diest sand ridges that were lined up parallel to the former ENE–WSW trending coastline. Finally, the Boom Clay took over the role of erosion-resistant formation. The 450 ka event triggered the formation of the Flemish Valley, probably in different steps, with extensions towards the south and the east. Low-altitude terraces, up to 20 m above the current floodplain, can be found along, e.g., the rivers Scheldt and Dijle. In the Nete basin, several planation surfaces have been recognized at different altitudes. These are interpreted as cryopediments that were formed during periglacial conditions (Vandenberghe & De Smedt 1979). The western edge of the Campine Plateau is also considered a cryopediment connected with the development of the north–south oriented river network that gradually disappeared in the course of the Middle Pleistocene. Deep incisions are also known from the River Nete, where a deep fossilized gully, more than 10 m below the current surface, has been recognized on geoelectrical soundings near the village of Westerlo (Vandenberghe & De Smedt 1979). The incision is believed to be Early Weichselian in age.

During warm interglacials (Holsteinian, Eemian, Holocene), estuarine and marine sedimentation resumed in some parts in northern Belgium together with the deposition of sands and clays, testifying to important marine transgressions in the southern North Sea basin (Vandenberghe *et al.* 2004). Middle and Late Pleistocene estuarine conditions are recorded in the Ijzer basin, the Flemish Valley and the Scheldt estuary, but not in the Campine area itself. The widespread occurrence of aeolian deposits, coversands in the north and loess in the south suggests that during some intervals, climatic conditions were very dry. Most aeolian deposits are Weichselian in age, although older sediments can be found as well, especially in the loess area in the southern part of Flanders (Gullentops *et al.* 2001). During the Holocene, under a climate comparable to the present one, soils started developing, which led to the stabilization of the Weichselian land surface. In the northern, central and eastern Campine area, podzol soils developed on sandy substrate, whereas loamy and clayey soils with distinct texture B horizons developed in the southern–southeastern Campine and Lower Scheldt basin, respectively.

Evolution of the Flemish Valley. The long-term evolution of the surface environment of the Campine area is strongly linked with the evolution of the Flemish Valley. The Flemish Valley is defined as a broad and relatively deep depression below current sea-level that formed as a result of erosion during sea-level lowstands (Tavernier 1946; Tavernier & De Moor 1974) (Fig. 8). At present, it is completely filled with sediment. The deepest parts are situated in the Ghent area and

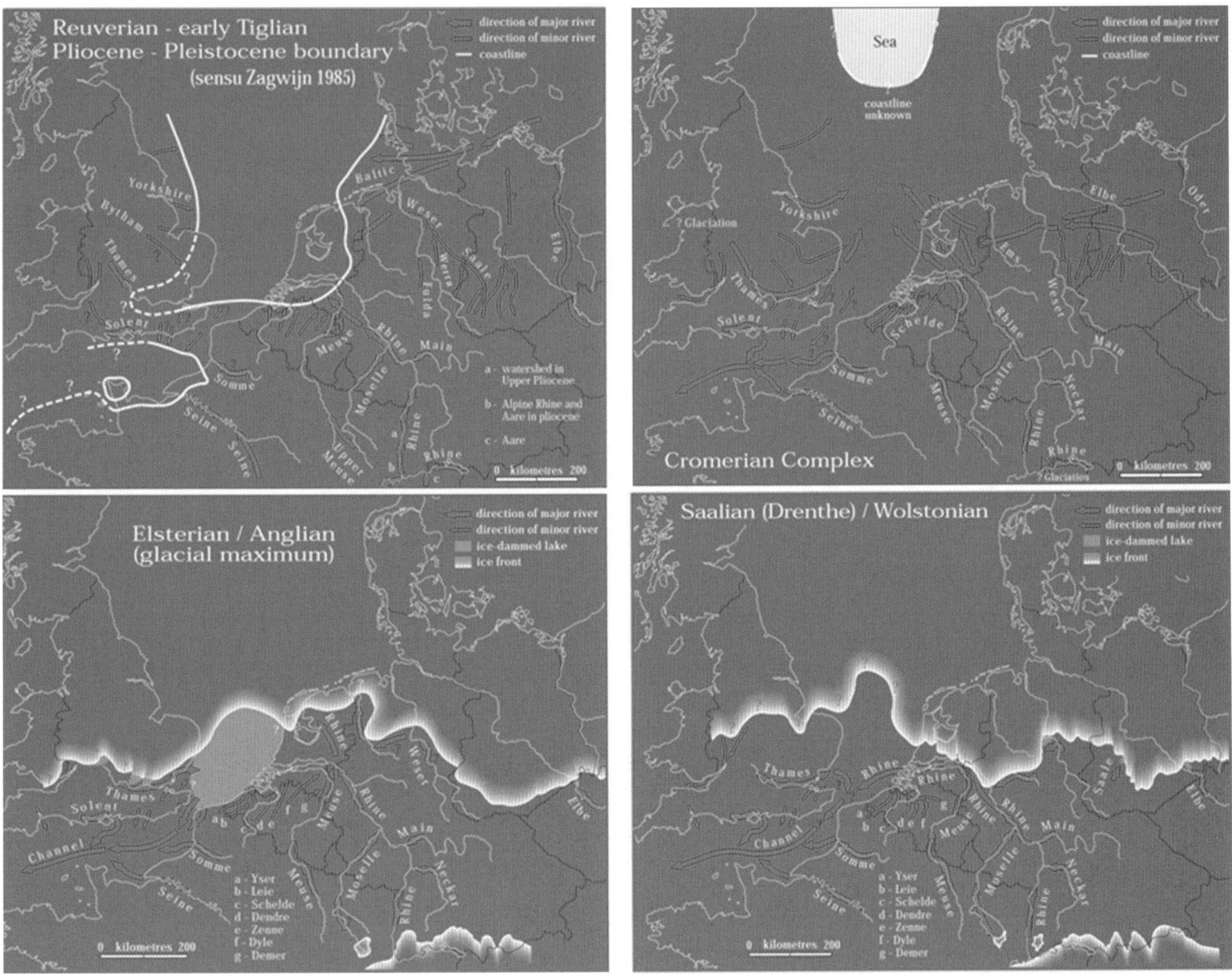

Fig. 7. Evolution of the river network in NW Europe from the Tiglian to the Saalian, according to Gibbard (1988; http://www.dov.vlaanderen.be, consulted in 2010). Reuverian–Early Tiglian, *c.* 2.5–2.0 Ma ago; Cromerian Complex, *c.* 0.9–0.5 Ma ago; Elsterian, *c.* 0.45 Ma ago; Saalian (Drenthe), *c.* 0.15 Ma ago. See also Hijma *et al.* (2012).

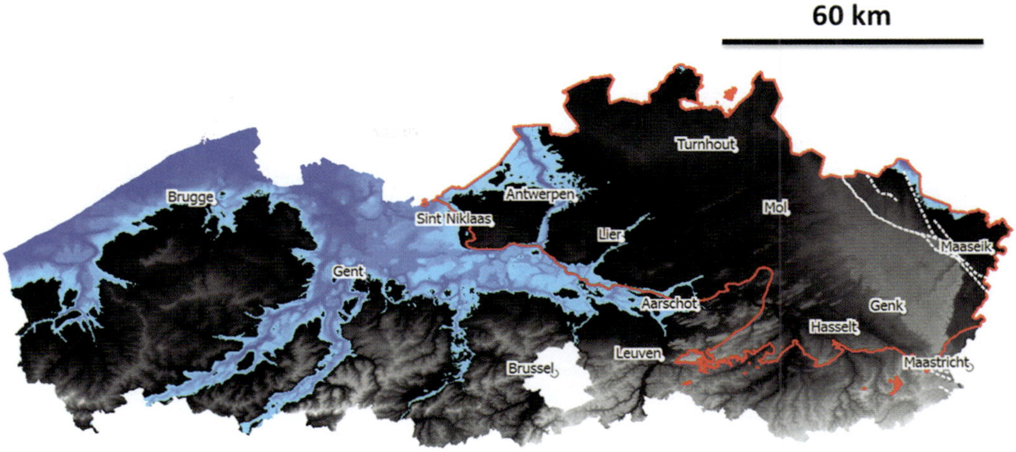

Fig. 8. Map with base of Quaternary deposits in Flanders: areas with base >0 m are in greyscale (between 0 and 150 m) and <0 m in colour scale (between −40 m and 0 m). The Boom Clay outcrop and subcrop zone is indicated by the red line, while major Quaternary faults are in white. The Flemish Valley can be defined as the zone in the western part of Belgium with Quaternary deposits below sea-level. Note that in the deepest parts of the Roer Valley Graben, the base of the Quaternary deposits is also below 0 m (east of the major faults). From Geologisch 3D Model Vlaanderen (v1.2011).

along the coastline, where it reaches depths below −15 m, down to −30 m. Extensions of the Flemish Valley can be found to the south and east of the Ghent area. The evolution of the Flemish Valley during the Quaternary has been described differently, according to various authors. The age of the Flemish Valley has been assigned to the Cromerian (Paepe *et al.* 1981) and, alternatively, as a result of polycyclic development during the Holsteinian, Saalian, Eemian and Weichselian (De Moor 1963). The southern and eastern extensions or branches are thought to have been developed first in the Saalian, but the deepest incisions are attributed to the Eemian and Weichselian (Bogemans 1993). In Figure 9, a schematic topographical transect is shown from the Campine down to the deepest part of the North Sea off the Belgian coast (see Fig. 4 for location of the profile). Also shown in the figure is the projected depth of the Flemish Valley, including its eastern branch. A strong gradient is present just off the current coastline, down to the bottom of the southern North Sea, with the presence of a wide submerged valley. This valley has developed in response to consecutive sea-level lowstands posterior to 450 ka when a huge offshore fluvial system drained the bottom of the North Sea, and/or the catastrophic drainage of proglacial lakes. The projected Flemish Valley can be identified as a subsequent hanging valley of this wide offshore valley system. With its present altitude of −30 m to −40 m, this offshore valley bottom seems to have served as the baseline for fluvial development in the Scheldt basin. As can be seen in Figure 8, the Flemish Valley has developed in subsequent position to the Boom Clay cuesta. This observation has already been described by Vandenberghe & De Smedt (1979), for example, and is attributed to the erosion resistance of the consolidated and very cohesive Boom Clay relative to the under- and overlying Tertiary sands. Note that dipping layers appear adjacent to each other on a geological map as shown in Figure 2. This means that the Boom Clay outcrop zone effectively hampers fluvial erosion and expansion of the Flemish Valley to the north. Here, we introduce the concept of the self-protecting role of Boom Clay against erosion. During the Pleistocene, two rivers succeeded in breaking through the Boom Clay cuesta: the rivers Nete and Scheldt. This resulted in *c.* 30 m of localized linear erosion.

Future evolution of the surface environment in the Campine area

Representative climates and derivation of relevant timeframes

A useful source of information on future climates within the next 1 Ma, in relation to radioactive waste disposal, is available from the BIOCLIM project (Modelling sequential BIOsphere systems under CLIMate change for radioactive waste disposal; BIOCLIM 2001, 2004). The main objective of the BIOCLIM project was to provide a scientific

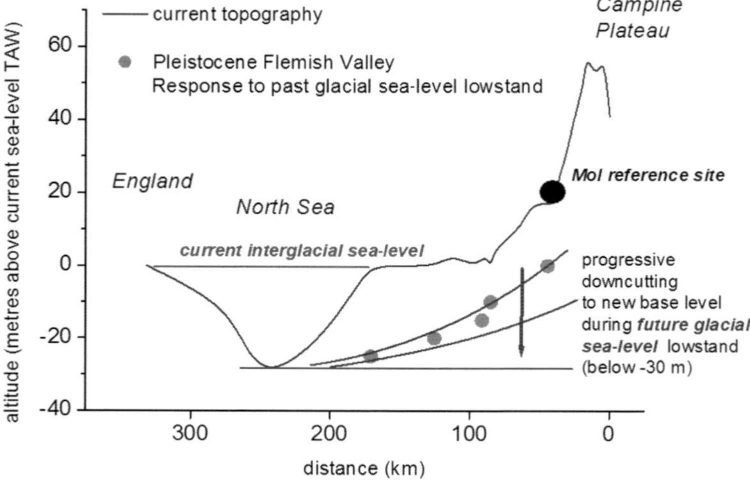

Fig. 9. General topography of the eastern and central Scheldt basin down to the bottom of the North Sea, along an east–west profile from the Campine area to the coast of England (see Fig. 1 for location of the profile). Also shown is the projected depth of the presumed valley floor of the fluvial Flemish Valley system, at discrete locations (bathymetry based on Google Ocean). The projected valley floor of the Flemish Valley system is based on the Quaternary geological map of Flanders, map sheets 16 (Lier), 23 (Mechelen), 24 (Aarschot), 13 (Brugge), 14 (Lokeren) and 15 (Antwerpen) (www.dov.vlaanderen.be). See Bogemans (2005) for a compilation of these map sheets.

basis and practical methodology for assessing possible long-term impacts of climate and environmental change on the safety of radioactive waste disposal facilities. Climate modelling for the future 1 Ma was based on natural insolation variations together with various assumptions on atmospheric CO_2 concentrations. This resulted in two suites of climate models (expressed in time after present (AP) v. northern hemisphere ice volume): those with natural CO_2 variations (A scenarios) and those with anthropogenically forced CO_2 concentrations (B scenarios). In the A scenarios, conditions as warm as the present day persist for a considerable time in northwestern Europe (up to *c.* 50 ka AP). In two of three subscenarios, a glacial period is then predicted at *c.* 53 ka AP.

In the B scenarios, the situation is completely different. The most striking result of the climate predictions, taking into account the anthropogenic scenarios, is that the next glaciation will be delayed and be less severe. The first glacial period is simulated to appear at 178 ka AP for the three subscenarios, with much smaller ice sheets than in the A scenario simulations. It appears that the anthropogenic fossil fuel contribution will have an impact on the future climate for at least the next 400 ka. The first important glacial period with the northern hemisphere ice volume exceeding 30×10^6 km^3 appears after 400 ka AP in one subscenario. After 500 ka AP, the impact of the fossil fuel contribution becomes smaller in favour of natural variations (i.e. the return to 'typical' Quaternary glacial–interglacial alternations).

The future climate scenarios are extremely dependent on the CO_2 forcing assumptions that are used in the calculations. Hence, the precise timing of glacial–interglacial periods and the volume of ice sheets differ considerably for the various climate simulations. Therefore, a spectrum of potential and representative climate conditions are identified, instead of deriving a single climatic evolution scenario for the next 1 Ma (De Craen *et al.* 2012; Van Geet *et al.* 2012). The conceivable climate types for our region include a warm climate (i.e. the result of continued global warming; Fichefet *et al.* 2007) with and without a marine transgression affecting the Campine area, and a cold climate with and without permafrost. The presence of an ice sheet advancing over northern Belgium is very unlikely but cannot be ruled out completely (Huybrechts 2010).

Next, relevant timeframes are defined. They either relate to the safety assessment of the geological disposal of high-level nuclear waste, or they are derived from the BIOCLIM modelling results that are thought to be representative for specific climate conditions (De Craen *et al.* 2012). The first timeframe from 0 to 10 ka AP is based on specific requirements from the viewpoint of performance assessment (see De Craen *et al.* 2012). The next timeframe, between 10 and 50 ka AP, is considered to be representative for a warm climate with a marine transgression, while the period 50–170 ka AP is considered to be representative for a warm climate without a marine transgression. The timeframes 170–400 ka AP and 400–1000 ka AP are considered to be representative for cold climatic conditions without and with permafrost, respectively.

Climatic changes and tectonic movements together will largely determine the evolution of individual components of the physical geographical system, such as the orohydrography, soils, vegetation and hydrology. The impact of human activities will not be considered in this work. As is the case with climatic changes, the future tectonic evolution is impossible to predict in a precise way. Therefore, in this exercise we will refer to tectonics in a qualitative way, to indicate whether the specific climatic conditions are combined with uplift or subsidence.

Timeframe 0–10 ka AP

Tectonics are not considered in this timeframe given the very limited amount of potential vertical movement. Global warming is expected to occur, with or without a marine transgression. Note that the global warming climatic condition without a marine transgression in the Campine area does not necessary exclude global sea-level rise.

Global warming without marine transgression in the Campine area. Various studies regard global warming in northwestern Europe as a transition from a temperate (present-day) to a subtropical climate (BIOCLIM 2004). Average temperatures are expected to increase by several degrees as a result of anthropogenic greenhouse gas emissions (possibly between +1.1 °C and +6.4 °C by the end of the twenty-first century already; IPCC 2007). Owing to the long residence time of CO_2 in the atmosphere, the impact of human activities on global climate may be maintained for several centuries and millennia. The landscape in the Campine area has proven relatively stable during the present interglacial (Holocene, from 11.7 ka up to the present; see Beerten & Leterme 2012), mainly as a result of the dense vegetation cover. The density of the vegetation cover is not expected to change significantly during this timeframe, relative to the current situation (BIOCLIM 2004). Soil formation processes may resume in areas where the soil has been destroyed, leading to enhanced stabilization of the surface over several thousand years (Beerten *et al.* 2012). It is therefore assumed that landscape stability will not decrease drastically (Fig. 10). Instead, human activities and changes in land use could have more impact on the landscape than global warming itself.

With regard to the water balance, several studies are in favour of increased seasonality with reduced summer precipitation and slight to moderate increases in winter precipitation (BIOCLIM 2004; Ozer *et al.* 2008; Loutre *et al.* 2010). Using climatic analogue stations, Leterme *et al.* (2012) found an increase in the annual groundwater recharge on grassland by 3% for a 5% increase in annual precipitation. However, in the case of warm and dry summers, which are to be expected in a subtropical climate, the groundwater pumping rate may increase and lower the groundwater table significantly (BIOCLIM 2004). Indeed, today, the water table may be 0.25 m lower during very dry summers in the central Campine region, relative to 'normal' summers (Labat 2008).

Global warming with marine transgression. Global warming with a full marine inundation of large parts of the Boom Clay outcrop and subcrop zone may well occur during the next 10 ka (Fichefet *et al.* 2007) (Figs 11 & 12). As a consequence of a marine transgression, a marine abrasion surface will be formed due to coastal erosion processes. During the transgression, the pre-existing topography with distinct highs and lows will be flattened out, in particular if the transgressed surface consists of unconsolidated sediments as is the case in the Campine area. However, marine erosion is limited and considered to be within several metres only (Vandenberghe *et al.* 1998). During a marine transgression, estuarine and marine sedimentation will resume in large parts of the Boom Clay area. The present-day fluvial environment of the Nete basin will evolve into an estuarine and marine environment. Furthermore, rising sea-level is thought to bring about several changes, first transforming the low-lying areas in a coastal zone. A sea-level rise of 20 m for instance will cause a full marine inundation of the Nete basin. Seepage zones will shift from inundating valleys to the edges of topographical ridges within the basin. The pore water composition of the aquifers will gradually change into that of seawater.

Timeframe 10–50 ka AP

Long-term future climate projections from BIOCLIM (2001) do not foresee major changes towards colder climates within this timeframe. Therefore, it is expected that one of the two climatic conditions described for the timeframe 0–10 ka AP will persist (subtropical with or without marine transgression), or will alternate with a temperate climate comparable to the current one. There might be episodes with marine inundations of the Campine area (see previous timeframe), but also episodes of sea-level comparable to the present-day

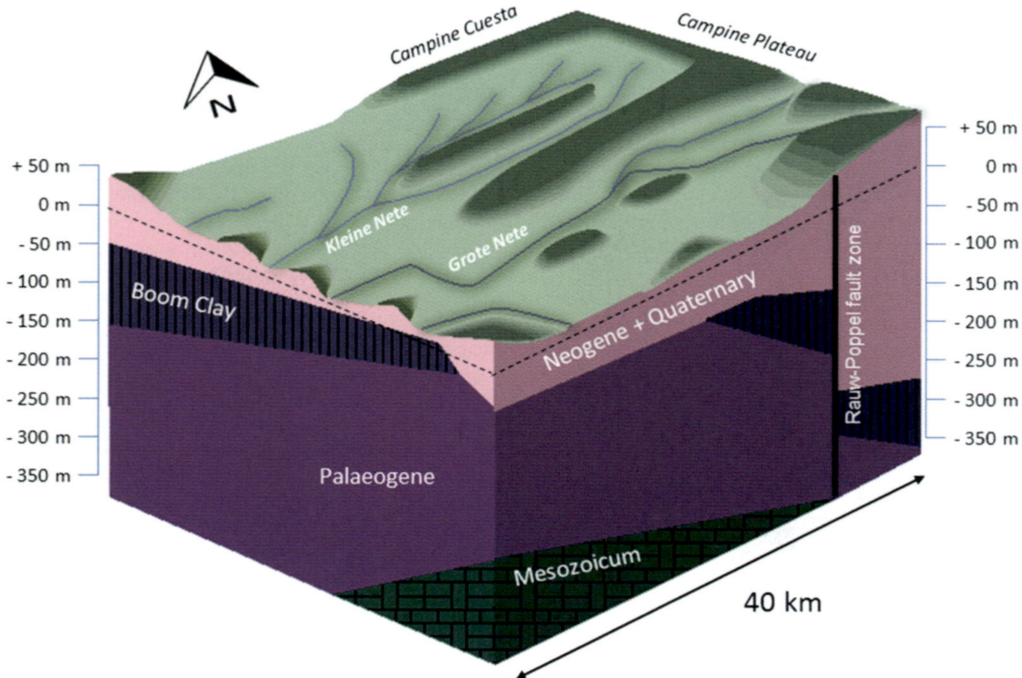

Fig. 10. Illustration of the present-day landscape and subsurface geology of the upstream parts of the Kleine and Grote Nete basins. Dark green indicates areas with mainly forest, and light green indicates meadow and arable land. This present-day stable situation is thought to be representative of a future warm climate without marine transgression. See Figure 4 for location.

situation. In such cases, ongoing soil formation processes may further transform the mostly sandy substrate into podzol soils with thick leached and deep accumulation horizons (Thompson 1981). Pronounced argillic horizons may develop on the glauconite-rich parent material (the Diest and Kattendijk formations and, to a lesser extent, the Kasterlee and Poederlee formations) that is typical for the Campine area (van Ranst & De Coninck 1983). These may significantly alter the hydrological characteristics of the unsaturated zone. Fault reactivation at the eastern border of the Campine Plateau will probably produce morphologically visible fault steps in the landscape (Vanneste *et al.* 2001). Tectonic movements on the order of several metres may be expected such that erosional processes become more or less important in the case of uplift or subsidence. It will also influence the position of the coastline.

Timeframe 50–170 ka AP

Within this timeframe, relatively warm conditions are expected to dominate the climate. Cold conditions with a significant sea-level lowering are not expected but cannot be ruled out completely. In fact, conditions described for the previous timeframes may persist in the case of the B scenarios (anthropogenically forced CO_2 contributions), while cold climatic conditions may already appear in the case of the A scenarios (natural CO_2 variations). In any case, erosion and denudation is not expected to become significant, not even with tectonic uplift. However, an extended period of high sea-level may lead to salinization of the aquifers above and below the Boom Clay, and also the Boom Clay itself. An inverse interpretation of modelled natural tracer profiles in Boom Clay has indicated that the chloride content at repository depth (*c.* 230 m) may already reach 2000 mg l^{-1} after 100 ka (Mazurek *et al.* 2008).

Timeframe 170–400 ka AP

Within this timeframe, cold climatic conditions will probably appear in the investigation area (for the first time, following the B scenarios). A cold climate in the Campine region without permafrost development may be identical to a subarctic (BIOCLIM 2004) and a boreal (Leterme *et al.*

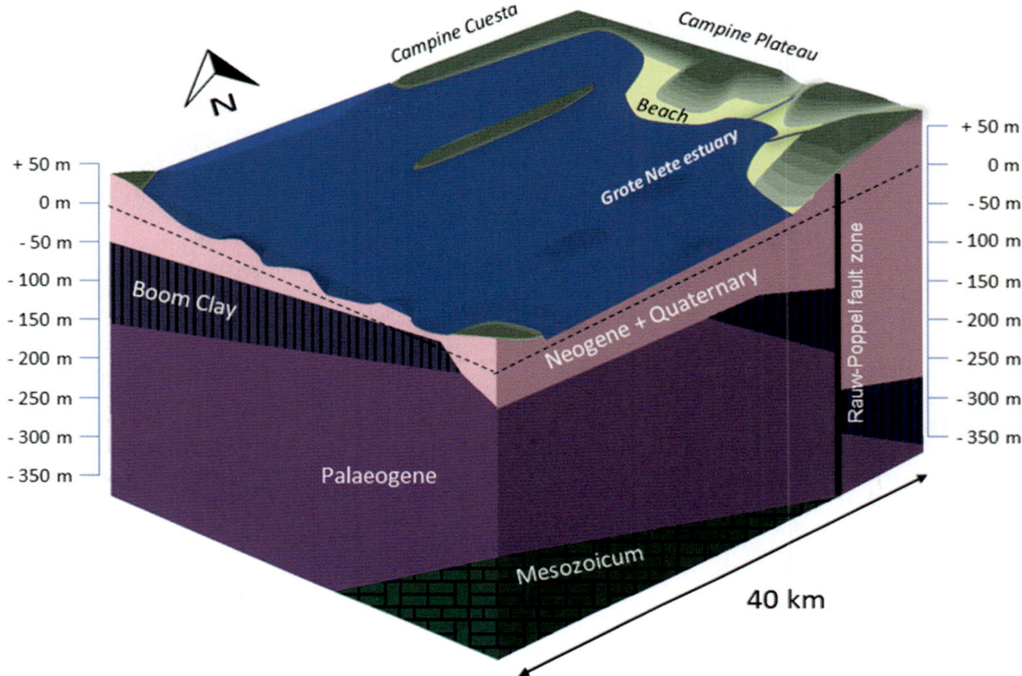

Fig. 11. Illustration of the landscape and subsurface geology of the upstream parts of the Kleine and Grote Nete basins with sea-level rising to *c.* +30 m. Dark green indicates areas with dense vegetation, and yellow indicates sandy beaches. See Figure 4 for location.

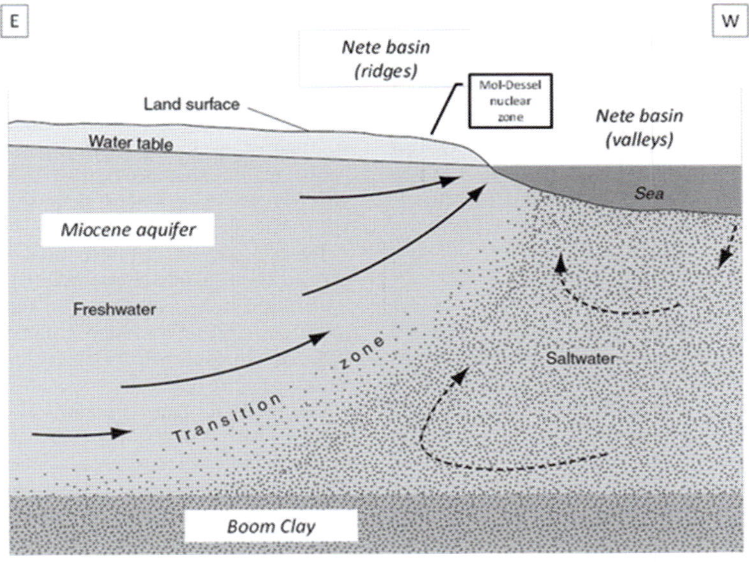

Fig. 12. Saltwater–freshwater boundary in a coastal aquifer, translated to the Nete basin (modified from Barlow 2003).

2010) climate. The main features of such a climate are lower temperatures, lower precipitation, different precipitation (more snow) and a seasonally frozen ground. Vegetation density is likely to remain unchanged, although the type of vegetation will drastically change: coniferous (boreal) forest is expected to be the dominant landscape element. Soil formation processes on a sandy soil, such as podzolization, will probably continue. The hydrographical network may show the appearance of braided rivers. Recharge conditions may be less favourable as a result of reduced infiltration rates (Leterme & Mallants 2012). The baseline for erosion will be lowered drastically due to a sea-level drop. It is thought that the dry North Sea valley bottom at *c.* 30–40 m below sea-level will be the limit, resulting in potential erosion of *c.* 60 m in the upper reaches of the Kleine Nete basin and *c.* 30–40 m near the River Scheldt estuary (see Fig. 9). Denudation of interfluvia and fluvial erosion may be enhanced because of the increased importance of frost weathering and increased erosivity of precipitation (i.e. snowmelt on a frozen substrate in spring). Uplift and/or subsidence may modify this pattern, where incision and denudation will be promoted by uplift and counterbalanced by subsidence. Estuarine conditions that may have developed during the previous timeframes may significantly alter the soil associations and surficial geology of the Campine area. Deposition of estuarine clays will probably change the hydrology and vegetation after the marine regression associated with a global sea-level drop. Runoff may increase drastically and enhance fluvial incision. Even though cold climatic conditions will likely occur during this timeframe, it is expected that they will alternate with warmer interglacial conditions. Sea-level rise following a sea-level lowstand may cause the incised valleys to fill up again with marine and/or estuarine sediments (sands and clays) and will change the porewater composition of the aquifers, depending on the magnitude and duration of the inundation. Given the duration of this timeframe, perhaps one or two climatic cycles may occur.

Timeframe 400 ka–1 Ma AP

During this timeframe, the typical Quaternary glacial–interglacial cycles of *c.* 100 ka will probably resume or continue, while permafrost development during strong glaciations becomes very likely (Fig. 13). A cold climate in the Campine region with permafrost development may be identical to an arctic (BIOCLIM 2004) or a tundra (Leterme *et al.* 2010) climate, while interglacials will be characterized by a climate that is comparable to the current one. The main features of an arctic or tundra climate are lower temperatures (mean annual temperature below -4 °C for discontinuous permafrost and below -8 °C for continuous permafrost), a lower amount of precipitation, different precipitation (more snow) and a permanently frozen ground. The characteristics of the Earth surface during successive glacial–interglacial cycles will shift from a stable landscape with dense vegetation, thick soils, meandering rivers and sufficient connection between the surface and subsurface hydrological system during interglacials to a generally instable landscape with less vegetation (densely to sparsely vegetated tundra), weakly developed (frozen) soils, braided river systems arranged in a dense network, strong erosion phenomena and temporarily isolated surface hydrological systems during glacial stages. In the presence of water, cryosols will develop that are characterized by cryoturbated horizons, frost heave, thermal cracking, ice segregation and patterned ground microrelief (IUSS Working Group WRB 2007). Depending on the degree of permafrost development, the hydrological system at the surface may become largely independent of the subsurface groundwater system. In this respect, we note that ^{14}C age distributions of the Ledo-Paniselian aquifer below the Boom Clay show a hiatus in the timeframe between *c.* 14 ka BP and *c.* 20 ka BP, which may well be related to the strongly reduced recharge conditions (Blaser *et al.* 2010).

Permafrost depth in the central Campine area during a glacial period that is analogous to the Weichselian glaciation is estimated to be *c.* 200 m during the coldest intervals considering snow and vegetation (Govaerts *et al.* 2011). This value of 200 m is significantly larger than previous permafrost calculations for a 'standard' western European context (*c.* 120 m; Delisle 1998) and estimations based on the thickness of now filled-up Weichselian pingo ruins in the Netherlands (*c.* 25 m; Berendsen 1998). However, the latter should probably be regarded as minimum depth estimates (Berendsen 1998). Continuous permafrost conditions would probably not last longer than several millennia (Huijzer & Vandenberghe 1998).

Denudation and erosion rates will drastically increase because of a combination of frost weathering, a vulnerable active layer in summer and peak runoff on a frozen and barren underground. Landscape development will be characterized by important slope retreat due to cryopedimentation. Peak discharge will cause periodical activation of braided river systems that will show alternations of deposition and erosion. The maximum amount of erosion is constrained by the altitude of the dry North Sea bottom at *c.* 30–40 m below current sea-level. A total number of ten sea-level lowstands may cause the relief in the Campine area to be lowered by 35 m in the west (Scheldt area), and

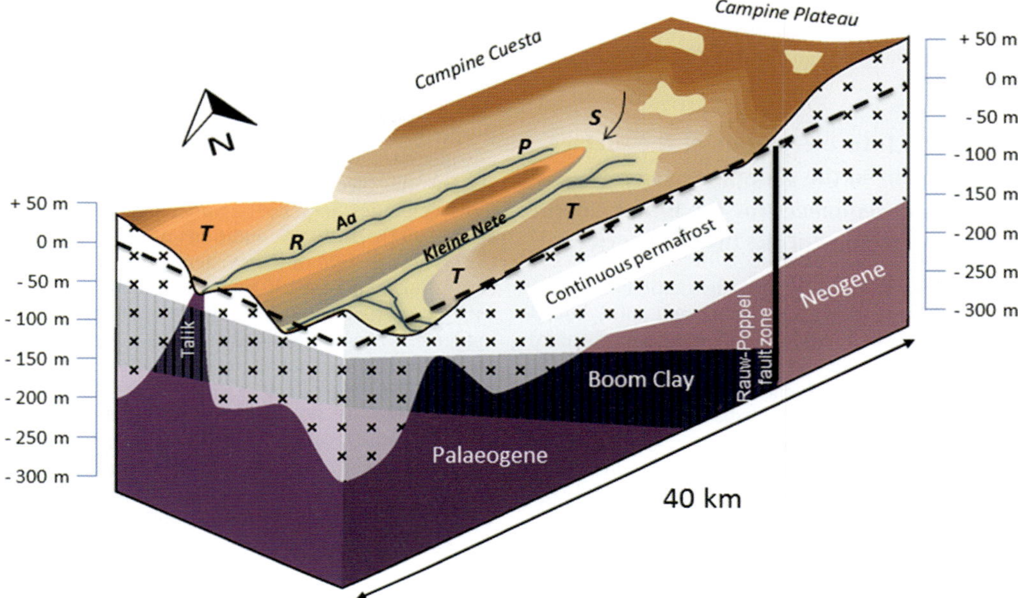

Fig. 13. Illustration of the landscape and subsurface geology of the upstream part of the Kleine Nete basin during a cold climate with permafrost, after several glacial–interglacial cycles. Brown areas refer to tundra vegetation with few or no trees at all, while yellow areas are sandy plains (plateaus and fluvial braided plains). As a result of fluvial erosion (*R*), terraces will be formed (*T*), while slope retreat (*S*) and cryopedimentation (*P*) will further widen the braided river plains and cause large-scale denudation. Note that some valleys are cut down below present sea-level, a situation similar to the core area of the Flemish Valley during the Late Pleistocene. The Campine Cuesta and Campine Plateau remain relatively stable due to their erosion-resistant composition (cohesive clay and gravel, respectively). Permafrost depth (for illustration purposes only) may be *c.* 200 m during the coldest stages assuming snow and vegetation, but will be significantly less in the valleys. See Figure 4 for location.

more than 60 m in the east (Nete area), as a result of the reactivation and headward expansion of the Flemish Valley. Warm/cold transitions are known to produce very unstable conditions at the Earth's surface (Vandenberghe 2008). During a warm–cold transition, deep fluvial incisions (10 m) are to be expected, followed by massive sediment input from denudating interfluves and retreating slopes. Vertical movements may modify this pattern because they influence the potential energy of fluvial processes and the relative position of the baseline for denudation and erosion. When looking at the last 1 Ma, uplift was more important than subsidence (see section 'Quaternary geological and geomorphological development'). Uplift of several tens of metres seems very likely for the next 1 Ma, especially for the southernmost areas of the Campine region (outside the Roer Valley Graben). Glacio-isostacy may further contribute to climate-induced uplift of 5–10 m and subsequent forebulge downwarping during an ice sheet advance over northwestern Europe (Peltier 2004; Steffen 2006; both cited in Busschers *et al.* 2007).

So far, the Pleistocene ice sheets did not reach the Campine area. Furthermore, ice sheet simulations using the coldest BIOCLIM climatic scenario (BIOCLIM 2001) do not predict a glacial advance over northern Belgium during the next 1 Ma, so this event is considered to be very unlikely. However, should it occur, the landscape will be completely covered with ice, and another type of erosion process – glacial erosion – will start playing a determining role in the evolution of the landscape. The geological record in northern Europe (minimum 100 km north of the Campine area) shows that ice sheets may drastically change the surface and subsurface environment with glaciotectonic deformation with the creation of ice-pushed ridges up to 100 m high and glacial basins up to 150 m deep (Zagwijn 1974), the formation of subglacial tunnel valleys that may reach depths of up to 200 m (Jørgensen & Sandersen 2006), and/or the development of ice-marginal valleys and outwash plains. Furthermore, with an advancing ice sheet, the groundwater system will become pressurized, with increased hydraulic heads and flow velocities, and discharge through meltwater channels, proglacial lakes and seepage zones in discontinuous permafrost (van Weert *et al.* 1997).

Summary: future evolution of the surface environment in the Campine area and possible consequences for the burial depth and hydrogeology of the disposal system

Sea-level variations, climatic fluctuations between glacial and interglacial conditions and uplift/subsidence have been recorded in the geological record from the Campine area and surroundings (e.g. Saalian ice sheet advance in the Netherlands). Differential movements of the land surface and climate changes are expected to occur in the next 1 Ma and profoundly change the surface environment. In this section, possible consequences on the burial depth and the hydrogeological system will be discussed, as these are expected to be rather sensitive to changes in the surface environment.

Those combinations of features, events and processes that would govern a marine inundation in the Campine area all have several changes in common. Typical examples are tectonic subsidence and/or sea-level rise as a result of an extended period of warm climatic conditions. The porewater composition of the aquifers (and eventually Boom Clay) will gradually change from meteoric to seawater, and the sedimentary overburden above the Boom Clay is likely to increase (sand and/or clay) by several metres, perhaps several tens of metres, depending on the uninterrupted duration of these conditions that may reach 100 ka and more in several climatic scenarios from BIOCLIM (2001). Subsidence may further contribute to these conditions, certainly in the Roer Valley graben where the subsidence rates will be higher. The flow system will also change drastically with strongly reduced flow velocities and changing flow paths.

Those combinations of features, events and processes that govern the persistence of continental conditions due to uplift, and the development of cold climatic conditions, also share several characteristics. Uplift over the next 1 Ma will promote fluvial incision and denudation of interfluvia, such that the thickness of the overburden decreases. Erosion-resistant landscape elements, such as the Campine Plateau, the Campine Cuesta and the Boom Clay Cuesta are expected to remain topographical highs, whereas the sands will become eroded more quickly, especially in the valleys. Reactivation and further development of the Flemish Valley during (extended) sea-level lowstands may further enhance the overburden reduction. Strong changes in relief will also bring about changes in the configuration of recharge and discharge zones. During episodes of continuous permafrost, the groundwater system may become completely disconnected from the surface hydrology, leading to strongly reduced recharge and discharge. However, it should be noted that if an ice sheet were to advance over northwestern Europe, the groundwater system may become pressurized, such that water flow velocities increase significantly (van Weert *et al.* 1997). No-flow conditions are thus not very likely, also in light of the existence of taliks (unfrozen ground underneath rivers and lakes). Even if such conditions should occur, they would not last longer than several millennia in our regions, based on palaeoenvironmental reconstructions of the Weichselian (Huijzer & Vandenberghe 1998). An ice sheet advance over northern Belgium may completely change the hydrogeological system as the overburden may be deformed and/or removed. Such an event is not expected during the next 1 Ma, but cannot be ruled out completely.

Conclusions

The large-scale geomorphological development of the Boom Clay outcrop and subcrop zone is generally well understood. The synthesis of Quaternary landscape evolution in this work has focused on the specific application of waste disposal, and lines up with the description of the current geomorphological and hydrographical configuration of the area of interest. Furthermore, it provides insights into the processes that might occur during the next 1 Ma, the timeframe under consideration in the long-term safety assessment of radioactive waste disposal in geological host rocks. At present, erosion estimates based on the geological archive of sea-level variations and uplift/subsidence are treated as extreme values and need further refinement for specific locations within the Boom Clay outcrop and subcrop zone from detailed geomorphological investigations. A full understanding of the future evolution of the surface environment also needs input from analogue approaches and modelling studies, for example with respect to the hydro(geo)logical evolution including seawater intrusion as a result of global sea-level rise, permafrost development during forthcoming glacial stages, and the advance of ice sheets in northwestern Europe. The evidence and reasoning presented in this paper clearly indicates that the surface environment of the Campine area will undergo significant changes during the next 1 Ma with respect to hydrological conditions. The current patterns of recharge and discharge will completely change during marine transgressions, permafrost development and ice sheet development (north of Belgium). Furthermore, uplift and erosion will change the aquifer configuration (infiltration areas) and the overburden thickness, while soil development and vegetation changes will have an impact on landscape stability and infiltration rates.

This work was performed in close cooperation with, and with the financial support of, ONDRAF/NIRAS, the Belgian Agency for Radioactive Waste and Fissile Materials, as part of the programme on geological disposal of high-level/long-lived radioactive waste that is carried out by ONDRAF/NIRAS. The views expressed in the paper do not necessarily correspond to those of ONDRAF/NIRAS. We are grateful to J. Marivoet and K. Vandersteen for comments and suggestions on earlier versions of the manuscript. We thank J. Van Hemelryck from artlab-11 for drawing Figures 10, 11 and 13.

References

AGIV (AGENTSCHAP GEOGRAFISCHE INFORMATIE VLAANDEREN) 2006. *DHM Vlaanderen, raster, 100 m.* Departement Mobiliteit en Openbare Werken en Vlaamse Milieumaatschappij.

AGIV (AGENTSCHAP GEOGRAFISCHE INFORMATIE VLAANDEREN) 1998. *Bodemassociaties, versie 1992.* Vlaamse Landmaatschappij.

BARLOW, P. M. 2003. *Groundwater in Freshwater-Saltwater Environments of the Atlantic Coast.* U.S. Geological Survey, Circular 1262.

BATELAAN, O. & DE SMEDT, F. 2007. GIS-based recharge estimations by coupling, surface–subsurface water balances. *Journal of Hydrology*, **337**, 337–355.

BEERTEN, K. 2010. *Geomorphological Evolution of the Nete Basin: Identification of Past Events to Assess the Future Evolution.* Belgian Nuclear Research Centre External Report SCK•CEN-ER-137.

BEERTEN, K. & LETERME, B. 2012. *Physical Geography of North-Eastern Belgium – The Boom Clay Outcrop and Subcrop Zone.* Belgian Nuclear Research Centre External Report SCK•CEN-ER-202.

BEERTEN, K., DEFORCE, K. & MALLANTS, D. 2012. Landscape evolution and changes in soil hydraulic properties at the decadal, centennial and millennial scale: a case study from the Campine area, northern Belgium. *Catena*, **95**, 73–84.

BEERTEN, K., DE CRAEN, M. & WOUTERS, L. 2013. Post-Rupelian patterns and estimates of erosion and burial in the Campine area, north-eastern Belgium. *Physics and Chemistry of the Earth*, **64**, 12–20.

BERENDSEN, H. J. A. 1998. *De vorming van het land.* Inleiding in de geologie en de geomorfologie. Van Gorcum & Comp., Assen.

BERGER, A. 1977. Support for the astronomical theory of climate change. *Nature*, **269**, 44–45.

BIOCLIM 2001. *Modelling Sequential BIOsphere Systems Under CLIMate Change for Radioactive Waste Disposal, Deliverable 3: Global Climatic Features Over the Next Million Years and Recommendation for Specific Situations to be Considered*, http://www.andra.fr/bioclim/documentation.htm

BIOCLIM 2004. *Modelling Sequential BIOsphere Systems Under CLIMate Change for Radioactive Waste Disposal, Deliverable D10-12: Development and Application of a Methodology for Taking Climate-Driven Environmental Change into Account in Performance Assessments*, http://www.andra.fr/bioclim/documentation.htm

BLASER, P. C., COETSIERS, M., AESCHBACH-HERTIG, W., KIPFER, R., VAN CAMP, M., LOOSLI, H. H. & WALRAEVENS, K. 2010. A new groundwater radiocarbon correction approach accounting for palaeoclimate conditions during recharge and hydrochemical evolution: the Ledo-Paniselian aquifer, Belgium. *Applied Geochemistry*, **25**, 437–455.

BOGEMANS, F. 1993. Quaternary geological mapping on basis of sedimentary properties in the eastern branch of the Flemish Valley. *Toelichtende Verhandeling voor de Geologische en Mijnkaarten van België*, **35**, 1–49.

BOGEMANS, F. 2005. *Technisch verslag bij de opmaak van de Quartairgeologische kaart van Vlaanderen.* Vlaamse Overheid, Dienst Natuurlijke Rijkdommen.

BUSSCHERS, F. S., KASSE, C. ET AL. 2007. Late Pleistocene evolution of the Rhine-Meuse system in the southern North Sea basin: imprints of climate change, sea-level oscillation and glacio-isostacy. *Quaternary Science Reviews*, **26**, 3216–3248.

CLARK, P. U., MIX, A. C. & BARD, E. 2001. Ice sheets and sea level of the Last Glacial Maximum. *Transactions of the American Geophysical Union*, **82**, 241 & 246–247.

CLOETHING, S. A. P. L., ZIEGLER, P. A., BOGAARD, P. J. F., 30 CO-WORKERS AND TOPO-EUROPE WORKING GROUP 2007. TOPO-EUROPE: the geoscience of coupled deep Earth–surface processes. *Global and Planetary Change*, **58**, 1–118.

DE CRAEN, M., BEERTEN, K., HONTY, M. & GEDEON, M. 2012. *Geo-Scientific Evidence to Support the I2 Isolation Function (Geology & Long-Term Evolution) as Part of the Safety and Feasibility Case 1 (SFC1).* Belgian Nuclear Research Centre External Report SCK•CEN-ER-184.

DELISLE, G. 1998. Numerical simulation of permafrost growth and decay. *Journal of Quaternary Science*, **13**, 325–333.

DE MOOR, G. 1963. Bijdrage tot de kennis van de fysische landschapsvorming in Binnen-Vlaanderen. *Bulletin de la Société Belge d'Etudes Géographiques*, **32**, 329–433.

DEMOULIN, A. & HALLOT, E. 2009. Shape and amount of the Quaternary uplift of the western Rhenish shield and the Ardennes (western Europe). *Tectonophysics*, **474**, 696–708.

DE SMEDT, F. 1992. De hydrologie. *In*: DENIS, J. (eds) *Geografie van België.* Gemeentekrediet, Brussel, 217–239.

EHLERS, J., GIBBARD, P. L. & HUGHES, P. D. (eds) 2011. Quaternary glaciations – extent and chronology. A closer look. *Developments in Quaternary Sciences*, **15**, 2–1108.

FICHEFET, T., DRIESSCHAERT, E., GOOSSE, H., HUYBRECHTS, P., JANSSENS, I., MOUCHET, A. & MUNHOVEN, G. 2007. *Modelling the Evolution of Climate and Sea Level During the Third Millennium (MILMO)*, Scientific Support Plan for a Sustainable

Development Policy, Belgian Science Policy, Brussels.

GEOLOGISCH 3D MODEL VLAANDEREN (V.1) 2011. Opgemaakt door VITO in opdracht van de Vlaamse overheid, ALBON.

Gibbard, P. L. 1988. The history of the great north-west European rivers during the past three million years. *Philosophical Transactions of the Royal Society of London*, **B318**, 559–602.

Gibbard, P. L. 2007. Europe cut adrift. *Nature*, **448**, 259–260.

Govaerts, J., Weetjens, E. & Beerten, K. 2011. *Numerical Simulation of Permafrost Depth at the Mol Site*. Belgian Nuclear Research Centre External Report SCK•CEN-ER-148.

Gullentops, F., Paulissen, E. & Vandenberghe, J. 1981. Fossil periglacial phenomena in NE Belgium (excursions in the Kempen on 26 and 27 September 1978). *Biuletyn Periglacjalny*, **28**, 345–365.

Gullentops, F., Bogemans, F., De Moor, G., Paulissen, E. & Pissart, A. 2001. Quaternary lithostratigraphic units (Belgium). *Geologica Belgica*, **4**, 153–164.

Gupta, S., Collier, J. S., Palmer-Felgate, A. & Potter, G. 2007. Catastrophic flooding origin of shelf valley systems in the English Channel. *Nature*, **448**, 342–345.

Hays, J. D., Imbrie, J. & Shackleton, N. J. 1976. Variations in the Earth's orbit: pacemaker for the ice ages. *Science*, **194**, 1121–1132.

Hijma, M. P., Cohen, K. M., Roebroeks, W., Westerhoff, W. E. & Busschers, F. S. 2012. Pleistocene Rhine–Thames landscapes: geological background for hominin occupation of the southern North Sea region. *Journal of Quaternary Science*, **27**, 17–39.

Hoek, W. Z. 2001. Vegetation response to the 14.7 and 11.5 ka cal. BP climate transitions: is vegetation lagging climate? *Global and Planetary Change*, **30**, 103–115.

Huijzer, B. & Vandenberghe, J. 1998. Climatic reconstruction of the Weichselian Pleniglacial in northwestern and central Europe. *Journal of Quaternary Science*, **13**, 391–417.

Huybrechts, P. 2010. *Vulnerability of an Underground Radioactive Waste Repository in Northern Belgium to Glaciotectonic and Glaciofluvial Activity During the Next 1 Million Year*. Departement Geografie VUB Report 10/01.

Imbrie, J., Hays, J. D. *et al.* 1984. The orbital theory of Pleistocene climate: support from a revised chronology of the marine $\delta^{18}O$ record. *In*: Berger, A., Imbrie, J., Hays, J. D., Kukla, G. & Salzman, B. (eds) *Milankovitch and Climate*. Reidel, Dordrecht, 269–306.

IPCC 2007. *In*: Solomon, S., Qin, D. *et al.* (eds) *Climate Change 2007: The Physical Science Basis. Contribution of Working Group I to the Fourth Assessment Report of the Intergovernmental Panel on Climate Change*. Cambridge University Press, Cambridge/New York.

IUSS WORKING GROUP WRB 2007. *World Reference Base for Soil Resources 2006, first update 2007*. World Soil Resources Reports, FAO, Rome, **103**.

Jørgensen, F. & Sandersen, P. B. E. 2006. Buried and open tunnel valleys in Denmark – erosion beneath multiple ice sheets. *Quaternary Science Reviews*, **25**, 1339–1363.

Kasse, C. & Bohncke, S. 2001. Early Pleistocene fluvial and estuarine records of climate change in the southern Netherlands and northern Belgium. *In*: Maddy, D., Macklin, M. G. & Woodward, J. C. (eds) *River Basin Sediment Systems: Archives of Environmental Change*. Balkema Publishers, Lisse, 171–193.

Labat, S. 2008. *Piezometric Measurements at the Mol-Dessel Site*. Site Characterisation for Disposal of Category A Waste. Belgian Nuclear Research Centre External Report SCK•CEN-ER-57.

Lambeck, K., Smither, C. & Ekman, M. 1998. Tests of glacial rebound models for Fennoscandinavia based on instrumented sea- and lake-level records. *Geophysical Journal International*, **135**, 375–387.

Leterme, B. & Mallants, D. 2012. *Simulation of Evapotranspiration and Groundwater Recharge in the Nete Catchment Accounting for Different Land Cover Types and for Present and Future Climate Conditions*. Belgian Nuclear Research Centre External Report SCK•CEN-ER-192.

Leterme, B., Hooker, P., Jacques, D., Mallants, D., De Craen, M. & Van den Hoof, C. 2010. *Long-Term Climate Change and Consequences for Near-Field, Geosphere and Biosphere Parameters*. NIROND TR 2009-07E V1.

Leterme, B.,, Mallants, D. & Jacques, D. 2012. Sensitivity of groundwater recharge using climatic analogues and HYDRUS-1D. *Hydrology and Earth System Sciences*, **16**, 2485–2497.

Loutre, M. F., Crucifix, M., Berger, A. & Fichefet, T. 2010. *Simulation of Extreme Climate Changes Under Interglacial Boundary Conditions for Radioactive Waste Disposal. A Study Performed with the LOVE-CLIM Model in a Belgian Perspective*. Georges Lemaître Centre for Earth and Climate Research (TECLIM), Université Catholique de Louvain, Scientific Report **2010/01**.

Louwye, S., De Schepper, S., Laga, P. & Vandenberghe, N. 2007. The Upper Miocene of the southern North Sea Basin (northern Belgium): a palaeoenvironmental and stratigraphical reconstruction using dinoflagellate cysts. *Geological Magazine*, **144**, 33–52.

Matthijs, J., Buffel, P. & Leroi, S. 2003. *Geologische dwarsprofielen doorheen het Tertiair in Vlaanderen, Schaal 1:100 000 – Technische Toelichting*. Geological Service Company bvba (GCS) in opdracht van het Ministerie van de Vlaamse Gemeenschap – Afdeling Natuurlijke Rijkdommen en Energie, Brussels.

Mazurek, M., Alt-Epping, P., Bath, A., Gimmi, T. & Waber, H. N. 2008. *Natural Tracer Profiles Across Argillaceous Formations: Review and Synthesis*. CLAYTRAC Project, NEA No. **6253**.

Meyus, Y., Woldeamlak, S. T., Batelaan, O. & De Smedt, F. 2004. *Opbouw van een Vlaams Groundwatervoedingsmodel: Eindrapport*. Rapport in opdracht van AMINAL, afdeling Water, Brussels.

Miller, K. G., Kominz, M. A. *et al.* 2005. The Phanerozoic record of global sea-level change. *Science*, **310**, 1293–1298.

Ozer, J., Van den Eynde, D. & Ponsar, S. 2008. *Evaluation of Climate Change Impacts and Adaptation Responses for Marine Activities: CLIMAR. Trend Analysis of the Relative Mean Sea Level at Oostende (Southern North Sea – Belgian Coast)*. Report of the

CLIMAR project for Belgian Federal Science Policy Office.

PAEPE, R., BAETEMAN, C., MORTIER, R. & VANHOORNE, R. 1981. The marine Pleistocene sediments in the Flandrian area. *Geologie en Mijnbouw*, **60**, 321–330.

PELTIER, W. R. 2004. Global glacial isostasy and the surface of the ice-age Earth: the ICE-5G (VM2) model and GRACE. *Annual Review of Earth and Planetary Sciences*, **32**, 111–149.

RMI 2011. Royal Meteorological Institute of Belgium (RMI). www.meteo.be

STEFFEN, H. 2006. *Determination of a consistent viscosity distribution in the Earth's mantle beneath Northern and Central Europe*. PhD thesis, Institut für Geologische Wissenschaften der Freie Universität Berlin.

STRAHLER, A. H. & STRAHLER, A. N. 1992. *Modern Physical Geography*, 4th edn. Wiley, New York.

SVENDSEN, J. I., ALEXANDERSON, H. *ET AL.* 2004. Late Quaternary ice sheet history of northern Eurasia. *Quaternary Science Reviews*, **23**, 1229–1271.

TAVERNIER, R. 1946. L'évolution du Bas Escaut au Pleistocène supérieur. *Bulletin de la Société Belge d'Etudes Géographiques*, **55**, 106–125.

TAVERNIER, R. & DE MOOR, G. 1974. L'évolution du Bassin de l'Escaut. *Centenaire de la Société Géologique de Belgique*, 159–233.

THOMPSON, C. H. 1981. Podzol chronosequences on coastal dunes of eastern Australia. *Nature*, **291**, 59–61.

TOUCANNE, S., ZARAGOSI, S. *ET AL.* 2009. A 1.2 Ma record of glaciation and fluvial discharge from the Western European Atlantic margin. *Quaternary Science Reviews*, **28**, 2974–2981.

VAN BALEN, R. T., HOUTGAST, R. F., VAN DER WATEREN, F. M., VANDENBERGHE, J. & BOGAART, P. W. 2000. Sediment budget and tectonic evolution of the Meuse catchment in the Ardennes and the Roer Valley Rift System. *Global and Planetary Change*, **27**, 113–129.

VAN DEN BERG, M. W. & VAN HOOF, T. 2001. The Maas terrace sequence at Maastricht, SE Netherlands: evidence for 200 m of late Neogene and Quaternary surface uplift. *In*: MADDY, D., MACKLIN, M. G. & WOODWARD, J. C. (eds) *River Basin Sediment Systems: Archives of Environmental Change*. Balkema, Rotterdam, 45–86.

VAN GEET, M., DE CRAEN, M. *ET AL.* 2012. Climate evolution in the long-term safety assessment of surface and geological disposal facilities for radioactive waste in Belgium. *Geologica Belgica*, **15**, 8–15.

VAN LANDUYT, W., VANHECKE, L. & HOSTE, I. 2006. Geografische factoren in de verspreiding van planten – Hoofdstuk 6. *In*: VAN LANDUYT, W., HOSTE, I., VANHECKE, L., VAN DEN BREMT, P., VERCRUYSSE, W. & DE BEER, D. (eds) *Atlas van de flora van Vlaanderen en het Brussels Gewest*. Nationale Plantentuin van België en Instituut voor Bosbouw en Wildbeheer (INBO), Impressum Meise, Brussel, 83–93.

VAN MOURIK, J. M., NIEROP, K. G. J. & VANDENBERGHE, D. A. G. 2010. Radiocarbon and OSL dating based chronology of a polycyclic driftsand sequence at Weerterbergen (SE Netherlands). *Catena*, **80**, 170–181.

VAN MOURIK, J. M., SLOTBOOM, R. T. & WALLINGA, J. 2011. Chronology of plaggic deposits; palynology, radiocarbon and optically stimulated luminescence dating of the Posteles (NE-Netherlands). *Catena*, **84**, 54–60.

VAN RANST, E. & DE CONINCK, F. 1983. Evolution of glauconite in imperfectly drained sandy soils of the Belgian Campine. *Zeitschrift für Pflanzenernährung und Bodenkunde*, **146**, 415–426.

VAN WEERT, F. H. A., VAN GIJSSEL, K., LEIJNSE, A. & BOULTON, G. S. 1997. The effects of Pleistocene glaciations on the geohydrological system of Northwest Europe. *Journal of Hydrology*, **195**, 137–159.

VANDENBERGHE, J. 2008. The fluvial cycle at cold–warm–cold transitions in lowland regions: a refinement of theory. *Geomorphology*, **98**, 275–284.

VANDENBERGHE, J. & DE SMEDT, P. 1979. Palaeomorphology in the eastern Scheldt basin (Central Belgium) – the Dijle–Demer–Grote Nete confluence area. *Catena*, **6**, 73–105.

VANDENBERGHE, N., LAGA, P., STEURBAUT, E., HARDENBOL, J. & VAIL, P. R. 1998. Tertiairy sequence stratigraphy at the southern border of the North Sea basin in Belgium. *In*: DE GRACIANSKY, P.-C., HARDENBOL, J., JACQUIN, T. & VAIL, P. R. (eds) *Mesozoic and Cenozoic Sequence Stratigraphy of European Basins*, Society of Economic Paleontologists and Mineralogists, Special Publications, **60**, 119–154.

VANDENBERGHE, N., VAN SIMAEYS, S., STEURBAUT, E., JAGT, J. W. M. & FELDER, P. J. 2004. Stratigraphic architecture of the Upper Cretaceous and Cenozoic along the southern border of the North Sea Basin in Belgium. *Netherlands Journal of Geosciences*, **83**, 155–171.

VANDERSTEEN, K., GEDEON, M. & LETERME, B. 2013. *Hydrogeology of North-East Belgium*. Belgian Nuclear Research Centre External Report SCK•CEN-ER-236.

VANNESTE, K., VERBEECK, K. *ET AL.* 2001. Surface-rupturing history of the Bree fault scarp, Roer Valley graben: evidence for six events since the late Pleistocene. *Journal of Seismology*, **5**, 329–359.

WATERBOUWKUNDIG LABORATORIUM 2012. *Hydrologisch Jaarboek 2011, HIC Meetstations*. Vlaamse Overheid, Departement Mobiliteit en Openbare Werken, WL Report **12_072**.

WESTAWAY 2001. Flow in the lower continental crust as a mechanism for the Quaternary uplift of the Rhenish Massif, north-west Europe. *In*: MADDY, D., MACKLIN, M. G., WOODWARD, J. C. (eds) *River Basin Sediment Systems: Archives of Environmental Change*. Balkema, Rotterdam, 87–167.

ZAGWIJN, W. H. 1974. The Paleogeographic evolution of The Netherlands during the Quaternary. *Geologie en Mijnbouw*, **53**, 369–385.

Earth tidal and barometric responses observed in the Callovo-Oxfordian clay Formation at Andra Meuse/Haute-Marne Underground Research Laboratory

M. DELCOURT-HONOREZ[1]* & E. SCHOLZ[2]

[1]*Dr ès Sciences, rue Sainte Barbe, 1A, BE 5537 Anhée-sur-Meuse, Belgium*

[2]*Andra, Centre de Meuse/Haute-Marne, RD960, 55290 Bure, France*

**Corresponding author (e-mail: fb874195@skynet.be)*

Abstract: Fluid pressures were recorded over 6.5 years in borehole EST207 located at the Andra Meuse/Haute-Marne Underground Research Laboratory in Bure (France). The borehole is equipped with a multipacker system monitoring 11 intervals, including 8 in the Callovo-Oxfordian clay Formation, 2 in the Dogger Formation and 1 at the Oxfordian base. Pressure data were analysed for responses to Earth tides and barometric pressure in the 11intervals. Fourier analyses on the pressure data revealed some of the Earth tidal components. Use of ETERNA software determined a better estimate of Earth tidal wave parameters (amplitudes, phases and their standard deviations) and barometric efficiency. Estimates of the tidal parameters K1, O1, P1, S1 and M2, N2, S2, K2 were made before and after barometric correction. It is suggested that in the Callovo-Oxfordian clay, the greater the clay content of the formation, the greater the barometric efficiency values. A barometric and tidal responses classification using four groups as a function of the 11 borehole intervals is in good agreement with the geological configuration. This preliminary geological interpretation will allow both the estimation of poroelastic and hydrogeological parameters of the clay formation from Earth tidal and barometric responses without classical pump test experiments, and the monitoring of these properties over a long period.

At the Andra Meuse/Haute-Marne Underground Research Laboratory (URL) site located in Bure (France), water levels and fluid pressures are measured in several boreholes in the Callovo-Oxfordian clay Formation and in overlying geological formations.

Fluid pressure or hydraulic head measured in wells in geological formations can respond to Earth tidal forces and atmospheric pressure variations. Earth tidal and barometric responses identified in wells can be used to estimate poroelastic and hydrogeological parameters, without classical pump tests, and to monitor these parameters over a long period (Hsieh *et al.* 1987; Delcourt-Honorez 1988, 1995, 1998, 2001).

In 2003, Andra investigated Earth tidal and barometric responses in boreholes EST201, EST203 and EST104 screened in the Oxfordian Formation at the URL site (Delcourt-Honorez 2003). The hydrogeological interpretation of data from boreholes EST203 and EST104 produced transmissivity and specific storage estimates consistent with the pump test results (Klubertanz *et al.* 2004).

Eleven absolute pressure data series recorded in a further borehole at the URL site (EST207) have also been analysed for possible Earth tidal and barometric pressure responses. This paper concerns the analytical results obtained so far (Delcourt-Honorez 2011) which suggest that in future it may be possible to use Earth tidal and barometric pressure responses to estimate poroelastic and hydrogeological parameters in the Callovo-Oxfordian clay Formation and in other formations at the URL site.

Borehole EST207 is located about 200 m away from the URL access shaft, and is drilled from the land surface down to 575 m depth. The borehole (Fig. 1) is equipped with a multipacker dividing the borehole into11 isolated absolute pressure monitoring intervals: 2 intervals are in the Dogger Formation (intervals 1 and 2), 8 intervals in the Callovo-Oxfordian clay Formation (intervals 3 to 6 in the lower, and intervals 7 to 10 in the upper Callovo-Oxfordian formations), and 1 interval (interval 11) in the Oxfordian base. The pressure sensor number 11 effectively records pressure in the interval 11, from 417.80 to 406.44 m depth. Atmospheric pressure is recorded at the surface. In borehole EST207, the fluid pressure and the atmospheric pressure were recorded every fifteen minutes from 2 June 2004 to 31 December 2010 (6.5 years' recording). The 215 802 pressure datasets at 215 802 time points in each interval (1–11) were analysed for possible responses to Earth tides and atmospheric pressure variations (Delcourt-Honorez 2011).

From: Norris, S., Bruno, J., Cathelineau, M., Delage, P., Fairhurst, C., Gaucher, E. C., Höhn, E. H., Kalinichev, A., Lalieux, P. & Sellin, P. (eds) 2014. *Clays in Natural and Engineered Barriers for Radioactive Waste Confinement*. Geological Society, London, Special Publications, **400**, 53–62.
First published online April 7, 2014, http://dx.doi.org/10.1144/SP400.17

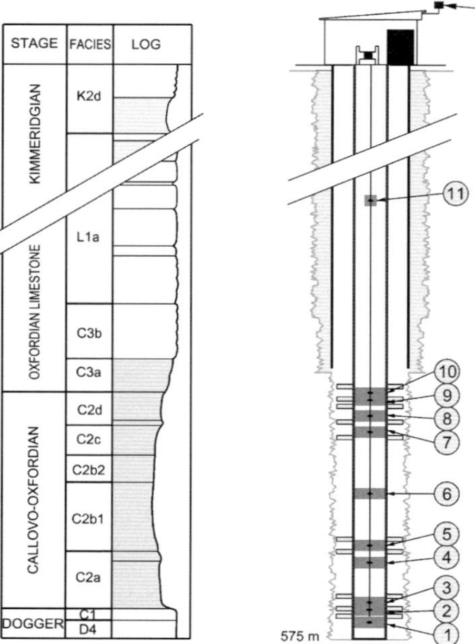

Interval	Interval depth measurement (m/BGL)	Lithological unit
11	406.44 – 417.80	C3a
10	418.30 – 424.97	C3a – C2d
9	425.97 – 429.57	C2d
8	430.57 – 439.31	C2d
7	440.31 – 448.93	C2d – C2c
6	449.93 – 512.24	C2c à C2b1
5	512.24 – 519.92	C2b1
4	520.92 – 547.59	C2a
3	548.59 – 555.26	C2a
2	556.26 – 559.86	C1
1	560.86 – 575.00	D4

Fig. 1. Geological log, equipment and configuration of borehole EST207.

Earth tides

Earth tides can be defined in a simplified way as phenomena resulting from an elastico-viscous deformation of the terrestrial globe caused mainly by the gravitational action of the Moon, the Sun, and other planets to a lesser extent. The tidal force derives from a 'generating potential', called 'Earth Tidal Potential' (Melchior 1983) (Table 1). Some principal diurnal waves are K1, O1, P1, S1, and semi-diurnal waves are M2, N2, S2, K2 (Table 2).

Data processing

The data used in Earth tides programs must meet several criteria such as time converted into Universal Time and Julian Time, and data being in a specific tidal format, including no gaps or a minimum of gaps. In some cases, original data had to be shifted so as to fit the regular quarter-hour time frame 00, 15, 30, 45. Various perturbed data such as gaps, drift and abnormal data had to be corrected. Data interpolation and filtering processes were performed to have data available every 15 min so that hourly data series could be generated: pressure data PAi (tj) (i = 1, 2, . . . , 11, tj = on the hour), and atmospheric pressure data atmP (tj), labelled atmP or AtmP in some figures. The processing of the pressure PAi and of the atmospheric pressure data series are detailed in Delcourt-Honorez (2011).

Spectral analyses

Spectral analysis (Fast Fourier Transform, FFT) was performed on the atmospheric pressure data

Table 1. *Earth tidal potential and Earth tidal dilatation, equations and explanations*

Earth tidal potential	$W = G(\mu/r)\sum_{n=2}^{\infty}(a/r)^n P_n(\cos z)$	a is the radius of the Earth, μ the disturbing body mass, r the vector radius, z the geocentric zenithal distance of the external body, G the gravitational constant, Pn the Legendre's polynomial of order n.
Earth tidal dilatation	$\varepsilon_T = \dfrac{1-2\nu_P}{1-\nu_P}(2h-6l)\dfrac{W_2}{ag}$	ν_p is the Poisson's coefficient, h and l Love numbers at the Earth's surface, a radius of the Earth, g acceleration due to gravitational force and W_2 lunisolar potential calculated for each tidal component.

Table 2. *Principal tidal waves (Melchior 1983)*

Wave name	Wave type	Wave period
M2	Lunar semi-diurnal	12 h 25 min 15 s
N2	Lunar semi-diurnal	12 h 39 min 30 s
S2	Solar semi-diurnal	12 h
K2	Lunisolar semi-diurnal	11 h 58 min 2 s
K1	Lunisolar diurnal	23 h 56 min 4 s
O1	Solar diurnal	25 h 49 min 10 s
P1	Solar diurnal	24 h 3 min 57 s
S1	Solar diurnal	24 h

and revealed only one amplitude peak at S2 solar semi-diurnal wave frequency. Figure 2a shows the S2 atmospheric pressure spectral amplitude peak.

Spectral analyses (FFT) carried out on all the pressure data series in the 11 intervals clearly revealed amplitude peaks at some of the various Earth tidal diurnal and semi-diurnal wave frequencies. Although in some pressure series, several frequencies cannot be clearly discriminated, the results of the spectral analyses show the presence of a response to Earth tidal forces as pressures in the Callovo-Oxfordian clay Formation and in the other formations. As an example, Figure 2b (in grey) shows spectral amplitude peaks at O1, K1, M2 and S2 frequencies in interval 9 in the Callovo-Oxfordian clay (PA9 data series from 7 October 2004 to 31 December 2010). At a second stage, spectral analyses were performed on each pressure data series after having removed the effect of the atmospheric pressure variations using barometric efficiency (BE) (see next paragraph 'Earth tides analyses'); because pressures in the clay can respond to atmospheric pressure variations, these analyses were done to show the effect of atmospheric pressure variations on the spectra (Fig. 2b, in black).

Table 3. *Barometric efficiency (BE) and standard deviation (σ (BE)) in each of the 11 intervals of borehole EST207*

Intervals	Barometric efficiency (BE)	σ (BE)
11	0.49805	0.00729
10	0.52570	0.00577
9	0.52211	0.00615
8	0.51566	0.00597
7	0.52254	0.00606
6	0.63001	0.00586
5	0.62976	0.00592
4	0.59549	0.00935
3	0.62022	0.01340
2	0.33212	0.00599
1	0.21432	0.00638

Earth tides analyses

Based on a different method from FFT spectral analysis, ETERNA (Wenzel 1996, 1998), software targeted at Earth tides analysis, supplies Earth tidal parameters (amplitude and phase), as well as providing useful statistical information and, in several cases, a better wave separation. ETERNA uses a least squares fit between the observed data and a

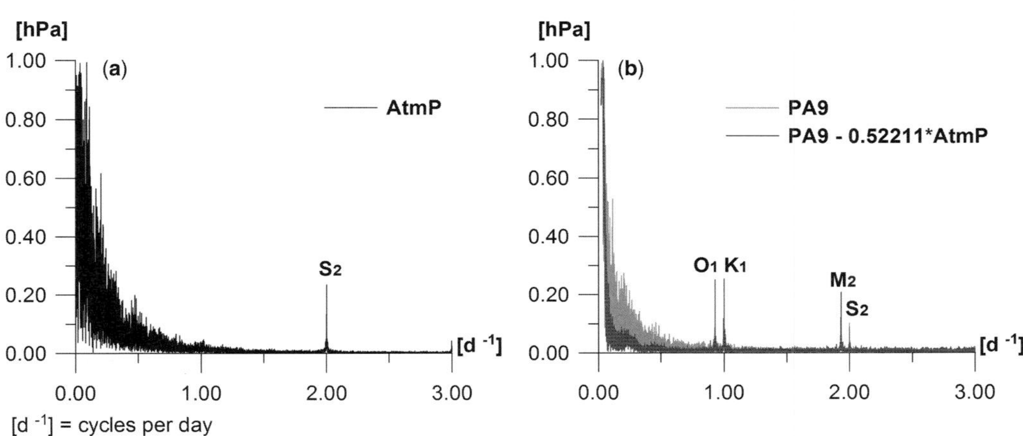

Fig. 2. Fourier analyses (Fast Fourier Transform) results: an example from borehole EST207, interval 9 in the Callovo-Oxfordian clay, from 10 July 2004 to 31 December 2010. (**a**) Amplitude spectrum of atmospheric pressure (AtmP). (**b**) Amplitude spectrum (in grey) of fluid pressure PA9 [hPa]; amplitude spectrum (in black) after having removed the effect of atmospheric pressure variations (0.52211= barometric efficiency (BE) calculated at a second stage by linear regression factor in Earth tidal analyses to show effect of barometric correction on the spectrum).

Table 4. *Earth tidal amplitude A and phaseϕ after barometric correction in each series PAi (hourly data) and their standard deviations: amplitudes and phases of the clearly identified waves are typed in bold grey cells, the less clearly identified waves and the not clearly identified waves are typed in black and in grey characters, respectively*

Wave		K1				O1				P1				S1			
Interval	Hourly data number	A[hPa]	± σ(A)	φ[°]	± σ(φ)	A[hPa]	± σ(A)	φ[°]	± σ(φ)	A[hPa]	± σ(A)	φ[°]	± σ(φ)	A[hPa]	± σ(A)	φ[°]	± σ(φ)
11	37850	0.237	±0.010	169.7	±2.2	0.227	±0.008	176.6	±2.1	0.131	±0.010	164.9	±4.4	0.076	±0.015	−151.7	±11.1
10	52075	0.349	±0.010	172.1	±1.5	0.336	±0.008	179.3	±1.4	0.166	±0.010	169.9	±3.4	0.099	±0.015	−135.3	±8.4
9	45326	0.366	±0.008	174.9	±1.3	0.337	±0.008	176.3	±1.2	0.184	±0.009	168.4	±2.7	0.055	±0.013	−124.8	±13.4
8	52091	0.394	±0.008	169.8	±1.3	0.351	±0.008	172.7	±1.3	0.172	±0.010	163.9	±3.3	0.060	±0.014	−144.4	±13.8
7	52268	0.391	±0.008	173.7	±1.2	0.353	±0.008	174.1	±1.2	0.190	±0.009	171.1	±2.7	0.054	±0.013	−145.6	±13.9
6	52308	0.451	±0.010	172.7	±1.2	0.410	±0.008	179.0	±1.2	0.203	±0.010	167.2	±2.8	0.088	±0.015	−155.5	±9.6
5	52293	0.458	±0.010	172.8	±1.2	0.423	±0.008	177.8	±1.1	0.207	±0.010	166.3	±2.8	0.124	±0.015	−135.9	±6.9
4	29303	0.447	±0.018	166.1	±2.3	0.410	±0.017	174.2	±2.3	0.247	±0.020	161.1	±4.6	0.057	±0.029	−129.9	±28.8
3	16269	0.426	±0.021	163.5	±2.8	0.392	±0.018	172.8	±2.6	0.248	±0.022	165.0	±5.1	0.099	±0.033	175.5	±19.1
2	52226	0.788	±0.007	169.3	±0.6	0.710	±0.007	170.8	±0.6	0.336	±0.008	168.1	±1.4	0.084	±0.012	−132.8	±8.5
1	49347	0.939	±0.007	−177.7	±0.5	0.845	±0.007	−177.5	±0.5	0.400	±0.009	−178.9	±1.2	0.053	±0.013	−123.8	±13.7
Wave		M2				S2				N2				K2			
Interval	Hourly data number	A[hPa]	± σ(A)	φ[°]	± σ(φ)	A[hPa]	± σ(A)	φ[°]	± σ(φ)	A[hPa]	± σ(A)	φ[°]	± σ(φ)	A[hPa]	± σ(A)	φ[°]	± σ(φ)
11	37850	0.281	±0.009	−168.7	±1.9	0.121	±0.009	−175.3	±4.3	0.061	±0.009	−173.8	±8.5	0.034	±0.007	171.2	±11.9
10	52075	0.406	±0.007	−167.5	±1.0	0.165	±0.007	−176.4	±2.6	0.085	±0.007	−161.0	±4.9	0.051	± 0.006	−174.5	±6.4
9	45326	0.377	±0.008	−172.3	±1.3	0.160	±0.008	−179.1	±3.0	0.056	±0.008	−175.8	±8.3	0.042	± 0.006	−178.9	±8.6
8	52091	0.401	±0.007	−177.4	±1.0	0.176	±0.007	178.9	±2.3	0.079	±0.007	−174.5	±4.9	0.047	±0.005	169.9	±6.5
7	52268	0.410	±0.008	−172.4	±1.1	0.183	±0.008	−177.6	±2.5	0.069	±0.008	−163.2	±6.6	0.030	±0.006	163.8	±11.6
6	52308	0.632	±0.007	−160.7	±0.7	0.271	±0.007	−166.0	±1.6	0.109	±0.008	−162.6	±3.9	0.067	±0.006	−161.4	±5.0
5	52293	0.633	±0.007	−160.3	±0.7	0.266	±0.007	−168.2	±1.5	0.110	± 0.007	−155.3	±3.7	0.067	±0.006	−174.3	±4.7
4	29303	0.580	±0.010	−163.3	±1.0	0.271	±0.010	−165.7	±2.1	0.110	± 0.010	−158.2	±5.1	0.057	±0.008	−170.7	±8.1
3	16269	0.618	±0.017	−163.2	±1.6	0.269	±0.017	−164.0	±3.7	0.104	±0.017	−150.7	±9.6	0.063	±0.014	−179.1	±12.3
2	52226	0.751	±0.007	176.1	±0.6	0.368	±0.007	165.6	±1.2	0.130	±0.007	177.9	±3.3	0.098	±0.006	162.2	±3.4
1	49347	0.991	±0.010	−171.5	±0.6	0.475	±0.009	−179.4	±1.2	0.180	±0.009	−166.4	±3.0	0.127	±0.007	−179.2	±3.3

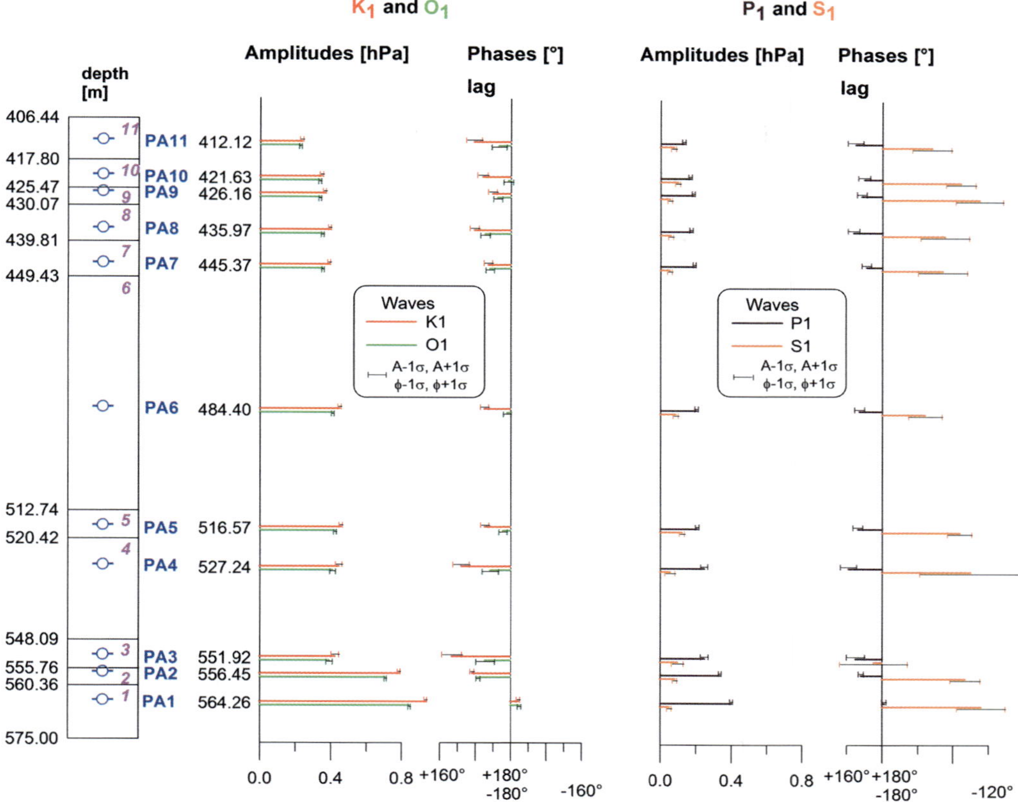

Fig. 3. Diurnal waves' (K1, O1, P1, S1) amplitudes A [hPa] and phases ϕ [°] (and error bars corresponding to $\pm\sigma$): diagram obtained from the results of Earth tidal analyses (data from 2004 to 2010) in 11 intervals from borehole EST207 after barometric correction.

tidal model with the possibility of additional environmental input channels, for example, atmospheric pressure records. In these Earth tidal analyses, a high pass filter is applied limiting the spectrum to the Earth tidal ranges. Using ETERNA for Earth tidal responses in the 11 pressure data series PAi, Earth tidal wave parameters, amplitude A (in hectopascal [hPa]) and its standard deviation σ (A) [hPa], phase ϕ (in degrees [°]) and its standard deviation σ (ϕ) [°], were estimated in diurnal and semi-diurnal frequency bands before and after having removed the atmospheric pressure variation effect from pressure data PAi (in this paper, removing the atmospheric pressure variation effect from pressure data PAi is called 'barometric correction').

The atmospheric pressure data were analysed with ETERNA for atmospheric pressure wave parameters. The solar S2 wave amplitude value can be used as a reference point (Delcourt-Honorez 2002). For the atmospheric pressure data series of borehole EST207, the estimated amplitude value (A = 0.367 hPa $\pm$ 0.005 hPa), is in good agreement with calculated values at other sites (Delcourt-Honorez 2002) and at the URL site (Delcourt-Honorez 2003, 2011). This result validates the atmospheric pressure dataset. A standardization of various BE definitions has been discussed in Delcourt-Honorez (1999). The BE and its standard deviation (σ(BE)) is calculated as a linear regression coefficient of the atmospheric pressure as an auxiliary channel in the ETERNA analyses. The calculated BE values for the 11 intervals (1–11) are presented in Table 3.

In the 11 data series PAi, high quality data series and multi-year pressure datasets made it possible to separate the most possible tidal waves in the diurnal band and in the semi-diurnal band. Among all the separated waves, K1, O1, P1, S1 and M2, N2, S2, K2 waves were first taken into account in each series. Table 4 contains the numerical results of ETERNA Earth tidal analyses after barometric correction in each series PAi (hourly data), that is,

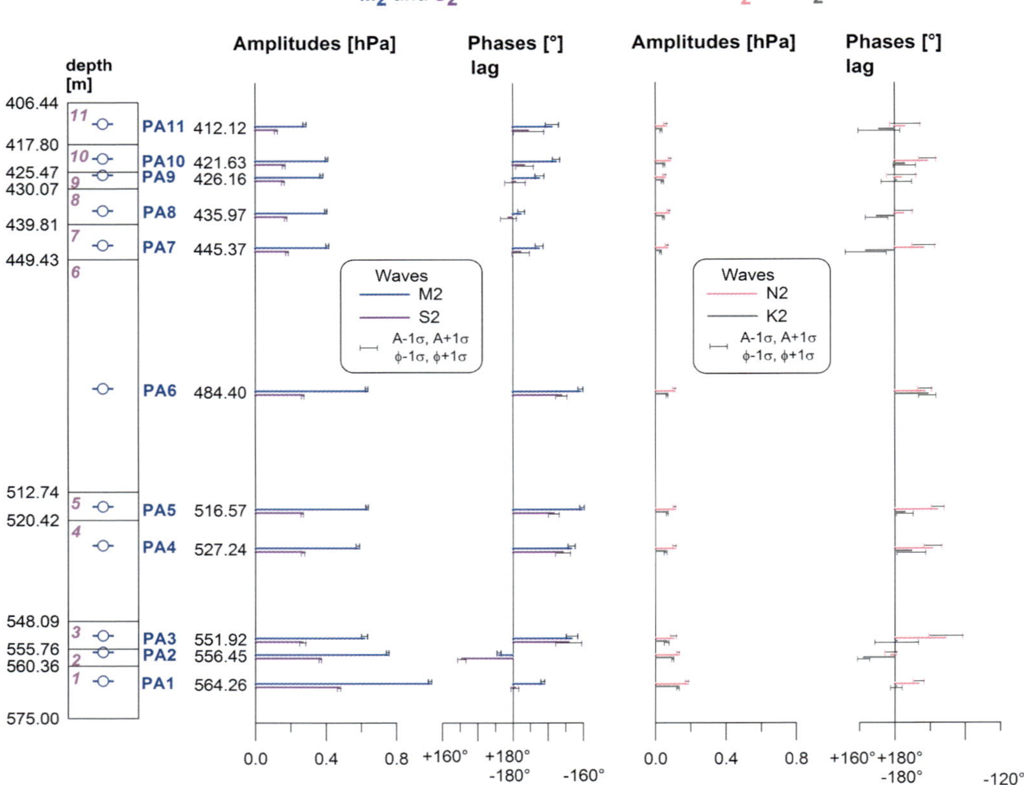

Fig. 4. Semi-diurnal waves' (M2, S2, N2, K2) amplitudes A [hPa] and phases ϕ [°] (and error bars corresponding to $\pm\sigma$): diagram obtained from results of Earth tidal analyses (data from 2004 to 2010) in 11 intervals from borehole EST207 after barometric correction.

amplitude A and phase ϕ, and their standard deviations. Figures 3 and 4 show diagrams illustrating diurnal and semi-diurnal amplitudes [hPa] and phases [°] and their error bars corresponding to $\pm\sigma$ (ϕ) after barometric correction.

To validate those results of the Earth tidal analyses that are in good agreement with the spectral analyses results, the significance of tidal parameters was investigated for each wave in the 11 data series. In Table 4, amplitudes and phases of the clearly identified waves are typed in bold grey cells, the less clearly identified waves and the not clearly identified waves are typed in black and in grey characters, respectively. The results can be summarized as follows, taking into account in this paper only the most clearly identified waves (details in Delcourt-Honorez 2011).

In the Callovo-Oxfordian clay (interval 3 (PA3) to interval 10 (PA10)), K1, O1, M2 and S2 are the most clearly identified waves; N2 is clearly identified in intervals 4 (PA4), 5 (PA5), 6 (PA6), 8 (PA8) and 10 (PA10); K2 is clearly identified in intervals 5 (PA5) and 6 (PA6). In the Dogger Formation (in the interval 1 (PA2) and 2 (PA2)), K1, O1, P1 and M2, N2, S2, K2 waves are clearly identified; and in the Oxfordian Formation (in interval 11 (PA11)), K1, O1 and M2, N2, S2 are clearly identified.

In the 11 intervals, while the M2 amplitude is not (or only slightly) perturbed by atmospheric pressure variations and the M2 phase is not, the S2 wave amplitude and phase are very sensitive to atmospheric pressure, as has been concluded in several studies of Andra and other sites, in which the effects of Earth tides on water level and pressure data series were analysed (e.g. Delcourt-Honorez 1988, 1998, 2001, 2003).

Preliminary geological interpretation

After barometric correction, the Earth tidal amplitudes generally decrease from the bottom of the

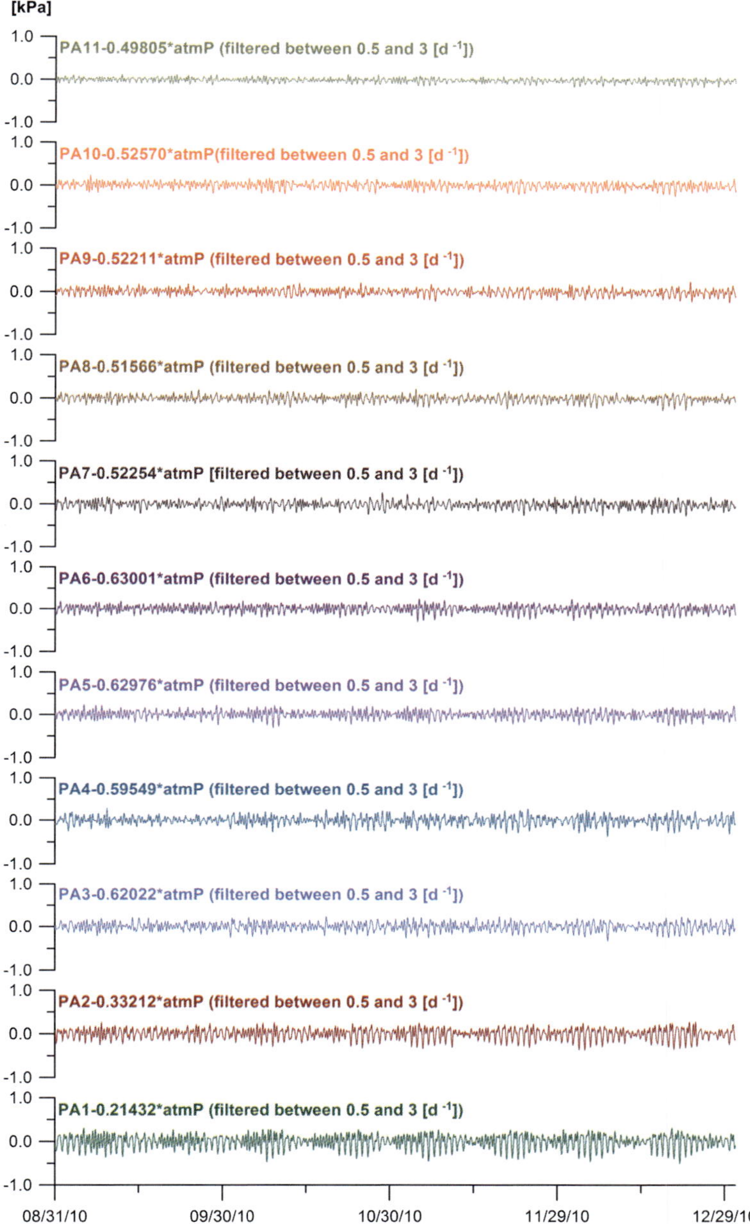

Fig. 5. Graphs showing time series of the pressure records in the 11 intervals from borehole EST207 after barometric correction and after filtering between 0.5 d^{-1} and 3 d^{-1} (atmP = atmospheric pressure data): the amplitudes are generally decreasing from the bottom of the borehole to the top (from PA1 to PA11).

borehole to the top (Fig. 5). The largest amplitudes are observed in PA1 and PA2, in intervals 1 and 2. In the lower Callovo-Oxfordian Formation, PA3 to PA6 show smaller amplitudes which are of the same order of magnitude; in the upper Callovo-Oxfordian Formation, PA7 to PA10 show slightly smaller amplitudes than in the lower Callovo-Oxfordian Formation; at the base of Oxfordian Formation, the smallest amplitude is detected in PA11.

Table 5. *Barometric and Earth tidal responses classification: barometric efficiency (BE) and Earth tidal amplitude in the four groups in borehole EST207*

Groups*	Intervals	Barometric efficiency (BE)	Amplitudes [hPa] (after barometric correction)						
			K1	O1	P1	M2	S2	N2	K2
4	11	≈0.50	≈0.24	≈0.23	≈0.13	≈0.28	≈0.12	≈0.06	
3	7 to10	≈0.52	≈0.38	≈0.35	≈0.17	≈0.40	≈0.17	≈0.07	≈0.047
2	3 to 6	≈0.60 to ≈0.63	≈0.45	≈0.40	≈0.22	≈0.62	≈0.27	≈0.11	≈0.064
1	2	≈0.33	≈0.79	≈0.71	≈0.34	≈0.75	≈0.37	≈0.13	≈0.098
	1	≈0.21	≈0.94	≈0.85	≈0.40	≈0.99	≈0.48	≈0.18	≈0.127

*Groups 2 and 3 in Callovo-Oxfordian clay; group 1 in the Dogger Formation; group 4 at the Oxfordian base (K2 is not clearly identified in interval 11).

In a preliminary geological interpretation, the detailed intercomparison of the results of tidal analyses according to the barometric correction effect, the BE values and the quality of the wave identification, result in a classification of the barometric and tidal responses as a function of the various intervals. The barometric and Earth tidal responses were classified into four groups based on the monitoring intervals: in the Dogger Formation (group 1, i.e. intervals 1 and 2), the lower Callovo-Oxfordian Formation (group 2, i.e. intervals 3 to 6), the upper Callovo-Oxfordian Formation (group 3, i.e. intervals 7 to 10) and the base of Oxfordian Formation (group 4, i.e. interval 11). This classification of the responses matches the geological configuration. Table 5 shows BE values

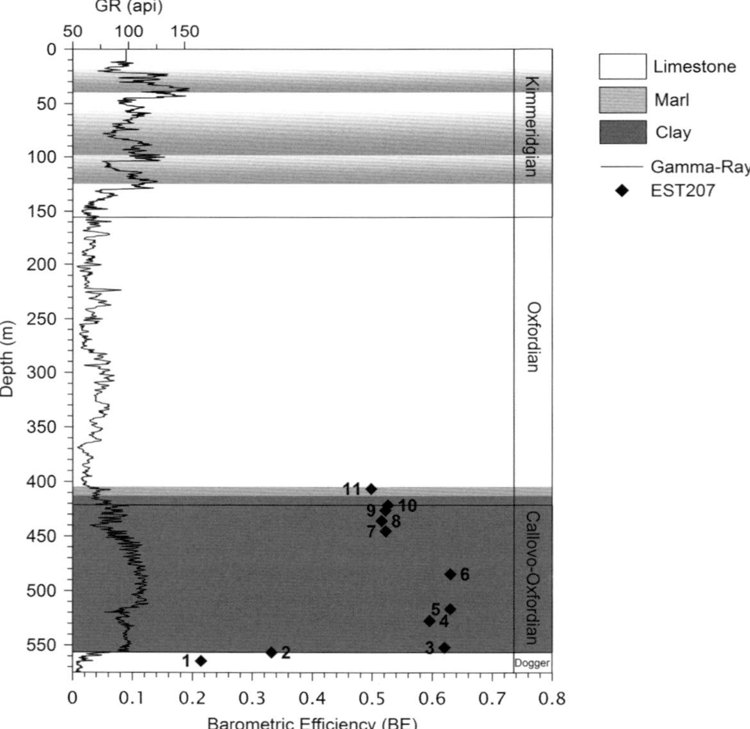

Fig. 6. Gamma ray (GR) log and barometric efficiency (BE) calculated during Earth tidal analyses and plotted together on the same graph using data from borehole EST207.

and amplitudes of Earth tidal waves (K1, O1, P1, M2, S2, N2 and K2) in the four groups (averaged values for several intervals); S1 is not mentioned as this wave is less clearly identified in intervals 5 and 10 and not clearly identified in the others; K2 is not mentioned in interval 11 as it is not clearly identified.

Figure 6 shows the gamma ray log and BE values plotted together using data from borehole EST207. A first glance could lead one to deduce that there is a relationship between gamma ray and BE values: the figure may suggest that in the Callovo-Oxfordian clay the greater the clay content of the formation (in fact gamma ray measurements are taken as a measure of clay content), the greater the BE values. At this stage, no detailed analysis was carried out on the data from borehole EST207.

Conclusion and prospects for future research

The pressures in the 11 intervals of borehole EST207 respond to Earth tidal forces and to atmospheric pressure variations. In responses to Earth tides, several waves were clearly identified, in amplitude and phase. From the responses analyses, a preliminary geological interpretation was carried out. At this stage, the data from the Callovo-Oxfordian clay Formation suggest that the greater the clay content of the formation, the greater the barometric efficiency (BE) values; this will lead to further numerical investigation of the relationship between BE values and clay content. The classification of the Earth tidal and barometric responses into four groups has shown that barometric and Earth tidal responses are dependent on the formations' geology. This preliminary geological interpretation of the Earth tidal and barometric responses is in good agreement with the geological configuration.

Earth tidal and barometric responses in wells can be used to estimate poroelastic and hydrogeological parameters (e.g. Hsieh *et al.* 1987; Delcourt-Honorez 1988, 1995, 1997, 1998, 2001). In 2004, Andra investigated Earth tides and barometric responses in boreholes EST201, EST203 and EST104 screened in the Oxfordian Formation at its Meuse/Haute-Marne URL site (Delcourt-Honorez 2003). From Earth tidal responses that were analysed using ETERNA software, the hydrogeological interpretation of data from boreholes EST203 and EST104 produced transmissivity and specific storage estimates consistent with the pump test results (Klubertanz *et al.* 2004).

The results of the analyses of Earth tidal and barometric responses, obtained from the pressure data series recorded during 6.5 years in the 11 intervals of borehole EST207, and the preliminary geological interpretation of the responses, are the first stage in continuing studies. At a later stage, the responses in amplitude and phase to Earth tidal pressures, and the barometric responses as identified by use of ETERNA and as explained in this paper, could be used to estimate the poroelastic and hydrogeological parameters of the whole Callovo-Oxfordian clay Formation (and of the other formations). The estimation of parameters in this way, without classical pump test experiments, allow monitoring of geological properties over a long period. The preliminary interpretation could be refined by analysing the tidal parameters and barometric responses combined with a tidal dilatation model. This tidal dilatation model could also be developed from data recorded by appropriate equipment, such as three-directional extensometers.

References

Delcourt-Honorez, M. 1988. *Sur la contribution des variations de la pression atmosphérique et des variations du niveau des nappes aux variations de la pesanteur.* Dissertation doctorale. Faculté des sciences, Université Catholique de Louvain, Belgium.

Delcourt-Honorez, M. 1995. Earth tides in wells: poroelastic aquifer parameter estimates. *In*: Dubois, C. (ed.) *Gas Geochemistry.* Sciences Reviews, Northwood, 245–256.

Delcourt-Honorez, M. 1997. Earth tides, hydrogeology and radon. *In*: Virk, H. S. (ed.) *Applications in Earth and Environmental Science.* Guru Nanak Dev University, Amritsar, 34–40.

Delcourt-Honorez, M. 1998. *Interprétation hydrogéologique des phénomènes de marées terrestres observés dans les puits de la Vienne. Années 1997–1998.* Andra report D.RP.0INU.98.001/A.

Delcourt-Honorez, M. 1999. *Analyse détaillée des données de pression dans les forages en vue d'une interprétation hydrogéologique.* Andra report D.RP. 0INU.99.001/A.

Delcourt-Honorez, M. 2001. *Marées terrestres, séismes et barométrie dans les forages: apports à la connaissance des paramètres poro-élastiques.* Syllabus des Cours enseignés au DEA de génie civil. École doctorale de génie civil de Lille. USTL/EUDIL. France.

Delcourt-Honorez, M. 2002. Contribution à l'étude des marées atmosphériques. *In*: Flick, J. A. & Stomp, N. (eds) *Sciences de la Terre au Grand-Duché de Luxembourg: Réminiscences.* Musée National d'Histoire Naturelle, Centre Européen de Géodynamique et de Séismologie, Luxembourg, 196–203.

Delcourt-Honorez, M. 2003. *Etude des marées terrestres dans les forages EST201, EST203 et EST104.* Andra report D.RP.0INU.03.001/A.

Delcourt-Honorez, M. 2011. *Etude des pressions mesurées dans le forage EST207. Recherche du signal de marées terrestres.* Andra report D.RP.0INU. 11.0002/B.

HSIEH, P. A., BREDEHOEFT, J. D. & FARR, J. M. 1987. Determination of aquifer transmissivity from Earth tide analysis. *Water Resources Research*, **23 (10)**, 1824–1832.

KLUBERTANZ, G., SCHWARZ, R., LAVANCHY, J. M. & DELCOURT-HONOREZ, M. 2004. *Interprétation hydrogéologique des marées terrestres sur la base des pressions aux forages EST201, EST203 et EST104.* Andra report D.RP.0CPE.04.019/A.

MELCHIOR, P. J. 1983. *The Tides of the planet Earth.* 2nd edn. Pergamon Press, Oxford.

WENZEL, H. G. 1996. *Eterna 3.30 and 3.34.help in ETERNA 3.30.* Black Forest Observatory, D-Karlsruhe.

WENZEL, H. G. 1998. The nanogal software: earth Tides Data Processing Package Eterna 3.30. *In*: *Proceedings of the 13th International Symposium on the Earth Tides.* Brussels, July 22–25, 1997. Observatoire Royal de Belgique, Série Géophysique, Brussels, 487–494.

Enhanced Sealing Project (ESP): evolution of a full-sized bentonite and concrete shaft seal

D. A. DIXON[1], D. G. PRIYANTO[1]*, J. B. MARTINO[1], M. DE COMBARIEU[2], R. JOHANSSON[3], P. KORKEAKOSKI[4] & J. VILLAGRAN[5]

[1]*Atomic Energy of Canada Limited, Pinawa, Canada*

[2]*Agence Nationale pour la gestion des déchets radioactifs ANDRA, France*

[3]*Svensk Kärnbränslehanertring AB SKB, Stockholm, Sweden*

[4]*Posiva Oy, Eurajoki, Finland*

[5]*Nuclear Waste Management Organization, Toronto, Canada*

**Corresponding author (e-mail: priyantd@aecl.ca)*

Abstract: A full-scale shaft seal was designed and installed in the 5 m-diameter access shaft at Atomic Energy of Canada Limited's (AECL's) Underground Research Laboratory at the point where the shaft intersects an ancient water-bearing, low-angle thrust fault at a depth of *c.* 275 m in granitic rock. The seal consists of a 6 m-thick bentonite-based component sandwiched between 3 m-thick, keyed upper and lower concrete components. This design was adopted in order to limit the mixing of saline groundwater from the deeper regime with the fresher, near-surface groundwater regime. Construction of the shaft seal was done as part of Canada's Nuclear Legacy Liabilities Program. A jointly funded monitoring project, called the Enhanced Sealing Project (ESP), was developed by AECL (Canada) and jointly funded by NWMO (Canada), SKB (Sweden), Posiva Oy (Finland), and ANDRA (France), and since mid 2009 the thermal, hydraulic and mechanical evolution of the seal has been constantly monitored. The evolution of the type of seal being monitored in the ESP is of relevance to repository closure planning by demonstrating the functionality of shaft seals. Although constructed in a crystalline rock medium, the results of the ESP are relevant to the performance of seals in a variety of host rock types.

Description of the shaft seal

The seal being monitored by the Enhanced Sealing Project (ESP) is so far as possible a full-scale simulation of the type of structure that could be used in an actual repository. Although this seal does not have a specifically defined or required life-span in this particular installation, monitoring of its evolution and performance provides valuable information regarding the type of installation that will ultimately be needed in repository closure and how their initial performance might be monitored. No radioactive materials are stored at the AECL's Underground Research Laboratory (URL). The seal's primary goal in this installation is to limit the mixing of more saline groundwater below a major fracture zone with the fresher, near-surface groundwater regime. In this respect, it is intended to function in much the same manner as a shaft seal in a repository containing radioactive materials. These goals dictated the design of the seal and the materials used in constructing it.

The shaft seal installed in the access shaft of the URL is shown schematically in Figure 1. The 3 m-thick keyed upper and lower concrete components sandwich a 6 m-long bentonite-based component, which is an *in situ*, densely compacted mixture of bentonite clay (40% dry mass) and quartz/feldspar sand (60% dry mass). Construction of the shaft seal was done as part of Canada's Nuclear Legacy Liabilities Program (NLLP). A jointly funded monitoring project, called the Enhanced Sealing Project, was developed by AECL (Canada) and funded by NWMO (Canada), SKB (Sweden), Posiva Oy (Finland) and ANDRA (France) and since mid 2009 the thermal, hydraulic and mechanical evolution of the seal has been constantly monitored.

The concrete components consist of a heavily reinforced lower concrete component and a non-reinforced upper concrete component. These concrete components are keyed into the rock to a maximum depth of *c.* 0.5 m, providing a mechanical restraint to any upwards or downwards movement of the seal as the result of swelling pressure generated by the bentonite-based component. Reinforcement of the lower concrete component was necessary for safety reasons as it was required

From: Norris, S., Bruno, J., Cathelineau, M., Delage, P., Fairhurst, C., Gaucher, E. C., Höhn, E. H., Kalinichev, A., Lalieux, P. & Sellin, P. (eds) 2014. *Clays in Natural and Engineered Barriers for Radioactive Waste Confinement*. Geological Society, London, Special Publications, **400**, 63–70.
First published online May 1, 2014, http://dx.doi.org/10.1144/SP400.33

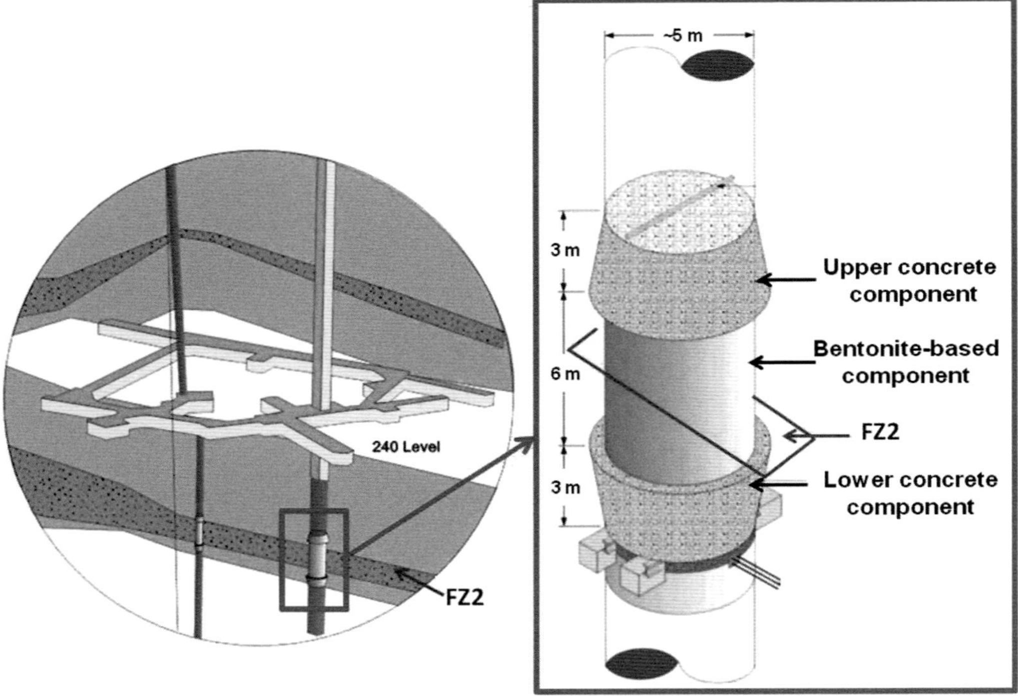

Fig. 1. Location and geometry of access shaft seal.

to be capable of supporting the entire mass of material above it as well as a fully flooded upper shaft. The upper concrete component has no load-bearing requirement and so did not require installation of reinforcement. The type of concrete selected for use is a low-heat, low-pH, high-strength formulation described by Martino *et al.* (2010). The low-heat was needed to limit cooling shrinkage and potential thermally induced cracking in the period immediately following its heat-generating curing phase. Limiting thermally induced volume change will also limit the size of any shrinkage-induced gaps along the rock-concrete contacts. The low pH will limit the environmental influence of the concrete on the nearby groundwater as well as minimizing potentially adverse chemical interactions with the adjacent bentonite-based materials. The internal shrinkage of the concrete and its thermal evolution are monitored as part of the ESP. Following installation of the upper concrete component and its initial curing, a displacement monitor was installed at its upper surface in order to determine if any subsequent movement of the plug occurred.

The 6 m-thick bentonite-based component was *in situ* compacted to a dry density of 1810 kg/m^3 as a series of layers and spans the entire width of the fracture zone, extending at least 1 m above or below its identified extent. As the bentonite-based component takes on water from its surroundings, it swells and forms a tight, low-permeability seal, restricting any vertical movement of water via the shaft.

Monitoring of shaft seal

The main objective of the ESP is to monitor the performance of the shaft seal installed at the URL. The installation of the monitoring system in and around the seal needed to be done in parallel with its construction, without unduely slowing its completion. Sensors needed to be selected and installed in such a manner as to allow them to be monitored from a distance of approximately 350 m (280 m of this was vertically up an ultimately water-filled shaft). As the seal was not accessible following its construction, sensor durability was critical, as was recognition of the limitations of some monitoring (e.g. psychrometers and thermistors had maximum cable lengths, *c.* 40 m). There was also a need to minimize risk in monitoring through the use of a variety of sensor types and technologies. Details of the design and construction of the shaft seal are provided by Dixon *et al.* (2009) and Martino *et al.* (2010).

In total there were 100 sensors installed in 68 instruments (32 instruments had dual sensors installed in them, e.g. temperature and pressure or temperature and suction). In planning sensor locations and selecting technologies, operational lifespan was critical. Sensors which required nearby loggers (e.g. psychrometers and thermistors) were installed in locations where logger flooding and subsequent failure would not compromise the overall monitoring plan. For example, psychrometers were installed in locations where early saturation of the bentonite-based component was expected and once saturation was achieved they had served their function and subsequent flooding of the logger used to monitor them was therefore not an issue.

As of mid 2012 (after 3 years of monitoring), approximately 90% of the sensors having a longer-term monitoring function were still functioning. This survival/failure rate is fairly typical for such installations in harsh environments, highlighting one of the challenges to long-term monitoring. The failed sensors do not, however, show any discernible technological or locational pattern to their failure that would lead one to avoid any particular technology, or justify terminating further monitoring of the installation. A detailed description of the sensors used, their calibration, installation and monitoring results is provided by Dixon *et al.* (2009), Martino *et al.* (2010), Holowick *et al.* (2011) and Dixon *et al.* (2012).

Results of monitoring

Flooding of underground excavations

The URL has extensive horizontal development at the 240 m (above the seal) and 420 m (below the seal) levels. The regions below 275 m were artificially flooded in 2009 prior to construction of the shaft seal in order to provide a safer working location during installation as well as to avoid producing a large, isolated air pocket that would pressurize as water ingress occurred. The region above the shaft seal is being allowed to flood naturally as water entered the upper regions (*c.* 4.3 m^3/day). As water reached the 240 m level the extensive horizontal excavations has resulted in a very slow rise in the head of water above the shaft plug. This has resulted in an essentially constant head condition existing since mid 2010, aiding in interpretation of the sensor output. By approximately 2017 the 240 m level is expected to be completely filled with water, but this date will depend on the actual water inflow rate experienced since shaft closure. Shaft flooding above the 240 m level will take approximately one year based on its measured volume and current water inflow rate. As the main inflow fractures above the 240 m level are reached, hydraulic head in the shaft will slow the rate of inflow, causing the final head recovery to occur slowly, potentially delaying completion of flooding until after 2018.

Concrete components

The concrete components were monitored to determine the influence of the thermal pulse induced during concrete curing as well as to determine what, if any, internal strain or bulk movement has occurred since initial curing was complete. Thermal evaluation has been accomplished using an array of thermocouples in and around the concrete components, while internal shrinkage induced by concrete curing and cooling has been monitored using fibre-optic strain gauges. The physical location of the top of the upper concrete component has been monitored using displacement transducers.

Figure 2 shows the limited magnitude and duration of the thermal transient associated with the concrete curing in the lowermost concrete component. In order to facilitate description of sensor locations, a descriptive labelling approach was adopted and is included in the legends in Figure 2 and subsequent figures. For example 2.5 m/CL refers to a sensor located 2.5 m above the lower concrete–clay contact at the vertical centre-line (CL) of the shaft. Owing to the large number of temperature sensors installed, Figure 2 does not include a sketch showing the physical location of the sensors; a detailed description and plot of the sensor locations can be found in the report by Dixon *et al.* (2009). Temperature rise was limited to *c.* 17 °C, much lower than for traditional concretes where *c.* 60 °C rise would be anticipated. The upper concrete component showed a similarly small temperature rise. These low temperature rises translate into small cooling shrinkages for these seals (Fig. 3).

If the net radial shrinkage of the concrete within the shaft is uniform and if no bonding to the adjacent rock is assumed, a gap of 0.6–0.9 mm would be present between the concrete and rock. It is known from visual observation that such a gap exists in the lower concrete component (water was observed draining past the lower concrete component prior to bentonite-based component installation). Monitoring of porewater pressures in both concrete components also confirms that these interfaces are hydraulically open, as anticipated in the initial design. The swelling, bentonite-based material component, confined by the concrete components, is expected to supply the sealing capability to the construction.

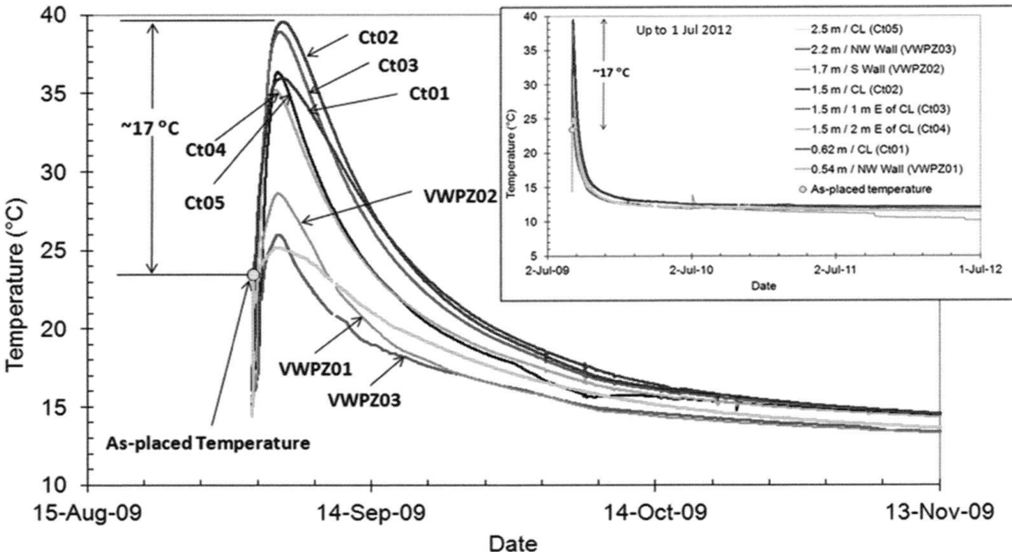

Fig. 2. Temperature rise associated with curing of lowermost concrete component.

The hydraulic pressure at the rock–concrete interface and in the shaft above the seal is constantly monitored. Figure 4 shows the hydraulic pressures measured and clearly shows that the rock–concrete interfaces are hydraulically open and are recording the water pressures present in the nearby shaft.

The displacement sensors installed at the upper surface of the concrete component show that the seal has not moved discernibly since its installation. At most, a 0.2 mm (downwards) movement has occurred at the upper surface of the seal, confirming its physical stability.

Water movement past seal

Of particular interest in the results of monitoring the ESP is determining what degree of isolation the seal provides to the regions above and below it in the shaft. As can be seen in Figure 4, there is a substantial difference in the water pressures present above and below the seal. This pressure difference peaked in September 2011 at approximately 480 kPa (or *c.* 49 m of water head). The proportion of this differential that can be explained by elevation head is only *c.* 60 kPa (6 m), which

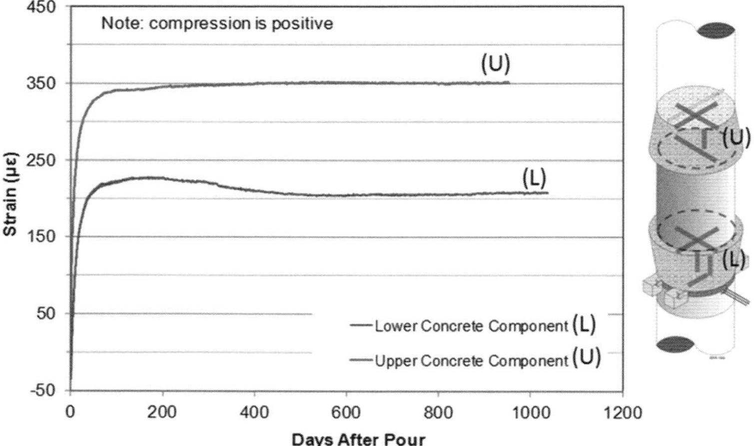

Fig. 3. Shrinkage of concrete components.

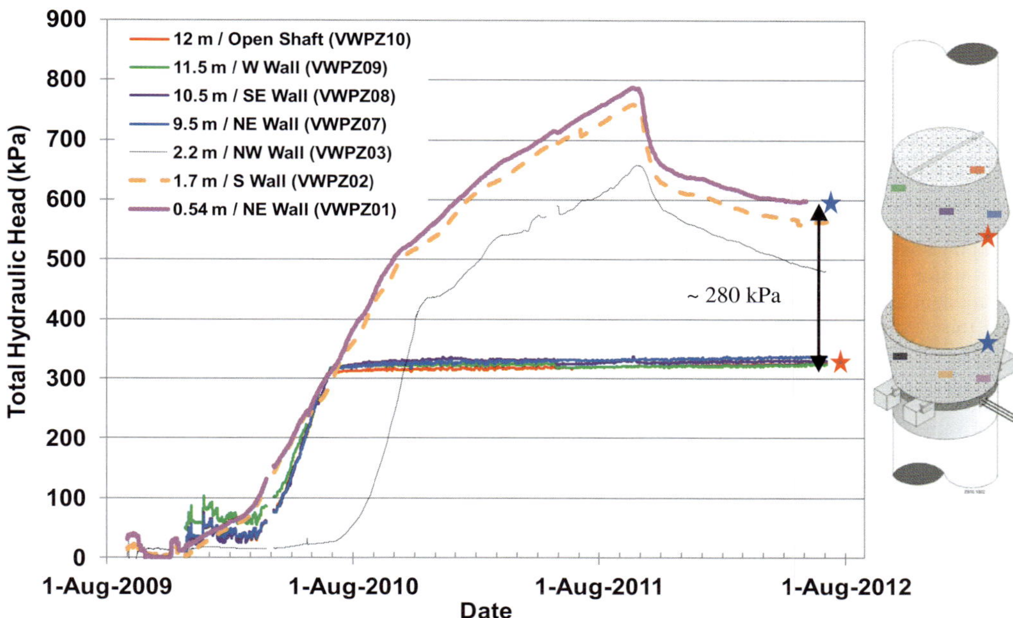

Fig. 4. Hydraulic pressures in upper shaft and at rock–concrete interfaces. Elevations in legend refer to distance from the base of the lower concrete component.

would exist if the concrete–bentonite interfaces were open. This leaves a 420 kPa (42 m) pressure differential to be accounted for. Since pressure below the shaft peaked in September 2011, the pressure differential has decreased somewhat to *c.* 280 kPa (*c.* 22 m more than can be attributed to elevation change) and appears to be stabilizing as of mid 2012.

The excess water pressure at the base of the bentonite-based component is strong evidence that the bentonite-based component is limiting vertical water movement and the Excavation Damaged Zone (EDZ) has been quite effectively cut by the keying of the concrete components. Only with a hydraulic disconnect should such a pressure differential develop and persist. The decay in the differential in late 2011 to mid 2012 indicates that there is likely a very limited/restricted connection between the lower shaft and the Fracture Zone (FZ), which has allowed the excess porewater pressure to decrease. The maintainance of the differential porewater pressure across the entire seal indicates that this connection is poor to non-existent between the FZ and the upper shaft. Were it not disconnected, there would be at most a 60 kPa difference between the base and top of the bentonite-based component (the result of 6 m of water head). The most likely explanation for the observed differential is the presence of a poorly connected EDZ in the region below the FZ, which allows for a very limited movement of water from the lower shaft to the FZ. This means that the shaft seal is fulfilling its function in terms of limiting water movement vertically.

The pressure differential across the shaft seal will ultimately decrease towards zero once the shaft is fully flooded and the regional groundwater regime recovers. At that time there will also be no pressure differential across the seal and no mechanism trying to induce water movement across the seal.

Conditions in the bentonite-based component

Monitoring of central bentonite-based component includes recording of the temperature, porewater pressures, total pressures and water content. Temperature within this region has changed little since the installation of the upper concrete component, which induced a brief, localized increase in temperature as the result of heat transfer from the curing concrete. Most of the temperature sensors are associated with other measurements (e.g. piezometric pressure, total pressure and water content (psychrometers)), but their output is monitored as a matter of course. By mid 2012 the temperature of the bentonite-based component had stabilized at 11–12 °C, which is close to the current ambient temperature of the surrounding rock mass. There is still a small influence of seasonal variations in the water-filled upper shaft, but this is no more than *c.* 1 °C.

Water pressure within the bentonite-based component is an indicator of two things: first, achieving saturation and, second, equilibration of the bentonite-based component with its surroundings. The water pressures within the bentonite-based component are shown in Figure 5, clearly showing that this region is far from equilibrium. There are regions immediately adjacent to the upper and lower bentonite–concrete interfaces that are showing substantial porewater pressures (indicating local saturation), but their magnitude is still well below that of the nearby water-filled shaft. Other, more centrally located piezometers are not indicating any substantive pore pressures (saturation not yet achieved).

Total pressure (sum of swelling/mechanical pressure and water pressure) measurements within the bentonite-based component and at the contact with its confinement (rock and concrete) are presented in Figure 6. These data show the effects of both the as-yet unsaturated conditions within the bentonite-based component and the hydraulic pressure decrease in the lower shaft in the variability of the total pressure readings. Ultimately when fully saturated and flooding of the overlying shaft has occurred, the total pressures within the bentonite-based component should reach approximately 3.5 MPa (*c.* 0.8 MPa swelling pressure plus approximately 2.7 MPa hydraulic head).

Water uptake by the bentonite-based component of the seal is monitored using thermocouple psychrometers and time domain reflectometers (TDRs). The psychrometers were installed in the region closest to the bentonite–rock interface while the TDRs were installed in the central region. This was done because the psychrometers provide a better measure of localized saturation state and because of their shorter design life (the logger at 240 m level is expected to be lost when flooding of the level is completed). The TDRs measure conditions in a larger volume of soil and their design life is much longer since their logger is pressure protected. The psychrometer readings indicate that, by early 2012, the seal had saturated to a distance of approximately 1–1.5 m in from the contact with the rock and concrete. The central bentonite-filled region is still unsaturated and water uptake is continuing much more slowly in these regions than it did near the rock and concrete contacts. This is a feature that has been observed in numerous other large-scale sealing demonstrations involving bentonite-based materials (see for example Dixon *et al.* 2001; Martino *et al.* 2008). Given that this phenomenon is related to the bentonite-based component, it is also of relevance to seals installed in other geological media.

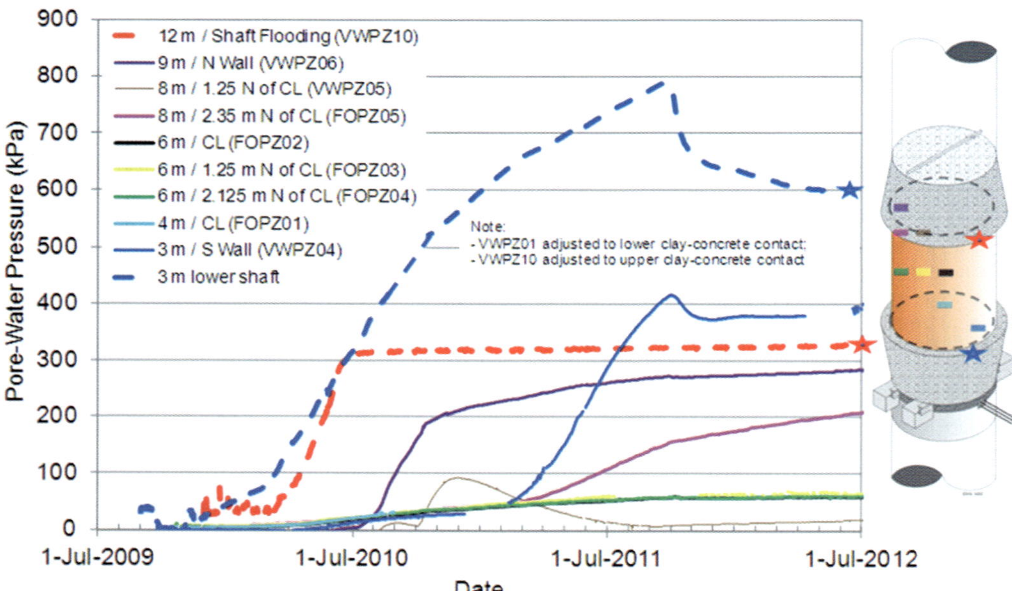

Fig. 5. Porewater pressures in bentonite-based component along rock–bentonite and concrete–bentonite contacts.

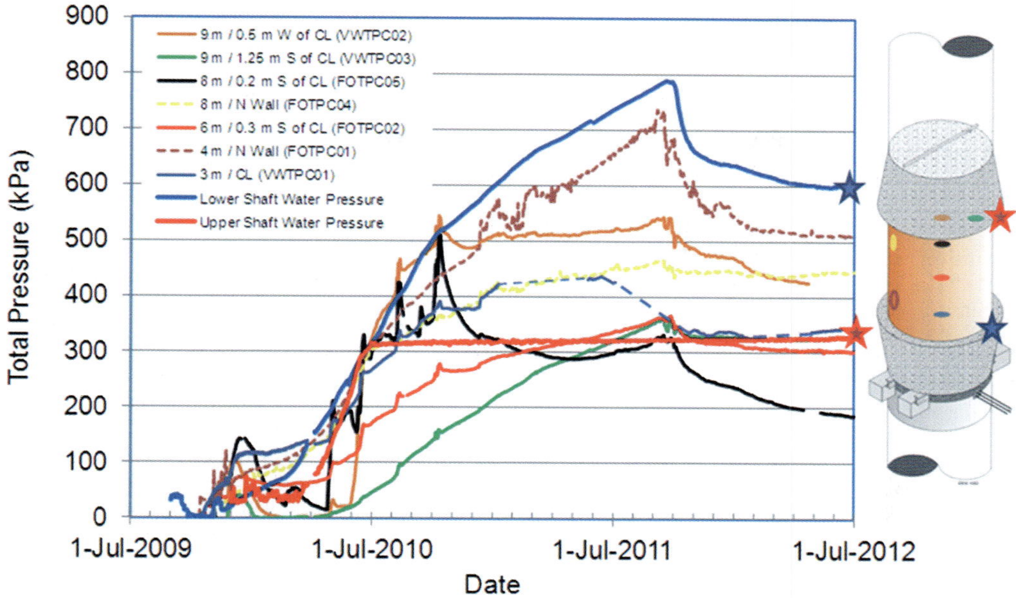

Fig. 6. Total pressures measured in bentonite-based component.

Figure 7 shows the water contents in the central volume of the bentonite-based component as recorded by the TDRs, with the volumetric water content rising from *c.* 23% (*c.* 68 ± 3% degree of saturation) to approximately 25% (*c.* 74 ± 1% degree of saturation), over the three years of water

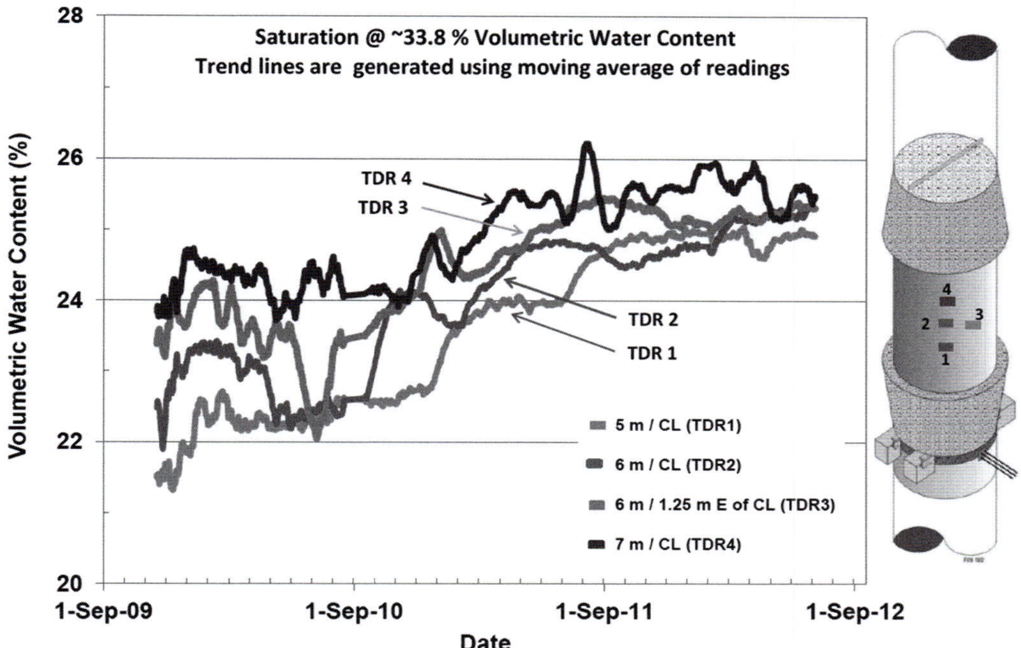

Fig. 7. Water uptake in central regions of bentonite-based component of shaft seal.

uptake as well as a trend towards uniform moisture conditions.

Summary

The shaft seal installed in the access shaft of the URL provides valuable experience related to the construction and performance of the type of massive sealing structures that will be needed in the closure of an actual repository for nuclear fuel waste. The ESP has provided the opportunity to monitor a full-scale shaft seal installed 275 m below the surface of a granitic rock mass for the first 3-years after its installation. The monitoring results have provided valuable information regarding the evolution of such a structure, especially the complexity of the pressure conditions within it as it takes on water and interacts with its surroundings.

It is expected that, as the upper shaft of the URL floods and the regional groundwater conditions are re-established, the system will trend towards less variability in the hydraulic and total pressure conditions within the bentonite-based component. It is also expected that, once flooding of the underground excavations is completed and regional groundwater conditions are re-established, the currently observed hydraulic pressure difference across the seal will trend towards a value attributable to elevation difference only. The swelling pressures within the confined bentonite-based component of the seal should ultimately reach approximately 0.8 MPa (based on laboratory measurements of this material) and remain at that level indefinitely. The hydraulic pressure at the midpoint of the seal will be approximately 2.7 MPa. These conditions will not, however, develop until after 2018 when flooding is essentially completed and regional groundwater pressures have recovered. In the period between 2012 and 2018, ongoing monitoring of the seal at the URL will provide valuable information regarding the evolution of a key component of the engineered barriers system proposed for use in sealing a repository for used nuclear fuel. The observation of the degree to which swelling pressures continue to show non-uniformity and how long the hydraulic pressure differential across the bentonite-based component persists will be of particular interest and relevance to other repository programs in other geological media. Both the swelling pressure and hydraulic pressure measurements provide an indication of the seal's effectiveness in isolating the regions above and below it, because maintaining of a hydraulic pressure differential across the seal is an indicator of restricted mass transport.

Although installed in a granitic medium, the design of the URL shaft seal and its evolution are relevant to a variety of geological media, specifically with respect to its construction and monitoring. The use of keys to disconnect the shaft EDZ, mass-poured low pH concrete, monitoring of water uptake and stress development in the bentonite-based component are largely medium-independent, making the results of the ESP relevant to other geological environments. The ESP also provides a valuable demonstration of the methods needed and the results of constructing of both shaft and tunnel/ramp seals in a disturbed rock environment. Comparison of the ESP results with other tests is recommended for future study and may help further assess relevancy of the results of this study with other geological environments.

The Enhanced Sealing Project (ESP) was jointly funded by the Nuclear Waste Management Organization (NWMO), Agence nationale pour la gestion des déchets radioactifs (ANDRA), Svensk Kärnbränslehantering AB (SKB), and Posiva Oy to monitor the seal at the Atomic Energy of Canada Limited (AECL)'s Underground Research Laboratory from 2009–2013. The construction of the seals was funded by NRCan through the Nuclear Legacies Liabilities Program (NLLP). The participation and support of all of these organizations is gratefully acknowledged.

References

Dixon, D. A., Martino, J. B. & Onagi, D. P. 2009. *Enhanced Sealing Project (ESP): design, construction and instrumentation plan.* Technical Report, Nuclear Waste Management Organization, **APM-REP-01601-0001**.

Dixon, D. A., Chandler, N. A., Stroes-Gascoyne, S. & Kozak, E. 2001. *The isothermal buffer–rock–concrete plug interaction test: results, issues, synthesis, Ontario Power Generation*, Supporting Technical Report, **06819-REP-01200-10056-R00**, Toronto, Ontario.

Dixon, D. A., Priyanto, D. G. & Martino, J. B. 2012. *Enhanced Sealing Project (ESP): project status and data report for period ending 31 December 2011.* Technical Report, Nuclear Waste Management Organization, **APM-REP-01601-0005**.

Holowick, B. E., Dixon, D. A. & Martino, J. B. 2011. *Enhanced Sealing Project (ESP): project status and data report for period ending 31 December 2010.* Technical Report, Nuclear Waste Management Organization, **APM-REP-01601-0004**.

Martino, J. B., Dixon, D. A. & Stroes-Gascoyne, S. 2008. *Project Report: The Tunnel Sealing Experiment 10 Year Summary Report.* Atomic Energy of Canada Limited, **URL-121550-REPT-001**, Chalk River.

Martino, J. B., Dixon, D. A., Holowick, B. E. & Kim, C.-S. 2010. *Construction of full scale shaft seals and Enhanced Sealing Project (ESP) monitoring equipment installation.* Technical Report, Nuclear Waste Management Organization, **APM-REP-01601-0003**.

EBS behaviour immediately after repository closure in a clay host rock: HE-E experiment (Mont Terri URL)

I. GAUS[1]*, K. WIECZOREK[2], K. SCHUSTER[3], B. GARITTE[1], R. SENGER[4], R. VASCONCELOS[5] & J. C. MAYOR[6]

[1]*Nagra, Wettingen, Switzerland*

[2]*GRS, Braunschweig, Germany*

[3]*BGR, Hannover, Germany*

[4]*Intera Swiss Branch, Ennetbaden, Switzerland*

[5]*CIMNE, Barcelona, Spain*

[6]*Enresa, Madrid, Spain*

**Corresponding author (e-mail: irina.gaus@nagra.ch)*

Abstract: The evolution of the clay-based engineered barrier system (EBS) of geological repositories for radioactive waste has been the subject of many research programmes during the last decade. The early post-closure thermal behaviour is elucidated by the HE-E experiment, a 1:2 scale heating experiment (at the Mont Terri Rock Laboratory), which was implemented in the first semester of 2011, with the initiation of the heating phase in June 2011. A maximum temperature of 140 °C was reached in June 2012. After 15 months of heating, the temperature evolution in the EBS and the Opalinus Clay reflects the design calculations, and thermally induced porewater overpressures are being measured at a few metres' distance in the Opalinus Clay. Seismic methods proved to be a sensitive tool for the continuous characterization of changes of EBS and Opalinus Clay properties. Design modelling and predictive modelling based on the as-built parameter dataset with established coupled codes (TOUGH, CODE_BRIGHT; using various geometries) are described. The results indicate that the models are generally in agreement with the observations and capable of capturing the evolution of the experiment.

The evolution of the clay-based engineered barrier system (EBS) of geological repositories for radioactive waste has been the subject of many research programmes during the last decade (e.g. the 6th Framework NF-PRO project). The emphasis of the research activities was on understanding the complex thermo-hydro-mechanical-(chemical) processes that are expected to take place in the early post-closure period in the near-field. It is important to understand the coupled thermo-hydro-mechanical (-chemical) (THM(-C)) processes and their evolution occurring in the EBS during the early post-closure phase in order to confirm that the safety functions will be fulfilled in the later phases when the radionuclides will be released after the breaching of the canisters. In particular, it is necessary to ensure that interactions during the resaturation phase (heat pulse, gas generation, non-uniform water uptake from the host rock) do not affect the near-field in terms of its safety-relevant parameters (e.g. swelling strain and pressure, hydraulic conductivity, retention behaviour).

The 7th Framework EURATOM PEBS project (Long Term Performance of Engineered Barrier Systems) aims at providing in-depth process understanding for constraining the conceptual and parametric uncertainties in the context of long-term safety assessment. As part of the PEBS project, a series of laboratory and *in-situ* experiments are envisaged to describe the EBS behaviour after repository closure when the resaturation and heating take place. In this paper the very early post-closure period is investigated when the EBS is subjected to high temperatures and unsaturated conditions with low but increasing moisture content. So far the detailed thermo-hydraulic behaviour of a bentonite EBS in a clay host rock has not been evaluated at a large-enough scale in response to temperatures of up to 140 °C at the canister surface, produced by high-level waste (and spent fuel), as anticipated in some of the designs considered. Furthermore, earlier THM experiments (Zhang *et al.* 2007) have shown that upscaling of thermal conductivity and its dependency on water content and/or humidity

From: Norris, S., Bruno, J., Cathelineau, M., Delage, P., Fairhurst, C., Gaucher, E. C., Höhn, E. H., Kalinichev, A., Lalieux, P. & Sellin, P. (eds) 2014. *Clays in Natural and Engineered Barriers for Radioactive Waste Confinement*. Geological Society, London, Special Publications, **400**, 71–91.
First published online March 7, 2014, http://dx.doi.org/10.1144/SP400.11

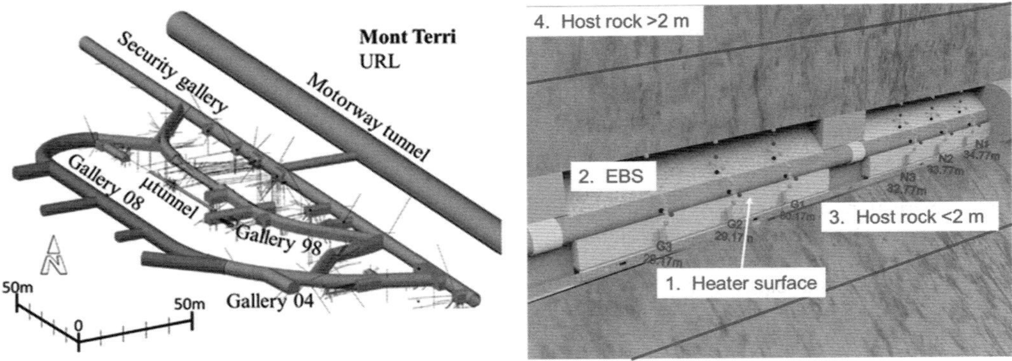

Fig. 1. Left: location of the HE-E experiment in the microtunnel (µtunnel) in the Mont Terri URL (Switzerland). Right: instrumentation concept for the HE-E experiment consisting of four monitoring zones: 1, heater surface; 2, engineered barrier system; 3, Opalinus Clay <2 m from the HE-E microtunnel; 4, Opalinus Clay >2 m from the HE-E microtunnel.

from the laboratory scale to a field scale need further attention.

Objectives and setup of the experiment

The HE-E experiment is a 1:2 scale heating experiment considering natural resaturation of the EBS at a maximum heater surface temperature of 140 °C. The experiment is planned to run at least until 2014. The experiment is located at the Opalinus Clay (OPA) of the Mont Terri Underground Research Laboratory (URL; Switzerland) in a 50 m-long microtunnel of 1.3 m diameter (Fig. 1). The test section of the microtunnel has a length of 10 m and has been characterized in detail during the Ventilation Experiment, which took place in the same test section (Mayor *et al.* 2007). The heating started in June 2011, whereby the maximum temperature was reached in June 2012. Since then, the temperature is being held constant.

The aims of the HE-E experiment are elucidating the early non-isothermal resaturation period and its impact on the thermo-hydro-mechanical behaviour, namely: (a) to provide the experimental database required for the calibration and validation of existing THM models of the early resaturation phase; and (b) to upscale thermal conductivity of the partially saturated EBS from laboratory to field scale (pure bentonite and bentonite-sand mixtures).

The experiment (Fig. 2; Gaus (2011), Teodori & Gaus (2011)) consists of two independently heated sections of 4 m length each, whereby the heaters are placed in a steel liner supported by MX80 bentonite blocks (dry density 1.8 g cm^{-3}, water content 11%). The two sections are fully symmetric apart from the granular filling material. While section 1 is filled with a 65/35 granular sand–bentonite mixture, section 2 is filled with pure MX80 bentonite pellets. This allows comparison of the thermo-hydraulic behaviour of the two EBS materials under almost identical conditions. Both materials have been characterized in the laboratory with respect to their thermo-hydraulic properties. The MX80 materials (blocks and pellets) are similar to those materials considered for the repository

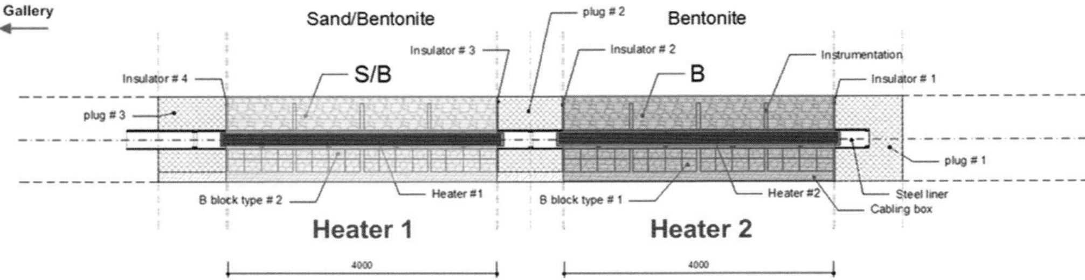

Fig. 2. Schematic layout of the HE-E experiment showing the section in the back of the tunnel filled with bentonite pellets and the section in the front of the tunnel filled with sand–bentonite.

EBS in Switzerland. The sand–bentonite mixture is under consideration as an alternative EBS material in Germany.

Construction and instrumentation of the experiment

The construction of the HE-E experiment took place between December 2010 and June 2011. In a first step, a U-profile railway was installed in the 50 m-long tunnel. Subsequently the host rock was instrumented. The EBS instrumentation modules were then towed into the tunnel, leading to the connection of the heater liner elements into one continuous liner. An auger, adapted to the 1.3 m diameter of the tunnel, was used to emplace the granular EBS material. Emplacement densities, established during off-site tests for the MX80 ranged around 1.45 g cm^{-3} while for the sand–bentonite mixtures the densities were estimated to be 1.50 g cm^{-3}. The test sections in the tunnel were separated by three concrete plugs also containing thermal insulation and a vapour barrier. Finally, in the last step, the two 4 m-long heaters were emplaced in the central liner.

The instrumentation concept is targeting four zones (Fig. 1): (a) the heater surface where the temperature is measured; (b) the EBS itself and the interface with the Opalinus Clay with very dense measurements of temperature and relative humidity; (c) the Opalinus Clay close to the microtunnel, which was under the influence of the ventilation before and during construction where temperature, humidity, hydraulic pressure and displacement are monitored; and (d) the Opalinus Clay at several metres from the microtunnel, where hydrostatic conditions were less disturbed by the activities in the microtunnel and where hydraulic pressures are monitored. In the design phase it was assumed that no significant swelling pressure would develop in the EBS. During the early resaturation phase the permeability of the granular-like buffer materials is still high, but resaturation rate is limited by the low permeability of the rock, resulting in low inflow rates. Differential thermal expansion of water and solid skeleton is likely to induce hydraulic overpressures in the saturated Opalinus Clay at some distance from the EBS–rock interface. The measurement and assessment of these overpressures became an additional objective of the experiment (see the section 'Model calculations').

Heater control and heater surface instrumentation

Two 4 m-long electrical heaters (each providing a maximum of 2400 W) were installed in a central steel liner. Heaters are designed to be operated by either power or temperature control and the heaters function independently for the two sections. Twenty-four thermocouples (12 for each section), placed on six radial planes at the inner part of the 8 mm-thick heater liner allow for monitoring of the heater surface temperature in four radial directions. Heat transmission between the sections and through the front and back plugs of the test section is limited by thermal insulation.

EBS instrumentation

A total of 18 humidity/temperature sensors are emplaced at the Opalinus Clay/EBS interface and another 60 in the EBS materials (24 sensors in the blocks and 36 in the granular materials). The sensor location in the bentonite blocks and in the granular material is indicated in Figure 3 (left). Instrumentation rings were designed with insulating material to avoid thermal bridges. The high sensor density ensures an accurate spatial characterization of the thermal behaviour and how it is affected by the saturation in the Opalinus Clay and vapour formation owing to the heating.

Host rock (Opalinus Clay) instrumentation up to 2 m from the microtunnel

Direct measurements. In the immediate vicinity of the microtunnel (up to 2 m distance) a dense instrumentation configuration in 11 planes perpendicular to the tunnel axis is installed consisting of 34 mini-piezometers (hydraulic pressure); 20 capacitive humidity/temperature sensors (humidity, temperature); eight extensometers (displacement); and 17 psychrometers (humidity; Fig. 3, right).

Indirect measurements. A seismic array consisting of five piezoelectric transducers that serve as emitters and 10 transducers that serve as receivers was installed in three 1 m-deep boreholes. The details and interpretation of the observations are given in the section 'Seismic characterization of Opalinus Clay and sand–bentonite mixture'.

Host rock (Opalinus Clay) instrumentation at larger distances from the microtunnel

Design calculations (Czaikowski *et al.* 2012) indicated that hydraulic overpressures owing to the thermal pulse will not occur close to the microtunnel but can occur at several metres' distance in the OPA. Therefore, on top of two existing boreholes, two additional boreholes were drilled from the main gallery (indicated in Figure 1 as Gallery 98) and each equipped with a quadruple-packer system allowing the monitoring of hydraulic pressure at

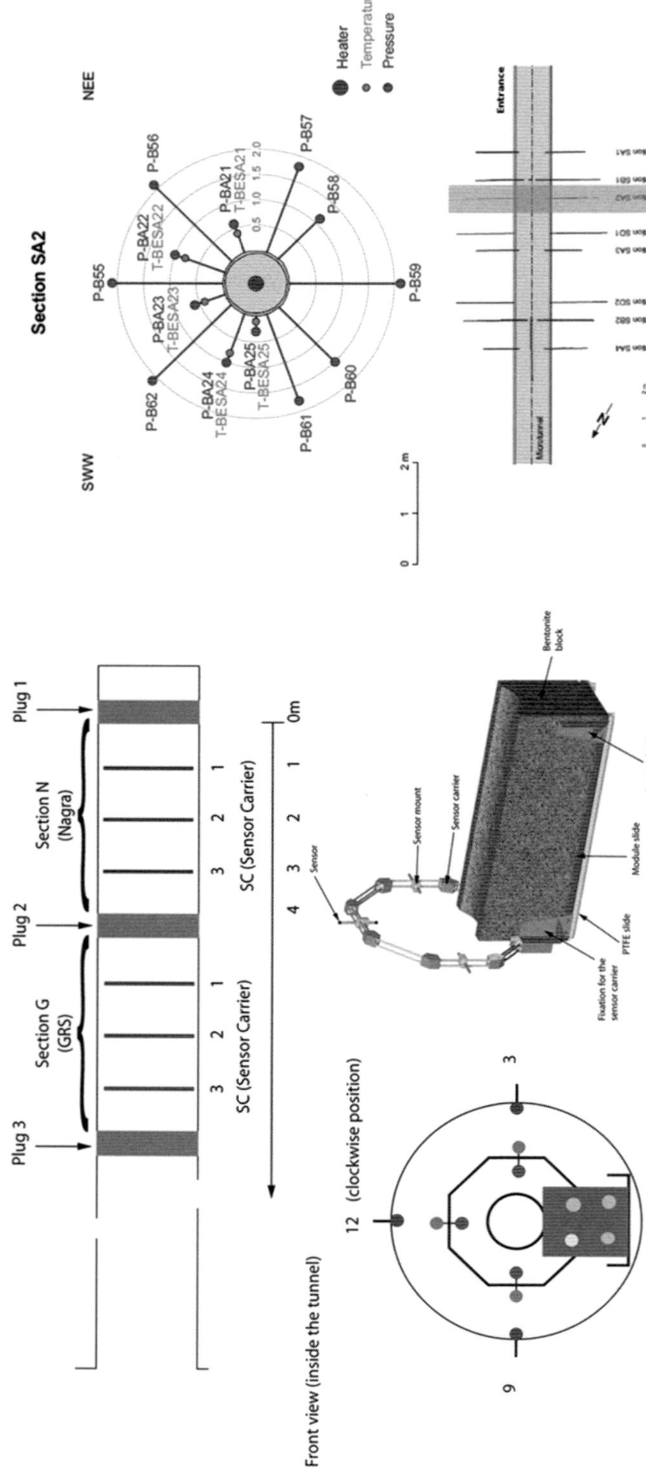

Fig. 3. Instrumentation in the Engineered Barier System and design of the modules (blocks and instrumentation rings) on which the heater liner was placed (left). Instrumentation in the Opalinus Clay at distances <2 m from the microtunnel surface, section SA2, located in the middle of the sand–bentonite section, shown as an example (right).

relevant distances. This allowed measurements up to 6 m distance from the microtunnel.

Observations during the first 15 months after heating start

The impact of the thermo-hydraulic response on both the EBS and the Opalinus Clay was monitored, first by establishing the initial conditions before heating started and then during the heating phase. The results are discussed per zone as indicated in Figure 1.

Heating system

In the first year of the experiment, the heating power was gradually increased to reach the target temperature of 140 °C after 1 year of operation. While the initial steps were power-controlled, after a few months a linear temperature controlled increase was imposed. This strategy of slow increase was based on the temperatures derived from repository-scale modelling including estimated heat output from the canisters (Johnson *et al.* 2002), and this avoided strong temperature gradients at the beginning of the experiment. The power evolution of the heaters and temperature evolution at the heater surface is shown in Figure 4. The power needed to maintain the heater surface temperature in the bentonite section is significantly higher than in the sand–bentonite section, indicating an overall higher thermal conductivity in the bentonite section.

The temperature at the surface of the steel liner in each section is not entirely homogenous: the basis of the liner resting on the bentonite blocks has a slightly lower temperature than the top and the sides in both sections. The difference is likely to be caused by a better contact with the EBS materials at the base and the higher thermal conductivity of the blocks compared with the granular material, indicating that the heat profile is not radially symmetric in the early stages. Convection within the heater system might be an additional factor as hot air in the heater system has a tendency to move upward.

Engineered barrier (bentonite and sand–bentonite) and EBS–Opalinus Clay interface

The temperatures in both engineered barrier sections mirror the heater surface temperatures closely (Fig. 5, top). The temperature in the sand–bentonite section is slightly lower, possibly owing to some heat loss through the front of the tunnel. In both cases a strong temperature gradient is established going from 140 °C at the heater surface to 40 °C at the EBS–Opalinus Clay interface over a distance of 50 cm. The difference in measured relative humidities between the two sections is larger than for the thermal field (Fig. 5, bottom). At the start of the heating the relative humidities are almost identical in both sections and above 60% at the EBS–Opalinus Clay interface owing to the water intake from the OPA. Water uptake and redistribution continues during the heating phase, leading to 100% humidity in the sand–bentonite section after 4 months while after 15 months the bentonite interface has not reached 100% yet. The relative humidities at smaller distances from the heater are almost identical after 4 months and continue to decrease owing to evaporation and subsequent vapour diffusion. Even after the maximum temperature was reached in June 2012, this trend continues.

The difference in thermal conductivity between the bentonite blocks and the pellets is illustrated in Figure 6. The temperatures at similar distances from the heater surface are higher in the blocks than in the pellets (Fig. 6), although the heater surface temperature is lowest where it is in contact with the blocks. These observations indicate a lower temperature gradient in the blocks and thus, as a direct consequence of Fourrier's law, a larger thermal conductivity in the blocks. The larger thermal conductivity of the blocks can be attributed to their higher initial water content and higher dry density. In the early stages of the experiment the blocks thus act as a thermal bridge compared with the pellets. Although the relative humidity in the blocks is initially higher than in the pellets, it has become identical at similar distances from the heater surface after 15 months (Fig. 6, bottom). Heating seems to have a determining influence on the homogenization of the humidity distribution inside the EBS based on the currently available observations. At 10 cm from the heater surface the relative humidity reduces to around 10% after 15 months.

An illustration of the longitudinal homogeneity of the observations in the EBS is given in Figure 7, which shows all sensor readings parallel to the heater axis, at a distance of 25 cm from the heater surface, in the 3 o'clock direction when facing the front of the experiment. The temperature variations (Fig. 7, top) are small, of the order of <5 °C (<8% relative difference), and the variation between the two sections is larger than within the same section. The relative humidities within the same section are slightly different between the two sections but almost identical within each section (Fig. 7, bottom).

After 15 months of heating, of which the last 3 months are at 140 °C, heating is the dominant

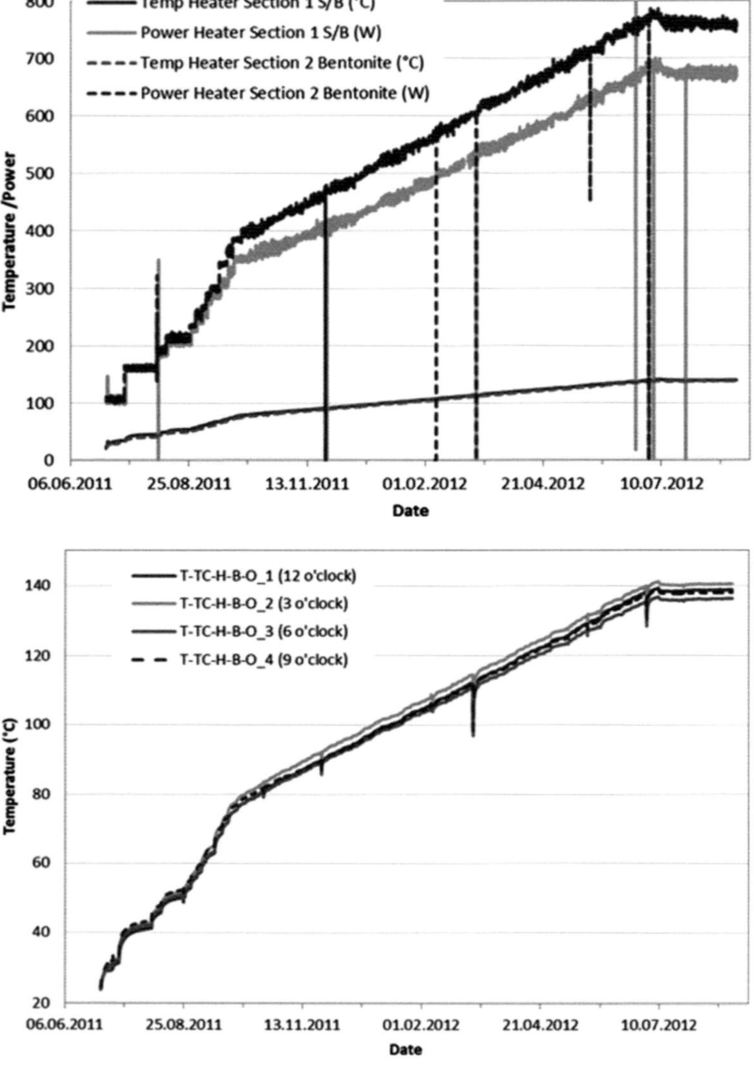

Fig. 4. Temperature at the heater surface and applied power in the two sections of the HE-E experiment (top; vertical lines indicate short power interruptions). Heater surface temperature in section 2 at four locations (bottom).

driving force behind the hydraulic processes within the EBS, although water uptake is probably not negligible. The latter is dependent on the permeability of the Opalinus Clay on the one hand and the initial suction in the EBS on the other. The permeability of the Opalinus Clay being very low; it is likely to be the determinant factor. This assumption is supported by the observation of relative humidity increase observed at the OPA/EBS interface. Additionally, part of this increase can also be induced by vapour diffusion from the hotter inner part of the EBS. The role of vapour diffusion cannot be understood from the observations only and requires numerical modelling.

Opalinus Clay close to the microtunnel (<2 m)

Before and during the construction of the experiment the microtunnel was ventilated, inducing changes in hydraulic pressure, saturation and humidity close to the tunnel surface. At the start of the experiment the hydraulic pressure up to 2 m from the surface was atmospheric and humidities

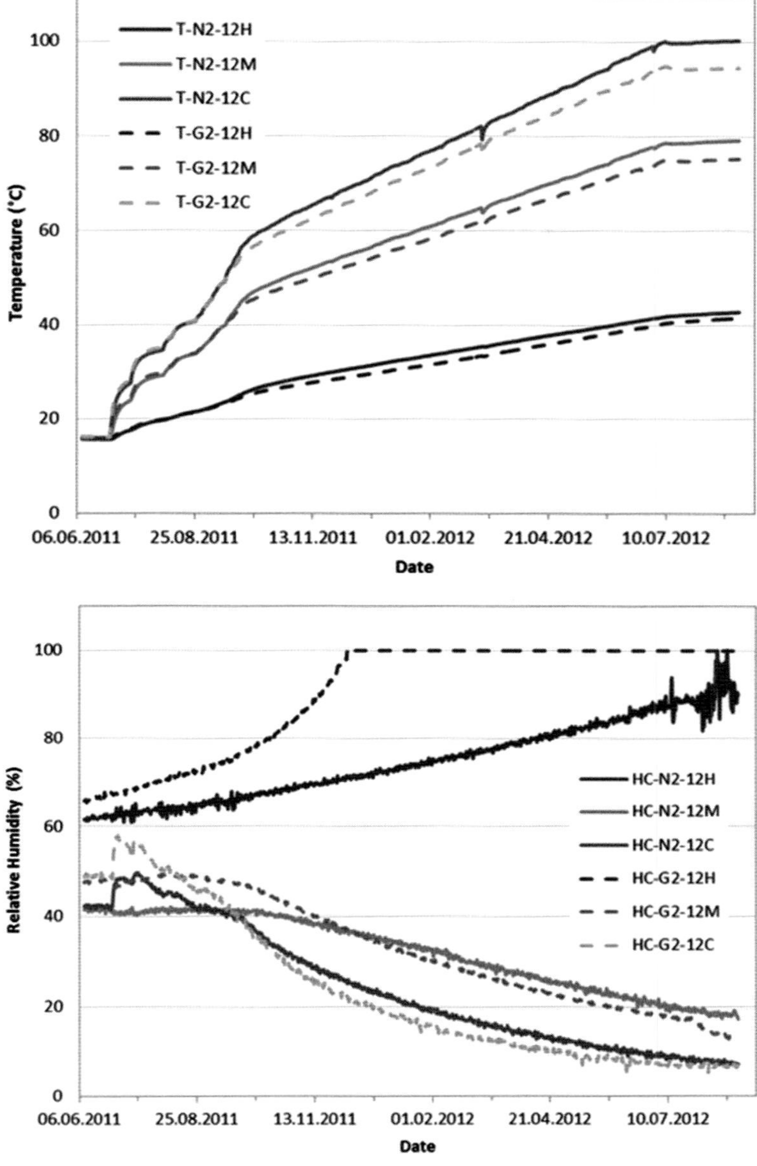

Fig. 5. Temperature and relative humidity evolution in the middle of the bentonite section (N2) and the sand–bentonite section (G2) (Temperature (top) and relative humidity (bottom), in the vertical upward direction (12 o'clock) at 10 cm (C), 25 cm (M) and 45 cm (H) from the heater surface.

at the surface were approximately 98% (not shown). Recording of atmospheric pressure indicates that this area is in suction. Piezometers kept recording values near atmospheric during the heating phase, indicating that, owing to the low permeability of the Opalinus Clay, no significant flux is taking place. Figure 8 (top) shows sensors up to about 2 m from the microtunnel wall in the radial section in the middle of the sand–bentonite part. It illustrates how slowly the porewater pressure increase in the Opalinus Clay occurs. Sensors close to the microtunnel are still indicating suction after 15 months; however, sensors at 2 m from the microtunnel register a porewater pressure increase 5–10

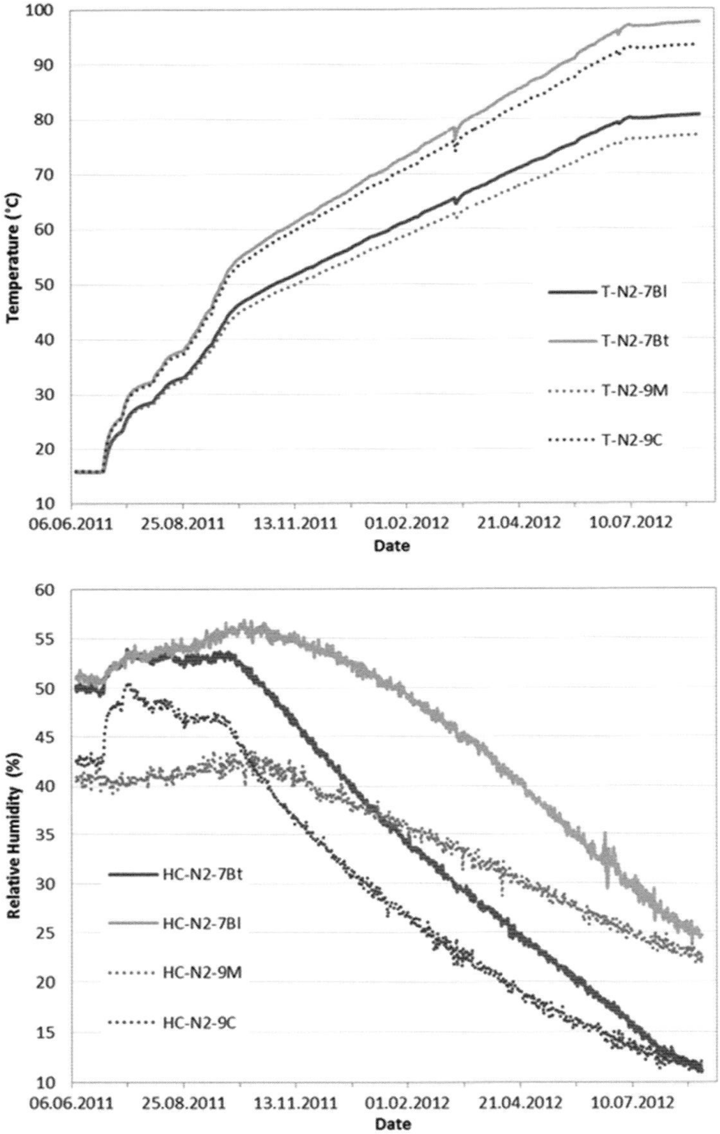

Fig. 6. Temperature and relative humidity evolution in the middle of the bentonite section: comparison of blocks at 11.6 cm (Bt) and 21.1 cm (Bl) from the heater surface at 7 o'clock v. the pellets at 10 cm (C) and 25 cm (M) at 9 o'clock from the heater surface. Top, temperature; bottom, relative humidity.

months after the heaters started. Temperatures in the Opalinus Clay in the same section (Fig. 8, bottom) reflect the steady increase after heaters were switched on, although temperatures remain fairly low (35 °C at a depth of 0.4 m). While heat dispersion in the bentonite and sand–bentonite is fairly homogeneous, in the Opalinus Clay an anisotropy regarding the thermal conductivity can be suspected with higher temperatures in the plane horizontal to the microtunnel and lower temperatures towards the top (no measurements at the bottom are available).

Opalinus Clay at larger distances from the microtunnel (>2 m)

At a few metres from the microtunnel, positive pore-water pressure conditions were maintained, even

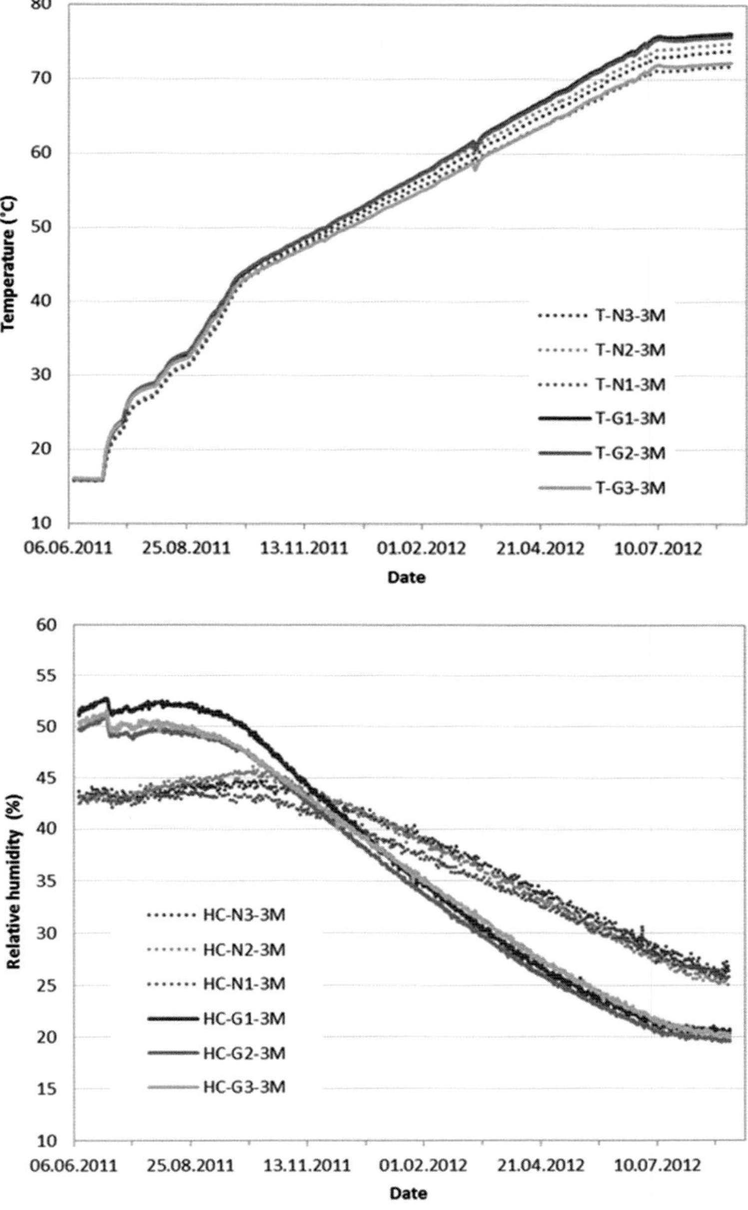

Fig. 7. Assessment of the homogeneity of the observations: all measurement in the two sections at 25 cm from the heater surface along the 3 o'clock axis in both the bentonite and the sand–bentonite sections. Top, temperature; bottom, relative humidity.

during the operations in the microtunnel. This was monitored through multipacker systems in four boreholes drilled from the main gallery perpendicular to the direction of the microtunnel. The hydraulic pressures before the start of the experiment are somewhat different from those assuming pure hydrostatic control as there is an impact from the presence of the main gallery (see Fig. 1) and the laboratory as a whole. However, almost immediately after switching on the heaters, the progression of

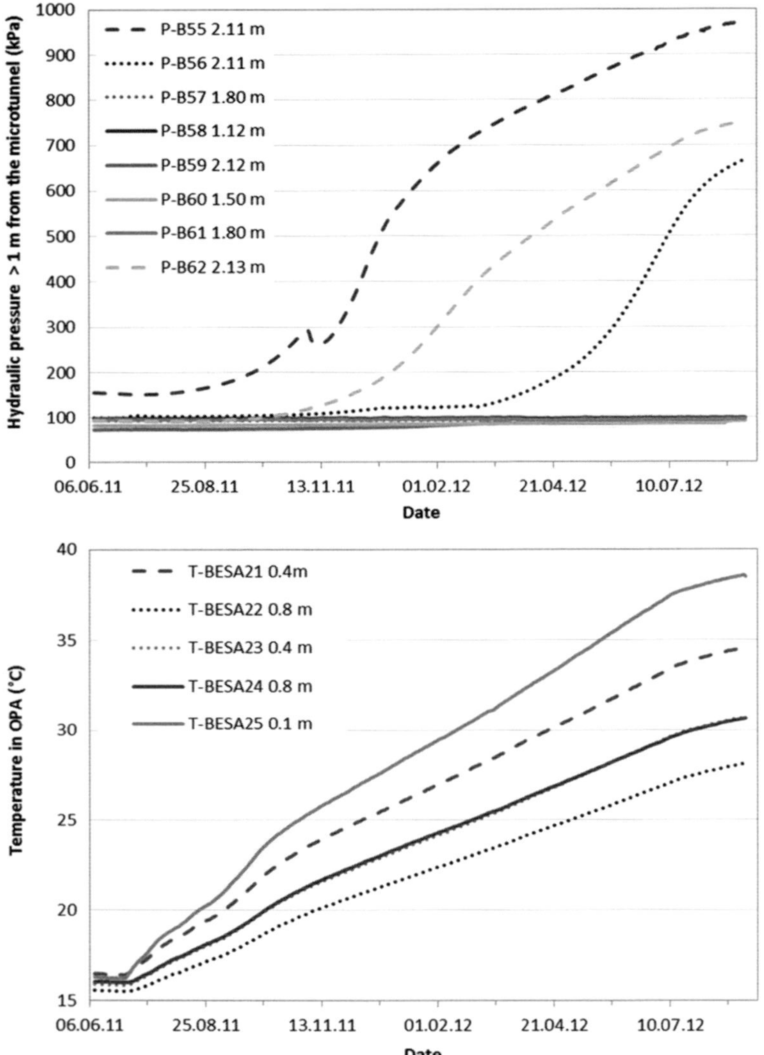

Fig. 8. Top: hydraulic pressures in the Opalinus Clay at distance between 1.0 and 2.0 m in the middle of the sand–bentonite section (section SA2) since the start of the heating; distances are with respect to the microtunnel surface (top). Bottom: temperatures in the Opalinus Clay up to 0.8 m from the microtunnel surface in the middle of the sand–bentonite section (section SA2) since the start of the heating.

the temperature front leads to increasing hydraulic pressures in all intervals owing to the difference in thermal expansion coefficients of the Opalinus Clay and the porewater (Fig. 9, example of borehole BVE91). The maximum increase is of the order of 0.7 MPa at 3.5 m distance, reducing to 0.4 MPa at 5.6 m distance from the microtunnel, indicating that the area around the microtunnel in which overpressures are generated is currently (after 15 months of heating) approximately <10 m radius (based on linear extrapolation). While the heater temperature was stable from June to August 2012 close to the heater, ongoing temperature increase further away causes pressures to increase following the same trend as before heater temperature stabilization. At this stage the maximum overpressures that will be reached owing to thermal expansion can only be estimated by modelling.

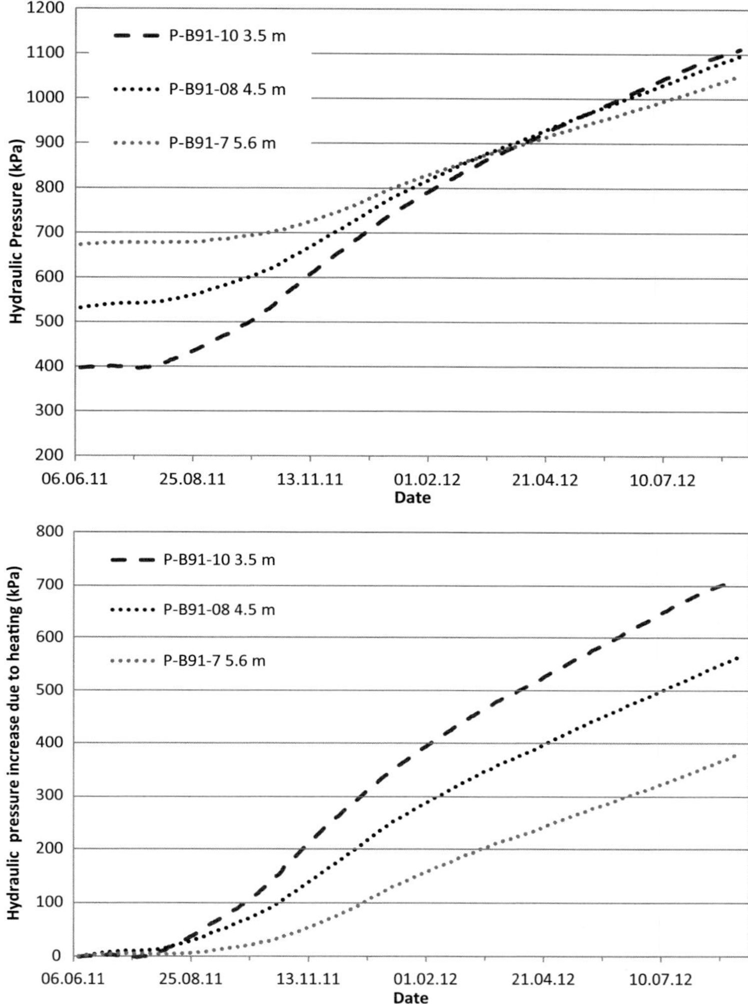

Fig. 9. Hydraulic pressures measured after starting the heating at larger distances from the microtunnel where no prior impact on hydraulic pressures occurred (top, absolute hydraulic pressure; bottom, relative increases since the start of the heaters).

Seismic characterization of Opalinus Clay and sand–bentonite mixture

Layout of the seismic experiment

Seismic transmission measurements aim at the characterization of the sand-bentonite (S/B) material and the OPA with the help of seismic parameters. A change in rock properties in the vicinity of the microtunnel and especially in the S/B is expected owing to the de- and resaturation processes, stress redistribution and the temperature impact, all caused by the installation work, the closure of the section and finally the heating process. Seismic parameters like P-wave velocity (v_p) and the amplitudes of first arrival phases react very sensitively to appropriate changes. These parameters characterize the rock in an integral way over distances between 0.26 and 0.86 m depending on the emitter–receiver locations. Not all correlations and dependencies between varying seismic parameters and related rock property changes are completely understood.

The evolution of both materials has been monitored since 12 March 2011, approximately 50 days before the S/B was emplaced and the section was closed, and 108 days before the heater was switched on. A daily automatic seismic measurement is performed using a small-scale seismic array consisting of five piezoelectric transducers which serve as emitters and 10 transducers which serve as receivers. Three existing 1 m-long boreholes in the test section of the microtunnel were used for the installation (boreholes BVE-112, BVE-113 and BVE-114; see Fig. 10). The boreholes are located in the S/B-section of the experiment. The installation of all prefabricated components was performed by the Gesellschaft für Materialprüfung und Geophysik GmbH while data processing and interpretation are being done by BGR.

Details of the layout of the seismic array are given in Figure 10. The boreholes are located on the eastern wall of the test section around 3 m from plug 3 between instrumentation G1 and G2 (cf. Fig. 1). All three boreholes are subhorizontal and nearly parallel. With respect to the tunnel axis they are oriented *c.* 30° south. Three transducers are located outside the boreholes in order to observe the evolution in the backfill material. According to the experience from former experiments (VE, Schuster 2007 and EB, Schuster & Alheid 2004) the observation concentrates on the first 50 cm of the rock, where the main seismically detectable changes in rock properties are expected.

The chosen layout of the seismic transducers allows seismic waves travelling roughly preferably normal, parallel and with 45° towards the bedding planes of the OPA. Concerning the coverage of different depths levels and orientations of the ray paths towards the bedding, a compromise had to be made. Not all depth levels for all three orientations could be realized. The automatic daily measurement between all emitter–receiver combinations results in 50 different ray paths. The regular measuring phase started on 12 March 2011 with one measurement phase every night at about 1 a.m. During this measuring phase of approximately 50 min, the 10 seismic receivers record successively seismic signals emitted by the five emitters. Each recording is repeated 2048 times in order to improve the signal-to-noise ratio. In total, 50 different seismic traces are recorded every night. From these data, seismic parameters are derived for further interpretation. In this stage the focus is on the P-wave arrival phases. More data will be analysed in a later stage.

Seismic data and derived parameters

With the derived travel times (t) and the calculated distances (L) between seismic emitters and receivers, seismic P-wave velocities are calculated

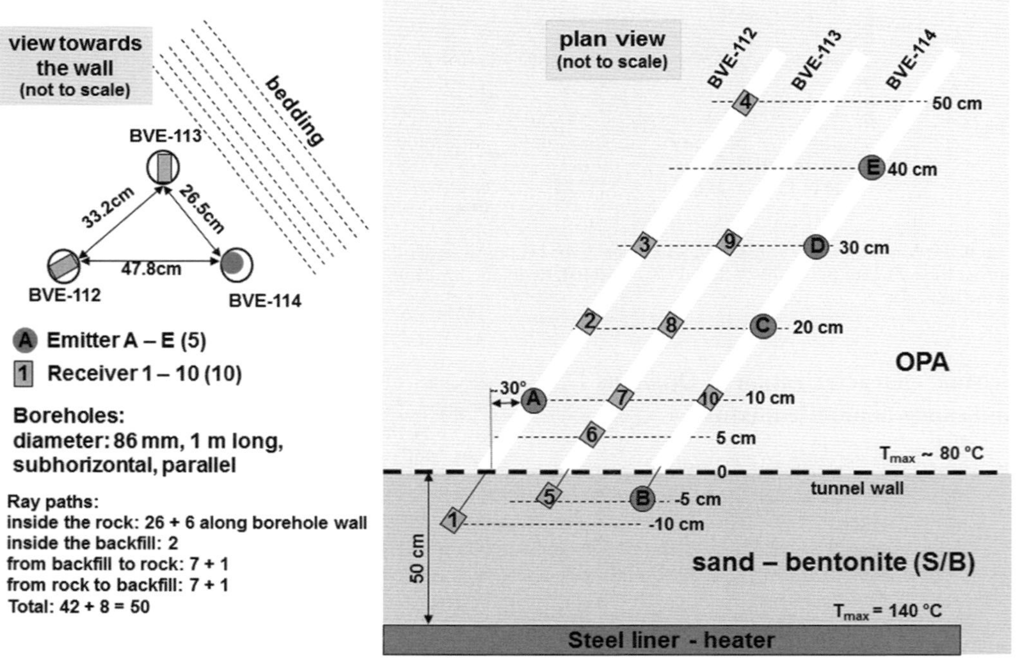

Fig. 10. Layout of the seismic transmission experiment.

($v_p = L/t$). To optimize comparison, all velocities are normalized to the values derived for day 1 of the measurements. Furthermore, owing to minor uncertainties in the distance estimation with a v_p normalization, 'over-interpretation' of the absolute v_p-values is avoided. All derived absolute v_p values are lower than the values obtained in our former *in-situ* investigations outside the excavation damaged/disturbed zone. However, the relative differences between the absolute velocities for the three orientations are in line with the anisotropic structure of the OPA (v_p-normal $< v_p$−45° $< v_p$-parallel). The frequencies of the P-wave phases are in the range between 12 and 20 kHz, which is at the lowermost edge of ultrasonic signals.

A rather qualitative explanation for the v_p variation can be given according to our experience gained at several ultrasonic *in-situ* experiments at the Mont Terri Rock Laboratory. In general, a drop in v_p can be explained by a loosening of the rock and an increase in v_p by a consolidation of the material. Loosening can be caused by the creation of microcracks, even when they are macroscopically not visible, or an increase in porosity. Furthermore, a v_p increase can be caused by an increase in saturation, involving the swelling and consolidation of the rock (sealing of microcracks) Interpretations concerning the influences of pore pressure and saturation (gas and fluids) on signal variations and v_p, especially for *in-situ* investigations in argillaceous rocks, have to be enlarged.

As an example, seismic sections from two different emitter–receiver pairs are shown in Figure 11 covering the time 12 March 2011 (day 1) to 14 September 2012 (day 553). Only every second trace is plotted in order to obtain a clearer visualization. Some recordings are defective, for example around day 305. Both sections are ensemble normalized. The P-wave arrival phases are indicated. On the left side in Figure 11 seismic traces for emitter B (E02) and receiver 1 (R01), both located in the S/B at 5 and 10 cm distance from the interface S/B–OPA, are shown (cf. Fig. 10). Between day 1and day 51 no seismic signals can be correlated because no buffer material was emplaced and consequently no seismic energy could propagate. The emplacement of the sand–bentonite mixture started on day 52. Starting from this day the seismic signals become stronger and seismic P-wave phases can be correlated clearly. A decrease in travel times corresponds to an increase in P-wave velocity (v_p). This process continues until day 180 followed by constant values until day 300. Then travel times decrease in two different steps, between day 300 and day 500 with a higher gradient as between day 500 and 553. The corresponding v_p graph is shown in Figure 12.

Figure 11 (right) shows the seismic section for ray path running in the OPA at approximately 45° towards the bedding at a distance of 10 cm from the interface S/B–OPA (cf. Fig. 10). The P-wave arrivals can be identified very clearly at around 210 μs (day 1). The pronounced variation of these first arrival phases (travel time and amplitude) with time is a clear indication for changes of petrophysical properties of the OPA along the travel path. Around day 25 the travel times start to increase (v_p drops) and the strength of amplitudes drops considerably. During that time the installation work and the emplacement of EBS in the rear section of the tunnel started. The test section was closed at day 53. The heater elements were switched on at day 109. From day 130 on the amplitudes start to become stronger and v_p increases. Around day 300 travel times and amplitudes are very close to the starting situation. At day 350 travel times start to increase again (v_p drops). This process continues until the end of the observations. The P-wave phases change at the same time towards a lower frequency-content ('broader signals') and lower amplitudes (not visible in this data representation). This could be an indication of an increase in saturation of the OPA and/or porosity increase. The appropriate derived and normalized seismic P-wave velocities are plotted in Figure 12. Seismic trace plots from different distances (5–45 cm, not displayed here) show a clear qualitative trend: the further the travel paths from the S/B–OPA

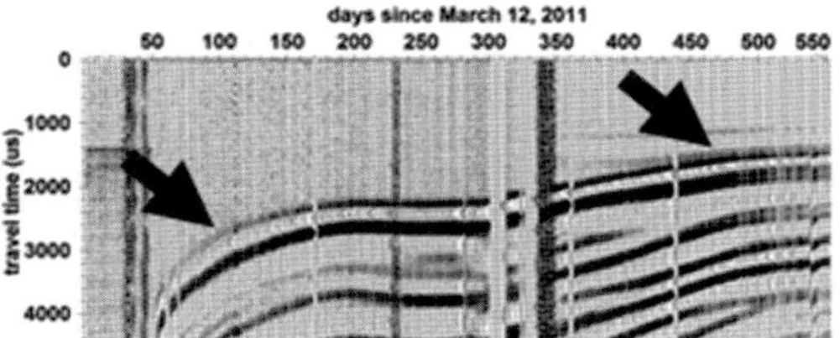

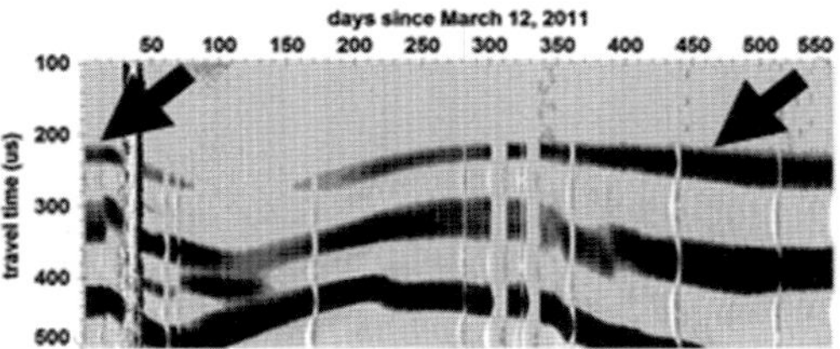

Fig. 11. Seismic sections recorded between 12 March 2011 and 14 September 2012. Left: seismic traces from the sand–bentonite mixture, between 5 and 10 cm from the S/B–OPA interface (B-R1). Right: seismic traces derived from ray paths running 45° to bedding in the OPA at a distance of 10 cm from the S/B–OPA interface (A-R10). P-wave arrivals are marked.

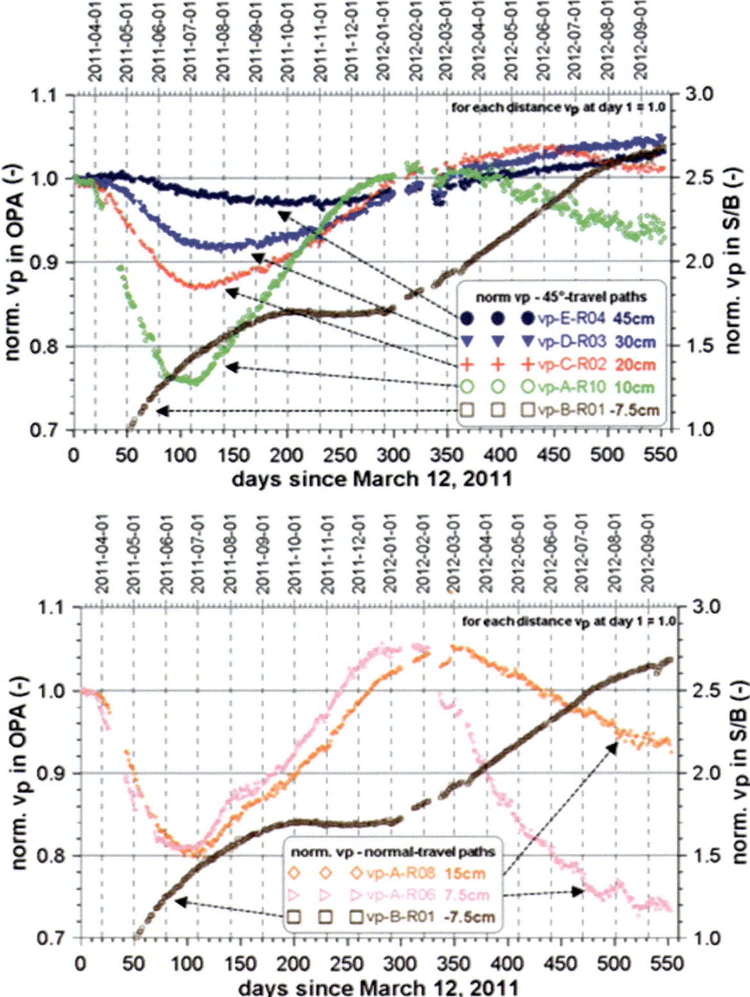

Fig. 12. Derived normalized seismic P-wave velocities at different depth levels. Top: for ray paths preferably running 45° to bedding. Bottom: for ray paths preferably running normal to bedding. For the v_p normalization, v_p at day 1 is set to 1.0. Note: the right *y*-axis in both graphs applies to the normalized v_p derived from the S/B.

interface, the less the P-wave phase variation is present (travel time and amplitudes).

First interpretation of seismic data

In Figure 12 several v_p graphs from different distances of the ray paths from the S/B–OPA interface and two orientations towards the bedding are shown. All velocities are normalized ($v_p = 1$ for day 1). Additionally, the v_p evolution between 5 and 10 cm distance from the S/B–OPA interface of the sand–bentonite mixture is given in both figures.

In the S/B material a rather stepwise v_p increase can be observed that indicates a clear stepwise consolidation of the S/B mixture. The plateau is remarkable between days 200 and 300, as well as the bend in the graph around day 500 several days after the heaters were held at constant temperature (cf. Fig. 6). The notch in the v_p graph at day 540 results most probably from a power failure that resulted in a similar notch in the temperature profiles several days before.

Comparing the four v_p-graphs for the 45° orientation of travel paths (Fig. 12, top) a similarity in the general trend can be seen, but it is obvious that the rock material closer to the interface S/B–OPA (10 cm) is more affected by microcrack creation than at greater distances (20, 30 and 45 cm). The decline of v_p starts between day 20 (v_p at 10 cm) and day 50 (v_p at 45 cm). Likewise, the absolute minima are reached at different times and different intensities, varying between 25% (v_p at 10 cm, day 112) and 3% (v_p at 45 cm, day 240). The ventilation (desaturation near the microtunnel) during the installation is most probably responsible for that. Between days 280 and 380 the four v_p -graphs reach their start value (1), with a greater but comparable delay as they decline between days 20 and 50. The start values are exceeded by several per cent. A hydration of the OPA (gradual closing of microcracks) owing to vapour coming from the drying S/B and a moderate radial pressure buildup towards the OPA would support this. The v_p values derived near the interface (10 and 20 cm) start to decline again at days 410 and 480, although the relative humidity RH stays constant at 100% from day 250 onwards (cf. Fig. 5). An expansion of the OPA into the still unconsolidated S/B could explain this v_p reduction. From nearby locations (instrumentation sections SD1 and SD2) displacements in the range of up to −20 mm were measured between days 110 and 350. A porewater pressure increase involved with a porosity increase, as predicted in the models (cf. Fig. 16), could also lead to a v_p decrease in certain situations.

The v_p evolution in the OPA at two distances from the interface for the normal travel paths are shown in Figure 12, bottom. The general trend is comparable to the 45° oriented travel path data but the first minima are reached for both distances at the same time (day 105) and the same intensity (−20%). The later decline comes earlier and is more pronounced than in the 45° orientation plot. It is worth mentioning that several tens of days after the v_p in the S/B starts to rise (day 300), the v_p for the short distances in the OPA starts to decline.

It seems that a 'pulse' is entering the rock, and at later stages coming from the rock, but with a pronounced dependency on the orientation to the bedding (anisotropy of OPA). Which driving forces (thermal impact, vapour pressure, degree of saturation) generate this 'pulse' and what process interactions define its origin needs further attention. It is shown, however, that seismic monitoring is a useful tool for the characterization of continuously ongoing changes of rock properties.

Model calculations

The early post-closure phase is characterized by high temperature gradients and low saturation that could potentially have an impact on the further evolution of the system canister–EBS–rock. A profound understanding of the ongoing THM processes is necessary to assess the performance of the system. In this framework, model simulations are an essential part of the HE-E. Modelling is structured in different phases:

- Scoping calculations were performed during the planning phase of the experiment in order to make sure that the experiment layout meets the requirements regarding temperature evolution and that instrumentation is adequate.
- Interpretation modelling is performed in prediction–evaluation cycles. First predictive calculations have been performed and are currently compared with the data measured so far. A second cycle is in preparation and will be validated on the measurements towards the end of the experiment. In comparison to the scoping calculations, the interpretation modelling is going more into depth, because it can use not only the *in-situ* measured data, but also results of parallel laboratory testing for determination of material behaviour and parameters.
- Long-term extrapolation: the models and parameters validated and calibrated in the prediction–evaluation cycles will be used for long-term extrapolation until final saturation state and negligible temperature gradients in the EBS. The final state and the remaining uncertainties will be assessed and the impact on the safety functions evaluated.

Three modelling teams are involved in the simulation calculations: One team is performing coupled thermal–hydraulic simulations with a full 3D model using the TOUGH2 code (Pruess *et al.* 1999). The other two (CIMNE and GRS) use CODE_BRIGHT (CODE_BRIGHT User's Manual 2009) with full THM coupling and different 2D symmetries (CIMNE, axial symmetry; GRS, plane strain model) and material models.

Scoping calculations

The THM coupled scoping calculations (Czaikowski *et al.* 2012) showed that, taking into account the actual test field history, maximum porewater pressures would occur in the Opalinus Clay a few metres away from the EBS–rock interface (Fig. 13). This result led to a rearrangement of instrumentation, involving additional packer probes at a larger distance from the tunnel (see the section 'Host rock (Opalinus Clay) instrumentation at larger distances from the microtunnel'). Gens *et al.* (2007) showed that the increase in porewater pressure is mainly due to the differential thermal expansion of

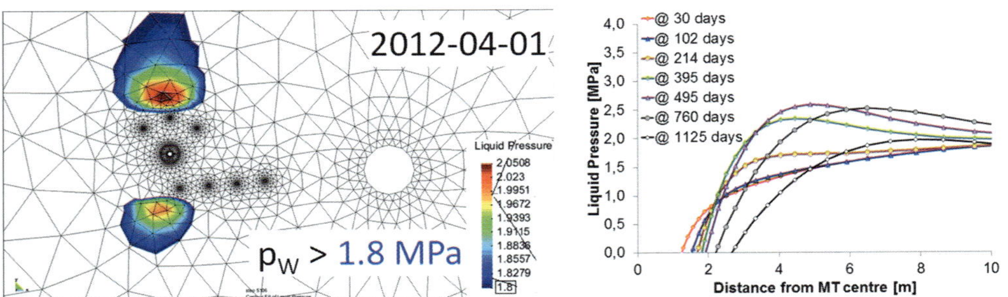

Fig. 13. THM design calculation results (CODE_BRIGHT): calculated pore pressures above 1.8 MPa after one year of heating for the GRS model (left) and pore pressure as a function of distance from the microtunnel axis for different heating times for the CIMNE model (right).

the water and the pores in combination with the relatively low water permeability that prevents full drainage of the generated overpressure.

Another interesting result of TOUGH2's calculations is the difference in the two types of EBS. Even where both EBS materials have equal

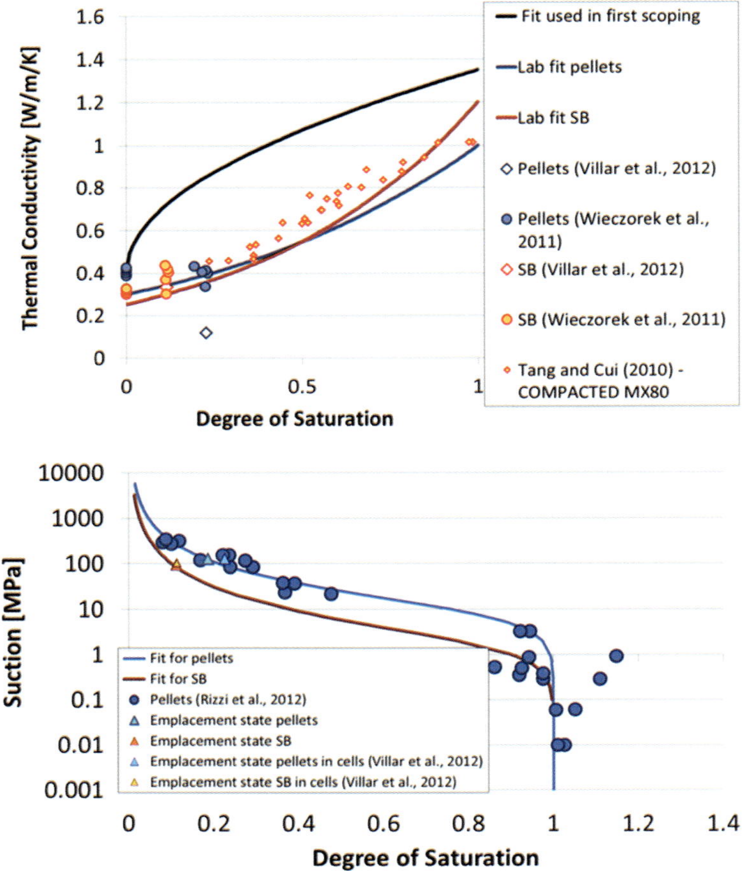

Fig. 14. CIMNE model: thermal conductivity (top) and suction (bottom) as a function of saturation for the bentonite pellets (blue) and granular sand–bentonite (red). Original fit of thermal conductivity for the scoping calculations in black.

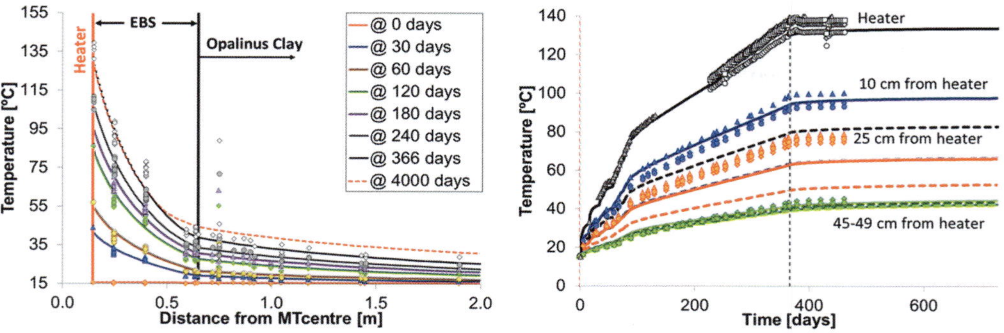

Fig. 15. CIMNE model: measured (dots) and calculated (lines) temperatures (based on updated parameters) v. distance to the microtunnel centre during heating (left) and temperature evolution v. time at different distances from the heater surface (right). In the right figure, the dotted lines show the results from the scoping computation (assuming actual power input). Time is referenced to heating start. Measurements are included up to 450 days (end of September 2012).

permeabilities, the lower initial suction in the sand–bentonite section leads to a slower resaturation of this section (not shown).

Predictive calculations

During the ongoing prediction/evaluation modelling cycles the behaviour of the system heater–EBS–rock is analysed in more detail. Special progress has been made with respect to the material parameters of the EBS components which have been determined in parallel laboratory tests (Wieczorek *et al.* 2011; Villar *et al.* 2012). Particularly critical parameters are the thermal conductivity and the suction as functions of saturation (Fig. 14). To date, thermal conductivity measurements were only available for the low saturation range (up to 20%). This is the likely working range in the hot part of the EBS for the moment as heating induced drying. In the colder part, higher degrees of saturation are expected and the fit was realized using measurements from a study considering compacted MX80 bentonite (Tang & Cui 2010).

Simulation results for temperature using full THM coupling (CIMNE's model) were compared with measurements during the first 450 days of the experiment as a function of the distance to the microtunnel centre and as a function of time (Fig. 15). The measurements from about 36 sensors in the EBS and 15 sensors in the Opalinus Clay are depicted. Simulation results (assuming heater power rather than heater temperature as the boundary condition) are given for the fit of thermal conductivity used in the first scoping computation (dotted lines) and for the laboratory updated fit (full lines). The sensitivity of the temperature to changes of thermal conductivity in the EBS is illustrated and the match is satisfactory. As mentioned in the previous sections, the variability of the temperature measurements in the EBS at the same distance from the heater surface is probably

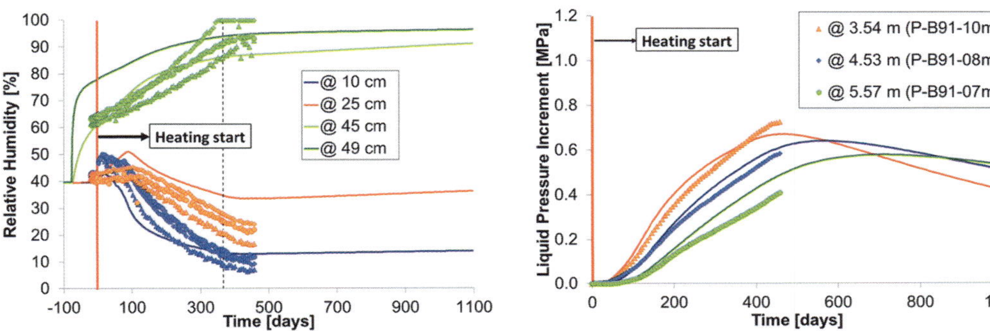

Fig. 16. CIMNE model: measured (dots) and simulated (lines) relative humidity evolution in the EBS (left) and porewater pressure increments in the Opalinus Clay at 3.5, 4.5 and 5.5 m from the EBS–rock interface (right).

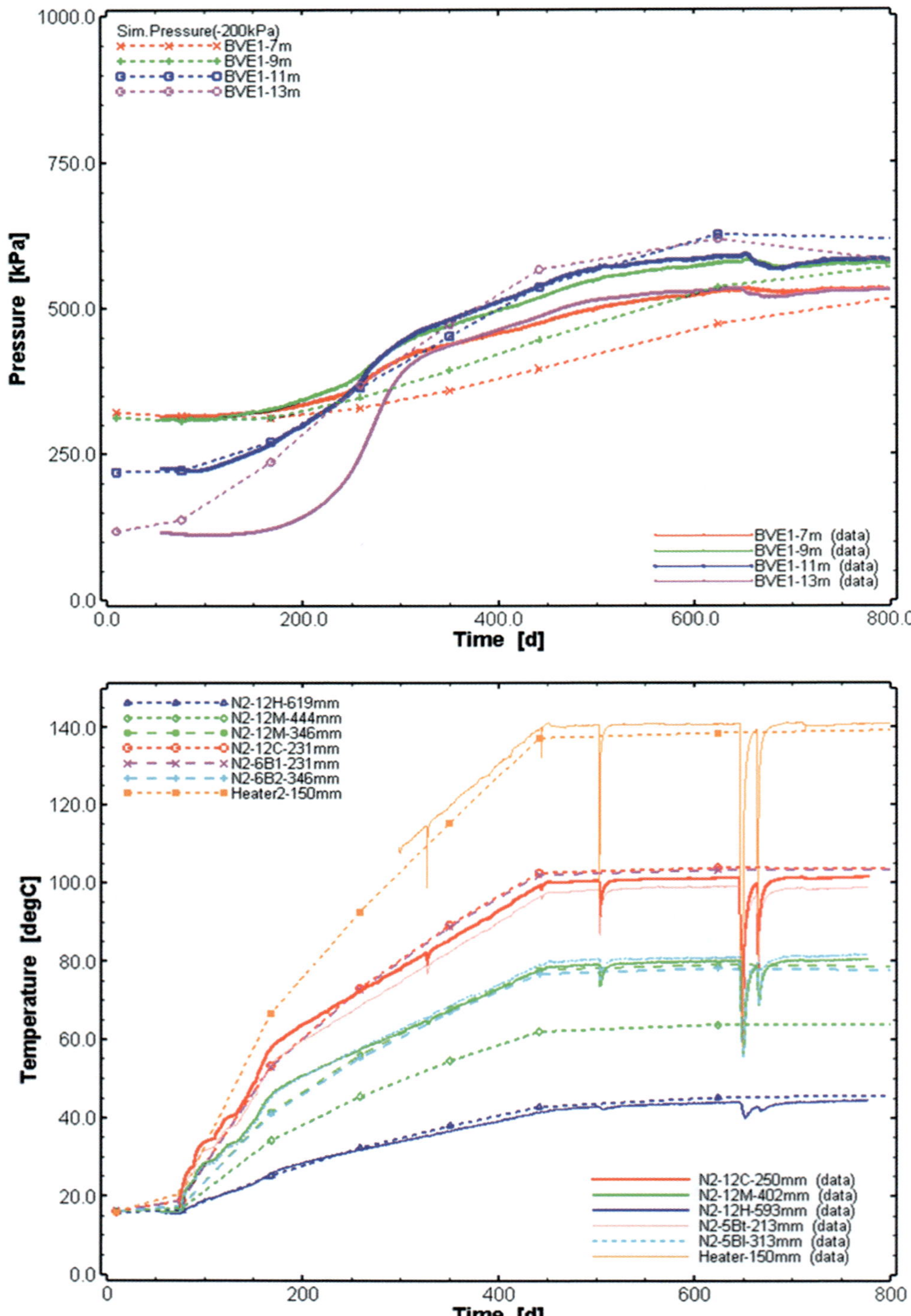

Fig. 17. TOUGH2 simulation. Simulated (dotted) and measured (lines) pressures in the Opalinus Clay (top), simulated and measured temperatures in the bentonite pellets (bottom).

due to: (a) the impact of the ends of the instrumented section and the gap between the two heaters; and (b) the presence of the more conducting bentonite blocks supporting the heaters. Both aspects are not taken into account in the axisymmetric simulation. The high temperature gradient in the EBS in comparison to the much lower gradient observed in Opalinus Clay is a direct consequence of their large difference in thermal conductivity values.

The evolutions of relative humidity (RH) in the EBS and the liquid pressure in the Opalinus Clay at different distances from the heater are illustrated in Figure 16 (CIMNE model). Sensors near the heater register first an increase in RH that is attributed to diffusion of vapour coming from the zones between the sensor considered and the heater where liquid water evaporates as a consequence of the high temperature. Water evaporation and diffusion towards the colder zone dries the material near the heater and results in a subsequent decrease in the RH. Near the rock, most of the increase in RH is attributed to water intake from Opalinus Clay, although part of the increase could also be attributed to the diffusion of water vapour from warmer zones. The simulated RH reproduces reasonably well the observed trend; the same can

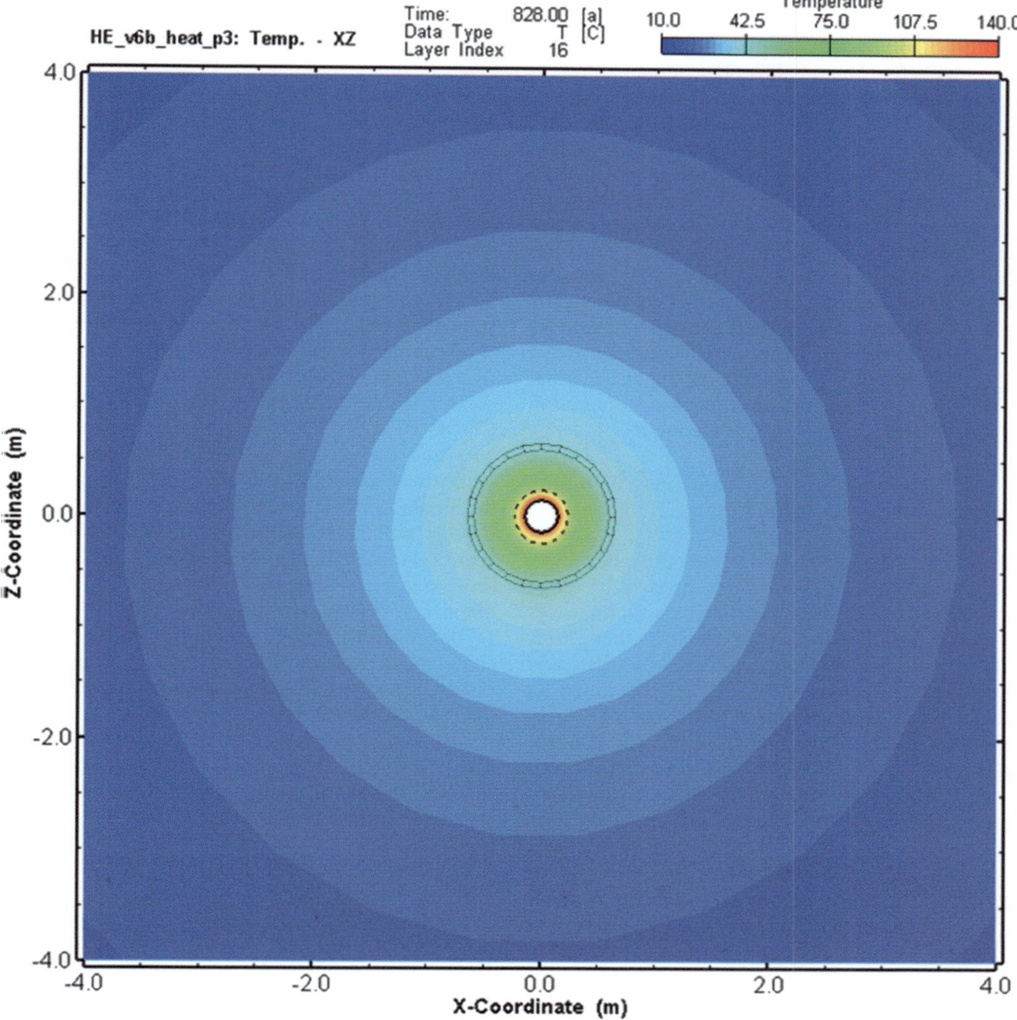

Fig. 17. (*Cont.*) Spatial distribution of temperatures in a vertical section through the microtunnel and Opalinus Clay one year after the start of heating.

be stated for the liquid pressure evolution in the Opalinus clay.

Results of a full 3D TH simulation with TOUGH2's model are shown in Figure 17. The mechanical effect on pore pressure, which is especially important with respect to the test field history (excavation), can be simulated only in a simplified manner in the TH model; as a consequence, the pore pressure in the rock at start-up of heating is offset by about 200 kPa from the measured values. The relative pressure increase during heating is quite well reproduced in the simulation, considering that some of the node locations differ somewhat from the actual BVE-1 monitoring intervals owing to the mesh discretization.

The modelled temperature field after one year of heating shows the effect of the higher thermal conductivity of the bentonite blocks compared with the bentonite pellets: temperatures in the floor below the microtunnel are higher than in the walls and at the top. The detailed temperature histories in the bentonite pellets section show that the simulated temperatures reproduce the overall trend, whereby some of the offset can be related to the differences in distance from the heater for the corresponding node location. However, it is noticeable that the measured temperatures in the bentonite blocks below and the bentonite pellets above the heater show relative small differences, whereas the simulated temperature distribution shows somewhat higher temperatures owing to the higher thermal conductivities at the higher initial water saturation in the bentonite blocks used in the model.

Conclusions and outlook

The HE-E experiment was successfully designed, implemented and initiated in 2011. After 1 year of gradually increasing power, the maximum heater temperature of 140 °C was reached in June 2012. The observed temperature increases in EBS and Opalinus Clay are in line with those predicted by the design calculations (slight variations are attributed to differences in model setup and conceptualization).

The EBS is characterized by a very strong temperature gradient owing to its low thermal conductivity in the current dry state. Drying in the inner part of the EBS continues while a complex development of the humidity profiles takes place. The latter is strongly determined by the different water contents and densities of the materials at installation, high sensitivity to changing two-phase flow parameters and the impact of vapour diffusion in a changing porous matrix.

The Opalinus Clay is at relatively low temperature while still partially unsaturated close to the microtunnel. The point at which hydraulic pressures are registered, influenced by the anisotropy of the Opalinus Clay, is converging slowly toward the tunnel wall. After 15 months, it is still located at more than 1.5 m from the microtunnel wall. As predicted by the models, a hydraulic pressure increase, associated with the differential thermal expansion of the Opalinus Clay and the porewater, is observed in the saturated Opalinus Clay at a larger distance from the tunnel. The porewater pressure increase, which started developing shortly after switching on the heaters, is still continuing and the maximum value has not been reached yet (although the heaters' temperature has been maintained constant for the last three 3 months).

Based on the seismic transmission measurements, a variation of the derived P-wave velocity evolution was observed with pronounced dependencies on the orientation of the travel paths to the bedding (anisotropy) and their distances from the S/B–OPA interface. A sequence of decreasing and increasing P-wave velocities points to the creation and sealing of invisible microcracks within the first 50 cm from the S/B–OPA interface, most probably caused by desaturation, the thermal pulse and the saturation and porewater pressures changes. Seismic methods have been shown to be a very sensitive tool for the continuous characterization of changes in rock properties.

Outcomes of the design modelling and predictive modelling based on the as-built parameter dataset with established coupled codes (TOUGH, CODE_BRIGHT using various geometries) indicate that the models are generally in agreement with the observations and are capable of capturing the evolution of the experiment. In a next step, models will be calibrated on 18 months of data (period June 2011 to December 2012). Furthermore, the parameter estimations derived from ongoing 50 cm column tests carried out on the EBS materials (Villar *et al.* 2012) will be incorporated. The refined calculations will be compared with observations (up to December 2013) in order to further assess the predictive capability of the models.

Research leading to these results has received funding from the European Atomic Energy Community's Seventh Framework Programme (FP7/2007-2011) under grant agreement no. 249681 and from the German Bundesministerium für Wirtschaft und Technologie under contract No. 02E10689.

References

CODE_BRIGHT USER'S MANUAL 2009. UPC Geomechanical Group.

Czaikowski, O., Garitte, B., Gaus, I., Gens, A., Kuhlmann, U. & Wieczorek, K. 2012. *Design and predictive modeling of the HE-E test*. PEBS Deliverable D3.2-1. NAB 12-003. Nagra, Wettingen.

Gaus, I. (ed.) 2011. *Long Term performance of Engineered Barrier Systems (PEBS). Mont Terri HE-E experiment: detailed design report*. NAB 11-001. Nagra, Wettingen.

Gens, A., Vaunat, J., Garitte, B. & Wileveau, Y. 2007. In situ behaviour of a stiff layered clay subject to thermal loading: observations and interpretation. *Géotechnique*, **57**, 207–228.

Johnson, L. H., Niemeyer, M., Klubertanz, G., Siegel, P. & Gribi, P. 2002. *Calculations of the temperature evolution of a repository for spent fuel, vitrified high-level waste and intermediate level waste in Opalinus Clay*. Nagra Technical Report NTB 01-04. Nagra, Wettingen.

Mayor, J. C., García-Siñeriz, J. L. *et al.* 2007. Ventilation experiment in Opalinus Clay for the disposal of radioactive waste in underground repositories. *In*: Bossart, P. & Nussbaum, C. (eds) *Mont Terri Project – Heater Experiment, Engineered Barrier Emplacement and Ventilation Experiment Rep*. Swiss Geological Survey, Bern, **1**, 182–240

Pruess, K., Oldenburg, C. & Moridis, G. 1999. *TOUGH2 User's Guide*. LBNL-43134, Earth Sciences Division. Lawrence Berkeley National Laboratory, University of California, Berkeley, CA.

Schuster, K. 2007. *High resolution seismic investigations within the VE-Experiment*. Mont Terri Technical Report TR 07-06. Swisstopo, Wabern.

Schuster, K. & Alheid, H.-J. 2004. *EB: Engineered Barrier Experiment; Seismic Long Term Observation in the EB Niche*. Mont Terri Technical Report TR 04-02. Swisstopo, Wabern.

Tang, A. M. & Cui, Y. J. 2010. Effects of mineralogy on thermo-hydro-mechanical parameters of MX80 bentonite. *Journal of Rock Mechanics and Geotechnical Engineering*, **2**, 91–96.

Teodori, S.-P. & Gaus, I. (ed.) (2011). *Long term performance of engineered barrier systems (PEBS)*. Mont Terri HE-E experiment: as-built report. NAB 11-025. Nagra, Wettingen.

Villar, M. V., Martín, P. L., Gómez-Espina, R., Romero, F. J. & Barcala, J. M. (2012). *THM cells for the HE-E test: setup and first results*. PEBS Report D2.2.7a. CIEMAT Technical Report CIEMAT/DMA/2G210/02/2012, Madrid.

Wieczorek, K., Miehe, R. & Garitte, B. 2011. *Measurement of Thermal Parameters of the HE-E Buffer Materials*. PEBS Deliverable D2.2-5.

Zhang, C.-L., Rothfuchs, T. *et al.* 2007. *Thermal effects on the Opalinus Clay. A joint heating experiment of ANDRA and GRS at the Mont Terri URL (HE-D Project)*. Final Report, **GRS-224**.

Self-sealing experiments and gas injection tests in a backfilled microtunnel of the Mont Terri URL

G. W. LANYON[1]*, P. MARSCHALL[2], T. TRICK[3], R. DE LA VAISSIÈRE[4], H. SHAO[5] & H. LEUNG[6]

[1]*Fracture Systems Ltd, St Ives, Cornwall, UK*

[2]*National Cooperative for the Disposal of Radioactive Waste, 5430 Wettingen, Switzerland*

[3]*Solexperts AG, 8617 Mönchaltorf, Switzerland*

[4]*French National Agency for Radioactive Waste Management, 55290 Bure, France*

[5]*Federal Institute for Geosciences and Natural Resources, 30655 Hannover, Germany*

[6]*Nuclear Waste Management Organization, Toronto, Ontario, Canada*

**Corresponding author (e-mail: bill@fracture-systems.co.uk)*

Abstract: This paper describes a large-scale experiment on gas transport and hydromechanical processes around underground structures as part of a long-term geoscientific research programme at the Mont Terri Underground Rock Laboratory in the Jura Mountains of Switzerland. A horizontal microtunnel with a diameter of 1 m and a length of 13 m was drilled in an overconsolidated claystone formation. After installing monitoring instruments in the open tunnel, the end of the tunnel was backfilled with sand (test section) and a large hydraulic packer was emplaced in the seal section. The packer was inflated and subsequently the test interval was saturated with a synthetic pore-water. Following saturation an extended programme of hydraulic testing was performed over a two year period. A series of gas injection tests was then performed over a period of approximately 1.5 years. Following this first series of gas injections, a long post-gas hydraulic test has been initiated. The paper presents data and interpretation of the gas injections and subsequent hydraulic testing. The ability of the excavation damage zone to transport gas at pressures below fracturing is demonstrated. The post-gas hydraulic performance is considered and related to the self-sealing of the damage zone observed during saturation and hydraulic testing.

The investigation of damage zones around excavations such as seal sections in tunnels or shafts and their impact on gas migration are key issues in the field of underground waste disposal. The experiment ('Gas path through host rock and along seal sections/HG-A') was designed as a long-term gas experiment in a backfilled microtunnel, to investigate both leak-off rates and gas release paths from a sealed tunnel section in an ultra-low permeability host rock (Opalinus Clay). The aims of the HG-A experiment are to:

- provide evidence for barrier function of the Opalinus Clay on the tunnel scale (scale effects in rock permeability);
- investigate self-sealing of the excavation damage zone (EDZ) after tunnel closure (mechanical self-sealing in response to packer inflation and pore pressure changes);
- provide evidence for gas transport capacity of Opalinus Clay (intact host rock and EDZ).

The Opalinus Clay

The Opalinus Clay in Northern Switzerland has been identified as a potential host rock formation for the disposal of radioactive waste (Nagra 2002). The formation is part of a thick Mesozoic–Cenozoic sedimentary sequence which was deposited 180 Ma ago in a shallow marine environment.

At Mont Terri the Opalinus Clay formation reached a maximum depth of about 1000 m and can be classified as a slightly overconsolidated rock, the estimated overconsolidation ratio varying between 2.5 and 3.5 (estimates derived from both laboratory tests and burial history). The principal stress σ_1 at laboratory level is in the order of 6.5 MPa, subvertically oriented, and it reflects the overburden; σ_3 is NE–SW-oriented (2.5 MPa) and σ_2 runs in a NW–SE direction (4.5 MPa) parallel to the Security Gallery (Fig. 1; Martin *et al.* 2002). There is some uncertainty on the stress tensor, in particular relating to the magnitude of

From: Norris, S., Bruno, J., Cathelineau, M., Delage, P., Fairhurst, C., Gaucher, E. C., Höhn, E. H., Kalinichev, A., Lalieux, P. & Sellin, P. (eds) 2014. *Clays in Natural and Engineered Barriers for Radioactive Waste Confinement*. Geological Society, London, Special Publications, **400**, 93–106.
First published online March 5, 2014, http://dx.doi.org/10.1144/SP400.8

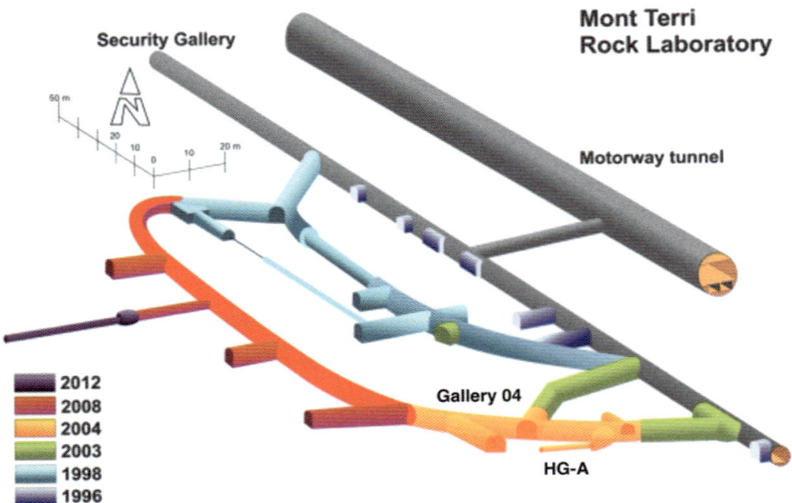

Fig. 1. Layout of the Mt Terri Rock URL. Tunnels are coloured by year of excavation. The HG-A experiment niche and microtunnel are visible at the bottom right of the figure.

σ_3 (see Martin & Lanyon 2003 and Corkum & Martin 2007).

Quantitative laboratory analyses of core samples from Mont Terri showed a total mass fraction of clay minerals of 47–60%, a quartz content of 14–30% and 16–22% carbonates. The fraction of swelling clay minerals of 23–27% (illite, illite/smectite mixed layers) is of particular relevance for the consolidation behaviour of the rock. Further minerals are siderite, pyrite and feldspar (Pearson *et al.* 2003).

Intact Opalinus Clay at Mont Terri exhibits a very low hydraulic conductivity, with a mean hydraulic conductivity of 2×10^{-13} m s^{-1}, and a moderate spatial variability, which is less than an order of magnitude (Marschall *et al.* 2004). Microscopic observation of the fabric of the Opalinus Clay at Mont Terri suggests that there may be a significant core-scale hydraulic anisotropy. The ratio between bedding-parallel and bedding-normal permeability is thought to lie between 1 and 10. The very fine pore network is saturated with a Na–Cl–SO_4 connate pore-water of marine origin with a mean content of dissolved solids of about 12 g l^{-1}. Even though the rock is fractured, a distinct fracture transmissivity has not been observed, suggesting that the fractures are generally tight for the given stress conditions.

Rock mechanical characterization of the Opalinus Clay is challenging owing to its ultra-low permeability, over-consolidation and distinct bedding. Typical values for key geotechnical parameters are shown in Table 1 (Bock 2000).

EDZ experiments at Mont Terri

Since its inception, EDZ experiments have been a focus of the Mont Terri programme (Thury & Bossart 1999; Bossart *et al.* 2004; Blümling *et al.* 2007). Experience from the experiments suggests that the damage zone geometry and properties are a function of:

- excavation method (drill and blast or mechanical methods);
- orientation of the tunnel relative to the *in situ* stress field and to the bedding fabric of the rock;
- excavation geometry (size and shape);
- presence of pre-existing faults and fractures;
- environmental conditions within the excavation (ventilation/humidity).

The HG-A experiment supplements results from previous EDZ experiments in terms of scale *c.* 1 m (intermediate between boreholes and tunnels) and orientation (bedding-parallel).

The response to excavation of the microtunnel and the associated creation and development of the EDZ is discussed in Marschall *et al.* (2006, 2008). The EDZ around the microtunnel is formed by the interaction of the rock and stress anisotropy with significant breakout zones ('notches') at 3 o'clock and 9–11 o'clock (see Fig. 10a).

The test section saturation and hydraulic testing prior to gas injection are presented in Lanyon *et al.* (2009). This paper presents the results of gas leak-off testing and subsequent post-gas hydraulic

Table 1. *Geotechnical reference parameters of the Opalinus Clay at the Mont Terri Underground Laboratory (after Bock 2000)*

Parameter	Value	Remarks
Bulk density (mg m^{-3})	2.45	Water saturated
Grain density (mg m^{-3})	2.71	
Porosity (%)	13.7	Range: 10–16%
Water content (% wt)	6.1	Range: 6–7%
Young's modulus (GPa)	104	Parallel to beddingNormal to bedding
Shear modulus	1.2	
Poisson's ratio (−)	0.27	
Uniaxial compressive strength (MPa)	1016	Parallel to beddingNormal to bedding
Tensile strength (MPa)	21	Parallel to beddingNormal to bedding
P-wave velocity (m s^{-1})	34102620	Parallel to beddingNormal to bedding
S-wave velocity (m s^{-1})	19601510	Parallel to beddingNormal to bedding
Fracture toughness K_{IC} (MN m$^{-1.5}$)	0.530.12	Parallel to beddingNormal to bedding

testing together with an overview of the rock's response to testing.

Experimental layout and sequence

The HG-A experiment is located in the southern part of the Mont Terri Rock Laboratory off Gallery 04 (see Fig. 1). The 1 m-diameter, 13 m-long microtunnel was excavated during February 2005 using a steel augur from a niche in Gallery 04. The microtunnel was excavated parallel to bedding strike and bedding parallel features run along the tunnel, replicating the expected relationship between bedding and emplacement tunnel orientation in a deep repository (Nagra 2002), where bedding is expected to be flat-lying and emplacement tunnels to be subhorizontal. The excavation was monitored by a borehole array containing piezometers and deformation gauges (clino-chain and chain deflectometers). The borehole array was subsequently augmented with additional piezometer boreholes and borehole stress meters (Fig. 2a).

The first 6 m of the microtunnel was lined with a steel casing immediately after excavation to stabilize the opening. The gap behind the liner was then cement-grouted, but not sealed. The purpose-built hydraulic megapacker (diameter 940 mm and sealing section length 3000 mm) was installed in 2006. The sealing section was located at 6–9 m with a 1 m grouted zone containing the non-sealing part of the packer and retaining wall from 9–10 m. The final 3 m of the microtunnel from 10–13 m forms the test section which was instrumented and backfilled prior to packer emplacement (see Fig. 2b).

Instrumentation

The test section was instrumented with piezometers, extensometers, strain-gauges and time domain reflectometers (TDRs) to measure pressure, deformation and water content. After instrumentation the test section was backfilled with sand behind a retaining wall (Fig. 2b).

The seal section was instrumented with piezometers, total pressure cells and TDRs prior to the installation of the megapacker. Following the installation of the megapacker the volume between the retaining wall and the megapacker was filled with a cement grout. Table 2 lists the instruments in the geosphere, test and sealing sections.

Saturation and long-term hydraulic testing

Saturation of the test section and surrounding rock was started in November 2006 following emplacement of the megapacker (June 2006) and subsequent grouting of the section between the megapacker and test section (see Fig. 3). A variety of saturation tests was performed using a synthetic pore-water (Pearson *et al.* 2003) until January 2008, when a long-term multirate hydraulic test was initiated (see Lanyon *et al.* 2009). This test continued until February 2010 and involved a series of constant rate injection steps. During the test the applied injection rate was reduced from *c.* 10 to 0.1 ml min^{-1} (144 ml per day). The results from the hydraulic testing indicated progressive self-sealing (Lanyon *et al.* 2009; Bock *et al.* 2010).

During hydraulic testing the effective stress conditions in the seal zone were altered by changing the megapacker pressure. At low megapacker pressures (*c.* 2000 kPa) the minimum measured radial stress was *c.* 1600 kPa, which was close to the then test section pressure, resulting in very low effective stress conditions and apparently higher EDZ permeability (Lanyon *et al.* 2009). Since June 2009 the megapacker pressure has been maintained above 2600 kPa with measured total pressures in

(a)

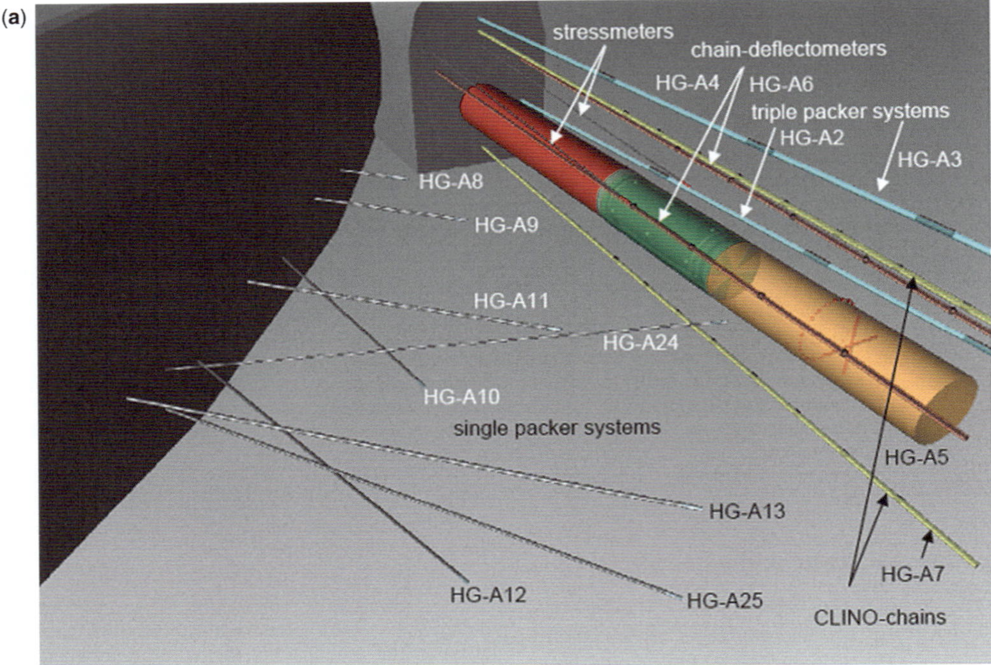

(b)

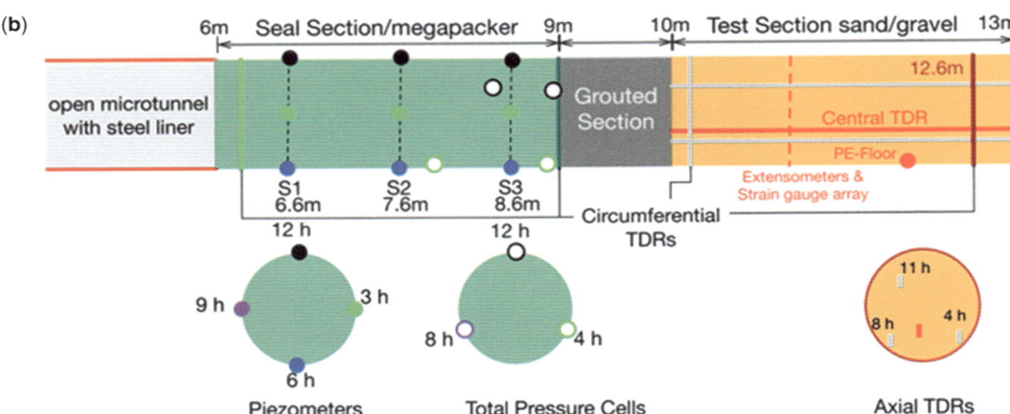

Fig. 2. Schematic drawing of the microtunnel and the site instrumentation: (**a**) layout and borehole instrumentation (colour coding refers to the steel liner, red; the seal section, green; and the backfilled test section, orange); (**b**) seal section and test section instrumentation.

the sealing section always above 1950 kPa, considerably above the test section pressure.

Immediately prior to the start of the gas injection phase the water injection was reduced from 0.1 to 0.04 ml min^{-1} to lower the initial pressure in the test section. This resulted in a test section pressure prior to gas injection of 300 kPa absolute.

Gas injection phase

The gas injection phase included three separate nitrogen gas injections, listed in Table 3; microtunnel sensor responses to gas injection are shown in Figure 4. After each gas injection, following a shut-in period, water was extracted from the test section and depressurized to remove trapped gas

Table 2. *HG-A experiment instrumentation*

Instrument	Count	Measurement
Test section instrumentation		
Piezometers	2	Pore pressure
Extensometers	2	Horizontal/vertical deformation
Strain gauges	22	Circumferential deformation
TDRs	8	Volumetric water content
Geophones	8	Acoustic emission
Seal section instrumentation		
Piezometers	12	Pore pressure
Total pressure cells	6	Load on tunnel wall + temperature
TDRs	2	Volumetric water content
Geosphere instrumentation		
Piezometers	14	Pore pressure
Deflectometers	2	Deformation (8-point)
Clino-chains	2	Deformation (8-point)
Stressmeters	3	Deformation and stress + temperature

(gas–water exchange). The degassed water was then re-injected into the test section. This procedure provided a well-defined initial gas saturation in the test section pore-water for the subsequent gas injection. During gas injection a low constant rate (*c.* 0.02 ml min^{-1}), water injection was maintained in the test section.

Pressure response to gas injection

Gas pressure during GI1 was limited to 1200 kPa, significantly below the minimum stress, to avoid coupled mechanical effects. After an initial 20 ml N min^{-1} injection when pressure rose quickly, the injection rate was reduced to about 10 ml N min^{-1}

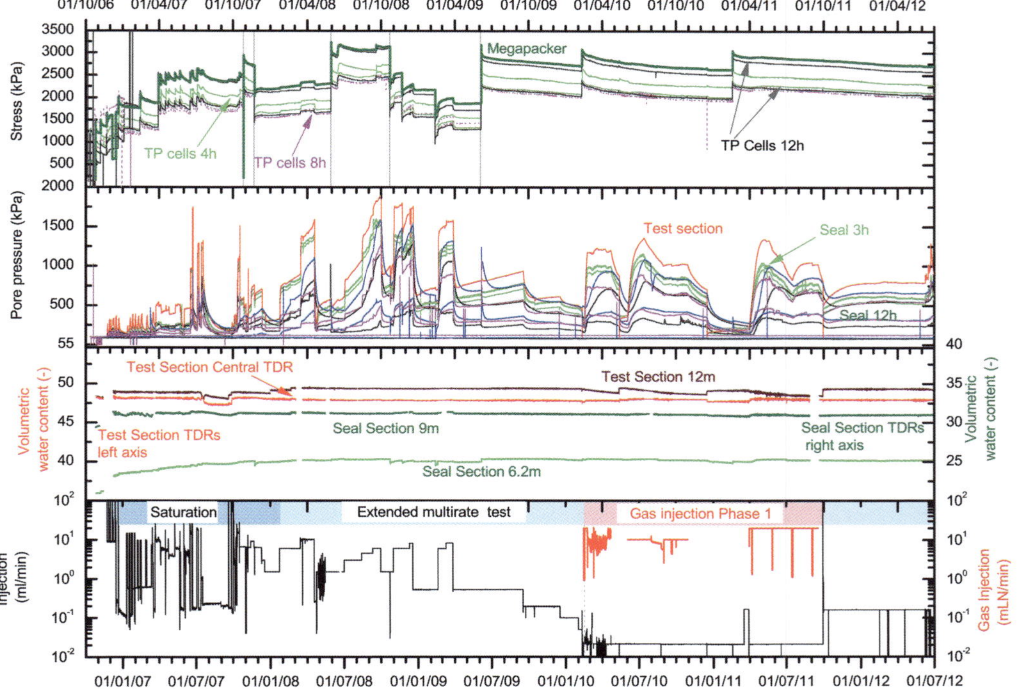

Fig. 3. Microtunnel stress, pore pressure, water content and flow rate measurements from start of saturation.

Table 3. *Gas injection steps*

	Injection Rate (ml N min^{-1})	Duration (days)	P_0 (kPa)	P_{max} (kPa)	Total gas volume (l N)
GI1	20 switching to constant pressure *c.* 1200 kPa	65	290.2	1234	1028
GI2	10.3	149	361.6	1363	2096
Gi3	20.3	171	204.2	1347	4864

to maintain an approximately constant test section pressure.

During GI2 test section pressure rose more slowly than in GI1 and then peaked at 1363 kPa on 12 July 2010. After the pressure breakdown, the test section pressure dropped over about a month by about 350–975 kPa and then stabilized at about 1040 kPa.

During GI3, test section pressure again peaked at 1347 kPa but with a broader peak than in GI2 and then dropped to about 840 kPa owing to an interruption in injection before recovering (after resumption of gas injection) and stabilizing at about 1040 kPa.

Pressures in the sealing section during gas injection were lower and lagged those in the test section, as would be expected. Pressure response was highly heterogeneous, indicating a sparsely connected system of flow paths. Pressure along the 3 o'clock sensors reacted most strongly and quickly to the test section suggesting a high diffusivity connection. However, the pressure at section 1 (6.6 m) in PES-S1–3h (close to the rear of the sealing element) was high (in fact higher than PES-S2-3h a sensor in the centre of the seal section at 7.6 m) and comparable to PES-S3-3h (8.6 m) closest to the test section. This suggests that this piezometer is not well connected to the open tunnel. PES-S1-6h shows a small response but the other piezometers in section 1 show no response to gas injection and remain at close to atmospheric pressure.

After shut-in and depressurization, remnant high pressures were observed at some piezometers in the sealing section, indicating possible closure after gas injection or some other loss of connectivity, perhaps owing to water invasion and blocking.

Total pressure cells

The measured total pressures in the sealing section largely followed the applied megapacker pressure.

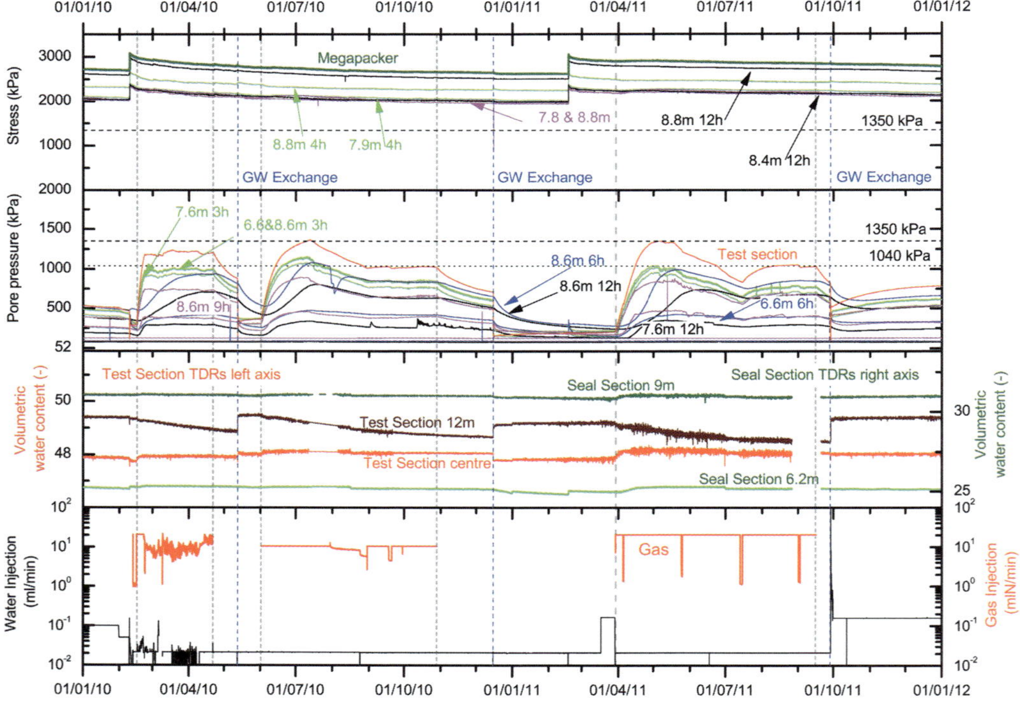

Fig. 4. Microtunnel stress, pore pressure, water content and flow rate measurements during gas injection. Gas–water exchange marked as vertical blue dashed lines. Sensor locations shown in Figure 2b.

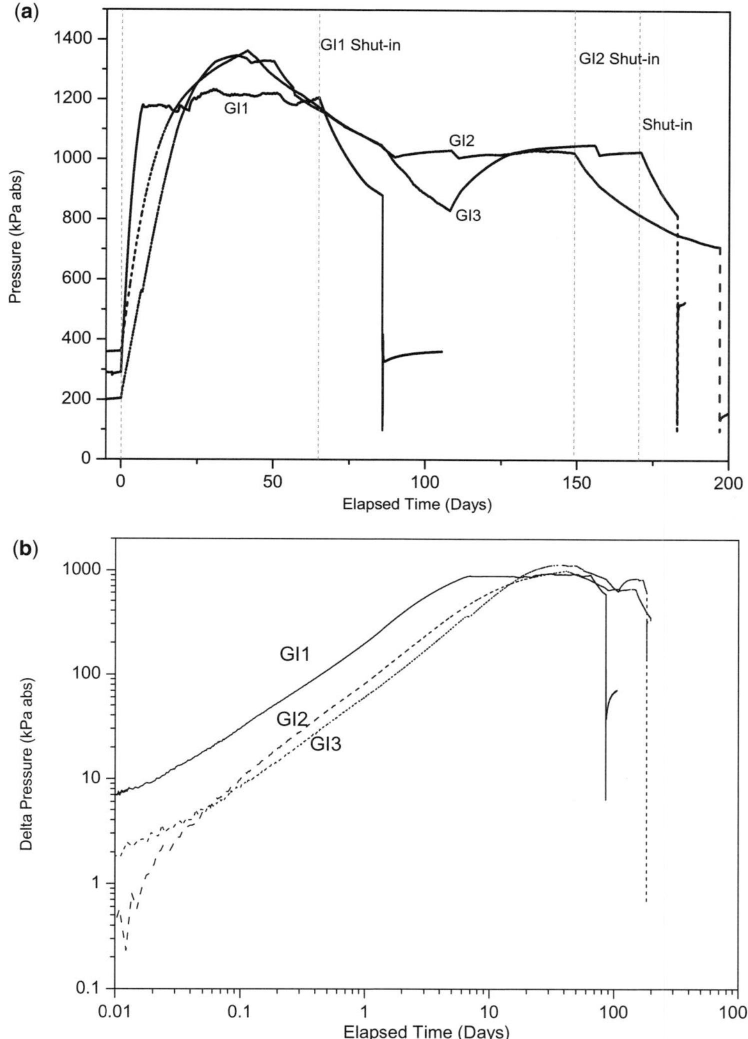

Fig. 5. (**a**) Linear and (**b**) log–log plots of delta pressure from GI1–3.

Total pressures were typically 75–95% of the mega-packer pressure. Stresses were more heterogeneous in section 1 (7.85–8.36 m) than section 2 (8.84–8.87 m), although this could have been due to instrumentation offsets. The influence of test section pressure can be seen in the data, with load increasing slightly as test section pressure increases.

Temperature

Temperature was monitored within the laboratory, in the test section and geosphere throughout the experiment. The laboratory temperature showed a variable seasonal trend between 10 and 15 °C. Borehole temperature data showed an attenuated seasonal response between 12 and 13 °C with a slight upward trend. Temperatures measured in the test section in a circumferential strain gauge array showed an overall slow increase similar to that measured in the geosphere. Two temperature gauges located at the top of the microtunnel showed irregular temperatures occurring after the start of each gas injection.

TDRs

The three circumferential TDRs showed a small reduction (*c.* 1%) in water content during gas

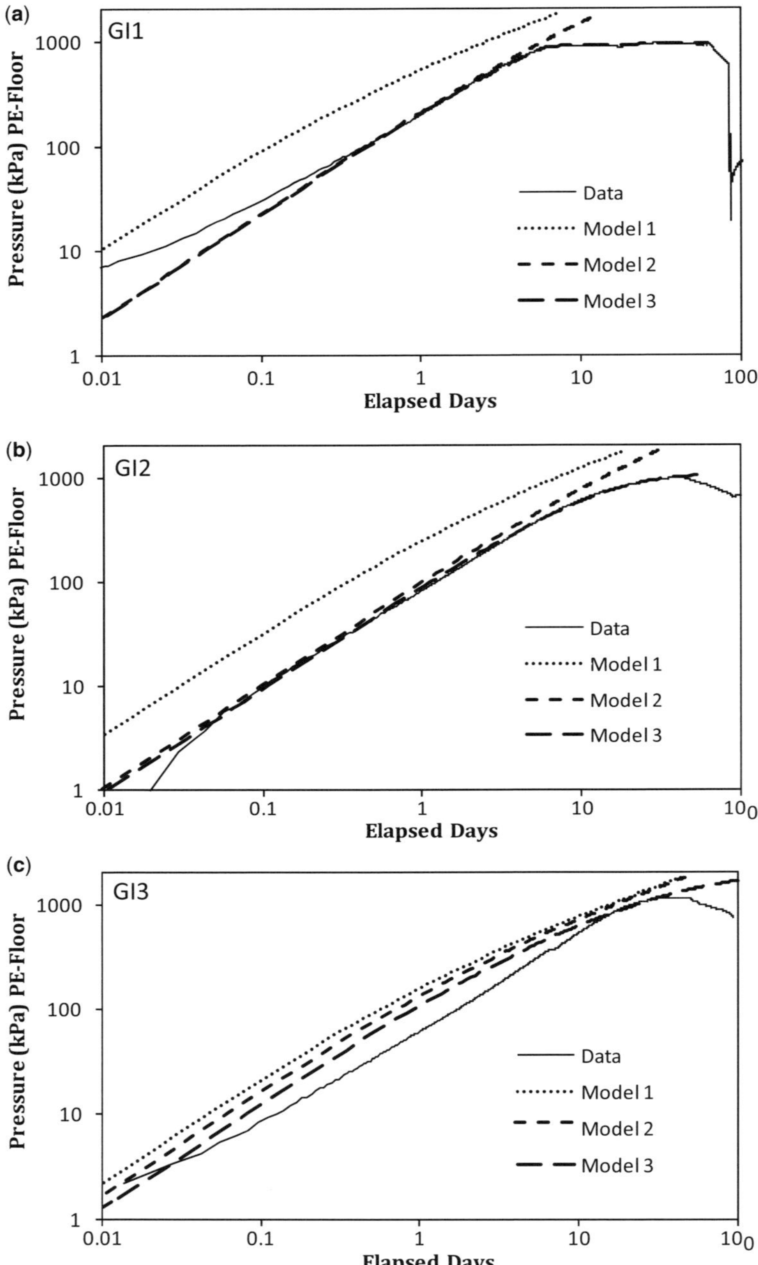

Fig. 6. Log–log plots of delta pressure (sensor PE-Floor) for each gas injection for models 1–3: (**a**) GI1; (**b**) GI2; (**c**) GI3.

injection which recovered after the gas–water exchange. The lack of recovery during shut-in and pressure drop and subsequent recovery after gas–water exchange suggest that this was a response to the presence of gas rather than a coupled response owing to test section pressure. The responses were, however, small and would indicate that a totally desaturated zone might have existed over a small

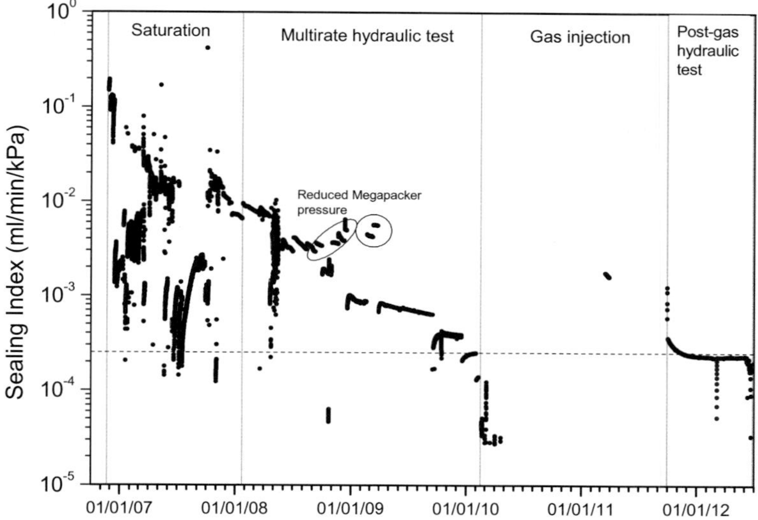

Fig. 7. Sealing index S during water injection. Periods of reduced megapacker pressure (resulting in low effective stress) are marked.

part of the circumference owing to a gas layer in the top of the test section. If the desaturation was only partial, the desaturated layer could have been thicker. The onset of response was slightly delayed from the start of gas injection, indicating possible dissolution or trapping of the first gas injected and the delay was greatest during GI2. The TDR at the rear of the test section showed a more delayed response as it did not extend to the top of the microtunnel.

The axial TDRs showed a different pattern with increased water content after the first gas–water exchange and small reductions midway through GI2. Water content dropped after the gas–water exchange and then increased at the start of GI3. This behaviour is difficult to explain and requires further consideration. It is possible that it reflects both changes in porosity within the sand/gravel as well as saturation. Alternatively it may have been an instrumentation artefact.

The sealing section TDRs showed significantly lower water content than the test section (contrast between porosity of rock and backfill) together with a minor reduction in water content after the GI2 gas–water exchange in both the 9.0 and 6.2 m section TDRs. This drop in water content recovered at the start of GI3. Again it is difficult to interpret these small responses.

Geosphere response

Pore pressure and strain responses (clinometer, deflectometer and borehole stress gauge) in the geosphere showed responses largely related to the test section pressure similar to those observed during long hydraulic testing. No clear evidence of gas transport was obvious from the geosphere data.

Post-gas hydraulic testing

Following the final gas–water exchange, a constant rate 0.17 ml min^{-1} water injection into the test section was initiated to determine any effect of the gas testing on the water leak-off characteristics of the EDZ (via the sealing index). The test started on 30 September 2011 and is ongoing (September 2012).

Analysis

The HG-A experiment is being modelled by several groups both within the HG-A experiment and as part of the EU FORGE project (FORGE 2009). Numerical models including two-phase flow and hydromechanical coupling are being used to assess the complete experimental sequence. Here simple models are presented to consider pre-breakdown behaviour of the test section and self-sealing.

Pre-breakdown behaviour

Here we present an analytical model of the pre-breakdown response of the test section pressure illustrating the different processes. Figure 5 shows a comparison between the test section pressures in

(a)

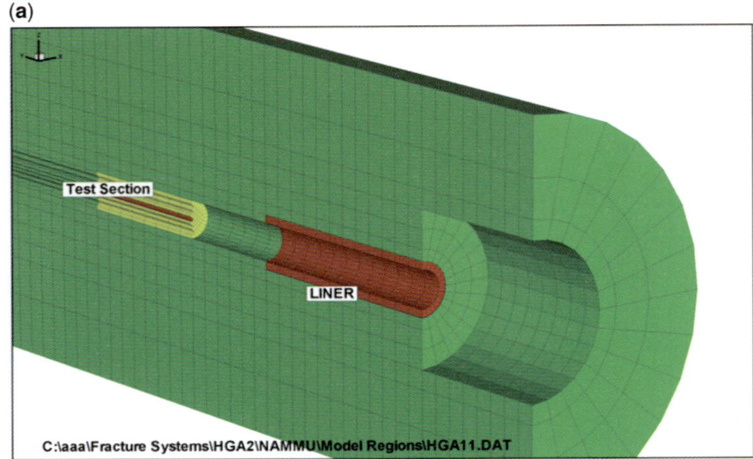

(b)

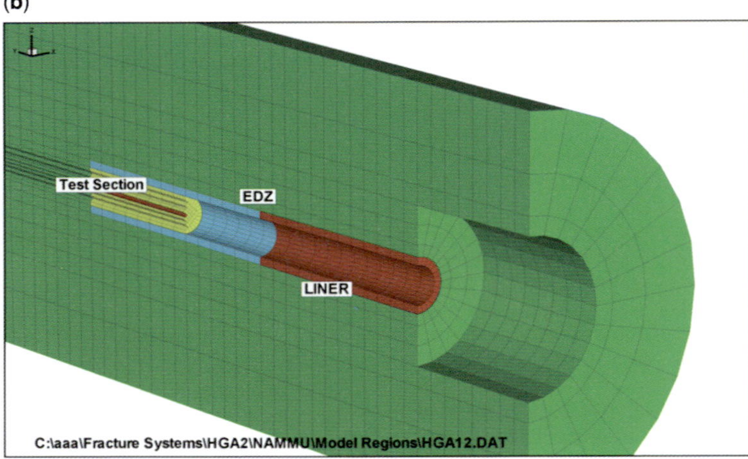

(c)

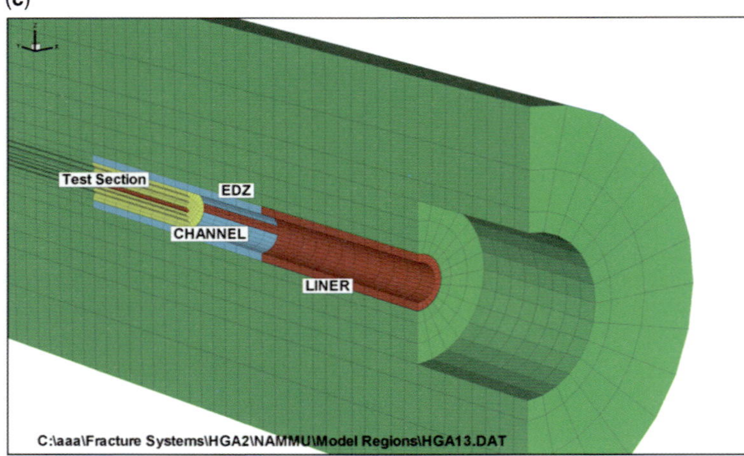

Fig. 8. Continuum models A, B and C from Lanyon *et al.* (2009).

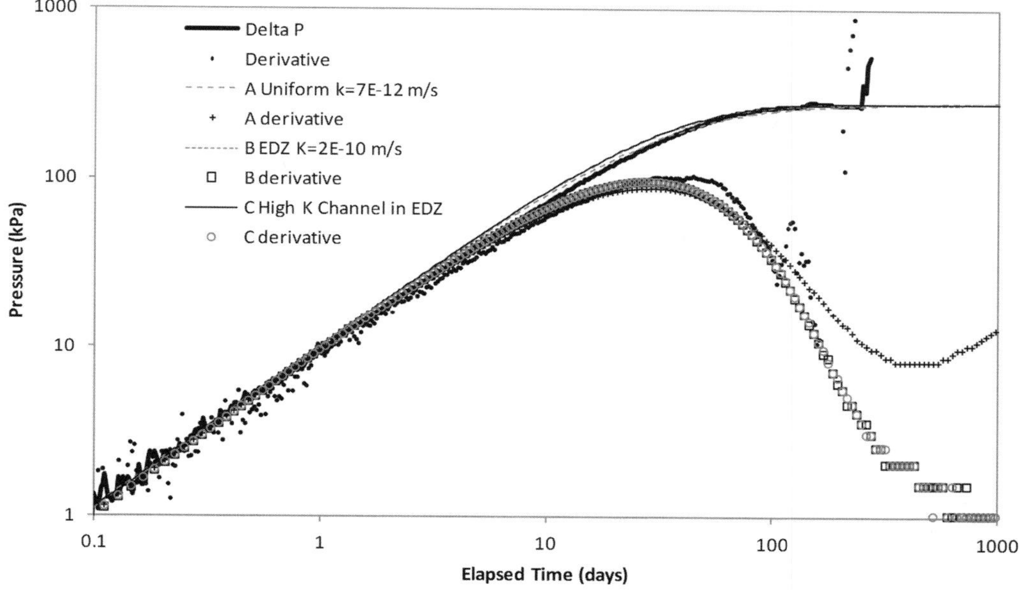

Fig. 9. Diagnostic plot from post-gas hydraulic test. Measured and calculated delta pressure and derivative.

GI1, GI2 and GI3. Inspection of the log–log plots (Fig. 5b) shows an approximately linear behaviour for the first 20 days prior to the subsequent breakdown in GI2 and GI3. In GI1 pressure is held constant and no breakdown was observed.

Predicted test section delta-pressures for GI1–3 from three simple models are shown in Figure 6a–c. The models describe the relationship between test section pressure P (kPa) and gas volume V (m^3).

Model 1: test section and gas compressibility only

$$\Delta P = C.\Delta V; \quad \Delta V = q\Delta t$$

Model 2: compressibility and dilution

$$\Delta V = \left(q\Delta t - \frac{HM_w \Delta P}{P_0 \rho_{g0}} \right) \frac{P_0}{P}$$

Table 4. *Rock hydraulic conductivity (m s^{-1}) fitted to post-gas hydraulic test for models A–C: (see Fig. 8)*

Model	Rock	EDZ	Channel
A– no EDZ	7×10^{-12}		
B – cylindrical EDZ	10^{-13}	2×10^{-10}	
C– EDZ + channel	10^{-13}	10^{-11}	6×10^{-9}

Model 3: compressibility, dilution and leakage

$$\Delta V = \left(q\Delta t - \frac{HM_w \Delta P}{P_0 \rho_{g0}} \right) \frac{P_0}{P} - q_w \Delta t$$

The models are characterized by:

- q, gas injection rate (ml min^{-1} @ standard temperature and pressure);
- C, test section compressibility (m^3 kPa^{-1});
- M_w, mass of water in test section (kg);
- q_w (ml min^{-1}) water leakage calculated from S, sealing index (ml min^{-1} kPa^{-1}).

Standard values for nitrogen properties (density ρ, Henry's law constant H) were used. Analysis of the long-term multirate test suggested a C of 10^{-5} m^3 kPa^{-1} and S of 5×10^{-4} ml min^{-1} kPa^{-1} (see Fig. 7). M_w was estimated as 700 kg from the tunnel volume and an estimated porosity of 30%. These values together with the applied gas flow rate were used in models for GI1 and GI2, while for GI3 it was necessary to significantly increase the compressibility, possibly indicating that there was a significant amount of free gas in the test section prior to the start of GI3, despite the gas–water exchange.

Model 3, incorporating test section compressibility, gas dissolution and water leakage using parameters derived from the long term multirate test, is able to reproduce the pre-peak behaviour of

GI1 and GI2 (see Fig. 6a, b). The model is not able to replicate the response in GI3, suggesting that conditions had changed prior to GI3, probably owing to the presence of free gas. It is possible that leakage from the test section may also have changed after GI2, but such a change would be difficult to determine given uncertainty in initial conditions.

Self-sealing

Self-sealing of fractures in claystone rocks has been observed at Mont Terri and other sites (Bock *et al.* 2010). Lanyon *et al.* (2009) define a simple measure of the flow resistance across the seal section as the sealing index, S (ml min^{-1} kPa^{-1}) based on the injection rate and test section pressure as:

$$S = \frac{Q}{P_{\mathrm{TestSection}} - 100\ \mathrm{kPa}}$$

where Q is the flow into the test section in ml min^{-1} and $P_{\mathrm{TestSection}}$ is the test section pressure in kPa absolute. This measure was chosen as it is not possible to determine equivalent permeability or hydraulic conductivity without assumptions concerning the geometry of the flow paths in the EDZ. Assuming linear flow and a flow-path length of 3 m (seal section), the sealing index can be converted into an EDZ conductance (m^3 s^{-1}) by multiplying by a factor 5×10^{-7} (considering only resistance along the seal section). Figure 7 shows the calculated sealing index after filtering data for flow rate changes affected by storage.

In the year prior to gas injection, the sealing index S reduced from 10^{-3} to almost 10^{-4} (ml min^{-1} kPa^{-1}), equivalent to a hydraulic conductance of $1–10 \times 10^{-11}$ m^3 s^{-1}. After gas injection during the post-gas hydraulic testing, S quickly reduced to about 2.5×10^{-4} ml min^{-1} kPa^{-1}, only slightly higher than that prior to gas injection after a relatively short recovery.

A simple groundwater flow continuum model as described in Lanyon *et al.* (2009) has been used to model the post-gas hydraulic test for three different cases (see Fig. 8):

(1) uniformly permeable rock, no EDZ;
(2) cylindrical EDZ (40 cm thick) in uniform permeability rock;
(3) channel in EDZ in uniformly permeable rock.

The hydraulic conductivity of the different rock types has been adjusted to provide a match to the observed delta-pressure as shown in Figure 9 and Table 4. It can be seen that the models with an EDZ (creating a 'linear' flow geometry) are a

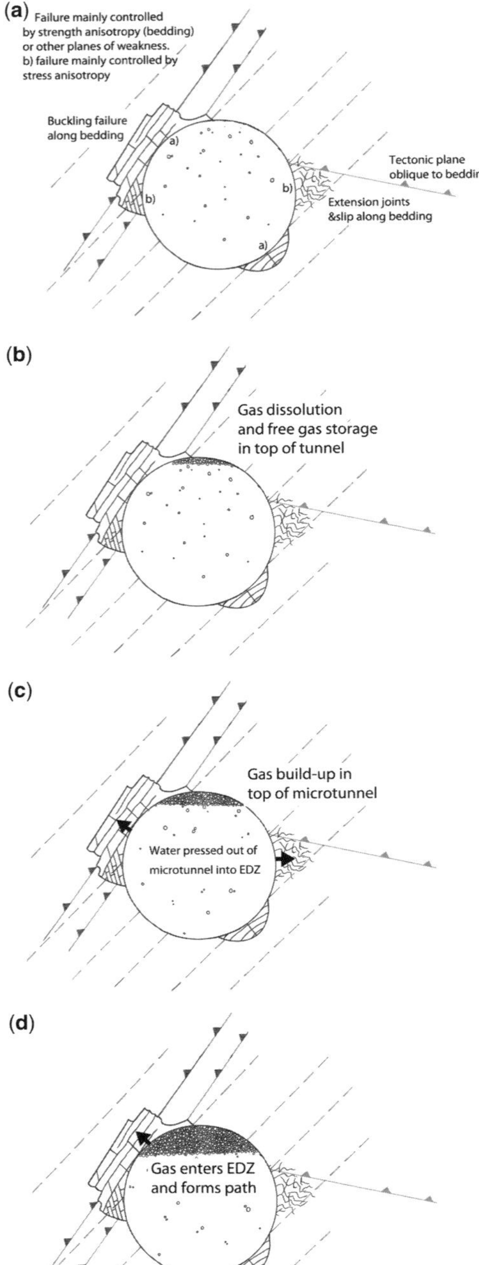

Fig. 10. EDZ schematic (from Marschall *et al.* 2008) with stages of gas injection: (**a**) EDZ structure; (**b**) storage and dissolution with linear pressure response (*c.* 20 days); (**c**) storage and leakage, pressure flattens prior to breakdown; and (**d**) breakdown and gas flow into the EDZ. View looking towards end of microtunnel.

better match than the uniformly permeable model, but that it is not possible to discriminate between the two EDZ models (2 and 3), both with an effective EDZ conductance of 3×10^{-10} m^3 s^{-1}. This slightly higher value of conductance, compared with that from the sealing index, may reflect the inclusion of the resistance to flow out of the test section and to the constant head boundary imposed at the liner.

Within the models the permeability of the cylindrical EDZ (model B) is 1–2 orders of magnitude greater than the undisturbed host rock, while that of a narrow channel would be higher but of smaller cross-section area.

Discussion

Test section behaviour

The observed response of the test section to gas injection suggests that the dominant pre-breakdown processes are: test section compressibility, gas dilution in pore-water and water leakage from the test section. These result in (a) an initial storage period with linear log–log response, followed by (b) a transitional period when leakage becomes important as test section pressure rises and (c) a final breakdown when the gas level in the microtunnel has been driven down to the point where the gas column is in contact with a permeable feature and gas overpressure is sufficient to enter the feature, as illustrated in Figure 10.

Independent of the gas injection rate, the peak pressures during gas injection are in the order of 1.3–1.4 MPa, suggesting that gas breakthrough is controlled by the percolation process in the sparse channel network along the EDZ. No evidence of gas-'fracturing' was observed. Test section pressures at peak (*c.* 1350 kPa) and during continuing gas injection (*c.* 1025 kPa) were well below the measured radial stress along the seal section (minimum of eight sensors, 1950 kPa) and no unusual responses to the breakdown were observed in the seal section.

Gas flow in the EDZ

The pressure data support a model of a heterogeneous EDZ/contact zone along the sealing section with a potential channel associated with the 3 o'clock position (close to one of the notches observed immediately after excavation). However, this channel appears to be poorly connected to any outflow from the system (atmospheric boundary condition). The 6 o'clock sensor at 8.6 m shows high pressures after degassing, suggesting that it is not well connected to the test section. The small oscillations and build-ups also observed suggest either a meta-stable or developing gas flow field rather than steady flow through an established pathway. The analysis of the pore pressure measurements in response to the gas injections gives clear evidence for localized gas leak-off along the EDZ.

Self-sealing and post-gas hydraulics

The water injections following the gas injection sequence confirmed a long-term sealing tendency of the EDZ. The effective conductance showed an ongoing reduction during saturation and hydraulic testing. Following gas injection the effective conductance swiftly reduced to values comparable to those prior to gas injection. The test zone response suggests a relatively linear flow (e.g. in a channel or some part of a cylindrical shell with conductance *c.* $>5 \times 10^{-10}$ m^3 s^{-1}).

This work was supported by ANDRA, BGR, Nagra and NWMO via the Mont Terri Consortium. The authors would like to thank the management and staff of the consortium together with the many contractors who have supported the experiment for their invaluable assistance in the technical planning and performance of the experiment.

References

Blümling, P., Bernier, F., Lebon, P. & Martin, C. D. 2007. The excavation damaged zone in clay formations time-dependent behaviour and influence on performance assessment. *Chemistry of the Earth*, **32**, 588–599.

Bock, H. 2000. *RA Experiment – Rock Mechanics Analyses and Synthesis: Data Report on Rock Mechanics.* Mont Terri Technical Report TR 2000-02.

Bock, H., Dehandschutter, B., Martin, C. D., Mazurek, M., De Haller, A., Skoczylas, F. & Davy, C. 2010. *Self-sealing of Fractures* in Argillaceous Formations in the Context of Geological Disposal *of* Radioactive Waste Review and Synthesis *Report*. OECD NEA 6184. Waste Management. NEA, OECD, Paris.

Bossart, P., Meier, P. M., Moeri, A., Trick, T. & Mayor, J.-C. 2004. Structural and hydrogeological characterisation of the excavation-disturbed zone in the Opalinus Clay (Mont Terri Project, Switzerland). *Applied Clay Science*, **26** (Clays in Natural and Engineered Barriers for Radioactive Waste Confinement), 429–448.

Corkum, A. G. & Martin, C. D. 2007. Modelling a mine-by test at the Mont Terri rock laboratory, Switzerland. *International Journal of Rock Mechanics and Mining Sciences*, **44**, 846–859, http://dx.doi.org/10.1016/j.ijrmms.2006.12.003

FORGE. 2009. Fate of Repository Gases. Collaborative Project of the European Commission, part of the 7th EURATOM framework (2009–2013), http://www.bgs.ac.uk/forge/

Lanyon, G. W., Marschall, P., Trick, T., de La Vaissière, R., Shao, H. & Leung, H. 2009.

Hydromechanical evolution and self-sealing of damage zones around a microtunnel in a claystone formation of the Swiss Jura Mountains. *In*: *43rd U.S. Rock Mechanics Symposium & 4th U.S. – Canada Rock Mechanics Symposium*, 28 June to 1 July 2009, Asheville, NC, 652–663, paper 09-333.

Marschall, P., Croisé, J., Schlickenrieder, L., Boisson, J.-Y., Vogel, P. & Yamamoto, S. 2004. Synthesis of hydrogeological investigations at the Mont Terri site (phases 1 to 5). *In*: Heitzmann, P. (ed.) *Mont Terri Project- Hydrogeological Synthesis, Osmotic Flow*. Reports of the Federal Office for Water and Geology, FOWG, Geology Series No. **6**, Federal Office for Water and Geology, FOWG, Bern.

Marschall, P., Distinguin, M., Shao, H., Bossart, P., Enachescu, C. & Trick, T. 2006. Creation and evolution of damage zones around a microtunnel in a claystone formation of the Swiss Jura Mountains. *In*: *SPE International Symposium and Exhibition on Formation Damage Control*, SPE Paper 98537, Society of Petroleum Engineers, Richardson, Texas.

Marschall, P., Trick, T., Lanyon, G. W., Delay, J. & Shao, H. 2008. *Hydro-mechanical Evolution of Damaged Zones around a Microtunnel in a Claystone Formation of the Swiss Jura Mountains*. ARMA, San Francisco, CA.

Martin, C. & Lanyon, G. 2003. Measurement of in-situ stress in weak rocks at Mont Terri Rock Laboratory, Switzerland. *International Journal of Rock Mechanics and Mining Sciences*, **40**, 1077–1088, http://dx.doi.org/10.1016/S1365-1609(03)00113-8

Martin, C. D., Lanyon, G. W., Blümling, P. & Mayor, J.-C. 2002. The excavation disturbed zone around a test tunnel in the ppalinus clay. *In*: *Proceedings of 5th North American Rock Mechanics Symposium and 17th Tunnelling Association of Canada Conference: NARMS/TAC 2002*, University of Toronto Press, Toronto, **2**, 1581–1588.

NAGRA. 2002. *Projekt Opalinuston – Synthese der geowissenschaftlichen Untersuchungsergebnisse. Entsorgungsnachweis für abgebrannte Brennelemente, verglaste hochaktive sowie langlebige mittelaktive Abfälle*. Nagra Technical Report **NTB 02-03**. Nagra, Wettingen, Switzerland.

Pearson, F. J., Arcos, D. et al. 2003. *Geochemistry of Water in the Opalinus Clay Formation at the Mont Terri Laboratory*. Report of the Federal Office for Water and Geology (Bern, Switzerland), Geology Series, **5**.

Thury, M. & Bossart, P. 1999. The Mont Terri Rock Laboratory, a new international research project in a mesozoic shale formation, in Switzerland. *Engineering Geology*, **52**, 347–359.

Installation of the PRACLAY Seal and Heater

P. VAN MARCKE[1]*, X. L. LI[2], G. J. CHEN[2], J. VERSTRICHT[2], W. BASTIAENS[2] & X. SILLEN[1]

[1]*ONDRAF/NIRAS, Avenue des Arts 14, 1210 Brussels, Belgium*

[2]*EURIDICE, Belgian Nuclear Research Centre (SCK•CEN), Boeretang 200, 2400 Mol, Belgium*

**Corresponding author (e-mail: p.vanmarcke@nirond.be)*

Abstract: In 2011 the last phase in the installation of the PRACLAY In-Situ Experiment in the underground research facility HADES (Mol, Belgium) was completed. The main goal of the experiment is to perform a large-scale *in-situ* Heater Test. The Heater Test will examine the effect of the thermal load generated by heat-emitting waste on Boom Clay, currently considered as a potential host rock in the Belgian R&D programme for geological disposal. In 2007 the PRACLAY gallery was constructed to host the Heater Test. In 2010 a bentonite-based hydraulic seal was installed in this gallery, isolating the heated part from the non-heated part of the PRACLAY gallery. The primary objective of the seal is to provide undrained hydraulic boundary conditions for the Heater Test. As the performance of the seal is crucial to the Heater Test, it has been instrumented accordingly. The seal also provides an opportunity to gather additional information on the *in-situ* behaviour of bentonite-based repository structures. Finally the placement of a heating system and water-saturated sand in the heated section of the PRACLAY gallery completed the experiment installation. The water-saturated backfill sand has to assure undrained hydraulic boundary conditions at the interface between the clay and the gallery lining.

The PRACLAY In-Situ Experiment fits into the framework of the safety and feasibility assessment of the Belgian reference concept as developed by NIRAS/ONDRAF for the geological disposal of high-level waste (HLW) in Boom Clay (ONDRAF/NIRAS 2001). An important aspect in this assessment is the effect of the heat generated by the high-level waste on the clay host rock. It is important to verify that this heat will not affect the contribution of the host rock to the safety functions of the disposal system. This will be examined by the PRACLAY Heater Test in which a 30 m-long gallery section will be heated for 10 years.

The PRACLAY setup is intended to be representative of a generic drift or gallery for the disposal of heat-emitting waste. To anticipate possible future changes in the repository design and because it is not possible to fully reproduce the time scale, the spatial scale and the boundary conditions of a real repository, the Heater Test was conceived to be as independent as possible of the final repository design by conducting it under reasonably conservative thermal, hydraulic and mechanical boundary conditions (Li *et al.* 2009).

This implies, among other requirements, undrained conditions. These conditions are realized by the installation of a hydraulic seal at the intersection between the heated and the non-heated part of the gallery and installing a water-saturated backfill sand in the heated part of gallery. The installation of the hydraulic seal constitutes the Seal Test.

The Heater Test will be hosted in the PRACLAY gallery in the underground laboratory HADES (Fig. 1). The PRACLAY gallery is 45 m long and has an external diameter of 2.5 m. The construction of this gallery and its crossing with the already existing gallery connecting the two shafts of HADES (the so-called 'Connecting gallery') constitutes the Gallery and Crossing Test. The Gallery and Crossing Test, the Seal Test and the Heater Test make up the PRACLAY In-Situ Experiment, the layout of which is shown in Figure 2.

The PRACLAY gallery was constructed in 2007 (Van Marcke & Bastiaens 2010). The successful realization of the gallery crossing and the PRACLAY gallery itself meant an important milestone in the Belgian research programme for the geological disposal of HLW. The design proved to be adequate, although it appeared that, from a practical point of view, the gallery diameter of 2.5 m was at the low end of what is feasible using this construction technique. The same technique was successfully used for the construction of a 4.8 m gallery at HADES. Since the future disposal galleries have an intermediate diameter, their construction feasibility can be considered proven.

During the excavation, the response of the clay formation was monitored. The results were in line with previous observations and confirmed the highly coupled and anisotropic hydro-mechanical behaviour of the Boom Clay and known fracturing processes.

From: Norris, S., Bruno, J., Cathelineau, M., Delage, P., Fairhurst, C., Gaucher, E. C., Höhn, E. H., Kalinichev, A., Lalieux, P. & Sellin, P. (eds) 2014. *Clays in Natural and Engineered Barriers for Radioactive Waste Confinement*. Geological Society, London, Special Publications, **400**, 107–115.
First published online March 5, 2014, http://dx.doi.org/10.1144/SP400.4

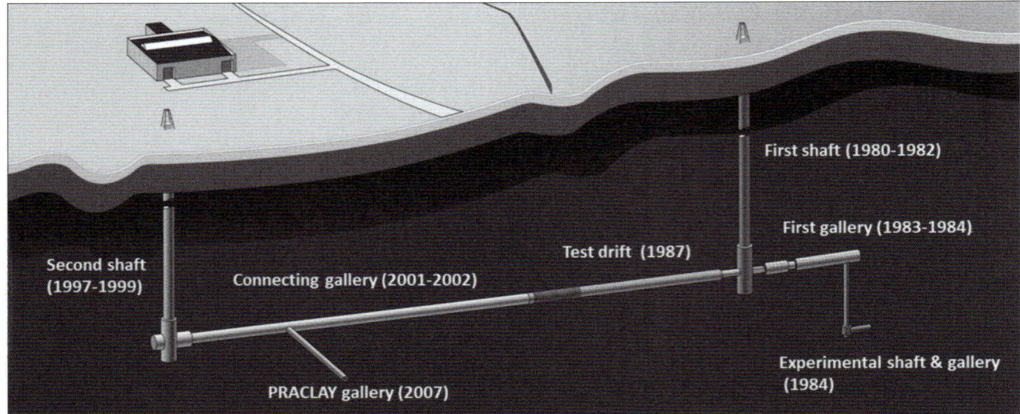

Fig. 1. Layout of the underground laboratory HADES in Mol (Belgium).

Subsequently the hydraulic seal was fitted in the PRACLAY gallery in 2010, separating the heated and the non-heated gallery sections. Finally, in 2011, the heating system and backfill material were installed in the heated section, completing the setup of the PRACLAY In-Situ Experiment.

The hydraulic seal

The main objective of the hydraulic seal is to close off the heated part of the PRACLAY gallery and to hydraulically cut off the preferential pathway through the gallery and the excavation damaged zone around the gallery. In that way an undrained hydraulic boundary at the intersection between the non-heated and the heated section is created. As the heated section is backfilled and saturated with water before the start of the heating phase, this also imposes an undrained boundary condition at the interface between the gallery lining and the clay. Such an undrained boundary is chosen to achieve the most penalizing conditions that are reasonably achievable during the Heater Test. The seal also provides an opportunity to gather additional information on the *in-situ* behaviour of bentonite-based repository structures.

Finally, the hydraulic seal has to allow watertight feed-through of the instrumentation placed in the upstream part of the PRACLAY gallery (this is the side of the heated part of the PRACLAY Gallery; the downstream side is the side towards the Connecting gallery) and feed-through of the heater cables placed in a later phase.

Seal design

The seal is installed in the PRACLAY gallery at 10 m from the crossing with the Connecting Gallery to limit mutual interactions between the Heater Test and the Connecting Gallery. Scoping calculations indicated that a seal length of 1 m was

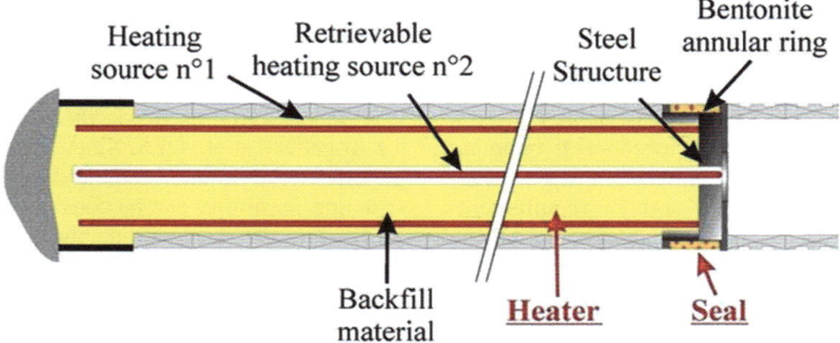

Fig. 2. Layout of the PRACLAY In-Situ Experiment, which consists of the Gallery and Crossing Test, the Seal Test and the Heater Test.

sufficiently effective and that no significant gain was obtained by further increasing its length. A bentonite-based hydraulic seal was chosen instead of, for example, a technical seal consisting of inflatable packers, as bentonite is generally considered as a suitable material for seals in repository designs for the disposal of radioactive waste. Although the design of the seal is primarily driven by the requirements of the Heater Test and it is not mimicking the design of a repository seal, it can provide lessons on the feasibility of installing such a seal and on the behaviour of the bentonite.

The hydraulic seal consists of a stainless steel structure closing off the heated part of the gallery from the rest of the underground infrastructure and an annular ring of bentonite placed against the clay (Fig. 2). The bentonite is naturally hydrated by water coming from the surrounding Boom Clay and artificially by filters included in the steel structure supporting the bentonite ring.

As the hydraulic seal was installed after the construction of the PRACLAY gallery, an alternative lining in the zone of the future hydraulic seal was needed instead of the concrete wedge block lining to be able to place the bentonite blocks against the clay sidewall. The alternative lining consisted of four steel rings and wooden plates placed between these rings (Fig. 3). Before the installation of the hydraulic seal, the wooden plates were removed to allow the bentonite blocks to be placed against the Boom Clay. The steel rings remained in place.

The design and installation of the hydraulic seal are illustrated in Figure 4, where a cross-section of the PRACLAY gallery at the location of the hydraulic seal is shown for the different steps of the seal installation. A steel cylinder was placed in the gallery at the upstream side of the location for the hydraulic seal (Fig. 4a). The cylinder was closed in its middle by a plate with a manhole. In the next step the temporary part of the alternative lining (i.e. the wooden plates; Fig. 3) at the location of the hydraulic seal was removed (Fig. 4b). Subsequently a flange was placed against the concrete lining at the downstream side (Fig. 4c) and another flange was placed against the concrete lining at the upstream side (Fig. 4d). Because the flanges were too large to be installed in one piece, they were constructed in four segments and assembled on-site. After both flanges were installed, bentonite blocks were placed between the flanges and against the clay massif (Fig. 4e). Finally the cylinder was pushed in against the downstream flange and assembled to the two flanges (Fig. 4f).

Fig. 3. Alternative gallery lining at the location of the hydraulic seal consisting of steel rings and removable wooden plates in between.

The bentonite had to meet the following specifications:

- Its swelling pressure was larger than 2.5 MPa to avoid the creation of negative effective stresses around the hydraulic seal during the Heater Test (the maximum pore water pressures in the Boom Clay around the hydraulic seal during the Heater Test were estimated at 2.5 MPa).
- Its maximal swelling pressure was 6 MPa to avoid fracturing the clay and jeopardizing the integrity of the stainless steel structure of the hydraulic seal.
- Its hydraulic conductivity at saturated state was as low as possible (at least lower than the conductivity of undisturbed of Boom Clay (*c.* 10^{-12} m s^{-1}) and preferably one order of magnitude lower).

It was decided to use MX80 bentonite compacted into blocks. The bentonite was chosen as seal material for its good swelling capacity upon hydration and for its low permeability. Relevant experience and information with this type of bentonite exists from its use in other experiments in underground research facilities (Mont Terri, Bure, ASPO, AECL's underground research laboratory) and in the laboratory (by CEA, CIEMAT, CERMES and SKB). Furthermore it is a Na-bentonite, which makes it chemically compatible with Boom Clay water. The initial dry density of the bentonite (1.8 t m^{-3}), which affects its swelling pressure and its saturated permeability, was determined by scoping calculations. The scoping calculations took account of a void volume of 7% in the bentonite ring before hydration. This void was necessary to allow the installation of the bentonite ring. The final dry density of the bentonite, after swelling, was estimated at 1.6 t m^{-3}.

Seal installation

The seal installation sequence is explained in Figure 4. First the wood from the alternative lining was removed, after which the two flanges

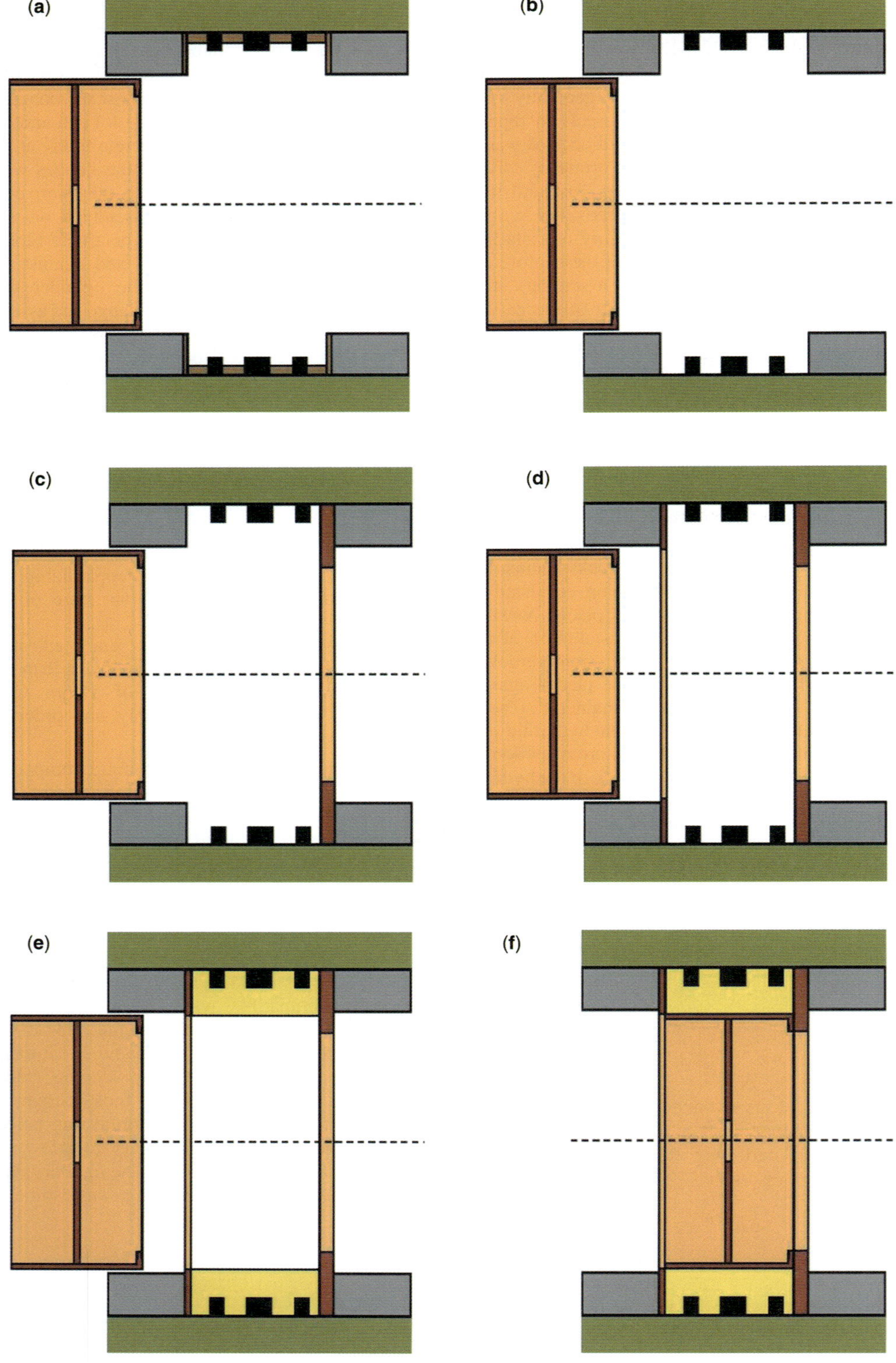
(a)
(b)
(c)
(d)
(e)
(f)

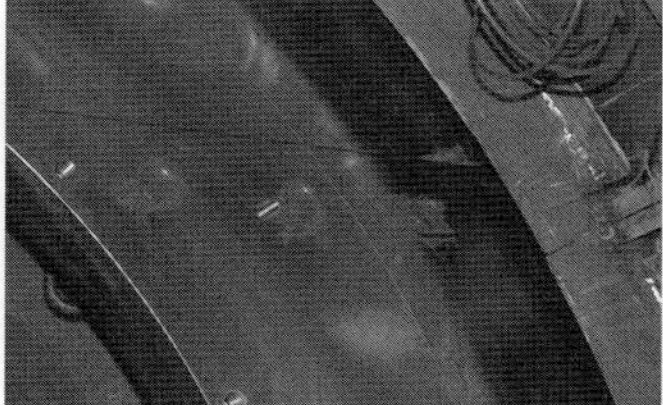
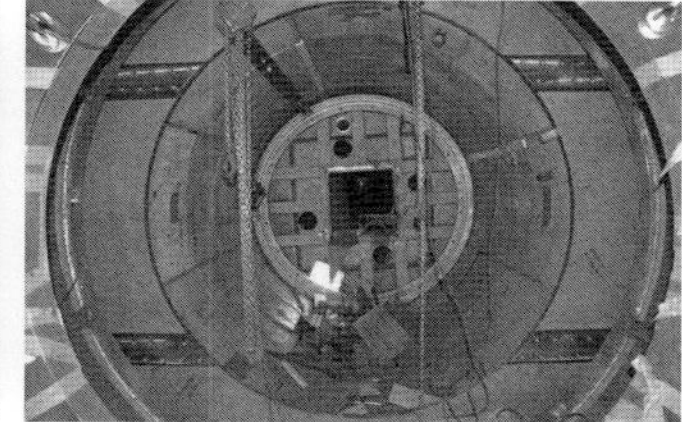

Fig. 5. Downstream flange (left); assembly of the flange segments by bolts and filling of the bolts with a sealing resin (middle); upstream flange with cylinder (waiting to be pushed in place after installation of the bentonite) in the background (right).

were placed against the concrete lining (Fig. 5). Because the flanges were too large to be installed in one piece, they were composed of four segments that are assembled on-site. Initially it was foreseen to assemble the segments by welding them together, but the risk of large deformations owing to the welding was considered too high. Therefore the assembly was done by bolting the segments together. Their connections were made watertight by placing a sealing resin on the segment interfaces.

After both flanges were installed, an annular ring of bentonite blocks was built between the flanges and the clay sidewall (Fig. 6). The bentonite ring was composed of in total 236 blocks or *c.* 3200 kg bentonite. An anticipated technological void of 8% of the total volume was needed to allow the placement of the bentonite blocks. Much instrumentation was placed in the bentonite blocks to obtain information on the bentonite hydration and to be able to evaluate the performance of the hydraulic seal.

Subsequently the steel cylinder was placed inside the annular bentonite ring and between the two flanges. In that way the bentonite was enclosed between the two flanges, the cylinder and the Boom Clay (Fig. 7). The hydraulic seal also provides feedthrough for the instrumentation and heater cables in the heated section of the PRACLAY gallery. Because this part of the gallery has to remain accessible after the hydraulic seal installation for the placement of heater, backfill sand and instrumentation, the seal has a central manhole. After the complete installation of the PRACLAY Experiment, a plate was welded onto the manhole to completely close off the heated section of the PRACLAY gallery.

The assembly of the heavy steel components (the heaviest single part weighed *c.* 1.6 t and the total weight of the assembled downstream flange was *c.* 2.3 t) in the very limited working space was not at all straightforward. This demonstrates the necessity of limiting underground operations as they are complex and imply high risks.

The heater

The heating system has to heat the clay at the gallery extrados from its *in-situ* temperature of 16 °C to a temperature of 80 °C and keep the temperature at the gallery–Boom Clay interface at this temperature for *c.* 10 years. In the current reference design of a disposal gallery in Boom Clay containing heat-emitting waste, the maximum temperature at the interface between the gallery and the clay should be reached after 10–25 years. The thermal boundary condition imposed in the Heater Test is conservative, with a faster temperature increase rate and a slightly higher maximum temperature.

The heating system consists of a primary heater close to the gallery intrados and a secondary heater inside a central tube (Fig. 8). Both heating systems consist of electrical heaters. The primary heater is inaccessible during the Heater Test and therefore twice as many primary heater cables than are necessary were installed (100% redundancy). The secondary heater is a backup and will remain accessible and replaceable at all times during the test. It was placed in a central tube.

The primary heating system is made up of three sections: a front-end section (the first metres behind the hydraulic seal), a middle section of *c.* 30 m and a far-end section (the last metres of the PRACLAY gallery). To obtain a constant temperature along the extrados of the 30 m-long middle section, a higher power is applied at the front-end and the

Fig. 4. Schematic overview of the different steps of the hydraulic seal installation in a cross-section of the PRACLAY gallery at the location of the hydraulic seal (the left side of the drawings is the upstream side). Green, Boom Clay; grey, neighbouring concrete lining; black, steel rings of the permanent part of the gallery lining at the location of the hydraulic seal; brown, wood making up the temporary part of the gallery lining at the location of the hydraulic seal; red-orange, steel components of the hydraulic seal; yellow, bentonite ring.

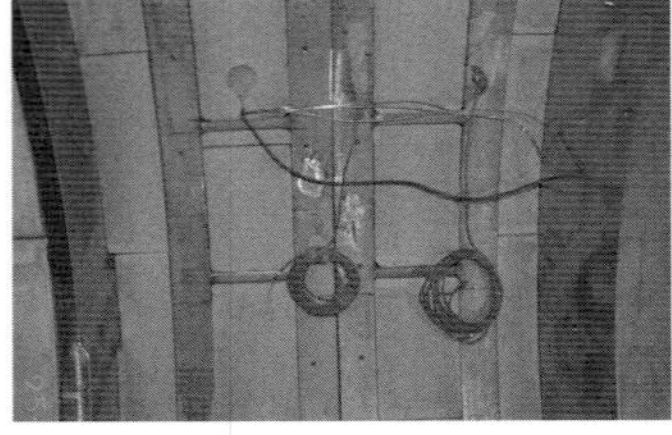
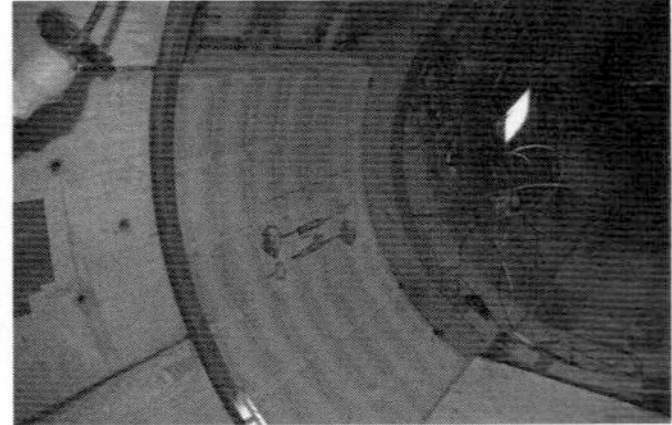

Fig. 6. Installation of the annular bentonite ring consisting of two layers of bentonite made from bentonite blocks.

far-end ends to compensate for the larger heat dissipation at these ends.

In addition to these heating elements, a control system regulating the heating power as a function of measured and target temperatures makes up part of the heating system. During the start-up phase the temperature is increased at a limited rate to limit the thermal gradient over the gallery lining, and consequently the thermal stresses in the lining. This heating rate is nevertheless faster than the one that will be imposed by the heat-emitting wastes in the current reference design of a repository in Boom Clay.

The backfill sand

A water-saturated backfill material was installed in the heated part of the PRACLAY gallery to achieve an undrained hydraulic boundary condition at the interface between Boom Clay and gallery lining. The material has to have a sufficient thermal conductivity to efficiently transfer the heat generated by the heating system to the gallery lining and the clay and as such limit the temperature of the heating elements. It also has to be sufficiently hydraulically conductive to allow a rapid homogenization of the water pressures in the backfill material, thus ensuring a uniform hydraulic boundary condition along the gallery interface with the clay.

Mol Sand M34 was selected as backfill material. The calculated saturated thermal and hydraulic conductivities of the sand are respectively 2.9 $W\ m^{-1}\ K^{-1}$ and $5 \times 10^{-4}\ m\ s^{-1}$ (for a dry density of $1.60\ t\ m^{-3}$), which were considered to be sufficiently high. Furthermore, the Mol Sand also has a narrow grain size distribution, which limits the density differences between the top and bottom of the backfill sand owing to segregation. Segregation could lead to an insulating water layer at the top of the gallery and a less homogeneous temperature profile around the gallery.

Approximately 145 t Mol sand was required to fill the gallery (Fig. 9). It was installed by blowing the sand in a dry state through an injection hose on top of the PRACLAY gallery. As the sand front progressed towards the hydraulic seal, the injection hose was pulled back. A dry installation was preferred over the injection of a sand–water mixture as the latter would be a relatively complex operation. The possible abrasion of the heating elements by the injected sand was checked beforehand on a surface. No damage to the cable in the surface test could visually be detected. Also, during the backfill installation, the cables in the gallery were regularly inspected.

In a later phase the backfill sand was saturated through six saturation filters placed at the bottom of the heated part of the PRACLAY gallery. Five vent filters at the top of the gallery allowed venting of the air of the gallery during the backfill saturation.

Approximately 43 m^3 water was injected between November 2011 and June 2012, leading

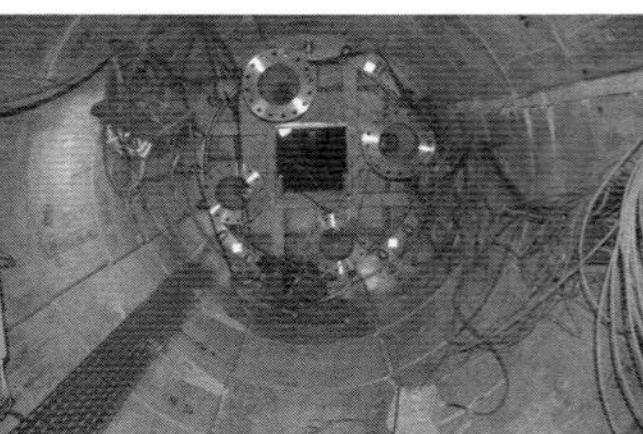
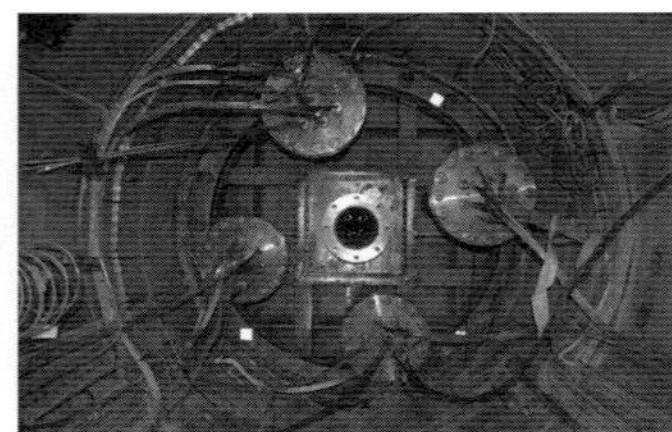

Fig. 7. Moving the steel cylinder against the downstream flange (left); manhole in the structure for the later installation of the heater and backfill sand (middle); closure of the manhole after the heater and backfill sand installation (instrumentation feed-throughs are included in the four flanges; right).

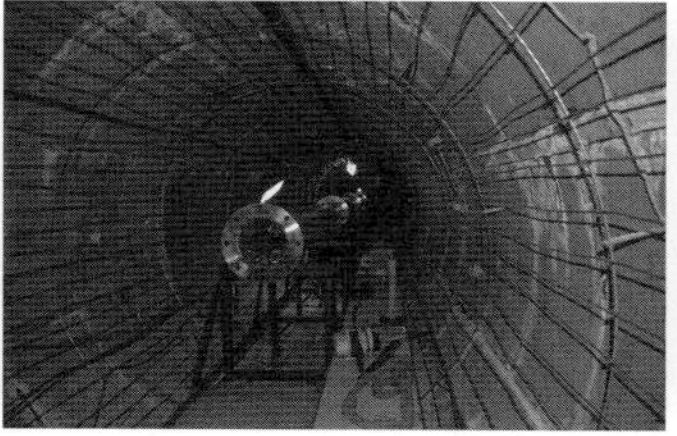
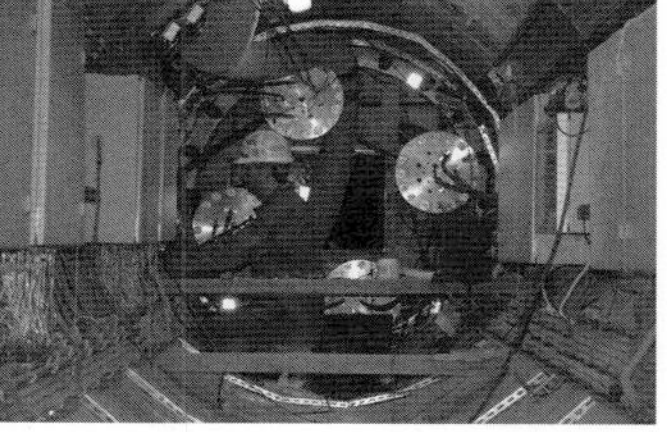

Fig. 8. The primary heating system consisting of heater cables placed close to the gallery lining (left); the secondary heating system consisting of heater cables placed in a central tube (middle); connection of the heater cables to the control system in the non-heated part of the PRACLAY gallery (right).

to a water pressure of 5 bar in the backfill sand. The gallery was pressurized stepwise to increase the saturation degree of the gallery (by compression and dissolution of the air in the backfill) and to already test the whole seal performance and stability under cold conditions. During the saturation and pressurization period, no water leakage around the seal and no significant seal movements were observed.

The PRACLAY Heater Test

The next step in the completion of the PRACLAY Experiment is the switch-on of the heater and the start of the PRACLAY Heater Test. The switch-on timing depends on the progress of the bentonite hydration as a sufficiently high bentonite swelling pressure is needed for the seal to perform its function. The pore water pressures in the clay are expected to increase up to a value of about 2.5 MPa during the thermal period if no drainage towards the galleries takes place. To ensure that this pressure can be maintained for the Heater Test, the bentonite has to reach a swelling pressure larger than 2.5 MPa. Scoping calculations suggest that, after the start of the heating, the swelling pressures will increase in the seal as a result of the temperature increase. Hence, the current requirement on the swelling pressure value at switch-on could possibly be relaxed in the future. A decision on the moment for heater switch-on is thus based on a good understanding of the state of the bentonite hydration and a sufficient confidence in the performance of the hydraulic seal.

The bentonite is naturally hydrated by pore water coming from the Boom Clay and artificially by water injected through filters at the outside of the cylinder of the seal. The natural hydration started immediately after the bentonite installation (February 2010). The artificial hydration was started 2 months later (April 2010).

The swelling, total pressure and pore pressure of the bentonite and the surrounding clay are continuously measured and analysed. Also the hydraulic conductivity at the bentonite–Boom Clay interface and the surrounding clay are being measured systematically. After 2 years of hydration the measured hydraulic conductivity at the interface with the Boom Clay was in the order of the conductivity of undisturbed Boom Clay (*c.* 10^{-12} m/s). Furthermore, numerical simulations of the seal hydration are made to obtain a better understanding of its evolution (Chen & Li 2011).

The radial swelling pressure exerted by the bentonite on the Boom Clay sidewall is shown in Figure 10. The pressures started increasing a few days after the start of the artificial hydration, which indicates that in the first days of artificial hydration a closure of the technological voids in the bentonite ring took place. After all voids were closed, the stresses in the bentonite ring started

Fig. 9. Injection tube for the blowing of the backfill sand (left); backfilled part of the PRACLAY gallery (middle); on-surface testing of the abrasion of the injected sand on cables (right).

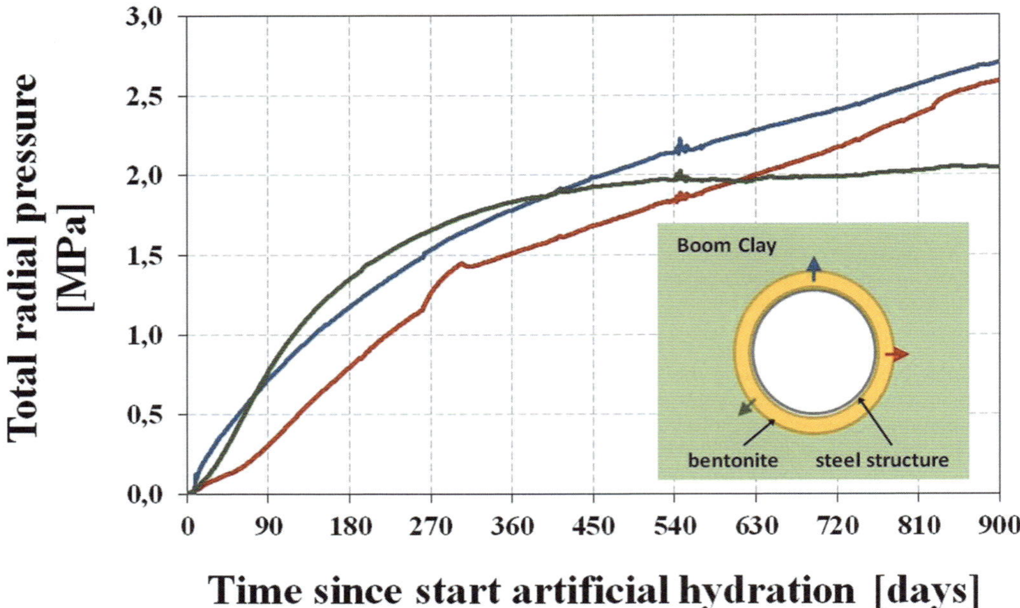

Fig. 10. Evolution of the radial pressure exerted by the bentonite on the Boom Clay since the start of the artificial bentonite hydration.

increasing. After almost 3 years of bentonite hydration the measured radial swelling pressures range from 2.0 to 2.7 MPa. The measured pressures in the top and middle section still clearly show an increasing trend with a current stress increase rate of *c.* 0.6 MPa/year. However, the increase in pressure measured in the lower section is very limited. The stress has increased less than 0.1 MPa in the last year.

The bentonite hydration is clearly still on-going and, as most bentonite is still unsaturated, there is still large potential for further bentonite swelling as the hydration goes on. The hydration however occurs much slower than initially expected and the early scoping calculations overestimated the hydration rate. This reveals the difficulty in modelling the bentonite behaviour and in particular in accurately modelling the hydration rate.

Conclusions

By the installation of the heater and backfill sand in the PRACLAY gallery in 2011, the last phase in the setting up of the PRACLAY In-Situ Experiment was completed. The experiment fits into the framework of the safety and feasibility assessment of the Belgian reference concept as developed by NIRAS/ONDRAF for the geological disposal of HLW in Boom Clay (ONDRAF/NIRAS 2001). An important aspect in this assessment is the effect of the heat from the HLW on the key clay properties that support the safety functions of the disposal system. This will be examined by the PRACLAY Heater Test, in which a 30 m long gallery section will be heated for 10 years.

The PRACLAY setup is intended to be representative of a generic drift or gallery for the disposal of heat-emitting waste. To anticipate possible future changes in the repository design and because it is not possible to fully reproduce the time scale, the spatial scale and the boundary conditions of a real repository, the Heater Test was conceived to be as independent as possible of the final repository design and will be conducted under a well-controlled and reasonably conservative combination of thermal, hydraulic and mechanical boundary conditions (Li *et al.* 2009).

This implies, among other requirements, undrained conditions that are realized by the installation of a hydraulic seal separating the heated part from the non-heated part of the gallery. A bentonite based hydraulic seal was chosen, instead of, for example, a technical seal consisting of inflatable packers, as bentonite is an often proposed material in the design for radioactive waste repository. The bentonite seal can also provide lessons on the feasibility of installing such a seal and on the behaviour of the bentonite.

The hydraulic seal was installed in 2010. Subsequently a system to heat up the clay at the gallery extrados from its *in-situ* temperature of 16 °C to a temperature of 80 °C was placed in the PRACLAY gallery. The system has to keep the temperature at the interface gallery–Boom Clay at 80 °C for *c.* 10 years.

Finally the heated section of the PRACLAY gallery was backfilled with sand. The sand was blown into the gallery in dry state and later saturated with water to achieve quasi-undrained hydraulic boundary conditions at the interface between Boom Clay and gallery lining.

The switch-on of the heating system, and thus the start of the Heater Test, depends on the progress of the bentonite hydration as a sufficiently high bentonite swelling pressure is needed for the seal to perform its function. As the pore water pressures in the clay are expected to increase up to a value of about 2.5 MPa during the Heater Test, the bentonite has to reach a minimal swelling pressure of 2.5 MPa.

The radial swelling pressures the bentonite exerts on the Boom Clay, as measured almost 3 years after the start of the hydration, range from 2.0 to 2.7 MPa. The bentonite hydration occurs more slowly than expected.

References

Chen, G. & Li, X. 2011. Numerical Study of PRACLAY Seal Test in Mol, Belgium. *In*: *2nd International Symposium on Computational Geomechanics*, Cavtat–Dubrovnik, Croatia, 27–29 April.

Li, X., Bastiaens, W., Weetjens, E. & Sillen, X. 2009. The THM boundary conditions control in the design of the large scale in-situ PRACLAY heater test. *In*: *International Conference and Workshop, 'Impact of Thermo-Hydro-Mechanical–Chemical (THMC) Processes on the Safety of Underground Radioactive Waste Repositories'*, Luxembourg, 29 September to 1 October.

ONDRAF/NIRAS 2001. *SAFIR-2, Safety Assessment and Feasibility Interim Report 2*. ONDRAF/NIRAS, Brussels.

Van Marcke, P. & Bastiaens, W. 2010. Construction of the PRACLAY Experimental Gallery at the HADES URF. *In*: *Clays in Natural and Engineered Barriers for Radioactive Waste Confinement*, Nantes, 29 March to 1 April.

Hydro-chemical modelling of *in situ* behaviour of bituminized radioactive waste in Boom Clay

N. MOKNI[1]*, S. OLIVELLA[1], E. VALCKE[2], N. BLEYEN[2], S. SMETS[2], X. LI[3] & X. SILLEN[4]

[1]*Department of Geotechnical Engineering and Geosciences, Universitat Politècnica de Catalunya, Barcelona, Spain*

[2]*Waste and Disposal Expert Group, The Belgian Nuclear Research Centre (SCK•CEN), Boeretang 200, 2400 Mol, Belgium*

[3]*EIG EURIDICE, Boeretang 200, 2400 Mol, Belgium*

[4]*ONDRAF/NIRAS, Kunstlaan 14, 1210 Brussel, Belgium*

**Corresponding author (e-mail: nadia.mokni@upc.edu)*

Abstract: The hydro-chemical (CH) interaction between swelling Eurobitum bituminized radioactive waste (BW) and Boom Clay was investigated to assess the feasibility of geological disposal for the long-term management of this waste. First, the long-term behaviour of BW in contact with water was studied. A CH formulation of chemically and hydraulically coupled flow processes in porous materials containing salt crystals is discussed. The formulation incorporates the strong dependence of the osmotic efficiency of the bitumen membrane on porosity and assumes the existence of high salt concentration gradients that are maintained for a long time and that influence the density and motion of the fluid. The impacts of temporal and spatial variations of key transport parameters (i.e. osmotic efficiency (σ), intrinsic permeability (k), diffusion, etc.) were investigated. Porosity was considered the basic variable. For BW porosity varies in time because of the water uptake and subsequent processes (i.e. dissolution of salt crystals, swelling of hydrating layers, compression of highly leached layers). New expressions of σ and k describing the dependence of these parameters on porosity are proposed. Several cases were analysed. The numerical analysis was proven to be able to furnish a satisfactory representation of the main observed patterns of the behaviour in terms of osmotic-induced swelling, leached mass of $NaNO_3$ and progression of the hydration front when heterogeneous porosity and crystal distributions have been assumed. Second, the long-term behaviour of real Eurobitum drums in disposal conditions, and in particular its interaction with the surrounding clay, was investigated. Results of a CH analysis are presented.

Final disposal in deep geologically stable formations is considered worldwide as the preferred option for the long-term management of long-lived intermediate- and high-level radioactive waste. The disposal concepts proposed by many countries consist of a system of natural and engineered barriers to separate the waste from the biosphere. The Belgian Agency for the Management of Radioactive Waste and Fissile Materials (ONDRAF/NIRAS) envisages geological disposal of Eurobitum bituminized radioactive waste (BW) (ONDRAF/NIRAS 2009). Eurobitum is an intermediate-level long-lived radioactive waste form that consists of *c.* 60 w% of hard bitumen Mexphalt R85/40 and *c.* 40 wt% of waste. The waste originates from the chemical reprocessing of spent nuclear fuel and from the cleaning of high-level waste storage tanks, and contains mainly $NaNO_3$ and $CaSO_4$. These salts are present with a concentration of, respectively, 20–30 and 4–6 wt% of Eurobitum. In Belgium, the Boom Clay, which is a 30–35 million years old and *c.* 100 m thick marine sediment, is being studied as a potential host formation because of its favourable properties to limit and delay the migration of any leached radionuclides to the biosphere over extended periods of time. The current disposal concept foresees that several 220 l drums of Eurobitum would be grouped in a thick-walled cement-based secondary container, which in turn would be placed in concrete-lined disposal galleries that are excavated at mid-depth in the clay layer (Mariën *et al.* 2013). Only 80–90% of the total volume of the drum would be filled with Eurobitum. The remaining voids between the drums would be backfilled with a cement-based material. The emplacement of Eurobitum BW in the Boom Clay is expected to result in a geomechanical and a physico-chemical perturbation of the clay (Valcke *et al.* 2010). The large amounts of dehydrated and hygroscopic salts in the BW, with

From: Norris, S., Bruno, J., Cathelineau, M., Delage, P., Fairhurst, C., Gaucher, E. C., Höhn, E. H., Kalinichev, A., Lalieux, P. & Sellin, P. (eds) 2014. *Clays in Natural and Engineered Barriers for Radioactive Waste Confinement*. Geological Society, London, Special Publications, **400**, 117–134.
First published online June 16, 2014, http://dx.doi.org/10.1144/SP400.40

bitumen acting as a highly efficient membrane, will provoke an osmosis-driven uptake of pore water, resulting in swelling of the BW and/or increased stress on the clay (i.e. a geomechanical perturbation of the clay). In addition, diffusion of large amounts of salts, mainly $NaNO_3$, will result in physico-chemical perturbation of the clay. After closure of the disposal gallery, two different phases can be distinguished (Valcke *et al.* 2010; Mariën *et al.* 2013). In the first, free swelling phase, an osmotically induced swelling of the BW will take place until all free volume is filled with swollen BW. Afterwards, the continued water uptake and swelling of the BW will induce an increasing stress on the clay. The resultant deformation of the clay creates additional space for the BW to swell, but compared with the free swelling phase, the swelling rate is limited by the increasing stress exerted by the clay. This second phase is the restricted swelling phase. The number of drums per gallery cross-section must be limited to prevent unacceptable geomechanical perturbations of the clay.

An experimental programme aiming at understanding the swelling and leaching behaviour of BW samples in contact with alkaline solutions is in progress at the Belgian Nuclear Research Centre SCK•CEN. The set of tests that has been performed consists of water uptake tests under restricted (i.e. constant stress and constant volume) and free swelling conditions (Valcke *et al.* 2010; Mariën *et al.* 2013). In a previous paper (Mokni *et al.* 2011) a hydro-chemical (CH) formulation of chemically and hydraulically coupled flow processes in porous materials containing salt crystals was discussed. The formulation incorporates the strong dependence of the osmotic efficiency of the membrane on porosity, and assumes the existence of high concentration gradients that are maintained for a long time and that influence the density and motion of the fluid. The CH formulation was used to analyse the influence of osmosis on the swelling of Eurobitum BW owing to water uptake under laboratory conditions. The experimental results of pressure increase, swelling deformation and formation of low permeability compressed outer layers were reproduced successfully by Mokni *et al.* (2011) and Mariën *et al.* (2013). However, some differences were observed between the model results and the experimental observations in terms of progression of the hydration front and leached mass of $NaNO_3$. Indeed, the modelled time to dissolve all salt crystals was shorter than observed in the experiments. According to the model, all crystals in the sample are dissolved after about 2000 days, while in reality only 12–21% of the initial $NaNO_3$ content was leached after about 4 years of hydration and only the outer layers with a thickness of 1–2 mm were hydrated after this time period (Mariën *et al.* 2013). Mariën *et al.* (2013) stated that the difference is related to the simple relationship between porosity and membrane efficiency proposed by Mokni *et al.* (2011), which shows a too rapid decrease in efficiency with porosity increase (see Fig. 2 further in the text).

In order to better reproduce the experimental observations in terms of progression of the hydration front and leached mass of $NaNO_3$, a new expression of the osmotic efficiency coefficient (σ), including the dependence of σ on porosity and on crystal volume fraction, is proposed in this paper. Because of the interdependency of most of the key transport functions (i.e. permeability, diffusion coefficient and osmotic efficiency) that control the magnitude of coupled fluxes (principally chemical osmosis and ultrafiltration), a new expression of the intrinsic permeability (κ), including the dependence of κ on porosity, is proposed in this paper. The water uptake tests under constant vertical stress performed on BW were modelled using the proposed expressions and several cases were analysed. A sensitivity analysis was performed in order to study the effect of the variation rates of σ and κ with porosity on swelling, leaching and progression of the hydration front of BW. The CH formulation was used to model the *in situ* behaviour of BW, to obtain insights into the kinetics of water uptake by BW, dissolution of the embedded $NaNO_3$ crystals, solute leaching and maximum generated pressure under disposal conditions.

Long-term swelling behaviour of BW: phenomenology and main processes observed in laboratory experiments

A research programme involving the testing of the swelling behaviour of specimens of BW under constant volume and constant stress conditions was performed at the Belgian Nuclear Research Centre SCK•CEN. The experimental results are reported in Valcke *et al.* (2010) and Mariën *et al.* (2013). Swelling under various constant stresses, osmosis-induced pressure in constant volume conditions and leached mass of $NaNO_3$ were measured during the tests. After dismantling, the hydrated samples were characterized with micro-focus X-ray computer tomography (μCT) and environmental scanning electron microscopy (ESEM). Modelling results of those tests were presented by Mokni *et al.* (2011); Mokni (2011) and Mariën *et al.* (2013). The phenomenology of the water uptake, swelling and salt leaching processes in restricted swelling conditions is now reasonably well understood (Valcke *et al.* 2010; Mokni *et al.* 2011; Mariën *et al.* 2013). The reported results and observations show that the uptake of water by and the subsequent swelling and/or pressure increase of BW (composed of 40 wt% of waste containing 28.5 wt% soluble salts $NaNO_3$) are largely controlled by osmosis. The bitumen surrounding the soluble salts and other inorganic compounds behaves as a highly efficient semipermeable membrane, especially when tested under restricted swelling conditions. After almost four years of hydration of small BW samples (38 mm diameter, 10 mm thickness), osmostic-induced pressures of up to 20 MPa and swelling deformations ranging between 10 and 20% are being measured in constant volume and constant stress water uptake tests, respectively. A major characteristic of the medium, which plays an important role in the process of water uptake, is the presence of salt crystals. In fact, the large amount of hygroscopic salt incorporated into the bitumen matrix will attract water from the surrounding reservoir, resulting in dissolution of the soluble crystals, diffusion of the dissolved salts and volume increase of the leaching BW. When the external applied stress is lower than the osmotic pressure, water is driven into the pores of the medium because of the gradient of concentration existing between the reservoir and the solution initially filling the pores. The pore water concentration will decrease until it reaches a value corresponding to an osmotic pressure equal to the external applied stress. The diffusion of $NaNO_3$ out of the BW, combined with the external stress and the osmosis-driven transport of water from the outermost pores with a lower $NaNO_3$ concentration (higher water activity) towards deeper pores with a higher $NaNO_3$ concentration (lower water activity), contributes to a local consolidation and the formation of a re-compressed, low-porosity layer at the outer boundary of the leached material (Fig. 1b). Compression of the outermost layers is considered here as a limiting or controlling process for the water uptake and the release of $NaNO_3$. Valcke *et al.* (2010) characterized some leached samples with μCT. This technique permitted the hydrated area of the leached sample to be visualized. Figure 1a shows the μCT image of samples S0 and S4 after 887 and 1472 days of hydration, respectively. The red parts represent the stainless steel filters. The thin green–yellow layers near the filters are the hydrated layers. The central part in red–yellow corresponds to the dry Eurobitum. For both samples, only a thin layer had been hydrated. The central part of the samples appeared to be not hydrated, which suggested that it would take several years before the hydration front reached the centre of the samples, and it would require even more time before all $NaNO_3$ had dissolved. During that time, the sample was expected to continue to swell. In fact, size reduction of the outer pores induces a decrease in the permeability and the diffusion coefficient, and an increase in the osmotic efficiency in these compressed layers, while deeper in the sample these parameters vary in the opposite way. Therefore, the outer layers act as a highly efficient semipermeable membrane surrounding the swollen material with less efficient membrane properties. This maintains a very slow inflow of water towards the sample and outflow of the solute over a long time.

Coupled CH formulation

Mokni *et al.* (2010*a*, 2011) presented a CH formulation of chemically and hydraulically coupled flow processes in porous materials containing salt crystals, which incorporates the strong dependence of the efficiency of the membrane on porosity. The formulation assumes the existence of high concentration gradients that are maintained for a long time and that influence the density and motion of the fluid. The CH formulation is presented briefly below.

Balance equations

The medium is composed of three phases:

- solid phase – solid matrix and salt crystals;
- liquid phase (l) – dissolved salt in water, dissolved air;
- gas phase (g) – mixture of dry air and water vapour.

It also comprises four species:

- solid matrix (m) – matrix embedding the crystals;
- salts – as crystal (c) and as dissolved salt (s) resulting from the dissolution of the crystals;
- water (w) – as liquid or evaporated in the gas phase;
- air (a) – dry air as gas or dissolved in the liquid phase.

The total volume of the medium (V_t) can be decomposed into the volume occupied by the crystals (V_c), the pores (voids, V_v) and the bitumen (V_b), that is, $V_t = V_c + V_v + V_b$. The volume of the pores varies as a consequence of dissolution of the crystals and may be occupied by water or air.

These volumes permit the volume fraction of the different components (ϕ_c and ϕ_b) and the volume fraction of pores, that is, the porosity (ϕ_f), to be defined. These variables are defined as:

$$\phi_c = \frac{V_c}{V_t}; \quad \phi_f = \frac{V_v}{V_t}; \quad \phi_b = \frac{V_b}{V_t};$$
$$V_t = V_c + V_v + V_b; \; 1 = \phi_c + \phi_f + \phi_b = \phi_c + \phi_m$$
$$\phi_m = \phi_f + \phi_b \tag{1}$$

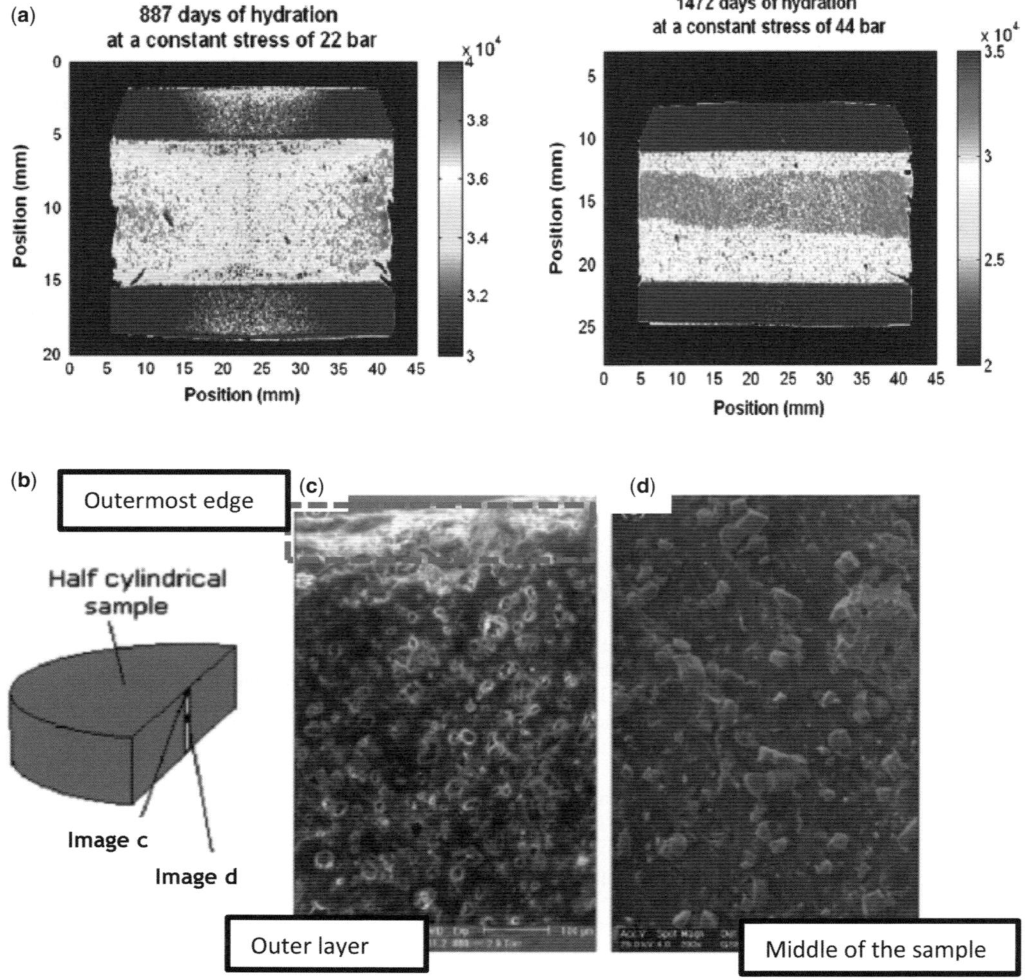

Fig. 1. (**a**) Micro-focus X-ray computer tomography (μCT) images of two Eurobitum samples: visualization of the hydration front in Eurobitum samples with 28.5 wt% $NaNO_3$ after 887 and 1472 days (Valcke *et al.* 2010). (**b**) ESEM photos of the leached layer of an Eurobitum sample (sample S0) after 887 days of contact with 0.1 M KOH in constant stress conditions (Valcke *et al.* 2010).

The sum of the porosity (ϕ_f) plus the volume fraction of the bitumen (ϕ_b) is defined as the volume fraction of the porous bitumen matrix ϕ_m, that is, $\phi_m = \phi + \phi_b$. In this way, $1 - \phi_m = \phi_c$ is the volumetric fraction occupied by the crystals.

An initial porosity ϕ_{f_0} is assumed, corresponding to the volume initially occupied by the gas and water in contact with salt. As a consequence of the dissolution of the crystals, the pore volume available for the fluids (ϕ_f) will increase, resulting in a deformation of the sample.

To establish the balance equations, the compositional approach is adopted, consisting of balancing the species rather than the phases. The mass balance equations for water, dissolved salts, crystals and solid phase, including the coupled flows, namely osmotic flow and ultrafiltration, and the dissolution/precipitation of salts, are described in detail in Mokni *et al.* (2011). The formulation outlined below has been discretized in space and time in order to be used for numerical analysis and has been included in the computer code CODE_BRIGHT (Olivella *et al.* 1996).

Constitutive equations

The following assumptions are made in the CH formulation:

- The medium constitutes a single soluble salt ($NaNO_3$) plus an inert solid phase.
- Multiphase flow including phase change is considered (dissolution/precipitation of the salt crystals).
- Only chemical and hydraulic gradients are considered.
- The medium is assumed not to be saturated with water.
- Binary diffusion of salt and water at high concentration is assumed.
- Permeability, diffusion and osmotic efficiency coefficients are considered to be variable with porosity.

The constitutive equations that relate the mass-average flow of fluid and the mass flow of solute to chemical and pressure gradients are as follows (Mokni *et al.* 2011):

(1) Flux of liquid phase ($\mathbf{J}_l$) induced by both pressure and concentration gradients:

$$\mathbf{J}_l = -\frac{\mathbf{k}k_{rl}}{\mu}(\nabla p + \rho_l \mathbf{g} \nabla z) + \frac{\alpha_s}{\mu}\mathbf{k}\sigma\nabla(w_l^s);$$
$$\alpha_s = \frac{k_{rl}RT\rho_l}{M_s} \tag{2}$$

This is an advective flux in the sense that it transports both the solute and the water.

(2) Nonadvective flux of solute:

$$\mathbf{J}_l^s = -\sigma w_l^s \rho_l \mathbf{J}_l - D\rho_l(1-\sigma)(1+\gamma w_l^s)\nabla w_l^s;$$
$$\gamma = \frac{1}{\rho_l}\frac{\partial \rho_l}{\partial w_l^s} \tag{3}$$

(3) Nonadvective flux of water including ultrafiltration is written as:

$$\mathbf{J}_l^w = \frac{\beta}{\alpha}\sigma w_l^s \rho_l \mathbf{J}_l - D\rho_l(1-\sigma)(1-\gamma w_l^w)\nabla w_l^w$$
$$\alpha = 1 + \frac{1}{\rho_l}\frac{\partial \rho_l}{\partial w_l^s} w_l^s = 1 + \gamma w_l^s$$
$$\beta = 1 + \frac{1}{\rho_l}\frac{\partial \rho_l}{\partial w_l^w} w_l^w = 1 - \gamma w_l^w \tag{4}$$

where $\mathbf{k}$ (m^2) denotes the intrinsic permeability tensor, k_{rl} the relative permeability of the liquid, $\mathbf{g}$ the gravity vector, μ (MPa s) the dynamic viscosity, p (MPa) the hydraulic pressure, w_l^s the mass fraction of solute in the liquid phase, w_l^w the mass fraction of water in the liquid phase, R (J (mol K)$^{-1}$) the gas constant, T (K), the temperature, ρ_l (kg m^{-3}) the liquid density, M_s (kg mol^{-1}) the solute molar mass, σ the osmotic efficiency coefficient, and D ($m^2 s^{-1}$) the diffusion coefficient.

The total fluxes of water and solutes in the liquid phase are expressed as:

$$\mathbf{j}_l^w = -w_l^w \rho_l \frac{\mathbf{k}}{\mu}\nabla p + D(1-\sigma)\rho_l(1-\gamma w_l^w)\nabla w_l^s + \frac{\mathbf{k}\sigma}{\mu}(w_l^w \rho_l)\nabla\pi + \frac{\mathbf{k}\sigma^2}{\mu}\left(\frac{\beta}{\alpha}w_l^w \rho_l\right)\nabla\pi - \frac{\mathbf{k}\sigma}{\mu}\left(\frac{\beta}{\alpha}w_l^w \rho_l\right)\nabla p + w_l^w \rho_l \phi_f \frac{d\mathbf{u}}{dt} \tag{5a}$$

$$\mathbf{j}_l^s = -w_l^s \rho_l \frac{\mathbf{k}}{\mu}\nabla p - D(1-\sigma)\rho_l(1+\gamma w_l^s)\nabla w_l^s + \frac{\mathbf{k}\sigma}{\mu}(w_l^s \rho_l)\nabla p - \frac{\mathbf{k}\sigma^2}{\mu}(w_l^s \rho_l)\nabla\pi + \frac{\mathbf{k}\sigma}{\mu}(w_l^s \rho_l)\nabla\pi + w_l^s \rho_l \phi_f \frac{d\mathbf{u}}{dt} \tag{5b}$$

The first and second terms in equations (5a) and (5b) represent Darcy's and Fick's laws, respectively. The solute diffusion law is dependent upon the osmotic efficiency coefficient (σ). This is an additional coupling effect included in order to model diffusion for semipermeable membrane materials. For the total flux of water (equation 5a) the third term represents chemical osmosis. The fourth and the fifth terms are coupled fluxes representing respectively the flow of water induced by an osmotic ($\nabla\pi$) and a hydraulic (∇p) pressure gradient. For the total flux of the solute, the coupled fluxes terms include ultrafiltration (third term). The solute flow is also driven by an osmotic pressure gradient (fourth and fifth terms in equation 5b). The sixth terms in both equations describe the advective flux owing to solid motion.

Constitutive relationships

Important parameters that affect the water uptake and salt release rates and hydration front are the bitumen membrane efficiency, the permeability of the BW, and the diffusion coefficient of dissolved salt in the BW. These parameters depend on the porosity of the BW, which varies in time because of the water uptake and subsequent processes, that is, dissolution of salt crystals, swelling of hydrating layers and compression of highly leached layers.

The diffusivity of solutes and water ($m^2 s^{-1}$). The diffusion coefficient controls the diffusion of dissolved salt and water. The parameter D is considered as an effective diffusion coefficient equal to $D = \tau\, \phi_f\, D_0$ where τ is the tortuosity and D_0 is a Fickian diffusion coefficient for a solute in free water. The effective diffusivity for semipermeable membranes is defined as $D^* = (1 - \sigma)D$. Therefore, for media presenting ideal membrane properties, the diffusion of the solute and water is completely hindered.

The osmotic efficiency σ. The osmotic efficiency coefficient controls the flow of water and solutes owing to osmotic effects. This parameter describes the nonideality of membrane behaviour. It ranges from 1 for an ideal membrane to 0 for porous media having no membrane properties. For clays, it has been shown that the efficiency of the clay membrane depends on several factors such as cation exchange capacity, solute concentration or porosity. For a composite porous medium such as BW, composed of a porous bituminized matrix embedding crystals, Mokni *et al.* (2011) proposed a simple relationship that describes the decrease of σ when porosity increases:

$$\sigma = \sigma_0\left(\frac{\phi_0}{\phi_f}\right); \quad \phi_0 \leq \phi_f \tag{6a}$$

This equation shows a rapid decrease in osmotic efficiency with porosity increase (Fig. 2a).

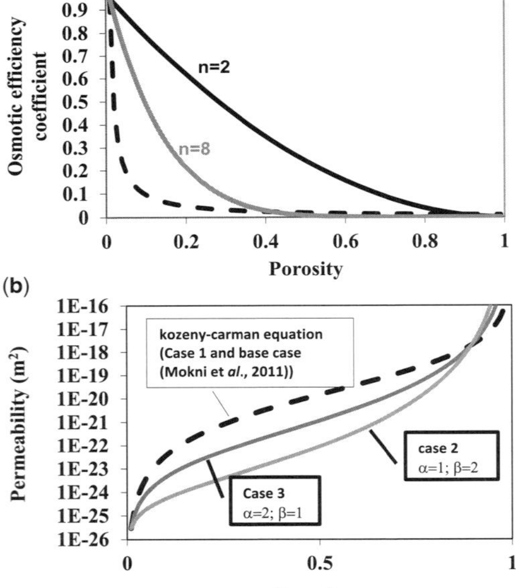

Fig. 2. Evolution of (**a**) osmotic efficiency coefficient for different values of *n* in equation (6b) and for previous function used in Mokni *et al.* (2011) and (**b**) intrinsic permeability used in the different analysed cases and Kozeny–Carman function used in Mokni *et al.* (2011).

In the present paper, the osmotic efficiency coefficient is assumed to be dependent on porosity and on crystal content. The latter dependency has been reported by Valcke *et al.* (2010), who mentioned the existence of a threshold salt content at which BW loses its membrane properties. Indeed, for high crystal contents, the bitumen membrane around the salt crystals becomes too thin to induce any osmotically driven swelling.

The osmotic efficiency coefficient σ is assumed to be a nonlinear function of the porosity (ϕ_f) and of the crystal volume fraction (ϕ_c):

$$\sigma = \sigma_0\left(\frac{1-\phi_f}{1-\phi_0}\right)^n(1-\phi_c-\phi_f); \quad \phi_0 \leq \phi_f \tag{6b}$$

where σ_0 is the reference osmotic efficiency for the reference porosity ϕ_0, and n is a material parameter that controls the rate of decrease of σ with increase of ϕ_f and ϕ_c. Figure 2a shows the evolution of the osmotic efficiency coefficient as a function of porosity for several values of n. The choice of this expression for the membrane efficiency is based on experimental observations. Indeed, there is experimental evidence that, even when the layers are partially leached, the membrane efficiency is still very high (Valcke *et al.* 2010).

The intrinsic permeability k. The intrinsic permeability controls the flow of water owing to hydraulic and osmotic pressure gradients. It also controls the flow of solutes by advection. Mokni *et al.* (2011) used a Kozeny–Carman-type equation (Fig. 2b) to describe the evolution of permeability with porosity expressed as:

$$k = k_0\frac{\phi_f^3}{(1-\phi_f)^2}\frac{(1-\phi_0)^2}{\phi_0^3} \tag{7a}$$

This equation is useful for soil materials. Nevertheless, for a composite porous media such as BW, experiencing a structure evolution induced by dissolution and swelling, an expression that is specifically applicable to that particular material is needed. Mokni *et al.* (2011) performed a sensitivity analysis on the effect of the initial values of intrinsic permeability, osmotic efficiency, and diffusion coefficients on the magnitude and duration of swelling deformation of BW. It was demonstrated that these three parameters, dependent on porosity, evolve in a coupled way. Indeed, a material having a high value of permeability and diffusion has a low value of the efficiency coefficient (less efficient semi permeable membrane).

In this paper, in order to investigate the effect of the rate of change of permeability with porosity, a generalized form of equation (7a) is proposed for BW.

$$k = k_0 \frac{\phi_f^{\alpha}}{(1-\phi_f)^{2\beta+2}} \frac{(1-\phi_0)^{2\beta+2}}{\phi_0^{\alpha}} \quad (7b)$$

where k_0 is the reference permeability at the reference porosity ϕ_0, and α and β are material parameters between 1 and 2 that control the rate of increase of k with ϕ_f. Figure 2b shows the evolution of the intrinsic permeability as a function of porosity for several values of α and β.

Sensitivity analysis: effect of the rate of variation of σ and k with porosity on swelling, leached mass of $NaNO_3$ and on leached thickness of BW

In this section, results from the modelling work of the water uptake tests under constant stress conditions described in Valcke *et al.* (2010) using the proposed expressions of σ and k (equations 6b & 7b) are presented. Several cases are analysed considering different values of n, α and β, parameters controlling the rate of change of osmotic efficiency and intrinsic permeability with porosity, swelling, leached mass of $NaNO_3$ and leached thickness.

For the experimental aspects of these water uptake tests, the reader is referred to Valcke *et al.* (2010). Specifically for this paper, it is important to mention that, prior to contact with water, the samples in the water uptake cells were submitted to repeated compression–decompression steps (‘compressibility test’) to determine the mechanical properties of BW. Mokni *et al.* (2008, 2010*a*, *b*) proposed an elasto-viscoplastic constitutive model to simulate these compression tests in oedometric conditions. The CH formulation, combined with the elasto-viscoplastic mechanical constitutive model (Mokni *et al.* 2010*b*) was used to simulate the water uptake tests. The result of the different analyses performed will be compared with the result of the Base case described in Mokni *et al.* (2011).

Model domain, boundary conditions and (initial) parameter values

The model domain consists of two regions (Fig. 3). One region corresponds to a reservoir representing a filter that is saturated with a low $NaNO_3$ concentration solution. The initial $NaNO_3$ mass fraction for the reservoir is $w = 0.001$ kg $NaNO_3$ kg^{-1} solution (*c.* 0.01 M). Within the reservoir, the permeability is set to high values ($k = 10^{-16}$ m^2). The other region corresponds to the BW containing $NaNO_3$ crystals occupying 16% of the total volume of the BW (corresponding to 28 wt%).

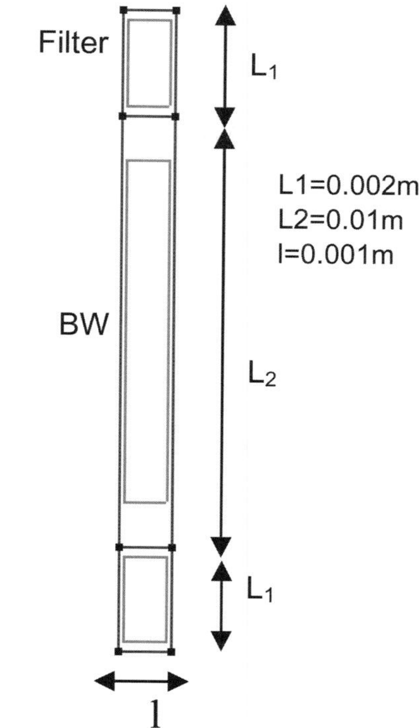

Fig. 3. Schematic representation of the model domain.

Pure water inflow, brine outflow and salt diffusion through the reservoir are allowed at the upper and lower sides of the sample. The medium is considered to be saturated ($S_r = 1$) with $NaNO_3$ brine (initial mass fraction $w = 0.47$ kg $NaNO_3$ kg^{-1} solution (*c.* 7.46 M)).

The different cases were simulated under a constant vertical stress. Along the vertical boundaries of the domain, horizontal displacements are restricted to represent the oedometric conditions. Material properties are summarized in Table 1. The initial values for permeability, osmotic efficiency and diffusion coefficient are the same reference values used by Mokni *et al.* (2011) and Mariën *et al.* (2013). The mechanical parameters were derived from the calibration of oedometer compressibility tests that are presented in Mokni *et al.* (2010*b*).

Effect of α and β on swelling and leached mass of $NaNO_3$

Table 2 summarizes the different cases analysed. The transport functions (permeability and osmotic efficiency) used for the different cases are displayed in Figure 2b.

Table 1. *Material properties*

Parameter (initial values)	Symbol	Value	
Crystal volume fraction	ϕ_{c_0}		0.16
Porosity	ϕ_{f_0}		0.01
Intrinsic permeability	k_0 (m^2)	BW	3×10^{-26}
	k (m^2)	Filter	10^{-16}
Efficiency coefficient	σ_0		0.95
Diffusion coefficient	D^* (m^2 s^{-1})		1.6×10^{-16}
Dissolution rate constant	κ (kg s^{-1} m^{-3})		10^{-5}
Solute mass fraction	w_l^s (kg $NaNO_3$ kg^{-1} solution)	BW	0.47
		Filter	0.001

Initial values of transport parameters used for the different cases.
BW, Eurobitum bituminized radioactive waste.

Table 2. *Transport functions used for the different cases*

Analysed cases	Constitutive relationships
Case 1*	$k = k_0 \frac{\phi_f^3}{(1-\phi_f)^2} \frac{(1-\phi_0)^2}{\phi_0^3}$; $\sigma = \sigma_0 \left(\frac{1-\phi_f}{1-\phi_0}\right)^n$; $n = 2,3,5$
Case 2	$k = k_0 \frac{\phi_f}{(1-\phi_f)^6} \frac{(1-\phi_0)^6}{\phi_0}$; $\sigma = \sigma_0 \left(\frac{1-\phi_f}{1-\phi_0}\right)^n$; $n = 2,3,8$
Case 3	$k = k_0 \frac{\phi_f^2}{(1-\phi_f)^4} \frac{(1-\phi_0)^4}{\phi_0^2}$; $\sigma = \sigma_0 \left(\frac{1-\phi_f}{1-\phi_0}\right)^2$; $n = 2$

*Kozeny–Carman function used in Mokni *et al.* (2011).

Case 1. In this case the Kozeny–Carman type permeability function (equation 7a) used by Mokni *et al.* (2011) is maintained, while for the osmotic efficiency coefficient the new expression (equation 6b) that describes the dependency of σ on ϕ_f and ϕ_c is used. Several values of n are considered. The evolution of volumetric deformation (i.e. swelling) is displayed in Figure 4. The plot shows

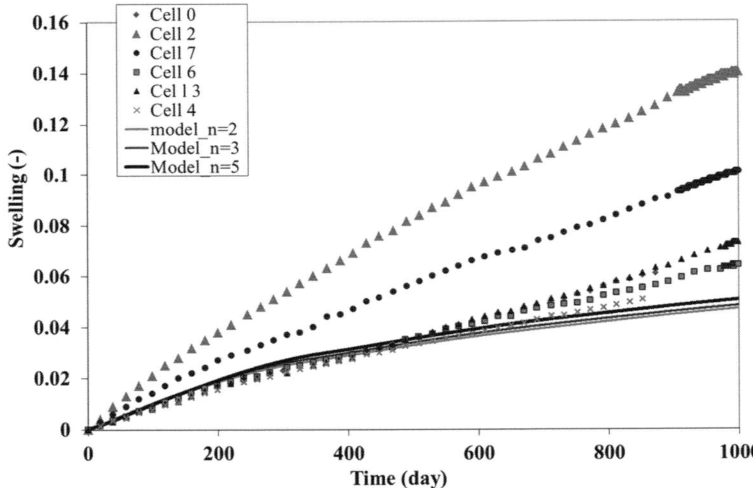

Fig. 4. Case 1: swelling (or volumetric deformation) for different values of n (equation 6b).

comparisons between the experimental data and the numerical simulations (at constant vertical applied stress, 2.2 MPa). It is observed that the model predicts lower swelling rates in comparison to the experimental results. Indeed, in the CH formulation, water uptake, swelling and leaching are considered to be the result of several transient fully coupled processes (diffusion, advection, dissolution, involving a low permeability material with evolving thickness and properties) that are strongly nonlinear. The overall time-dependence of the swelling and leaching rates and rate of progression of the hydration front depend on the relative contribution of all these processes and on their time constants. In fact, the variation of the intrinsic permeability and osmotic coefficient plays an important role in the coupled migration of water and solute. While the direct advective fluxes are proportional to k, the coupled fluxes are proportional to the product $k\sigma$ (equation 5). It is convenient to define at this point the following transport functions: $k\pi/\mu$, $k\pi\sigma/\mu$ and $k\pi\sigma^2/\mu$. These functions will be compared with the diffusion coefficient D^* to identify the major processes that control the transport of water and solute. Figure 5 shows the evolution with porosity of the above-defined functions together with the diffusion coefficient D^* for case 1 (Fig. 5a) and for the base case presented by Mokni *et al.* (2011) (Fig. 5b). The plots indicate that, for case 1, the transport functions controlling the swelling and leaching rates are significantly different from the base case. Indeed, for case 1 the transport functions are similar and evolve with porosity at similar rates. In addition, higher values and

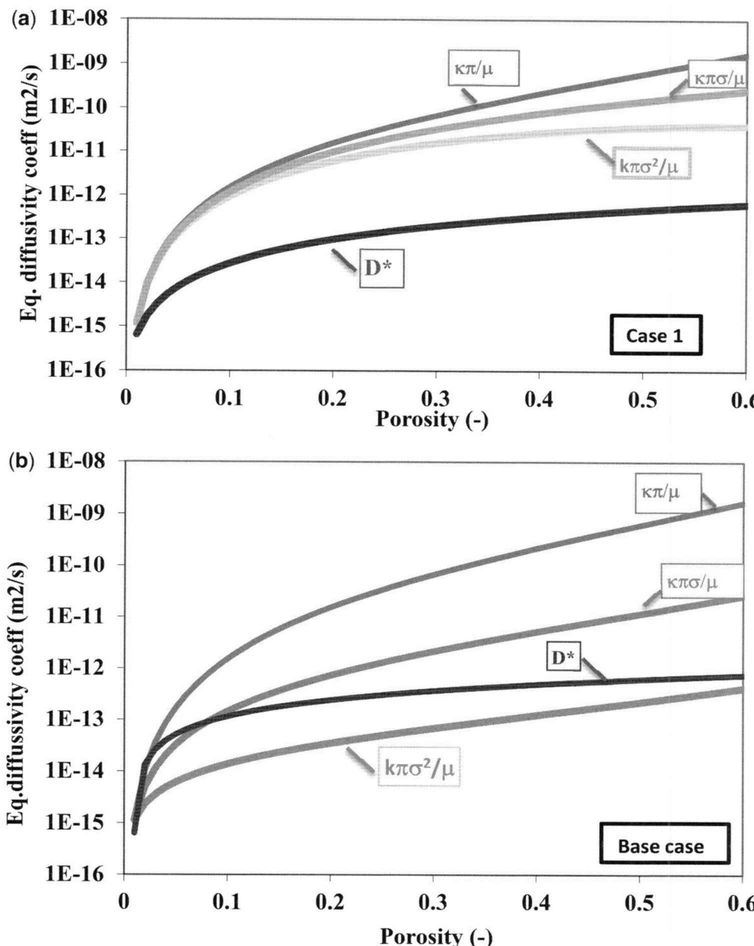

Fig. 5. Variation with porosity of the transport functions $k\pi/\mu$, $k\pi\sigma/\mu$, $k\pi\sigma^2/\mu$ and D^* for (**a**) case 1 and (**b**) base case (Mokni *et al.* 2011).

evolution rate are observed for the transport function $k\pi\sigma^2/\mu$ in comparison to the base case. As a result there is rapid leaching of the outermost layers of the BW sample and a rapid decrease in $NaNO_3$ concentration and therefore in osmotic pressure in these layers owing to the higher water and solute flow rates. The transport of $NaNO_3$ solutes out of the BW, combined with the external stress and the osmosis-driven transport of water from the outermost pores with a lower $NaNO_3$ concentration (higher water activity) towards deeper pores with a higher $NaNO_3$ concentration (lower water activity), contributes to a local consolidation of the hydrated layers. Swelling of the deeper layer is then counteracted by compression of the leached layers leading to a fast levelling off of the leaching and swelling rates.

Case 2. In this case the new permeability (equation 7b) and osmotic efficiency (equation 6b) functions are used (Table 2). The used permeability function predicts a slower increase of the permeability with increasing porosity compared with the base case (Fig. 2b). Several values of the parameter n, which controls the rate of decrease of the osmotic efficiency coefficient with porosity, are considered. The same initial values of the transport function as for the previous case have been used (Table 1). The evolution of volumetric deformation is displayed in Figure 6. Experimental and numerical results concerning the sodium nitrate release are shown in Figure 7. In this case the model reproduces well the swelling and leaching rates. Several observations can be made. First, it is seen that, after more than 4 years of hydration, the samples

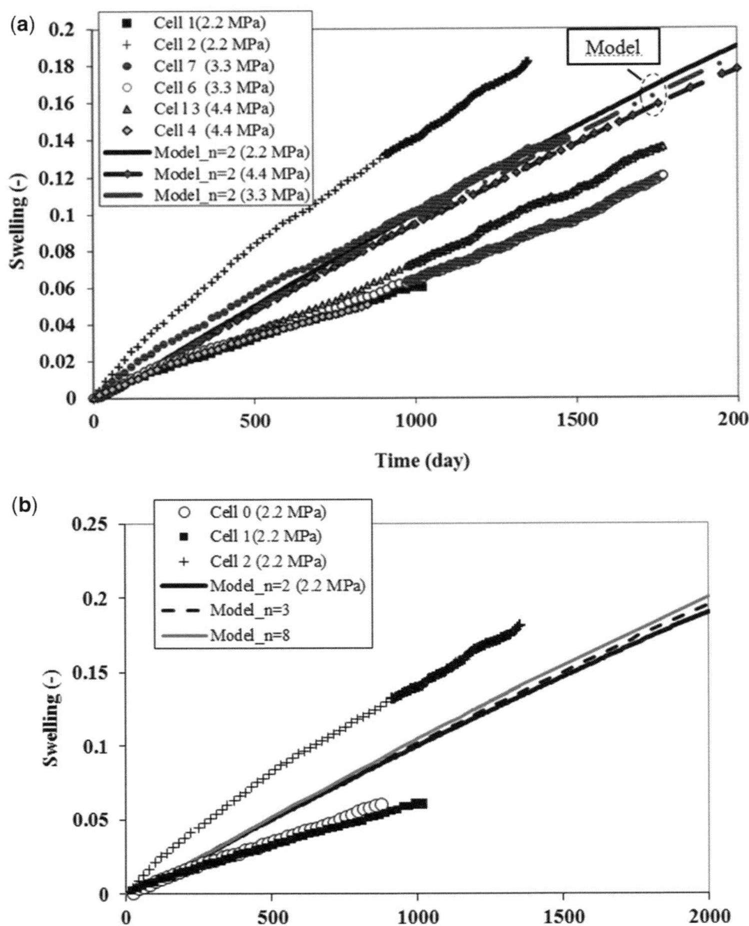

Fig. 6. Case 2: swelling (or volumetric deformation) for (**a**) different applied stresses and (**b**) different values of n. Experimental results and model predictions.

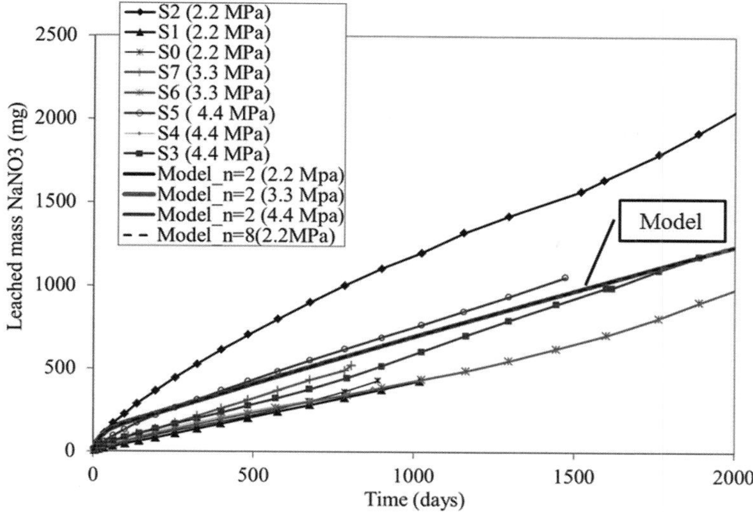

Fig. 7. Case 2: leached mass of $NaNO_3$. Experimental results and model predictions.

continued to swell with no tendency to level off. Both modelling and experiments were performed under various vertical applied stresses (2.2, 3.3, and 4.4 MPa). For this range of stresses, the effect on swelling and leaching rates is moderate according to the model. The effect of stress level on swelling was discussed by Mokni *et al.* (2011) in regard to the maximum swelling pressure achievable taking into account the high osmotic pressure associated with saturated sodium nitrate pore solution. Second, no significant effect of the variation of the value of n is observed on the swelling and leaching rates (Figs 6b & 7). Figure 8 shows the evolution with porosity of the above-defined transport functions together with the diffusion coefficient D^* for several values of the parameter n. The plot shows that, in both cases (i.e. cases of $n = 8$ and $n = 2$) for low porosity, diffusion is the dominant process for the transport of water and solute. As porosity increases, permeability increases and the direct and coupled advective fluxes become the dominant processes. For both cases, the transport parameters controlling the swelling and leaching rates are significantly different at high porosity values, while more similar values are observed for low porosities. The fact that no effect of n on the swelling

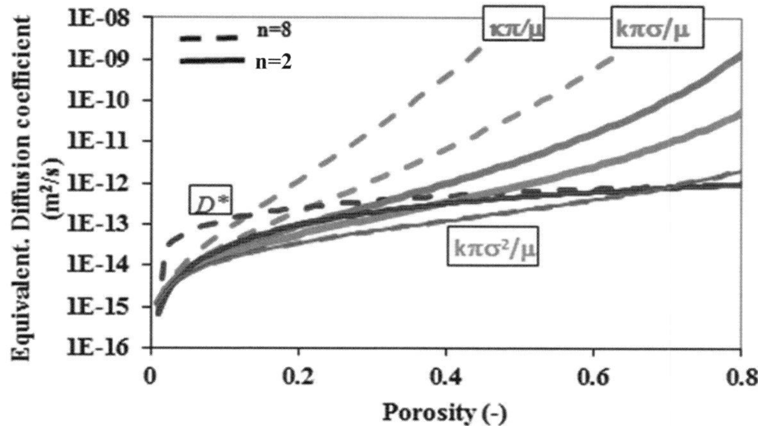

Fig. 8. Case 2: variation with porosity of the transport functions $k\pi/\mu$, $k\pi\sigma/\mu$ and $k\pi\sigma^2/\mu$ and D^*, which control the magnitude of the direct and coupled fluxes of water and solute (for a given osmotic and hydraulic pressure gradients; $\pi = 42$ MPa and $\mu = 10^{-9}$ MPa s).

and leaching rates is observed demonstrates that both are controlled by the fluxes in the outermost low porosity layer. Compression of the outermost layers is then a limiting or controlling process for the water uptake and the release of $NaNO_3$.

Case 3. In this case, the rate of increase of permeability with porosity is higher than in case 2 but lower than in case 1 (Fig. 2a; Table 2). The same material parameters as for the previous cases have been used (Table 1). Modelling results of swelling and $NaNO_3$ leaching of the samples in the constant stress test ($\sigma_v = 2.2$ MPa) leads to lower swelling and leaching rates in comparison to the experimental results (Fig. 9). In fact, in this case the outermost layers are predicted to leach quite rapidly owing to the higher water and solute flow rates inducing a fast decrease in $NaNO_3$ concentration and in osmotic pressure in these layers. The osmotic pressure would then rapidly approach the value of the externally applied stress, which would lead to compression of the hydrated layers.

Effect of α *and* β *on leached thickness of BW and hydration front*

The progression of the water front is the result of the relative contribution of several transient processes (diffusion, advection and dissolution) involving a low porosity and low permeability material with evolving thickness and properties and that are strongly coupled. Valcke *et al.* (2010) and Mariën *et al.* (2013) observed that almost no pores can be seen on ESEM images of the dry BW. Indeed,

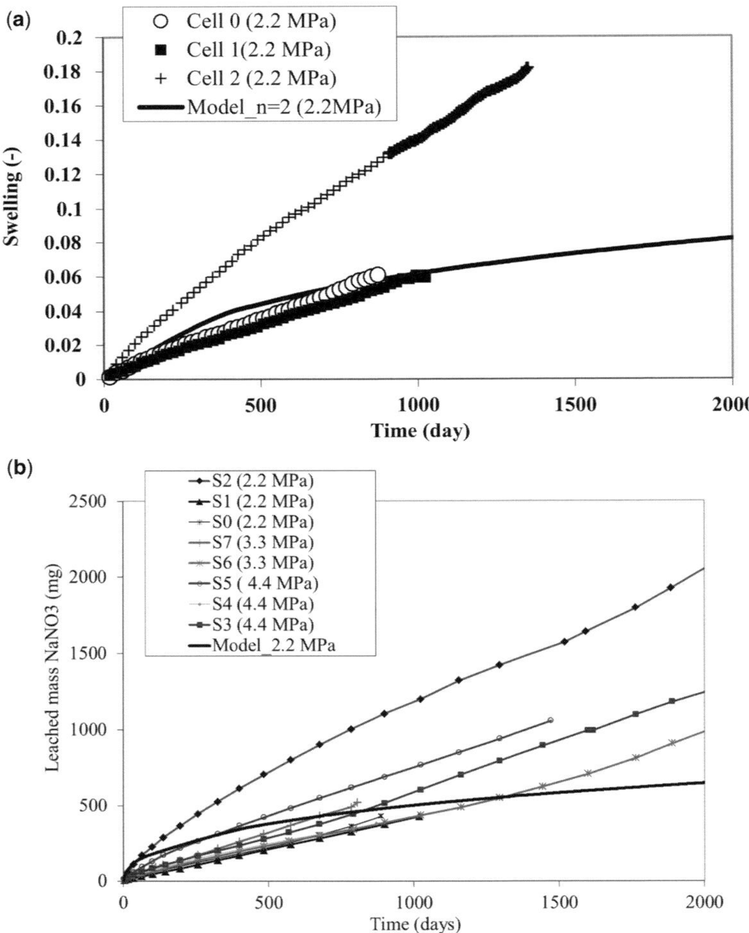

Fig. 9. Case 3. (**a**) Volumetric deformation v. time. (**b**) Leached mass of $NaNO_3$. Experimental results and model predictions.

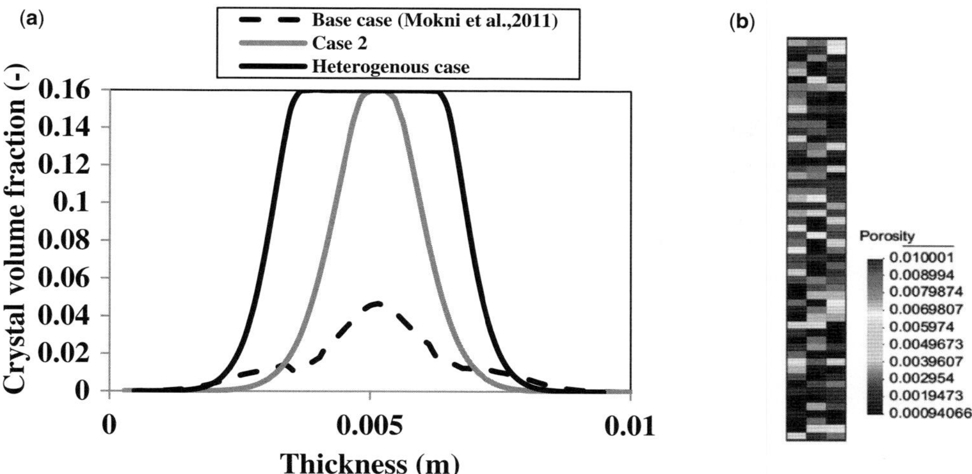

Fig. 10. (**a**) Profiles of crystal volume fraction after 1500 days of hydration for case 2, heterogeneous case and base case (Mokni *et al.* 2011). (**b**) Heterogeneous distribution of porosity (each finite element is characterized by a single value of porosity).

in contact with water the $NaNO_3$ salt crystals at the outer boundary of the material will dissolve rapidly. This results in the formation of pores filled with a highly saturated $NaNO_3$ solution. As a consequence permeability and diffusivity increase but most of the water that is taken up further by the BW stays in the first layer of pores to dissolve the remaining crystals and in a next strep to dilute the saturated $NaNO_3$ solution. Profiles of crystal volume fraction (Fig. 10a) allow the thickness of the leached zones to be calculated. Table 3 summarizes the calculated and measured leached thicknesses (obtained by μCT and ESEM analyses; Valcke *et al.* 2010) at different times for the different analysed cases. For case 2, the model predicts a slower progress of the water front in comparison to the base case presented in Mokni *et al.* (2011). However, even for this slower progress of the water front, there is still a difference between the modelled results and the μCT analyses, which show that after about *c.* 1500 days of hydration the water front had not yet reached the centre of the sample (Table 3). For case 3, the model predicts a slower dissolution of the crystals in the centre of the sample and a slower progression of the hydration front compared with the base case and case 2 (Table 3). Indeed, except for the outermost layers, the largest part of the sample is not hydrated after 2000 days. This is a direct consequence of the size reduction of the outer pores, which induces a decrease in the permeability and the diffusion coefficient, and an increase of the osmotic efficiency in these compressed layers, while deeper in the sample these parameters are varying in the opposite way. This maintains a very slow inflow of water towards the sample and outflow of the solute over a long time.

Analysis of the effect of heterogeneous distribution of porosity on leached thickness of BW and hydration front

Visual examination of the μCT images after different hydration periods shows irregular shape of

Table 3. *Calculated and measured leached thickness (note that these data relate to the total thickness, that is, the sum of the thicknesses of the two leached layers per sample)*

Duration (days)	Leached thickness (mm) μCT	Leached thickness calculated (mm) (case 2)	Leached thickness calculated (mm) (case 3)	Leached thickness calculated (mm) (heterogenous case)	Leached thickness calculated (base case) (mm) (Mokni *et al.* 2011)
c. 1500	6.29	9	6.4	*c.* 6.62	*

*Hydration front has already reached the centre of the sample.

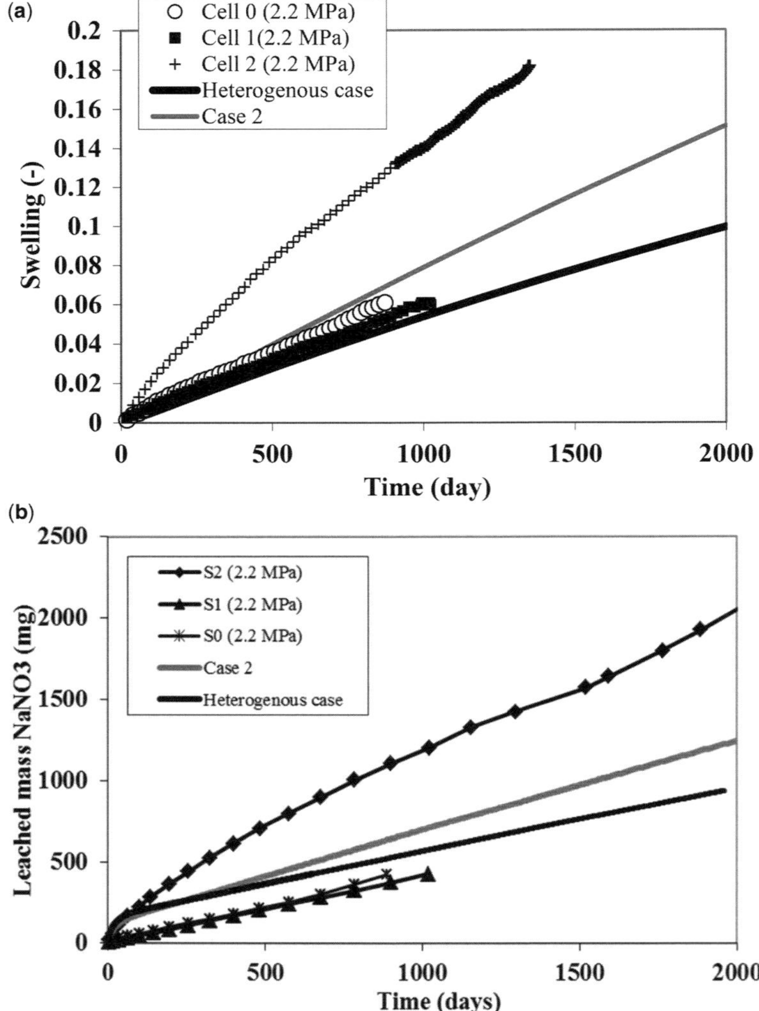

Fig. 11. Heterogeneous case. (**a**) Volumetric deformation v. time. (**b**) Leached mass of $NaNO_3$. Experimental results and model predictions.

the hydration front (Fig. 1a), which can be attributed to pre-existing heterogeneities inside the BW leading to local variation of permeability, diffusivity and osmotic efficiency of the material. This heterogeneity may originate from the distribution of crystals, which are often grouped together in larger crystal clusters, generating zones with variable salt content and porosity (Fig. 1b).

In an additional effort to more realistically simulate the progression of the wetting front, some heterogeneity of the transport properties has been considered. The same transport function as in case 2 are used (Table 2, $n = 2$) since they lead to good estimates of swelling and leaching rates. Figure 10b shows the initial porosity field considered. The mean porosity is $\phi_{\text{mean}} = 0.01$ (same initial porosity as the homogenous case: case 2) and the variance is in the order of 10^{-5}. This random field has no spatial correlation. The range of initial porosity considered is justified by the existence of only a few isolated pores with different sizes observed in ESEM images of dry BW (Valcke *et al.* 2010; Mariën *et al.* 2013). Note that, owing to the small variations in porosity considered, initial values of permeability, osmotic efficiency and diffusivity are different for each

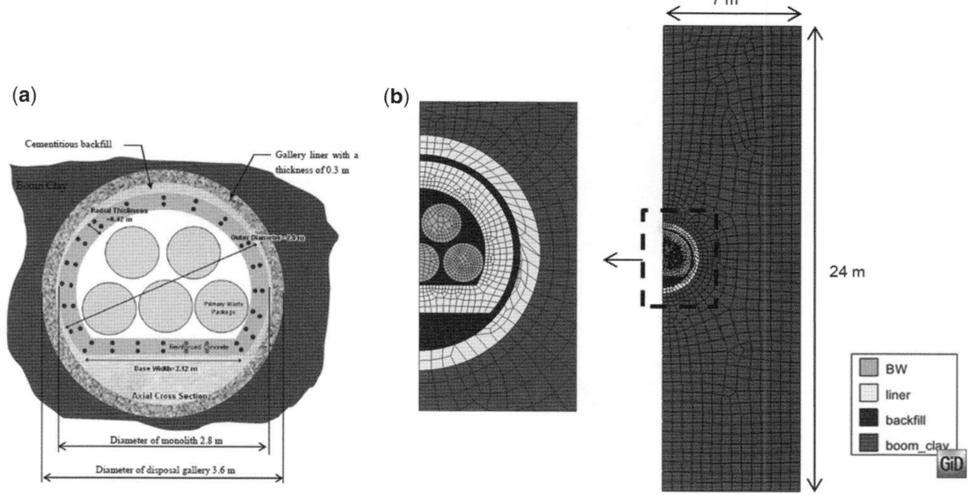

Fig. 12. (**a**) Axial cross section of the monolith for Eurobitum (Wacquier & Van Humbeeck, 2009). (**b**) 2D Finite element model.

finite element. The profile of crystal volume fraction drawn after 1500 days for the heterogeneous case is shown in Figure 10a. When comparing the thickness of the hydrated layer with the experimental results, the heterogeneous case shows a comparable result (Table 3). Figure 11 shows the evolution of swelling and leached mass of $NaNO_3$ with time. Although the model predicts a slower swelling and leaching rates as compared with case 2, induced by the marked heterogeneity in the initial values of the transport properties, the results are still in the range of the experimental measurements.

CH modelling of the *in-situ* disposal of BW

According to the present disposal design of ONDRAF/NIRAS, ten 220 l drums, each filled with on the average 170 l of Eurobitum, would be grouped in two arrays of five drums in cement-based secondary containers, which are to be placed in concrete lined disposal galleries that are excavated at mid depth in the host formation (Fig. 12a; Wacquier & Van Humbeeck 2009). The void space between the primary waste packages and the secondary concrete container, and between the secondary concrete container and the gallery liner, will

Table 4. *Initial values of transport functions for the different materials considered in the in-situ case*

Parameters (initial values)		Symbols	Values
BW	Crystal volume fraction	ϕ_{c0}	0.16
	Porosity	ϕf_0	0.01
	Intrinsic permeability	k_0 (m^2)	3×10^{-26}
		α	1
		β	2
	Efficiency coefficient	σ_0	0.95
		n	2
	Diffusion coefficient	D (m^2 s^{-1})	1.6×10^{-16}
	Dissolution rate constant	κ (kg s^{-1} m^{-3})	10^{-5}
	Solute mass fraction	w_l^s	0.47
Boom Clay	Intrinsic permeability	K (m^2)	10^{-19}
	Solute mass fraction	w_l^s	0.001
Concrete container+liner	Intrinsic permeability	K (m^2)	10^{-18}
	Solute mass fraction	w_l^s	0.001
Backfill	Intrinsic permeability	K (m^2)	10^{-16}
	Solute mass fraction	w_l^s	0.001

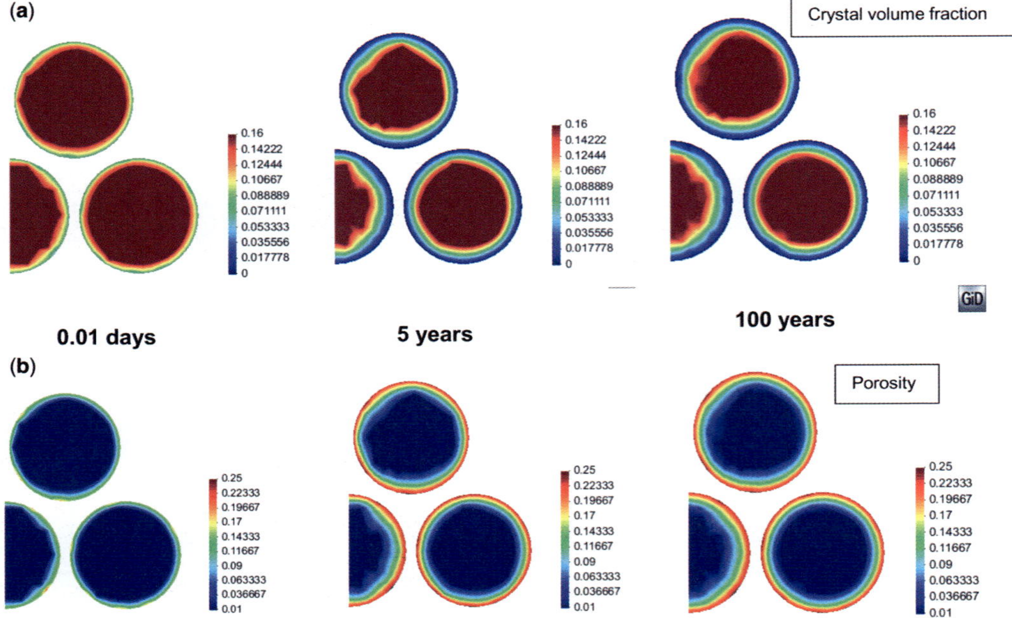

Fig. 13. Contours of crystal volume fraction (**a**) and porosity in BW drums (**b**) after 0.01 days, and 5 and 100 years

be backfilled with cement-based materials. It is important to recognize that several aspects of the Eurobitum disposal system are not yet finalized, in particular the composition of the concrete secondary container, the presence or the absence of reinforcement bars in the secondary concrete contained and the composition of the cement based backfill material. The design provided in Figure 12a is based on the latest information and assumptions.

In the frame of the study of the long-term behaviour of Eurobitum bituminized radioactive waste in final disposal conditions, a preliminary hydro-chemical analysis is presented here. The objective is to obtain first insights into the kinetics of water uptake, dissolution, solute leaching and maximum generated swelling and pressure under disposal conditions. A 2D finite element model is used considering a section of the gallery taken perpendicularly to the axial direction (Fig. 12b). The cementitious backfill, the concrete container, the concrete liner and the Boom Clay are considered to be saturated with water with low $NaNO_3$ concentration (initial mass fraction is $w = 0.001$ kg kg^{-1}). The BW filling the drums contains $NaNO_3$ crystals occupying 16% of the total volume of the BW (corresponding to 28% in weight) and is considered to

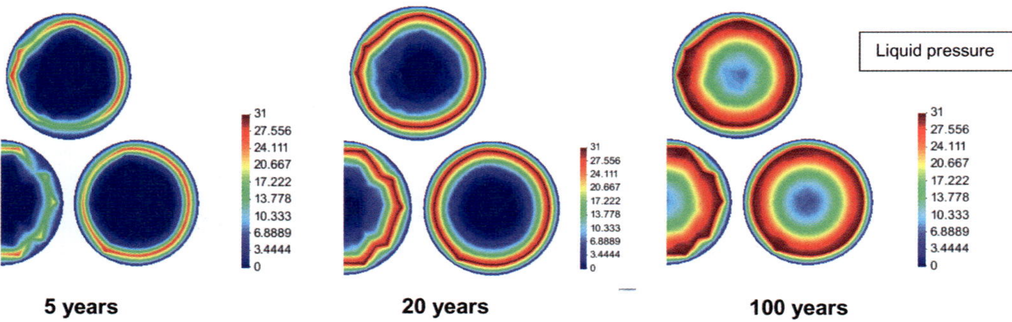

Fig. 14. Contour field of liquid pressure in BW drums after 5, 20 and 100 years.

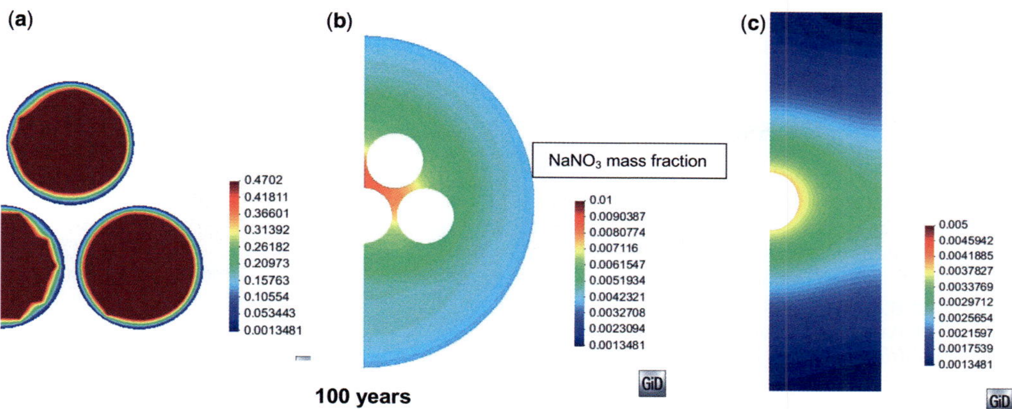

Fig. 15. Contour field of NaNO3 mass fraction in the liquid phase after 100 years in (**a**) BW, (**b**) backfill, concrete container and liner and (**c**) Boom Clay.

be saturated with $NaNO_3$ brine (initial mass fraction: $w = 0.47\ \mathrm{kg\ kg^{-1}}$). The same initial values for permeability, osmotic efficiency and diffusion coefficient for BW as for case 2 are used for this analysis. Detailed materials parameters are summarized in Table 4.

Figure 13 shows the contour field of crystal volume fraction after, 0.01 days, and 5 and 100 years together with contours of porosity. At early time, owing to high concentration gradient, water is attracted from the backfill to the BW. After 5 years, the crystal volume fraction (ϕ_c) decreases and tends to zero at the outer boundary of the drums, while no change in ϕ_c is observed in the middle of the drum after 100 years (Fig. 13a). This suggests that it will take several hundred years before the hydration front has reached the centre of the drums. As a consequence of crystal dissolution, porosity increases at the hydrated layers (Fig. 13b). No compression of the outer hydrated boundary is observed since no mechanical coupling is considered. Figure 14 shows contour field of liquid pressure after 5, 20 and 100 years. The figure shows that there is a pressure front that progresses slowly towards the centre of the drums. After 5 years the maximum value is attained at the outer boundary of the BW drums and is around 31 MPa. A decrease of the maximum value of liquid pressure is also observed after 100 year at the outer hydrated layers. In fact, the increase of porosity at this zone induces an increase of permeability and diffusion coefficient and a decrease of efficiency of the membrane. Drainage and solute diffusion out of the BW are favoured, resulting in a decrease in osmotic pressure and therefore in pore pressure. Figure 15a shows contours field of concentration after 100 years in BW. Except for the outer boundary where the crystals have dissolved, the concentration of the $NaNO_3$ solution inside the BW drums remains constant and equal to that of a saturated $NaNO_3$ solution after 100 years. In fact, the crystals provide a source term of solutes that prevent the decrease of the concentration. Contours field of concentration through the backfill material, liner, concrete container and Boom Clay after 100 years are shown in Figures 15b and c. Because the BW is not a perfect semipermeable membrane, the solute is allowed to diffuse out of BW and the $NaNO_3$ concentration increases in the surrounding materials.

Conclusion

In this paper, equations of the key transport variables in the CHM formulation proposed by Mokni *et al.* (2011) are improved in order to better reproduce some of the experimental results. In this context a new expression of the osmotic efficiency coefficient (σ), including the dependence of σ on porosity and on crystal volume fraction, has been proposed. A new expression describing BW permeability evolution with porosity has been presented as well. The modelling results illustrate that α and β, parameters controlling the rate of change of intrinsic permeability with porosity, have a significant influence on swelling and leaching rates and on progression of the hydration front. The proposed functions are interdependent and control the magnitude of the coupled fluxes, namely chemical osmosis and ultrafiltration. In addition, it was demonstrated that assuming a simple realization of random fields of porosity that is of transport properties (permeability, osmotic efficiency and diffusivity) improves the modelling results in terms of progression of the hydration front. Indeed, in this case, the calculated and the measured thicknesses of the leached layers, after 1500 days of hydration, were quite similar. This approach implies different magnitude of direct and coupled fluxes at each finite element in the domain. Accounting for spatial and temporal variations of the key transport parameters was essential to reproduce all the aspects that were observed in the experiments (swelling, leaching of sodium nitrate and thickness of the leached layer). Finally, the hydro-chemical analysis of the behaviour of Eurobitum drums under disposal conditions provided first insights into the kinetics of water uptake, dissolution, solute leaching and maximum generated pressure under disposal conditions.

This work was undertaken in close cooperation with, and with the financial support of ONDRAF/NIRAS, the Belgian Agency for the Management of Radioactive Waste and Enriched Fissile Materials, as part of its programme on geological disposal of medium-level long-lived waste.

References

Mariën, A., Mokni, N., Valcke, E., Olivella, S., Smets, S. & Li, X. 2013. Osmosis-induced water uptake by Eurobitum bituminized radioactive waste and pressure development in constant volume conditions. *Journal of Nuclear Materials*, **432**, 348–365.

Mokni, N. 2011. *Deformation and Flow Induced by Osmotic Processes in Porous Materials*. PhD thesis, Universitat Politècnica de Catalunya, España.

Mokni, N., Olivella, S., Li, X., Smets, S. & Valcke, E. 2008. Deformation induced by dissolution of salts in porous media. *Physics and Chemistry Earth, Parts A/B/C*, **33**(Supplement 1), S436–S443.

Mokni, N., Olivella, S. & Alonso, E. E. 2010*a*. Swelling in clayey soils induced by the presence of salt crystals. *Applied Clay Science*, **47**, 105–112.

Mokni, N., Olivella, S., Li, X., Smets, S., Valcke, E. & Mariën, A. 2010*b*. Deformation of bitumen based porous material: experimental and numerical analysis. *Journal of Nuclear Materials*, **404**, 144–153.

Mokni, N., Olivella, S., Valcke, E., Mariën, A., Smets, S. & Li, X. 2011. Deformation and flow driven by osmotic processes in porous materials: application to bituminized waste materials. *Transport in Porous Media*, **86**, 665–692.

Olivella, S., Gens, A., Carrera, J. & Alonso, E. E. 1996. Numerical formulation for a simulator (CODE_BRIGHT) for the coupled analysis of saline media. *Engineering with Computers*, **13**, 87–112.

ONDRAF/NIRAS 2009. *The Long-term Safety Assessment Methodology for the Geological Disposal of Radioactive waste*. ONDRAF/NIRAS report NIROND **TR-2009-14 E**, www.nirond.be

Valcke, E., Mariën, A., Smets, S., Li, X., Mokni, N. & Olivella, S. 2010. Osmosis-induced swelling of Eurobitum bituminized radioactive waste in constant stress conditions. *Journal of Nuclear Materials*, **406**, 304–316.

Wacquier, W. & Van Humbeeck, H. 2009. *B&C Concepts and Open Questions*. ONDRAF/NIRAS Technical note, 2009-0146.

Studies of construction methods for Bentonite engineered barrier systems for sub-surface disposal: vibratory compaction

ATSUO YAMADA[1*], YOSHIHIRO AKIYAMA[1], MASAKI NAKAJIMA[1], TSUTOMU YADA[1], MASAKAZU CHIJIMATSU[2] & TAKAHIRO NAKAJIMA[2]

[1]*Radioactive Waste Management Funding and Research Center, 1-15-7 Tsukishima, Chuo-ku, Tokyo 104-0052, Japan*

[2]*Hazama Corporation, 515-1 Karima, Tsukuba City, Ibaraki Prefecture 305-0522, Japan*

**Corresponding author (e-mail: atsuo.yamada@rwmc.or.jp)*

Abstract: Bentonite buffer is a part of the engineered barrier systems (EBS) for sub-surface disposal. The EBS is required to meet two quality standards at completion of construction. One is very low permeability of less than 5×10^{-13} m s^{-1} and the other is very low diffusibility of less than 1×10^{-12} m^2 s^{-1}. The bentonite buffer is required for the first quality standard of low permeability and must be constructed with high density in order to make the buffer a low-permeability layer. To confirm that the bentonite buffer is actually constructed with high density, we have carried out a full-scale mock-up test for part of the Demonstration Test of Underground Cavern-Type Disposal Facilities commissioned by the Japanese government in fiscal year 2005. This paper covers the test results from the construction of the bentonite buffer.

Classification of radioactive waste disposal methods in Japan

Radioactive waste in Japan can be roughly classified into two groups: high-level and low-level. There are four disposal concepts for radioactive waste, which depend on depth and differences in the barrier's ability to retard leakage of radioactive substances. These concepts are shown in Figure 1 (Government of Japan 2008*a*).

Sub-surface disposal system

The concept of a sub-surface disposal with engineered barrier systems (EBS) is illustrated in Figure 2 (Government of Japan 2008*b*). EBS components for sub-surface disposal consist of filler, a concrete pit, a low-diffusion layer, a buffer, backfill materials and a tunnel support. The design goals are:

(1) to maintain the waste vessel in stable condition for a long period of about 1000 years or more;
(2) to reduce groundwater inflow and limit release of radionuclides by advection;
(3) to form a diffusion regime in accordance with the low-permeability requirement and to limit radionuclide release by diffusion.

The main EBS components include:

(1) a bentonite buffer that reduces groundwater inflow through the facilities;
(2) a cementitious low-diffusion layer of cement material, which is inside the low-permeability layer and contains no groundwater that may seep from the inner diffusion area.

Objectives of this study

Successful completion of the first main EBS component, the bentonite buffer, requires the construction of a low-permeant bentonite buffer. Because previous studies have never done a full-scale mock-up test in actual conditions, such as a cavern, it is necessary to confirm that the bentonite buffer is actually constructed to its stated quality (having low permeability). In this paper, we present the results of a confirmation test of construction methods for a low-permeant bentonite buffer.

Assigned goals and solutions

Required performance and initial value

The required performance of the bentonite buffer as a part of an EBS is as follows:

(1) lower permeability than the bedrock surrounding the facility;
(2) provision of a diffusion field inside the facility's concrete pit.

It is expected that this required performance will deteriorate over time after completion of facility construction. In this paper, the 'initial value' of the

From: Norris, S., Bruno, J., Cathelineau, M., Delage, P., Fairhurst, C., Gaucher, E. C., Höhn, E. H., Kalinichev, A., Lalieux, P. & Sellin, P. (eds) 2014. *Clays in Natural and Engineered Barriers for Radioactive Waste Confinement*. Geological Society, London, Special Publications, **400**, 135–144.
First published online March 5, 2014, http://dx.doi.org/10.1144/SP400.2

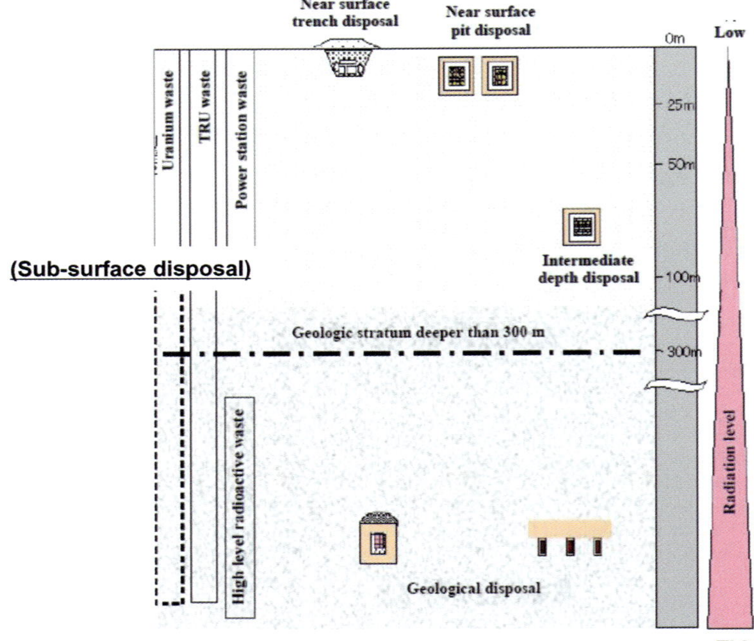

Fig. 1. Radioactive waste disposal methods in Japan.

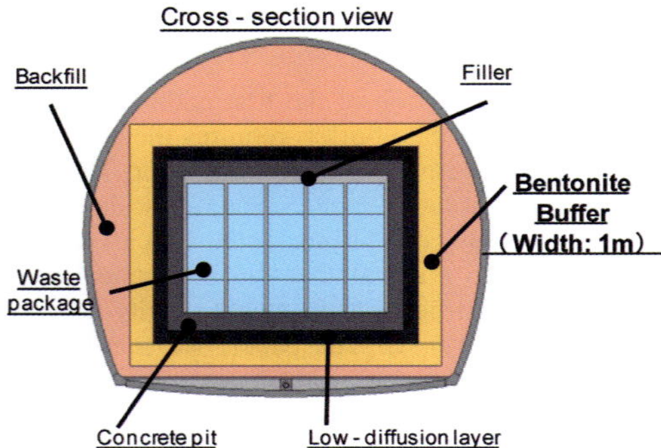

Fig. 2. Sub-surface disposal system concept.

bentonite buffer is defined as the performance when construction is completed, and is set as shown in Table 1. This initial value indicates very low permeability, and was set in consideration of the effects of long-term deterioration from predictable changes (such as chemical and mechanical changes) and the effects of uncertain phenomena.

Density of the bentonite buffer for low permeability

The required performance of the bentonite buffer is low permeability. Results of many previous studies about bentonite will be adopted as a part of the EBS in Japan.

Figure 3 shows the correlation between permeability and the dry density of granular bentonite mined in Tsukinuno, Yamagata Prefecture, Japan (Radioactive Waste Management Funding and Research Center 2006). According to this figure, permeability is well correlated with dry density, and thus values for dry density can be adopted as target values for construction management. As

Table 1. *Required performance and initial value of bentonite buffer*

Required performance	Status	Initial value
1. Lower permeability than the bedrock surrounding the facility	Permeability	5×10^{-13} m s^{-1}
2. To provide a diffusion field inside the facility's concrete pit	Thickness	1 m

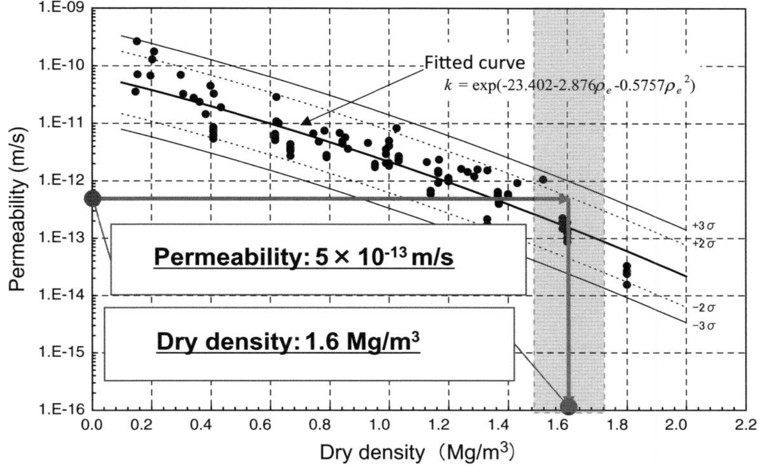

Fig. 3. Correlation between permeability and dry density.

shown in the previous section, the initial permeability of the bentonite buffer is set to 5×10^{-13} m s^{-1}, which corresponds to a dry density value of about 1.4 mg m^{-3}. In the present study, the dry density value used in construction management was set to a larger value than 1.4 mg m^{-3} owing to the effects of uncertain phenomena in construction.

We took into account previous studies on the effects of uncertain phenomena in construction and experience with construction of similar bentonite buffers to set the construction management values as follows:

(1) a median of 1.6 mg m^{-3}, equivalent to 1.2 times the dry density value (about 1.35 mg m^{-3}, which corresponds to the initial value for permeability (5×10^{-13} m s^{-1});

(2) a density range between 1.5 and 1.7 mg m^{-3}, which is within $\pm 5\%$ of the median (1.6 ± 0.1 mg m^{-3});

(3) a mean value of each layer's dry density that is within the set range (1.6 ± 0.1 mg m^{-3}).

Setting of assigned goals

Assigned goals to construct a low permeant bentonite buffer are set as follows:

(1) the initial value for permeability of the bentonite buffer is 5×10^{-13} m s^{-1};

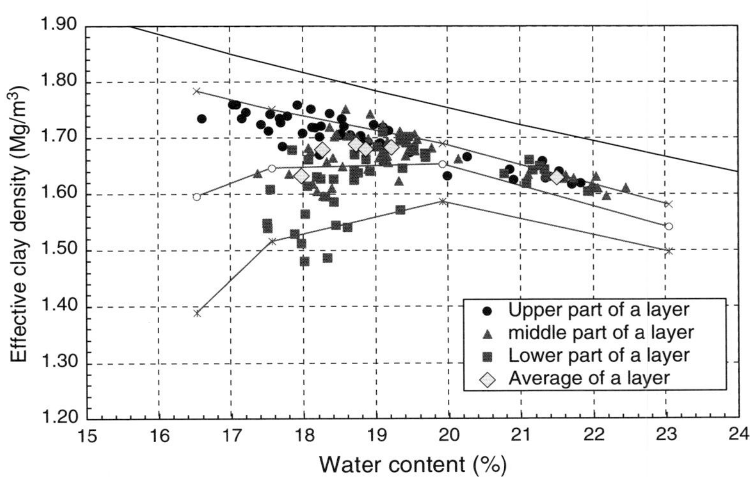

Fig. 4. Correlation between water content and dry density of divided core into vertical three parts.

Table 2. *Quality check test*

Item	Unit	Management value	Standard for check test	Frequency
Maximum grain size	mm	10	JIS A 1204	Every batch
Natural water content	%	<10	JIS A 1203	
Plastic limit	%	<30	JIS A 1206	
MBC (Methylene Blue Capacity)	mmol/100 g	63–77	JBAS-107-91	
Saturation power	ml/2 g	≥10 or greater	JBAS-104-77	

Table 3. *Remarks about quality conditions*

Category	Item	Details
Management of quality	Bentonite	Kunigeru GX (100%)
	Packaging	Flexible container bag
	Target range of water content	21 ± 2%
	Storage method	Temporary storage
	Frequency of checking water content	Three locations per batch, about 150 g × 3 samples

(2) the bentonite buffer must be constructed with high density (1.6 mg m^{-3}) in order to make the buffer a low-permeability layer;

(3) the distribution of dry density shall show as little variation as possible.

Solutions

The first assigned goal concerning permeability was solved by solving the second assigned goal concerning dry density. The second assigned goal, concerning dry density, can be solved by two methods: (a) selecting bentonite materials that are easy to make compact; and (b) selecting a construction method that attains sufficient compaction energy. The bentonite material selected was granular bentonite ore from Tsukinuno, Yamagata Prefecture (Kunigeru GX, maximum particle size 10 mm), which was selected on the basis of a previous study. In the present study, the confirmation test for compaction using Kunigeru GX confirmed that a buffer with sufficiently high dry density of over 1.6 mg m^{-3} can be constructed if a rammer is used in the concrete pit construction and the layer thickness is 10 cm.

The third assigned goal can be solved by preparing bentonite with the proper water content. A previous study showed that, as water content increases, the variation in the dry density becomes smaller (Kudo *et al.* 2004). Figure 4 shows that the dry density distribution starts to narrow at values around 21%. We therefore set the target range for the preparation of water content at 21 ± 2%.

Construction methods and management

Construction conditions

In the present study, the bentonite buffer was constructed in the side area of the test facility's concrete pit (the lower part of the bentonite buffer underneath the pit is discussed in more detail in a previous study; Akiyama *et al.* 2011) The side area of the concrete pit has a trench-like form, 1 m in width with walls on both sides having a maximum height of 8 m. Under this narrow condition, the construction methods and available equipment that can be adapted are limited (see Fig. 5).

Overview of construction

Granular bentonite (maximum diameter: 10 mm) with fixed water content was used. The *in situ* construction procedure was as follows:

(1) Spread the bentonite manually using an original spreading machine referred to as an 'asphalt finisher'.

Table 4. *Mixer specifications*

Main equipment	Specification	Remarks
Mixer	Granular mixer (pan model) (Eirich Mixer)	Rotation container with stirring wings

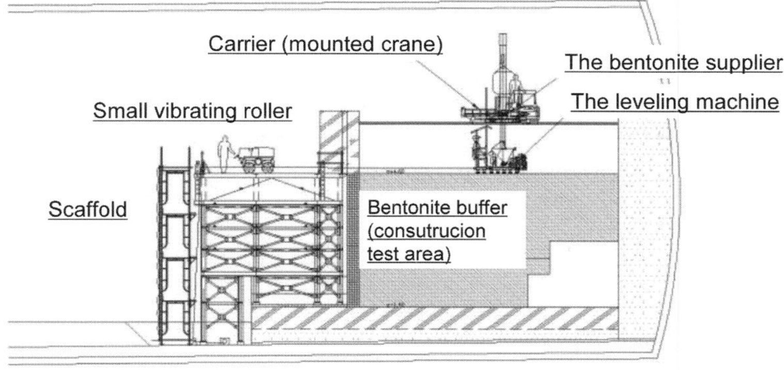

Fig. 5. Construction conditions (cross-section view).

(2) In the main area, compact the bentonite with a small vibrating roller without the use of vibration.
(3) At the far end, compact the bentonite with tamping rammer.

Preparation of water content

It is very important to prepare the bentonite water content so that the workability of the compaction and the specified quality can be ensured. Also, the selected preparation method must take into account manufacturing capacity and economical cost. In this study, bentonite was mixed with the specified amount of added water by using an Eirich Mixer to yield mixed bentonite with water content of $21.0 \pm 2\%$. First, the natural water content of the bentonite was measured, and then the specified volume of water was calculated based on the quality of material in Table 2 (Ito *et al.* 2007). Additional remarks about quality conditions are shown in Table 3. Specifications and remarks about the preparation equipment are shown in Table 4 and Figure 6.

Spreading, levelling and compaction of bentonite

In consideration of our construction track record, construction efficiency and other factors (Chijimatsu *et al.* 2006), we chose to perform on-site compaction of bentonite with a vibratory roller. A previous study showed an example of attaining a dry density of 1.6 mg m^{-3} or more with the use of Kunigeru GX 100% (Chijimatsu *et al.* 2006). In the present study, we find that using a vibratory roller is an outstanding construction method from the viewpoint of construction efficiency and achieving the required density.

The spreading, levelling and compaction of bentonite must be performed in a narrow space of 1 m in width, so a construction machine with this width was used. A spreading machine with the same operation design as an asphalt finisher was manufactured as a prototype. This machine includes a hopper and a screw feeder that spreads and levels bentonite in the transverse direction. Because this

Fig. 6. Eirich mixer.

Fig. 7. Spreading and levelling of bentonite with the spreading machine.

Fig. 8. Supplying bentonite to the spreading machine.

machine is a prototype, the drive system was made into a traction system by use of a winch (see Fig. 7).

The bentonite supplier that supplies bentonite to the spreading machine uses a mechanism in which the bentonite is fed into the spreading machine hopper from the upper part of the concrete pit through a conveyor belt and chute. This supplier enables bentonite to be supplied while the

Fig. 9. Vibrating roller (1.5 ton class).

Fig. 10. Sampling a core with a drill.

Table 5. *Vibrating roller, main details*

Item	Details
Length	1500 mm
Height	1200 mm
Width	610–850 mm
Body mass	1548 kg
Engine	Two cylinders, air-cooled, diesel
Speed	1.2 km h^{-1}
Width of rim	278 × 2 to 398 × 2
Operation	Remote control

machine moves alongside the spreading machine (see Fig. 8).

Bentonite compaction was performed by using a vibrating roller designed for 1.5 tons (Fig. 9). This vibrating roller is used for compaction in trench-like and similar areas and is highly adaptable to the construction area in the present research project. In order to compact bentonite in the concrete wall case, it is necessary to protect the concrete surface. For this protection, a shock-absorbing material was attached to the sidewall of the wheels. The vibration amplitude was then adjusted in order to adapt the machine to compaction of thin-layered bentonite. The main details about the vibrating roller are summarized in Table 5.

After the bentonite was compacted, the volume of each layer was calculated from the layer thickness surveyed during levelling and the area of the layer surveyed before compaction. The dry density of each layer was calculated from the volume of the layer, the wet density of the spread and levelled bentonite, and the average water content of the bentonite. For each layer, we judged construction to be complete based on the dry density calculation. To ascertain the final quality, we measured the density using core samples collected by a core drill in each layer (see Fig. 10). The dry densities of the cores were then calculated from the mass and volume measured with a caliper. After density was calculated, the cores were ground and the water content was measured by the furnace drying method.

Results of construction test

Preparation of water content

The distribution of the water content of the bentonite is shown in Figure 11. The water content has a mean of 22%, a median of 21.5%, a maximum of 22.9%, a minimum of 19.1% and a standard deviation of 0.9%. After adjustment of the bentonite water content, its values could fit in the target range of 19.0–23.0% for the estimated suitable moisture

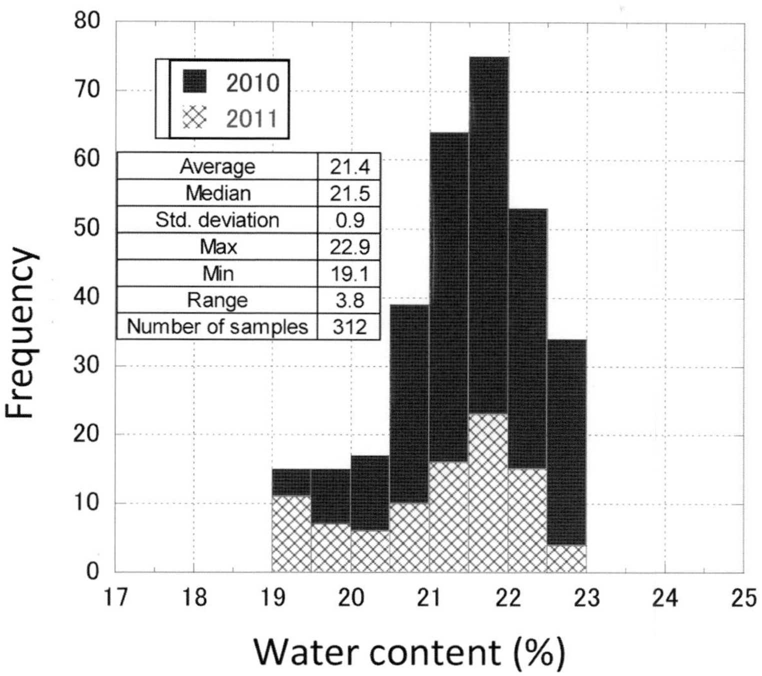

Fig. 11. Bentonite water content after adjustment.

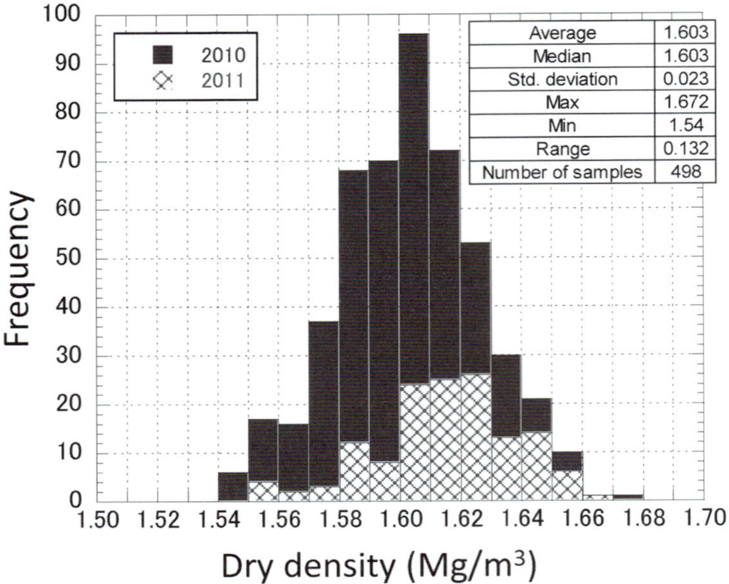

Fig. 12. Dry density of the bentonite buffer estimated from sample cores.

input calculated by measuring the water content of the raw materials for every batch from this result. As stated above, these results confirm that water content adjustment of the bentonite allows the water content to fit within the target range of values when using the selected mixer.

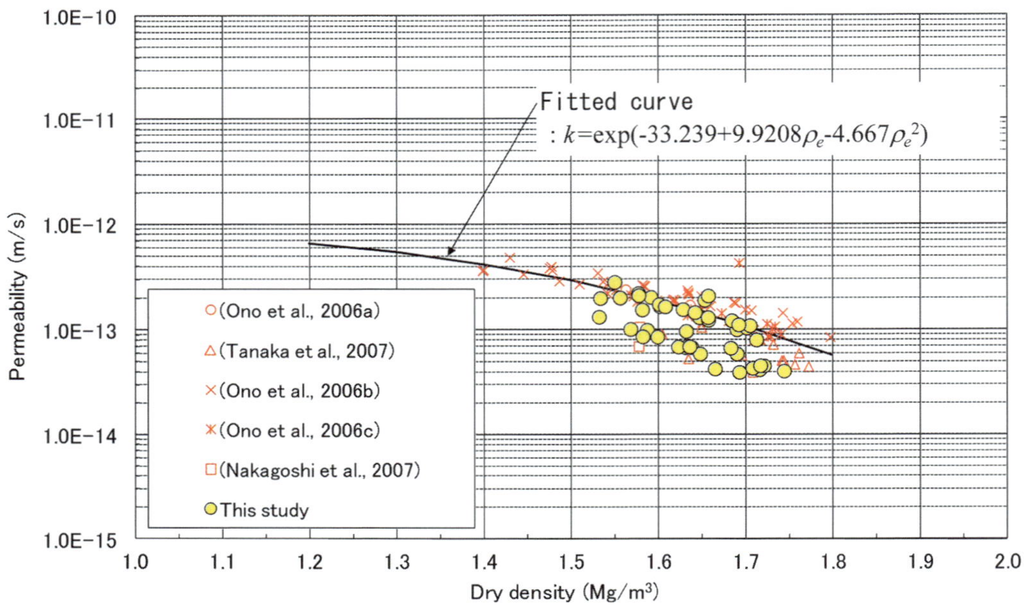

Fig. 13. Correlation between dry density and permeability (Ono *et al.* 2006*a–c*; Nakagoshi *et al.* 2007; Tanaka *et al.* 2007).

Spreading, levelling and compaction of bentonite

The dry density measurements of the core extracted after construction are shown in Figure 12. The bentonite buffer in the side area was constructed with a total of 83 layers within two years, 2010 and 2011. The total number of samples was 498. The dry density has a mean of 1.603 mg m^{-3}, a median of 1.603 mg m^{-3}, a maximum of 1.672 mg m^{-3}, a minimum of 1.540 mg m^{-3} and a standard deviation of 0.023 mg m^{-3}. As a result of the spreading, levelling and compaction of bentonite whose water content had been adjusted to within the target range, we also confirmed that the dry density of the construction settled in the target range of 1.50–1.70 mg m^{-3}.

Because the predetermined density was attained, we expect the constructed bentonite buffer to have a low permeability that satisfies the performance requirements. In the present study, the permeability was measured directly from the core. The correlation between permeability and dry density is shown in Figure 13, where the yellow points indicate the permeability values obtained in the present study. This data is mostly in agreement with the trend line (fitted curve) acquired from previous studies. Complete fulfillment of the performance requirements is shown by the fact that the dry density clears the management target values.

Conclusions

In order to confirm that the bentonite buffer is actually constructed at its stated quality (low permeability), we carried out confirmation tests of the construction method for a low-permeability bentonite buffer. Kunigeru GX, a granular bentonite ore, was used for the bentonite ore. Initial performance values were set to a permeability of 5×10^{-13} m s^{-1} or less for the tychopotamic performance requirement for the EBS. As a substitute for permeability, dry density of 1.60 ± 0.1 mg m^{-3} and water content of $21.0 \pm 2\%$ were set as the target range of values for construction based on the correlation between permeability and dry density and taking into account the variation that occurs during construction.

For the construction method, the water content of the bentonite was adjusted through the method of granulation mixture with water by use of a mixer. Also, the bentonite was compacted on-site using a vibrating roller.

The results of the construction confirmation test confirm that the bentonite values can fit in the target range when the water content is adjusted using the selected mixer. It was also confirmed that, after the bentonite with water content adjusted to be within the target range is spread, levelled and compacted, the dry density of the construction will settle in the target range of 1.50–1.70 mg m^{-3}. Because the dry density has cleared the management target values, it has been shown that the low-water-permeability performance requirements are completely fulfilled.

This report is a part of the report on a demonstration test for a cavern-type disposal facility in 2010 and 2011 at the consignment of the Agency of Natural Resources and Energy, the Ministry of Economy Trade and Industry of Japan. This study received the guidance of the committee of the Demonstration Test for Cavern-Type Disposal Facility. The authors wish to express their appreciation.

References

AKIYAMA, Y., TERADA, K., ODA, N., YADA, T. & NAKAJIMA, T. 2011. Demonstration test of a underground cavern-type disposal facilities fiscal 2010 status. *Proceedings of the 14th International Conference on Environmental Remediation and Radioactive Waste Management*, ICEM 2011-59180.

CHIJIMATSU, M., YOSHIKOSHI, I., NAKAGOSHI, A. & AMEMIYA, K. 2006. Examination of on-site compaction of bentonite (Part 1) – construction confirmation test results of the narrow area. *In*: *Collection of Summaries of the 2006 Annual Meeting of the Atomic Energy Society of Japan*, **B26** (in Japanese).

GOVERNMENT OF JAPAN 2008*a*. *Joint Convention on the Safety of Spent Fuel Management and on the Safety of Radioactive Waste*. National Report of Japan for the Third Review Meeting, Fig. L6-1.

GOVERNMENT OF JAPAN 2008*b*. *Joint Convention on the Safety of Spent Fuel Management and on the Safety of Radioactive Waste*. National Report of Japan for the Third Review Meeting, Fig. L6-3-2.

ITO, H., CHIJIMATSU, M. & MURAKAMI, T. 2007. The development of bentonite material for on-site construction. *In*: *Collection of Summaries of the 62nd Annual Academic Lecture Meeting of the Japan Society of Civil Engineers*, CS5-001, 161–162 (in Japanese).

KUDO, K., TANAKA, Y. *ET AL*. 2004. Compaction test of bentonite ore in a concrete pit. *In*: *Collection of Summaries of the 59th Annual Academic Lecture Meeting of the Japan Society of Civil Engineers*, CS1-050, 99–100 (in Japanese).

NAKAGOSHI, A., CHIJIMATSU, M., NIWASE, K. & TANI, T. 2007. Study on uniformity of density of a bentonite block gap. *In*: *Collection of Summaries of the 62nd Annual Academic Lecture Meeting of the Japan Society of Civil Engineers*, CS5-008, 175–176 (in Japanese).

ONO, F., NIWASE, K., TANI, T., NAKAGOSHI, A. & CHIJIMATSU, M. 2006*a*. Permeability measurements in the boundary of the layer of compaction. *In*: *Collection of Summaries of the 61st Annual Academic Lecture Meeting of the Japan Society of Civil Engineers*, CS5-09, 323–324 (in Japanese).

ONO, F., NIWASE, K., TANI, T., NAKASHIMA, H. & ISHII, T. 2006*b*. Fast permeability test of bentonite ore – press saturation method. *In*: *Collection of Summaries of the 2006 Fall Meeting of the Atomic Energy Society of Japan* (in Japanese).

ONO, F., NIWASE, K., TANI, T., NAKASHIMA, H. & ISHII, T. 2006*c*. Fast permeability test of bentonite ore by saturation method of compacted bentonite specimens. *In*: *Collection of Summaries of the 2006 Fall Meeting of the Atomic Energy Society of Japan* (in Japanese).

RADIOACTIVE WASTE MANAGEMENT FUNDING AND RESEARCH CENTER 2006. *TRU Related Waste Disposal Technology Survey, Cavern-Type Disposal Facility Performance Confirmation Test Report, a Geological Isolation Technology Survey in the Fiscal Year 2005*, 4–133.

TANAKA, Y., NAKAMURA, K. ET AL. 2007. Evaluation of the permeability of compacted bentonite ground considering heterogeneity by geostatistics. *Doboku Gakkai Ronbunshuu C*, **63**, 207–223 (in Japanese, abstract in English).

Analysis of the ambient conditions in an IL-LLW storage cell in a deep clay repository during the waiting closure period

LUC-VINCENT BÉNET[1]*, CHARLAINE BOUILLET[1] & JACQUES WENDLING[2]

[1]*SOCOTEC SA/Projets Industriels/AME, 1 Avenue du Parc, 78640 Montigny-le-Bretonneux, France*

[2]*ANDRA, 92298 Châtenay-Malabry Cedex, France*

**Corresponding author (e-mail: luc-vincent.benet@socotec.com)*

Abstract: In the 2009 Andra repository concept, Intermediate Level Long Lived Wastes are stored in several 100-m-long storage cells. Concrete over-packs are stacked in piles before being emplaced by row. The large sections opened in the slab floor to enable emplacement by the conveyor could have a major effect on the temperature distribution in the storage cell. The aim of the study is to assess the ability of the ventilation to regulate temperatures along the cell. We focus on the period before closure, when the ventilation rate is at a minimum. The issue has been addressed by undertaking numerical simulations. A specific modelling approach based on head-loss correlations has been used to calculate the distributions of temperature and air velocity at a decimetre scale along the storage cell. The heat transfer and the air flow problems have been solved on a 3D mesh representing 225 rows of waste packages, the concrete walls and the surrounding rock. Different scenarios have been considered about the air flow rate, the heat release and the closure of the conveyor sections at several locations. Results have been analysed in terms of flow patterns, temperature distributions and thermal gradients in air and in concrete.

Since the end of the 1990s, the French agency for the management of nuclear wastes, Andra, has been studying the feasibility of locating a repository in the Callovo-Oxfordian argillites on the borders of the Meuse and Haute-Marne departments at a depth of about 500 metres. In Andra's design (Andra 2009) the Intermediate Level Long Lived Wastes (IL-LLW) are stored in 0.5-km-long galleries named 'modules'. The radioactive wastes are packaged in steel canisters, with eight canisters placed into a concrete over-pack. In some IL-LLW modules, the over-packs are stacked before being emplaced into the storage cell. The piles are carried along the storage cell by a conveyor underneath the gallery floor.

During the emplacement period, the ventilation air will flow through the gaps which remain between the piles, the pile and the wall, the pile and the ceiling, and under the over-packs through the conveyor sections that remain open. All these longitudinal gaps and apertures are connected with the transverse gaps between the rows.

During the period before closure, the ventilation rate will be reduced; however, it must be sufficient to prevent hydrogen concentrations and air temperatures from getting too high, as some IL-LLW produce hydrogen and heat. The presence of hydrogen within the storage area and the subsequent issue of the ATEX risk (i.e. explosive atmosphere) are not addressed here (Lajoie *et al.* 2009). This study focuses on the thermal issue.

The thermal issue

The heat released in the waste packages will be partly removed by the air ventilation, and partly stored in the over-packs and in the walls; thermal conduction in the surrounding rock will also occur over time. So the temperature in the air and in solid media depends not only on the air ventilation rate but also on the material properties: the thermal conductivity and capacity of concrete and rock, and the emissivity for infrared radiant fluxes. But the distribution of air temperatures also depends on the distribution of the air flow through the network of gaps between two piles, pilles and ceiling, pile and wall, and pile and conveyor-section floor. The air flow will be dispersed into these gaps depending on the balance of surface and volume forces: (i) with frictional forces, the air will tend to flow from narrow apertures (pile-to-pile and pile-to-wall gaps) into large apertures (conveyor section); (ii) with inertial forces, the air will tend to flow straight rather than bend; and (iii) with buoyancy forces which generate free convection, the hotter air will tend to flow upwards. Thus the contrasts in hydraulic diameters between the conveyor section and other longitudinal gaps, and free convection in row-to-row transverse gaps due to temperature variations, are of such a nature as to amplify spatial variations in air flow rate and air temperature within the cross sections of the cell. These variations, which often take the form of thermal

From: Norris, S., Bruno, J., Cathelineau, M., Delage, P., Fairhurst, C., Gaucher, E. C., Höhn, E. H., Kalinichev, A., Lalieux, P. & Sellin, P. (eds) 2014. *Clays in Natural and Engineered Barriers for Radioactive Waste Confinement*. Geological Society, London, Special Publications, **400**, 145–161.
First published online April 9, 2014, http://dx.doi.org/10.1144/SP400.18

stratification, prevent an efficient homogenization of the air temperature and so promote the build-up of high temperature zones which generally set under the ceiling. So the specific design of IL-LLW storage considered in this study is potentially exposed to such an issue during the operational period.

High temperatures could be damaging to the concrete infrastructure. Thus, it is desirable that the temperature in concrete does not rise above 65 °C to prevent the structure from being ruined by chemical reactions. Regarding the risk of thermal cracks in concrete, it is desirable that the thermal gradient does not rise above 7 °C m^{-1} in non-reinforced concrete (see the concrete without reinforcement in brown in Fig. 1).

General purpose of the study

The main objective of the study is to determine which zone of the storage cell will be exposed to high temperatures or vulnerable to the formation of cracks in the concrete structure for this specific design of IL-LLW storage cell (see the cross section in Fig. 1). Another objective is to determine which operational conditions will promote high temperatures locally, and which conditions will prevent them.

The temperature assessment in the storage zone has been addressed by undertaking numerical simulation. The thermal transient has been simulated over a 4-year-long period after the emplacement of waste packages. The numerical simulations have been carried out on a 3D model representing a storage cell composed of 225 waste package rows, the concrete walls and the surrounding rock. The simulations are of the evolution in time of: (i) the temperature in the air and in solid media; (ii) the air flow rate; and (iii) the thermal gradients in the concrete structures (wall, ceiling and over-packs) throughout the storage cell. The analysis of the results enables a good understanding of how the coupling of air flows and heat transfers in such a long cell depend on the various operating options considered: the air ventilation rate, the age of the waste package or changes in the air pathway within the cell.

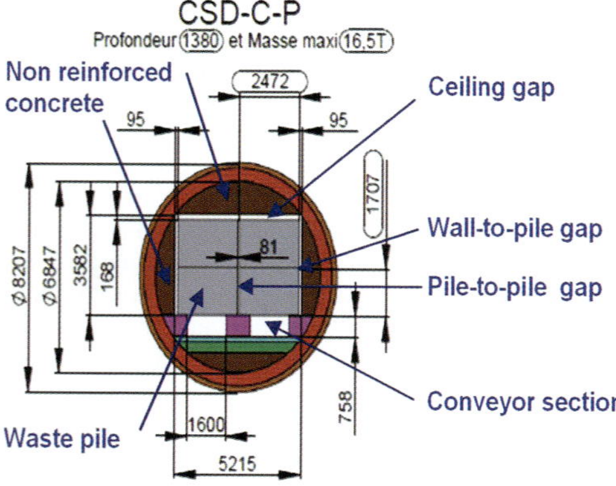

Fig. 1. Cross-section of an IL-LLW storage cell (Andra 2009).

Air flow modelling

Modelling approach

In the storage cell, the air flow and the conveyed heat are dispersed through a network of gaps between the waste packages, the walls and the ceiling. In this context, the head losses have a major effect on the flow because of the huge contrast between hydraulic diameters at a decimetre scale and the length of the cell (about 325 m in the model). So a Darcy-like model based on the head pressure modelling has been set up and applied to a mesh composed of about 100 000 elements, with only one element in the gap thickness the size of which equals the size of an aperture. Head-loss correlations (Idel'cik 1979) have been implemented to simulate the effects of the air friction on walls and of the gathering and splitting streams at the junction between lateral and longitudinal gaps. Thus, in each element of the mesh, the velocity and the temperature calculated by the model have to be considered as the mean values over the thickness of the gap.

Furthermore, this model is aimed at simulating a thermal transient over 4 years. So the thermal transfers through the walls and the rock have been modelled, and the calculation domain extended to the host rock. Consequently, the advective timescales (about 10 s for the heat to cross a gap, and an hour to cross the storage cell) are negligible compared to the specific time of thermal conduction through the wall (about 7 days to cross 1 m of concrete): the time step used to compute the 4-year-long thermal transient is around a week.

This model has been used instead of a classical Computational Fluid Dynamic (CFD) model based on the Navier-Stokes equations with a turbulence-like viscosity which would have required several million elements for such a geometry.

$$\overbrace{\underbrace{\nabla(P) - \rho g}_{\text{Volume forces}} + \underbrace{\rho\left(\frac{\partial V}{\partial t} + V\cdot\nabla(V)\right) - \nabla.\left((\mu + \mu_T)\nabla(V)\right)}_{\text{Inertial and frictional forces with turbulence}} = 0}^{\text{Navier-Stokes equations}}$$

$$\underbrace{\overbrace{\nabla(P) - \rho g} - \overbrace{[\zeta]\rho V_m} = 0}_{\text{Equivalent Darcy-like equation}}$$

where P is the air pressure, V the local velocity and V_m the mean velocity in the gap thickness, g the gravity acceleration, ρ the air density, $[\zeta]$ the head-loss matrix, μ the dynamic viscosity, and μ_T the turbulence-like viscosity.

The Darcy-like equation is usually adopted to model a flow in a network of ducts (1½D problem); in some industrial boilers, for example, the flow is generated by buoyancy forces due to mixed water and steam, and moderated by head losses.

Similar approaches are adopted on 3D problems to model the effect of a network of obstacles on the flow without meshing any obstacle (Ritz & Bénet 2002; Bellivier 2004; Pinson 2006). The discontinuous zone is meshed as continuous medium whose properties are adapted through spatial homogenization methods to take into account the various effects of the obstacles on the heat and mass transfers (volume, dispersion and turbulence changes, head losses, and so on.). However, such problems are modelled with Navier-Stokes equations in order to better discretize the inertial forces in 3D.

The model used in this study consists of an extension of the Darcy-like approach to a 2½D problem. Its main advantage is to focus on the right timescale and the right space scale to deal with the issue of the study. However, this model has limitations that a CFD model does not have. These limitations are due to: (i) the decimetre scale of the geometric resolution (the flow patterns under a decimetre scale such as boundary layers, local vortexes or free convection plumes cannot be captured by the model); (ii) the time step used which is much longer than the specific time period of air velocity fluctuations; and (iii) the validation field of the Darcy-like model which is smaller than the CFD one (in particular, head-loss correlations used are defined for steady state flows without any disturbance from upstream, and the modelling of inertial forces is less accurate between gap junctions than on them).

Buoyancy modelling and analysis

The temperature variations in the air generate free convection due to the buoyancy forces because of variations in space of the air density due to thermal dilatation. In this study, the buoyancy forces are modelled with Boussinesq's hypothesis. The variations of the air density have been considered only in the buoyancy forces:

$$\rho g \cong \rho_0 g \beta \Delta T$$

where ρ_0 is the reference density, β the thermal dilatability, and ΔT the temperature variation from the reference value T_0.

Richardson's number is defined as the ratio of the potential energy (gravity) to the kinetic energy. It is used to evaluate whether free convection or forced convection will dominate within a horizontal flow. In the example of the modelling of a vault storage system (Ritz & Bénet 2002), the transition of regime from forced convection to free convection is observed for Richardson's numbers from 3 to 7:

$$\text{Richardson} = \frac{g \cdot \beta \cdot \Delta T \cdot h}{U^2}$$

where $h = 3.58$ m, being the height of vertical gaps, and U (m s^{-1}) the mean horizontal air flow velocity.

In the storage cell, the air temperature will increase with distance from the entrance due to the heat released by the radioactive waste. Consequently, the lower the flow rate, the shorter the distance without significant free convection.

Head-loss modelling

The head pressure losses are modelled via an orthotropic matrix $[\zeta]$ locally defined on each cell of the fluid mesh.

$$[\zeta] = \frac{yV_m\zeta_f}{2D_h} + \begin{bmatrix} \zeta_{11} & & \\ & \zeta_{22} & \\ & & \zeta_{33} \end{bmatrix}$$

where D_h is the hydraulic diameter, V_m the velocity in the mesh cell, ζ_f the head-loss factor due to the air friction through a cylindrical duct, and y the shape factor (1.5 for a flat rectangular duct). This parameter is calculated as a function of the Reynolds' number (Idel'cik 1979; diagram 2.2) for laminar flows (Reynolds $<$ 2000), turbulent flows (Reynolds $>$ 4000) and intermediate flows (see the kink in the curve in Fig. 2).

The ζ_{ii} factors take both lateral ($\zeta_{\text{lat}} = \zeta_{22}$ or ζ_{33}) and straight ($\zeta_{\text{straight}} = \zeta_{11}$) head losses into account in order to model the splitting streams. They are applied as a frictional head loss on the lateral direction and on the longitudinal direction respectively (Idel'cik 1979; diagrams 7.21 and 7.23).

$$\zeta_{\text{lat}}^{\text{split}} = \frac{V_m^2}{2L_{\text{lat}}V_{\text{lat}}}\left(1 + \left(\frac{V_{\text{lat}}}{V_m}\right)^2 - 2\frac{V_{\text{lat}}}{V_m}\right)$$

$$\zeta_{\text{straight}}^{\text{split}} = \frac{0.4V_m^2}{2L_{\text{long}}V_{\text{straight}}}\left(1 - \frac{V_{\text{straight}}}{V_m}\right)^2$$

where L_{long} is the size of the junction cell in the longitudinal direction, L_{lat} the size of the junction cell in the lateral direction, V_{straight} and V_{lat}, the

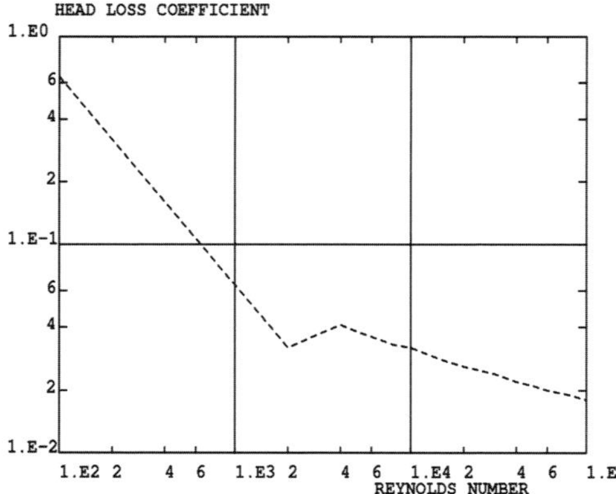

Fig. 2. Frictional head-loss law for a cylindrical duct ($y = 1$).

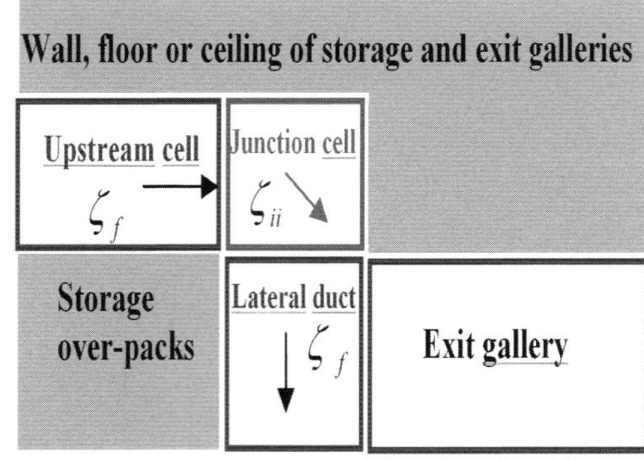

Fig. 4. Locations of the head-loss model in the mesh cells at the exit of the storage cell.

components of the air velocity in the longitudinal direction and in the lateral direction respectively.

Only the straight head loss (Idel'cik 1979; diagram 7.7) is implemented in order to model the gathering streams:

$$\zeta_{\text{straight}}^{\text{gath}} = \frac{V_m^2}{2L_{\text{long}}V_{\text{straight}}}\left(1.55\frac{Q_{\text{lat}}}{Q_m} - \left(\frac{Q_{\text{lat}}}{Q_m}\right)^2\right)$$

where Q_{lat} is the flow rate from the lateral direction, $Q_m = Q_{\text{lat}} + Q_{\text{straight}}$, and Q_{straight} the flow rate from the longitudinal direction.

These head-loss models are implemented at the junctions (Figs 3 & 4) between the transverse row-to-row gaps and the longitudinal gaps (i.e. the gaps along both the wall and the ceiling, along the conveyor sections and between the package piles).

At the end of the storage cell, the air exits from the conveyor section, the pile-to-wall and the pile-to-ceiling gaps, and turns to enter a smaller cylindrical gallery of same direction through a 10-cm-thick transverse gap between the wall section of this gallery and the last row. The head loss generated is modelled as a splitting stream with a zero flow rate downstream in the longitudinal duct. The lateral duct gives access to the exit gallery (see the scheme in Fig. 4).

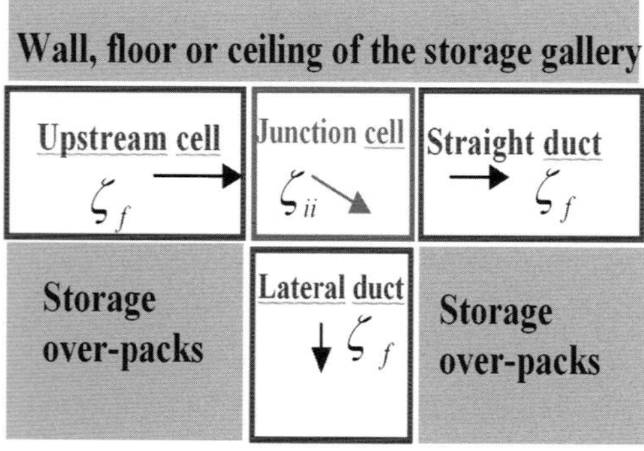

Fig. 3. Locations of the head-loss model in the mesh cells in a splitting flow.

Consistency between the Darcy-like model and the Navier-Stokes model

Two test cases have been achieved in order to verify the consistency between the Darcy-like model and the Navier-Stokes model. The numerical simulations have been performed with the Cast3M software on a 2D problem. The Navier-Stokes equations, combined with the so-called 'k-ε' equations and the standard wall functions for the turbulence modelling are discretized with a linear Finite Element method (Cast3M Gounand 2012) (see (Pinson 2006) for the validation of the method on results from the Comte-Bellot experimentation (Comte-Bellot 1965)). The steady state flow is reached by solving a pseudo-transient problem with an implicit algorithm. The head-loss model is discretized with a Mixed Hybrid Finite Element method (Bénet 1996; Dabbene 1998) and the steady state problem solved using an implicit method. However, an iterative algorithm is used to enable an implicit assessment of head-loss coefficients because of the non-linearity of these coefficients.

The consistency of the two models is evaluated on a 2D problem derived from the real geometry of an IL-LLW storage cell. The longitudinal extent of the mesh is reduced to 30 package rows. The longitudinal and vertical gaps such as pile-to-pile

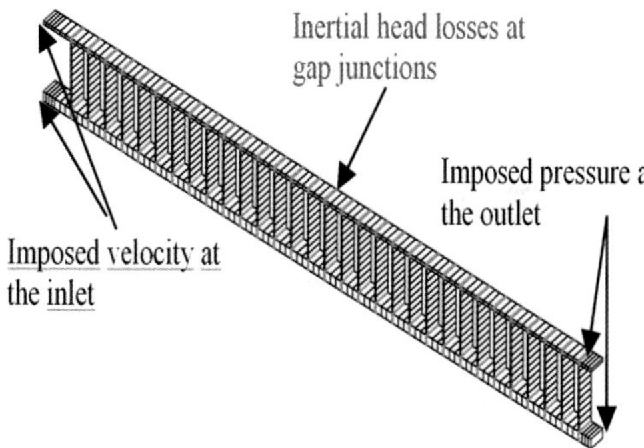

Fig. 5. Mesh of the Darcy-like model used for the comparison with the Navier-Stokes model.

and pile-to-wall gaps are not modelled, which forces the air to bypass the transverse gaps when flowing from a longitudinal gap to another gap. The mesh of the Navier-Stokes model has 39 000 elements, and the mesh of the Darcy-like model 634 elements.

The numerical simulations are performed without buoyancy forces in the first test case. In the second test case, the buoyancy forces have been taken into account considering a 14 °C gradual increase in air temperature over the first 12 rows; the heat advection has not been modelled.

In the two test cases, the same velocity is imposed at the entrance of the conveyor section and in the pile-to-ceiling gaps. The air flow redistribution between the ceiling and the conveyor sections is generated by (i) higher frictional forces above over-pack rows than below (the hydraulic diameters of the ceiling gap and conveyor sections equal 34 mm and 1 m respectively), and (ii) inertial head losses at each gap junction. Consequently, the flow rate through each longitudinal gap evolves with the row number. In the second test case, the flow rate gradually decreases in the ceiling gap (see Fig. 5) and increases in the conveyor sections. In the third test case, the flow rate sharply decreases in the ceiling gap due to buoyancy forces as the temperature gets higher with distance from the entrance, and then slightly increases beyond the twelfth row as the temperature remains uniform and the head losses are higher in the conveyor sections.

It should be noted that the gathering streams into the conveyor sections generate turbulence which tends to increase the frictional stresses downstream. This phenomenon is well simulated by the Navier-Stokes model but not taken into account in the Darcy-like model. In the former model, the frictional head loss is defined for flows in long ducts (Idel'cik 1979) that are not affected by upstream disturbance; in contrast the gathering streams from the ceiling gap disturb the air flow by producing and then advecting turbulence through the conveyor sections. The higher head losses and the lower flow redistribution observed in the Navier-Stokes results are probably due to this effect, which cannot be captured by the Darcy-like model without turbulence modelling.

Two Darcy-like simulations have been performed with different values of the shape factor of the frictional head-loss model (see the flow redistributions in Fig. 6a, b for the two shape factors). A shape factor equal to 0.75 has been considered as the reference value in order to better match the results of the Navier-Stokes model.

Boundary conditions

In the period before closure, the air renewal rate is supposed to be in the range of 0.3–3 $m^3\ s^{-1}$ per module. The air flow enters the first row through four different apertures: (i) the conveyor sections and the gaps between; (ii) package piles (81 mm

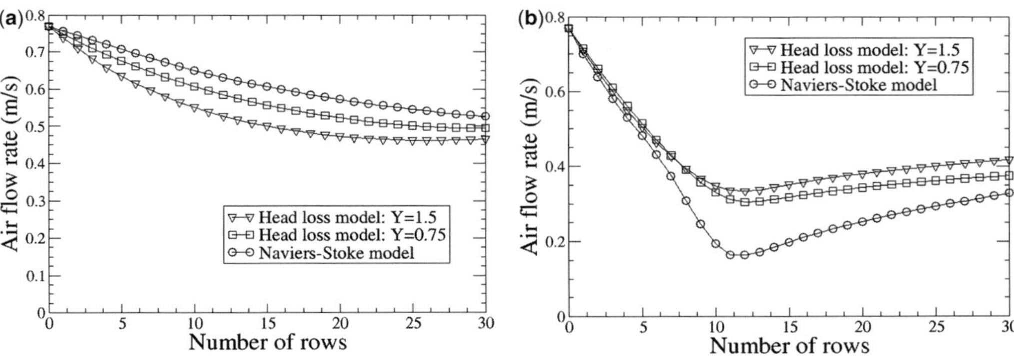

Fig. 6. Evolution along the storage cell of the mean velocity in the apertures between the piles and the ceiling for the Darcy-like model and the Navier-Stokes model (**a**) without buoyancy forces, and (**b**) with buoyancy forces.

thick); (iii) piles and wall (95 mm thick); and (iv) piles and ceiling (170 mm thick).

At the entrance, the air velocity is imposed equally in all apertures. Indeed, the entrance head losses induced by the huge differences of hydraulic diameter between the upstream gallery (6 m) and the gaps (from 0.34 m to 1 m) are high enough to equalize the inlet velocity. The relevance of this scenario has been verified by means of numerical simulation based on the Navier-Stokes equations.

A uniform pressure is imposed on the outlet boundary at the end of the gallery. The spatial velocity profile is smoothed by modelling a high head loss just upstream of the outlet boundary. All other boundary conditions are no-friction and no-heat fluxes.

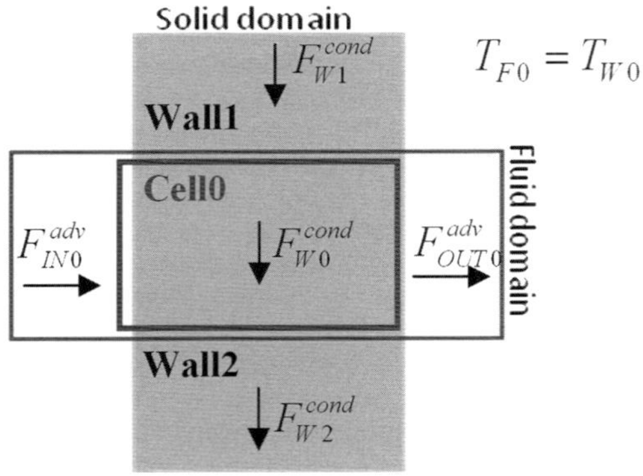

Fig. 7. Scheme of the interface between the fluid and solid media for the one-temperature model.

Heat transfer modelling

Physical modelling

The physical modelling of thermal transfers includes:

(1) The advection flux F^{adv};
(2) The wall-to-wall radiated heat exchanges F^{rad} through the gaps;
(3) The air-to-wall heat exchanges E^{conv} due to forced or free convection;
(4) The heat release Q^{w} from nuclear waste packages;
(5) The thermal conduction flux F^{cond} through the air, the concrete structures and the rock.

The air is assumed to enter the storage cell at the *in situ* temperature of 22 °C (i.e. the geothermal temperature at the repository level) because of convective heat exchange with the walls throughout the upstream part of the ventilation network. The air entering the cell by bypassing the first row is named 'fresh air' in the text. The fresh air is heated by the radioactive wastes by crossing the storage cell, and conveys the heat out of the cell when exiting by bypassing the last row.

Wall-to-wall radiated heat exchange. The face-to-face radiated heat exchange flux F^{rad} between *Wall1* and *Wall2* (see the scheme on Fig. 7) derives from the Stefan-Boltzman's law:

$$F^{\text{rad}} = \frac{\sigma \cdot (T_{W1}^4 - T_{W2}^4)}{(1/\varepsilon_{W1} + 1/\varepsilon_{W2}) - 1} = h_{\text{rad}}(T_{W1} - T_{W2})$$

with

$$h_{\text{rad}} = \frac{\sigma\left(T_{W1}^2 + T_{W2}^2\right)(T_{W1} + T_{W2})}{(1/\varepsilon_{W1} + 1/\varepsilon_{W2}) - 1} \quad \text{and} \quad \varepsilon_{W1} = \varepsilon_{W2} = 0.9$$

where ε is the emissivity, σ the Stefan-Boltzman's parameter, and h_{rad} the heat exchange between *Wall1* and *Wall2*.

It should be noted that the radiated heat exchanges are significant even under geothermal conditions due to the scenario of a high emissivity of concrete surfaces ($h_{\text{rad}} = 4.7\ \text{W m}^{-2}\ \text{K}^{-1}$ for $T_{W1} = T_{W2} = 22\ °\text{C}$).

Considering that the gap thickness e_0 is far smaller than the other dimensions of the fluid domain, the radiated heat flux through gaps is modelled as an equivalent conductivity which is locally applied on *W0*, the gap mesh of the solid thermal problem:

$$F^{\text{rad}} = e_0 h_{\text{rad}} \vec{\nabla}_{W0}(T) = F^{\text{cond}}_{W0}$$

Air-to-wall convective heat exchange. The heat exchange coefficient h_c is deduced from Nusselt's number which depends on the specific length $L = D_h$, and the thermal conductivity of the air λ:

$$\text{Nusselt} = \frac{h_c \cdot \lambda}{L}$$

For a turbulent forced convection (Reynolds > 10^4), Nusselt's number is calculated with Colburn's formula (Sacadura *et al.* 1980) according to the Reynolds' (Re) number:

$$\text{Nusselt} = 0.023\text{Pr}^{1/3}\text{Re}^{0.8}$$

For a turbulent-free convection (Grashof (Gr) > 10^9), Nusselt's number is defined with McAdams' formula according to the temperature variation between wall and air (McAdams 1942; Kern 1950):

$$\text{Nusselt} = 0.12\,\text{Gr}^{1/3}$$

The exchange coefficient used is the highest of both for a mixed convection.

Table 1. *Material properties*

Medium	Conductivity ($W\ m^{-1}\ K^{-1}$)	Mass density ($kg\ m^{-3}$)	Thermal capacity ($J\ kg^{-1}\ K^{-1}$)
Concrete	2.3	2460	1070
Rock(x, y)	2.2	2390	1015
Rock(z)	1.7	–	–
Air(22 °C)	0.025	1.2	1000

Heat release. The heat release value considered depends on the age of the radioactive waste at the date it is emplaced, because of the radioactive decay. Two scenarios have been studied: the storage of (i) just-made packages and (ii) 10-year-old waste packages (reference option).

Eight steel canisters are placed into a concrete over-pack before being emplaced in the storage cell. Each package row is supposed to be composed of four over-packs. The heat released per steel canister is 14 and 7.5 W respectively, which corresponds to 448 and 240 W per row or 100.8 and 54 kW per storage cell respectively.

It should be noted that the heat released by nuclear wastes is assumed to remain constant over the thermal transient in the model even though it is not the case. This modelling choice tends to overestimate the temperatures.

Thermal properties. The thermal properties of concrete (waste packages, concrete liner) and host rock are supposed to be independent of temperature and water saturation (Table 1).

Air-and-wall thermal coupling

Two numerical approaches have been used to force the air-and-wall thermal coupling. In both approaches, the temperature in solid media is defined and solved on the fluid mesh. The solid and fluid temperatures are the same in the first approach (see details in 'The one-temperature model' below) while they are distinct in the second approach (see details in 'The two-temperature model' below). Both methods have been tested and their results compared to the reference calculation case (see the definition of case 1 in Table 2; see also 'Comparison between the two thermal models' below).

The one-temperature model. The air-and-wall coupling is modelled by defining an artificial conductivity in the air that takes into account the convective heat exchange between the air and the walls and the radiated heat exchange between face-to-face walls:

$$F_{W0}^{\mathrm{cond}} = e_0(h_{\mathrm{rad}} + h_{\mathrm{conv}})\overrightarrow{\nabla}_{W0}(T)$$

where e_0 is the thickness of the gap that is meshed with only one cell (see the cell C0 in Fig. 7, noted *W*0 in the solid thermal problem).

Heat transfers are discretized with a Finite Volume method in the fluid and in the solid domain. The implicit solver enforces the heat balance without an iterative process in contrast to the two-temperature model.

The two-temperature model. Heat transfers through the fluid and solid media are split to be solved. Heat transfers are discretized with a Finite Element method in the fluid domain, and with a Finite Volume method in the solid domain. Both resolutions are coupled via an iterative process to assess convective heat exchanges between air and walls while ensuring the heat balance enforcement.

The real interface between solid and fluid domains (see the surface of *Wall1* and *Wall2* in Fig. 8) has been transposed into the fluid between the walls (see the cell C0 in Fig. 8, noted *F*0 in the fluid thermal problem) to define a volume density

Table 2. *Reference case and selected sensitivity cases*

Case number	Air flow ($m^3\ s^{-1}$ per cell)	Heat release (kW per can)	Section closure (every 10 rows)
1	3.0	7.5	no
2	0.3	7.5	no
3	0.3	7.5	yes
41	3.0	14.0	no
42	3.0	14.0	yes
43	0.3	14.0	no

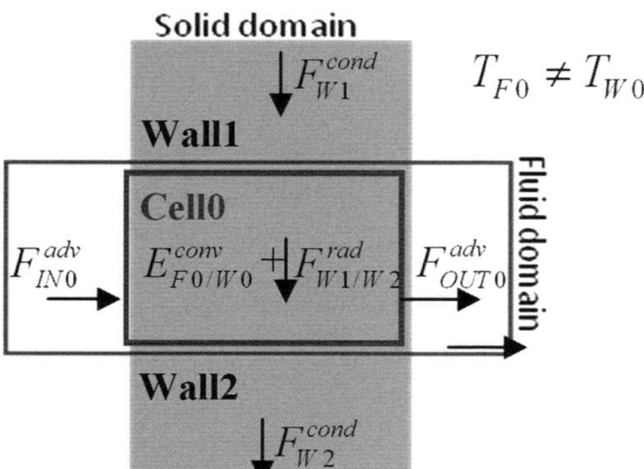

Fig. 8. Scheme of the interface between the fluid and solid media for the two-temperature model.

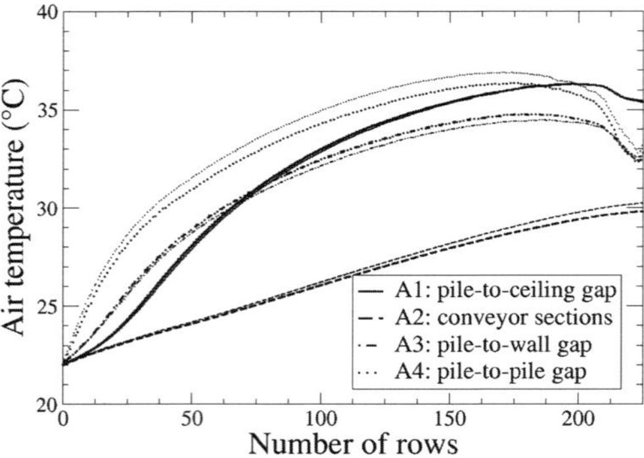

Fig. 9. Evolution along the storage cell of air temperatures through the longitudinal gaps and the conveyor sections from the results of the one-temperature model (fine curves) and the two-temperature model (thick curves) obtained after 4-year-long transient.

of heat exchange E_{F0}^{conv} between the two media. The elementary surfaces δS_{W1} and δS_{W2} of the cell interface between walls and air are expressed as a surface density related to the elementary volume δV_0 of the cell Cell0:

$$\frac{\delta S_{W1}}{\delta V_0} = \frac{\delta S_{W2}}{\delta V_0} = \frac{1}{e_0}$$

The convective heat exchange E_{F0}^{conv} between the air and walls is defined on the interface cell Cell0 as follows:

$$E_{F0}^{\text{conv}} \delta V_0 = h_{\text{conv}} \delta S_{W1}(T_{F0} - T_{W1}) + h_{\text{conv}} \delta S_{W2}(T_{F0} - T_{W2})$$

E_{F0}^{conv} can be also written as follows:

$$E_{F0}^{\text{conv}} = \frac{2h_{\text{conv}}}{e_0}(2T_{F0} - T_{W0})$$

where T_{F0} is the mean air temperature, T_{W0} the mean wall temperature at the interface of the cell Cell0, and T_{W1} and T_{W2} the temperatures on the *Wall*1 and the *Wall*2 respectively.

Note that a zero thermal inertia is considered in the gap mesh Cell0 of the solid domain (W0).

Comparison between the two thermal models. The reference case 1 (see the definition in 'Operating conditions' below) has been calculated with the two thermal model in order to compared the results. Figure 9 shows the evolutions of the air temperature through the longitudinal gaps and the conveyor sections obtained with the one-temperature model (see the fine curves from A1 to A4) and the two-temperature model (see the thick curves from A1 to A4) at the end of the 4-year-long thermal transient. The temperature evolutions assessed by the two models are very similar. The maximum temperature differential between both models is observed in the pile-to-pile gap; the one-temperature model slightly overestimates the air temperature by 0.5 °C.

Given that the results are very similar, the one-temperature model has been chosen to perform the other thermal simulations of the study because of its better numerical robustness.

Spatial modelling

A 325-m-long cell of 225 waste package rows within an 80-m-thick host rock layer has been represented by 358 517 finite elements, with 98 100 elements in the concrete walls, 40 500 in the waste packages, 42 300 in the air and 135 900 in the rock (see the mesh in the Fig. 10). Only the storage cell of the module has been meshed; neither the exit gallery nor the access gallery of the module has been included. However, a 10-m-long mesh has been added at the end of the cell in order to prevent the hydrostatic pressure from interfering with the exit boundary conditions (see 'Boundary conditions' above).

It should be noted that the storage cells of the Andra 2009 design is either 260 or 400 m long. So the length of the modelled storage cell is an intermediate between the two storage cell designs.

The calculation domain is reduced to a width of 25 m by taking into account the two longitudinal and vertical planes of symmetry (i) between two modules and (ii) along the axis of the storage cell. Each pile of two over-packs is discretized with 180 elements. The steel canister geometry is not represented. The heat released by the nuclear wastes is uniformly introduced into the piles.

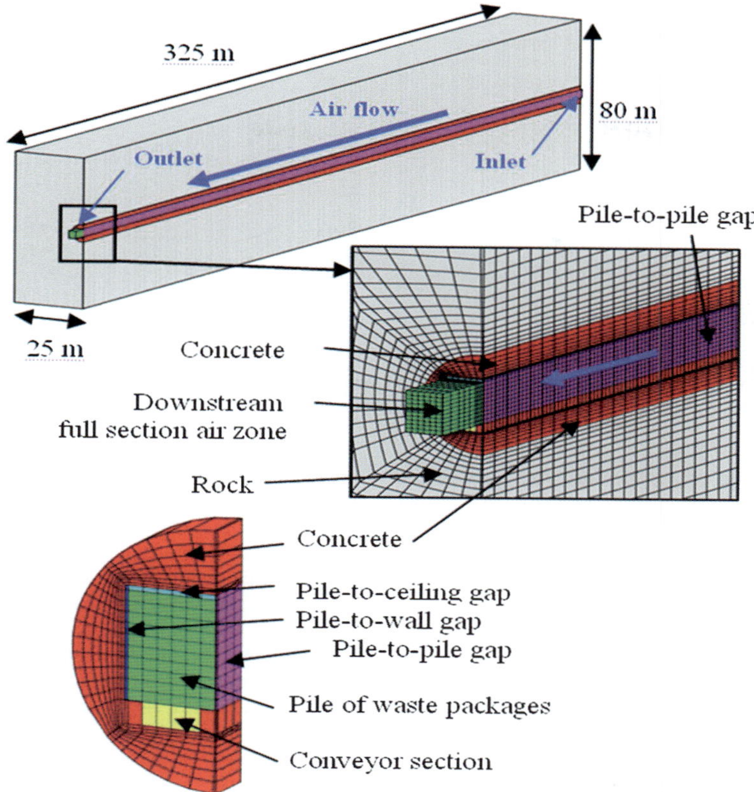

Fig. 10. Mesh of the coupled heat transfer and air flow problem.

The gap thickness is meshed with only one cell in thickness (see the modelling approach for heat transfers in 'Physical modelling' above) in order to reduce the geometrical problem to a network of thick planes (2½D problem). Thus, the velocity gradient across the gap thickness cannot be represented as the frictional forces on walls are not discretized in space but are taken into account by equivalent head losses (see 'Modelling approach' above).

Numerical methods

The flow problem has been discretized with a Mixed Hybrid Finite Element method (Dabbene 1998) and the thermal problem in the air with a Finite Volume method (Le Potier 2005). In solid media, the thermal problem has been solved with a linear Finite Element method for the two-temperature model, and with the Finite Volume method for the one-temperature model. The thermal transient has been solved with an implicit method with a first-order time scheme.

The physical models and the algorithmic process coupling the air flow and the heat transfers have been set up with the GIBIANE language. All the modelling, the meshing and problem solving have been achieved with the Cast3M software (Cast3M).

Operating conditions

The operating conditions considered in the reference case are: (i) a ventilation rate of 3 $m^3 s^{-1}$ per cell; (ii) the storage of 10-year-old waste packages; and (iii) no closure of the conveyor sections.

Other scenarios have been tested through five sensitive cases where either ventilation rate, or the age of the stored radioactive waste, or the air pathway through the storage cell have been changed:

(1) A ventilation rate of 0.3 $m^3 s^{-1}$ per cell: a low ventilation rate will reduce the cost of operating conditions which will be maintained over tens of years until the end of the reversibility period of the storage.

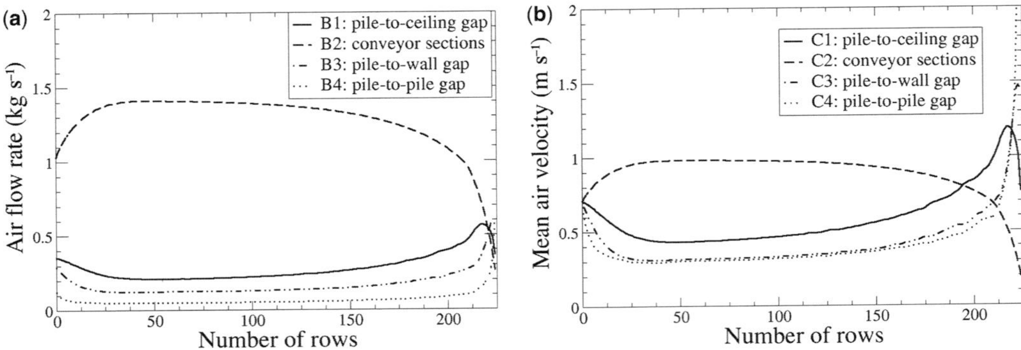

Fig. 11. Evolution along the storage cell of (**a**) air flow rates and (**b**) mean velocities in the longitudinal gaps; results after 4 years from case 1 ($D_{\text{air}} = 3\ \text{m}^3\ \text{s}^{-1}$).

(2) The storage of fresh waste packages which induces a higher heat release: 14 kW per waste canister instead of 7.5 kW with 10-year-old waste packages.

(3) The stepwise closure of a conveyor section after every 10 or 20 rows: the closure of a conveyor section will divert the air flow upwards to the transverse row-to-row gaps, and then to the other longitudinal gaps. The air flow redistribution among the longitudinal gaps will improve the air mixing and possibly the cooling of waste packages.

The operating conditions for the six calculation cases are shown in Table 2.

Results and analysis

The following analyses are based on results obtained after 4 years.

Case 1: reference scenarios

This calculation case is based on the higher flow rate scenario (3 $\text{m}^3\ \text{s}^{-1}$ per module) and the storage of 10-year-old waste packages (about 254 kW per module). The air flow enters the storage cell by bypassing the first row with a 0.75 m s^{-1} uniform velocity (Fig. 11b).

The flow distribution between the gaps changes along the first 30 rows: 78% of the ventilation flow bypasses row 30 through the conveyor section while only 55% bypasses row 1 by this pathway (see curve B1 in Fig. 11a). This increase in the air flow rate through the conveyor section is due to (i) its higher hydraulic diameter (Reynolds analysis), and (ii) the buoyancy forces due to the temperature rise with distance from the entrance: these forces pull the fresh air downwards when entering the cell.

Along the last ten rows, the air flow distribution drastically changes because of the peripheral reduction of the free cross-section area from the storage zone to the exit gallery. The air flow is gradually diverted, from the ceiling gap and the conveyor sections which are partially obstructed, to the pile-to-wall and pile-to-pile gaps (see the curves B1, B2, B3 and B4 respectively in Fig. 11).

At the end of the storage cell, the air velocity lowers to 0.2 m s^{-1} in the conveyor sections, and reaches 3.5 m s^{-1} in the pile-to-pile gap (see the curves C2 and C4 in Fig. 11). The air under the ceiling becomes warmer than in the conveyor section beyond row 10 (see the curves A1 and A2 respectively in Fig. 12). The air flow distribution remains quite uniform until row 100, even though the temperature variation from the conveyor sections to the ceiling increases with distance from the entrance. Beyond row 100, the air flow decreases through the conveyor sections and increases in the

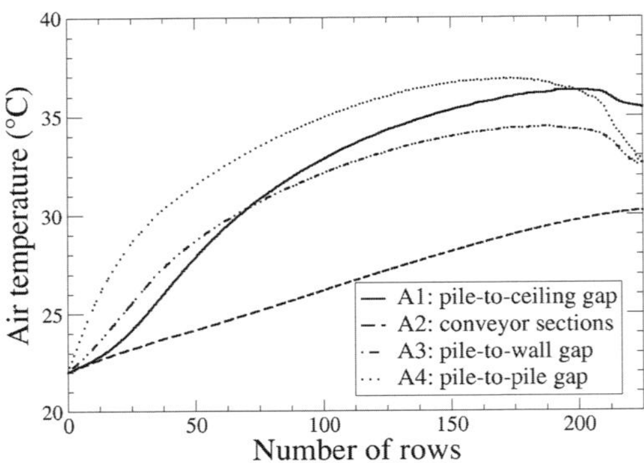

Fig. 12. Evolution along the storage cell of mean air temperatures in the longitudinal gaps; results after 4 years from case 1 ($D_{\text{air}} = 3\ \text{m}^3\ \text{s}^{-1}$).

other longitudinal gaps due to free convection in vertical gaps that generates a thermal stratification. Around row 175, the temperature difference between the bottom and the ceiling rises to 7 °C.

The coefficients of radiated heat exchange between piles and walls equal about 4.3 to 5.2 W m^{-2}. The radiated heat exchanges are overall more efficient than the convective heat exchanges because of the assumption of a high emissivity of concrete surfaces (0.9): the convective exchanges account for about 70% of radiated heat exchanges through the pile-to-wall gaps and the pile-to-ceiling gap, and 85% through the conveyor sections.

After 4 years, two-thirds of the heat release has been stored in rock and concrete while one-third has been removed by the ventilation (see Fig. 13). However, the thermal inertia of walls and packages has a major effect during the first year by delaying the increase of the air temperature throughout the cell. Thus, after 4 months, the air temperature reaches a maximum value of 30 °C, and 35 °C after a year. After 4 years the highest temperature (37 °C) is reached in the pile-to-pile gap (see the curve A4 in Fig. 12).

The maximum thermal gradients in the concrete walls are reached in the upper corner of the cell cross section. They reach 7 °C m^{-1} at the end of the storage cell where the air temperature is the highest (see curve D2 in Fig. 14).

The thermal gradient in the waste packages reaches 10 °C m^{-1} at the bottom of the piles (see curve D1 in Fig. 14). High thermal gradients are induced by the convective heat exchanges along the conveyor sections. Actually, the conveyor sections bring fresh air farther than other longitudinal gaps because the flow rate and the air velocity over the first 200 rows are higher (see curve D2 in Fig. 14). The thermal gradient in the pile bottom tends to be lower along the first 20 rows and the last 100 rows because the air flow redistribution between gaps that happens at both ends of the cell improves the cooling of the waste packages along the other gaps, especially through the transverse row-to-row gaps.

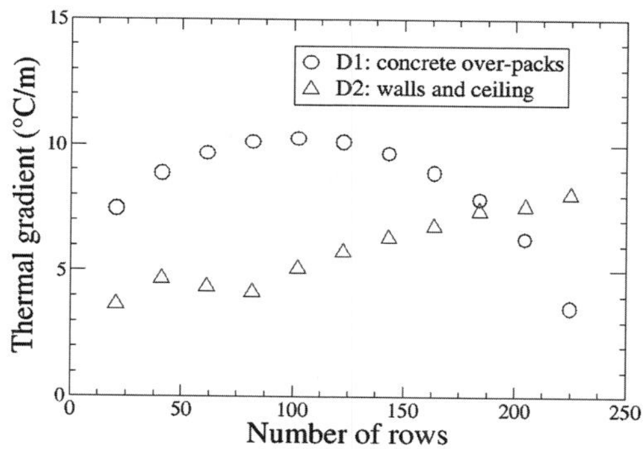

Fig. 14. Evolution along the storage cell of maximum thermal gradients in the concrete; results after 4 years from case 1 ($D_{air} = 3\ m^3\ s^{-1}$).

Case 2: low air-flow-rate scenario

In this calculation case, a ventilation rate of a factor 10 lower than the reference case, and the storage of 10-year-old waste packages have been considered without conveyor-section closure. Figure 15 shows the evolution of the flow rates, and Figure 16 the evolution of air temperatures throughout the storage cell and the maximum thermal gradients in the wall and the concrete over-packs after 4 years.

The flow pattern is drastically different from the reference case since a reverse flow spreads in the ceiling gap over the first 150 rows (see the

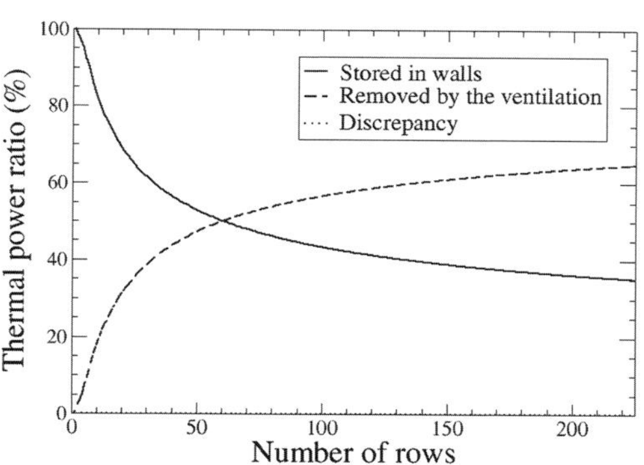

Fig. 13 Evolution in time of the heat stored in the walls and removed by the air from the storage cell; results after 4 years from case 1 ($D_{air} = 3\ m^3\ s^{-1}$).

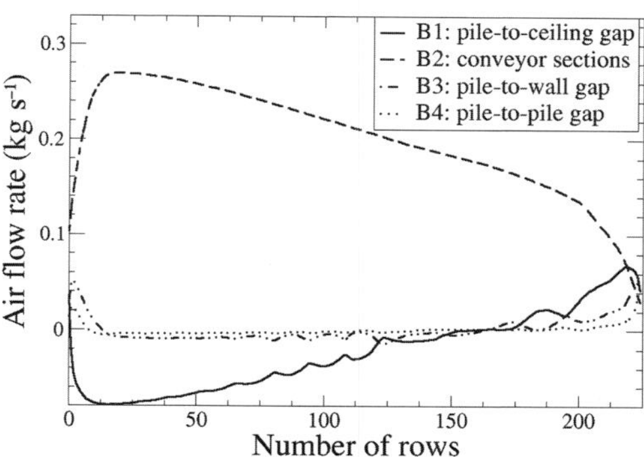

Fig. 15. Evolution along the storage cell of flow rates in the longitudinal gaps for half a cell; results after 4 years from case 2 ($D_{air} = 0.3\ m^3\ s^{-1}$).

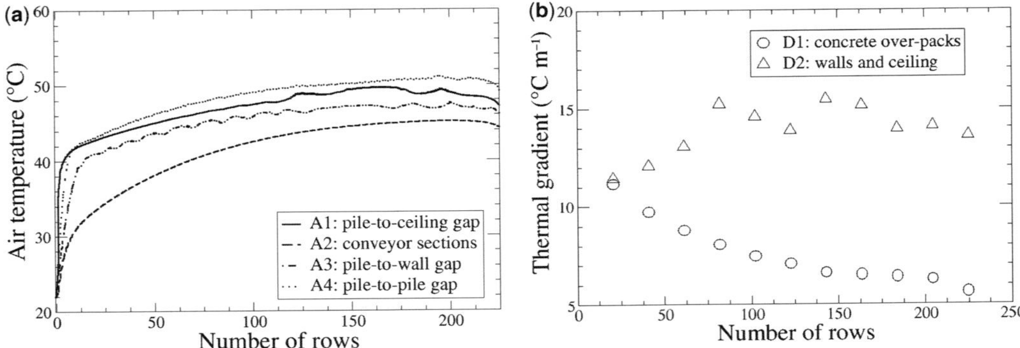

Fig. 16. Evolution along the storage cell of (**a**) mean air temperatures in the longitudinal gaps and (**b**) maximum thermal gradients in concrete; results after 4 years from case 2 ($D_{air} = 0.3$ m^3 s^{-1}).

negative flow rate on curve B1 in Fig. 15). These results are consistent with a high Richardson's number of about 700, a value which suggests that the air flow is governed by the buoyancy forces. This recirculation is enhanced by larger horizontal gaps (ceiling 0.17 m and conveyor section 0.76 m) than vertical ones (wall 0.095 m and pile-to-pile 0.08 m).

The air flow through the conveyor sections rises up to 144% of the ventilation rate over the first 15 rows where the reverse flow from the ceiling gap returns into the conveyor section. Then, the air from the conveyor section flows up through the transverse row-to-row gaps along the following 150 rows because of the free convection, and feeds the reverse flow until row 150 and beyond the equal-current flow in the ceiling gap. Consequently the air flow in the conveyor section gradually decreases down to 50% at row 200 (see curve B2 in Fig. 15).

This low ventilation rate generates an increase in air temperature up to 51 °C instead of a maximum value of 37 °C in the reference case (see curves A1 in Figs 16 & 12). The maximum air temperatures are found in the transverse row-to-row and in the longitudinal pile-to-pile gaps at the end of the storage gallery. The low ventilation generates a thermal stratification all along the cell. The difference in air temperature between the conveyor section and the ceiling evolves gradually from 12 °C around row 10 to 5 °C around row 315 (see the difference between curves A1 and A2 in Fig. 16a).

The heat removed by the ventilation is significantly lower than the heat stored in the rock (13% v. 87%). However, the reverse flow through the upper part of the cell conveys heat up to the first rows. Moreover, the air temperature in the peripheral gaps exceeds 40 °C from the very first rows while it only exceeds 30 °C after row 60 in the reference case (see curves A1 in Figs 16a & 12). Consequently, this reverse flow spreads efficient heat exchange throughout the storage cell in contrast to case 1.

The thermal gradient rises up to 11 °C m^{-1} at the bottom of piles in the first rows. It decreases with distance from the air entrance down to 6 °C m^{-1} at the end of the storage cell. However, the thermal gradient levels reached in both cases 1 and 2 are the same (see curves D1 in Figs 14 & 16b). In contrast, the maximum thermal gradients in the concrete walls are higher than in case 1: the maximum thermal gradients in the concrete walls exceed 15 °C m^{-1} around some rows (see curves D2 in Figs 14 & 16b), probably due to a higher temperature differential between the rock and the air (51 °C reached in case 2 instead of 37 °C in case 1).

Case 3: low air-flow-rate scenario with stepwise closure of conveyor sections

The scenarios considered in case 3 are: (i) a ventilation rate of 0.3 m^3 s^{-1}; (ii) the storage of 10-year-old waste packages; and (iii) stepwise closure of conveyor sections.

The stepwise closure of some conveyor sections aims at preventing a thermal stratification within the storage cell. The stepwise approach to closure has been studied through two calculation cases considering a closure after either every 10 rows or every 20 rows. Both scenarios lead to similar results. Only the results obtained with a ten-row closure are shown here.

The stepwise closure of conveyor sections after every ten rows forces the air to cross the transverse row-to-row gaps before being distributed among the other longitudinal gaps. The first closure diverts half the overall flow to the two pile-to-wall gaps and a quarter to each of the pile-to-pile and pile-to-ceiling gaps (see curves B1 to B4 in Fig. 17a). Between the two first closures, only about a

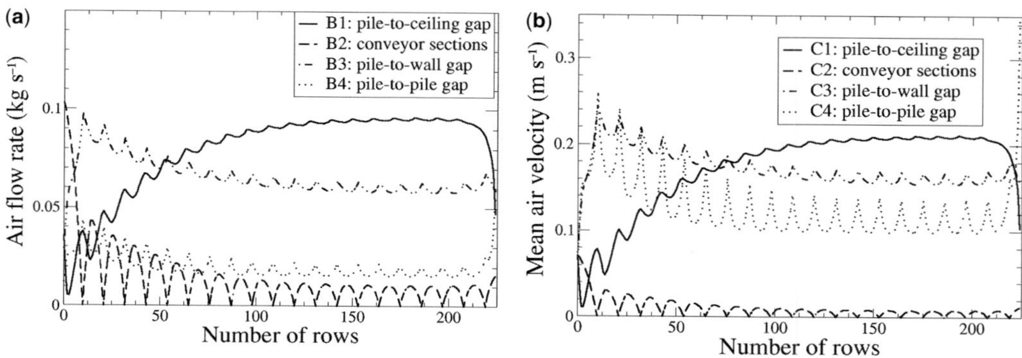

Fig. 17. Evolution along the storage cell of mean air flows in the longitudinal gaps for half a cell; results after 4 years from case 3 ($D_{air} = 0.3$ m^3 s^{-1} and 'section closures').

quarter of the overall flow returns into the conveyor sections. Along the 100 following rows, the flow is mainly diverted to the pile-to-ceiling gap, and the flow rates in the other longitudinal gaps decrease. Beyond 200 rows and between two closures, the air flow distribution in the cross section is 50% through the pile-to-ceiling gap, 35% through the pile-to-wall gaps, 10% between piles, and 5% through the conveyor sections. Moreover, the variations in flow rate between two closures reach about 5% of the overall ventilation flow through the conveyor sections, 3% through the pile-to-pile gap, 2% through the pile-to-wall gaps, and 1% through the pile-to-ceiling gap. It should be noted that the variation in flow rate is higher with the scenario of a closure after every 20 rows.

The air temperature increases along the first 100 rows and reaches a maximum of 53.5 °C in the pile-to-pile gaps beyond 150 rows. In this gap and in the conveyor sections, the temperature slightly fluctuates by 1 °C and 2 °C respectively with the closure of conveyor sections (see curves A1 and A2 in Fig. 18). Temperatures remain quite uniform in the other gaps (around 48 °C) where the air is cooled by the walls or the ceiling.

There is no significant thermal stratification: the variation in air temperature between the conveyor section and the gap does not exceed 3 °C. But, paradoxically, this flow pattern generates a 2 °C increase of the maximum air temperature compared to results from the case with no section closure (case 2). Effectively, the zone of efficient heat exchange with walls and ceiling are less extended in case 3 (about 80% of the storage-cell length according to a criterion of a 40 °C minimum air temperature) than in case 2 (about 100%), because there is no reverse flow under the ceiling in case 3.

The section closure succeeds in preventing a thermal stratification in the cell, which proves the efficiency of the section closures in mixing the air temperatures. Nevertheless, the air temperatures are higher than in case 2 and the maximum of thermal gradients in the walls or the ceiling remains higher than 10 °C m^{-1} over the three-quarters of the cell length with a maximum value of 13 °C m^{-1} (see curve D2 in Fig. 19).

The thermal gradients in the waste piles stay under 7 °C m^{-1} (see curve D1 in Fig. 19) which is a better result than in case 1 and 2.

Case 41: storage of fresh waste packages

The scenarios considered in case 41 are: (i) a ventilation rate of 3 m^3 s^{-1}; (ii) the storage of fresh waste packages; and (iii) without closure of conveyor sections.

The flow pattern is very similar to reference case 1. The air temperature along the storage cell rises to 50 °C (see curve A4 in Fig. 20) instead of 37 °C

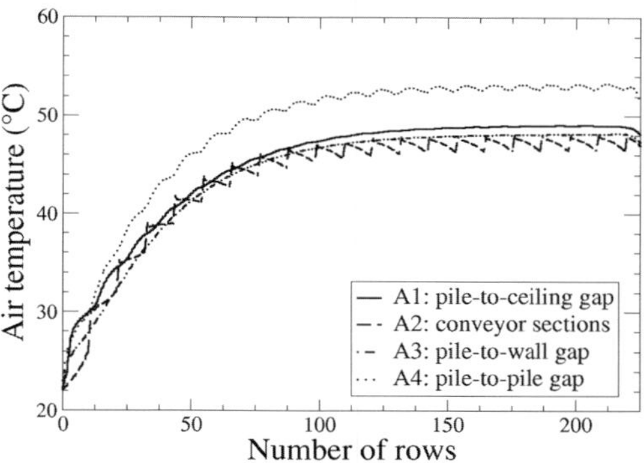

Fig. 18. Evolution along the storage cell of mean air temperatures in the longitudinal gaps; results after 4 years from case 3 ($D_{air} = 0.3$ m^3 s^{-1} and 'section closures').

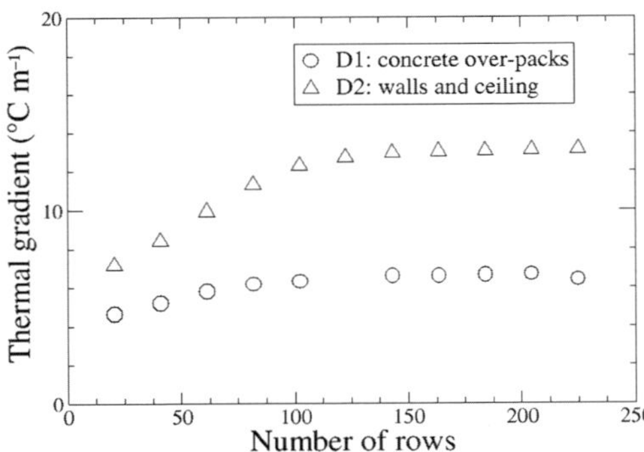

Fig. 19. Evolution along the storage cell of maximum thermal gradients in concrete; results after 4 years from case 3 ($D_{air} = 0.3$ m^3 s^{-1} and 'section closures').

for case 1 (see curve A4 in Fig. 12), and the thermal stratification rises to a 15 °C temperature differential between the conveyor-section floor and the ceiling around row 140, instead of 8 °C in case 1 (see curves A1 and A2 in Figs 20 & 12).

The maximal thermal gradient into the concrete packages rises to 20 °C m^{-1} (8 °C m^{-1} in case 1). The highest values are reached below piles along the second third of the cell (see curve D1 in Fig. 20b) where the air distribution between the longitudinal gaps is uniform (see as an indication, curve B from case 1 in Fig. 11a): a minimum air flow crosses the transverse row-to-row gap which reduces the convective heat exchanges on the pile interface.

The maximum of thermal gradients in walls increases from 6 to 15 °C m^{-1} with distance from the entrance (see curve D2 in Fig. 20b). The evolution of the thermal gradients along the cell is consistent with the increase of air temperature along the cell through the pile-to-wall and the pile-to-ceiling gaps (see curves A3 and A1 in Fig. 20a).

Case 42: storage of fresh waste packages with section closure

The scenarios considered in case 42 are: (i) a ventilation rate of 3 m^3 s^{-1}; (ii) the storage of fresh waste packages; and (iii) stepwise closure of conveyor sections after every ten rows.

The maximum air temperature is almost the same as in case 41 (50 °C) even though the flow pattern and the temperature distribution are different. The stepwise closure of conveyor sections after every ten rows seems to reverse the thermal stratification, as the temperature in the pile-to-ceiling gap is about 6 °C lower than in the conveyor sections (see curves A1 and A3 in Fig. 21a). However, this result is consistent with the flow distribution through the longitudinal gaps. Indeed, the fresh air which enters the pile-to-ceiling gap flows faster (2 m s^{-1}) through the pile-to-ceiling than through the other gaps, and low temperatures from the entrance are spread farther through the cell for this reason.

The thermal gradients in concrete rise up to 11 °C m^{-1} into the walls or the ceiling, and to 8.5 °C m^{-1} into the waste packages (see curves D2 and D1 respectively in Fig. 21b). The maximum thermal gradients into the waste packages are located on the top of piles, instead of below in case 41, because the convective heat exchanges are boosted in the pile-to-ceiling gap by a high flow rate and low air temperatures.

Case 43: storage of fresh waste packages with low ventilation

The scenarios considered in case 43 are: (i) a ventilation rate of 0.3 m^3 s^{-1}; (ii) the storage of fresh waste packages; and (iii) without closure of conveyor sections.

The flow pattern and the temperature profiles along the storage cell are very similar to those

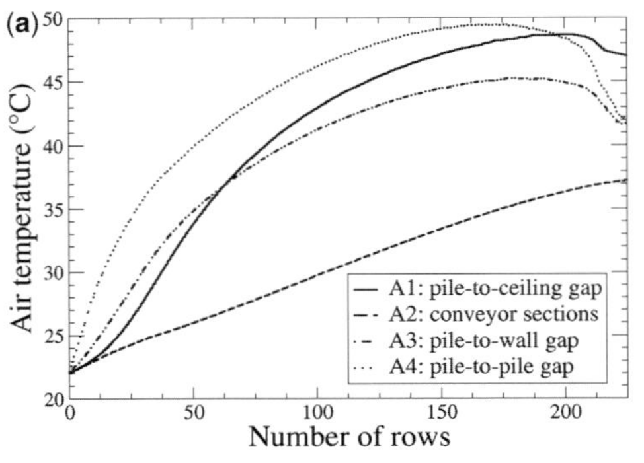

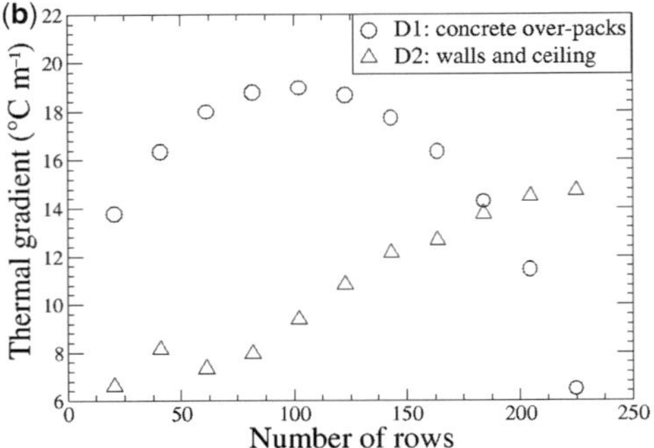

Fig. 20. Evolution along the storage cell of (**a**) mean air temperatures in the longitudinal gaps and (**b**) maximum thermal gradients in concrete; results after 4 years from case 41 ($D_{air} = 3$ m^3 s^{-1} and maximum heat release scenario).

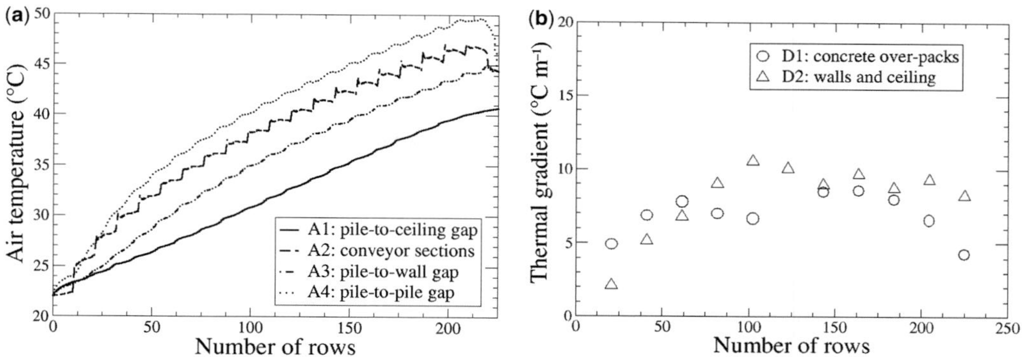

Fig. 21. Evolution along the storage cell of (**a**) mean air temperatures in the longitudinal gaps and (**b**) maximum thermal gradients in concrete; results after 4 years from case 42 ($D_{\text{air}} = 3\ \text{m}^3\ \text{s}^{-1}$, maximum heat release and 'section closure' scenarios.

obtained for the storage of 10-year-old waste packages (case 2), but the temperatures and the thermal gradients are higher. The air temperature reaches 74 °C (see curve A4 in Fig. 22a) and the thermal gradient rise up to 24 °C m^{-1} at the top of walls and to 20 °C m^{-1} at the bottom of the first rows above the conveyor sections (see curves D2 and D1 in Fig. 22b).

Comparative analysis

Maximum air temperature (see Table 3). The operating conditions in the reference case lead to a maximum air temperature of 37 °C (case 1: scenarios of a high ventilation rate and a low heat release). The air temperature only exceeds the threshold of 65 °C in case 43 with a maximum of 74 °C. This case adds up two scenarios favourable to high air temperatures: the high heat release and the low ventilation rate. Moreover, these operating conditions also maximize the thermal gradients in the walls (24 °C m^{-1}) and in the concrete over-packs (20 °C m^{-1}). In the other sensitive cases (cases 2, 3, 41 and 42), the range of the maximum air temperatures is 50–54 °C, because these calculation cases associate a scenario favourable to low temperatures (i.e. high ventilation or low heat release) with another favourable to high temperatures (i.e. low ventilation or high heat release).

The Richardson's analysis suggests the prominence of free convection whatever the operating conditions simulated (see the definition of Richardson's number in 'Buoyancy modelling and analysis' above), which is consistent with the thermal stratifications observed without 'section closure'.

Maximum thermal gradient in the concrete over-packs (see Table 3). Low thermal gradients in concrete over-packs are promoted by the air-flow redistribution between the longitudinal gaps and the conveyor sections because the ventilation of the transverse row-to-row gaps increases convective

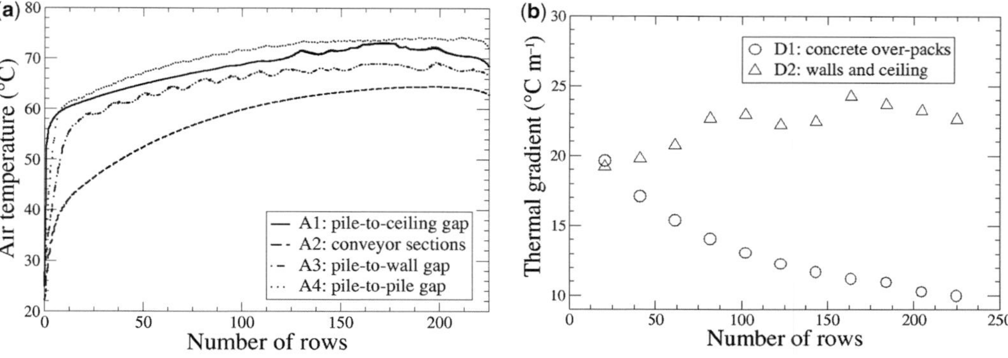

Fig. 22. Evolution along the storage cell of (**a**) mean air temperatures in the longitudinal gaps and (**b**) maximum thermal gradients in concrete; results after 4 years from case 43; ($D_{\text{air}} = 0.3\ \text{m}^3\ \text{s}^{-1}$ and maximum heat release scenario).

Table 3. *Maximum air temperature and thermal gradient reached and specific flow pattern for each selected calculation case*

Case number	$\vec{\nabla}(T)_{max}$ in piles ($^{\circ}C\ m^{-1}$)	$\vec{\nabla}(T)_{max}$ in walls ($^{\circ}C\ m^{-1}$)	T_{max} in the air ($^{\circ}C$)	Richardson's No. (−)	Reverse flow
1	10.2	8	37	8	no
2	11.2	15	51	1750	yes
3	6.7	13	53.5	1900	no
41	19	15	50	17	no
42	8.6	11	50	17	no
43	20	24	74	3100	yes

heat exchanges with the waste over-packs: the lowest maximum thermal gradients in over-packs are observed along the last rows in case 1 (4 $^{\circ}C\ m^{-1}$) and case 41 (6 $^{\circ}C\ m^{-1}$) because of the increase of the air-flow redistribution at the end of the cell, and in case 3 (6.7 $^{\circ}C\ m^{-1}$) and case 42 (8.6 $^{\circ}C\ m^{-1}$) because of the 'section closures'. In case 41, in contrast, the maximum gradient reaches 19 $^{\circ}C\ m^{-1}$ because of the low redistribution of the air flow in the second third of the cell.

Maximum thermal gradients in the walls and the ceiling (see Table 3). Low thermal gradients in walls and ceiling are mainly promoted by low air temperatures because the temperature potential through the wall mainly depends on the *in situ* rock temperature (22 °C) and on the air temperature.

The lower maximum thermal gradient is observed in case 1 (8 $^{\circ}C\ m^{-1}$) with the reference operating conditions: a 3 $m^3\ s^{-1}$ ventilation rate for storage of 10-year-old waste packages. In contrast, the higher maximum thermal gradient is observed in case 43. The 'section closures' slightly reduces the maximum thermal gradients by 2 $^{\circ}C\ m^{-1}$ for storage of 10-year-old waste packages with a low ventilation rate (case 3 and case 2), and by 4 $^{\circ}C\ m^{-1}$ for storage of fresh waste packages with a high ventilation rate (case 42 and case 41).

The maximum thermal gradients are potentially damaging to the non-reinforced concrete of the walls or the ceiling whatever the operating conditions simulated.

Synthesis and conclusions

In some IL-LLW storage modules from the 2009 Andra repository design (Andra 2009), the large sections that will remain in the slab floor of the storage cell after the emplacement of the waste over-packs will have a key role in the ventilation of the storage cell, and therefore on the cooling of waste packages. The aim of the study has been to determine to what extent some operating conditions could prevent the air temperatures and the thermal gradients in concrete from getting too high (i.e. temperature in the air under 65 °C and gradients under 7 $^{\circ}C\ m^{-1}$ in the non-reinforced concrete structure). The temperatures in the air, in the concrete structure of the storage cell and in the rock have been evaluated by numerical simulation, considering various scenarios about the operating conditions: the ventilation rate, the age of the radioactive waste when emplaced and the possible stepwise closure of some conveyor sections.

A specific modelling approach has been adopted to assess the distribution of temperatures and air flows at a decimetre scale (size of the smallest mesh elements). This modelling choice has been motivated by the huge contrast between the hydraulic diameter of apertures (some decimetres) and the length of the storage cell (325 m). Thus, the overall number of elements of the mesh has been kept under 360 000. Head-loss correlations have been implemented to calculate the air-flow distribution through the network of gaps between package piles, walls, ceiling and conveyor-section floor. The numerical calculations have been performed over a 4-year-long thermal transient.

Two air ventilation rates have been considered: 3 $m^3\ s^{-1}$, the reference value (case 1); and 0.3 $m^3\ s^{-1}$, a lower ventilation rate which is envisaged because it is more cost-effective (case 2). The maximum air temperature reaches 37 °C and 51 °C respectively for the two cases. The heat released in waste packages is partly removed from the storage cell by the ventilation, and partly transferred to the walls and ceiling by radiated and convective heat exchange. After 4 years, the ventilation removes from a storage cell two-thirds of the heat released in the reference case, and only a seventh with the lower ventilation rate scenario.

The flow patterns obtained with both ventilation rate scenarios are very different. With the higher ventilation scenario, the air mainly flows through the conveyor sections, below the package rows. In the other longitudinal gaps, the heat is conveyed by forced convection but free convection is locally active in the transverse gaps between package rows. Free convection generates a 10 °C thermal stratification that spreads along two-thirds of the storage cell (case 1). With the lower ventilation rate (case 2), free convection dominates, and a reverse flow under the ceiling spreads over the first two-thirds of the storage cell. The stepwise closure of conveyor sections after every ten rows prevents the reverse flow formation by forcing the flow to be partly redistributed into the longitudinal gaps by crossing the transverse gaps between rows.

The air temperature exceeds the threshold of 65 °C only in case 43 with a high heat release (i.e. storage of fresh waste packages) and a low ventilation rate. Other operating conditions tested (cases 3, 41 and 42) lead to a maximum air temperature of between 50 °C and 54 °C.

With regards to thermal gradients, the comparative analysis of results shows that: (i) low maximum temperatures in the air promote low maximum thermal gradients in the walls and the ceiling; and (ii) an efficient ventilation of transverse gaps between the package rows promotes low maximum thermal gradients in the concrete over-packs by improving convective heat exchanges all around them. Therefore, the stepwise closure of conveyor sections significantly reduces the thermal gradient in the over-packs because the air redistribution following every conveyor-section closure improves the ventilation between rows all along the cell.

The over-packs are less exposed to damage by thermal gradient because their concrete structure is reinforced. The maximum thermal gradient is reached either on the top or on the bottom of the pile. In contrast, the non-reinforced concrete of the walls and the ceiling could be exposed to damage by a high thermal gradient. The maximum thermal gradient is located at the corner of the cell and exceeds 7 $^{\circ}C\ m^{-1}$, whatever the operating conditions simulated. However, it should be noted that taking account of the decrease of the heat release due to radioactive decay should lead to a downward reassessment of this result, whereas considering a 400-m-long storage cell should have the opposite effect.

Among operating conditions tested in the study, the optimal conditions to consider in order to minimize thermal gradients in walls are probably the higher ventilation rate combined with the lower heat release and the stepwise closure of conveyor sections after every ten rows: the stepwise closure of some conveyor sections is an effective approach to reducing the thermal gradient in the concrete structure.

This work was funded by Andra, and the models presented are based on data provided by Andra. The authors are grateful to S. Watson and an anonymous reviewer for their helpful and constructive remarks.

References

ANDRA 2009. *Dimensionnement et architecture générale d'un stockage. Site de Meuse/Haute-Marne.* Note n °C.NSY.ASTE.080171. Andra, Châtenay-Malabry, France.

Bellivier, A. 2004. *Modélisation numérique de la thermo-aéraulique du bâtiment: des modèles CFD à une approche hybride volumes finis /zonale.* PhD thesis, Université de La Rochelle.

Bénet, L. V. 1996. *Résolution des équations de Darcy exprimées en fonction de la pression et de la vitesse.* Prise en compte des effets de densité. Technical Report CEA/DRN/DMT/96/250.

CAST3M http://www-cast3m.cea.fr/

Comte-Bellot, G. 1965. *Ecoulement turbulent entre deux parois parallèles.* Publications scientifiques et techniques du Ministère de l'Air, France. http://hal.archives-ouvertes.fr/docs/00/63/02/13/PDF/Comte-Bellot-1965-FrancaisOCR.pdf

Dabbene, F. 1998. Mixed hybrid finite elements for transport of pollutants by underground water. *In*: Hafez, M. & Heinrich, J.-C. (eds) *Proceedings of the 10th International Conference on Finite Elements in Fluids.* Tucson, Arizona, 456–461.

Gounand, S. 2012. *Introduction à la méthode des éléments finis en mécanique des fluides incompressible*, Cours ENSTA B2-1. http://www-cast3m.cea.fr/

Idel'cik, I. E. 1979. *Mémento des pertes de charges, collection de la direction des études et recherches d'Electricité De France*, Introduction ála méthode des eéléments finis en mécanique des fluides incompressible. Eyrolles, Paris.

Kern, D. Q. 1950. *Process Heat Transfer.* McGraw-Hill Classic Textbook Reissued Series (1989). McGraw-Hill Book Company, Inc., New York.

Le Potier, C. 2005. Schéma volumes finis pour des opérateurs de diffusion fortement anisotropes sur des maillages con structurés. *Comptes Rendus de l'Académie des sciences Series I*, **340**, 921–926.

McAdams, W. H. 1942. *Heat Transmission.* 2nd edn. McGraw-Hill Book Company, Inc., New York.

Pinson, F. 2006. *Modélisation à l'échelle macroscopique d'un écoulement turbulent au sein d'un milieu poreux.* PhD thesis, Institut National Polytechnique de Toulouse.

Ritz, J.-B. & Bénet, L.-V. 2002. *Faisabilité d'un outil de prédimensionnement thermique des halls d'entreposage de combustibles usés.* Technical Report HI-82/01/51/A, EDF R&D.

Sacadura, J. F., Barrand, J. F. *et al.* 1980. *Initiation aux transferts thermiques*, Technique et Documentation, Lavoisier, Paris.

Regional groundwater flow modelling of the confined aquifers below the Boom Clay in NE Belgium

K. VANDERSTEEN[1]*, M. GEDEON[1], J. MARIVOET[1] & L. WOUTERS[2]

[1]*SCK•CEN, Belgian Nuclear Research Centre, Boeretang 200, B-2400 Mol, Belgium*

[2]*ONDRAF/NIRAS, Belgian Agency for Radioactive Waste and Enriched Fissile Materials, Kunstlaan 14, B-1210 Brussels, Belgium*

**Corresponding author (e-mail: katrijn.vandersteen@sckcen.be)*

Abstract: For more than 35 years, SCK•CEN has been investigating the possibility of high-level radioactive waste disposal in the Boom Clay in northeastern Belgium. This research, defined in the long-term management programme for high and medium long-lived waste of ONDRAF/NIRAS, includes the regional hydrogeological modelling of the aquifer systems surrounding the Boom Clay. In this paper, the most recent update of the Deep Aquifer Pumping model (DAP) is described, which is capable of quantifying the regional groundwater flow in the complex confined aquifers lying below the Boom Clay in NE Belgium. The DAP model was successfully calibrated using an automated inverse optimization algorithm and is able to reproduce satisfactorily the general trends in the observed groundwater heads. This model can be used as a tool for planning future characterization efforts and reducing the predictive model uncertainty.

In the framework of the Belgian research programme on the long-term management of high-level and/or long-lived radioactive waste coordinated by ONDRAF/NIRAS, the Boom Clay is considered as a reference host rock for the geological disposal of such radioactive waste in NE Belgium. The hydrogeological programme at SCK•CEN supports the long-term performance assessment studies of the geological disposal of radioactive waste by performing phenomenological research of the aquifer systems surrounding the studied disposal system. One of the important components of this programme is the regional hydrogeological modelling, which aims at:

(a) understanding the hydrogeology in the aquifers surrounding the Boom Clay;
(b) providing boundary conditions for the local flow and transport models that are used in the performance assessment studies which in turn aim at determining the long-term radiological impact due to a hypothetical radioactive waste repository in the Boom Clay (Marivoet *et al.* 2002);
(c) simulating the hydrogeological system behaviour for changing boundary conditions, for example, due to climatic changes.

The regional hydrogeology is studied using two main models, the steady-state Neogene Aquifer Model (NAM) (Gedeon 2008) and the transient Deep Aquifer Pumping model (DAP) (Gedeon & Wemaere 2009), developed to characterize and quantify the regional groundwater flow in, respectively, the aquifers lying above (NAM) and below (DAP) the Boom Clay in NE Belgium. This paper summarizes the most recent update of the DAP model, which comprises implementing a much more detailed 3D geological model, a modified parameterization and an updated groundwater extraction dataset. For more details, the reader is referred to Vandersteen *et al.* (2012).

Conceptual model assumptions

The DAP model includes the confined parts of the Oligocene and Ledo-Paniselian-Brusselian aquifer systems and their respective confining layers, the Bartoon aquitard and the Boom aquitard (Fig. 1a). The assumed bottom boundary of this deep aquifer system is formed by a sequence of two aquitard systems, the Paniselian aquitard and the Ypresian aquitard.

Figure 1b gives the detailed hydro- and lithostratigraphy of the subsurface from the Miocene to the Ledo-Paniselian-Brusselian aquifer, as implemented in the current update of the DAP model. We mainly based the geometry of the hydrostratigraphic units of the deep aquifer system on the studies by Welkenhuysen *et al.* (2012) and Houthuys (1990). In total, 17 different hydrogeological units were implemented in the model. The horizontal extent of the DAP model is given in Figure 2. The model extends in the south as far as the outcrops of the major aquitards: the Asse/Ursel Clay (lower part of the Bartoon aquitard)

From: Norris, S., Bruno, J., Cathelineau, M., Delage, P., Fairhurst, C., Gaucher, E. C., Höhn, E. H., Kalinichev, A., Lalieux, P. & Sellin, P. (eds) 2014. *Clays in Natural and Engineered Barriers for Radioactive Waste Confinement*. Geological Society, London, Special Publications, **400**, 163–177.
First published online April 30, 2014, http://dx.doi.org/10.1144/SP400.22

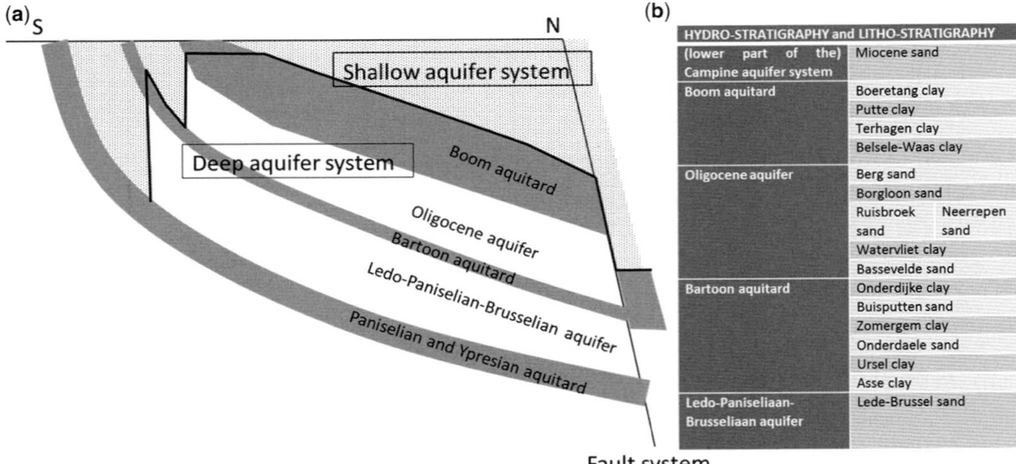

Fig. 1. (**a**) Conceptual model cross-section; (**b**) hydrostratigraphic units and corresponding lithological layers incorporated in the DAP model. The extent of the deep aquifer system is given by the thick black line.

confining the Ledo-Paniselian-Brusselian aquifer, and the Boom Clay confining the Oligocene aquifer. The northern and western boundaries of the model are chosen arbitrarily and extend nearly as far as the Oosterschelde and Hollands Diep to the north, and as far as Zeeland towards the west. The eastern boundary is placed at the most western fault of the Roer Valley Graben fault system, in the east of Flanders.

Since the studied system as a whole is buried under very low-permeable Boom Clay (vertical hydraulic conductivity *c.* 10^{-12} m s^{-1}), the groundwater sources are limited, as the flow percolating through the Boom Clay is negligible. The main recharge sources are the outcrops of the formations forming the studied system, located south of the Boom Clay outcrop. Groundwater extraction is the most important process that affects the natural flow. Because of the imbalance between the groundwater sources and sinks, the groundwater heads are continuously decreasing at most of the piezometers with time series of more than 30 years (Labat 2011). As opposed to the studied deep aquifer system, the aquifer system overlying the Boom Clay is dominated by surface hydrological processes. Here, the heads fluctuate seasonally without any apparent long-term trend and we can consider these groundwater heads as constant in time for the considered time period of about 30 years. Since the processes in the two aquifer systems are assumed to be independent, the groundwater flow processes occurring outside the model domain (in the unconfined parts of the studied aquifers and in the aquifer system located above the Boom Clay) can be replaced by fixed head boundary conditions.

Two other boundary types are defined at the eastern, western and northern limits of the model. At the eastern border, the massive fault system related to the Roer Valley Graben structure displaces the entire stratum vertically, possibly causing an interconnection between the Miocene and the Oligocene aquifers. Here, we assume that a zero-flux boundary condition is the most appropriate as suggested by estimated ages based on measured ^{14}C concentrations (Marivoet *et al.* 2000) and piezometric data near the fault (Labat 2011).

In the north and west, the deep aquifer system continues to deepen underneath Dutch territory (north) and the North Sea (west). Here, an equilibration of groundwater heads towards zero (sea-level) can be assumed at a far distance from the model boundary. Appropriate boundary conditions (general head) reflect this assumption (Gedeon & Wemaere 2003).

Field measurements suggest a decrease with depth of the hydraulic conductivity of the confined aquifers in NE Belgium (Vandersteen *et al.* 2013). We adopt the approach presented by Whittemore *et al.* (1993), that is, the permeability decreases exponentially with increasing depth. We further assume that the specific storage of the aquifers decreases with increasing depth according to the formula of van der Gun (1979).

Groundwater extraction is at present the most important process that affects the groundwater flow in the confined aquifer system below the Boom Clay. Therefore, it is essential to reconstruct the groundwater extraction history in these aquifers as far as possible into the past. Unfortunately, the information about groundwater extraction in the

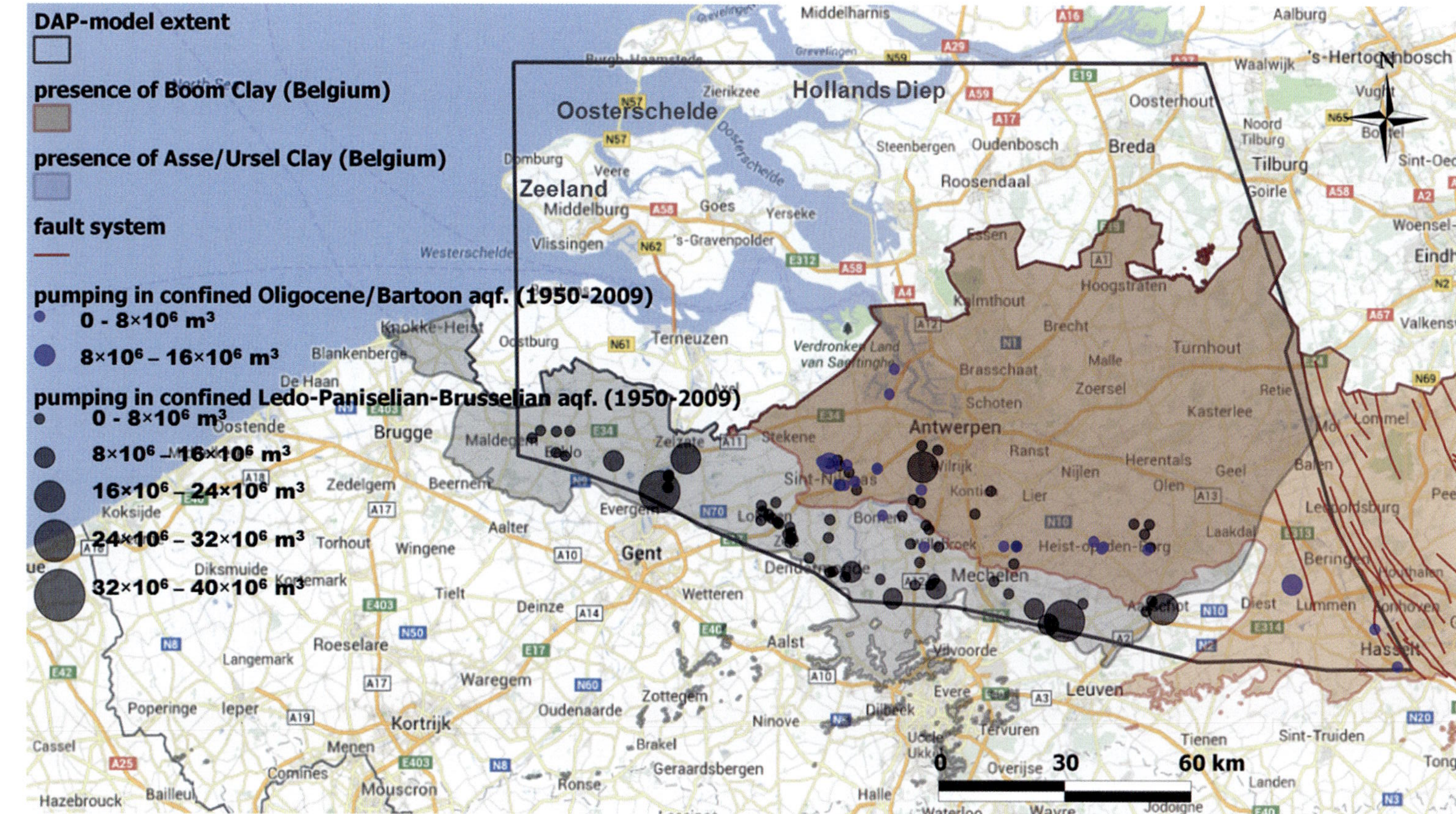

Fig. 2. Extent of the DAP model (in black) in relation to the presence of Boom Clay, Asse/Ursel Clay and the fault system related to the Roer Valley Graben, including location of large groundwater extraction installations in the confined part of the Oligocene aquifer and Bartoon aquitard (blue) as well as in the confined part of the Ledo-Paniselian-Brusselian aquifer (black).

deep aquifers is limited to the last decennia. In this study, the year 1950 was chosen as a starting date for groundwater extraction because we assume that the majority of the industries started to recover from the World War II economic depression and because of the introduction of legislation concerning the use of groundwater in 1946.

Data collection and analysis

Three types of data for use in the groundwater model were collected and analysed: groundwater extraction data, groundwater head time series and hydraulic property measurements.

In order to build a transient model of the aquifers below the Boom Clay, detailed information (going as far back as possible into the past) concerning the groundwater extraction rates in the aquifers was gathered.

Figure 2 shows the position of the large groundwater extraction installations in the confined part of the Oligocene aquifer, the Bartoon aquitard and the Ledo-Paniselian-Brusselian aquifer. Most of the large installations are situated near the outcrop of the Boom Clay (for the wells in the Oligocene aquifer) or the outcrop of the Asse/Ursel Clay (Bartoon aquitard and Ledo-Paniselian-Brusselian aquifer). Much less groundwater extraction occurs from the Oligocene aquifer and the Bartoon aquitard than from the Ledo-Paniselian-Brusselian aquifer.

From the available piezometers (Databank Ondergrond Vlaanderen 2011; Labat 2011) with at least a 10-year record, both the interpolated groundwater heads and the temporal groundwater head trends for the Oligocene aquifer, the sandy parts of the Bartoon aquitard and the Ledo-Paniselian-Brusselian aquifer were analysed. The latter were calculated as the slope of the regression line through the yearly mean groundwater heads.

The interpolated groundwater level measurements in the Oligocene and the Ledo-Paniselian-Brusselian aquifers for the year 2010 are visualized in Figures 3 and 4. The aquifers are recharged from the east and south, where the water is partly removed by pumping. The groundwater flow is mainly in a northwesterly or (more to the north of Belgium) westerly direction and is heavily influenced by groundwater extraction south of Antwerp. Towards the northern part of Belgium, the heads in the Ledo-Paniselian-Brusselian aquifer are considerably (>5 m) lower than the heads in the Oligocene aquifer.

Figure 3 also shows the trend in the head observation data for the Oligocene aquifer and the sandy parts of the Bartoon aquitard. Towards the northern Belgian border, neither increase nor decrease in heads is observed. A groundwater head decrease is observed in the eastern part of the model (near the city of Mol). Towards the southern model boundary, near the outcrop of the Boom Clay, an increase in groundwater heads is present in the data. More to the SE of the model near the outcrop, the groundwater heads are constant in time or have a negative trend. Figure 4 shows the trends in the head observation data in the Ledo-Paniselian-Brusselian aquifer. In the northern part of Belgium, a negative trend is prevalent. Although the groundwater extraction is concentrated in the south, it causes the groundwater heads to decrease, even far northwards. Near the outcrop of the Boom Clay and the Bartoon aquitard, the trend is mainly upward.

It is most likely due to a combination of the pumping in the south, which is greater for the Ledo-Paniselian-Brusselian aquifer than for the Oligocene aquifer, and the larger transmissivity of the Ledo-Paniselian-Brusselian aquifer, that the heads in the Ledo-Paniselian-Brusselian aquifer are considerably lower than the heads in the Oligocene aquifer. This was also pointed out by Olsthoorn (2011).

The characteristic time for pressure changes to propagate a given distance through an aquifer is known as the aquifer hydraulic diffusion time, D_{H} $[L^2T^{-1}]$, which is defined as follows:

$$D_{\mathrm{H}} = \frac{T}{S} = \frac{K}{S_{\mathrm{s}}} \tag{1}$$

where T is transmissivity $[L^2T^{-1}]$, S is storage coefficient [-], K is hydraulic conductivity $[LT^{-1}]$ and S_{s} is specific storage $[L^{-1}]$.

The characteristic time for diffusion t_D over distance L is given by Harrar *et al.* (2001):

$$t_{\mathrm{D}} = \frac{L^2}{D_{\mathrm{H}}} = \frac{L^2 S}{T} = \frac{L^2 S_{\mathrm{s}}}{K} \tag{2}$$

The time it takes for a pressure change in the Oligocene or Ledo-Paniselian-Brusselian aquifers due to groundwater abstraction in the south, to be seen in the north of the area, is thus dependent on the distance, the specific storage of the aquifer and its hydraulic conductivity. On the assumption that the storage coefficient in both aquifers is the same, only the difference in an aquifer's transmissivity will determine the difference in characteristic time for both aquifers. The larger the aquifer's transmissivity, the earlier a pressure change in the south will be observed in the north. In case of the Oligocene and Ledo-Paniselian-Brusselian aquifers, the heads in the latter are going down in the north, while the heads in the former are staying constant in the north. This most likely implies that the transmissivity of the Oligocene aquifer is lower than the transmissivity of the Ledo-Paniselian-Brusselian aquifer. This effect is reinforced by the greater

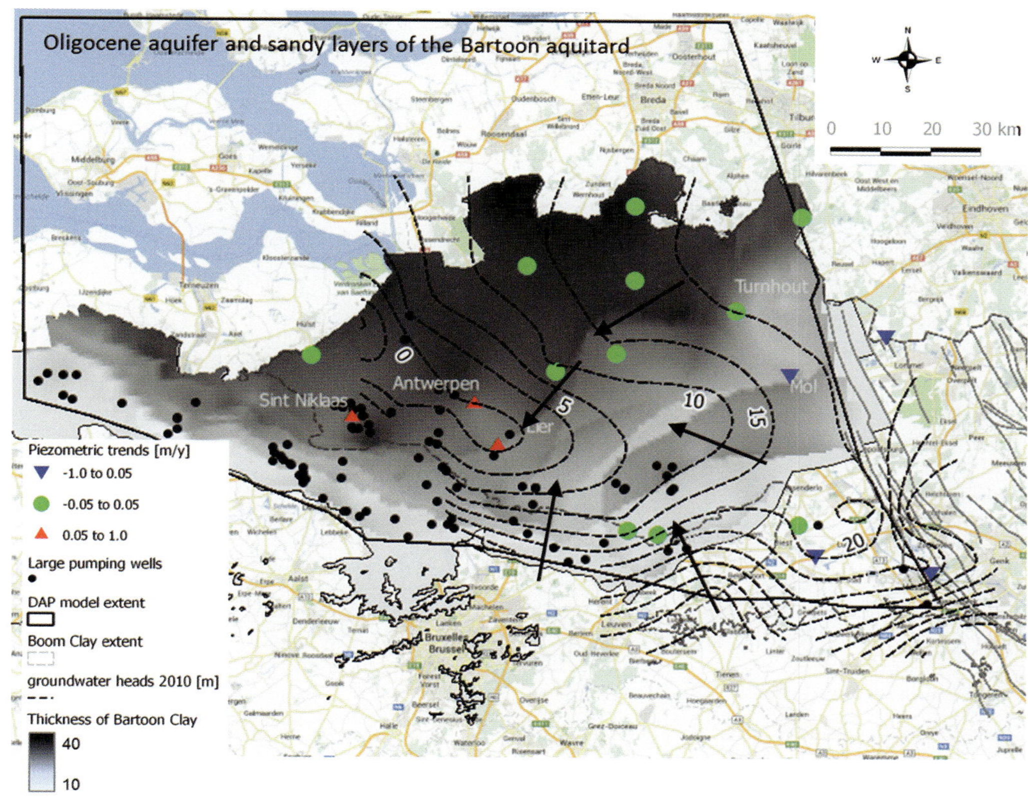

Fig. 3. Main trends in groundwater heads (negative = decrease; positive = increase of the heads) observed in the Oligocene aquifer for the period 2000–2010, together with measured interpolated groundwater heads for 2010. The trends in the Oligocene aquifer are related to the thickness of the Bartoon Clay (shaded). The black arrows indicate the groundwater flow direction.

amount of pumping from the Ledo-Paniselian-Brusselian aquifer (causing larger pressure drops).

However, in the northeastern part of Belgium (Mol), far from the groundwater abstractions in the south, the groundwater head in the Oligocene aquifer is also decreasing, The explanation here is most likely the limited thickness and/or possibly variations in the hydraulic conductivity of the Bartoon aquitard in this region, which separates the Oligocene aquifer from the underlying Ledo-Paniselian-Brusselian aquifer, compared to the greater thickness of the Bartoon aquitard more to the NW. Analogous to the previous reasoning, the characteristic time for a pressure head change in the over/underlying aquifer to propagate through the Bartoon aquitard can be determined from equation (2). Around Mol, the Bartoon aquitard's thickness is *c.* 10 m, while more to the NW, the thickness is *c.* 50 m. Assuming that the storage coefficient, *S*, of the Bartoon Clay is the same in both regions, the difference in characteristic time in both regions will only be influenced by the aquitard's thickness and/or its hydraulic conductivity: the larger the thickness of the Bartoon aquitard and/or the smaller the hydraulic conductivity, the larger t_D and the longer it takes for a pressure change in the Ledo-Paniselian-Brusselian aquifer to be seen in the Oligocene aquifer.

A recent decrease in the amount of groundwater extraction from the Oligocene and Ledo-Paniselian-Brusselian aquifers, mainly towards the western part of the Belgium, can explain the upward trends in the groundwater heads in both aquifers near the outcrop of the Boom Clay.

Detailed hydraulic conductivity data of the Boom Clay in NE Belgium are available from five boreholes (Mol, Zoersel, Weelde, Essen and Doel), the results of which are summarized by Yu *et al.* (2013). At these five boreholes, we determined upscaled hydraulic conductivity values for use in the hydrogeological model (Table 1), which is described in detail in Vandersteen *et al.* (2012).

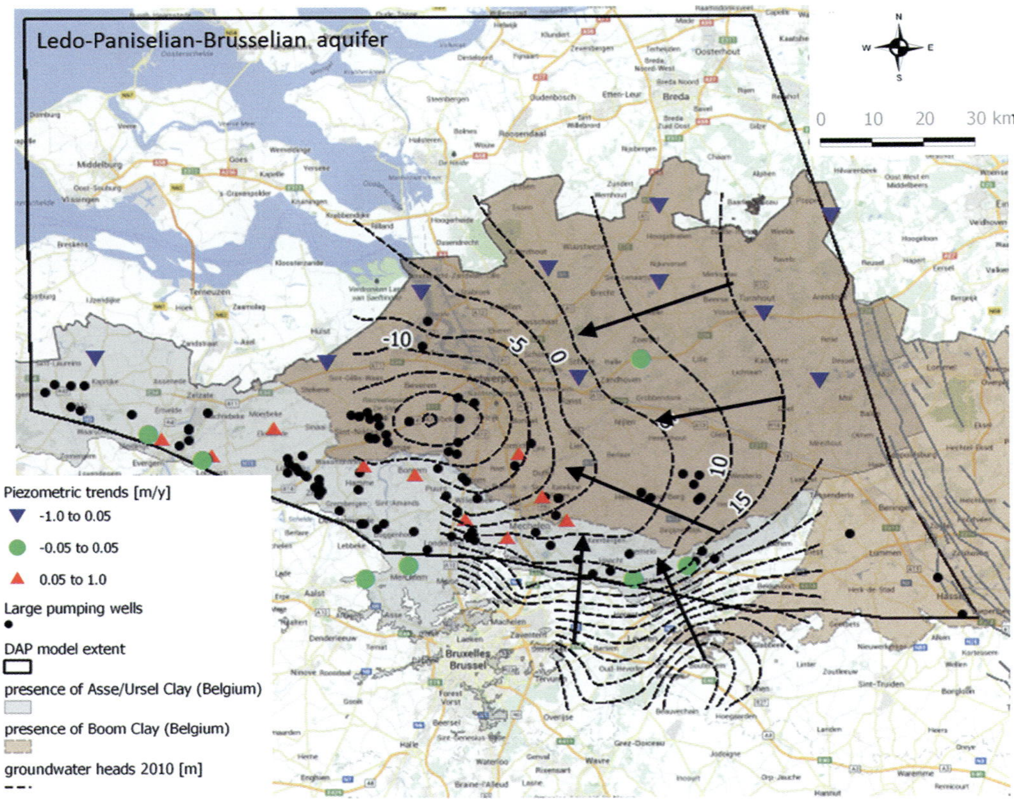

Fig. 4. Main trends in groundwater heads (negative = decrease; positive = increase of the heads) observed in the Ledo-Paniselian-Brusselian aquifer for the period 2000–2010, together with measured interpolated groundwater heads for 2010. The groundwater flow direction is indicated by black arrows.

From the detailed sequence of hydraulic conductivity values, the upscaled vertical and horizontal conductivity were calculated as the harmonic mean and the arithmetic mean respectively of all the samples. Based on the experimental data for the Boom Clay, it was found by Yu *et al.* (2013) that the vertical hydraulic conductivity of the Boom Clay in the Putte-Terhagen Member is depth-dependent following an exponential relationship. At three regional boreholes (Mol, Zoersel and Weelde), hydraulic conductivity measurements on the Asse Clay (Bartoon aquitard) are available. Upscaled horizontal and vertical hydraulic conductivity values are calculated in the same way as for the Boom Clay. Results are given in Table 1.

Table 1. *Conductivity of Boom Formation and Asse Clay*

	Upscaled (regional) value	
Layer	K_v (m s^{-1})	K_h (m s^{-1})
Boeretang	6.6×10^{-12}	4.2×10^{-11}
Putte + Terhagen	4.1×10^{-12}	1.2×10^{-11}
Belsele-Waas	2.2×10^{-11}	1.8×10^{-9}
Asse Clay	2.8×10^{-12}	1.2×10^{-11}

A limited number of measurements of the hydraulic conductivity (i.e. laboratory measurements on cores, slug test measurements and pump test measurements) of the Oligocene aquifer, the sandy layers of the Bartoon aquitard and the Ledo-Paniselian-Brusselian aquifer are available (Table 2). We used these values as guidance for an initial estimate of the groundwater model parameters.

Model design

The updated version of the 3D DAP model is implemented in the MODFLOW-2005 code

Table 2. *Aquifer conductivity measurements*

Borehole	K_h Oligocene aquifer [m s^{-1}]		K_h Ledo-Paniselian-Brusselian aquifer [m s^{-1}]		
	Lab. meas.	Slug test	Lab. meas.	Slug test	Pump test
Doel (Wemaere *et al.* 2008)	1.4×10^{-8}				
Zoersel (Wemaere *et al.* 2004*b*)	4.8×10^{-9}	3×10^{-8}	3.86×10^{-7}	9.0×10^{-7}	
Mol (Wemaere *et al.* 2002)	2.4×10^{-8}				
Weelde (Wemaere *et al.* 2005*a*)	2.8×10^{-9}				
Rijkevorsel (Wemaere *et al.* 2004*a*)		1×10^{-7}			2.9×10^{-6}
Turnhout (Wemaere *et al.* 2005*b*)		4×10^{-8}			3.1×10^{-6}
Herenthout (Labat *et al.* 2008*a*)	1.52×10^{-7}, 1.58×10^{-8}*		5.11×10^{-7}		
North of Leuven (Ministerie Vlaamse Gemeenschap 2004)				2.8×10^{-5}	
Lembeke (Carreres 1993)					4.9×10^{-5}

*Measurements at different depths.

(Harbaugh 2005). The transient nature of the overexploitation problem in the deep aquifers requires a transient simulation. The time frame of the transient model extends from 1950 up to the end of 2009. We divided the total simulation time of 59 years into 29 stress periods, which were further divided into yearly time steps. Four stress periods were used for the time span from 1950 to 1985, for which little groundwater extraction data and no observation data are available. Twenty-five stress periods were defined for the years from 1985 to 2009, for which more detailed information on groundwater extraction and head monitoring data exist.

The horizontal schematization is irregular with cell dimensions ranging from 500 to 4000 m. The most detailed representation is at locations with a high density of groundwater extraction wells. In the vertical direction, the model includes 11 numerical layers: one layer representing the Miocene; three layers the Boom Formation; five layers the heterogeneous Oligocene aquifer and Bartoon aquitard; and two layers the Ledo-Paniselian-Brusselian aquifer. We used the Hydrogeologic-Unit Flow (HUF) package of MODFLOW-2005 (Anderman & Hill 2000) given the complexity of the sandy and clayey sediments of the aquifers, to be able to define the numerical finite difference grid separately from the geological layer geometry.

The definition of hydraulic parameters for the DAP model represents a challenging task because of its 17 different hydrogeological units. Parameterizing each of these units separately can lead to problems in the parameter estimation, that is, non-sensitivity and parameter correlation, and would require a large number of model runs within an appropriate global inversion framework, which is outside the goal of the current work. To be able to apply the local optimization methods classically used in groundwater flow modelling, we used a number of assumptions, reducing the number of parameters:

(1) In the complex Oligocene aquifer and the Bartoon aquitard, the hydraulic conductivity parameters of all sandy layers were combined in one parameter (OB_sand) and all clayey layers in another parameter (OB_clay).
(2) For the Ledo-Paniselian-Brusselian aquifer, horizontal variation in the hydraulic conductivity was incorporated based on the lithostratigraphy described in Houthuys (1990). Three different zones for the horizontal hydraulic conductivity of the Ledo-Paniselian-Brusselian aquifer were implemented using zone arrays in MODFLOW-2005.
(3) The vertical anisotropy of all sandy layers was taken as equal to one.
(4) The specific storage parameters were combined into two parameter clusters: one cluster with all the parameters of the sandy layers and one cluster with all the parameters of the clayey layers. We assumed that the specific storage for sandy aquifers decreases with increasing depth, according to van der Gun (1979), through the use of multiplier arrays. We applied one multiplier array for the Miocene aquifer, one for the sandy layers of the Oligocene aquifer and one for the Ledo-Paniselian-Brusselian aquifer. Each multiplier array contained values of the specific storage for every cell in the layer The parameter

included in the model represents a scaling factor for the multiplier arrays.

(5) The groundwater extraction sinks were parameterized by including one single well multiplication factor for all groundwater extraction rates, which was used to assess the global effect of groundwater extraction rates on the simulated heads.

Calibration

The DAP model was calibrated using the universal automated calibration program UCODE 2005 (Poeter *et al.* 2005). The method used by this code consists of finding the model parameter values that minimize a misfit criterion, in our case, the sum of squared residuals (SSR) based on observed and predicted groundwater heads. UCODE 2005 uses a local optimization procedure. If the objective function includes several local optima, convergence of local search procedures will depend on the starting point of the search. To minimize the chances of finding a wrong optimum of the objective function, the automated calibration was preceded by a manual calibration step aimed at finding good starting values for the optimization procedure.

Throughout the calibration procedure, the parameter sensitivities of the parameters in Table 3 were evaluated using the composite scaled sensitivities (Hill 1998), indicating the total amount of information provided by the observations for the estimation of one parameter, as well as the correlation between the different parameters. The composite scaled sensitivity for the j_{th} parameter (css_j) was calculated using the scaled sensitivities for all observations (ss_{ij}) as follows (Hill 1998):

$$css_j = \sqrt{\frac{\left[\sum_{i=1}^{ND} (ss_{ij})^2\right]_{\bar{b}}}{ND}} \tag{3}$$

where

$$ss_{ij} = \sum_{k=1}^{ND} \left[\left(\frac{\partial y_i'}{\partial b_j} \right) b_j \sqrt{\omega_{ii}} \right] \tag{4}$$

and ND is the number of observations; y_i' is a simulated value; b_j is the j_{th} estimated parameter; $(\partial y_i'/\partial b_j)$ is the sensitivity of the simulated value with respect to the j_{th} parameter evaluated at the set of parameters in $\bar{b}$; $\bar{b}$ is a vector which contains the parameter values at which the sensitivities are evaluated; and ω_{ii} is the weight of the i_{th} observation.

Insensitive parameters (with composite scaled sensitivities that are more than a factor of 100 less than the maximum composite scaled sensitivity for a set of estimated parameters) were excluded from the parameter estimation process. As the well multiplication factor was highly correlated with the hydraulic conductivity of the Ledo-Paniselian-Brusselian aquifer, we excluded the well multiplication factor (WELL_MLT) from the calibration. Estimating the groundwater extraction rates at individual wells is beyond the scope of this study, which aims at modelling the global effects of overexploitation. Also, changes in the individual groundwater extraction rates are difficult to justify. Different values of the global groundwater extraction amount (corresponding to a potential total error in data) were, however, investigated within the framework of the sensitivity analyses. Table 3 shows the parameter values calibrated using the automated calibration procedure. The time series charts showing the observed versus calculated groundwater heads at selected observation points (Fig. 5) indicate that at most observation points, the general trend in evolution of the measured groundwater heads is reproduced by the model; however, the transient heads at some piezometers in the Oligocene aquifer are not well reproduced, especially at the SCK_20d piezometer. This misfit had already been observed in the previous version of the model (Gedeon & Wemaere 2009), where it was attributed to the use of a simplistic representation of the Oligocene and Bartoon sediments. The fact that this misfit appears in the current version of the DAP model using a complex geometry of the Oligocene and Bartoon sediments, points out that there are still conceptual issues to resolve. Plausible reasons for the misfits are the lack of detail in the parameterization or issues in the geological representation of the layers.

Evaluation of estimated parameters

After the parameter estimation, we performed a backward check of the estimated hydraulic conductivity values by comparing them with measured values. In case of the Putte-Terhagen Member of the Boom Formation, estimated values are in the range of the measured values (Table 4). In particular, for the Essen, Doel and Zoersel boreholes, the estimated values are very close to the measured values. The greatest difference between measured and estimated values occurs in Mol, where the measured value is overestimated by a factor 3. These, so far unexplained, differences in Mol were also described by Yu *et al.* (2013) and Jeannée (2012). This anomaly also occurs in the DAP model, since it uses a single depth-dependency equation to distribute the hydraulic conductivity values spatially in the modelling domain, and no adjustments at the Mol area were made. In case of the Bartoon aquitard, the estimated vertical hydraulic conductivity

Table 3. *Parameter values and fit statistics for the automatically calibrated model*

Parameter name	Hydrostratigraphic unit		Material type	Starting value	Automatically calibrated value	Unit
HK_mio	Miocene		Sand	1.16×10^{-10}	Not estimated	ms^{-1}
VANI_mio				2×10^{-6}	Not estimated	–
HK_BmBo	Boom–Boeretang		Clay	4.17×10^{-11}	Not estimated	ms^{-1}
VANI_BmBo				6.4	Not estimated	–
HK_BmPuTe	Boom–Putte		Clay	1.85×10^{-11}	Not estimated	ms^{-1}
VANI_BmPuTe	Boom–Terhagen			2.9	Not estimated	–
KD_BmPuTe				0.004	1×10^{-4}	m^{-1}
HK_Bmbw	Boom–Belsele-Waas/Bilzen–Berg		Clay/sand	1.85×10^{-9}	Not estimated	ms^{-1}
VANI_Bmbw				84	Not estimated	–
HK_OB_sand	Oligocene–Borgloon		Sand	1.74×10^{-8}	Not estimated	ms^{-1}
	Oligocene–Ruisbroek					
	Oligocene–Neerrepen					
	Oligocene–Bassevelde					
VANI_OB_sand	Bartoon–Buisputten			1	Not estimated	–
KD_OB_sand	Bartoon–Onderdaele			0.001	Not estimated	m^{-1}
HK_OB_clay	Bartoon–Watervliet		Clay	5.23×10^{-11}	Not estimated	ms^{-1}
	Bartoon–Onderdijke					
	Bartoon Zomergem					
VANI_OB_clay	Bartoon–Ursel			4.52	0.546	–
KD_OB_clay	Bartoon–Asse			0.004	0.004488	m^{-1}
HK1_LPB	Ledo-Paniselian-Brusselian–Vlierzele-Aalter	Zone1		2.3×10^{-5}	9.8×10^{-6}	ms^{-1}
HK2_LPB	Ledo-Paniselian-Brusselian –Brussels	Zone2		2.3×10^{-5}	1.7×10^{-5}	
HK3_LPB	Ledo-Paniselian-Brusselian –Wemmel-Lede	Zone3		2.3×10^{-5}	7.2×10^{-6}	
VANI_LPB	Ledo-Paniselian-Brusselian			1	Not estimated	–
KD_LPB				0.0058	0.005299	m^{-1}
SS_sand (scaling factor)	All sandy units		Sand	1	0.03893	–
SS_clay	All clayey units		Clay	1×10^{-4}	3.2×10^{-5}	m^{-1}
WELL_MLT	–		–	1	–	–
Fit statistics						
Sum squared weighted residuals					1392	
Mean residual (ME)					0.12	
Mean absolute residual (MAE)					1.29	
Root mean sq. residual (RMSE)					1.64	
R^2 (measured v. simulated)					0.985	

HK, hydraulic conductivity; VANI, vertical anisotropy; KD, depth-dependence parameter; SS, specific storage, WELL_MLT, well multiplication factor.

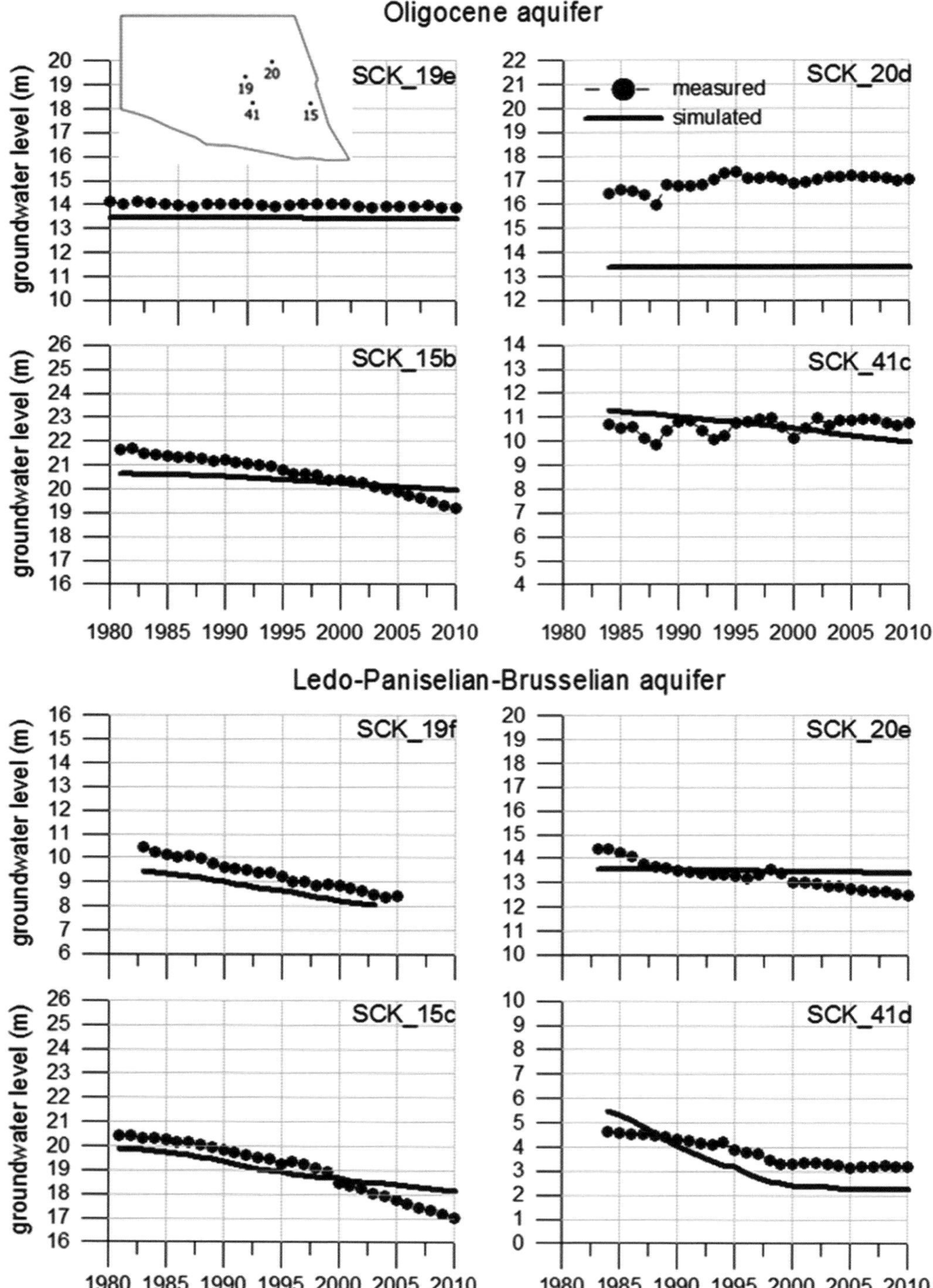

Fig. 5. Time series of simulated and yearly mean observed values at four observation points in the Oligocene and Ledo-Paniselian-Brusselian aquifers. The model extent locating the observation points is shown in the upper left corner.

Table 4. *Comparison between observed and estimated aquitard conductivity values*

Borehole	K_v Putte-Terhagen member [m s^{-1}]		K_v Asse Clay [m s^{-1}]	
	Laboratory measurement	Estimated value	Laboratory measurement	Estimated value
Mol	2.0×10^{-12}	6.0×10^{-12}	1.0×10^{-11}	3.4×10^{-12}
Zoersel	4.6×10^{-12}	6.2×10^{-12}	1.9×10^{-12}	7.1×10^{-12}
Weelde	3.3×10^{-12}	5.9×10^{-12}	1.0×10^{-12}	7.5×10^{-13}
Essen	5.6×10^{-12}	6.1×10^{-12}		
Doel	6.5×10^{-12}	6.2×10^{-12}		

values fall within the range of the measured values for the Asse Clay (Table 4). The greatest difference between measured and estimated values occurs in Zoersel, where the measured value is overestimated by almost a factor 4.

In case of the Oligocene aquifer, Table 5 shows that the differences between the estimated and measured hydraulic conductivity values are within an order of magnitude. The greatest differences between measured and estimated values occur in Rijkevorsel and Turnhout (slug test measurements). In Rijkevorsel, the measured value is underestimated by a factor 10 and in Turnhout by a factor 4.

In case of the Ledo-Paniselian-Brusselian aquifer, the measured values are underestimated in the northern part of the model by a factor 14 (at Turnhout) and 30 (at Rijkevorsel) (Table 5). At the groundwater extraction site at Lembeke-Oosteeklo the values are underestimated by a factor 7.5. Near Zoersel and Herenthout, differences between measured and observed values are rather small (up to a factor 4). For the measurement point north of Leuven, the estimations are in good agreement with the measurements.

Sensitivity analysis

The composite scaled sensitivity (*css*) (equation 3) for the final estimated parameters of the reference model (Fig. 6) was calculated. The well multiplication factor, which was not estimated, but which is known to be an influential parameter, was also included. From Figure 6, we see that the most sensitive parameters of the model are the well multiplication factor (WELL_MLT), the depth-dependence coefficient of the Ledo-Paniselian-Brusselian aquifer (KD_LPB) and the depth-dependence coefficient of the clayey layers of the Oligocene aquifer/Bartoon aquitard (KD_OB_clay).

We optimized the model for different values of the well multiplication factor. Two alternative values were tested, representing a 20% and a 40% increase in the total groundwater extraction. As pointed out previously, only an increase in the

Table 5. *Comparison between observed and estimated aquifer conductivity values*

Borehole	K_h Oligocene aquifer [m s^{-1}]		K_h Ledo-Paniselian-Brusselian aquifer [m s^{-1}]	
	Laboratory measurement (Slug test measurement)	Estimated value	Laboratory measurement (Slug test measurement) [pump test]	Estimated value
Doel	1.4×10^{-8}	1.24×10^{-8}		
Zoersel	4.8×10^{-9} (3×10^{-8})	1.1×10^{-8}	(9.0×10^{-7})	3.43×10^{-7}
Mol	2.4×10^{-8}	9.2×10^{-9}		
Weelde	2.8×10^{-9}	7.3×10^{-9}		
Rijkevorsel	(1×10^{-7})	9.4×10^{-9}	(2.9×10^{-6})	9.52×10^{-8}
Turnhout	(4×10^{-8})	8.9×10^{-9}	(3.1×10^{-6})	2.1×10^{-7}
Herenthout			5.11×10^{-7}	2.36×10^{-6}
Lembeke-Oosteeklo			[4.9×10^{-5}]	6.5×10^{-6}
North of Leuven			2.8×10^{-5}	1.25×10^{-5}

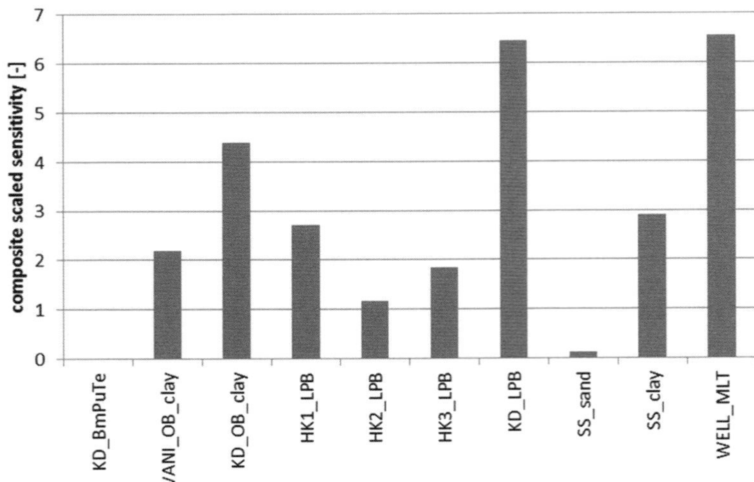

Fig. 6. Composite scaled sensitivity of the final estimated parameters of the reference model.

total groundwater extraction was considered, as we assume that the current groundwater extraction data represent an underestimation of reality. This assumption is based on two facts: (1) only the wells with the highest groundwater extraction rates were selected in our modelling analysis and (2) the groundwater extraction dataset analysis makes us suspect that groundwater extraction activities had already started before 1950, which we consider as a starting date in our model. Table 6 and Figure 7 summarize the results of this analysis.

An increase in the well multiplication factor results in an increase in both the hydraulic conductivity of the Bartoon aquitard (decrease of VANI_OB_clay) and the hydraulic conductivity of the Ledo-Paniselian-Brusselian aquifer (HK1-2-3_LPB). In the central and northern part of the model, this hydraulic conductivity increase is compensated by higher values of the depth-dependence coefficients of the clayey Oligocene and Bartoon Formations (KD_OB_clay) and of the Ledo-Paniselian-Brusselian aquifer (KD_LPB), such that the differences between the hydraulic conductivity values of the reference model and the alternative models are not more than 6% (Fig. 7, Turnhout and Rijkevorsel). The global increase in groundwater extraction rates results in higher estimates of the hydraulic conductivity values VK_OB_clay and HK_LPB, in the southern part of the model (Fig. 7, Lembeke and Leuven), where all groundwater extraction wells are located. Here, differences between the hydraulic conductivity values of the reference model and the alternative models are up to 75%. The model fit (SSR value) is almost identical for all three evaluated models. These results point out the robustness of the estimated hydraulic conductivity values for the areas located further from the groundwater extraction wells (or outcrops).

Table 6. *Conceptual model sensitivity analysis showing the most sensitive estimated parameters and fit statistics of the different tested conceptual models*

Parameter	Well_MLT = 1.2	Well_MLT = 1.4	Reference
Well_MLT [−]	1.2	1.4	1
HK1_LPB [m s^{-1}]	1.103	1.382	0.8491
HK2_LPB [m s^{-1}]	2.065	2.754	1.477
HK3_LPB [m s^{-1}]	0.865	1.136	0.624
VANI_OB_clay [−]	0.399	0.303	0.546
KD_OB_Clay [m^{-1}]	4.897×10^{-3}	5.279×10^{-3}	0.4488×10^{-2}
KD_LPB [m^{-1}]	5.697×10^{-3}	6.03×10^{-3}	0.5299×10^{-2}
SSR	1389	1387	1393

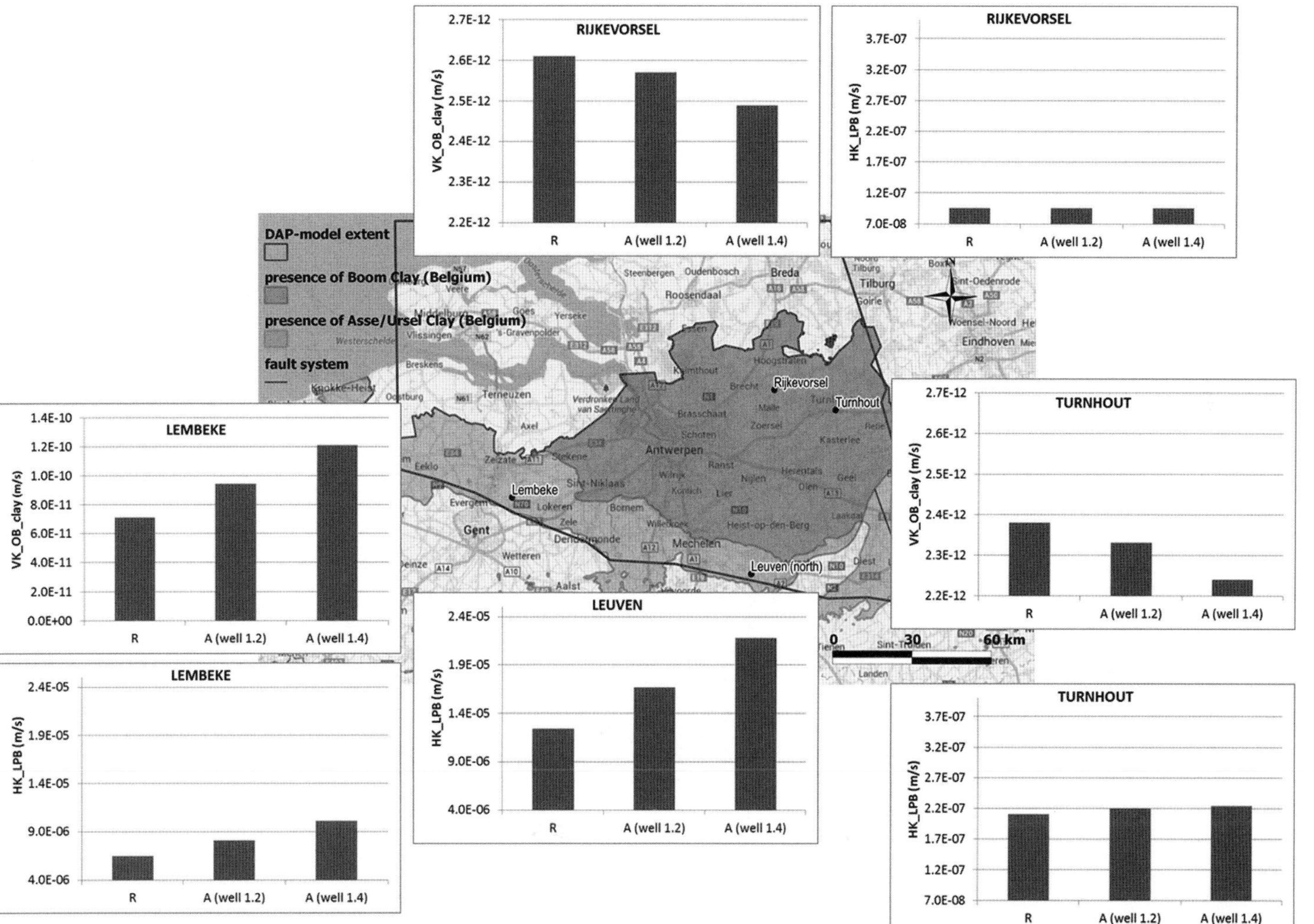

Fig. 7. Comparison between the vertical hydraulic conductivity of the clayey layers of the Oligocene and Bartoon sediments (VK_OB_clay) and the horizontal hydraulic conductivity of the Ledo-Paniselian-Brusselian aquifer (HK_LPB) estimated by the reference model (R) and the alternative models (A (well 1.2) and A (well 1.4)) at different locations in the model.

Conclusions

A conceptual model for the confined parts of the aquifers located below the Boom Clay was developed. A detailed 3D geological model and parameterization were implemented as well as detailed groundwater data. We believe that the current groundwater extraction dataset reflect the accessible and available knowledge about actual groundwater extraction rates in this area. However, uncertainty over both the actual groundwater extraction rates and the groundwater extraction history remains.

The analysis of the hydraulic heads of the confined aquifers below the Boom Clay enabled further insight to be obtained into the effects of intensive groundwater extraction in these aquifers. Since the Oligocene aquifer has a significantly lower transmissivity compared to the Ledo-Paniselian-Brusselian aquifer, the Oligocene groundwater extraction triggers only local effects on groundwater heads. Hence, the regional effects (constant decrease of groundwater heads) in the Oligocene aquifer are presumably caused by groundwater extraction in the Ledo-Paniselian-Brusselian aquifer, whereby the clayey layers of the Bartoon aquitard dampens these effects. The amount of this dampening is given by the spatial distribution of the hydraulic properties of the Bartoon aquitard and/or its variable thickness. For the piezometers located in the Ledo-Paniselian-Brusselian aquifer, we see that although groundwater extraction is concentrated in the south, it causes the water level to decrease even far northwards, which implies that this aquifer has a higher transmissivity than the Oligocene aquifer.

The developed model was successfully calibrated using an automated calibration algorithm. The model is able to reproduce satisfactorily the general trends in the observed groundwater heads. The most sensitive parameters of the model are the amount of groundwater extraction, the depth-dependence coefficient of the Ledo-Paniselian-Brusselian aquifer and the depth-dependence coefficient of the clayey layers in the Oligocene aquifer and the Bartoon aquitard. Changing the amount of groundwater extraction considerably influences the estimated hydraulic conductivity of the Ledo-Paniselian Brusselian aquifer and the Bartoon aquitard in the southern part of the model.

In order to further decrease the level of uncertainty of the DAP model, incorporation of more detail in the parameterization and/or geology of the Ledo-Paniselian-Brusselian aquifer and implementation of the detailed geology of the Netherlands should be performed. As a preparatory task in this context, additional calibration and sensitivity analyses using a global inverse model algorithm may enable more insight to be gained in the model behaviour.

We acknowledge the Flemish Environmental Agency (VMM) for providing groundwater extraction and groundwater head data. We are grateful to two anonymous reviewers for their valuable comments on the manuscript.

References

ANDERMAN, E. R. & HILL, M. C. 2000. *MODFLOW-2000, the U.S. Geological Survey Modular Ground-Water Model – Documentation of the Hydrogeologic-Unit Flow (HUF) Package*. USGS Open File Report **00-342**, Denver, Colorado.

CARRERES, E. L. 1993. The groundwater pumping station at Lembeke. MSc thesis, Faculty of Applied Sciences, Free University Brussels.

DATABANK ONDERGROND VLAANDEREN 2011. Head measurements and filters: Table made on 08/04/2011. http://dov.vlaanderen.be

GEDEON, M. 2008. *Neogene Aquifer Model*. External Report of the Belgian Nuclear Research Centre SCK•CEN-ER-48, Mol, Belgium.

GEDEON, M. & WEMAERE, I. 2003. *Updated Regional Hydrogeological Model for the Mol Site (The North-Eastern Belgium Model)*. External Report of the Belgian Nuclear Research Centre SCK•CEN-R-3751, Mol, Belgium.

GEDEON, M. & WEMAERE, I. 2009. *Transient Model of the Confined Aquifers Located Below the Boom Clay. Regional Hydrogeological Model for the Mol Site*. External Report of the Belgian Nuclear Research Centre SCK•CEN-ER-72, Mol, Belgium.

HARBAUGH, A. W. 2005. *MODFLOW-2005, The U.S. Geological Survey modular ground-water model – the Ground-Water Flow Process*. U.S. Geological Survey Techniques and Methods 6-A16, USGS, Reston, Virginia.

HARRAR, W. G., WILLIAMS, A. T., BARKER, J. A. & VAN CAMP, M. 2001. Modelling scenarios for the emplacement of palaeowaters in aquifer systems. *In*: EDMUNDS, W. M. & MILNE, C. J. (eds) *Palaeowaters in Coastal Europe: Evolution of Groundwater Since the Late Pleistocene*. Geological Society, London, Special Publications, **189**, 213–229.

HILL, M. C. 1998. *Methods and Guidelines for Effective Model Calibration*. U.S. Geological Survey Water-Resources Investigations Report **98-4005**, USGS, Denver, Colorado.

HOUTHUYS, R. 1990. *Vergelijkende studie van de afzettingsstructuur van getijdenzanden uit het Eoceen en van de huidige Vlaamse banken*. Aardkundige Mededelingen 5. University Press, Leuven, Belgium.

JEANNÉE, N. 2012. *Boom Clay: Integrating data and upscaling using geostatistical techniques. Transferability of regionalized variables*. ONDRAF/NIRAS report NIROND-TR 2012-01.

LABAT, S. 2011. *Overview and Analysis of 30 Years Piezometric Observations in North-East Belgium*. External Report of the Belgian Nuclear Research Centre SCK•CEN-ER-163, Mol, Belgium.

Labat, S., Wemaere, I., Marivoet, J. & Maes, T. 2008*a*. *Herenthout-1 and Herenthout-2 Borehole of the Hydro/05neb Campaign: Technical Aspects and Hydrogeological Investigations*. External Report of the Belgian Nuclear Research Centre ER-59, Mol, Belgium.

Marivoet, J., Van Keer, I. *et al.* 2000. *A Palaeohydrogeological Study of the Mol site (PHYMOL Project)*. EC report EUR 19146 EN, Luxembourg.

Marivoet, J., Sillen, X., Mallants, D. & De Preter, P. 2002. Performance assessment of geological disposal of high-level radioactive waste in a plastic clay formation. *In*: McGrail, P. & Cragnolino, G. (eds) *Scientific Basis for Nuclear Waste Management XXV, 26–29 November 2001*, Boston, Massachusetts. Materials Research Society, Symposium Proceedings, **713**, 189–200.

MINISTERIE VLAAMSE GEMEENSCHAP. DEPARTEMENT LEEFMILIEU EN INFRASTRUCTUUR, ADMINISTRATIE MILIEU-, NATUUR-, LAND- EN WATERBEHEER, AFDELING WATER 2004. *Ontwikkelen van regionale modellen ten behoeve van het Vlaams Grondwater Model (VGM) in GMS/MODFLOW. Perceel nr. 3: Brulandkrijtmodel.*

Olsthoorn, T. N. 2011. *Geologische Berging van Radioactief afval in België: Mogelijke Gevolgen voor Nederland.* Rapport opgesteld in opdracht van Provincie Noord-Brabant, TUDelft, Delft University of Technology, The Netherlands.

Poeter, E. P., Hill, M. C., Banta, E. R., Mehl, S. & Christensen, S. 2005. *UCODE_2005 and Six other Computer Codes for Universal Sensitivity Analysis, Calibration and Uncertainty Evaluation Constructed using the JUPITER AVI.* U.S. Geological Survey Techniques and Methods, 6-A11, USGS, Reston, Virginia.

Vandersteen, K., Gedeon, M. & Rogiers, B. 2012. *Transient Model of the Confined Aquifers Below the Boom Clay: 2011 Update*. External Report of the Belgian Nuclear Research Centre SCK•CEN-ER-199, Mol, Belgium.

Vandersteen, K., Gedeon, M. & Leterme, B. 2013. *Hydrogeology of North-East Belgium. Status Report 2012*. External report of the Belgian Nuclear Research Centre SCK•CEN-ER-236, Mol, Belgium.

van der Gun, J. 1979. *Schatting van de elastische bergingscoëfficiënt van zandige watervoerende paketten.* Jaarverslag 1979, TNO dienst waterverkenning, Delft, The Netherlands.

Welkenhuysen, K., Vancampenhout, P. & De Ceukelaire, M. 2012. Quasi-3D model van de Formatie van Maldegem, de Groep van Tongeren en de Groep van Rupel. *Professional Papers of the Belgian Geological Survey*, 2012/1 – No.311, 46.

Wemaere, I., Marivoet, J., Beaufays, R., Maes, T. & Labat, S. 2002. *Core Manipulations and Determination of Hydraulic Conductivities in the Laboratory for the Mol-1 Borehole (April-May 1997).* External Report of the Belgian Nuclear Research Centre SCK•CEN-R-3590, Mol, Belgium.

Wemaere, I., Marivoet, J., Labat, S., Maes, T. & Beaufays, R. 2004*a*. *Rijkevorsel Borehole of the Hydro/96neb Campaign: Technical Aspects and Hydrogeological Investigations.* External Report of the Belgian Nuclear Research Centre SCK•CEN-R-3930, Mol, Belgium.

Wemaere, I., Marivoet, J., Labat, S., Maes, T. & Beaufays, R. 2004*b*. *Zoersel Borehole of the Hydro/96neb Campaign: Technical Aspects and Hydrogeological Investigations.* External Report of the Belgian Nuclear Research Centre SCK•CEN-R-3892, Mol, Belgium.

Wemaere, I., Marivoet, J., Labat, S., Beaufays, R. & Maes, T. 2005*a*. *The Weelde Boreholes of the Hydro/96neb Campaign: Technical Aspects and Hydrogeological Investigations.* External Report of the Belgian Nuclear Research Centre SCK•CEN-R-4187, Mol, Belgium.

Wemaere, I., Marivoet, J., Labat, S., Maes, T. & Beaufays, R. 2005*b*. *Turnhout Borehole of the Hydro/96neb Campaign: Technical Aspects and Hydrogeological Investigations.* External Report of the Belgian Nuclear Research Centre SCK•CEN-R-4124, Mol, Belgium.

Wemaere, I., Marivoet, J. & Labat, S. 2008. Hydraulic conductivity variability of the Boom Clay in north-east Belgium based on four core drilled boreholes. *Physics and Chemistry of the Earth*, **33**, 24–36.

Whittemore, D. O., Macfarlane, P. A. *et al.* 1993. *The Dakota Aquifer Program.* Annual Report, FY92, Kansas Geological Survey Open-file Report 93-1.

Yu, L., Rogiers, B., Gedeon, M., Marivoet, J., De Craen, M. & Mallants, D. 2013. A critical review of laboratory and in-situ hydraulic conductivity measurements for the Boom Clay in Belgium. *Applied Clay Science*, **75–76**, 1–12.

Microbial processes relevant for the long-term performance of high-level radioactive waste repositories in clays

ARTUR MELESHYN

Gesellschaft für Anlagen- und Reaktorsicherheit, Theodor-Heuss-Str. 4, D-38122 Braunschweig, Germany (e-mail: artur.meleshyn@grs.de)

Abstract: The primary purpose of the present work was to qualitatively evaluate the relevance of microbial activity for the long-term performance of a deep geological repository for high-level radioactive waste and spent nuclear fuel utilizing clay and to identify which safety-relevant processes and properties can be potentially influenced by this activity. Deterioration of clay properties accompanying destabilization and destruction of clay mineral structure as a result of microbial actions can be considered as the primary microbial impact on clay. The present analysis identified eight clay properties essential for maintaining safety functions of containment and retardation of the disposal system – swelling pressure, specific surface area, cation exchange capacity, anion sorption capacity, porosity, permeability, fluid pressure and plasticity – which can potentially be influenced by microbial processes in clay-based materials and claystone within a repository. Radioactive waste canisters and over-packs made from cast metal or steel represent a further component of the engineered barrier system which can be strongly affected by microbial activity in the clay buffer or in the adjacent host rock. This work should provide a basis for a follow-up quantitative estimation of the maximum possible effects of microbial processes on the barrier system of a deep geological repository in clay.

A number of investigations on occurrence and viability of microbes in compacted clays have been aimed at studying possible microbial effects on long-term performance of a deep geological repository (further in the text: repository) for high-level radioactive waste (HLW) and spent nuclear fuel (SF). Clays are considered in current repository designs either as a buffer material (compacted bentonite) or as a host rock (claystone). Accordingly, those investigations have been carried out in countries pursuing concepts of final HLW/SF disposal utilizing not only a claystone formation as a host rock for a repository (Switzerland, Belgium, France; Nagra 2002; ONDRAF 2004; ANDRA 2005), but also a granitic formation (Canada, Finland, Sweden), for which the repository performance largely relies on confinement properties of copper canisters amended by the favourable chemical and mechanical properties of an enveloping clay buffer (NWMO 2005; Neall *et al.* 2007; SKB 2011). Therefore, a survey of investigations of microbes in clays in the current analysis provides a compilation of issues relevant for the long-term performance of a repository in granitic and claystone formations.

Recent research presented conclusive evidence that living microbes are present in very old (16 million up to 111 million years) sediments in an abundance of more than 10^6 cells/cm^3 as deep as 400 m up to 1.6 km below the seafloor (Schippers *et al.* 2005; Roussel *et al.* 2008). Based on the abundance of phospholipid fatty acids indicative of living microbes, a 170-million-year-old sediment from the Opalinus Clay – a candidate host rock for a repository in Switzerland – has been estimated to contain 5×10^6 living microbes per gram of dry claystone (Mauclaire *et al.* 2007). An *in situ* experiment in the Opalinus Clay formation has demonstrated that, even if the undisturbed clay might be devoid of some microbial species, they would most probably be introduced there by human activity during repository construction and operation (Stroes-Gascoyne *et al.* 2011*a*; Wersin *et al.* 2011*b*). The question is therefore not whether microbes will be present in a repository in the post-closure phase, but rather whether they will exert a negative impact on the repository performance in the long term and if so, to what extent.

Even if the process of microbial reduction may be extremely latent in the host rock owing to a limited pore space or supply of nutrients, the excavation of a repository and placement of radioactive waste will inevitably, although possibly only partially and temporarily, lift these limitations. Backfilling and sealing of a repository will not lead to a complete restoration of the pristine host rock conditions, at least within a sizeable period of time, and leave microenvironments conducive to microbes, such as waste–buffer and buffer–host rock interfaces or those between blocks of clay buffer (De Cannière *et al.* 2004; Kursten 2004; Stroes-Gascoyne *et al.* 2011*a*, *b*; Wersin *et al.* 2011*a*).

A systematic evaluation of the relevance of microbial activity for the long-term performance

From: Norris, S., Bruno, J., Cathelineau, M., Delage, P., Fairhurst, C., Gaucher, E. C., Höhn, E. H., Kalinichev, A., Lalieux, P. & Sellin, P. (eds) 2014. *Clays in Natural and Engineered Barriers for Radioactive Waste Confinement*. Geological Society, London, Special Publications, **400**, 179–194.
First published online March 5, 2014, http://dx.doi.org/10.1144/SP400.6

of a repository requires, as a prerequisite, the identification of (a) the safety-relevant processes, which can be accelerated (e.g. a conversion of smectite to illite) or aggravated (e.g. a change from uniform corrosion to pitting corrosion) by microbial activity, and (b) the safety-relevant properties (e.g. swelling pressure) of clay, which can be deteriorated as a result of microbial activity. This evaluation represents the primary purpose of the present analysis and is in turn an essential prerequisite for a quantitative estimation of the maximum possible effects of microbial processes in a repository to be made in follow-up studies.

Such quantitative estimation should additionally take into account (c) the changes in the repository's physical and geochemical conditions during the operation and post-closure phases, which can control microbial activity (e.g. increases in temperature or in water content upon re-saturation of clay). This issue deserves a special attention, as limitations on water and space in a repository in clay can strongly influence the size and activity of microbial populations (Stroes-Gascoyne *et al.* 2007; Stroes-Gascoyne 2010; Stroes-Gascoyne *et al.* 2010). Note that ionizing radiation, on the contrary, is not expected to control microbial activity, as microbes can survive in such highly irradiated environments, as in the damaged Three Mile Island reactor, with a dose rate of 100 Gy h^{-1} (Wolfram & Dirk 1997). A further subject of essential importance for assessing long-term performance of a repository in clay concerns (d) the impact of microbial activity on radionuclide transport and radionuclide redox chemistry in clay environments. Detailed consideration has been given to the latter two issues (c) and (d) in reviews published elsewhere (Pedersen 1997; Pedersen 2000; Lloyd 2003; Humphreys *et al.* 2010; Sherwood Lollar 2011; Wolfaardt & Korber 2012), and the present analysis focuses on the issues (a) and (b) only.

Microbes utilize redox processes as a general way to gain energy for metabolism and growth by directing the flow of electrons from the donor substrates to an intermediate, intracellular electron acceptor and from that to the terminal, extracellular electron acceptor (Fig. 1). With respect to the limiting condition of energy accessibility, subsurface sediments can be divided into two broad categories of electron-donor limited and electron-acceptor limited ones (Chapelle 2000). Pristine sediments with the content of organic carbon, representing an essential source of electrons for microbial metabolism, being lower than that of electron acceptors (O_2, Fe(III), SO_4^{2-} and CO_2) are regarded thereby as limited in electron donors.

In deep subsurface systems with limited supply of electron donors and under anaerobic conditions, which can be expected to establish in a repository several years after its closure, the Fe(III)-reducing microbes are expected to outcompete sulphate-reducing microbes, which in turn outcompete methane-producing microbes for electron donors and thus effectively inhibit sulphate reduction and/or methane production (Chapelle 2000). Therefore, the processes of microbial reduction and microbial dissolution of clay minerals related to the activity of Fe(III)-reducing microbes are dealt with first in this overview, followed by the processes of microbially influenced corrosion and of microbial gas production and conversion, which is not limited to, but is strongly influenced and contributed by, methanogenic microbes. Biofilm formation, a frequent precursor to microbially influenced corrosion, is considered before the latter. A discussion on subsurface systems limited in electron acceptors, which can be characterized by the reverse dominance order of microbial reduction processes, precedes the final section evaluating the relevance of these issues for the long-term performance of a repository for HLW/SF.

Microbial reduction of clay minerals

It is now accepted that microbial metabolism – and not abiotic mechanisms – primarily controls iron redox chemistry in most environments (Weber *et al.* 2006). A wide variety of microbes has been found to reduce structural Fe(III) in iron-containing clay minerals, such as montmorillonite, nontronite, illite, chlorite or biotite, to maintain their metabolism (Dong *et al.* 2009). Three primary mechanisms have been proposed to explain the transfer of electrons to extracellular Fe(III) incorporated in the mineral structure (Weber *et al.* 2006; Dong *et al.* 2009): (a) the direct contact between the microbe and the clay mineral; (b) the production of endogenous/use of exogenous soluble electron shuttles; and (c) the production of chelating ligands to facilitate the mineral dissolution providing soluble Fe(III).

Two modes of the direct microbe–mineral interaction providing electron transfer are currently known (Weber *et al.* 2006). One mode requires a very close microbe approach to the mineral surface, so that electron transfer can occur via metalloproteins associated with the outer cell membrane. In another mode, microbes form extracellular appendages with a diameter of up to 0.2 μm and a length of several tens of microns – called bacterial nanowires owing to their electrical conductivity – in response to electron-acceptor limitation in order to transport electrons to solid-phase Fe(III) (El-Naggar *et al.* 2010).

In the second mechanism, microbes reduce such redox-active organic compounds as humic acid,

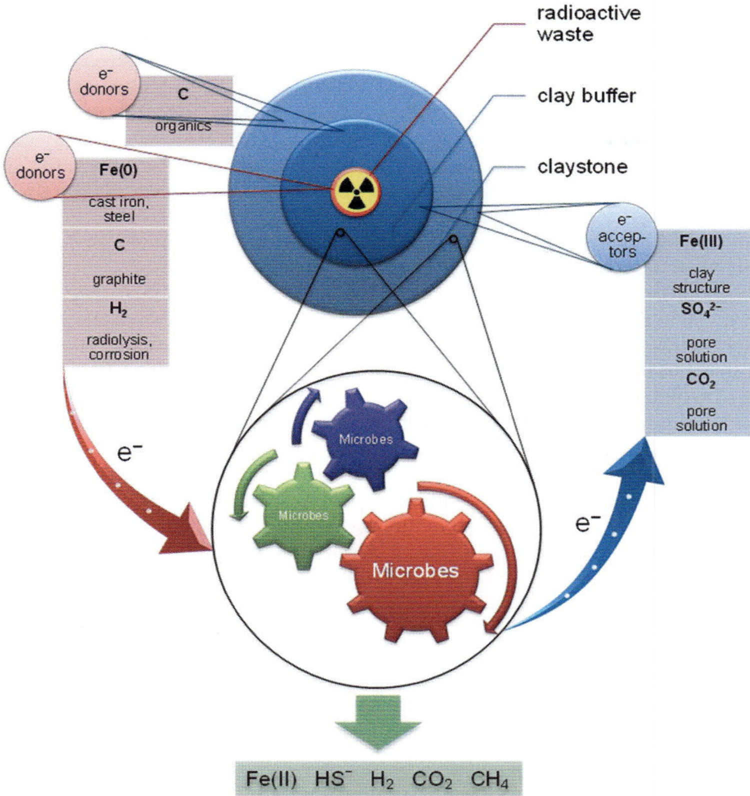

Fig. 1. Schematic summary of sources of electron donors and acceptors for microbial activity in a repository in clay as well as of microbial metabolism products of major importance for its long-term performance.

which diffuse to the mineral surface and donate electrons to the structural Fe(III) (Lovley *et al.* 1998). Model humic acid compounds have been confirmed to increase the extent of clay mineral reduction, that is the amount of structural Fe(III) ions reduced to Fe(II) within the clay mineral matrix, by a factor of 2–4 as well as the rate of this reaction (Lovley *et al.* 1998). Microbes can also produce endogenous electron shuttles, such as flavins, which have been shown to transfer two electrons per molecule to metals separated by more than 50 μm from microbes (Marsili *et al.* 2008). The third mechanism involves the use of exogenous or endogenous chelating compounds, which in addition to a solubilization of Fe(III) from the clay structure can also increase the extent of clay mineral reduction (Kostka *et al.* 1999*a*).

Within several days to weeks, up to 30% of Fe(III) in the mineral structure can be reduced by Fe(III)- and sulphate-reducing as well as methane-producing microbes in clay suspensions at anaerobic conditions, as shown for example for nontronite by Jaisi *et al.* (2008) and Liu *et al.* (2011, 2012). Interestingly, even although facultative anaerobic Fe(III)-reducing microbes do not reduce Fe(III) at aerobic conditions, they still can change hydration properties of compacted clays. Indeed, a bentonite powder containing such microbes and compacted to a dry density of 1.3 g cm^{-3} took up about 60% more water than a sterile control, presumably owing to microbially induced textural changes leading to an increase in the pore space (Perdrial *et al.* 2009). Note that no nutrients were added to clay in this experiment in a resemblance to repository conditions, but the intended anaerobic conditions were apparently not reached after initial contact with air, as evidenced by an increasing formation of a Fe(III) oxyhydroxide lepidocrocite in the presence of a Fe(II,III) hydroxycarbonate green rust (Perdrial *et al.* 2009), which is characteristic of oxidation reactions (Legrand *et al.* 2004; Génin *et al.* 2006; Pantke *et al.* 2012).

An increase in negative layer charge as a result of a microbial smectite reduction proceeding by a solid-state transformation and limited to a production of *c.* 2 mmol structural Fe(II) ions per gram of clay can be accompanied by a decrease in swelling pressure (*safety-relevant property*) by *c.* 40% (as a general rule, an increase in layer charge results in decreased crystalline swelling of smectites, as discussed by Laird (2006)), an increase in cation exchange capacity (*safety-relevant property*) by *c.* 20–30%, and a decrease in specific surface area (*safety-relevant property*) by *c.* 30–50% as determined by Kostka *et al.* (1999*b*) for a pure smectite. As a further consequence, microbial Fe(III) reduction can significantly improve anion sorption capacity (*safety-relevant property*) of clay minerals (note that the sorption of metal ions on mineral surfaces involves adsorption, surface precipitation and co-precipitation phenomena (Charlet & Manceau 1992)). Indeed, owing to the presence of Fe(II) in their structure as a result of Fe(III) reduction, clay minerals become able to reduce and effectively immobilize, for example, highly mobile Tc(VII) existing as pertechnetate (TcO_4^-) anion in oxidizing environments (Jaisi *et al.* 2009). Tc(VII) appears to be stable in nuclear waste glass at the relevant moderately reducing conditions ($Fe(II)/\Sigma Fe < 0.95$) (Lukens *et al.* 2007). In addition, Tc(VII) is present in spent nuclear fuel, and a part of it occurs in interstices between uranium oxide grains and is readily mobilized upon contact with solution (Laverov *et al.* 2009). Importantly, Tc(VII) is slowly reduced below the reduction threshold, except when biogenic Fe(II) is present as a sorbed species, because its reduction depends more critically on the availability and steric distribution of electron donors than on the overall redox potential of the system (Icenhower *et al.* 2008).

In summary, although the process of microbial reduction of clay minerals does not necessarily require a direct contact between microbes and clay minerals, the interaction of electron shuttles or chelating ligands with clay can be expected to limit its influence zone in a repository to proximities of the interfaces conducive to microbes.

Microbial dissolution of clay minerals

The processes of microbial dissolution and reduction of clay minerals are intimately related to each other as they both are triggered by the microbial demand for Fe(III) as terminal electron acceptor. A recent study has suggested that, upon a solid-state microbial reduction of *c.* 1.2 mmol Fe(III) per g clay, smectite structure destabilizes and a release of structural Fe(II) becomes possible (Jaisi *et al.* 2008). Accordingly, a progressive reduction of up to *c.* 3 mmol of structural Fe(III) per gram of clay within several days resulted in a progressively increasing release of up to *c.* 1.5 mmol of Fe(II) per gram of clay, which partitioned to surface complexation sites followed by ion-exchangeable sites and, upon the saturation of these sorption sites on the smectite surfaces, into the aqueous solution. Lower Miocene mudstones from Madrid Basin provide a conclusive mineralogical record of microbial reduction of Fe(III) and release of Fe(II) from biotite and chlorite that were depleted of Fe by up to 24 and 40 wt%, respectively (Sanz-Montero *et al.* 2009).

In the presence of sufficient dissolved potassium, microbial Fe(III) reduction and clay dissolution result in the irreversible conversion of smectite to illite. Without microbial activity, this most important diagenetic clay reaction occurs in mudstones and shales over a temperature range of 50–180 °C and geological times of 0.5–300 myr (Pollastro 1993). Owing to microbial activity, however, a strong acceleration of this reaction takes place, which results in the formation of illite by the reductive dissolution of smectite and simultaneous precipitation of illite in 14 days at 25 °C and 0.1 MPa (Kim *et al.* 2004). A recent study of Middle Cambrian mudstones has presented a strong evidence of microbial smectite dissolution accompanied by precipitation of illite, which occurred at temperatures in the range from 26 to 69 °C during early diagenesis (Vorhies & Gaines 2009).

Importantly, illite neo-formation from dissolution of smectites and other clay minerals occurs by precipitation in pore space and is accompanied by precipitation of silica there (Nadeau *et al.* 2002; Dong *et al.* 2009; Vorhies & Gaines 2009). Such neo-formed illite and silica particles are submicrometre-sized and can readily fill the void spaces of a comparable size, which control the permeability in mudstones and claystones (Nadeau *et al.* 2002; Mauclaire *et al.* 2007). It is argued that relatively minor amounts of neo-formed illite, less than 5%, within mudstones can greatly reduce permeability by blocking the effective pore network, and thereby render them prone to overpressure (Nadeau *et al.* 2002). Microbial dissolution of clay minerals is argued to be responsible for much of the early silica cementation occurring within shales and adjacent rocks (Vorhies & Gaines 2009).

Low-molecular-weight organic acids or their conjugates, such as nitrilotriacetic acid, oxalate, citrate or malate, which are weak Fe(III)-chelators, have been shown to facilitate microbial dissolution of smectite (Kostka *et al.* 1999*a*). An even stronger effect has been observed for siderophores (low-molecular-weight compounds with a very high and specific affinity for Fe(III)) produced by aerobic and facultative anaerobic microbes (Cheah *et al.* 2003). Moreover, the presence of, for example, oxalate was observed to increase the dissolution of solid-state Fe(III) by siderophores up to four-fold in a synergistic effect (Cheah *et al.* 2003; Brantley *et al.* 2004). Formation of a Fe(III)–chelator surface complex is considered to weaken the bonds between Fe(III) and the mineral lattice and to result in a solubilization of chelated Fe(III) (not of reduced Fe(II)) in agreement with the observation of dissolved Fe(III) concentrations of up to 8 $\mu mol\ l^{-1}$ at pH of 7.4 as compared with *c.* 10^{-4} $\mu mol\ l^{-1}$ in the absence of chelating compounds (Brantley *et al.* 2004).

It can be concluded that microbial dissolution of clay minerals can result in decreases in porosity (*safety-relevant property*) and permeability (*safety-relevant property*). These decreases are accompanied by an increase of fluid pressure (*safety-relevant property*) up to a development of overpressure that can eventually lead to a temporal loss of clay plasticity (*safety-relevant property*) and formation of fractures (Ortiz *et al.* 2002). Microbial dissolution of swelling smectites and neoformation of non-swelling illites result in decreases in swelling pressure (*safety-relevant property*), cation exchange capacity (*safety-relevant property*) and specific surface area (*safety-relevant property*) of clay. Furthermore, a decrease in Fe(II) content in the clay mineral structure as a result of its dissolution should be considered as detrimental to anion sorption capacity (*safety-relevant property*) of clay. At the same time, a decrease in porosity and permeability as a result of cementation can be expected to eventually limit the microbial activity to some background level because of decreasing living space as discussed, for example, for the clayey formation of Bure by Lerouge *et al.* (2011).

Biofilm formation

Microbes produce various extracellular polymers as attachment structures to anchor to the surface (Barker *et al.* 1998). Provided that nutrients are available, this initial colonization of the surface can proceed through division of attached microbes, attachment of further planktonic ones, and continuing secretion of extracellular polymers to eventually develop a biofilm, which can be one up to thousands of micrometres thick and completely cover the original solid surface (Brown *et al.* 1994; Barker *et al.* 1998). Extracellular polymers represent a matrix embedding individual microbes within a biofilm and protecting them from desiccation and other physical, chemical and biological stresses. A further protection is provided through a thin boundary structure, presumably a lipid bilayer similar to that constituting the microbial cell membrane, enveloping the entire biofilm (Palsdottir *et al.* 2009).

These protective components give a strong advantage to microbes organized in a biofilm over those in planktonic form. They allow microbes, for example, to control the level of toxic metal ions to which they are exposed (Brown *et al.* 1994; Barker *et al.* 1998). Biofilms often contain certain metals at concentrations 2–4 orders of magnitude greater than those in the contacting aqueous solution by accumulating them in a highly insoluble form as a detoxification measure (Templeton & Knowles 2009). A natural anaerobic biofilm collected from a borehole *c.* 1.5 km below the land surface has been reported to maintain even *c.* 2×10^7 times higher Zn^{2+} concentrations (in the form of zinc sulphide) than the contacting borehole water (MacLean *et al.* 2007). The net effect of biofilm formation on cation and anion sorption capacities of the solid substrate (*safety-relevant properties*) depends, however, on the solute identity. Indeed, the adsorption of Co(II), Th(IV) and Np(V) on the granite rock surface has been observed to decrease and that of Pm(III) and Am(III) not to change significantly or even to increase upon the formation of a biofilm under *in situ* subsurface conditions at the Äspö Hard Rock Laboratory (Anderson *et al.* 2007).

Moreover, biofilm formation can influence chemical conditions of at least parts of the adjacent bulk solution. Biofilms have been observed to plug the few-micrometre-thick pore space between exfoliation planes of biotite grains within 2 weeks and to decrease the pH from the bulk value of 7 to the value of 3 inside the biofilm and the confined pore space, which corresponds to a *c.* 20-fold increase in dissolution rate of biotite (Barker *et al.* 1998).

Furthermore, biofilm formation can substantially influence mass transport and hydrodynamics in porous media. Packed columns of crushed granitic rock from Äspö Hard Rock Laboratory have been reported to become impermeable within 2 days because of formation of biofilms of Fe(III)-reducing microbes indigenous to the Äspö rocks (Tuck *et al.* 2006). In another case, filter canister designed to remove 0.5–800 μm small fine debris suspended during the defuelling operations in the damaged Three Mile Island reactor have been plugged at abnormally fast rates with very little total solids entrained owing to biofilm formation (Wolfram & Dirk 1997).

In conclusion, biofilm formation on a mineral surface, which can occur within a few days, can facilitate clay mineral reduction and dissolution and thus aggravate – at least locally – microbial impact on safety-relevant properties discussed above. In an immediate effect, biofilm formation and growth can modify cation exchange or anion sorption capacities (*safety-relevant properties*) of the underlying mineral surface and strongly reduce porosity and permeability (*safety-relevant properties*) of the adjacent pore space. Availability of void spaces is, however, a prerequisite for formation of biofilms, which can therefore represent an important issue predominantly at the interfaces in a repository.

Microbially influenced corrosion and activity of sulphate-reducing bacteria

The ability of sulphate-reducing bacteria (SRB) to accelerate corrosion of waste canisters or overpack materials in the reference designs of engineered barrier systems adopted in different countries (carbon steel in Belgium, France and Switzerland or copper in Sweden, Finland and Canada) has been so far a major concern of microbial research related to final disposal of radioactive waste in geological formations (Nagra 2002; ONDRAF 2004; ANDRA 2005; NWMO 2005; Neall *et al.* 2007; Landolt *et al.* 2009; SKB 2011). The Canister Materials Review Board, convened by Nagra to deal with the concerns of the Swiss Federal Government over the corrosion issues in a repository in the Opalinus Clay, has recommended a focus on the sustainability of the microbially influenced corrosion (MIC) in the repository environment and the on the maximum extent of possible MIC damage instead of trying to demonstrate the total absence of SRB activity in unsaturated bentonite (Landolt *et al.* 2009). Since then, Nagra has started an ongoing research effort on microbial processes in the host rock settings (Schwyn *et al.* 2012).

A transfer of electrons from zero-valent metal to an external electron acceptor, which results in release of the metal ions into the contact solution and deterioration of the metal, is termed corrosion. An acceleration of this electrochemical process owing to microbial activity is referred to as MIC. While rates of abiotic, anaerobic iron corrosion by water attack (primary dissolution $Fe \leftrightarrows Fe^{2+} + 2e^-$ followed by water dissociation and reduction of derived protons $2e^- + 2H^+ \rightarrow 2H_{(adsorbed)} \rightarrow H_{2(adsorbed)} \rightarrow H_{2(aqueous)}$) are expected to be <10 μm/a, anaerobic corrosion rates as high as 700 μm/a can be observed in the case of MIC for non-clay environments (Sherar *et al.* 2011).

Anaerobic MIC is most closely identified with SRB and is currently considered to be able to proceed according to the following major mechanisms in dependence on the microbial species involved or environmental conditions: (a) an indirect mechanism by hydrogen sulphide attack; (b) an indirect mechanism by adsorption of microbially produced extracellular polymers at the metal–water interface; (c) a direct mechanism by scavenging of hydrogen atoms accumulating on the metal surface as a result of corrosion; and (d) a direct mechanism by an uptake of metal electrons into microbial cells. Note, however, that the extent of the influence of the mechanisms (b) and (c) is still much debated and the occurrence of the mechanism (d) in the deep biosphere remains to be confirmed, as discussed elsewhere (Beech & Sunner 2004; Meleshyn 2011).

Thus, a direct contact between SRB and the metal surface is not generally required. Indeed, results of a recent study strongly indicate that corrosion of copper plates buried within a clay buffer compacted to a density of 2.0 g cm^{-3} occurred as a result of production of sulphide by SRB at the compacted clay–groundwater interface and its diffusion to the copper surface with a diffusion coefficient of 0.2×10^{-11} m^2 s^{-1} (Pedersen 2010).

SRB activity has been concluded to have doubled the corrosion rate of carbon steel to *c.* 30 μm/a in 20% suspensions of Callovo-Oxfordian clay in synthetic pore water as compared with sterile experiments (El Hajj *et al.* 2010). This activity led to the formation of na SRB biofilm and corrosion pits in the carbon steel within 1 month. With undisturbed cores of Callovo-Oxfordian claystone and at the *in situ* pressure of 12 MPa, the corrosion rate increased from a value of 11 μm/a in the sterile control to 27 μm/a in the presence of SRB.

A recent study with Wyoming MX-80 bentonite revealed that the clay itself is a source of SRB and that a heating at 120 °C for 15 h did not sterilize the bentonite, although it reduced the number of viable SRB (Masurat *et al.* 2010). The reduction of SRB population resulted in a decrease in copper corrosion rate by 4 and 16 times for clay densities of 1.8 and 2.0 g cm^{-3}, respectively. In a further example of heat tolerance, SRB have been found to actively reduce sulphate in the hydrothermal deep-sea sediments with optima at 80–90 and 103–106 °C (Jørgensen *et al.* 1992).

Compelling evidence of SRB activity in a disturbed deep clay formation has been obtained very recently in an *in situ* experiment in the Opalinus Clay, in which synthetic pore water had been circulating in a borehole for five years (Wersin *et al.* 2011*a*). A substantial SRB activity developed in the borehole solution and adjacent clay within a few months after the start of the experiment. This activity led to a more than three-fold decrease in sulphate concentration from the value of 14.7 mmol l^{-1} – despite a continuous in-diffusion of sulphate from the surrounding claystone – and an increase in sulphide concentration to 1.0 mmol l^{-1}, which did not exceed this value because of the observed precipitation of ferrous sulphides (Stroes-Gascoyne *et al.* 2011*a*; Wersin *et al.* 2011*a*). This was a result of an unintentional placement of a degradable source of organic carbon in the borehole solution, presumably the highly soluble glycerol from gel-filling of reference electrodes used for a continuous monitoring of redox potential and pH. Furthermore, the graphite tip of an electrode has been almost completely leached in this experiment, which may indicate microbial utilization of graphite (Gregory *et al.* 2004) as well.

The activity of SRB in Boom Clay was evidenced by a sulphide concentration of 4.7 μmol l^{-1} in the pore water strongly exceeding the 0.01–0.1 μmol l^{-1} expected from equilibrium with pyrite (De Cannière *et al.* 2004). Moreover, extensive amounts of SRB cells have been observed, and a high sulphide concentration as of 5.6 mmol l^{-1} has been measured in the pore water extracted from the interstitial space between lining cement blocks of the HADES tunnel and surrounding Boom Clay (Lydmark & Persson 2010).

A compelling demonstration of a potential SRB impact on performance of metals under repository conditions has been provided by a five-year-long mock-up experiment, OPHELIE, simulating the waste disposal architecture pursued at that time in Belgium. In this experiment, a heat-generating stainless steel tube was surrounded by a layer of pre-compacted blocks of clay-based backfill material with a dry density of *c.* 1.8–2.0 g cm^{-3} and placed within a stainless steel lining (Verstricht & Dereeper 2003). Despite the high density and backfill temperatures above 100 °C for about four years, strong indications of MIC have been observed on multiple locations on the lining and the heat-generating tube (Kursten 2004) with, for example, chromium sulphides formed at the interface between the lining and the backfill, and up to 150 μm-deep pits underneath. A sulphide concentration of *c.* 0.6 mmol l^{-1} has been measured in the hydration circuit water, which is comparable to the value measured in the *in situ* experiment in the Opalinus Clay.

Microbial gas production and conversion

The microbial reduction of Fe(III) or sulphate considered in the preceding sections leads to the production and/or consumption of gases. For example, Fe(III)-reducing microbes produce 1 mol of CO_2 per 1 mol of carbon atoms of the oxidized organics and 4 mol of the reduced Fe(III). Similarly, 1 mol of gaseous hydrogen sulphide (H_2S) and 2 mol of CO_2 are produced by SRB per 1 mol of reduced sulphate with acetate as electron donor. Note, however, that the inventory of gases in a repository will depend on the amount of the available solution, as at temperatures and pressures relevant for a repository after the re-establishing of clay saturation (which can take a couple of centuries; Johnson & King 2008), about 1 mol of H_2S and 1 mol of CO_2 can be dissolved per kilogram of pore solution (Duan *et al.* 2006, 2007).

A build-up of gas overpressure around a repository for HLW/SF because of a disparity between the gas diffusion and gas production rates is considered to be a possible scenario that can result in fracturing of the clay rock (Nagra 2002; Ortiz *et al.* 2002). If occurs, this fracturing, being preceded by the loss of clay plasticity (*safety-relevant property*), will be accompanied by gas discharge to relieve the excess pressure. Overpressure is, however, mostly related to issues regarding the formation of H_2 as a consequence of abiotic metal corrosion in a repository; the oxidation of H_2 by microbes, which is considered in more detail in this section, would reduce the risk of overpressure. To account for further potential microbial sources or sinks of gas in a repository for HLW/SF, this section proceeds by considering two microbial processes of high importance in the deep subsurface – fermentation and methane production (methanogenesis).

Fermentation refers to microbial metabolic processes that do not employ external electron acceptors for oxidation of organics and result in the accumulation of reduced compounds such as acetate (CH_3COO^-), formate ($CHOO^-$) and H_2 as well as oxidized carbon (CO_2; McMahon *et al.* 1992). In this process, up to 2 mol of H_2 and up to 1 mol of CO_2 are produced per mole of carbon atoms of the oxidized organics, for example, for glucose: $C_6H_{12}O_6 + 6H_2O \rightarrow 6CO_2 + 12H_2$. Some microbial species often co-existing with SRB, which require acetate as a carbon source (Muyzer & Stams 2008), are able to metabolize these gaseous products in a reverse reaction consuming 4 mol of H_2 and 2 mol of CO_2 per mole of produced acetate: $4H_2 + 2CO_2 \rightarrow CH_3COOH + 2H_2O$.

Fermentation has been shown to occur at geologically significant rates in buried sediments of the Black Creek formation in South Caroline, USA, characterized by clay layers with a thickness of up to *c.* 10 m and organic carbon content of up to *c.* 0.7 wt% interbedded with sands at depths of *c.* 50–900 m (McMahon *et al.* 1992). As a result of fermentation, a net diffusive flux of low-molecular-weight organic acids and CO_2 from clay into sands has been occurring during the post-depositional period, which has favoured the activity of SRB at clay–sand interfaces. As high acetate and formate concentrations as 1.8 and 6.4 mmol l^{-1} in pore water of a clay layer were decreased by the SRB activity to lower than 2×10^{-3} mmol l^{-1} in pore waters of the adjacent sand. In a further example, a microbial community consisting of fermentative and sulphate-reducing microbes was shown to be active in consolidated Cretaceous rock at the interface between shale with an organic carbon content of *c.* 1.5 wt% and sandstone owing presumably to a diffusion of organic material from shale (Krumholz *et al.* 1997).

These two cases and the ability of fermentative microbes to effectively produce or consume gases exemplify a possible impact of microbial fermentation in a repository. Importantly, increasing temperature to 50–90 °C has been shown to provide energy for making recalcitrant organic matter more degradable for microbes (Parkes *et al.* 2007). Furthermore, microbes can strongly facilitate H_2 formation in sediments added by common minerals or rocks and heated to 40–100 °C (Parkes *et al.* 2011). As a result, dissolved H_2 concentrations can increase by up to 325 times (to 1.6 mmol l^{-1}) and that of CO_2 by 2.6 times (to 7.7 mmol l^{-1}) as compared with their background production levels in sediments. To set the former figure in context, H_2 pressure inside a corroding canister in a repository is expected to quickly reach several megapascals, with for example, 5 MPa corresponding to a dissolved H_2 concentration of 50 mmol l^{-1} (Carbol *et al.* 2005). In a further effect, the previous presence of microbial activity (at temperatures below 100 °C) was observed to enhance abiotic gas-generating reactions at higher temperatures of 125–155 °C (Parkes *et al.* 2011).

Methanogenesis occurs by the three major pathways through (a) reduction of CO_2 (e.g. from oxidation of H_2: $4H_2 + CO_2 \rightarrow CH_4 + 2H_2O$), (b) fermentation of acetate ($CH_3COO^- + H^+ \rightarrow CH_4 + CO_2$) and (c) fermentation of methanol or methylamines ($4CH_3OH \rightarrow 3CH_4 + CO_2 + 2H_2O$). Extensive core samplings of deep (down to *c.* 300 m below seafloor) marine sediments off Vancouver Island have recently revealed that methane production by CO_2 reduction pathway is one to three orders of magnitude larger than that by acetate fermentation in silty clay and clay layers (Yoshioka *et al.* 2010). This deep subsurface methane production has been concluded to probably be responsible for formation of one of the largest deposits of methane hydrate worldwide.

In the *in situ* experiment in the Opalinus Clay, methane concentration has been observed to steadily increase for about 4 years and to reach a value of 0.7 mmol l^{-1}, which was three orders of magnitude higher than that at the start of the experiment (Wersin *et al.* 2011*a*) and resulted from microbial methanogenic activity (Stroes-Gascoyne *et al.* 2011*a*). This study concluded, based on a reactive transport modelling effort, that methanogenesis played only a minor role in degradation of the available organic carbon (Wersin *et al.* 2011*b*). However, the observed methane concentration is comparable to those of *c.* 1.1 and *c.* 1.6 mmol l^{-1} measured in the groundwater of shallow sandy aquifers contaminated by gas condensate in Colorado and by crude oil in Minnesota, respectively, where methanogenesis has been identified as a major microbial process (Baedecker *et al.* 1993; Gieg *et al.* 1999).

In extension of their ability to reduce structural Fe(III) in clay minerals (Liu *et al.* 2011), methanogens have been recently revealed to produce hydrogen and methane in a mixture of metallic iron and montmorillonite (Chastain & Kral 2010). The observation of higher methane production and corrosion rates by at least three orders of magnitude in a mixture of a fine powder of metallic iron with Boom Clay–water slurry as compared with a mixture containing steel instead of metallic iron (Ortiz *et al.* 2002) can be considered as a further demonstration for this methanogen's ability.

In addition to the rather diversified mechanisms of gaining energy for metabolism and growth, methanogens also exhibit extraordinary survival capabilities. The most heat-resistant microbe currently known is a methanogen that can grow and produce methane at 122 °C under the *in situ* pressure of 20 MPa, whereas the maximum temperature of its proliferation decreases to 116 °C at 0.4 MPa (Takai *et al.* 2008).

Electron-acceptor limited subsurface settings

In the early 1990s, growing experimental evidence has resulted in a formulation of criteria allowing prediction of the sequence in which microbial reduction processes will occur along the direction of groundwater flow in a pristine or contaminated subsurface environment based solely on information about the availability of key terminal electron acceptors and electron donors (Chapelle 2000). It became increasingly clear that, for example the dominance order of microbial reduction processes – $O_2 > NO_3^- > Fe(III) > SO_4^{2-} > CO_2$ with the activity of O_2-reducing microbes prohibiting that of NO_3^--reducing microbes until O_2 is consumed, and so on – characteristic of a pristine subsurface environment with a limited supply of electron donors can be reversed upon chemical contamination by human activities.

A further complication to the model of rather clearly discriminated zones of predomination of single microbial reduction processes comes about with the possibility that microbes responsible for different reduction processes are simultaneously active in the same parts of subsurface environments. The *in situ* borehole experiment in the Opalinus Clay has revealed that a concomitant activity of methanogens, SRB and fermentative microbes can occur at the interfaces in a disturbed clay formation in the excess of electron-donors, which resulted in considerable increases of methane, sulphide and acetate concentrations of up to 0.7, 1.0, and 13.2 mmol l^{-1}, respectively (Wersin *et al.* 2011*a*). The degradation of the electron-donor source within first two years of the experiment was followed by a consumption of acetate, which was accompanied by a *c.* 30-fold increase in methane concentration and a *c.* 20-fold decrease in sulphide concentration.

In Late Eocene and Late Paleocene clay formations in Georgia, USA, methane production has been inferred to be the predominant microbial reduction process at four out of seven sampled locations despite available Fe(III) (Shelobolina *et al.* 2005). These locations were characterized by organic carbon contents of 0.6–2.5 wt%. Yet at another location with organic carbon content of 5.3 wt%, sulphate reduction predominated. Fe(III)-reducing microbes were present at all these locations, as has been demonstrated by a stimulation of their activity using a Fe chelator. On the contrary, two remaining locations with comparable amounts of Fe(III), but characterized by at least one order of magnitude lower organic carbon contents of 0.06–0.07 wt%, have featured Fe(III) reduction as the predominant microbial process. According to these observations, the local availability of electron donors in excess of that of electron acceptors or, alternatively, the decreased accessibility of the latter as compared with that of the former can be a principal factor governing which microbial reduction process predominates in a geological setting.

Evaluation and summary

The present analysis identified eight clay properties essential for maintaining safety functions of containment and retardation (SKB 2011) of the disposal system – swelling pressure, specific surface area, cation exchange capacity, anion sorption capacity, porosity, permeability, fluid pressure and plasticity – which can potentially be influenced by microbial processes in clay buffer and claystone within a repository for HLW/SF. Radioactive waste canisters and over-packs made from cast metal, carbon steel or stainless steel represent a further component of the engineered barrier system that can be strongly affected by microbial activity in the clay buffer or in the adjacent host rock.

According to the current state of knowledge, iron(III)-reducing, sulphate-reducing, fermentative, methane-producing and methane-oxidizing microbes, either indigenous or introduced there by human activity during repository construction and operation, can be considered to be present in any clay formation to be utilized either as a source of clay buffer material or as a host rock for a repository for HLW/SF. Moreover, the growing body of observations suggests that each habitat includes a massive number of microbial niches with perhaps only a small proportion of the species being metabolically active under the habitat's conditions, the remainder not becoming extinct – which makes microbes discontinuously different from larger organisms (Patterson 2009). Thus, for example, hyperthermophilic microbial species are available in currently rather cold environments, such as deep clay formations (Stroes-Gascoyne *et al.* 2011*a*), and can become active as soon as temperatures in a repository increase as a result of the placement of HLW/SF. Although the corresponding thermal phase in a repository will be limited to several centuries, the effect of their activity (e.g. production of H_2 or low-molecular-weight organic acids) should be assessed with respect to its relevance for barrier performance.

It can further be concluded that clays contain electron donors and electron acceptors in amounts sufficient for these microbes to remain active, even though perhaps at low metabolic rates, during geological times. The excavation of a repository, placement of radioactive waste backfilling and sealing of a repository will add further sources of electron donors and acceptors. The ability of microbes to use sophisticated systems, either nanowires or shuttles, to transfer electrons to clay Fe(III) will additionally ease the limitation of local availability of terminal electron acceptors imposed on the activity of microbial population. A schematic summary of possible sources of electron donors and acceptors and of microbial metabolism products of major importance for the long-term performance of a clay-utilizing repository for HLW/SF is given in Figure 1.

Deterioration of clay properties accompanying destabilization and destruction of clay mineral structure as a result of microbial actions at the interfaces in a repository can be considered as the primary microbial impact on clay. Structural Fe(III) of clay mineral layers represents the primary electron-accepting reactant in redox reactions driven by Fe(III)-reducing and methane-producing microbes as well as by sulphate-reducing microbes in the case of limited supply of sulphate. However, even in the case that the latter microbes predominantly use sulphate as the terminal electron acceptor, the product of their metabolism, hydrogen sulphide, reacts with structural Fe(III) in an abiotic redox reaction, which is facilitated by further, organic products of SRB metabolism.

Furthermore, almost all hyperthermophilic microbes, characterized by optimum growth temperatures in the range of *c.* 80–121 °C, isolated so far from deep subsurface sediments have been found to reduce Fe(III) (Lovley *et al.* 2004).

These microbial species have the ability to maintain metabolic activity with H_2 as the only electron donor, which may be supplied in the vicinity of waste canisters not only by corrosion or radiolysis but also by microbially mediated generation from minerals, as revealed recently (Parkes *et al.* 2011). These considerations and the ability of Fe(III)-reducing microbes to potentially directly influence eight safety-relevant properties of clay point up the necessity of accounting for their possible impact when assessing long-term performance of a clay-utilizing repository for HLW/SF.

Microbial production of chelating compounds in the repository environment was recently judged to be particularly important in the safety analysis of radioactive waste disposal while being one of the least understood processes (Pedersen 2005). The possible production of siderophores during the oxygen availability in a repository and synergistic effects of organic acids and siderophores (either available in the clay formation or secreted by microbes introduced into a repository) are other critical, yet unresolved research issues. For example, pore waters of the Opalinus Clay and Callovo-Oxfordian claystone (rocks characterized by organic carbon contents of up to 1 wt%, which are typical for deep clay formations) contain lactate and acetate in concentrations of 8–9 and 203–1865 $\mu mol\ g^{-1}$, respectively (Courdouan *et al.* 2007*a*, *b*). Even if concentrations of these chelators and of siderophores in a host formation or a buffer material of a repository might be several orders of magnitude lower, their synergistic effect on clay should conservatively be assumed based on observations made for Fe(III) (hydr)oxides. Indeed, comparable rates of, for example, goethite dissolution have been observed in the presence of either 0.5 $\mu mol\ ml^{-1}$ oxalate or >0.1 $\mu mol\ ml^{-1}$ siderophore, or of only 0.04 $\mu mol\ ml^{-1}$ oxalate combined with 0.01 $\mu mol\ ml^{-1}$ siderophore (Cheah *et al.* 2003).

Although biofilms form on clay minerals under laboratory conditions, the present study could not find any *in situ* evidence of biofilm formation in workings excavated in claystone. A probable reason for that may be rock desiccation by the venting, as water availability has been identified as the limiting factor for biofilm formation in granitic environments (Brown *et al.* 1994). After the repository closure, however, disturbed host rock and clay buffer will inevitably be saturated with water. A decrease of available pore space as a result of swelling of clay upon water saturation will pose another principal limitation on biofilm formation, growth or survival, which however can be eased to some extent by the availability of interfacial microenvironments conducive to microbes as discussed above. In addition, changes in these microenvironments occurring after repository backfilling may exert stress on indigenous or inoculated anaerobic microbial communities and foster their organization into biofilms. Although general effects of biofilm formation on properties of mineral surfaces and pore solutions in the deep subsurface become increasingly known, the specific impact of biofilms on repository components and host material remains to be identified and quantified in future research.

Owing to the ability of SRB to reduce structural Fe(III) of clay directly or through production of hydrogen sulphide, SRB activity can negatively influence up to eight safety-relevant properties of clay. This aspect of SRB activity was strongly under-represented in previous experimental investigations in the field of geological disposal of radioactive waste and clearly requires a proper consideration in the future research activity. Clay-containing sediments at depths of 1–80 m below the surface have been concluded to be characterized by clay-Fe(III) half-lives of *c.* 100 000–2 400 000 years depending on the deposition rate and mineralogy when exposed to a 1 $mmol\ l^{-1}$ solution of SRB-produced hydrogen sulphide (Raiswell & Canfield 1996). The first value, which should be conservatively used if no value specific to the clay host rock of a repository is available, is significantly lower than the timeframe of 1 million years after repository closure assigned in accordance with regulatory requirements in many countries for the assessment of the safety of potential repository for HLW/SF. This underlines the necessity of a quantification of the maximum potential effect of SRB activity in a repository not only for canister corrosion but also for clay buffer and clay host rock at the interfaces, especially considering that sulphide concentrations of 0.5 and 1.0 $mmol\ l^{-1}$ have been observed in the groundwater and borehole solution at the Äspö and Mont Terri rock laboratories, respectively (Masurat *et al.* 2010; Wersin *et al.* 2011*a*).

In its report to Nagra, the Canister Materials Review Board postulates that it is reasonable to assume that corrosion of steel canisters in saturated bentonite is uniform, so that the corroded area equals the canister surface (Landolt *et al.* 2009). However, recent mock-up and laboratory studies of SRB-induced MIC in a clay environment (Kursten 2004; El Hajj *et al.* 2010) suggest that such an assumption might not be generally valid. Provided that the observation made in the latter laboratory experiment with undisturbed Callovo-Oxfordian claystone will be reproduced in future studies, a localized, pitting corrosion with an initial rate of up to *c.* 700 $\mu m/a$ can be assumed for MIC of steel in contact with clay as perhaps the worst corrosion scenario instead. The latter value corresponds well to the maximum corrosion rate as a result of MIC under anaerobic conditions reported for non-clay environments (Sherar *et al.* 2011).

Therefore, the ability of SRB to induce pitting corrosion even at low levels of biomass (Sherar *et al.* 2011) has to be taken into account when evaluating possible MIC effects in a repository for HLW/SF. It cannot currently be concluded from the available experimental data whether, and if so to what extent, the change to pitting corrosion is favoured by which one of the identified four mechanisms of microbially influenced corrosion. This exposes a clear research demand on this topic in order to be able to evaluate the possibility of occurrence of pitting corrosion and its maximum possible effect in a repository for HLW/SF. Most importantly, since a safety assessment of a repository requires information, not only on the initial corrosion rate, but also on that applying to the subsequent corrosion stage, during which corrosion products can provide protection to metal surfaces, research effort is needed to determine the according rates of MIC relevant for clay environment.

Furthermore, in no case should the potential impact of SRB be underestimated based on a possible argument of comparably low biomass of the microbes in contact with metal surfaces or dissolved metals. The timeframe of their potential activity must be considered concomitantly, as in a space of several years an SRB population confined in a relatively small biofilm (which within several years occupied a large fraction of the volume of a 30 mm-diameter and 25 m-long borehole at 1.474 km depth) encountering submicrometre concentrations of dissolved metal in a deep subsurface environment is able to precipitate metal amounts that, if continued over geological time, would produce an economic deposit (MacLean *et al.* 2007).

Production and conversion of gases in reactions involved in microbial reduction of Fe(III), sulphate or CO_2 and microbial fermentation represent a further important issue for the consideration of the overall microbial impact on the long-term performance of radioactive waste repositories in clays. Since an *in situ* experiment in the Opalinus Clay measured microbial methane concentrations comparable to those in hydrocarbon-contaminated subsurface settings where methanogenesis was a major microbial process, future research effort should be aimed at providing data necessary for a quantification of the potential impact of methanogens on the repository safety.

While the occurrence of either of the discussed microbial process in the repository environment, even though perhaps at only low metabolic rate, appears to be backed by the available experimental data, the question of their interplay at repository conditions should be considered still open. Contrary to the earlier, simplistic conception of the dominance order of microbial processes, which was based solely on energetic considerations, the present view recognizes that their occurrence critically depends on local heterogeneity of microbial habitats with respect to, most importantly, the availability of electron donors and acceptors.

Quantifying inventories of electron donors and electron acceptors both available in a geological formation chosen to host a repository and to be introduced there during the excavation, operation and closure of a repository is indispensable for predicting which microbial processes may be active or predominate at which locations within the repository and at which times after its closure. Such an approach can build on the sound ground given by the extensive previous work, for example by West *et al.* (2006), which is a result of a more than two decades of research effort started as early as in 1982 owing to the realization that microbial processes should be accounted for in the safety assessment of a repository (West *et al.* 1982). It is also essential that quantitative estimations give proper attention to a possible interplay of single microbial processes in a repository and not only to their separate contributions. Future studies should develop a methodological basis and identify parameters and data required for implementation of such an approach in safety analyses of deep geological repositories.

Finally, as clay materials are characterized by very low permeability, the issue of the access to space for a significant biological development should be addressed. Indeed, the compaction of a claystone can strongly decrease microbial activity because of decreasing living space, as discussed for example for sulphate-reducing bacteria in the Callovian-Oxfordian formation by Lerouge *et al.* (2011). According to Krumholz *et al.* (1997), the microbial activity in shales appears to be greatly reduced, presumably because of their restrictive pore size. Moreover, for another type of rocks (limestone), Bottrell *et al.* (2000) demonstrated that the activity of sulphate-reducing bacteria was possible in the fissures but not in the limestone matrix because of its characteristic pore throat size of <0.5 μm, which is too small for bacteria to pass.

The highly compacted bentonite buffer emplaced in a repository can also be considered to suppress microbial activity as long as it maintains a uniform dry density ≥ 1600 kg m^{-3} (Stroes-Gascoyne *et al.* 2011*b*). Such high densities do not, however, eliminate the viable microbial population in bentonite, and the latter study showed that, upon expansion of compacted bentonite into a void, the resulting reduction in dry density stimulated or restored microbial culturability. Therefore, localized enhanced microbial activity at buffer interfaces or in sealed desiccation fractures within the buffer – if they were not to regain the original density of the emplaced bentonite upon buffer saturation – was concluded by Stroes-Gascoyne *et al.* (2011*b*) to remain of concern.

It should be stressed that the volume of the voids in contact with a bentonite buffer in a repository can be considerable. For example, the French concept of HLW disposal does not incorporate a bentonite buffer between disposal packages with a diameter of 60 cm and 2.5 cm thick metallic sleeves separating 30 m-long disposal tunnels with a diameter of 70 cm from the Callovian-Oxfordian claystone (ANDRA 2005). Thus, *c.* 13% of the tunnel volume represents a void space around disposal packages, which contacts 3 m-thick swelling clay plugs on the head side of disposal tunnels. Further voids are potentially available in canisters with vitrified HLW. For example, the void space in CSD-V canisters produced in La Hague and to be disposed of in France, Germany, Japan, Switzerland, Belgium and The Netherlands amounts to 20 l per canister or *c.* 12% of the canister volume (Meleshyn & Noseck 2012). Similarly, the potential void space in a spent fuel container, which may be filled by water upon the loss of the container integrity, can have a volume in the range 22–46% of the spent fuel volume (Gmal *et al.* 2004).

Although the biochemical perturbation in the borehole of the *in situ* experiment in the Opalinus Clay was intense during five years, the chemical disturbances were strongly buffered in the claystone and the zone of the influence of the borehole perturbation appeared to be limited to the intermediate interface region (Koroleva *et al.* 2011). For example, the concentration of dissolved Fe(II) in the borehole showed an intermediate increase from *c.* 2 $\mu mol\ l^{-1}$ expected in the clay pore water to a rather high value of *c.* 100 $\mu mol\ l^{-1}$ (Wersin *et al.* 2011*a*), and a release from clay minerals was suggested as one of the most likely sources of Fe(II) in the borehole solution (Koroleva *et al.* 2011). Indeed, the increase in dissolved Fe(II) may be supposed to result from either a direct microbial dissolution of clay or an interaction of biogenic hydrogen sulphide with Fe(III) in clay matrix. However, no corresponding mineralogical changes could be revealed within 15 cm of the claystone adjacent to the borehole.

Thus, it appears that the microbial activity can only be significant when some void space (interfaces, borehole, fissures and so on) is available. The issue of the extension of the microbiological perturbation inside the clay matrix is, however, still an open question and should be addressed by future works. Reactive transport/chemistry calculations could help to solve this issue, as has been done for other chemical perturbations (see e.g. Marty *et al.* 2009).

It is further important to keep in mind that a safety assessment of a repository is not restricted to the 'normal evolution scenario', but should consider also, for example, 'disruptive event scenarios', which account for for example buffer failure (see e.g. Villagran 2012). In the latter case, water can move through the buffer because of void space availability, which also means that microbial activity and growth are possible there. Because of the relevance of such scenarios for safety assessments, there is a clear demand for thorough understanding and proper consideration of microbial processes in a repository environment, taking into account different kinds of void spaces probably or only potentially available there.

I thank P. De Cannière (Federal Agency for Nuclear Control, Belgium) for his extensive substantive and editorial suggestions on the original draft of the report published earlier (Meleshyn 2011). Comments and discussions with U. Noseck (GRS Braunschweig) are gratefully acknowledged. This manuscript also benefited from helpful comments from two anonymous reviewers and the corresponding editor S. Norris. This study was supported by the Federal Ministry of Economics and Technology (BMWi) under the identification number 02 E 10548 titled Wissenschaftliche Grundlagen zum Nachweis der Langzeitsicherheit von Endlagern (WiGru).

References

Anderson, C. R., Jakobsson, A. M. & Pedersen, K. 2007. Influence of in situ biofilm coverage on the radionuclide adsorption capacity of subsurface granite. *Environmental Science and Technology*, **41**, 830–836.

ANDRA 2005. *Dossier 2d005 Argile SYNTHESIS. Evaluation of the feasibility of a geological repository in an argillaceous formation.*

Baedecker, M. J., Cozzarelli, I. M., Eganhouse, R. P., Siegel, D. I. & Bennet, P. C. 1993. Crude oil in a shallow sand and gravel aquifer – III. Biogeochemical reactions and mass balance modeling in anoxic groundwater. *Applied Geochemistry*, **8**, 569–586.

Barker, W. W., Welch, S. A., Chu, S. & Banfield, J. F. 1998. Experimental observations of the effects of bacteria on aluminosilicate weathering. *American Mineralogist*, **83**, 1551–1563.

Beech, I. B. & Sunner, J. 2004. Biocorrosion: towards understanding interactions between biofilms and metals. *Current Opinion in Biotechnology*, **15**, 181–186.

Bottrell, S. H., Moncaster, S. J., Tellam, J. H., Lloyd, J. W., Fisher, Q. J. & Newton, R. J. 2000. Controls on bacterial sulphate reduction in a dual porosity aquifer system: the Lincolnshire Limestone aquifer, England. *Chemical Geology*, **169**, 461–470.

Brantley, S. L., Liermann, L. J., Guynn, R. L., Anbar, A., Icopini, G. A. & Barling, J. 2004. Iron isotope fractionation during mineral dissolution with and without bacteria. *Geochimica et Cosmochimica Acta*, **68**, 3189–3204.

Brown, D. A., Kamineni, D. C., Sawicki, J. A. & Beveridge, T. J. 1994. Minerals associated with biofilms occurring on exposed rock in a granitic underground research laboratory. *Applied and Environmental Microbiology*, **60**, 3182–3191.

Carbol, P., Cobos-Sabate, J. et al. 2005. *The Effect of Dissolved Hydrogen on the Dissolution of ^{233}U-doped $UO_2(s)$, High Burn-up Spent Fuel and MOX Fuel.* Technical Report **TR-05-09**. Svensk Kärnbränslehantering AB, Stockholm.

Chapelle, F. H. 2000. The significance of microbial processes in hydrogeology and geochemistry. *Hydrogeology Journal*, **8**, 41–46.

Charlet, L. & Manceau, A. A. 1992. X-ray absorption spectroscopic study of the sorption of Cr(III) at the oxide-water interface: II. Adsorption, coprecipi-tation, and surface precipitation on hydrous ferric oxide. *Journal of Colloid and Interface Science*, **148**, 443–458.

Chastain, B. K. & Kral, T. A. 2010. Zero-valent iron on Mars: an alternative energy source for methanogens. *Icarus*, **208**, 198–201.

Cheah, S. F., Kraemer, S. M., Cervini-Silva, J. & Sposito, G. 2003. Steady-state dissolution kinetics of goethite in the presence of desferrioxamine B and oxalate ligands: implications for the microbial acquisition of iron. *Chemical Geology*, **198**, 63–75.

Courdouan, A., Christl, I., Meylan, S., Wersin, P. & Kretzschmar, R. 2007*a*. Characterization of dissolved organic matter in anoxic rock extracts and in situ pore water of the Opalinus Clay. *Applied Geochemistry*, **22**, 2926–2939.

Courdouan, A., Christl, I., Meylan, S., Wersin, P. & Kretzschmar, R. 2007*b*. Isolation and characterization of dissolved organic matter from the Callovo–Oxfordian formation. *Applied Geochemistry*, **22**, 1537–1548.

De Cannière, P., De Boever, P. & Daumas, S. 2004. *Microbiological analyses and detection of sulphides in the OPHELIE mock-up.* Presentation on EIG EURIDICE' OPHELIE Day, 10 June, http://www.euridice.be/eng/06publicaties2004agendaO.shtm (accessed 25 May).

Dong, H., Jaisi, D. P., Kim, J. & Zhang, G. 2009. Microbe–clay mineral interactions. *American Mineralogist*, **94**, 1505–1519.

Duan, Z., Sun, R. & Zhu, C. 2006. An improved model for the calculation of CO_2 solubility in aqueous solutions containing Na^+, K^+, Ca^{2+}, Mg^{2+}, Cl^-, and SO_4^{2-}. *Marine Chemistry*, **98**, 131–139.

Duan, Z., Sun, R. & Zhu, C. 2007. An accurate thermodynamic model for the calculation of H_2S solubility in pure water and brines. *Energy & Fuels*, **21**, 2056–2065.

El Hajj, H., Abdelouas, A., Grambow, B., Martin, C. & Dion, M. 2010. Microbial corrosion of P235GH steel under geological conditions. *Physics and Chemistry of the Earth*, **35**, 248–253.

El-Naggar, M. Y., Wanger, G. et al. 2010. Electrical transport along bacterial nanowires from Shewanella oneidensis MR-1. *Proceedings of the National Academy of Sciences of the United States of America*, **107**, 18 127–18 131.

Génin, J.-M. R., Ruby, C., Géhin, A. & Refait, P. 2006. Synthesis of green rusts by oxidation of Fe(OH)2, their products of oxidation and reduction of ferric oxyhydroxides; Eh–pH Pourbaix diagrams. *Comptes Rendus Geoscience*, **338**, 433–446.

Gieg, L. M., Kolhatkar, R. V., McInerney, M. J., Tanner, R. S., Harris, S. H., Sublette, K. L. & Suflita, J. M. 1999. Intrinsic bioremediation of petroleum hydrocarbons in a gas condensate-contaminated aquifer. *Environmental Science and Technology*, **33**, 2550–2560.

Gmal, B., Hesse, U., Hummelsheim, K., Kilger, R., Krzykacz-Hausmann, B. & Moser, E. F. 2004. *Untersuchungen zur Kritikalitätssicherheit in der Nachbetriebsphase eines Endlagers für ausgediente Kernbrennstoffe in unterschiedlichen Wirtsformationen.* Report **GRS-A-3240**, BfS-FKZ WS 1005/8488-2. GRS, Garching, (http://www.bfs.de/de/endlager/publika/AG_2_Einzelaspekte_Kriti_Text.pdf)

Gregory, K. B., Bond, D. R. & Lovley, D. R. 2004. Graphite electrodes as electron donors for anaerobic respiration. *Environmental Microbiology*, **6**, 596–604.

Humphreys, P. N., West, J. M. & Metcalfe, R. 2010. *Microbial Effects on Repository Performance*. Technical Report. UK NDA.

Icenhower, J. P., Qafoku, N. P., Martin, W. J. & Zachara, J. M. 2008. *The Geochemistry of Technetium: A Summary of the Behavior of an Artificial Element in the Natural Environment.* Report PNNL-18139. Pacific Northwest National Laboratory, Richland, WA.

Jaisi, D. P., Dong, H. & Morton, J. P. 2008. Partitioning of Fe(II) in reduced nontronite (NAu-2) to reactive sites: reactivity in terms of Tc(VII) reduction. *Clays and Clay Minerals*, **56**, 175–189.

Jaisi, D. P., Dong, H. L., Plymale, A. E., Fredrickson, J. K., Zachara, J. M., Heald, S. & Liu, C. X. 2009. Reduction and long-term immobilization of technetium by Fe(II) associated with clay mineral nontronite. *Chemical Geology*, **264**, 127–138.

Johnson, L. & King, F. 2008. The effect of the evolution of environmental conditions on the corrosion evolutionary path in a repository for spent fuel and high-level waste in Opalinus Clay. *Journal of Nuclear Materials*, **379**, 9–15.

Jørgensen, B. B., Isaksen, M. F. & Jannasch, H. W. 1992. Bacterial sulfate reduction above 100 °C in deep-sea hydrothermal vent sediments. *Science*, **258**, 1756–1757.

Kim, J., Dong, H., Seabaugh, J., Newell, S. W. & Eberl, D. D. 2004. Role of microbes in the smectite-to-illite reaction. *Science*, **303**, 830–832.

Koroleva, M., Lerouge, C., Mäder, U., Claret, F. & Gaucher, E. 2011. Biogeochemical processes in a clay formation in situ experiment: Part B – results from overcoring and evidence of strong buffering by the rock formation. *Applied Geochemistry*, **26**, 954–966.

Kostka, J. E., Haefele, E., Viehweger, R. & Stucki, J. W. 1999*a*. Respiration and dissolution of iron(III)-containing clay minerals by bacteria. *Environmental Science and Technology*, **33**, 3127–3133.

Kostka, J. E., Wu, J., Nealson, K. H. & Stucki, J. W. 1999*b*. The impact of structural Fe(III) reduction by bacteria on the surface chemistry of smectite clay minerals. *Geochimica et Cosmochimica Acta*, **63**, 3705–3713.

Krumholz, L. R., McKinley, J. P., Ulrich, G. A. & Suflita, J. M. 1997. Confined subsurface microbial communities in Cretaceous rock. *Nature*, **386**, 64–65.

Kursten, B. 2004. *Results from corrosion studies on metallic components and instrumentation installed in the OPHELIE mock-up*. Presentation on EIG EURIDICE' OPHELIE Day, 10 June, http://www.euridice.be/eng/06publicaties2004agendaO.shtm (accessed 25 May 25).

Laird, D. A. 2006. Influence of layer charge on swelling of smectites. *Applied Clay Science*, **34**, 74–87.

Landolt, D., Davenport, A., Payer, J. & Shoesmith, D. 2009. *A Review of Materials and Corrosion Issues Regarding Canisters for Disposal of Spent Fuel and High-Level Waste in Opalinus Clay*. Technical Report **02-05**. Nagra, Wettingen.

Laverov, N. P., Yudintsev, S. V. & Omel'yanenko, B. I. 2009. Isolation of long-lived technetium-99 in confinement matrices. *Geology of Ore Deposits*, **51**, 259–274.

Legrand, L., Mazerolles, L. & Chaussé, A. 2004. The oxidation of carbonate green rust into ferric phases: solid-state reaction or transformation via solution. *Geochimica et Cosmochimica Acta*, **68**, 3497–3507.

Lerouge, C., Grangeon, S. et al. 2011. Mineralogical and isotopic record of biotic and abiotic diagenesis of the Callovian–Oxfordian clayey formation of Bure (France). *Geochimica et Cosmochimica Acta*, **75**, 2633–2663.

Liu, D., Dong, H. et al. 2011. Reduction of structural Fe(III) in nontronite by methanogen Methanosarcina barkeri. *Geochimica et Cosmochimica Acta*, **75**, 1057–1071.

Liu, D., Dong, H. et al. 2012. Microbial reduction of structural iron in interstratified illite-smectite minerals by a sulfate-reducing bacterium. *Geobiology*, **10**, 150–162.

Lloyd, J. R. 2003. Microbial reduction of metals and radionuclides. *FEMS Microbiology Reviews*, **27**, 411–425.

Lovley, D. R., Fraga, J. L., Blunt-Harris, E. L., Hayes, L. A., Phillips, E. J. P. & Coates, J. D. 1998. Humic substances as a mediator for microbially catalyzed metal reduction. *Acta Hydrochimica et Hydrobiologica*, **26**, 151–157.

Lovley, D. R., Holmes, D. E. & Nevin, K. P. 2004. Dissimilatory Fe(III) and Mn(IV) reduction. *Advances in Microbial Physiology*, **49**, 219–286.

Lukens, W. W., McKeown, D. A., Buechele, A. C., Muller, I. S., Shuh, D. K. & Pegg, I. L. 2007. Dissimilar behavior of technetium and rhenium in borosilicate waste glass as determined by X-ray absorption spectroscopy. *Chemistry of Materials*, **19**, 559–566.

Lydmark, S. & Persson, J. 2010. *Production of Gas and Sulphide by Bacteria in Boom Clay*. Report NIROND-TR **2010-16E**.

MacLean, L. C. W., Pray, T. J., Onstott, T. C., Brodie, E. L., Hazen, T. C. & Southam, G. 2007. Mineralogical, chemical and biological characterization of an anaerobic biofilm collected from a borehole in a deep gold mine in South Africa. *Geomicrobiology Journal*, **24**, 491–504.

Marsili, E., Baron, D. B., Shikhare, I., Coursolle, D., Gralnick, J. A. & Bond, D. R. 2008. Shewanella secretes flavins that mediate extracellular electron transfer. *Proceedings of the National Academy of Sciences of the United States of America*, **105**, 3968–3973.

Marty, N. C. M., Tournassat, C., Burnol, A., Giffaut, E. & Gaucher, E. C. 2009. Influence of reaction kinetics and mesh refinement on the numerical modelling of concrete/clay interactions. *Journal of Hydrology*, **364**, 58–72.

Masurat, P., Eriksson, S. & Pedersen, K. 2010. Microbial sulphide production in compacted Wyoming bentonite MX-80 under in situ conditions relevant to a repository for high-level radioactive waste. *Applied Clay Science*, **47**, 58–64.

Mauclaire, L., McKenzie, J. A., Schwyn, B. & Bossart, P. 2007. Detection and cultivation of indigenous microorganisms in Mesozoic claystone core samples from the Opalinus Clay formation (Mont Terri rock laboratory). *Physics and Chemistry of the Earth*, **32**, 232–240.

McMahon, P. B., Chapelle, F. H., Falls, F. W. & Bradley, P. M. 1992. Role of microbial processes in linking sandstone diagenesis with organic-rich clays. *Journal of Sedimentary Petrology*, **62**, 1–10.

Meleshyn, A. 2011. *Microbial Processes Relevant for the Long-Term Performance of Radioactive Waste Repositories in Clays*. Report **GRS-291**, BMWi-FKZ 02 E 10548. GRS, Braunschweig, http://www.grs.de/publi cation/GRS-291

Meleshyn, A. & Noseck, U. 2012. *Radionuclide Inventory of Vitrified Waste after Spent Nuclear Fuel RePro-cessing at La Hague*. Report **GRS-294**. BMWi-FKZ 02 E 10548. GRS, Braunschweig, http://www.grs.de/publication/grs-294-radionuclide-inventory-vitrified-waste-after-spent-nuclear-fuel-reprocessing-la-hague

Muyzer, G. & Stams, A. J. M. 2008. The ecology and biotechnology of sulphate-reducing bacteria. *Nature Reviews*, **6**, 441–454.

Nadeau, P. H., Peacor, D. R., Yan, J. & Hillier, S. 2002. I-S precipitation in pore space as the cause of geopressuring in Mesozoic mudstones, Egersund Basin, Norwegian continental shelf. *American Mineralogist*, **87**, 1580–1589.

NAGRA. 2002. *Project Opalinus Clay. Safety Report. Demonstration of Disposal Feasibility for Spent Fuel, Vitrified High-Level Waste and Long-Lived Intermediate-Level Waste (Entsorgungsnachweis)*. Technical Report **02-05**. Nagra, Wettingen.

Neall, F., Pastina, B., Smith, P., Gribi, P., Snellman, M. & Johnson, L. 2007. *Safety Assessment for a KBS-3H Spent Nuclear Fuel Repository at Olkiluoto – Complementary Evaluations of Safety Report*. Report POSIVA **2007-10**, Olkiluoto.

NWMO. 2005. *Choosing a Way Forward. The Future Management of Canada's Used Nuclear Fuel (Final Study)*. Nuclear Waste Management Organization, Toronto.

ONDRAF/NIRAS, 2004. *Multi-Criteria Analysis on the Selection of a Reference EBS Design for Vitrified High Level Waste*. Report NIROND **2004-03**.

ORTIZ, L., VOLCKAERT, G. & MALLANTS, D. 2002. Gas generation and migration in Boom Clay, a potential host rock formation for nuclear waste storage. *Engineering Geology*, **64**, 287–296.

PALSDOTTIR, H., REMIS, J. P. ET AL. 2009. Three-dimensional macromolecular organization of cryofixed Myxococcus xanthus biofilms as revealed by electron microscopic tomography. *Journal of Bacteriology*, **191**, 2077–2082.

PANTKE, C., OBST, M., BENZERARA, K., MORIN, G., ONA-NGUEMA, G., DIPPON, U. & KAPPLER, A. 2012. Green rust formation during Fe(II) oxidation by the nitrate-reducing Acidovorax sp. strain BoFeN1. *Environmental Science and Technology*, **46**, 1439–1446.

PARKES, R. J., WELLSBURY, P., MATHER, I. D., COBB, S. J., CRAGG, B. A., HORNIBROOK, E. R. C. & HORSFI ELD, B. 2007. Temperature activation of organic matter and minerals during burial has the potential to sustain the deep biosphere over geological time scales. *Organic Geochemistry*, **38**, 845–852.

PARKES, R. J., LINNANE, C. D., WEBSTER, G., SASS, H., WEIGHTMAN, A. J., HORNIBROOK, E. R. C. & HORSFIELD, B. 2011. Prokaryotes stimulate mineral H_2 formation for the deep biosphere and subsequent thermogenic activity. *Geology*, **39**, 219–222.

PATTERSON, D. J. 2009. Seeing the big picture on microbe distribution. *Science*, **325**, 1506–1507.

PEDERSEN, K. 1997. The microbiology of radioactive disposal. *In*: WOLFRAM, J. H., ROGERS, R. D. & GAZSÓ, L. G. (eds) *Microbial Degradation Processes in Radioactive Waste Repository and in Nuclear Fuel Storage*. Kluwer Academic, Budapest, 189–209.

PEDERSEN, K. 2000. *Microbial Processes in Radioactive Waste Disposal*. Technical Report TR-00-04. Svensk Kärnbränslehantering AB, Stockholm.

PEDERSEN, K. 2005. Microorganisms and their influence on radionuclide migration in igneous rock environments. *Journal of Nuclear and Radiochemical Sciences*, **6**, 11–15.

PEDERSEN, K. 2010. Analysis of copper corrosion in compacted bentonite clay as a function of clay density and growth conditions for sulfate-reducing bacteria. *Journal of Applied Microbiology*, **108**, 1094–1104.

PERDRIAL, J. N., WARR, L. N., PERDRIAL, N., LETT, M.-C. & ELSASS, F. 2009. Interaction between smectite and bacteria: implications for bentonite as backfill material in the disposal of nuclear waste. *Chemical Geology*, **264**, 281–294.

POLLASTRO, R. M. 1993. Considerations and applications of the illite/smectite geothermometer in hydrocarbon-bearing rocks of Miocene to Mississippian age. *Clays and Clay Minerals*, **41**, 119–133.

RAISWELL, R. & CANFIELD, D. E. 1996. Rates of reaction between silicate iron and dissolved sulfide in Peru Margin sediments. *Geochimica et Cosmochimica Acta*, **60**, 2777–2787.

ROUSSEL, E. G., CAMBON BONAVITA, M.-A., QUERELLOU, J., CRAGG, B. A., WEBSTER, G., PRIEUR, D. & PARKES, R. J. 2008. Extending the sub-sea-floor biosphere. *Science*, **320**, 1046.

SANZ-MONTERO, M. E., RODRÍGUEZ-ARANDA, J. P. & PÉREZ-SOBA, C. 2009. Microbial weathering of Fe-rich phyllosilicates and formation of pyrite in the dolomite precipitating environment of a Miocene lacustrine system. *European Journal of Mineralogy*, **21**, 163–175.

SCHIPPERS, A., NERETIN, L. N., KALLMEYER, J., FERDELMAN, T. G., CRAGG, B. A., PARKES, R. J. & JØRGENSEN, B. B. 2005. Prokaryotic cells of the deep sub-seafloor biosphere identified as living bacteria. *Nature*, **433**, 861–864.

SCHWYN, B., LEUPIN, O. X., BAGNOUD, A. & BERNIER-LATMANI, R. 2012. Microbiologically mediated processes in a repository sited in a clay host rock. *In*: *Clays in Natural & Engineered Barriers for Radioactive Waste Confinement, O/06A/1KP*. ANDRA, Montpellier.

SHELOBOLINA, E. S., PICKERING, S. M., JR. & LOVLEY, D. R. 2005. Fe-cycle bacteria from industrial clays mined in Georgia, USA. *Clays and Clay Minerals*, **53**, 580–586.

SHERAR, B. W. A., POWER, I. M., KEECH, P. G., MITLIN, S., SOUTHAM, G. & SHOESMITH, D. W. 2011. Characterizing the effect of carbon steel exposure in sulfide containing solutions to microbially induced corrosion. *Corrosion Science*, **53**, 955–960.

SHERWOOD LOLLAR, B. 2011. *Far-field Microbiology Considerations Relevant to a Deep Geological Repository – State of Science Review*. Report NWMO TR-2011-09. Nuclear Waste Management Organization, Toronto.

SKB. 2011. *Long-Term Safety for the Final Repository for Spent Nuclear Fuel at Forsmark*. Main report of the SR-Site project. Technical Report TR-11-01. Svensk Kärnbränslehantering AB, Stockholm.

STROES-GASCOYNE, S. 2010. Microbial occurrence in bentonite-based buffer, backfill and sealing materials from large-scale experiments at AECL's underground research laboratory. *Applied Clay Science*, **47**, 36–42.

STROES-GASCOYNE, S., SCHIPPERS, A. ET AL. 2007. Microbial community analysis of Opalinus Clay drill core samples from the Mont Terri underground research laboratory, Switzerland. *Geomicrobiology Journal*, **24**, 1–17.

STROES-GASCOYNE, S., HAMON, C. J., MAAK, P. & RUSSELL, S. 2010. The effects of the physical properties of highly compacted smectitic clay (bentonite) on the culturability of indigenous microorganisms. *Applied Clay Science*, **47**, 155–162.

STROES-GASCOYNE, S., SCHIPPERS, C. ET AL. 2011*a*. Biogeochemical processes in a clay formation in situ experiment: Part D – microbial analyses – synthesis of results. *Applied Geochemistry*, **26**, 980–989.

STROES-GASCOYNE, S., HAMON, C. J. & MAAK, P. 2011*b*. Limits to the use of highly compacted bentonite as a deterrent for microbiologically influenced corrosion in a nuclear fuel waste repository. *Physics and Chemistry of the Earth, Parts A/B/C*, **36**, 1630–1638.

TAKAI, K., NAKAMURA, K. ET AL. 2008. Cell proliferation at 122 °C and isotopically heavy CH4 production by a hyperthermophilic methanogen under high-pressure cultivation. *Proceedings of the National Academy of Sciences of the United States of America*, **105**, 10949–10954.

TEMPLETON, A. & KNOWLES, E. 2009. Microbial transformations of minerals and metals: recent advances in geomicrobiology derived from synchrotron-based

X-ray spectroscopy and X-ray microscopy. *Annual Review of Earth and Planetary Sciences*, **37**, 367–391.

Tuck, V. A., Edyvean, R. G. J., West, J. M., Bateman, K., Coombs, P., Milodowski, A. E. & McKervey, J. A. 2006. Biologically induced clay formation in subsurface granitic environments. *Journal of Geochemical Exploration*, **90**, 123–133.

Verstricht, J. & Dereeper, B. 2003. The OPHELIE mock-up experiment: first step in the demonstration of the feasibility of HLW disposal. *Proceedings of the International Conference 'Waste Management '03'*, Tucson, AZ.

Villagran, J. E. 2012. *Development of a Monitoring Program for a Deep Geological Repository for used Nuclear Fuel*. Report NWMO TR-2012-18, Toronto.

Vorhies, J. S. & Gaines, R. R. 2009. Microbial dissolution of clay minerals as a source of iron and silica in marine sediments. *Nature Geoscience*, **2**, 221–225.

Weber, K. A., Achenbach, L. A. & Coates, J. D. 2006. Microorganisms pumping iron: anaerobic microbial iron oxidation and reduction. *Nature Reviews Microbiology*, **4**, 752–764.

Wersin, P., Leupin, O. X. *et al.* 2011*a*. Biogeochemical processes in a clay formation in situ experiment: part A – overview, experimental design and water data of an experiment in the Opalinus Clay at the Mont Terri Underground Research Laboratory, Switzerland. *Applied Geochemistry*, **26**, 931–953.

Wersin, P., Stroes-Gascoyne, S., Pearson, F. J., Tournassat, C., Leupin, O. X. & Schwyn, B. 2011*b*. Biogeochemical processes in a clay formation in situ experiment: part G – key interpretations and conclusions. Implications for repository safety. *Applied Geochemistry*, **26**, 1023–1034.

West, J. M., McKinley, I. G. & Chapman, N. A. 1982. Microbes in deep geological systems and their possible influence on radioactive waste disposal. *Radioactive Waste Management and the Nuclear Fuel Cycle*, **3**, 1–15.

West, J. M., McKinley, I. G., Neall, F. B., Rochelle, C. A., Bateman, K. & Kawamura, H. 2006. Microbiological effects on the Cavern Extended Storage (CES) repository for radioactive waste – a quantitative evaluation. *Journal of Geochemical Exploration*, **90**, 114–122.

Wolfaardt, G. M. & Korber, D. R. 2012. *Near-field Microbiological Considerations Relevant to a Deep Geological Repository for used Nuclear Fuel – State of Science Review*. Report NWMO TR-2012-02. Nuclear Waste Management Organization, Toronto.

Wolfram, J. H. & Dirk, W. J. 1997. Biofilm development and the survival of microorganisms in water systems of nuclear reactors and spent fuel pools. *In*: Wolfram, J. H., Rogers, R. D. & Gazsó, L. G. (eds) *Microbial Degradation Processes in Radioactive Waste Repository and in Nuclear Fuel Storage Areas*. Kluwer Academic, Dordrecht, 139–147.

Yoshioka, H., Maruyama, A., Nakamura, T., Higashi, Y., Fuse, H., Sakata, S. & Bartlett, D. H. 2010. Activities and distribution of methanogenic and methane-oxidizing microbes in marine sediments from the Cascadia Margin. *Geobiology*, **8**, 223–233.

Two-phase-flow pore-size simulations in Opalinus clay by the Lattice Boltzmann Method

M. DYMITROWSKA[1]*, A. PAZDNIAKOU[2] & PIERRE M. ADLER[2]

[1]*IRSN, B.P.17, 92262 Fontenay-aux-Roses Cedex, France*

[2]*Sisyphe-UPMC, 4, place Jussieu, 75252-Paris, France*

**Corresponding author (e-mail: magdalena.dymitrowska@irsn.fr)*

Abstract: The experimental determination of transport properties of low permeability clay rocks, especially of relative permeabilities and capillary pressure curves for water and gas, is a very challenging issue, in particular at high water saturation (very low gas permeability resulting in long equilibration times) and for gaseous hydrogen (due to the high pressures involved and the resulting explosion risk). Navier-Stokes equations are solved inside a porous medium on the pore scale, so as to derive the absolute and relative (two-phase-flow) permeabilities. For this purpose microtomography data of Opalinus clay samples acquired in the Mont Terri Ventilation Experiment are used to visualize the pore space in 3D at a micrometric scale (porosity size >0.7 μm). The corresponding percolating porosity is mainly composed of micrometric cracks parallel to the bedding and attributed to shrinkage. Two-phase flow is calculated in the percolating cracks by an immiscible Lattice Boltzmann (LBM) code. In addition, some validation results of the LBM model for three-phase systems (liquid-gas-solid) are presented.

Clays have been intensively studied as potential host rocks for geological disposal of radioactive wastes, especially because of their very low permeability (Andra 2005). Significant production of hydrogen (and incidentally other gases) is expected within such a facility mainly due to anaerobic corrosion of steel elements. A gas phase may appear and affect the hydraulic charge, the mechanical conditions and the potential transport of radionuclides in the disposal facility components, including the host rock. Thus, the containment capacity of the disposal system could be altered due to the generation and migration of gas. This problem must be thoroughly understood when assessing the safety of a deep repository. Several recent national and international programmes have been partially or totally devoted to this topic because of its importance (theoretical: French GNR MoMaS; modelling and experimental: Mont Terri Project, 7th PCRD FORGE Project).

However, regarding the flow of water and hydrogen in very low permeability materials such as clays, only a limited number of specific measures of relative permeability and capillary pressure are available (Marschall *et al.* 2005; Zhang & Rothfuchs 2007; Cariou *et al.* 2012). The experimental determination of these properties is a challenging issue especially at high water saturation (the very low gas permeability implies long equilibration times and requires using high gas pressures) and with hydrogen (the high pressures involved increase explosion risk).

An alternative path has been used to acquire the data necessary to carry out gas migration simulations at the disposal scale. Recent developments in 3D imaging, such as X-ray tomography (Wildenschild *et al.* 2002), have been used to reconstruct the 3D pore space of opaque solids with a submicron resolution. On the one hand, these reconstructions can be used to perform simulations of multiphase flow through the pores. On the other hand, the generalized Darcy equation which is commonly used and easier to solve numerically, introduces a bias and its applicability to very low permeability media is uncertain. Numerous methods based on mesoscopic approaches (Dissipative Particles' Dynamics (Henrich *et al.* 2007), Lattice Boltzmann Method (LBM: for a comprehensive review see Succi (2001), Volume of Fluid (Fukai *et al.* 1995)) exist for efficient simulations of multiphase flow in pore spaces with complex topologies. Here, the Lattice Boltzmann Model (Rothman & Keller 1988) is solved inside a realistic pore network. The results can be processed to obtain the macroscopic transport parameters used in large-scale calculations such as permeabilities and the diffusion coefficient. Actually, by fitting the Darcy or generalized Darcy equation to the results, it is possible to extract the absolute and relative two-phase permeabilities and capillary pressure. In this paper, the geometrical properties of the pore space as measured from tomography experiments on Opalinus clay samples are described. The LBM is briefly presented as well as its applicability to simulate three-phase flow with capillarity effects. Then, it is applied to two types of two-phase flow simulations with LBM. The first group consists of fixed saturation

From: Norris, S., Bruno, J., Cathelineau, M., Delage, P., Fairhurst, C., Gaucher, E. C., Höhn, E. H., Kalinichev, A., Lalieux, P. & Sellin, P. (eds) 2014. *Clays in Natural and Engineered Barriers for Radioactive Waste Confinement*. Geological Society, London, Special Publications, **400**, 195–206.
First published online April 7, 2014, http://dx.doi.org/10.1144/SP400.20

simulations and the second one consists of desaturation (drainage) simulations. Some sensitivity results as well as the analysis in terms of relative permeabilities and suction curve are presented.

Properties of the Opalinus clay samples

High resolution X-ray microtomography (0.7 μm) data of Opalinus samples extracted during the Ventilation Experiment of the Mont Terri Project (Bossart & Thury 2008) were used. The average total porosity obtained by mercuro-porosimetry is equal to 18% for all samples. Further description of the structural organization of minerals and porosity in the samples is given by Matray *et al.* (2007).

From the initial data, after an appropriate segmentation into pore and solid voxels, the pore voxels which do not belong to the percolating connected component were eliminated. This procedure revealed that the connected (or percolating) porosity above 0.7 μm was mainly composed of micrometric cracks roughly parallel to the bedding and attributed to shrinkage. Only 8 samples among 14 were percolating. Thus, the average aperture and the roughness of the microcracks were determined with the concepts and tools used for fractures. These properties are summarized in Table 1 (this analysis was conducted only for six selected samples). Figure 1 gives the images of the connected porosity of the corresponding samples.

Lattice Boltzmann method

Introduction to LBM

The Navier-Stokes equations for two-phase flow are solved using a LBM. Historically, the 'classical' Lattice Boltzmann equation (LBE) has been developed empirically, with basic ideas borrowed from cellular automata; the physical space of interest is discretized by a regular lattice populated by discrete particles that 'jump' from one site of the lattice to another with discrete velocities and collide/interact with each other at the lattice nodes. It has been shown that it is possible to choose the lattice geometry and the discrete velocities so as to ensure the rotational invariance of the momentum flux tensor at the macroscopic level. Using Chapman-Enskog analysis, one can recover the governing continuity and Navier-Stokes equations from the LBE. The simulation code developed by Pazdniakou (2012) implements recent improvements of the classical multiphase LBM (Rothman & Keller 1988) due to Gunstensen *et al.* (1991). The model consists of two LBEs, one for each fluid phase. These fluids are named red and blue. For a multiphase flow of two immiscible fluids, all the points of the lattice occupied by fluid can be divided into three categories, namely purely red fluid points, purely blue fluid points, and interface points (lattice points where the red and blue fluids coexist). The evolution of the fluid in purely red or purely blue regions is described by the standard multirelaxation time (MRT) LBE. The relaxation parameters are tuned by using the standard two relaxation times (TRT) model. At the fluid-fluid interface, the collision matrix is defined by the majority rule, that is, it is based on the dominant density fluid (D'Ortona *et al.* 1995). The bounce back rule is applied at the solid-fluid interface (Ginzburg & Adler 1994).

In the present model, a continuum method for modelling surface tension is used. The main idea of the method is to use a volumetric surface tension force at the fluid-fluid interface in order to obtain the surface tension effects. The force direction is normal to the interface and is proportional to the curvature of the interface at the point of application. Since the interface between the two fluids is not a surface, the main difficulty is to find a way to calculate the curvature of the interface. For this purpose, the colour field is introduced:

$$C(r, i) = \frac{\rho_r(r, i) - \rho_b(r, i)}{\rho_r(r, i) + \rho_b(r, i)} \quad (1)$$

where $\rho_\alpha(r, i)$ is the local density of the α-th fluid. In the present model, the colour field $C(\mathbf{r}, t)$ is fixed at

Table 1. *Pore properties of some Opalinus samples: <b> mean aperture, σ−/σ+ roughness of the lower and upper surface (the standard deviation of the Z-position), open (percolating) porosity Φ and initial porosity Φinit*

Sample	<b>	σ−/σ+	Φinit	Φ	Size in voxels
96-2_1	3.3	42/43	3	0.69	1171 × 1130 × 201
96-2_2	3.1	48/48	3	0.69	1171 × 1166 × 1024
96-4	3.1	24/24	2.9	2.4	1163 × 1050 × 1024
103-1_1	2.1	20/20	3.8	0.68	500 × 325 × 198
103-1_2	29.6	18/26	3.8	0.68	324 × 325 × 198
103-4B	20	15/5	6	1.8	339 × 450 × 115

The unit length is the size of a voxel (0.7 μm).

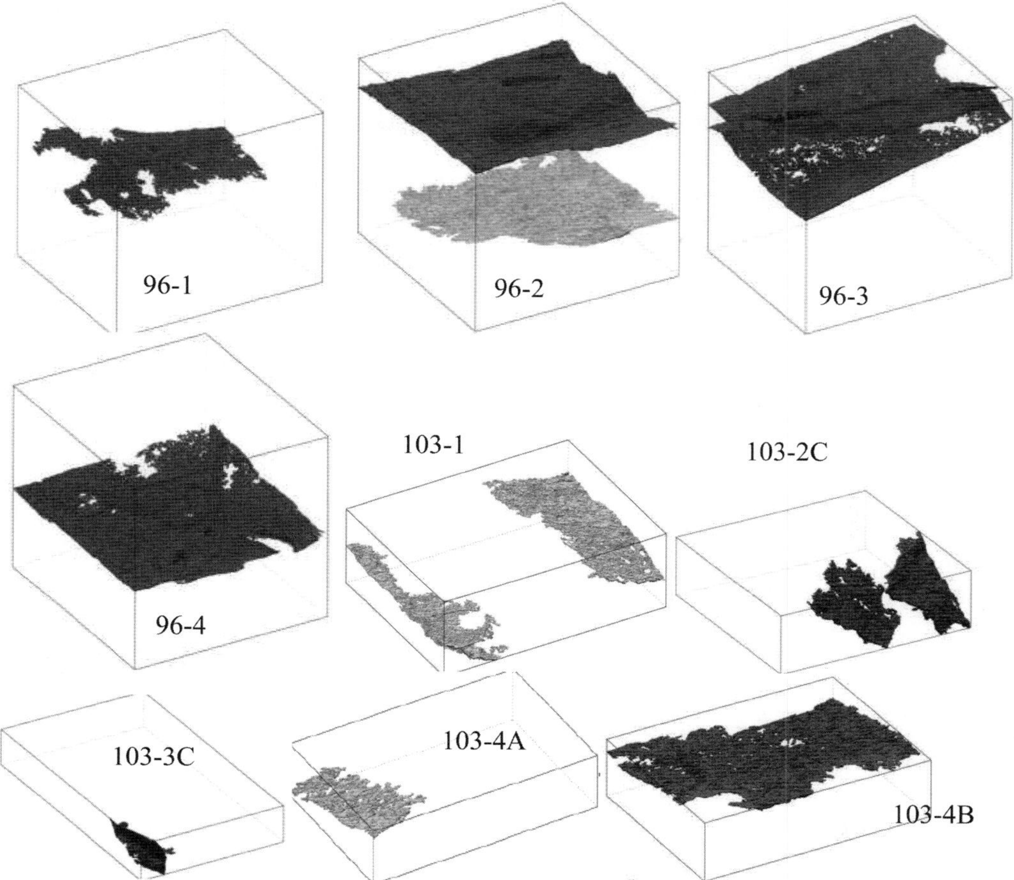

Fig. 1. The percolating component of porosity in Opalinus samples. Voxel size: 0.7 μm. Some of the samples are composed of two disjoint pore volumes and are subdived into two parts for calculations.

the solid wall point **r** in order to model contact angle. It can be demonstrated that

$$\cos(\theta) = C(\mathbf{r}, t) \tag{2}$$

where θ is the contact angle.

Validation of LBM code on three-phase flow tests

The first part of this work has been devoted to the validation of the two-phase Lattice Boltzmann code since it involves many physical features such as surface tension and wetting properties. The validation has been conducted in several different cases including static properties, simple dynamic properties and complex dynamic properties. In all cases, good agreement was obtained between the simulations and the existing experimental and/or theoretical results. As an example, capillary rise of a wetting liquid between parallel (infinite) plates was studied as illustrated in Figure 2.

The evolution of the height h(t) of a liquid in a plane channel (i.e. in 2D) can be approximated by the Washburn equation (Washburn 1921). The computational domain is presented in Figure 2 and results for one test case are shown in Figure 3. For this particular set of physical variables, very good agreement with the theoretical prediction is found, not only for the asymptotic regime, but also for early times (Attar & Körner 2009). The simulations were performed for a large set of physical parameters and the numerical results obtained by LBM are in good agreement with theoretical predictions when the following conditions are verified:

- the tube radius R is sufficiently large (>10);
- the rise height is large with respect to R;
- the rise velocity is low ($Ca \ll 1$).

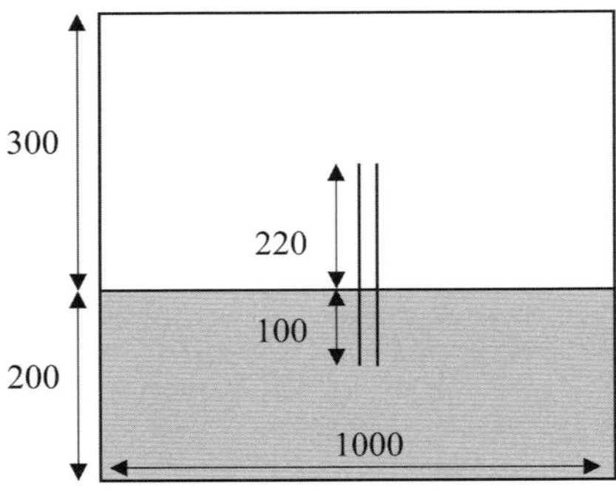

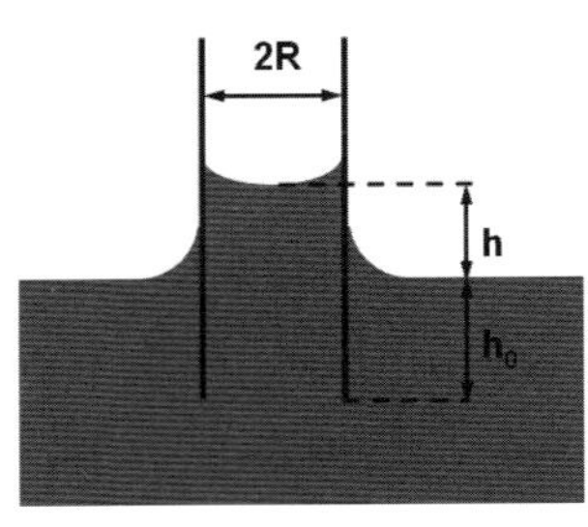

Fig. 2. Geometry and dimensions (in lattice units) of the simulation domain for capillary rise between plates.

It should be noted that in order to obtain an even better agreement between the simulations and the analytical prediction, the latter should be enriched by including the non-established velocity profile and the dynamic change in contact angle during the rise (Hoffman 1975; Bonn *et al.* 2009).

Further validation tests including drop spreading are given in Pazdniakou (2012).

Results of LBM simulations in Opalinus samples

Absolute permeability

Single-phase flow was calculated in the percolating cracks by solving the Stokes equations of motion with no slip at the solid interface. The absolute permeability was also determined for two perpendicular directions in the crack plane whenever it percolates along these two directions (Table 2). The values of permeabilities form two groups, one of the order of 10^{-20} m^2 and the second one of the order of 10^{-15} m^2. These results are within the interval of values of absolute permeability measured for sound Opalinus clay (*c.* 10^{-20}–10^{-21} m^2) and that of the EDZ of the Opalinus clay in Mont Terri, estimated by Bossart *et al.* (2004) at about 10^{-14} m^2. It should be noted also that the vertical (perpendicular to the bedding) permeability is much lower than in the bedding plane, as expected.

Relative permeability and capillary pressure

Calculations were performed in symmetric configurations: in this approach, the initial sample is

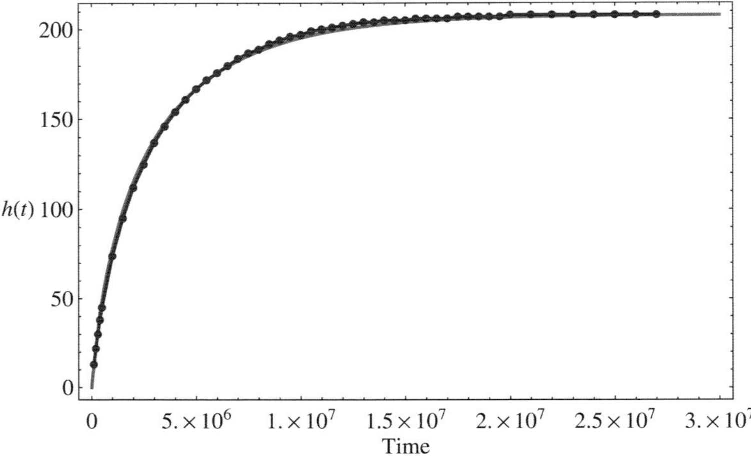

Fig. 3. Verification of the Washburn equation (continuous line) by the LBM code (line with dot labels): domain size 1000 × 1 × 500, surface tension $\gamma = 0.004$, gravitational constant g = 5e-7, fluid density ratio 1, red/blue fluid viscosity ratio 0.002, contact angle cos $\Phi = -0.52$, tube radius 20, $h(\infty) = 208$. All physical properties are dimensionless.

Table 2. *Absolute permeability of all tested samples of Opalinus clay (in m^2)*

Sample	Kx	Ky	Kz
96-1	0	1.2e-16	0
96-2_1	1.0e-19	4.2e-15	0
96-2_2	4.0e-20	0	0
96-3	1.0e-15	2.2e-15	0
96-4	3.0e-15	3.1e-15	0
Mean(96)	1.0e-15	2.0e-15	0
103-1_1	1.5e-20	0	0
103-1_2	3.0e-20	0	0
103-2A	1.6e-20	0	0
103-3C	0	0	1.0e-20
103-4A	1.2e-16	0	0
103-4B	1.3e-15	1.1e-15	0
Mean(103)	1.2e-16	5.9e-17	1.4e-21

doubled by its mirror image in order to obtain periodic boundary conditions. In all simulations, the parameters are chosen so as to maintain a low Reynolds number. The pressure gradient (imposed via a body force) is equal to 10^{-6} for both fluids. The interfacial tension is chosen to be equal to 10^{-4} in order to keep the capillary number Ca in the range of 10^{-4}–10^{-3}. The blue fluid is supposed to wet the solid surface. The contact angle used in the simulations is 0°, that is, the solid surface is completely wetted by the blue fluid. Both fluids in pure state have a density equal to 1 and the square of the sound speed equal to 1/3. The kinematic viscosity is equal to 0.00185 and 1 for red and blue fluids, respectively. All the values listed above are given in lattice units. Particular attention was given to obtain reliable results for water saturations close to 1, and for this purpose the calculations were repeated at several spatial resolutions.

Two types of simulations were performed. In the first one, saturation is imposed via the initial conditions (random distribution of both phases within the pore space) and a constant pressure gradient is applied to the system in order to reach a stationary flow after a transient phase. First, the fluids coalesce under the action of the interfacial tension and of the flow; second, the spatial distribution of fluids tends to a steady state. In the second class of simulations, the fully saturated pore volume is put in contact with a gas saturated space; a pressure gradient is also applied in order to force the desaturation of the pore space.

In all cases, the flow results were analyzed in terms of equivalent relative permeabilities and capillary pressure. The fluxes of both fluids are measured across a predefined surface perpendicular to the pressure gradient and the relative permeability k_{ri} for fluid i is deduced by the generalized Darcy equation:

$$V_\alpha(i) = \frac{k_{r\alpha}(i)K}{\mu_\alpha}\overline{\nabla p} \tag{3}$$

Where V_α is the Darcy velocity of the fluid α and μ_α its viscosity, K is the absolute permeability of the medium, $\overline{\nabla p}$ the applied pressure gradient and i the i-th time step. The relative permeability value was calculated as a mean over N (usually 1000) time steps.

If we want to calculate the capillary pressure with sufficient precision, we need to consider only points which could represent pure fluids. In practice, it means that they are distant from the interfaces and have $C(r, t)$ close to 1 or -1. The points with $|C(r, t)| \geq 0.99$ are kept for the calculation of capillary pressure. For each of such points, the local pressure is calculated as follows:

$$P_L(r, i) = \begin{cases} c_r^2\rho_r(r, i) : C(r, i) \geq 0 \\ c_b^2\rho_b(r, i) : C(r, i) < 0 \end{cases} \tag{4}$$

where c_α is the sound velocity of fluid α. In order to obtain the capillary pressure assimilated to the difference of pressures between the two fluids, the averages of P_L are calculated over the points belonging to the fluid α $\{\}_\alpha$ where $|C(r, t)| \geq 0.99$:

$$P_c(i) = \{P_L(r, i)\}_r - \{P_L(r, i)\}_b \tag{5}$$

The final value of the capillary pressure is obtained by averaging over N time steps; N is usually equal to 1000.

Steady state calculations of two-phase flow

The 103-4B sample was selected since it presents one of the largest mean apertures (20 voxels) and percolates along the X and Y directions. This relatively large fracture is appropriate to represent in the original discretization a system which may be composed of up to two interfaces (usually of about 5–6 voxels wide) and three single fluid regions (wetting/non-wetting/wetting fluid layers inside the fracture). However, with the initial sample size of 339 × 450 × 115 voxels, doubled by its mirror image in order to obtain periodic boundary conditions, it would be difficult to study the influence of the discretization, and the calculations are very long. For this reason, a subsample was extracted as shown in Figure 4. The initial subsample was discretized by 200 × 200 × 66 voxels.

The mean aperture of this subsample is only 2.6 voxels. Thus, it was sub-discretized twice resulting in sizes of 400 × 400 × 132 and 800 × 800 × 264.

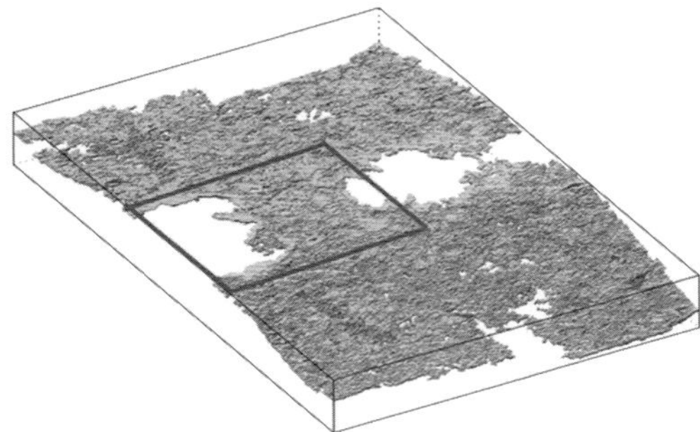

Fig. 4. Subsample of the 103-4B used for the two-phase flow calculations. Its size is 200 × 200 × 66 voxels.

The relative permeabilities were calculated for the three discretizations. Results are displayed in Figure 5. The relative permeability of the gas phase goes rapidly to zero when the saturation S_b (saturation of the blue fluid) increases. The curves for different discretizations are seen to be relatively similar except for, rather surprisingly, the intermediate values 0.4 and 0.6 of water saturation. In order to understand the origin of the differences, some phase configurations obtained for the two most refined discretizations and for two saturations (0.6 and 0.8) are represented in Figure 6.

It is clear especially for the saturation equal to 0.8 that discretization has a strong effect on the phase configurations. However, one might suspect that further refining of the discretization down to the level where the gas phase can become continuous may result in a significant increase in gas permeability, as happened for a saturation of 0.6 when passing from 400 × 400 × 132 to 800 × 800 × 264 discretization. The problem of high saturation is more obvious when looking at capillary pressure curves. They are given in Figure 7 for the three discretizations. Again the results are seen to be qualitatively similar. A problem appears for high saturations where the capillary pressure is seen to be an increasing function of S_w. This problem was specifically studied and it was concluded that the capillary pressure rise for high saturations of blue fluid (water) is possibly due to the radius decrease of the red fluid (gas) microdrops (capillary pressure is proportional to the interface curvature and thus inversely proportional to the radius of the droplets).

In order to overcome this difficulty, several tests have been undertaken by modifying the initial conditions or the wetting properties (contact angle of 30° instead of 0°). In all these cases, similar results were obtained. Unfortunately, further refinement of discretization cannot be considered as an immediate and appropriate solution, since the memory limits of the available computers are reached. On the other hand, the description of the pore surface cannot be refined to the extent needed for the overall consistency of the simulations.

Drainage calculations

In this section, we consider how the non-wetting fluid (red) invades a sample which is fully saturated with the wetting fluid (blue), which allows us to explore water saturations very close to 1. Calculation was restricted to the subsample of 103-4B as in the previous section (though without the mirror symmetric configuration), using the following method. A reservoir of gas fluid is created on one side of the sample as well as an exit for the water fluid on the other side. This is done by adding a slit of length L_s along the *YZ* border of the sample, as illustrated in Figure 8. The drainage calculations are done as follows. Initially, the fracture is filled with water and the slit with gas. Then, a constant pressure drop is applied across the whole set-up and spatially periodic boundary conditions are imposed. Therefore, the gas is forced into the fracture and water is pushed out of the fracture on the other side. In the slit, at a distance L_t of the fracture outlet, a transmutation plane is defined; any water which goes through this plane is transformed into gas and therefore no water is forced back into the fracture because of the spatial periodicity.The body force is fixed for both fluids at 10^{-4}; all other parameters are the same as in the *Steady state flow* section.

Several configurations with varying L_s and L_t were tested since they should be chosen in order to not perturb the simulations ($L_s = (5,10,20,40)$, $L_t = (3,5,10,20)$). It can be concluded that the slit size must be sufficiently large to obtain realistic results. The transmutation plane must be located at a reasonable distance from the sample outlet in order to avoid red fluid returning to the fracture. Based on the considered configurations, the following parameters can be recommended: $L_t > 20$ and $L_s > 10$ lattice steps. To prevent computational overload, the slit should not be too large. The results obtained with $L_t = 40$ and $L_s = 20$ lattice steps are shown below.

Starting from the initial discretization 100 × 200 × 66, the discretization is doubled twice, as in the previous section. Since the physical parameters used in simulations are kept constant in unit length (the unit length is equal to the voxel size), they are not physically equivalent for the three cases. For this type of calculation, a strong discretization effect is also observed, see Figure 9. For the first discretization, the sample is only partially drained, even after 4.7×10^7 iterations, and there is almost no change during the last 1.2×10^7 iterations. For the doubled discretization, the sample

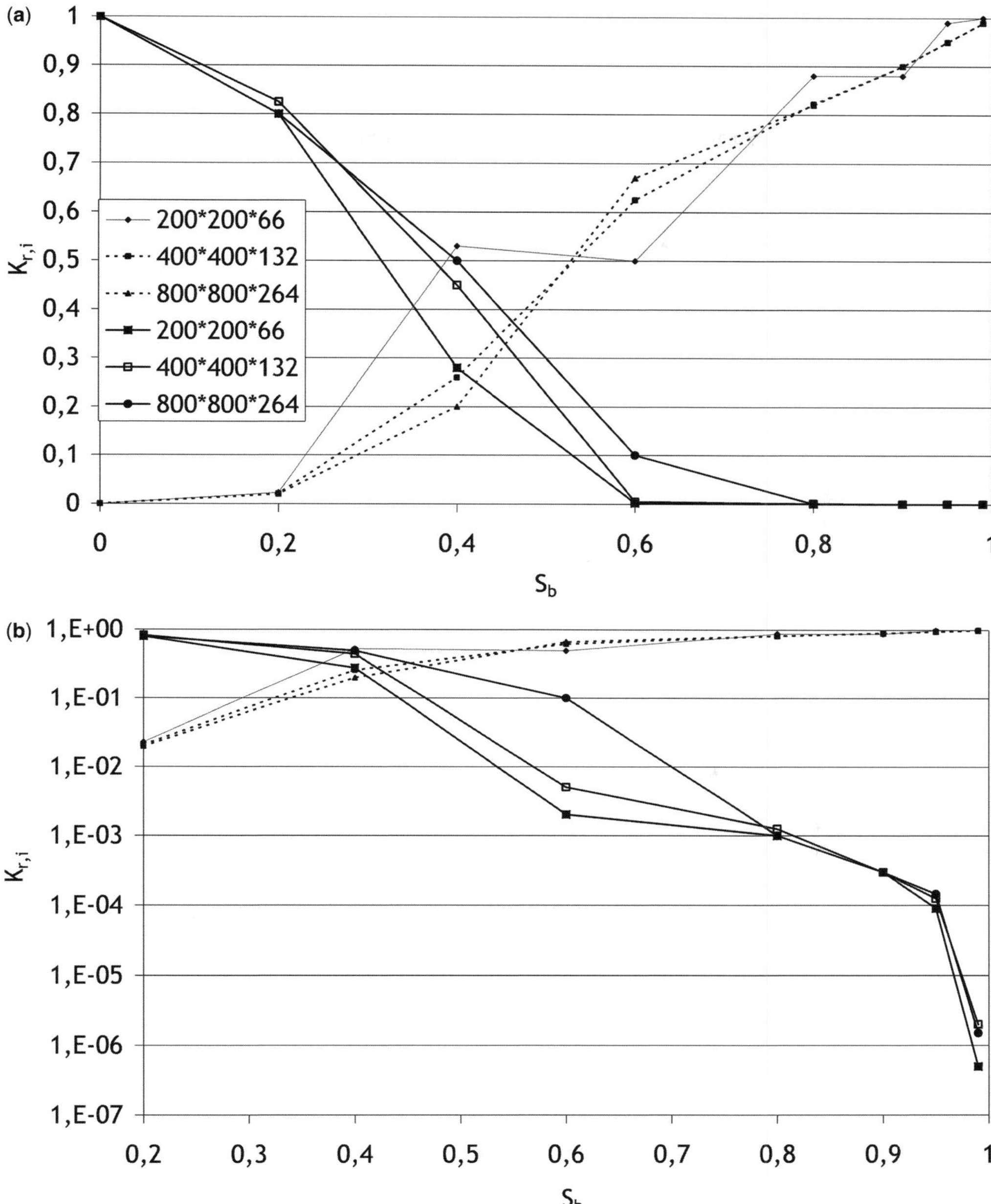

Fig. 5. Relative permeabilities for three discretizations of the subsample of 103_4B in arithmetic (left) and semilog (right) coordinates.

is almost completely drained after 1.3×10^7 iterations. For the finest discretization, the red phase propagates inside the blue phase which wets the fracture surface; moreover, the red phase percolates through the fracture after 3.6×10^6 iterations.

Relative permeabilities and capillary pressure for the three cases are plotted in Figure 10. They were calculated using the fluxes of both fluids averaged over the pore volume (excluding the added slit). It can be concluded that the discretization of

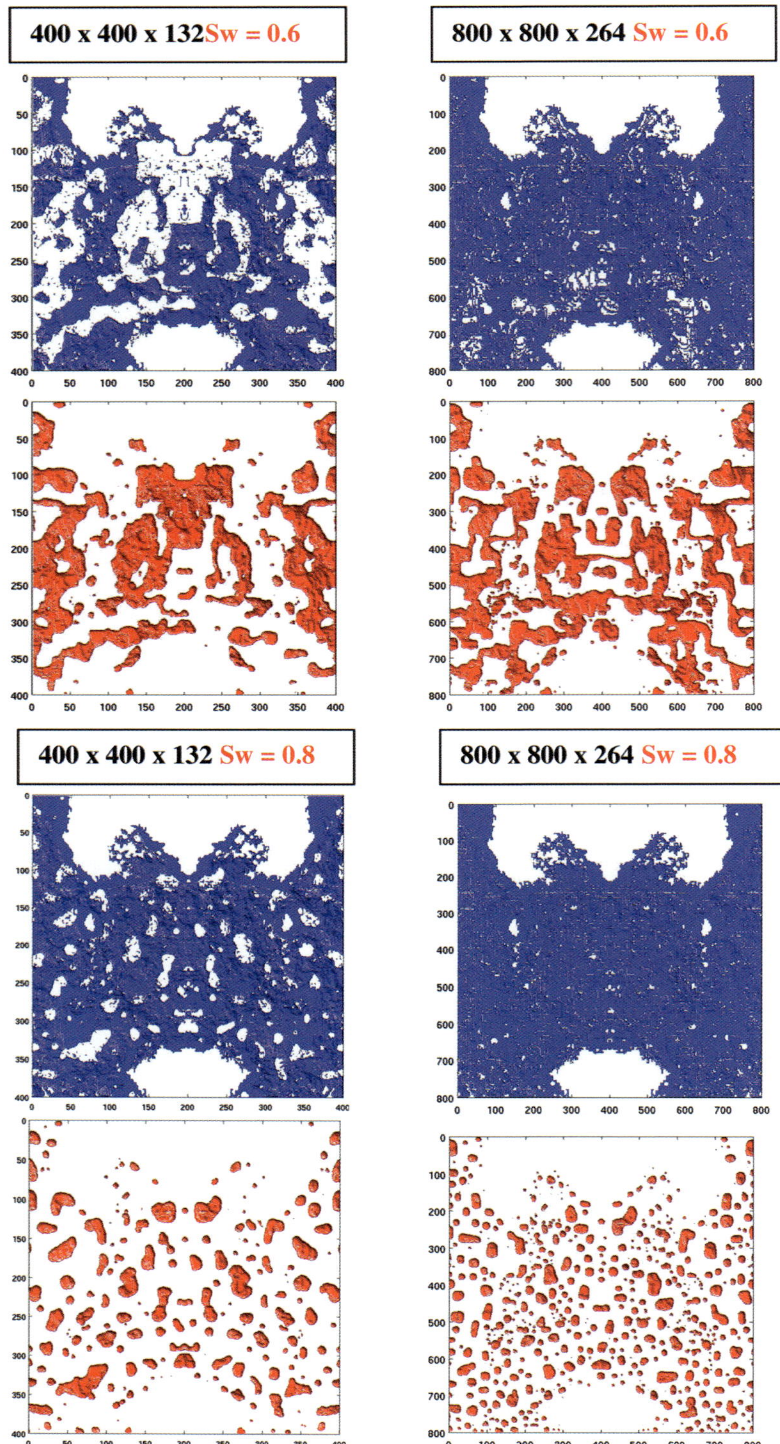

Fig. 6. Phase configurations at several discretizations for Sw = 0.6 and 0.8.

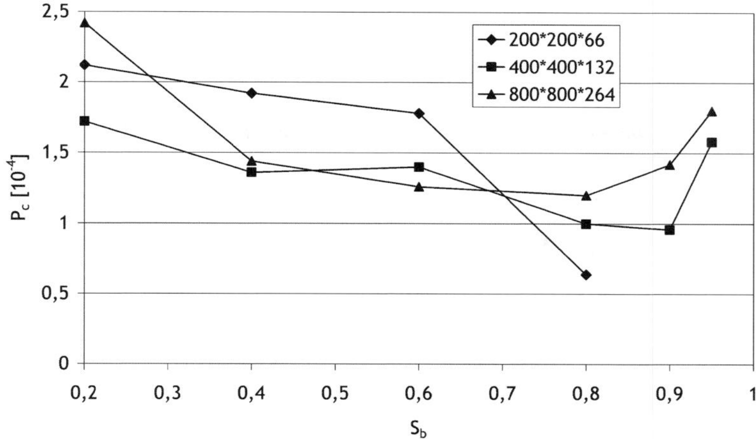

Fig. 7. The capillary pressures for the three discretizations of the subsample of 103-4B.

the sample can have a crucial influence on the drainage results. The finest discretization gives the most realistic results; the red phase propagates inside the blue phase which wets the fracture surface. However the oscillations present in Figure 10 show that the volume of the studied sample is too small in comparison with the relevant elementary volume (REV) for this material. Large computational resources were necessary because of the sample size; all calculations presented here were run over several months and it can be seen that for the two most refined discretizations only a relatively small number of time steps could have been done.

Conclusions and perspectives

Flow properties of a low permeability natural medium, namely Opalinus clay from the Mont Terri Laboratory, were studied by application of the Lattice Boltzmann method (LBM).

A series of X-ray tomography images with 0.7 μm resolution were used to perform one- and two-phase flow simulations inside a real pore space. The absolute permeability was calculated for all (8 of 14) percolating samples and was found to be very close to the permeability of the excavation damaged zone in Mont Terri. In order to

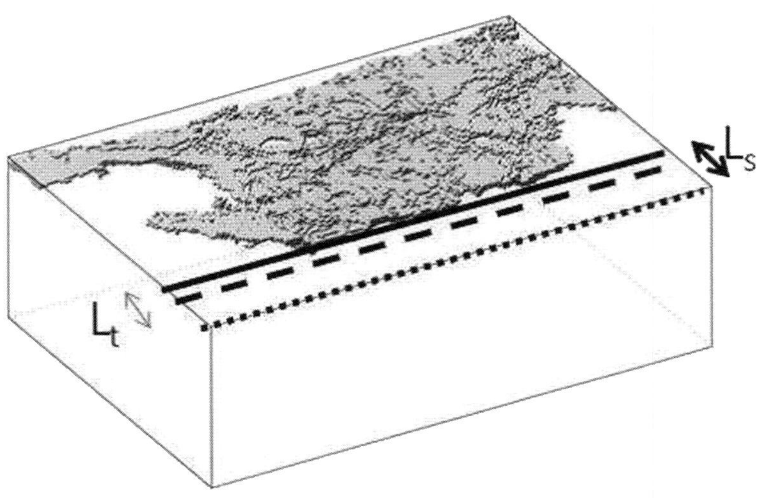

Fig. 8. The geometrical configuration of the fracture with a slit of length L_s. L_t is the location of the transmutation plane.

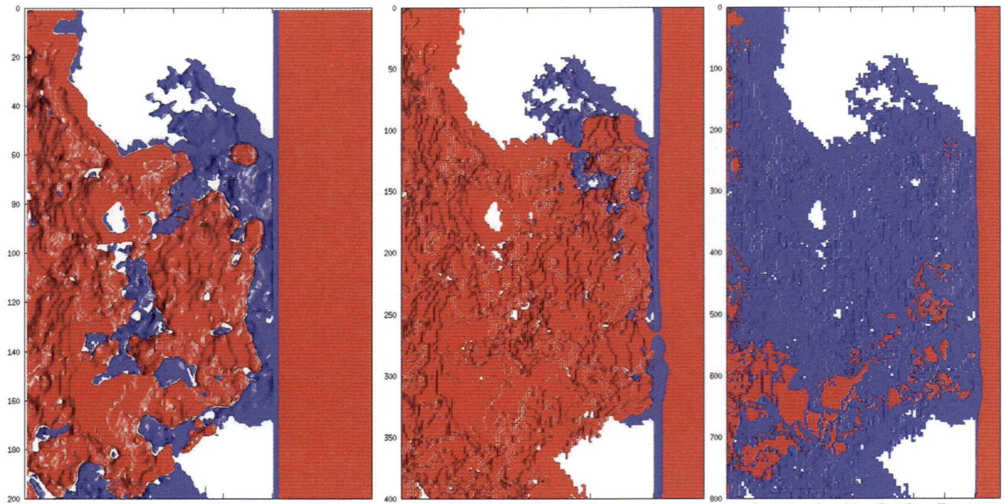

Fig. 9. Saturation distributions at final times for the drainage simulations for three different discretizations. The initial discretization of $100 \times 200 \times 66$ at 4.7×10^7 iterations (left), $200 \times 400 \times 132$ at 1.3×10^7 iterations (middle), $400 \times 800 \times 264$ at 3.6×10^6 iterations (right).

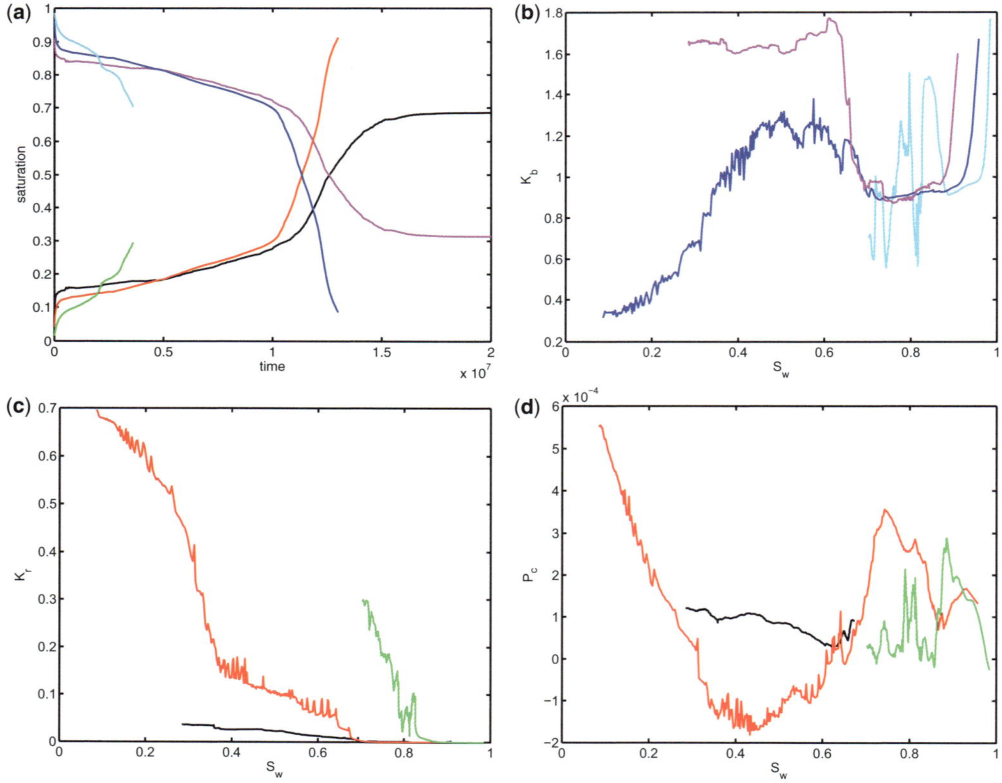

Fig. 10. Comparison between drainage results for three discretizations. Black and magenta lines correspond to initial discretization, red and blue lines to $200 \times 400 \times 132$, and cyan and green lines to $400 \times 800 \times 264$.

demonstrate the ability of the LBM to accurately simulate two-phase flows in porous media, the simulation results were compared with some analytical results for special cases (capillary rise, droplet spreading, etc.). Under reasonable flow conditions (capillary and Reynolds numbers not too large), very good agreement was found with theoretical predictions.

Two types of two-phase flow simulations were then conducted on one of the X-ray tomography subsamples. Since the main purpose of this work was to reach saturation states close to one, we studied steady state flow at increasing saturation. It has been shown that it is possible to fit the obtained steady state fluxes to the generalized Darcy equation and extract physically coherent relative permeability curves at water saturation lower than 0.6. A second type of simulation consisted of drainage tests, where a fully saturated porous medium was put in contact with a reservoir of non-wetting fluid which expelled the wetting fluid. This kind of simulation was performed for various configurations, and the influence of several parameters on the final results was studied. It was found that the surface tension σ had a small influence on the results for the considered configurations and for the chosen physical parameters. The effect of the sample discretization was also studied for both types of simulation. As expected, it was found that the finest discretization gave the most realistic results, but there was a significant difference between the results obtained for the two levels of greatest refinement. This means that we cannot exclude further modification to results on finer grids. This hypothesis is reinforced by the fact that with the finest resolution tested, the aperture of the pore space was at some places only 12 elementary voxels. This is not enough to represent correctly a multilayer system composed of three fluid layers and two interfaces, especially with the D'Ortona model (D'Ortona *et al.* 1995) which is known to present a diffuse interface region between the two fluids. Such a diffuse interface is definitely a realistic feature with respect to gas–liquid systems, but it appears to be a drawback of the method for studying two-phase flow in disordered tight porous media. The simplest solution to the problem would be refinement of the meshes of the samples down to the level where the resolution of the three-layer system in the smallest pores would be acceptable. However, this would lead to very big meshes and would drastically increase the time needed for the simulations.

In order to be able to conduct LBM calculations in argillite samples, it appears necessary to use a completely immiscible variant LBM method resulting in a very sharp interface. The best candidate is the original Rothman-Keller method with Gunstensen extensions (see the analysis in Yang & Boek (2013)). In order to be able to treat whole samples as obtained from the microtomography, it will also be essential to make extensive use of parallel computing. Another very interesting aspect is the search for a way to include in the analysis of transport properties the majority of the porosity which was detected by X-ray tomography, but which has then beendisregarded due to the lack of visible connectivity (Moctezuma-Berthier *et al.* 2004).

The research leading to these results has received funding from the European Atomic Energy Community's Seventh Framework Programme (FP7/2007-2011) under Grant Agreement no230357, the FORGE project.

References

ANDRA, 2005. *Dossier 2005 Argile.* http://www.andra.fr/international/pages/en/dossier-2005-1636.html

ATTAR, E. & KÖRNER, C. 2009. Lattice Boltzmann method for dynamic wetting problems. *Journal of Colloid and Interface Science*, **335**, 84–93.

BONN, D., EGGERS, J., INDEKEU, J., MEUNIER, J. & ROLLEY, E. 2009. Wetting and spreading. *Reviews of Modern Physics*, **81**, 739–805.

BOSSART, P. & THURY, M. (eds) 2008. *Mont Terri Rock Laboratory. Project*, Programme 1996 to 2007 and Results, Rapports du Service géologique national, n° 3.

BOSSART, P., TRICK, T., MEIER, P. M. & MAYOR, J. C. 2004. Structural and hydrogeological characterization of the excavation-disturbed zone in the Opalinus Clay (Mont Terri Project, Switzerland). *Applied Clay Science*, **26**, 429–448.

CARIOU, S., SKOCZYLAS, F. & DORMIEUX, L. 2012. Experimental measurements and water transfer models for the drying of argillite. *International Journal of Rock Mechanics & Mining Sciences*, **54**, 56–69.

D'ORTONA, U., SALIN, D., CIEPLAK, M., RYBKA, R. B. & BANAVAR, J. R. 1995. Two-color nonlinear Boltzmann cellular automata: surface tension and wetting. *Physical Review E*, **51**, 3718.

FUKAI, J., SHIIBA, Y. ET AL. 1995. Wetting effects on the spreading of a liquid droplet colliding with a flat surface: experiment and modeling. *Physics of Fluids*, **7**, 236.

GINZBURG, I. & ADLER, P. M. 1994. Boundary flow condition analysis for three-dimensional lattice Boltzmann model. *Journal de Physique II France*, **4**, 191–214.

GUNSTENSEN, A. K., ROTHMAN, D. H., ZALESKI, S. & ZANETTI, G. 1991. Lattice Boltzmann model of immiscible fluids. *Physical Review A*, **43**, 4320.

HENRICH, B., CUPELLI, C., MOSELER, M. & SANTER, M. 2007. An adhesive DPD wall model for dynamic wetting. *Europhysics Letters*, **80**, 60004.

HOFFMAN, R. L. 1975. A study of the advancing interface. I. Interface shape in liquid-gas systems. *Journal of Colloid and Interface Science*, **50**, 228–241.

MARSCHALL, P., HORSEMAN, S. & GIMMI, T. 2005. Characterisation of gas transport properties of the Opalinus Clay, a potential host rock formation for radioactive waste disposal. *Oil & Gas Science and Technology – Revue de L'Institut Français du Petrole*, **60**, 121–139.

Matray, J. M., Parneix, J. C., Tinseau, E., Prêt, D. & Mayor, J. C. 2007. Structural organization of porosity in the Opalinus Clay at the Mont Terri Rock Laboratory under saturated and unsaturated conditions. *In*: *Clays in Natural & Engineered Barriers for Radioactive Waste Confinement*. Andra, Châtenay-Malabry, 445–446

Moctezuma-Berthier, A., Vizika, O., Thovert, J. F. & Adler, P. M. 2004. One- and two-phase permeabilities of vugular porous media. *Transport in Porous Media*, **56**, 225–244. .

Pazdniakou, A. 2012. *Lattice models in porous media studies*. PhD thesis, University of Pierre and Marie Curie, Paris.

Rothman, D. H. & Keller, J. M. 1988. Immiscible cellular-automaton fluids. *Journal of Statistical Physics*, **52**, 1119–1127.

Succi, S. 2001. *The Lattice Boltzmann Equation for Fluid Dynamics and Beyond*. Oxford University Press, Oxford.

Washburn, E. W. 1921. The dynamics of capillary flow. *Physical Review*, **17**, 273–283.

Wildenschild, D., Hopmans, J. W., Vaz, C. M. P., Rivers, M. L., Rikard, D. & Christensen, B. S. B. 2002. Using X-ray computed tomography in hydrology: systems, resolutions, and limitations. *Journal of Hydrology*, **267**, 285–297.

Yang, J. & Boek, E. S. 2013. A comparison study of multicomponent Lattice Boltzmann models for flow in porous media applications. *Computers and Mathematics with Applications*, **65**, 882–890.

Zhang, C.-L. & Rothfuchs, T. 2007. Moisture effects on argillaceous rocks. *In*: Schanz, T. (ed.) *Proceedings of the 2nd International Conference of Mechanics of Unsaturated Soils*. Springer Proceedings in Physics, **112**, 319–326.

Oxidation front and oxygen transfer in the fractured zone surrounding the Meuse/Haute-Marne URL drifts in the Callovian–Oxfordian argillaceous rock

AGNÈS VINSOT[1]*, FRANÇOIS LEVEAU[1,2], ALAIN BOUCHET[3] & APOLLINE ARNOULD[2]

[1]*Andra, Centre de Meuse/Haute-Marne, RD 960, 55290 Bure, France*

[2]*GEOTER, Pôle géoenvironement, 3 rue Jean Monnet, 34830 Clapiers, France*

[3]*ERM, Centre Régional d'Innovation du Biopôle, 4 rue Carol Heitz, 86000 Poitiers, France*

**Corresponding author (e-mail: agnes.vinsot@andra.fr)*

Abstract: Deep argillaceous rocks are reducing environments. When exposed to air, reduced minerals of these rocks react with oxygen, modifying the surrounding chemical conditions. Thus, oxidation is an issue in studies about the confining properties of such rocks in the framework of geological disposal projects for radioactive waste. Previous studies in several underground research laboratories (URLs) in argillaceous rocks have shown that oxidation reactions mainly occurred in the excavation-induced fracture network surrounding the drifts. In the Callovian–Oxfordian argillaceous rock, at −490 m in drifts from the Meuse/Haute-Marne URL, oxidized features were systematically looked for in 115 borehole cores. The concerned drifts were of various ages, from a few days to 6.5 years. After 5 months, oxidized features were encountered in numerous excavation-induced extensional fractures. In excavation-induced shear fractures, oxidized features were observed in a few borehole cores after 2 years, and they became frequent after 6 years. In all cases, the oxidized features observed were found on the fracture walls or were connected to them, and were less than 1.8 m from the drift walls. These observations about the oxidation front and its evolution over time provide insights regarding the properties of excavation-induced fractures with respect to oxygen transfer.

The Callovian–Oxfordian clay mineral-rich sedimentary layer has been under investigation for the past 20 years in the eastern part of the Paris Basin (France) as a potential host rock for the installation of a deep geological disposal for radioactive waste (Andra 2005, 2013). In 2000, Andra, the French national radioactive waste management agency, began the construction of the Meuse/Haute-Marne underground research laboratory (MHM URL) in the southern area of the Meuse district in order to study this rock *in situ* and test the construction of repository elements. In 2004, the two URL vertical shafts reached the top of the target layer at a depth of 420 m. The URL main level now consists of horizontal drifts over a total length of more than 1 km and lying at a depth of 490 m, in the middle of the Callovian–Oxfordian clay-rich layer.

An interdisciplinary programme has been carried out in the MHM URL to study the geological, mechanical, hydrogeological and geochemical properties of this clay-rich rock as well as its responses to the mechanical and chemical disturbances induced by a disposal facility (Delay *et al.* 2007; Delay *et al.* 2014). Among the scientific questions addressed in this programme are those concerning the consequences of rock oxidation by atmospheric air. Indeed, the chemical reactions involved in the oxidation of this type of rock may affect the pH and redox conditions in the rock and thus influence the durability of the disposal materials (especially metals) and radionuclide mobility (Baeyens *et al.* 1985; Charpentier *et al.* 2001; Descostes 2001; Mazurek *et al.* 2003). Studies addressing these questions have been carried out in several URLs in clay-rich rocks over the past 20 years. Their objectives were to identify the changes induced on the rock by oxidation, the extension of the rock domain affected by these changes, and their evolution over time.

For instance, at URL HADES in Mol (Belgium), within the Boom Clay, piezometers were installed in two drifts of different ages (6 and 20 years). Mineralogical and leaching water analyses were performed on the core samples, and water was regularly sampled in six piezometer intervals located at distances between 0.4 and 8.2 m from the drift wall, making it possible to monitor its composition over months to years (De Craen *et al.* 2008). Dissolved sulphate concentration was shown to increase in the core and borehole water samples issuing from

From: Norris, S., Bruno, J., Cathelineau, M., Delage, P., Fairhurst, C., Gaucher, E. C., Höhn, E. H., Kalinichev, A., Lalieux, P. & Sellin, P. (eds) 2014. *Clays in Natural and Engineered Barriers for Radioactive Waste Confinement*. Geological Society, London, Special Publications, **400**, 207–220.
First published online June 2, 2014, http://dx.doi.org/10.1144/SP400.37

the first metre from the drift wall when compared to the pristine porewater concentration, whatever the drift age. In addition to this experiment, oxidation effects were observed on numerous Boom Clay cores and around excavation-induced fractures (Baeyens *et al.* 1985; De Craen *et al.* 2004; De Craen *et al.* 2008). From the results, these authors concluded that (1) the oxidation process was related to excavation-induced fractures and occurred over the same extension from the drift walls as occupied by these fractures, that is, *c.* 1 m (Bastiaens *et al.* 2007); (2) the dominating reaction was pyrite (FeS_2) oxidation, often associated with calcite ($CaCO_3$) dissolution and gypsum ($CaSO_4$, $2H_2O$) precipitation, or with jarosite ($KFe_3(SO_4)_2(OH)_6$) precipitation when calcite is absent, and that these reactions might involve microorganisms; (3) mineralogical changes were only observed within the first 4.5 cm of rock away from the lining concrete/rock interface. They did not report any significant difference depending on drift age between the effects of oxidation observed. Moreover, they mentioned that even if oxidation significantly affects the seepage water composition in air-drilled boreholes, this influence may become negligible after a few months of continuous water extraction. Other than the mineralogical changes induced by rock oxidation, slight but significant effects of oxidation on the Boom Clay organic matter were also described, for example around extruded rock cores from drift walls that were exposed to atmospheric air (Blanchart *et al.* 2012).

At the Mont Terri rock laboratory in Switzerland, within the Opalinus Clay, excavation-induced air-filled fractures are present in the 1 m zone surrounding the drifts (Bossart *et al.* 2002). Rock oxidation, as in Boom Clay, is linked to these fractures, with the main reactions described being pyrite oxidation, calcite dissolution and gypsum precipitation (Mäder & Mazurek 1998; Pearson *et al.* 2003; Zheng *et al.* 2008; Techer *et al.* 2009). The effects of oxidation on this rock were also studied in the 140-year-old Old Hauenstein railway tunnel (Mäder & Mazurek 1998). In this tunnel, the visible oxidized features observed on drill cores have been associated with excavation-induced fractures. Based on the sulphur loss deduced from rock sample chemical analyses, the maximum thickness of the oxidized zone at the fracture walls is *c.* 3 cm, whereas the thickness of the visible brownish rims is less than 1.5 cm (Mäder & Mazurek 1998).

At Tournemire (France), detailed mineralogical and chemical analyses were performed on two sets of Toarcien marls core samples: one taken at less than 1 m from the 4- and 100-year-old tunnel walls and the other taken at 14–14.5 m from the 100-year-old tunnel wall (Charpentier *et al.* 2004). The first set of core samples were oxidized, whereas the second set were preserved. This study aimed to identify the impact of oxidation on the rock clay minerals. It mentioned that pyrite oxidation occurred, as expected, and that gypsum, jarosite, celestite ($SrSO_4$) and Fe-hydroxides precipitated. Regarding the clay minerals, tenuous and localized alteration effects due to oxidation were observed in the mixed-layer illite-smectite clay minerals: their relative Si content and Fe(III)/Fe ratio increased slightly statistically, and illite layers were dissolved. As a complement, the excavation-induced and tectonic fractures were described in six boreholes drilled from three of the Tournemire URL tunnels (Matray *et al.* 2007).

In a drift of the MHM URL, systematic observations were performed on the cores of several boreholes up to 7 m in length that were drilled at various dates (between 1 day and 18 months after drift excavation; Belcourt 2009). This study showed that pyrite oxidation occurred within the first metre from the drift wall, where the excavation-induced air-filled fractures were encountered. Further evolution in this zone led to the formation of gypsum and bassanite ($2CaSO_4$, H_2O), jarosite and natrojarosite ($NaFe_3(SO_4)_2(OH)_6$), celestite, iron hydroxide and native sulphur. The location of these secondary minerals progressed over time. For instance, gypsum, which serves as a mark of the oxidation process, was found only on the drift wall after 3 months, whereas the whole zone affected by excavation-induced air-filled fractures contained gypsum after 12 months. On the other hand, no gypsum was detected outside this zone, even after 18 months.

Stemming from these studies, it appears that the locations of the oxidized rock zones surrounding the URL drifts in clay-rich rocks rely first on existing structures that display a greater hydraulic conductivity than the pristine rock: in HADES, Mont Terri and MHM URLs, these structures are fractures generated by drift excavation, whereas at Tournemire, fractures with both a tectonic origin and induced by excavation are present, and may display a greater hydraulic conductivity than the pristine rock. Furthermore, oxidized pyrite minerals were observed on each rock surface that had been directly in contact with atmospheric air. Oxidation was not observed within the rock matrix at more than a few centimetres from a surface exposed to atmospheric air, even after more than 100 years of exposure.

In this context, oxidized features indicate that oxygen interacted with the rock at this location. They prove that a transfer path for gaseous or dissolved oxygen has existed between the fracture and the drift. As a consequence, they give insights on gas transfer in the fracture network. The first aim of the study presented here is to identify the location of the oxidation front surrounding the MHM URL drifts, its relationships with the excavation-induced fracture geometry, and its possible evolution over time. The oxidation front is defined as the set of deepest points away from the drift walls where oxidation reactions have taken place. The second aim of this study is to contribute to the characterization of gas transfer in the excavation-induced fracture network. Indeed, this fractured zone constitutes the exchange interface between the drift atmosphere and the pristine rock. Transfer of water vapour and other gases like O_2 or CO_2 through this interface plays an important role in the long-term evolution of hydraulic and chemical conditions in the drifts and may affect near-field evolution of a geological disposal for radioactive waste.

Since 2010, 115 new boreholes have been cored from the MHM URL drifts, and macroscopic oxidized features have been systematically sought in the cores. The maximum distance from the drift walls at which oxidized features were encountered was recorded, and some of the oxidized features were sampled for mineralogical analyses. The results were analysed taking into account borehole direction, drift orientation and the lag time between the drift excavation and borehole coring. This analysis was based on the recent detailed description of excavation-induced fractures by Armand *et al.* (2014).

Environmental conditions surrounding the URL drifts

The Callovian–Oxfordian argillaceous rock

Sediments of the Callovian–Oxfordian unit consist of a dominant clay fraction associated with carbonate, quartz with minor feldspars, and accessory minerals (Lerouge *et al.* 2011). At the MHM URL main level (−490 m depth), the clay fraction is high (40–60%), and clay minerals consist of illite, ordered illite/smectite mixed layers with dominant illite, kaolinite, tri-octahedral, iron-rich chlorite, and minor biotite (Lerouge *et al.* 2011). Accessory diagenetic minerals represent *c.* 1–2% of the rock mass and include pyrite, sphalerite, celestite, siderite, glauconite and silica. Accessory detrital minerals comprise chalcopyrite, rutile and titanomagnetite (Lerouge *et al.* 2011).

The Callovian–Oxfordian clay-rich rock porosity lies between 14% and 20% at the MHM URL site and is close to 18% at the URL main level (Yven *et al.* 2007), and natural water content ranges between 5% and 8%. Hydraulic conductivity is between 10^{-14} and 10^{-12} m s^{-1} (Delay *et al.* 2006), corresponding to a permeability between 10^{-21} and 10^{-19} m^2, and tritiated water (HTO) effective diffusion coefficient is in the order of 10^{-11} m^2 s^{-1} (Altmann *et al.* 2012). Pore pressure at the URL main level was measured at *c.* 4.7 MPa (Armand *et al.* 2014).

The excavation-induced fracture network in the MHM URL

The morphology of the excavation-induced fractures in the MHM URL (Fig. 1) has been described in detail by Armand *et al.* (2014), who showed that the excavation-induced fracture pattern and extent depend on the drift orientation with respect to the *in situ* stress field axes:

- Drifts parallel to the major horizontal stress axis σ_H (N155°E) show extensional and shear fractures up to 0.4 diameters and one diameter from the lateral walls, respectively. Beyond the floor and the ceiling, shear and extensional fractures are present at up to 0.15 diameters.
- Drifts parallel to the minor horizontal stress axis σ_h (N65°E) show extensional and shear fractures up to 0.5 diameters and one diameter, respectively, beyond the floor and ceiling. Beyond the lateral walls, shear and extensional fractures are present at up to 0.2 diameters.

The zone including extensional fractures exhibits the highest hydraulic conductivity, with local values greater than 10^{-10} m s^{-1}. This zone is called the traction zone in the following text. Further from the wall, the zone including only shear fractures exhibits average hydraulic conductivity values similar to those of pristine rock. This zone is called the shear zone in the following text. Neither natural joint nor fault has been observed in the Callovian–Oxfordian layer at the MHM URL site (Armand *et al.* 2014). Taking into account the Tsang *et al.* (2005) definition of the 'excavation damaged zone' (EDZ), the excavation-induced fractured zone extent in the MHM URL is greater than the EDZ, which corresponds to the traction zone.

Materials and methods

Borehole cores

The 115 boreholes studied were cored in several directions and in various types of drift (Fig. 2) in terms of orientation with respect to the stress field axes and age: the drifts dated back from a few days to 6.5 years ago, when the boreholes were drilled. The borehole cores from the MHM URL were examined in detail by geologists. Each borehole core was reconstituted on a dedicated table, which included a fixed decametre used as a measuring reference. The core structural analyses make it

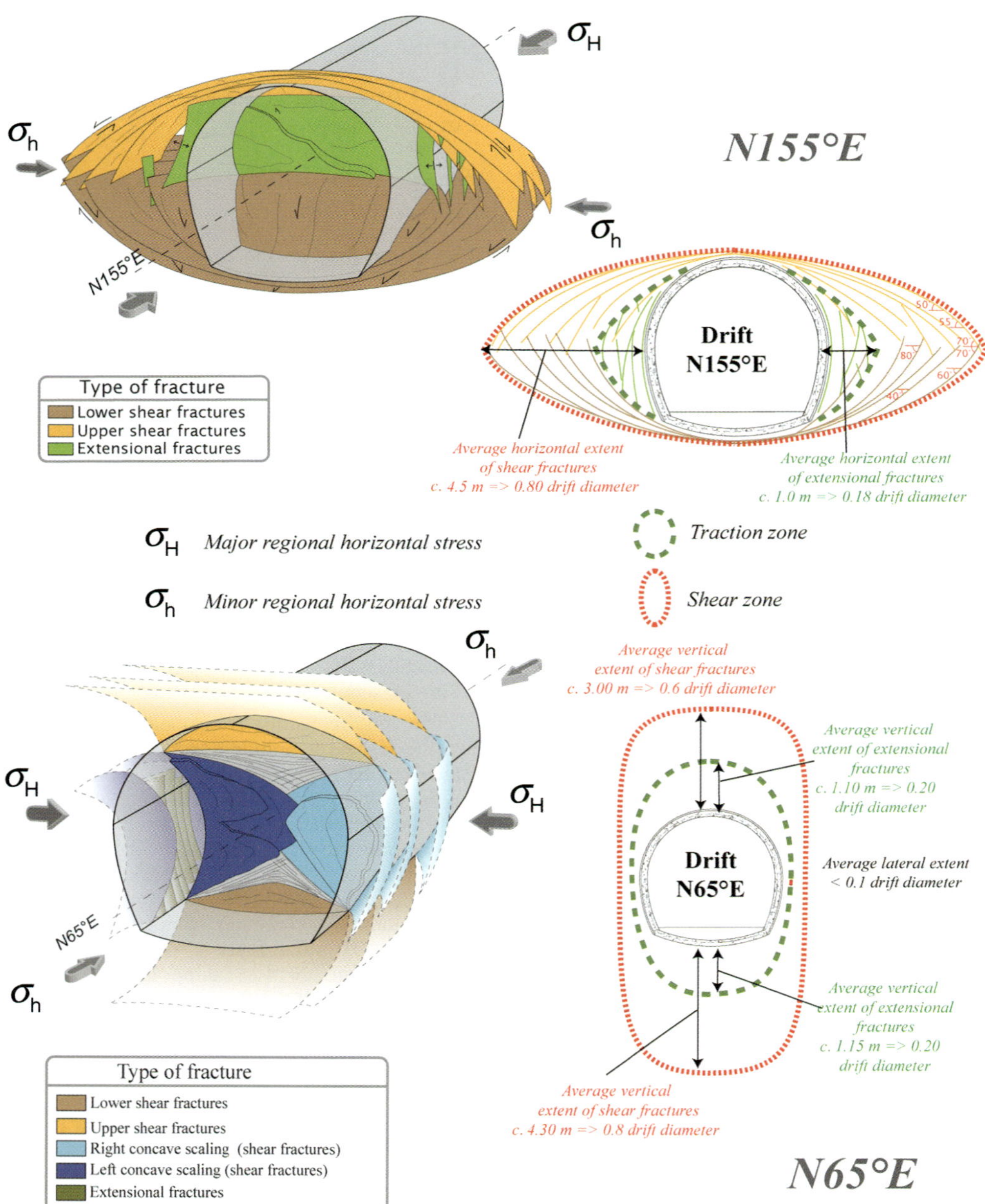

Fig. 1. Conceptual model of the excavation-induced fracture network surrounding the MHM URL drifts.

possible to characterize each fracture and to classify most of them either as extensional or shear (Armand *et al.* 2014). Moreover, since 2010, the geologists have systematically inventoried the visible oxidized features on the cores and recorded their maximum depth, as well as the lag time between the drift excavation and borehole coring. They noted the elapsed time between the core drilling and the geological survey, and cracked open some of the rock samples in order to look for oxidation traces within the rock matrix. All results were compiled in a dedicated database.

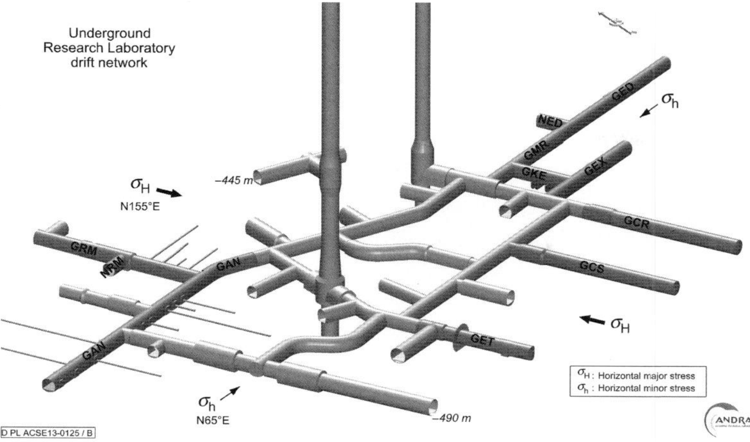

Fig. 2. Schematic view of the MHM URL, showing drift location and orientation with respect to the *in situ* stress field axes.

For each borehole core, a parameter was calculated to compare the positions of both the deepest oxidized feature and the extensional fracture. This parameter, termed the 'traction ratio', is the ratio of their depths (i.e. deepest oxidized feature depth over deepest extensional fracture depth). Traction ratio values smaller than 1 indicate that the deepest oxidized features are located in the traction zone, whereas values greater than 1 imply that they are located in the shear zone. When no fracture was found in a borehole core, the traction ratio had no value. More than 20 cores containing oxidized features were sampled for laboratory analyses. These samples were conditioned under an anoxic atmosphere within a few hours of coring.

Analyses

A total of 30 thin sections were made in core samples from eight boreholes. The thin sections were chosen to intersect visible oxidized features or apparently preserved pyrite minerals. They were orientated either perpendicular or parallel to the core sample surfaces. Thin sections perpendicular to core sample surfaces including oxidized features made it possible to evaluate their thickness, whereas thin sections parallel to these surfaces helped to identify the mineralogical characteristics of the oxidized elements.

A petrographic study of these core-sample thin sections was performed at various scales, using a polarizing optical microscope and a scanning electron microscope (SEM). The SEM (JEOL 5600 LV) was equipped both with an Everhart–Thornley type detector and a Centaurus backscattered electron detector (KE Development IC, Cambridge), which made it possible to observe the rock surfaces. An analysis system (Bruker detector) allowed the detection of chemical elements with atomic mass greater than 5. The Quantax software (provided by Bruker) was used to perform chemical analyses on small surfaces and for the acquisition of X-ray maps.

Results

Characteristics of the oxidized features observed

Three main types of oxidized features were observed on cores at a macroscopic scale during the geological survey (Fig. 3):

- oxidized sedimentary elements marked by brown rust colours;
- oxidized patina identified by a rust colour observed on fracture walls;
- white gypsum spots.

Oxidized sedimentary elements were mainly bioturbations and sometimes fossils (ammonites or bivalves). Bioturbations were induced by burrowing animals. Their typical size observed on the cores was measured in centimetres for the length and in millimetres for the diameter. When preserved from oxidation, bioturbations were visible because they created a contrast of lighter or darker area within the rock matrix. Their filling was mainly pyrite of diagenetic origin (Lerouge *et al.* 2011). When they were oxidized, they displayed a brown rust colour. The rare fossils encountered were of centimetre size. As for the bioturbations, they were mainly filled by pyrite and had a brown rust colour when oxidized.

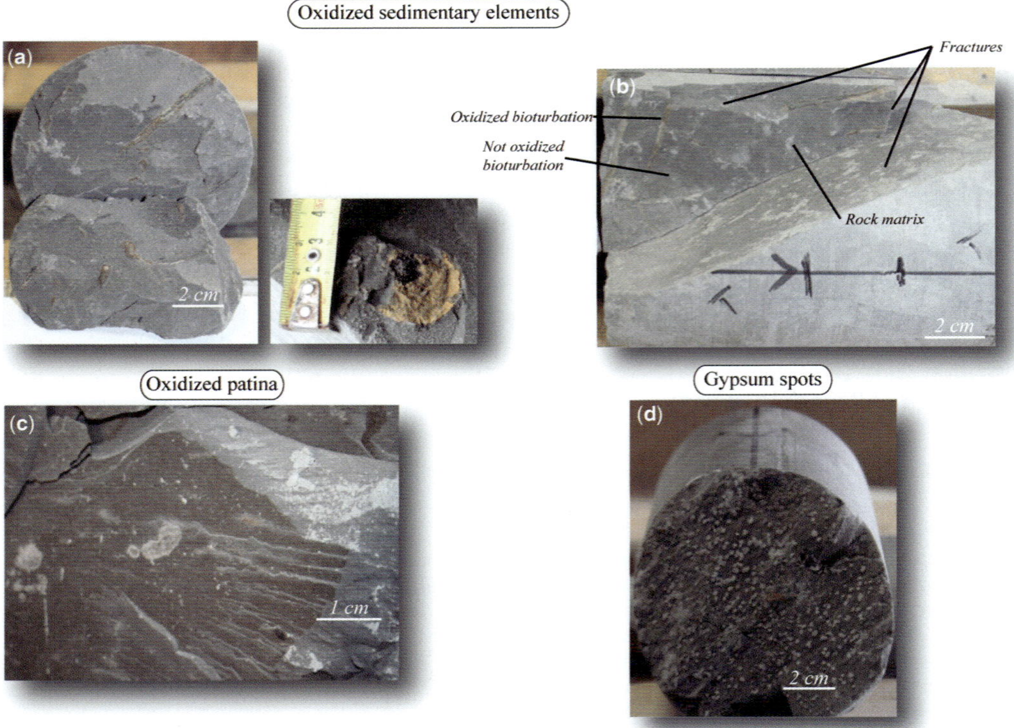

Fig. 3. Types of oxidized feature observed in borehole cores from the MHM URL: (**a**) bioturbations and fossil on fracture walls; (**b**) oxidized elements in the rock matrix connected to fractures; (**c**) rust-coloured oxidized patina on a fracture wall; (**d**) gypsum spots on a fracture wall.

The oxidized sedimentary elements observed generally intersected fractures. However, a few oxidized sedimentary elements were also found in the rock matrix after having cracked open core pieces. In all the cases encountered, these features were connected to a fracture by a macroscopic structure like a tube from a bioturbation or crack. The microscopic observation of oxidized sedimentary elements confirmed the presence of pyrite minerals surrounded by white halos of gypsum, themselves in contact with outer iron oxide and hydroxide brown rims (Fig. 4). On the other hand, millimetre- to centimetre-sized cracks filled with automorphic gypsum crystals were often observed, and were connected to pyrite minerals in all the cases encountered (Fig. 4). At a larger scale, X-ray maps helped to locate weathering minerals smaller than 10 μm. For instance, Figure 4 shows gypsum, and iron oxide minerals associated with calcite, within the weathering halo surrounding the pyrite mineral. This observation suggests that gypsum replaced calcite, while iron oxides invaded the clay-rich matrix.

Within the rock matrix, non-weathered pyrite minerals were frequently found at less than 1 cm from weathered ones. In such cases, no sign was found of a structure connecting the minerals to a fracture. By contrast, when oxidation marks were observed in the rock matrix, a connecting structure was found in all the cases encountered.

Oxidized patina observed on fracture walls consist of a thin layer of iron oxides and hydroxides. Cross-sections cut perpendicular to the fracture surfaces made it possible to confirm that the oxidized patina was composed of iron oxide and hydroxide, and also to estimate the thickness of that layer at less than 50 μm. Chemical analyses performed on thin-section surfaces confirmed that the macroscopic millimetre-sized white spots observed on fracture walls were made of gypsum.

Oxidation front

Of the 115 borehole cores studied, 21 did not show any fractures at all nor oxidized features. These 21 boreholes, which were mainly cored vertically

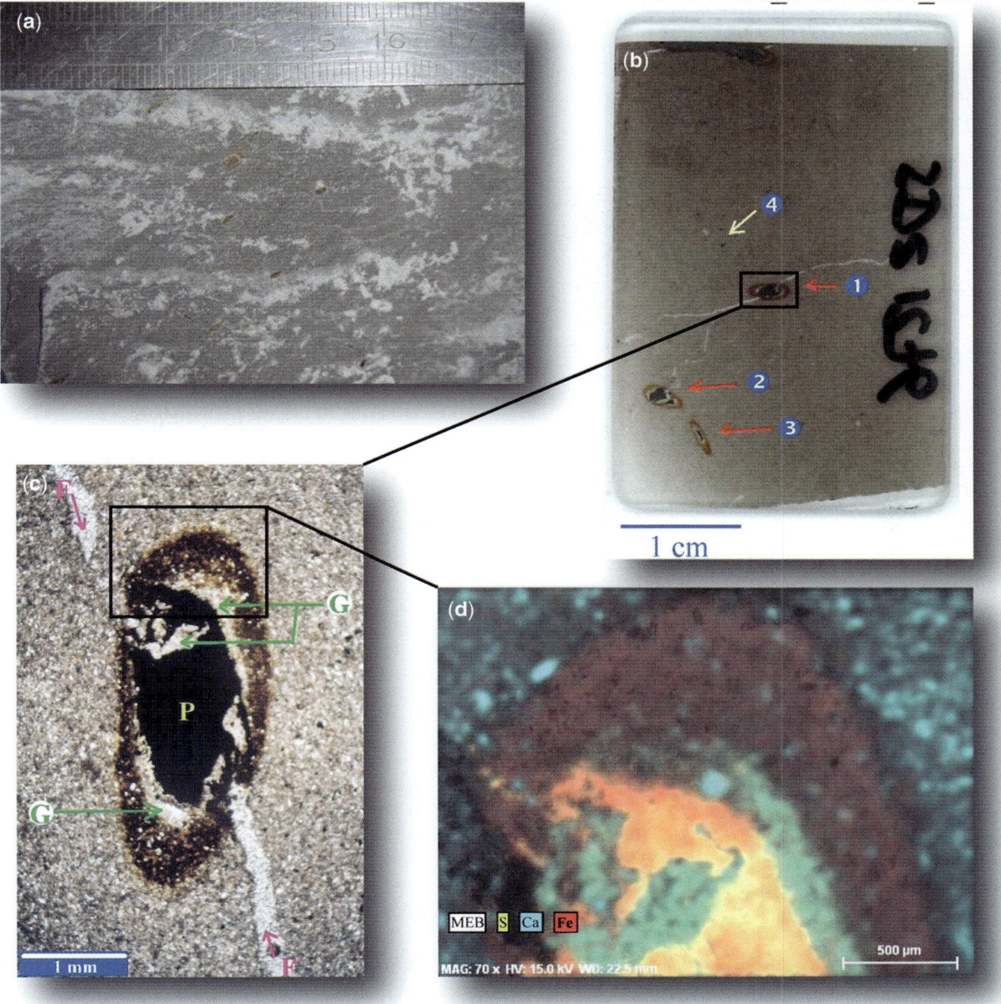

Fig. 4. Microscopic observations of oxidized features: (**a**) the macroscopic sample; (**b**) thin section with both weathered (1, 2 and 3) and non-weathered (4) pyrite; (**c**) detail of weathered pyrite 1; (**d**) SEM X-ray map (S, Ca, Fe) of weathered pyrite detail.

from drifts along the major horizontal stress axis, are not taken into account in Tables 1 and 2, nor in Figures 5–9.

The deepest oxidized features of the three types described above were all found at about the same distance from the drift walls (Table 1), although oxidized patina were sometimes encountered at a shorter distance than the other features. However, this oxidized feature was also the most difficult to identify and could have been missed during the geological survey from time to time. The different oxidized feature types are considered together in the following.

All the oxidized features were encountered on excavation-induced fracture walls or were connected to them. As a consequence, their locations depend on the fracture network geometry. The extension of the excavation-induced fractured zone surrounding the MHM URL drifts varied from less than 0.5 m below or above the drifts orientated along the major horizontal stress axis, up to 4.5 m horizontally from the same drifts or vertically below the drifts orientated perpendicular to the major horizontal stress axis (Armand *et al.* 2014) (Fig. 1). Thus, it is necessary to relate the deepest oxidized feature depths to the drift orientations in

Table 1. *Depth range of the deepest oxidized features of the three types identified in relationships with drift orientation and age (maximum values in bold)*

Borehole direction	Type of oxidized feature	Time duration since drift excavation (day)	Borehole number	Deepest oxidized feature depth (m)			Traction ratio (m m^{-1})			Drifts
				Min.	Ave.	Max.	Min.	Ave.	Max.	
Laterally from N155°E drifts	Gypsum	3–31	5	0.00	0.00	**0.00**	0.00	0.00	**0.00**	GCS/GCR
		122–305	32	0.00	0.29	**0.97**	0.00	0.30	**1.00**	GCS/GET
		488–1067	6	0.40	0.65	**0.85**	0.30	0.69	**1.33**	NED/GRM
	Oxidized elements	3–31	5	0.00	0.00	**0.00**	0.00	0.00	**0.00**	GCS/GCR
		122–305	32	0.00	0.36	**0.90**	0.00	0.38	**1.05**	GCS/GET
		488–1067	6	0.41	0.41	**1.20**	0.34	0.79	**1.41**	NED/GRM
	Oxidized patina	3–31	5	0.00	0.00	**0.00**	0.00	0.00	**0.00**	GCS/GCR
		122–305	32	0.00	0.23	**0.16**	0.00	0.24	**1.05**	GCS/GET
		488–1067	6	0.00	0.40	**0.90**	0.00	0.37	**0.96**	NED/GRM
Vertically from N155°E drifts	Gypsum	3–31	1	0.00	0.00	**0.00**	0.00	0.00	**0.00**	GCR
		122–305	3	0.32	0.37	**0.39**	0.84	0.95	**1.00**	GCS/GET
		488–1067	4	0.00	0.01	**0.03**	0.00	0.02	**0.67**	NED
	Oxidized elements	3–31	1	0.00	0.00	**0.00**	0.00	0.00	**0.00**	GCR
		122–305	3	0.00	0.26	**0.39**	0.00	0.67	**1.00**	GCS/GET
		488–1067	4	0.00	0.24	**0.45**	0.00	0.75	**1.00**	NED
	Oxidized patina	3–31	1	0.00	0.00	**0.00**	0.00	0.00	**0.00**	GCR
		122–305	3	0.00	0.13	**0.39**	0.00	0.33	**1.00**	GCS/GET
		488–1067	4	0.00	0.11	**0.45**	0.00	0.25	**1.00**	NED
Laterally from N65°E drifts	Gypsum	3–31	2	0.00	0.00	**0.00**	0.00	0.00	**0.00**	GAN/NRM
		122–305	13	0.00	0.13	**1.00**	0.00	0.22	**1.05**	GAN
		488–1067	4	0.00	0.28	**0.56**	0.00	0.61	**1.40**	GED/GAN
		1922–2379	9	0.30	0.67	**1.11**	0.91	1.19	**1.86**	GEX/GMR
	Oxidized elements	3–31	2	0.00	0.00	**0.00**	0.00	0.00	**0.00**	GAN/NRM
		122–305	13	0.00	0.19	**0.68**	0.00	0.50	**1.05**	GAN
		488–1067	4	0.51	0.58	**0.71**	1.00	1.10	**1.40**	GED/GAN
		1922–2379	9	0.30	0.66	**1.11**	0.84	1.17	**1.86**	GEX/GMR
	Oxidized patina	3–31	2	0.00	0.00	**0.00**	0.00	0.00	**0.00**	GAN/NRM
		122–305	13	0.00	0.08	**1.00**	0.00	0.08	**1.00**	GAN
		488–1067	4	0.00	0.00	**0.00**	0.00	0.00	**0.00**	GED/GAN
		1922–2379	9	0.00	0.39	**0.95**	0.00	0.65	**1.86**	GEX/GMR
Vertically from N65°E drifts	Gypsum	3–31	2	0.00	0.00	**0.00**	0.00	0.00	**0.00**	GAN
		122–305	2	0.36	0.43	**0.50**	0.30	0.32	**0.33**	GAN
		488–1067	9	0.51	0.92	**1.45**	0.47	0.84	**1.38**	GED
		1922–2379	2	0.93	1.02	**1.10**	0.94	1.03	**1.12**	GED/GKE
	Oxidized elements	3–31	2	0.00	0.00	**0.00**	0.00	0.00	**0.00**	GAN
		122–305	2	0.42	0.46	**0.50**	0.30	0.34	**0.38**	GAN
		488–1067	9	0.72	1.04	**1.66**	0.45	0.93	**1.38**	GED
		1922–2379	2	0.93	1.35	**1.77**	0.94	1.37	**1.81**	GED/GKE
	Oxidized patina	3–31	2	0.00	0.00	**0.00**	0.00	0.00	**0.00**	GAN
		122–305	2	0.00	0.00	**0.00**	0.00	0.00	**0.00**	GAN
		488–1067	9	0.72	1.05	**1.70**	0.55	0.97	**1.62**	GED
		1922–2379	2	1.13	1.21	**1.28**	1.15	1.22	**1.29**	GED/GKE

order to analyse the evolution of their absolute distances extending from the drift walls over time. Two representations were used (Figs 5–8): (1) the absolute distances of the deepest oxidized features were plotted v. the lag time between the drift excavation and the borehole coring, separately for each drift orientation; (2) the traction ratio was plotted v. the lag time, separately for each drift orientation.

Boreholes cored up until a month (31 days) after drift excavation did not show any oxidized feature in

Table 2. *Summary of the deepest oxidized feature depth ranges in relationships with drift orientation and age (maximum values in bold)*

Borehole direction	Time duration since drift excavation (day)	Borehole number	Deepest oxidized feature depth (m)			Traction ratio ($m\ m^{-1}$)			Drifts
			Min.	Ave.	Max.	Min.	Ave.	Max.	
Laterally from N155°E drifts	3–31	5	0.00	0.00	**0.00**	0.00	0.00	**0.00**	GCS/GCR
	122–305	32	0.00	0.48	**1.16**	0.00	0.49	**1.05**	GCS/GET
	488–1067	6	0.41	0.75	**1.20**	0.34	0.80	**1.41**	NED/GRM
Vertically from N155°E drifts	3–31	1	0.00	0.00	**0.00**	0.00	0.00	**0.00**	GCR
	122–305	3	0.38	0.39	**0.39**	1.00	1.00	**1.00**	GCS/GET
	488–1067	4	0.00	0.24	**0.45**	0.00	0.75	**1.00**	NED
Laterally from N65°E drifts	3–31	2	0.00	0.00	**0.00**	0.00	0.00	**0.00**	GAN/NRM
	122–305	13	0.00	0.21	**1.00**	0.00	0.52	**1.05**	GAN
	488–1067	4	0.51	0.58	**0.71**	1.00	1.10	**1.40**	GED/GAN
	1922–2379	9	0.30	0.70	**1.11**	0.95	1.21	**1.86**	GEX/GMR
Vertically from N65°E drifts	3–31	2	0.00	0.00	**0.00**	0.00	0.00	**0.00**	GAN
	122–305	2	0.42	0.46	**0.50**	0.30	0.34	**0.38**	GAN
	488–1067	9	0.81	1.16	**1.70**	0.61	1.05	**1.62**	GED
	1922–2379	2	1.28	1.53	**1.77**	1.29	1.55	**1.81**	GED/GKE

the cores. The next boreholes were cored 3 months later (122 days) (Figs 5, 7 & 8) and numerous oxidized features were encountered in the cores from 153 days. Most were located within the traction zone, no matter what the borehole or drift orientations were. Between 122 and 305 days, the deepest oxidized features reached the deepest extensional fractures in a few of the borehole cores (Table 2). Within this set, a small number of oxidized features were encountered within the shear zone, close to the external limit of the traction zone (traction ratio = 1.05) (Table 2). In terms of absolute distance, the deepest oxidized feature was found at *c.* 1.2 m from the drift wall when the lag time between the drift excavation and the borehole drilling was in that range (Table 2). This depth was reached in a horizontal borehole from a drift along the major horizontal stress axis (Fig. 5). Between 488 and 1067 days after drift excavation, a significant portion of the deepest oxidized features were found in the shear zone with traction ratios close to 1.4 (Table 2). The deepest oxidized feature was found at 1.7 m from the drift wall, in a vertical borehole from a drift perpendicular to the major horizontal stress axis (Fig. 7). Between 1922 and 2379 days after drift excavation, boreholes were cored only from drifts orientated perpendicular to the major horizontal stress axis. Laterally and vertically, more than half of the deepest oxidized features observed on these cores were found in the shear zone, with maximum traction ratios above 1.8 (Table 2, Figs 7 & 8).

Finally, over 6.5 years, the oxidation front has propagated deeper into the fractured zone and has developed in the shear zone. Figure 9 illustrates this general tendency. Four lag time ranges were defined, and these ranges were also used to aggregate the data in Tables 1 and 2. The ranges were chosen in order to group together boreholes that were drilled within 6 months of each other, and to separate boreholes for which the time between drilling exceeded 6 months. In addition, the boreholes cored over the first 31 days were put in a specific group, and the last isolated data (2379 days) were joined to the fourth group. For each range, two values were calculated: the proportion of boreholes containing oxidized features located at the external traction zone limit or deeper (traction ratio ≥ 1) and the proportion of boreholes with oxidized features located in the shear zone (traction ratio > 1). Both values increased over time: the first exceeded 0.8 after 5.2 years (1900 days), and the second exceeded 0.5 at the same time. Nevertheless, all the oxidized features were encountered at less than 1.8 m from the drift walls (Table 2), and none were found within the deepest part of the shear zone over the maximum lag time of the study (6.5 years).

Discussion

The results of this study showed that, in the MHM URL, oxidized features were only found in the excavation-induced fractured zone. This result is coherent with those obtained previously in similar URLs. Within the excavation-induced fractured zone, it has been shown that the oxidation front evolved over the 6.5-year time period of the study. This evolution was considered in relation to the

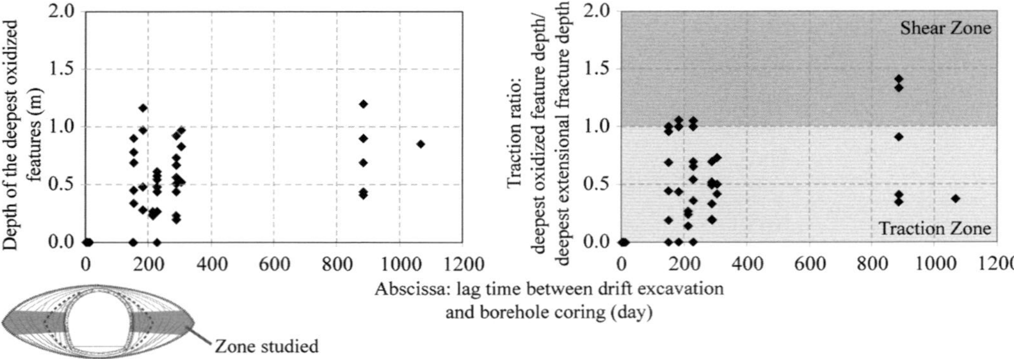

Fig. 5. Deepest oxidized features in horizontal boreholes from N155°E drifts.

site-specific configuration of this fractured zone. The main processes involved in the formation of oxidized features are examined below before discussing what can be learned from the results.

Reactions involved in oxidized feature formation

Observation of the oxidized features confirmed that the following reactions occurred, as previously described by the authors cited above:

- pyrite oxidation: $4\ FeS_2 + 15\ O_2 + 10\ H_2O \longrightarrow 4\ FeO(OH) + 8\ SO_4^{2-} + 16\ H^+$
- calcite dissolution: $8\ CaCO_3 + 16\ H^+ \longrightarrow 8\ Ca^{2+} + 8\ H_2O + 8\ CO_2$
- gypsum precipitation: $8\ Ca^{2+} + 8\ SO_4^{2-} + 16\ H_2O \longrightarrow 8\ CaSO_4{\cdot}2\ H_2O$

Indeed, pyrite minerals react and oxidize when exposed to oxygen, inducing local acid conditions. Calcite minerals locally dissolve and gypsum precipitates due to the increase of calcium and sulphate concentrations in the porewater, while the released iron forms oxides and hydroxides. Gypsum may crystallize either in fissures created in the clay rock or in the vicinity of pyrite minerals. Other reactions should also take place, such as cation exchange at the clay mineral surfaces and reactions involving iron and carbonates. Microorganisms should contribute to these processes and can also affect the natural organic matter.

Role of water-filling

The water-filling state of fractures plays a role in the formation of oxidized features. For instance, in gas-filled fractures, gaseous oxygen can come into contact with fracture wall surfaces, whereas only dissolved oxygen can reach the wall surfaces of water-filled fractures. Gas circulation tests were performed in both the traction and shear zones of a drift orientated along the major horizontal stress axis (de La Vaissière *et al.* 2014). These tests showed that fractures may be connected to one another and allow gas circulation only in the traction zone.

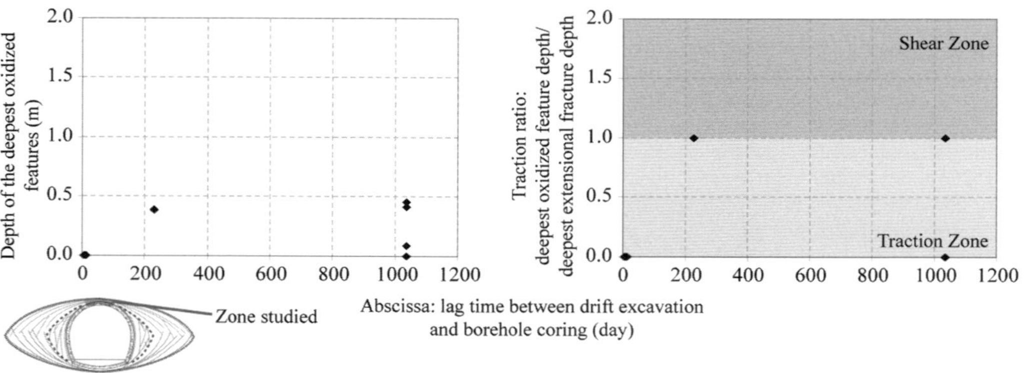

Fig. 6. Deepest oxidized features in vertical boreholes from N155°E drifts.

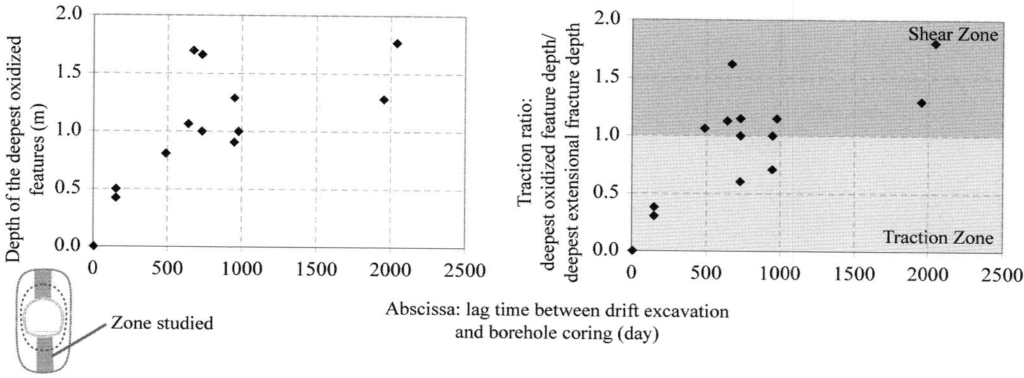

Fig. 7. Deepest oxidized features in vertical boreholes from N65°E drifts.

When drifts age, water vapour exchange between the drift atmosphere and the rock could contribute to partially drying the fractures and favouring gas penetration. Indeed, in the MHM URL, water vapour fluxes mainly pass from the rock to the ventilating air, because the latter is, most of the time, under saturated with respect to liquid water. Nevertheless, study of excavation-induced fractured zone evolution in the MHM URL (Armand *et al.* 2013) suggests that, over the years following drift excavation, some extensional fractures in the traction zone may remain filled by gas, whereas shear fractures should remain closed and mainly filled by water.

Oxygen transfer

Oxidized features appeared in borehole cores drilled *c.* 5 months after drift excavation. Belcourt (2009) reported that oxidation marks became visible after 3 months of rock exposure to the atmospheric air. Correlatively, when the MHM URL rock is exposed to atmospheric air for about 3 months (or to the equivalent oxygen global quantity), macroscopic oxidation marks should form. Conversely, a rock surface that does not display any macroscopic oxidation sign can be surmised to have been exposed to atmospheric air for less than 3 months.

The results for oxidation front propagation suggest that the deepest fractures displaying oxidized features after long lag times (>1 year) had not been exposed enough to oxygen to produce visible oxidation marks earlier. The oxygen mass that potentially came into contact with the fracture when the drift was excavated was not sufficient to induce oxidized features. As a consequence, oxygen fluxes from the drift to the location of these deepest oxidized features are necessary to explain the development of these delayed oxidized features.

Two processes were possibly responsible for the transfer of oxygen: advection and diffusion. Based on the discussion above, gas advection should

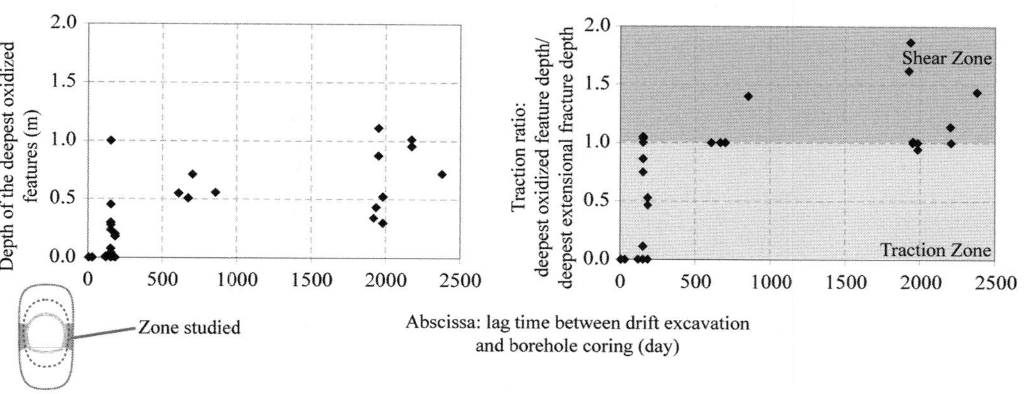

Fig. 8. Deepest oxidized features in horizontal boreholes from N65°E drifts.

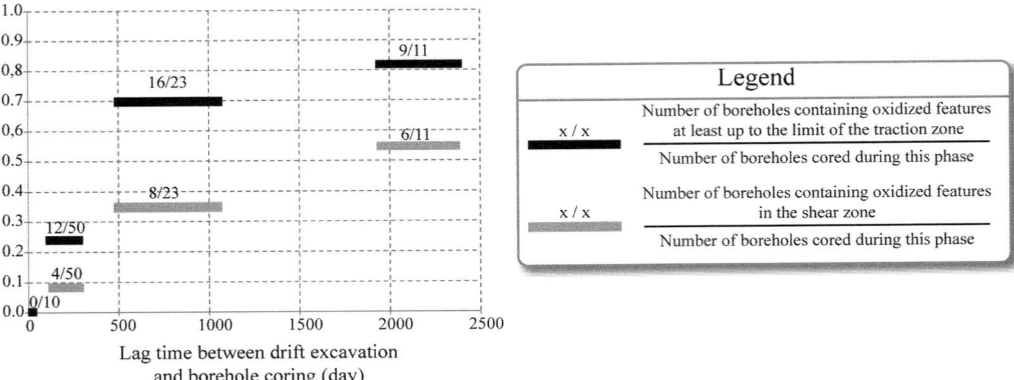

Fig. 9. Evolution over time of the proportion of boreholes having (1) oxidized features in the whole traction zone (black bars) and (2) oxidized features in the shear zone (grey bars).

occur in some extensional fractures, but not in shear fractures, where dissolved oxygen could be transferred by diffusion. Previous studies (cited already) have shown that oxygen diffusion in a pristine rock matrix needs a hundred years to produce visible oxidation marks over a few centimetres from a surface exposed to atmospheric air. Here, the deepest oxidized features are located at several tens of centimetres further away from the traction zone external limit. As a consequence, the diffusion properties of some of the shear fractures have to be different from those of the pristine rock.

Similar results have already been obtained at Mont Terri and Tournemire. At Mont Terri, measured profiles of dissolved helium over up to 10 m perpendicular from a drift wall showed that this gas was able to diffuse more easily through the 3 m zone from the drift wall than in pristine rock (Rübel *et al.* 2002; Pearson *et al.* 2003). This result suggests that the rock diffusion properties were modified up to a depth of 3 m, that is, deeper than the zone affected by excavation-induced fractures and displaying a greater hydraulic conductivity than the pristine rock. The same type of result based on helium measurements was also obtained at Tournemire (Lavielle *et al.* 2012).

It may be deduced from these results that some of the excavation-induced shear fractures display oxygen diffusion properties different from the pristine rock over the time period studied. In these fractures, dissolved gas diffusion may be facilitated for several years after drift excavation. On the other hand, no oxidized features were found in the deepest excavation-induced shear fractures. Further observation will help to determine if this is due to an insufficient study duration or if these fractures have diffusion properties close to those of the pristine rock.

Consequences for drift evolution

The excavation-induced fractured zone constitutes the interface between the gaseous atmosphere and the rock in drifts. Water vapour and oxygen are transferred through this interface when the drifts are ventilated with atmospheric air. Other gases, such as CO_2, H_2S and H_2, could also be transferred through this interface when present. This study has provided insights on the gas diffusion properties of the rock in this fractured zone over a few years after drift excavation. It may help to evaluate the gas fluxes through this interface in the framework of studies on the evolution of a geological disposal for radioactive waste. However, further investigations are needed to determine how these properties will evolve over time.

In the Callovian–Oxfordian rock, porewater will seep into the drifts when ventilation with dry air is stopped. This water will dissolve the soluble secondary mineral phases produced by rock oxidation in the excavation-induced fracture zone. In this zone, it will also dilute the salt-concentrated porewater induced by water evaporation (Vinsot *et al.* 2013) and rock oxidation. As a consequence, study of the oxidation front surrounding the MHM URL drifts will help to evaluate the mass of salt produced by rock oxidation. Evaluation of this mass will help in discussions about the composition of the water that will fill the drifts after their closure.

Conclusion

Oxidized features were systematically sought in 115 borehole cores drilled in various directions from drifts of the Meuse/Haute-Marne URL. The studied drifts were of varying ages, from a few days to 6.5 years, and with various orientations with respect to the stress field axes. The oxidized features encountered mainly consisted of weathered pyrite minerals associated with gypsum, as well as iron oxides and hydroxides. All the oxidized features observed were found on excavation-induced fracture walls or were connected to them, and were located less than 1.8 m from the drift walls. Absolute distances of oxidized features from drift walls depend on the excavation-induced fractured zone morphology, as described in detail by Armand *et al.* (2014). Two zones are distinguished in the excavation-induced fractured zone: a traction zone displaying a greater hydraulic conductivity than the pristine rock and including extensional fractures, and a shear zone including only shear fractures and displaying hydraulic conductivity values similar to those of the pristine rock.

The oxidation front was defined as the set of deepest points away from the drift walls where oxidation reactions have occurred. This front propagated in the traction zone and started to invade the shear zone over the first few years after drift excavation. The evolution of the oxidation front over time implies that gas diffusion was facilitated in the shear zone when compared to the pristine rock.

The survey of oxidation front evolution may help to evaluate the various gas fluxes expected in the excavation-induced fractured zone surrounding the drifts of an underground facility in the Callovian–Oxfordian rock. Furthermore, it could help to estimate the composition of the water that will fill the drifts after their closure.

We gratefully thank A. Dottesi, A. Toussaint (Geoter), C. Righini, P. Landrein, C. Aurière, the operating and facilities management department team (Andra) and the Cofor drilling team for their involvement in the realization of this study, J. Delay, A. Noiret, R. de La Vaissière and M. Lundy (Andra) for their internal review, K. Fournier for English advice, and the two anonymous reviewers and the editor for their very helpful comments.

References

Altmann, S., Tournassat, C., Goutelard, F., Parneix, J.-C., Gimmi, T. & Maes, N. 2012. Diffusion-driven transport in clayrock formations. *Applied Geochemistry*, **27**, 463–478.

ANDRA 2005. Synthesis–evaluation of the feasibility of a geological repository in an argillaceous formation. http://www.andra.fr/international/download/andra-international-en/document/266va.pdf

ANDRA 2013. The Cigeo project: Meuse/Haute-Marne reversible geological disposal facility for radioactive waste. http://www.andra.fr/international/download/andra-international-en/document/editions/504va.pdf

Armand, G., Leveau, F. *et al.* 2014. Geometry and properties of the excavation-induced fractures at the Meuse/Haute-Marne URL drifts. *Rock Mechanics and Rock Engineering*, **47**, 21–41.

Armand, G., Noiret, A., Zghondi, J. & Seyedi, D. M. 2013. Short- and long-term behaviors of drifts in the Callovo-Oxfordian claystone at the Meuse/Haute-Marne Underground Research Laboratory. *Journal of Rock Mechanics and Geotechnical Engineering*, **5**, 221–230.

Baeyens, B., Maes, A., Cremers, A. & Henrion, P. N. 1985. Aging effects in Boom Clay. *Radioactive Waste Management and the Nuclear Fuel Cycle*, **6**, 409–423.

Bastiaens, W., Bernier, F. & Li, X. L. 2007. SELFRAC: experiments and conclusions on fracturing, self-healing and self-sealing processes in clays. *Physics and Chemistry of the Earth, Parts A/B/C*, **32**, 600–615.

Belcourt, O. 2009. *La perturbation chimico-minéralogique (hydratation, état d'oxydation, eau interstitielle) de la zone perturbée excavée et ses relations avec la perturbation texturale et mécanique: application aux argilites des galeries de Bure*. PhD thesis, Université Henri Poincaré, Nancy I.

Blanchart, P., Faure, P., Bruggeman, C., De Craen, M. & Michels, R. 2012. *In situ* and laboratory investigation of the alteration of Boom Clay (Oligocene) at the air–geological barrier interface within the Mol underground facility (Belgium): consequences on kerogen and bitumen compositions. *Applied Geochemistry*, **27**, 2476–2485.

Bossart, P., Meier, P. M., Moeri, A., Trick, T. & Mayor, J.-C. 2002. Geological and hydraulic characterisation of the excavation disturbed zone in the Opalinus Clay of the Mont Terri Rock Laboratory. *Engineering Geology*, **66**, 19–38.

Charpentier, D., Cathelineau, M., Mosser-Ruck, R. & Bruno, G. 2001. Évolution minéralogique des argilites en zone sous-saturée oxydée: exemple des parois du tunnel de Tournemire (Aveyron, France). *Comptes Rendus de l'Académie des Sciences–Series IIA–Earth and Planetary Science*, **332**, 601–607.

Charpentier, D., Mosser-Ruck, R., Cathelineau, M. & Guillaume, D. 2004. Oxidation of mudstone in a tunnel (Tournemire, France): consequences for the mineralogy and crystal chemistry of clay minerals. *Clay Minerals*, **39**, 135–149.

De Craen, M., Van Geet, M., Wang, L. & Put, M. 2004. High sulphate concentrations in squeezed Boom Clay pore water: evidence of oxidation of clay cores. *Physics and Chemistry of the Earth, Parts A/B/C*, **29**, 91–103.

De Craen, M., Van Geet, M., Honty, M., Weetjens, E. & Sillen, X. 2008. Extent of oxidation in Boom Clay as a result of excavation and ventilation of the HADES URF: experimental and modelling assessments. *Physics and Chemistry of the Earth, Parts A/B/C*, **33**(Suppl. 1), S350–S362.

de La Vaissière, R., Morel, J. *et al.* 2014. Excavation-induced fracture network surrounding tunnel: properties and evolution under loading. *In*: Norris, S., Bruno, J. *et al.* (eds) *Clays in Natural and Engineered Barriers for Radioactive Waste Confinement*. Geological Society, London, Special Publications, **400**. First published online May 6, 2014, http://dx.doi.org/10.1144/SP400.30

Delay, J., Trouiller, A. & Lavanchy, J.-M. 2006. Propriétés hydrodynamiques du Callovo-Oxfordien dans l'Est du bassin de Paris: comparaison des résultats obtenus selon différentes approches. *Comptes Rendus Geosciences*, **338**, 892–907.

Delay, J., Vinsot, A., Krieguer, J.-M., Rebours, H. & Armand, G. 2007. Making of the underground scientific experimental programme at the Meuse/Haute-Marne Underground Research Laboratory, North Eastern France. *Physics and Chemistry of the Earth, Parts A/B/C*, **32**, 2–18.

Delay, J., Bossart, P. *et al.* 2014. Three decades of Underground Research Laboratories. What have we learned? *In*: Norris, S., Bruno, J. *et al.* (eds) *Clays in Natural and Engineered Barriers for Radioactive Waste Confinement*. Geological Society, London, Special Publications, **400**. First published online March 5, 2014, http://dx.doi.org/10.1144/SP400.1

Descostes, M. 2001. *Evaluation d'une perturbation oxydante en milieu argileux: mécanisme d'oxydation de la pyrite*. PhD thesis, Université Paris 7.

Lavielle, B., Matray, J.-M., Thomas, B., Dauzères, A., Bensenouci, F. & Gilabert, E. 2012. Stages of evolution of a Toarcian compacted claystone around galleries excavated between 1 and 124 years ago by the study of noble gases dissolved in pore water at the Tournemire Underground Research Laboratory (France). *Applied Geochemistry*, **27**, 1403–1416.

Lerouge, C., Grangeon, S. *et al.* 2011. Mineralogical and isotopic record of biotic and abiotic diagenesis of the Callovian–Oxfordian clayey formation of Bure (France). *Geochimica et Cosmochimica Acta*, **75**, 2633–2663.

Mäder, U. K. & Mazurek, M. 1998. Oxidation phenomena and processes in Opalinus clay: evidence from the excavation disturbed zones in Hauenstein and Mont Terri tunnels, and Siblingen open clay pit. *Material Research Society Symposium Proceedings*, **506**, 731–739.

Matray, J. M., Savoye, S. & Cabrera, J. 2007. Desaturation and structure relationships around drifts excavated in the well-compacted Tournemire's argillite (Aveyron, France). *Engineering Geology*, **90**, 1–16.

Mazurek, M., Pearson, F. J., Volckaert, G. & Bock, H. 2003. *Features, Events and Processes Evaluation Catalogue for Argillaceous Media*. OECD/NEA Report **4437**.

Pearson, F. J., Arcos, D. *et al.* 2003. Mont Terri Project–geochemistry of water in the Opalinus Clay Formation at the Mont Terri Rock Laboratory. *Rapport de l'OFEG, Série Géologie*, **5**, 143.

Rübel, A. P., Sonntag, C., Lippmann, J., Pearson, F. J. & Gautschi, A. 2002. Solute transport in formations of very low permeability: profiles of stable isotope and dissolved noble gas contents of pore water in the Opalinus Clay, Mont Terri, Switzerland. *Geochimica et Cosmochimica Acta*, **66**, 1311–1321.

Techer, I., Clauer, N. & Liewig, N. 2009. Ageing effect on the mineral and chemical composition of Opalinus Clays (Mont Terri, Switzerland) after excavation and surface storage. *Applied Geochemistry*, **24**, 2000–2014.

Tsang, C.-F., Bernier, F. & Davies, C. 2005. Geohydromechanical processes in the excavation damaged zone in crystalline rock, rock salt, and indurated and plastic clays–in the context of radioactive waste disposal. *International Journal of Rock Mechanics and Mining Sciences*, **42**, 109–125.

Vinsot, A., Linard, Y., Lundy, M., Necib, S. & Wechner, S. 2013. Insights on desaturation processes based on the chemistry of seepage water from boreholes in the Callovo–Oxfordian argillaceous rock. *Procedia Earth and Planetary Science*, **7**, 871–874.

Yven, B., Sammartino, S., Geraud, Y., Homand, F. & Villieras, F. 2007. Mineralogy, texture and porosity of Callovo–Oxfordian argillites of the Meuse/Haute-Marne region (eastern Paris Basin). *Mémoires de la Société géologique de France*, **178**, 73–90.

Zheng, L., Samper, J., Montenegro, L. & Mayor, J. C. 2008. Multiphase flow and multicomponent reactive transport model of the ventilation experiment in Opalinus Clay. *Physics and Chemistry of the Earth, Parts A/B/C*, **33**(Suppl. 1), S186–S195.

Effect of montmorillonite content on mechanical and hydraulic properties of bentonite and its numerical modelling

YUSUKE TAKAYAMA[1]*, SHUHEI TSURUMI[1], ICHIZO KOBAYASHI[2], HITOSHI OHWADA[3], TOMOKO ISHII[3], RYOJI YAHAGI[3] & ATSUSHI IIZUKA[4]

[1]*Graduate School of Engineering, Kobe University, Kobe City, 657-8501, Japan*

[2]*Kajima Corporation, 2-19-1, Tobitakyu, Chofu-shi, Tokyo, 182-0036, Japan*

[3]*Radioactive Waste Management Funding and Research Center, 1-15-7, Tsukishima, Chuo-ku, Tokyo, 104-0052, Japan*

[4]*Research Center for Urban Safety and Security, Kobe University, Kobe City, 657-8501, Japan*

**Corresponding author (e-mail: 091t123t@stu.kobe-u.ac.jp)*

Abstract: Recently, the hydraulic/mechanical/chemical (HMC) analytical method has been studied with the aim of evaluating the long-term performance of Trans-Uranium (TRU) geological repositories. In this particular research, the hydraulic/mechanical modelling of bentonite materials for HMC analyses has been studied, with bentonite materials considered as the engineered barrier. It is said that the bentonite material in the TRU disposal facilities is altered chemically by cement leachate. Moreover, there is a concern that chemical alteration changes the expansion characteristics and water permeability of the bentonite material. In this paper, using past test results, the influence of montmorillonite content and the exchange of sodium and calcium ions on mechanical and hydraulics behaviour is examined. Hydraulic/mechanical mathematical modelling is also presented, describing the mechanical behaviour of bentonite from an initially unsaturated state while also taking into account the hydraulic and mechanical characteristics of bentonite mentioned above. We also simulate the saturation process from an initially unsaturated state in a TRU disposal facility, which is modelled one-dimensionally.

Bentonite materials are considered to be the engineered barrier in Trans-Uranium (TRU) disposal facilities (shown in Fig. 1). It is known that the mechanical properties of bentonite materials change depending on the content of montmorillonite and the degree of water saturation. In particular, the compressibility expansion characteristics of bentonite materials are governed by their montmorillonite content under loading/unloading or water dissipation/absorption. It is not easy, however, to quantitatively grasp their mechanical properties because of the intermixing of conditions during loading/unloading and water dissipation/absorption. Furthermore, alteration of the mechanical properties due to the exchange of sodium and calcium ions in the montmorillonite complicates the interpretation of the mechanical behaviour of bentonite materials. In order to mechanically model the compression behaviour of fully saturated bentonite materials under monotonic loading, all the preceding factors must be experimentally examined. On the other hand, from a hydraulics viewpoint, disintegration of the soil skeleton due to swelling seems to increase the specific surface area, apparently because of water molecules adsorbing onto the surface of montmorillonite particles. Accordingly, a method for evaluating the coefficient of permeability (taking into consideration any variation in the specific surface area) in order to express hydraulic performance was investigated. Following this, numerical modelling was carried out and we simulated the saturation process from the initially unsaturated state. The mechanical behaviour of bentonite material in the TRU waste disposal facility during the saturation process was examined.

Mechanical behaviour of fully saturated bentonite materials under monotonic loading

According to Kobayashi *et al.* (2007), swelling characteristics such as normally consolidated lines, the relationship between equilibrium swelling pressure and equilibrium swelling deformation, can be obtained from a full saturation line as shown in Figures 2 and 3. The full saturation line can be obtained by one-dimensional air-exhausting

From: Norris, S., Bruno, J., Cathelineau, M., Delage, P., Fairhurst, C., Gaucher, E. C., Höhn, E. H., Kalinichev, A., Lalieux, P. & Sellin, P. (eds) 2014. *Clays in Natural and Engineered Barriers for Radioactive Waste Confinement*. Geological Society, London, Special Publications, **400**, 221–235.
First published online April 7, 2014, http://dx.doi.org/10.1144/SP400.13

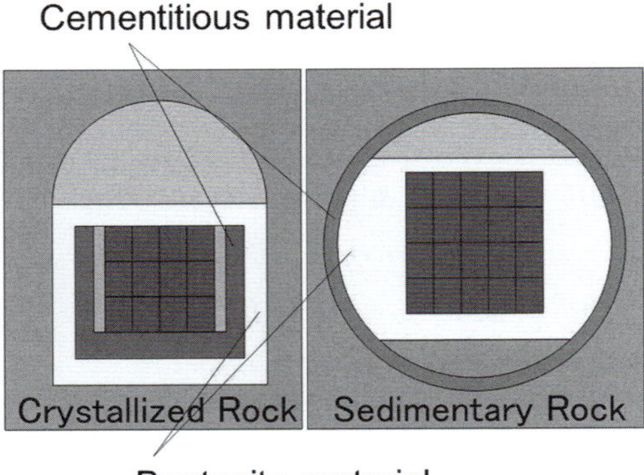

Fig. 1. Example of layout of TRU disposal facilities.

compression testing on the bentonite-engineered barrier material. The test is important for verifying the mechanical model of the bentonite-engineered barrier material and for evaluating the change in mechanical behaviour due to chemical alteration. As shown in Figure 4, a full saturation line becomes linear when plotted against logarithms of dry density and compression pressure, and when dry density is converted to a void ratio it becomes a curved line. The full saturation line is a straight line of constant saturation where the degree of saturation is 100% and the intersection of the constant water content line and the constant saturation line is the saturation point.

One-dimensional air-exhausting compression tests were conducted by Kobayashi *et al.* (2011) to obtain full saturation lines for Na montmorillonite, Ca montmorillonite, Na bentonite and manmade Na bentonite constituted by a mixture of fine silica sand and montmorillonite that takes on the same montmorillonite content as KuniGel V1. The test procedure is as follows:

(1) Materials with various initial water contents as a parameter were prepared.
(2) Each material was filled into the compaction mould and statically compressed one-dimensionally at a constant rate. Although the compression rate differed according to the air-exhaust rate of the materials, the air-exhaust condition of the mould, and so on, during this test, 0.016 mm/min was used for Na montmorillonite and air was exhausted from the top.
(3) Loading pressure and wet density were recorded during compression.
(4) Dry density was calculated from wet density and initial water content. The results were summarized as the relationship between dry density and static compression pressure.

Figure 5 shows the theoretical development flow diagram of the mechanical model obtained by one-dimensional compression tests. In the diagram,

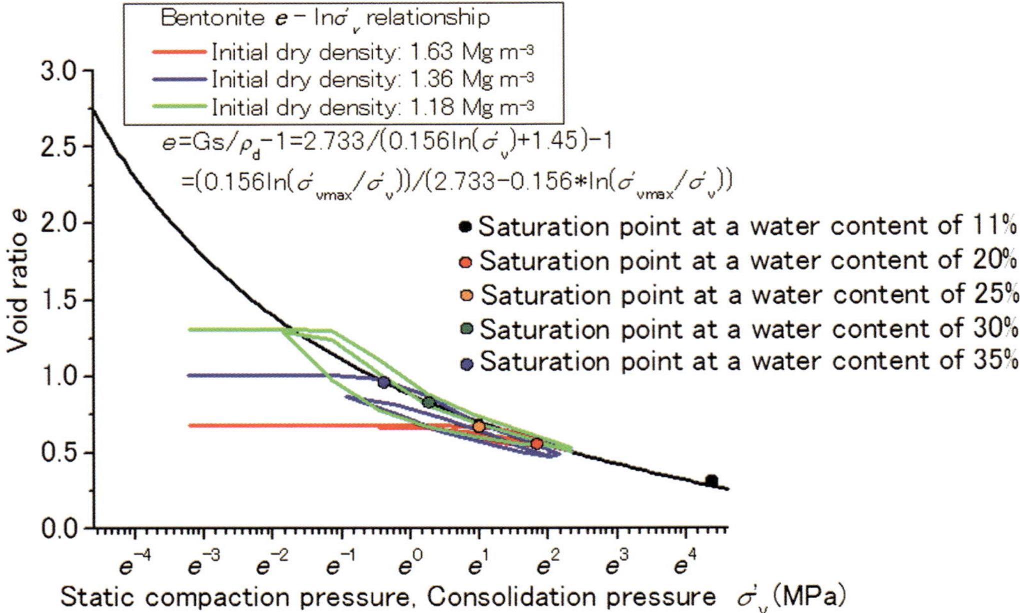

Fig. 2. Consistency between full saturation line and normalized consolidation line (Kobayashi *et al.* 2007).

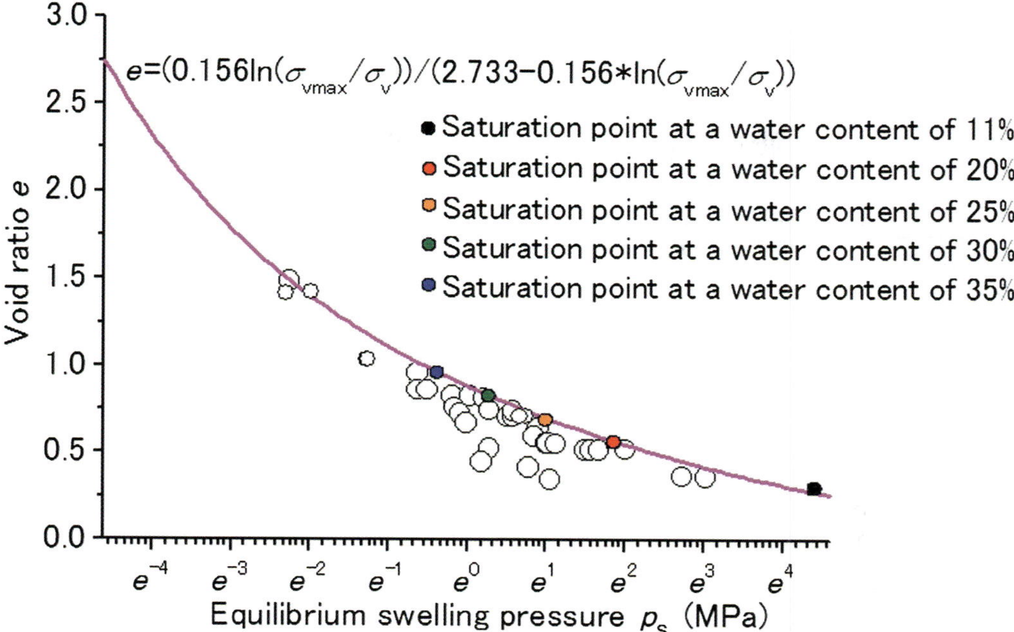

Fig. 3. Consistency between full saturation line, swelling pressure in equilibrium and constant swelling volume in equilibrium (Kobayashi *et al.* 2007).

lettering in red represents material properties. When one-dimensional air-exhausting compression tests were conducted for the respective materials and the results were summarized according to the flow diagram, full saturation lines for individual materials could be obtained. When the full saturation lines of various materials were obtained, consideration of chemical alteration based on the theory of equilibrium could be incorporated into the mechanical model.

An example of the relationship between saturation points and full saturation lines in one-dimensional air-exhausting compression tests is shown in Figure 6. As shown in the figure, dry density

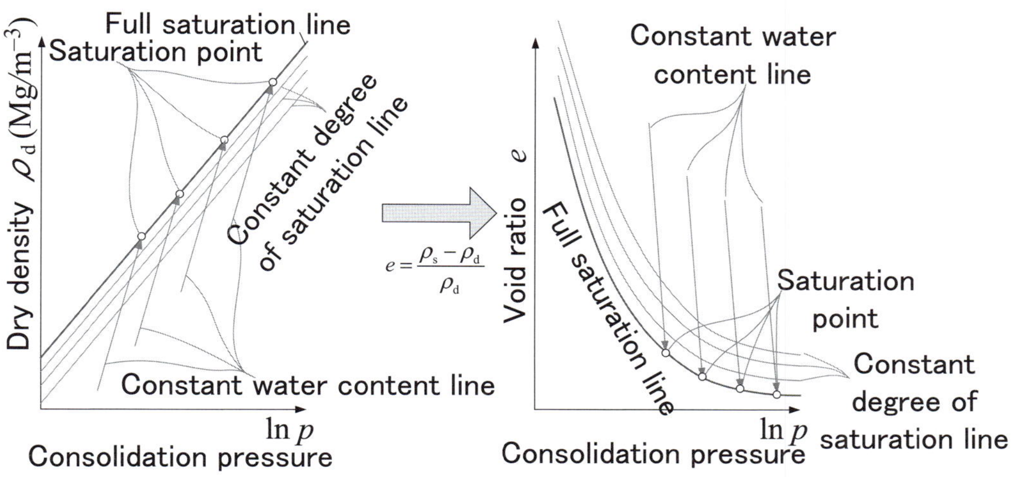

Fig. 4. Relationship between full saturation line, constant saturation line and constant water content line.

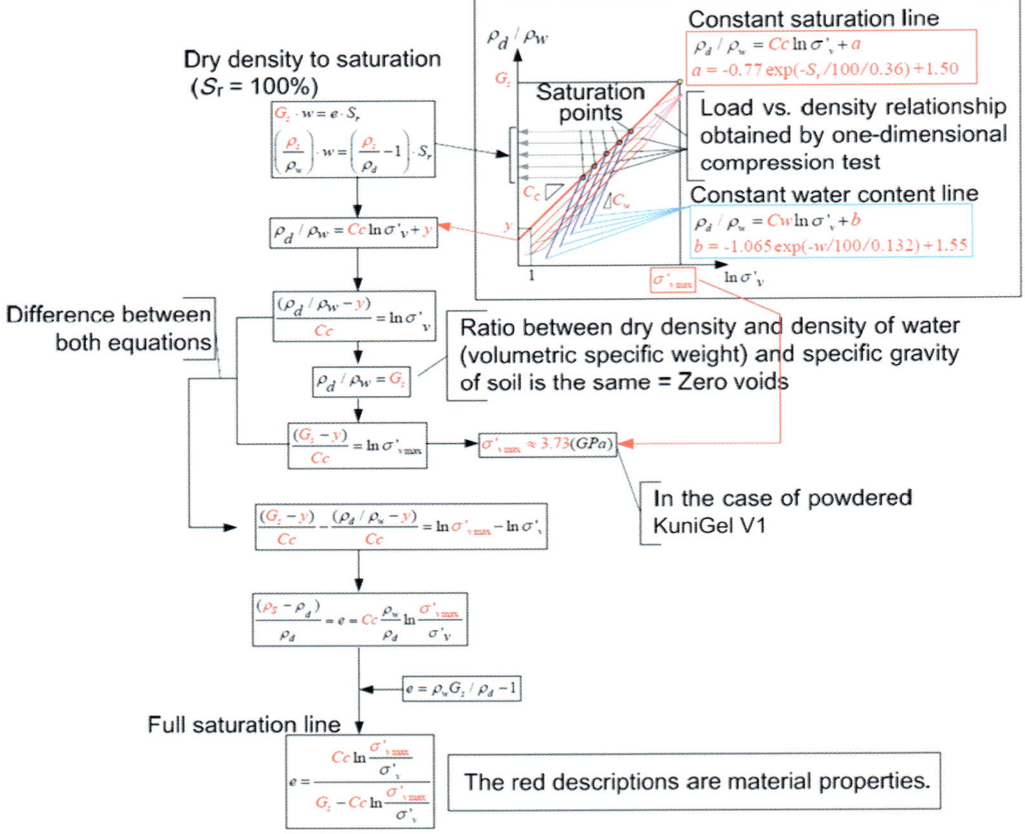

Fig. 5. Theoretical development of the one-dimensional compression test.

increases with an increase in the compression pressure when the materials were compressed with exhaust air. Because water is hardly drained during the process of air-exhausting compression, this process can be regarded as having constant water content. It is also shown that the gradient of the constant water content line in the straight-line section is the same, and independent of the initial water content. Figure 7 shows full saturation lines obtained from one-dimensional air-exhausting compression tests for Na montmorillonite, Ca montmorillonite, Na bentonite and manmade Na bentonite. The test results indicated the following:

(1) The full saturation lines of Na montmorillonite and Ca montmorillonite are consistent with each other. Furthermore, the three saturation lines for Na bentonite, Ca bentonite and manmade Na bentonite, which have the same montmorillonite content as the Na bentonite, all coincided.
(2) When the full saturation lines were plotted against the logarithms of dry density and consolidation pressure, the lines were straight in all cases. Also, the gradients of the full saturation lines for montmorillonite and bentonite were the same.
(3) The position of full saturation lines for bentonite moved to the right with increasing montmorillonite content.
(4) The full saturation lines were not dependent on any type of cation between the layers.

As mentioned above, when the montmorillonite content of both manmade bentonite and Na bentonite is the same, they have the same full saturation lines. Accordingly, it can be considered that the consolidation and swelling behaviours of materials with the same montmorillonite content are the same. In other words, montmorillonite content is one of the key parameters to determine the mechanical behaviour of bentonite.

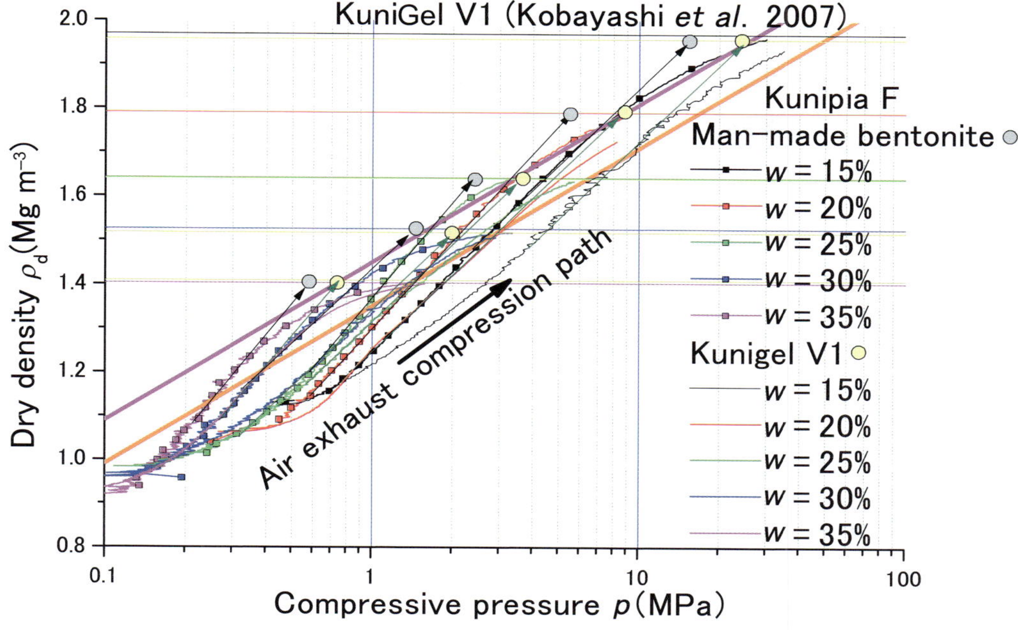

Fig. 6. Relationship between saturation points and full saturation lines in air-exhausting one-dimensional compression tests of Na bentonite and manmade Na bentonite (Kobayashi *et al.* 2011).

Permeability of the bentonite materials

The coefficient of permeability is a parameter that indicates how easily liquid flows through a porous medium and it is an important parameter in the evaluation of impermeability of the bentonite-engineered barrier in the geological repositories of radioactive waste. The coefficient of permeability is usually determined by the relationship between the hydraulic gradient and the flow rate obtained through experiments, assuming that Darcy's law applies. Although a variety of estimation equations for hydraulic conductivity are proposed, most of the equations are a function of the total volume of voids, the types of liquid and the microscopic structure of the materials. In the equations, the voids are represented by the void ratio and the dry density of the soil, the types of liquid are represented by

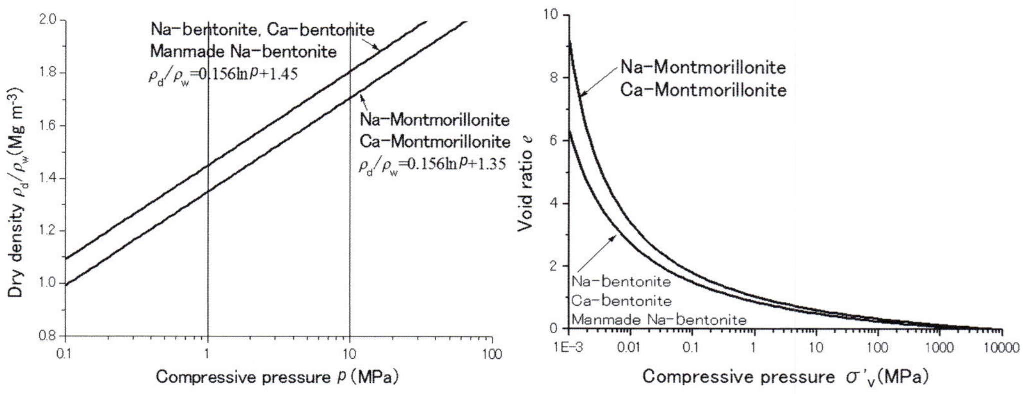

Fig. 7. Relationship between saturation points and full saturation lines in one-dimensional air-exhausting compression tests.

viscosity, and the microscopic structures of the materials are represented by specific surface area and particle size.

In the case of bentonite-engineered barriers, montmorillonite, which characterizes impermeability, exists in the form of secondary particles constituted by the laminate form of primary particles in dry conditions. Such secondary particles constitute a microscopic structure. When water permeates into montmorillonite, the hydration of the interface between the primary particles and the osmosis of water occurs, and the spacing between the primary particles that constitute secondary particles increases. The increase of space between the primary particles causes the decay of the microscopic structures. Because a change in the microscopic structure is dependent on dry density and degree of saturation, the microscopic structure required to estimate the coefficient of permeability should be quantitatively evaluated with constant dry density and in wet conditions. Because quantitative evaluation of the microscopic structure of bentonite under a constant dry density and in wet conditions is difficult, most of the estimation equations for the coefficient of permeability of the bentonite-engineered barrier are based on the inductive method, and very few equations are based on the deductive method. Accordingly, in this study, the method of estimating the hydraulic conductivity of the bentonite-engineered barrier according to the Kozeny–Carman law, which is conventional but derived deductively, has been investigated to obtain the coefficient of permeability. The Kozeny–Carman law is stated as follows:

$$k = k(e, S_v) = \frac{1}{c}\frac{\rho_w g}{\mu}\frac{1}{S_v{}^2}\frac{e^3}{1+e} \qquad (1)$$

where S_v is the specific surface area and e is the void ratio.

It is therefore necessary to determine the specific surface area in order to apply this Kozeny–Carman law. However, it is very difficult to measure directly the specific surface area of the compressed bentonite block in wet conditions. Kobayashi *et al.* (2011) therefore focused on the water content in a two-layer hydration state, and developed a method for calculating the specific surface area based on this concept.

Water content control X-ray diffraction (XRD) test

Definition of two-layer hydration water content

The method formulated by Kawamura *et al.* (1999) was used to determine a two-layer hydration condition. The condition was defined as a state where

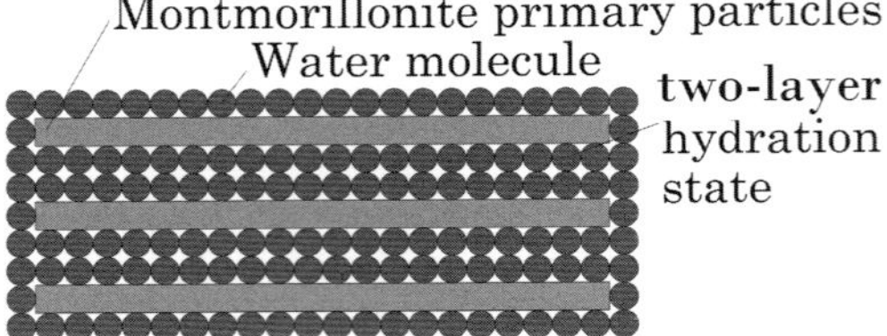

Fig. 8. Two-layer hydration state of montmorillonite.

one water molecule is uniformly adsorbed onto the surface of each primary particle (see Fig. 8). In this method, the two-layer hydration condition can be measured as basal spacing by XRD. The two-layer hydration water content was also defined as the water content in a two-layer hydration state, and the specific surface area of the compressed bentonite block in the swelling state was obtained from this two-layer hydration water content.

According to Sudo (1967), a projected area of one water molecule on the plane of the montmorillonite particle is given as 10.8 Å^2 (m^2), the surface area of montmorillonite per unit weight S_g can be expressed as

$$s_g = 10.8\text{Å}^2 \times 6.02 \times 10^{23}/18$$
$$= 3.6 \times 10^3 (\text{m}^2/\text{g}) \qquad (2)$$

where Avogadro's number (6.02×10^{23}) and the molecular weight of water (18 g mol^{-1}) were used.

Because water content is defined as the ratio of porewater weight and soil particle weight, the specific surface area per unit weight of soil particles can be derived by the substitution of water content in the two-hydration state w^*, where a water molecule is uniformly adsorbed without overage or shortage on the surface of each primary particle for equation (2):

$$s_v = 3.6 \times 10^3 \times (w^*/100) = 36w^*(\text{m}^2/\text{g}) \qquad (3)$$

Thus, the specific surface area of the compressed bentonite block in wet conditions can be determined when two-layer hydration water content w^* is obtained. This is because the water molecule adsorbs onto the surface of the primary particle and the surface then becomes a water path in a montmorillonite block, and the specific surface area obtained from equation (3) will be the specific surface area required for the Kozeny–Carman law.

It should be noted that the two-layer hydration water content can be regarded as a constant, regardless of the dry density, when it is smaller than the saturated water content (see Fig. 9).

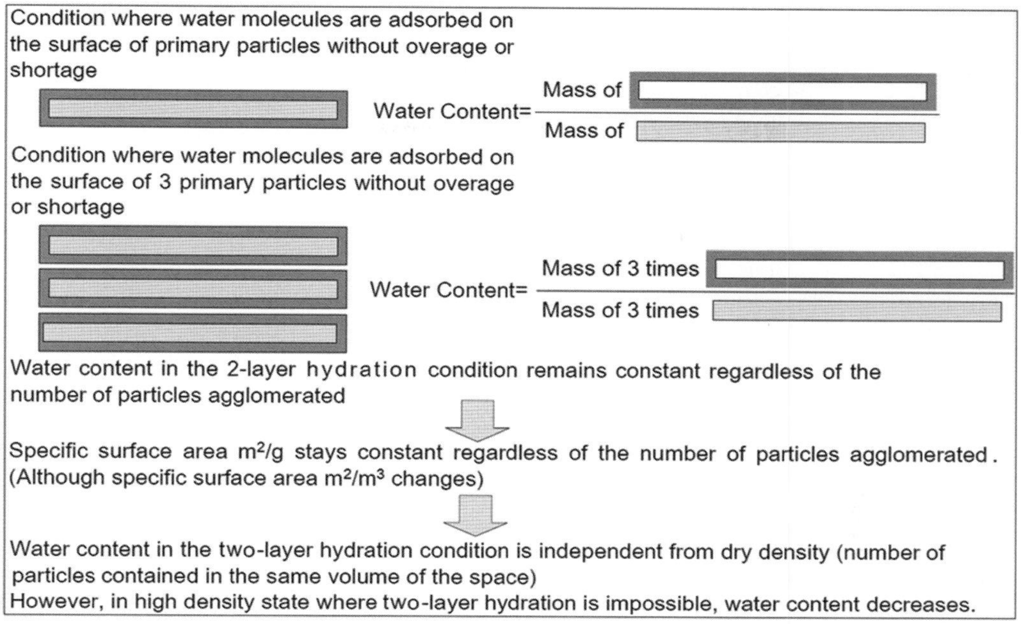

Fig. 9. Schematic diagram of two-layer hydration water content.

Measurement of two-layer hydration water content using XRD

Kobayashi *et al.* (2011) conducted XRD tests on five types of material: Na montmorillonite, Ca montmorillonite, Na bentonite (KuniGel V1), Ca bentonite (Ca KuniGel V1) and manmade Na bentonite, which is composed of a mixture of fine silica sand and montmorillonite and has the same montmorillonite content as KuniGel V1. The montmorillonite used here is of Kunigel V1 origin. The two-layer hydration water content w^* was determined with the following process based on the logical approach explained previously:

(1) Each material was chilled to about $-20°$.
(2) Powdered ice chilled by liquid nitrogen was prepared.
(3) The chilled material and powder ice were mixed as powder at a temperature of $-20°$ to prevent defrosting during mixing.
(4) The weight ratio of the chilled material and powder ice becomes the water content.
(5) The structure of the XRD cell consists of a polycarbonate pipe with a thickness of 1.5 mm and a stainless-steel sheath to prevent volumetric changes, with a window for X-ray transmission. The chilled mixtures were compressed into the XRD cell and thawed to obtain the XRD specimen under constant water content and dry density conditions (see Fig. 10).
(6) Specimens with various water contents as a parameter were subjected to an XRD test, and basal spacing was measured for these samples. The water content (two-layer hydration water content w^*), that is, the threshold between the two-layer hydration and three-layer hydration (a state where three water molecules are present between primary particles), was determined based on the measured basal spacing. During this test, the X-ray tube used for XRD was molybdenum (MoKα).
(7) Determination of the two-layer or three-layer hydration was made by peak separation of the diffraction pattern based on the position of the peaks of two-layer and three-layer hydration.

Test results

Figure 11 presents XDR diffraction patterns for all cases. As a general trend, the peak value shifts from the peak of two-layer hydration (2.6792°) to the peak of three-layer hydration (2.2623°) with increasing water content. Because the peak intensity of three-layer hydration is higher than that of two-layer hydration, peak intensity tends to increase with an increase in water content. Ca montmorillonite had been thought to have a smaller swelling

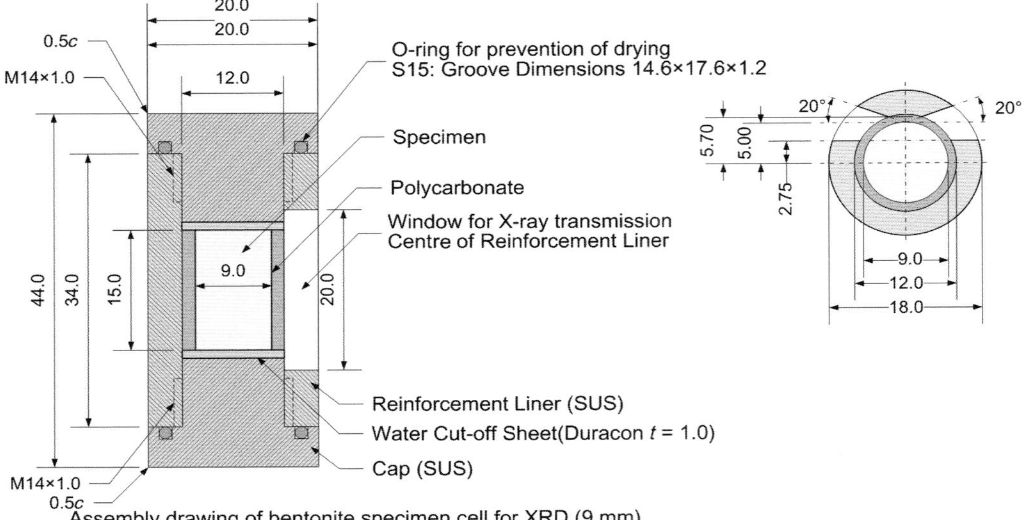

Fig. 10. XRD cell for measurement of two-layer hydration.

capacity and to be unable to absorb water endlessly, with two- or three-layer hydration representing its limit (Morodome 2009). In this test, however, as shown in Figure 11, we were able to observe four-layer hydration. In existing research (Morodome 2009), the adsorbed water was supplied by controlling the relative humidity around the specimen. It may be that the difference in how the water supplement is adsorbed might cause a difference in the peak properties of Ca montmorillonite. In terms of the materials, it was found that the peak diffraction positions of the cases are consistent when the montmorillonite partial water content (defined as the weight ratio of porewater and montmorillonite) is the same. This means that most of the porewater in the bentonite was absorbed by montmorillonite, because the specific surface area of non-swellable accessory minerals for bentonite was notably smaller than that of montmorillonite. Montmorillonite partial water content w_{mon} is expressed by the following equation:

$$w_{mon} = \frac{W_w}{W_{\mathrm{smon}}} = \frac{W_w}{W_s \alpha_{\mathrm{mon}}} = \frac{w_{\mathrm{ben}}}{\alpha_{\mathrm{mon}}} \tag{4}$$

where W_w is the weight of the porewater, W_{smon} is the weight of the montmorillonite, W_s is the weight of the soil particles in bulk, w_{ben} $(=W_w/W_s)$ is water content for the entire bentonite and α_{mon} $(=W_{\mathrm{mon}}/W_s)$ is the montmorillonite content (0.59 in this study).

As the above relationship also applies to the two-layer hydration state, the two-layer hydration water content of bentonite w^*_{ben} can be expressed as $w^*_{\mathrm{ben}} = \alpha_{\mathrm{mon}} w^*$, where the two-layer hydration water content of montmorillonite alone is w^*. Because the montmorillonite content of the bentonite used in this study was 0.59, the two-layer hydration water content of bentonite was 0.59 times the two-layer hydration water content of Na montmorillonite.

Relationship between effective bentonite dry density and two-layer hydration water content

The summarized results regarding the relationship between effective bentonite dry density and two-layer hydration water content for the five bentonite materials mentioned above are shown in Table 1. The effective bentonite dry density is defined as the ratio of the mass of bentonite to the volume of bentonite plus void in the mixture. In the case of pure bentonite, effective bentonite dry density corresponds to dry density. When Table 1 is summarized further, the equation for two-layer hydration water content is as follows:

$$w^* = \frac{\alpha_{\mathrm{mon}}}{c_{\mathrm{mon}} + a_{\mathrm{mon}} {\rho_d}^{b_{\mathrm{mon}}}} \tag{5}$$

where a_{mon}, b_{mon} and c_{mon} are given as $a_{\mathrm{mon}} = 10^{-1.16\mathrm{CR}-5.42}$, $b_{\mathrm{mon}} = 3.5\mathrm{CR} + 12.1$ and $c_{\mathrm{mon}} = 0.014\mathrm{CR} + 0.052$ for Na montmorillonite, and for Na bentonite they are given as $a_{\mathrm{mon}} = 10^{-2.75\mathrm{CR}-7.89}$, $b_{\mathrm{mon}} = 6.75\mathrm{CR} + 17.7$ and $c_{\mathrm{mon}} = 0.014\mathrm{CR} + 0.052$. α_{mon} is montmorillonite content and a_{mon}, b_{mon} and c_{mon} are functions of CR,

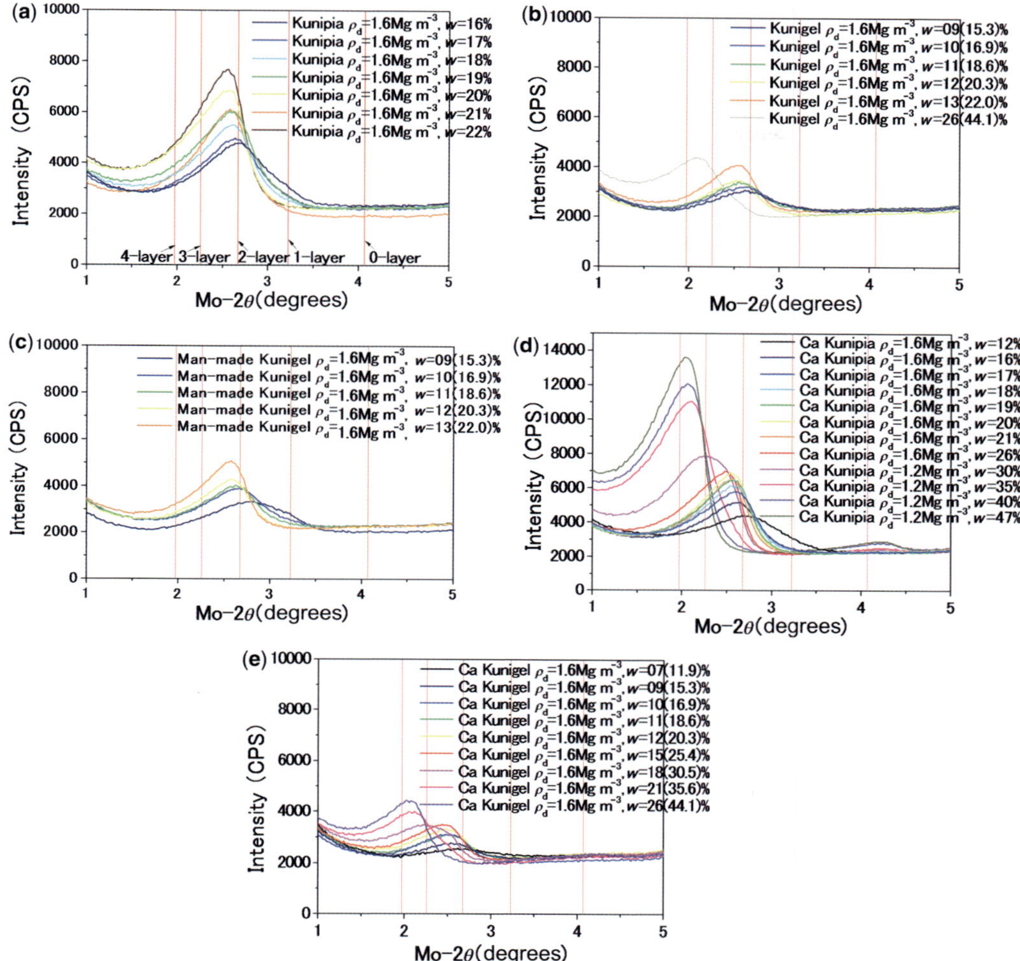

Fig. 11. XRD diffraction patterns for all types of material tested (Kobayashi *et al.* 2011): Na montmorillonite, Na bentonite, manmade KuniGel V1, Ca montmorillonite, Ca bentonite.

where CR is defined as the variation in the ratio of exchangeable Ca ions and the total amount of exchangeable Na and Ca ions in the interlayer of bentonite. The reference state for the definition of CR is the intact state, which is before the cation exchange reaction. CR is termed the ‘Ca ions ratio’.

Figure 12 shows the relationship between effective bentonite dry density and water content in a two-layer hydration state, the surface area per unit weight, and the surface area per unit volume. As shown in the figures, the surface area per unit volume required in the Kozeny–Carman equation peaks at *c.* 1.9 Mg m^{-3} dry density; when the density is larger, the specific surface area decreases due to the lack of physical space able to become the two-layer hydration state.

The relationship between the saturated hydraulic conductivity and the effective bentonite dry density calculated by the Kozeny–Carman law using these relationships is shown in Figure 13, which also shows the relationship between hydraulic conductivity in a saturation state and effective bentonite dry density proposed by existing research. It is clear that the trend in the relationship between the hydraulic conductivity in a saturation state and effective bentonite dry density calculated by the Kozeny–Carman equation in this study is consistent with the JAEA Database and the existing study results by Pusch (1994), unlike Åkesson *et al.* (2010) and Ichikawa *et al.* (2002), when compared with the equation proposed in the past. The results of the present study are particularly

Table 1. *Relationship between two-layer hydration water content and effective bentonite dry density for different materials*

Material	w^* (%)	Effective bentonite dry density and two-layer hydration state water content relation
Na montmorillonite	19	$w^* = 1/(0.052 + 3.8E - 6\rho_d{}^{12.1})$
Na bentonite KuniGel V1	11	$w_{ben} = 1/(0.091 + 2.2E - 8\rho_d{}^{17.57})$ $w^* = \alpha_{mon}/(0.052 + 3.8E - 6\rho_d{}^{12.1})$
Manmade KuniGel V1	11	$w_{ben} = 1/(0.091 + 2.2E - 8\rho_d{}^{17.57})$ $w^* = \alpha_{mon}/(0.052 + 3.8E - 6\rho_d{}^{12.1})$
Ca montmorillonite	15	$w_{ca}{}^* = 1/(0.066 + 2.6E - 7\rho_d{}^{15.16})$
Ca bentonite Ca KuniGel V1	9	$w_{caben}{}^* = 1/(0.11 + 3.9E - 11\rho_d{}^{24.5})$ $w_{ca}{}^* = \alpha_{mon}/(0.066 + 2.6E - 7\rho_d{}^{15.16})$

w^*, two-layer hydration state water content; α_{mon}, montmorillonite content; ρ_d, effective bentonite dry density.

consistent within the practical range of dry density for the bentonite-engineered barrier. Figure 13 also shows the results for Na montmorillonite, Ca montmorillonite and Ca bentonite. The observed variation in hydraulic conductivity due to differences in the materials, by up to tenfold times can be explained by the change that takes place in the specific surface area only (change in two-layer hydration water content).

Mathematical modelling

The final objective of this study was the establishment of a long-term performance evaluation system that can take the chemical reaction of a cementitious engineered barrier and the bentonite-engineered barrier into account. In current research, the coupled analysis of hydraulic/mechanical/chemical (HMC) properties has been examined. For example, in the HMC coupled analysis method proposed by Ishii *et al.* (2013), fully saturated state after saturation process is regarded as the initial condition. When we conduct the HMC coupled simulation we must appropriately determine the initial condition, which is the density and permeability distribution in the saturated buffer after the saturation process. It is therefore necessary to simulate the saturation process. Furthermore, a numerical model that can consistently express the mechanical behaviour from the unsaturated to saturated states is required. In this research, the mechanical and hydraulic properties of bentonite were assumed to be determined by the montmorillonite content, dry density and CR. Accordingly, numerical modelling that reflects these mechanical

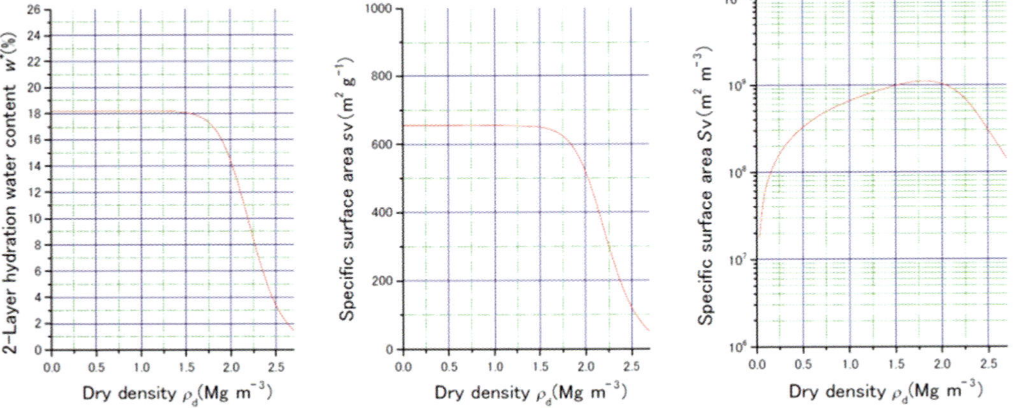

Fig. 12. Relationship between effective bentonite dry density and two-layer hydration water content, specific surface area.

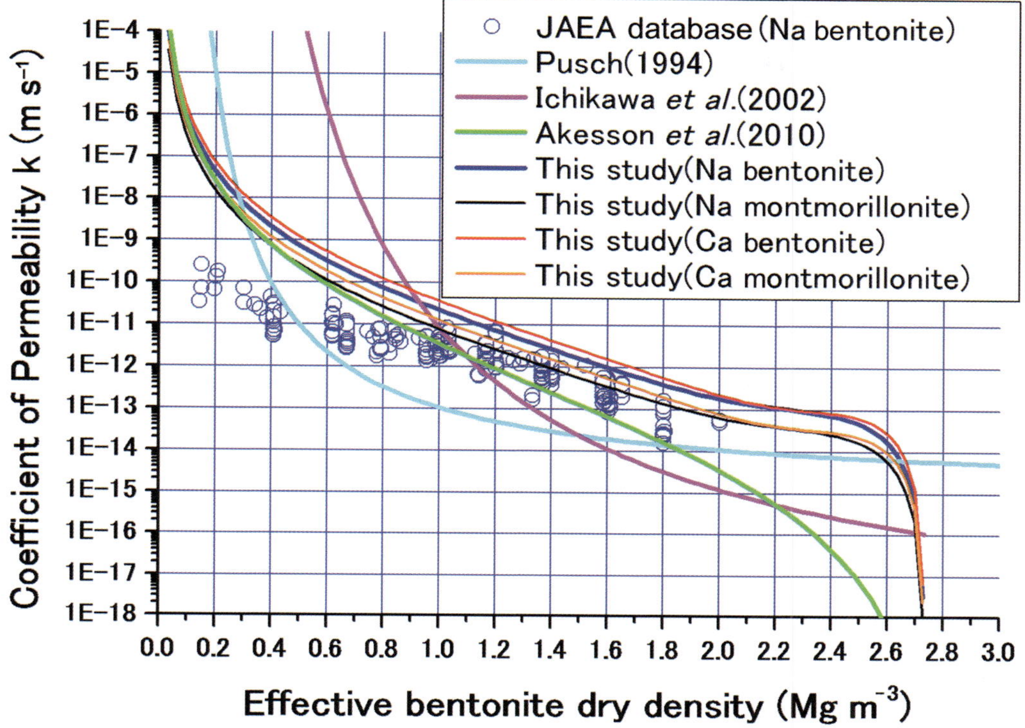

Fig. 13. Comparison of relationships between effective bentonite dry density and hydraulic conductivity.

and hydraulic properties can be easily related to any chemical alteration. In this section, the hydraulic and mechanical mathematical model for HMC analyses based on the concept of the full saturation line and the coefficient of permeability is established, taking into account variation in specific surface area. We then simulate the saturation process.

Bentonite can be modelled as an elasto-plastic material. Tachibana *et al.* (2011) extended the elasto-plastic constitutive model for unsaturated soils, proposed by Ohno *et al.* (2007), to express

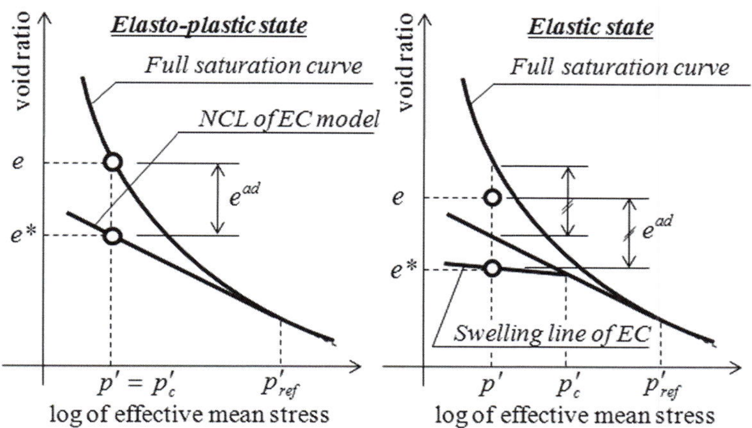

Fig. 14. Schematics of relationship between void ratios under isotropic stress conditions. NCL, normally consolidation line.

the expansion characteristics of bentonite based on the concept of the full saturation line proposed by Kobayashi *et al.* (2007). The model is implemented in the finite element (FE) program DACSAR-U (Iizuka *et al.* 2000; Kawai *et al.* 2009), which was coded based on a three-phase mixture theory consisting of soil, porewater and pore air.

According to Kobayashi *et al.* (2007), the normal compression line of saturated bentonite corresponds to the full saturation line. It can be expressed as (see Fig. 5)

$$e = \frac{\lambda_\rho \ln(\sigma'_{v\max}/\sigma'_v)}{G_s - \lambda_\rho \ln(\sigma'_{v\max}/\sigma'_v)} \quad (6)$$

where G_s is the density of soil particles, λ_ρ is the gradient of the normal compression line and $\sigma_{v\max}$ is the compression stress equivalent to the state with no void. Regardless of montmorillonite content or type of interlayer ion, $\lambda_\rho = 0.156$. For KuniGel V1, $G_s = 2.733$ and $\sigma_{v\max} = 3.73$ GPa. In the constitutive formulation, the effective vertical stress of equation (6) is interpreted to be the effective mean stress. In order to consider the expansion characteristics of bentonite, the following decomposition of void ratio is also introduced:

$$e = e^* + e^{ad} \quad (7)$$

Herein, e^* is assumed to be a component that can be described by the usual elasto-plastic constitutive model for unsaturated and saturated soils. In this paper, the Exponential Contractancy (EC) model proposed by Ohno *et al.* (2006), which is a version of *Se* (effective degree of saturation) hardening models for unsaturated and saturated soils by Ohno *et al.* (2007), is used to describe the component e^*. In this EC model, contractancy characteristics during shearing of soils is expressed by an exponential function, and the change in void ratio due to isotropic compression is described as

$$e^* = e_0 - \lambda \ln \frac{p'}{p'_0} \text{ in NC state} \quad (8)$$

$$e^* = e_0 - \kappa \ln \frac{p'}{p'_0} \text{ in OC state} \quad (9)$$

The yielding function is then expressed as

$$f = \frac{\lambda - \kappa}{1 + e_0^*} \ln \frac{p'}{p'_c} + \frac{\lambda - \kappa}{1 + e_0^*} \frac{1}{n_E} \left(\frac{q}{p'M} \right)^{n_E} - \varepsilon_v^p = 0 \quad (10)$$

where ε_v^p is the plastic component of volumetric strain, NC is normally consolidation, OC is over consolidation and n_E is a fitting parameter to express the contractancy characteristics during shearing by the exponential curve.

In applying the above model to the concept of the full saturation line employed by Kobayashi *et al.* (2007), it is assumed that a line, tangential to the full saturation line of bentonite at a reference stress p'_{ref}, is the normally consolidated line that determines the change in void ratio e^*. Its gradient as a normally compressive index is expressed as

$$\lambda = -\frac{de}{d(\ln p')}\bigg|_{p'=p'_{\text{ref}}} = G_s\lambda_\rho \bigg/ \left(G_s - \lambda_\rho \ln \frac{p'_{\max}}{p'_{\text{ref}}} \right)^2 \quad (11)$$

On the other hand, the component e^{ad} (additional void ratio), newly introduced into equation (7) to take into account the expansion characteristics of

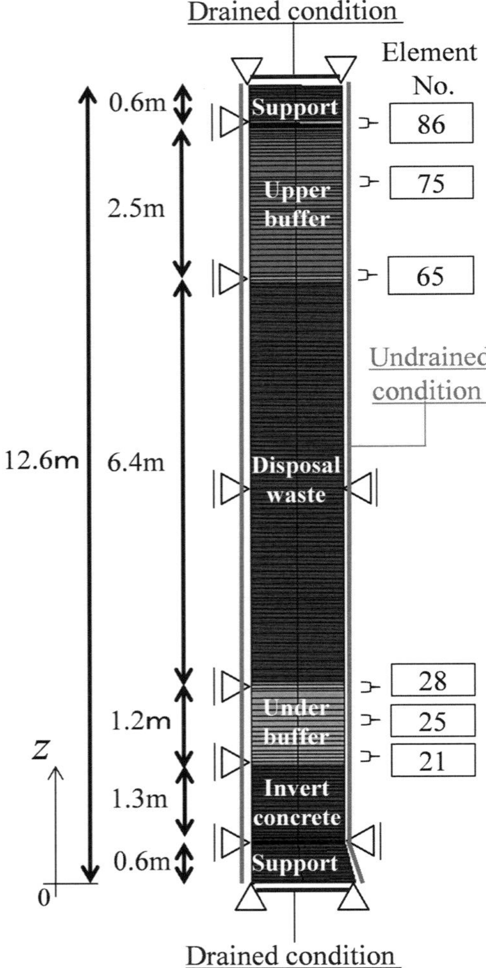

Fig. 15. Finite element model.

Table 2. *Input parameters of upper buffer material*

Gs	$\Lambda\rho$	$p'_{\max}$ (GPa)	p'_{ref} (GPa)	Λ	μ	ν	n
2.733	0.156	3.73	0.05	0.85	0.49	0.43	1
a	n_E	m	S_{ra}	A_w	B_w	A_d	B_d
100	1.3	0.023	0.1	−8	1.25	−12	1.58

bentonite, can be expressed from equations (6–8) as (see Fig. 14):

$$e^{ad} = \lambda_\rho \ln \frac{p'_{\max}}{p'} \Big/ \left(G_s - \lambda_\rho \ln \frac{p'_{\max}}{p'} \right) - e_0 + \lambda \ln \frac{p'}{p'_0} \quad (12)$$

When the transition from the unsaturated state to the saturated state is considered during simulation for the expansion of bentonite, it is necessary to introduce the water retention curve under the unsaturated state. In this study, the water retention characteristics of bentonite are assumed to be expressed by the following logistic curve, following the work of Sugii & Uno (1995) (see Kawai *et al.* (2002)):

$$S_r = S_{ra} + \left(S_{rf} - S_{ra}\right)/\left(1 + s^B \exp A\right) \quad (13)$$

where S_r is the degree of saturation, S_{ra} is the residual degree of saturation when $s \to \infty$, S_{rf} is the degree of saturation when $s = 0$, s is the suction, and A and B are fitting parameters.

The constitutive expression under the unsaturated state is summarized as follows, with the definition of effective stress given by

$$\boldsymbol{\sigma}' = \boldsymbol{\sigma}_{\text{net}} + p_s \mathbf{1} \quad (14)$$

where $\boldsymbol{\sigma}_{\text{net}}$ $(=\boldsymbol{\sigma} - p_a\mathbf{1})$, p_s $(=sS_e)$, $s = p_a - p_w$, $S_e = (S_r - S_{ra})/(1 - S_{ra})$, $\boldsymbol{\sigma}'$ is the effective stress tensor, $\boldsymbol{\sigma}_{\text{net}}$ is the net stress tensor, $\mathbf{1}$ is the second-order unit tensor, $\boldsymbol{\sigma}$ is the total stress tensor, s is the suction, p_s is the suction stress, p_a is the pore air pressure, p_w is the porewater pressure and S_e is the effective degree of saturation.

The hardening law of the elasto-plastic constitutive model used in the simulation is expressed as

$$p'_c = p'_{\text{sat0}} \zeta \exp \frac{\varepsilon_v^p}{(\lambda - \kappa)/(1 + e_0)} \quad (15)$$

$$\zeta = \exp\left[(1 - S_e)^n \ln a\right] \quad (16)$$

where n is fitting parameter to control the shape of the yield line. The evaluation expression to determine e^{ad} in equation (7) is assumed to be a function of the degree of saturation, consistent with equation (18):

$$e^{ad} = (S_e)^m e_{ad}^{\text{sat}} \quad (17)$$

$$e_{\text{sat}}^{ad} = \lambda_\rho \ln \frac{p'_{\max}}{p'} \Big/ \left(G_s - \lambda_\rho \ln \frac{p'_{\max}}{p'} \right) - e_0 + \lambda \ln \frac{p'}{p'_0} \quad (18)$$

where m is a fitting parameter.

The permeability coefficient of the bentonite material used in this work is estimated from the Kozeny–Carman relation, which is expressed as

$$k = k(e, S_v) = \frac{1}{c} \frac{\rho_w g}{\mu} \frac{1}{S_v{}^2} \frac{e^3}{1 + e} \quad (19)$$

where c is the shape coefficient of the soil particle ($c = 5$) and μ is the viscosity coefficient of porewater ($\mu = 0.001$ Pa s)). The specific surface area S_v is derived from equations (3) and (5).

Table 3. *Input parameters of material under buffer*

Gs	$\Lambda\rho$	$p'_{\max}$ (GPa)	p'_{ref} (GPa)	Λ	μ	ν	N
2.705	0.156	1.73	0.05	0.8	0.63	0.33	1
a	n_E	m	S_{ra}	A_w	B_w	A_d	B_d
100	1.3	0.23	0.1	−8	1.25	−12	1.58

Numerical simulation of the saturation process

Figure 15 indicates the geometric configuration employed in this paper, in which a TRU waste disposal facility is modelled one-dimensionally. The total water head up to the height of the top surface of the facility was imposed on the permeable boundary as the hydraulic condition to simulate the saturation process.

The upper buffer comprises compacted KuniGel V1. Under the buffer material is the compacted KuniGel V1/sand mixture material, with a KuniGel V1 content of 70% (dry mass). The montmorillonite content of the material under the buffer is therefore less than that of the upper buffer material, so it is considered that the position of the full saturation lines for the material under the buffer move to the right in comparison to those of the upper buffer material. The input parameters used in the simulation are summarized in Tables 2 and 3, where M is the critical state stress ratio, ν is Poisson's ratio for the elastic part of bentonite, Λ is the irreversibility ratio (defined as $\Lambda = 1 - \kappa/\lambda$), p'_{ref} is the reference mean stress defined in equation (11), a and n are fitting parameters defined in equation (15) and S_{ra}, A_w, B_w, A_d and B_d are parameters that specify the water retention curve defined in equation (13). The initial conditions used in the simulation are summarized in Table 4.

Figure 16 shows the development of the degree of saturation with water absorption over time. Figure 17 shows the change in void ratio with water absorption over time. In these figures, the behaviour of element nos 21, 25, 28, 65, 75 and 86 are shown. Element nos 86 and 21 are the elements located outside the facility. Element nos 75 and 25 are the elements located in the middle of the buffer material. Element nos 65 and 28 are the elements located inside the facility. According to numerical simulations, the bentonite buffer reaches its fully saturated state over quite a long period of time, as shown in Figure 16. Also, Figure 17 shows that the bentonite buffer close to the boundaries shows remarkable expansion in the early stage of the wetting process and then reduces its void ratio up to a fully saturated equilibrium state.

Table 4. *Initial conditions*

	Upper buffer	Under buffer
Initial dry density (Mg m^{-3})	1.3	1.6
Initial degree of saturation	0.29	0.29
Initial suction (kPa)	1960	1960

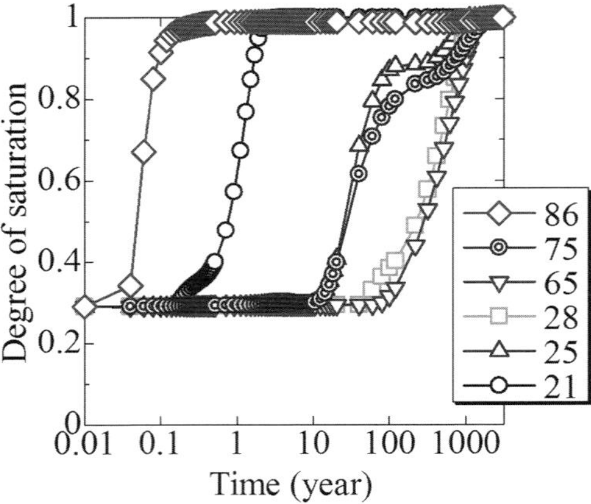

Fig. 16. Change of degree of saturation over time.

Concluding remarks

This paper discusses the influence of montmorillonite content and types of interlayer ions on mechanical and hydraulic properties. In terms of the mechanical property, the gradients of the full saturation lines of montmorillonite and bentonite were the same. The positions of full saturation lines for the bentonite moved to the right when montmorillonite content increased. The full saturation lines did not depend on the type of interlayer exchangeable cations, but on montmorillonite content. In terms of the hydraulic property, the variation in hydraulic conductivity due to differences in the materials, by up to tenfold times can be explained by change in the specific surface area only. We established the HM analytical method, considering

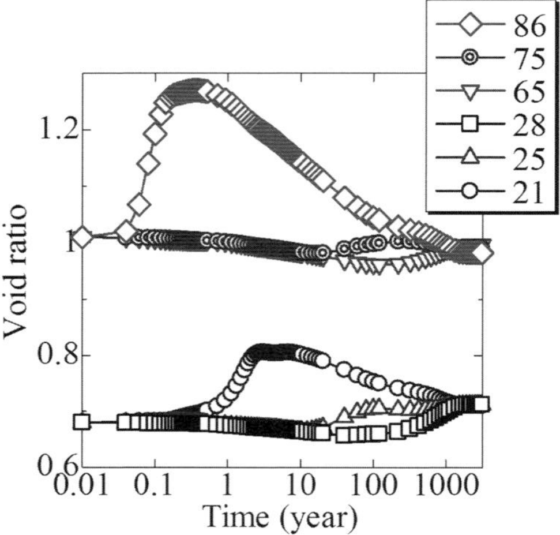

Fig. 17. Change of void ratio over time.

these properties. This mathematical modelling can describe the mechanical behaviour of bentonite from its initially unsaturated state, taking into account the above characteristics of bentonite. Using this mathematical model, numerical simulations for the saturation process from the initially unsaturated state can be demonstrated.

Beyond this, the authors intend to establish a thermal/hydraulic/mechanical/chemical (THMC) coupled analysis method that introduces these models in order to establish a long-term performance evaluation system for the bentonite-engineered barrier. Moreover, the establishment of a long-term performance evaluation system that is able to take into account the chemical reaction of a cementitious engineered barrier with a bentonite-engineered barrier is the final objective of this study.

This research is part of 'Development of Techniques for Evaluating Long-term Performance of EBS, FY2011' under a grant from the Agency of Natural Resources and Energy, the Ministry of Economy Trade and Industry of Japan (METI).

References

Åkesson, M., Börgesson, L. & Kristensson, O. 2010. *SR-Site Data Report, THM Modelling of Buffer, Backfill and Other System Components*. SKB Report **TR-10-44**, Swedish Nuclear Fuel and Waste Management Company.

Ichikawa, Y., Kawamura, K., Fujii, N. & Theramast, N. 2002. Molecular dynamics and multiscale homogenization analysis of seepage/diffusion problem in bentonite clay. *International Journal of Numerical Methods in Engineering*, **54**, 1717–1749.

Iizuka, A., Honda, M., Nishida, H., Kawai, K. & Karube, D. 2000. Soil/water coupled analysis considering unsaturated pore water distribution. *JSCE Journal of Geotechnical Engineering*, **659**, 165–178 (in Japanese).

Ishii, T., Yahagi, R. et al. 2013. Coupled chemical-hydraulic-mechanical modeling of long-term alteration of bentonite. *Journal of Fine Particle Science*, **48**, 331–341.

JAEA, BUFFER MATERIAL DATABASE, Japan Atomic Energy Agency. https://bufferdb.jaea.go.jp/bmdb/

Kawai, K., Wang, W. & Iizuka, A. 2002. The expression of hysteresis appearing on water characteristic curves and the change of stress in unsaturated soils. *JSCE Journal of Applied Mechanics*, **5**, 777–784 (in Japanese).

Kawai, K., Shibata, M., Kanazawa, S., Tachibana, S., Ohno, S., Iizuka, A. & Honda, M. 2009. Simulation of static compression with unsaturated soil water coupled FE code. *JSCE Journal of Applied Mechanics*, **12**, 429–436 (in Japanese).

Kawamura, K., Ichikawa, Y., Nakano, M., Kiayama, K. & Kawamura, H. 1999. Swelling properties of smectite up to 90°C: *in situ* X-ray diffraction experimental and molecular dynamics simulation. *Engineering Geology*, **54**, 75–59.

Kobayashi, I., Toida, M., Sasakura, T. & Ohta, H. 2007. Interpretation of swelling behavior of compacted bentonite based on elasto-plastic mechanics. *Journal of JSCE*, **63**, 1065–1078.

Kobayashi, I., Owada, H. & Ishii, T. 2011. Hydraulic/mechanical modeling of smectitic materials for HMC analytical evaluation of the long term performance of TRU geological repository. *Proceedings of the 14th International Conference on Environmental Remediation and Radioactive Waste Management*, ICEM2011-59090.

Morodome, S. 2009. *On Swelling Behavior of Montmorillonite With Various Exchangeable Cations and Structure of Montmorillonite–Water System – in situ Observation and Numerical Simulation of X-ray Diffraction and Small Angle X-ray Scattering Method.* **Doctoral thesis, TIT**.

Ohno, S., Iizuka, A. & Ohta, H. 2006. Elastoplastic model for soil by using the descriptive nonlinear function of contractancy. *JSCE Journal of Applied Mechanics*, **9**, 407–414 (in Japanese).

Ohno, S., Kawai, K. & Tachibana, S. 2007. Elasto-plastic constitutive model for unsaturated soil applied effective degree of saturation as a parameter expressing stiffness. *JSCE Journal of Geotechnical Engineering*, **63**, 1132–1141 (in Japanese).

Pusch, R. 1994. *Waste Disposal in Rock*. Developments in Geotechnical Engineering, **76**, Elsevier, The Netherlands.

Sudo, S. 1967. On the specific surface of soil and its measurement by BET methods. *Journal of the JSSP*, **16**, 39–42 (In Japanese).

Sugii, T. & Uno, T. 1995. A new modeling of moisture characteristics curve. *Proceedings of the Annual Conference of the JSCE*, **III**, 130–131 (in Japanese).

Tachibana, S., Takayama, Y., Iizuka, A., Kawai, K., Ohno, S. & Kobayashi, I. 2011. Elasto-plastic constitutive model for expansive soils with a concept of fully saturation curve. *Proceedings of the 5th Asia-Pacific Conference on Unsaturated Soils*, AP-UNSAT2011.

Reactive transport modelling of iron–bentonite interaction within the KBS-3H disposal concept: the Olkiluoto site as a case study

PAUL WERSIN[1]* & MARTIN BIRGERSSON[2]

[1]*Institute of Geological Sciences, University of Bern, Baltzerstrasse 1 + 3, 3012 Bern, Switzerland*

[2]*Clay Technology AB, Ideon, 223 70 Lund, Sweden*

**Corresponding author (e-mail: paul.wersin@geo.unibe.ch)*

Abstract: The interaction of steel with bentonite used as buffer material in high-level waste repositories may result in changes to the properties of the buffer. One of the repository designs (KBS-3H) developed by Posiva and SKB foresees the horizontal emplacement of so-called supercontainers, consisting of copper canisters surrounded by compacted bentonite and an outer perforated steel shell. The corrosion of the steel shell and the interaction of iron with the clay may impair the long-term safety functions of the buffer.

The corrosion and iron–clay interaction processes within the KBS-3H concept were assessed with a kinetically based reactive transport model and a comprehensive thermodynamic database. The large uncertainty related to precipitation rates of corrosion products and iron silicates was considered by defining a series of test cases. The results generally indicate a limited effect on the stability of montmorillonite, thus affecting only a few centimetres next to the iron source. Upon complete corrosion only insignificant changes are predicted. These results are explained by (i) the diffusional constraint of mass transfer, (ii) low solubility of corrosion products and (iii) slow transformation kinetics of montmorillonite. Model results further suggest that the largest impact arises from 'indirect' processes, such as microbial sulphate reduction, which may lead to a strong increase in pH.

In most designs of geological repositories for high-level and spent-fuel radioactive waste, emplaced swelling-clay materials constitute an important safety barrier. Thus, in the Swedish and Finnish repository concept, termed KBS-3V, compacted bentonite surrounding the copper canister is foreseen to be emplaced in vertical deposition holes (SKB 2011). As an alternative, a horizontal emplacement concept, termed KBS-3H, is envisioned (SKB/Posiva 2008), where modules of canister-bentonite are surrounded by a perforated metal shell (termed the 'supercontainer') and emplaced in deposition drifts (Fig. 1). Carbon steel is regarded as option for this shell material.

Steel is unstable in the repository environment and will corrode anaerobically producing hydrogen gas and oxidized iron species once the residual oxygen is depleted by redox processes (Wersin *et al.* 2003*a*). Upon repository closure, the bentonite will be saturated by ingress of the groundwater from the crystalline host rock and extrude through the porous steel shell by swelling. Saturation times will depend on groundwater flow rates and will also be affected by the decay heat of the waste contained in the canister and the consumption of water via the corrosion reaction; on the basis of thermo-hydraulic modelling they are expected to be in the range of tens to hundreds of years (Gribi *et al.* 2007). Corrosion of the steel shell will already have been initiated during saturation. Initially, molecular oxygen from the residual air will react with the metal and Fe(III) (oxyhydr)oxides will be formed. However, oxygen depletion will occur fairly rapidly, within a few years or less, due to the corrosion of the steel shell and other redox reactions in the near-field (Wersin *et al.* 2003*b*). Subsequently, anaerobic corrosion of the steel shell, mostly under saturated and close-to-ambient temperature conditions (with maximum temperatures of about 50 °C), will take place generating hydrogen gas and Fe(II):

$$Fe(0) + 2H_2O \rightarrow Fe^{2+} + H_2 + 2OH^- \quad (1)$$

Therefrom, magnetite may be formed, but also other corrosion products, such as green rust (GR) phases (which may also be precursors to magnetite), and, depending on solution conditions, siderite or iron sulphides may precipitate (Gribi *et al.* 2007; Wersin *et al.* 2007). Reduced iron also interacts with the bentonite clay (predominantly montmorillonite) thus further complicating the corrosion process. The details of this iron–clay interaction process are still only poorly understood although progress

From: Norris, S., Bruno, J., Cathelineau, M., Delage, P., Fairhurst, C., Gaucher, E. C., Höhn, E. H., Kalinichev, A., Lalieux, P. & Sellin, P. (eds) 2014. *Clays in Natural and Engineered Barriers for Radioactive Waste Confinement*. Geological Society, London, Special Publications, **400**, 237–250.
First published online May 1, 2014, http://dx.doi.org/10.1144/SP400.24

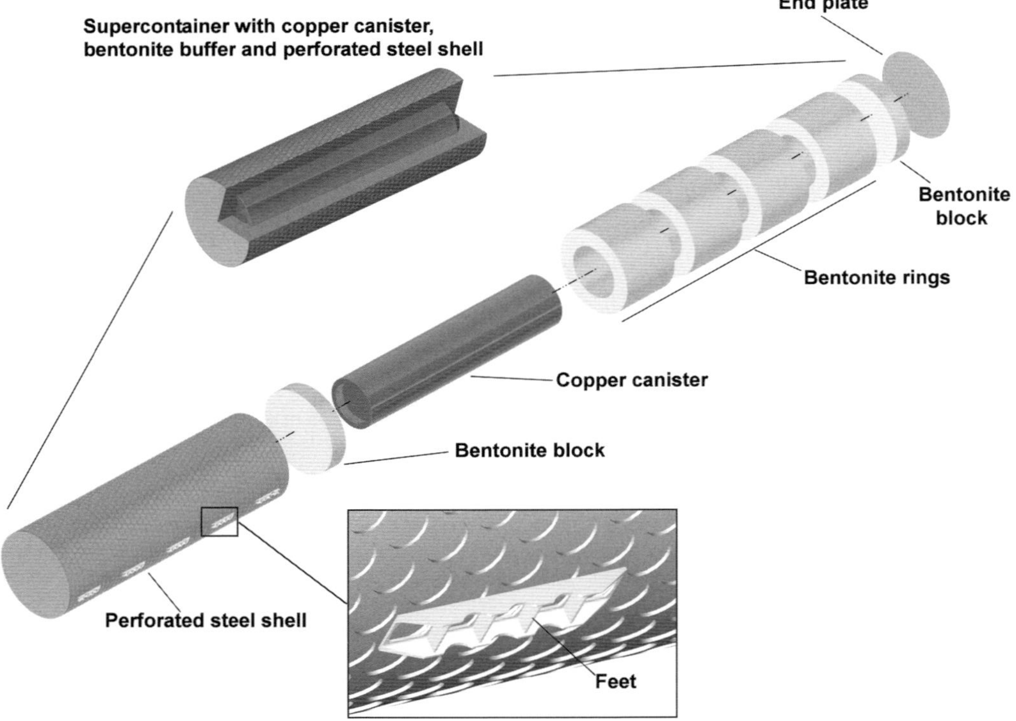

Fig. 1. Supercontainer shell design in the KBS-3H concept (*SKB/Posiva 2008*).

has been made in recent years. In principle, sorption of Fe(II) to the clay (Géhin *et al.* 2007), reduction of structural Fe(III) (Lantenois *et al.* 2005) in the clay and/or neoformation of Fe phyllosilicates (e.g. berthierine, cronstedtite) have been observed in different experimental studies (Lantenois 2003; Wilson *et al.* 2006; Kumpulainen *et al.* 2010; Mosser-Ruck *et al.* 2010; Lanson *et al.* 2012). Furthermore, the increase in pH induced by the corrosion reaction (equation 1) may enhance dissolution of montmorillonite and neoformation of iron silicate phases (Marty *et al.* 2010).

The interaction of the steel shell with the clay may impair the safety function of the bentonite buffer (by reducing the swelling pressure), especially in the case of montmorillonite dissolution and transformation to a non-swelling phyllosilicate. From a mass balance perspective, the amount of iron in the 8 mm thick steel shell is sufficient to transform a maximum of 30% of montmorillonite in the buffer to a non-swelling iron phyllosilicate (Wersin *et al.* 2006) which would result in significant reduction of the swelling pressure in the buffer. However, the iron–clay interaction process will be limited by mass transfer and kinetic constraints. The scope of this study is to assess the impact of iron shell corrosion on the bentonite buffer, in the context of the KBS-3H disposal concept, by a reactive transport model. The Olkiluoto site in Finland is taken as a case study. The model builds on a previous preliminary modelling exercise (Wersin *et al.* 2006, 2007). The focus is to consider recent thermodynamic data of clays and other potentially forming silicate minerals, and advancements in experimental (e.g. Kumpulainen *et al.* 2010) and modelling (Bildstein *et al.* 2006; Hunter *et al.* 2007; Marty *et al.* 2010; Savage *et al.* 2010) efforts. The simulations consider both the initial corrosion stage (expected to last for about 5000 years, see 'Results and discussion' below) and the subsequent stage of up to 50 000 years after repository closure.

Model description

Representation of geochemical conditions in and around the repository

The geochemical conditions of the crystalline host rock around the repository at Olkiluoto, foreseen at about 400 m depth, are described in Pitkänen *et al.* (2004). Groundwater flow is variable, in the range of <0.01–10 ml min^{-1} per deposition drift (Hellä *et al.* 2006). The groundwater composition is of saline NaCl-type, but is predicted to evolve towards more dilute waters as a result of land uplift at the site. The differences in groundwater composition over time at repository conditions have been represented by 'reference' and 'limiting' waters, comprising 'saline', 'limiting saline' and 'limiting dilute' waters (e.g. Wersin *et al.* 2007). In general, deep groundwaters at Olkiluoto display conditions of microbial sulphate reduction and may contain measurable levels of dissolved sulphide up to *c.* 1 mg l^{-1}. For the purpose of this study, we consider a modified dilute water type throughout the modelled time span, which is the most unfavourable in terms of Fe fluxes from the iron shell to the buffer, as is indicated from previous modelling (Wersin *et al.* 2006, 2007).

The emplaced buffer is represented by a simplified composition of MX-80 bentonite including Na-montmorillonite, Ca-montmorillonite, quartz, and small amounts of gypsum and calcite (Table A1, SI). The buffer's dry density is 1450 kg m^{-3} (saturated density 1920 kg m^{-3}) leading to a porosity of 47.6%. The selected density is slightly lower than the target saturated density of 2000 kg m^{-3} and accounts for the possible density decrease by the initial gap between the steel shell and the drift wall. As part of initial conditions in the modelling, the bentonite is assumed to be water-saturated by equilibration with the surrounding groundwater before corrosion is initiated. The assumption of initially saturated conditions rather than unsaturated ones is conservative with regard to the migration of corrosion-derived iron into the buffer. For the derivation of the initial porewater composition, a simple thermodynamic bentonite model, as described in Wersin (2003), was applied, but with some modifications. These included a lowering of the CO_2 partial pressure (pCO_2) and sustaining calcite undersaturation in order to increase pH effects induced by corrosion. Also, sulphate concentrations were lowered in the porewater because scoping calculations showed substantial gypsum precipitation and clogging at the buffer/rock interface. However, the effect of both higher pCO_2 and sulphate were evaluated in separate model runs (see 'Results and discussion'). Moreover, close-to-zero concentrations for Fe(II) and Si in the porewater were conservatively assumed in order to enhance the effect of montmorillonite dissolution. The resulting initial porewater composition is shown in Table A2. Because of the generally acknowledged limitation of microbial activity in compacted bentonite (e.g. Stroes-Gascoyne *et al.* 2010), no sulphate reduction or methane production reactions were considered, which was achieved by suppressing these redox reactions in the model. But the possibility of microbial sulphate reduction at the bentonite/host rock boundary was assessed in separate test cases (see below 'Results and discussion').

For the anaerobic corrosion of the 8 mm thick perforated steel shell, a uniform corrosion rate of 1 μm a^{-1} was assumed. This value is based on the long-term corrosion study of Smart *et al.* (2004) which indicated a strong decrease in corrosion rates with time due to the formation of a protective iron oxide layer. These rates, however, are higher than those obtained in water (Smart *et al.* 2001) which was explained by slower build-up of the protective oxide film because of the incorporation of the iron in the clay. The application of a constant corrosion rate of 1 μm a^{-1} leads to the complete corrosion of the iron source after approximately 5000 years.

Selection of thermodynamic data

The thermodynamic data used is based on the THERMODDEM database (Blanc *et al.* 2007; www.thermoddem.fr, release 2009-09-04) which is designed for waste management applications, in particular for clay and cement barriers, and includes a large range of clay, zeolite and other silicate data.

Before implementing this database in the reactive transport model a screening procedure was adopted in order to decrease the number of phases and to check the geochemical consistency of the dataset. In a first step, the high-temperature phases were screened out. Then scoping calculations under 'static' conditions using the PHREEQC code (Parkhurst & Appelo 1999) were performed to check for the stability of the different clay phases in the porewaters adjacent to a corroding iron source (modelled as zero-order reactions). This revealed an obvious inconsistency between the 2:1 clay mineral log K data which originated from different sources. For example, the implemented saponite and vermiculite minerals showed considerable oversaturation relative to montmorillonite under all tested conditions. Because of the rarely observed direct transformation of dioctahedral smectite to these trioctahedral smectites and also because the focus of the modelling exercise was to assess transformation to non-swelling phases, saponite and vermiculite data were excluded. The Fe(III)-rich phyllosilicates, such as Fe(III)-montmorillonite and nontronite, were excluded from the database because of their strong undersaturation under reducing conditions. Furthermore, the rare minerals celadonite (showing strong oversaturation) as well minnesotaite and lizardite were excluded. Regarding zeolites, the most common low-temperature minerals philippsite, analcime, chabazite and heulandite (Gaucher & Blanc 2006) were retained. At elevated pH, which may result from the iron corrosion process (see equation 1), smectites may transform into zeolites, although this reaction is commonly observed at higher temperatures (e.g. Mosser-Ruck & Cathelineau 2004). For the same reason, Calcium Silicate Hydrate (CSH) aluminate phases were considered; hence the commonly described minerals tobermorite, gyrolite and hydrotalcite were included.

Concerning potentially forming iron corrosion products, all phases included in THERMODDEM were retained, except for hematite in view of its large activation energy at low temperature. In addition, green rust (GR) phases which have been shown to form in natural environments (Génin *et al.* 2001) and as iron corrosion products (Castermant *et al.* 2008) were considered. Equilibrium calculations were performed to test the stability of mixed hydroxo-GR phases (fougerite) from the thermodynamic data of Bourrié *et al.* (1999). These indicated that the $Fe(II)_2Fe(III)(OH)_7$ was the most stable GR in Olkiluoto-type waters and therefore was added to the thermodynamic data. The calculations also revealed unrealistically low solubilities of berthierine with Fe(II) $<10^{-25}$ mol l^{-1}. Therefore, the solubility constant was adapted to yield Fe(II) concentrations of *c.* 10^{-6} mol l^{-1}. For one test case, however, the effect of low solubility berthierine was evaluated (see 'Definition of base case and alternative cases'). The considered final mineral database is shown in Table A3.

No explicit ion exchange model was included. The reasons for this omission are the difficulty of a consistent thermodynamic description of both ion exchange and solubility for montmorillonite (e.g. Kittrick 1979) and the prevention of the 'double-counting' of exchangeable cations when montmorillonite dissolves.

Model set-up and kinetic description

The rather complicated model geometry was represented in the simplest possible manner considering a 1D linear model with an iron source embedded in a bentonite column (Fig. 2). This geometry approximately describes the mid-section of the KBS-3H supercontainer (Fig. 1). The model column is divided into 2 cm wide cells and the iron source is represented by one such cell which initially contains only iron and water. The amount of iron was adjusted to give an iron/bentonite mass ratio which agrees with the KBS-3H design. The total column length is 42 cm.

The code CrunchFlow, a software package for multicomponent transport in porous media (Steefel 2006), was applied for implementing the model. A specific feature of the model is the kinetic formulation of mineral reactions based on the transition state theory (Lasaga 1989). It allows for parallel kinetic rate laws for any mineral which are summed up to a give an overall rate law for a given mineral. In this model, the dissolution/precipitation rate (mol per m^3 porous media and per second) for a given mineral is typically expressed as:

$$\text{Rate} = \frac{\mathrm{d}m}{\mathrm{d}t} = S \times k_{T_0} \times \left(1 - \frac{Q}{K}\right) \quad (2)$$

where S is the reactive mineral surface area (m^2 m^{-3} porous media), k_{T_0} a kinetic rate parameter (mol m^{-2} s^{-1}) evaluated at $T_0 = 25$ °C (the assumed temperature at all times in the model), Q the ion activity product for the mineral reaction and K its equilibrium constant.

Kinetic rates and reactive surface areas are generally rather sparse and medium-specific. This particularly holds for precipitation reactions where far fewer experimental data are available than for dissolution reactions. This uncertainty was addressed by treating the product of k_{T_0} and S as adjustable parameter for the relevant iron mineral precipitation reactions, as detailed in 'Results and discussion'. For minerals with rather well-established rate data, these were taken from the compilation of Palandri & Kharaka (2004) or from the CrunchFlow database (www.csteefel.com/CrunchPublic/DownloadCrunchflow.html). For silicates with less well-established data, notably for those phases which might form upon montmorillonite dissolution (smectites, zeolites, CSH phases) a constant rather high k_{T_0} of 10^{-11} mol m^{-2} s^{-1} was assumed together with roughly estimated surface areas. This

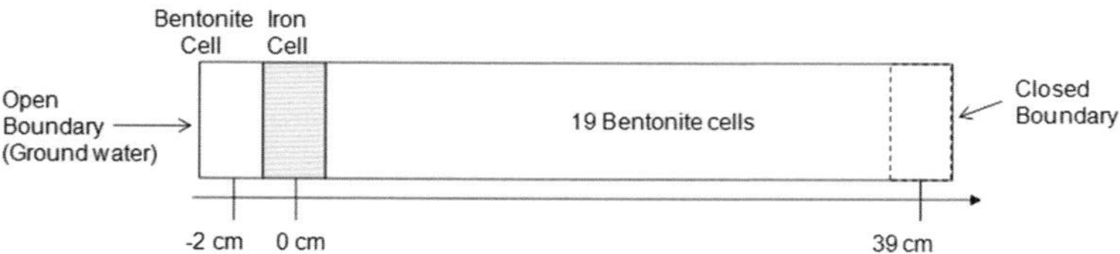

Fig. 2. Schematic view of model set-up.

somewhat subjective selection was done in order to not underestimate montmorillonite transformation. The listing of the rate parameters for all considered minerals is given in Table A3.

The corrosion reaction (assumed to be 1 μm a^{-1}) was modelled as first-order rate expression, since completely constant rates are difficult to implement in CrunchFlow, with:

$$\text{Rate} = \frac{dn}{dt} = S \times k_{T_0,\text{Iron}} = s_{\text{Iron}} \times M_{\text{Iron}} \times n(t) \times k_{T_0,\text{Iron}} \quad (3)$$

where $n(t)$ is the amount of iron (mole) at time t. The selected parameters are $s_{\text{Iron}} = 1.68 \times 10^{-4}$ m^2 g^{-1} and $k_{T_0,\text{Iron}} = 10^{-8.35}$ mol m^{-2} s^{-1}, which together with the iron molar mass $M_{\text{Iron}} = 55.845$ g mol^{-1} gives the loss of iron as

$$n(t) = n(0) \times e^{-t/(756\,\text{years})} \quad (4)$$

From this expression it is seen that approximately only 0.1% of the initial iron is present after 5000 years.

Transport through compacted bentonite is dominated by diffusion which according to Fick's first law:

$$j_{\text{diff}} = \phi^m \times D_0 \times \nabla c \quad (5)$$

where j_{diff} is the diffusive flux, ϕ denotes the porosity, D_0 the pore diffusion coefficient, c the porewater concentration of the considered species and m the so-called cementation factor, which allows for a formation factor of the form of Archie's law. In CrunchFlow, updating the porosity as a result of mineral precipitation or dissolution is implemented by coupling porosity to the effective diffusivity as:

$$D_e = D_0 \times \phi^m \quad (6)$$

Here a cementation factor of one was assumed throughout.

A fixed D_0 of 4.5×10^{-11} m^2 s^{-1} was selected for all species assumed to diffuse in the full (physical) porosity. This is a strong simplification of the diffusive process in compacted clays with permanent negative-layer charge which are affected by multicomponent diffusion and diffuse double-layer effects (Bourg *et al.* 2006; Appelo & Wersin 2007). Transport in compacted bentonite was shown to be strongly influenced by diffusion in the interlayers (Glaus *et al.* 2007, 2010), but there is controversy concerning both bentonite microstructure and the conceptualization of its porosity in terms of accessibility to diffusion (Birgersson & Karnland 2009; Tournassat & Appelo 2011). In view of the incomplete understanding of the diffusion process and the experimentally demonstrated importance of interlayer diffusion, we deem our simple model assumptions justified although we acknowledge the significant conceptual uncertainties involved. The selected diffusion coefficient is numerically equivalent to the so-called 'apparent' diffusivity deduced by Birgersson & Karnland (2009) from tracer diffusion tests in compacted bentonite.

Definition of base case and alternative cases

The initial geochemical conditions have been described above. In the base case (BC), all selected iron minerals (Table A1) were allowed to form except for the iron sulphides (see 'Model description') and chlorites. The latter phases generally form at higher temperatures.

The effect of the precipitation kinetics of the corrosion products (Table 1) was studied in cases A1–A3 by lowering the rate constants of the different iron (oxyhydr)oxide minerals (fougerite, magnetite, goethite, $Fe(OH)_2$).

In case B1, the effect of microbial sulphate reduction and formation of FeS amorphous (FeS(am)) and pyrite was tested.

Chlorites as potential transformation products were included in case C1 (Table 1). The influence of the highly uncertain thermodynamic data on berthierine (see above) was tested in case E1.

Cases C1–C3 explored the different geochemical conditions of the surrounding groundwater. Specifically, higher values of sulphate and variable CO_2 partial pressure conditions were evaluated.

Results and discussion

Base case

The calculations were performed for a timescale of 50 000 years. The instant conversion of Ca-montmorillonite to Na-montmorillonite was observed in the entire buffer. This is a priori not expected from the groundwater composition, it rather suggests some inconsistency in the thermodynamic montmorillonite logK data. In fact, these data would suggest the solubility of Na/Ca/Mg-montmorillonite to be a strong function of the variation in charge-compensating cations, which is not supported by experimental or natural observations. For the purpose of this study, however, this 'internal' montmorillonite conversion process has virtually no effect on the iron–bentonite interaction.

With regard to the corrosion process, the major part of the released Fe precipitates in the iron cell

Table 1. *Rate constants log (k_{T_0}) (mol m^{-2} s^{-1}) of iron corrosion products for the different test cases (for siderite, a value of −8.9 was used for all cases)*

Case	$Fe_3(OH)_7$	Magnetite	Goethite	$Fe(OH)_2$	Pyrite	FeS(am)	Berthierine	Chlorites*
Base	−11	−10.78	−7.94	−7.9			−8	
A1	−31	−10.78	−7.94	−7.9			−8	
A2	−31	−30.78	−7.94	−7.9			−8	
A3	−31	−30.78	−27.94	−7.9			−8	
B1	−11	−10.78	−7.94	−7.9	−5.0	−5.0	−8	
C1	−11	−10.78	−7.94	−7.9			−8	−11
D1	−11	−10.78	−7.94	−7.9			−8	
D2	−11	−10.78	−7.94	−7.9			−8	
D3	−11	−10.78	−7.94	−7.9			−8	
E1†	−11	−10.78	−7.94	−7.9			−8	

* Chamosite, Clinochlore, "Al_Free_Chlorite", "Chlorite_Cca-2" selected from the THEMODDEM database.
†The extremely low solubility constant for berthierine given in THERMODDEM was used for this case (see text).

as fougerite (mixed Fe(II)/Fe(III) hydroxide) which is the most insoluble corrosion product under the conditions of interest:

$$3Fe + 7H_2O \rightarrow Fe_3(OH)_7 + 3.5H_2 \quad (7)$$

This reaction preserves pH which is also represented in the pH evolution of the iron cell with a pH of about 8; pH buffering is also effective in the rest of the bentonite column, but a slight pH increase from about 8 to 8.5 is noted. This indicates that some of the OH^- ions produced by the corrosion reaction (equation 1) diffuse into the buffer. The corrosion produces rather large amounts of H_2 (equation 7) leading to maximum hydrogen pressures of 40 bar similar to the expected swelling pressures (data not shown). In reality, mechanical interaction with clay may be due to this gas pressure which is not included in the model.

Initially, there is a strong redox gradient between the iron source and the clay which is assumed to be oxic before the interaction process (Table A2). Reducing conditions are rapidly established, thus after 30 years almost constant Eh conditions of about -200 mV_{SHE} (where SHE is Standard Hydrogen Electrode) throughout the clay column are simulated (Fig. A1). In the long run, after the depletion of the iron source, conditions become less reducing but stable (*c.* -140 mV_{SHE}) and are controlled by fougerite equilibrium.

The mineral distribution in the column after 5000 years, that is, at the end of the corrosion stage, is depicted in Figure 3. It illustrates that almost no montmorillonite is transformed under conditions of the BC and indicates the generally low mineral conversion rates apart from the corrosion reaction (Fig. 3a). Small amounts of gibbsite and zeolite phases (chabazite, Na-phillipsite) are formed (Fig. 3b). Some calcite at the expense of gypsum is precipitated in the iron cell, but further out the calcite remains fairly constant with time. After the corrosion stage very little further reaction occurs. Thus after 50 000 years, the mineral distribution in the bentonite column is very similar to that after 5000 years.

Precipitation kinetics and stability of iron corrosion products

Suppression of GR formation leads to formation of magnetite (case A1). The overall changes relative to the BC are small, the main differences are the slightly higher pH, Fe(II) and H_2 concentrations (data not shown). Table 2 illustrates the spatial extent of montmorillonite dissolution and nature of transformation for the different cases.

If magnetite precipitation is also suppressed, goethite is formed (case A2). This results in a further increase in pH (up to 9.5), and Fe(II) in the iron cell because of the higher solubility of this Fe(III) phase at the low redox potentials generated. The higher pH induces more montmorillonite dissolution and zeolite precipitation. Nevertheless, the amount of montmorillonite alteration and its spatial extent remain limited (Table 2).

In the case that precipitation of all iron (oxyhydr)oxides except for $Fe(OH)_2$ is suppressed (case A3) the pH and Fe(II) concentrations increase even more, leading to considerable montmorillonite alteration and neoformation of zeolites. Also, considerably more siderite than in the previous cases precipitates in the iron cell. This case is unrealistic because $Fe(OH)_2$ will transform into more insoluble iron (oxyhydr)oxides, but it is presented here for illustrative purposes.

If an extended set of chlorite phases from the THERMODDEM database are included in the model (case D1), the changes relative to the BC

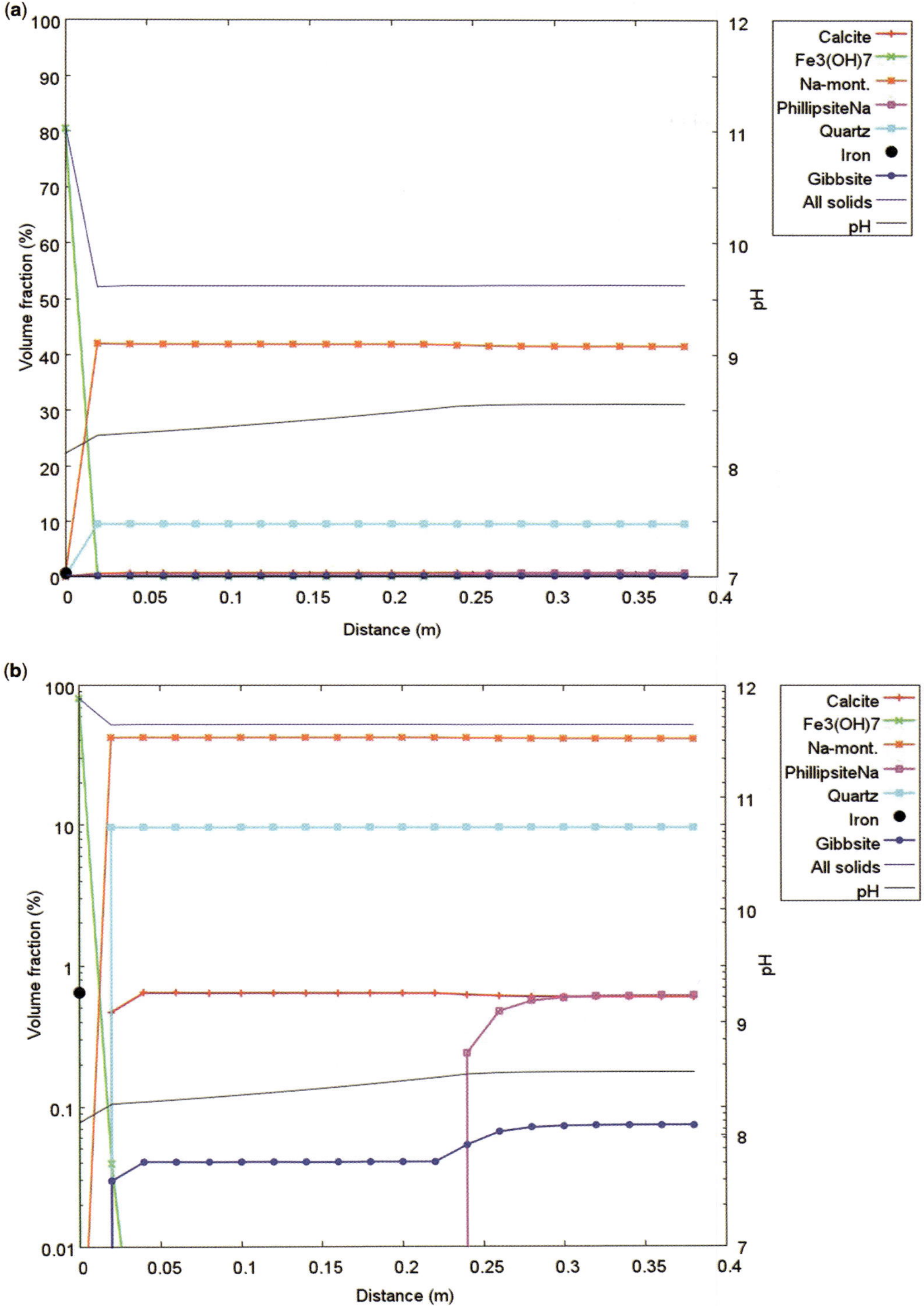

Fig. 3. Distribution of minerals and pH after 5000 years in the base case: (**a**) linear scale; (**b**) logarithmic scale.

Table 2. *Selected modelling results after 5000 years*

Case	Main Fe mineral formed		Second Fe mineral formed		Other minerals formed	Distance alteration (cm)
	Name	% of Fe_{tot}	Name	% of Fe_{tot}	Mineral names	(mont. loss >5%)
Base	Fougerite	100			Calcite, zeolites	0–2
A1	Magnetite	100			Calcite, zeolites	0–2
A2	Goethite	>99	Siderite	<1	Calcite, zeolites	2–4
A3	$Fe(OH)_2$	>99	Siderite	<1	Calcite, zeolites	4–6
B1	Fougerite	86	Pyrite	14	Calcite, zeolites, hydrotalcite	6–8
C1	Fougerite	>99	Chamosite	<1	Calcite, zeolites, Chlorites (minor)	0–2
D1	Fougerite	100			Zeolites, gypsum	
D2	Fougerite	100			Zeolites	2–4
D3	Fougerite	70	Siderite	30	Calcite	0–2
E1	Fougerite	>99	Berthierine	<1	Calcite	0–2

are small. A small amount of (<0.1 vol%) of chamosite is formed in the bentonite adjacent to the iron source. The small amount of chlorite formed is mainly the consequence of the slow kinetics of montmorillonite dissolution.

The berthierine logK data are highly uncertain as pointed out previously. Implementing the original logK (case E1) from the THERMODDEM database results in unrealistically low Fe(II) values, and a minor contribution of this phase to the corroded products. The transformation is a direct consequence of the iron corrosion process and almost stops after exhaustion of the iron source.

Influence of CO_2 and sulphate input conditions

To assess the conditions favourable for siderite precipitation, pCO_2 conditions at the source were varied. Increase in pCO_2 from $10^{-2.71}$ to 10^{-2} bar (case D3) induces calcite and some siderite precipitation next to the iron cell in spite of slight lowering of pH. The increased flux of dissolved inorganic carbon flux thus outweighs the pH effect. In contrast, lowering pCO_2 leads to pH increase and some calcite dissolution and slightly increased montmorillonite dissolution relative to the BC (Table 2, case D2).

Increase in sulphate concentration at the source leads to substantial gypsum precipitation and clogging in the initial cell (case D1). This effect is triggered by the release of Ca^{2+} from the exchanger resulting from Ca-montmorillonite–to–Na-montmorillonite conversion. As outlined above, this conversion process is probably the result of the inconsistent thermodynamic montmorillonite data and hence such a clogging effect is not expected to occur in reality.

Effect of microbial sulphate reduction

The impact of microbially mediated sulphate reduction on the iron–clay interaction process was tested. No kinetic constraint was imposed and sulphate reduction was assumed to occur instantaneously according to thermodynamic constraints. Sulphide was assumed to be controlled either by FeS(am) or pyrite (cases B1 and B2). The results for both cases are similar, indicating a significant dissolution and transformation of montmorillonite compared to the BC. This is primarily caused by the strong pH increase (up to 11.5) at the corroding source (Fig. 4a):

$$\frac{31}{7}Fe + 4H_2O + 4H^+ + 2SO_4^{2-} \longrightarrow FeS_2 + \frac{12}{7}Fe_3(OH)_7 \quad (8)$$

Note that this reaction occurs without hydrogen production and accordingly, no hydrogen pressure build-up is predicted by the model. The high pH destabilizes the clay, and montmorillonite transforms rapidly to zeolite minerals and hydrotalcite, but no iron silicate is formed. The zone of montmorillonite alteration is predicted to reach about 7 cm (Fig. 4b).

This test case represents the most extreme scenario in terms of montmorillonite alteration which is entirely induced by the indirect effect of pH increase from the sulphate reduction process. It highlights the importance of pH buffering reactions in the clay which are only partly accounted for in the modelling. Thus, proton exchange reactions occurring at the clay surface are neglected. Also, calcite undersaturation in the source fluid is imposed (see 'Model description').

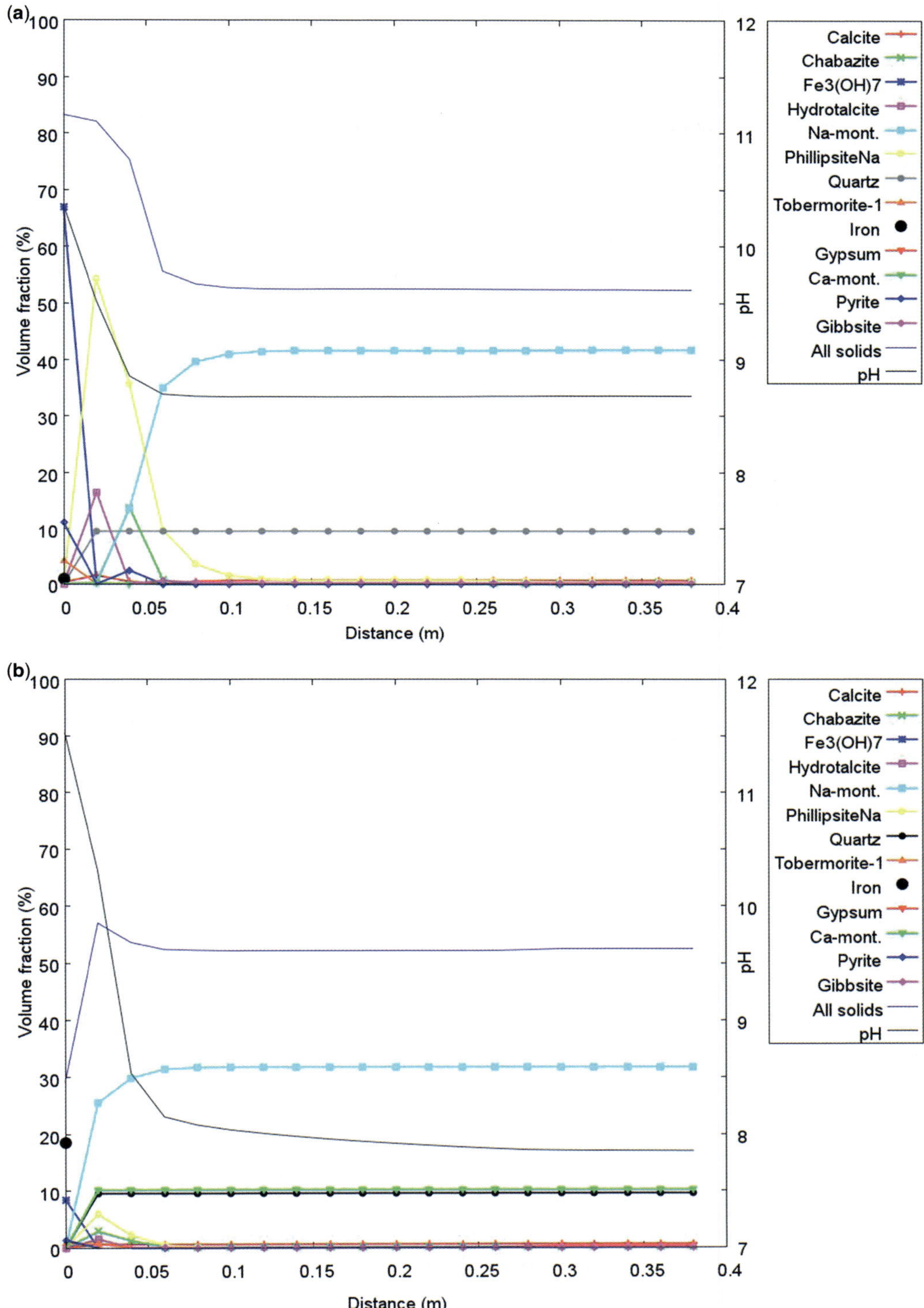

Fig. 4. Distribution of minerals and pH for sulphate reduction case B1 after (**a**) 100 years, and (**b**) 5000 years.

Uncertainties

There are a number of both conceptual and data uncertainties inherent in the modelling approach. The main conceptual uncertainties concern the lack of mechanistic understanding of the iron–clay interaction and the description of the diffusion process itself. This has been addressed by the generally conservative assumptions with regard to maximizing the effect of montmorillonite alteration.

The thermodynamic data for the clay minerals constitute a major data uncertainty which has also been underpinned in this modelling exercise in identifying inconsistencies in logK data. In the thermodynamic database applied for this modelling study only a restricted dataset of potential iron silicate alteration products was available. Moreover, as outline above, the logK data for berthierine are doubtable. Thus, the predicted very limited formation of iron silicates, in particular of 1:1 clays, bears considerable uncertainty. From a qualitative viewpoint, however, it should be born in mind that the kinetics of smectite transformation are slow and that for the considered KBS-3H system the depletion of the Fe source is rather rapid (*c.* 5000 years).

Another important uncertainty relates to the biogeochemical conditions at the buffer/rock interface. This particularly concerns the relevance of microbially induced sulphate reduction in this zone, possibly enhanced by corrosion-derived hydrogen acting as an electron donor (Pedersen 2008). This redox process may lead to high pH conditions, but as stated above, not all pH buffering reactions have been considered in the model. Moreover, mineral reaction kinetics constitute a general uncertainty which have been evaluated by the selection of the different test cases and by exploring a wide parameter space.

Overall, in spite of all uncertainties inherent in the modelling exercise, the applied model is considered to be robust with regard to the estimated loss of smectite and the spatial extent of the affected buffer material.

Conclusion

A kinetically based reactive transport modelling exercise has been conducted to assess the impact of the anaerobic corrosion of the steel shell on the bentonite buffer within the Swedish and Finnish KBS-3H disposal concept. The results generally indicate a limited effect on the stability of montmorillonite, thus affecting only a few centimetres next to the iron source. Upon complete corrosion only insignificant changes are predicted. These results, which are qualitatively in line with previous modelling (Bildstein *et al.* 2006; Wersin *et al.* 2007; Savage *et al.* 2010) and experimental studies (Kumpulainen *et al.* 2010), are explained by (i) the diffusional constraint of mass transfer, (ii) low solubility of corrosion products and (iii) slow transformation kinetics of montmorillonite. Model results further suggest that the largest impact arises from 'indirect' processes, such as microbial sulphate reduction processes, which may lead to a strong increase in pH.

In the applied thermodynamic database only a restricted dataset for potential iron silicate alteration products was available. The predicted very limited formation of iron silicates therefore bears considerable uncertainty.

In general, the model results have highlighted the need for characterizing biogeochemical conditions in the near-field of KBS-3H-type repositories. Furthermore, they have shown the necessity for consistent thermodynamic data of clay minerals (in particular smectites, and 1:1 Fe-rich clays).

We express our thanks to Margit Snellman for her support and fruitful discussions. Daniel Svensson is acknowledged for helpful comments. Financial support by Posiva and SKB within the KBS-3H programme is acknowledged.

Appendix

Table A1. *Simplified composition of MX-80 bentonite used for initial conditions in model*

Mineral	Volume (%)
Na-montmorillonite	30
Ca-montmorillonite	12
Quartz	9.6
Calcite	0.6
Gypsum	0.2
(Porosity)	47.6

Table A2. *Initial and boundary water composition, T = 25 °C (see text)*

	Groundwater/ Initial bentonite pore water	Comments
pH	7.391	
Eh	750 mV	
pCO_2	$10^{-3.5}$ bar	Undersaturation with calcite assumed
DIC	$2.05 \cdot 10^{-4}$ M	
Al^{3+}	10^{-12} M	
Mg^{2+}	$1.65 \cdot 10^{-3}$ M	
Ca^{2+}	$1.045 \cdot 10^{-2}$ M	
Fe^{2+}	10^{-15} M	Conservatively assumed to be low
Na^{+}	$1.668 \cdot 10^{-1}$ M	
K^{+}	$5.3 \cdot 10^{-4}$ M	
Cl^{-}	$1.765 \cdot 10^{-1}$ M	Obtained by charge balance
SO_4^{2-}	$7.42 \cdot 10^{-3}$ M	Decreased by a factor of 10
$SiO_2(aq)$	10^{-15} M	Conservatively assumed to be low

DIC: total inorganic carbon

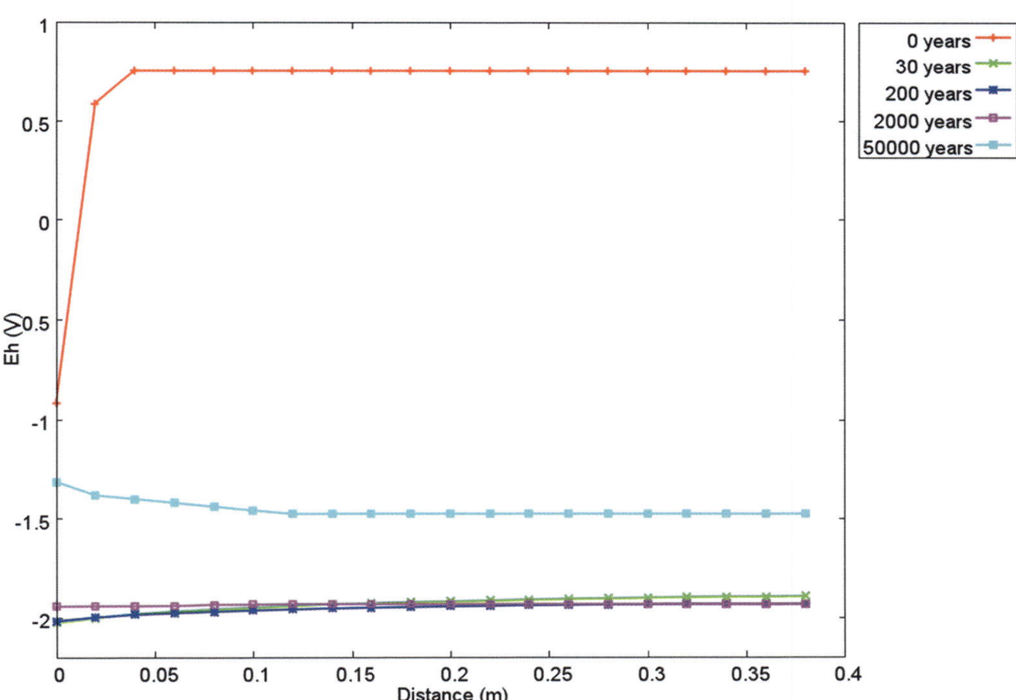

Fig. A1. Evolution of redox potential as a function of distance from iron source and of time for base case.

Table A3. *All considered mineral phases with kinetic parameters (see text)*

Mineral	Mineral type	Log k_25 ($mol/m^2/s$)	Specific surface area (m^2/g if not otherwise stated)
Analcime	Zeolite	−11	0.1
Beidellite Ca	Smectite	−11	1.0
Beidellite-FeCa	Smectite	−11	1.0
Beidellite-FeK	Smectite	−11	1.0
Beidellite-FeMg	Smectite	−11	1.0
Beidellite-FeNa	Smectite	−11	1.0
Beidellite Mg	Smectite	−11	1.0
Beidellite Na	Smectite	−11	1.0
Berthierine	1:1 clay	−8	1.0
Calcite	Carbonate	−5.81/CF*/	100 m^2/m^3 porous media
Chabazite	Zeolite	−11	0.1
Chalcedony	Tectosilicate	−11	100 m^2/m^3 porous media
CSH_0.8	CSH phase	−11	0.01
CSH_1.2	CSH phase	−11	0.01
CSH_1.6	CSH phase	−11	0.01
Dolomite	Carbonate	−11	0.01
Fe(OH)2	Hydroxide	−7.9/CF/	0.01
Fe3(OH)7	Hydroxide, green rust	−11	0.001
Goethite	Oxy-hydroxide	−7.94/CF*/	100 m^2/m^3 porous media
Gyrolite	CSH phase	−11	0.01
Heulandite Ca	Zeolite	−11	1.0
Heulandite Na	Zeolite	−11	1.0
Hydrotalcite	LDH	−11	1.0
Hydrotalcite-CO3	LDH	−11	1.0
Iron**	Metal	−8.35	$1.68 \cdot 10^{-4}$
Magnetite	Oxide	−10.78/P&K/	0.01
Mg-Montmorillonite-Ca	Smectite	−14.41/P&K/	37.5
Mg-Montmorillonite-Mg	Smectite	−14.41/P&K/	37.5
Mg-Montmorillonite-Na	Smectite	−14.41/P&K/	37.5
Phillipsite Ca	Zeolite	−11	1.0
Phillipsite Na	Zeolite	−11	1.0
Quartz, alpha	Tecto-silicate	−13.39/CF/	100 m^2/m^3 porous media
Siderite	Carbonate	−8.9/CF/	0.01
Tobermorite_11A	CSH phase	−11	0.01
Tobermorite_14A	CSH phase	−11	0.01
FeS, am	Sulphide	−8.9/CF/	0.001
Pyrite	Sulphide	−5.0/CF*/	0.001
Chamosite	Chlorite	−11	1.0
Clinochlore	Chlorite	−11	1.0
Al_free_chlorite	Chlorite	−11	1.0
Chlorite_Cca-2	Chlorite	−11	1.0
Illite_IMt-2	Illite	−11	1.0
Illite-Al	Illite	−11	1.0
Illite-FeII	Illite	−11	1.0
Illite-FeIII	Illite	−11	1.0
Illite-Mg	Illite	−11	1.0

/CF/ refers to CrunchFlow database /P&K/ refers to Palandri and Kharaka (2004); *modified according to Wersin *et al.* (2007), **calculated to fit corrosion rate of 0.1 μm/a.

References

APPELO, C. A. J. & WERSIN, P. 2007. Multicomponent diffusion modeling in clay systems with application to the diffusion of tritium, iodide and sodium in Opalinus Clay. *Environmental Science and Technology*, **41**, 5002–5007.

BILDSTEIN, O., TROTIGNON, L., PERRONNET, M. & JULLIEN, M. 2006. Modelling iron–clay interactions in deep geological disposal conditions. *Physics and Chemistry of the Earth*, **31**, 618–625.

BIRGERSSON, M. & KARNLAND, O. 2009. Ion equilibrium between montmorillonite interlayer space and an external solution – Consequences for diffusional transport. *Geochimica et Cosmochimica Acta*, **73**, 1908–1923.

BLANC, P., LASSIN, A. & PIANTONE, P. 2007. *THERMODDEM – A Database Devoted to Waste Minerals*. BRGM Orléans, France, http://thermoddem.brgm.fr

BOURG, I. C., SPOSITO, G. & BOURG, A. C. M. 2006. Tracer diffusion in compacted, water-saturated bentonite. *Clays and Clay Minerals*, **54**, 363–374.

BOURRIÉ, G., TROLARD, F., GENIN, J. M. R., JAFFREZIC, A., MAITRE, V. & ABDELMOULA, M. 1999. Iron control by equilibria between hydroxy-green rusts and solutions in hydromorphic soils. *Geochimica et Cosmochimica Acta*, **63**, 3417–3427.

CASTERMANT, J., MENDONÇA, C. A., REVIL, A., TROLARD, F., BOURRIÉ, G. & LINDE, N. 2008. Redox potential distribution inferred from self-potential measurements associated with the corrosion of a burden metallic body. *Geophysical Prospecting*, **56**, 269–282.

GAUCHER, E. & BLANC, P. 2006. Cement/clay interactions – a review: experiments, natural analogues, and modeling. *Waste Management*, **26**, 776–788.

GÉHIN, A., GRENÈCHE, J.-M., TOURNASSAT, C., BRENDLÉ, J., RANCOURTE, D. G. & CHARLET, L. 2007. Reversible surface-sorption-induced electron-transfer oxidation of Fe(II) at reactive sites on a synthetic clay mineral. *Geochimica Cosmochimica Acta*, **71**, 863–876.

GÉNIN, J.-M. R., REFAIT, P., BOURRIÉ, G., ABDELMOULA, M. & TROLARD, F. 2001. Structure and stability of the Fe(II)-Fe(III) green rust 'fougerite' mineral and its potential for reducing pollutants in soil solutions. *Applied Geochemistry*, **16**, 559–570.

GLAUS, M. A., BAEYENS, B., BRADBURY, M. H., JAKOB, A., VAN LOON, L. R. & YAROSHCHUK, A. 2007. Diffusion of ^{22}Na and ^{85}Sr in montmorillonite: evidence of interlayer diffusion being the dominant pathway at high compaction. *Environmental Science and Technology*, **41**, 478–485.

GLAUS, M. A., FRICK, S., ROSSÉ, R. & VAN LOON, L. R. 2010. Comparative study of tracer diffusion of HTO, $^{22}Na^{+}$ and $^{36}Cl^{-}$ in compacted kaolinite, illite and montmorillonite. *Geochimica et Cosmochimica Acta*, **74**, 1999–2010.

GRIBI, P., JOHNSON, L., SUTER, D., SMITH, P., PASTINA, B. & SNELLMAN, M. 2007. *Safety Assessment for a KBS-3H Nuclear Fuel Repository at Olkiluoto*. Process report. Posiva report 2007-09, Olkiluoto, Finland.

HELLÄ, P., AHOKAS, H., PALMÉN, J. & TAMMISTO, E. 2006. *Analysis of Geohydrological Data for Design of KBS-3H Repository Layout*. SKB report R-08-27, Stockholm, Sweden.

HUNTER, F., BATE, F., HEATH, T. & HOCH, A. 2007. *Geochemical Investigation of Iron Transport into Bentonite as Steel Corrodes*. SKB Technical Report TR-07-09, Stockholm, Sweden.

KITTRICK, J. A. 1979. Chapter 20: Ion exchange and mineral stability: are the reactions linked or separate? *In*: *Chemical Modeling in Aqueous Systems*, ACS Symposium Series, **93**, 401–412.

KUMPULAINEN, S., CARLSSON, T. ET AL. 2010. *Long-Term Alteration of Bentonite in the Presence of Metallic Iron*. SKB Report R-10-52, Stockholm, Sweden.

LANSON, B., LANTENOIS, S., VAN AKEN, P. A., BAUER, A. & PLANÇON, A. 2012. Experimental investigation of smectite interaction with metal iron at 80 °C: structural characterization of newly formed Fe-rich phyllosilicates. *American Mineralogist*, **97**, 864–871.

LANTENOIS, S. 2003. *Réactivité fer métal/smectites en milieu hydraté à 80 deg*. PhD thesis, Université d'Orléans, France.

LANTENOIS, S., LANSON, B., MULLLER, F., BAUER, A., JULLIEN, M. & PLANCON, A. 2005. Experimental study of smectite interaction with metal Fe at low temperature: 1. smectite destabilization. *Clay and Clay Minerals*, **53**, 597–612.

LASAGA, A. C. 1989. *Kinetic Theory in the Earth Sciences*. Princeton University Press, Princeton, NJ.

MARTY, N. C. M., FRITZ, B., CLÉMENT, A. & MICHAU, N. 2010. Modelling the long term alteration of the bentonite barrier in an underground radioactive waste repository. *Applied Clay Science*, **47**, 82–90.

MOSSER-RUCK, R. & CATHELINEAU, M. 2004. Experimental transformation of Na,Ca-smectite under basic conditions at 150 deg. C. *Applied Clay Science*, **26**, 259–273.

MOSSER-RUCK, R., CATHELINEAU, M., GUILLAUME, D., CHARPENTIER, D., ROUSSET, D., BARRES, O. & MICHAU, N. 2010. Effects of temperature, pH, and iron/clay and liquid/clay ratios on experimental conversion of dioctahedral smectite to berthierine, chlorite, vermiculite, or saponite. *Clays and Clay Minerals*, **58**, 280–291.

PALANDRI, J. L. & KHARAKA, Y. K. 2004. *A Compilation of Rate Parameters of Water-Mineral Interaction Kinetics for Application to Geochemical Modeling*. U.S. Geological Survey, Open File Report **2004–1068**.

PARKHURST, D. L. & APPELO, C. A. J. 1999. User's Guide to PHREEQC (Version 2) – *A Computer Program for Speciation, Batch Reaction, One-Dimensional Transport, and Inverse Geochemical Calculations*. U.S. Geological Survey, Water Resources Investigations Report **99-4259**.

PEDERSEN, K. 2008. *Microbiology of Olkiluoto Groundwater 2004–2006*. Posiva Report 2008-02, Olkiluoto, Finland.

PITKÄNEN, P., PARTAMIES, S. & LUUKONEN, A. 2004. *Hydrogeochemical Interpretation of Baseline Groundwater Conditions at the Olkiluoto Site*. Posiva Report 2003-07, Olkiluoto, Finland.

SAVAGE, D., WATSON, C., BENBOW, S. & WILSON, J. 2010. Modelling iron-bentonite interactions. *Applied Clay Science*, **47**, 91–98.

SKB 2011. *Long-Term Safety for the Final Repository for Spent Nuclear Fuel at Forsmark*. SKB Report TR-11-01, Stockholm, Sweden.

SKB/POSIVA 2008. *Horizontal Deposition of Canisters for Spent Nuclear Fuel – Summary of the KBS-3H Project 2004–2007*. SKB Technical Report TR-08-03. Swedish Nuclear Fuel and Waste Management Co (SKB), Stockholm, Sweden and also published as Posiva Report 2008-03, Posiva Oy, Olkiluoto, Finland.

SMART, N. R., BLACKWOOD, D. J. & WERME, L. 2001. *The Anaerobic Corrosion of Carbon Steel and Cast Iron in Artificial Groundwaters*. SKB Technical Report TR-01-22, Stockholm, Sweden.

SMART, N. R., RANCE, A. P. & WERME, L. O. 2004. Anaerobic corrosion of steel in bentonite. *Materials Research Society Symposia Proceedings*, **807**, 441– 446.

STEEFEL, C. I. 2006. *CrunchFlow. Software for Modeling Multicomponent Reactive Flow and Transport. User's manual*. Lawrence Berkeley National Laboratories Report, Berkeley, USA.

STROES-GASCOYNE, S., HAMON, C. J., MAAK, P. & RUSSELL, S. 2010. The effects of the physical properties of highly compacted smectitic clay (bentonite) on the culturability of indigenous microorganisms. *Applied Clay Science*, **47**, 155–162.

TOURNASSAT, C. & APPELO, C. A. J. 2011. Modelling approaches for anion-exclusion in compacted Na-bentonite. *Geochimica et Cosmochimica Acta*, **75**, 3698–3710.

WERSIN, P. 2003. Geochemical modelling of bentonite porewater in high-level waste repositories. *Journal of Contaminant Hydrololgy*, **61**, 405–422.

WERSIN, P., JOHNSON, L. H. & SCHWYN, B. 2003*a*. Assessment of redox conditions in the near field of nuclear waste repositories: application to the Swiss high-level and intermediate level waste disposal concept. *Materials Research Society Proceedings*, **807**, 539–544, http://dx.doi.org/10.1557/PROC-807-539

WERSIN, P., JOHNSON, L. H., SCHWYN, B., BERNER, U. & CURTI, E. 2003*b*. *Redox Conditions in the Near Field of a Repository for SF/HLW and ILW in Opalinus Clay*. Nagra Technical Report **NTB 02-13**, Nagra, Wettingen, Switzerland.

WERSIN, P., JOHNSON, L. H. & SNELLMAN, M. 2006. Impact of iron released from steel components on the performance of the bentonite buffer: a preliminary assessment within the framework of the KBS-3H disposal concept. *Materials Research Society Proceedings*, **932**, 117.1, http://dx.doi.org/10.1557/PROC-932-117.1

WERSIN, P., BIRGERSSON, M., OLSSON, S., KARNLAND, O. & SNELLMAN, M. 2007. *Impact of Corrosion-Derived Iron on the Bentonite Buffer within the KBS-3H Disposal Concept – The Olkiluoto Site as Case Study*. Posiva Report **2007-11**, Olkiluoto, Finland.

WILSON, J., CRESSEY, G., CRESSEY, B., CUADROS, J., RAGNARSDOTTIR, K. V., SAVAGE, D. & SHIBATA, M. 2006. The effect of iron on montmorillonite stability: (II) experimental investigation. *Geochimica et Cosmochimica Acta*, **70**, 323–336.

Thermo-hydraulic modelling of the bentonite buffer in deposition hole 6 of the Prototype Repository

D. MALMBERG* & O. KRISTENSSON

Clay Technology AB, IDEON Science Park, S-223 70 Lund, Sweden

**Corresponding author (e-mail: dm@claytech.se)*

Abstract: The Prototype Repository is an on-going full-scale trial of a final disposal of spent nuclear fuel designed in close agreement with the Swedish KBS-3V design. It provides an excellent opportunity to test our modelling capabilities and our ability to predict evolution of conditions in the planned repository for nuclear waste in Sweden. Here we describe our current efforts to simulate the experiment and compare results with available measured data. To reduce the computational demand, a modelling strategy was adopted in which large-scale low-resolution models (including the entire experiment and a large volume of surrounding rock) were used to determine boundary conditions for local-scale high-resolution models (including one deposition hole). The results obtained so far generally show reasonably good agreement with measurements available from the field experiment.

The Prototype Repository is a full-scale trial designed in close agreement with the Swedish concept (KBS-3V) for final storage of spent nuclear fuel. According to the KBS-3V design, the spent fuel is placed within a copper canister (5 m high and 1.05 m in diameter) which is installed into a surrounding buffer of bentonite clay into an 8 m deep deposition hole with a diameter of 1.75 m. The deposition holes are drilled within deposition tunnels, which are to be excavated in the Swedish crystalline bedrock *c.* 500 m below sea-level.

The Prototype Repository experiment, installed at 450 m depth in the Äspö Hard Rock Laboratory in Sweden, consists of six deposition holes (DHs), each equipped with a full-scale bentonite buffer and canister (see Fig. 1 for a schematic overview). The deposition tunnel is backfilled with a mixture of 70% crushed rock and 30% bentonite. The heat generation of the spent fuel is simulated by heaters installed inside the canisters. The experiment is divided into two sections, the inner (I) consists of four DHs (DH1 to DH4) and the outer (II) consists of two DHs (DH5 and DH6). When fully installed, the sections were sealed with a concrete plug. Sensors were installed in the rock, tunnel backfill and DH buffer to monitor thermal, hydraulic and mechanical (THM) processes taking place during the operational phase.

Section I was installed during the autumn of 2001 and section II during the spring and autumn of 2003. In 2011, section II was excavated, the buffers in DH5 and DH6 were sampled, and their properties (density and water content) determined.

In this paper we describe the modelling of the Prototype Repository, as performed at Clay Technology AB within the framework of the Task Force on Engineered Barrier Systems before the excavation data were available. Special focus is put on modelling the thermo-hydraulic (TH) evolution in the buffer in DH6.

Numerical strategy

The modelling presented here was performed using the finite-element program Code_Bright, which was developed with simulating unsaturated flow in porous media in mind (Olivella *et al.* 1996).

Fractured rock

Presently when using Code_Bright, fractures in the rock domain have to be represented by volumes (when modelling in 3D, as is the case here). Generally, this would lead to complicated meshes if a large number of fractures were to be included. To represent the hydraulic transport properties of the rock without including a complex fracture network, the rock around the repository is treated as a homogeneous material in the models, except for just around the DHs, where a more detailed description is used (as explained in section 'Representing the rock near the DHs').

Step-wise solution

To accurately model the evolution in the DHs, the interaction between all DHs must be included, as well as the response in the rock due to the presence of both the heater and the unsaturated clay. One solution would be to model the entire Prototype Repository in a coupled TH model, including a very large

From: Norris, S., Bruno, J., Cathelineau, M., Delage, P., Fairhurst, C., Gaucher, E. C., Höhn, E. H., Kalinichev, A., Lalieux, P. & Sellin, P. (eds) 2014. *Clays in Natural and Engineered Barriers for Radioactive Waste Confinement*. Geological Society, London, Special Publications, **400**, 251–263.
First published online May 1, 2014, http://dx.doi.org/10.1144/SP400.25

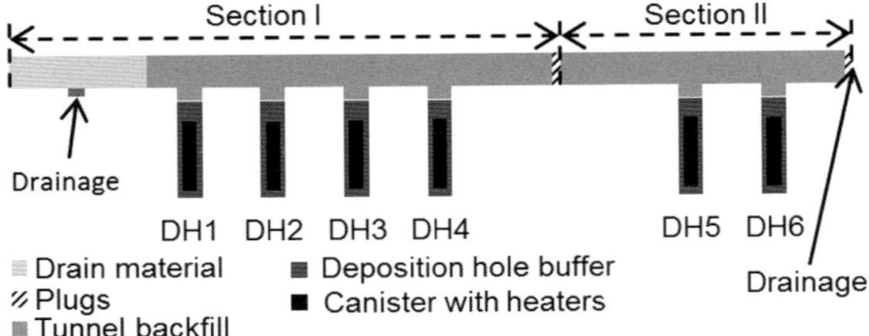

Fig. 1. Schematic overview of the backfill tunnel and the six deposition holes (DHs) in the Prototype Repository.

volume of rock (such that constant temperature and hydrostatic pressure can be assumed at the boundary). However, to include fracture flow in the rock near the DHs, as well as to resolve the bentonite buffer well enough to achieve a grid-independent solution, a Code_Bright model of the entire experiment would require a mesh with more than 10^6 nodes. This would lead to exceedingly long computation times, and would most likely lead to numerical problems (especially if mechanical processes were included, which is the long-term goal of the modelling presented here).

To avoid such problems, the modelling was done in steps. Two types of geometries were used: (1) a large-scale 3D model of the entire experiment and a large volume of surrounding rock; and (2) a local-scale 3D model of one DH, with corresponding tunnel backfill and a small part of the surrounding rock. To capture the evolution, three sets of models were used:

(1) A large-scale model solving for the liquid pressure before installation of the buffer (but after excavation of the rock). This was used to calibrate the rock material with respect to measured tunnel and DH inflows.
(2) Large-scale models after installation of the buffer, solving for temperature and liquid pressure separately, to determine boundary conditions for the local model.
(3) A set of local-scale, coupled TH models, used to study the wetting process in the buffer from installation to excavation. The temperature and liquid pressure at the boundary of this local-scale model was obtained from the large-scale solution.

The set-up and results of the large-scale models are briefly discussed in section 'Large-scale models', while the set-up and results of the local-scale models are presented in section 'Local-scale models'. Before this, in section 'Pre-installation characterization', an overall description of field measurements performed before installation of the canisters and buffer is given, as these are used to calibrate our models.

Pre-installation characterization

The prototype tunnel and deposition boreholes were extensively characterized during boring and before installation of the buffer (see e.g. Forsmark *et al.* (2001) for a comprehensive summary). For example, the inflow into the tunnel and all DHs were measured and mapped. A large uncertainty was present in these measurements. As an example, the total inflow into DH6 was measured three times, resulting in three significantly different values (Rhen & Forsmark 2001): 6.1 ml min^{-1} (December 1999), 2.7 ml min^{-1} (March 2000) and 7.4 ml min^{-1} (June/July 2000). Of these, the value of 2.7 ml min^{-1} was considered to be most representative of the true inflow (see Rhen & Forsmark (2001) for further details), and was thus used here. It is important to remember that if the true inflow was closer to the two other measurements, our models would underestimate the rate of hydration in the buffer.

Aside from the total inflow into the DH it is also essential to know how the water enters it. Two extreme cases (which with some modification were used in the modelling, see section 'Representing the rock near the DHs') can be used as outer bounds on the real situation:

- *Diffuse inflow*: the water enters evenly over the entire DH wall, or
- *Local inflow*: the water enters through discrete fractures intersecting the DH, hence on the DH wall local zones with high inflow can be identified, while all other parts of the wall are essentially dry.

As part of the characterization of the DH several local inflow zones were identified through visual inspection of the wall of DH6 (Forsmark *et al.* (2001), see also section 'Representing the rock near the DHs'). To estimate the water flux in these zones, plastic bags were glued to the rock wall.

These collected the water entering the DH through the local zones, and by measuring the volume after a given period of time the flux was measured.

According to plastic-bag measurements, the local inflow zones account for approximately 20% of the total inflow into DH6. If correct, the majority (80%) of the inflow enters the DH as a diffuse flow.

A second type of flow mapping was done by covering the DH wall with diapers (Forsmark & Rhen 2005). These were left in place for approximately one week; the water that had accumulated in them was then determined by weighing. This technique gave a rather different result from the plastic-bag measurements, as the diaper measurements indicated that almost all the inflow enters DH6 in the identified local zones.

Large-scale models

As described above, a step-wise solution strategy was adopted when analysing the Prototype Repository. Coarse, 3D large-scale models, including the entire experiment with all six DHs, were used to determine representative thermal and hydraulic boundary conditions, which were prescribed on the boundary of the more detailed local-scale models. Two sets of large-scale simulations were carried out, modelling thermal and hydraulic processes separately.

In the hydraulic model, prior to simulating the operational phase after buffer installation, the case where water flows into the empty tunnel and DHs was simulated. By comparing the calculated tunnel and DH inflows with experimental measurements before buffer installation (see section 'Pre-installation characterization'), the permeability in the material model, describing the water transport (Darcy's law), was calibrated.

This calibration was then used when simulating the hydraulic evolution during the operational phase, that is, after installation of the buffer. One example of the results from the large-scale models is shown in Figure 2; here, the liquid pressure is shown in a vertical plane which is perpendicular to the prototype-experiment-tunnel axis and taken at the position of DH6. At the top is the pressure field just after buffer installation in DH6, while at the bottom the situation 1000 days after buffer installation is shown. As can be seen in the picture, the large-scale models include a tunnel which is situated parallel to the Prototype Repository tunnel. This side tunnel was partly used to monitor the evolution in the experiment, as the cables from most sensors were led there and connected to data collectors. The side tunnel has little effect on the temperature field in the rock which is elliptical in shape and centred on the mid-point of the canister. However, as it is open to the atmosphere, it significantly perturbs the liquid-pressure field around the prototype-experiment tunnel, which would otherwise have been elliptical in shape and centred on the mid-point of the tunnel floor.

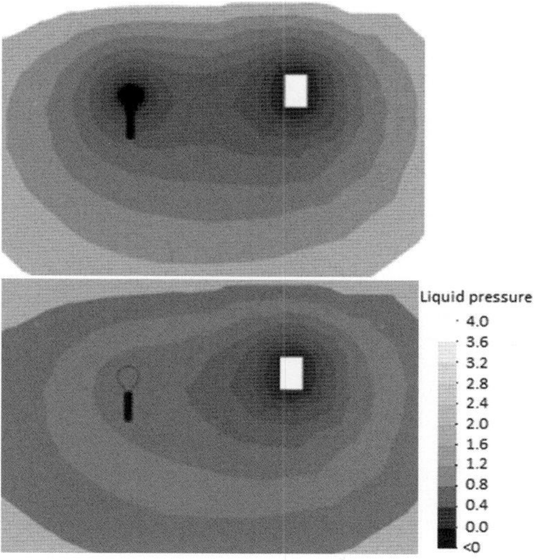

Fig. 2. Liquid pressure in the rock from the large-scale model just after the installation of bentonite (top) and 1000 days after installation (bottom).

Local-scale models

Geometry

The local-scale models were done in full 3D. The geometry is shown in Figure 3; the oval shape and dimensions were chosen so as to agree with the liquid-pressure isolines shown in Figure 2. This was done in order to simplify the hydraulic boundary conditions (described in section 'Boundary conditions').

Representing the rock near the DHs

Except for a 1 m thick layer around the DHs, the rock was treated as a homogeneous material, thereby effectively averaging the flow over both the rock matrix and fractures. The transport properties of this homogeneous material were calibrated in the large-scale models, where the aim was to reproduce the measured inflows in the experiment tunnel. The same material properties were used in the local-scale model. However, as mentioned, near the DHs, the homogenized representation was not used, as it would have resulted in much too high inflows.

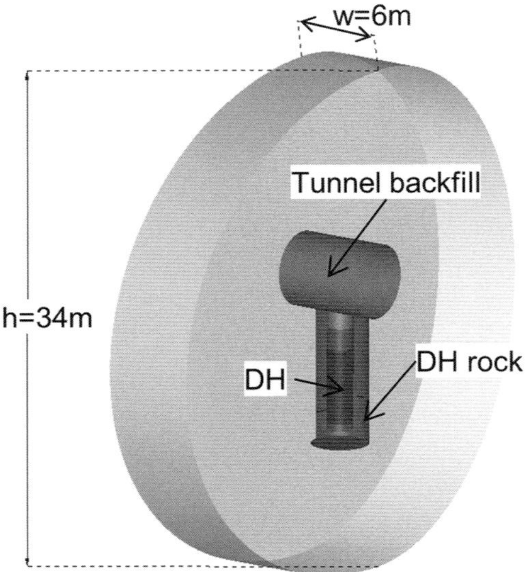

Fig. 3. The geometry of the local model. The shape was selected so as to agree reasonably well with the temperature and liquid pressure (see Fig. 2) isolines in the large-scale models.

As discussed in section 'Pre-installation characterization', the DHs were extensively characterized before buffer installation, with the total inflow measured, as well as local inflows on the wall identified, and the flow in these estimated using two methods. These flow measurements in the field were used to construct the model conceptualization of the rock close to the DHs.

A cylindrical volume with radius 1.875 m and depth 9 m was placed around each DH (which has a radius of 0.875 m and depth of 8 m), and assigned a material called 'DH Rock'. In this volume five horizontal 45° arcs with inner radius of 0.875 m, outer radius of 1.875 m and a thickness of 0.1 m were cut out at the positions where local inflows had been identified in the field and assigned materials called Local Zones 1–5. In Figure 4 the position of these zones is indicated in the left panel, which shows a schematic of the wall in DH6. The right panel shows the identified zones of high inflows (black squares) and is taken from Forsmark & Rhen (2005).

The materials (DH Rock and Local zones 1–5) were then calibrated under atmospheric conditions in two different ways:

- *'Diffuse-flow model'* in which the permeabilities were calibrated such that the total inflow equaled 2.7 ml min^{-1}, while the flow distribution corresponded to that estimated using the plastic-bag measurements: 20% of total inflow in local zones, 80% spread evenly over the rest of the DH wall.

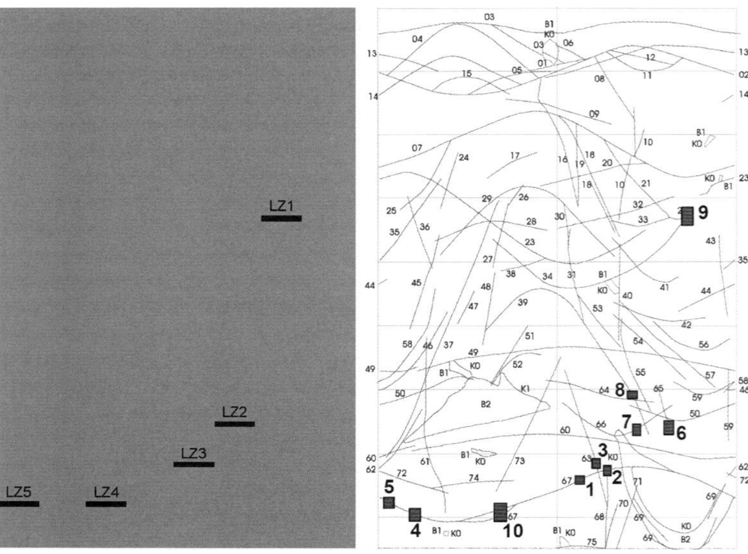

Fig. 4. A schematic overview of the position of the local inflow zones (LZ1–LZ5) in the model (left) compared to those identified in the field (black squares in the right panel) using visual inspection of the deposition-hole wall (from Forsmark & Rhen 2005).

- '*Local-flow model*' in which the permeabilities were calibrated such that the total inflow equaled 2.7 ml min^{-1}, and all the inflow was concentrated in the identified local zones. The ratio between the inflows in these zones is, however, taken from the plastic-bag measurements.

Material parameters

In the modelling completed so far, only thermal and hydraulic processes have been considered, while mechanical processes have been neglected. As bentonite clay is strongly swelling and its hydraulic properties change with porosity, neglecting mechanical processes means that we cannot fully capture the real evolution. To quantify the uncertainty caused by this simplification, two sets of models were constructed: (1) installation models, in which material properties simulated those just after installation, and (2) homogenized models, where the buffer was assumed to be fully radially homogenized. The evolution and final state of liquid saturation in the field experiment was then expected to be bounded by the results from these two sets of models.

The bentonite blocks used to backfill the DHs in the Prototype Repository have very similar properties to the blocks which are to be used in the spent fuel repository. Thus, their TH properties have been extensively characterized in laboratory experiments. Suitable constitutive laws and parameters to use when modelling the buffer using Code_Bright are available in Åkesson *et al.* (2010). A summary is shown in Table 1 but for further details the reader is directed to Åkesson *et al.* (2010). The material parameters describing the tunnel backfill, a 70:30 mixture of crushed rock and bentonite from Milos, are also included in Table 1. The permeability and retention parameter values were obtained from studying experimental evaluations of the material, reported in Johannesson (2005).

The rock was represented by the material parameters listed in Table 2. To reproduce the thermal evolution of the large-scale model, a thin-layer (1 dm thick) material with calibrated thermal properties had to be applied on each side of the local-scale geometry. This was needed due to the heat load asymmetry around DH6 which was neglected when constructing the local model.

Initial conditions

The solid density of all the materials used to represent the rock was set to 2770 kg m^{-3}, and the porosity to 0.003. The solid density of the bentonite was set to 2780 kg m^{-3}, and its heat capacity to 800 J kg^{-1} K^{-1}.

The initial temperature was set to 15 °C in the entire model. The prescribed porosities/void ratios and initial liquid pressures are listed in Table 3.

The initial conditions of the buffer in the installation model were based on the characterization of the materials done at installation of the buffer in the Prototype Repository (see installation report in Johannesson *et al.* (2004)). As the experiment is a trial version of the KBS-3V design, these initial conditions agree well with those identified in Åkesson *et al.* (2010).

A schematic picture of the two modelled states of the buffer (the installation and homogenized states respectively) can be seen in Figure 5. The parameters were chosen such that the total available pore volumes in the two models were identical.

Boundary conditions

As discussed in section 'Geometry', the shape of the local model was chosen so as to agree with the shape of the liquid-pressure isolines seen in the large-scale models (see Fig. 2). However, the presence of the neighbouring parallel tunnel (on the north side of the Prototype Repository tunnel) perturbs the liquid-pressure field, making it impossible to prescribe the same liquid pressure around the entire boundary. It was therefore divided into several segments with slightly different hydraulic boundary conditions (see lower part in Fig. 6).

The temperature isolines are not perturbed by the neighbouring tunnel. However, they are offset with respect to the liquid-pressure isolines. While the latter are centred on a point approximately level with the tunnel floor, the temperature isolines are centred on the vertical mid-point of the heater. Thus, the temperature on the boundary also had to be prescribed in segments. The boundary conditions used are shown in Figure 6.

It should be noted that here we only prescribe the liquid pressure and temperature on the boundary of the local model. To verify that the evolution in the local model follows that in the large-scale model we have verified that the liquid and thermal fluxes on the local-model boundary are similar to those in the large-scale model.

Local model: results

In total, four local models were used; they will hereafter be identified as: Inst_Diff; Inst_Local; Hom_Diff; Hom_Local (Table 4). They represent the two different states (installation and homogenized) as well as the two different inflow calibrations (diffuse flow and local flow). As only the TH evolution is simulated, the quantities which can be analysed are the temperature and relative humidity

Table 1. *Thermal and hydraulic constitutive laws and parameters of the buffer materials*

		Parameter	Rings	Cylinders	Pellets	Tunnel backfill
Liquid advective flux	$\mathbf{q}_l = -\frac{k \times k_{rl}}{\mu_l}[\nabla P_l - \rho_l \mathbf{g}]$	k [m^2]	1.2×10^{-21}/ 5.1×10^{-21}*	2.0×10^{-21}/ 5.9×10^{-21}*	5.2×10^{-19}	5×10^{-18}
	$k_{rl} = S^3$					
Liquid density	$\rho_l = 1002.6 \exp[4.5 \times 10^{-4}(P_l - 0.1) - 3.4 \times 10^{-4}\,T]$					
Liquid viscosity	$\mu_l = 2.1 \times 10^{-12} \exp\left[\frac{1808.5}{273.15 + T}\right]$					
Vapour non-advective flux	$\mathbf{i}_g^w = -[n\rho_g(1 - S)D_m^w]\nabla\omega_g^w$					
	$D_m^w = 5.9 \times 10^{-6}\frac{(273.15 + T)^{2.3}}{P_g}$					
	ρ_g: ideal gas law					
Retention curve	$S(\Psi) = \left[1 + \left(\frac{\Psi}{P_0}\right)^{\frac{1}{1-\lambda}}\right]^{-\lambda}$	P_0 [MPa]	67.2/8.93	43.5/15.2	0.508	0.332
	$\Psi = P_g - P_l$	λ	0.48/0.22*	0.38/0.25*	0.26	0.23
Conductive flux of heat	$\mathbf{i}_l = -\lambda_i(S)\nabla\mathbf{T}$					
	$\lambda_i(S) = 0.7 \times (1 - S) + 1.3 \times S$					

*These values are used for the homogenized models.

Table 2. *Thermal and hydraulic properties of the rock material used in the local-scale model*

		Parameter	Rock	TL* material on side towards other DHs	TL* material on side towards open tunnel
Liquid advective flux	$\mathbf{q}_l = -\frac{k \times k_{rl}}{\mu_l}[\nabla P_l - \rho_l \mathbf{g}]$	k [m^2]	10^{-17}	10^{-17}	10^{-17}
	$k_{rl} = -\sqrt{S}[1-(1-S^{1/\lambda})^{\lambda}]^2$	λ	0.24	0.24	0.24
Liquid viscosity	$\mu_l = 2.1 \times 10^{-12} \exp\left[\frac{1808.5}{273.15 + T}\right]$				
Retention curve	$S(\Psi) = \left[1 + \left(\frac{\Psi}{P_0}\right)^{\frac{1}{1-\lambda}}\right]^{-\lambda}$	P_0 [MPa]	0.6	0.6	0.6
	$\Psi = P_g - P_l$	λ	0.24	0.24	0.24
Conductive flux of heat	$\mathbf{i}_l = -\lambda_i \nabla \mathbf{T}$	λ_i	2.685	1	60
Spec. heat capacity		C [J kg^{-1} K^{-1}]	770	1500	15 000

*TL: the Thin-Layer (1 dm thick) rock material applied on both sides of the local-scale model.

Table 3. *Initial conditions of the materials*

Material	Void ratio/ Porosity	Liq. pressure [MPa]
Rings	0.571/0.363	−45.9
Cylinders	0.623/0.384	−45.9
Pellets	1.78/0.64	−99.9
Hom. rings	0.757/0.431	−45.9
Hom. cylinders	0.773/0.436	−45.9
Backfill	0.580/0.367	−2.33
Canister	0.001/0.001	−45.9
Rock	0.003/0.003	3.3

(RH), as well as the final degree of liquid saturation. As the main focus is on the buffer, here we will only discuss, and compare the model results with, data from sensors placed inside the buffer, and will not consider the evolution in the tunnel backfill or nearby surrounding rock.

Temperature

The agreement between model data and measurements of the temperature in the experiment is in general very good in all models, although there are some small discrepancies, in particular above and below the heater. The temperature evolution in the buffer is (weakly) coupled to the hydraulic evolution since the thermal conductivities of the buffer materials are functions of the degree of liquid saturation. As such, even though all models have identical thermal boundary conditions as well as thermal material properties, the temperature evolution differs slightly.

Hydraulic evolution

The hydraulic evolution in the field experiment is monitored using both RH and suction sensors (for more details, see for example Goudarzi & Johannesson (2009)). The data from the RH sensors are only reliable when RH <95%, corresponding to a suction of about 8 MPa at a temperature of 60 °C. To monitor the evolution for higher values of RH (>95%), the installed suction sensors can instead be used, as the two quantities are related through Kelvin's law:

$$\mathrm{RH} = 100\ \exp\left[\frac{-\Psi \times M_{\mathrm{W}}}{R(273.15 + T)\rho_l}\right] \quad (1)$$

where Ψ identifies the suction in the material, R the universal gas constant, M_{w} the molar mass of water, T the temperature (measured in °C) and ρ_l the liquid density.

When comparing the modelled evolution with experimental data it is important to keep in mind that RH measurements in the field are difficult to make and that there are significant uncertainties in the signal from the sensors. Thus, the model data should be compared with the general trend seen in the measured RH, not the exact values.

In Figure 7 (installation models) and Figure 8 (homogenized models) we plot the RH evolution at 13 different positions in the buffer of DH6, each corresponding to the position of a RH sensor in the experiment. The experimental data are represented by solid lines (RH sensors) and squares (suction-sensor data converted to RH using Kelvin's law), while dashed lines represent model

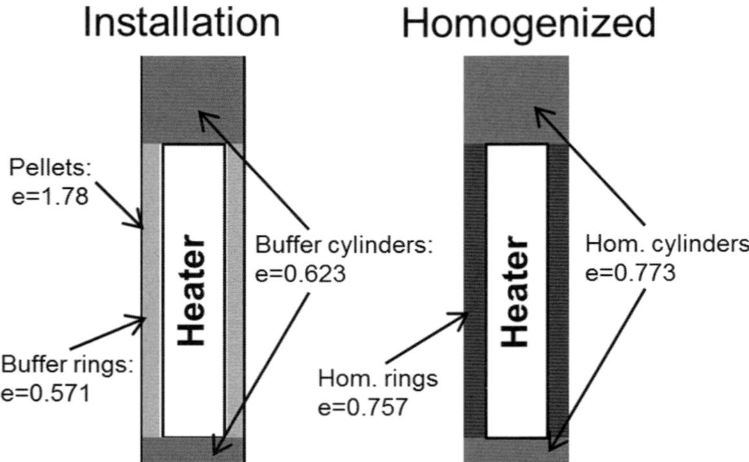

Fig. 5. Schematic overview of the two states modelled. To the left is the state of the buffer just after installation, before any swelling has occurred. In the right part is the state of the buffer assuming that only radial swelling occurs, and that the final, homogenized state has been reached.

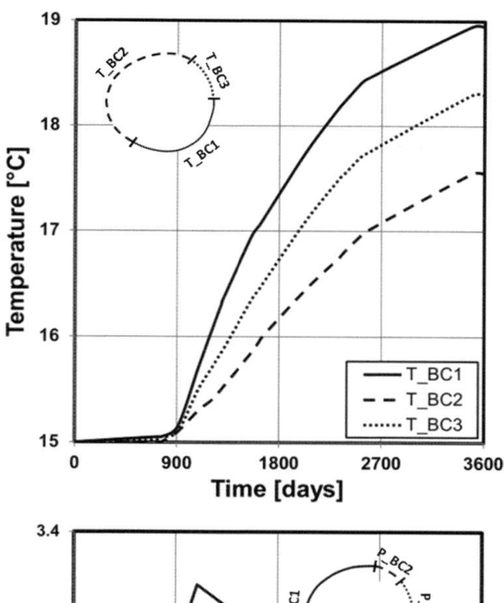

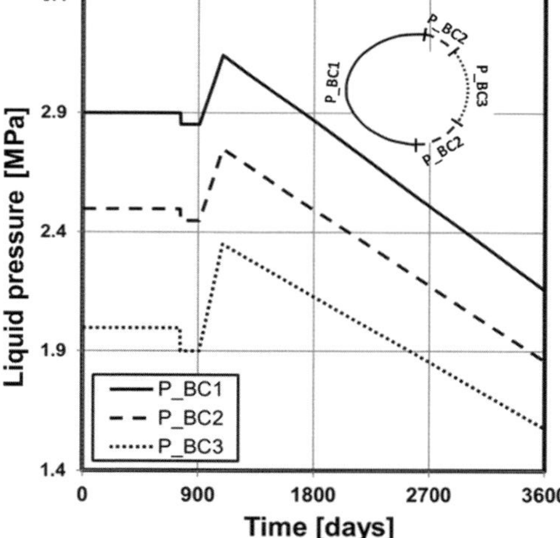

Fig. 6. Temperature (above) and liquid pressure (below) prescribed at the outer boundary of the local model. The three different lines in each panel correspond to different segments (BC1–BC3) on the boundary as illustrated in the overlaid figure in the upper corner of each graph.

data. The left column in both figures contains data from the models where the water enters as a diffuse flow, while the right column contains data from the models with only local inflow. In general, the diffuse-flow models agree better with experimental data than the local-flow models, except for perhaps during the first few 100 days.

It should be pointed out that one important feature of the experiment has been left out of the modelling: the closing of both the inner and outer tunnel drainage (see Fig. 1). This occurred approximately 700 days after the installation of the buffer

Table 4. *Overview of the four models done*

Model	Initial state	Flow calibration
Inst_Diff	Installation	Diffuse flow
Inst_Local	Installation	Local flow
Hom_Diff	Homogenized	Diffuse flow
Hom_Local	Homogenized	Local flow

in DH6. As can be seen in the RH data from the experiment in Figures 7 and 8, at this time, the RH in the sensors placed around the heater almost instantaneously rose to 100%; furthermore, pore and total pressure sensors in both the DH and in the tunnel backfill just above DH6 showed a significant spike. Most likely, the closing of the drainage led to a rapid increase in the liquid pressure in the rock around the experiment, which pushed water in between the blocks around the heater in DH6. This might have filled the gap between the heater and the buffer rings with water. Furthermore, since all the sensors are connected with the outside world using cables/tubes, it is possible that water could be transported directly to the sensors via the cable/tube pathways, causing the instantaneous increase to 100% RH.

Had this effect been included in the model it is possible that the RH evolution in the models with only local inflow would have come much closer to measured evolution, and as such they cannot be discarded by only looking at the comparison in Figures 7 and 8.

If considering only the results shown in Figures 7 and 8 it is, however, clear that the diffuse-flow calibration reproduces the hydraulic evolution in the field better than the local-flow calibration, in particular in the buffer rings around the canister. The general trend is well reproduced. The perhaps most important discrepancy, that the buffer rings around the canister reach full saturation too early in the models, arises when including suction-sensor data. This indicates that taking the buffer from 95% RH up to 100% RH was very slow in the experiment, a behaviour which is not reproduced in the models.

Final state. Measuring the RH during the experiment is complicated and many uncertainties exist. However, measuring the water content after excavation is a straightforward task where high accuracy can be achieved. As such, perhaps the most important result of the models is the final state of the buffer in terms of the degree of saturation. Unfortunately, at the time of writing this paper, no data were available from the excavation of the buffer in DH6, and therefore we cannot present a comparison with experimental results.

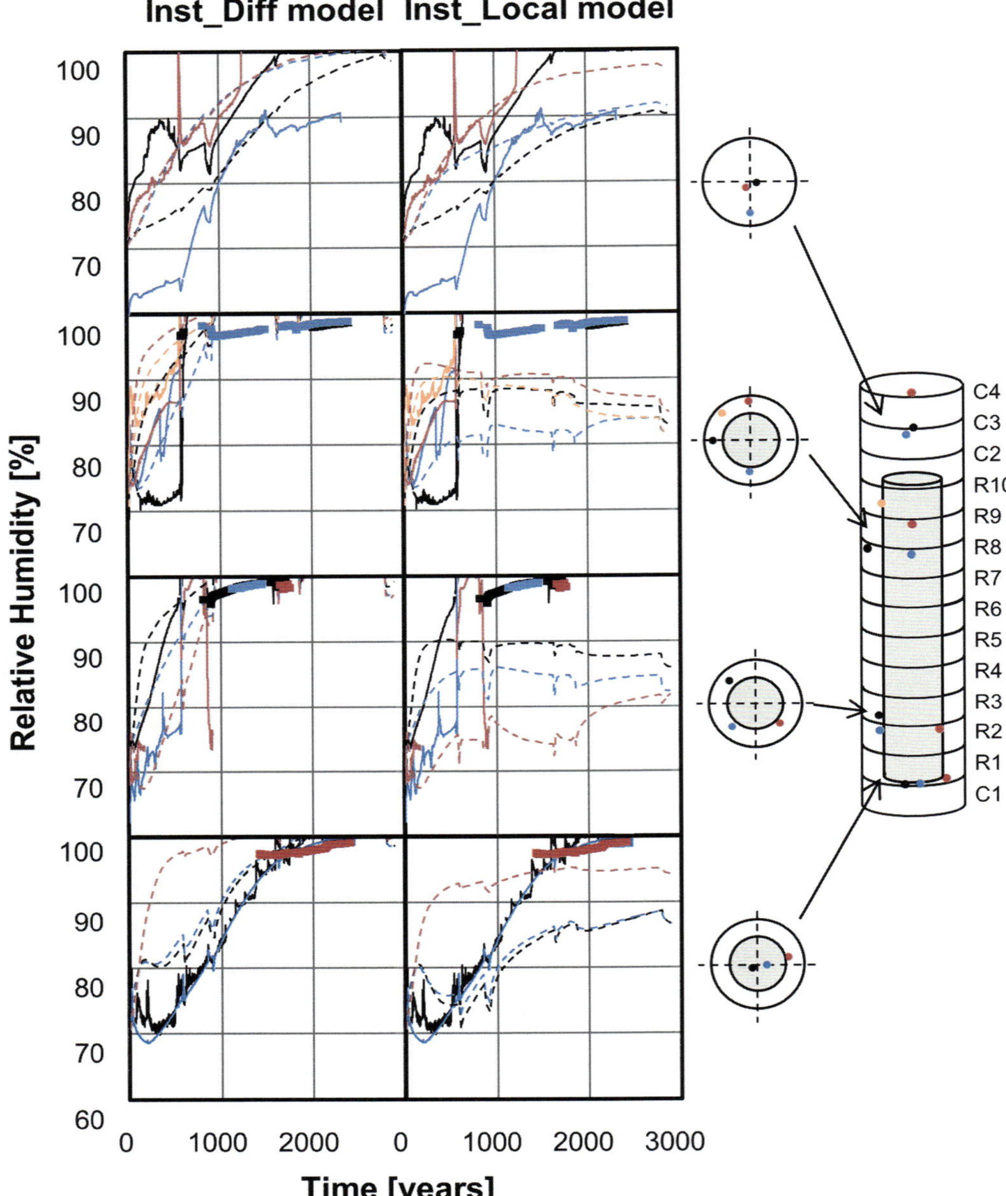

Fig. 7. The evolution of relative humidity at different depth and radii in DH6. Model data (dashed lines) are taken from the two models with the buffer in the installation configuration (see also Table 4). The solid lines and squares represent experimental data.

Contour graphs of the degree of saturation in the buffer from the four models can be seen in Figure 9. The contour graphs show both a vertical cut through the entire DH, as well as horizontal cuts at four different depths. The contour graphs shows the state of the buffer at the time when the dismantling excavation of the buffer in DH6 was completed.

Several important features can be observed. If the plastic-bag measurements are most accurate (i.e. that the inflow is mostly diffuse, models

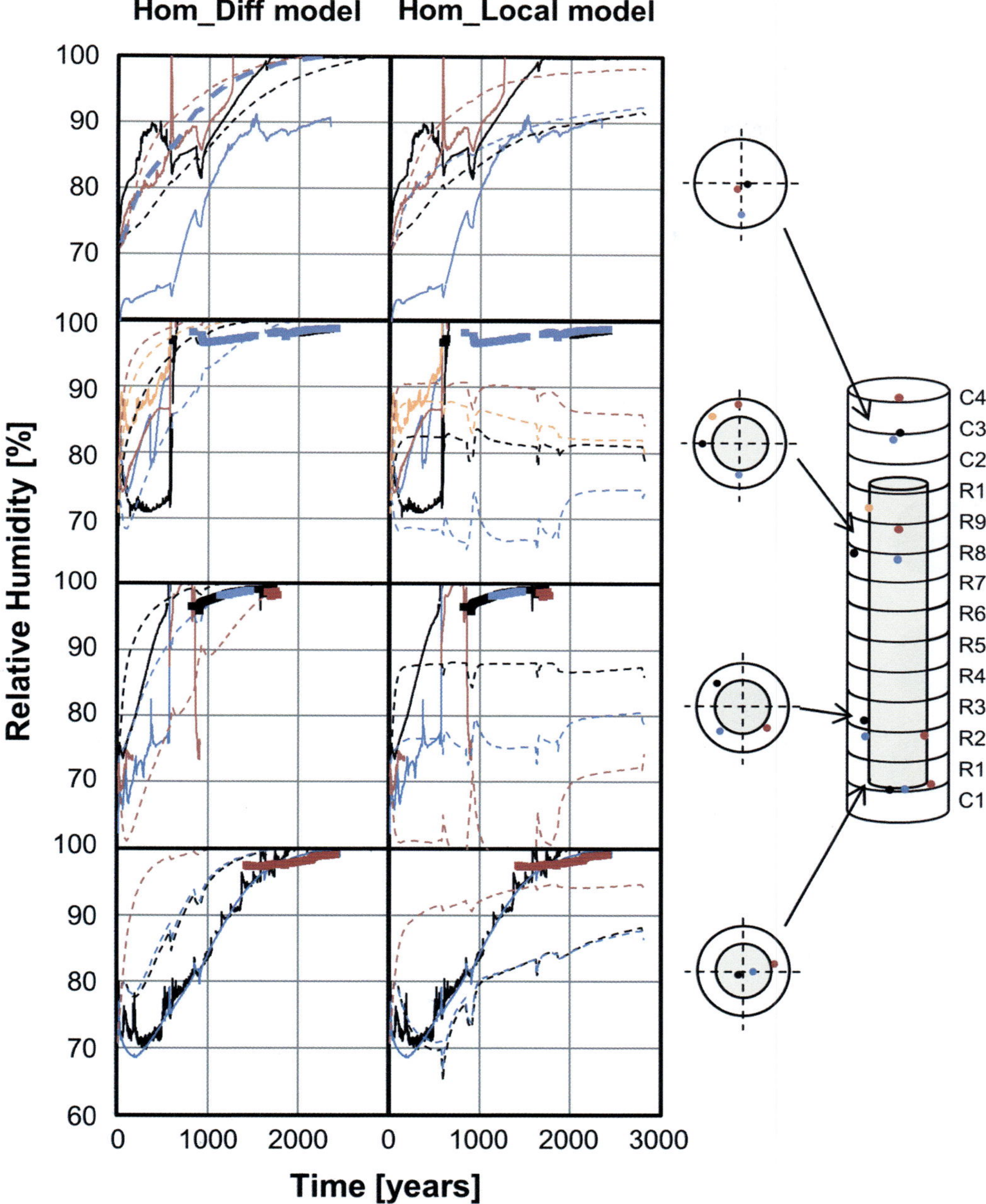

Fig. 8. The evolution of relative humidity in the buffer at several different depths and radii in DH6. Model data (dashed lines) are taken from the two models with the buffer in the homogenized state (see also Table 4). The solid lines and squares represent experimental data.

Inst_Diff and Hom_Diff) the models indicate that buffer should have been almost fully saturated at excavation of the experiment; the only unsaturated part is found on top of the heater. Furthermore, the degree of saturation should be essentially axisymmetric; hence no effect can be seen from the local inflow zones at this point.

On the other hand, if the diaper measurements are most accurate the excavated buffer should be relatively dry, with full saturation reached only

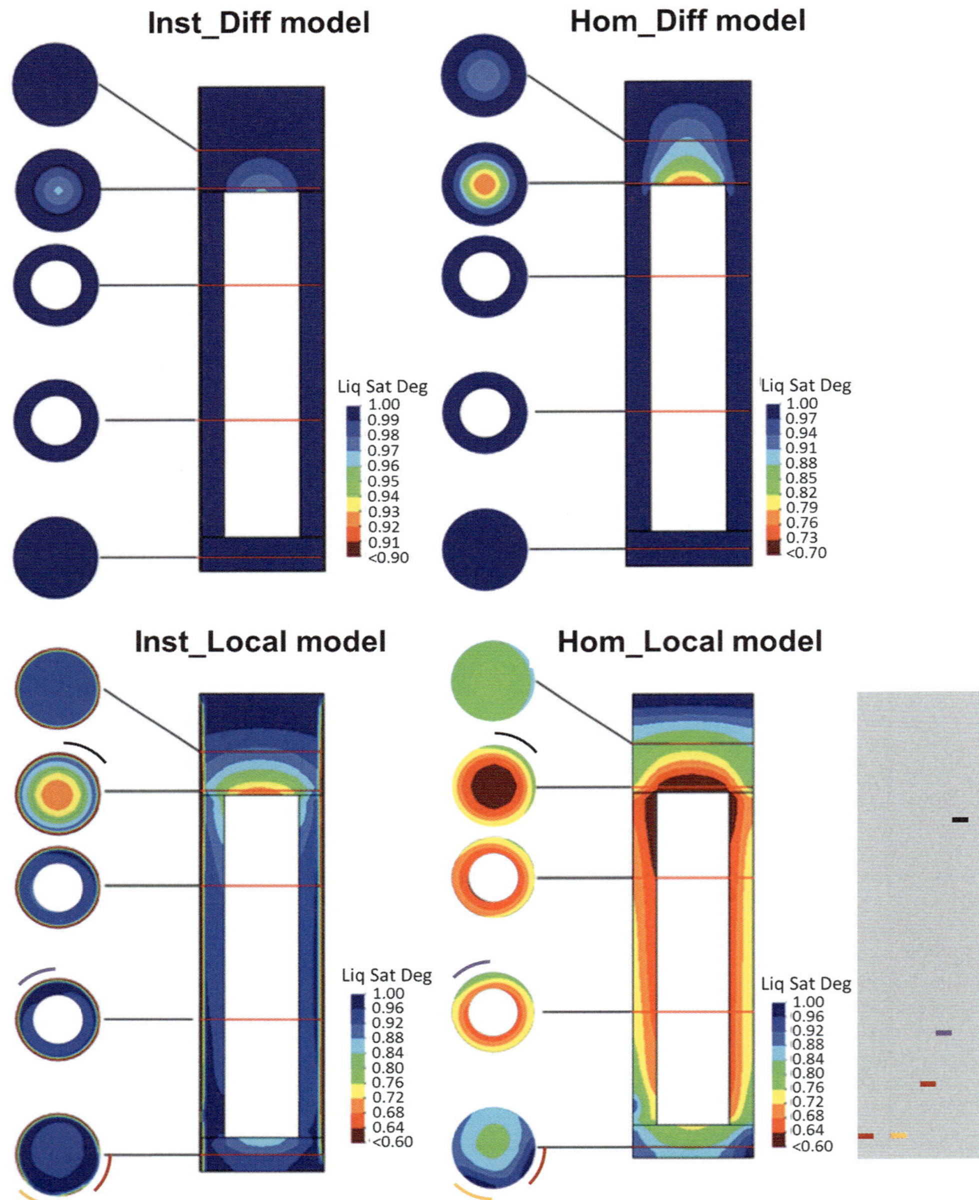

Fig. 9. Contour plots showing the liquid saturation in all four models presented in this paper (see Table 4 for further details) at the same time as when the buffer was excavated in the experiment. In the lower-right corner a schematic drawing, showing the position of the local-flow zones on the deposition-hole wall, is included. Where appropriate, the orientation of these zones is also shown around the horizontal cuts in the figure.

just in front of the local inflow zones, as well as in the top buffer cylinder. In the latter case, the water has entered the DH via the tunnel backfill instead of through any of the local inflow zones. One important result in the local-flow models is that the wetting is far from axisymmetric.

In the models with mostly diffuse inflow (Inst_Diff and Hom_Diff), the difference between the installation state and the homogenized state is small; these models are, however, somewhat drier on top of the heater. In contrast, in the local-flow models there is a clear difference between the installation and homogenized models, as the latter are considerably drier.

From comparing, for example, Inst_Diff and Inst_Local, it is clear that any prediction of the final state from the models done here is extremely sensitive to how the water enters the DH, whether it is through diffuse flow, or if it is concentrated to a few high-flowing zones.

Conclusions

Some of the models presented are able to reproduce the general trends seen in relative humidity sensor data from the field experiment. This gives confidence in that the modelling strategy used, with large, coarse-meshed models used to calibrate smaller, well-resolved models, works well when studying the evolution of conditions in the Prototype Repository.

From the characterization of the rock around both the prototype tunnel, as well as the deposition holes (DHs), it is clear that the hydraulic properties are quite complex. In the DHs, a significant part of the inflow, as measured during atmospheric conditions in the field, comes through local zones on the walls. However, measuring the exact volume of water entering is not trivial, and large uncertainties are present. Here, these uncertainties were handled using two different inflow calibrations, one with all the flow into DH6 through local zones, and one with a more homogeneous (diffuse) inflow over the DH wall.

The sharp difference in the results from these two different inflow calibrations shows that the model results are extremely sensitive to the pre-installation characterizations done in the field. Given the uncertainties in the characterizations, it is very hard to predict the evolution in the buffer, as well as its final state.

Summary

The strategy of using associated models at different scales to simulate the Prototype Repository enables the capture of the main features of the hydraulic evolution seen in the experiment. At the same time, this modelling approach, using Code_Bright, is significantly more numerically efficient than doing well-resolved models of the entire repository at once.

How well the final state in the models agrees with the actual final state in the field is not known at this point, as data from the excavation of the buffer in one of the deposition holes (DH6) have not yet been made available. However, the reasonable agreement with sensor data in the *diffuse-flow* models indicates that the final state in these models should best match the final state in the experiment. It is also clear that in order to make a more precise prediction of the final state of the experiment, a more detailed characterization of the transport properties of the rock, with much smaller uncertainties, would be needed.

The financial support from SKB AB within the Task Force on Engineered Barrier Systems project and the Prototype Repository project are acknowledged. We also would like to thank the reviewers for their comments, which improved the manuscript considerably.

References

Åkesson, M., Börgesson, L. & Kristensson, O. 2010. *SR-Site Data Report. THM modelling of buffer, backfill and other system components*. SKB TR-10-44, Stockholm, Sweden.

Forsmark, T. & Rhen, I. 2005. *Äspö Hard Rock Laboratory. Prototype Repository. Hydrogeology – diaper measurements in DA3551G01 and DA3545G01, flow measurements in section II and tunnel G, past grouting activities*. SKB IPR-05-03, Stockholm, Sweden.

Forsmark, T., Rhen, I. & Andersson, C. 2001. *Äspö Hard Rock Laboratory. Prototype repository. Hydrogeology – Deposition and lead-through boreholes: inflow measurements, hydraulic responses and hydraulic tests*. SKB IPR-00-33, Stockholm, Sweden.

Goudarzi, R. & Johannesson, L. E. 2009. *Äspö Hard Rock Laboratory. Prototype Repository. Sensors data report (Period 010917-091201)*. Report No: **22**, SKB IPR-10-05, Stockholm, Sweden.

Johannesson, L. E. 2005. *Äspö Hard Rock Laboratory. Prototype repository. Laboratory tests on the backfill material in the prototype repository*. SKB IPR-05-11.

Johannesson, L. E., Gunnarsson, D., Sandén, T., Börgesson, L. & Karlzén, R. 2004. *Äspö Hard Rock Laboratory. Prototype repository. Installation of buffer, canisters, backfill, plug and instruments in Section II*. SKB IPR-04-13, Stockholm, Sweden.

Olivella, S., Gens, A., Carrera, J. & Alonso, E. E. 1996. Numerical formulation for a simulator (CODE_BRIGHT) for the coupled analysis of saline media. *Engineering Computations*, **13**, 87–112, http://dx.doi.org/10.1108/02644409610151575

Rhen, I. & Forsmark, T. 2001. *Äspö Hard Rock Laboratory. Prototype repository. Hydrogeology. Summary report of investigations before the operation phase*. SKB IPR-01-65, Stockholm, Sweden.

Coupled hydromechanical modelling of the mine-by experiment at Meuse-Haute-Marne underground rock laboratory France

K. YILDIZDAG[1], H. SHAO[2], J. HESSER[2], A. NOIRET[3] & J. SOENNKE[2]*

[1]*Shepton-Mallet Ring 10, 30655 Hannover, Germany*

[2]*Federal Institute for Geosciences and Natural Resources, Stilleweg 2, 30655 Hannover, Germany*

[3]*National Radioactive Waste Management Agency (ANDRA) 1/7, rue Jean Monnet Parc de la Croix-Blanche, 92298 Châtenay-Malabry cede, France*

**Corresponding author (e-mail: juergen.soennke@bgr.de)*

Abstract: At the Meuse-Haute-Marne underground rock laboratory, France, a mine-by experiment was performed in the niche GCS by ANDRA. A 3D coupled hydraulic and mechanical (HM) continuum model was applied to understand the coupled HM mechanisms in the Callovo–Oxfordian claystone. Features such as transverse isotropy, shrinkage phenomena and permeability change induced by mechanical deformation were considered using the numerical code RockFlow. Intensive parameter studies in comparison with measured deformation and pore pressure data reduced model uncertainties. Fully saturated models showed very precise calculation results for the far-field zone sensors, but rapidly decreased pore pressure and continuously increased deformation in the near-field zone cannot be interpreted adequately even considering shrinkage and partially saturated flow models. A 2D isotropic damage model taking stiffness degradation into account was applied and acceptable calculation results were obtained for pore pressure evolution. However, the time-dependent deformation cannot be evaluated by the current model.

In the framework of the project OHZ (Observation and Monitoring of the Excavation Damaged Zone – EDZ), an excavation experiment (mine-by) was performed in the niche GCS at Meuse-Haute-Marne (MHM) underground rock laboratory (URL) (Fig. 1). This project was initiated in 2008 by ANDRA. The aims of this project are: to investigate hydromechanical response of Callovo-Oxfordian claystone (COx) during drift excavation (short- and long-term material behaviour), to characterize EDZ and its evolution over time, and to develop numerical models by comparing their results with *in situ* measurements. The project consists of experiments and analyses concerning different niches located 490 m below the surface level. Experimental designs of these niches were realized considering excavation method, support system and *in situ* stress orientation (Armand *et al.* 2007). During the excavation of the GCS, a soft support system (yieldable wedges combined with shotcrete) is used to allow long-term deformation of the shotcrete shell (Noiret *et al.* 2011). Radii of the drift GCS and adjacent niche GAT are 2.6 and 2.85 m, respectively (Fig. 1).

Callovo-Oxfordian claystone

The argillaceous rock COx is of a special interest because of its very favourable characteristics for being a possible host rock for the disposal of radioactive waste. Argillaceous rocks can be characterized by relatively high stiffness (highly consolidated), high potential for sealing cracks (swelling capacity and visco-plasticity), favourable radionuclide retention properties (high sorption capacity for most radionuclides) and very low water permeability, with low hydraulic gradients (diffusion dominated mass transport).

Bedding layers in the COx at the URL are nearly flat with a slight 1–2% regional dip towards NW (Distinguin & Lavanchy 2006; Delay *et al.* 2006). These bedding layers cause anisotropy of hydraulic and mechanical behaviour of the rock.

Analysis of the experimental data

Hydromechanical response of COx during the mine-by experiment was observed by monitoring deformation and pore water pressure changes around the GCS. Pore water pressures (P_L) around the excavated tunnel are measured using multi-packer systems (Fig. 2). *In-situ* pore pressure evolution is presumed to be influenced by excavation-induced pore volume variations owing to compaction, expansion and fracturing (short-term behaviour), as well as shrinkage-induced desaturation processes during ventilation (long-term behaviour). In the near-field zone of the chambers

From: Norris, S., Bruno, J., Cathelineau, M., Delage, P., Fairhurst, C., Gaucher, E. C., Höhn, E. H., Kalinichev, A., Lalieux, P. & Sellin, P. (eds) 2014. *Clays in Natural and Engineered Barriers for Radioactive Waste Confinement*. Geological Society, London, Special Publications, **400**, 265–278.
First published online June 16, 2014, http://dx.doi.org/10.1144/SP400.42

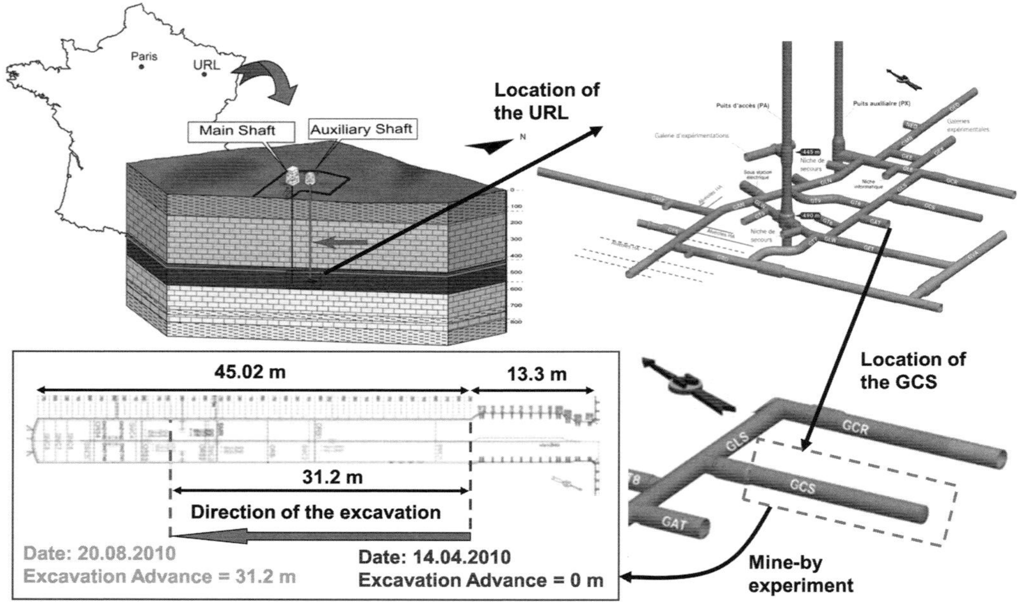

Fig. 1. Location of the drift GCS (mine-by experiment) in the Meuse-Haute-Marne (MHM) underground rock laboratory (URL).

(distance to the drift $\leq$ radius of the drift = 2.6 m) and especially parallel to the bedding (along OHZ1521), compression-induced pore space compaction is predominant in the horizontal plane. A mean stress increase owing to a stress redistribution after an excavation might cause these pore space compactions, which lead to overpressures ($P_L > P_{L,\,initial}$ = 50 bar). Subsequent to these increases, accelerated drop to atmospheric pressure (1 bar) was observed (Fig. 2), which means that water drained immediately. This incident indicates a possible damaged zone or an elastic pore volume expansion. Generally, pore pressure results show anisotropic pattern owing to the bedding layers (Fig. 12). In the sidewall overpressures up to 75 bar in Chamber 4-OHZ1521 (distance to the drift wall $\approx$4.8 m) were observed, unlike those in the roof and floor.

Deformations around the excavated tunnel are measured using extensometers and deflectometers (Fig. 3). Measured relative displacement values during the mine-by experiment in the adjacent rock show diverse magnitudes owing to anisotropic material behaviour and coupling phenomena. Relative displacements remained within a range of millimetres during the mine-by experiment (14 April 2010 to 31 August 2010), and they never exceed 35 mm in the radial (σ_h) direction and 3.5 mm in the axial (σ_H) direction.

Moreover, higher displacements in near-field zone anchors (distance to the drift $\leq$ radius of the drift) are observed than those located in the far-field zone (distance to the wall > radius of the drift). These results obtained from near-field zone anchors may indicate a possible damaged zone. In OHZ1501 the magnitude of relative displacements increases systematically with distance to the drift after the excavation passage (Fig. 3).

Considering all of these complex phenomena caused by anisotropy and fluid–solid coupling during the excavation in the COx, coupled hydro-mechanical (HM) process modelling was conducted by the Federal Institute for Geosciences and Natural Resources to understand the underlying processes. Modelling calculations are divided into two main phases: 3D coupled HM modelling without damage and 2D isotropic damage modelling.

Numerical background and material parameters of the Callovo-Oxfordian claystone

The code RockFlow is based on the finite element method. Liquid-phase flux was calculated using a modified Darcy's law (Bear & Bachmat 1990). Richards's approximation was used to define the partially saturated medium by means of

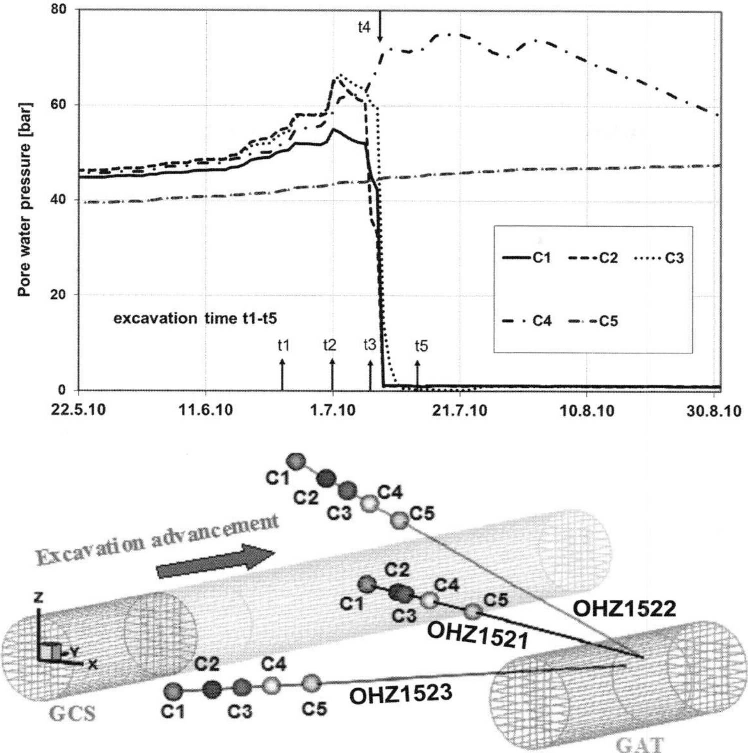

Fig. 2. Measured pore water pressures in OHZ1521 and the multipacker arrangement with chambers. NB. The date on which the excavation face reached the level of the corresponding sensor is shown by the arrow on the figure.

single-phase formulation (Richards 1931). The dependency between relative permeability and liquid saturation was defined by employing the Mualem–van Genuchten model (Mualem 1976; van Genuchten 1980) and capillary pressure–liquid saturation dependency was defined by employing the van Genuchten model (van Genuchten 1980). Generalized Hooke's law was employed to describe the constitutive equation for linear elastic material.

Transverse isotropic material properties were assumed to model the hydraulic and mechanical anisotropy caused by the bedding in COx (Table 1). Therefore COx was assumed to exhibit different hydraulic and mechanical properties along the bedding layer and in the layer normal to the bedding. Coupling effects between hydraulic and mechanical parameters were realized by Terzaghi's effective stress concept, temporal change of displacements implemented in the mass balance equation, Biot coefficient and specific storage coefficient. Terzaghi's effective stress concept describes the effects of a liquid phase on volumetric part of stress (von Terzaghi 1954). Total stress was recalculated by employing effective stress in a solid skeleton, pore water pressure, Biot coefficient and Bishop parameter (Bishop & Blight 1963). Moreover, Biot coefficient and specific storage coefficient were implemented in the mass balance equation to simulate the impacts of compressibility of elements (pores, solid and liquid) in a porous medium. The time derivative of solid displacement in the mass balance equation stands for the consolidation effect on a liquid flow (Kohlmeier 2006). Swelling/shrinkage and deformation-dependent porosity models (Massmann 2009) were also applied for a better understanding of HM processes. The effect of shrinkage and swelling phenomena in claystone was modelled by connecting swelling strain change directly with liquid saturation evolution. Like thermal strain calculations, strains caused by swelling are added into the elastic strains as a stress-independent part of strain tensors. By using deformation-dependent porosity models, porosity can be determined depending on its initial value (φ_0), elastic volumetric

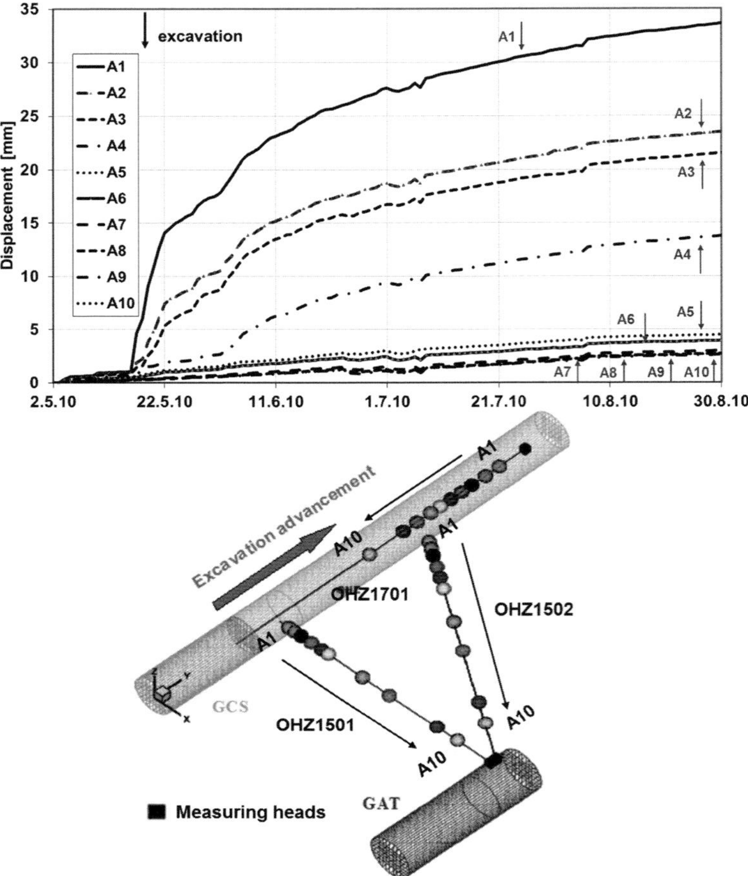

Fig. 3. Measured relative displacements in OHZ1501 and borehole arrangement with extensometers. NB. The date on which the excavation face reached the level of the corresponding sensor is shown by the arrow on the figure.

strain and swelling/shrinkage volumetric strain of a material.

3D coupled hydromechanical modelling of the mine-by experiment

3D coupled HM model setup

The uniform mesh used for 3D simulations with the applied *in-situ* stresses and the HM anisotropy in the COx for the base case is depicted in Figure 4. The *in-situ* stress state was set to $\sigma_h = -11.5$ MPa $= \sigma_v$, $\sigma_H = -14.9$ MPa (Morel *et al.* 2010), and fixed (zero) displacement boundary conditions were applied to the outer model boundaries. Initial pore water pressure ($P_{L,\ initial}$) was set to 50 bar (Su 2007) in the model domain. The value of the hydraulic boundary condition, along the excavated zone of the GCS ($P_{L,\ Drift}$), was assigned a value of atmospheric pressure (1 bar), thus fully saturated flow ($P_L > 0$ bar) was realized. Excavation advance is simulated as a time-dependent deactivation of elements. The excavation interval between 13.3 m (14 April 2010) and 44.5 m (20 August 2010) was used for the modelling of mine-by experiment (Fig. 1).

Parameters in the base case were stepwise varied based on the measured data to understand the underlying physical processes. Analyses showed that parametric variations of Young's modulus and intrinsic permeability have the dominant influence on computed pore pressure and displacement evolution compared with variations of initial stress and Poisson's ratio. Additional modelling concepts (e.g. EDZ as a separate material group, shotcrete lining) and parameter variations were performed based on literature research, laboratory tests and

Table 1. *Hydromechanical material properties of Callovo–Oxfordian claystone based on literature research*

Parameter	Value	Remarks
Intrinsic permeability (m^2)	$k_{Parallel} = 2 \times 10^{-20}$	Parallel to the bedding
	$k_{normal} = 2 \times 10^{-21}$	Normal to the bedding
Initial porosity	$\Phi_0 = 0.16$	
Young's modulus (MPa)	$E_{Parallel} = 6000$	Parallel to the bedding
	$E_{normal} = 4000$	Normal to the bedding
Poisson's ratio	$\nu = 0.3$	
Cross shear modulus (MPa)	$G_{cross} = 1540$	
Biot coefficient	$\alpha = 0.6$	Coupling parameter
Specific storage coefficient (Pa^{-1})	$S_s = 0$*	Coupling parameter

*In a fully water saturated medium storage will be caused only by skeleton strain.

in-situ measurements. Important parameter values of selected best-fit model settings (nos. 32, 39 and 43) are given in Table 2.

Simulation results of selected sensors

Model variations show diverse results depending on sensor locations, anisotropy and HM coupling mechanism. In this paper, hydromechanical responses of the rock at four (out of 38) sensors are discussed. These sensors are divided into two spatial categories (near- and far-field zone) with respect to their measurement results and influence of the excavation.

Results in the near-field zone of the drift during the mine-by experiment

Displacement values obtained at the Anchor 2 of borehole OHZ1501 with the base case model were

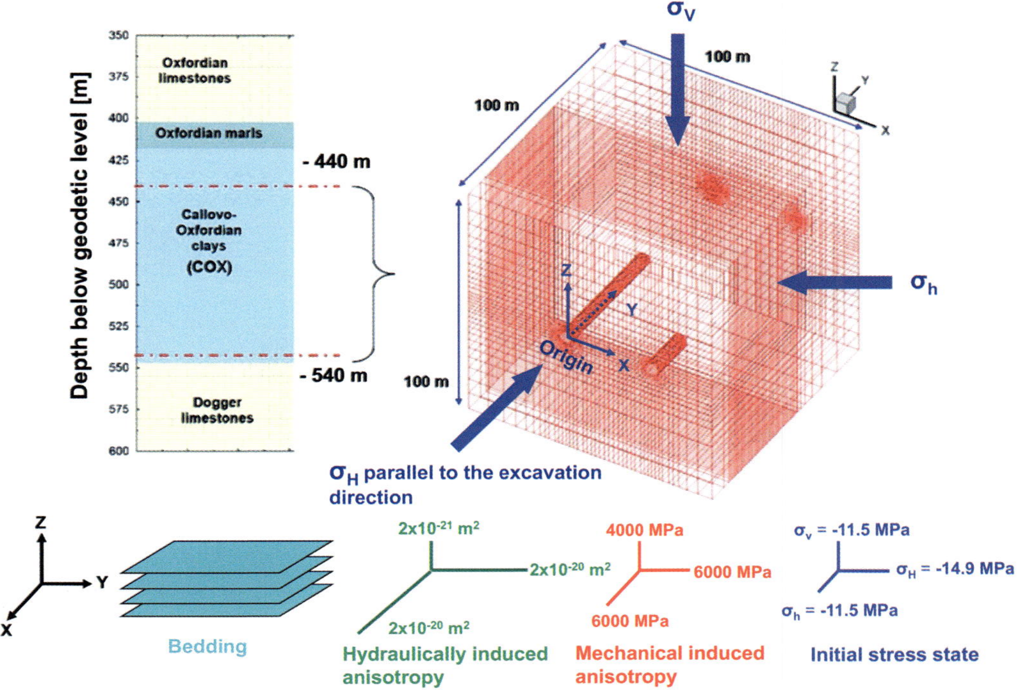

Fig. 4. Stratification at the MHM URL (Distinguin & Lavanchy 2006, p. 9, fig. 6) and the modelling mesh with *in-situ* stress state, anisotropy and bedding in the Callovo–Oxfordian claystone. The drift GCS and the niche GAT are showed on the left side and the right side in the mesh, respectively.

Table 2. *Parameters of selected model settings*

Parameter	No. 32	No. 39	No. 43
$k_{parallel}$ (m^2)	2×10^{-19}	2×10^{-19}	2×10^{-19}
k_{normal} (m^2)	2×10^{-20}	2×10^{-20}	2×10^{-20}
$E_{parallel}$ (MPa)	4000	4000	3795
E_{normal} (MPa)	1500	1500	2530
G_{cross} (MPa)	1540	682	973
S_s (Pa^{-1})	0	2.04×10^{-10}	6.7×10^{-10}
$P_{L, Drift}$ (bar)	0	−998	−998

lower than measured ones, which is mainly due to the fully saturated flow condition. Displacement of Anchor 2 indicates the convergence towards the GCS during the mine-by experiment, which means that this anchor might locate in the convergence zone at the rock. A partially saturated flow and swelling/shrinkage model was used to obtain an increase in compressive stresses by coupling and an increase in displacement towards the GCS as well. The coupling effect on the mechanical part was then maximized with applying partially saturated flow ($P_{L, Drift} = -998$ bar) and using the swelling/shrinkage material model in model no. 43. However oscillations of computed displacements under partial saturation were observed owing to the coupling in the near-field zones. Specific storage value was then increased to become more compressible and less stiff, which makes the liquid less sensitive to the gradient changes (caused by boundary conditions) and to the coupling effects. Higher displacements were obtained by model no. 43 with increases up to 6 mm at Anchor 2-OHZ1501 (Fig. 5) in comparison with the base case. When the excavation face reached the level of this sensor ($t = 19.05.10$), rock displaced 3.8 mm towards the drift and displacements of 2.6 and 1 mm were obtained by model no. 43 and the base case model, respectively. Both computed and measured displacement rates reached their maximum values owing to the influence of the advancing excavation face (steep curves); afterwards they started to decrease with the time (less steep more flat curves). Long-term relative displacement evolution could not be reproduced by any of these model variations despite using partially saturated flow, coupling and swelling/shrinkage, because in these models visco-plastic material behaviour was not considered. At the end of the mine-by experiment ($t = 20.08.12$) rock was displaced 23 mm towards the drift; however, in model no. 43 and the base case model rock was displaced only 11.2 and 5.5 mm, respectively. During the mine-by test, the evolution tendency of computed relative displacement values at this anchor was in agreement with

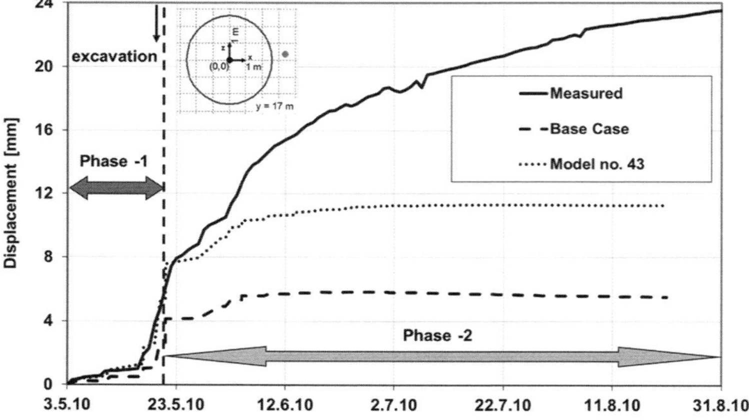

Fig. 5. Displacement evolutions at Anchor 2-OHZ1501, comparison of measured and computed values. NB. The date on which the excavation face reached the level of the corresponding sensor is shown by the arrow on the figure. Dot and circle on the sketch represent a location of a corresponding sensor and the drift GCS, respectively. Number in a bracket is a distance of a sensor to the model origin.

the measured ones, which indicates convergence towards the drift.

The computed displacement evolution in the near-field zone can be divided into two phases (Fig. 5) taking into account the coupling mechanism (effective stress calculations), mechanical anisotropy (Young's modulus) and initial stress anisotropies. In phase 1, anisotropy was more dominant because of elastic stress rearrangement owing to the excavation. This phase started from the beginning of calculations ($t = 03.05.10$) until one day after the excavation passage at the anchor ($t = 20.05.10$). Model no. 43 showed very precise fittings during this phase. At the location of this anchor, effective stress was strongly decreasing owing to the boundary condition change (partially saturated flow) during the excavation and increasing overpressure.

In phase 2, coupling effects induced by partially saturated flow became more dominant. The computed displacement rate started to decrease and finally stayed constant, because coupled deformation in the *elastic medium* was not capable of reproducing long-term material behaviour (possible visco-plastic response of the rock). Until the end of the simulation, effective stress recovered with decreasing pore pressure during drainage.

Volumetric strain evolution at the level of Anchor 2-OHZ1501 was analysed (Fig. 9). Before and after the excavation passage at this anchor's level, expansion (tensile volumetric strain) was observed above the roof and below the floor of the drift, and compaction (compressive volumetric strain) was observed on the side wall zone. Tensile volumetric strain continued to develop around the drift until the end of the mine-by experiment.

Computed and measured pore water pressure curves at the Chamber 2-OHZ1521 are shown in Figure 6. Under fully saturated flow condition (base case) water could not drain towards the drift after the excavation passage. Measured pore pressure reached its maximum value, $P_L = 65.2$ bar ($t = 01.07.10$), owing to compression and then started to decrease rapidly during the excavation passage. The effect of the excavation face was within a distance of around two diameters of the drift. After 10 days it was equal to atmospheric pressure. Computed pore pressure in the base case model reached its maximum value $P_L = 93.5$ bar subsequent to excavation ($t = 09.07.10$) and then started to decrease slowly, unlike the measured value. During 43 days, pore pressure ($P_L = 41.2$ bar) drained too slowly and could not decrease to atmospheric pressure.

In order to obtain this accelerated pressure drop in the near-zone, partially saturated flow with high suction pressure ($P_{L,\,Drift} = -998$ bar) was applied in model no. 39. Effective stress calculation (HM coupling) in partially saturated media led to oscillations and amplifications (overpressures) of computed pore pressure at the near-zone. The effect of pore pressure on stress calculations was then prevented via disabling the effective stress model (no influence of H on M). This modelling approach is adapted from Martin & Lanyon (2003) and Massmann (2009). According to Martin & Lanyon (2003), the applicability of Terzaghi's effective stress law on calculations for claystone is questionable. In a claystone, part of natural water is not free to migrate and it is bonded to clay particles (adhesion). The specific storage was then

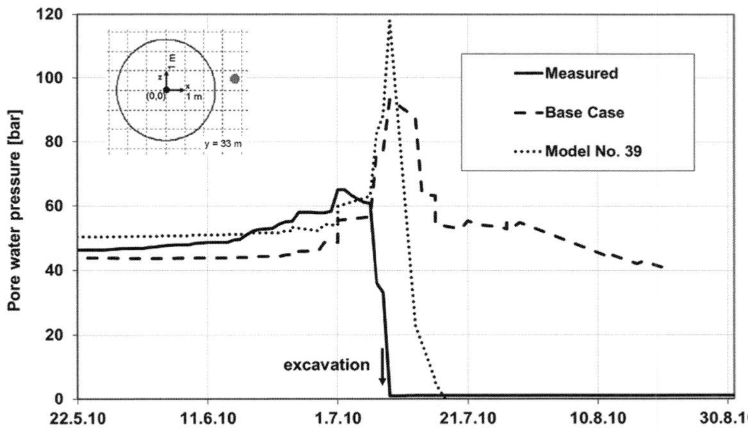

Fig. 6. Pore-water pressure evolutions at Chamber 2-OHZ1521, comparison of measured and computed values. NB. The date on which the excavation face reached the level of the corresponding sensor is shown by the arrow on the figure. Dot and circle on the sketch represent a location of a corresponding sensor and the drift GCS, respectively. Number in a bracket is a distance of a sensor to the model origin.

increased to adjust fluid compressibility in pore space. Computed pore pressure in model no.39 could drain faster than in the base case, but high suction caused overpressure despite modelling modifications. Computed pore pressure reached its maximum value ($P_L = 118.1$ bar) subsequent to the excavation ($t = 09.07.10$) and after 11 days water could drain ($P_L = -8.2$ bar under $S_L = 0.9$).

Accelerated pressure drops in the near-zone cannot be accurately reproduced using elastic models even using partially saturated conditions, material anisotropy, coupling and swelling/shrinkage. This might be due to the lack of a damage mechanism in the 3D models. According to the experimental results obtained by ANDRA around the drift GCS, this chamber is located inside of a possible EDZ, wherein connected fracture networks were dominant, thus, drastic permeability enhancement was measured and then water was assumed to drain within this fracture network immediately. In an elastic model, accelerated pressure drop subsequent to excavation may be better predicted by a partially saturated model in comparison to a fully saturated one. On the other hand, compaction induced pressure increase before the excavation passage was accurately reproduced by anisotropic elastic models using HM coupling (Biot coefficient).

Results in the far-field zone of the drift during the mine-by experiment

Back analysis of the base case (fully saturated) results showed that the adjacent niche GAT (Fig. 3) had a diminishing influence on displacement calculations of anchors at the extensometer OHZ1501 and on pore pressure calculations of chambers at the multipacker OHZ1521 owing to the lack of a support system around the GAT. This influence became more dominant with decreasing distances to GAT. In order to compensate for this effect, the displacement boundary condition value along GAT was assigned a value of zero (fixed type) in the (fully saturated) model no. 32 instead of using a shotcrete lining in the model. Therefore possible coupling effects of mechanical parameters on hydraulic subcalculations around GAT were disabled.

Computed rock displacements at the Anchor 6-OHZ1501 with the base case were lower than the measured ones (Fig. 7), which were then successfully increased in model no. 32. At the end of the mine-by test ($t = 20.08.10$) rock was displaced 3.8 mm towards the drift GCS. It was displaced 2.4 and 1.7 mm in model no. 32 and the base case, respectively. The increasing trend of rock displacement during the experiment was reproduced by both models.

Pore pressure measured at Chamber 4-OHZ1521 (Fig. 8) started to increase ($t = 23.05.10$) and reached its maximum value, $P_L = 75.1$ bar ($\Delta P \approx$ 29 bar), 12 days after the excavation passage ($t = 18.07.10$). At the end of the mine-by test ($t = 20.08.10$) pressure decreased up to 64.3 bar ($\Delta P \approx 11$ bar). In the base case pore pressure started to increase ($t = 09.06.10$) and reached its maximum value $P_L = 47.3$ bar ($\Delta P \approx 3$ bar) on the excavation passage date ($t = 06.07.10$). At the end of the mine-by test pressure decreased up to 46.5 bar ($\Delta P \approx 1$ bar). Pressure evolution in the

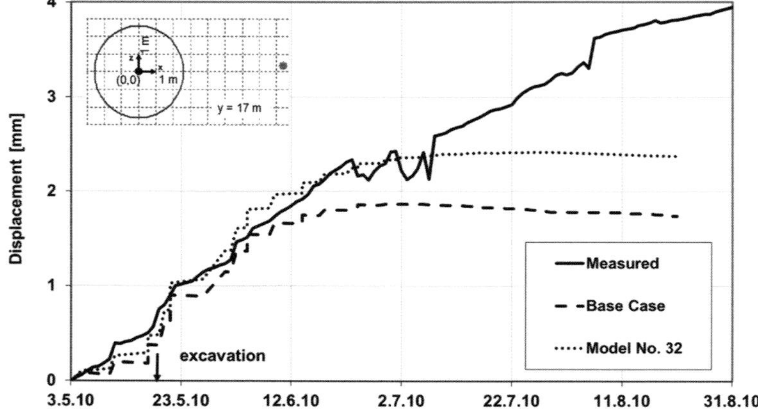

Fig. 7. Displacement evolutions at Anchor 6-OHZ1501, comparison of measured and computed values. NB. The date on which the excavation face reached the level of the corresponding sensor is shown by the arrow on the figure. Dot and circle on the sketch represent a location of a corresponding sensor and the drift GCS, respectively. Number in a bracket is a distance of a sensor to the model origin.

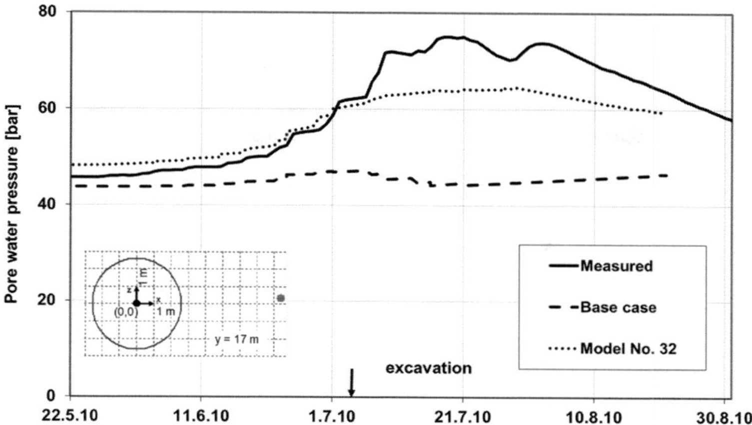

Fig. 8. Pore-water pressure evolutions at the Chamber 4-OHZ1521, comparison of measured and computed values. NB. The date on which the excavation face reached the level of the corresponding sensor is shown by the arrow on the figure. Dot and circle on the sketch represent a location of a corresponding sensor and the drift GCS, respectively. Number in a bracket is a distance of a sensor to the model origin.

near-field zone in model no. 32 was corrected by preventing the water drainage towards the GAT through the boundary condition modification mentioned before. In model no. 32 pressure started to increase ($t = 22.05.10$) and reached its maximum value $P_L = 64.6$ bar ($\Delta P \approx 16$ bar) 23 days after the excavation passage ($t = 29.07.10$). At the end of the mine-by test pressure decreased by up to 59.6 bar ($\Delta P \approx 5$ bar).

Comparative evaluation of several model results and sensitivity analyses showed that, in comparison to the anisotropy of *in-situ* stress state in the URL ($\sigma_H/\sigma_h = \sigma_H/\sigma_v \approx 1.3$), mechanical anisotropy plays a more important role in explaining these overpressures (induced by compaction) during the experiment than the hydraulic anisotropy (Massmann 2009). Generally it can be concluded that in the far-field zone relative displacements and pore pressures can be very precisely reproduced by coupled fully saturated elastic model settings. For this reason, the rock in the far-field zone of the drift (distance to the wall > radius of the drift) might remain fully saturated and might exhibit an elastic deformation during the mine-by experiment.

Projected volumetric strain (model no. 43) and pore water pressure evolutions (model no. 39) are depicted in Figure 9. During the mine-by experiment compressive volumetric strains (with negative sign) mostly along the bedding and volumetric extension strains (with positive sign) mostly along the direction normal to the bedding were observed in model no. 43. Water could drain faster along the bedding than along the direction normal to the bedding in model no. 39. The extent of the partially saturated zone was therefore larger in the bedding direction. Overpressures were observed mainly along the bedding direction in the far-field zone. Inconsistent patterns of volumetric strains and pore pressures occurred owing to the coarser elements located in the excavation direction (y).

2D coupled damage modelling of the mine-by experiment

Long-term material behaviour (possibly induced by inelasticity) was observed in the near-field zone of the drift. Such responses of the rock – accelerated pressure drop and continuously increasing displacement – cannot be calculated even considering partially saturated flow, coupling and/or swelling/shrinkage. Therefore a damage model was applied at the locations of Chamber 2-OHZ1521 and Anchor 2-OHZ1501 to reproduce long-term material behaviour in the near-field zone during the mine-by experiment.

Damage model concept

This damage model was based on the assumptions that the material is brittle and the damage mechanism is purely tensile with an isotropic pattern (Massmann 2009), which means failure in tension rather than in shear regime.

The damage mechanism is based on a *scalar damage variable*, D (Kachanov 1958). In the chosen zone P within the representative elementary volume (REV), the damage variable (D) is

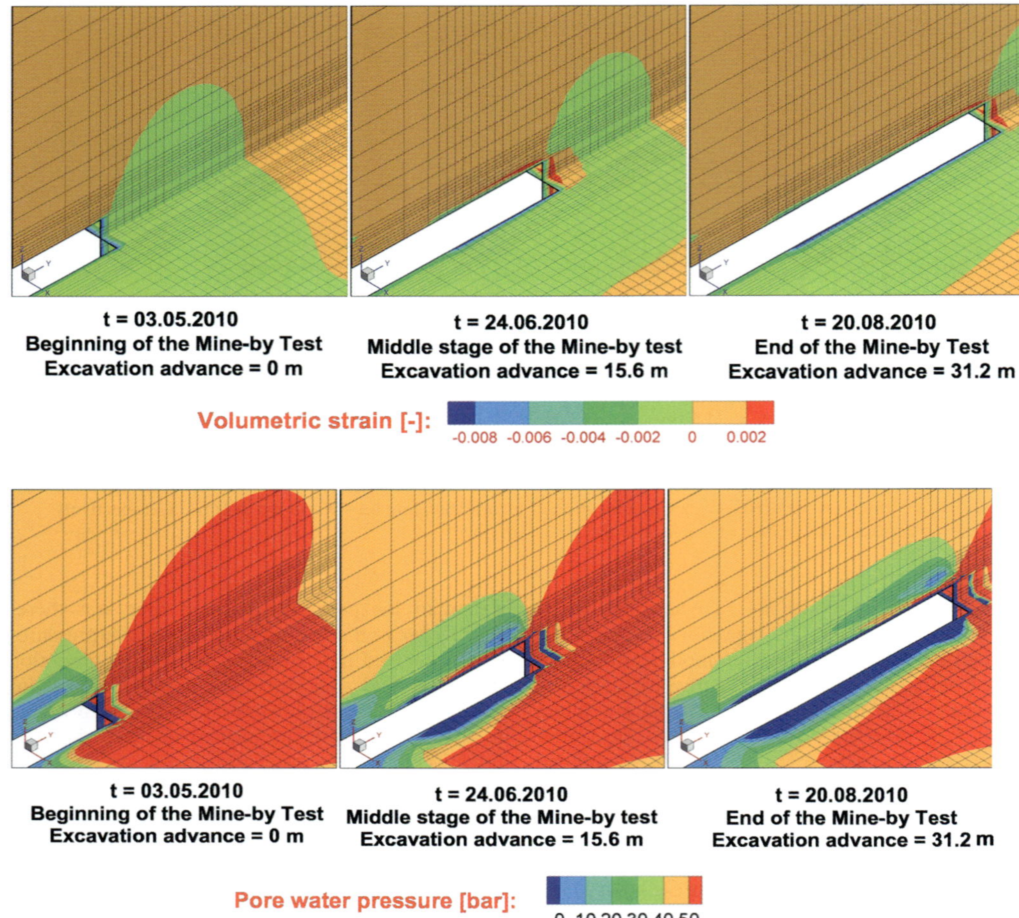

Fig. 9. Results of model no. 43 for the mechanical calculations (above) and results of model no.39 for the hydraulic calculations (below).

related to the area of a plane (dA (m^2)) and the area of intersection of all microcracks or cavities in dA (dA_D (m^2)):

$$D = \frac{dA_D}{dA} \quad (1)$$

Post-condition: 0 (undamaged-virgin element) $\leq D \leq 1$ (fully damaged – ruptured element).

Progressive degradation of material stiffness was realized via updating of the initial elastic material tensor (C_{init}) with evolving damage. The linear decrease of the initial Young's modulus (E) in a damage model can be simply presented in the one-dimensional case. Change in Young's modulus during damage (E_D) is given by:

$$E_D = E \cdot (1 - D) \quad (2)$$

The evolution function, which allows the coupling of elasticity with damage mechanism, is given according to Marigo (1981):

$$D_t = \left(\frac{r_t - r_{init}}{A}\right)^B, \quad \text{if } r_t > r_{init} \quad (3)$$

where r_{init} is the damage threshold (energy barrier) in the initial (undamaged) state (Pa), A is the damage consolidation modulus (Pa) and B is the exponent. After parametric studies, damage

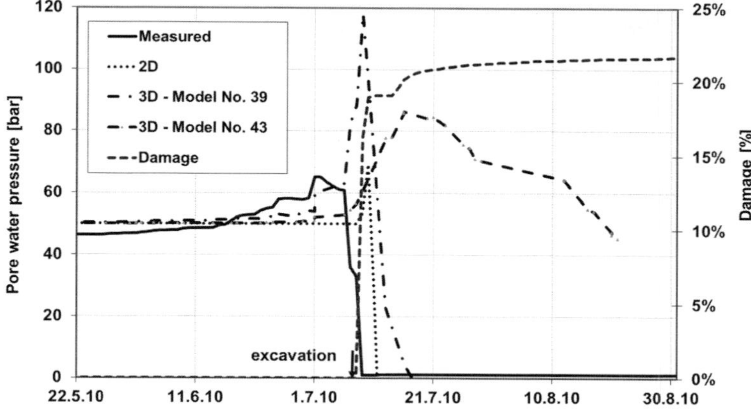

Fig. 10. Pore-water pressure and damage variable evolutions at the Chamber 2-OHZ1521, comparison of measured and computed (2D and 3D) values. NB. The date on which the excavation face reached the level of the corresponding sensor is shown by the arrow on the figure.

parameters in the model were assigned values of: $r_{init} = 0.02$ MPa, $A = 5$ MPa and $B = 0.3$.

2D coupled damage model setup

The 2D mesh ($100 \times 100\ m^2$) within the *xz*-plane was used for damage modelling of the mine-by experiment. *In-situ* stress state was set to $\sigma_h = -11.5\ \text{MPa} = \sigma_v$ and fixed (zero) displacement boundary conditions were applied on outer model boundaries. Initial pore water pressure ($P_{L,\ initial}$) was set to 50 bar in the model domain. Evolution of damage was realized by HM coupling only with a partially saturated flow development. Therefore, $P_{L,\ Drift}$ was assigned a value of -998 bar. According to previous results obtained by model nos. 39 and 43, the effect of pore pressure on stress calculations was minimized to prevent the influence of suction on stress and specific storage was then increased. Reference values of the stiffness and the intrinsic permeability (base case) were adapted as initial (undamaged) rock properties. Advancement of an excavation face in a 3D space cannot be represented in 2D, so damage simulation in 2D was realized by applying a hydraulic boundary condition on an open drift contour (free displacement) during the mine-by experiment and, thus, deformation and pore pressure evolution before the excavation passage could not be considered in 2D.

Simulation results of selected sensors

Results of computed pore pressure evolutions by 2D (damage) and 3D calculations (without damage, model nos. 39 and 43) are depicted with measured data at the Chamber 2-OHZ1521 (Fig. 10). Both magnitude and trend of pressure evolution owing to excavation could be adequately reproduced using the 2D damage model. Pore pressure and displacement in the damage model stayed unchanged until one day before the excavation passage date ($t = 08.07.10$). This occurred because of the geometrical limitations as mentioned before. Pore pressure increased up to 69.3 bar one day after the excavation passage on 10.07.10, which corresponds to the measured pore pressure (65.2 bar) and then it decreased rapidly to -20.9 bar (under the suction conditions).

Rock started to be damaged with a value of 16% ($D = 0.16$) on the excavation passage date ($t = 09.07.10$) at Chamber 2-OHZ1521. One day after the excavation passage 3% more damage occurred. Maximum damage (21%) was obtained 15 days after the passage. At the end of the experiment 22% of rock was damaged.

Results of rock displacement in the near-zone (Fig. 11) show that long-term deformation behaviour observed at Anchor 2-OHZ1501 could not be reproduced by the damage model. In the 2D model, rock had started to displace towards the drift (7.7 mm) after the excavation passage ($t = 20.05.10$) and at the same time 18% of rock was damaged (*in-situ* case: 0.9 mm on $t = 20.05.10$). Rock continued to displace and also continued to be damaged but with decreasing rates. At the end of the experiment rock was displaced 10.3 mm, which is less than measured value (19.2 mm), and 22% of it was damaged.

According to Massmann (2009), the damage criterion is a function of strain energy, considering tensile strains only. In the 2D model, using

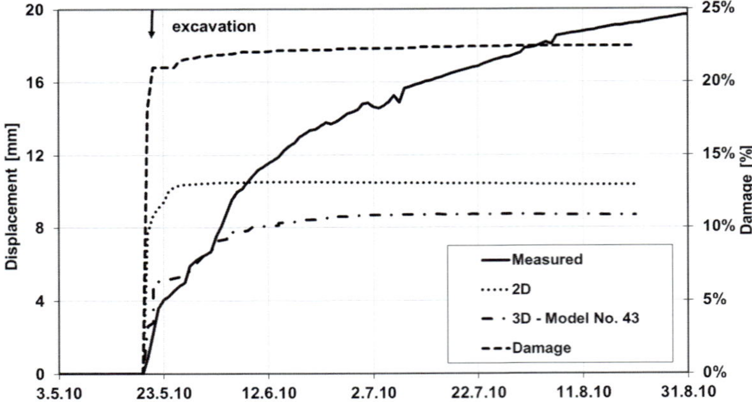

Fig. 11. Displacement and damage variable evolutions at the Anchor 2-OHZ1501, comparison of measured and computed (2D and 3D) values. NB. The date on which the excavation face reached the level of the corresponding sensor is shown by the arrow on the figure.

relative higher suction pressure around the drift causes rapid desiccation and this leads to overpressures (tensile strains) via HM coupling and change of strain energy. When this strain energy exceeds the threshold then damage occurs and this causes an enhancement of intrinsic permeability (approximately 5 orders of magnitude here). This increase in intrinsic permeability causes the accelerated pressure drop immediately after the excavation passage. Accurate modelling of measured pore pressure was hence obtained only considering damage. However, all of these processes led to poor estimation of rock mass displacement because desiccation occurred rapidly within a short time interval and then slightly evolved until the end of the experiment. Therefore displacements also reached their maximum values with decreasing Young's modulus via coupling within a short time interval, and slightly evolved during the rest of the time.

Excavation damaged zone assessment

Computed and measured pore water pressure distributions are depicted in Figure 12. All diagrams in this figure are isochronic ($t = 20.08.10$ = End of the mine-by experiment and simulations) and the cross-section in 3D model no. 39 is taken within a plane $y = 33.53$ m, which is the excavation date of the chosen near-field zone sensor (Chamber 2-OHZ1521) level. These partially saturated zones are considered to be an indicator of the hydraulically

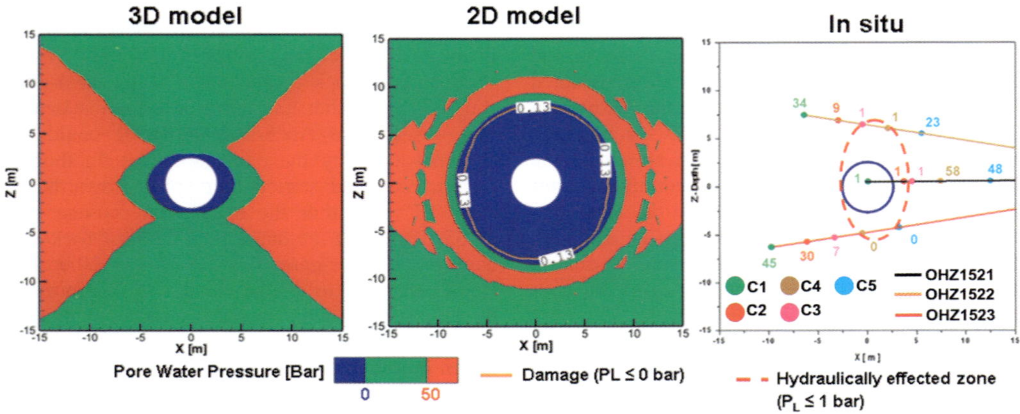

Fig. 12. Spatial extent of hydraulically effected zone obtained in 2D damage and 3D models in comparison with *in-situ* data at the time step $t = 20.08.2010$ (end of the mine-by experiment and simulations).

effected zone owing to the excavation. The value of the damage variable within the hydraulically affected zone in the 2D calculations ($D = 0.13$) is given in the white box in the figure. This value is assumed to be the lower limit that leads to partially saturation in the near-field zone. Values lower than this limit are assumed not to cause partial saturation ($S_L = 1$).

At the end of the mine-by experiment, partially saturated rock zone around the drift (blue zone confined by red circle in middle diagram) was at least 13% damaged. The zone beyond was fully saturated with water and was damaged less then 13%. Overpressures were mostly observed near to the side walls in both models as well as in the *in-situ* case (Fig. 12). The maximum extents of the partially saturated zone in the *in-situ* case were $\Delta x \approx 1.9$ m and $\Delta y \approx 4.4$ m. The extent of the computed partially saturated zone was increased using damage in the 2D model ($\Delta x \approx 5.4$ m, $\Delta y \approx 5.9$ m) in comparison with the 3D model ($\Delta x \approx 1.9$ m, $\Delta y \approx 0.4$ m).

Considering the isotropic stress state in 2D and the isotropic mechanism of damage based primarily on the stiffness degradation, mechanical anisotropy is more decisive than permeability anisotropy for coupled damage evolution. As mentioned before, this inference is also valid for a 3D model without damage. Mechanical anisotropy ($E_{\text{Parallel}}/E_{\text{normal}} = 1.5$ in 2D model) led to damage differences around the drift. Rock zones above the roof and below the floor were more damaged than those near to the sidewalls of the drift. Pore pressures above the roof and below the floor decreased faster than those near to the sidewalls ($\Delta y > \Delta x$) owing to the different damage zones. This fact was also addressed by Massmann (2009) for an anisotropic rock, where mechanical anisotropy is more significant than permeability anisotropy for possible amplification (e.g. overpressure) or attenuation (e.g. pressure drop) of pore pressure near to the drift after excavation.

In order to explain the *in-situ* partially saturated zone development during the mine-by experiment, further parametric studies should be performed with the 2D damage model. Absence of a 3D stress state and constant pressure boundary condition around the drift may account for the inconsistency between computed and measured data. The effect of a shotcrete lining should be taken into account too. During the mine-by experiment the shotcrete lining emplacement might have stabilized the partially saturated flow development. Induced fractures preferably propagate perpendicular to the direction of the minimum principal stress (σ_3). However, it is not possible to represent this incident in the 2D plane. Furthermore hydraulic boundary condition around the drift should be modified for excavation face advancement. Analyses considering an equivalent pressure change in the drift to simulate the front advancement in a 2D damage model are still pending.

Conclusions

A better understanding of complicated hydro-mechanical responses to the excavation in COx has been gained through fully coupled HM modelling with stepwise inclusion of influencing properties, for example, transverse isotropic properties, swelling-shrinkage phenomena of clay material, partially saturated flow conditions and damage mechanism. In order to deepen the knowledge of the excavation-induced hydromechanical behaviour of the COx, long-term *in situ* measurements are being conducted at the MHM URL by ANDRA.

Special emphasis was put on the study of the specific storage coefficient (an important coupling term), partially saturated flow, mechanical anisotropy and permeability anisotropy to explain the increasing rock mass displacement towards the drift and pore water pressure changes measured during the excavation. In 3D modelling intrinsic permeability values should be increased to one order of magnitude and Young's modulus values should be reduced to 40–66% in comparison with the base case parameters from the literature to obtain better agreement with the measurement results. Using the 2D damage model, which considers stiffness degradation, hydraulic near-zone observations could be reproduced reasonably. Generally, far-field zone observations of both displacement and pore water pressure could be accurately reproduced by a coupled elastic and fully saturated model considering material anisotropy. This may imply that partial saturation zone and inelastic deformation have less influence on the rock part in the far-field zone.

Under partially saturated conditions, suction with high negative pressure may play an important role. Long-term mechanical response of rock in the near-field zone could not be reproduced by the current coupled HM anisotropic models owing to limitations. However, spatial extent and temporal development of EDZ has been explained by desiccation-induced coupled HM damage models for transverse isotropic materials.

It is recommended to focus on the following aspects for further work for better understanding of a mine-by experiment in a transversely isotropic medium: determination of capillary pressure on drift wall during the excavation; analysis of influence of temperature; revision of effective stress concept for claystones; and activation of support elements during an excavation.

Researches on the degradation of mechanical properties under partially saturated conditions are needed (e.g. Young's modulus change with liquid saturation). Moreover further parametric analyses including laboratory study on the constitutive law from coupled hydromechanical point of view are necessary to reproduce long-term mechanical response of rock in the near-field zone.

We would like to thank the German Federal Ministry for Economic Affairs and Energy (BMWi = Bundesministerium für Wirtschaft und Technologie) for funding the research and development activities described in this paper.

References

Armand, G., Rebours, H., Richard, L. & Delay, J. 2007. *Specifications de L'Experimentation OHZ Laboratoire de Recherche Soutterain de Meuse/Haute-Marne*, ANDRA-France Technical Report **D.SP.ALS.07.1074**.

Bear, J. & Bachmat, Y. 1990. *Introduction to Modeling of Transport Phenomena in Porous Media*. Kluwer Academic, Dordrecht.

Bishop, A. W. & Blight, G. E. 1963. Some aspects of effective stress in saturated and partly saturated soils. *Geotechnique*, **13**, 177–197.

Delay, J., Wileveau, Y. & Lesavre, A. 2006. *The French Underground Research Laboratory at Bure as a Precursor for Deep Geological Repositories*. ANDRA France, Technical Report **D.NT.ADPE.06.0290**.

Distinguin, M. & Lavanchy, J. M. 2006. Determination of hydraulic properties of the Callovo–Oxfordian argillite at the Bure site: synthesis of the results obtained in deep boreholes using several in situ investigation techniques. *Physics and Chemistry of the Earth*, **32**, 379–392, ScienceDirect.

Kachanov, L. M. 1958. Time of the rupture process under creep conditions, Leningrad University. *Izvestia Akademii Nauk SSSR, Otdelenie tekhnicheskich. Nauk.*, **8**, 26–31.

Kohlmeier, M. 2006. *Coupling of thermal, hydraulic and mechanical processes for geotechnical simulations of partially saturated porous media*. PhD dissertation, 72/2006, Hannover Leibniz University, Germany.

Marigo, J. -J. 1981. Formulation d'une loi d'endommagement d'un matériau élastique. *Comptes rendus de l'Académie des sciences Paris Séries II*, **292**, 1309–1312.

Martin, C. D. & Lanyon, G. W. 2003. Measurement of in-situ stress in weak rocks at Mont Terri Rock Laboratory. *International Journal of Rock Mechanics and Mining Sciences*, **40**, 1077–1088.

Massmann, J. 2009. *Modelling of excavation induced coupled hydraulic-mechanical processes in claystone*. PhD dissertation, 77/2009, Hannover Leibniz University, Germany,.

Morel, J., Delay, J., Armand, G., Rebours, H. & Conil, N. 2010. In situ experiments for the determination of rock properties and behaviour at the Meuse/Haute Marne Centre. *In*: *ICEM 14 – 14th International Conference on Experimental Mechanics*, Poitiers, France, EPJ Web of Conferences, **6**, id.17003.

Mualem, Y. 1976. A new model for predicting the hydraulic conductivity of unsaturated porous media. *Water Resource Research*, **12**, 513–522.

Noiret, A., Armand, G., Cruchaudet, M. & Coinl, N. 2011. Mine by experiments in order to study the hydromechanical behavior of the Callovo-Oxfordian claystone at the Meuse Haute-Marne Underground research laboratory (France). *In*: *8th International Symposium on Field Measurements in Geo-Mechanics*, Berlin, 12–16 September 2011.

Richards, L. A. 1931. Capillary conduction of liquids through porous media. *Physics*, **1**, 318–333.

Su, K. 2007. *Development of hydro-mechanical models of the Callovo–Oxfordian argillites for the geological disposal of radioactive waste (MODEX-REP)*, European Commission Nuclear Science and Technology and ANDRA, Final Report **FIKW-CT-2000-00029**.

Von Terzaghi, K. 1954. *Theoretical Soil Mechanics*, 7th edn. John Wiley & Sons, New York.

Van Genuchten, M. Th. 1980. A closed-form equation for predicting the hydraulic conductivity of unsaturated soils. *Soil Science Society of America Journal*, **44**, 892–898.

Excavation-induced fractures network surrounding tunnel: properties and evolution under loading

R. DE LA VAISSIÈRE[1]*, J. MOREL[1], A. NOIRET[1], P. CÔTE[2], B. HELMLINGER[3], R. SOHRABI[4], J.-M. LAVANCHY[4], F. LEVEAU[5], C. NUSSBAUM[6] & J. MOREL[7]

[1]*Agence nationale pour la gestion des déchets radioactifs (ANDRA), RD960 55290 Bure, France*

[2]*Institut Français des Sciences et Technologies des Transports, de l'Aménagement et des Réseaux (IFSTTAR), Route de Bouaye CS4, 44344 Bouguenais, France*

[3]*Solexperts France, Technopôle Nancy-Brabois 3, rue du Bois de la Champelle, 54500 Vandœuvre-lès-Nancy, France*

[4]*AF Consult, Taefernstrasse 26, 5405 Baden, Switzerland*

[5]*GEOTER, RD960 55290 Bure, France*

[6]*SWISSTOPO, Fabrique de Chaux, CH-2882 St-Ursanne, Switzerland*

[7]*SOLDATA, Parc de l'île, 21 rue du port, 92022 Nanterre, France*

**Corresponding author (e-mail: remi.delavaissiere@andra.fr)*

Abstract: The investigation of the induced fractures network around seals in drifts or shafts, and in particular its evolution, is a key issue for the performance assessment of an underground waste repository. Within this framework, a specific experiment was designed and implemented in the Meuse/Haute-Marne Underground Research Laboratory (URL). This experiment, called CDZ (Compression of the Damaged Zone), is dedicated to studying the effect of mechanical compression within the induced fractures zone of the Callovo-Oxfordian claystone (COx). An unequalled level of knowledge in the 3D structure of the fractures network has been attained. A multidisciplinary approach was applied to observe not only the initial state of the induced fracture zone but also its evolution during a loading cycle. The investigations show that the fracture network which composed the Excavation Damaged Zone (EDZ) was initially interconnected and open for flow and then partially closed progressively following the increasing mechanical stress applied on the drift wall. Moreover, the evolution of the EDZ after unloading indicates a self-sealing process.

During the excavation of a shaft or drift, the rock properties around the underground openings could be altered owing to the level of the ratio of in situ stresses to the rock strength. In some case, the rock mass could even be fractured. One concern regarding waste disposal is that the associated disturbance and damage created in the area around these excavations might change the favourable properties of such rocks and thus negatively impact the repository performance. Many authors speak about this zone, which is commonly called the Excavation Damaged Zone (EDZ). At the main level of the Meuse/Haute-Marne Underground Research Laboratory (URL; −490 m), an induced fracture network has been observed (Armand *et al.* 2014) around the drifts and it is made up of two types of fractures (shear and extensional fractures).

Taking into account Tsang *et al.*'s (2005) definition of the EDZ and safety considerations from Bauer *et al.* (2003), Armand *et al.* (2014) discussed the EDZ v. the observed induced fracture network. Based on the hydraulic conductivity results, they showed that the excavation-induced fracture zone extent is larger than the EDZ, and they concluded that the EDZ is included in the zone with extensional fracture in the Callovo-Oxfordian claystone at the Meuse/Haute-Marne URL.

To limit radionuclide migration along drifts and through the EDZ, seals will be implemented in drifts and shafts. These seals will be composed of swelling clay (mainly bentonite). After natural hydration from the surrounding rock mass, the bentonite will swell and apply radial pressure against the drift wall. The investigation of the EDZ, and in particular its evolution around seals in drifts and shafts, is a key issue for the performance assessment of an underground waste repository.

Within this framework, an experiment (called CDZ, Compression of the Damaged Zone), dedicated to studying the effect of mechanical compression within the induced fractures of the Callovo-Oxfordian claystone, was designed and

From: NORRIS, S., BRUNO, J., CATHELINEAU, M., DELAGE, P., FAIRHURST, C., GAUCHER, E. C., HÖHN, E. H., KALINICHEV, A., LALIEUX, P. & SELLIN, P. (eds) 2014. *Clays in Natural and Engineered Barriers for Radioactive Waste Confinement*. Geological Society, London, Special Publications, **400**, 279–291.
First published online May 6, 2014, http://dx.doi.org/10.1144/SP400.30

implemented in the Meuse/Haute-Marne URL. One of the objectives of this experiment is to study the effects of mechanical loading and unloading on selected properties of the EDZ. The procedure intends to reproduce the sealing effect of bentonite swelling in a drift as well as to mimic the effect of a liner being removed before the establishment of a sealing element. A further objective, not discussed in this paper, is to study the effect of the hydration on the EDZ. *In situ* sealing of a damaged clay formation owing to hydration and associated swelling was demonstrated earlier in one experiment at the Mont-Terri URL (Bossart *et al.* 2002). However this paper concerns only the effect of purely

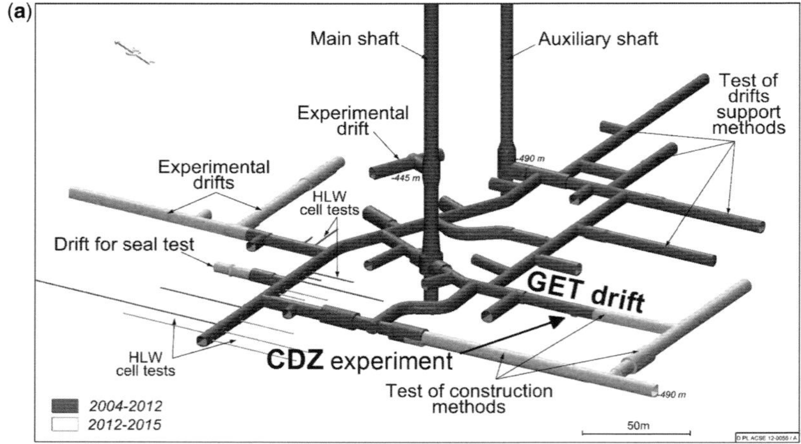

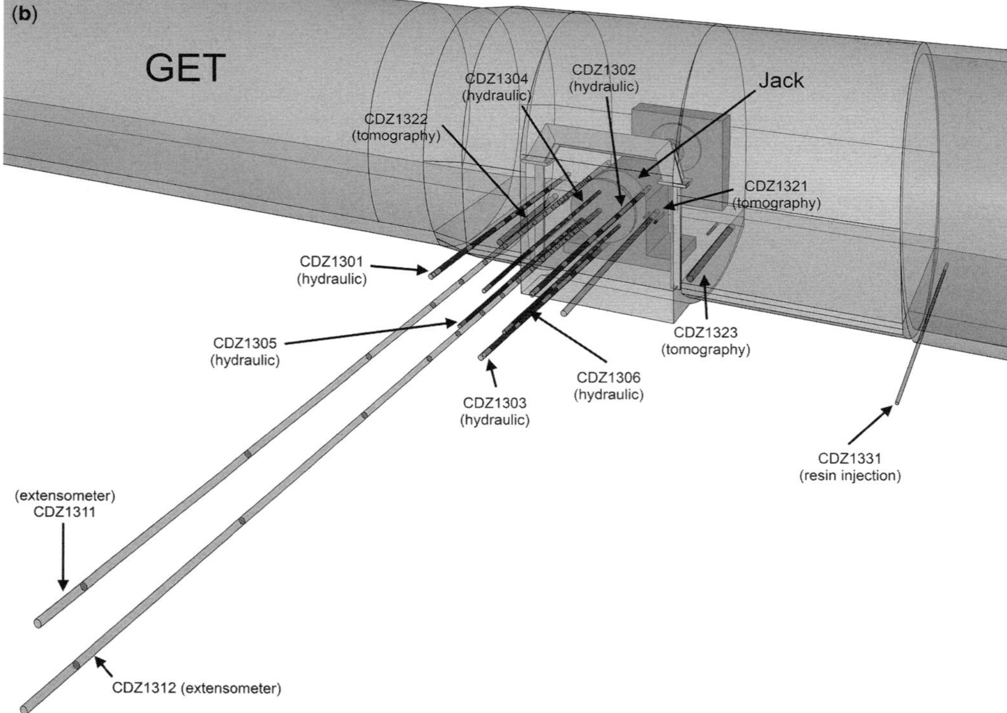

Fig. 1. Schematic diagrams of the CDZ experiment: (**a**) view of the drifts in the Meuse/Haute-Marne Underground Rock Laboratory showing the location of the CDZ experiment; (**b**) CDZ experimental setup.

mechanical loading on the EDZ without artificial hydration.

Experimental setup and test sequence

The CDZ experiment was installed in the Meuse/Haute-Marne URL in the GET (Galerie d'Expérimentation Trois) drift. This drift is oriented parallel to the horizontal major stress direction. The vertical stress is nearly equal to the horizontal minor one (Wileveau *et al.* 2007). Figure 1a, b provides a general overview of the URL and the CDZ setup.

The CDZ experimental setup is similar to that of the SELFRAC-1 experiment which was performed at the Mont Terri Rock Laboratory (Meier *et al.* 2000; Bernier *et al.* 2007). The loading apparatus consists of a single end plate, with a cross section of 1 m^2, transferring a hydraulically induced load to the drift wall. Displacement of the loading plate and the force applied to the move of the plate are monitored. The loading apparatus was oriented in such a way as to apply a load in more or less normal direction to extensional EDZ fractures in the sidewall rock (Fig. 1b).

Multipacker systems (Fierz *et al.* 2007) for hydraulic measurements and test performances were installed into six boreholes: three behind the loading plate (CDZ1304 to CDZ1306) and three in the immediate surrounding (CDZ1301 to CDZ1303).

Table 1 details the locations of the intervals of the six multipacker systems. Behind the loading plate, the multipacker systems are composed of six intervals with lengths each equal to 20 cm, excepting the last interval at the end of each borehole. In the surroundings zone, the multipacker systems are composed of five intervals with different lengths. The intervals were filled with gas (nitrogen) at atmospheric pressure with the exception of the deep end intervals of all boreholes, which were filled with synthetic water. The minimum distance between the closest boreholes is about 0.6 m.

Two extensometers were placed into boreholes, one in the middle of the loading area (CDZ1312) and one (CDZ1311) at a distance of 21 cm from the loading area. The density of the extensometers' measurement points decreases gradually with distance from the drift wall. Three boreholes were also equipped for seismic tomography (CDZ1321 to CDZ1323), CDZ1321 being only used for acoustic reception. To complete the seismic tomography system, very short boreholes were drilled (CDZ1324 to CDZ1327) to get additional measurement near the drift wall. All these boreholes are located in one horizontal plane leading through the centre of the loading axis. Furthermore, one sub-horizontal borehole was dedicated to resin injection (CDZ1331) and is located a few metres from the others in the GET drift. A total of 12 boreholes were drilled which, to date, represents the highest local density of boreholes in the Meuse/Haute-Marne URL.

All boreholes were drilled between December 2010 and February 2011 and the jack was in place before the end of February 2011. A first loading cycle was conducted between March and May 2011, applying three levels of loading (2, 3 and 4 MPa), prior to releasing stress rapidly.

The duration of loading steps was 2 weeks, except for the last level, which lasted 3 weeks. Five measurement campaigns took place: before (January 2011) and during the loading as well as after releasing stress (August 2011). Each measurement campaign included seismic tomography and hydraulic tests.

Methodology

Gas tests methods

In the intervals filled with gas, two types of gas tests are performed: (a) gas tests to determine the permeability around the intervals (they are composed of pulse or constant pressure tests); and (b) gas tracer tests with helium. They are composed of a qualitative injection test to detect connections between intervals and a semi-quantitative injection test to further describe helium migration within a selected flow path. The first type of test is repeated

Table 1. *Location of the intervals of the multipacker systems in metres*

	CDZ1301	CDZ1302	CDZ1303	CDZ1304	CDZ1305	CDZ1306
Interval 6 (PRE06)				0.48–0.68	0.53–0.73	0.38–0.58
Interval 5 (PRE05)	0.59–1.29	0.32–1.02	0.29–0.99	0.88–1.08	0.93–1.13	0.78–0.98
Interval 4 (PRE04)	1.49–2.19	1.22–1.92	1.19–1.89	1.28–1.48	1.33–1.53	1.18–1.38
Interval 3 (PRE03)	2.49–2.89	2.22–2.62	2.19–2.59	1.88–2.08	1.93–2.13	1.78–1.98
Interval 2 (PRE02)	3.89–4.09	3.62–3.82	3.59–3.79	3.08–3.28	3.13–3.33	2.98–3.18
Interval 1 (PRE01)	5.09–5.4	4.82–5.07	4.79–5.15	4.98–5.23	5.03–5.24	4.88–5.22

Distances are given relative to the drift wall (claystone).

for each hydraulic campaign, whereas the second one is done before loading, at the maximum loading level and after unloading.

Note that, in the intervals filled with water, water permeability tests are performed but results will not be discussed in this paper. Indeed no evolution of the water permeability at the deep borehole end is observed during loading.

Gas permeability tests. Gas permeability measurements are performed through pulse tests (instantaneous increase or decrease of the pressure) or constant rate injection tests. Constant rate injection tests are performed in partially saturated zones whereas pulse tests are carried out in saturated zones. The evaluation of saturation around intervals filled with gas is done by a short gas extraction period before the gas permeability tests.

The tests are interpreted with the software MULTISIM (Tauzin & Johns 1997), a flexible borehole simulator for the analysis of single-phase hydraulic tests in low-permeability formations. This simulator can account for the following situations: liquid or gas flow; non-ideal test conditions owing to variable wellbore storage, wellbore temperature variations and pre-test pressures; single and dual porosity, homogeneous or composite formations, fractional dimension flow, no-flow or constant pressure boundaries; and any sequence of slug tests, pulse tests, constant or variable pressure tests, and constant or variable rate tests. The general model theory and application for hydraulic permeability test was described by Bourdarot (1996) and Horne (1997). Pickens *et al.* (1987) provided an application of the method for a radial composite flow.

The analysis of all of the pulse tests shows that there is no influence of the outer zone on the simulations, even with an outer boundary at 0.1 m (outside the borehole damaged zone). This means that no boundary is imposed near the borehole. Moreover, the imposed differential gas pressure for the pulse is certainly below the capillary pressure, so the gas cannot penetrate the rock by advection. For all these tests, only an upper bound for equivalent gas permeability can be estimated at 10^{-12} m s^{-1}. This value is the permeability value of the sound claystone (Delay *et al.* 2006).

For gas constant flow rate tests, typically performed in partially saturated zones, a homogeneous model with a uniform boundary at constant pressure, set at 0.1 m from the borehole wall, is necessary to simulate the measurements. This unique model is used to simulate the test sequence composed of constant rate and pressure recovery for each test campaign. This test analysis methodology implies that the relative evolution of permeability with respect to the initial values is possible. The uncertainty around the permeability reference values is in the range 2–5.

Gas tracer tests

Qualitative helium injection test. The aim of this test is, through migration of helium, to detect the hydraulic connections within the fracture network. Binary results are obtained, that is, helium is detected or not. A gas mixture (95% N_2 and 5% He) is injected into one interval over a duration of a few minutes and helium detectors measure the helium migration through the EDZ into the other intervals. During the test, all of the other intervals are submitted to a gas extraction flow rate. At the end of the test, a flush-out is performed from the injection interval during the following few hours before initiating the next injection. This sequence is repeated for all gas-filled intervals. These qualitative results will not be detailed in this paper but results are used to choose several pairs of intervals for the semi-quantitative helium detection test.

Semi-quantitative helium detection test. This type of test involves the injection of a nitrogen–helium mixture into one interval and the extraction of gas in one another interval. The helium concentration is measured at the extraction interval. The expected result is the tracer breakthrough curve with the first

Table 2. *Maximal distance from drift wall of tensile and shear fractures in CDZ boreholes*

	CDZ1301	CDZ1302	CDZ1303	CDZ1304	CDZ1305	CDZ1306	CDZ1311	CDZ1312	CDZ1321	CDZ1322	CDZ1323
Borehole length (m)	5.4	5.07	5.15	5.23	5.24	5.22	15.84	15.76	4.4	2.22	2.61
Maximal extension of tensile fractures (m)			1.32	0.85	0.85	1.46	0.8	1.53	1.13	1.25	1.94
Maximal extension of shear fractures (m)	4	3.83	3.65	3.91	3.96	3.82	3.67	4.1	4.3	1.49	2.46

arrival of gas tracer and the relative helium concentration. A simple comparison of the breakthrough curves of the tracer between the successive test campaigns reveals any possible change inside the fracture network during loading. A purely quantitative analysis was not performed as it would require taking into account a large number of assumptions about the conceptualization of the fractured medium (homogeneous porous medium, fracture network with different flow properties).

Seismic tomography methods

Geophysical methods provide an effective way to characterize the damage around underground structures, including the extent and degree of the damage (Schuster *et al.* 2001). Seismic tomography is based on measuring travel times of the compression wave (P wave) between source and receiver points between boreholes (see Fig. 1b). From these measurements, it is then possible to derive a compression wave velocity map. In the case of *a priori* homogeneous material, the appearance of a zone with lower velocity than sound claystone indicates that the material has been damaged. The same algorithm (see Abraham *et al.* 1998) is applied to the travel times and wave amplitudes for all seismic tomography campaigns in order to ensure that velocity maps can be compared with each other.

Initial excavation damaged zone

Armand *et al.* (2014) describe the different methods that are used in the Meuse/Haute-Marne URL to assess the extension of induced fracture networks and the EDZ around drifts. Two methods of direct observation are used to characterize the induced fractures network in the CDZ experiment. The first method is based on structural analysis of the core samples from CDZ boreholes to characterize the type of fracture and to define the fracture density and orientation according to depth. This structural analysis is coupled with camera logging within CDZ boreholes. The second method is based on the overcoring of resin-filled EDZ fractures in a dedicated borehole CDZ1331. The extension of the damaged zone before loading could also be depicted by seismic tomography (see Fig. 6).

Drill core observation on 11 CDZ boreholes

Cores are reconstituted and orientated, that is, determination of the original position in a borehole is achieved using sedimentary stratification (nearly horizontal in laboratory drifts). For horizontal drillings like in CDZ, orientation relative to north direction (given by orientation of the drill) is obvious, but the top of the stratification is not defined and two possibilities exist. In CDZ boreholes,

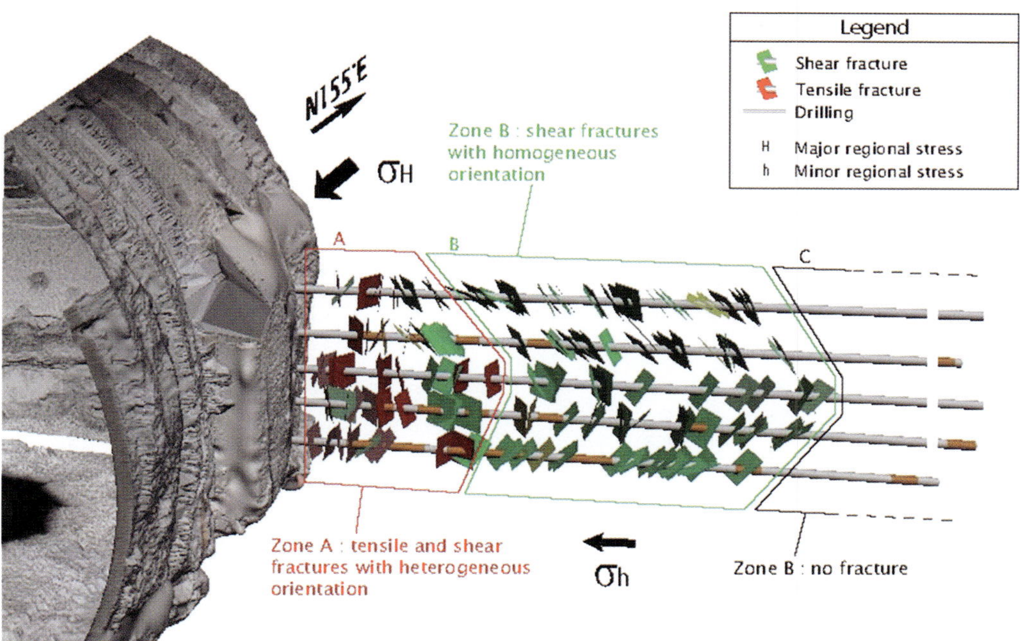

Fig. 2. 3D visualization of the fracture network in the GET drift (CDZ boreholes).

oriented borehole images are available and the cores could be oriented beyond doubt.

The geologist measures all fractures and slickensides and classifies them according to the type of break (shear fracture or unloading tensile fractures). Some fractures which were induced by the drilling of the borehole are identified and are not taken into account during analysis. Some other fractures in which the break mode could not be seen are not classified. Maximal extensions for the two types of break are given in Table 2.

Based on the data obtained, a 3D representation of the fracture plan in the GET drift is proposed (Fig. 2). Two different zones of the excavation-induced fracture can be clearly distinguished. The first zone (zone A), near the drift, is made up of tensile and shear fractures. Their orientation (both for extensional and shear fractures) is very heterogeneous with steeply dipping fractures. The maximal extension of tensile fractures is 1.94 m at CDZ1323.

The second zone (zone B), further from the drift, is made up of shear fractures only. Their orientation

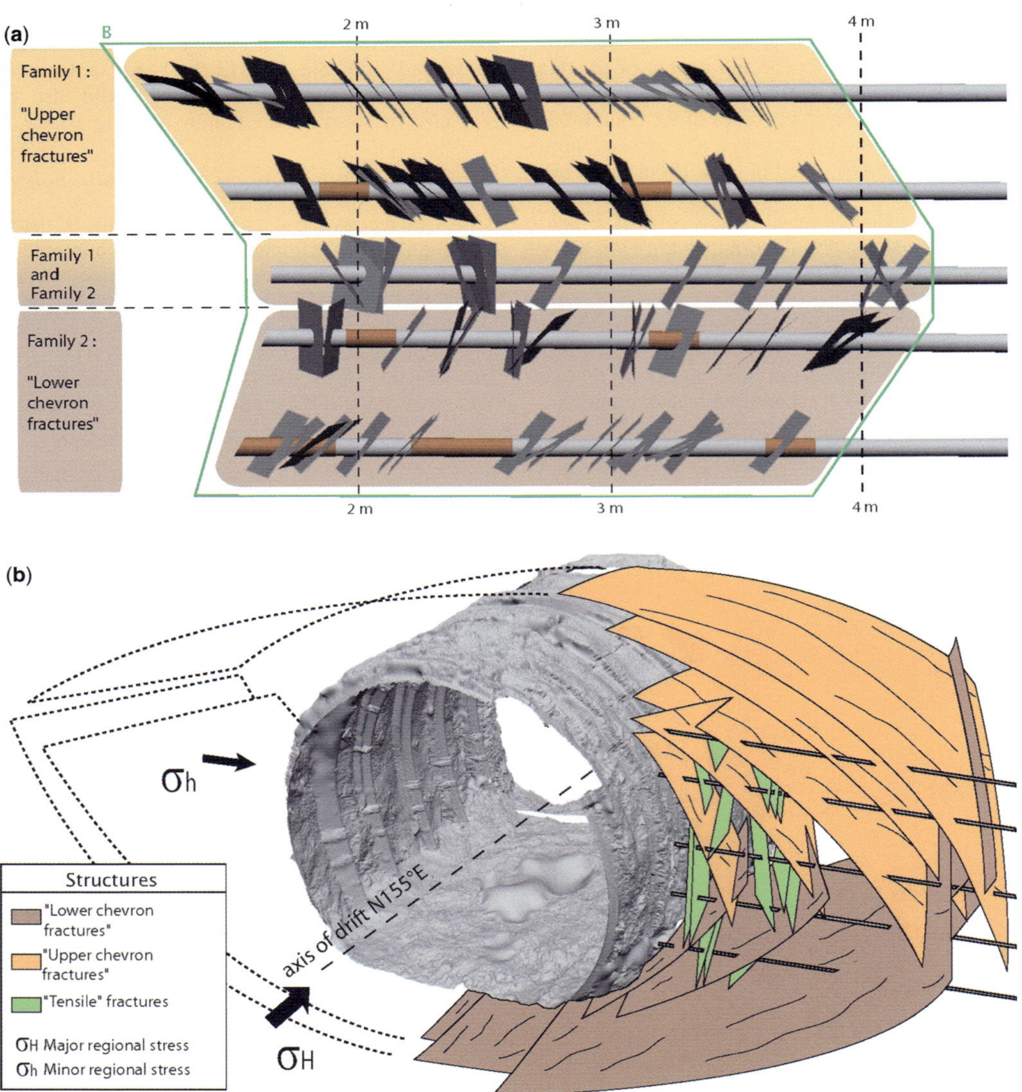

Fig. 3. Setup and orientation of the induced fractures: (**a**) close-up of the homogeneous zone; (**b**) proposed conceptual model.

is more uniform with gently dipping fractures. In this zone, we can extrapolate fractures and associate them to the fractures observed in drifts. The strike of these fractures is nearly parallel to the drift axis (0–10°). Fractures can be classified into two major families (Fig. 3a). In the drift's upper part, fractures

(a)

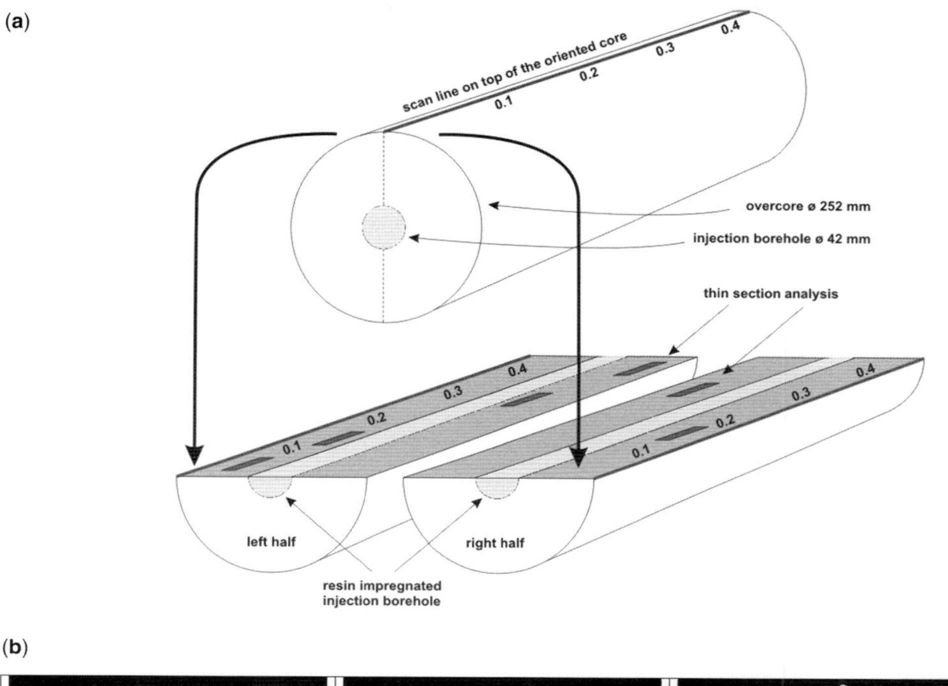

(b)

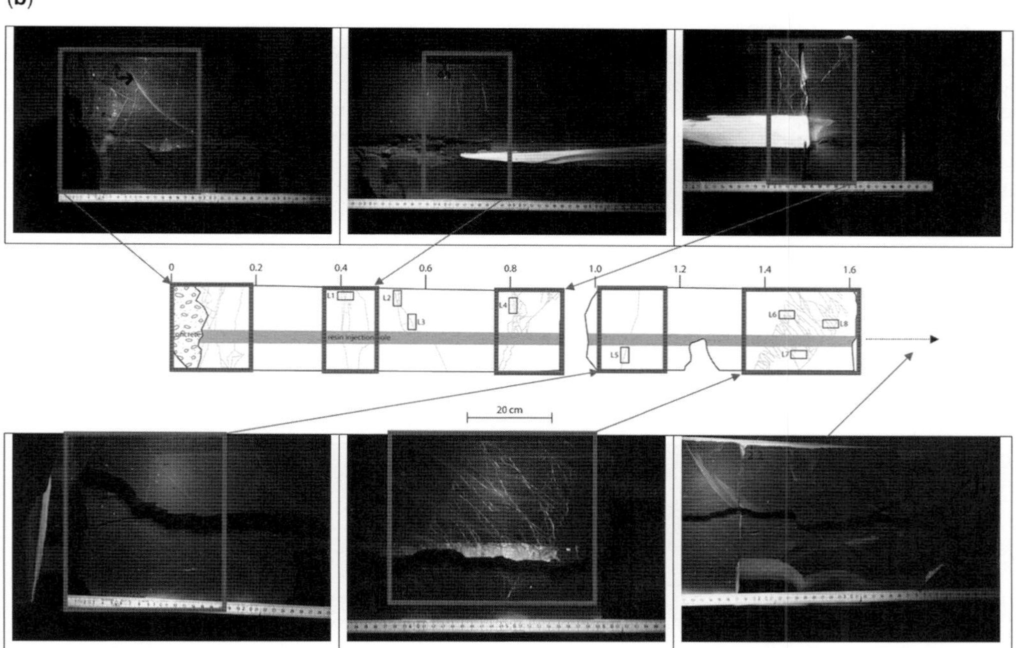

Fig. 4. Resin injection in CDZ1331 borehole: (**a**) sketch of the cut overcore; and (**b**) most important types of excavation-induced fractures found in the CDZ1331 core material.

have a dip oriented towards the block. They correspond to the extension of 'upper chevron fractures' in the block. In the drift's lower part, fractures have a dip oriented towards the drift. They correspond to the extension of 'lower chevron fractures' in the block. At middle height these structures rejoin or intersect each other. Figure 3b shows the extrapolation of the observed fractures to determine a model of the fracture network. The maximal extension of shear fractures is 4.3 m at the CDZ1321 borehole.

Resin injection

The principle of resin injection for studying the fractured rock mass in a borehole wall was first developed and applied by Möri & Bossart (1999) and it is also illustrated in Bossart *et al.* (2002). The detailed method is described in Armand *et al.* (2014). First an injection borehole is filled with resin doped with a fluorescent dye (fluorescein). A few days afterwards, this borehole is overcored and cores are analysed under UV light (Fig. 4a).

Based on the resin-impregnated fracture analysis, the core can be divided into three main parts (Fig. 4b). The first 1.1 m section is behind the gallery wall, where the fracture network consists of sub-vertical extensional fractures oriented oblique to the drift axis. These fractures developed during the excavation and are strongly controlled by the direction of excavation and the distance to the drift front ('onion skin'), thus they are mimicking former tunnel fronts during excavation. The fracture network in this zone is heterogeneous. The second section, located between 1.4 and 1.6 m from the gallery wall, consists of a 0.1 m-thick fracture network zone with an envelope dipping towards the tunnel front at an angle of about 60° and striking parallel to the tunnel axis. According to its strike and dip, this fracture zone may correspond to the lower part of an en-chevron shaped fracture. This section may be interpreted as Riedel structures within a sinistral shear zone. Riedel structures are networks of shear bands, commonly developed in zones of simple shear during the early stages of faulting (Passchier & Trouw 1995). The third section, between 1.9 and 4.5 m from the drift wall, contains a rather homogeneous network of non-impregnated fractures dipping steeply (83°) towards the gallery. The fractures occur every 5–10 cm and are oriented parallel to the tunnel axis. Fractures within the homogeneous zone more than 1.9 m behind the tunnel wall are of shearing origin only and thus not impregnated. The observations from CDZ1331 are well in agreement with core analysis on the other CDZ boreholes.

EDZ evolution during loading

Displacement

The extensometer in the borehole CDZ1312 gives the displacement of the rock at different depths during the loading. Measurements are not presented but they show that displacement is maximal at the wall surface and decreases with depth. This diminution of the displacement with depth is due to the decrease in normal stress with depth. The displacement of the loading plate at the drift wall is also recorded. Figure 5 presents the relative loading in respect to the relative displacement of the loading plate.

During the first loading step (at 2 MPa), most of the displacement measured at the wall (approximately 10 mm) is due to the adjustment of the plate on the drift wall. The rest of the displacement (approximately 9 mm) is due to horizontal settlement

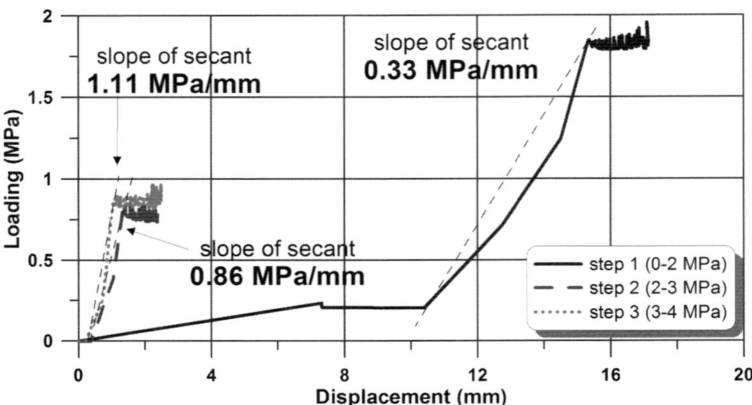

Fig. 5. Loading v. displacement. Slope of secant.

in the rock. Based on the Boussinesq solution (Goodman 1989), a strain modulus could be calculated through the ratio between loading and displacement (also called subgrade reaction coefficient K_s). Strain modulus E (in Pa) is given in Equation (1):

$$E = \left(1 + \frac{2}{\pi}\right) \times (1 - \upsilon^2) \times R \times K_s \quad (1)$$

where υ is the Poisson ratio (taken equal to 0.3) and R is the radius of the loading plate (equal to 0.56 m). The Boussinesq solution implies a pure elastic behaviour. In reality, owing to inelasticity and secondary stress field, this formula is valid as a first approximation only.

The calculated strain modulus is equal to 279, 724 and 938 MPa for the loading steps 0–2, 2–3 and 3–4 MPa, respectively, and increases with loading. The only way to explain this result is to consider sound blocks of claystone (Young's modulus equals to 4 GPa for sound claystone see Andra 2010), with open fractures that absorb strain by closing. The strain modulus evolution depicts the

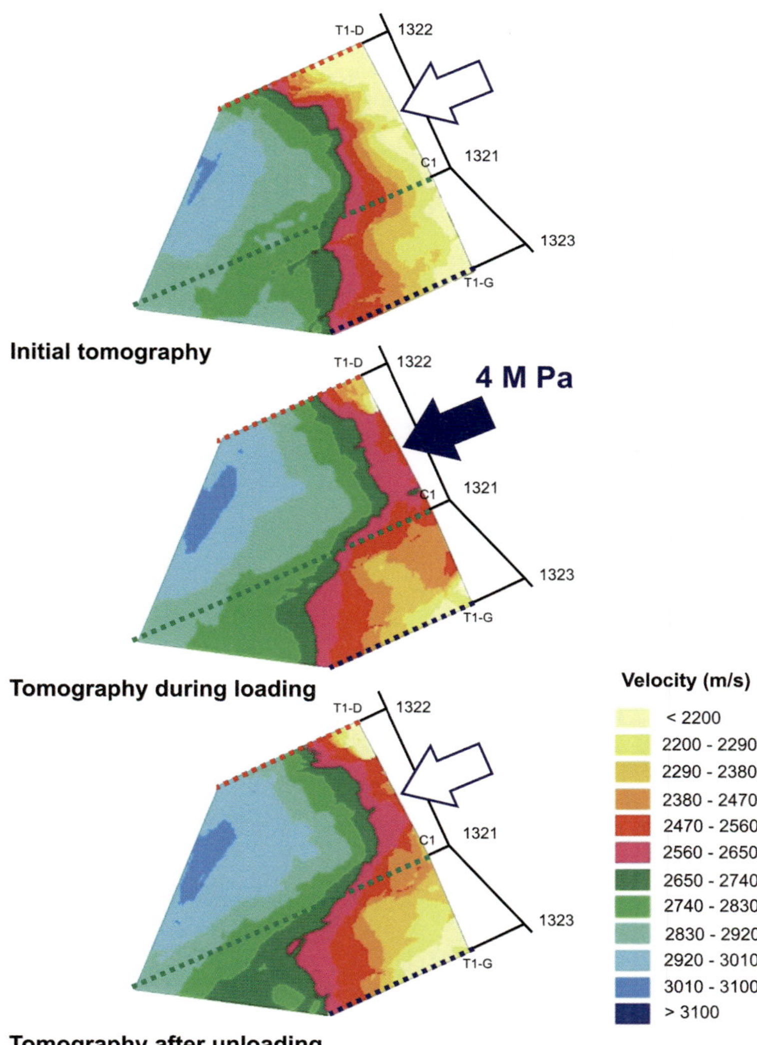

Fig. 6. Seismic tomography before loading, during loading at 4 MPa and after unloading. Arrow: location of the jack. Velocity in m s^{-1}. Grid mesh equal to 0.5 m.

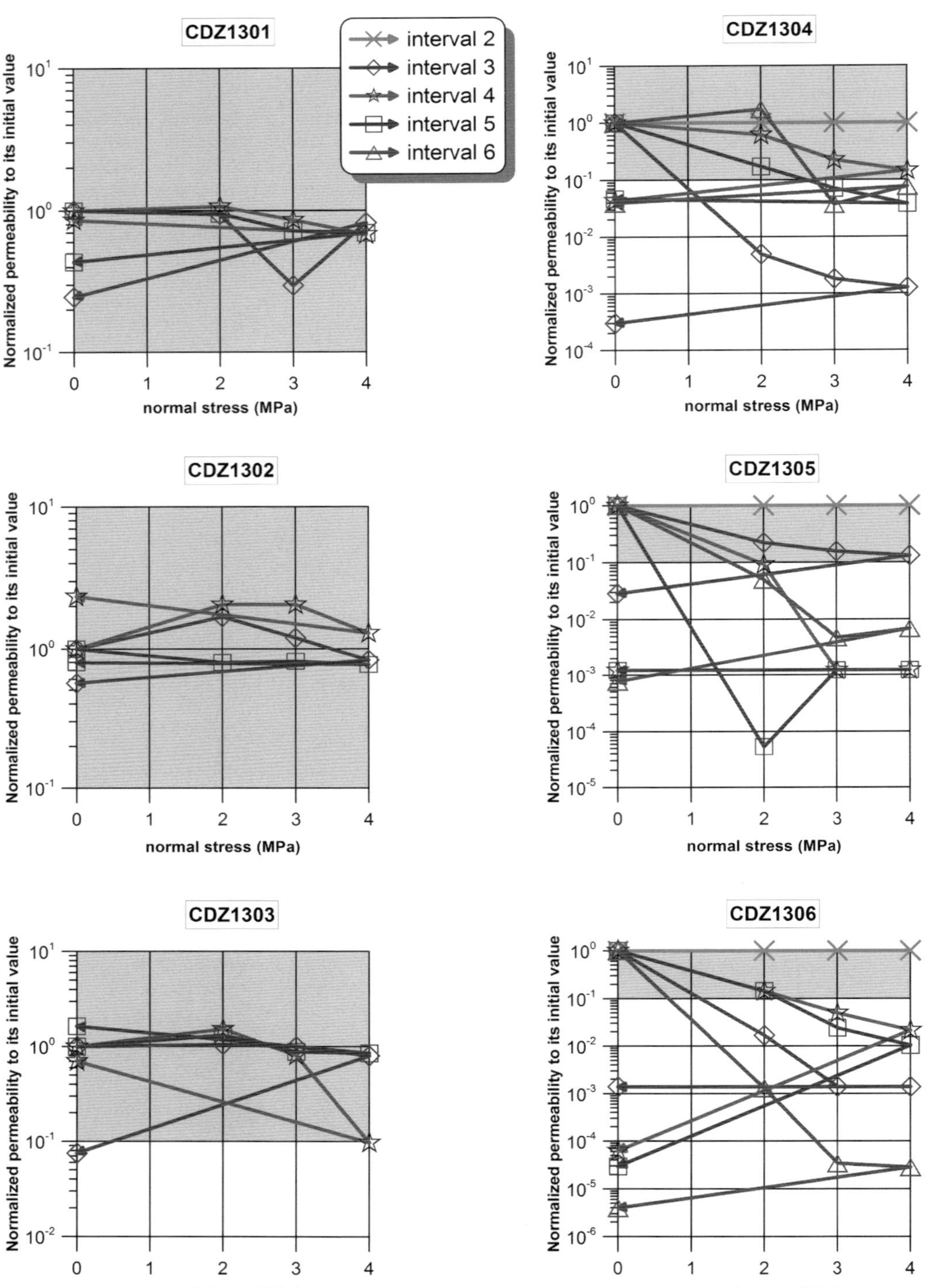

Fig. 7. Normalized permeability v. normal stress during loading cycle. Left: outside loading area. Right: below loading area. Grey rectangle: normalized permeability evolution is not significant.

geometry of the open fracture network (depth and orientation) in respect to the generated stress at different depths.

After unloading, the displacements measured at different depths do not return to their initial values (not presented). Some extensional fractures were closed during loading but they do not reopen after unloading.

Seismic tomography

Figure 6 shows the P wave velocity evolution between the two emission boreholes CDZ1322 and CDZ1323 and the reception borehole CDZ1321 before and during loading at 4 MPa as well as after releasing stress. The initial velocity map can be compared with the direct methods of observation. Velocities below 2500 m s^{-1} coincide well with the maximal extension of the extensional fractures seen by drill core observation and resin injection (zone A). The zone characterized by these velocities is indeed deeper near borehole CDZ1323. It is also in this borehole that the deepest extensional fracture is observed (see Table 2). Beyond this zone, the shear fracture zone is characterized by velocities above 3100 m s^{-1}.

During loading at 4 MPa, the velocities increase by several hundreds of metres per second below the loading area. After unloading, there is no significant change in the velocity map as compared with the map at 4 MPa.

Permeability

Figure 7 details the evolution of the normalized permeability, from its initial value before loading, in response to the normal stress during the loading cycle. Owing to the permeability's uncertainty, ratios of permeability at time *n* v. initial permeability lower (greater) than $1/10$ (or $\times 10$) are not considered significant. This insignificant ratio is depicted by a grey rectangle in Figure 7.

The three plots on the left of Figure 7 correspond to the permeability measurements performed in the boreholes outside the loading area. These plots show that there is no evolution of permeability owing to the loading. Conversely, a strong permeability drop during loading can be observed for all the intervals except those beyond 3 m depth (see Table 1). The limit of the loading influence is between 2 and 3 m from the drift wall. The normalized permeability drop at the loading step of 4 MPa ranges between 1 and 4 orders of magnitude.

After unloading, most of normalized permeabilities are still well below the initial values. Moreover, some of them show a strong drop in comparison with the load at 4 MPa.

Gas tracer injection

Four pairs of intervals were chosen in order to inject helium into one interval and extract gas in another (see section on 'gas tracer tests'). Figure 8 shows one pair consisting of the fifth intervals of CDZ1302 and CDZ1305 boreholes. The distance between the two intervals is 0.8 m. It is the only pair in which the gas tracer is detected during the loading step at 4 MPa as well as after unloading.

At the loading step at 4 MPa, the breakthrough curve shows a clear delay in the first arrival of helium and also a drop in relative helium concentration. This implies a restriction of the advection

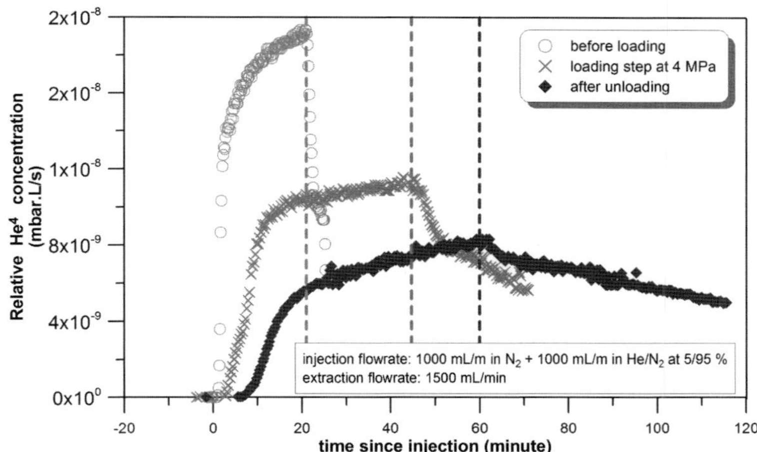

Fig. 8. Helium breakthrough curve at CDZ1304, interval 5, before loading, at the loading step at 4 MPa and after unloading. Vertical dotted lines: gas injection stop.

of the tracer in the fracture network at the loading step at 4 MPa compared with the initial state. This gas flow restriction further increases after unloading.

Discussion

The reduction of the fracture transmissivity following mechanical normal loading of fractures is well documented in the literature (see Gentier *et al.* 2000). In clay material, gas permeability reduction owing to the increase in the confining pressure was largely demonstrated in Callovo-Oxfordian or Opalinus clay fractured samples (Zhang & Rothfuchs 2008; Zhang 2011). It was also demonstrated on samples that after unloading the permeability remains lower than during the initial stage.

When fractures are induced on samples, smectite minerals are released in the fracture walls. The confining pressure redistributes water saturation inside the sample and increases the relative humidity in the fracture, thus inducing smectite mineral swelling. A self-sealing process according to the definition given by Bock *et al.* (2010) occurs within the clay sample. It should be noted that the self-sealing process could be only partial because permeability could still be higher than the permeability of a sound clay sample.

In situ EDZ sealing owing to compaction was investigated in the Mont-Terri URL in the SELFRAC project (Bernier *et al.* 2007). The two *in situ* experiments SELFRAC-1 and SELFRAC-2 have shown the reduction of the water permeability in respect to the confining pressure, but the two physical–chemical phenomena of the loading and self-sealing of the fracture owing to water redistribution have not been distinguished. In the case of the CDZ experiment in which permeability tests with gas are performed, it is the first time that gas permeability reduction during loading has been observed *in situ*. Gas permeability reduction is associated with an increase of the P-wave velocity and a compression measured by the extensometer behind the loading plate. Gas tracer advection into the induced fracture network drops during loading owing to the closure of the fracture under load.

All aforementioned results show that it is the extensional fractures in 'zone A' (see Fig. 2) that are closed during loading. Conversely, 'zone B' is not affected by the loading, or by the fact that stress increase remains small at this depth.

After unloading, the gas permeability and gas tracer advection further continue to decrease. The velocity map remains high and confirms that some fractures remain closed. Thanks to having gas-filled intervals, the results clearly show an *in situ* self-sealing process in the damaged zone around a drift without saturation of the fracture network.

Conclusions

The high density of boreholes in the CDZ experiment has led to an unique level of knowledge of the three-dimensional structure of the induced fractures network around the GET drift in the Meuse/Haute-Marne URL. A multidisciplinary approach was taken to observe not only the initial state of the induced fractures network but also its evolution during a loading cycle. The results show that it is the first 'zone A', where there are extensional fractures, which is impacted during loading. This zone corresponds to the exact definition of the EDZ described in Armand *et al.* (2014).

From a hydraulic point of view, loading closes extensional fractures and reduces fracture transmissivities with a drop in the EDZ permeability by 1–4 orders of magnitude at a normal stress of 4 MPa. This result confirms the interest in seals based on swelling clay which will reduce the EDZ permeability. Furthermore the closure of the fracture under load exhibits a significant self-sealing by the redistribution of water from rock matrix to fractures, thus inducing clay mineral swelling within the flow paths.

From a mechanical point of view, CDZ results provide information on what would be the 'stiffness' of the EDZ. The EDZ acts as a buffer material which absorbs part of the strain from the creep of the sound claystone with very low strain kinetics. This determines the impact of mechanical loading for key components of the future repository, such as the concrete liner and casings of high level waste cells. The CDZ experiment will continue with a new stage which is focused on the effect of hydration on the properties of the induced fracture network.

We gratefully acknowledge M. Piedevache and F. Bréhin (Solexperts France) for the experimental setup. We also gratefully acknowledge G. Armand (Andra) and J. Smuteck (PhD at Czech Technical University in Prague) for their internal reviews, C. Fairhust for his external review and the two anonymous reviewers for their improving comments.

References

ABRAHAM, O., BEN SLIMANE, K. & COTE, P. 1998. Factoring anisotropy into iterative geometric reconstruction algorithms for seismic tomography. *International Journal of Rock Mechanics and Mining Sciences*, **35**, 31–41.

ANDRA 2010. Référentiel du site Meuse/Haute-Marne – Tome 2. C.RP.ADS.09.0007.

ARMAND, G., LEVEAU, F. ET AL. 2014. Geometry and properties of the excavation induced fractures at the meuse/haute-marne URL drifts. *International Journal of Rock Mechanics and Mining Sciences*, **47**, 21–41. http://dx.doi.org/10.1007/s00603-012-0339-6

Bauer, C., Pépin, G. & Lebon, P. 2003. EDZ in the performance assessment of the Meuse/Haute-Marne site: conceptual model used and questions addressed to the research. *In*: *Proceedings of the EC Cluster Conference on Impact of the EDZ on the Performance of Radioactive Waste Geological Disposal*, 3–5 November 2003, Luxembourg. European Commission Report **EUR21028EN**.

Bernier, F., Li, X. L. *et al.* 2007. Fractures and *self-healing within the excavation disturbed zone* in *clays* (SELFRAC). Final Report. European Commission Report, **EUR 22585**, 62.

Bock, H., Dehandschutter, B., Martin, C. D., Mazurek, M., de Haller, A., Skoczylas, F. & Davy, C. 2010. Self-sealing of fractures in argillaceous formations in the context of geological disposal for radioactive waste – review and synthesis. Report no. **6184**, Nuclear Energy Agency.

Bossart, P., Meier, P. M., Moeri, A., Trick, T. & Mayor, J.-C. 2002. Geological and hydraulic characterization of the excavation disturbed zone in the Opalinus Clay of the Mont Terri. *Engineering Geology*, **66**, 19–38.

Bourdarot, G. 1996. *Essais de puits: méthodes d'interprétation*. Éditions Technip, Institut français du pétrole, Paris.

Delay, J., Trouiller, A. & Lavanchy, J.-M. 2006. Propriétés hydrodynamiques du Callovo-Oxfordien dans l'Est du bassin de Paris: comparaison des résultats obtenus selon différentes approches. Comptes Rendus. *Geoscience*, **338**, 892–907. http://dx.doi.org/10.1016/j.crte.2006.07.009

Fierz, T., Piedevache, M., Delay, J., Armand, G. & Morel, J. 2007. Specialized instrumentation for hydromechanical measurements for hydromechanical measurements in deep argillaceous rock. *In*: *FMGM 2007: Seventh International Symposium on Field Measurements in Geomechanics*, Boston.

Gentier, S., Hopkins, D. & Riss, J. 2000. Role of fracture geometry in the evolution of flow path under stress. Geophysical monograph **122**: 'Dynamics of fluids in fractured rocks' AGU edition, 169–184.

Goodman, E. R. 1989. *Introduction to Rock Mechanics*. 2nd edn., Wiley, Singapore.

Horne, R. N. 1997. *Modern Well Test Analysis. A Computer-Aided Approach*. 2nd edn., Petroway, Palo Alto.

Meier, P., Trick, Th., Blümling, P. & Volckaert, G. 2000. Self-healing of fractures within the EDZ at the Mont Terri Rock Laboratory: results after one year of experimental of work. *In*: *Proceedings of International Workshop on Geomechanics, Hydromechanical and Thermohydro-mechanical Behaviour of Deep Argillaceous Rocks: Theory and Experiment*, 11–12 October, Paris.

Möri, A. & Bossart, P. 1999. Visualisation of Flow Paths (FM-B experiment). *In*: Thury, M. & Bossart, P. (eds) *Results of the Hydrogeological, Geochemical and Geotechnical Experiments (performed in 1996 and 1997)*. Geological Report no. **23**. Swiss National Geological Survey and Hydrogeological Survey, Ittigen-Berne, 113–121.

Passchier, C. W. & Trouw, R. A. J. 1995. *Microtectonics*. Springer, Heidelberg.

Pickens, J. F., Grisak, G. E., Avis, J. D., Belanger, D. W. & Thury, M. 1987. Analysis and interpretation of borehole hydraulic tests in deep boreholes: principles, model development, and applications. *Water Resources Research*, **23**, 1341–1375.

Schuster, K., Alheid, H.-J. & Böddener, D. 2001. Seismic Investigation of the EDZ in Opalinus Clay. *Engineering Geology*, **61**, 189–197.

Tauzin, E. & Johns, R. T. 1997. A new borehole simulator for well test analysis in low permeability formations. *In*: *Proceedings of the IAMG'97 the Annual Conference of the International Association for Mathematical Geology*, Barcelona.

Tsang, C. F., Bernier, F. & Davies, C. 2005. Geohydromechanical processes in the excavation damaged zone in crystalline rock, rock salt, and indurated and plastic clays in the context of radioactive waste disposal. *International Journal of Rock Mechanics and Mining Science*, **42**, 109–125.

Wileveau, Y., Cornet, F. H., Desroches, J. & Blumling, P. 2007. Complete in situ stress determination in an argillite sedimentary formation. *In*: *International Meeting 'Clays in Natural and Engineered – Barriers for Radioactive Waste Confinement'*, 17–20 September 2007, Lille. *Physics and Chemistry of the Earth*, **32**, 866–878.

Zhang, C. L. 2011. Experimental evidence for self-sealing of fractures in claystone. *In*: *International Meeting 'Clays in Natural and Engineered – Barriers for Radioactive Waste Confinement'*, 28 March to 1 April 2010, Nantes. *Physics and Chemistry of the Earth*, **36**, 1972–1980.

Zhang, C.-L. & Rothfuchs, T. 2008. Damage and sealing of clay rocks detected by measurements of gas permeability. *In*: *International Meeting 'Clays in Natural and Engineered – Barriers for Radioactive Waste Confinement'*, 17–20 September 2007, Lille. *Physics and Chemistry of the Earth*, **33**, S363–S373.

Measuring hydraulic conductivity and swelling pressure under high hydraulic gradients

LUCIE HAUSMANNOVA* & RADEK VASICEK

Centre of Experimental Geotechnics, Faculty of Civil Engineering, Czech Technical University in Prague, Thákurova 7, 166 29 Prague 6, Czech Republic

**Corresponding author (e-mail: lucie.hausmannova@fsv.cvut.cz)*

Abstract: The concept of a deep geological repository for high-level waste in the Czech Republic is based on disposal in a granite environment. The construction of an engineered barrier between the rock and canister is considered, and compacted swelling clay (bentonite) is chosen as the most suitable material for such a barrier. The most important geotechnical properties of bentonite in this context are low hydraulic conductivity and high swelling pressure. Measuring these parameters in such a low-permeable material requires much time. This study focuses on combined measurements of hydraulic conductivity and swelling pressure and on the impact of using high hydraulic gradients that are consistent with possible porewater pressures in a deep repository. High hydraulic gradients may also quicken the determination of these parameters. Results show a decrease in hydraulic conductivity with increasing saturation pressure up to a threshold dry density, with no further change at higher dry densities. The influence of saturation pressure on measured pressure decreases with dry density until a certain value is reached. For samples with lower density, the slope of the increase is close to unity.

A deep geological repository is widely considered the safest way to ensure the isolation of spent nuclear fuel from the biosphere. Packages containing radioactive waste will be emplaced in the deep repository and protected by a so-called multi-barrier system consisting of engineered barriers surrounding a natural barrier, the surrounding rock. The multi-barrier system will prevent or at least substantially delay the migration of radionuclides over an extremely long period of time.

The Czech concept of a deep repository is based on disposal in granite with engineered barriers. One part of the engineered barrier, the geotechnical barrier, will include a buffer placed between the waste package and the natural barrier. Swelling clay (bentonite) is generally regarded to be the optimal material for the buffer, because this clay material meets the requirements of geological disposal. The important geotechnical properties of bentonite are its low hydraulic conductivity, high swelling pressure, plasticity, high thermal conductivity and rheological stability or, rather, the long-term stability of these parameters.

This paper focuses on the experimental determination of hydraulic conductivity and swelling pressure. Both parameters are determined after full saturation is achieved. In such impermeable materials, this process requires much time. It can be accelerated by increasing the pressure under which water is injected into the sample (i.e. the saturation pressure, σ_{sat}), resulting in a high hydraulic gradient. Therefore, the influence of changes in hydraulic gradient on hydraulic conductivity and swelling pressure is tested in order to improve the measurement capabilities. Another important aim of this work is the study of possible long-term changes of these parameters in response to the changes in hydraulic gradients in a deep repository that can be caused by changes in the environment during its lifetime. These changes can be caused, for example, by a glacial episode (Graham *et al.* 2012).

Darcy's law is used for the calculation of hydraulic conductivity, which should be independent of the hydraulic gradient once a critical hydraulic gradient is exceeded (Villar & Gómez-Espina 2009).

Materials

Two Czech bentonites were used in the experiments–bentonite 75 (B75) and Na-activated Sabenil 65 (Sab65). Both were industrially prepared (crushed and sieved; Keramost Ltd) with an initial water content of *c.* 10%. The chemical compositions of both materials are provided in Table 1.

The material was uniaxially compacted in the laboratory to reach the required dry density (1.2–1.75 Mg m^{-3}). The sample diameter was 30 mm, and its height *c.* 20 mm. The height was measured accurately after compaction. The initial values of hydraulic conductivity (see Fig. 1) and swelling pressure (see Fig. 2) were evaluated using a saturation pressure $\sigma_{sat} = 1$ MPa (corresponding to a

From: Norris, S., Bruno, J., Cathelineau, M., Delage, P., Fairhurst, C., Gaucher, E. C., Höhn, E. H., Kalinichev, A., Lalieux, P. & Sellin, P. (eds) 2014. *Clays in Natural and Engineered Barriers for Radioactive Waste Confinement*. Geological Society, London, Special Publications, **400**, 293–301.
First published online May 15, 2014, http://dx.doi.org/10.1144/SP400.36

Table 1. *Chemical composition of both tested materials*

Component	Weight %	
	B75	Sab65 (Na-activated)
SiO_2	51.91	49.52
Al_2O_3	15.52	14.50
Fe_2O_3	8.89	11.92
TiO_2	2.28	3.08
CaO	4.60	4.21
MgO	2.22	2.18
Na_2O	1.21	2.03
K_2O	1.27	0.59
P_2O_5	0.40	0.44
MnO	0.11	0.08
FeO	2.95	0.65
SO_3	0.09	0.08
$CaCO_3$	11.71	8.84
CO_2	5.15	3.89
Lost–drying	10.65	9.85

hydraulic gradient of $i = 5000$; see equations (1) and (2)). Swelling pressure values were evaluated at zero saturation pressure.

Device and method

A special device (see Fig. 3 & 4) was used to simultaneously measure the hydraulic conductivity and the swelling pressure. The set-up of this device was inspired by the work of Dixon and colleagues (Dixon *et al.* 1999). An internal calibration procedure was performed before testing. No axial pressure was applied. Distilled water was used as a saturating fluid and was pushed by compressed argon.

The sample was placed between two permeable plates and enclosed in a stainless steel cell. No change in volume of the sample was allowed during the experiment. Distilled water was pushed from the bottom of the cell at different pressures (1–6 MPa) corresponding to hydraulic gradients i ranging between 5000 and 30 000. No backpressure was applied; that is, the saturation pressure directly determined the hydraulic gradient according to equations (1) and (2):

$$i = \frac{\Delta h}{\Delta l} \tag{1}$$

where i is the hydraulic gradient [–], Δh is the hydraulic head [m] and Δl is the sample length [m],

$$\Delta h = \frac{\Delta \sigma_{SAT}}{\rho \times g} \tag{2}$$

where Δh is the hydraulic head [m], $\Delta \sigma_{SAT}$ is the hydrostatic (saturation) pressure [Pa], ρ is the fluid density (water) [Mg m^{-3}] and g is gravitational acceleration [m s^{-2}].

Evaluation of hydraulic conductivity

Once constant flow was reached, the volume of water passing through the sample (ΔV) was determined over a given period of time (Δt). The

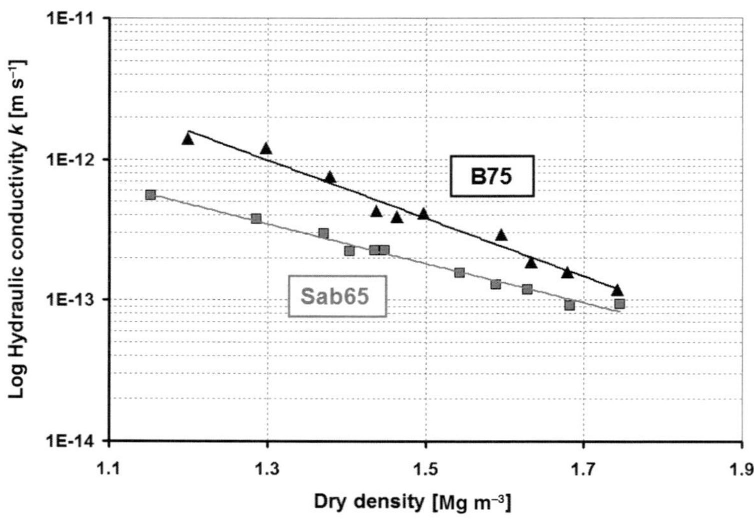

Fig. 1. Initial values of hydraulic conductivity of B75 and Sab65 dependent on dry density.

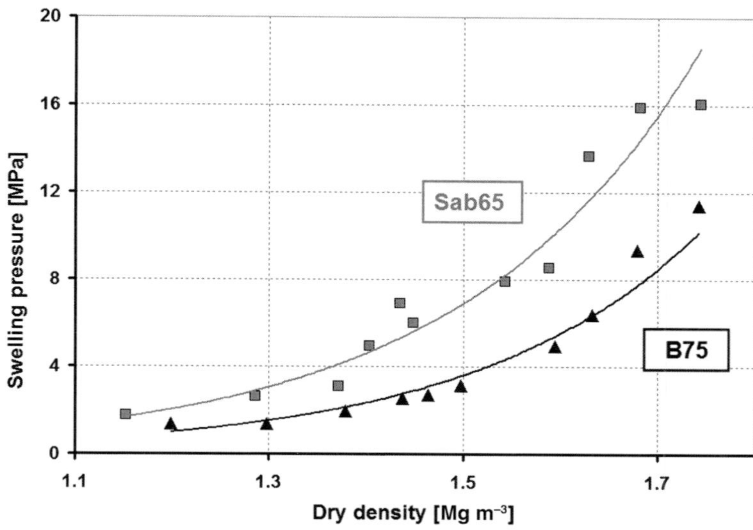

Fig. 2. Initial values of swelling pressure of B75 and Sab65 dependent on dry density.

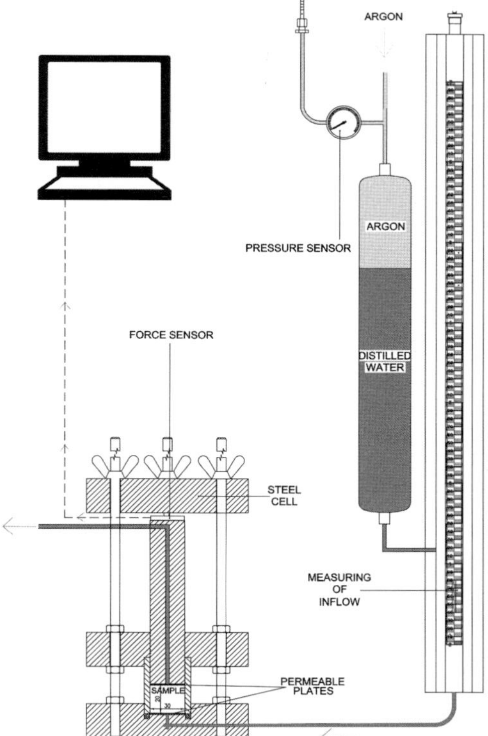

Fig. 3. Schematic of device for measuring hydraulic conductivity and swelling pressure.

hydraulic conductivity was evaluated according to Darcy's law (3):

$$k = \frac{\Delta V \times l}{A \times \Delta t \times \Delta h} \quad (3)$$

where k is hydraulic conductivity [m s^{-1}], l is sample length [m], A is surface area [m^2], Δh is the hydraulic head [m], Δt is time [s] and ΔV is the volume of water [m^3].

For low-permeability materials, the applicability of Darcy's law is limited, in particular regarding the threshold hydraulic gradient J_0 and the critical hydraulic gradient J_C (Dixon *et al.* 1992).

J_0 is the hydraulic gradient below which no flow occurs in the porous medium, and J_C is the lowest

Fig. 4. Device for measuring hydraulic conductivity and swelling pressure.

value of the gradient for which Darcy's law applies. Although these gradients are not determined in this study, tests on Spanish bentonite (FEBEX experiment) concluded that the critical gradient was in this case close to 2000 (ENRESA 2006). Because the lowest gradient used in the present study was 5000, we can therefore expect that all used gradients are larger than the critical gradient, and that equation (3) can be used for the evaluation of hydraulic conductivity.

Evaluation of swelling pressure

The swelling pressure was calculated from the continuously recorded axial force F measured by force sensors (Honeywell) (equation (4)):

$$\sigma_{sw} = \frac{F}{A} \tag{4}$$

where σ_{sw} is the swelling pressure [Pa], F is the recorded force [N] and A is the surface area [m^2].

This force is influenced by the injection pressure. Accordingly, the value of the force used for the evaluation of swelling pressure must be registered when the injection pressure is zero and the measured force is settled; when no injection pressure is applied, the measured force is only caused by swelling of the sample.

Testing procedure

A total of 21 samples with various dry densities (the material was compacted at 1.2–1.75 Mg m^{-3}) were tested. All samples were subjected to the same testing procedure, which consisted of loading and unloading phases achieved by increasing and decreasing the saturation pressure. Both phases consisted of a number of step-by-step increases in saturation pressure (in steps of 1 MPa, i.e. $i =$ 5000; or in steps of 2 MPa, i.e. $i = 10\,000$). The loading phase started when the saturation pressure reached 1 MPa ($i = 5000$). Following stabilization of the inflow, the saturation pressure was increased in steps of 1 MPa up to a maximum of 6 MPa ($i = 30\,000$). Unloading of the samples started thereafter, with a decrease in saturation pressure in 2 MPa intervals down to 0 MPa. Both hydraulic conductivity and swelling pressure (from the recorded force) were calculated for each step of the loading/unloading process.

Results for hydraulic conductivity

The evolution of hydraulic conductivity with respect to the hydraulic gradient is shown in Figure 5 (for B75) and Figure 6 (for Sab65). Only the smallest, middle and highest dry densities are plotted. Arrows in the graphs indicate the beginning of the loading phase.

Hydraulic conductivity decreases with increasing saturation pressure (resp. hydraulic gradient), and hysteresis is observed after unloading. The decrease in hydraulic conductivity upon increasing the hydraulic gradient from 5000 to 30 000 is shown in Figure 7 for all samples. For Na-activated Sab65, the decrease seems independent of the dry density and is *c.* 25%. Conversely, the results for non-activated Mg–Ca bentonite B75 at low densities show a dependency on dry density (with a

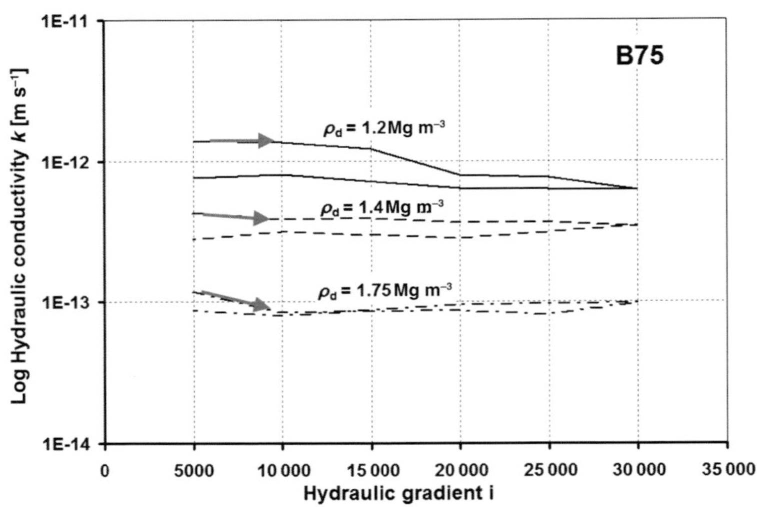

Fig. 5. Dependence of hydraulic conductivity on the hydraulic gradient of bentonite B75.

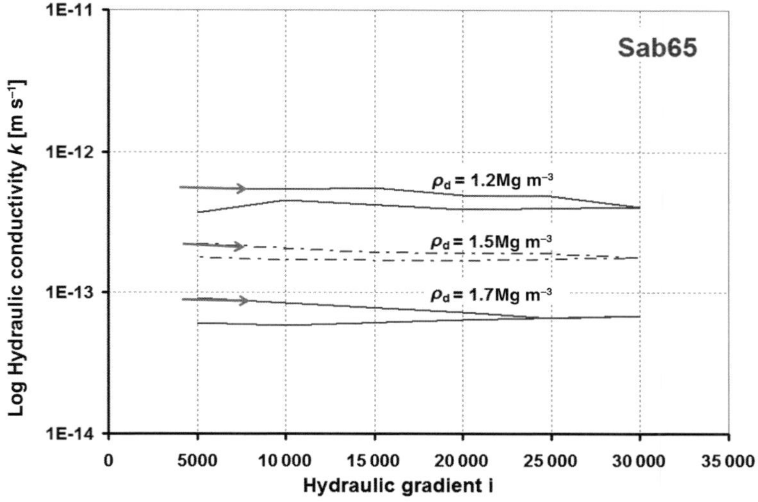

Fig. 6. Dependence of hydraulic conductivity on the hydraulic gradient of bentonite Sab65.

decrease from 60% to 20%) up to a threshold dry density of *c.* 1.45 Mg m^{-3}. Above this density, the hydraulic conductivity decrease stabilizes at *c.* 30%, independent of the dry density.

At the end of the experiment, that is, after the unloading phase, an irreversible decrease in hydraulic conductivity with respect to the hydraulic gradient was observed for all samples (see Fig. 8). These results are comparable to those obtained after the loading phase: (1) the decrease depends on dry density only for the non-activated Mg–Ca bentonite B75 until a threshold dry density close to 1.45 Mg m^{-3}; (2) for other values and for the Na-activated Sab65, the hydraulic conductivity decrease does not depend on dry density.

Results for swelling pressure

Swelling pressure, continuously monitored by the force sensor placed on top of the steel cylinder (see Fig. 3), was the second parameter observed

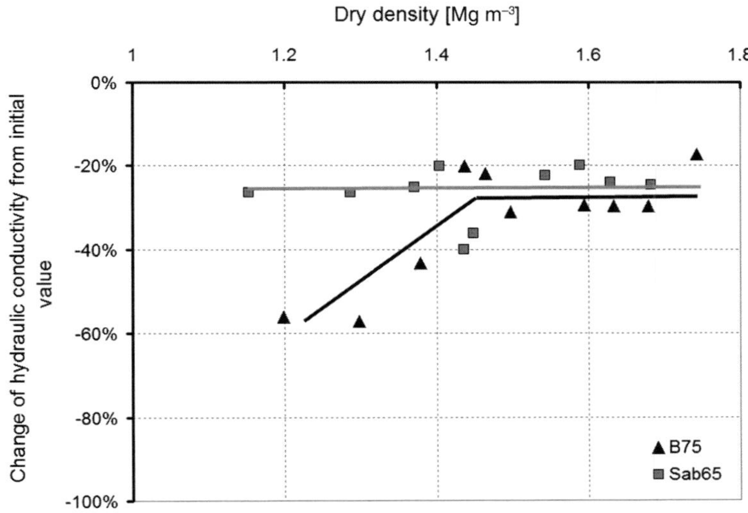

Fig. 7. Decrease of hydraulic conductivity after increasing the hydraulic gradient from 5000 to 30 000.

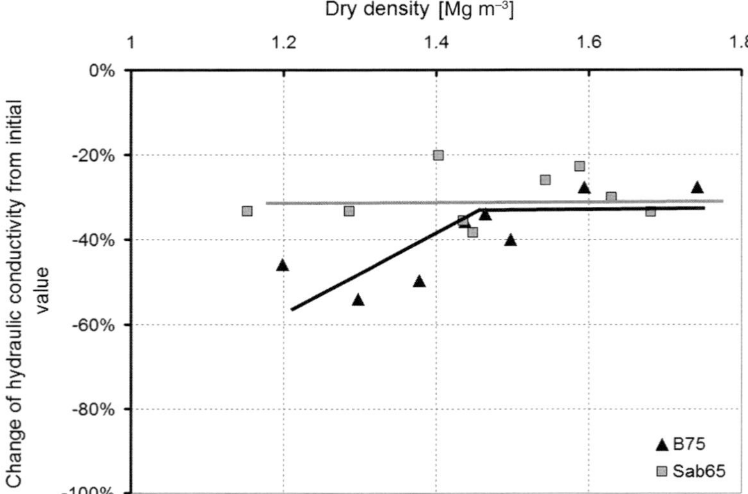

Fig. 8. Decrease of hydraulic conductivity after increasing and decreasing the hydraulic gradient from 5000 to 30 000 and back to 5000.

during the testing procedure. The results obtained from samples of B75 and Sab65 at three different dry densities are presented in Figure 9 (for B75) and Figure 10 (for Sab65). It seems that additional compaction of the sample during loading may have caused the slight increase in axial pressure σ observed after unloading. This hypothesis should, however, be examined in further tests.

The pore pressure was set to zero at one end of the sample with no backpressure applied. Assuming the validity of Terzaghi's effective stress, the final measured pressure σ should be the sum of the effective pressure (σ_{sw}, swelling pressure in this case) and the saturation pressure (σ_{sat}). When the saturation pressure is increased by 1 MPa, the measured pressure should also increase by 1 MPa.

However, it is known from the literature (Vanicek 2000) that this general assumption must be modified in the case of swelling materials. The impact of the saturation pressure is reduced by a

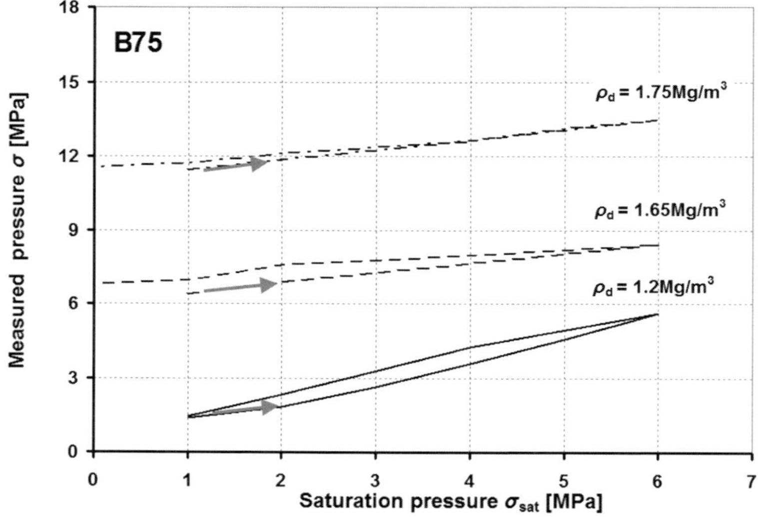

Fig. 9. Increase of measured pressure related to increasing saturation pressure for B75.

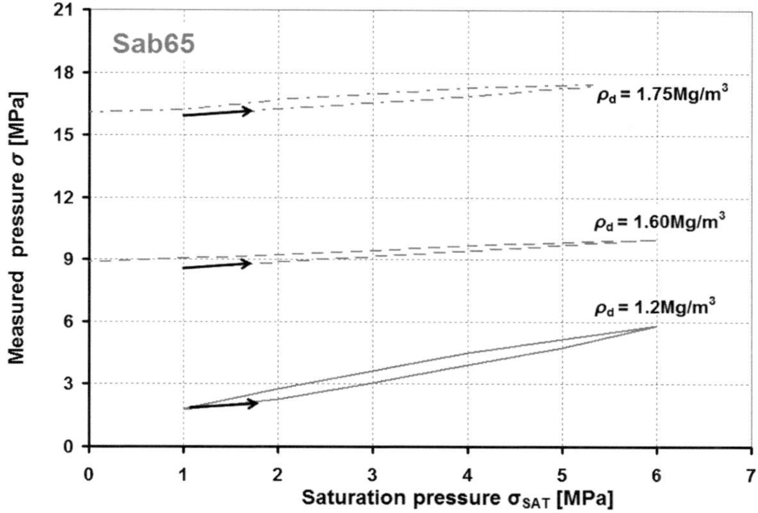

Fig. 10. Increase of measured pressure related to increasing saturation pressure for Sab65.

coefficient α, defined by the following equation:

$$\sigma = \alpha \times \sigma_{\mathrm{sat}} + \sigma_{\mathrm{sw}} \tag{5}$$

where σ is the measured pressure [Pa], α is a coefficient [−], σ_{sat} is the saturation pressure [Pa] and σ_{sw} is the swelling pressure [Pa].

All the results obtained during the loading phase are shown in Figures 11 (B75) and 12 (Sab65). Each line represents a single dry density and is presented with its own formula.

The increase in saturation pressure has a slope equal to unity. It is obvious that all loading lines for samples above this saturation pressure line are roughly parallel with a coefficient α close to 0.4. The lines intersecting the saturation pressure line have slopes progressively closer to the saturation pressure line.

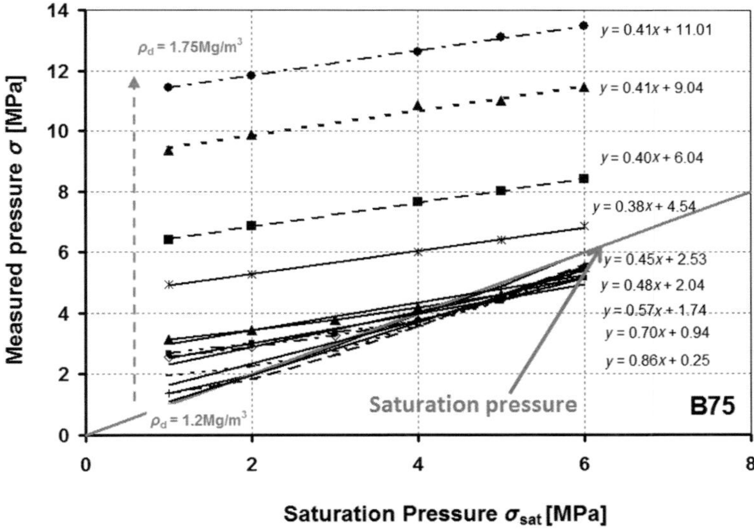

Fig. 11. Increase of measured pressure from the loading phase for B75 with different dry densities (1.2–1.75 Mg m^{-3}), with values of saturation pressure indicated.

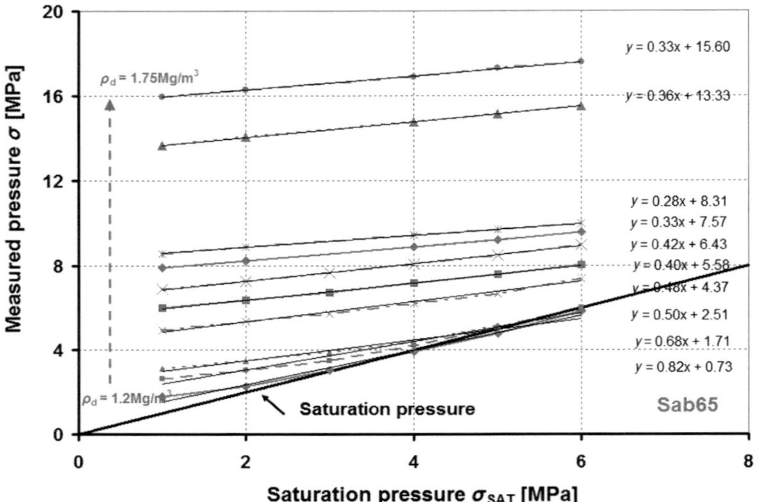

Fig. 12 Increase of measured pressure from the loading phase for Sab 65 with different dry densities (1.2–1.75 Mg m^{-3}), with values of saturation pressure indicated.

In a previous study performed on samples with a swelling pressure much lower than the applied saturation pressure, Harrington and Birchall (2007) obtained coefficient α close to 0.8.

In the present study, the values of coefficient α determined from the formulas are plotted against the initial swelling pressure (pressure measured when the saturation pressure is equal to 1 MPa) in Figure 13. An obvious outcome is that for samples in which the initial swelling pressure is higher than the saturation pressure, coefficient α is constant and close to 0.4. For samples with swelling pressure lower than the saturation pressure this coefficient ranges from 0.4 to 1, because the decrease in initial swelling pressure corresponds to an increase in coefficient α towards a value of 1.

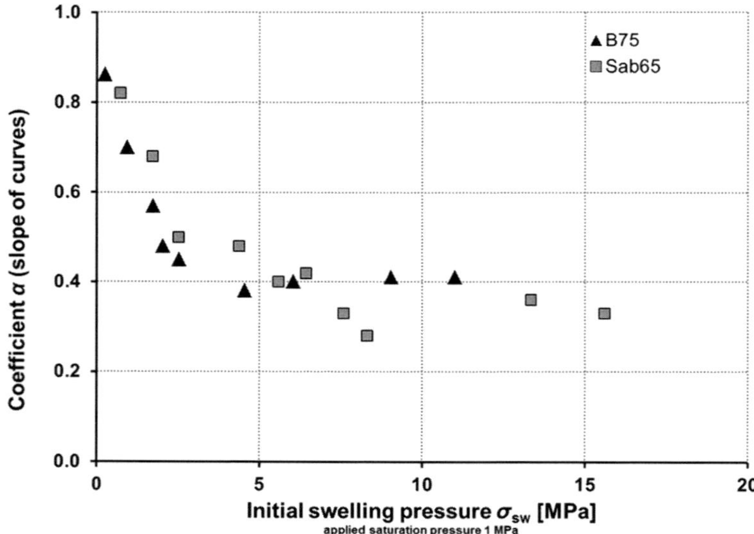

Fig. 13. Dependence of coefficient α on initial swelling pressure for B75 and Sab65, with saturation pressure increasing from 1 to 6 MPa.

Conclusion

A series of hydraulic conductivity tests with concurrent measurement of swelling pressure have been carried out on two types of Czech bentonite–Bentonite 75 and Sabenil 65 (Ca–Mg type v. Na-activated).

A saturation pressure of 1–6 MPa was applied in steps of 1 or 2 MPa. Both loading and unloading (by saturation pressure) phases were carried out.

For the Ca–Mg material B75, the impact of saturation pressure on hydraulic conductivity (the rate of sample compression during the test) decreases with increasing dry density up to a threshold dry density value of *c.* 1.45 Mg m^{-3}. At higher values, the hydraulic conductivity does not change with increasing saturation pressure (resp. hydraulic gradient); this means that above a certain value of dry density, the impact of saturation pressure on the hydraulic conductivity decrease is constant at *c.* 30% of the hydraulic conductivity value. Na-activated Sabenil 65 exhibits a constant decrease of 25% for all studied dry density ranges. After unloading from 6 to 1 MPa (with corresponding hydraulic gradient from 30 000 to 5000 resp.), the value of hydraulic conductivity does not increase back to its initial value. It can be caused by compaction of the sample, but this effect should be examined in further tests.

The influence of saturation pressure on measured pressure decreases with dry density (resp. initial swelling pressure) until a certain value is reached. It can be concluded that when the saturation pressure is lower than the swelling pressure, the measured pressure increases linearly, with coefficient α close to 0.4. The recorded pressures were slightly higher during unloading than in the loading phase. This could also correspond to sample compaction during the loading procedure, but this hypothesis should also be examined in further tests.

Although changes in these parameters should be taken into account, high saturation pressure can be used for the testing of hydraulic conductivity and swelling pressure.

This research was supported by the Czech Ministry for Trade and Industry (project no. FR-TI1/362 and institutional research plan MSM6840770031–'Complex System of Methods for Directed Design and Assessment of Functional Properties of Building Materials'). This support is gratefully appreciated.

References

Dixon, D. A., Gray, M. & Hnatiw, D. 1992. Critical gradients and pressures in dense swelling clays. *Canadian Geotechical Journal*, **29**, 1113–1119.

Dixon, D. A., Graham, J. & Gray, M. N. 1999. Hydraulic conductivity of clays in confined tests under low hydraulic gradients. *Canadian Geotechical Journal*, **36**, 815–825, http://dx.doi.org/0008-3674 (available at http://hdl.handle.net/1993/2792)

ENRESA 2006. *Full-Scale Engineered Barriers Experiment*. Updated Final Report **1994–2004**, ISBN (ISSN) 1134-380X.

Graham, C. C., Harrington, J. F., Cuss, R. J. & Sellin, P. 2012. Impact of pore–pore pressure cycling on bentonite in constant volume experiments. *Clays in Natural and Engineered Barriers for Radioactive Waste Confinement*. Montpellier, France, 22–25 October 2012. Andra, Dostupnéz, 572–573, www.montpellier2012.com/meeting-program.php

Harrington, J. F. & Birchall, D. J. 2007. *Sensitivity of Total Stress to Changes in Externally Applied Water Pressure in KBS-3 Buffer Bentonite [online]*. SKB, Stockholm, 26 [cit. 2012-11-11], http://nora.nerc.ac.uk/9983/

PROUDOVÝ TLAK, Vanicek, I. 2000. *Geomechanika 10: Mechanika zemin*. 3. vyd., ČVUT, Praha, 46–47, ISBN 80-01-01437-1

Villar, M. V. & Gómez-Espina, R. 2009. *Report on Thermo-Hydro-Mechanical Laboratory Tests Performed by CIEMAT on Febex Bentonite 2004–2008*. CIEMAT, Madrid.

Pore-pressure cycling experiments on Mx80 Bentonite

C. C. GRAHAM[1]*, J. F. HARRINGTON[1], R. J. CUSS[1] & P. SELLIN[2]

[1]*British Geological Survey, Keyworth, Nottingham NG12 5GG, UK*

[2]*Svensk Kärnbränslehantering AB, Stockholm, Sweden*

**Corresponding author (e-mail: caro5@bgs.ac.uk)*

Abstract: The Swedish concept for geological disposal of radioactive waste involves the use of bentonite as part of an engineered barrier system. A primary function of the bentonite is its ability to swell when hydrated by its surroundings. One particular uncertainty is the impact on this function, resulting from deviations in pore-water pressure, p_w, from expected *in situ* hydrostatic conditions. We present results from a series of laboratory experiments designed to investigate the form of the relationship between swelling pressure and p_w, for compacted Mx80 bentonite, from low to elevated applied water pressure conditions. The experiments were conducted using constant volume cells, designed to allow the total stresses acting on the surrounding vessel to be monitored (at five locations) during clay swelling. The results demonstrate that swelling pressure reduces non-linearly with increasing p_w, becoming less sensitive to changes at elevated pressures. After cyclic loading a marked hysteresis was also observed, with swelling pressure remaining elevated after a subsequent reduction in applied water pressure. Such behaviour may impact the mechanical and transport properties of the bentonite and its resulting performance. However, such hysteric behaviour was not always observed. Further testing is required to better understand the causes of this phenomenon and the controls on such behaviour.

In the Swedish concept for disposal of radioactive waste, copper canisters containing the vitrified waste material are emplaced within a crystalline host-rock. The canisters are isolated from the surrounding rock by blocks and pellets of pre-compacted bentonite. The bentonite provides a number of functions within the design, including: (a) structural support, preventing collapse and protecting the canister from excessive stresses; (b) a diffusional barrier surrounding the canister; and (c) a swelling characteristic on hydration, which aids in the closure of voids and joints within the deposition hole. As such, the swelling properties of the bentonite are of primary importance to the design of the system. In order for the material to act as a successful barrier, it must be able to generate and maintain its expected swelling pressure, Π (when constrained by its surroundings), over long timescales.

Early experimental studies in bentonite demonstrated that the development of swelling pressure occurs rapidly in the initial stages of swelling, before notably reducing in rate as the final pressure is approached (Pusch 1980; Börgesson 1985; Madsen & Müller-Vonmoss 1989; Bucher & Müller-Vonmoos 1989). The rate of this development may be affected by factors such as the initial water content of the material and the applied water pressure. However, the final pressure achieved by the swelling bentonite, for a given pore water chemistry, is strongly dependent upon the initial dry density (Pusch 1980; Bucher & Müller-Vonmoos 1989; Oscarson *et al.* 1990). A number of test geometries are commonly used for testing the swelling properties of clays; for example, a sample is allowed to swell under a specific load, or sample loading is achieved by constraining the sample from swelling. This latter approach, where sample volume is maintained constant, is the most representative of compacted bentonite blocks swelling in an enclosed engineered barrier system (EBS). As such, a constant volume arrangement was utilized in this study, in order to provide the most appropriate boundary condition.

The SKB safety case for a Swedish radioactive waste repository highlights the potential importance of long-term fluctuations in local pore-water pressures on repository functions (SKB 2011). For example, during future glaciation events, pore-water pressures are likely to be significantly elevated for considerable periods of time. One particular uncertainty is the likely effect of elevated pore-water pressures on the safety functions of the engineered barrier. Over the repository lifetime such changes in pore-water pressure may well be cyclic in nature, as successive glacial episodes lead to loading and unloading of the engineered barrier. It is therefore crucial to have a full understanding of the impact of such changes in pore-water pressure on the swelling properties of the bentonite. There are a considerable number of studies focussing on the impact of cyclic drying and wetting of clays and argillaceous rocks within the

From: Norris, S., Bruno, J., Cathelineau, M., Delage, P., Fairhurst, C., Gaucher, E. C., Höhn, E. H., Kalinichev, A., Lalieux, P. & Sellin, P. (eds) 2014. *Clays in Natural and Engineered Barriers for Radioactive Waste Confinement*. Geological Society, London, Special Publications, **400**, 303–312.
First published online May 7, 2014, http://dx.doi.org/10.1144/SP400.32

literature (Osipov *et al.* 1987; Basma *et al.* 1996; Pejon & Zuquette 2002; Doostmohammadi *et al.* 2008). However, there is a paucity of data relating to the impact of water pressure cycling on fully saturated clays at applied pressures in the vicinity of expected *in situ* conditions for a radioactive waste repository. The focus of this paper is on the effects of such deviations from hydrostatic conditions on bentonite, within the context of the Swedish radioactive waste disposal programme.

Bucher & Müller-Vonmoos (1989) investigated the development of stress generated by swelling of Mx80 and Montigel bentonite. They observed a strong correlation between swelling stress, σ, and increased dry density. By applying stepped increases in the applied water pressure, the consequent increase in σ was determined in order to delineate the relationship between the two. For samples of dry density below 1.40 mg mm^{-3}, this relationship was seen to be approximately linear, although the full range of potential water pressures expected in the repository environment was not examined; instead, low to moderate pressures were considered only. However, for higher dry densities, swelling stress was observed to be less sensitive to increasing p_w, with a generated value of only 60–70% of the applied water pressure, for a dry density of 1.90 mg m^{-3}. The authors suggested that this observed deviation from the standard effective stress law (Terzaghi 1943) may be at least partly explained by the exceptionally small pore volume in highly compacted bentonite, which clearly limits the availability of free water within the material. While Terzaghi's principal requires that the effective stress, or 'swelling pressure' for clays, results from the difference between generated stress and the applied pore pressure, Bucher and Müller-Vonmoos suggested the addition of a proportionality constant to allow for this deviation from linear behaviour.

Harrington & Horseman (2003) and Harrington & Birchall (2007) also observed this deviation for Mx80 bentonite, under a constant volume boundary condition. When investigating the change in total stress, $d\sigma$, resulting from a change in applied water pressure, dp_w, their observations also required the introduction of a proportionality constant, α (equal to $d\sigma/dp_w$), to adequately describe them. As such, for a clay–water system with the pore-water in thermodynamic equilibrium with an external reservoir of water at pressure, p_w, the total stress acting on the surrounding vessel can be expressed as:

$$\sigma = \Pi + \alpha p_w \quad (1)$$

where Π is the swelling pressure. This is the definition for swelling pressure used in this study and throughout the remainder of this paper.

While it has been previously suggested that, with ncreasing applied water pressure, the swelling pressure may decline to the point where liquefaction of the bentonite occurs, the findings from a number of studies contradict this (Harrington & Horseman 2003), instead indicating that, for higher-density bentonite, swelling pressure becomes increasingly insensitive to changes in applied water pressure at elevated values of p_w. The same observation is made in this paper, where the form of this relationship is examined at higher applied water pressures.

We present observations from constant volume experiments on four bentonite samples (Mx80-10, -11, -13 and -14), focussed on elucidating the relationship between p_w and Π. The samples tested were prepared from blocks of pre-compacted Mx80 bentonite, which is a candidate clay for the Swedish repository concept. The data from testing specimens Mx80-10 and Mx80-11 have been previously presented in technical reports by Harrington & Horseman (2003) and Harrington & Birchall (2007). However, here we present these results in combination with those from two further tests (Mx80-13 and Mx80-14), providing a larger evidence base from which to compare findings. Our intention is to present laboratory results of the impact of pore-pressure on swelling behaviour in Mx80 bentonite and, in particular, the influence of cyclic applied pressure changes. This is intended as an observational paper, as further expansion of the dataset is required in order to reliably interpret the findings.

Methodology

Experimental set-up

The experiments described here were carried out using custom-designed constant volume and radial flow (CVRF) apparatuses, constructed initially to examine the sensitivity of gas flow in buffer bentonite to test boundary conditions (Harrington & Horseman 2003). The CVRF systems consist of: (a) a thick-walled stainless steel pressure vessel; (b) a fluid injection system; (c) three independent backpressure systems, each consisting of an array of four filters acting as fluid sinks; (d) five total stress sensors to measure radial and axial stress; and (e) a logging system. The pressure vessel (Fig. 1) comprises a stainless-steel steel, dual-closure, tubular vessel whose end-closures are secured by 12, high-tensile cap-screws that can also apply a small pre-stress to the specimen if required. The positions of the sink arrays ([1], [2] and [3]) and the stress sensors (labelled PT1, 2, 3, 5 and 6) are shown in Figure 1.

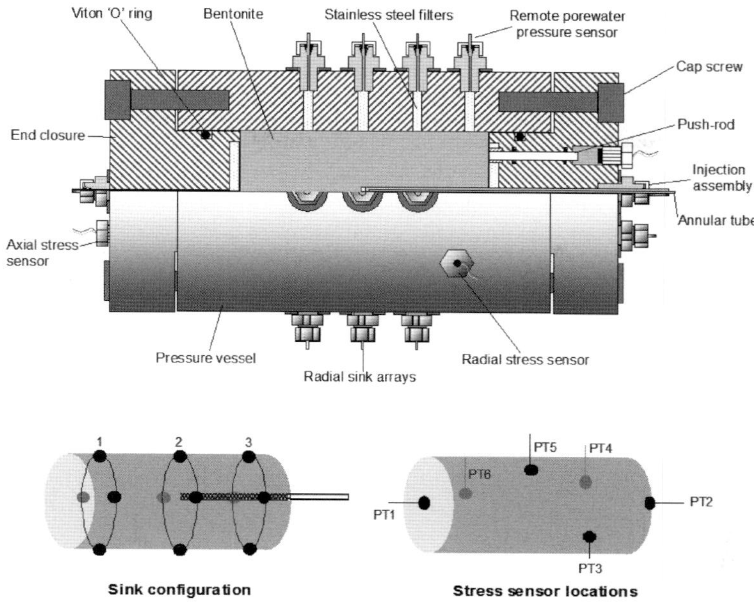

Fig. 1. Cut-away section of the constant volume radial flow (CVRF) cell, showing the two end-closures with their embedded drainage filters, the central fluid injection filter, the 12 radial sink filters, the five total stress sensors and the independent pore pressure sensor.

Two CVRF apparatuses were used for the tests described in this paper (Fig. 1), which are both based on a similar design. However, a few notable differences are described below. In CVRF1, all ports, except those for the direct measurement of stress, contain sintered stainless steel porous filters which are profiled to match the bore of the pressure vessel. The stress gauges are an in-house design, using a steel push-rod fitted with an 'O'-ring seal, to compress a small volume of liquid contained within a chamber at the front face of a miniature Sensotec Model pressure transducer. This rig also has a port with an independent pore-water pressure sensor (PT4, Fig. 1), where the push-rod is replaced by a sintered stainless steel porous filter, enabling water pressure to act on the front face of the transducer.

CVRF2 has no pore-pressure sensor directly connected to the pressure vessel itself, but the three pore-pressure arrays can be isolated in connection to a pressure transducer, allowing independent monitoring of the pore-water pressure, or connected to the backpressure pump in order to monitor outflow. While sintered stainless steel filters are embedded in the end closures and the central injection rods of both apparatuses, in CVRF2 the radial arrays contain sintered high-density polyethylene plugs. In this set-up, stress measurements are made using push-rods which are each in direct contact with a Burster miniature load cell (model 8402-6005). In both cases, the central filter is embedded at the end of a 6.4 mm-diameter stainless steel tube which can be used to inject permeant, for gas flow testing purposes. Results from the gas

Table 1. *Pre-test (calculated with an assumed grain density of 2.77 mg m^{-3})*

Sample no.	Moisture content (%)	Bulk density (mg m^{-3})	Dry density (mg m^{-3})	Void ratio	Saturation (%)	Test apparatus
Mx80-10	26.7	2.005	1.582	0.751	98.6	CRVF1
Mx80-11	25.6	2.016	1.605	0.726	97.6	CVRF1
Mx80-13	20.1	2.064	1.718	0.612	91.1	CVRF2
Mx80-14	26.6	1.999	1.579	0.754	97.7	CVRF2

Table 2. *Post-test*

Sample no.	Saturation (%)	Maximum p_w experienced (MPa)	Duration of test (days)
Mx80-10	≥100*	7	93
Mx80-11	≥100*	46	340
Mx80-13	≥100*	42	140
Mx80-14	≥100*	41	152

*Measured value indicates sample was fully saturated, within uncertainty limits of the measurement.

testing phase of these experiments are not the focus of this paper and are in preparation for publication or presented elsewhere (Graham *et al.* 2012).

The pressure and flow rate of test fluid is controlled using either an ISCO Teledyne-100 or an ISCO Teledyne-260, Series D syringe pump, operated by an ISCO pump controller. These units have an RS232 serial port, which allows volume, flow rate and pressure data from each pump to be transmitted to a bespoke logging system. Additional test parameters are logged simultaneously by the same system and the typical acquisition

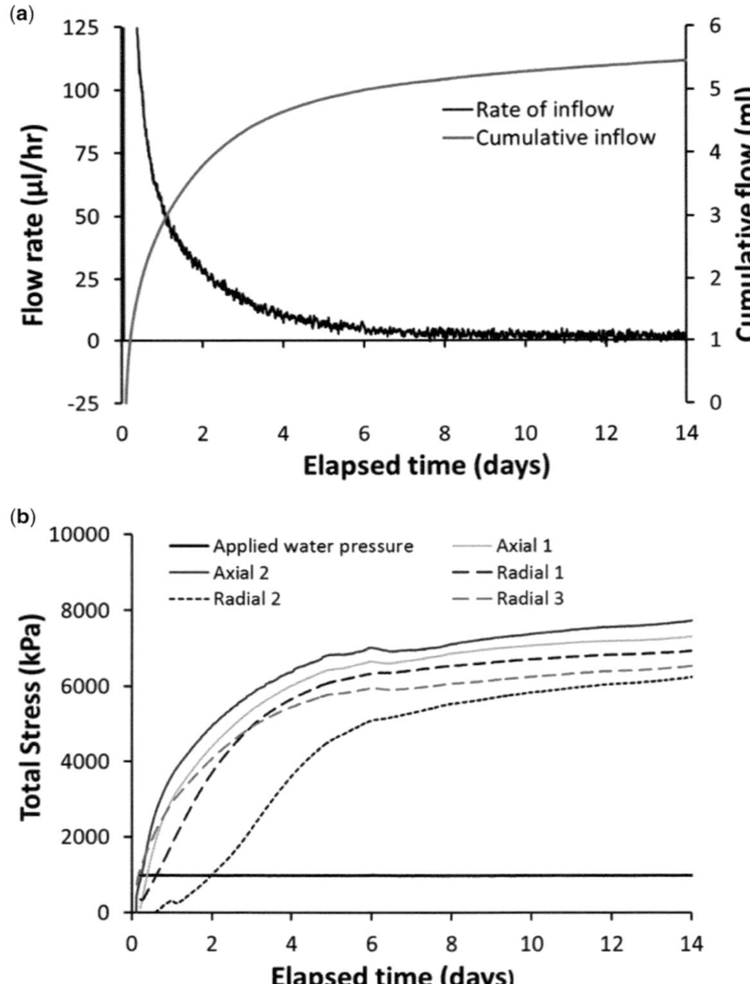

Fig. 2. All samples were initially hydrated at a constant applied pore-water pressure of 1 MPa and allowed to re-equilibrate. (**a**) A typical inflow during the hydration phase test Mx80-14. (**b**) Both axial and radial stresses were clearly observed to increase during sample swelling (test Mx80-14).

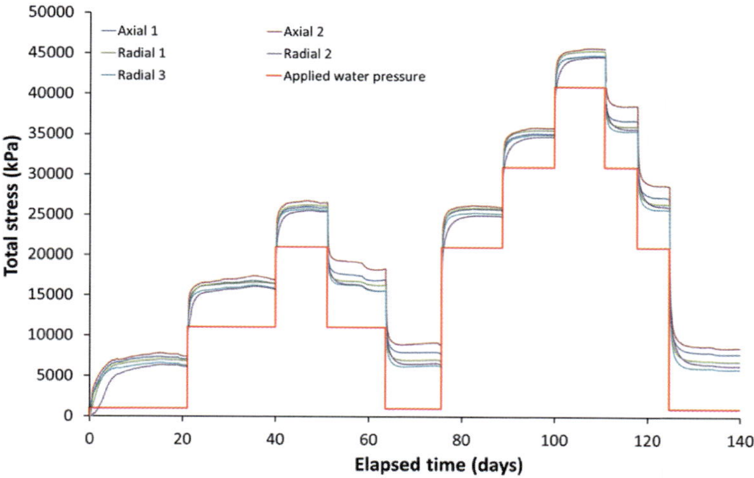

Fig. 3. A typical applied water pressure loading history, applied to sample Mx80-14. In the test shown, two loading and unloading cycles were applied to the specimen over the course of 152 days. The stress response observed in this test is atypical, in that no hysteresis was observed (see Fig. 6a).

rate is one scan every 2 min. All stress and pore pressure sensors were calibrated against laboratory standards by applying incremental steps in pressure, from atmospheric pressure to a pre-determined maximum value. This was followed by a descending history to quantify any hysteresis.

Sample preparation and properties

Testing was carried out on samples of Mx80 bentonite, which is the selected buffer material for the Swedish radioactive waste repository concept. Mx80 bentonite is a fine-grained sodium bentonite, from Wyoming, which contains around 80% montmorillonite (Karnland 2010). Blocks of pre-compacted bentonite were manufactured by Clay Technology AB (Lund, Sweden), by rapidly compacting bentonite granules in a mould under a one dimensionally applied stress (Johannesson *et al.* 1995). In this paper, results are presented from experiments carried out on four cylindrical test specimens (with a diameter = 60 mm and length = 120 mm), sub-sampled from the bentonite. Samples Mx80-10 and Mx80-11 were manufactured by hand-trimming using a tubular former with a sharpened leading edge. The upper and lower surfaces were finished using a scraping action with a flat-bladed knife, leaving the end surfaces flat and parallel. The former was mounted in a lathe and a 6.4 mm-diameter hole drilled in the clay to accommodate the source filter and tubing assembly. The specimen was then extruded from the former into the pressure vessel using a screw-driven press. Rather than using the former, samples Mx80-13 and Mx80-14 were instead turned to the same dimensions on a lathe, producing a high quality finish on the cylindrical surface, and a hole drilled into the clay as before.

Standard geotechnical properties for each sample pre- and post-test are shown in Tables 1 and 2, respectively. The water content of the specimens was determined by weighing them pre-test, then oven-drying them post-test before weighing again. The void ratio, porosity and degree of saturation are based on a grain density for the bentonite of 2.77 mg m^{-3} (taken from an average of the values measured by Karnland 2010). Dry densities of the samples were *c.* 1.55–1.75 mg m^{-3} and, therefore, above the dry density of 1.4 mg m^{-3} where Bucher & Müller-Vonmoos (1989) noted a change in the relationship between applied water pressure and the resulting stress generated.

Pore-pressure cycling

For each sample, hydration was carried out with de-ionized water, applied through all filters. A high-precision syringe pump was used to maintain a constant applied water pressure, while the rate of inflow, and consequent stress development, were monitored. Each test was begun by applying a constant applied water pressure of 1 MPa. An example showing the typical stress development and rate of inflow (for sample Mx80-14) is given in Figure 2. The monitored stresses are shown as measured in three radial locations, as well as

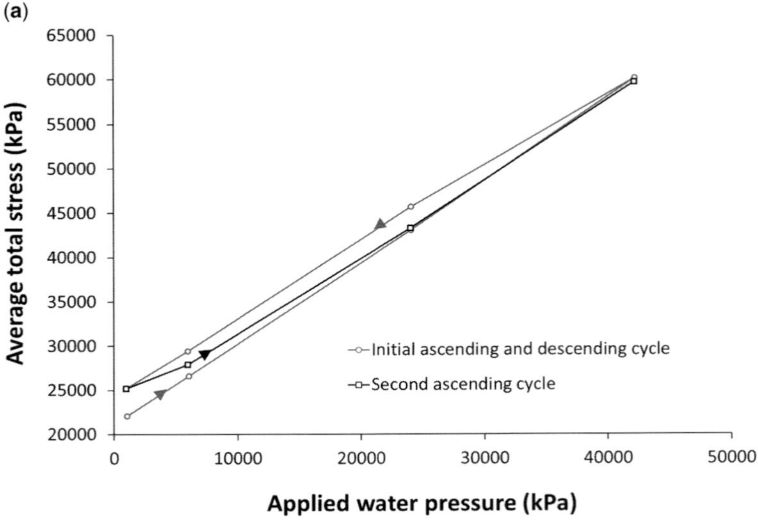

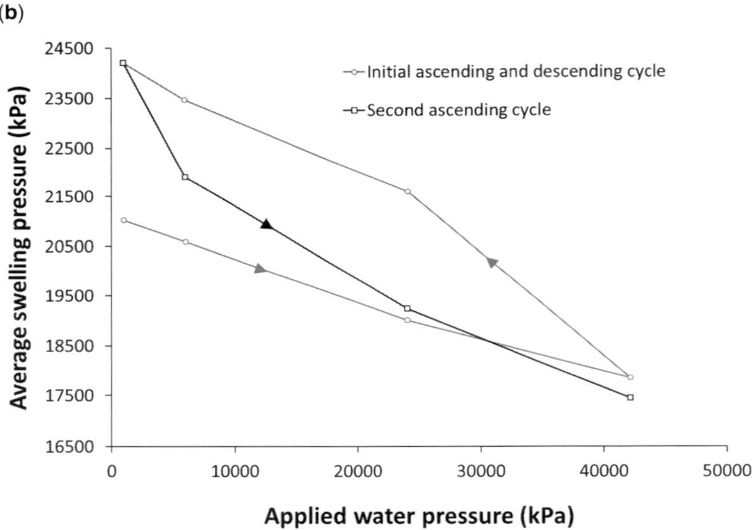

Fig. 4. (**a**) An example stress response history, resulting from applied water pressure cycling (for sample Mx80-13). (**b**) An example swelling pressure response history, resulting from applied water pressure cycling (for sample Mx80-13).

at the samples ends (axial). The drop-off in water inflow, as the rate of sample swelling slowly reduces and the clay equilibrates with the applied water pressure, is clearly shown. For all water pressure increments, the sample was only subjected to a new applied pressure once this stage had been clearly reached.

Each sample was subjected to a series of incremental constant pore pressure steps, with all samples experiencing at least one loading and unloading cycle. Samples Mx80-10 and Mx80-14 were subjected to one and two full cycles, respectively. In the case of specimen Mx80-11, three cycles were carried out (the last two with smaller pressure steps) and specimen Mx80-13 was subject to one full cycle, plus additional loading increments. In an ideal system (where alpha equals one), swelling pressure, Π, is equal to the effective stress (Harrington & Birchall 2007). In order to examine the validity of this relationship, swelling pressure was calculated using this assumption, as shown in the following section.

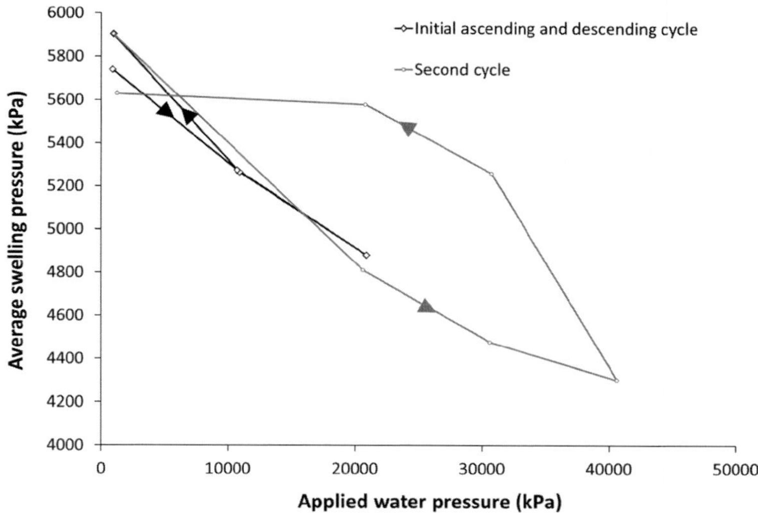

Fig. 5. An example swelling pressure response history, resulting from applied water pressure cycling (for sample Mx80-14).

Results

In this section we present results from all four experiments. A typical loading history is shown in Figure 3 (for sample Mx80-14), which demonstrates the significant timescales required to carry out testing of this nature. The measured dry and bulk densities for each sample are given in Table 1. All samples were found to be fully saturated after testing was completed (Table 2). An example loading history is shown in Figure 4 (for Mx80-14), along with generated total stresses (at three radial and two axial locations) and calculated swelling response. For all tests, the swelling pressure was clearly observed to decrease with increasing pore-water pressure (e.g. Fig. 5 – Mx80-14). This behaviour was observed by Harrington & Horseman (2003) at lower water pressures. However, by investigating the form of this relationship over a more extensive range of pressures, it is clear that the sensitivity of the swelling pressure to changes in the applied water pressure decreases asymptotically as p_w is increased (Fig. 5). This implies that, within the range of expected repository conditions, there is a physical limit beyond which swelling pressure becomes insensitive to further increases in applied water pressure, most likely owing to the reduced mobility of water within the clay under these conditions. As such, care must clearly be taken when attempting to extrapolate swelling pressure behaviour at higher pore-water pressures, based on laboratory data measured at lower applied pressures.

In addition, the results from cycling loading and unloading of the bentonite indicate that pore-pressure cycling may lead to a persistent elevation in the average swelling pressure of the clay, in spite of a consequent reduction in the applied pore-water pressure. By plotting the measured average total stress and the calculated swelling pressure at each applied pressure step (Fig. 6), a significant degree of hysteresis is made apparent for samples Mx80-10, Mx80-11 and Mx80-13. Higher density bentonite was used for test Mx80-13, explaining the noticeably higher total stresses and swelling pressures observed. However, hysteresis was still clearly observed at these higher pressures. In the case of sample Mx80-14, the observed behaviour is anomalous: while significant non-linearity in the behaviour of the bentonite (Fig. 5) was apparent, the resulting swelling pressure after the first and second pore-pressure cycles did not significantly deviate from the initially measured values. Incremental values for alpha were calculated, by fitting to the slope of the measured stress v. the applied water pressure between each step (i.e. calculating $d\sigma/dp_w$). The average total stress data for all tests yield incremental alpha values generally within the range 0.8–1.0 (Fig. 7), which is in agreement with previous results for Mx80 bentonite (Bucher & Müller-Vonmoos (1989). It should be noted that the initial elevated swelling pressures observed in sample Mx80-13 were the result of a significantly higher starting dry density of the material (Table 1).

However, the observed differences in behaviour for Mx80-14 do not appear to be explained

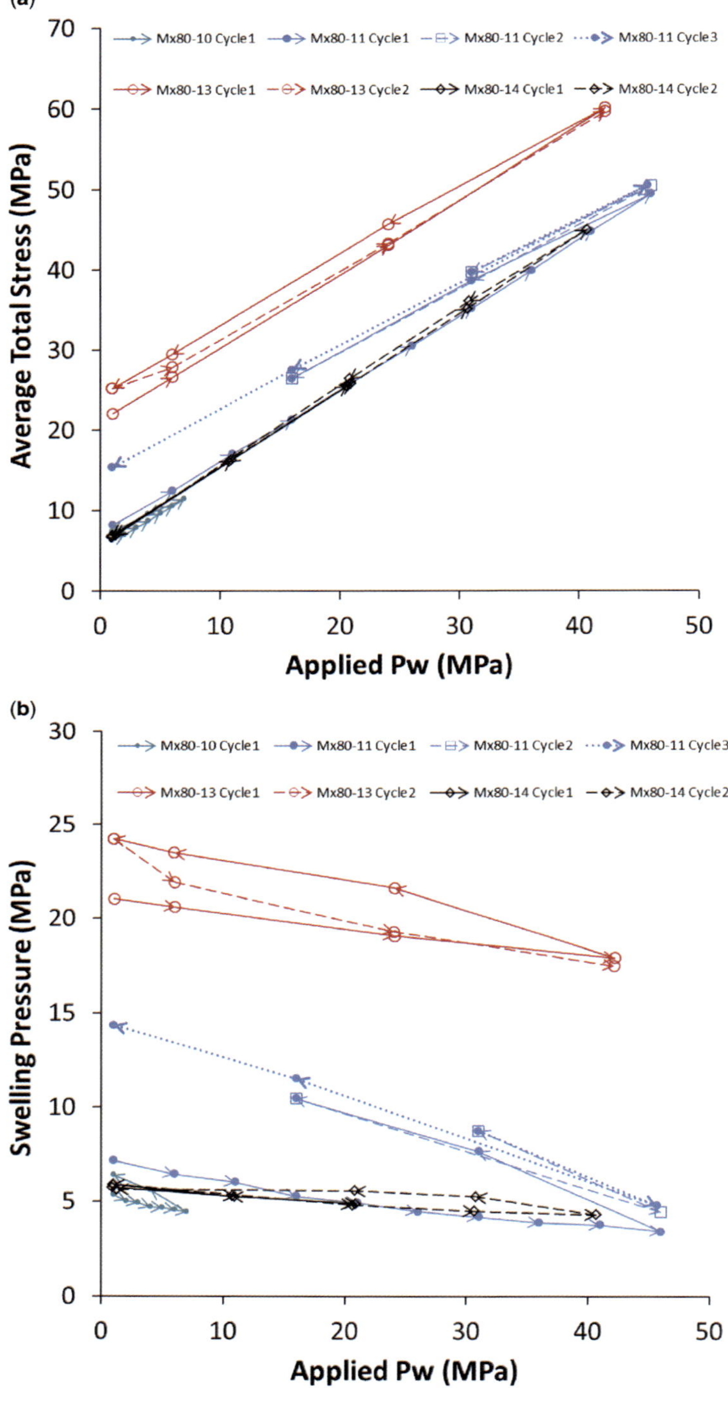

Fig. 6. (**a**) Average total stress plotted against externally applied water pressure (backpressure). (**b**) Swelling pressure plotted against externally applied water pressure (backpressure). Samples Mx80-11 and Mx80-13 show a clear departure from the predicted ideal behaviour, as does Mx80-10 (as a percentage of the initial swelling pressure).

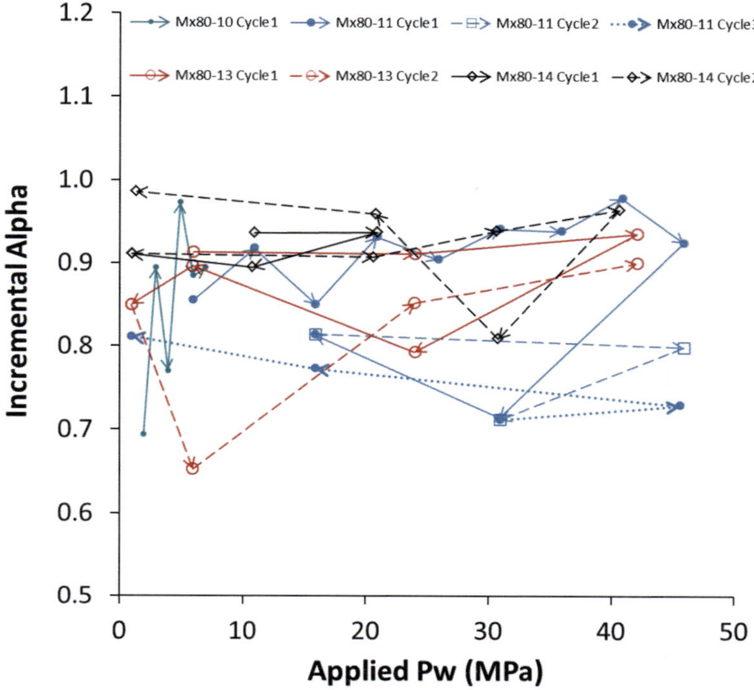

Fig. 7. Calculated incremental values of α (a proportionality constant equal to $d\sigma/dp_w$), resulting from applied water pressure cycling (for all four samples of Mx80 bentonite).

simply by dry density alone. While a different stress measurement method was utilized in test Mx80-14, hysteric behaviour was observed for sample Mx80-13, which was conducted with the same experimental apparatus. Care was taken to ensure that samples were at hydraulic equilibrium before calculating the resulting swelling pressure. Time may be another potential influence on the observed behaviour though, if a degree of time-dependency exists in the response of the bentonite. However, no evidence for such time dependency could be observed when comparing the length of pressure steps for experiments within this dataset. Instead, it seems more likely that the observed hysteresis is an artefact of the underlying physics governing bonding of water to the clay and its coupling to the stress state of the sample. However, further expansion of the data set is required in order to better elucidate the cause of and controls on this observed behaviour.

Conclusions

In order to examine the form of the relationship between swelling pressure and applied water pressure, a series of pressure cycling experiments were conducted on samples of pre-compacted Mx80 bentonite. The experiments were conducted using a specially designed constant volume cell, which allows the evolution of the total stresses acting on the surrounding vessel to be monitored during clay swelling. The results clearly demonstrate a significant decrease in swelling pressure with increase applied water pressure, for all four Mx80 samples. However, it is also apparent that this sensitivity to changes in applied water pressure is significantly reduced at more elevated values of p_w.

In addition to this observed non-linearity in behaviour, test observations also indicate that an elevated swelling pressure may be retained at significant magnitudes within the clay, as a result of a cyclic water pressure loading. While the observed behaviour is not extreme in magnitude, the effect is notable and suggests that our understanding of the relationship between applied water pressure and swelling pressure is not yet complete. As such, a better understanding of the causes of this behaviour may be needed when assessing repository performance in response to periods of elevated or decaying pore-water pressure (e.g. during interglacial and glacial events). However, it should also be noted that such hysteric behaviour is not always observed and the cause of this remains, as yet, unclear. Further testing will help to clarify the controls on this behaviour and to elucidate the nature of the relationship between pore-water pressure, swelling pressure and the resulting total stress in such systems.

The research leading to these results has received funding from SKB and the European Atomic Energy Community's Seventh Framework Programme (FP7/2007-2011) under Grant Agreement no. 230357, the FORGE project. We would like to thank the reviewers for their comments and improvements. This paper is published with the permission of the Executive Director of the British Geological Survey (NERC).

References

Basma, A. A., Al-Homoud, A. S., Husein Malkawi, A. I. & Al-Bashabsheh, M. A. 1996. Swelling-shrinkage behaviour of natural expansive clays. *Applied Clay Science*, **11**, 211–227.

Börgesson, L. 1985. Water flow and swelling pressure in non-saturated bentonite-based clay barriers. *Engineering GeologyEngineering Geology*, **21**, 229–237.

Bucher, F. & Müller-Vonmoos, M. 1989. Bentonite as a containment barrier for the disposal of highly radioactive wastes. *Applied Clay Science*, **4**, 157–177.

Doostmohammadi, R., Moosavi, M. & Araabi, B. N. 2008. A model for determining the cyclic swell-shrink behaviour of argillaceous rock. *Applied Clay Science*, **42**, 81–89.

Graham, C. C., Harrington, J. F., Cuss, R. J. & Sellin, P. 2012. Gas migration experiments in bentonite: implications for numerical modelling. *Mineralogical Magazine*, **76**, 3279–3292.

Harrington, J. F. & Birchall, J. D. 2007. *Sensitivity of total stress to changes in externally applied water pressure in KBS-3 buffer bentonite*. SKB Silver Series Technical Report **TR-06-38**, Svensk Kärbränslehantering AB, Stockholm.

Harrington, J. F. & Horseman, S. T. 2003. *Gas migration in KBS-3 buffer bentonite: sensitivity of test parameters to experimental boundary conditions*. Report **TR-03-02**, Svensk Kärbränslehantering AB, Stockholm.

Johannesson, L.-E., Börgesson, L. & Sandén, T. 1995. *Compaction of bentonite blocks: development of technique for industrial production of blocks which are manageable by man*. TR 95-19, Svensk Kärnbränslehantering AB.

Karnland, O. 2010. *Chemical and mineralogical characterization of the bentonite buffer for the acceptance control procedure in a KBS-3 repository*. Report **TR-10-60**. Svensk Kärbränslehantering AB (SKB), Stockholm.

Madsen, F. T. & Müller-Vonmoss, M. 1989. The swelling behaviour of clays. *Applied Clay Science*, **4**, 143–156.

Oscarson, D. W., Dixon, D. A. & Gray, M. N. 1990. Swelling capacity and permeability of an unprocessesd and a processed bentonitic clay. *Engineering Geology*, **28**, 281–289.

Osipov, V. I., Bik, N. N. & Rumjantseva, N. A. 1987. Cyclic swelling of clays. *Applied Clay Science*, **2**, 363–374.

Pejon, O. J. & Zuquette, L. V. 2002. Analysis of cyclic swelling of mucrocks. *Engineering Geology*, **67**, 97–108.

Pusch, R. 1980. *Swelling pressure of highly compacted bentonite*. Report **TR-80-13**. Svensk Kärbränslehantering AB, Stockholm.

Svensk Kärnbränslehantering AB 2011. *Long-term safety for the final repository for spent nuclear fuel at forsmark: main report of the SR-site project*. TR-11-01.

Terzaghi, K. 1943. *Theoretical Soil Mechanics in Engineering Practice*. John Wiley, New York.

Mechanical interpretations of the homogeneous nature of bentonite due to swelling

I. KOBAYASHI[1]*, K. SUZUKI[2], H. ASANO[2], P. SELLIN[3], C. SVEMAR[3] & M. HOLMQVIST[4]

[1]*Kajima Corporation, Tobitakyu 2-19-1, Chofu, Tokyo 182-0036, Japan*

[2]*RWMC, Pacific Marks Tsukishima, 1-15-7, Tsukishima, Chuo-ku, Tokyo 104-0052, Japan*

[3]*SKB, Box 250, SE-101 24 Stockholm, Sweden*

[4]*SKB International, Box 1091, SE-101 39 Stockholm, Sweden*

**Corresponding author (e-mail: koba13@kajima.com)*

Abstract: To achieve the high sealing capability required for the bentonite engineered barriers used in geological disposal facilities for radioactive waste, the bentonite should be highly compacted. Previous investigation of methods to enable such compaction has revealed that the construction methods used have an impact on the density distribution. However, these investigations focused on achieving the required average dry density in the bentonite barriers and little attention has been paid to the distribution of dry density. As a result, there has been a lack of understanding of how the construction method used affects this distribution, which is highly pertinent to the evaluation of the long-term performance of bentonite engineered barriers.

This paper first describes the residual density distribution after resaturation and then highlights how swelling behaviour can be studied and evaluated based on elasto-plastic theory. The conceptual understanding is that the buffer does not swell to achieve an entirely homogeneous density, but will reach an equilibrium state with a remaining gradient.

Clay-based buffers in geological disposal facilities for radioactive high level waste (HLW) are normally installed as a combination of blocks and pellets with voids in between these components. The swelling of these components to form a homogeneous and watertight buffer is an important process in ensuring that the voids are filled and that long-term buffer performance is achieved after resaturation. As part of a bentonite resaturation programme, Japan's Radioactive Waste Management Funding and Research Center (RWMC) has initiated and performed studies to evaluate the homogenization of installed buffer components as observed during the resaturation period. One of the project's objectives is continuous evaluation of long-term buffer performance using quantitative methods. Although a variety of processes associated with the resaturation period are investigated in the project, this paper focuses on the evolution of, and changes in, the dry density of the bentonite buffer during the process of swelling.

Review of previous research

In the current concept of geological disposal for radioactive waste (e.g. JAEA 2000), high sealing capability is required for bentonite engineered barriers to ensure the feasibility of the approach. To achieve the high compaction properties required for this sealing capability, various methods such as *in situ* compacting, bentonite block placement and the filling and spraying of pellets into gaps between the blocks have been investigated. The results show that density distribution in such barriers depends on the construction method used. However, previous investigations have focused on achieving the required average dry density in buffer materials, while little attention has been paid to the changes in the distribution of dry density based on microscopic investigation of how bentonite swells homogeneously due to water permeation. As a result, there has been a lack of study on how the construction method used affects this distribution, which is highly pertinent to the evaluation of long-term performance of bentonite engineered barriers.

According to Kobayashi *et al.* (2007), the method used for the construction of buffer materials and their long-term performance are not independent. In the study, methods were investigated in terms of density distribution as shown in Figure 1. The average dry density of the buffer materials was 1.6 Mg m^{-3} in all cases. In a case involving block placement, the region including the gaps between blocks was considered as having a low dry density of 1.22 Mg m^{-3} and width of 10 mm.

From: NORRIS, S., BRUNO, J., CATHELINEAU, M., DELAGE, P., FAIRHURST, C., GAUCHER, E. C., HÖHN, E. H., KALINICHEV, A., LALIEUX, P. & SELLIN, P. (eds) 2014. *Clays in Natural and Engineered Barriers for Radioactive Waste Confinement*. Geological Society, London, Special Publications, **400**, 313–321.
First published online April 2, 2014, http://dx.doi.org/10.1144/SP400.21

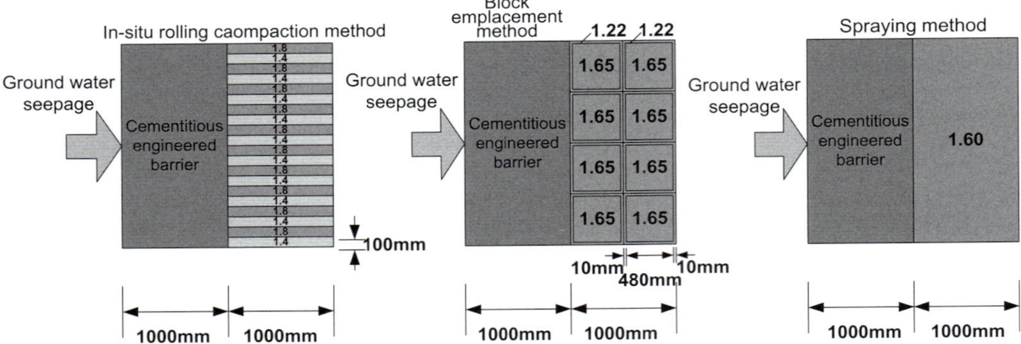

Fig. 1. Density distributions for different construction methods (Kobayashi *et al.* 2007).

Analytical investigation of how the construction method affects long-term performance of buffer materials (as shown in Fig. 2) revealed a significant influence in terms of dry density distribution for bentonite engineered barriers. The study of Kobayashi *et al.* (2007) involved analysis of trans-uranium (TRU) disposal facilities. This showed that high-pH cement leachate rich in Ca ions and with high

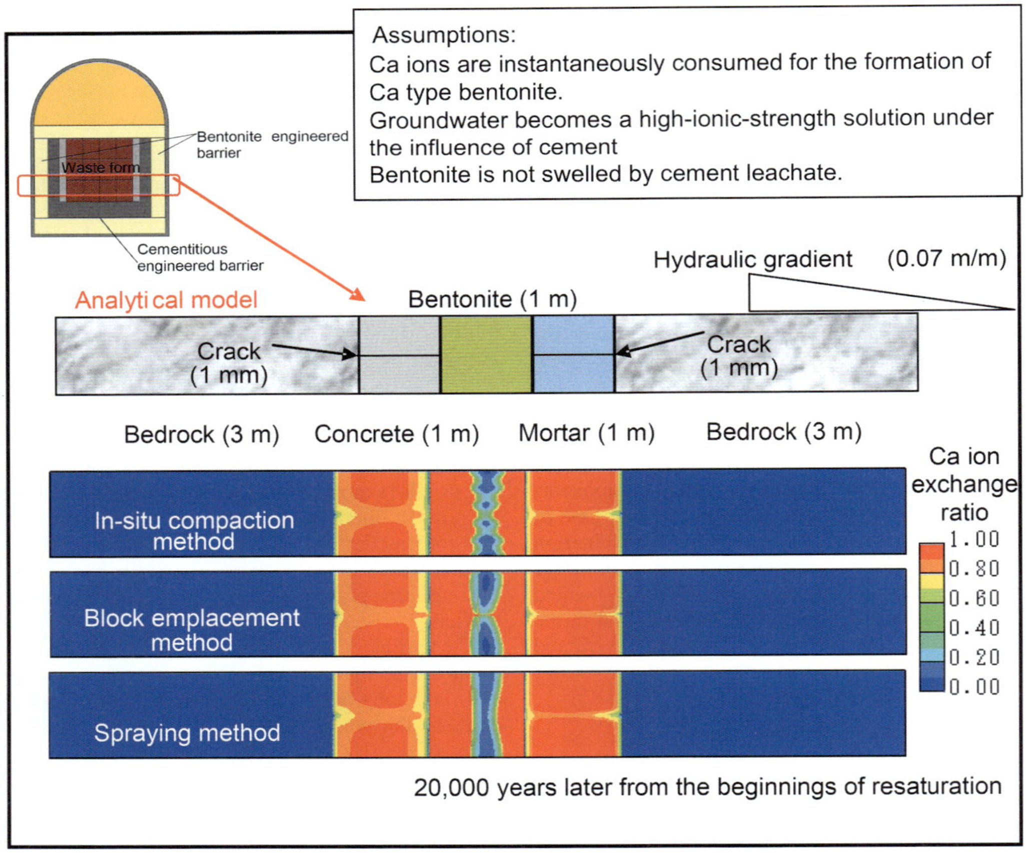

Fig. 2. Effects of the construction method on the long-term performance of bentonite engineered barriers (Kobayashi *et al.* 2007).

ion strength was found to permeate buffer materials. As a result, overly conservative analysis conditions were used involving the assumption of no bentonite swelling. If homogenization caused by buffer material swelling had been investigated in further detail, more realistic analytical conditions could have been used.

Earlier research on the development of density distribution in buffers during swelling indicated that the buffer did not reach a homogeneous state and density distribution variations remained (e.g. Lars-Erik 2007). In addition, it has proven difficult in previous studies to estimate theoretically the extent to which residual density distribution changes over time.

In other studies (e.g. RWMC 2009), the effects of construction methods on long-term performance were investigated in swelling deformation tests of specimens with a focus on dry density distribution. In the RWMC study, a single specimen containing high- and low-density regions was tested and the interface between these regions observed to enable calculation of residual density distribution. In this conventional method, the swelling deformation test cell often has a window for observation of the interface area. However, the position determined in such an observation does not always accurately represent swelling deformation because contact between the interface and the window causes friction.

This paper first describes the density distribution after resaturation and then highlights how swelling behaviour can be studied based on elasto-plastic theory. The conceptual understanding is that the buffer does not swell to an entirely homogeneous density, but will reach an equilibrium state with a remaining (residual) gradient.

Experimental investigation of buffer material homogenization

In this study, a series swelling deformation test apparatus was developed as shown in Figure 3. In the test buffer swelling behaviour was investigated with two ordinary apparatus set-ups connected in series via pistons. In this apparatus specimens with high dry density will swell and compress

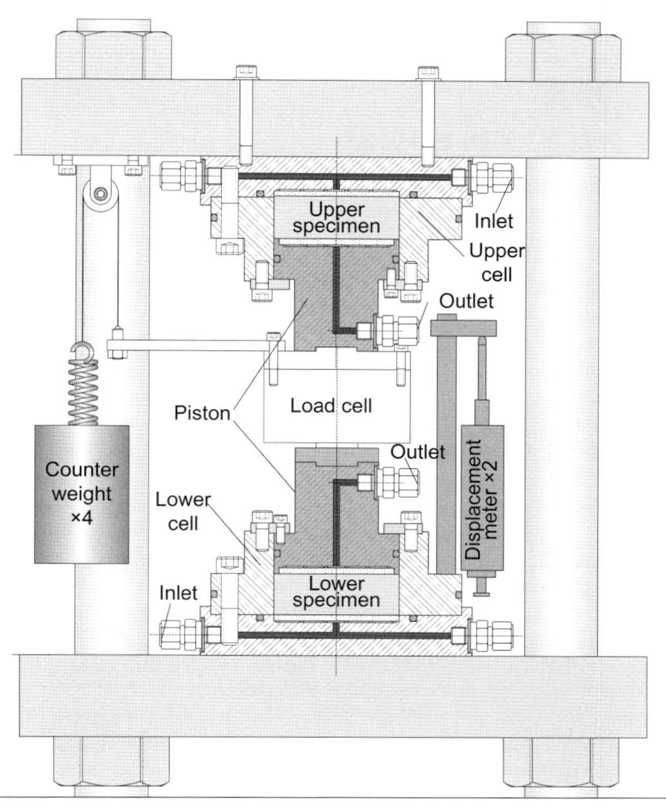

Fig. 3. Series swelling deformation test apparatus.

low-density specimens. Swelling deformation is measured from the position of the piston and swelling pressure is measured using a load cell located at the piston centre. It should be noted that the load cell can measure only the lower load when two different loads are applied to the top and bottom of the load cell, and that the load cell moves due to the difference between such loads.

A counterweight was used to offset the weight of the piston and load cell acting on the lower specimen. The buffer material used in this project was a mixture of Kunigel V1 (a Japanese sodium-type bentonite) and 30 wt% silica sand. The montmorillonite content of Kunigel V1 is approximately 60 wt%.

In order to adjust water content for uniformity, chilled dry bentonite and powdered ice were mixed in an amount equivalent to 10% of the weight of the dry bentonite. The resulting mixture was set in the cell after defrosting and compacted to achieve the planned dry density. The compacted bentonite was subjected to testing without being removed from the cell. After static compaction , the pedestal was locked to prevent unloading-related deformation. As mentioned above, efforts were made to maintain a stress state.

The experimental investigation was conducted to verify the applicability of elasto-plastic theory to the predicted residual density distribution in buffers. Series swelling deformation tests were conducted for three cases as shown in Table 1. Cases 1 and 2 show the evaluated homogenization process of specimens with different dry densities. In Case 1, a lower specimen with a dry density of 1.8 Mg m^{-3} and an upper specimen with a dry density of 1.4 Mg m^{-3} were connected in series. In Case 2, a lower specimen with a dry density of 2.0 Mg m^{-3} and an upper specimen with a dry density of 1.2 Mg m^{-3} were connected in series. The density difference in Case 1 was smaller than that in Case 2, but the average dry density for both was the same at 1.6 Mg m^{-3}. In both cases, just after the lock restricting the volume change was released, water was supplied from the upper part of the upper specimen and from the lower part of the lower specimen. Both specimens would have swollen in isolation due to the water supply, but the high-density specimen compressed the low-density specimen due to the higher pressure caused by its swelling.

Case 3 was investigated to evaluate the effects of stress history on homogenization. The dry densities of the specimen pair in this case were the same, but their stress histories were different. As shown in Figure 4, one specimen was made by causing swelling from a dry density of 1.8 Mg m^{-3} to a value of 1.6 Mg m^{-3} (called the overconsolidated (OC) specimen). Water was supplied from the bottom of this specimen, whose test apparatus had a lock to prevent the dry density from falling below 1.6 Mg m^{-3} in relation to swelling. The other specimen was created by applying saturation with the dry density maintained at 1.6 Mg m^{-3} (called the normally consolidated (NC) specimen). Consequently, the average dry density was almost the same in all test cases. When the dry density of the OC specimen reached 1.6 Mg m^{-3}, both specimens (OC and NC) were connected via a piston and the locks were released. Both specimens would have swollen in isolation due to the water supply. Thus, if the swelling pressure of bentonite is uniquely determined by its dry density, swelling deformation can be avoided.

Experimental results

The swelling deformation variations observed in Case 1 and Case 2 are shown in Figure 5, where positive values indicate low-density specimen compression by the high-density specimen. It can be seen that the high-density specimen compressed the low-density one in both cases. In other words, the dry density difference decreased and homogenization progressed. Homogenization appears to have been stable for more than 800 days in Case 2

Table 1. *Test cases*

Case	Objective	Dry density (Mg m^{-3})	Dry density (Mg m^{-3})	Bentonite partial dry density (Mg m^{-3})		Average dry density (Mg m^{-3})	Liquid supplied	Bentonite
		Lower	Upper	Lower	Upper			
1	Homogenization	1.798 (1.8)	1.399 (1.4)	1.581	1.164	1.599 (1.6)	Distilled water	Na bentonite with 30 wt% silica sand
2	Homogenization	1.984 (2.0)	1.190 (1.2)	1.792	0.963	1.587 (1.6)		
3	Inhomogenization due to stress history	1.794 → 1.596 (1.8 → 1.6)	1.589 (1.6)	1.576 → 1.368	1.357	1.593 (1.6)		

Numbers in brackets represent design values.

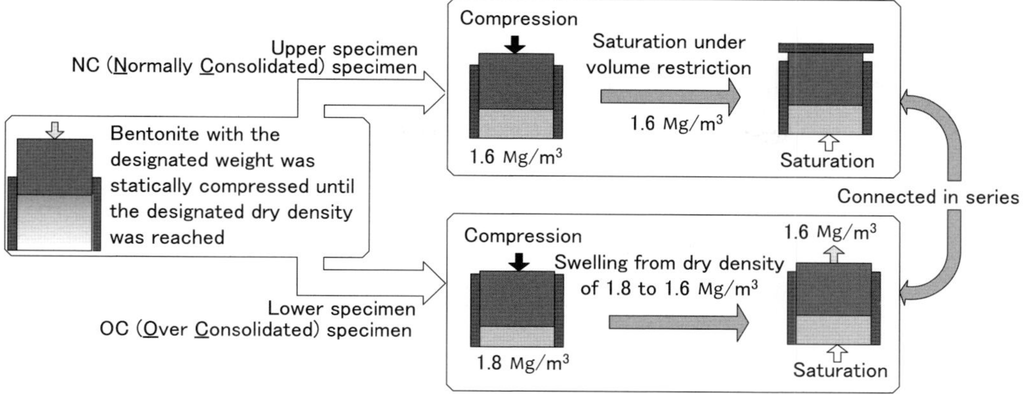

Fig. 4. Illustration of the specimens for Case 3.

and progressed gradually in Case 1. Temporal variations in dry density are plotted in Figure 6, which indicates that density differences decreased due to swelling in both cases. It appears that homogenization had almost finished and a density difference remained in Case 1, while a residual density difference remained in Case 2.

Temporal variations in swelling deformation of the OC specimen (Case 3) are plotted in Figure 7. The series swelling deformation test was conducted after the specimen had swollen from a dry density of approximately 1.8 Mg m^{-3} to 1.6 Mg m^{-3}. Compression was observed in the specimen just after the beginning of the test. Temporal variations in dry density (Case 3) are shown Figure 8, which indicates that although the value for the OC specimen was consistently higher than that for the NC specimen during the test, the latter compressed the former.

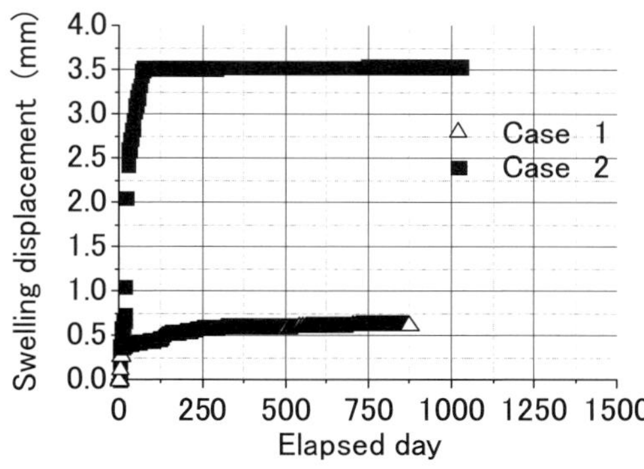

Fig. 5. Temporal variations in swelling deformation (Case 1 and Case 2).

Mechanical interpretation of the experimental results

The natural interpretation of these results is that the relationship between the dry density of bentonite and swelling pressure is not unique, as the latter is affected by stress history. This explains why a density difference arose in Case 3 despite the specimens in the series swelling deformation test having almost the same dry density.

In soil mechanics, clay-type materials are treated as elasto-plastic. Mechanically speaking, materials deform until a state of equilibrium (rather than a state of homogeneous density) is reached. A typical stress/strain relationship for elasto-plastic materials is shown in Figure 9, where the stress state of the high-dry-density specimens in Case 1 and Case 2 is plotted as H (white dot) and those of the low-dry-density specimen is plotted as L (black dot). At the beginning of the series swelling test, the stress state of H was greater than that of L. Accordingly, the stress state of H traced the unloading line with swelling and reached the elastic region during the test. The stress state of L traced the loading line with compression and remained in the elasto-plastic region. When both stress values equalize, or reach an equilibrium state, the necessary strain for homogenization does not always occur. The term *stress state* here refers to the total pressure, as the swelling pressure that can be measured in swelling pressure tests includes variations of effective stress, pore water pressure, void air pressure, pressure caused by osmotic swelling, and so on.

The behaviour observed in Case 3 can be similarly explained. The initial stress states of the NC and OC specimens are plotted as NC (white

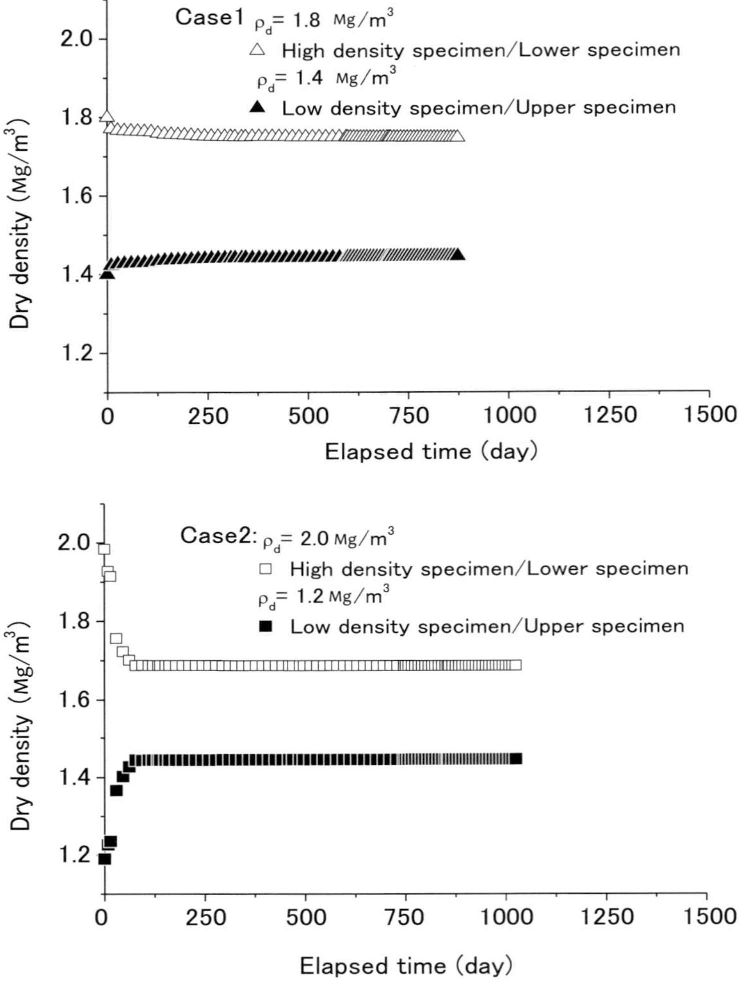

Fig. 6. Temporal variations in dry density (Case 1 and Case 2).

square) and OC (black square), respectively, in Figure 9. There is no strain difference at the beginning of the test, but the stress state of the specimens is not in equilibrium; that at NC is larger. This shows that the OC specimen was compressed and the NC specimen swelled in Case 3 until an equilibrium state was reached despite the specimens having almost the same dry density.

Mechanical judgment of residual density differences

As mentioned previously, differences in stress history affected the homogenization of buffer specimens. This residual density difference relating to stress history can be explained mechanically based on the relationship between dry density (ρ_d) and logarithmic pressure (log$P$) as shown in Figure 10. This relationship is equivalent to that between volumetric strain and stress. It should be noted that dry density is used in Figure 10, while the explanation was based on strain in Figure 9. The ρ_d–log$P$ relationship for Kunigel V1 as shown in Figure 10 is given in Sasakura *et al.* (2004). An explanation in mechanical terms of residual density difference is described below.

First, the dry densities of the high- and low-density specimens (corresponding to the swelling pressure P as determined in the series swelling

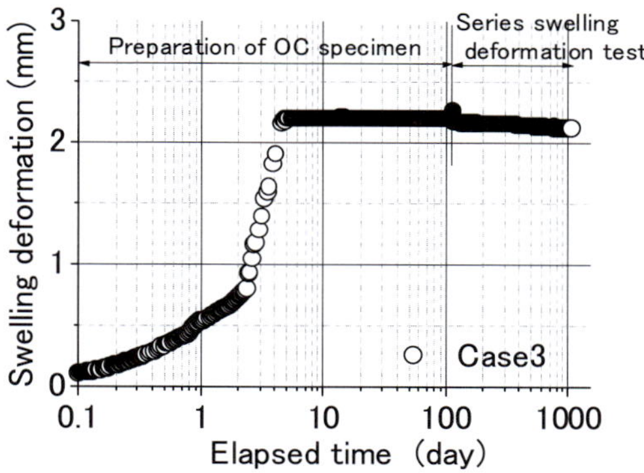

Fig. 7. Temporal variations in swelling deformation of the OC specimen (Case 3).

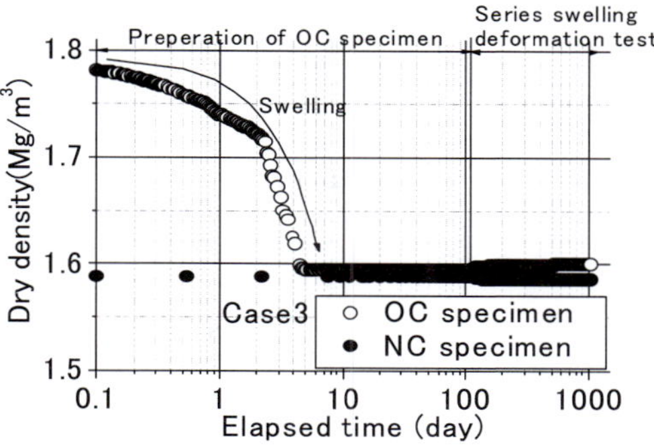

Fig. 8. Temporal variations in dry density (Case 3).

deformation tests) were ascertained from the ρ_d–log$P$ relationship for Kunigel V1 as points on the unloading and loading lines respectively. As a load cell set between two different loads can measure only the lower load, that measured is due to the swelling pressure of the low-density specimen. Second, the dry densities calculated from the swelling deformation measurement data were plotted on the ρ_d–log$P$ relationship line. Third, the dry density values corresponding to pressure measurements were compared with dry density measurement data. Essentially, both sets of values should correspond; if they do not, an equilibrium state is not reached in the series swelling deformation test. Finally, when the stress state reaches equilibrium, the difference between points corresponding to the swelling pressure on the unloading and loading lines must represent the residual density difference (i.e. the density difference in a state of equilibrium).

The test results for all cases are plotted in Figure 10 based on the procedure outlined above. It can be seen that the measured pressure values were consistent with those of dry density for the low-density specimen in all cases. However, in Case 1 and Case 3, the dry density corresponding to the swelling pressure differed significantly from the measured dry density. Accordingly, it can be inferred that the stress state in these two cases had not yet reached a state of equilibrium. As gradual swelling still seemed to be in progress as shown in Figures 5 and 7, this judgment is consistent with the swelling deformation test results for Case 1 and Case 3. Meanwhile, the dry density corresponding to the swelling pressure in Case 2 was almost the same as the measured dry density. Accordingly, it can be judged that the stress state had almost reached equilibrium. As swelling behaviour seems to have settled for more than 800 days as shown in Figure 5, this judgment is consistent with the swelling deformation test results for Case 2. It can thus be inferred that the residual dry density difference would have remained if the series swelling test had been continued.

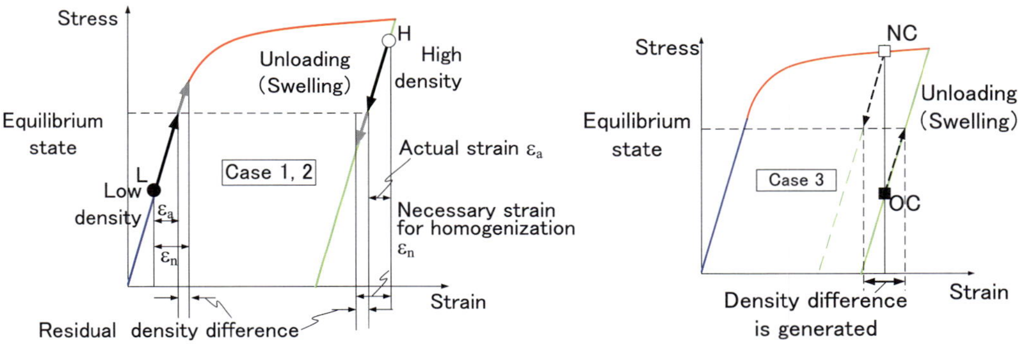

Fig. 9. Stress-strain relationships of elasto-plastic materials.

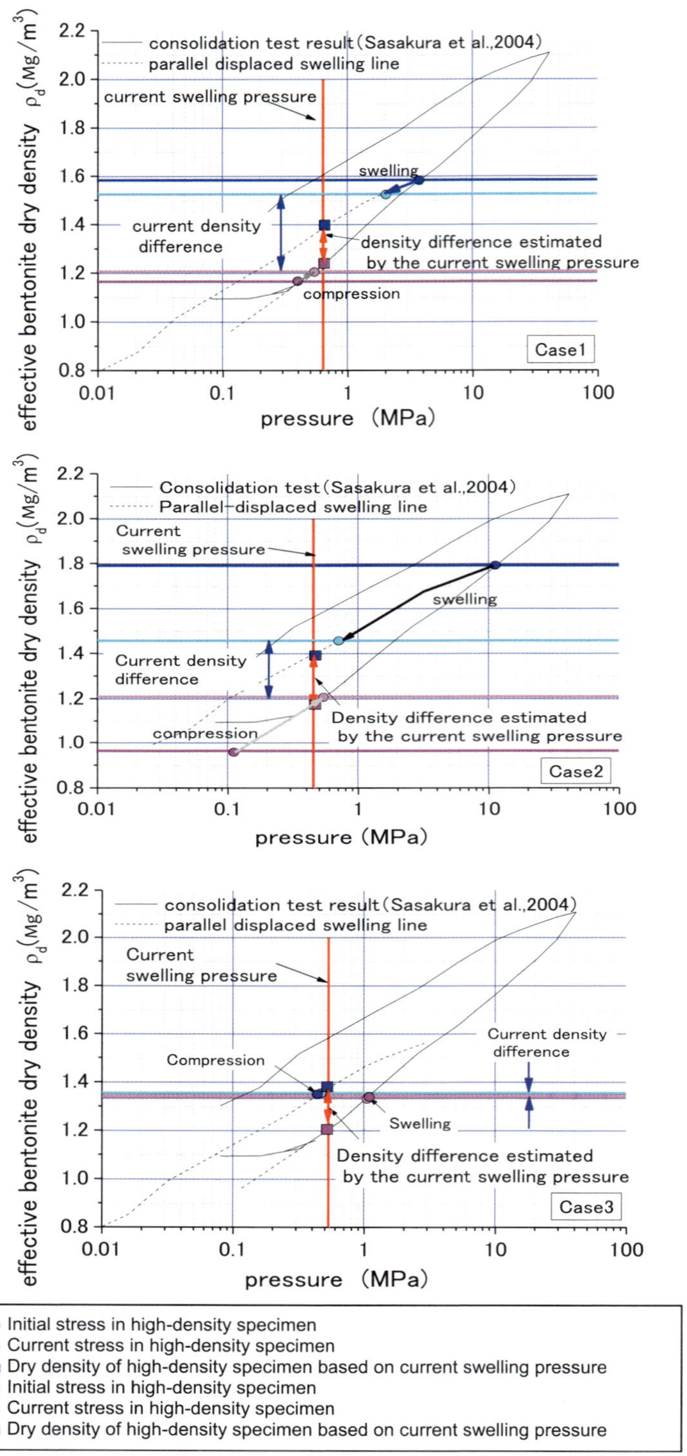

Fig. 10. Projection of residual dry density differences.

From a mechanical viewpoint, it is therefore a rational interpretation that bentonite with a density distribution does not become homogenized. This implies that the construction method should be carefully evaluated for long-term buffer performance. In future work, this research will be expanded to determine the permissible range of density distribution resulting from the approach to construction.

Conclusions

Experimental investigation showed that there is a density distribution present in installed buffer components due to stress history. This means that the residual density difference should be considered in the evaluation of long-term performance based on the given initial conditions. It is therefore necessary to define the requirements for the manufacture and installation of buffer components in consideration of long-term performance because the residual dry density is dependent upon the mechanical characteristics of the buffer components immediately after installation.

In addition, the shotclay method (Kobayashi *et al.* 2010) and other construction approaches that minimize the density distribution in bentonite engineered barriers are expected to be effective in avoiding uncertainty relating to residual density difference with respect to the long-term performance of such barriers.

The research and development work reported here was partially financed by the FY 2011 Development of Advanced Engineering for Disposal Systems initiative of the Agency of Natural Resources and Energy under Japan's Ministry of Economy, Trade and Industry.

References

JAPAN ATOMIC ENERGY AGENCY (JAEA, former JNC) 2000. *H12 report: Supporting Report 2*, Repository Design and Engineering Technology, JNC TN1410 2000-003. http://jolissrch-inter.tokai-sc.jaea.go.jp/pdfdata/JNC-TN14 10-2000-003.pdf

KOBAYASHI, I., TOIDA, M., NAKAJIMA, M., TANAKA, T., TERADA, K., NONAKA, K. & YABE, J. 2007. Development of construction methods for bentonite engineering barriers using the high-density shotclay system. *In*: PUSCH, R. (ed.) *Proceedings of the Workshop on the Long-term Performance of Smectitic Clays Embedding Canisters with Highly Radioactive Waste*, Lund, Sweden, 52–59.

KOBAYASHI, I., FUJISAWA, S., NAKAJIMA, M., TOIDA, M., NAKASHIMA, H. & ASANO, H. 2010. Design options for HLW repository operation technology, (IV) Shotclay technique for seamless construction of EBS. *In*: *Proceedings of the 13th International Conference on Environmental Remediation and Radioactive Waste Management, ICEM*, Tsukuba, Japan.

LARS-ERIK, J. 2007. *Äspö Hard Rock Laboratory. Canister Retrieval Test. Dismantling and Sampling of the Buffer and Determination of Density and Water Ratio.* SKB IPR-07-16, 21–26, http://www.skb.se/upload/publications/pdf/ipr-07-16.pdf

RADIOACTIVE WASTE MANAGEMENT FUNDING AND RESEARCH CENTER (RWMC) 2009 *Research and Development on Disposal Technology FY 2009. Study on Remote Operation Technology for HLW Repository.* (Research conducted under a grant from Japan's Ministry of Economy, Trade and Industry).

SASAKURA, T., KOBAYASHI, I., SAHARA, F., MURAKAMI, T., OHI, T., MIHARA, M. & ITOH, H. 2004. Studies on the mechanical behavior of bentonite for development of an elasto-plastic constitutive model, DisTec 2004. *In*: International Conference on Radioactive Waste Disposal, 498–507.

Characterization of excavated claystone and claystone–bentonite mixtures as backfill/seal material

CHUN-LIANG ZHANG

Gesellschaft für Anlagen- und Reaktorsicherheit (GRS) mbH, PO Box, 38122 Braunschweig, Germany (e-mail: chun-liang.zhang@grs.de)

Abstract: Crushed claystone produced by excavation of repository openings has been investigated as backfill/seal material. The raw coarse-grained claystone can be used for backfilling repository openings and, in mixture with bentonite, for sealing boreholes, drifts and shafts. The investigation programme focused on characterizing the thermo-hydro-mechanical properties of the excavated Callovo-Oxfordian claystone and the compacted claystone–bentonite mixtures, including (1) mechanical compaction, (2) gas and water permeability as a function of porosity, (3) water retention and saturation, (4) swelling capacity and (5) thermal properties of the materials. The major results are presented in this paper.

The disposal concepts developed by many countries for repositories in clay formations are based on the principle of a multi-barrier system, which comprises the natural geological barriers provided by the host rock and its surroundings, and an engineered barrier system (EBS). After emplacement of radioactive waste containers, the disposal boreholes, drifts and shafts must be backfilled and sealed with suitable materials to prevent a release of radionuclides into the biosphere. The geotechnical properties of each EBS component must be specified with regard to their functional requirements. In most concepts, bentonite-based materials are considered for use in filling and sealing disposal boreholes. For instance, the Swiss concept (Nagra 2002) provides for waste containers being emplaced on compacted bentonite blocks positioned on the drift floor and the remaining space being backfilled with granular bentonite. Meanwhile, according to the French concept (Andra 2005), the drifts are backfilled with excavated claystone and sealed with compacted bentonite-based seal core confined between concrete plugs.

Crushed claystone produced by the excavation of repository openings may be a favourable alternative as backfill/seal material; its use has many advantages, including chemical-mineralogical compatibility with the host rock, availability in sufficient amounts, low cost of material preparation and transport, and there is no need for it to occupy surface space if it is used as backfill for the excavated claystone. Indeed, crushed claystone with coarse grains can be considered for use for backfilling repository openings and, in mixture with bentonite, for sealing boreholes, drifts and shafts. Recently, a research programme has been conducted at the GRS geotechnical laboratory to characterize the geotechnical properties of the excavated Callovo-Oxfordian claystone and the compacted claystone–bentonite mixtures under relevant repository conditions. The following properties of the backfill/seal materials are most important in terms of their barrier functions and have thus been investigated:

(1) The compaction behaviour of the backfill material in underground openings. This controls mechanical interactions with support linings (if existing) and with the surrounding rock and, in turn, determines the evolution of backfill porosity and permeability.
(2) The permeability of the backfill/seal materials in relation to porosity, which dominates fluid transport and radionuclide migration in the backfill and seal.
(3) Water retention and saturation of the backfill/seal materials, which determines the water saturation process and storage capacity and influences the mechanical behaviour and thermal conductivity of the backfill and seal.
(4) The swelling capacity of the seal material in boreholes, drifts and shafts, which is required to seal any remaining gaps within the seals and at the seal/rock interfaces and to support the surrounding rock against damage propagation.
(5) The thermal conductivity of the buffer material surrounding heat-emitting high-level radioactive waste containers, which controls heat transfer and the temperature distribution in the multi-barrier system.

The main results of the investigations are summarized in this paper.

Testing materials

The crushed claystone was produced by drift excavation in the Callovo-Oxfordian argillaceous

From: Norris, S., Bruno, J., Cathelineau, M., Delage, P., Fairhurst, C., Gaucher, E. C., Höhn, E. H., Kalinichev, A., Lalieux, P. & Sellin, P. (eds) 2014. *Clays in Natural and Engineered Barriers for Radioactive Waste Confinement*. Geological Society, London, Special Publications, **400**, 323–337.
First published online April 30, 2014, http://dx.doi.org/10.1144/SP400.28

formation (COx) at the main level of −490 m in the Underground Research Laboratory at Bure in France. On average, the COx claystone contains 25–55% clay minerals, 20–38% carbonates, 20–30% quartz, and small amounts of others (Andra 2005). The claystone has been overconsolidated through specific geological histories over hundreds of millions of years to low porosities of 14–18% and low permeabilities of 10^{-20} to 10^{-21} m^2. The initially saturated material with a water content of *c.* 7.2% was somewhat desaturated to residual values of 4–5% due to exposition to air after excavation.

It is convenient to use the excavated material immediately to backfill the repository openings without any further treatment. Therefore, raw crushed claystone with grain sizes up to a diameter of $d = 32$ mm was studied experimentally. Considering that coarse grains are further crushed and segregated during transport and backfilling, relatively small grains of $d < 20$ mm and $d < 10$ mm were also prepared for testing by sieving the raw material. The grain size distribution curves used for testing are depicted in Figure 1. Additionally, fine-grained claystone powder with grain sizes of $d < 0.5$ mm was mixed with the expansive MX80 bentonite of $d < 0.5$ mm in different ratios. The mixtures were considered for use in the construction of seals in disposal boreholes, drifts and shafts. The COx powder has grain sizes similar to those studied by others (Tang *et al.* 2011). The delivered MX80 bentonite was relatively dry with a water content of $w = 9.6\%$.

Compaction and permeability of crushed claystone backfill

Sample preparation and testing method

Coarse claystone aggregates with grain sizes of $d < 32$ mm and $d < 20$ mm were tested. The partly saturated material was first wetted by spraying with synthetic COx clay water and then mixed in a mixing drum to a homogeneous aggregate. The water content of the samples varied between $w = 6.6\%$ and 7.8%, comparable to the original water content of the claystone. The aggregate was used to fill a rubber jacket of 280 mm diameter in seven layers of 100 mm each. Each layer was compacted by different approaches, including stamping by hand, vibration by a vibrator, or compression by load piston. The following initial dry densities were obtained: $\rho_d = 1.45$ g cm^{-3} by vibration, $\rho_d = 1.60$ g cm^{-3} by stamping and

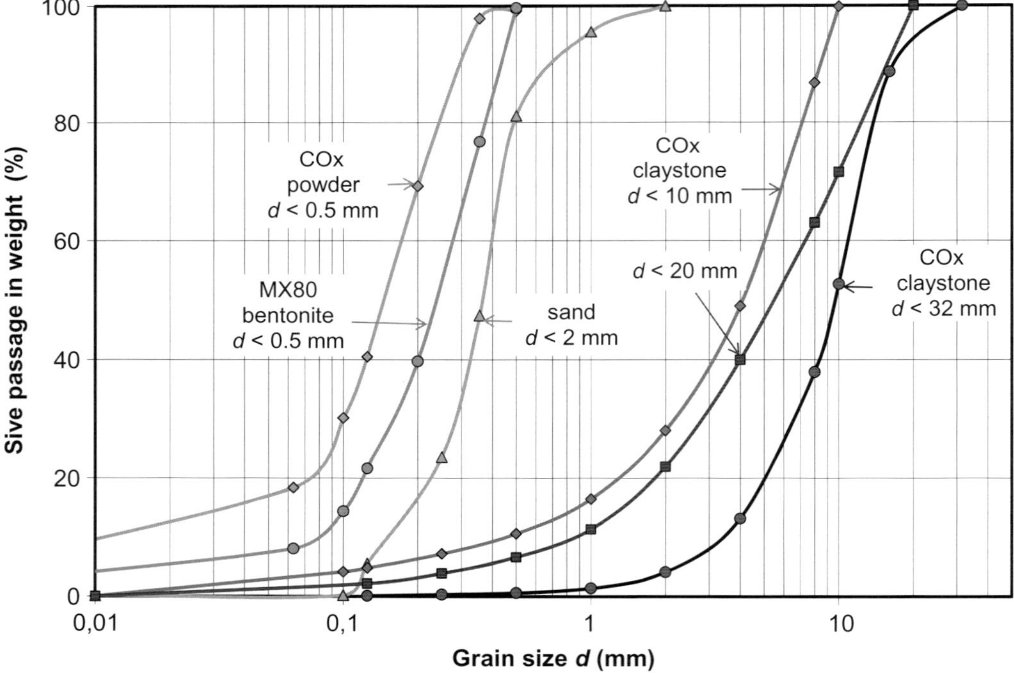

Fig. 1. Grain size distribution curves of studied crushed claystone, bentonite and sand.

Table 1. *Characteristics of the initial state of prepared backfill samples*

Parameter	Sample 1	Sample 2	Sample 3	Sample 4
Grain size d (mm)	<20	<32	<32	<32
Precompaction by	Stamping	Compression	Compression	Vibration
Diameter D (mm)	281.4	262.5	282.2	273.4
Height H (mm)	637.6	661.5	644,4	680.6
Mass G (kg)	68.24	69.44	78.54	62.01
Dry density ρ_d (g cm^{-3})	1.595	1.819	1.826	1.456
Water content w (%)	7.82	6.62	6.62	6.62
Porosity ϕ (%)	40.9	32.6	32.4	46.1

$\rho_d = 1.82$ g cm^{-3} by mechanical compression at 1 MPa. The characteristics of the prepared samples are presented in Table 1.

The prepared samples were installed in a big triaxial cell, as illustrated in Figure 2. The test system allows the testing of large samples of 280 mm diameter and 700 mm length under coupled thermo-hydro-mechanical (THM) boundary conditions. The upper limitations are the maximum lateral stress $\sigma_{r\text{-max}} = 50$ MPa and axial stress $\sigma_{a\text{-max}} = 75$ MPa, the maximum temperature $T_{max} = 150$ °C and the maximum pore fluid pressure $p_{max} = 15$ MPa. External stress is applied to the sample by regulating the outer axial load and lateral pressure in the cell. Axial strain is measured by a deformation transducer mounted on the lower piston outside the cell, while radial strain is recorded by two circumferential extensometers at the middle and lower positions of quarter sample length, respectively. From the axial and radial deformation, the resulting volume change is calculated. The permeability is measured in the axial direction by flowing nitrogen gas through at a constant injection pressure.

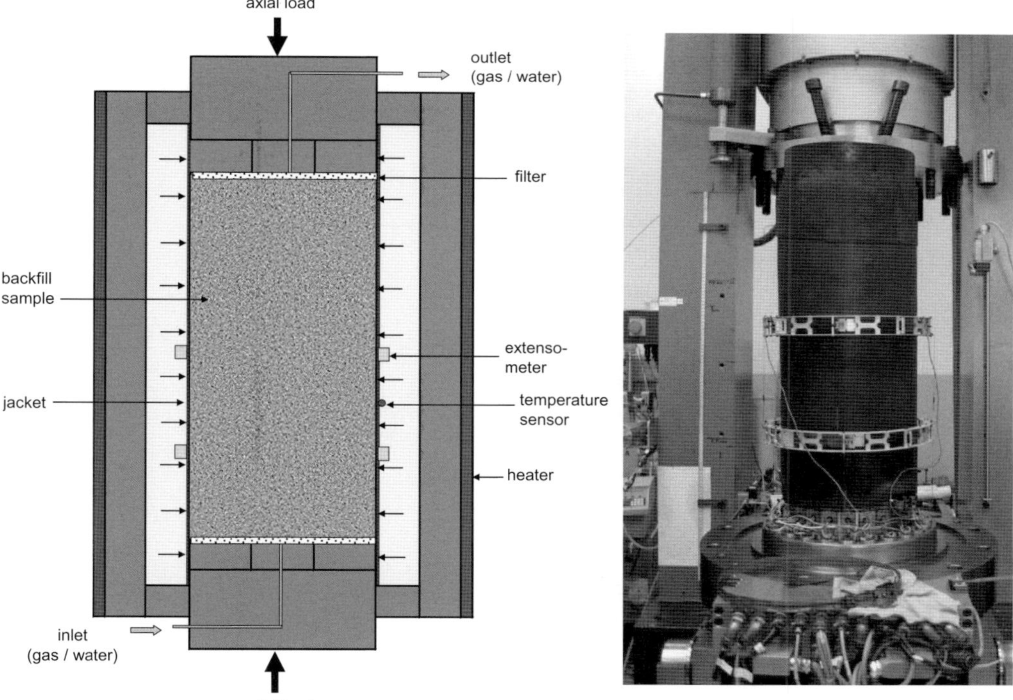

Fig. 2. Schematic of coupled compaction–permeability tests on backfill material in a big triaxial testing apparatus.

Results

Figure 3 presents a test example with sample 1. It was compacted by increasing the lateral stress step by step from 0.5 MPa to 1, 4 and 6 MPa at fixed axial strain ($\Delta\varepsilon_a = 0$). Each step was carried out over a period of 7–14 days. The applied load path is similar to the compression conditions applied to backfill in long boreholes and drifts. At $\Delta\varepsilon_a = 0$, the increase in radial stress up 6 MPa caused a radial compression of $\varepsilon_r \approx 6\%$ and a slight increase in axial stress of $\sigma_a = 0.1–0.4$ MPa. At each stress level, the radial strain increased steadily with time at a decreased rate. The corresponding reduction in porosity was dominated by irreversible viscoplastic deformation. Compaction led to a decrease in gas permeability from $k_g = 1 \times 10^{-12}$ m^2 at an initial porosity of $\phi = 37\%$ to $k_g = 5 \times 10^{-19}$ m^2 at the final porosity of $\phi = 24.5\%$. A sharp permeability drop occurred in the low porosity range below $\phi = 26\%$.

Figure 4 summarizes the mean stress v. porosity curves obtained on three crushed claystone samples. As mentioned above, sample 1 ($d < 20$ mm) was compressed by increasing σ_r at $\Delta\varepsilon_a = 0$ while the other two samples with coarse grains ($d < 32$ mm) were compressed quasi-hydrostatically. It is obvious that the three curves are close to each other, indicating little influence of the grain sizes and load paths. The mean stress v. porosity relation

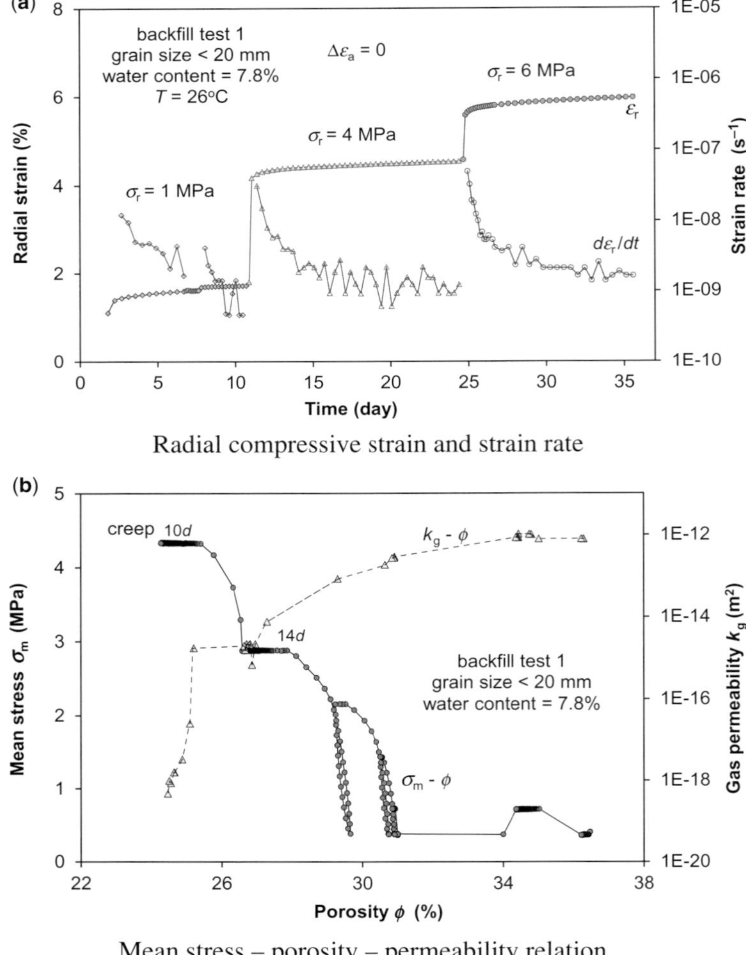

Fig. 3. Results of a compaction and permeability test on crushed claystone by increasing the lateral confining stress at fixed axial strain.

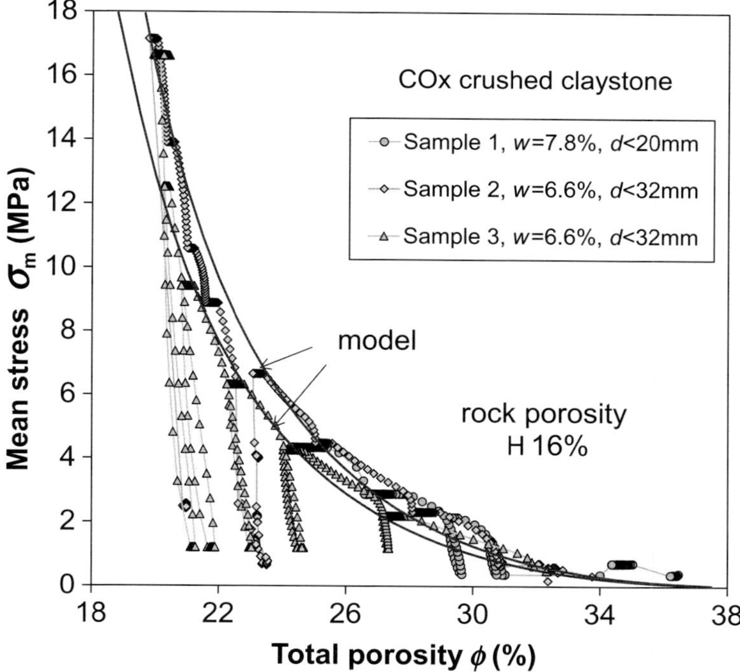

Fig. 4. Mean stress–porosity relation of crushed claystone with different grain sizes.

is non-linear and may be expressed by an exponential function. The compressibility of the material is relatively high; that is, its resistance against external load is relatively low. At mean stresses of 12–16 MPa, corresponding to the lithostatic stress at depths of 500–600 m, the backfill can be compacted to a low porosity of *c.* 20% which is close to the rock porosities of 16–18%. The stiffness of the porous material increases with decreasing porosity. The elastic bulk modulus increases exponentially with decreasing porosity from $K = 710$ MPa at $\phi = 27\%$ to $K = 1820$ MPa at $\phi = 19\%$.

The compacted backfill can sustain certain deviatoric loads. Figure 5 shows the stress v. strain curves obtained on a compacted sample of 21% porosity under multiple lateral stresses of 2–7 MPa. Each stress v. strain curve shows typical elastoplastic behavior, but with evolution of dilatation ($\varepsilon_v < 0$) during deviatoric stressing. From the linear part of each curve we can determine the elastic parameters, which are depicted in Figure 6a as a function of the lateral stress. Young's modulus varies from $E = 430$ to 1030 MPa, while Poisson's ratio lies in a range between $\nu = 0.50$ and $\nu = 0.64$. The values of $\nu > 0.5$ indicate dilatation of the compacted backfill under the deviatoric stresses. The peak strength values determined for each lateral stress are depicted in Figure 6b. The strength of the compacted backfill can be described by the Mohr–Coulomb criterion with an inherent cohesion of 3.7 MPa and an internal friction angle of 12°.

The gas permeability values obtained on the crushed claystone backfill with grain sizes of $d < 20$ mm and $d < 32$ mm are illustrated in Figure 7 as a function of porosity and also compared with the water permeability of the backfill with small grains of $d < 10$ mm. It can be seen that (1) the water permeability is much lower than the gas permeability and (2) the permeability of the coarse-grained backfill is higher than that of the fine-grained material at a given porosity. The low water permeability is mainly attributed to the effects of water-induced swelling and slaking of the clay grains into the pores. The permeability decreases much faster at low porosities below *c.* 25% for gas flow and below *c.* 30% for water flow. Very low permeability values of less than 10^{-19} to10^{-20} m^2 were measured with water flow at a porosity of *c.* 27% and with gas flow at lower porosities of 20–22% depending on the grain size. These values are close to that of the intact clay rock.

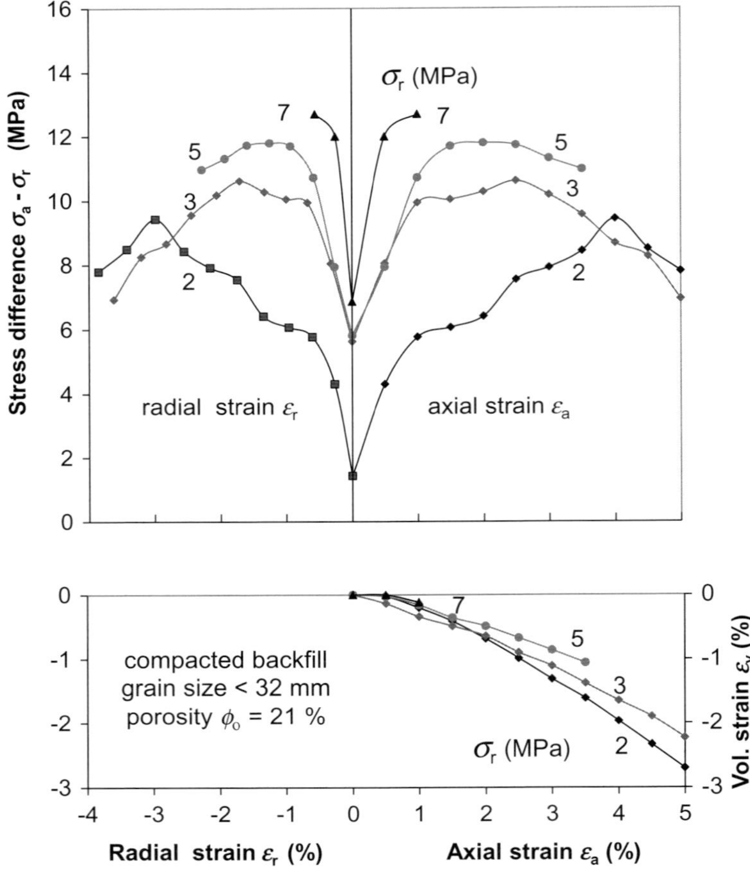

Fig. 5. Deviatoric stress–strain behaviour of compacted claystone backfill.

Sealing properties of compacted claystone–bentonite mixtures

Compacted bentonite and bentonite–sand mixtures have been widely investigated for the sealing of underground repositories in crystalline and clay formations. Crushed claystone is also considered as a main component of seals in repositories. In order to meet the specific requirements dictated by the geotechnical properties of seals to be constructed in different locations (for instance, sealing boreholes, drifts and shafts), crushed claystone is mixed with bentonite and compacted to particular densities. Commonly, the seals must have sufficient supporting capacity against damage propagation from the surrounding rock, a low hydraulic conductivity against migration of radionuclides with fluids, and a certain swelling capacity for sealing gaps and interfaces between compacted blocks and the surrounding rock. The sealing properties of compacted claystone–bentonite mixtures with various ratios were determined by performing various kinds of experiment, some of which were designed to investigate particular issues.

Compaction

Fine-grained COx claystone powder with grains of $d < 0.5$ mm and coarse-grained claystone of $d < 10.0$ mm were mixed with MX80 bentonite powder of $d < 0.5$ mm in different ratios. Both fine-grained and coarse-grained aggregates had a water content of $w = 4.3\%$, while the delivered MX80 bentonite had a water content of $w = 9.6\%$. The claystone–bentonite mixtures were compacted in oedometer cells at an axial strain rate of 1×10^{-6} s^{-1} to a maximum load of 30 MPa. The compaction curves are summarized in Figure 8, while the key properties of the compacted samples are given in Table 2. The mixtures exhibit different initial

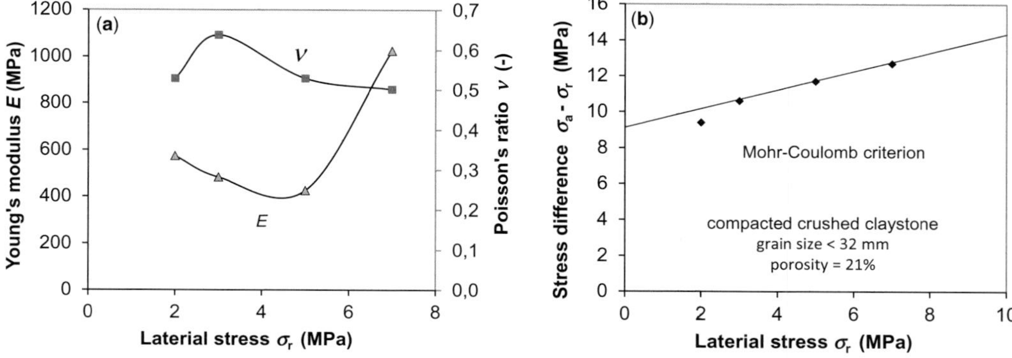

Fig. 6. Elastic parameters and peak strength of compacted claystone backfill.

porosities due to the different claystone/bentonite ratios, grain size distributions and water contents. This leads to different densities being achieved by application of the same energy, from one mixture to another. At the final load of 30 MPa, the achieved dry density increases (and accordingly the porosity decreases) with increasing fraction of claystone from $\rho_d = 1.56$ g cm^{-3} ($\phi = 44\%$) for pure bentonite to $\rho_d = 2.02$ g cm^{-3} ($\phi = 25\%$) for pure claystone powder. The macropores in the coarse claystone aggregate ($d < 10$ mm) can be more densely filled with bentonite powder to a low porosity of $\phi = 23\%$ at a mixture ratio of 50/50. The compacted mixtures exhibit a high stiffness (the slope of each σ–ϕ curve) and hence a high supporting capacity.

Water retention

Resaturation of an unsaturated sealing material is mainly controlled by the relationship between water content and suction (usually called the water retention curve). The water retention curves of four compacted claystone–bentonite mixtures were determined for unconfined samples using the vapour equilibrium technique. They were mixed with COx claystone and MX80 bentonite to the following ratios: COx, 80COx + 20MX80, 60COx + 40MX80, and MX80. The mixtures were compacted to 50 mm diameter and 33 mm length at a load of 30 MPa. The samples were unconfined and placed in desiccators at different relative humidity values of RH = 15–100% and at an ambient temperature

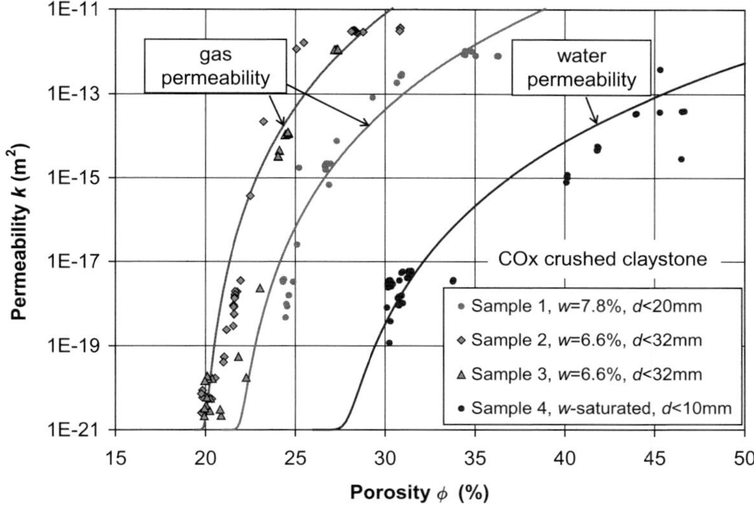

Fig. 7. Water and gas permeability of crushed claystone backfill as a function of porosity.

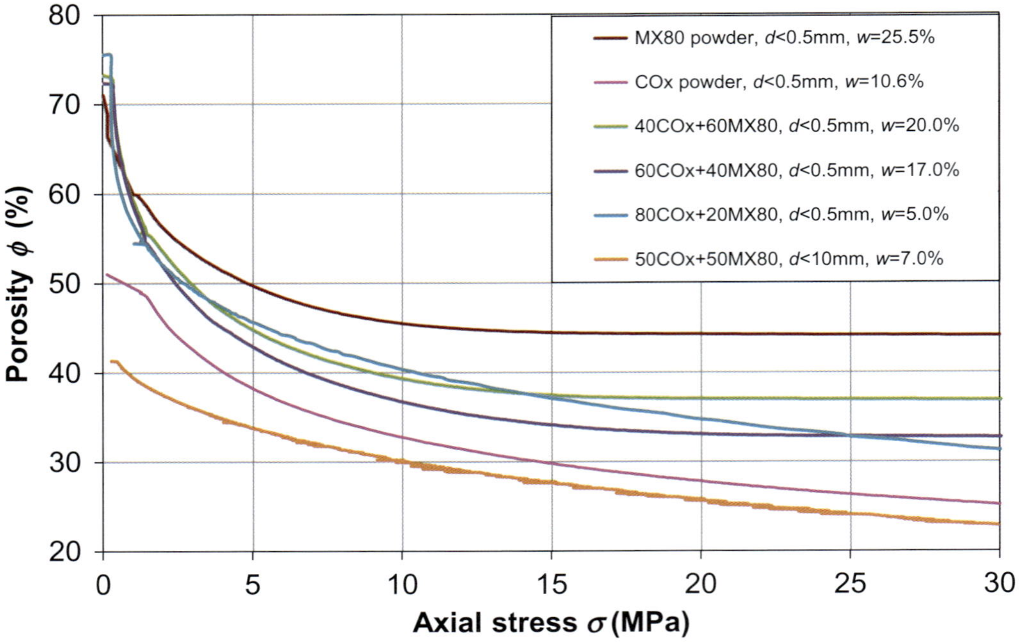

Fig. 8. Compaction behaviour of crushed claystone–bentonite mixtures.

of 25 °C. The corresponding suctions lie between 0.0 and 258 MPa. The water content of the samples evolved with time towards an equilibrium with the preset humidity. The tests continued for more than six months until equilibrium was reached.

The measured equilibrium water contents for all samples at all humidity conditions are depicted in Figure 9 as a function of suction. One can recognize that (1) the water content of each mixture increases with decreasing suction or increasing humidity, and (2) moisture uptake at a given suction is proportional to the bentonite content of the mixture. Note that in the wet environment at zero suction or RH = 100%, all the mixtures can take up large amounts of water. Water contents were measured up to $w = 12\%$ for the crushed claystone and $w = 48\%$ for the bentonite.

Swelling capacity

Swelling strain. In correspondence with the change in water content, the compacted claystone–bentonite mixtures expanded or contracted during the abovementioned tests for water retention. The

Table 2. *Characteristics of compacted claystone–bentonite mixtures*

Parameter	COx powder	MX80 powder	40COx + 60MX80	60COx + 40MX80	80COx + 20MX80	50COx + 50MX80
Diameter D (mm)	46.0	46.0	46.0	46.0	46.0	100.0
Height H (mm)	48.2	49.5	52.2	50.6	20.6	91.0
Grain size d (mm)	<0.5	<0.5	<0.5	<0.5	<0.5	<10.0
Water content w (%)	10.6	25.5	20.0	17.0	5.0	7.0
Bulk density ρ_b (g cm^{-3})	2.25	1.95	2.08	2.14	1.97	1.91
Dry density ρ_d (g cm^{-3})	2.02	1.56	1.67	1.79	1.86	1.79
Grain density ρ_s (g cm^{-3})	2.70	2.78	2.75	2.73	2.72	2.74
Total porosity ϕ (%)	25.0	44.0	39.0	35.0	31.0	22.7
Water saturation S_w (%)	85.6	90.4	85.6	87.0	30.0	55.2
Swelling pressure (MPa)	3.0	>7.6	6.0	6.0	6.5	9.5
Water permeability (m^2)	–	$<1 \times 10^{-20}$	2×10^{-20}	3×10^{-20}	2×10^{-19}	–

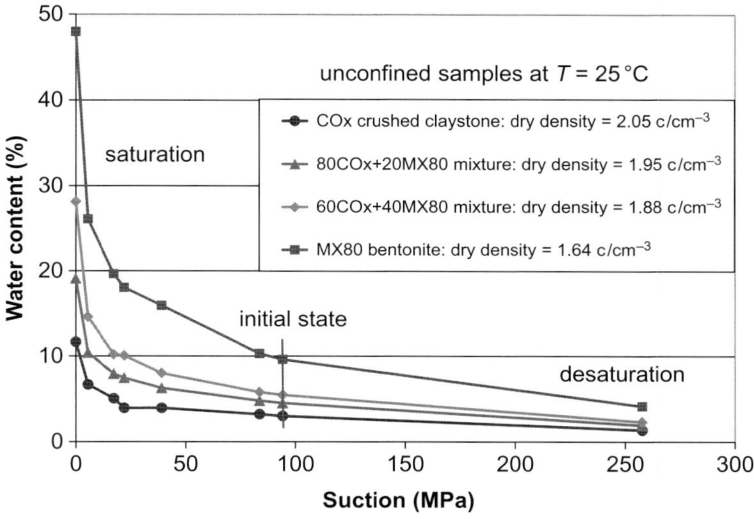

Fig. 9. Water retention curves of compacted claystone–bentonite mixtures.

volumetric strains measured at equilibrium are illustrated in Figure 10 as a function of water content. The volume of each compacted mixture increases almost linearly with increasing water content. Under a high humidity of RH = 100%, the compacted COx claystone expanded to $\varepsilon_v = 12\%$, while the others under a slightly lower humidity of RH = 96% expanded even further to $\varepsilon_v = 21\%$ (80COx + 20MX80), $\varepsilon_v = 27\%$ (60COx + 40MX80) and $\varepsilon_v = 38\%$ (MX80) due to the presence of the highly expansive bentonite. All the compacted claystone–bentonite mixtures exhibit large free swelling capacities.

In addition to the free swelling, significant swelling strains were also observed on the compacted mixtures under certain loads. Figure 11 demonstrates the swelling strain curves of two compacted claystone–bentonite mixtures in ratios of 50/50 and 80/20, in correspondence to the water uptake during saturation at axial load of 1 MPa in

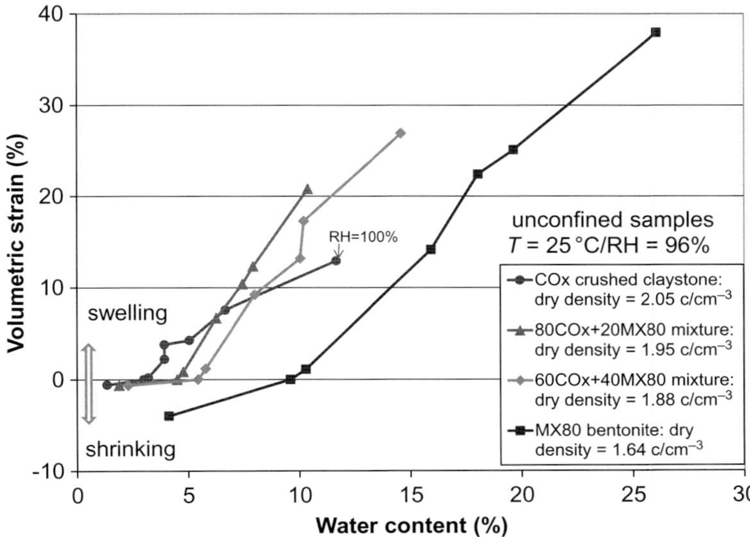

Fig. 10. Volumetric expansion of compacted claystone–bentonite mixtures with increasing water content.

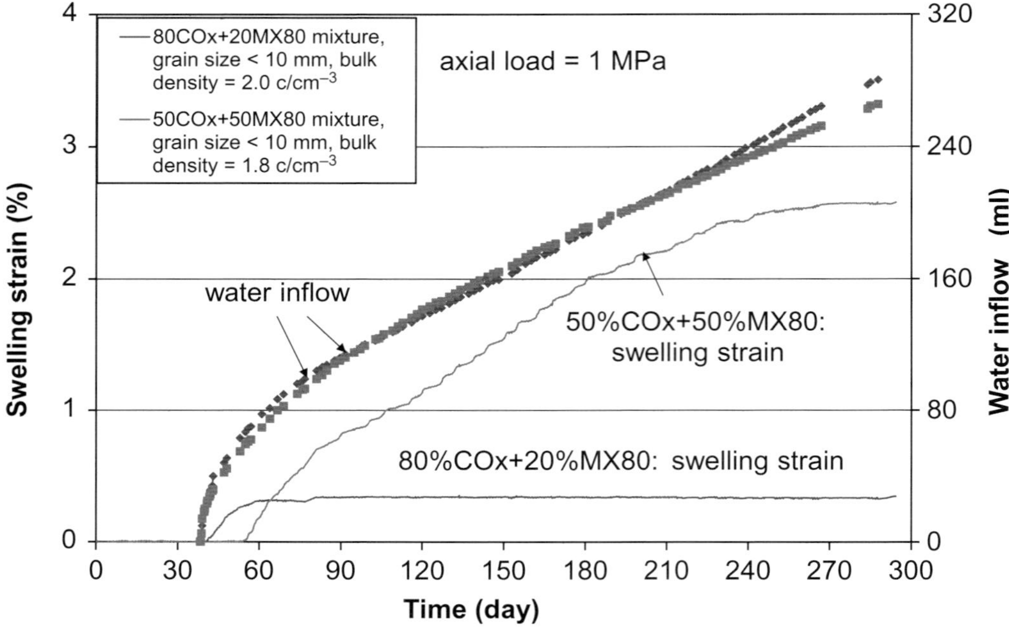

Fig. 11. Swelling strains of compacted claystone–bentonite mixtures in ratios of 50/50 and 80/20 during water saturation at an axial load of 1 MPa in oedometer cells.

oedometer cells. Whereas the 80COx + 20MX80 mixture expanded immediately on contacting the water and approached a maximum value of 0.35%, the swelling of the 50COx + 50MX80, with its higher bentonite component, was delayed by several days but reached a large value of 2.6% after seven months.

Swelling pressure. The swelling pressure of the compacted mixtures was measured on the samples in our own designed cells, as schematically shown in Figure 12. The samples had a length of 50 mm and a diameter of 46 mm, which was smaller than the inner cell diameter of 50 mm. The annulus between sample and cell was filled with fine-grained quartz sand of $d < 0.5$ mm. After installation, the sample was axially preloaded to *c.* 5 MPa and then fixed. To simulate realistic conditions with the unsaturated seal being first wet by humid air and then by liquid water from the surrounding rock, each sample was first ventilated by circulating air with controlled humidity through the surrounding sand layer and then flooded with synthetic clay water. The response of the axial stress was recorded by means of a pressure sensor installed at the top of the cell between load piston and cap. The variation of the axial stress with humidity is herein defined as the swelling pressure.

Figure 13 presents the swelling pressures measured on the compacted pure claystone and bentonite powder. The stressed samples were first dried by dry air, leading to a drop in the axial stress from 4.5 MPa to zero. The subsequent wetting by water vapour rapidly increased the stress to 7.6 MPa in the bentonite and to 3.0 MPa in the

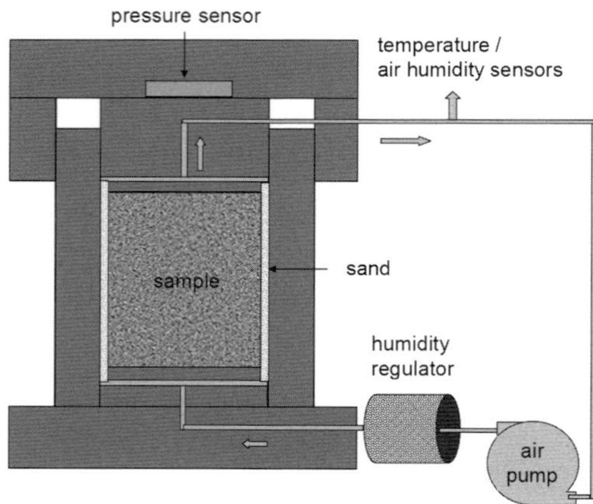

Fig. 12. Schematic of swelling pressure measurement on claystone–bentonite mixtures in our own designed cell.

compacted claystone powder. Because of the soft lateral confinement provided by the loosely filled sand layer, the highly increased axial swelling pressure exceeded the strength of the compacted bentonite, leading to damage and a residual stress of *c.* 4 MPa. The lateral swelling of the bentonite consolidated the sand layer and increased the stiffness of the confinement, thus allowing a further increase in the swelling pressure to 6 MPa. Finally, water flooding induced a further increase in swelling pressure in the bentonite to a high value of 7 MPa due to swelling of the remaining unsaturated particles, but a stress drop in the claystone powder down to 1.6 MPa due to wetting-induced weakness of particles and collapse of pore structures under the soft lateral confinement.

Figure 14 illustrates the swelling pressures observed on two compacted claystone–bentonite mixtures in COx/MX80 ratios of 40/60 and 60/40 during air drying/wetting and water flooding. Both samples had a very similar response of axial stress to moisture change. After vapour wetting over a month, the increased swelling pressures in both mixtures stabilized at a level of 4.2–4.7 MPa. Additional water flooding caused a rapid increase in swelling pressure to 6.0 MPa in both mixtures, which remained nearly constant over the flooding period.

In addition to the powdered claystone–bentonite mixtures, another mixture with coarse-grained COx aggregate of $d < 10$ mm and MX80 bentonite powder in a ratio of 50/50 was tested under confined conditions in an oedometer cell of 100 mm diameter and 90 mm length. An end face of the sample was in contact with synthetic clay water. Figure 15 shows the water uptake and the build-up of swelling pressure v. saturation time. In correlation with water uptake, the swelling pressure increased gradually to the maximum value of 9.5 MPa after seven months. Post-testing showed that the 90-mm-long sample was not yet fully saturated, and the pores between the coarse claystone grains were densely filled with bentonite powder.

Water permeability

The main function of seals in a repository is to prevent fluid access into the repository and the release of radionuclides from it. Accordingly, the seals are required to have very low hydraulic conductivities from the start and over long time periods of thousands of years. The permeability of the compacted claystone–bentonite mixtures as seal materials was determined with synthetic clay water.

The compacted samples were confined in oedometer cells. After water saturation over two weeks, the measurement of water permeability was carried out over several months. The measured water permeability values are plotted in Figure 16 against time. It is obvious that the permeability values of all compacted mixtures are quite low and keep

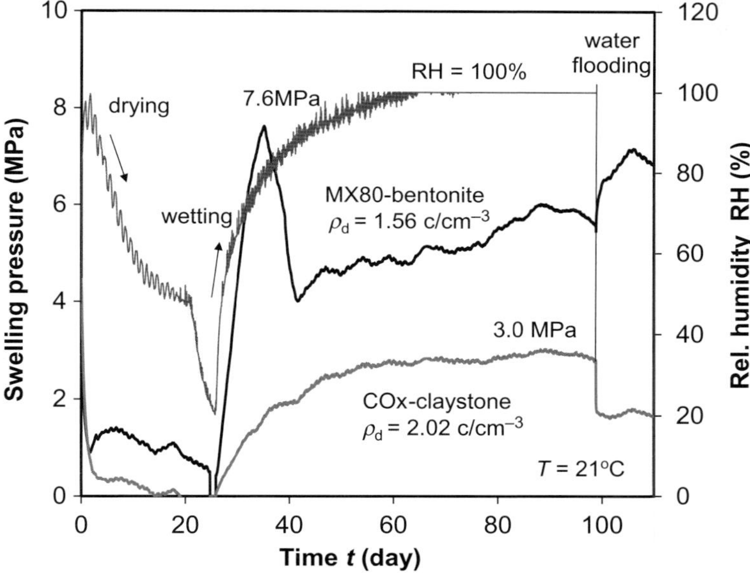

Fig. 13. Evolution of swelling pressures measured on compacted COx claystone powder and MX80 bentonite during air drying/wetting and water flooding.

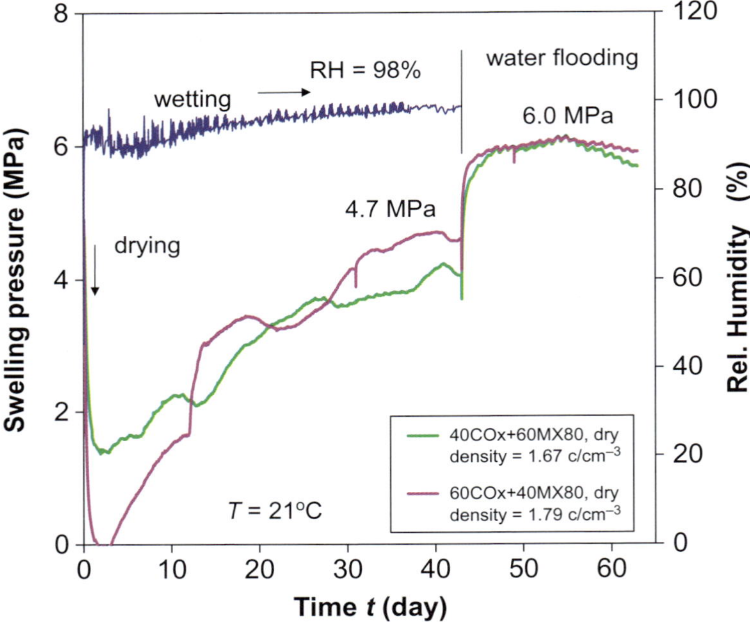

Fig. 14. Evolution of swelling pressures of two compacted claystone–bentonite mixtures in ratios of 40/60 and 60/40 during air drying/wetting and water flooding.

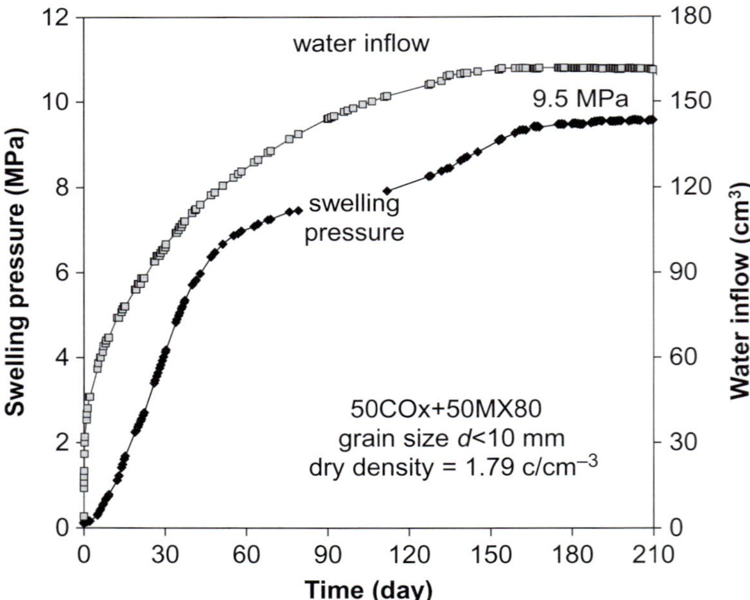

Fig. 15. Evolution of swelling pressure of compacted claystone–bentonite mixture in 50/50 ratio during wetting with synthetic clay water.

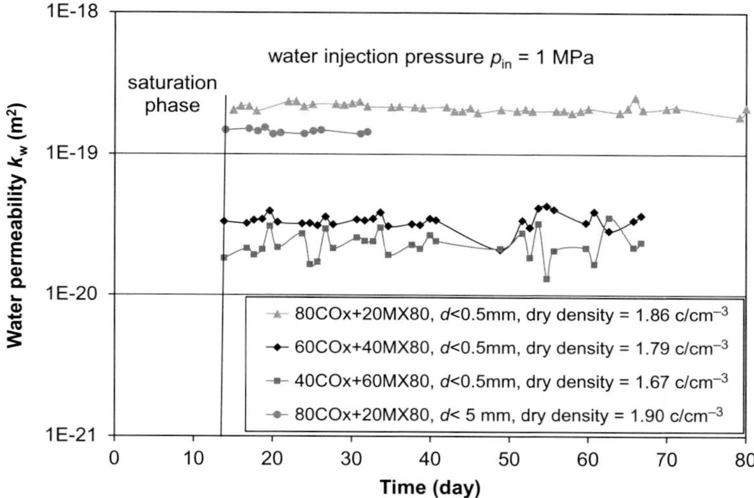

Fig. 16. Evolution of water permeability of compacted claystone–bentonite mixtures.

relatively constant: $k_w = 2 \times 10^{-19}$ m^2 at 80COx + 20MX80, $k_w = 3 \times 10^{-20}$ m^2 at 60COx + 40MX80 and $k_w = 2 \times 10^{-20}$ m^2 at 40COx + 60MX80. In the case of the crushed claystone with grains of $d < 10$ mm, the water permeability becomes very low too, $k_w < 1 \times 10^{-19}$ m^2, as the porosity is below 30% (cf. Fig. 7). The very low water permeability values of the compacted mixtures are comparable with that of the intact clay rock ($<10^{-20}$ m^2).

Thermal properties

As the excavated claystone and claystone–bentonite mixtures are considered to be used as buffer material surrounding high-level radioactive waste containers, knowledge of their thermal properties is required for analyses of the thermal processes in the buffer and in the whole repository system. In the investigation programme, the thermal properties of crushed COx claystone with grain sizes of $d < 10$ mm were characterized as a function of porosity and water saturation.

The measured data for specific heat capacity and thermal conductivity are depicted in Figure 17 as a function of water saturation at different porosities of 28.5, 30.5 and 32.5%. The heat capacity varies from 800 to 1000 J kg^{-1} K^{-1} in the dry and saturated states, respectively, and is only slightly dependent upon porosity. A mean value of 900 J kg^{-1} K^{-1} can be determined. The thermal conductivity increases with water saturation on average from 0.8 W m^{-1} K^{-1} under dry conditions to 1.6 W m^{-1} K^{-1} under saturated conditions. Because of the relatively large data scatter, the dependence on porosity could not be clearly identified. The thermal conductivities obtained on compacted claystone aggregate at a dry density of 1.80–1.93 g cm^{-3} are somewhat higher than the 0.4–1.3 W m^{-1} K^{-1} of FEBEX bentonite at a dry density of 1.57–1.72 g cm^{-3} (Huertas *et al.* 2000) and 0.4–1.0 W m^{-1} K^{-1} of MX80 bentonite at a dry density of 1.45–1.80 g cm^{-3} (Tang & Cui 2010).

Summary

Excavated claystone as backfill material and compacted claystone–bentonite mixtures as seal material have been comprehensively investigated. The main findings are summarized as follows.

As backfill material, the raw coarse-grained claystone with grain sizes up to 32 mm can be compacted from initial porosities of 32–46% to a low level of *c.* 20% as the applied stress increases to 12–16 MPa, mainly depending on the applied load rate or time and water content. The stiffness and support capacity of the backfill increases with compaction. The compaction leads to a decrease in porosity and permeability. It was found that, at a given porosity, (1) the permeability of the coarse-grained backfill is higher than that of the fine-grained material and (2) the water permeability is much lower than the gas permeability. The low water permeability is mainly attributed to the effects of water-induced swelling of clay grains with a subsequent clogging of the pores. The permeability decreases much faster at porosities

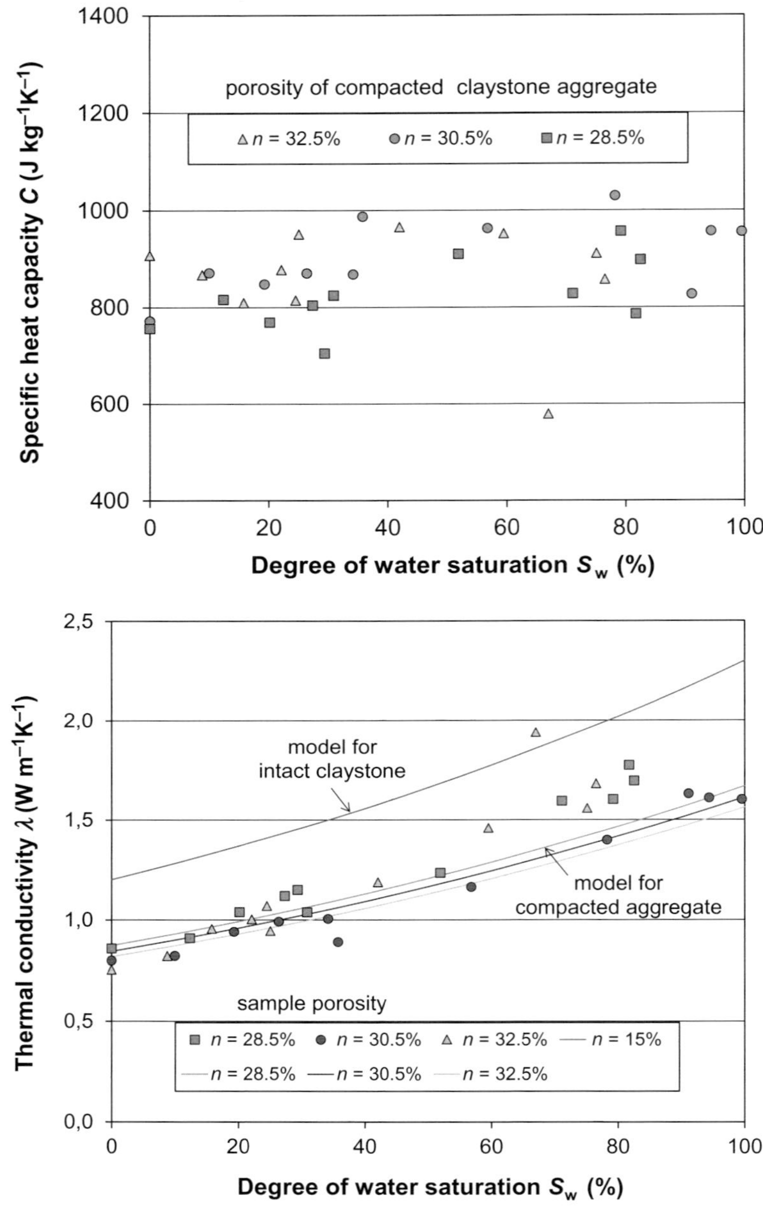

Fig. 17. Measured specific heat capacity and thermal conductivity of compacted claystone aggregate as a function of water saturation and porosity.

below *c.* 25% for gas flow and below *c.* 30% for water flow due to an increasing disconnection of the flow channels in the pore network. In low-porosity regions, the measured permeability values of less than 10^{-19} m^2 are close to that of the intact clay rock.

As a seal material, compacted claystone–bentonite mixtures with different COx/MX80 ratios of 0/100, 40/60, 50/50, 60/40, 80/20 and 100/0 were prepared by compaction at a pressure of 30 MPa. By application of the same energy, the dry density increases with increasing claystone content in the mixtures from 1.56 g cm^{-3} for pure bentonite to 2.00 g cm^{-3} for the claystone aggregate. The most important characteristics of the compacted mixtures are as follows:

(1) The water retention capacity of the claystone–bentonite mixtures increases with the fraction of bentonite. All the mixtures can take up large amounts of water from a humid environment, ranging up from 12% for the pure claystone and *c.* 50% for the pure bentonite under unconfined conditions.

(2) With water uptake, the unconfined mixtures can expand with a volumetric increase of 12% of the compacted claystone and 40% of the compacted bentonite. The swelling capacity is directly proportional to the bentonite content of the mixture.

(3) In axially fixed and laterally soft confined conditions, hydration causes a swelling pressure increase in the rigid axial direction. The maximum swelling pressures recorded vary between 3 MPa for the compacted claystone and 8 MPa for the compacted bentonite. Under constant-volume conditions, a rather high swelling pressure of 9.5 MPa was found for a compacted COx/MX80 mixture with a ratio of 50/50. All studied mixtures exhibit sufficiently high swelling capacities to meet the requirements for sealing repository openings.

(4) All compacted claystone–bentonite mixtures show very low water permeabilities, with a determined range of 10^{-19} to 10^{-20} m^2, which is close to that of the intact clay rock. With regard to hydraulic conductivity, all tested mixtures are suitable as a seal material.

(5) The thermal properties of compacted claystone as buffer are dependent on porosity and water saturation. On average, the specific heat capacity is *c.* 900 J kg^{-1} K^{-1}, the thermal conductivity increases with water saturation from 0.8 W m^{-1} K^{-1} under dry conditions to 1.6 W m^{-1} K^{-1} under saturated conditions, values that are comparable with those of compacted bentonite.

Generally, all the compacted claystone–bentonite mixtures exhibit favourable geotechnical properties with respect to their barrier functions in terms of preventing the release of radionuclides from a repository into the biosphere.

This research work was funded by the German Federal Ministry of Economics and Technology (BMWi) under contract number 02E10377. Support from ANDRA in providing the test material and for fruitful discussions is also gratefully acknowledged.

References

ANDRA: DOSSIER 2005. *Synthesis – Evaluation of the feasibility of a geological repository in an argillaceous formation.*

HUERTAS, F., FUENTES-CANTILLANA, J. L. ET AL. 2000. *Full-Scale Engineered Barriers Experiment for a Deep Geological Repository for High-Level Radioactive Waste in Crystalline Host Rock (FEBEX Project).* Final report EUR 19147 EN, European Commission.

NAGRA 2002. *Project Opalinus Clay, Models, Codes and Data for Safety Assessment – Demonstration of Disposal Feasibility for Spent Fuel, Vitrified High-Level Waste and Long-Lived Intermediate-Level Waste.* Technical Report 02-05, Nagra.

TANG, A. M. & CUI, Y. J. 2010. Effects of mineralogy on thermo-hydro-mechanical parameters of MX80 bentonite. *Journal of Rock Mechanics and Geotechnical Engineering*, **2**, 91–96

TANG, C. S., TANG, A. M., CUI, Y. J., DELAGE, P., SCHROEDER, C. & LAURE, E. D. 2011. Investigating the swelling pressure of compacted crushed Callovo-Oxfordian claystone. *Physics and Chemistry of the Earth*, **36**, 1857–1866.

Equivalent upscaled hydro-mechanical properties of a damaged and fractured claystone around a gallery (Meuse/Haute-Marne Underground Research Laboratory)

RACHID ABABOU[1]*, ISRAEL CAÑAMÓN[2] & ADRIEN POUTREL[3]

[1]*Institut de Mécanique des Fluides de Toulouse, Unité Mixte de Recherche 5502 (CNRS-INPT-UPS), 1 Allée du Professeur Camille Soula, 31400 Toulouse, France*

[2]*Departamento de Matemática Aplicada y Médodos Informáticos, Universidad Politécnica de Madrid, ETSI Minas y Energía, C./Ríos Rosas 21, 28003 Madrid, Spain*

[3]*Agence Nationale pour la Gestion des Déchets Radioactifs (ANDRA), 1/7 rue Jean Monnet, Parc de la Croix-Blanche, 92298 Châtenay-Malabry cedex, France*

**Corresponding author (e-mail: ababou@imft.fr)*

Abstract: In this work, we present calculations and analyses of equivalent continuum (upscaled) coefficients describing the damaged, fissured and fractured claystone around an underground gallery. We focus here on mechanical and coupled hydro-mechanical properties of the damaged claystone (the upscaled Darcy permeability of the same claystone was studied in a previous paper focused on hydraulics without mechanical deformations). Concerning the geometric structure of the damaged clay stone around the cylindrical excavation, we use a hybrid 3D geometric model of fissuring and fracturing, comprising (a) a set of 10 000 statistical fissures with radially inhomogeneous statistics (size, thickness and density increasing near the wall), and (b) a deterministic set of large curved 'chevrons' fractures, periodically spaced along the axis of the drift according to a 3D chevron pattern. The hydro-mechanical coefficients calculated here are second- and fourth-rank tensors, which are displayed using ellipsoids. For simplicity, we also calculate equivalent isotropic coefficients extracted from these tensors: Young's modulus (E), bulk modulus (K), Lamé shear modulus (μ), Poisson's ratio (ν), Biot coefficient (B, stress–pressure coupling) and Biot modulus (M, pressure–fluid production coupling). All of these coefficients are affected by the degree of damage and fracturing, which increases near the wall of the gallery. Both 3D and '2D transverse' distributions are analysed, on grids of 3D cubic voxels and 2D pixels, respectively. Global coefficients upscaled over the entire damaged and fractured zone are also analysed. Other types of averages are presented, for example, upscaled values over a cylindrical annular shell at various radial distances from the gallery wall. The relation to the degree of fracturing is discussed, including for instance the effect of fracturing on bulk and shear stiffnesses, and on the hydro-mechanical coupling coefficients of the damaged claystone.

The main objective in this paper is to determine by *homogenization*, or *upscaling* from fine to coarser scales, the coupled hydro-mechanical (H-M) behaviour of a damaged (fissured and fractured), water-saturated porous clay rock around a cylindrical excavation. Here, the term 'upscaling' is used in the following sense: obtaining equivalent continuum properties for the fractured porous rock at selected scales (averaging volumes). The excavation considered in this work is a subhorizontal cylindrical 'drift' (named the GMR gallery), located at about 525 m depth in the Underground Research Laboratory (URL) 'CMHM' (Centre de Meuse/Haute-Marne, Bure, France) operated by ANDRA. This URL is used to develop research studies on the stability and isolation properties of a deep geological repository for radioactive waste in Callovo-Oxfordian claystone.

In this context, a programme was developed towards the simulation of hydraulic, mechanical and coupled hydro-mechanical processes in the near field, in the damaged and fractured zone around a cylindrical underground excavation. Our objective is to obtain a set of 'upscaled' continuum equivalent equations and coefficients, based on the geometry and properties of the discrete fractures, and based on the properties of the porous rock matrix. The zone of interest is the so-called excavation damaged zone (EDZ), a cylindrical annular zone around the gallery. The next section will present the study site briefly, but we also refer the interested reader to Ababou *et al.* (2011) for more

From: Norris, S., Bruno, J., Cathelineau, M., Delage, P., Fairhurst, C., Gaucher, E. C., Höhn, E. H., Kalinichev, A., Lalieux, P. & Sellin, P. (eds) 2014. *Clays in Natural and Engineered Barriers for Radioactive Waste Confinement*. Geological Society, London, Special Publications, **400**, 339–358.
First published online July 16, 2014, http://dx.doi.org/10.1144/SP400.44

details concerning the overall objectives, the programmatic context, the URL site and some of the relevant data.

To sum up, the present work focuses on obtaining a continuum equivalent model (and the corresponding 'upscaled' coefficients) for coupled hydro-mechanical processes in the EDZ around a cylindrical excavation (gallery).

Specifically, we are interested in achieving this continuum characterization in three dimensions (3D) by a fast method that does not require detailed numerical equations for the processes taking place in the fractured porous rock to be solved. On the other hand, we require that the method is able to take into account in some way the 3D geometric structure and the hydro-mechanical properties of the fractures (embedded discontinuities) and of the 'intact' porous rock (embedding matrix).

The present paper elaborates on a previous presentation by the authors at the Clay 2012 conference (Ababou *et al.* 2012).

Basis of the superposition method for upscaling

With this in mind, we have selected a computationally fast upscaling method that can be considered as a *generalized superposition method*. Superposition methods have been used for upscaling permeability in fractured media (quite frequently), and also sometimes (more rarely) for upscaling hydro-mechanical processes in fractured media, but without including the full role of the water-filled matrix porosity in these processes. Thus, in Oda (1986) and Stietel *et al.* (1996), the rock matrix is impervious hydraulically, and does not participate in pressure–stress coupling mechanically. Similarly, the recent paper by Sævik *et al.* (2013) includes a section on superposition methods, where the fractured medium is assumed to be made up of fractures imbedded in a totally impervious matrix.

Note: in addition to the superposition methods (effective medium approximation), there are several other groups of upscaling methods, such as the Self-Consistent approximations and their variants (e.g. Sævik *et al.* 2013), and there are many other upscaling methods, some quasi-analytical and some others numerical, which are all largely developed in the literature. Even if we do not aim to discuss these other methods in this paper, we briefly discuss their main limitations to state the benefits of the superposition method in the case of fractured porous media. For example, the Self-Consistent approximation has well-known limitations for high density of fracturing (while the superposition approach, on the contrary, performs well for high-density fracturing but has drawbacks for low-density or poorly connected systems). There are also methods that are more heavily numerical, where the upscaled coefficients are computed by solving a detailed solution of partial differential equations (PDEs) that are similar in complexity to the non-homogenized PDEs themselves. Such methods are therefore more computationally intensive, and expensive, than the superposition approach. Finally, there are also numerical solvers that aim at obtaining detailed simulations of the behaviour of the porous matrix, the discrete fractures and their interactions. In some cases, these solvers use a set of local-scale effective coefficients, such as the parallel and orthogonal effective permeability of a joint embedded in a porous matrix (this concept was also used in Cañamón (2006, 2009) and Ababou *et al.* (2011) as a building block for implementing the hydraulic superposition method on larger scales). For instance, Mourzenko *et al.* (2010) numerically solve single-phase flow in a matrix–fracture system with thousands of fractures, tens of wells and millions of elementary volumes, with local effective permeability attributed locally to the discrete joints surrounded by porous matrix.

In the present paper, on the other hand, the aim is to upscale, not to simulate, and the method for upscaling is to extend the superposition approach to obtain a fast evaluation of global and spatially distributed upscaled coefficients describing the coupled H-M properties of the fractured porous rock on chosen support scales. In this upscaled description, the rock is viewed as an equivalent H-M continuum. The resulting coefficients can potentially be used as inputs for numerical simulations using more or less 'standard' H-M solvers and codes accepting tensorial coefficients, and this without the need for a detailed representation of fractures. In this paper, we calculate and analyse the upscaled coefficients without implementing the simulations themselves.

The *generalized* superposition method considered in this paper goes as follows:

- For hydraulic upscaling, the local Darcy and/or Poiseuille fluxes in both the permeable matrix and the cracks are superposed, and the final result is a tensorial flux/gradient law with macro-permeability K_{ij} (this problem is fully treated in Ababou *et al.* (2011) for the same URL site and the same GMR gallery).
- For mechanical upscaling, local strains are superposed, in a manner that involves both the isotropic elastic porous matrix and the strongly anisotropic elastic cracks. When the coupling is ignored, this results in continuum equivalent

tensorial coefficients of compliance (C_{ijkl}) and stiffness (R_{ijkl}).

- When taking into account fully coupled H-M processes (the object of this paper), the local strains in the matrix and cracks are still being superposed but, in addition, the interactions with variations of fluid pressure and with fluid production are also taken into account. Moreover, these interactions occur both in the water-filled cracks and in the saturated porous matrix. At least two new coefficients are thus obtained, the Biot coefficient, B_{ij}, and the Biot modulus, M (described in more detail below).

Our implementation of the effective medium/superposition method, to date, does not deal explicitly with fracture–fracture interactions. On the other hand, by various refinements of the method, we account here for the following types of interactions in 3D: (a) pressure–stress coupling and fluid production within the matrix; (b) pressure–stress coupling and fluid production within the fracture system; and (c) hydro-mechanical matrix–fracture interactions.

The tensorial superposition calculations leading to C_{ijkl}, R_{ijkl}, B_{ij} and M, are rather involved. Some relevant theoretical relations thus obtained will be presented in this paper, along with the quantitative results obtained specifically for the EDZ around the gallery of our study site.

In practice, for our specific application, the upscaling method is implemented as follows:

- Implementation starts with the definition of a permeable and deformable porous rock matrix, and with the specification of several sets of discrete embedded cracks or fractures. In the present case, we have two main sets: (a) a statistical set of small planar disc fissures (joints, cracks); and (b) a deterministic set of large curved fractures organized in a periodic 'chevron' pattern. The latter fractures are then numerically discretized into triangular patches before applying the generalized superposition method (see Ababou *et al.* 2011). The statistical cracks are disc-shaped planar objects that are generated only once (single realization). In the end, all planar 'fractures' or 'cracks' resulting from these procedures are stored in a unique object database, with 3D geometric parameters and other properties attributed to each crack. It should be emphasized that the cracks are hydro-mechanical as well as geometric objects: they are hydraulic conductors, and they are deformable objects. In particular, they all contribute to pressure–stress coupling, as does the porous rock matrix itself (further details on this are given below).
- Finally, given the geometric, hydraulic and mechanical properties of these materials at the local scale (the intact claystone matrix and the various types of cracks), the superposition method yields an upscaled system of continuum laws (Darcy, Hooke, Biot) with equivalent continuum coefficients (tensors), such as mechanical stiffnesses and H-M coupling coefficients, as mentioned earlier.

Concerning the equivalent Darcy law, and the calculation of macropermeability, we recall again that this was addressed separately in a previous work (Ababou *et al.* 2011). The macroscale permeability tensor K_{ij} was calculated in the EDZ using the flux superposition method. Therefore, in this work, we need only be concerned with the upscaling of coupled H-M properties, using the generalized strain superposition method sketched earlier.

Finally, let us discuss the 'scale' at which the equivalent continuum is calculated. In our procedure, the 'equivalent continuum' coefficients are evaluated at some specified 'homogenization scale' (meso- or macroscale), which is analogous to the support scale or measurement scale of a field measurement (its volume of influence). We will see in this paper how the upscaled coefficients obtained in the EDZ depend on the chosen scale of homogenization. When the selected scale is sufficiently 'local' (e.g. with respect to gallery diameter), we obtain not only the numerical values but also the spatial distributions of these upscaled coefficients within the damaged zone around the gallery (EDZ).

Several types of spatial distributions will be analysed, depending on the selected scales and supports of averaging:

- 3D spatial distributions obtained by upscaling locally on small cubic voxels (0.5 m);
- 2D transverse distributions of H-M properties – these are obtained by upscaling along the entire axis of the gallery as well as transversely on a grid square pixels (0.5 m);
- 1D radial distributions – these are obtained by upscaling on annular regions, that is, over cylindrical annular shells (thickness 0.5 m) located at various radial distances from the drift wall;
- 0D global coefficients ate the scale of the entire EDZ – these are obtained by upscaling over the entire damaged and fractured zone around the cylindrical drift (here the drift radius is 2 m and the total thickness of global annular shell is taken to be 4.0 m).

We expect that these types of results, taken together, will lead to a simplified quantification of the effects of fracturing on the claystone stiffness, and on the

Hydro-Mechanical coupling coefficients, at various scales of analyses.

Application site (Underground Research Laboratory)

The parameters used in the upscaling calculations presented later on in this paper correspond to the CMHM (Centre de Meuse/Haute-Marne, Bure, France), the Underground Research Laboratory (URL) operated by ANDRA – the French national agency for the management of radioactive waste.

We focus here on the damaged zone around the 'GMR' gallery, a subhorizontal gallery, located at about 525 m depth underground, within the Callovo-Oxfordian formation, a thick 130 m claystone layer between depths 400 and 600 m. The GMR gallery is oriented parallel to the minor horizontal principal stress (σ_h). For more details, we refer the interested reader to Ababou *et al.* (2011) and references therein, concerning the overall scientific objectives, the programmatic context, the URL site and some of the available *in-situ* data and observations.

For instance, see Armand & ANDRA (2007) concerning measurements of radial permeability profiles, and observations of rock fracture traces in the EDZ around galleries at the URL site (including the GMR gallery and other galleries).

The reader may consult other publications (Vincké *et al.* 1997; Coste *et al.* 1999; Cosenza *et al.* 2002) concerning various evaluations of mechanical and hydro-mechanical coefficients in the Callovo-Oxfordian claystone. We will refer more precisely to these works when discussing results (see section on 'Inputs, results, analyses: equivalent H-M coefficients').

Finally, the reader may consult ANDRA (2005*b*) for an overall description of ANDRA's URL site and its objectives.

Based on *in-situ* observations like those of Armand & ANDRA (2007) and others, it was decided that the geometric structure of fissures and fractures around the gallery should be represented by combining two sets: a statistical set of planar disc cracks, and a deterministic set of large curved chevron fracture surfaces (as explained in the previous section on objectives and methods). In summary, the small disc fractures (which contribute essentially to near-wall damage) are represented statistically, while the large curved fractures (which have a larger extent that the diameter of the drift) are modelled as deterministic parametric surfaces, periodically arranged along the drift axis. These two synthetic sets of fractures constitute essentially our 'model' for the internal geometric structure of the fractured and damaged EDZ. This is defined more quantitatively in the next section.

Geometric structure of fissured and fractured claystone

Concerning the geometric structure of the damaged claystone around the cylindrical excavation (gallery), we use a hybrid statistical and deterministic geometric model of 3D fissuring and fracturing, comprising:

(1) A statistical set of 10 000 randomly distributed and randomly oriented, disc-shaped planar fissures. This set has radially inhomogeneous statistics: in particular, the fractures are more densely distributed near the wall. Some of the statistical/geometric parameters of this random set of cracks were calibrated by comparing upscaled permeabilities with measured permeability profiles along radial boreholes (Ababou *et al.* 2011).

(2) A deterministic set of large curved 'chevron' fractures, periodically distributed along the axis of the gallery (inter-spacing 0.5 m) and forming a 3D chevron pattern. The curved surface of each large 'chevron' fracture was represented using a parametric surface model (a modified *conoidal* surface resembling the pinched end of a toothpaste tube). These curved fractures surfaces were further discretized into triangular planar cracks (patches).

The complete 3D system of fractures is shown in Figure 1.

The reader is referred to Ababou *et al.* (2011) for more details on the procedures that allowed us to generate and represent each subset of fractures, and for a description of the complete set of geometric parameters, statistical as well as deterministic.

Theory and equations of the upscaling method

Equivalent continuum equations, tensorial coefficients and variables

The upscaled behaviour of the claystone is described by a system of coupled equivalent continuum hydro-mechanical PDEs that enforce mass conservation and (quasi-static) momentum conservation. The variables involved are total stress (σ_{ij} or $\Delta\sigma_{ij}$), total strain (ε_{ij}), and water pressure (p or Δp). Other variables involved are the water flux density vector (q_i) expressed by Darcy's law (Darcy 1856), and the water production term $\Delta\xi$

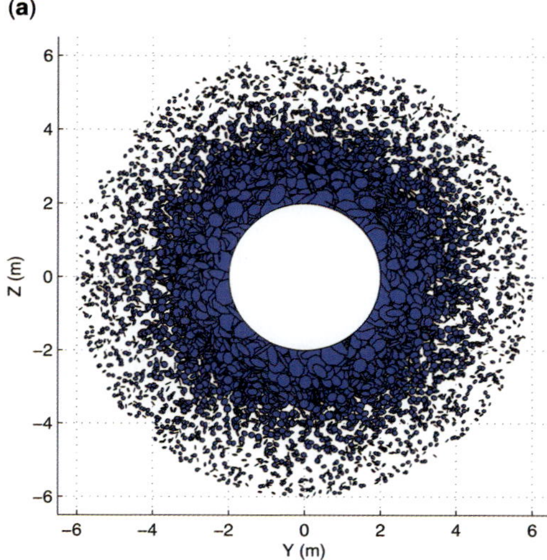

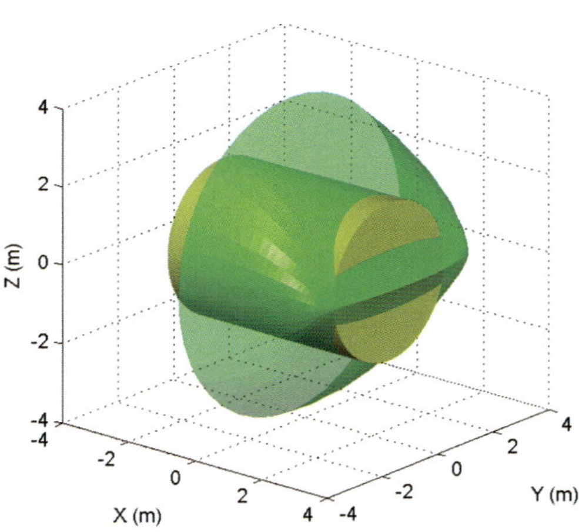

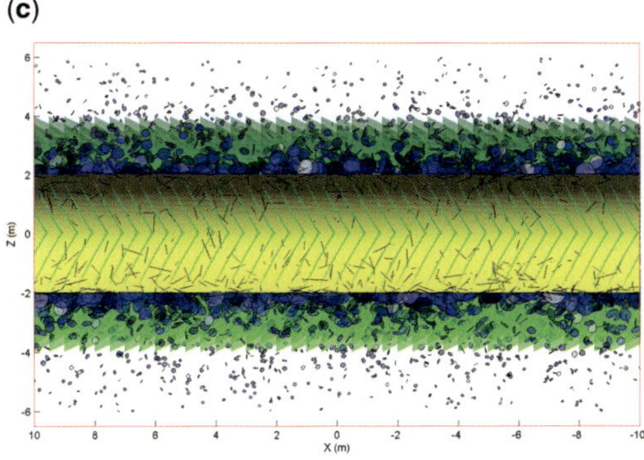

Fig. 1. (**a**) 2D transverse view of the statistical set of fissures (small fractures); (**b**) 3D view of a single curved chevron fracture surface; (**c**) 2D longitudinal section of the fissured and fractured system.

(defined further below). All these variables depend on 3D space and time. The symbol Δ designates a variation in time at any given point (x, y, z).

The mesoscale equivalent continuum equations obtained by our generalized superposition method take the form of linear *tensorial laws involving tensorial coefficients*, all the coefficients being calculated in terms of the given geometric structure and properties of the fractures and of the porous rock matrix (*note: we use everywhere Einstein's notation of implicit sum on repeated indices, unless mentioned otherwise*).

- Hooke's stress–strain law (without H-M coupling) contains tensorial stiffness coefficients R_{ijkl} (Pa),

$$\Delta\sigma_{ij} = R_{ijkl}^{\mathrm{MC}}\,\varepsilon_{kl} \tag{1}$$

where the superscript MC denotes 'matrix and cracks' (i.e. the equivalent homogenized medium). Some classical scalar coefficients can be derived by various contractions of the upscaled R_{ijkl} tensor, a process we call '*isotropization*' (a neologism described in Appendix 1). In this way, we obtain, for instance, isotropic equivalent values of Young's modulus E, the bulk modulus K, the shear modulus μ and Poisson's ratio ν. This will be useful when calculating the equivalent tensor R_{ijkl} on annular regions or at the global scale of the entire EDZ; however, the full anisotropic R_{ijkl} tensor is of more interest on local scales (i.e. half-metre-scale cubic voxels).

- The hydro-mechanical Hooke/Biot stress–strain–pressure law contains a pressure term that couples stress to pressure via a tensorial Biot coefficient B_{ij} as follows:

$$\sigma_{ij} = R_{ijkl}^{\mathrm{MC}}\Delta\varepsilon_{kl} + B_{ij}^{\mathrm{MC}}\Delta p \tag{2a}$$

$$\varepsilon_{ij} = C_{ijkl}^{\mathrm{MC}}\Delta\sigma_{kl} + \bar{B}_{ij}^{\mathrm{MC}}\Delta p \tag{2b}$$

In equation (2a), σ_{ij} is the stress tensor, R_{ijkl} is the stiffness tensor, ε_{ij} is the strain tensor, B_{ij} is a dimensionless 'Biot' coupling coefficient and p is the water pressure. All stresses are taken as negative under compression (thus, fluid stress is $-p\delta_{ij}$). Note that Terzaghi's (1936) 'effective stress' model corresponds to $B_{ij} = 1 \times \delta_{ij}$ (effective stress $\sigma_{ij} + p\delta_{ij}$). Biot's (1941, 1956) theory generalizes Terzaghi by considering a scalar coupling coefficient B such that $0 < B \leq 1$ (effective stress $\sigma_{ij} + B\,p\delta_{ij}$). In the more general model here, B_{ij} is a tensorial coupling coefficient obtained by upscaling the saturated porous rock with its water-filled cracks. The tensor $\bar{B}_{ij}$ appearing in equation (2b) is called in this work the 'reciprocal Biot coefficient', and it corresponds to the anisotropic form of the classical 'poro-elastic expansion coefficient' ($\bar{B}_{ij}$ is not the inverse of B_{ij}). To sum up, equation (2a) expresses the stress–strain relation under pressure coupling, while equation (2b) is a formulation of the reciprocal strain–stress relation under the same pressure coupling.

- The hydro-mechanical relation expressing water production $\Delta\xi$ (m^3 of water m^{-3} of clay rock) in terms of water pressure variations Δp (via the Biot modulus M) and in terms of the global strain (ε_{ij}) of the poro-elastic medium.

$$\Delta\xi = B_{ij}^{\mathrm{MC}}\varepsilon_{ij} + \frac{1}{M_{\mathrm{MC}}}\Delta p \qquad (3a)$$

where strain tensor is defined as the symmetric part of the gradient of displacement:

$$\varepsilon_{ij} = \left(\frac{\partial u_i}{\partial u_j} + \frac{\partial u_j}{\partial u_i}\right)\Big/2 \qquad (3b)$$

Equation (3a) relates water production $\Delta\xi$ to strain and pressure, at any fixed point (x, y, z) in space. The variations in time from an initial reference state are represented by the symbol Δ (as in $\Delta\xi$ and Δp). Note: the strain (ε_{ij}) also contains implicitly a time variation of the displacement vector (u_i), which could also be written equivalently (Δu_i).

- Finally, the water production $\Delta\xi$ also appears in the water flow equation that enforces mass conservation of the moving fluid. This conservation equation takes the form:

$$\frac{\partial m_{\mathrm{W}}}{\partial t} = -\frac{\partial}{\partial x_j}\{\rho_{\mathrm{W}} Q_j\} \qquad (4a)$$

where $\Delta m_{\mathrm{W}} = \rho_{\mathrm{W}}\Delta\xi$ (w representing water).

The equation on the left involves the negative divergence of the mass flux of water ($\rho_{\mathrm{W}} Q_i$). The water flux density vector Q_i (($m^3\,s^{-1}$) m^{-2}) is expressed by an equivalent Darcy's law, which relates linearly water flux to pressure gradient at macroscopic scale via an equivalent permeability tensor k_{ij} (m^2) (and the dynamic viscosity of water μ_{W}):

$$Q_i = -\frac{k_{ij}^{\mathrm{MC}}}{\mu_{\mathrm{W}}}\frac{\partial(p + \rho_{\mathrm{W}} g z)}{\partial x_j} \qquad (4b)$$

This can be simplified in some cases as follows, in terms of hydraulic conductivity K_{ij} ($m\,s^{-1}$), using a reference value for ρ_{W}:

$$Q_i = K_{ij}^{\mathrm{MC}} J_j = -K_{ij}^{\mathrm{MC}}\frac{\partial((p/\rho_{\mathrm{W}}^0 g) + z)}{\partial x_j} \qquad (4c)$$

Inserting the water production term $\Delta\xi$ of equation (3a) and the macroscale Darcy law (equation 4b, c) in the mass conservation equation of the moving fluid (equation 4a) finally closes the entire system of equivalent (upscaled) continuum equations for hydraulics, mechanics and hydro-mechanics. Concerning fluid production, see also Coussy (1991), among other authors.

In the remainder of this work, we focus on mechanics and coupled hydro-mechanics, and we assume that the hydraulics upscaling has already been calculated. However, for the sake of completeness, we will also give a brief comparative outline of various upscaling methods for hydraulics and hydro-mechanics in fractured porous media (see below).

Upscaling method for hydro-mechanics: superposition and coupling

As explained in a previous section, all the equivalent tensorial coefficients (stiffnesses, compliances, coupling coefficients) were calculated in this work by the superposition method, an upscaling technique that takes into account the given geometric structure of the damaged fractured rock (geometry of cracks), as well as the local microscale properties of the fractured rock under various assumptions (quasi-elastic crack with given apertures and normal/shear stiffnesses, isotropic elastic porous matrix, etc.). Figure 2 shows a rough schematic of the fractured porous matrix at the microscale (both the pores of the rock matrix and the cracks are assumed to be filled with water).

It is not the purpose of this paper to give a full detailed account of the mathematical upscaling theory used in this work; however, the interested reader may consider the following discussion and the cited references concerning the mathematical bases of this approach.

Hydraulic upscaling and equivalent macro-permeability tensor. Concerning *hydraulic upscaling*, the reader is referred to Ababou *et al.* (2011): that paper contains a detailed description of the flux superposition method for obtaining an equivalent Darcy permeability, or hydraulic conductivity (K_{ij}). The permeability upscaling method was applied to the same site as in the present work (gallery GMR of the Meuse/Haute-Marne URL), using the same synthetic geometric structure of fractures (the near-wall statistical fractures were fitted to hydraulic conductivity measurements).

For the sake of completeness, let us give a brief outline of permeability upscaling calculations by the flux superposition method. The principle of

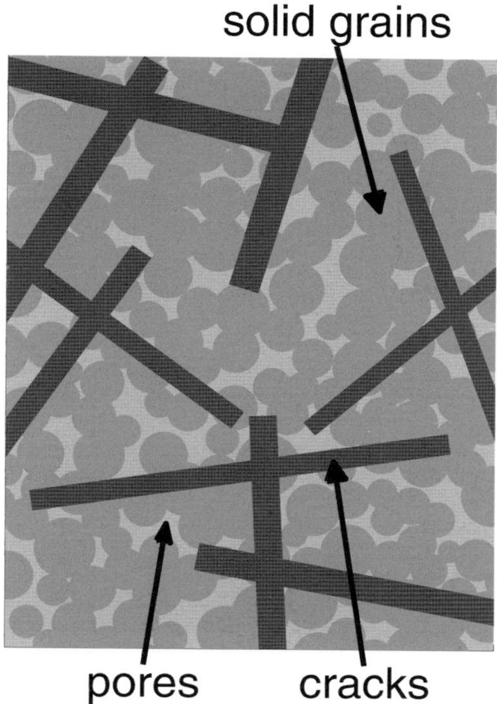

Fig. 2. Micro-scale schematic of the matrix solids (orange circles (grey in print version)), matrix pores (light blue background (light grey in print verison)) and planar cracks (dark blue bars (dark grey in print version)).

the methods is similar for hydro-mechanics, where a tensorial strain superposition is applied (with specific procedures for accounting for stress–pressure–fluid production coupling).

In the fractured porous medium, the local flow is assumed to be governed by the isotropic form of Darcy's law in the porous matrix, and by Poiseuille law in the plane of the cracks. This yields for each crack a 'cubic law', that is, the flow rate in each crack is proportional to the power 3 of its aperture. Upscaling the dual medium (the permeable matrix and cracks) based on a generalized form of the flux superposition method yields a tensorial Darcy law (equation 4a) with an 'equivalent conductivity tensor' (K_{ij}). The latter incorporates the matrix permeability, the hydraulic effects and geometric structure of cracks (apertures, planar sizes, etc.): see Ababou *et al.* (2011). When matrix permeability goes to zero, the resulting K_{ij} is similar to that of Oda & Hatsuyama (1985) and Oda (1986) without Oda's connectivity factor.

Mechanical upscaling. Mechanical upscaling (without hydro-mechanical coupling) is essentially based on the superposition of local strain values ε_{ij}, similar to the superposition of local fluxes in hydraulics. This approach leads, first, to upscaling the compliance coefficients. The equivalent compliance tensor (resulting from strain superposition) contains, essentially, weighted arithmetic averages of local matrix and crack compliances (although this is a rather simplified description of the full expressions).

Second, the upscaled stiffness tensor is obtained as a fourth-rank tensorial inverse of the upscaled compliance tensor. As a consequence, upscaled stiffnesses can be viewed essentially as weighted harmonic averages of local matrix and crack stiffnesses (in a general tensorial sense).

Coupled hydro-mechanical upscaling. The method is again based on the superposition of local strains, but we assume this time that there is a local stress–pressure coupling in each crack, and at each location within the porous matrix. The local H-M coupling in the water-filled cracks is based on Terzaghi's (1936) effective stress concept (equivalent to taking a Biot coefficient equal to one in each crack: $B_C = 1$). The local H-M coupling within the saturated porous rock matrix is either disregarded (cf. hypothesis 'iii.a' in Table 3), or fully taken into account through a matrix Biot coefficient B_M between 0 and 1 ($B_M \approx 0.5$–0.6 from experimental evidence).

The latter result, taking into account Biot coupling in the porosity of the matrix as well as in the cracks, is the more interesting one. Nevertheless, we want to conserve the results obtained both with and without H-M coupling in the matrix for the purpose of analysis (decomposition of coupling effects owing to the cracks v. those owing to the porous matrix).

Technically, and mathematically, the upscaling of hydro-mechanical couplings in the dual matrix–crack medium is a little complex; the resulting upscaled coefficients (B_{ij}, M) contain contractions of fourth-rank tensors, and cannot be described as simple averages except for special cases (the details of the general calculation remain beyond the scope of this paper).

For more details on the strain superposition method for both mechanics and coupled hydro-mechanics under various hypotheses, the reader is referred to, Cañamón *et al.* (2007, 2009); Cañamón (2009) and Ababou *et al.* (1994*a*, *b*). The present paper constitutes a natural generalization, and also a different application, of the latter works. Let us briefly compare the above-cited works with the present work, and particularly, let us point out what simplifying hypotheses have now been relaxed in the present work: compared with the previous cited works, the hydro-mechanical role of the permeable porous matrix is fully taken into account. The pressure coupling terms calculated in the present paper can fully take into account the contribution of the water-filled porous matrix as well as the water-filled cracks.

However, for simplicity and for comparison purposes, when displaying the spatial distributions of ellipsoids representing the coupling coefficients (B_{ij}, M), these are calculated for the sole contribution of the cracks, and in that case, they are labelled ($B_{ij}^{(0)}$, $M^{(0)}$). On the other hand, all the global and annular values of (B, M) given in the text and tables include the complete coupling effects of both cracks and matrix.

Spatial averaging procedures

It should be emphasized that the theoretical upscaling method can be used, technically, for obtaining different types of upscaled coefficients using various spatial averaging 'supports' and various scales (e.g. cubic or rectangular voxels of various sizes, annular regions of various thicknesses).

Indeed, in this work, several distinct versions of the upscaled continuum coefficients are calculated, as mentioned earlier. The equivalent coefficients are obtained either as 3D fields distributed in (x, y, z), or as 2D fields distributed in a transverse cross-section (y, z) orthogonal to the axis of the gallery (x). Other types of averages will also be presented, such as the annular average over a cylindrical annular shell at various radial distances from the wall of the gallery. Furthermore, 'global' values of the coefficients are calculated by taking the annular domain to be as thick as the entire damaged and fractured zone (about 4 m thickness around the wall of the gallery). These different versions of the upscaled continuum coefficients (3D, 2D, annular, global) can be computed by two possible alternative methods:

(1) *Direct upscaling* – the term 'direct upscaling' designates the usual upscaling procedure based on strain superposition (as described in previous sections), starting with discrete matrix–crack hydro-mechanical laws at the microscale, and upscaling them to equivalent continuum laws at the desired scale (meso or homogenization scale). The upscaling domain may be anything here: a 3D voxel, a parallelepiped box, an annular shell or a moving spherical window, etc.

(2) *Sequential upscaling* – here is an example of 'sequential upscaling' that was used in this work. The transverse distributions of '2D upscaled' quantities were obtained, first, by calculating the '3D upscaled' quantities on a grid of cubic voxels (each voxel is a subdomain), and second, by applying axial averaging operations (compatible with the upscaling theory) to obtain '2D upscaled' quantities over a transverse planar grid of pixels. Note: each pixel represents an elongated parallelepiped having the same length as the stretch of gallery considered (L = 20m here).

These different ideas are illustrated in Figure 3.

Equivalent isotropic scalar coefficients at annular and global scales

One more word of caution is needed concerning the interpretation of 'global scale' and 'annular scale' equivalent coefficients (see results in the next sections on global and annular upscaling, and see Tables 3–6).

- Annular and global scale coefficients are obtained from the tensorial superposition method applied to a cylindrical annular shell. When the annular shell has the thickness of the EDZ, we call the resulting tensorial coefficients 'EDZ-scale' coefficients. For annular shells of smaller thickness (e.g. 0.5 m), we obtain an annular tensorial coefficient for each various shell (the shells being located at various radial distances from the drift wall).
- At this stage, the calculated annular coefficients remain generally tensorial (more precisely, we obtain in this way annular or global values of R_{ijkl}, C_{ijkl}, B_{ij} and M).
- However, the point we want to make is that the choice of annular upscaling leads naturally to a loss of information on the anisotropic structure of the coefficients (except for 'M', which is in essence a scalar). For this reason, the global and annular results presented in the next section (Tables 3–6) are shown as scalars rather than tensors. The scalar values are isotropic equivalents obtained with the '*isotropization*' procedure defined in Appendix 1. There is only one exception: the shear compliance diagonal values S_{jj} are given for various annular shells, in Tables 4–6. This information is only used to show one example of annular coefficients without *isotropization*.

In summary, most annular and global quantities calculated in this work based on tensorial superposition are finally evaluated as equivalent isotropic scalar properties (Appendix 1). On the other hand, local scale continuum equivalent coefficients are still evaluated as full tensors, to be visualized in the next section as ellipsoids distributed around the gallery. The term 'local scale'

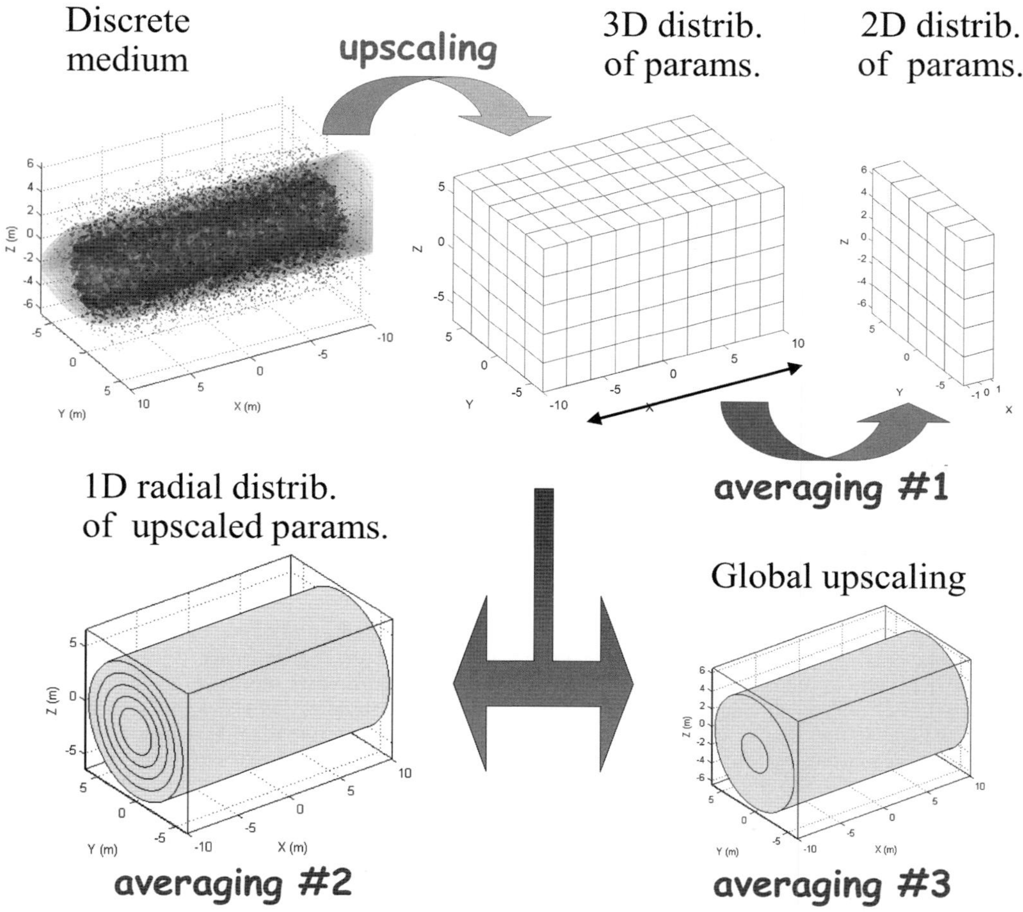

Fig. 3. Upscaling and averaging, leading to 3D, 2D or radially distributed, equivalent continuum upscaled coefficients (compliances, stiffnesses, coupling coefficients).

corresponds to the case where the two transverse dimensions of the averaging domain are smaller than drift diameter (e.g. cubic voxels of size $0.5 \ll 4.0$ m).

Inputs, results, analyses: equivalent 'H-M' coefficients

Inputs, outputs and post-treatment of hydro-mechanical upscaling

Inputs/outputs of upscaling and numerical values of input parameters. Tables 1 and 2 present two types of information:

(1) A list of inputs/outputs – the 'input' parameters (porous rock matrix and cracks) and the 'output' coefficients to be obtained by implementing the upscaling procedure (set of tensorial upscaled Hydro-Mechanical coefficients); and
(2) A list of numerical values of the basic hydro-mechanical and geometrical input parameters, such as matrix porosity, crack aperture, crack stiffnesses (normal and shear), matrix Young's modulus, Poisson's ratio, etc.

Hydro-mechanical coefficients (B, M) *for the rock matrix: experimental evidence.* Some of the parameters of Tables 1 and 2 (particularly for the rock matrix) were selected by comparison with the internal databank of ANDRA (see for instance ANDRA 2005*a*), and by comparison with other works in the literature, as explained below. The mean value of the Biot coefficient *B* over different layers in the Callovo-Oxfordian (between

Table 1. *Input and output parameters for the upscaling process: lists of input parameters and output coefficients (hydro-mechanical)*

Parameters	Crack (c)	Porous matrix	Equivalent continuum
Hydraulic	$a^{(c)}, n_1^{(c)}$ (Poiseuille)	K_{M}	K_{IJ}^{MC}
Mechanic	$K_N^{(c)}, K_S^{(c)}$	$E_{\mathrm{M}}, \nu_{\mathrm{M}}$	R_{ijkl}^{MC}
Hydro-mechanical	$B^{(c)}=1$ (Terzaghi)	$B_{\mathrm{M}}, M_{\mathrm{M}}$	$B_{ij}^{\mathrm{MC}}, M^{\mathrm{MC}}$

depths 400–500 m roughly) was calculated by Cosenza *et al.* (2002) in their table 1. They obtained the value $B \approx 0.4$ (range: $B = 0.37–0.42$) *and* observed that this value is close to that obtained directly by hydro-mechanical tests in other works, and they quote in particular the tri-axial tests of Coste *et al.* (1999), who obtained $B = 0.36$ *'for the same claystone'*, while other authors like Vincké *et al.* (1997) obtained higher values from $B = 0.4$ up to 0.8, also *'for the same claystone'*. Other data obtained *in-situ* at the Meuse/Haute-Marne URL seem to point to the value $B \approx 0.6$ for the 'intact' rock matrix (although small fissures cannot be avoided on rock matrix samples extracted *in-situ*). For these reasons, we have taken $B_{\mathrm{MATRIX}} \approx 0.5$ for the intact rock matrix in this study.

Concerning the Biot modulus '*M*', based on Cosenza *et al.* (2002), their calculated Biot modulus is about $M \approx 8$ GPa (range: $M = 7.69–8.36$ GPa in their table 1). There are few (if any) other more direct *in-situ* measurements of the Biot modulus of this claystone in the literature. We will take therefore $M \approx 8$ GPa or so, for the Biot modulus of the intact claystone matrix. We have verified that this value is close to that which can be obtained theoretically (from Biot's isotropic theory of granular materials) for the intact rock matrix using reasonable microscale parameters (final value: $M_{\mathrm{M}} = 8.68$ GPa in our Table 3).

Ellipsoidal representation and visualization of upscaled tensorial coefficients. Recall that most of the H-M coefficients are tensorial (except for the scalar Biot modulus '*M*'). For this reason, we use an algebraic interpretation that allows us to represent any positive definite symmetric second-rank tensor, such as K_{ij} and B_{ij}, as a 3D ellipsoid. For example, the tensorial version of Darcy's law (with hydraulic conductivity tensor K_{ij}) can be used to define a directional conductivity along the gradient direction (K_{grad}) and a directional conductivity along the flux direction (K_{flux}); it is then

Table 2. *Input and output parameters for the upscaling process: numerical values of input parameters used for obtaining upscaled hydro-mechanical coefficients*

Radius and length of the stretch of drift	Drift radius: $R = 2$ m, Axial length of drift stretch: $L = 20$ m
Apertures of planar disc fissures (small fractures)	$a_{\mathrm{FISSURES}} \approx 5 \times 10^{-5}$ m near the wall, decreasing rapidly away from wall
Diameter of planar disc fissures (small fractures)	$d_{\mathrm{FISSURES}} \approx 0.8$ m near the wall, decreasing rapidly away from wall
Density of planar disc fissures (small fractures)	$(\rho_{32})_{\mathrm{FISSURES}} \approx 6.11\ \mathrm{m}^2\ \mathrm{m}^{-3}$ near the wall, decreasing rapidly away from the wall
Aperture of curved 'chevron' fractures	$a_{\mathrm{FRACT/CHEVRON}} \approx 1 \times 10^{-4}\ \mathrm{m} = 100\ \mu$ constant
Size of curved 'chevron' fractures	Horizontal extension: 4 m Vertical extension: 4 m
Density of large curved 'chevron' fractures	$(\rho_{1\mathrm{D}})_{\mathrm{FRACT/CHEVRON}} = 2/\mathrm{m}$, periodic along the drift axis
Normal specific stiffness of planar cracks	$K_{\mathrm{N}} = 1 \times 10^{10}\ \mathrm{Pa} = 10$ GPa
Shear specific stiffness of planar cracks	$K_{\mathrm{S}} = 1 \times 10^{9}\ \mathrm{Pa} = 1$ GPa
Rock matrix Young's modulus	$E_{\mathrm{M}} = 5 \times 10^{9}\ \mathrm{Pa} = 5$ GPa
Rock matrix Poisson ratio	$\nu_{\mathrm{M}} = 0.30$
Rock matrix porosity	$\theta_{\mathrm{M}} = 0.14$

Table 3. *Global equivalent coefficients: hydro-mechanical coefficients of the fissured and fractured claystone, upscaled at the global scale of the 4 m thick damaged zone, and isotropized*

Table parameter (global)	Intact matrix	Matrix and fissures	Matrix and fissures and chevron fractures
Young's modulus, E (GPa)	5.0	3.0	1.31
Poisson ratio, ν	0.30	0.34	0.37
Shear modulus, G (GPa)	3.85	2.24	0.96
Biot coefficient, B (iii.a)	0	0.113	0.281
Biot coefficient, B (iii.b)	0.50	0.556	0.641
Biot modulus, M (GPa) (iii.a)	∞	36.8	14.8
Biot modulus, M (GPa) (iii.b)	8.68	7.47	6.25

Hypothesis 'iii.a' ignores H-M coupling within the matrix (coupling is due only to cracks), hypothesis 'iii.b' includes both the coupling role of the porous matrix and of the cracks
Note: 'matrix' values are also shown in this table. The 'matrix' is the intact porous rock; the 'fissures' are the statistical planar disc fissures (i.e. small fractures); the 'chevrons' are the large curved fractures organized in a chevron pattern along the drift axis.

possible to show that the polar plots of $\sqrt{(1/K_{\mathrm{grad}})}$ and of $\sqrt{(K_{\mathrm{flux}})}$ both describe an ellipse in 2D or an ellipsoid in 3D (Bailly 2009: chapter 10; Bailly *et al.* 2011). However, for simplicity, we have chosen here to represent with ellipsoids the matrix quantities A_{ij} (i.e. specifically K_{ij}, S_{ij} or S_{IJ}, B_{ij}, etc.), such that the three principal axes of an ellipsoid indicate the directions of the three eigenvectors of matrix A_{ij}, and the three principal radii of the ellipsoid are equal to the corresponding three eigenvalues of matrix A_{ij} (as in Ababou *et al.* 2011).

Accordingly, at any spatial position (or for any subdomain), the upscaled second-rank tensor B_{ij} (Biot coefficient) is represented directionally by an ellipsoid, just like the upscaled hydraulic conductivity K_{ij} in the work of Ababou *et al.* (2011). Thus, strongly anisotropic B_{ij}s are indicated by strongly elongated and/or flat ellipsoids (strongly prolate like a rugby ball, strongly oblate like a saucer, etc.). The Biot modulus M is plotted as a sphere, whose radius indicates the magnitude of M. Indeed, any scalar quantity like 'M' can be expressed equivalently as an isotropic (spherical) tensor, which can be represented by a sphere.

Finally, the purely mechanical coefficients of compliance (C_{ijkl}) and stiffness (R_{ijkl}) are higher-order tensors (fourth-rank). This makes it more difficult to represent their spatial distribution. The interested reader is referred to other possible representations in the literature, some of them based on Kelvin's decomposition of fourth-rank elastic tensors: Mehrabadi & Cowin (1990); Basser & Pajevic (2007); Pouya (2007, 2011); see also Charlez (1991).

In the present work, we have chosen to reduce the amount of information, and to display only some parts of the C_{ijkl} and R_{ijkl} tensors. For example, the stiffness tensor R_{ijkl} can be described (partially) via two 3×3 matrices – each of which can be represented by ellipsoids:

- 'normal stiffness' matrix N_{IJ} containing stiffness coefficients $R_{1111}, \ldots, R_{3333}, R_{1122}$, etc;
- 'shear stiffness' matrix S_{IJ} containing stiffness coefficients $R_{2323}, \ldots, R_{2312}, R_{1313}$,etc.

Therefore, the upscaled mechanical coefficients, such as 'normal stiffness' N_{IJ} and 'shear stiffness' S_{IJ}, can be represented by ellipsoids. In fact, the shear matrix ellipsoid (S_{IJ}) is easier to interpret, as it reduces to a sphere in the case of isotropic shear stiffnesses (whereas the normal ellipsoid N_{IJ} does not reduce to a sphere in the case of isotropic elasticity). We focus in this paper on the shear submatrix S_{IJ} (see also Table 6 and Appendix 1).

Post-calculation of equivalent 'isotropized' scalar coefficients. In order to remedy some possible interpretation problems with these ellipsoids (particularly for the so-called 'normal' ellipsoids of stiffnesses and compliances), it was decided to extract a set of more 'classical' mechanical coefficients (E, K, λ, μ, ν) from the full fourth-rank compliance and stiffness tensors and similarly for the H-M coupling coefficients (B_{ij}, M), which become (B, M) after *isotropization.* These calculations were performed using adequate contractions of second- and fourth-rank tensors, based on the definition of spherical quantities similar to the definition of a spherical bulk stiffness K in isotropic elasticity. This is summarized in Appendix 1 (a fully detailed algebraic implementation of these concepts for non-isotropic/non-orthotropic fourth-rank elasticity would be beyond the scope of this paper).

The resulting scalar coefficients (E, K, λ, μ, ν, B, M) should be considered as 'isotropized' versions of the upscaled tensorial coefficients (C_{ijkl}, R_{ijkl}). Each scalar can then be represented as a spatial distribution (if upscaled locally), or else it can be computed as a global quantity (if upscaled at the macroscale of the entire damaged/fractured zone). For example, numerical values of global isotropized coefficients (E, K, λ, μ, ν, B, M) is presented in Table 3.

Note: E = Young modulus; K = bulk spherical modulus; (λ, μ) are the Lamé moduli (in particular μ is the shear modulus); ν is the Poisson ratio; and finally (B, M) are the coupling coefficients (Biot coefficient B, and Biot modulus M).

3D upscaling: distribution of H-M coefficients on a cartesian grid (voxels)

In this section, we display graphically the 3D upscaled tensorial coefficients around the drift, as spatially distributed ellipsoids located on a cartesian grid of voxels (x_{IJK}, y_{IJK}, z_{IJK}).

Recall that the upscaled coefficients are generally second- and fourth-rank tensors, to be represented via 3×3 ellipsoids. The total 3D domain of damaged and fractured rock, in our case, is a 20 m stretch of gallery, with transverse size 13 × 13 m (comprising the 4 m diameter drift). The chosen size of the cubic voxels (subdomains) is 0.50 m. For these reasons, displaying the spatial distributions of tensorial coefficients in 3D space over the entire domain can be cumbersome. Several 3D views would be needed in order to clearly see all the 3×3 ellipsoids in (x, y, z) space. Therefore, to save space, we show only the *first transverse layer* of the 3D distribution, that is, the (Y, Z) plane with $X \in [-10.0\text{ m}, -9.5\text{ m}]$. This gives an idea of the 3D heterogeneity of upscaled coefficients inside the damaged/fractured zone around the drift.

The results are displayed in Figure 4, showing the first transverse layer of the 3D upscaled tensors (3×3 ellipsoids) for several types of coefficients (normal compliance N_{ij}, shear compliance S_{IJ}, Biot coefficient B_{ij}, and inverse Biot modulus $1/M$).

2D upscaling: distribution of H-M coefficients on a transverse grid of pixels

In this section, we display graphically the 2D upscaled tensorial coefficients around the drift, as spatially distributed ellipsoids located on a cartesian grid of pixels (y_{JK}, z_{JK}): see Figure 5. Note that, although their spatial distribution is 2D, these tensorial coefficients are still represented by 3×3 ellipsoids, so their '3D anisotropy' can still be seen by looking at the ellipsoids (e.g. ellipsoids of shear compliance S_{ij} and Biot coefficient B_{ij}).

In comparison with the 3D spatial distributions shown earlier, the 2D upscaled coefficients appear less anisotropic, and less variable in space (particularly near the wall). This effect was expected: it is due to the additional smoothing induced by averaging along the axis of the drift (compared with the 3D results which were not axially averaged). The advantage of the 2D transverse representation is that it offers a more synthetic view of the upscaled coefficients around the drift, but at the price of some loss of information (anisotropy and heterogeneity have been significantly smoothed out by axial averaging).

Global upscaling: H-M coefficients on a 4 m thick annular cylinder (EDZ)

The strain superposition upscaling procedure was used to calculate the tensorial coefficients at the global scale (macroscale) of the entire EDZ, that is, the entire damaged zone around the 20 m long stretch of the drift. The 'global' upscaling domain, in this case, is the EDZ, defined as the 20 m long cylindrical annular region of thickness 4 m around the drift (EDZ thickness is about one diameter around the drift, and the drift radius, R, is 2 m).

Once the global tensors have been calculated by the tensorial superposition method (R_{ijkl}, C_{ijkl}, B_{ij}, M), they are processed algebraically to obtain 'equivalent isotropic' scalar quantities (E, K, λ, μ, ν, B, M), as explained in Appendix 1. These final results, summarized in Table 3, provide a simplified synthetic view of the global properties of the fissured and fractured claystone around the drift. In particular, consider the Biot modulus (M). Note that 'M' can be interpreted as the *stiffness of the coupling* between pressure variations and fluid production. It can be seen, as expected, that the global value of the coupling stiffness 'M' is reduced as more cracks are 'added':

$$M_{\mathrm{M}} = 8.68\,\mathrm{GPa} > M_{\mathrm{M+Fiss}} = 7.47\,\mathrm{GPa} > M_{\mathrm{M+Fiss+FractChevr}} = 6.25\,\mathrm{GPa}$$

Annular upscaling: H-M coefficients on annular domains around the drift: 'near-wall' v. 'far-wall' values, and comparisons with global values

In this section, we present results obtained by upscaling the H-M coefficients over *annular shells*, rather than voxels or pixels. For simplicity, we present here not the full tensorial coefficients, but their equivalent isotropized values (as explained

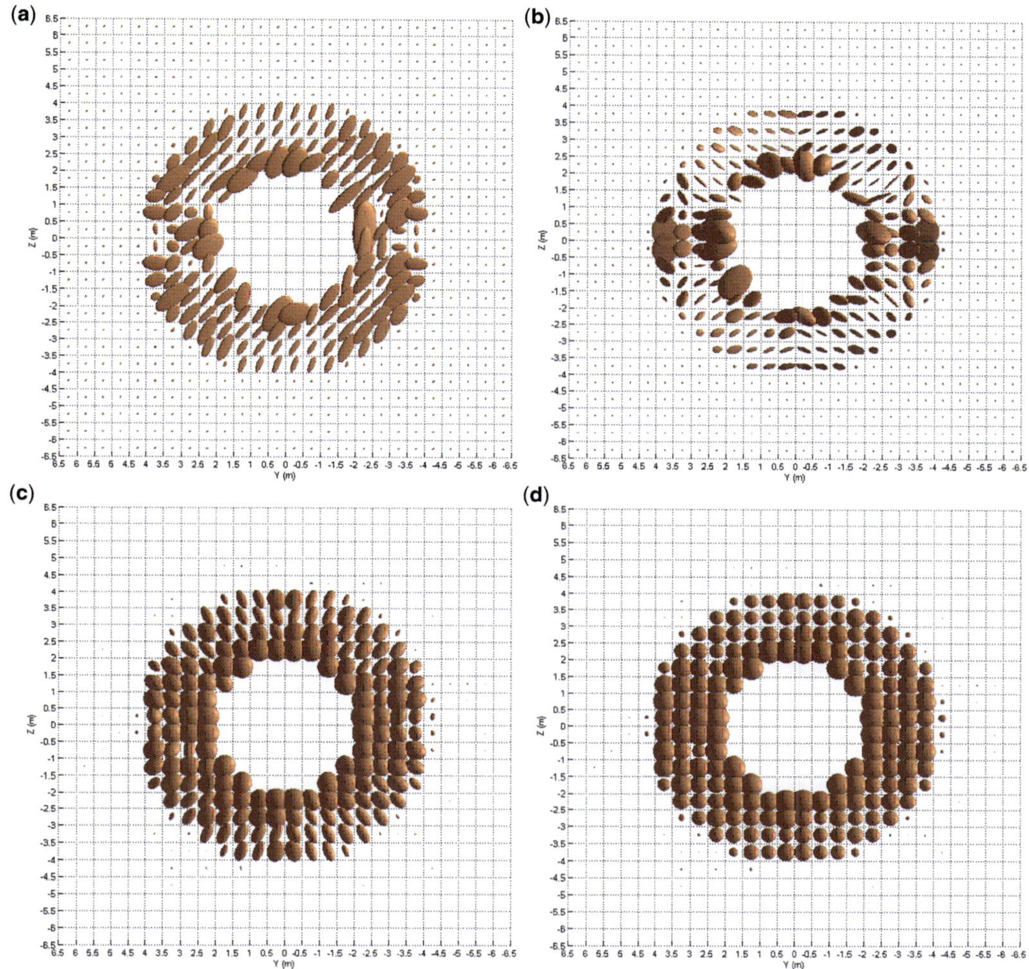

Fig. 4. Frontal views of the 3D upscaled coefficients: (**a**) normal compliance N_{ij}; (**b**) shear compliance S_{ij}; (**c**) Biot coefficient $B_{ij}^{(0)}$ (ellipsoids); and (**d**) inverse of the Biot modulus $1/M^{(0)}$ (spheres) in the first transverse layer (Y, Z) with $X \in [-10.0\ \text{m}, -9.5\ \text{m}]$. These upscaling calculations take into account the porous rock matrix (except for its coupling effects on B_{ij} and M), including the statistical set of fissures (small fractures), and the periodic set of large curved fractures (chevron pattern). For the coupling coefficients $B_{ij}^{(0)}$ and $M^{(0)}$, in this figure, hypothesis (iii.a) was used; superscript (0) indicates that the coefficient is calculated *relative* to the sole coupling effects of cracks (fissures and fractures).

in Appendix 1). The resulting scalar values represent coefficients upscaled on annular shells of various thicknesses.

For example, using this technique, we obtain $B \approx 0.90$ near the wall, that is, over a 0.5 m thick annular shell adjacent to the wall ($r \in [2.0\ \text{m}, 2.5\ \text{m}]$). This value $B \approx 0.90$ should be compared with the global value $B = 0.641$ given earlier at the global scale of the entire 4 m thick and 20 m long stretch of damaged/fractured rock around the drift. It should also be compared with the value $B = B_M = 0.50$ assumed for the intact rock matrix (Table 3).

We show a few more results along these lines in Tables 4–6. The tables describe the radial distributions of several mechanical and hydro-mechanical properties of the damaged claystone. These properties are obtained in two steps: (a) by upscaling the tensorial coefficients on annular cylindrical shells; and (b) by evaluating the resulting *equivalent isotropic* coefficients, at each radial distance from the centre of the drift.

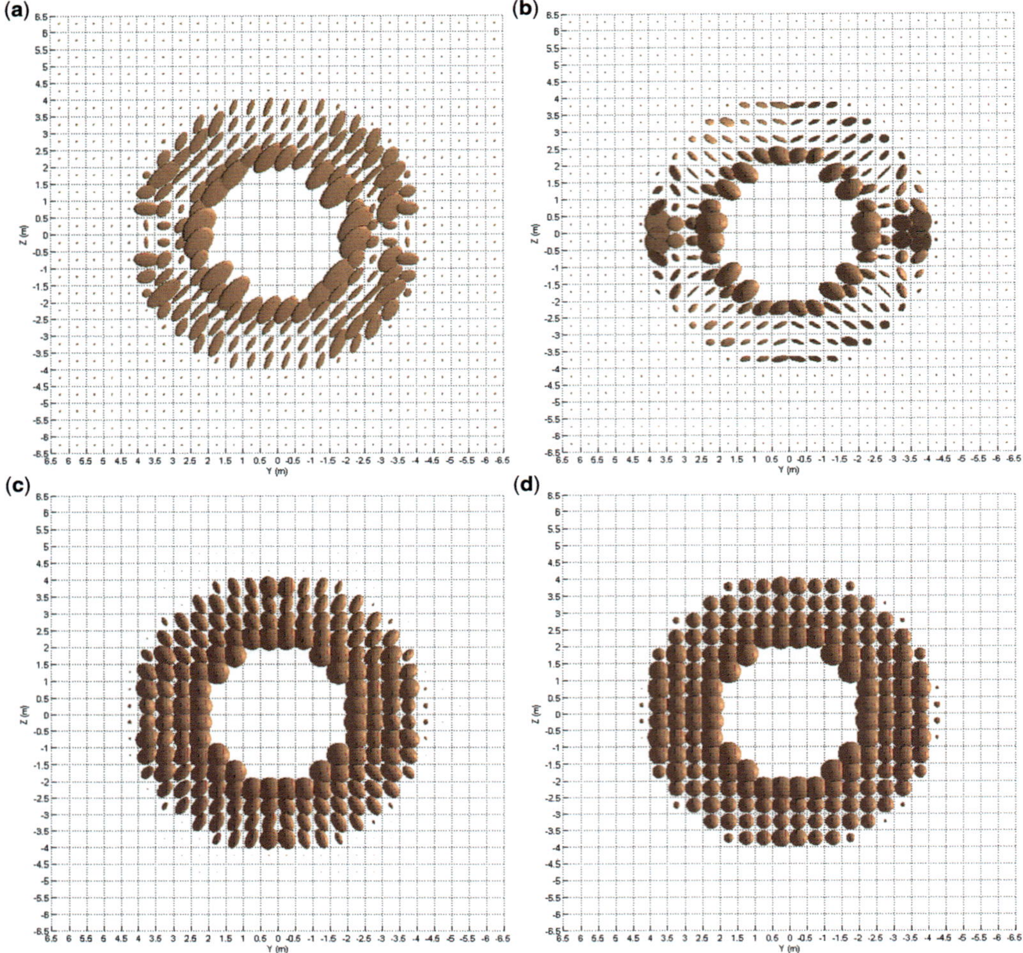

Fig. 5. Transverse 2D distributions (averaged along the X axis) of upscaled coefficients: (**a**) normal compliance N_{ij}; (**b**) shear compliance S_{ij}; (**c**) Biot coefficient $B_{ij}^{(0)}$ (ellipsoids); and (**d**) inverse of the Biot modulus $1/M^{(0)}$, taking into account the porous matrix, the small fractures (fissures) and the large chevron fractures. For the coupling coefficients B_{ij} and M, in this figure, hypothesis (iii.a) was used; superscript (0) indicates values *relative* to the sole coupling effects of cracks (fissures and fractures).

However, we also show the values of a tensorial coefficient (the shear compliance submatrix S_{IJ}) in order to examine its degree of anisotropy v. radial distance: looking at the diagonal values S_{II} (II = 44,55,66), it can be seen that the claystone is not only more compliant but also more anisotropic in the near-wall region (say within 0.5–1.0 m from the wall). On the other hand, the apparent anisotropy of S_{IJ} is not very strong, most probably because of the annular averaging process (see discussion in Section 3.4). The important thing is that shear compliance decreases by a factor 15 or so from near-wall to 4.5 m away from the wall. We will return to a brief discussion on the results of Tables 4–6 in the next section.

Conclusions, discussion and outlook

Conclusions and discussion

The results obtained so far for the upscaled hydromechanical properties of the damaged, fissured and fractured claystone around a gallery at the Meuse/Haute-Marne URL site are now summarized and discussed. (Recall that the upscaling considers the porous claystone matrix with embedded cracks

Table 4. *Radial distributions of mechanical and hydro-mechanical equivalent continuum coefficients upscaled on annular shells around the drift (the drift wall is located at R = 2.0 m from the drift centre): hydro-mechanical coupling coefficients v. radial distance R (for annular shells with R comprised between R1 and R2)*

R1 m	R2 m	*B* (Biot coefficient)	*M* (Biot modulus, GPa)
2.0	2.5	0.8717	6.5015
2.5	3.0	0.7812	6.9262
3.0	3.5	0.7696	6.9818
3.5	4.0	0.7519	7.0718
4.0	4.5	0.5219	8.5102
4.5	5.0	0.500	8.6778
5.0	5.5	0.500	8.6778
5.5	6.0	0.500	8.6778

comprising a set of statistical planar disc fissures, and a set of large curved 'chevron' fractures discretized into triangular patches.)

Global v. local rock properties. At the *global scale* of the damaged zone (4 m thick annular cylinder), the Young's and shear *moduli* are *reduced by a factor 4* compared with the intact rock matrix. The *global* Poisson ratio is not much changed with respect to the matrix value: it is only slightly increased, from 0.30 to <0.40.

On the other hand, at the *local scale*, the stiffness moduli can be *reduced by one order of magnitude* or more in the near-wall region, as shown for instance by comparing annularly upscaled values ($r \approx 2.0$ m–2.5 m) with global values.

Furthermore, looking at the '3D' coefficients, locally upscaled on a grid of cubic voxels (Figure 4 ellipsoids), it is clear that material compliances like the shear compliance matrix S_{IJ}, and the inverse Biot modulus $1/M$, increase significantly near the gallery wall.

This is confirmed by looking at Tables 4–6, where it is also seen that material *anisotropy* also increases significantly near the gallery wall (see the values of S_{II} in Table 6, and see also the '3D' ellipsoids of S_{IJ} and B_{ij} in Figure 4.

Hydro-mechanical coupling coefficients B_{ij} *and* M *(upscaled results).* The Biot coefficient B and the Biot modulus M are calculated first by taking into account the sole effect of the cracks (hypothesis 'iii.a'), and then by taking into account the complete coupling effects owing to porous matrix as well as cracks (hypothesis 'iii.b').

Looking at the *global* values shown in Table 3 under the more general hypothesis 'iii.b', and comparing these values with those of the intact rock matrix, it is clear that the upscaled Biot coefficient (B) *increases with damage* (while staying below unity, as it should), and that the Biot modulus (M) *decreases with damage.* These results are as expected physically, since damaging and fracturing should weaken the rock.

Extent of the hydro-mechanically damaged zone (based on upscaling results). It is seen here that, for all H-M coefficients (except the permeability, examined in Ababou *et al.* 2011), the 'far-wall' values obtained beyond $R = 4.5$ m coincide with the intact rock matrix properties. Since the radius of the gallery is $R = 2.0$ m, it can be concluded that, according to these upscaling calculations, the hydro-mechanically damaged rock lies in an annular region of thickness 2.5 m around the gallery wall.

Remarks on full tensorial coefficients v. simplified isotropic equivalents. From a theoretical point of view, the proposed upscaling method yields generally non-isotropic/non-orthotropic *tensorial*

Table 5. *Radial distributions of mechanical and hydro-mechanical equivalent continuum coefficients upscaled on annular shells around the drift (the drift wall is located at R = 2.0 m from the drift centre): mechanical coefficients (obtained from* C_{ijkl}*) v. radial distance R*

R1 m	R2 m	*E* (Young's modulus) (GPa)	μ (shear modulus) (GPa)	ν (Poisson's ratio)
2.0	2.5	0.42694	0.14971	0.4259
2.5	3.0	0.85225	0.30159	0.4129
3.0	3.5	0.69583	0.24379	0.4271
3.5	4.0	0.65671	0.22932	0.4319
4.0	4.5	3.8936	1.4592	0.3342
4.5	5.0	5.0000	1.9231	0.3000
5.0	5.5	5.0000	1.9231	0.3000
5.5	6.0	5.0000	1.9231	0.3000

Table 6. *Radial distributions of mechanical and hydro-mechanical equivalent continuum coefficients upscaled on annular shells around the drift (the drift wall is located at R = 2.0 m from the drift centre): shear compliance tensor S_{ij} (principal components) v. radial distance R*

R1 m	R2 m	S_{44} (Pa^{-1})	S_{55} (Pa^{-1})	S_{66} (Pa^{-1})
2.0	2.5	1.8859×10^{-9}	2.1661×10^{-9}	1.5341×10^{-9}
2.5	3.0	9.0955×10^{-10}	1.2262×10^{-9}	7.0979×10^{-10}
3.0	3.5	7.4276×10^{-10}	1.0856×10^{-9}	7.8636×10^{-10}
3.5	4.0	7.0106×10^{-10}	9.1224×10^{-10}	8.0620×10^{-10}
4.0	4.5	1.5526×10^{-10}	1.5485×10^{-10}	1.5815×10^{-10}
4.5	5.0	1.3000×10^{-10}	1.3000×10^{-10}	1.3000×10^{-10}
5.0	5.5	1.3000×10^{-10}	1.3000×10^{-10}	1.3000×10^{-10}
5.5	6.0	1.3000×10^{-10}	1.3000×10^{-10}	1.3000×10^{-10}

stress–strain–pressure relations, represented by tensorial coefficients of rank up to 4.

As explained earlier, the mechanical tensors of rank 4 (R_{ijkl}, C_{ijkl}) were reduced to 3×3 'normal' and 'shear' submatrices representing the upscaled stiffness and compliance tensors of rank 4. These submatrices were then visualized as spatially distributed ellipsoids on the top parts of Figures 4 and 5 (admittedly, shear compliance ellipsoids are easier to interpret than normal compliances). For second-rank tensors, their representation via ellipsoids is more direct and straightforward. The symmetric tensor B_{ij} is shown as an ellipsoid (like the permeability K_{ij} in other works), and the scalar M is shown as a sphere representing the spherical tensor $M\delta_{ij}$: see the bottom parts of Figures 4 and 5.

Finally, recall that the *scalar values* of upscaled coefficients, like those given in Tables 4 and 5, were obtained by 'post-processing' the *tensorial* stress–strain–pressure relations with their second- and fourth-rank tensorial coefficients. Indeed, we have exploited an algebraic theoretical formulation, which has allowed us to define consistently the *equivalent isotropic coefficients* for these laws. For a second-rank tensor like B_{ij}, one way to obtain the isotropic equivalent is to take the spherical part: $B = \mathrm{Tr}(B)/3 = B_{kk}/3$ (however the complete theory is more involved and will not be detailed here).

Remarks on the validity of results. The equivalent continuum H-M coefficients obtained in this work depend on two sets of assumptions and approximations: (a) assumptions on the geometric structure and local properties of the fractured rock (cracks and porous matrix); and (b) approximations and hypotheses linked to the method of upscaling for the matrix–crack medium (here the superposition method).

First, let us recall that the 'geometric model' used for generating the structure of the fractured rock around the gallery, was defined in a previous paper (Ababou *et al.* 2011) where hydraulic upscaling was studied for the same site and the same gallery. In particular, let us point out the following:

- The fracture system in the EDZ was composed of two subsets – the periodic set of large curved chevron fractures, and a statistical set of smaller planar fractures.
- The curved chevron fractures were modelled deterministically based on direct observations of large fracture traces on different planes (exploration ditches); a 3D parametric surface model was then adjusted (a generalized conoid having no axial symmetry).
- The probabilistic/geometric structure of the statistical set was validated based on comparisons of upscaled v. borehole permeability profiles (*in-situ*). This served in particular to adjust the radial inhomogeneity of statistical fracturing in the near-wall region.

In this paper, we assume that the geometric structure is reasonably well established from these previous tests (Ababou *et al.* 2011). The assumed geometric structure of the fractured rock in the EDZ was partially validated by hydraulic measurements and by observation of large fracture traces in that work. That is why we have conserved the same sets of fractures in the present study of hydro-mechanical upscaling.

Second, the superposition method of upscaling does not deliver exact results except in special cases. However, it is fast to implement, it is flexible in terms of the choice of averaging support, and it is able to convey the *anisotropic* and *coupled* hydro-mechanical response of a porous fractured rock. Other upscaling methods may not be as flexible and fast, and they too suffer from defects, approximations and limited range of applicability (as discussed briefly in previous sections).

On the other hand, as shown from the previous discussion of results, our equivalent continuum H-M coefficients match quantitatively some of the available observations, and they behave qualitatively as expected with the degree of fracturing.

Outlook

In future, we will present the mathematical upscaling theory which has led us to the above-discussed results. A brief description of the theoretical upscaling method based on strain superposition was provided earlier, along with appropriate references. The *upscaling hypotheses* were clearly listed, and it was noted that the current results on coupled hydro-mechanics were obtained under the most general hypothesis (i.e. under hypothesis 'iii.b' rather than 'iii.a'; see Section 3.2 and results of Table 3).

Note that we have obtained $B \approx 0.90$ for the 'near-wall' value of the Biot coefficient (upscaled over a 0.5 m thick annular shell adjacent to the wall). This value should be compared with the global value $B = 0.641$ for the entire 4 m thick global domain. More generally, the reader can compare the radial distributions of stiffnesses and coupling coefficients shown in Tables 4–6 with the global values shown in Table 3.

We plan to complete these results and analyses (*upscaled H-M coefficients of the damaged and fractured clay stone*) by implementing the upscaling method on a variety of other '*spatial supports*', for example, smaller size voxels in 3D and annular shells of smaller thicknesses. In addition, we plan to obtain continuously distributed upscaled parameters using *moving windows* (rather than fixed subdomains or partitions): see for instance the upscaled radial permeability profiles $K(r)$ obtained by Ababou *et al.* (2011). Similarly, it is expected that the current hydro-mechanical upscaling will allow us to obtain properties such as K (bulk stiffness) and B (spherical Biot coefficient) along 'numerical boreholes'.

Other *extensions* of our upscaling method are currently being considered for hydro-mechanical processes. The method, based on superposition of strains and fluxes, can be extended to account for: (a) weakly non-elastic/non-reversible deformation behaviour of the discrete cracks (a brief review of such effects can be found in Oda (1986) and references therein); and (b) retro-active 'feedback' effects of stress and strain on the upscaled properties themselves, such as hydraulic conductivity K_{ij} (which is very sensitive to crack apertures), but also other hydro-mechanical properties (compliances C_{ijkl}, and coupling coefficients B_{ij} and M). Our current theoretical developments and extensions indicate that the Biot modulus 'M' is indeed most sensitive to crack apertures and normal stiffnesses. Therefore it is expected that 'M' is quite sensitive to mechanical feedbacks owing to aperture variations. Note: the tensorial feedback effect of strain–stress on permeability k_{ij} was previously described theoretically in Ababou *et al.* (1994*b*) and Stietel *et al.* (1996), for 'Poiseuille cracks'.

Finally, we are continuing the extension of our theoretical frame for obtaining equivalent continuum H-M laws and coefficients for more general types of *dual porosity media*, including not only the current case of a fractured porous claystone comprising small statistical cracks as well as large curved fracture surfaces, but also other types of heterogeneous geological media as well, for example, with cavities and holes rather than thin cracks.

The first two authors wish to acknowledge financial support by Andra.

Appendix 1

Equivalent isotropic coefficients from fourth-rank stiffness R_{ijkl}

The equivalent continuum (upscaled) material coefficients obtained in this work are *not* assumed to be orthotropic at voxel scales (see discussion in Section 3.4). However, on large transverse scales of averaging (e.g. annular averaging), we have derived equivalent scalar quantities from the upscaled tensorial coefficients. These scalars are obtained from spherical equivalent tensors, through a procedure we have called '*isotropization*'. This is briefly explained in this Appendix. Let us first review the list of tensorial continuum equivalent coefficients in this paper:

- B_{ij} – *stress–pressure coupling Biot coefficient (second-rank tensor)*. In this case, the 'isotropized' value B is obtained by taking the spherical part of B_{ij}, that is: $B = (B_{kk})/3 = \mathrm{Tr}(\underline{\underline{B}})/3$ where $\mathrm{Tr}(\underline{\underline{B}})$ is the trace of tensor $\underline{\underline{B}}$.
- M – *fluid production–pressure coupling (scalar Biot modulus)*. This coefficient is by essence a scalar, that is, a zero-order tensor (it does not require 'isotropization').
- R_{ijkl}, C_{ijkl} – *stiffness and compliance coefficients (fourth-rank tensors)*. These upscaled coefficients are generally non-isotropic and non-orthotropic tensors, although they do have all the other symmetries of elasticity (this is one of the consequences of the linear superposition method).

We expect the upscaled elastic tensors R_{ijkl} and C_{ijkl} to become isotropic in the case of geometrically isotropic materials. We have verified that this is indeed the case via numerical tests with statistically isotropic and homogeneous fracturing (*not shown here*). In the isotropic case, the classical scalar coefficients of isotropic elasticity (E, ν, μ, etc.) can be directly extracted from the fourth-rank tensors (R_{ijkl} or C_{ijkl}). Our goal is to obtain similar scalar coefficients for the case of non-isotropic/non-orthotropic upscaled materials.

To simplify the analysis of upscaled tensorial coefficients, we can extract algebraic subsets from the fourth-rank stiffness and compliance tensors (R_{ijkl}, C_{ijkl}), as explained in the text (section 'Inputs, outputs and post-treatment of hydro-mechanical upscaling'). Let us focus in particular on 'shear stiffness' (Pa) or 'shear compliance' (Pa^{-1}) submatrix S_{IJ}, which contains the nine stiffness coefficients $R_{2323}, \ldots, R_{2312}, R_{1313}$ (similarly for compliance). Using Kelvin's well-known 6×6 index notation for R_{ijkl}, this 3×3×3×3 tensor can be expressed as a 6×6 matrix R_{IJ} with indices $I \in \{1, \ldots, 6\}$, $J \in \{1, \ldots, 6\}$, where $I = \{1, 2, 3, \text{and } 4, 5, 6\}$ corresponds to $(i, j) = \{(1, 1), (2, 2), (3, 3) \text{ and } (2, 3), (1, 3), (1, 2)\}$. The (3×3) shear submatrix S_{IJ} (analysed in Table 6) corresponds to the symmetric lower right part of the (6×6) R_{IJ} matrix. Note that the R_{IJ} matrix, with Kelvin's notation, does *not* represent a (6×6) second-rank tensor. For the same reason, S_{IJ} is a matrix rather than a tensor, *a priori*. Nevertheless, S_{IJ} being a symmetric (3×3) matrix which also appears to be positive-definite (or at least positive), it can be diagonalized, and then, it can be represented by an ellipsoid – which reduces to a sphere in the case of isotropic shear behaviour. Finally, a few words about notation: '*S*' stands for 'shear'; the diagonal components of S_{IJ} can be labelled S_{44}, S_{55}, S_{66} with Kelvin's (*I*, *J*) notation, but they could be labelled S_{11}, S_{22}, S_{33} if we revert to the standard notation S_{ij} for a 3×3 matrix with $i = \{1, 2, 3\}$ and $j = \{1, 2, 3\}$. Note also that S_{IJ} represents shear stiffness when it is extracted from R_{ijkl} (Pa), and shear compliance when it is extracted from C_{ijkl} (Pa^{-1}).

We now briefly describe how equivalent isotropic coefficients are defined from fourth-rank mechanical tensors, leading to bulk and shear moduli, Young's coefficient (*E*) and Poisson's ratio (ν):

Step 1. We define a scalar *bulk stiffness modulus* (*K*) for a non-isotropic/non-orthotropic material, in terms of the fourth-rank stiffness tensor R_{ijkl}, as follows:

$$3K = R_{ppll}/3 = (R_{1111} + R_{2211} + R_{3311} + R_{1122} + \cdots R_{3322} + R_{1133} + \cdots + R_{3333})/3 \quad (A.1)$$

Step 2. One can define as follows an equivalent tensor of volumetric stiffness (K_{ijmn}):

$$\begin{aligned} 3K_{ijmn} &= 3K\delta_{mi}\delta_{nj} = (R_{ppll}/3)\delta_{mi}\delta_{nj} \\ &= (R_{ppll}/3)(\delta_{mi}\delta_{nj} + \delta_{mi}\delta_{nj})/2 \end{aligned} \quad (A.2)$$

The last equality is obtained by exploiting the symmetries of elastic mechanics (these symmetries are conserved by our upscaling: they are applicable to macroscale coefficients).

Step 3. We interpret this relation in terms of the known elastic coefficients that characterize an (equivalent) isotropic material. We do this by inserting, in the K_{ijmn} and R_{ijkl} tensors, their known expressions in terms of Lamé moduli (λ, μ) for an isotropic material. This yields:

$$\begin{aligned} 3K_{ijmn} &= 3K(\delta_{mi}\delta_{nj} + \delta_{mj}\delta_{ni})/2 \\ &= (3\lambda + 2\mu)(\delta_{mi}\delta_{nj} + \delta_{mj}\delta_{ni})/2 \end{aligned} \quad (A.3)$$

where

$$K = (3\lambda + 2\,\mu)/3 \quad (A.4)$$

This macroscale result is as expected, and it is also consistent with the scalar *K* that we directly introduced in the first step above ($3K = R_{ppll}/3$).

Step 4. Continuing the previous comparison process with the case of isotropic materials, we introduce the Young modulus, *E*, and the Poisson ratio, ν, and we keep the shear modulus, μ, to describe the equivalent isotropic material. Thus:

$$\begin{aligned} &3K = (3\gamma + 2\mu);\ E = 3\ (1 - 2\nu); \\ &E = 2\mu(1 + \nu) \end{aligned} \quad (A.5)$$

Step 5. Defining for isotropic materials the shear stiffness modulus as $G = 2\mu$ (instead of the more usual $G = \mu$), we extend this notion to a shear stiffness tensor G_{ijmn} for the case of a non-isotropic/non-orthotropic material, again in terms of the fourth-rank stiffness tensor R_{ijkl}:

$$G_{ijmn} = R_{ijmn} - R_{ppmn}\delta_{ij}/3 \quad (A.6a)$$

Step 6. For isotropic materials, the shear stiffness tensor Gijmn reduces to:

$$G_{ijmn} = 2\mu(\delta_{mi}\delta_{nj} + \delta_{mj}\delta_{ni})/2 \quad (A.6b)$$

For verification, one can show that the scalar shear stiffness $G = 2\mu$ is indeed obtained upon introducing the isotropic version of the G_{ijmn} tensor (as given just above in equation A.6b) into the general fourth-rank tensor R_{ijkl}, and then assuming isotropy of the fourth-rank R_{ijkl}.

Step 7. Further contractions of the non-isotropic Gijmn of equation (A.6a) can lead to the equivalent isotropic form of equation (A.6b), and this eventually defines the scalar equivalent *G* or μ.

In summary, when dealing with fully non-orthotropic materials, we can use contracted forms of the fourth-rank stiffness tensor R_{ijkl} to define isotropic scalar equivalents of the classical coefficients of isotropic elasticity (*K*, *E*, ν, μ). We proceed similarly for the H-M coupling coefficients: thus, the spherical part of the second-rank Biot tensor B_{ij} yields the scalar *B*. We call this procedure '*isotropization*' (isotropic equivalents for fourth- and second-rank tensors).

References

Ababou, R., Millard, A., Treille, E., Durin, M. & Plas, F. 1994*a*. Continuum modeling of coupled thermo-hydro-mechanical processes in fractured rock. *In*: Peters, A. et al. (eds) *Computational Methods in Water Resources*. Kluwer Academic, Dordrecht, **1**, 651–658.

Ababou, R., Millard, A., Treille, E. & Durin, M. 1994*b*. *Coupled thermo-hydro-mechanical modeling for the near field benchmark test 3 (BMT3) of DECOVALEX phase 2 – progress report*. Rapport DMT/93/488, Commissariat à l'Energie Atomique, Saclay, France, 17 May.

Ababou, R., Cañamón, I. & Poutrel, A. 2011. Macro-permeability distribution and anisotropy in a 3D fissured and fractured clay rock: 'excavation damaged zone' around a cylindrical drift in Callovo-Oxfordian argilite (Bure). *In*: *Special issue on 'Clays in Natural and Engineered Barriers for Radioactive Waste Confinement' (4th International Meeting, CLAYS 2010*, Nantes, 29 March to 1st April 2010). *Journal of Physical and Chemistry of the Earth*, **36**, 1932–1948. http://dx.doi.org/10.1016/j.pce.2011.07.032

Ababou, R., Cañamón, I. & Poutrel, A. 2012. 3D Hydro-mechanical homogenization and equivalent continuum properties of a fractured porous claystone around a gallery: application to the damaged and fractured zone at the Meuse/Haute-Marne Underground Research Laboratory. *In*: *Proceedings of the International Conference CLAY 2012: Clays in Natural and Engineered Barriers for Radioactive Waste Confinement*, 22–25 October 2012, Montpellier, France, Poster Session 'Geomechanics/Numerical Modeling': poster Gm/NM/19, extended abstract.

ANDRA 2005*a*. *Dossier Andra 2005 (collective publication)/Rapport Technique ANDRA CRPADS040022 (collectif, 15 déc. 2005)*. Dossier 2005 – Référentiel du Site Meuse/Haute-Marne – Tome 2, **7**, chap. 32.

ANDRA 2005*b*. *Évaluation de la faisabilité du stockage géologique en formation argileuse. Dossier 2005 Argile. Synthèse: évaluation de la faisabilité du stockage géologique en formation argileuse*. ANDRA, December.

Armand, G. & ANDRA 2007. Analyse des perméabilités mesurées autour des ouvrages du LSMHM au niveau − 490 m pour déterminer des lois empiriques utilisables dans des calculs hydromécaniques couplés en milieu continu (Laboratoire de Recherche Souterrain de Meuse/Haute-Marne). Note Technique ANDRA D.NT.ALS.07.0453 A, 24 May.

Bailly, D. 2009. Vers une modélisation aux grandes échelles des écoulements dans les milieux très fissurés de type karst: étude morphologique, hydraulique et changement d'échelle. Doctoral thesis, Institut National Polytechnique de Toulouse, Université de Toulouse, France, 24 June, http://ethesis.inp-toulouse.fr/archive/00000975/

Bailly, D., Ababou, R. & Quintard, M. 2011. Macro-permeability K of highly fissured porous media: inertial and percolation effects. *In*: *EGU 2011: General Assembly 2011 of the European Geosciences Union (Session HS8.2.1: Stochastic Groundwater Hydrology)*, Vienna, Austria, 3–8 April 2011, poster and abstract. Geophysical Research Abstracts, **13**, EGU 2011-4718.

Basser, P. J. & Pajevic, S. 2007. Spectral decomposition of a 4th-order covariance tensor: applications to diffusion tensor MRI. *Signal Processing*, **87**, 220–236.

Biot, M. A. 1941. General theory of three-dimensional consolidation. *Journal of Applied Physics*, **12**, 155–164.

Biot, M. A. 1956. General solutions of the equations of elasticity and consolidation for a porous material. *Journal of Applied Mechanics*, **23**, 91–96.

Cañamón, I. 2009. *Coupled Phenomena in 3D Fractured Media Analysis and Modeling of Thermo-Hydro-Mechanical Processes)*. VDM Verlag.

Cañamón, I., Elorza, F. J. & Ababou, R. 2006. 3D fracture networks: optimal identification and reconstruction. *In*: *Proceedings of IAMG'06: International Association of Mathematical Geology, XIth International Congress*, University of Liège, Belgium, 3–8 September.

Cañamón, I., Ababou, R. & Elorza, F. J. 2007. A 3-dimensional homogenized model of coupled thermo-hydro-mechanics for nuclear waste disposal in geologic media. *In*: *Proceedings of ENC 2007 (European Nuclear Conference 2007): 'Nuclear Waste Modeling'*, Brussels, 16–20 September, http://www.euronuclear.org/events/enc/enc2007/home.htm.

Cañamón, I., Ababou, R. & Elorza, F. J. 2009. Numerical modeling and upscaling of a 3D fractured rock with Thermo-Hydro-Mechanical coupling. *In*: Amaziane, B. et al. (eds) *Proceedings of MAMERN09, 3rd International Conference on Approximation Methods and Numerical Modelling in Environment and Natural Resources*, Pau, France, 8–11 June, Editorial Universidad de Granada.

Charlez, Ph. A. 1991. *Rock Mechanics, Volume 1, Theoretical Fundamentals*. Editions Technip, Paris.

Cosenza, P., Ghoreychi, M., De Marsily, G., Vasseur, G. & Violette, S. 2002. Theoretical prediction of poroelastic properties of argillaceous rocks from in situ specific storage coefficient. *Water Resources Research*, **38**, 1207, http://dx.doi.org/10.1029/2001WR001201

Coste, F., Bounenni, A. & Chanchole, S. 1999. *Evolution des propriétés hydrauliques et hydromécaniques lors de l'endommagement des argilites de l'Est. Paper at Journées Scientifiques 'Etude de l'Est du Bassin Parisien'*. ANDRA and CNRS, Nancy, France, December.

Coussy, O. 1991. *Mécanique des Milieux Poreux*. Editions Technip, Paris.

Darcy, H. P. G. 1856. *Les Fontaines Publiques de la Ville de Dijon, Exposition et Application des Principes à Suivre et des Formules à Employer dans les Questions de Distribution d'Eau*. Victor Dalmont, Paris, Appendix D.

Mehrabadi, M. M. & Cowin, S. C. 1990. Eigentensors of linear anisotropic elastic-materials. *Quarterly Journal of Mechanics and Applied Mathematics*, **43**, 15–41.

Mourzenko, V. V., Bogdanov, I. I., Thovert, J. F. & Adler, P. M. 2010. Three-dimensional numerical simulation of single-phase transient compressible flows and well-tests in fractured formations. *Mathematics and Computers in Simulation*, **81**, 2270–2281.

Oda, M. 1986. An equivalent continuum model for coupled stress and fluid flow analysis in fractured rock masses. *Water Resources Research*, **22**, 1845–1856.

Oda, M. & Hatsuyama, Y. 1985. Permeability tensor for jointed rock masses. *In*: *Proceedings of International Symposium on Fundamentals of Rock Joints*, Bjorkliden, 15–20 September, 303–312.

Pouya, A. 2007. Ellipsoidal anisotropies in linear elasticity – extension of Saint Venant's work to phenomenological modelling of materials. *International Journal of Damage Mechanics*, **16**, 95–126.

Pouya, A. 2011. Ellipsoidal anisotropy in linear elasticity: approximation models and analytical solutions. *International Journal of Solids and Structures*, **48**, 2245–2254.

Sævik, P. N., Berre, I., Jakobsen, M. & Lien, M. 2013. A 3D computational study of effective medium methods applied to fractured media, *Transport in Porous Media*, **100**, 115–142.

Stietel, A., Millard, A., Treille, E., Vuillod, E., Thoraval, A. & Ababou, R. 1996. Continuum representation of coupled hydro-mechanical processes of fractured media: homogenisation and parameter identification. *In*: Stephansson, O., Jing, L. & Tsang, C-F. (eds) *Coupled Thermo-Hydro-Mech. Proceedings (Decovalex International Project)*. Developments in Geotechnical Engineering. Elsevier, Amsterdam, **79**, 135–164.

Terzaghi, V. K. 1936. The Shearing Resistance of Saturated Soils and the Angle between the Planes of Shear. *In*: *First International Conference on Soil Mechanics*. Harvard University Press, Cambridge, MA, **1**, 54–56.

Vincké, O., Longuemare, P., Boutéca, M. & Deflandre, J. P. 1997. *Etude du comportement poromécanique d'argilites dans les domaines élastiques et post-élastiques. Paper presented at Journées Scientifiques 'Etude de l'Est du Bassin Parisien*. ANDRA and CNRS, Bar-le-Duc, France, October.

Numerical modelling of moisture controlled laboratory swelling/shrinkage experiments on argillaceous rocks

W. J. XU[1,2*], H. SHAO[1], J. HESSER[1] & O. KOLDITZ[2,3]

[1]*Federal Institute for Geosciences and Natural Resorces, 30655 Hanover, Germany*

[2]*Technical University Dresden, 01062 Dresden, Germany*

[3]*Helmholtz Center for Environmental Research, 04318 Leipzig, Germany*

**Corresponding author (e-mail: wenjiexu84@googlemail.com)*

Abstract: Moisture-induced swelling and shrinkage deformation is a well-known phenomenon for clayey material. To quantify this deformation a fully coupled thermal-hydro-mechanical (TH2M) model based on the modified effective stress concept has been developed and implemented in the numerical code OGS. Three laboratory tests have been analysed using this concept. All specimens for the tests are argillaceous rocks, from different underground laboratory sites (Tournemire, Bure, France, and Mont Terri, Switzerland). The laboratory tests are performed under different conditions; the first test is free swelling/shrinkage and other two tests are with uni-axial loading. The relative humidity is also controlled. Only one test is carried out under varied temperature conditions. In this study the anisotropic swelling/shrinkage deformation wais simulated and analysed using this modified effective stress concept.

Moisture-induced swelling/shrinkage is a well-known and important issue for the final disposal of nuclear waste in argillaceous formations. After excavation of an underground repository, the saturated rock is contacted with the relatively dry air. Owing to the de-saturation process, shrinkage takes place and, as a result, fractures may occur. During the post-closure phase of a repository, the near field of the tunnel will be re-saturated with the pore water and, as a result of swelling, fractures will be closed. The opening and closing of fractures owing to shrinkage and swelling have a significant influence on the hydro-mechanical behaviour of argillaceous rocks and the sealing function of a repository.

Laboratory experiments have been carried out to investigate the swelling/shrinkage mechanism of argillaceous rocks. Different models have been developed to understand this coupled hydro-mechanical process. The most common way to simulate the swelling/shrinkage deformation is introducing an additional strain component, which depends on the saturation state. This model works well for an isotropic material, such as betonite. However, it is difficult to represent the anisotropic swelling/shrinkage behaviour of argillaceous rocks.

This work focuses on the numerical modelling of swelling and shrinkage induced anisotropic deformation process. A fully coupled thermal-hydro-mechanical (TH^{2}M) model is developed and implemented in the numerical code OpenGeoSys (OGS) to simulate mechanical processes during de- and re-saturation experiments on the argillaceous rock samples. In this model a water saturation-dependent effective stress concept is applied, taking both gas and capillary pressure in unsaturated porous medium into consideration. The Bishop coefficient is introduced (Bishop 1959; Bishop & Blight 1963; Gens 2010), which can be expressed as a function of effective saturation (Maßmann 2009). The anisotropic elastic behaviour of the medium is described by the generalized Hooke's law. Combining the unsaturated effective stress concept with an anisotropic stiffness model, the anisotropic swelling/shrinkage deformation can be automatically represented.

The hydraulic process of de- and re-saturation is simulated by a two-phase-flow model. Gas pressure and capillary pressure are defined as the primary variables. The relationship between capillary pressure and water saturation is described by the van Genuchten model, in which the parameters (gas entry pressure and shape factor) are fitted with laboratory suction experiment with controlled relative humidity. The relative permeability to water and gas is calculated by the Mualem approach. The flow process is described by the Darcy law. Similar to the mechanical model the intrinsic permeability is anisotropic to consider the bedding plane structure of argillaceous rocks. With the coupling of the thermal process, water vapour in gas phase is involved. Gas phase is assumed to be a mixture of two perfect ideal gases, dry air and water vapour. Diffusion of water vapour in the gas phase is described by the Fick law. In this TH^{2}M model only one parameter, the Bishop coefficient, is additionally introduced.

From: NORRIS, S., BRUNO, J., CATHELINEAU, M., DELAGE, P., FAIRHURST, C., GAUCHER, E. C., HÖHN, E. H., KALINICHEV, A., LALIEUX, P. & SELLIN, P. (eds) 2014. *Clays in Natural and Engineered Barriers for Radioactive Waste Confinement*. Geological Society, London, Special Publications, **400**, 359–366.
First published online May 8, 2014, http://dx.doi.org/10.1144/SP400.29

Three laboratory experiments on the specimens taken from the Tournemire site (France; Valès *et al.* 2004), the Bure site (France) and the Mont Terri Rock Laboratory (Switzerland; Zhang *et al.* 2004; Zhang & Rothfuchs 2007) have been analysed and recalculated. For the specimen from Tournemire site a free swelling/shrinkage experiment is performed with controlled relative humidity. Deformations in parallel and perpendicular directions of the bedding plane are recorded. The experiments of the specimens from Bure site are under a uniaxial load condition and moisture is not controlled. Two specimens are tested simultaneously with loading parallel and perpendicular to the bedding plane. The specimens are exposed to room air conditions and the axial shrinkage deformation is measured. The third experiment wais performed on the specimen from Mont Terri Rock Laboratory. Both moisture and temperature are controlled during the test. The specimen is also under uniaxial load conditions. Strong swelling/shrinkage capacity of all specimens is observed when the moisture changed. All of the experiments are analysed and the details are presented in the section 'Modelling of laboratory experiments'.

Governing equations

To model the laboratory and *in-situ* gas transport experiments, a coupled numerical TH^{2}M analysis is required. The primary variables of the problem are temperature T, capillary pressure p^c, gas pressure p^g and solid displacement $\mathbf{u}$. The porous medium is assumed to be a continuum homogenous.

Balance equations

Mass balance equations. In a deformable porous, medium flow of water and gas is considered. The water phase is assumed to be incompressible and the gas phase is compressible and obeys the ideal gas law. The water vapour in gas phase is considered. The mass conservation of the water and the vapour is expressed as (Sanavia *et al.* 2006)

$$\begin{aligned} & n[\rho^w - \rho^{gw}]\left[\frac{\partial S_w}{\partial p^c}\frac{\partial p^c}{\partial t}\right] + [1 - S_w]n\left[\frac{\partial \rho^{gw}}{\partial p^c}\frac{\partial p^c}{\partial t}\right] \\ & + [\rho^w S_w + \rho^{gw}[1 - S_w]]\mathrm{div}\left(\frac{\partial \mathbf{u}}{\partial t}\right) \\ & - \mathrm{div}\left(\rho^g \frac{M_a M_w}{M_g^2}\mathbf{D}_g^{gw}\mathrm{grad}\left(\frac{p^{gw}}{p^c}\right)\right) \\ & + \mathrm{div}\left(\rho^w \frac{\mathbf{k}k^{rw}}{\mu^w}[-\mathrm{grad}(p^g) + \mathrm{grad}(p^c) + \rho^w \mathbf{g}]\right) \\ & + \mathrm{div}\left(\rho^{gw}\frac{\mathbf{k}k^{rg}}{\mu^g}[-\mathrm{grad}(p^g) + \rho^g \mathbf{g}]\right) = 0 \qquad (1) \end{aligned}$$

where $\rho^{w,gw,g}$ is the density of water, vapour and gas phase. $\mathbf{k}$ is the material intrinsic permeability tensor, $k^{rw,rg}$ is the relative permeability of water and gas phase. $M_{a,w,g}$ is the molar mass of dry air, water and gas phase, which is mixed from vapour and dry air. $\mathbf{D}$ is the effective diffusivity tensor.

Similarly, the mass balance equation of the dry air is expressed as:

$$\begin{aligned} & -n\rho^{ga}\left[\frac{\partial S_w}{\partial p^c}\frac{\partial p^c}{\partial t}\right] + [1 - S_w]\rho^{ga}\mathrm{div}\left(\frac{\partial \mathbf{u}}{\partial t}\right) \\ & + [1 - S_w]n\left[\frac{\partial \rho^{ga}}{\partial p^c}\frac{\partial p^c}{\partial t} + \frac{\partial \rho^{ga}}{\partial p^g}\frac{\partial p^g}{\partial t}\right] \\ & - \mathrm{div}\left(\rho^g \frac{M_a M_w}{M_g^2}\mathbf{D}_g^{ga}\mathrm{grad}\left(\frac{p^{ga}}{p^c}\right)\right) \\ & + \mathrm{div}\left(\rho^{ga}\frac{\mathbf{k}k^{rg}}{\mu^g}[-\mathrm{grad}(p^g) + \rho^g \mathbf{g}]\right) = 0 \qquad (2) \end{aligned}$$

Energy balance equation. The energy balance equation for the heat transport process can be expressed as (Kolditz *et al.* 2012):

$$\begin{aligned} & C\rho\frac{\partial T}{\partial t} + (C^w\rho^w\mathbf{v}^w + C^g\rho^g\mathbf{v}^g)\mathrm{grad}T \\ & \quad - \mathrm{div}(\boldsymbol{\lambda}\mathrm{grad}T) = q_{th} \qquad (3) \end{aligned}$$

where C and ρ are the thermal capacity and density of the porous medium, $\rho = (1 - n)\rho^s + nS_w\rho^w + n(1 - S_w)\rho^g$; $\mathbf{v}^{w,g}$ is the advection velocity of water and gas phase; and $\boldsymbol{\lambda}$ is the thermal conductivity tensor. The evaporation and condensation are not considered.

Momentum balance equation. The water saturation is used as a weighting factor in the effective stress in unsaturated porous medium and the capillary pressure is modified with the Bishop coefficient (Bishop & Blight 1963; Kolditz *et al.* 2012). The linear momentum balance equation in the term of effective stress is expressed as:

$$\mathrm{div}(\boldsymbol{\sigma}' - \alpha(p^g - \chi p^c)\mathbf{I}) + \rho\mathbf{g} = 0 \qquad (4)$$

where $\boldsymbol{\sigma}'$ is the effective stress tensor, α is the Biot coefficient and χ is the Bishop coefficient, which can be described as a function of saturation (Maßmann 2009):

$$\chi = \begin{cases} 0 & S_w < S_{w,r} \\ \left(\dfrac{S_w - S_{w,r}}{1 - S_{w,r} - S_{g,r}}\right)^\kappa & S_{w,r} < S_w < 1 - S_{g,r} \\ 1 & S_w > 1 - S_{g,r} \end{cases} \qquad (5)$$

Figure 1 shows an example of the χp^c term in equation (4), showing its evolution with different κ. To this end, the saturation-induced deformation can be more flexibly controlled by introducing the Bishop coefficient.

This effective stress concept covers also the fully saturated situation. With fully saturated condition $p^c = 0$ and $p^g = p^w$, equation (4) has the same formation as the conventional effective stress concept for saturated flow.

Constitutive equations

Hydraulic process. In the two-phase flow process the relationship between water saturation and capillary pressure is required. In this paper the van Genuchten function (Van Genuchten 1980) is applied, which can be written as:

$$p^c = p_0\left(S_{ec}^{\frac{m}{1-m}} - 1\right)^{\frac{1}{m}} \tag{6}$$

with

$$S_{ec} = \frac{S_w - S_{wr}}{1 - S_{wr} - S_{gr}} \tag{7}$$

where p_0 is the gas entry pressure, m is a shape factor for van Genuchten model, S_{ec} is the effective saturation, and S_{wr} and S_{gr} are residual water and gas saturation.

The water and gas flow rate is described by the extended Darcy's law, considering the primary variables: capillary pressure and gas pressure. The water pressure is defined by the sorption equilibrium equation $p^w = p^g - p^c$ (Sanavia *et al.* 2006).

The relative permeability for water and gas phase is described with the approach of Mualem and has the following general form (Nagra 2008; Wang *et al.* 2011):

$$k_{r,g} = (1 - S_e)^{\frac{1}{2}}\left(1 - S_e^{\frac{m}{m-1}}\right)^{2(1-1/m)} \tag{8}$$

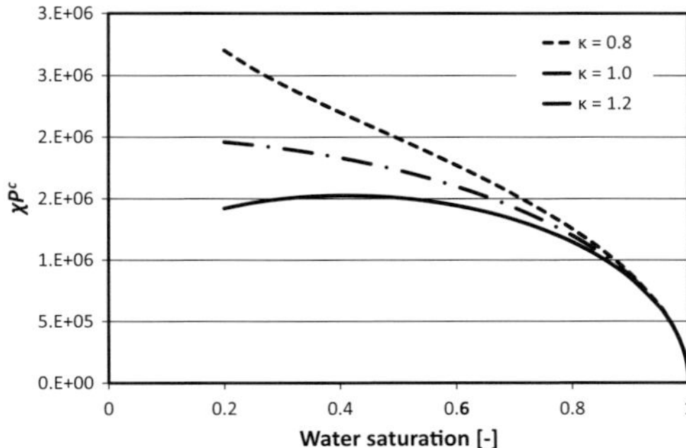

Fig. 1. Effective stress term χp^c with different κ.

and

$$k_{r,w} = S_e^{\frac{1}{2}}\left[1 - \left(1 - S_e^{\frac{m}{m-1}}\right)^{\frac{m-1}{m}}\right]^2 \tag{9}$$

The gas phase is a mixture of dry air and water vapour. The relative velocities of the diffusing species are defined by Fick's law (Sanavia *et al.* 2006):

$$\mathbf{v}_g^{ga} = -\frac{M_a M_w}{M_g^2}\mathbf{D}_g^{ga}\mathrm{grad}\left(\frac{p^{ga}}{p^g}\right) \tag{10}$$

$$\mathbf{v}_g^{gw} = -\frac{M_a M_w}{M_g^2}\mathbf{D}_g^{ga}\mathrm{grad}\left(\frac{p^{gw}}{p^g}\right) \tag{11}$$

where $\mathbf{v}_g^{ga,gw}$ is the relative velocity of dry air and water vapour and $\mathbf{v}_g^{ga} = \mathbf{v}_g^{gw}$.

Deformation process. The deformation process is described by the pure elastic model. The relationship between stress and strain is expressed with the general Hook's law and, considering the material anisotropic property, it has the following form (Valès *et al.* 2004):

$$\Delta\boldsymbol{\varepsilon} = \begin{bmatrix} \frac{1}{E_1} & \frac{-\upsilon_{12}}{E_2} & \frac{-\upsilon_{13}}{E_3} & & & \\ \frac{-\upsilon_{21}}{E_1} & \frac{1}{E_2} & \frac{-\upsilon_{23}}{E_3} & & & \\ \frac{-\upsilon_{31}}{E_1} & \frac{-\upsilon_{32}}{E_2} & \frac{1}{E_3} & & & \\ & & & \frac{1}{2G_{12}} & & \\ & & & & \frac{1}{2G_{13}} & \\ & & & & & \frac{1}{2G_{23}} \end{bmatrix}\Delta\boldsymbol{\sigma} \tag{12}$$

E is the Young's modulus, υ is Poisson ratio and G is the shear modulus. The indices (1, 2 and 3) indicate the three principal directions of the anisotropic material. Argillaceous rocks usually have transversely isotropic material properties owing to the bedding plane. The material properties in the bedding plane are isotropic and in the perpendicular direction to the bedding plane are anisotropic; $\boldsymbol{\varepsilon}$ is the total strain, which can be decomposed into elastic ($\boldsymbol{\varepsilon}^e$) and thermal ($\boldsymbol{\varepsilon}^{th}$) component,

$$\Delta\boldsymbol{\varepsilon} = \Delta\boldsymbol{\varepsilon}^e + \Delta\boldsymbol{\varepsilon}^{th} = \Delta\boldsymbol{\varepsilon}^e + \beta\mathbf{I}\Delta T \tag{13}$$

where β is the linear thermal expansion coefficient, $\mathbf{I}$ is the identity tensor and ΔT is the temperature change.

Modelling of laboratory experiments

Three laboratory experiments are presented and analysed in this section. The previously introduced TH^{2}M numerical model is used to simulate all

Table 1. *Laboratory experiment information (Valès* et al. *2004; Zhang* et al. *2004; Zhang & Rothfuchs 2007)*

Specimen origin	Tournemire site (France)	Bure site (France)	Mont Terri Rock Laboratory (Switzerland)
Loading condition	No loading	Uniaxial loading (15 MPa)	Uniaxial loading (1 MPa)
Moisture condition	Controlled (5–98%)	Room humidity (20–30%)	Controlled (20–95%)
Temperature	Room temperature (25 °C)	Room temperature (25 °C)	Controlled temperature (24–60 °C)
Specimen diameter	38 mm	45 mm	100 mm
Specimen length	80 mm	90 mm	220 mm

experiments with 3D. The summarized information of three experiments is listed in Table 1.

The loading and temperature conditions are applied as boundary conditions of the numerical model. The moisture condition cannot directly be used as boundary conditions. Therefore, it is converted into suction using the Kelvin relationship:

$$s = \rho^{w} RTM^{-1} \ln H \tag{14}$$

where R is the universal gas constant (8.314 J mol^{-1} K^{-1}), T is the absolute temperature (K), M is the molar mass of water (18.015 gmol^{-1}) and H is the relative humidity (%).

Free swelling/shrinkage experiment

A free swelling/shrinkage experiment is presented in Valès *et al.* (2004). During the experiment the specimen is situated in a tight box, in which the relative humidity is controlled by different saline solutions (Fig. 2). The weight sensor can measure the weight change during the experiment, and with that measured data the water saturation value can be also calculated.

The initial water saturation of the specimen is estimated at 90%. Five experiment stages with controlled relative humidity (98, 76, 50, 36 and <5%) are performed. The specimen is resaturated with 98% relative humidity condition at first, which leads to the increase of the water saturation of the specimen from 90% to 95%. During this resaturation stage the swelling of the specimen is observed. Then, shrinkage tests take place with decreased relative humidity. As a result, the water saturation of the specimen is decreased from 95 to 85, 60 and 45%. Deformation parallel and perpendicular to the bedding plane is measured with strain gauges and compared with the numerical results (Fig. 3).

The Young's modulus and Poisson ratio are assumed constant considering the measured values from the study of Valès *et al.* (2004). However, the laboratory investigation suggests that the stiffness properties of the specimen are varied with the saturation state. This phenomenon is not considered in this paper. The van Genuchten parameters are fitted with the suction and saturation relationship from the laboratory measurement. The temperature is constant and no heat transport

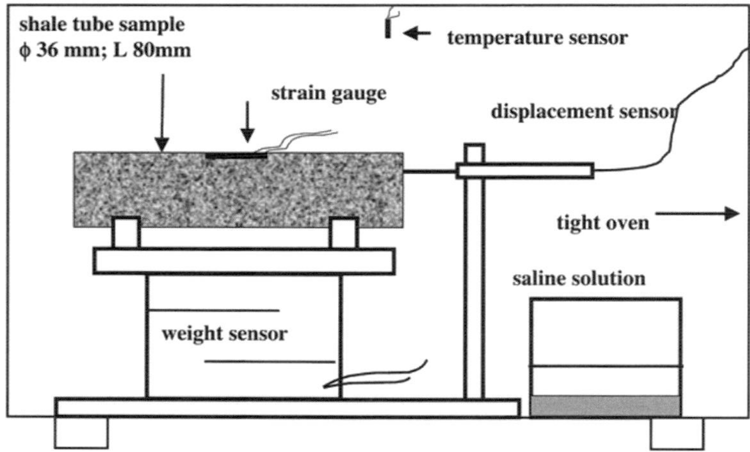

Fig. 2. Sketch of experiment layout (Valès *et al.* 2004).

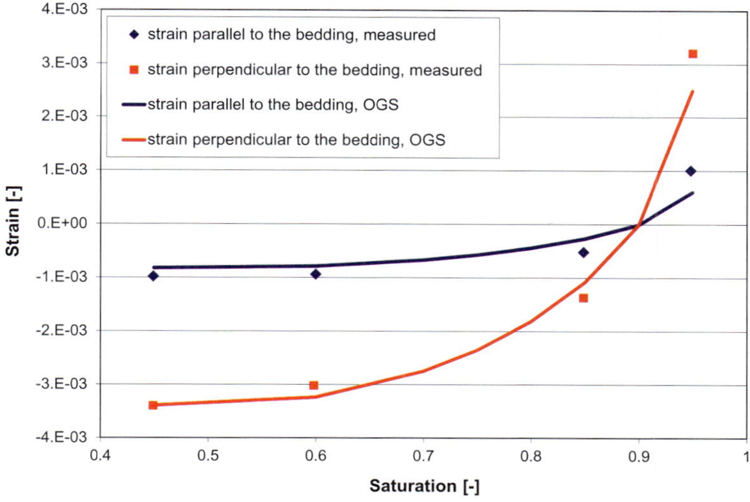

Fig. 3. Comparison of the measured data and calculated results of free swelling/shrinkage experiments.

process takes place. The summarized parameters are listed in Table 2.

The experimental measurements indicate that the moisture change-induced deformation is strongly anisotropic, which can be well represented with a numerical model. The anisotropic deformation behaviour is due to the bedding plane of the argillaceous rocks. The relationship between the deformation and saturation is nonlinear. This complex relationship can be automatically represented by the numerical simulation. According to the measured data, the most shrinkage deformation is induced with the saturation of the specimen from 95 to 60%. When the specimen is de-saturated from 60 to 30%, the shrinkage deformation is limited. Because of the low saturation in the dry porous medium, the capillary pressure makes relative little contribution to the effective stress.

Shrinkage experiments under uniaxial loading

Shrinkage experiments under uniaxial loading condition of two Callovo–Oxfordian specimens from Bure site are carried out by GRS (Gesellschaft für Anlagen-und Reaktorsicherheit; Zhang *et al.* 2004). The shrinkage experiments are performed immediately after a series of creep tests. During the creep tests the specimens are confined with rubber jacket under saturated condition. After four steps of creep tests the rubber jacket is removed and the specimens are exposed in the room air, which has a relative humidity about 25%. The uniaxial loading condition remains as the last creep step at 15 MPa. The loading direction is perpendicular (specimen 1) and parallel (specimen 2) to the bedding plane of two specimens. At the same the time mass evolution of another reference specimen is measured to investigate the water loss owing to the drying process.

Material parameters are listed in Table 3. The Young's moduli of two specimens in the loading direction are estimated from the creep tests, when the loading pressure is increased from 12 to 15 MPa. The parameter for the van Genuchten model is estimated by fitting the water retention curve, which is also measured by GRS.

Because there is no continually measured data of the relative humidity during the experiment, the saturation boundary condition is defined considering the water content evolution of the reference specimen. As in the free swelling/shrinkage experiment in the section 'Free swelling/shrinkage experiment', there is no heat transport process during the experiment. The system temperature remains constant.

Table 2. *Parameters for modelling of free swelling/shrinkage experiment*

Young's modulus $\perp/\|\|$ (GPa)	6.0/18.0	Poisson ratio	0.12
Permeability $\perp/\|\|$ (m^2)	$10^{-18}/2 \cdot 10^{-18}$	Porosity	0.07
p_0 for van Genuchten (MPa)	70	m for van Genuchten	1.96
κ for Bishop coefficient	1.4		

Table 3. *Parameters for modelling of shrinkage experiments under uniaxial loading*

Young's modulus ⊥/‖ (GPa)	4.2/6.0	Poisson ratio	0.3
Permeability ⊥/‖ (m^2)	$10^{-21}/10^{-20}$	Porosity	0.12
p_0 for van Genuchten (MPa)	20	m for van Genuchten	1.667
κ for Bishop coefficient	1.53		

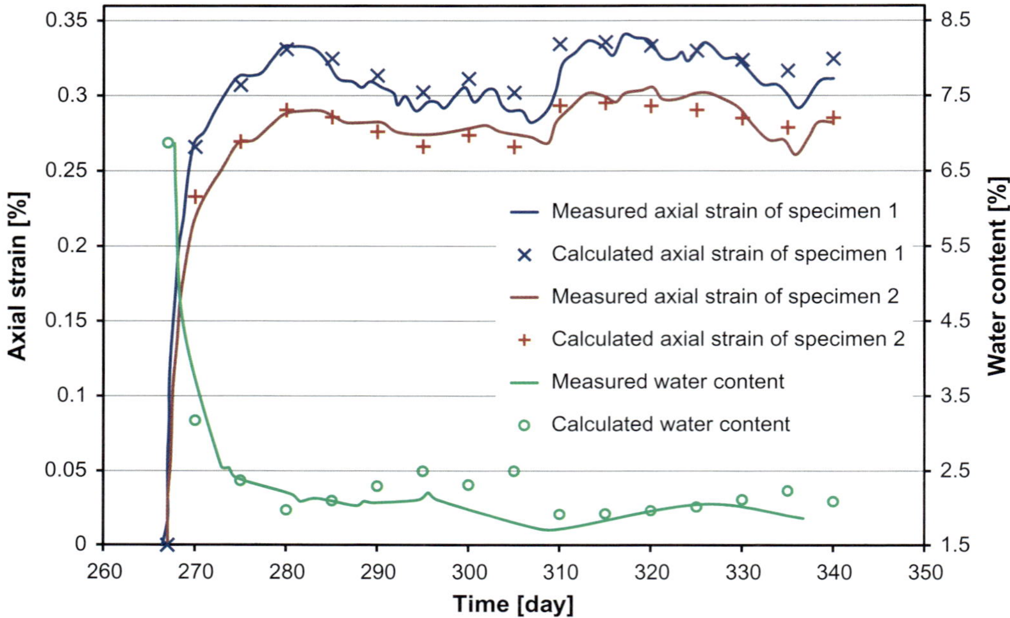

Fig. 4. Comparison of the measured data and calculated results of shrinkage experiment under uniaxial loading.

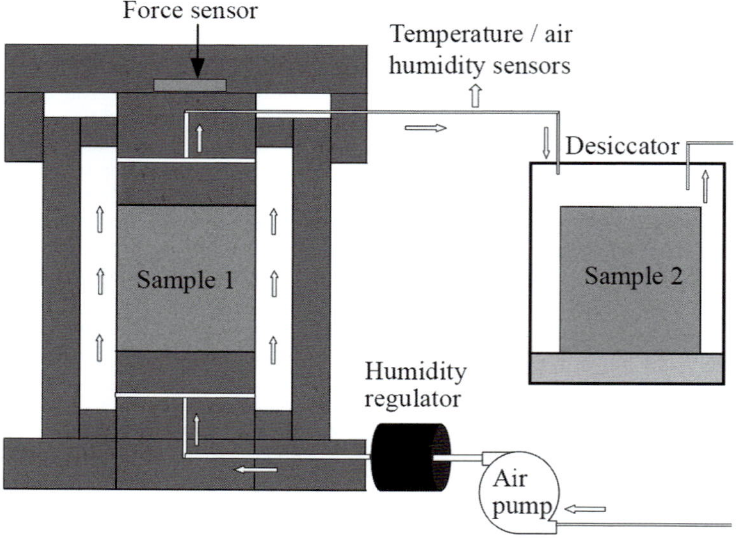

Fig. 5. Sketch of the laboratory experiment concept (Zhang & Rothfuchs 2007).

Table 4. *Parameters for modelling of experiments with controlled humidity and temperature*

Young's modulus ⊥/‖ (GPa)	3.2/8.0	Poisson ratio	0.28
Permeability ⊥/‖ (m^2)	$10^{-20}/10^{-19}$	Porosity	0.15
p_0 for van Genuchten (MPa)	25	m for van Genuchten	1.75
κ for Bishop coefficient	1.1	Thermal expansion	1.6E-5
Thermal capacity (J kg^{-1} K^{-1})	1000	Thermal conductivity	1.6

The measured axial deformation, as well as the numerical results, (Fig. 4) indicates that the shrinkage behaviour is also anisotropic under uniaxial loading conditions. The shrinkage in the bedding plane parallel direction is less than that in the perpendicular direction. The drying process needs about 15 days to reach a steady state. Because during the experiment the specimens are exposed to the room air, the relative humidity is changed and has a complex history, which leads to complicated deformation results, such as the axial strain increase around day 310.

Experiment with controlled humidity and temperature

An experiment with controlled relative humidity and temperature is carried out by GRS (Zhang & Rothfuchs 2007). The aim of this experiment is to investigate the swelling/shrinkage behaviour of Opalinus clay under non-isothermal conditions. The specimen is placed in an isolated chamber with uniaxial loading conditions (Fig. 5). The loading direction is parallel to the bedding plane with loading pressure constant at 1 MPa. Air is pumped into the chamber with controlled relative humidity and temperature. The air out of the chamber flows into a tight box, in which another similar specimen is situated without any mechanic loading. The water content of this reference specimen is measured to estimate the saturation state of the specimen in the chamber.

The whole experiment has three stages. In the first drying stage, air with relative humidity about 20–40% is pumped into the chamber. The temperature is slightly increased from 24 to 38 °C. In the

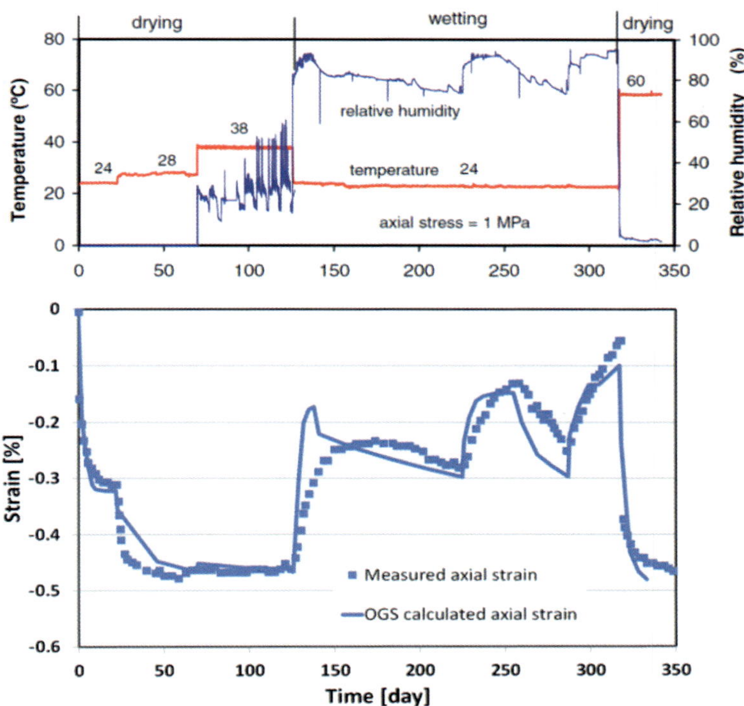

Fig. 6. Laboratory experiment conditions and comparison of measured and calculated axial strain.

second wetting stage the specimen is resaturated with relative humidity about 80–95% and the temperature is cooled from 38 to 24 °C. The last stage is the drying operation again. The relative humidity in the chamber is decreased to about 5% and the temperature is increased to 60 °C. The parameters for modelling are listed in Table 4.

The numerical results of the axial strain represent the experimental observations (Fig. 6). The shrinkage deformation can be reversed during the resaturation process. Owing to the low thermal expansion coefficient of the Opalinus clay (1.6×10^{-5} Table 4) the thermal deformation makes a smaller contribution than the swelling/shrinkage process to the total deformation during the whole experiment. At the first experiment stage, when the temperature is increased from 28 to 38 °C only slight expansion is observed. Even when the temperature is increased to 60 °C at the end of the experiment, the shrinkage process owing to the low relative humidity dominates the deformation behaviour. The last heating process has no significant influence on the shrinkage deformation.

The drying process is faster than the resaturation process. During both drying stages the shrinkage deformation reaches a steady state in a relatively short time, less than 10 days. The swelling process needs a longer time to reach a steady state when the specimen is resaturated. A possible explanation for this is the evolution of the intrinsic permeability owing to swelling and shrinkage. When the specimen is drying out, shrinkage can cause a permeability increase owing to the generation and opening of micro-fractures. With increased permeability the drying process is speeded up. Conversely, the wetting process is slowed down owing to micro-fracture closing during swelling with decreased permeability.

Conclusion

A TH^2M coupled numerical model is developed as a powerful tool to simulate the moisture change induced deformation. The Bishop coefficient is introduced into the effective stress model to control the swelling/shrinkage owing to the change in capillary pressure. The introduced effective stress concept takes both gas pressure and capillary pressure of the unsaturated porous medium into consideration. Three laboratory experiments with different conditions are simulated and well represented by the numerical model. The parameter sets for different rock types in this work can be used as calibrated values for modelling *in-situ* experiments.

Swelling/shrinkage behaviours of argillaceous rocks are strongly anisotropic owing to the bedding plane of argillaceous rocks. The moisture change-induced deformation in the perpendicular direction to the bedding plane is more than that in the parallel direction. The shrinkage process responds more quickly than the swelling process, possibly owing to the increased permeability. The numerical model can be further extended to involve the permeability evolution with the swelling/shrinkage deformation. More experimental measurements and evidence are needed to investigate the relationship between the permeability and moisture-induced deformation. For future work the plasticity of the argillaceous rocks should also be involved in this TH^2M model.

This work is supported by BMWi (Bundesministerium für Wirtschaft und Technologie, Berlin). The authors gratefully acknowledge Chun-Liang Zhang for scientific discussion.

References

Bishop, A. W. 1959. *The Principle of Effective Stress.* Teknisk Ukeblad, Oslo.

Bishop, A. W. & Blight, G. E. 1963. Some aspects of effective stress in saturated and partly saturated soils. *Géotechnique*, **13**, 177–197.

Gens, A. 2010. Soil-environment interactions in geotechnical engineering. *Géotechnique*, **60**, 3–74.

Kolditz, O., Görke, U-J., Shao, H. & Wang, W-Q. (eds) 2012. *Thermo-Hydro-Mechanical–Chemical Processes in Fractured Porous Media.* Springer, Heidelberg.

Maßmann, J. 2009. *Modelling of excavation induced coupled hydraulic–mechanical processes in claystone.* PhD thesis, Leibniz University Hanover.

NAGRA 2008. *Effects of post-disposal gas generation in a repository for low- and intermediate-level waste sited in the Opalinus Clay of Northern Switzerland.* Nagra Techenical Report **08-07**.

Sanavia, L., Pesavento, F. & Schrefler, B. A. 2006. Finite element analysis of non-isothermal multiphase geomaterials with application to strain localization simulation. *Computational Mechanics*, **37**, 331–348.

Valès, F., Nguyen Minh, D., Gharbi, H. & Rejeb, A. 2004. Experimental study of the influence of the degree of saturation of physical and mechanical properties in Tournemire shale (France). *Applied Clay Science*, **26**, 197–207.

Van Genuchten, M. Th. 1980. A closed-form equation for predicting the hydraulic conductivity of unsaturated soils. *Soil Science Society of America Journal*, **44**, 1–1.

Wang, W., Rutqvist, J., Gorke, U-J., Birkholzer, J. T. & Kolditz, O. 2011. Non-isothermal flow in low permeable porous media: a comparison of unsaturated and two-phase flow approaches. *Environ and Earth Sciences*, **62**, 1197–1207.

Zhang, C.-L. & Rothfuchs, T. 2007. Experimental study of the thermo-hydro-mechanical behavior of indurated clays. *Physics and Chemistry of the Earth*, **32**, 957–965.

Zhang, C.-L., Rothfuchs, T., Moog, H., Dittrich, J. & Müller, J. 2004. *Thermo-hydro-mechanical and geochemical behaviour of the Callovo-Oxfordian Argillite and the Opalinus Clay.* GRS-202.

Analytical modelling of a deep tunnel in a viscoplastic rock mass accounting for a simplified life cycle and extension to a particular case of porous media

F. DELERUYELLE[1]*, T. A. BUI[2], H. WONG[2] & N. DUFOUR[3]

[1]*IRSN, PRP-DGE, SEDRAN, BERIS, BP 17–92262 Fontenay-aux-Roses Cedex, France*

[2]*Université de Lyon, ENTPE/LGCB and LTDS UMR CNRS 5513, 3 rue Maurice Audin, 69518 Vaulx-en-Velin Cedex, France*

[3]*CETE Nord Picardie, RDT, MRG, MF, 42bis rue Marais, BP 99 Sequedin, 59482 Haubourdin, France*

**Corresponding author (e-mail: frederic.deleruyelle@irsn.fr)*

Abstract: An analytical approach to the viscoplastic behaviour of a rock mass around a deep tunnel in different stages of a simplified life cycle is presented. The viscoplasticity is modelled by means of a simple (linear) Norton–Hoff's law. First, the solution for a dry rock mass is presented, then an extension to a particular case of a porous saturated rock mass with a Poisson's ratio of 0.5 is examined. Some numerical examples are carried out to show the consistency and relevance of the solutions. Parametric studies illustrate the influence of key parameters such as rock mass viscosity and time of lining emplacement. This analytical approach constitutes a useful reference to test more complex numerical simulation and makes it possible to obtain practical estimations of the poromechanical behaviour of underground structures.

This paper deals with the time-dependent behaviour of rocks in the vicinity of underground cavities. Unlike other studies available in the literature, our approach is analytical and addresses different stages of the life cycle of the cavity. The theoretical framework is adopted from previous works (Wong *et al.* 2008*a*, *b*; Dufour *et al.* 2009, 2012) and completed to account for an irreversible viscoplastic behaviour, which has been experimentally shown (Boidy *et al.* 2002; Gatelier *et al.* 2002; Chiarelli *et al.* 2003; Gasc-Barbier *et al.* 2004; Zhou *et al.* 2011) to be more representative of *in situ* behaviour. In the literature, other attempts to account for creep with the use of viscoplasticity theory can be found (Pouya 1991; Giraud & Rousset 1996; Cosenza & Ghoreychi 1999; Kazmierczak *et al.* 2007; Carranza-Torres & Zhao 2009), but, most often, they require numerical tools and analytical approaches are more rare. Pouya (1991) provided an approximate analytical solution for the case of an unlined tunnel, accounting for a monophasic rock mass obeying Norton–Hoff's creep law and that was elastically incompressible (Poisson's ratio $\nu = 0.5$). Cosenza & Ghoreychi (1999) analytically dealt with the long-term behaviour of a spherical cavity inside a saturated poroviscoplastic rock mass in the limiting case of the stationary state, neglecting transitional stages.

In this paper, an analytical (resp. quasi-analytical) solution is developed for the case of a monophasic (resp. porous saturated) rock mass, taking into account a linear version of Norton–Hoff's creep law and a simplified life cycle. Owing to the complexity of the hydromechanical couplings encountered in the poromechanical part, only the elastically incompressible case ($\nu = 0.5$) is presented in this paper. The general case in which Poisson's ratio can take any value leads to much more complex analysis, which are still under progress and will be presented in a future publication.

General framework and main assumptions of the problem

We consider the plane-strain problem of a long, deep, cylindrical tunnel in a homogeneous initial stress field. Cylindrical symmetry reduces the problem to depend on only two variables: r (the distance to the tunnel axis) and t (time). Isothermal conditions and small displacements and strains are also assumed, and gravity is neglected, so volumetric forces are not taken into account.

The tunnel life cycle is idealized as three stages, as shown schematically in Figure 1. Initially, at $t = t_0 = 0$, the medium is supposed to be in equilibrium with the geostatic pressure P_∞ with zero displacements or strains. Depending on the considered case (monophasic or porous saturated), P_∞ denotes a monophasic or a total stress. The first stage

From: Norris, S., Bruno, J., Cathelineau, M., Delage, P., Fairhurst, C., Gaucher, E. C., Höhn, E. H., Kalinichev, A., Lalieux, P. & Sellin, P. (eds) 2014. *Clays in Natural and Engineered Barriers for Radioactive Waste Confinement*. Geological Society, London, Special Publications, **400**, 367–379.
First published online April 2, 2014, http://dx.doi.org/10.1144/SP400.14

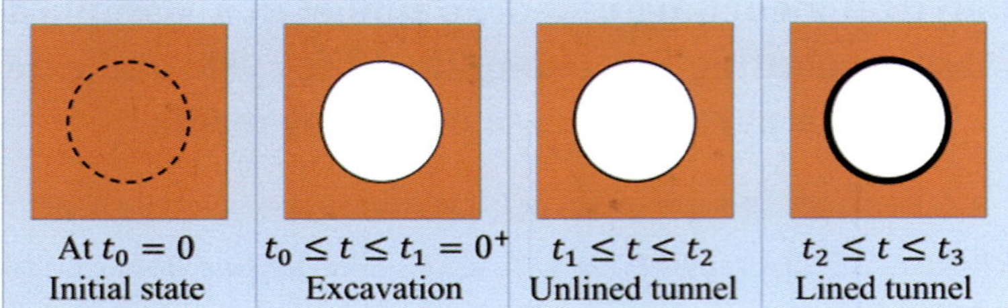

Fig. 1. Simplified life cycle of a typical tunnel.

corresponds to the excavation of a circular tunnel of radius a in an infinite viscoplastic medium characterized by Young's modulus E, Poisson's ratio v and rock mass viscosity η. Because the time required for the excavation is generally small compared to the characteristic creep time, this stage is assumed to be instantaneous and ends at $t = t_1 = 0^+$. It is then followed by the second stage, during which the tunnel remains unlined and a creep-induced convergence occurs. The second stage ends when a lining support is installed at $t = t_2$. In the third stage, from $t = t_2$ onwards, the lining reaction at the tunnel wall slows the creep-induced convergence. Indeed, to be complete, a fourth stage (post-closure stage) starting when the tunnel is backfilled ($t = t_3$) should be considered but leads to complex mathematical developments and thus remains under consideration.

Analytical solution in the case of a monophasic viscoplastic medium

In this section, the rock mass is supposed to be monophasic and to behave viscoplastically. Tensors and vectors are denoted in bold letters and scalars in normal letters. Because the problem is axisymmetrical, cylindrical coordinates are used. In such coordinate system, the stress field $\boldsymbol{\sigma}$ and strain field $\boldsymbol{\varepsilon}$ are diagonal and the displacement field $\boldsymbol{u}$ is purely radial:

$$\begin{aligned} &\boldsymbol{\sigma} = \mathrm{diag}[\sigma_r \;\; \sigma_\theta \;\; \sigma_z]; \;\; \boldsymbol{u} = \mathrm{diag}[u \;0 \;\; 0]; \\ &\boldsymbol{\varepsilon} = \mathrm{diag}[\partial_r u \;\; \frac{u}{r} \;\; 0]; \;\; U(t) \equiv \frac{u(a,t)}{a} \end{aligned} \tag{1}$$

where $U(t)$ is defined as the relative convergence (relative displacement at the tunnel wall). The volumetric strain is linked to the radial displacement by $\varepsilon = \mathrm{tr}(\boldsymbol{\varepsilon}) = \partial_r u + (u/r)$. Local perturbations due to excavation and the lining reaction in the near field involve a finite energy and therefore cannot propagate an infinite distance in a finite time. Hence, the boundary condition at an infinite distance from the tunnel is that all field quantities remain at their initial values at any time:

$$\begin{aligned} &u(\infty,\; t) = 0; \;\; \boldsymbol{\sigma}(\infty,\; t) = -P_\infty \mathbf{1} \\ &\boldsymbol{\varepsilon}(\infty, t) = 0; \;\; \varepsilon(\infty, t) = 0 \end{aligned} \tag{2}$$

As a result of the axisymmetrical and plane strain assumptions, the only non-trivial equilibrium equation is given by

$$r\frac{\partial \sigma_r}{\partial r} + (\sigma_r - \sigma_\theta) = 0 \tag{3}$$

The strain tensor $\boldsymbol{\varepsilon}$ is assumed to be the sum of elastic and viscoplastic parts, indicated by superscript 'e' and 'vp', respectively: $\boldsymbol{\varepsilon} = \boldsymbol{\varepsilon}^{\mathrm{e}} + \boldsymbol{\varepsilon}^{\mathrm{vp}}$. Elastic strain appears in the classic Hooke's law:

$$\dot{\boldsymbol{\varepsilon}}^e = \frac{1}{3K}\dot{\sigma}_m \mathbf{1} + \frac{1}{2G}\dot{\boldsymbol{s}} \tag{4}$$

where $\mathbf{1}$ is the second-order identity tensor, $\sigma_{\mathrm{m}} = \mathrm{tr}(\boldsymbol{\sigma})/3$ is the mean stress, $\boldsymbol{s} = \boldsymbol{\sigma} - \sigma_{\mathrm{m}}\mathbf{1}$ is the diagonal deviatoric stress tensor, E and v are, respectively, Young's modulus and Poisson's ratio, and $K = E/[3(1 - 2v)]$ and $G = E/[2(1 + v)]$ are, respectively, the bulk and shear moduli of the medium. Using thermodynamic methods (Lemaitre & Chaboche 2001), the viscoplastic strains defined by Norton–Hoff can be derived from a convex non-negative dissipation potential $\varphi^* \;(\boldsymbol{\sigma}) = q^{n+1}/[(n+1)\eta]$, where $q = \sqrt{(3/2)\boldsymbol{s} : \boldsymbol{s}}$ is the Von Mises equivalent stress, η is the viscosity of the medium, and n is a positive material constant. In the particular case of $n = 1$ (linear), the expression for the viscoplastic strain rate becomes

$$\dot{\boldsymbol{\varepsilon}}^{\mathrm{vp}} = \frac{\partial \varphi^*}{\partial \boldsymbol{\sigma}} = \frac{3}{2\eta}\boldsymbol{s} \tag{5}$$

This last equation, combined with equation (4) and the boundary conditions of equations (2), leads to

$$\dot{\sigma}_m = K\dot{\varepsilon}; \quad \sigma_m = K\varepsilon - P_\infty \tag{6}$$

Combining relations (4), (5) and (6), the expression of the total strain rate is obtained:

$$\dot{\boldsymbol{\varepsilon}} = \frac{1}{3}\dot{\varepsilon}\mathbf{1} + \frac{1}{2G}\dot{\boldsymbol{s}} + \frac{3}{2\eta}\boldsymbol{s} \qquad (7)$$

First stage: instantaneous excavation ($0 \le \boldsymbol{t} \le \boldsymbol{t}_1 = 0^+$)

The first stage assumes the instantaneous excavation of the tunnel. Because the dissipative mechanism has finite speed (material viscosity is finite), creep strains within an infinitesimal time must be infinitesimally small. Consequently, the behaviour of the medium can be considered *elastic*. The initial state, at $t_0 = 0$, under mechanical equilibrium, is characterized by a homogeneous field of geostatic stress. Furthermore, initial displacements and strains are null:

$$\begin{aligned} &\boldsymbol{\sigma}(r,\ 0) = -P_\infty \mathbf{1}; \\ &u(r,\ 0)0; \quad \boldsymbol{\varepsilon}(r, 0) = 0 \end{aligned} \qquad (8)$$

Boundary conditions at the far field are specified as in expressions (2). To simulate the excavation, the radial stress at the tunnel wall ($r = a$) is decreased from the geostatic overburden to zero; that is, in this stage $\sigma_r(a,\ t) = 0$. The above assumptions lead to the problem of a tunnel instantaneously excavated in an elastic medium. The solution of this problem is classical (e.g. Panet 1995). The main results at $t_1 = 0^+$ are as follows:

$$\begin{aligned} &u(r, 0^+) = -\frac{P_\infty}{2G}\frac{a^2}{r}; \\ &\varepsilon(r, 0^+) = \varepsilon_z(r, 0^+) = 0; \\ &\varepsilon_r(r, t) = -\varepsilon_\theta(r, t) = \frac{P_\infty}{2G}\frac{a^2}{r^2}; \\ &\sigma_r(r, 0^+) = -P_\infty\left[1 - \left(\frac{a}{r}\right)^2\right]; \\ &\sigma_\theta(r, 0^+) = -P_\infty\left[1 + \left(\frac{a}{r}\right)^2\right]; \\ &\sigma_z(r, 0^+) = -P_\infty \end{aligned} \qquad (9)$$

Second stage: free convergence of an unlined tunnel ($\boldsymbol{t}_2 = 0^+ \le \boldsymbol{t} \le \boldsymbol{t}_2$)

Stage 2 starts immediately after the excavation (at $t_1 = 0^+$). In this stage, the tunnel remains unlined while stresses, strains and displacements evolve. Due to time continuity, the initial conditions of this stage (at $t_1 = 0^+$) are defined by the solution of Stage 1 (see equations (9)). Concerning the boundary conditions, the radial stress remains at zero at the unlined tunnel wall ($\sigma_r(a, t) = 0$) while the stress and displacement fields remain undisturbed at the far field, as specified by expression (2). The zz-component of expression (7) is given by

$$\dot{s}_z + \frac{1}{T_0}s_z = -\frac{2G}{3}\dot{\varepsilon} \qquad (10)$$

where the characteristic time of relaxation, T_0, is defined by $T_0 = \eta/3G$. Taking into account the initial condition $s_z(r,0^+) = 0$ (deduced from expressions (9)), the solution of s_z is given by

$$\begin{aligned} &s_z = -\frac{2G}{3}\mathrm{e}^{-\frac{t}{T_0}}k_\varepsilon; \\ &k_\varepsilon = \int_0^t \mathrm{e}^{\frac{t'}{T_0}}\dot{\varepsilon}(r, t')\,\mathrm{d}t' \end{aligned} \qquad (11)$$

The same approach applied to the s_θ component, followed by $\mathrm{tr}(\boldsymbol{s}) = s_r + s_\theta + s_z = 0$, gives

$$\begin{aligned} &s_\theta = e^{-\frac{t}{T_0}}\left[2G\left(k_u - \frac{1}{3}k_\varepsilon\right) - P_\infty\left(\frac{a}{r}\right)^2\right]; \\ &k_u = \frac{1}{r}\int_0^t e^{\frac{t'}{T_0}}\dot{u}(r, t')\,\mathrm{d}t' \end{aligned} \qquad (12)$$

$$s_r = e^{-\frac{t}{T_0}}\left[2G\left(\frac{2}{3}k_\varepsilon - k_u\right) + P_\infty\left(\frac{a}{r}\right)^2\right] \qquad (13)$$

After substitution of equations (12) and (13) into equilibrium equation (3), derivation with respect to time and accounting for the boundary condition (2), we obtain, after some mathematical manipulations,

$$\frac{\partial\varepsilon(r, t)}{\partial t} + \frac{\varepsilon(r, t)}{T_1} = 0 \qquad (14)$$

where the characteristic time of creep, T_1, is defined by $T_1 = T_0(K + 4G/3)/K$. We notice that, obviously, $T_1 > T_0$ (i.e. creep is slower than relaxation), which is in accordance with classical results. The solution of equation (14), accounting for the initial condition $\varepsilon(r,0^+) = 0$, is

$$\varepsilon(r, t) = k_\varepsilon(r, t) = 0 \qquad (15)$$

Knowing that $\dot{\varepsilon}^{\mathrm{vp}} = 0$ from equation (5), the medium exhibits no volume change, whatever the value of Poisson's ratio. This can be explained by the assumptions of cylindrical symmetry and plane strain, whereby the stress field is essentially deviatoric. From expressions (15) and (6), we deduce that the mean stress in the medium remains constant at

$\sigma_m(r, t) = -P_\infty$. Injecting expression (15) into the compatibility relation $\varepsilon = \partial_r u + (u/r)$ and taking into account the boundary condition at the far field (expression (2)), we find the following important relation:

$$u(r, t) = U(t)\frac{a^2}{r} \qquad (16)$$

Hence, knowledge of the relative convergence $U(t)$ provides the entire displacement field inside the rock mass. To determine $U(t)$, we use the basic relation $\sigma_r = s_r + \sigma_m$ and combine it with the stress boundary condition ($\sigma_r(a, t) = 0$). A few steps of calculations provide the following solution:

$$U(t) = -\frac{P_\infty}{2G}\left(1 + \frac{t}{T_0}\right)$$
$$\dot{U}(t) = -\frac{P_\infty}{2G\,T_0} \qquad (17)$$

From this, the complete solution of this stage is derived below after a series of substitutions:

$$u(r, t) = -\frac{P_\infty}{2G}\left(1 + \frac{t}{T_0}\right)\frac{a^2}{r};$$
$$\varepsilon_r(r, t) = -\varepsilon_\theta(r, t) = \frac{P_\infty}{2G}\left(1 + \frac{t}{T_0}\right)\frac{a^2}{r^2};$$
$$\varepsilon_z(r, t) = 0; \qquad (18)$$
$$\sigma_r(r, t) = -P_\infty\left[1 - \left(\frac{a}{r}\right)^2\right];$$
$$\sigma_\theta(r, t) = -P_\infty\left[1 + \left(\frac{a}{r}\right)^2\right];$$
$$\sigma_z(r, t) = -P_\infty$$

It can be observed, by comparing expressions (18) with expressions (9), that the stresses are stationary during this stage, whereas, due to the creep, the displacements and strains evolve linearly with time, represented by the term t/T_0.

Third stage: operation of the tunnel after installation of the lining support ($t_2 \leq t$)

Stage 3 begins at $t = t_2$ with the installation of a linear elastic lining, which opposes ground convergence due to creep. Initially stress-free, the lining is assimilated to a thin cylindrical tube (thickness e), and its reaction is given by the following classic formulae (Panet 1995):

$$p_L(t) = -K_L[U(t) - U(t_2)];$$
$$K_L = \frac{E_L[a^2 - (a - e)^2]}{(1 + \nu_L)[(1 - 2\nu_L)a^2 + (a - e)^2]} \qquad (19)$$

E_L, ν_L and K_L indicate, respectively, Young's modulus, Poisson's ratio and the stiffness of the lining. The condition at infinity (expression (2)) is still valid for this stage, while in accordance with equations (19), the boundary condition on the radial stress at the tunnel wall is modified to

$$\sigma_r(a, t) = -p_L(t) = K_L[U(t) - U(t_2)] \qquad (20)$$

Due to time continuity, the initial conditions of this stage are given by the solution of Stage 2 in equations (18) at $t = t_2$. Following the same steps as from equations (10) to (13), the three principal deviatoric stresses are derived:

$$s_z = e^{-\frac{t}{T_0}}\left[-\frac{2G}{3}\hat{k}_\varepsilon\right]$$
$$s_\theta = e^{-\frac{t}{T_0}}\left[2G\left(\hat{k}_u - \frac{1}{3}\hat{k}_\varepsilon\right) - e^{\frac{t_2}{T_0}}P_\infty\left(\frac{a}{r}\right)^2\right] \qquad (21)$$
$$s_r = e^{-\frac{t}{T_0}}\left[2G\left(\frac{2}{3}\hat{k}_\varepsilon - \hat{k}_u\right) + e^{\frac{t_2}{T_0}}P_\infty\left(\frac{a}{r}\right)^2\right]$$

where $\hat{k}_\varepsilon(r, t)$ and $\hat{k}_u(r, t)$ are slightly different functions from their counterparts in the second stage :

$$\hat{k}_\varepsilon = \int_{t_2}^{t} e^{\frac{t'}{T_0}}\dot{\varepsilon}(r, t')d\tau;$$
$$\hat{k}_u = \int_{t_2}^{t} \frac{1}{r}e^{\frac{t'}{T_0}}\dot{u}(r, t')d\tau \qquad (22)$$

Following the same reasoning that led to equations (14) in Stage 2, the same conclusion (equation (15)) is drawn. This means that even after lining installation, which applies a compressive stress at the tunnel wall, the velocity field of the medium remains isochoric (no volume change). From this result, equation (16) remains valid for Stage 3. The same procedure as in Stage 2, expressing σ_r and applying stress boundary condition (20), leads to

$$\dot{U}(t) + \frac{1}{T_2}[U(t) - U(t_2)] = -\frac{P_\infty}{K_L T_2};$$
$$T_2 = \frac{K_L + 2G}{K_L}T_0 \qquad (23)$$

Accounting for the initial condition provided by expressions (18) at $t = t_2$, the solution of equations (23) becomes

$$U(t) = -P_\infty\left[\frac{1}{2G}\left(1+\frac{t_2}{T_0}\right)+\frac{1}{K_L}\left(1-\mathrm{e}^{-\frac{t-t_2}{T_2}}\right)\right] \tag{24}$$

The rate of convergence of the third stage (in absolute value) is therefore slower than in the second stage due to the presence of the lining. At large times, the relative convergence tends to stabilize to an asymptotic limit $U(\infty) = -P_\infty[(1/2G)(1+t_2/T_0) + 1/K_L]$. The term P_∞/K_L represents the convergence experienced by the lining. The displacement and lining reaction can then be determined:

$$u(r,t) = -\frac{a^2}{r}P_\infty\left[\frac{1}{2G}\left(1+\frac{t_2}{T_0}\right)+\frac{1}{K_L}\left(1-\mathrm{e}^{-\frac{t-t_2}{T_2}}\right)\right];$$

$$p_L(t) = P_\infty\left[1-\mathrm{e}^{-\frac{(t-t_2)}{T_2}}\right] \tag{25}$$

Subsequently, the stress fields can be expressed to complete the solution of this stage:

$$\sigma_r(r,t) = -P_\infty\left[1-\left(\frac{a}{r}\right)^2\mathrm{e}^{-\frac{(t-t_2)}{T_2}}\right];$$

$$\sigma_\theta(r,t) = -P_\infty\left[1+\left(\frac{a}{r}\right)^2\mathrm{e}^{-\frac{(t-t_2)}{T_2}}\right]; \tag{26}$$

$$\sigma_z(r,t) = -P_\infty$$

It is interesting to note that, due to the assumptions of cylindrical symmetry and plane strain, the deformation of the rock mass is isochoric, independent of the value of Poisson's ratio. In fact, in the final equations governing the evolution of stress, strain and displacement fields, neither Poisson's ratio nor the rock bulk modulus intervene.

Extension to saturated poro-viscoplastic media in the case of a Poisson's ratio of $\nu = 0.5$

In this section, the analytical approach presented above is extended to the case of a porous viscoplastic rock mass saturated with groundwater. Owing to the complexity of the analysis, only the particular case of an incompressible medium resulting from a particular choice of Poisson's ratio ($\nu = 0.5$) will be presented hereafter. An extension to the general case of a compressible medium is currently being completed.

The geometric, life cycle and general assumptions presented in above are still valid for this section. During Stage 1 (instantaneous excavation of the tunnel), the conditions are undrained and the pore pressure remains at its initial value p_{w0}. In Stage 2, the tunnel wall is in contact with ambient air at zero atmospheric pressure, so the pore pressure remains null at this location; that is, $p_w(a, t) = 0$. In Stage 3, the lining is supposed to be poro-elastic with infinite hydraulic conductivity. Consequently, the water pressure field is uniform through the lining and only depends on its value at the interior boundary (Dirichlet boundary condition). This assumption is justified for rocks that are much less permeable than the lining (e.g. argillite/concrete). Accordingly, the lining pore pressures have no effect on its mechanical behaviour. The pore pressure therefore remains null at the tunnel wall, and the mechanical boundary condition at this location does not change compared with the case of the monophasic medium, although the lining is now saturated. Hence, the mechanical conditions of the three stages are identical to those for the monophasic case described above.

We use the general framework proposed by Coussy (2004) to account for the presence of a saturating liquid phase (water) in the considered porous medium. Porosity ϕ is introduced as a new state variable and its variation ($\phi - \phi_0$) relative to its initial value ϕ_0 is assumed to be the sum of elastic and viscoplastic parts: $\phi - \phi_0 = \phi^e + \phi^{vp}$. The following state laws come are drawn from the classical thermodynamics framework (Lemaitre & Chaboche 2001; Coussy 2004):

$$\dot{\boldsymbol{\varepsilon}}^e = \frac{1}{3K}\dot{\sigma}'_m\mathbf{1} + \frac{1}{2G\dot{\boldsymbol{s}}} \tag{27}$$

$$\dot{\phi}^e = b\dot{\varepsilon}^e + \beta\dot{p}_w \tag{28}$$

where $b = 1 - K/K_m$ is Biot's coefficient, $\beta + (b - \phi_0)/K_m$ is pore compressibility, $\boldsymbol{\sigma}' = \boldsymbol{\sigma} + bp_w\mathbf{1}$ is the effective stress tensor, $\sigma_m{}' = \mathrm{tr}(\boldsymbol{\sigma}')/3 = \sigma_m + bp_w$ is the mean effective stress, and K_m is the bulk modulus of the solid matrix. K and G are the macroscopic bulk and shear moduli of the porous solid under drained conditions. Coefficients b and β represent hydro-mechanical coupling. Note that equation (27) is similar to equation (4), but uses the mean effective stress instead of the mean total stress, while equation (28) is specific to porous

Table 1. *Reference parameters used in the numerical applications*

$a = 5.0\,\text{m}$; $e = 0.5\,m$; $P_\infty = 12\,\text{MPa}$; $p_{w0} = 5\,\text{MPa}$;
$E = 4000\,\text{MPa}$; $\nu = 0.5$; $\eta = 10^{17}\,\text{Pa}\cdot\text{s}$;
$E_L = 1.5\times 10^4\,\text{MPa}$; $\nu_L = 0.2$; $\lambda_h = 10^{-17}\,\text{m}^4 N^{-1} s^{-1}$;
$b = 0.6$; $\beta_0 = 0\ \text{MPa}^{-1}$; $\phi_0 = 0.15$; $K_f = 2200\,\text{MPa}$;
$t_0 = 0$; $t_1 = 0^+$; $t_2 = 2$ months
Leading to
$G = 1333$ MPa; $1/M = 6.82\times 10^{-5}$ MPa^{-1}
$K_L = 1684$ MPa; $T_0 = 0.8$ years; $T_2 = 2.16$ years; $T_{hm} = 5.5$ years

media. Then, by adopting the simplifying assumption proposed by Coussy (2004),

$$\phi;^{\text{vp}} = b^{\text{vp}}\phi^{\text{vp}} \tag{29}$$

where b^{vp} is a material constant, we can postulate the dissipation potential $\varphi^* = \varphi^*(\boldsymbol{\sigma}'^{\text{vp}})$ under the Norton–Hoff form, which is formally similar to that of the monophasic case but using $\boldsymbol{\sigma}'^{\text{vp}} = \boldsymbol{\sigma} + b^{\text{vp}}p_w\mathbf{1}$ instead of $\boldsymbol{\sigma}$ (note that, generally, $\boldsymbol{\sigma}'^{\text{vp}} \neq \boldsymbol{\sigma}'$). Because $\mathbf{s}'^{\text{vp}} = \mathbf{s}$, the viscoplastic strains are still given by relation (5).

Accounting for $\dot{\varepsilon}^e = 0$ (because $\nu = 0.5$ implies $G = E/3$ and $K \to \infty$) and for $\dot{\varepsilon}^{\text{vp}} = 0$, the following equations can be derived:

$$\dot{\boldsymbol{\varepsilon}} = \frac{1}{2G}\dot{\mathbf{s}} + \frac{3}{2\eta_s}\mathbf{s};\quad \dot{\varepsilon} = 0 \tag{30}$$

Equation (16) still applies so that $u(r,t) = U(t)a^2/r$. In parallel, combining the mass balance equation of the fluid phase and Darcy's law, taking into account assumption (29) and noting that $\dot{\varepsilon}^e = \dot{\varepsilon}^{\text{vp}} = 0$, the following diffusion equation is obtained:

$$-\dot{p}_w + \lambda_h M\,\Delta\,\text{pw} = 0 \tag{31}$$

where M is Biot's modulus ($1/M = \beta + \phi_0/K_f$), β is pore compressibility, λ_h is hydraulic conductivity, K_f is the bulk modulus of the fluid phase, and $\Delta = \nabla\cdot\nabla$, the Laplace operator. We observe that the governing equations (30) and (31) are decoupled. Due to this uncoupling, the analytical resolution of the mechanical problem is similar to the previous section, expressed for each stage in expressions (9), (18), (25) and (26).

For the hydraulic problem, we chose to address it with the Laplace transform method, defined as $\bar{f}(r,s) = \mathcal{L}(f) = \int_0^\infty f(r,t)\text{e}^{-st}\text{d}t$, which transforms partial differential equations to ordinary ones. Once resolved in the Laplace-transformed space, the solution can be found in the time domain by using a numerical inversion algorithm such as that from Stehfest (1970):

$$f(r,t) = \mathcal{L}^{-1}[\bar{f}(r,s)] \cong \frac{\ln 2}{t}\sum_{n=1}^{\mathcal{N}} C_n\bar{f}\left(n\frac{\ln 2}{t}\right) \tag{32}$$

$$C_n = (-1)^{n+\frac{\mathcal{N}}{2}} \sum_{k=\text{Int}\left(\frac{n+1}{2}\right)}^{\min\left(n,\frac{\mathcal{N}}{2}\right)} \times \frac{k^{\frac{\mathcal{N}}{2}}(2k)!}{\left(\frac{\mathcal{N}}{2}-k\right)!k!(k-1)!(n-k)!(2k-n)!} \tag{33}$$

where Int(x) is the integer part of x, and N is an even positive integer. The validity of this numerical tool has already been verified (Wong *et al.* 2008*a*, *b*; Dufour *et al.* 2009, 2012). Because of this numerical inversion, the solution is called 'quasi-analytical'.

For Stage 2, the Laplace transform of the diffusion equation (31) is:

$$\frac{\partial^2\overline{p_w}(r,s)}{\partial r^2} + \frac{1}{r}\frac{\partial\overline{p_w}(r,s)}{\partial r} - \Omega^2(s)\overline{p_w}(r,s) = \frac{-\Omega^2(s)}{s}p_{w0} \tag{34}$$

Table 2. *Definition of dimensionless quantities*

$r^*; = \dfrac{r}{a}$;	$t^* = \dfrac{t}{T_0}$;	$t_1* = \dfrac{t_1}{T_0}$;	$t_2^* = \dfrac{t_2}{T_0}$;
$t_{hm} = \dfrac{t}{T_{hm}}$;	$t_{1hm} = \dfrac{t_1}{T_{hm}}$;	$t_{2hm} = \dfrac{t_2}{T_{hm}}$	
$\varepsilon_c = \dfrac{P_\infty}{2G}$;	$u^* = \dfrac{u}{a\varepsilon_c}$;	$\boldsymbol{\varepsilon}^* = \dfrac{\boldsymbol{\varepsilon}}{\boldsymbol{\varepsilon}_c}$;	$\boldsymbol{\sigma}^* = \dfrac{\boldsymbol{\sigma}}{P_\infty}$;
$p_L^* = \dfrac{p_s}{P_\infty}$;	$p_w^* = \dfrac{p_w}{p_{w0}}$		

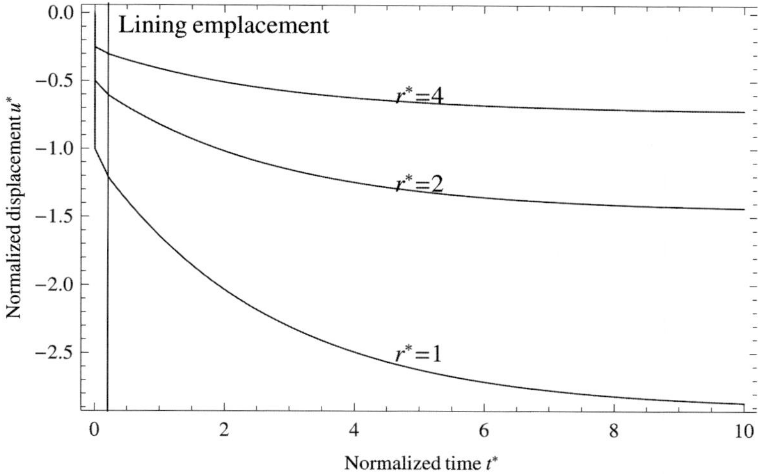

Fig. 2. Temporal evolution of normalized displacements at three different locations ($r = a$, $r = 2a$, $r = 4a$).

where $\Omega(s) = \sqrt{(s/\lambda_h M)}$. The general solution of equation (34) is known: $\overline{p_w}(r, s) = A(s)K_0[\Omega(s)r] + B(s)I_0[\Omega(s)r] + \frac{p_{w0}}{s}$
where I_0 and K_0 are the modified Bessel functions of order 0 and of the first and second kind, respectively. Accounting for the two hydraulic boundary conditions $p_w(a,\ t) = 0$ and $p_w(\infty,\ t) = p_{w0}$ and noting that $\lim_{x\to\infty} I_0(x) = \infty$ and $\lim_{x\to\infty} K_0(x) = 0$, the solution of the pore pressure in the Laplace space reduces to

$$\overline{p_w}(r, s) = \frac{p_{w0}}{s}\left[1 - \frac{K_0(\ \Omega r)}{K_0(\ \Omega a)}\right] \qquad (35)$$

The pore pressure is then obtained in the time domain using expressions (32) and (33) (Stehfest's algorithm).

For Stage 3, the pore pressure is still the solution of diffusion equation (31). Because of the hydro-mechanical uncoupling, this equation does not depend on the considered stage of the tunnel life cycle. As both Stages 2 and 3 have similar boundary conditions ($p_w(a,\ t) = 0$ and $p_w(\infty,\ t) = p_{w0}$), the evolution of the pore pressure during Stage 3 is described by the same expression (35) as during Stage 2 in the Laplace domain. It is then obtained in the time space using Stehfest's formula.

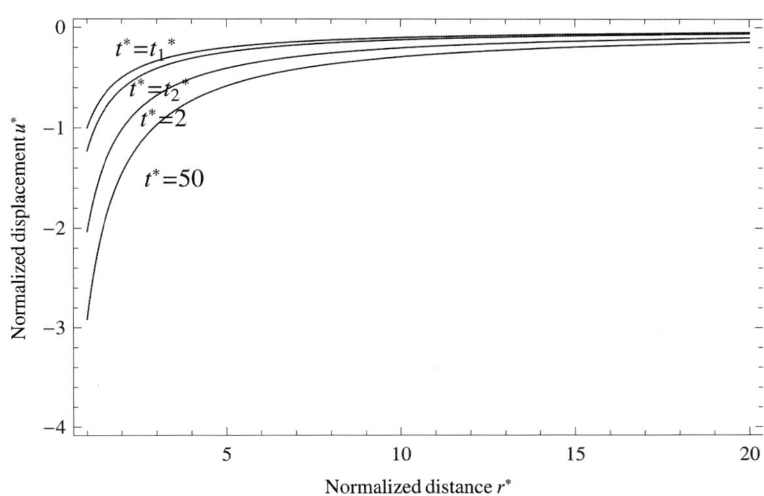

Fig. 3. Normalized displacement profiles at different times.

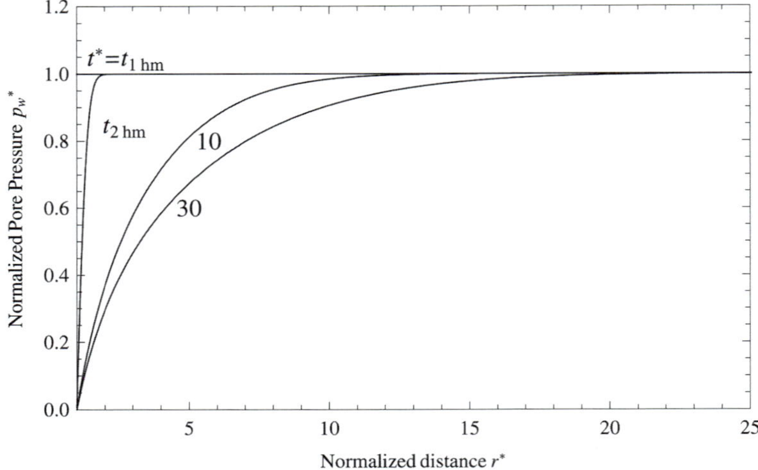

Fig. 4. Normalized pore pressure profiles at different times.

Numerical applications

In this section, a few numerical examples and parametric studies are carried out to illustrate the applicability of the analytical and quasi-analytical solutions. Table 1 summarizes the numerical data used in the calculations and provided by ANDRA (2005), Wileveau & Bernier (2008) and Zhang *et al.* (2010). These relate to the geological disposal project considered in France and will be referred to in the following as 'reference parameters'. Note that, according to Zhang *et al.* (2010), the linear case ($n = 1$) of Norton–Hoff's law considered in this paper seems to be acceptable for the following applications. Parameters are varied one by one in the following parametric study, while the others remain at their reference value. In Table 1, note that the elastic constants E and ν correspond to the drained condition. Moreover, the monophasic model is valid for a general value of Poisson's ratio. Notwithstanding this, in view of the poromechanical model to be presented in the next section, which is

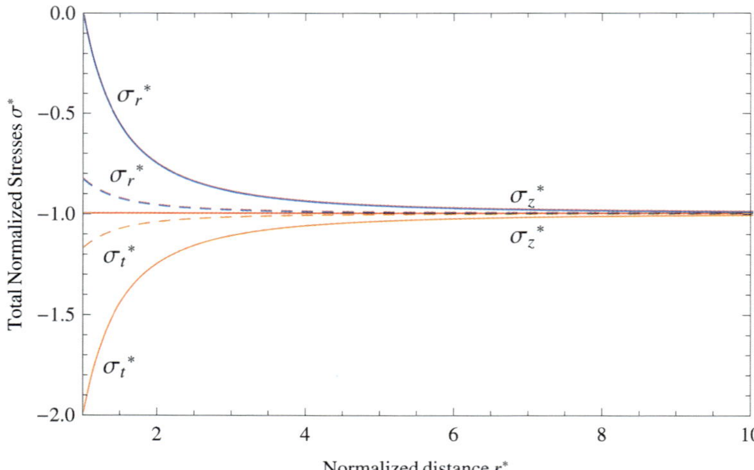

Fig. 5. Profiles of principal total stresses at t_2^* (lining emplacement, continuous lines) and $t^* = 5$ (lining settled, dashed lines).

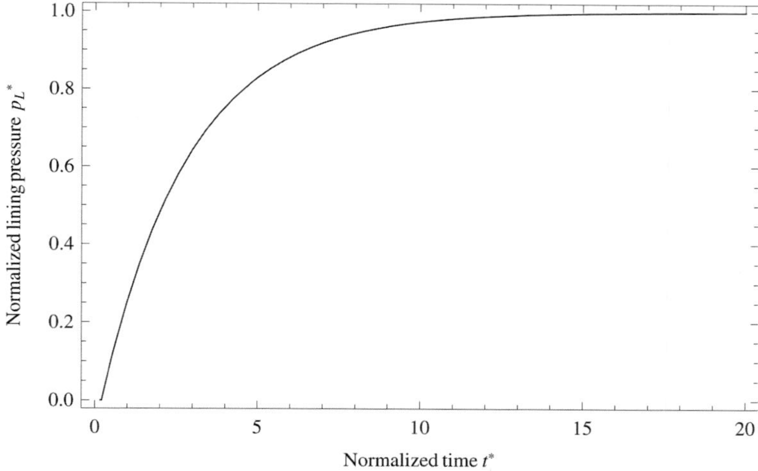

Fig. 6. Temporal evolution of lining pressure.

only valid for an elastically incompressible medium, we consider the particular value $\nu = 0.5$ in the numerical example.

Hydro-mechanical evolutions of normalized quantities

To obtain a more compact presentation and above all to better show the interplay between the various parameters and to underline their physical meaning, the physical quantities will be normalized to a dimensionless form (noted by a superscript *, or a subscript $_{\mathrm{hm}}$ for hydraulic quantities) by dividing them by their respective characteristic values. The radius of the tunnel a was chosen to normalize the coordinate variable. The characteristic time of relaxation, T_0, and of hydro-mechanical coupling, T_{hm}, will be used to normalize the time variable in the mechanical and hydro-mechanical evolutions, respectively. All physical quantities will depend on these variables instead of the dimensional variables. The mechanical stresses and pore pressures we will be normalized by the initial geostatic overburden P_∞ and the initial pore pressure p_{w0}, respectively. The ratio between the initial geostatic overburden and the shear modulus of the rock

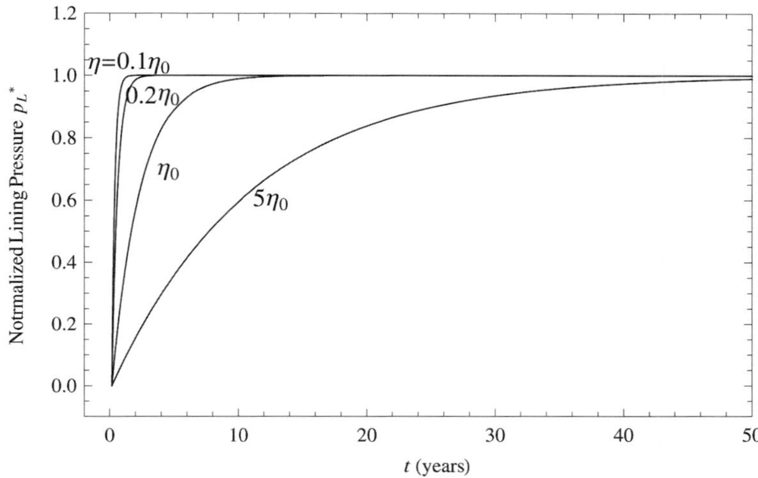

Fig. 7. Temporal evolution of normalized lining reaction with different values of rock viscosity (installation at $t_2 = 2$ months).

mass will be taken as the 'reference' strain ε_c. Table 2 summarizes the normalized quantities.

First, let us consider the reference case where all the parameters are set to their reference values (cf. Table 1). Figure 2 shows the temporal evolution of displacements at different positions in the host rock (at $r = a$, $r = 2a$ and $r = 4a$).

Negative values in this figure correspond to a converging movement towards the tunnel axis. At each point, the absolute value of the displacement first increases instantaneously due to the excavation. After excavation and before lining installation, the displacements continue to increase (in absolute value) more or less linearly with time. This convergence slows after lining installation and tends to an asymptotic value that represents the long-term mechanical equilibrium state. In contrast to the case of poro-elasticity treated in Dufour *et al.* (2012), where the tunnel wall was immobile after excavation, the displacements here evolve at each point before and after lining installation due to rock creep. Note that in the case of $v = 0.5$, hydraulic and mechanical decoupling makes Figure 2 valid for both monophasic and porous saturated rocks.

Figures 3 and 4 show the displacement and pore pressure profiles at different normalized times. In this reference case, from the data given in Table 1 and the definitions in Table 2, we have $t_1^* = 0^+$, $t_2^* = 0.21$, $t_{1hm} = 0^+$ and $t_{2hm} = 0.03$. As expected, the closer the tunnel wall, the greater the displacement. At any time, when r tends to infinity, the displacement tends to zero and the pore pressure tends to its initial hydrostatic value (its normalized value tends to 1). This is consistent with the assumption

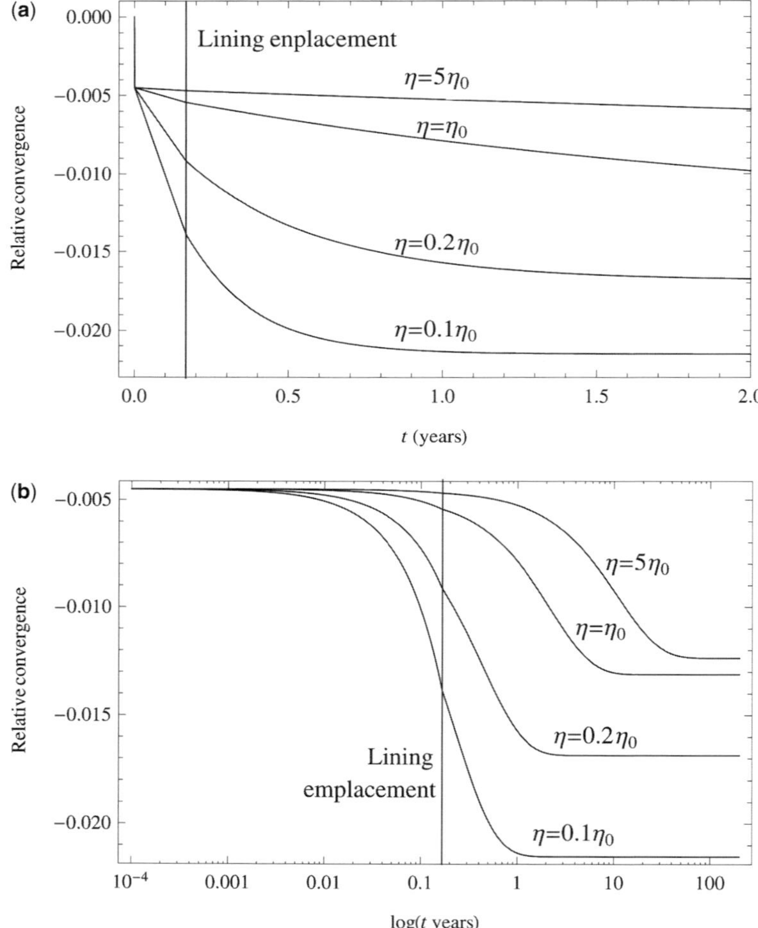

Fig. 8. Temporal evolution of relative convergence with different values of viscosity: (**a**) normal time scale and (**b**) logarithmic time scale.

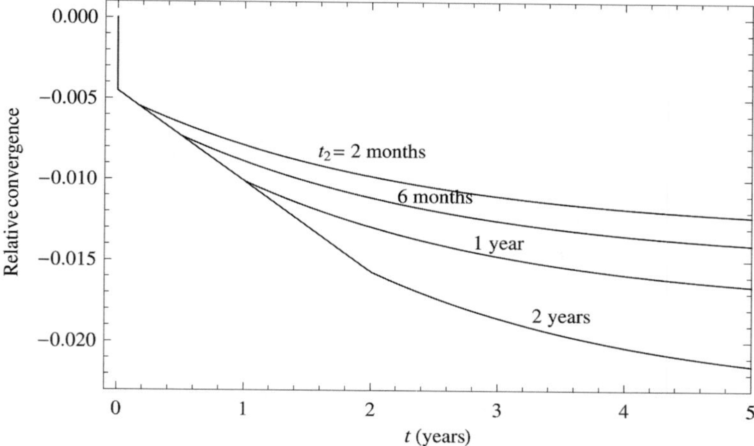

Fig. 9. Temporal evolution of relative convergence with different values of t_2 (time of lining emplacement).

that far from the tunnel, hydro-mechanical disturbances due to changes at the tunnel wall are negligible.

Figure 5 shows the profiles of the total stresses at $t^* = t_2^*$, exactly when the lining is emplaced (continuous lines) and at $t^* = 5$ (after lining emplacement, dashed lines) in the case of a porous saturated host rock. At $t = t_2$, the total radial stress is zero at the tunnel wall, which results from the condition of free surface boundary at this instant, but it becomes compressive with its absolute value increasing with time after lining installation, due to creep. Far from the tunnel and whatever the time, the principal total stresses tend to their initial geostatic value. The axial stress is observed to remain stationary whatever the time, and its value remains equal to the initial geostatic overburden. The circumferential and radial stress profiles are symmetrically distributed on both sides of the axial stress, which directly results from expressions (18) and (26). It can be seen that the stress field tends progressively to its initial isotropic state, equal to the geostatic pressure, due to the zero creep threshold.

Finally, the temporal evolution of the lining pressure is presented in Figure 6. As expected, starting from the moment of lining installation (t_2^*), this pressure increases with time and tends to the geostatic initial stress when t^* increases to infinity. At this limit, the rock mass reaches a final permanent

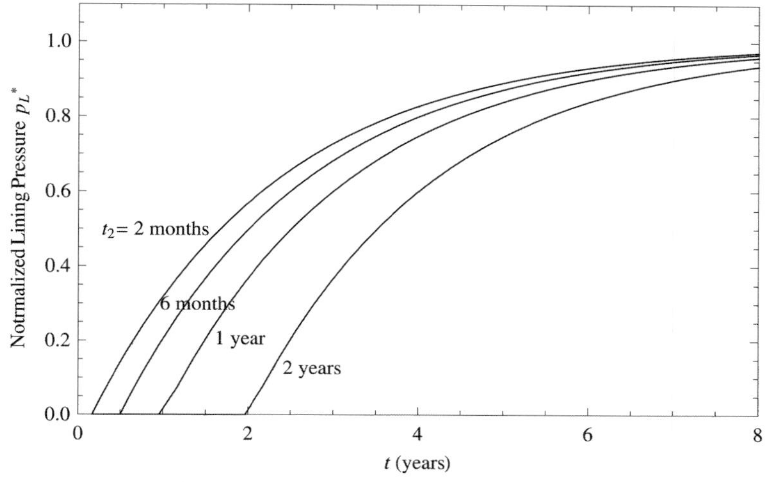

Fig. 10. Temporal evolution of normalized lining reaction with different values of t_2 (time of lining emplacement).

state of mechanical equilibrium, where all quantities remain constant.

Parametric studies

The rock mass viscosity η is a key parameter of the constitutive model used in this work. Therefore, a parametric study of the influence of this parameter is presented hereafter. It also seemed interesting to study the influence of a delay of lining installation on its loading. A parametric study on the instant of lining emplacement t_2 was also therefore carried out. However, because the rock mass viscosity η intervenes in the expression of the characteristic relaxation time T_0, and therefore also in the normalized time t^*, the use of a normalized time to present the results would be misleading in cases where the installation of the lining takes place at a fixed 'real' time. For this reason, in this section, the real time scale is preferred for plotting the figures.

The temporal evolution of the lining reaction is plotted in Figure 7. For any case, this load tends to the value of the initial geostatic pressure. As expected, the rate of this reaction is slow when the viscosity is large. When the viscosity tends to infinity (poro-elastic behaviour of the rock mass), this reaction reduces to zero, because the tunnel wall no longer moves in this case, as already mentioned in Dufour *et al.* (2012).

Figure 8a, b shows the influence of rock viscosity on the temporal evolution of relative convergence $U(t)$, with normal and logarithmic time scales. As expected, an increase in rock viscosity reduces the viscoplastic strains (equation (5)) and thereby decreases the convergence due to creep. An infinite viscosity corresponds to the limiting case of poro-elasticity. In any case, due to the presence of the lining, the relative convergence $U(t)$ always tends to an asymptotic value at large times.

Figures 9 and 10 show the influence of the time of lining emplacement t_2 on the radial displacement of the tunnel wall and on the lining reaction. Figure 9 shows that the more the lining installation is delayed (t_2 large), the larger would be the ground convergence, at any given time.

As expected, Figure 10 shows that, at any given time, the liner is less loaded the later it is put in place (t_2 large). However, at large times, this influence is no longer visible, which is consistent with expression (25). In other words, the benefit in terms of lining reaction gained from a delay of lining emplacement vanishes with time when $t \gg (t_2 - t_1)$.

Conclusion

This paper presents an analytical approach to the coupled behaviour of a deep tunnel inside a viscoplastic rock mass, accounting for a simplified life cycle. A fully analytical solution is provided for the case of a monophasic medium and is extended to porous saturated media in a quasi-analytical way with the assumption of a Poisson's ratio of 0.5. Viscoplasticity is modelled for a linear case (exponent $n = 1$) of Norton–Hoff's law. It is shown that the assumption of a Poisson's ratio for the medium of 0.5 decouples the mechanical and hydraulic behaviours. Some numerical examples were carried out to examine the consistency and relevance of the solutions. Parametric studies illustrate the influence of key parameters such as rock mass viscosity, which has a dominant influence on the evolution of convergence, and the delay of lining installation, which can substantially modify the loading it supports. Hence, this quasi-analytical solution constitutes a useful reference to comfort more complex numerical simulation, and makes it possible to obtain a practical estimation in the calculation of underground structures. A next step is to extend the quasi-analytical solution obtained for the case of a poro-viscoplastic medium to any value of Poisson's ratio. Research perspectives, such as accounting for partial saturation, anisotropic damage, or the backfilling and post-closure stage of the lifetime of the tunnel, are also under consideration.

References

ANDRA 2005. *Dossier 2005 argile: Evaluation de la faisabilité du stockage géologique en formation argileuse.* ANDRA, Paris.

Boidy, E., Bouvard, A. & Pellet, F. 2002. Back analysis of time-dependent behaviour of a test gallery in claystone. *Tunnelling and Underground Space Technology*, **17**, 415–424.

Carranza-Torres, C. & Zhao, J. 2009. Analytical and numerical study of the effect of water pressure on the mechanical response of cylindrical lined tunnels in elastic and elasto-plastic porous media. *International Journal of Rock Mechanics and Mining Sciences*, **46**, 531–547.

Chiarelli, A., Shao, J. & Hoteit, N. 2003. Modelling of elastoplastic damage behaviour of a claystone. *International Journal of Plasticity*, **19**, 23–45.

Cosenza, P. & Ghoreychi, M. 1999. Effects of very low permeability on the long-term evolution of a storage cavern in rock salt. *International Journal of Rock Mechanics and Mining Sciences*, **36**, 527–533.

Coussy, O. 2004. *Poromechanics*. John Wiley & Sons, Chichester.

Dufour, N., Leo, C. J., Deleruyelle, F. & Wong, H. 2009. Hydromechanical responses of a decommissioned backfilled tunnel drilled into a poro-viscoelastic medium. *Soils and Foundations*, **49**, 495–507.

Dufour, N., Wong, H., Deleruyelle, F. & Leo, C. J. 2012. Hydromechanical post-closure behaviour of a deep tunnel taking into account a simplified life cycle. *International Journal of Geomechanics*, **12**, 549–559.

Gasc-Barbier, M., Chanchole, S. & Berest, P. 2004. Creep behavior of Bure clayey rock. *Applied Clay Science*, **26**, 449–458.

Gatelier, N., Pellet, F. & Loret, B. 2002. Mechanical damage of an anisotropic porous rock in cyclic triaxial tests. *International Journal of Rock Mechanics and Mining Sciences*, **39**, 335–354.

Giraud, A. & Rousset, G. 1996. Time-dependent behaviour of deep clays. *Engineering Geology*, **41**, 181–195.

Kazmierczak, J., Laouafa, F., Ghoreychi, M., Lebon, P. & Barnichon, J. 2007. Influence of creep on water pressure measured from borehole tests in the Meuse/Haute-Marne Callovo-Oxfordian argillites. *Physics and Chemistry of the Earth*, **32**, 917–921.

Lemaitre, J. & Chaboche, J. L. 2001. *Mécanique des matériaux solides*. 2nd edn. Dunod, Paris.

Panet, M. 1995. *Le calcul des tunnels par la méthode convergence–confinement*. Presses de l'Ecole Nationale des Ponts et Chaussées ENPC, Paris.

Pouya, A. 1991. *Comportement rhéologique du sel gemme – application à l'étude des excavations souterraines*. PhD thesis, l'Ecole Nationale des Ponts et Chaussées ENPC.

Stehfest, H. 1970. Algorithm 368. *Communication of the Association for Computing Machinery*, **13**, 47–49.

Wileveau, Y. & Bernier, F. 2008. Similarities in the hydromechanical response of Callovo-Oxfordian clay and Boom clay during gallery excavation. *Physics and Chemistry of the Earth*, **33**, S343–S349.

Wong, H., Morvan, M., Deleruyelle, F. & Leo, C. J. 2008*a*. Analytical study of mine closure behaviour in a poro-elastic media. *Computers & Geotechnics*, **35**, 645–654.

Wong, H., Morvan, M., Deleruyelle, F. & Leo, C. J. 2008*b*. Analytical study of mine closure behaviour in a poro-visco-elastic medium. *International Journal of Numerical and Analytical Methods in Geomechanics*, **32**, 1737–1761.

Zhang, C. L., Czaikowski, O. & Rothfuchs, T. 2010. *Thermo-Hydro-Mechanical Behaviour of the Callovo-Oxfordian Clay Rock*. Gesellschaft für Anlagen- und Reaktorsicherheit (GRS) Final Report **12/2010**. http://www.grs.de.

Zhou, H., Zhang, K. & Feng, X. 2011. Experimental study on progressive yielding of marble. *Material Research Innovation*, **15**, S143–S146.

Self-sealing barriers of sand/clay mixtures – lessons learnt from *in situ* experiment and retrospective modelling

O. CZAIKOWSKI*, R. MIEHE & T. ROTHFUCHS

Gesellschaft für Anlagen und Reaktorsicherheit (GRS) mbH, PO Box 38122, Braunschweig, Germany

**Corresponding author (e-mail: oliver.czaikowski@grs.de)*

Abstract: In the framework of the SB experiment ('Self-sealing barriers of sand/clay mixtures'), a series of small *in situ* tests in several vertical boreholes were installed in October 2005 at the underground rock laboratory at Mont Terri. The boreholes were backfilled with sand/clay mixtures at different ratios as well as pure MX-80 bentonite. At the GRS laboratory the experimental setup was installed in advance and has run under nearly the same boundary conditions since March 2005. Sand/clay mixtures exhibit a high permeability to gas in the dry state and a low gas entry pressure and moderate permeability to gas in the saturated state, while the permeability to water is comparable to that of the undisturbed host rock. Therefore, gaseous corrosion products will be preferentially conducted through the seal, mitigating the risk of opening new pathways (fractures) in the host rock. The main objective of the SB experiment presented here is to test and demonstrate that the advantageous sealing properties of clay/sand mixtures determined in the laboratory can technically be achieved and maintained under repository-relevant *in situ* conditions.

As a result of the international experience gained in the field of geological radioactive waste disposal, it is widely agreed that the overall detection of geohydraulic conditions and their changes in response to geomechanical disturbances via *in situ* measurements is one of the most relevant prerequisites for a sound understanding of host-rock behaviour and its interaction with the sealing system. An adequate understanding of related coupled processes is needed for the development of reliable physical models, which in turn are required for numerical simulation of repository performance. To investigate disposal in clay formations, the GRS laboratory (hereafter GRS) carries out *in situ* testing under representative conditions at the Mont Terri underground Research Laboratory (MTRL), as described in Zhang *et al.* (2007), Miehe *et al.* (2010) and Rothfuchs *et al.* (2012).

Sand/clay mixtures exhibit a high permeability to gas in the dry state (between 1×10^{-13} and 1×10^{-15} m^2) and a low gas entry pressure and a moderate permeability to gas (between 1×10^{-16} and 1×10^{-17} m^2) in the saturated state, while the permeability to water in the saturated state of *c.* 1×10^{-18} m^2 is comparable to that of the undisturbed host rock. When used in repository seals, this material will preferentially conduct gaseous corrosion products, mitigating the risk of new pathways opening in the host rock.

The main objective of the SB experiment ('Self-sealing barriers of sand/clay mixtures') presented in this paper is to test and demonstrate that the advantageous sealing properties of clay/sand mixtures determined in the laboratory can technically be achieved and maintained under repository-relevant *in situ* conditions.

Laboratory investigations

Methodology

Before conducting *in situ* experiments, the installation techniques and the required saturation time for the material mixtures were investigated and optimized in a mock-up test on a 1:1 scale in the Geoscientific Laboratory of the GRS in Braunschweig. The principal layout of the mock-up test setup is shown in Figure 1.

The mock-up test was designed as a full-scale replica of the planned *in situ* experiments. The steel tube length in the laboratory was 2.5 m and the seal material was installed in thin layers of *c.* 5–10 cm thickness, similar to the planned *in situ* installation. Different techniques (hand stamping, vibrator technique) were tested and the maximum achievable density was determined.

The detailed objectives of the mock-up test were as follows:

(1) develop and test seal material installation techniques;
(2) determine the time needed to reach full seal saturation;

From: NORRIS, S., BRUNO, J., CATHELINEAU, M., DELAGE, P., FAIRHURST, C., GAUCHER, E. C., HÖHN, E. H., KALINICHEV, A., LALIEUX, P. & SELLIN, P. (eds) 2014. *Clays in Natural and Engineered Barriers for Radioactive Waste Confinement*. Geological Society, London, Special Publications, **400**, 381–397.
First published online April 7, 2014, http://dx.doi.org/10.1144/SP400.16

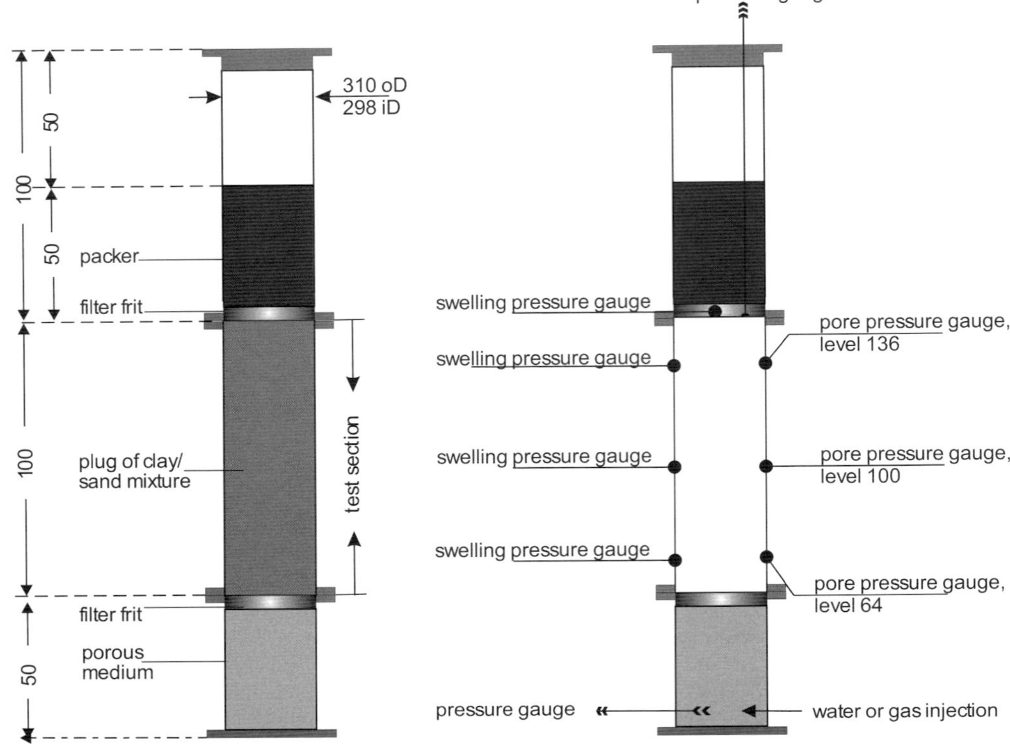

Fig. 1. Principle layout of the SB mock-up test setup with locations of measuring sensors (units in cm).

(3) test the measuring procedures for the determination of gas and water permeability as well as gas entry pressure for dry and saturated conditions, where the conditions were well defined, but slightly different from the *in situ* conditions.

Typical results and discussion

The mock-up test was started in October 2004. Based on experience from preliminary tests (Rothfuchs *et al.* 2004), the mock-up test was prepared carefully. A total amount of 148 728 g of the 65/35 sand/bentonite mixture was installed with an installation density of 2.07 g cm^{-3}, which is slightly better than the target value of 1.93 g cm^{-3}. The starting porosity was 27% and the water content was *c.* 5.8% with a corresponding degree of saturation of *c.* 42%. The initial gas permeability was determined to 6.4×10^{-14} m^2, which corresponds very well with the gas permeability determined in preceding investigations on small samples (see Table 1). Seal saturation in the mock-up test was started in April 2005 with an injection pressure of 1.1 MPa using a synthetic Opalinus clay porewater solution.

As shown in Figure 2, after *c.* 18 months of testing, the total pressure in the seal equalized in the lower and middle part of the seal (red and light green lines) at a value of *c.* 1.1 MPa, which corresponds to the applied water injection pressure. Surprisingly, one does not see a similar evolution of the pressure at the top of the seal (dark blue line).

The first water breakthrough, indicating a situation close to full seal saturation, was expected to occur within six months of the start of water injection. However, this was not observed until September 2007, after more than 29 months of injection. In July 2008, 39 months after start-up of testing, the water inflow and outflow rates equalized at *c.* 10 ml/day, yielding a calculated water permeability of 1.47×10^{-18} m^2, which is in very good agreement with the data obtained from the small samples used in the preceding laboratory tests. Water injection was stopped on 6 July 2008 by reducing the injection pressure to zero. The total amount of water injected was 28 400 g.

Table 1. *Overview of laboratory investigations and determined parameters*

Test parameter	Unit	Index	Initial state	Status at full saturation	Status after gas breakthrough	Comment
(a) Design values						
Installation density	g cm^{-3}	ρ	1.93	–	–	–
Fluid permeability	m^2	K_{fluid}	$>1.0 \times 10^{-15}$ (K_{gas})	1.0×10^{-17}–1.0×10^{-18} (K_{water})	$>1.0 \times 10^{-18}$ (K_{gas})	–
Pressure	MPa	P	–	<2.0 (P_{swelling})	<2.0 ($P_{\text{gas-entry}}$)	–
(b) Data from laboratory tests						
Installation density	g cm^{-3}	ρ	1.87–1.93	–	–	–
Fluid permeability	m^2	K_{fluid}	1.2×10^{-13} (K_{gas})	5.2×10^{-18} (K_{water})	1.4×10^{-17} (K_{gas})	–
Pressure	MPa	P	–	0.2–0.4 (P_{swelling})	0.4–1.1 ($P_{\text{gas-entry}}$)	–
(c) Data from mock-up test						
Density	g cm^{-3}	ρ	2.07/1.95/2.67 ($\rho_{\text{bulk}}/\rho_{\text{dry}}/\rho_{\text{grain}}$)	–	–	–
Porosity	–	ϕ_0	0.27		–	–
Saturation	–	S	0.42†	1.88†	1.47†	–
Mass of water	G	M_{w}	8150 ($M_{\text{w-initial}}$)	36 550 ($\Delta M_{\text{w-in}} = 28\ 400$)	28 630 ($\Delta M_{\text{w-out}} = 79\ 200$)	–
Water content	–	W	0.058	0.26	0.20	Related to $V_{\text{s}} = 140.58$ kg
Fluid permeability	m^2	K_{fluid}	6.4×10^{-14} (K_{gas})	1.5×10^{-18} (K_{water})	2.0×10^{-15} ($K_{\text{gas-break-through}} = 3.7 \times 10^{-17}$)	–
Fluid injection pressure	MPa	P_{fluid}	0.1–0.6 (P_{gas})	1.1 (P_{water})	1.4 ($P_{\text{gas-entry}} = 0.25$)	–
Swelling pressure	MPa	P_{sw}	–	–	0.25–0.35	–

†Related to the calculated porosity of 27% ($0.27 \times 72\ 000$ cm^3 = 19 440 cm^3).

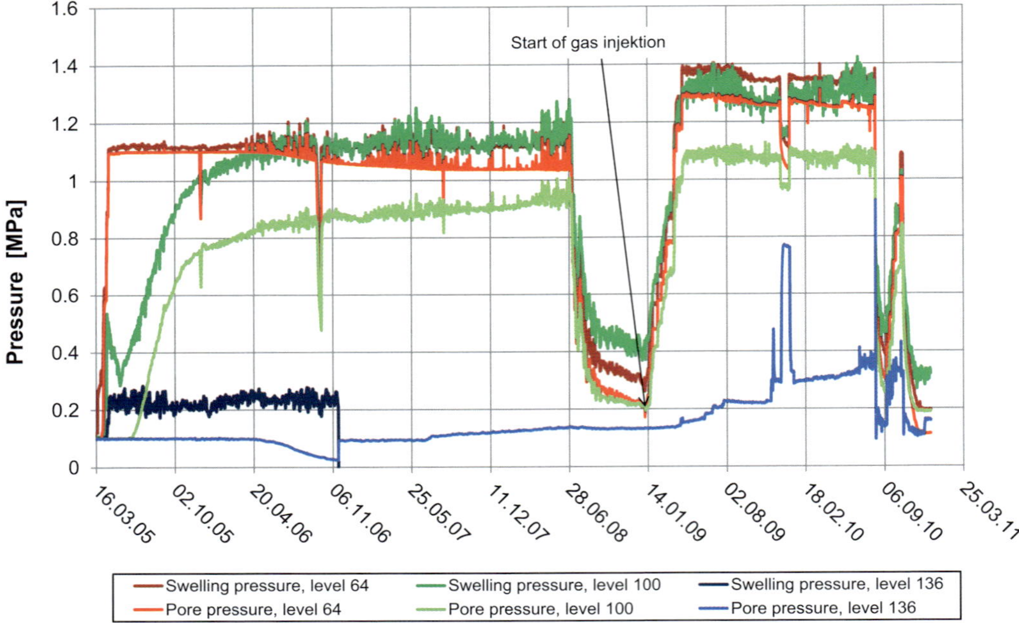

Fig. 2. Mock-up test. Evolution of pore and total pressure within the seal.

The swelling pressure was allowed to equalize for six months before starting the gas injection tests. In January 2009, after a testing period of *c*. 44 months, the total pressure stabilized at a final value of *c*. 0.35 MPa in the seal (Fig. 2) and 0.25 MPa at the packer bottom (sensor 1 in Fig. 3). These values agree with the swelling pressures determined on the small laboratory samples (compare Table 1) and thus confirm the expected seal properties.

The gas injection test was started on 7 January 2009 (Fig. 4). From the very beginning one can clearly see the gas entry and the corresponding water outflow from the seal, starting at a low pressure of only 0.25 MPa. This behaviour agrees very well with the design values given in Table 1 (part a).

By the end of November 2009, the inlet and outlet valves were closed (for technical reasons) for a period of time. An increase in the total pressure inside the seal could be observed during this period (Fig. 3). After opening the valves, the pressure returned to the previously monitored values. Over Christmas, the valves were closed again and a similar pressure increase was observed. The higher values are due to the longer period of time. After the valve was opened, pressures again reverted to the previous values.

A measurable continuous gas flow was only observed some months later in early March 2010. The variability of the data is due to the continuing mobilization of water in the seal material, which influences the gas flow. On the whole, an increase in gas flow rate was observed with a parallel decrease in water discharge rates. From 9 August 2010 onwards, 577 days after start-up of gas injection, no further water discharge was observed and the gas flow rate stabilized at 83 cm^3/min. The seal permeability to gas at this point of the experiment was determined to 3.7×10^{-17} m^2.

Overall, the observations of the mock-up test confirmed the seal properties of the selected 65/35 mixture, which are consistent with seal requirements. Table 1 summarizes the design values and compares them to earlier laboratory investigations and mock-up test results.

The porosity calculated taking into account the natural water content (sand 4.3%, bentonite 8.7%; Miehe 2007) at the seal installation (3986 g + 4166 g) and the amount of water (28 400 g) injected subsequently during the seal saturation phase had a value of 51%. In contrast, the porosity determined on the basis of the grain densities and the dry installation density gives a value of only 27%. Considering the total amount of water injected during seal saturation (28 400 g) and the initial water content of 8150 g, the determined porosity of 27% results in a theoretical saturation degree of 188%. This oversaturation might be explained by some water being adsorbed at a higher density to the interstitial

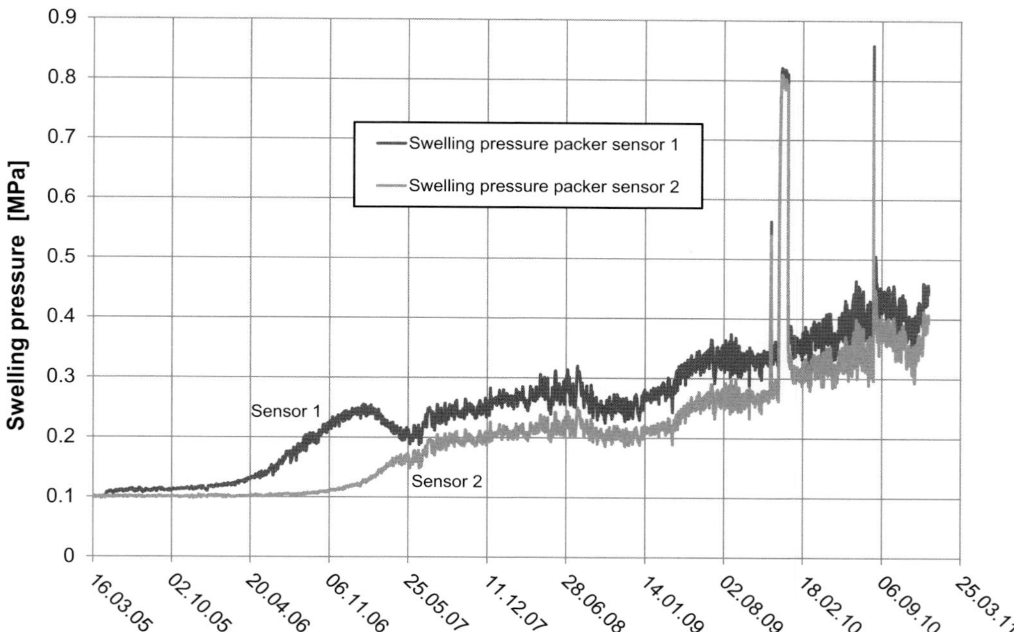

Fig. 3. Mock-up test. Evolution of swelling pressure below the packer.

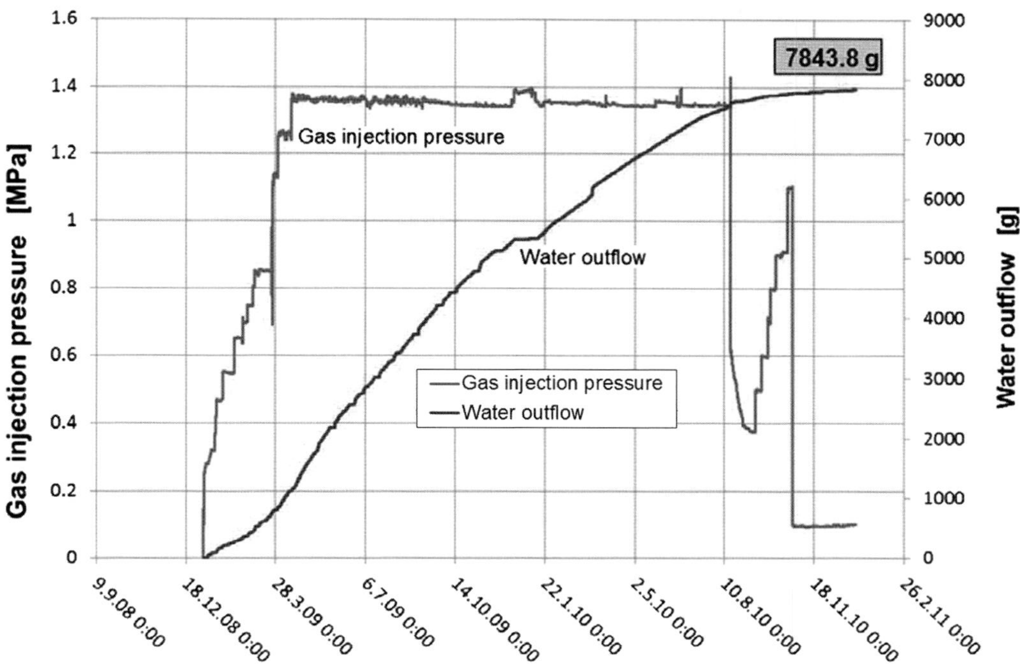

Fig. 4. Mock-up test. Evolution of gas injection pressure and water outflow.

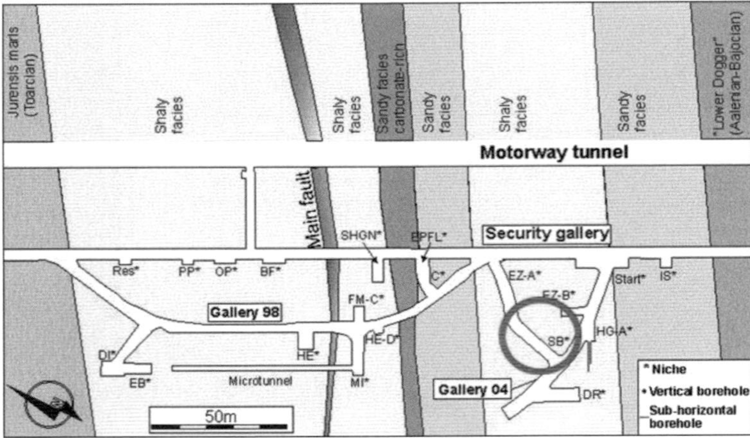

Fig. 5. Location of the SB experiment.

layers of the clay particles, a possible effect reported earlier by Hartge & Horn (1991).

In-situ investigations

Methodology

In the framework of the SB experiment, 65/35 and 50/50 sand/bentonite mixtures were selected to be investigated in the *in situ* experiments at the MTRL. In addition, one borehole was proposed by Nagra (the National Cooperative for the Disposal of Radioactive Waste, Switzerland) to be sealed with a pure MX-80 bentonite granulate, representing the buffer material in the Swiss high level waste (HLW) disposal concept. This experimental setup would be an excellent arrangement for a comparison of the sealing properties of the different sealing materials under representative *in situ* conditions.

The entrance part of a drift connecting Gallery 04 to the security gallery of the Mont Terri motorway tunnel has been excavated and prepared for the SB experiments (see Fig. 5). The test area has a length of *c*. 8 m, a width of 5 m and a height of 4 m. Four test boreholes, each 0.31 m in diameter, were drilled in the test room's floor to a depth of *c*. 3 m (Fig. 6).

The two boreholes SB1 and SB2 had seals of 1 m length consisting of 65/35 sand/bentonite mixtures, as originally planned and considered in the scoping calculations. The lengths of the seals in the two remaining boreholes in the foreground of the test area close to Gallery 04 were reduced to 0.5 m, because long saturation times were predicted in the scoping calculations for the 50/50 sand/bentonite mixture in borehole SB15 and the pure bentonite in borehole SB13 and it was desirable to keep the saturation time within the originally assessed testing duration of *c*. 2.5 years.

Instruments for measuring different hydro-mechanical parameters were also installed. In the lower part of the boreholes, the injection volume was filled with gravel as a porous medium. At the top of the porous medium, a filter frit was placed to ensure a homogeneous distribution of the injected water over the entire borehole cross-section. Above the filter frit, the seal was installed in several layers.

Above the seal, a further filter frit was installed for water and gas collection. This filter frit was connected via a measuring tube to the control panel in the test room's floor to allow measurement of water outflow from the seal after saturation. The entire borehole was sealed against the ambient atmosphere by a gas-tight packer. At the bottom of the packer, two swelling pressure sensors were

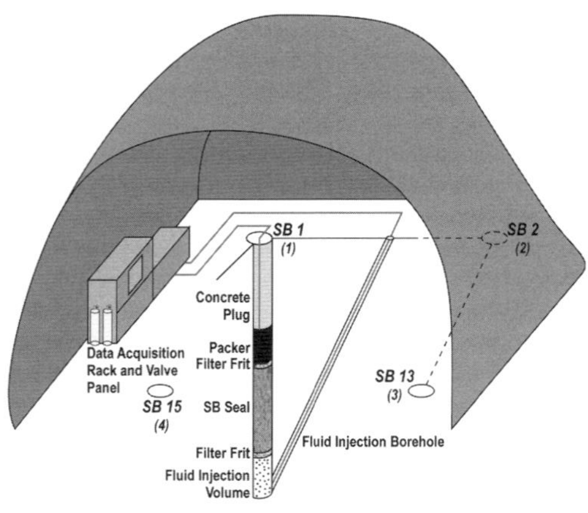

Fig. 6. Arrangement of SB test boreholes.

Fig. 7. Installation of seal material.

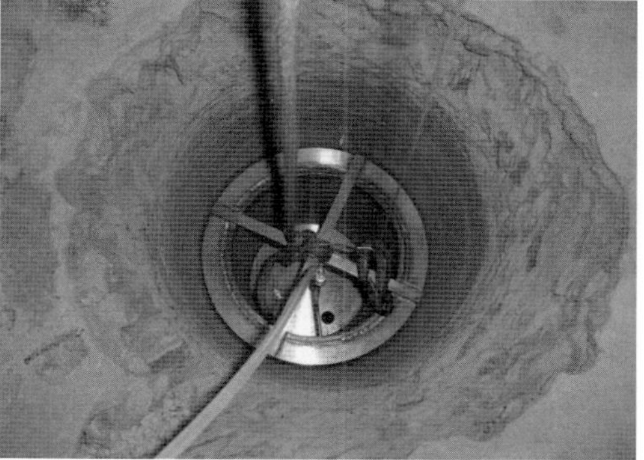

Fig. 9. Packer in borehole SB2.

installed. The uppermost part of the test borehole was grouted to keep the packer in place at the higher swelling pressures developed by the SB seal during water uptake.

For saturation or desaturation of the seal, water or gas could be injected through a tube running from the valve panel in the test room via an inclined borehole and into the lower fluid injection volume.

During the different test stages, as for the mock-up test, the following hydromechanical parameters of the borehole-sealing system were directly measured or determined:

- gas or water flow rate
- accumulated volume of injected water
- gas or water injection pressure
- swelling pressure of the seals
- seal permeability to water for the saturated seal
- gas entry pressure of the saturated seal.

The *in situ* experiments were installed at the MTRL between October 2005 and late October 2006. Figures 7–10 show a collection of photographs taken during the installation of test SB2.

Typical results and discussion

Two of the three *in situ* experiments (SB1 with a 65/35 sand/bentonite mixture and SB15 with a 50/50 sand/bentonite mixture) failed because of bypassing of the injected water through the borehole wall. Borehole SB13 with pure bentonite pellets, which was used for comparison, showed an increase of the swelling pressure up to the measuring limits of 3 MPa, but did not reach saturation within the testing period and also did not show any gas entry under the prevailing swelling pressure conditions.

Test SB2, in contrast, after the application of an improved testing procedure, demonstrated test results similar to the mock-up test. The main results are summarized as follows. The first installed setup of test SB2 with a 65/35 sand/bentonite seal was already started in October 2005 (Fig. 11). In contrast to the mock-up test, the pressure was initially kept at a comparably low level. There were two reasons for this: (1) to avoid greater loss of water into the surrounding rock and (2) because it was assumed that the surrounding rock would

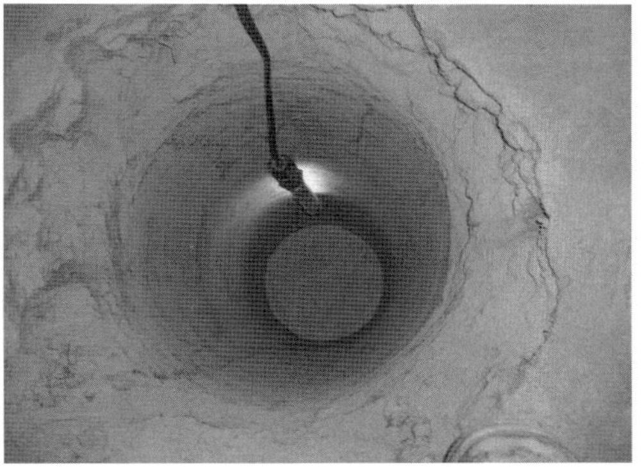

Fig. 8. Filter frit at top of seal.

Fig. 10. Grouted borehole cellar.

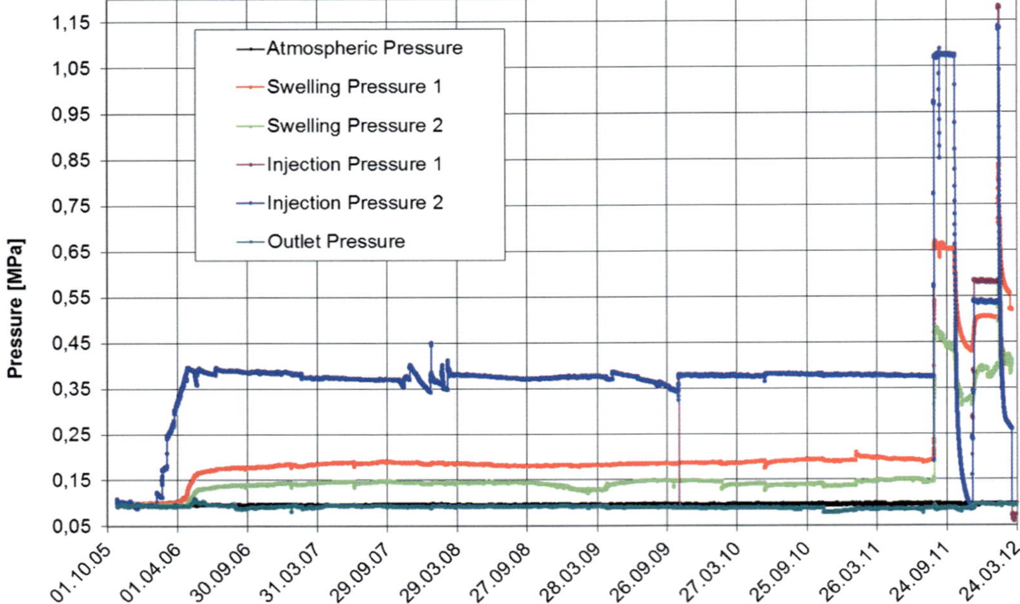

Fig. 11. Pressure evolution in test SB2 sealed with a 65% sand/35% bentonite mixture.

contribute to the seal saturation because of its natural water content and the high suction in the initially dry seal. Both factors could possibly lead to a comparably quick saturation of the seal.

Three and a half months later, from February to April 2006, the injection pressure was increased stepwise to a value of *c*. 0.38 MPa, similar to tests SB1 and SB15. The swelling pressure responded quickly, reaching near final values (between 0.15 and 0.19 MPa) less than one year after the start of water injection. This was a much faster response than in the mock-up test. The measured maximum swelling pressure values of between 0.15 and 0.19 MPa are similar to those determined on small laboratory test samples (compare Table 1 in section 'Typical results and discussion'). Therefore, sealing properties similar to those observed in small samples in the laboratory could be expected from the *in situ* experiment.

By the end of 2006 it was already known from the mock-up test results that the saturation time for the *in situ* experiment would most probably significantly exceed the initial calculated period of time. It was, however, expected that this *in situ* experiment with a 65/35 sand/bentonite mixture could be finished within a period of time similar to that observed in the mock-up test.

However, in June 2011, 68 months after the start of test SB2 (more than twice the duration and water volume of the mock-up saturation time), no water outflow was observed in this *in situ* experiment. Consequently, numerical modelling was performed to examine the interaction between the sealing material and the host rock with the goal of explaining the experimental results and, to the greatest extent possible, arriving at a prognosis on how to finalize the experiment successfully within a reasonable period of time.

Physical modelling and numerical simulation

Methodology

The idea was first to perform a sensitivity analysis on the basis of a variation of porosity and permeability data that could explain the mock-up saturation time adequately.

The computer code CODE_BRIGHT, developed by UPC, is used by GRS for the analysis of coupled thermo-hydro-mechanical (THM) phenomena in geological media (UPC 2002). The interaction between host rock excavation damaged zone (EDZ) evolution and backfill when submitted to desaturation processes can also be investigated. Within the interpretative modelling process, the numerically predicted change in fluid injection pressure achieved success. For detailed information, see Rothfuchs *et al.* (2012).

As far as possible, the simulation process agrees with the realistic time schedule, which means that after simulation of the excavation process in the SB test room, about five months (from mid November 2004 to the end of April 2005) of ventilation and associated desaturation of the whole system was simulated. As a consequence of the capabilities of the simplified elastic hydromechanical coupled approach that was used, which does not take into account any damaging process, the development of an EDZ at the gallery contour was not simulated. The time-dependent desaturation process of the surrounding rock mass induced by ventilation was, however, modelled by setting the hydraulic boundary condition to a negative pore pressure of *c.* −5 MPa, which is equivalent to a fluid saturation value of *c.* 95% according to the dataset and the corresponding retention curve for Opalinus clay. Consequently, the pore pressure field will vary with time. Swelling was not considered.

As a preliminary result, the parameter combination of $K = 1.2 \times 10^{-18}$ m^2 and $\phi = 0.4$ was selected for further calculations. The calibrated model was then used to recalculate the saturation process in both the mock-up test and the SB2 test. Figure 12 shows the evolution of the saturation process in the mock-up test and Figure 13 shows that of the SB2 test, neglecting any hydraulic impact of the surrounding rock. The calculation indicated good agreement of the measured and recalculated saturation time of *c.* 29 months for the mock-up and the saturation time of about four years for the *in situ* experiment with a 65/35 sand/bentonite seal, which in fact was not observed at the SB2 test at this time.

Because of this finding it was decided to take into account the interaction between the sealing material and the host rock in the modelling in order to come to a more reliable explanation of the current status and, as far as possible, make a prognosis as to how to successfully finalize the experiment in 2011.

For improved interpretative modeling, a rotation symmetrical model, 100 m in height and 50 m in length, was used (Fig. 14). In the centre of the model, the SB niche is located. The SB2 experimental layout was abstracted and modelled as a cylindrical borehole, 0.31 m in diameter, and without explicit modelling of the instrumentation in order to keep the calculation effort low. The anisotropic stress state and the hydraulic anisotropy were considered as follows.

At the MTRL, the maximum principal stress at laboratory level is on the order of 6.5 MPa, subvertically oriented, and its magnitude corresponds to the overburden of 250–300 m. One of the two subhorizontal principal stresses is roughly aligned with

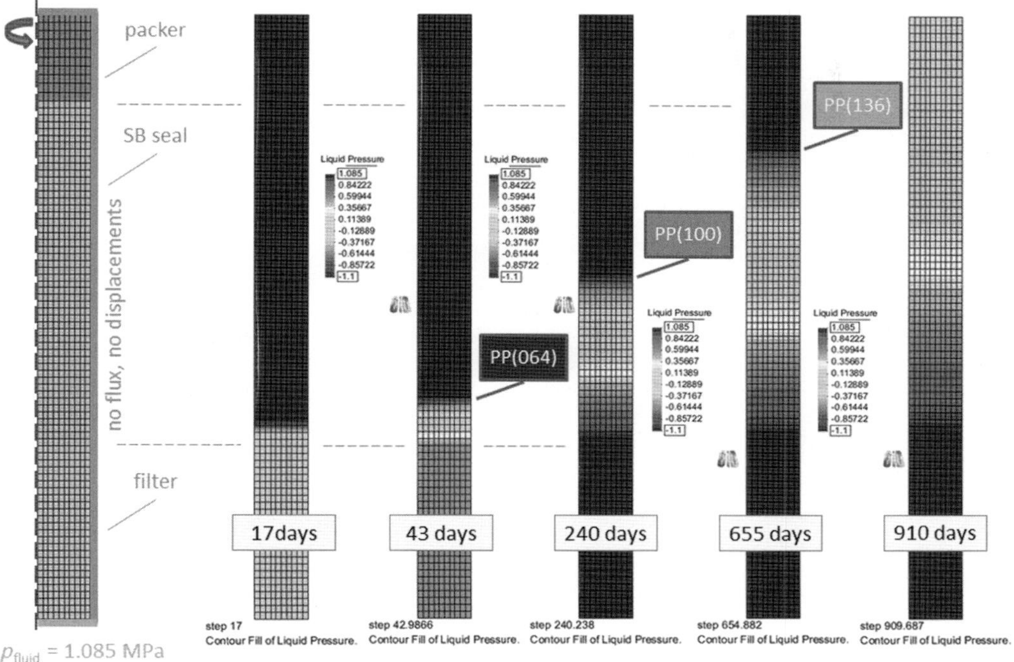

Fig. 12. Evolution of the fluid pressure over time in the mock-up test ($p_{injection} = 1.085$ MPa).

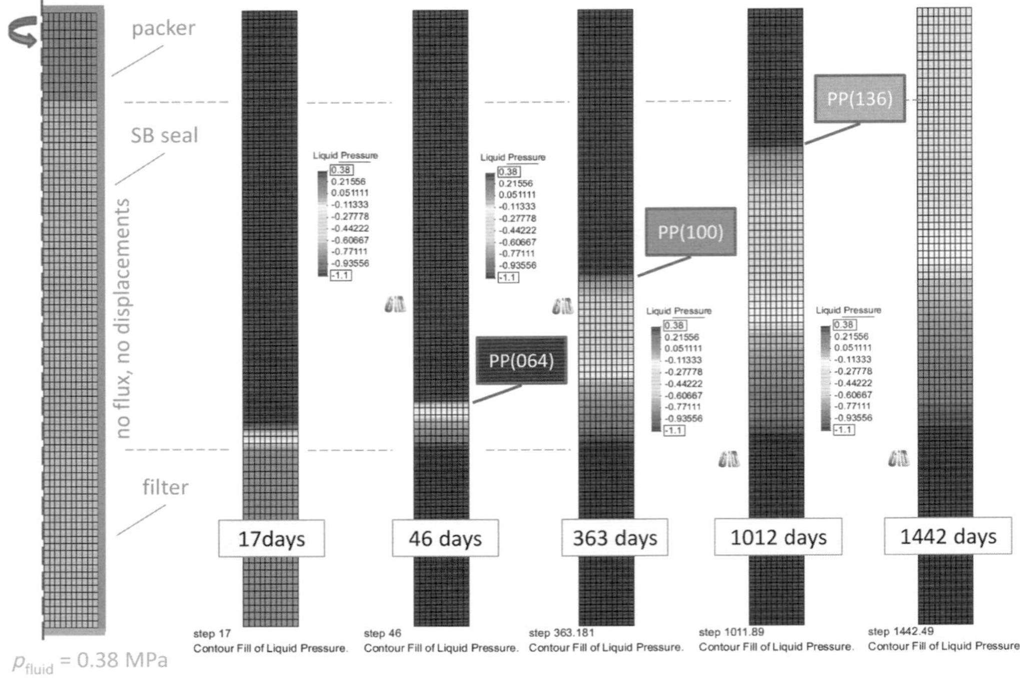

Fig. 13. Evolution of fluid pressure over time in the *in situ* experiment ($p_{injection} = 0.38$ MPa).

the orientation of the security tunnel, and the other is perpendicular to this drift. The described stress state is to be considered a best estimate, because stress measurements in clay-rich rocks are problematic because of the low strength, high anisotropy and swelling properties of the Opalinus clay. An extensive discussion of the stress state can be found in Martin *et al.* (2003). According to Vietor *et al.* (2006), the minor principal stress, which is perpendicular to the orientation of the security tunnel, is assumed to be 2.2 MPa, whereas the intermediate principal stress is oriented parallel and its

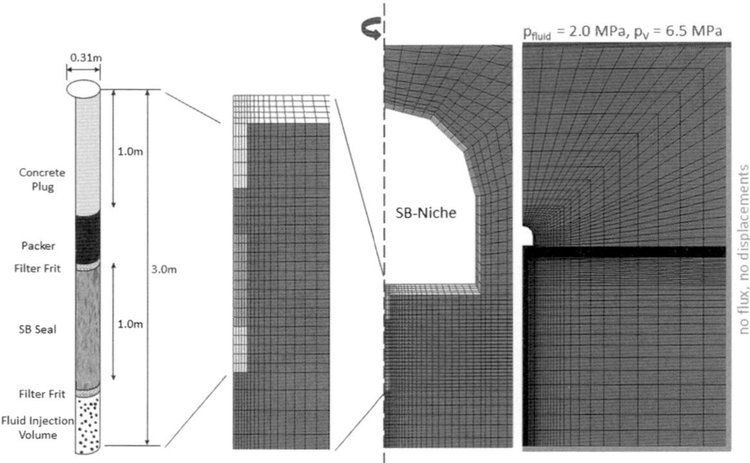

Fig. 14. Rotation symmetrical model.

value is *c*. 4.3 MPa. For the mechanical boundary conditions of the rotation symmetrical model a vertical stress of 6.5 MPa and a horizontal stress of 4.3 MPa are used.

Intact Opalinus clay exhibits a very low hydraulic conductivity, with a mean k_f of 2.0×10^{-13} m s^{-1} within the rock matrix and a much lower value in the direction perpendicular to the bedding planes ($k_f = 6.0 \times 10^{-14}$ m s^{-1}). The very fine pore network is assumed to be fully saturated, with an average porosity of 13.7%, and the pore pressure is assumed to be 2 MPa at laboratory scale (Bock 2008).

Typical results and discussion

On 29 April 2005, the borehole was excavated and prepared for installation. Figures 15 (excavation-induced stress distributions and decompression of the rock) and 16 (time-dependent pore pressure dissipation) show pore pressure propagation (obtained from hydromechanical coupled simulation) at different time steps during the installation sequence of the SB2 experiment. The spectrum of colours used for the water pressure is the same in Figures 15 and 16 and is limited by $p_{\text{fluid}} = 0.1$ MPa (in blue) and $p_{\text{fluid}} = 1.9$ MPa (in red). Values of pore pressure that are outside this range are not shown.

Figure 17 gives an impression on the impact of desaturation in a horizontal cross-section in the rock mass around the SB2 borehole at the buffer material mid-height at selected time steps. The results show that the level of porewater pressure of 0.7 MPa that was calculated before SB2 excavation is reached after 1 month of desaturation at *c*.1.5 m from the borehole contour.

Several porewater pressure sensors were installed in the vicinity of the experimental area between the BSB1 and BSB2 boreholes. Figure 18 shows the position of the corresponding boreholes BSB6–BSB10. The sensors are located at levels 1–2 m below the niche floor. Sensors SB-PWP9 and SB-PWP8 are located at a horizontal distance to the BSB2 axis of 0.36 m, whereas sensors SB-PWP7 and SB-PWP6 are located at distances of 0.75 and 1.5 m. According to Figure 18, the measurement results for SB-PWP7 show pore pressure values similar to atmospheric pressure. This shows that a circular area around the BSB2 borehole with a minimum radius of *c*. 0.75 m is highly influenced by the pre-installation process. Nevertheless, it could not be excluded that the fluid injection pressure itself has a certain influence on the near field rock mass.

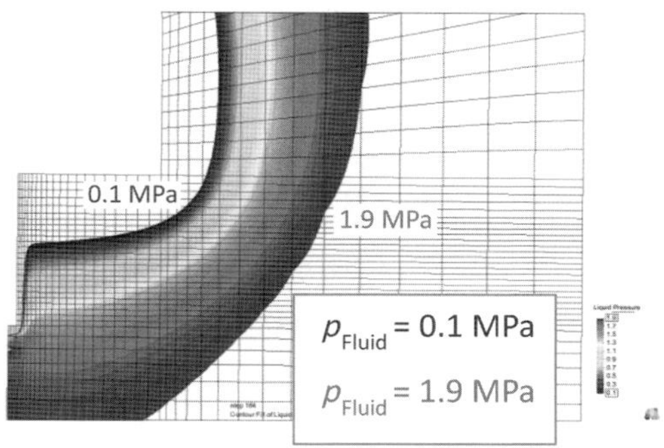

Fig. 15. Pore pressure distribution after drilling of borehole SB2, 29 April 2005.

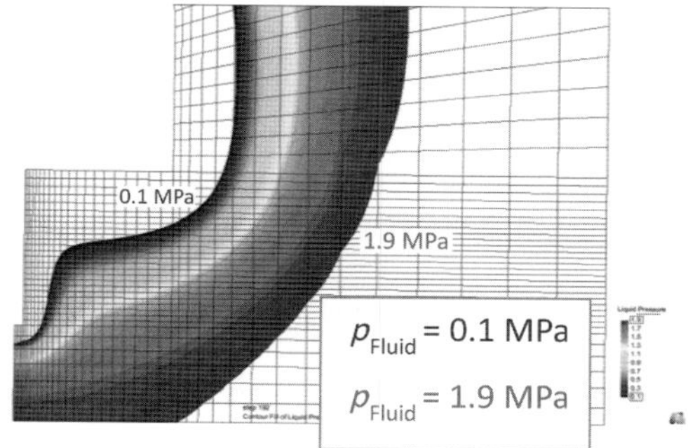

Fig. 16. Pore pressure distribution before SB2 installation, 27 May 2005.

In a second step, the existing model was extended to take into account the interaction between the sealing material and the host rock in order to come to a more reliable explanation of the current status. Taking advantage of earlier investigations made within the project, the finite element (FE) mesh used for the interpretative calculations of the mock-up test was extended with a rectangular body representing the influenced part of the near field rock mass with a radial extension of *c*. 1.5 m. The modified mesh and the new boundary conditions are shown in detail in Figure 19.

Figure 20 shows simulation results for the porewater distribution five years after experiment installation. Similar to Figure 16, the colour spectrum for water pressure is limited at a minimum value (here $p_{\text{fluid}} = 0.2$ MPa, in blue) and a maximum value (here $p_{\text{fluid}} = 0.38$ MPa, in brown). Values in the range of atmospheric pressure are not shown. The results show, furthermore, the importance of the consideration of the rock contour zone and its hydromechanical influence on the experimental trend. The simulation results show full saturation of the buffer after nearly 1.5 years (not shown in the Fig. 20).

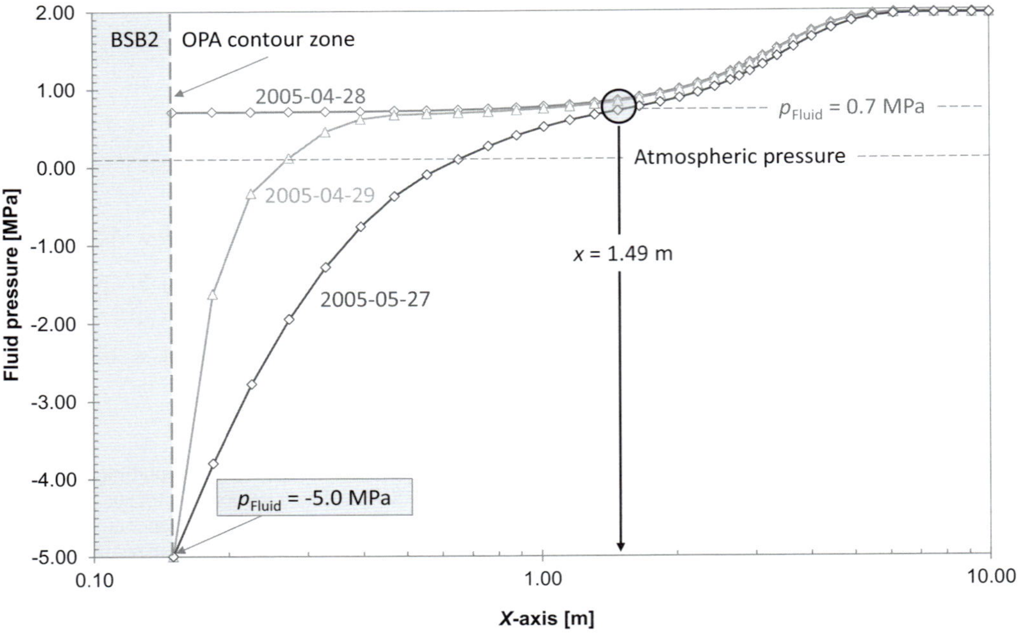

Fig. 17. Pore pressure propagation as a result of hydromechanical coupled simulation at different time steps in a horizontal cross-section.

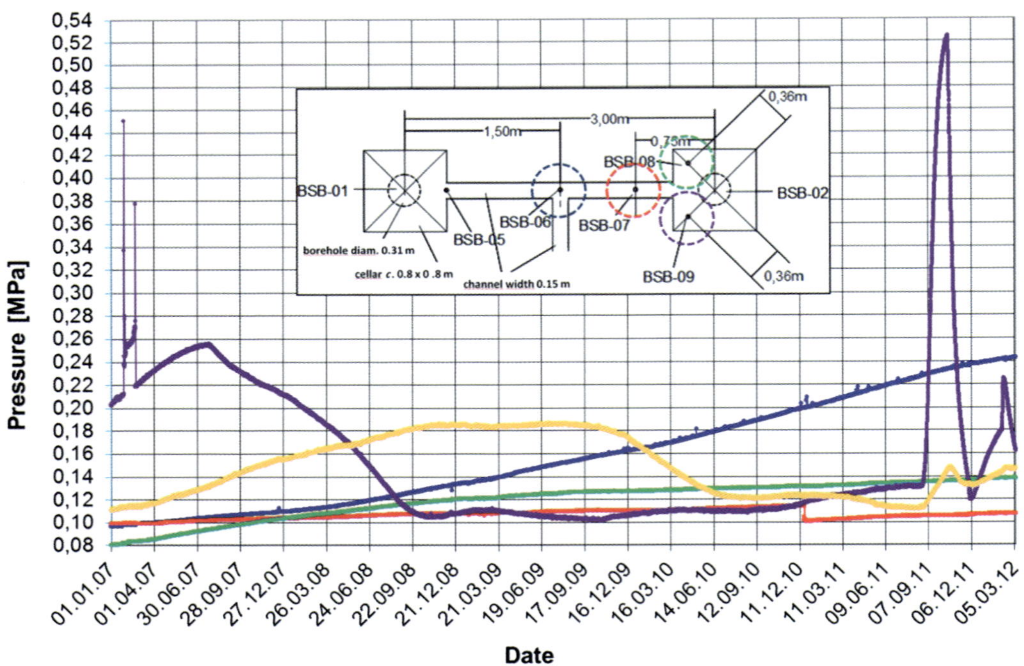

Fig. 18. Pore pressure measurement results in the near field of the BSB1 and BSB2 boreholes.

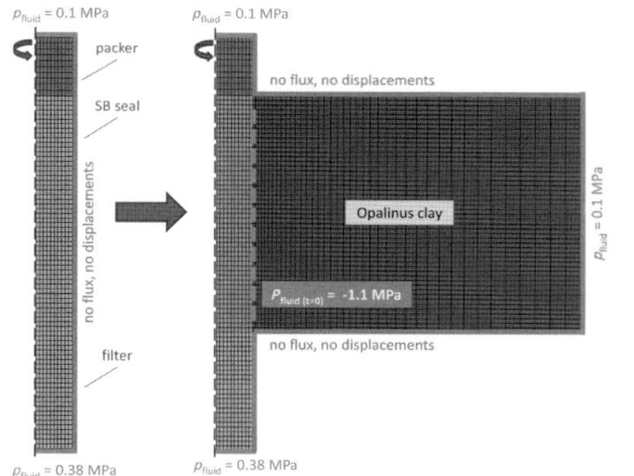

Fig. 19. Modified FE mesh with boundary conditions.

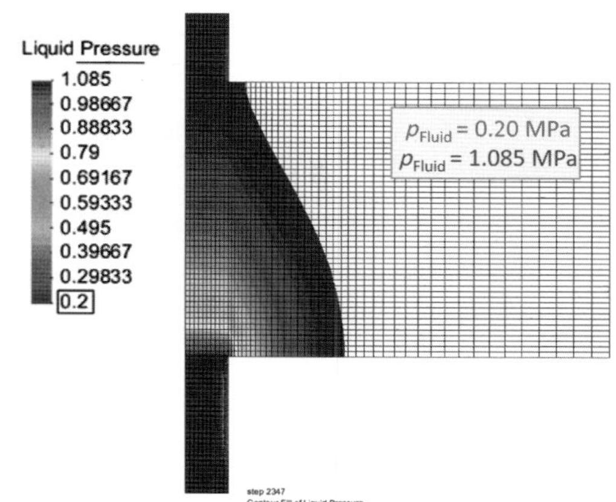

Fig. 21. Pore pressure distribution one day after injection pressure increase to 1.085 MPa (for colour see online version).

However, the water injection pressure of 0.38 MPa applied in the *in situ* experiment does not lead to a measurable pressure increase (<0.18 MPa) at the top of the SB2 sand/bentonite seal. In fact, the pressure distribution leads to equilibrium with the hydraulic boundary conditions of the near field rock mass. Vice versa, the porewater pressure flux towards the SB2 borehole leads to a faster resaturation than interpretatively calculated for the laboratory experiment.

In order to analyse how to finalize the experiment successfully in 2011, the water injection pressure was increased to the same level of *c*. 1.085 MPa that was applied in the mock-up test. Figure 21 shows the evolution of the porewater pressure in the system for 1 day after pressure increase and Figure 22 365 days later. The simulation results show that a distinct increase of the pore pressure to 0.37 MPa inside the seal and the adjacent rock might occur. This pressure level exceeds the porewater pressure of 0.2 MPa prevailing in the system before and could thus lead to a measurable water outflow at the top of the seal within a very short period of time of only 24 h, thereby enabling the pending final determination of the seal permeability.

Lessons learnt from retrospective modelling and final execution of the *in situ* test

Based on the modelling results, the fluid injection pressure was increased in August 2011. To enable the measurement of water flowing through the seal, a small measuring container was placed on a balance outside the borehole at the control valve panel and connected via a water-filled tube to the upper filter frit.

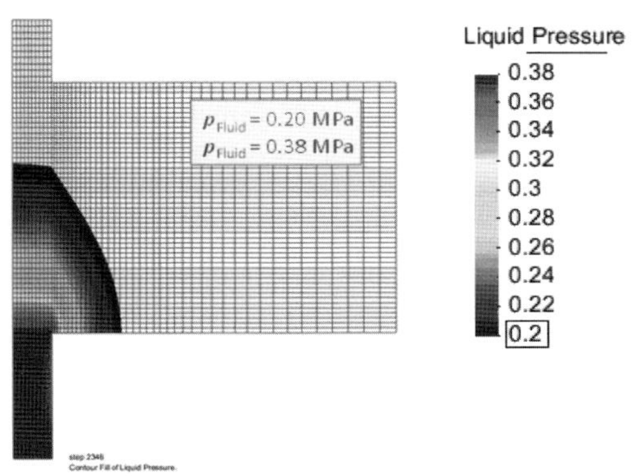

Fig. 20. Pore pressure distribution after five years at an injection pressure of 0.38 MPa (for colour see online version).

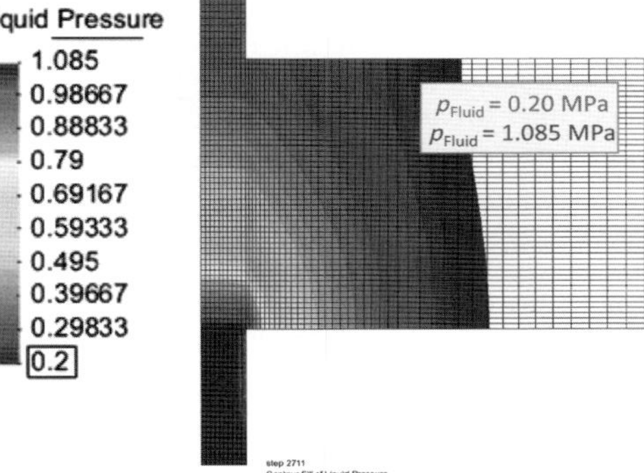

Fig. 22. Pore pressure distribution one year after injection pressure increase to 1.085 MPa (for colour see online version).

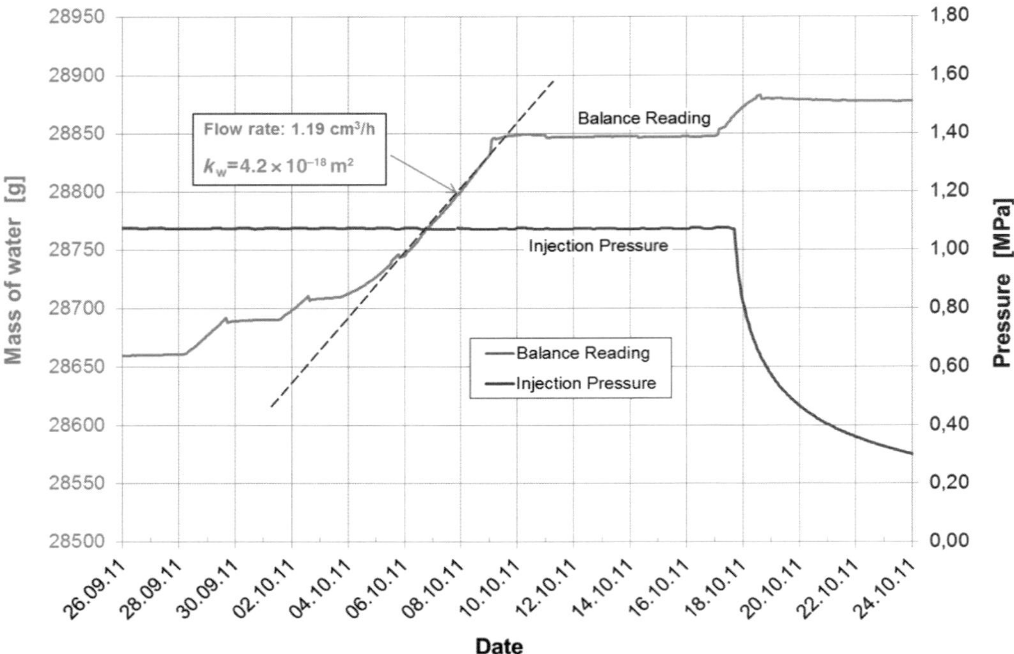

Fig. 23. Determination of seal permeability to water on basis of measured steady-state water outflow from the SB2 seal.

During the pressure increase phase, the water injection pressure was first increased to 1 MPa (absolute), resulting in a slight and soon stagnating water outflow. To enforce water outflow the pressure was again slightly increased to 1.1 MPa. In fact, from this point onwards – similar to the mock-up-test – a periodic water outflow could be seen together with an increase of the swelling pressure sensor signals. Figure 23 shows the evolution of the injection pressure and the measured cumulative mass of water. The steps in the balance reading curve can be explained by some gas volumes remaining in the seal. These gas volumes are displaced by the injected water and interrupt the continuous water flow from the seal to the upper filter frit and outlet tube.

Determination of the water permeability of the seal material

The water outflow rates in the different outflow phases were quite similar. Figure 23 shows the steady-state water outflow from the seal, which be used to determine the seal permeability to water in the saturated seal state. This was 4.2×10^{-18} m^2, a value that is in excellent agreement with the value determined on the small laboratory samples (see Table 1), hence confirming the expected seal properties.

Following the determination of the water permeability, the pending gas injection test started after the preceding pressure equalization phase already mentioned. The system was closed at the injection side, allowing the pressure to reduce and equalize via the seal and the surrounding rock (see Fig. 24). The relatively slow pressure reduction measured is obviously due to the tight contact of the seal to the rock in this test setup. The pressure reduction lasted until it faded almost completely.

Determination of the gas entry pressure of the seal material

The final gas injection test started at the end of November 2011 (see injection pressure curves in Figs 24 & 25). Two effects could be observed during the gas injection phase. First, the swelling pressure sensors' readings increased with the injection pressure, finally approaching values of 0.33 MPa and 0.43 MPa, respectively. This behaviour reflects the mechanical compaction of the saturated seal caused by the increasing gas pressure. Second, a decrease in the amount of water flowing back into the seal was observed. This reduction of the water backflow is most probably due to the closing of existing flow paths within the seal material because of its aforementioned compaction. Just 24 h after increasing the pressure in a second

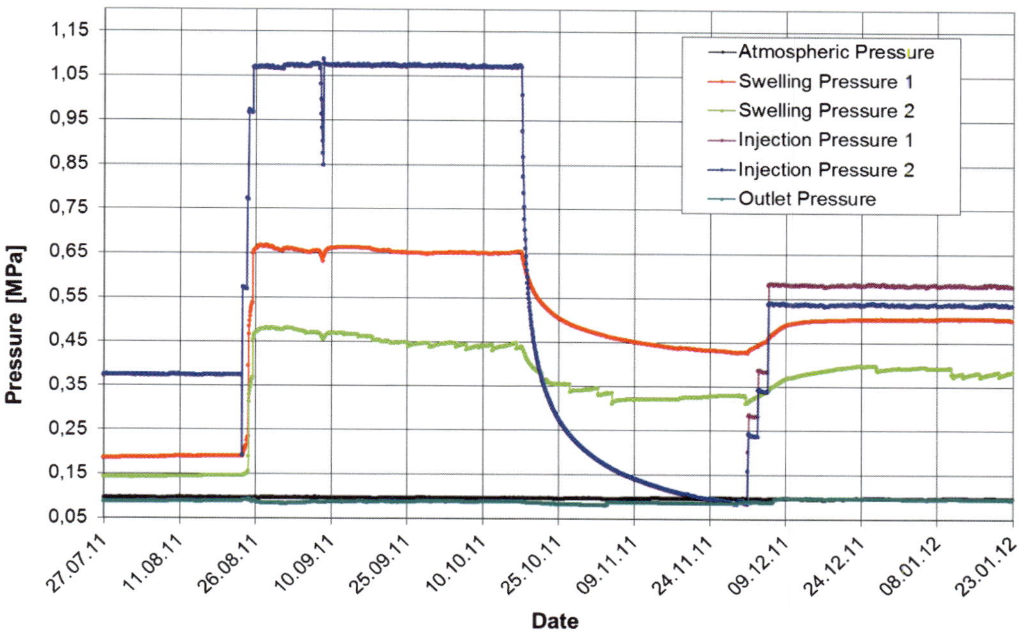

Fig. 24. Pressure signals of test SB2 during the water injection phase, the pressure equalization phase and the final gas injection phase.

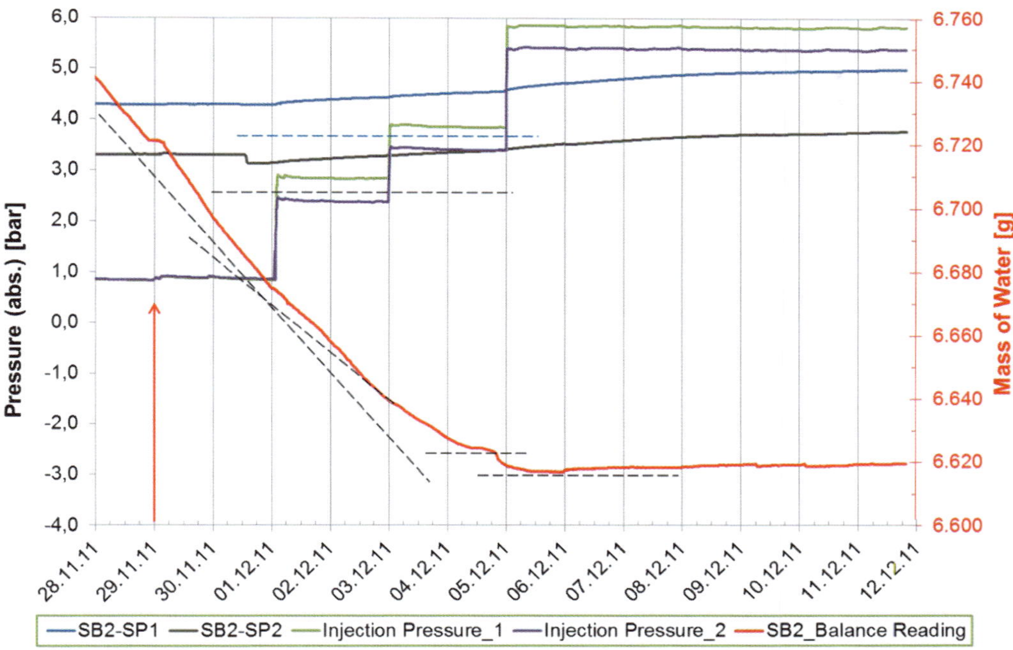

Fig. 25. Gas injection pressure (green and pink), swelling pressure readings (blue and black) and mass of water (red) monitored in the measuring container outside borehole SB2.

step to 0.4 MPa on 3 December 2011, the water backflow started to fade for the first time. A further pressure increase on 5 December 2011 finally led to complete stagnation of the water backflow. This behaviour reflects the starting gas entry into the seal material, which hinders further backflow of water from the outside measuring container.

Test SB2 was then shut in at the turn of 2011 into 2012. Gas injection was continued from 6 February 2012 onwards, with the injection pressure increasing to 1.2 MPa, the value that was also applied in the mock-up test. The awaited gas breakthrough was observed two days later on 8 February 2012 with continuously increasing gas flow rates.

Owing to the fact that the SB project was to be terminated in agreement with the funding institution before 30 June 2012, the GRS technicians stopped gas injection on 10 February 2012. The permeability to gas of the partly desaturated seal, as determined on the basis of the measured gas flow rates between 8 and 10 February, ranged between 9.3×10^{-17} m^2 and 4.1×10^{-16} m^2. This range of values agrees very well with the value determined for the mock-up test and also with the values determined for the small laboratory samples (see Table 1), and thus fulfils the requirements given in Table 1 and confirms the expected seal properties excellently.

Conclusion

The SB project represented a continuation of investigations on the suitability of sand/bentonite mixtures (Jockwer *et al.* 2000; Miehe *et al.* 2003) as an optimized sealing material for nuclear repositories containing gas-generating radioactive wastes. As found in the aforementioned projects, sand/bentonite mixtures exhibit a high permeability to gas in the unsaturated state, allowing the gases to migrate out of the repository. Even after water uptake from the host rock and compaction due to rock creep, these materials exhibit a comparably low gas entry pressure (with bentonite materials), so high gas pressures will not build up in the repository, neither in the unsaturated nor saturated states.

To test and demonstrate that the advantageous sealing properties of sand/bentonite mixtures can be realized technically and maintained *in situ* under repository-relevant conditions, the GRS started the SB project in January 2003 for the consideration of three major project phases: (1) preceding laboratory investigations for the selection of suitable material mixtures and the development of installation/emplacement techniques; (2) large-scale laboratory mock-up testing for the development of suitable material installation techniques and the determination of the time needed to reach full seal saturation; (3) *in situ* testing in boreholes under representative *in situ* conditions in the MTRL.

The preceding laboratory investigations for the determination of the petrophysical material parameters were performed on material mixtures containing 35%, 50% and 70% bentonite. The last of these was found unsuitable, but the other two were selected for further *in situ* testing.

The results of the mock-up test performed in the GRS Geoscientific Laboratory agree with the results of the preceding laboratory investigations determined on small-sized material specimens. However, the time period needed to reach saturation in the mock-up test was underestimated by the pretest modelling by a factor of about 5, indicating significant uncertainties in our understanding of the saturation processes.

Two of the three *in situ* experiments (one with a 65/35 sand/bentonite mixture and one with a 50/50 sand/bentonite mixture) failed because of a bypassing of the injected water through brittle borehole wall zones. The borehole with pure bentonite pellets, which was performed for comparison, showed an impressive evolution of the swelling pressure up to the measuring limits of 3 MPa, but did not reach saturation within the testing period and also did not show any gas entry under the prevailing swelling pressure conditions.

Table 2. *Overview of laboratory and* in situ *test results*

Test parameter	Design values (see Table 1)	Result of laboratory tests	Result of SB2 test
Installation density	1.93 g cm^{-3}	1.87–1.93 g cm^{-3}	1.91 g cm^{-3}
Gas permeability under dry conditions	High (>1×10^{-15} m^2)	1.2×10^{-13} m^2	3.29×10^{-14} m^2
Water permeability at full saturation	1×10^{-17}–1×10^{-18} m^2	5.2×10^{-18} m^2	4.2×10^{-18} m^2
Swelling pressure	<2 MPa	0.2–0.4 MPa	0.15–0.19 MPa
Gas entry pressure	<2 MPa	0.4–1.1 MPa	0.45 MPa
Gas permeability after gas breakthrough	High (>1×10^{-18} m^2)	1.4×10^{-17} m^2	9.3×10^{-17} –4.1×10^{-16} m^2

Test SB2 showed, after the application of an improved testing procedure, results similar to the mock-up test. The main results are given in the Table 2 as a comparison to the results obtained in the preceding laboratory investigations and in the mock-up test.

The data given in Table 2 show that the overall objective of the SB project 'to test and demonstrate that the sealing properties of sand/bentonite-mixtures determined in the laboratory can technically be realized and maintained *in situ* under repository relevant conditions' can be achieved, but there is more work to be done to successfully demonstrate the advantages of that kind of sealing system.

Items to be considered in pursuing R&D activities include the following:

(1) There is a discrepancy between the predicted and observed saturation times in the mock-up and *in situ* experiments. There may be two reasons for this. Either the models implemented in the codes used to design and predict the individual tests are not sufficiently representative of the physics of the water uptake, or the level of uncertainty of the parameter values used in the scoping calculations still needs to be reduced.
(2) Because of the settling behaviour of granular materials, the applicability of sand/bentonite mixtures as sealing materials in horizontal repository drifts needs to be confirmed by additional testing.
(3) So far, the suitability of sand/bentonite mixtures has been investigated at environmental temperature only. The suitability as sealing material under elevated temperature in disposal cells containing high-level radioactive waste is to be investigated.

Finally the following lessons have been learned:

(1) Model calibration on the basis of mock-up test data only was obviously not sufficient to explain the discrepancy concerning the saturation time of sand/bentonite mixtures.
(2) Only the consideration of additional information from the *in situ* test field, such as measured data for the porewater pressure distribution in the surroundings of the SB test area, enabled representative recalculation of important data such as adequate water injection pressure to achieve successful termination of the SB experiment.

The authors would like to thank the German Federal Ministry of Economics and Technology (BMWi) as well as the European Commission (EC) for the support given to the SB project under contract numbers 02E9894 and FI6W-CT-2003-508851. We also would like to thank the Mont Terri Project Management Team for their continuous and most welcomed assistance throughout the entire run time of the SB experiment.

References

Bock, H. 2008. *RA Experiment – Updated Review of the Opalinus Clay at the Mont Terri URL Based on Laboratory and Field Testing*. Technical Report **TR 2008-04.**

Hartge, K. H. & Horn, R. 1991. *Einführung in die Bodenphysik*. Ferdinant Enke Verlag, Stuttgart, 83–84.

Jockwer, N., Miehe, R. & Müller-Lyda, I. 2000. *Untersuchungen zum Zweiphasenfluss und diffusiven Transport in Tonbarrieren und Tongesteinen, Abschlussbericht, FZK-02 E 90170*. Gesellschaft für Anlagen- und Reaktorsicherheit (GRS) mbH, Köln, **GRS-167**.

Martin, C. D., Lanyon, G. W. & Blümling, P. 2003. Measurement of in-situ stress in weak rocks at Mont Terri Rock Lab. *International Journal of Rock Mechanics and Mining Sciences*, **40**, 1077–1088.

Miehe, R. 2007. *Bestimmung der Trockeneinbaudichten in den Bohrlöchern BSB1, BSB2, BSB13, BSB15, SB-Experiment im Mt. Terri Untertagelabor – Prüfbericht*. Gesellschaft für Anlagen- und Reaktorsicherheit (GRS) mbH, Braunschweig.

Miehe, R., Kröhn, P. & Moog, H. 2003. *Hydraulische Kennwerte tonhaltiger Mineralgemische zum Verschluss von Untertagedeponien (KENTON)*. Gesellschaft für Anlagen- und Reaktorsicherheit (GRS) mbH, Köln, **GRS-193**.

Miehe, R., Czaikowski, O. & Wieczorek, K. 2010. *BET – Barrier Integrity of the Isolating Rock Zone in Clay Formations, Final Report*. Gesellschaft für Anlagen- und Reaktorsicherheit (GRS) mbH, Köln, **GRS-261**.

Rothfuchs, T., Jockwer, N., Miehe, R. & Zhang, C.-l. 2004. *Self-Sealing Barriers of Clay/Mineral Mixtures in a Clay Repository SB Experiment in the Mont Terri Rock Laboratory, Final Report of the Pre-Project*. Gesellschaft für Anlagen- und Reaktorsicherheit (GRS) mbH, Köln, **GRS-212**.

Rothfuchs, T., Czaikowski, O., Hartwig, L., Hellwald, K., Komischke, M., Miehe, R. & Zhang, C.-L. 2012. *Self-sealing Barriers of Clay/Mineral Mixtures in a Clay Repository SB Experiment in the Mont Terri Rock Laboratory, Final Report*. Gesellschaft für Anlagen- und Reaktorsicherheit (GRS) mbH, Köln, **GRS-203**.

UPC 2002. *CODE_BRIGHT, a 3-D program for thermo-hydro-mechanical analysis in geological media, user's guide*. Technical University of Catalonia (UPC), Barcelona.

Vietor, T., Blümling, P. & Armand, G. 2006. Failure mechanisms of the Opalinus Clay around underground excavations. *EUROCK 2006 ISRM Regional Symposium on Multiphysics Coupling and Long-term Behaviour in Rock Mechanics*, Liège, Belgium, May 2006.

Zhang, C.-L., Rothfuchs, T. et al.. 2007. Thermal effects on the Opalinus Clay to heating – a joint heating experiment of ANDRA and GRS at the Mont Terri URL (HE-D Project), **GRS-224**, February 2007.

Preliminary modelling of the saturation of a full-sized clay and concrete shaft seal

D. G. PRIYANTO[1]*, D. A. DIXON[2], C.-S. KIM[1], P. KORKEAKOSKI[3] & J. E. VILLAGRAN[4]

[1]*Whiteshell Laboratories, Atomic Energy of Canada Limited, 1 Ara Mooradian Way, Pinawa, Manitoba, Canada R0E 1L0*

[2]*Golder Associates Ltd, Unit 1, 25 Scurfield Boulevard, Winnipeg, Manitoba, Canada R3Y 1G4*

[3]*Posiva Oy, Olkiluoto, FI-27160, Eurajoki, Finland*

[4]*Nuclear Waste Management Organization, 22 St Clair Avenue East, 6th Floor, Toronto, Ontario, Canada M4 T 2S3*

**Corresponding author (e-mail: priyantd@aecl.ca)*

Abstract: A shaft seal was installed at the intersection of the 5 m-diameter access shaft of the Atomic Energy of Canada Limited's Underground Research Laboratory with a major, moderately dipping fracture zone (FZ) to limit the mixing of the saline groundwater at the zone below the FZ with freshwater at the zone above the FZ. The shaft seal consists of 6 m-thick clay-based material that is sandwiched between 3 m-thick upper and lower concrete components, and its centre is located at *c.* 273 m depth below the surface. This paper presents the results of preliminary finite element analyses using 2D-axisymetric and 3D geometries simulating the saturation of the shaft seal. The components in the analyses include the clay, concrete, intact rock and FZ. Of particular interest are the field-observed differences in the hydraulic pressures above and below the shaft seal. The results of the numerical analyses were compared with the 3-year monitoring data. The results of the numerical models showed that the decrease in the pressure difference indicates that there was flow path that connects the zones above and below the shaft seal. A preliminary long-term estimation of the shaft seal hydraulic response (up to 1000 years) was done based on the hydraulic pressure and excavation volume data, combined with the assumptions that that the groundwater level will eventually recover to the initial conditions.

Monitoring of the full-sized clay and concrete shaft seal

The Atomic Energy of Canada Limited (AECL)'s Underground Research Laboratory (URL) was built to provide a facility where concepts for deep geological repositories for Canada's nuclear fuel waste could be studied (Fig. 1). In this document, the AECL's URL is referred to as the 'URL'. The URL operated since 1982 and provided much of the technical information used in developing the used fuel repository concept. In 2003, a decision was made to discontinue operation of the URL and close the underground portion. The vent raise and main shaft to access the URL intersect with the major fracture zone (FZ2; Fig. 1). The groundwater at the zone below the FZ2 is more saline than the groundwater above the FZ2 (Fig. 1). There is a need to preserve the freshwater at the surrounding area, after the closing of the URL's underground facility. Consequently, as part of the decommissioning of the URL, two full-sized seals were installed at the main shaft and ventilation raise where they intersect with a major fracture zone (FZ2; Fig. 1). Since there is no radioactive material stored at the URL, the configuration of the seal at the URL was different from a shaft sealing concept that would be installed at a deep geological repository containing radioactive materials. The shaft seals at the AECL's URL were designed to limit the mixing of the saline groundwater at the zone below the FZ2 with freshwater at the zone above the FZ2. In this respect, it is intended to function in much the same manner as a shaft seal in a repository containing radioactive materials.

Construction of the seals was done as part of the decommissioning of the URL and funded by Canada's Nuclear Legacies Liabilities Program. An Enhanced Sealing Project (ESP) was developed by AECL and jointly funded by NWMO (Canada),

From: Norris, S., Bruno, J., Cathelineau, M., Delage, P., Fairhurst, C., Gaucher, E. C., Höhn, E. H., Kalinichev, A., Lalieux, P. & Sellin, P. (eds) 2014. *Clays in Natural and Engineered Barriers for Radioactive Waste Confinement*. Geological Society, London, Special Publications, **400**, 399–412.
First published online May 15, 2014, http://dx.doi.org/10.1144/SP400.38

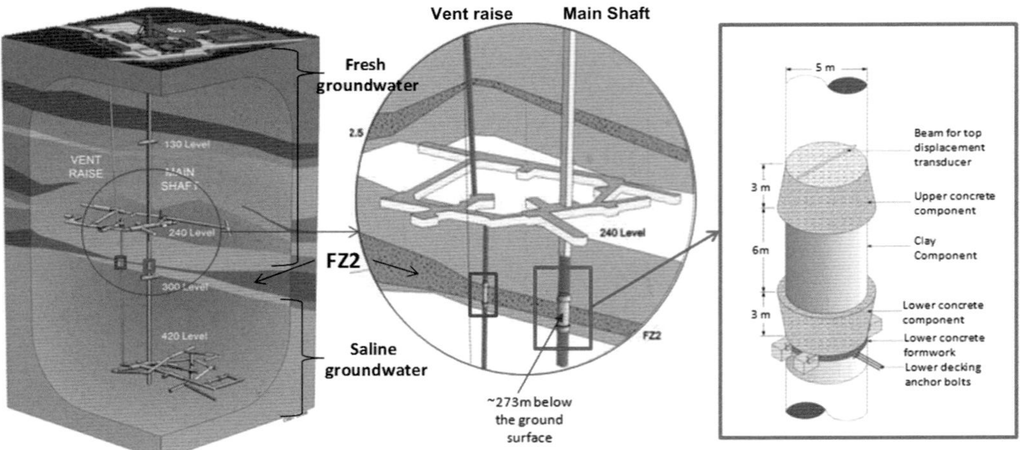

Fig. 1. Location and geometry of access composite shaft seal.

SKB (Sweden), Posiva Oy (Finland) and ANDRA (France). The ESP included: the installation of instruments in one of the larger seal in the main shaft (i.e. shaft seal) and the monitoring of thermal–hydraulic–mechanical responses from 2009 to 2013. The construction and instrument installation of the shaft seal were completed in 2009. The smaller seal in the vent raise was not part of the ESP and it was not monitored.

The shaft seal consists of a 6 m-thick clay component that is sandwiched between 3 m-thick upper and lower concrete components (Fig. 1). The mid-height of the shaft seal was located at a depth of *c.* 273 m below the ground surface. The shaft seal has a nominal diameter of *c.* 5 m.

The clay component was designed to span at least 1 m beyond the maximum identified vertical extend of the moderately dipping FZ2. The clay component was *in situ* compacted from a mixture of 40% bentonite and 60% quartz sand. The clay component provides the hydraulic sealing capability and assists in limiting saline groundwater transport through the seal.

The lower concrete component was made from reinforced concrete and conservatively designed to support the weight of all overlying materials, including the weight of groundwater if the shaft were completely filled. The upper concrete component was an unreinforced concrete structure to restrain the swelling of the clay component. Both upper and lower concrete components were made from high-performance concrete with a reduced heat of hydration. They are keyed approximately 0.5 m into the shaft wall at the elevation where they contact the clay component (Fig. 1). The concrete components provide mechanical support and confinement to the swelling of clay component, but they are not designed to serve a hydraulic sealing function, especially at the concrete–rock interface.

One-hundred sensors were installed at various locations within the shaft seal and nearby host rock to measure temperature, total pressure, hydraulic pressure, suction and volumetric water content, strains and displacements. Since the installation of the shaft seal in mid-2009, the thermal, hydraulic and mechanical evolution of the seal has been continuously monitored. Instrumentation and data monitoring are discussed in more detail in Dixon *et al.* (2012*a*–*c*, 2009), Holowick *et al.* (2011), Martino *et al.* (2010) and Martino & Kaatz (2010).

One of the most significant measurements that indicate the functionality of the shaft seal is the hydraulic pressure difference above and below the clay component (Fig. 2). Total hydraulic pressure measurements shown in Figure 2 were adjusted to the concrete–clay contact to determine the pressure difference above and below the clay component. The hydraulic pressure above the clay component was measured by means of vibrating wire piezometer sensors installed at the interface of the upper concrete component and the shaft wall (V7, V8 and V9) and at the top of the upper concrete component (V10). The hydraulic pressure below the clay component was measured by means of vibrating wire piezometers installed at the interface of the lower concrete component and the shaft wall (V1 and V2). As expected, the hydraulic pressures are relatively uniform at the upper concrete component (V7–V10) and lower concrete component (V1 and V2), because the concrete components were not designed to serve hydraulic sealing function in the rock–concrete interface. The clay component

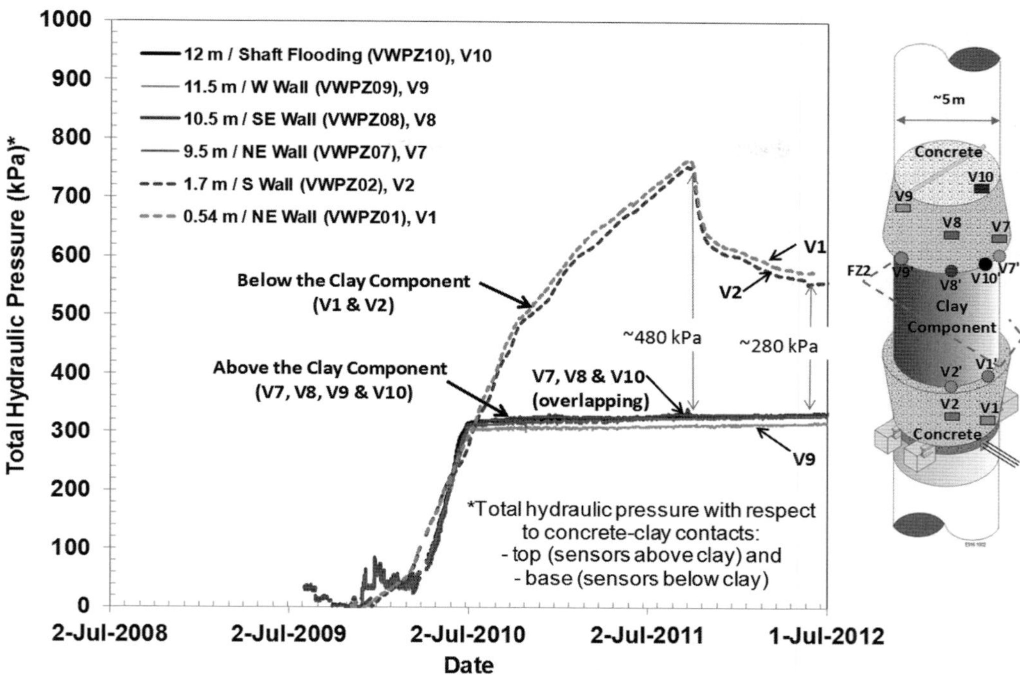

Fig. 2. Measurements showing total hydraulic pressure difference above and below the clay component. (Note for the diagram on the right: V1, V2, V7, V8, V9 and V10 are the original sensor locations; V1′, V2′, V7′, V8′, V9′ and V10′ are the reference locations to adjust total hydraulic pressure.)

provides isolation between the regimes above and below the FZ2. The degree of isolation provided by the clay component can be examined through the hydraulic pressure difference across the clay component (Fig. 2). The hydraulic pressure difference gradually increased and it peaked at approximately 480 kPa in September 2011. Afterwards, it gradually decreased to approximately 280 kPa in July 2012 and stabilized. This hydraulic pressure decrease was suspected owing to the existence of the flow path that connects the regions above and below the FZ2, which will be examined in this paper. This pressure difference is expected to gradually increase as the clay component starts to swell and seal this flow path. As of July 2012, the seal was still providing effective resistance to groundwater transport across the clay component and limiting the mixing of the saline groundwater below the FZ2 and fresh groundwater above the FZ2. This was indicated by the pressure difference, which was significantly greater than the hydrostatic pressure difference that would exist without the presence of the shaft seal (*c.* 60 kPa).

In addition to the ESP data, the groundwater levels have also been monitored from existing hydrogeology boreholes (>20 boreholes) around the URL site since 1984. These combined datasets can be used to calibrate the numerical model developed to describe the likely evolution of this system.

Preliminary numerical modelling of the shaft seal

Scope of work

This paper presents a preliminary numerical modelling to simulate the saturation of the shaft seal and the evolution of the groundwater pressure in the surrounding rock. The focus of this study is to simulate the observed differences in the hydraulic pressure above and below the clay component (Fig. 2) and to examine possible mechanism that can cause the gradual decrease of the hydraulic pressure (such as occurred in July 2012, as shown in Fig. 2). The effect of slight inclination of the FZ2 is also examined.

The study presented in this paper is a preliminary numerical modelling, in the sense that only limited comparisons between numerical modelling results and measurements are made. This work focuses on the hydraulic aspect. Limitations of the numerical models presented in this paper will be briefly mentioned in this paper whenever applicable.

The modelling of the conceptual shaft seals to store radioactive materials was previously done using COMSOL (e.g. Priyanto 2011; Priyanto *et al.* 2011*b*), and similar finite element software was used to complete analyses in this study. There is a difference in the seal configuration presented in this study compared with previous study. In the previous models of conceptual shaft seals to store radioactive materials, clay-based sealing materials were installed above and below the concrete components (e.g. Priyanto 2011; Priyanto *et al.* 2011*b*). In this study, there is no material above and below the upper and lower concrete components in the shaft seal (Fig. 1). Consequently, the groundwater pressure measurement above the upper concrete component of the seal (V10 in Fig. 2) is required as the boundary condition above the shaft seal.

The shaft seal is expected to function for a long time; consequently it is important to be able to simulate a long-term response of a shaft seal. Extrapolation of the groundwater pressure level was required to simulate long-term behaviour of the shaft seal. In this study, the extrapolation of the groundwater pressure above the shaft seal was calculated based on the excavation volume data, the assumption of constant flow and the assumption that the groundwater level will eventually recover to the pre-excavation conditions. Application of this extrapolated groundwater pressure (V10 in Fig. 2) as a boundary condition above the FZ2 allows the prediction of long-term hydraulic response of the shaft seal. This paper presents preliminary results of these models up to 1000 years. The assumption of constant flow rate was based on the data before July 2012; this assumption should be re-examined when more data is available in the future. The flow rate to flood the shaft is probably not constant owing to geological variations or other factors. The actual measurements of the groundwater level above the shaft should be used for the analysis whenever they are available in the future.

Since most of the saturation process has occurred under isothermal conditions, the thermal aspect was not considered in this study to simulate the saturation process. The temperature change was limited to the first 3 months owing to the curing of the concrete, after which the temperature in the shaft seal has remained relatively constant at 11–12 °C.

Three preliminary models

Three preliminary models to simulate saturation process of the shaft seal are discussed, in particular the groundwater pressure difference above and below the clay component of the shaft (Fig. 2). Model 1 utilized 2D-axisymmetric geometry and assumed no leakage seal at the rock–clay interface (Fig. 2). The flow path at the rock–concrete interface was assumed, as it was also observed by the hydraulic pressure measurements at the rock–concrete interface (Fig. 2). FZ2 was assumed to be perfectly horizontal.

Model 2 also utilized 2D-axisymmetric geometry. The ESP data show that the groundwater pressure below the shaft seal gradually decreased when the hydraulic pressure difference reached approximately 480 kPa after approximately 700 days (Fig. 2). Model 2 was developed in order to examine the hypothesis that the decrease of the hydraulic pressure difference was caused by the development of a flow path connecting the regions above and below the FZ2 when the pressure difference reached a certain level (*c.* 480 kPa in this case). There is no certain location where the flow path connects the regions below and above FZ2, whether the flow paths are located within the rock (e.g. the Excavation Damage Zone), within the clay component, in the rock–clay interface, or combination of various aspects. Figure 1 shows that the FZ2 intersects not only one main shaft, but also the vent raise of the URL. The purpose of model 2 was to examine what hydraulic response would be like if the flow path existed to connect the regions below and above FZ2. Model 2 assumed that the hydraulic conductivity of at the rock–clay interface increased after the pressure difference reached *c.* 480 kPa at 700 days. The thickness of the hypothetical flow path was assumed to be 20 cm. In this preliminary model, permeability was assumed as a function of time ($k_{\text{Rock-Clay}} = f$ (time)). This assumption may only be suitable for examining this particular hypothesis of decreasing hydraulic pressure, and not to predict the long-term behaviour of the shaft seal. The activation of this flow path may also be done by incorporating this feature within the constitutive models of the rock or clay component materials.

Model 3 (Fig. 4) utilized 3D geometry and assumed no leakage at the rock–clay interface, similar to the assumption used in model 1. The 3D model was constructed to take into account the moderate dip (*c.* 25°) of the FZ geometry. Modelling using 2D-axisymmetric geometry was faster than using the 3D geometry, but simplification was required in the 2D-axisymmetric model. The FZ was assumed to be perfectly horizontal and the dip of the FZ was not taken into account. This assumption resulted in perfectly axisymmetrical responses throughout the seal components, which was not observed in the field data for the shaft seal. In addition, if needed in the future, the actual geometry of the URL including both seals in the main shaft and vent raise should be included in the

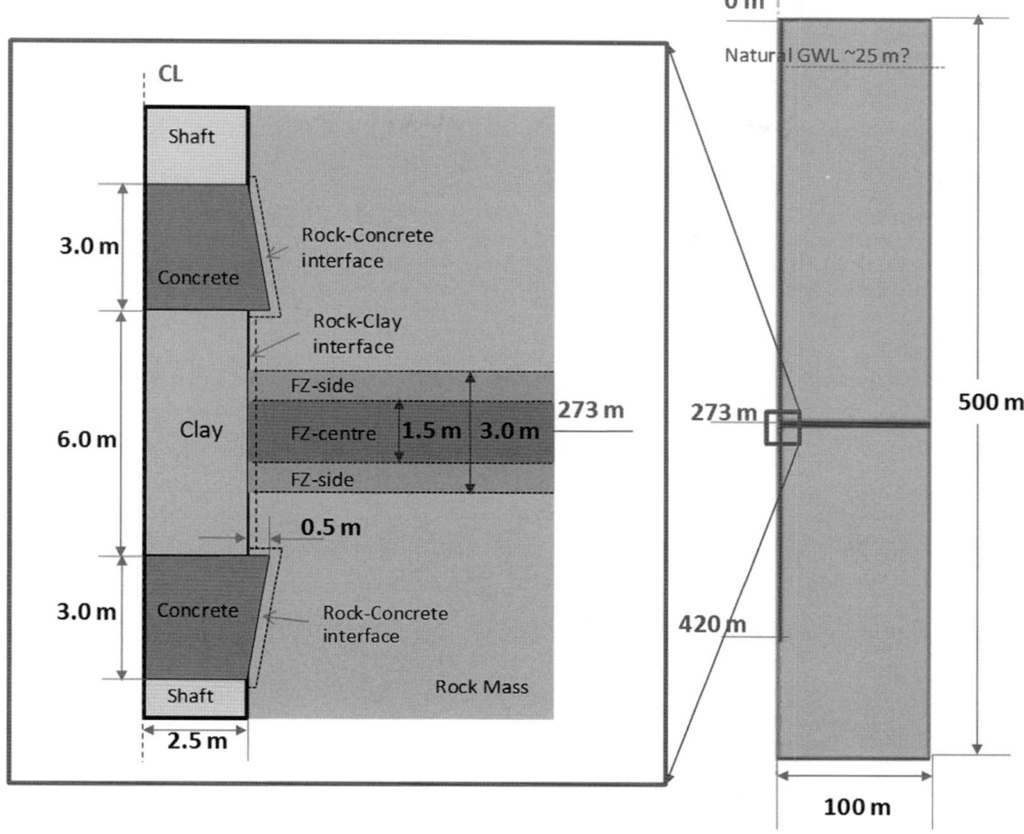

Fig. 3. 2D-axisymmetric geometry for numerical modelling of shaft seal.

numerical model, because they may be correlated. This can only be done using 3D geometry. Consequently, it is important to investigate using 3D geometry.

Model geometry, finite element mesh and parameters

Figure 3 shows the 2D-axisymmetrical geometry used for numerical analyses used in models 1 and 2. The entire domain of the models had a radius of 100 m and total thickness of 500 m. The radius of the shaft was equal to 2.5 m and the maximum depth of shaft was 420 m. The mid-height of the FZ2 was located at the depth of 273 m (the mid-height of the clay component in the shaft seal). The thickness of the clay component was 6 m, sandwiched between two 3 m-thick concrete components. The concrete components were keyed in the rock at 0.5 m depth. In the 2D-axisymmetric models, the FZ was assumed to be perfectly horizontal and had total thickness of 3 m.

Based on the observation at the FZ2 in the main shaft, the centre of the FZ tends to have more fracture and greater permeability (more groundwater seepage) than the regions closer to the intact rock. This indicates that the identified depth of FZ2 does not have uniform permeability. For simplicity, to incorporate this permeability variation along the depth of the FZ2, the FZ2 was considered as two zones in the analyses, including a 1.5 m-thick FZ-centre that was sandwiched between 0.75 m-thick FZ-parallel features.

The clay component of the shaft seal was designed to span at least 1 m beyond the maximum exposed vertical extent of the dipping FZ, since the 2D-axisymmetic geometry cannot take into account the inclination, as well as thickness variation of the FZ. In order to be more conservative,

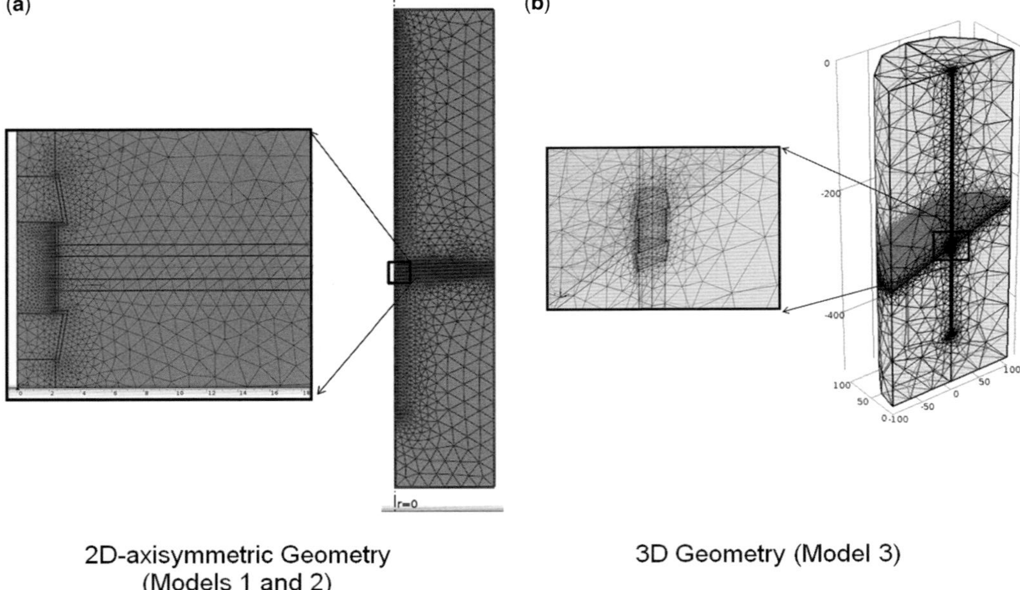

Fig. 4. Finite element meshes utilized in the numerical analyses.

a larger FZ was assumed for 2D-axisymmetrical geometry models. This also allowed the distance from the modelled FZ edge to edge of the clay component to be 1 m, as in the shaft seal. This assumption is sufficient for preliminary model that focuses on examining the mechanism, although it may result in faster saturation process.

Models 1–3 include clay and concrete components, intact rock mass, FZ and the rock–concrete interface. As the concrete components were not designed to provide a hydraulic seal, the rock–concrete interface zones were introduced in the models. For modelling purposes, the thicknesses of these zones were assumed equal to 20 cm and had a high permeability ($k_{\text{rock-concrete}} = 10^{-14}\ \text{m}^2$). Although in reality the interfaces are much smaller than 20 cm and of very high permeability, the use of this dimension for modelling serves to simulate an effectively open interface. This 20 cm thickness was selected in the model in order to avoid convergence problems in the finite element analyses owing to very small elements.

Model 2 considered the flow path connecting the regions below and above the FZ2. There was no certainty on the location of the flow path. A hypothetical flow path was assumed at the rock–clay interface. The permeability of the rock–clay interface increased when the groundwater pressure difference reached 480 kPa, after 700 days. For simplicity the permeability was assumed to be a function of time (i.e. $k_{\text{Rock-Clay}} = 10^{-17}\ \text{m}^2$ for $t < 700$ days, and $10^{-14}\ \text{m}^2$ for $t \geq 700$ days). This assumption was only suitable for examining whether the decrease in the hydraulic pressure difference was caused by the existence of a flow path connecting the regions above and below FZ2. For future prediction, the feature of increasing permeability should be included in the constitutive models of the clay or rock.

Figure 4 shows the finite element meshes using 2D-axisymmetric and 3D geometries, respectively. Denser meshes were utilized around the shaft seal locations. The inclination of the FZ in 3D geometry model was assumed equal to 25° and the thickness of the FZ was approximately 2 m; these assumptions more closely represent the observed geological structure of the FZ in the rock, although the thickness FZ variation (*c.* 1–2 m) was not considered in model 3.

Parameters and equations

The parameters used in the numerical modelling are summarized in Table 1. The numerical analyses were completed using COMSOL multiphysics combined with the subsurface flow and the structural mechanics modules. Overviews of the equations used in the COMSOL numerical modelling are described below.

Table 1. *Hydraulic–mechanical (H–M) parameters used for numerical modelling*

	Clay	Concrete	Rock, FZ and interface	Shaft (void)
Hydraulic parameters				
Water retention parameters, α (1/m)	0.00374	50	50	50
Water retention parameters, N	1.4	1.5	1.5	1.5
Water retention parameters, M	$1-1/N$	$1-1/N$	$1-1/N$	$1-1/N$
Water retention parameters, L	0.5	0.5	0.5	0.5
Saturated intrinsic permeability (m^2)	7×10^{-20}	1×10^{-19}	2×10^{-19} (intact rock) 10^{-17} (FZ-side) 10^{-16} (FZ-centre) 10^{-14} (rock–concrete interface) 10^{-14} (rock–clay interface) – model 2	1×10^{-4}
Bulk density, ρ_{bulk} (kg m^{-3})		2350	2650	1
Initial dry density, ρ_{dry} (kg m^{-3})	1810			
Total porosity, θ_s	0.33	0.009	0.003	0.99
Residual porosity, θ_r	0	0	0	0
Initial degree of saturation, S_w	0.66			
Initial water content, w_c	0.121			
Initial suction, s (Pa)	6×10^6	101 000 (atmospheric pressure)	NA	NA
Mechanical parameters				
Young's modulus, E (Pa)	$E=f$(vol) non-linear elastic $\kappa=0.065$	38×10^9	60×10^9	1
Poisson's ratio, ν	0.2	0.2	0.25	0.1

The hydraulic process in the unsaturated condition is simulated using Richard's equation:

$$\left[\frac{C}{\rho_f g}+S_e S\right]\frac{\partial u}{\partial t}+\nabla\cdot\left[-\frac{\kappa_s}{\eta}k_r\nabla(u+\rho_f g D)\right]=Q_s \quad (1)$$

where u is porewater pressure, C is the specific capacity, S_e denotes the effective saturation, S is the storage coefficient, κ_s is the intrinsic permeability, η is the fluid viscosity, k_r is the relative permeability, ρ_f is the fluid density (1000 kg m^{-3}), g is the gravitational acceleration (=9.81 m s^{-2}), D represents the vertical coordinate, and Q_s is the fluid source. The storage S is defined using the following equation:

$$S=\theta_s\chi_f+(1-\theta_s)\chi_p \quad (2)$$

This expression is equivalent to the Reuss average of the fluid and solid compressibility. Compressibility of fluid (χ_f) is 4×10^{-10} Pa^{-1} and compressibility of solid (χ_p) is $1/K_{\text{bulk}}$. The bulk modulus of solid (K_{bulk}) is calculated from Young's modulus (E) and poison's ratio (μ) (i.e. $K_{\text{bulk}}=E/(3\times(1-2\,\mu))$. In these models, K_{bulk} is constant for rock and concrete component and a function of specific volume, as it uses a non-linear elastic relationship for clay component; θ_s is the degree of saturation.

The relationship between the pressure and permeability with the degree of saturation is defined by the van Genuchten (1980) equations as follows:

$$\theta=\begin{cases}\theta_r+Se(\theta_s-\theta_r), & u_w<0\\ \theta_s, & u_w\ge 0\end{cases} \quad (3)$$

$$Se=\begin{cases}\dfrac{1}{\left[1+\left|\alpha\dfrac{p}{\rho_{f\cdot g}}\right|^N\right]^M}, & u_w<0\\ 1, & u_w\ge 0\end{cases} \quad (4)$$

$$C=\begin{cases}\dfrac{\alpha M}{1-M}(\theta_s-\theta_r)Se^{1/M}(1-Se^{1/M})^M, & u_w<0\\ 0, & u_w\ge 0\end{cases} \quad (5)$$

$$k_r=\begin{cases}Se^L[1-(1-Se^{1/M})^M]^2, & u_w<0\\ 1, & u_w\ge 0\end{cases} \quad (6)$$

where, θ is the volumetric water content or the volume of liquid per porous medium volume that ranges from a small residual value θ_r to the total porosity θ_s; C is the specific moisture capacity; k_r is the relative permeability; α, L and M are the van Genuchten parameters that describe the water retention curve.

The following formulation was used to take into account the mechanical aspect of the model. The mechanical process is solved by the quasi-static formulation with the governing equation as follows:

$$-\nabla \cdot \sigma = \rho \cdot g \tag{7}$$

where σ is the total stress tensor, ρ is the bulk density of the material.

Coupling of the hydraulic to mechanical (H–M) process is done for clay material by applying the Biot–Willis effective stress. Since the volume change of rock and concrete components does not significantly affect the H–M behaviour, the H–M coupling is not done for rock and concrete components. Coupling of the H–M processes and vice versa is done by applying body load ($F_{v,i}$) and mass flow (Q_m) as follows.

$$F_{v,i} = -\alpha_b \cdot Se \cdot \frac{\partial(p')}{\partial i} \tag{8}$$

$$Q_m = -\rho_f \cdot \alpha_b \cdot Se \cdot \frac{\partial}{\partial t}\varepsilon_{\mathrm{vol}} \tag{9}$$

where α_b is the Biot–Willis coefficient, which is assumed to be equal to 1, $\varepsilon_{\mathrm{vol}}$ is the volume strain, and $i = x, y, z$ for a cartesian coordinate system, or, r, z, θ for an axisymmetrical coordinate system.

For simplicity, the mechanical behaviour of the clay is described by a non-linear elastic model, which describes the specific volume (v) and effective stress (p') relationship as follows:

$$v = v_0 - \kappa \cdot \ln\left(\frac{p}{p_0}\right) \tag{10}$$

where v and v_0 are the current and initial specific volumes, respectively. Notation p' and p'_0 are the current and initial effective pressures, respectively. Effective stress p' used in the analyses is defined as:

$$p' = p + S_w \times (u_a - u_w) = p + S_w \times s \tag{11}$$

where S_w is the degree of saturation, u_a is pore air pressure, which is assumed to be constant (*c.* 0.101 MPa), u_w is pore water pressure, and s is the total suction, which is the difference between u_a and u_w. Since most of the volume changes would be due to the wetting process, parameter κ is assumed equal to 0.065, based on the results of wetting and drying tests of the bentonite–sand mixture (Tang 1999; Blatz 2000; Anderson 2003; Priyanto *et al.* 2011*a*). As the rock and concrete components have a much greater Young's modulus (E) than the clay component, the linear elastic model was used to describe the behaviour of the concrete and rock. For this study, the E of the rock is assumed to be constant throughout the depth for simplicity. The H–M coupling formulation used in this study only has a limited effect to couple hydraulic to mechanical processes and vice versa. Further investigation to improve the H–M coupling effect is still required for future study. Since this study focused the simulation on the hydraulic process, this limitation on the mechanical aspect will not significantly affect the hydraulic results presented in this study. Further analyses using different H–M coupling formulation or a more complex mechanical constitutive model to describe the clay component will be required to investigate the mechanical behaviour (e.g. Basic Barcelona Model (Alonso *et al.* 1990)).

Modelling stages, initial and boundary conditions

Two consecutive stages were required to simulate the saturation of the shaft seal. Stage 1 simulated the groundwater level during the operation of the URL and the changes of *in situ* stresses owing to excavation of the URL. The duration of stage 1 was 20 years, which represents the duration when the shaft was air-filled to 420 m depth during the operation of the URL. The initial groundwater level at the URL was approximately −25 m, based on the borehole measurements around the URL in 1984.

Stage 2 simulated the saturation of the shaft seal to 1000 years, recovery of the groundwater level and volume change (swelling and compression) of the clay components owing to an increase in the degree of saturation. The stress in the rock and concrete component also slightly changed owing to the changes in the porewater pressure during stage 2. The duration of the numerical model simulation for stage 2 was up to 1000 years in order to capture the recovery of the groundwater pressure of the host rock.

The boundary conditions applied in the models are as follows. During stage 1, the porewater pressure at the shaft wall was equal to 0 ($u_w = 0$ at $r \approx 2.5$ m). At the domain perimeter, the groundwater level remained constant at the depth (z) of 25 m, so that the hydraulic pressure was equal to:

$$u_{\mathrm{rock}} = -\rho_f \cdot g \cdot (z + 25) \quad \text{and} \quad u_{\mathrm{rock}} \geq 0 \tag{12}$$

The results of stage 1 analysis were used as the initial conditions of stage 2 at the host rock components, including the FZ. The initial porewater pressure (u_w) of the clay component was -6 MPa, corresponding to the initial suction (s) of 6 MPa. The model utilized Richard's formulation and assumed constant pore air pressure (u_a) to be equal to zero ($s = u_a - u_w$ and poreair pressure $(u_a) = 0 =$ constant). The S_e was 0.66, ρ_{dry} was 1810 kg m^{-3} and w_c was 12.1%. The initial porewater pressure of the concrete components was assumed equal to -0.101 MPa, representing the unsaturated initial condition of the concrete component and equal to the atmospheric pressure.

Prior to installation of the shaft seal, the water level on the shaft was intentionally flooded to approximately 1 m below the bottom of the shaft seal at the depth of 280 m for safety of the workers and to avoid buildup of the gas pressure below the shaft seal during flooding of the lower excavations (Dixon *et al.* 2012*a–c*). In order to represent this condition, the initial porewater pressure below the shaft seal ($u_{w,\text{below}}$) was set equal to

$$u_{w,\text{below}} = -\rho_f \cdot g \cdot (z + 280) \quad \text{and} \quad u_{w,\text{below}} \geq 0 \tag{13}$$

A time-dependent hydraulic pressure boundary condition was applied above the shaft seal ($u_{w,\text{above}} = f(t)$). This value ($u_{w,\text{above}}$) was based on the hydraulic pressure data from the vibrating wire piezometer (V10 in Fig. 2) installed at the top of the shaft seal that measures the groundwater level above the shaft seal. The extrapolation of the groundwater level above the shaft seal was calculated based on the first 3 year data up to July 2012, combined with known excavation volume of the URL, and an assumption that the flow rate will be relatively constant to flood the URL (Fig. 5). Please note that the extrapolation of the groundwater level shown in Figure 5 is still a preliminary result and this calculation was done solely for the purpose of the numerical modelling exercises to simulate the long-term hydraulic response of the shaft seal. This preliminary analysis (Fig. 5) indicated that the water level above the shaft seal will reach the depth of 25 m after approximately 7.5 years. This value should be re-analysed when more reliable data is available. An actual measurement of the groundwater level above the shaft (V10 Fig. 2) should be applied as the boundary condition whenever it is available. The assumption of constant flow rate to flood the shaft should also be examined, because this flow rate may depend on the geological variations along the shaft. At the time of reporting (July 2012) the groundwater level was still at the 240 level, where it covered a large cross section area (see Fig. 1). It is very difficult to generate an accurate groundwater flow rate to naturally flood the shaft. Much better accuracy of the groundwater flow rate to flood the shaft can be obtained when the 240 level is completely filled and the groundwater level is at the shaft and vent raise above the 240 level. At that time, the hydraulic pressure measurement will correspond to known cross sections of the main shaft and vent raise and a more accurate flow rate to flood the shaft can be determined and used for an input of the long-term behaviour of the shaft seal.

Numerical modelling results

Selected results from the numerical modelling results are discussed below. The discussion of the results is focused on the hydraulic processes.

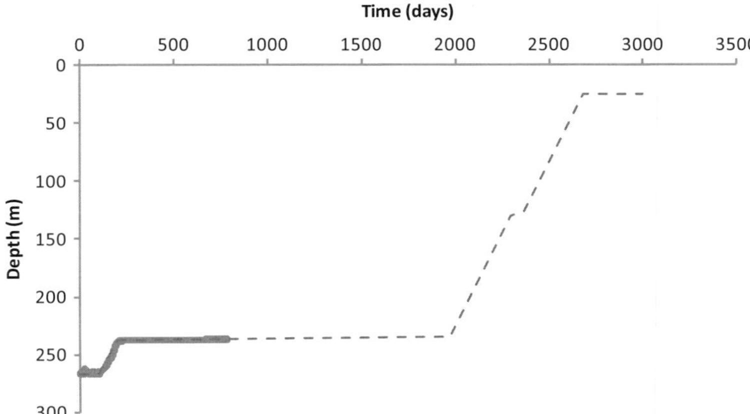

Fig. 5. Groundwater elevation above the shaft seal (solid line: measurements; dashed line: extrapolation based on constant flow rate and URL excavation volume).

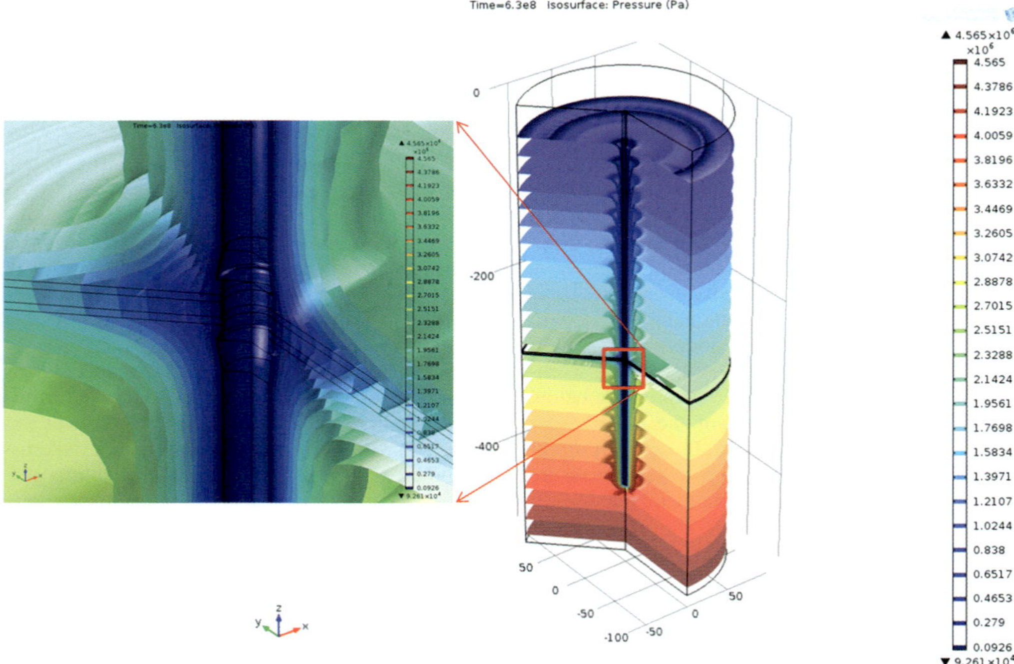

Fig. 6. Porewater pressure after 20 years of shaft operation (stage 1, 20 years).

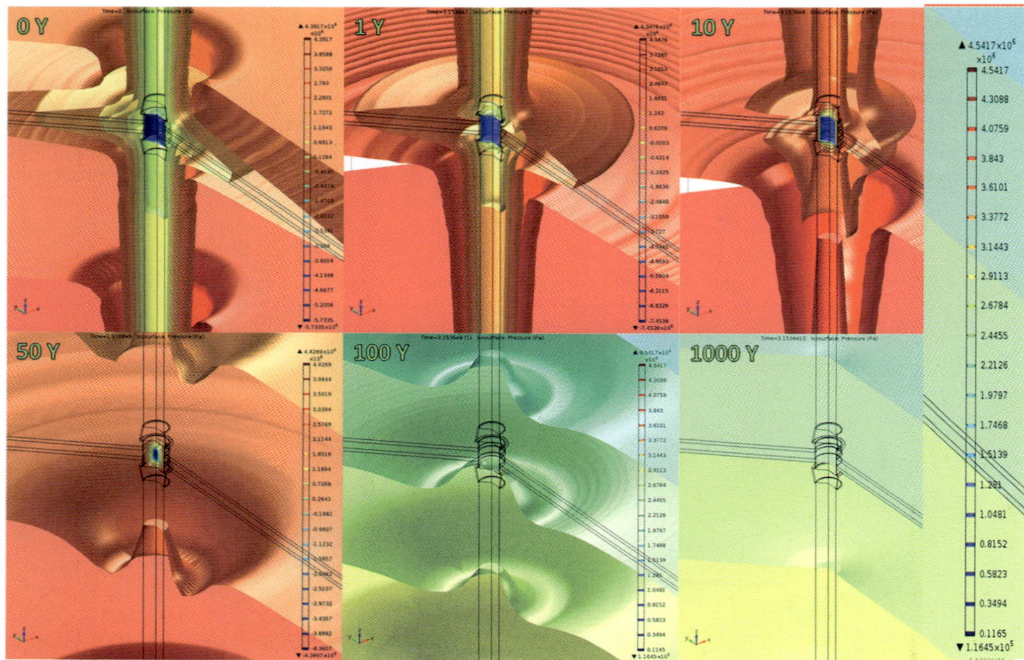

Fig. 7. Porewater pressure near shaft seal location at selected time after shaft seal installation (stage 2, t = 0, 1, 10, 50, 100 and 1000 years).

Figure 6 shows the porewater pressure of the host rock at stage 1, 20 years after the water level in the shaft was set at 420 m depth, during the URL operation. In the numerical model, the natural groundwater level was assumed to be located at the depth of 25 m. Zero porewater pressure was applied at the shaft wall. The porewater pressure near the shaft wall tended to decrease with time. This decrease was concentrated more around the FZ, as the FZ has greater permeability than the intact rock. The longer the shaft was kept open and drained, the wider the area affected by the drawdown of the groundwater was.

Figures 7 and 8 show the porewater pressure at stage 2 after seal installation at selected times up to 1000 years at a location near the shaft seal and for the entire domain, respectively. These models result in the porewater pressure of the rock recovering to the initial condition after 100–1000 years (Fig. 8).

Figure 9 shows calculated hydraulic pressures above and below shaft seal from models 1 and 2 for the first 3 years after seal installation compared with the first 3 years of field measurements. The hydraulic pressure above the seal was an input of the model and applied as the boundary condition, so they were similar for both models 1 and 2.

In the first year, models 1 and 2 overestimated the porewater pressure below the seal. The model assumed that the shaft seal worked perfectly to seal the shaft immediately after the installation. However, this was not the case in the actual shaft seal. During the seal construction, the groundwater was pumped at 1 m below the bottom of the concrete component for the safety of the workers. There was steel structure to support the shaft seal during the construction phase installed below the lower concrete component. At the beginning of the flooding of the shaft, there was empty space to be filled by the groundwater before the groundwater reached the seal. In addition, the clay–rock interface did not completely seal until the clay component started to swell after it had access to groundwater.

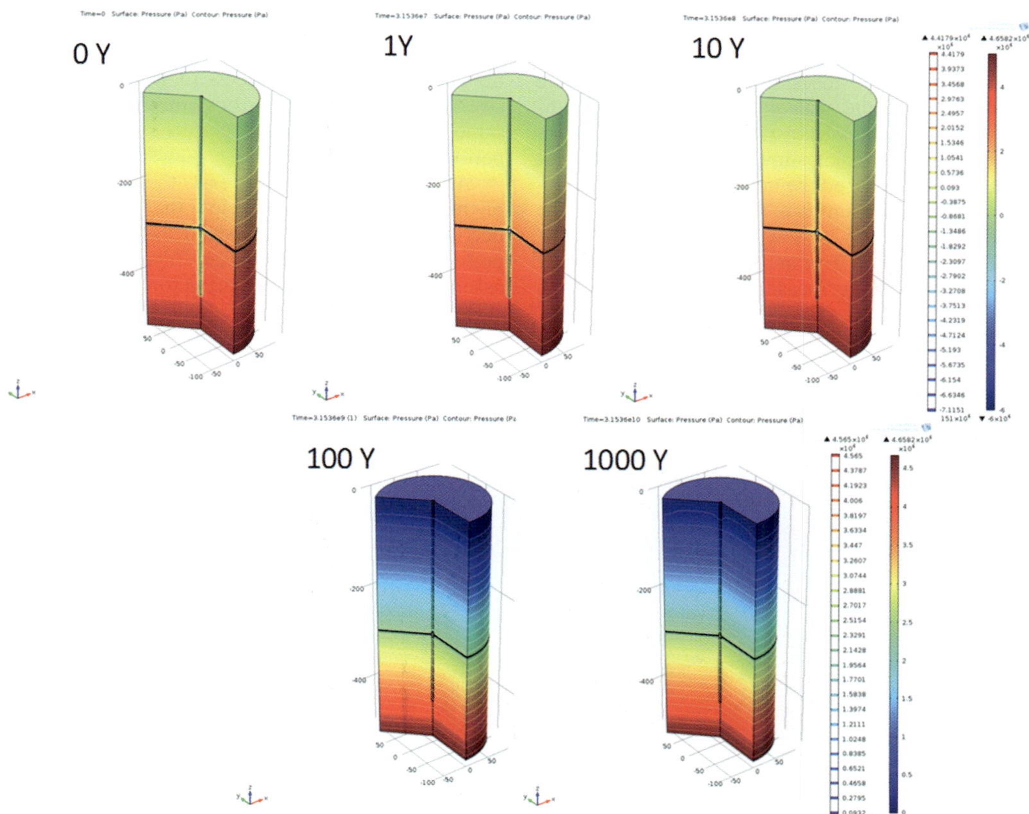

Fig. 8. Porewater pressure at entire domain at selected time after shaft seal installation (stage 2, $t = 0$, 1, 10, 100 and 1000 years).

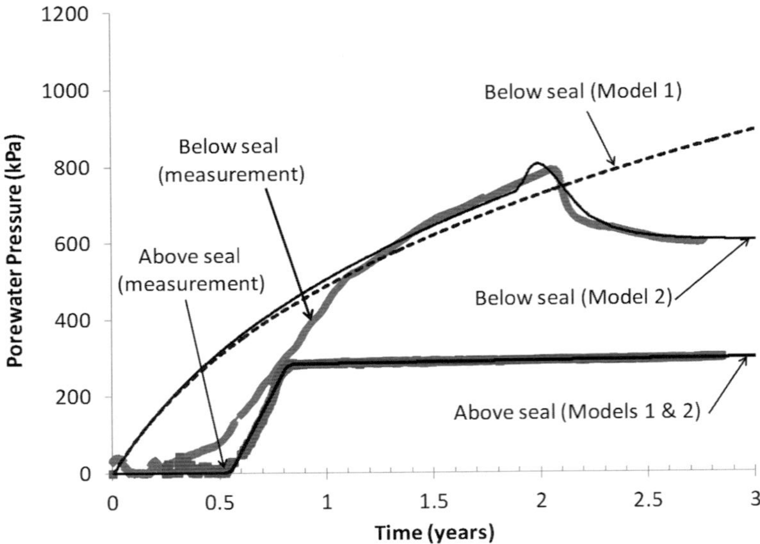

Fig. 9. Hydraulic pressures above and below shaft seal results from models 1 and 2 compared with the measurements for the first 6 years after seal installation (stage 2).

Consequently, the measurement of the hydraulic pressure below the seal did not increase immediately after the seal was installed (Fig. 9). This transition period was not simulated in the numerical modelling and caused the numerical model to overestimate the measurement at the first year of the measurement.

Between 1 and 2 years, the results of models 1 and 2 matched the measurement. In this period, the clay component has access to groundwater and swelled to seal the clay–rock interface, which is same as the assumptions of perfect seal during this period used in the model.

After approximately 2 years, the measurement of the hydraulic pressure below the seal gradually decreased (Figs 2 & 9). Model 1 assumed that the perfect seal still existed after 2 years. Model 2 assumed that a flow path started to exist when the

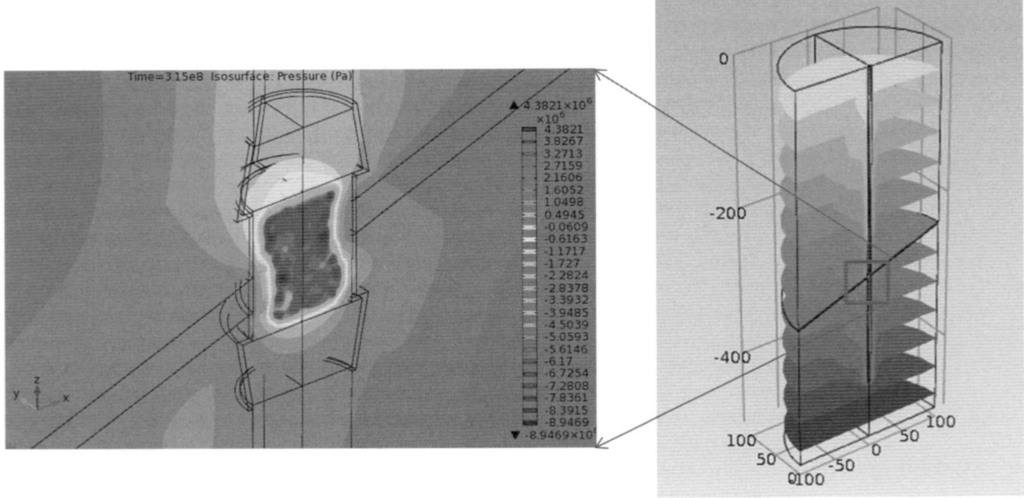

Fig. 10. Porewater pressure near shaft seal, 10 years after seal installation (stage 2, 10 years) (3D geometry).

pressure difference above and below the seal was greater than *c.* 480 kPa after *c.* 700 days to examine the reasons for the decrease in the hydraulic pressure in the measurement.

When assuming a perfect seal in the clay–rock interface (model 1), the hydraulic pressure below the shaft seal continued to increase, while it started to decrease if there was a flow path connecting the zone above and below the shaft seal (model 2). The hydraulic pressure above the seal analysed in model 2 was similar to what was actually measured. Inserting a flow path at the rock–clay interface connecting the host rock below and above the FZ improved the match between predicted and observed porewater pressures below the shaft seal. There was still uncertainty as to the locations of these flow paths, whether in the clay or rock components, or combinations of both. Other locations may also be possible since there were two different shafts at the URL (see Fig. 1). Despite this uncertainty of the location of the flow path, it can be concluded that the decrease in the pressure difference indicates that there was a flow path that connected the zones above and below the shaft seal. This pressure difference is expected to gradually increase as the clay component starts to swell and seal this flow path. The activation of this flow path may also be done by incorporating this feature within the constitutive models of the rock or clay component materials. Despite the hydraulic pressure decrease, the data show that the hydraulic pressure difference in the first 3-year period was significantly greater than the hydrostatic pressure difference that would exist without the presence of the shaft seal (*c.* 60 kPa). This indicates that shaft seal was still providing effective resistance to groundwater transport across the clay component and limiting the mixing of the saline groundwater below the FZ2 and fresh groundwater above the FZ2.

Modelling using 2D-axisymmetric geometry was beneficial as it was faster than 3D-geometry. However, the dip of the FZ was not taken into account. Modelling using 3D geometry has been started. Figure 10 shows the porewater pressure after 10 years of seal installation. As the ESP measurements of the suction and saturation were not perfectly axisymmetric, the effects of the FZ inclination can be observed in this figures and indicate the significance of considering 3D-geometry model to take into account the inclination of the FZ.

Summary and recommendation for future research

The preliminary results of the three different numerical models using 2D-axisymmetric and 3D-geometry have been presented in this study. Models 1 and 2 utilized 2D-axisymmetic geometries and model 3 utilized 3D geometry. Models 1 and 3 assumed an ideal hydraulic seal between rock–clay interfaces, while model 2 took into account a flow path connecting the regions above and below the FZ2 to examine the cause of the decrease in pressure difference observed in the measurements. The results of model 2 confirmed that taking into account a hypothetical flow path connecting between the regions above and below FZ can replicate the porewater pressure measurements below the seal. Although there is still uncertainty on the locations of the flow path, based on the results of this numerical modelling, it can be concluded that the decrease in the pressure difference indicates that there was a flow path that connected the zones above and below the shaft seal or FZ2.

As the measurements of the suction and saturation were not perfectly axisymmetric, it was important to take into account the inclination of the FZ in modelling, which was accomplished using a 3D-geometry model. Parameter calibrations, inclusion of the geological features and comparison with more measurement data as it becomes available are recommended for future modelling activities. Expanding modelling efforts to include other aspects of the shaft seal is also recommended (e.g. concrete shrinkage, mechanical behaviour, transfer of salt solution, etc.).

The authors would like to acknowledge financial support provided by NWMO, SKB, Posiva and ANDRA to the Enhanced Sealing Project to monitor the shaft seal at the Atomic Energy of Canada Limited's Underground Research Laboratory in 2009–2013.

References

Alonso, E. E., Gens, A. & Josa, A. 1990. A constitutive model for partially saturated soils. *Géotechnique*, **40**, 405–430.

Anderson, D. E. S. 2003. *Evaluation and comparison of mechanical and hydraulic behaviour of two engineered clay sealing material.* MSc thesis, Department of Civil Engineering, University of Manitoba, Winnipeg, Manitoba.

Blatz, J. A. 2000. *Elastic-plastic modelling of unsaturated high-plastic clay using results from a new triaxial test with controlled suction.* PhD thesis, Department of Civil Engineering, University of Manitoba, Winnipeg, Manitoba.

Dixon, D. A., Martino, J. B. & Onagi, D. P. 2009. *Enhanced Sealing Project (ESP): design, construction and instrumentation plan.* Nuclear Waste Management Organization (NWMO) Technical Report **APM-REP-01601-0001**.

Dixon, D. A., Priyanto, D. G. & Martino, J. B. 2012*a*. *Enhanced Sealing Project (ESP): project status and data report for period ending 31 December 2011.*

Nuclear Waste Management Organization (NWMO) Technical Report **APM-REP-01601-0005**.

Dixon, D. A., Martino, J. B., Priyanto, D. G. & Kim, C. S. 2012*b*. Three Years of Monitoring a Full-Scale Shaft Seal in a Granitic Rock Formation. *In*: *The 65th Canadian Geotechnical Conference*, 30 September to 3 October 2012. Winnipeg, Manitoba.

Dixon, D. A., Priyanto, D. G., Martino, J. B., De Combarieu, M., Johansson, R., Korkeakoski, P. & Villagran, J. 2012*c*. Enhanced Sealing Project (ESP): evolution of a full-sized concrete and bentonite shaft seal. Paper submitted to *The 5th International meeting on 'Clays in Natural and Engineered Barriers for Radioactive Waste Confinement'*, 22–25 October 2012, Montpellier.

Holowick, B. E., Dixon, D. A. & Martino, J. B. 2011. *Enhanced Sealing Project (ESP): project status and data report for period ending 31 December 2010*. Nuclear Waste Management Organization (NWMO) Technical Report **APM-REP-01601-0004**.

Martino, J. B. & Kaatz, R. 2010. *Enhanced Sealing Project (ESP): pre-construction grouting around seal locations*. Nuclear Waste Management Organisation (NWMO) Technical Report **APM-REP-01601-0002**.

Martino, J. B., Dixon, D. A., Holowick, B. E. & Kim, C.-S. 2010. *Construction of full scale shaft seals and Enhanced Sealing Project (ESP) monitoring equipment installation*. Nuclear Waste Management Organization (NWMO) Technical Report **APM-REP-01601-0003**.

Priyanto, D. G. 2011. Hydro-mechanical Modelling of a Shaft Seal in Crystalline and Sedimentary Media Using Comsol. *In*: *Canadian Nuclear Society (CNS) Conference, Waste Management, Decommissioning and Environmental Restoration for Canada's Nuclear Activities*, 11–14 September, 2011, Toronto.

Priyanto, D. G., Man, A. G., Blatz, J. A. & Dixon, D. A. 2011*a*. Hydro-mechanical constitutive model for unsaturated compacted bentonite–sand mixture (BSM): laboratory tests, parameter calibrations, modifications, and applications. *Physics and Chemistry of the Earth*, **36**, 1770–1782.

Priyanto, D. G., Dixon, D. A. & Man, A. G. 2011*b*. Interaction between clay-based sealing components and crystalline host rock. *Physics and Chemistry of the Earth*, **36**, 1838–1847.

Tang, G. X. 1999. *Suction characteristics and elastic-plastic modelling of unsaturated sand-bentonite mixture*. PhD thesis, Department of Civil Engineering, University of Manitoba, Winnipeg, Manitoba.

van Genuchten, M. Th. 1980. A closed form equation for predicting the hydraulic conductivity of unsaturated soils. *Soil Science Society of America Journal*, **44**, 892–898.

Evolution of temperature and humidity in an underground repository over the operation period

LUC-VINCENT BÉNET[1]*, CATALIN TULITA[1], LAURENT CALSYN[2] & JACQUES WENDLING[2]

[1]*SOCOTEC SA/Projets Industriels/AME, 1 Avenue du Parc 78640, Montigny-le-Bretonneux, France*

[2]*ANDRA, 92298 Châtenay-Malabry, Cedex, France*

Corresponding author (e-mail: luc-vincent.benet@socotec.com)

Abstract: The aim of the study is to describe the ambient conditions in an underground repository of nuclear waste over a 100-year-long operation period. The evolution over time of the moisture and the temperature in the ventilation network was assessed by means of numerical simulations. Condensation events are described in terms of location, frequency and flow rate. The physical conceptual model takes into account the heat and vapour advection and exchanges with walls, and heat conduction through the host rock. The results of simulations were analysed to highlight how the architectural design, the gradual extension of storage zones, the ventilation rate and the weather conditions are likely to influence the ambient conditions all along the shafts, the galleries and the storage modules. They illustrate the significant effects on ambient conditions of the wall thermal inertia, the variation in atmospheric pressure over the shaft height, and the ventilation partition in some galleries between incoming and outgoing air from storage modules.

Since the end of the 1990s, the French agency for the management of nuclear wastes, ANDRA, has been studying the feasibility of locating a repository in the Callovo-Oxfordian argillites on the borders of the Meuse and Haute-Marne departments at a depth of about 500 m. The ANDRA repository will be operated for about 70 years, and will stay open in pre-closure stage for about 100 years in order to enable the reversibility of waste storage.

During the operation period, the ventilation scheme will progress with the gradual construction of the different storage zones: two IL-LLW (intermediate-level long-life waste) storage zones (see in Fig. 1 the IL-LLW1 and IL-LLW2 zones) and three HL-LLW (high-level long-life waste) storage zones (see in Fig. 1 the HL-LLW0, HL-LLW1 and HL-LLW2 zones). The ventilation rate in the storage module depends on the radioactive waste stored (IL-LLW or HL-LLW) and on the ongoing operation stage: the construction, storage operation, sealing operation and pre-closure stages.

The ventilation system will ensure adequate air conditions for the staff in working zones and for the infrastructure. The ventilation air will undergo a set of temperature and moisture changes by interacting with the host environment throughout the repository: heat and vapour exchanges with walls and storage packages, pressure variations, evaporation and sometimes vapour condensation.

The aim of this study is to describe how the ambient conditions will evolve over the operation period all along the ventilation network. The ambient conditions have been defined in terms of temperature and relative humidity in the air flow and on contact with walls. They have been assessed by numerical simulation over 92 years, and analysed to determine to what extent the air-to-wall heat and vapour exchanges, the weather conditions, the gradual extension of the repository and the ventilation rate will change the ambient conditions in the repository, and possibly result in condensation.

This paper focuses on the description of the ambient conditions in the ventilation network of the storages zone HL-LLW1, HL-LLW2, IL-LLW1 and IL-LLW2 over the period of reversibility except for the construction stages.

Overview of the air-and-repository physical processes

The ventilation air will change in temperature and moisture by crossing the repository, because of heat and vapour exchanges with walls and storage packages, pressure variations along the ventilation network and sometime condensation. The fresh air flowing from the site surface down to the repository level will increase in temperature (+5 °C in adiabatic conditions) because of the atmospheric pressure increase from the site surface to the repository level, and will get dryer by getting hotter with depth. For the same reason, the air will get cooler

From: Norris, S., Bruno, J., Cathelineau, M., Delage, P., Fairhurst, C., Gaucher, E. C., Höhn, E. H., Kalinichev, A., Lalieux, P. & Sellin, P. (eds) 2014. *Clays in Natural and Engineered Barriers for Radioactive Waste Confinement*. Geological Society, London, Special Publications, **400**, 413–426.
First published online June 16, 2014, http://dx.doi.org/10.1144/SP400.41

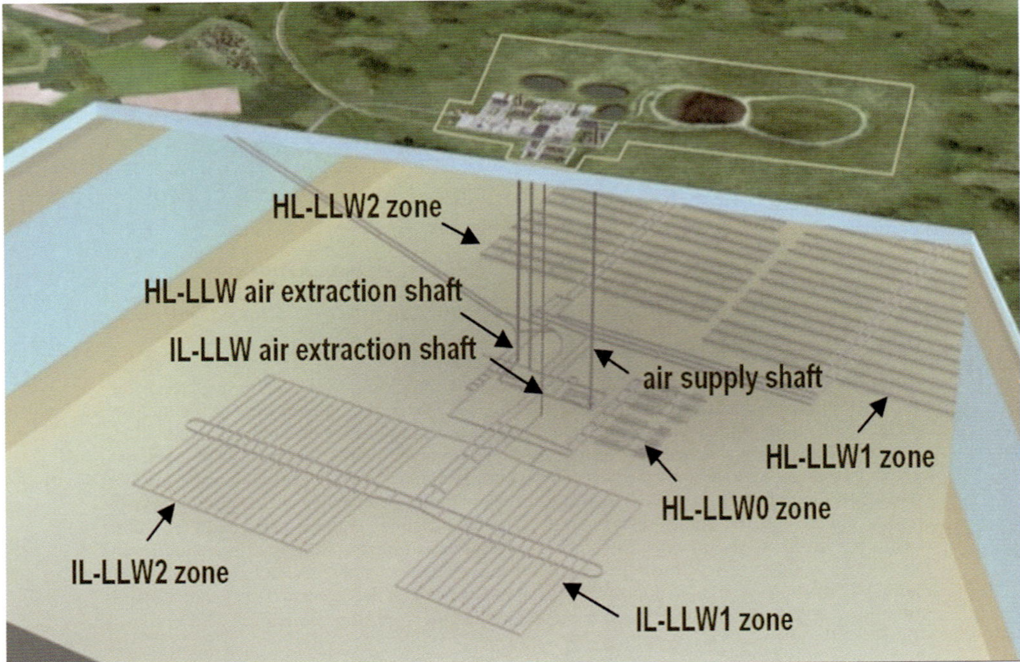

Fig. 1. Schematic view of the repository: architecture taken from ANDRA's 2009 design.

and wetter when rising in the air extraction shaft. The air pressure will also change throughout the repository, but to a lesser extent (a) with dynamic-pressure losses and (b) with the mechanical work of the ventilation system.

The ventilation air will transport weather fluctuations in temperature and moisture through the ventilation network. However, the heat exchanges with walls will partly buffer the weather temperature fluctuations: they will be progressively smoothed out as the distance from the inlet increases. In particular, this buffer effect depends on the time period of fluctuations (i.e. hourly, daily or seasonal variations). In contrast, the temperatures will slowly evolve on a longer time-scale in the concrete walls and in the host rock, because of the high thermal capacity of these materials.

Generally speaking, the features of the architectural design or the operation progress which have significant effects on the ambient conditions inside the repository are: (a) the extent of the air duct network – heat exchanges with the surrounding rock grow as the air duct network is getting longer; (b) the air ventilation rate – high rates of air renewal promote heat and vapour transport out of the repository; and (c) the air-duct hydraulic diameter. These two last parameters affect the efficiency of heat exchanges with walls. Another point is the heat released by some waste packages: the resulting increase in air temperature amplifies the thermal buffer effect on weather fluctuations and reduces the relative humidity by increasing the saturation vapour pressure.

The gradual construction of storage zones (i.e. one HL-LLW module and two IL-LLW modules opened every two years) lasts about 70 years. At the end of the 'active' operation period, the extension of the pre-closure stage to all the storage modules will significantly reduce the overall ventilation rate and consequently the transport of weather fluctuations throughout the repository. Thus, the time-scale and the spatial-scale are multiple:

In space,

(1) from 0.2 to 7 m for the diameters of the air cross-section;
(2) from 500 m to tens of kilometres for the depth of the storage and the length of the gallery network.

And in time,

(1) a few minutes – the air crossing the supply shaft;
(2) a few hours – the air crossing the repository, or a change in outdoor temperature over a day;
(3) a few days – a low-pressure period of weather, or the heat crossing 1 m of concrete;

(4) a few months – the duration of a season, or the heat crossing about 4 m of rock;
(5) 6–8 years – a storage module moving from the excavation stage to the pre-closure stage;
(6) a century – the duration of the repository operation, and the heat crossing about 75 m of rock.

Consequently, the system behaviour depends on both the specific durations of operation stages (construction, storage work, sealing work, pre-closure stage) and the characteristic times associated with the physical phenomena involved (weather changes, heat and mass transfers, etc.).

Ventilation and operation stage

The ventilation network

During the operation period, two separated ventilation networks coexist: the construction and the exploitation air networks. In the exploitation ventilation network, the ventilation air is named 'fresh air' upstream of a storage module and 'nuclear air' after crossing a storage module.

The untreated incoming air (i.e. fresh air) is characterized by significant variations of temperature and humidity between night and day and winter and summer. The air enters the repository through the air supply shaft (see Fig. 1), crosses the central zone through the main galleries, and enters the storage zones either through the construction air network or through the exploitation air network. In the storage zones, the air is distributed to the storage modules all along the secondary galleries networks. After crossing the storage module, the air (i.e. nuclear air) is confined in an air duct placed under the gallery vault, extracted from the repository through the HL-LLW or the IL-LLW air extraction shaft, and finally released into the atmosphere after being treated.

According to the 2009 ANDRA design, the maximum overall ventilation rate will reach about 400 $m^3 s^{-1}$. It will be reached after about 58 years of exploitation, just before the last HL-LLW storage module switches to the pre-closure stage, while only the last IL-LLW modules still remain in operation. The overall ventilation rate drops to 230 $m^3 s^{-1}$ once the 'active' operation stage ends and the pre-closure ventilation rate is generalized throughout the repository.

Operation and ventilation of the IL-LLW zone

The repository construction will start with the construction of the two IL-LLW storage zones. The construction of a module will last 2 years, the storage stage 4 years, and the following pre-closure stage until the closure of the repository at the end of the reversibility period (see some modules at different stages in Fig. 2b). The ventilation rate

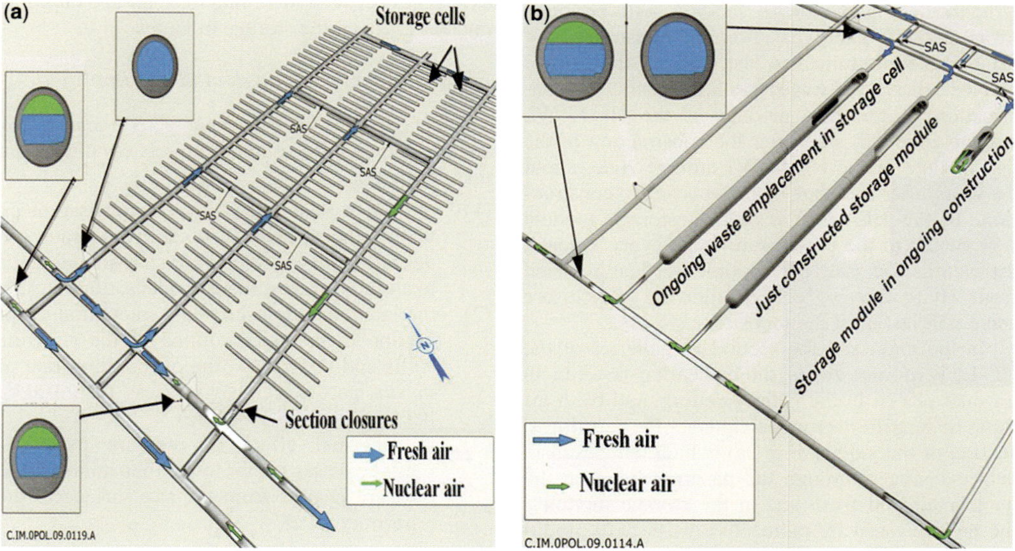

Fig. 2. Schematic view of the infrastructure and the ventilation network in a HL-LLW storage module (**a**), and in an IL-LLW storage zone (**b**) still in construction during the operation period; architecture taken from the ANDRA's 2009 design.

of an IL-LLW module is $10\ m^3\ s^{-1}$ during the 'active' operation period, and $3\ m^3\ s^{-1}$ during the pre-closure period (reference values).

The IL-LLW1 zone is composed of 21 storage modules including seven modules containing low exothermic wastes. The IL-LLW2 storage is composed of 36 storage modules. In ANDRA's 2009 design, an IL-LLW storage module is a 500 m-long gallery composed of a 400 m-long storage cell where rows of concrete over-packs are emplaced. The radioactive wastes are conditioned into four or six steel canisters per over-pack. In the exothermic module, the air temperature is expected to increase to about 50 °C during the first decade, and to rapidly decrease afterwards.

Operation and ventilation in The HL-LLW Zone

The HL-LLW1 and HL-LLW2 storage zones are each composed of 10 modules. An HL-LLW1 module is composed of 312 storage cells containing C5 type radioactive waste. An HL-LLW2 module is composed of 392 storage cells containing C6 type waste. The HL-LL waste is highly exothermic. The heat released by radioactive waste (350 W per canister at time of the emplacement) decreases in time. The storage cells will be sealed with a 7 m-long plug composed of concrete and bentonite (swelling clay).

The construction of the HL-LLW storage modules will start in the HL-LLW1 storage zone, about 20 years after the start of the repository construction. A new storage module will be open every 2 years. The excavation and construction of a module are assumed to last 4 years, the storage and the cell-closure stages will last 6 years. The construction of the first module of the HL-LLW2 storage zone will start after the construction of the last module of the HL-LLW1 storage zone, about 44 years after the start of the repository construction. In the HL-LLW storage, a storage module will switch to the pre-closure stage every 2 years; the ventilation rate per module is then reduced from 30 to $2\ m^3\ s^{-1}$. The following pre-closure stage will last until the repository closure.

In the main and the secondary galleries of the HL-LLW storage zones, the nuclear air flows in an air-duct placed in the gallery ceiling, and fresh air flows in the gallery cross-section (see the partitioned section of the gallery Fig. 2a), which will result in heat exchanges through the partition between the nuclear air and fresh air. In the storage modules, the heat released by radioactive waste will gradually diffuse through the surrounding rock toward the module gallery, of which only a small part will reach the gallery wall and be taken away by the ventilation. However, the air temperature is expected to reach 60 °C in some storage modules during the pre-closure stage.

Modelling approach

Physical processes modelled

The physical modelling has been designed in order to provide a good assessment of fluctuations in air temperature and vapour density throughout the ventilation network. The air temperatures and moistures are evaluated in the cross-section of air-ducts and on the wall surface using the saturation vapour pressure law. The relative humidity evolves with the heat and the vapour collected or lost all along the air network.

The physical modelling takes into account (a) air-to-wall heat exchanges owing to forced and free advection, and air-to-air heat exchanges through gallery partition, (b) heat advection by the air flow, (c) thermal storage and conduction in the concrete structure and the host rock, (d) vapour condensation on walls, (e) pore-water evaporation in walls through time, (f) weather fluctuations in temperature and in relative humidity and (g) heat released by radioactive waste through time.

In addition, the heat balance in the air is modelled considering (h) the air supply, the change in atmospheric pressure between the site surface and the repository level, (i) in the air extraction shafts, the thermal effect of mist formation and (j) in the HL-LLW partitioned galleries, air-to-air heat exchanges between incoming fresh air and outgoing nuclear air from the storage modules.

Physical processes neglected or simplified

Some phenomena have been neglected because they are considered to have minor effects upon heat and mass transfer:

(1) the latent heat released by condensation on walls – this mechanism tends to moderate condensation fluxes, so the physical model tends to slightly overestimate condensation rates;
(2) the latent heat consumed by interstitial evaporation in the pore volume of the concrete walls and the rock – the evaporation rate of *in situ* pore water is low (less than 100 kg per year per gallery metre).
(3) the thermal effect of pressure losses – pressure losses are far lower than atmospheric pressure change from the site surface to the repository level.

Moreover, some physical models have been simplified. As regards the vapour transfer, the evaporation rate of the *in situ* pore water along a gallery is defined as a function of time. These functions come from previous studies undertaken by the ANDRA with no consideration of the weather fluctuations in the ventilation. Consequently, the buffer effects on weather fluctuations in relative humidity and in vapour exchanges with porous media have not been simulated.

As regards the heat transfer, a variation-in-time laws of the air temperature at the exit of HA1 and HA2 modules have been used to model the heat collected by the air crossing the HL-LLW modules. These laws have been established by three-dimensional simulations of the wall-and-air thermal transient throughout a module. These simulations have shown that, during the pre-closure stage, the temperature in the incoming air has no significant effect on the outgoing temperature during the pre-closure stage. In fact, the residence time in a module is long enough for the air temperature to reach an equilibrium with the wall temperature before exiting the module given the low ventilation rate (2 $m^3 s^{-1}$ per module). The air temperature maximum reaches 60 and 55 °C respectively at the exit of HL-LLW1 and HL-LLW2 storage modules over 100 years (see Fig. 3). It should be noted that this modelling approach is also adopted during the first 6-year-long 'active' operation period, although the air flow is then high enough (30 $m^3 s^{-1}$ per module) to transport weather fluctuations beyond the module exit.

Modelling of the progressive extension of the repository

The air flow rate in a storage module depends on the ongoing operation stage. Consequently, the air-flow rate of the model is adapted every 2 years, as new storage modules open and other modules change into the pre-closure stage. Time-dependent parameters are calculated from the date of excavation of each gallery (i.e. ventilation rate, evaporation rate) or from the date of the waste package emplacement (power released by IL-LLW waste and air temperature outgoing from HL-LLW modules).

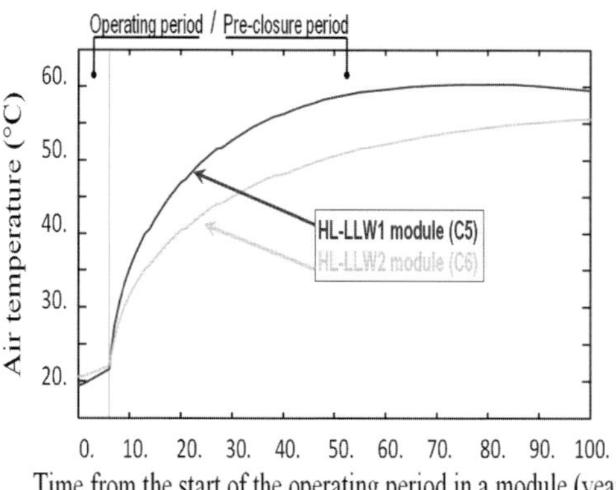

Fig. 3. Air temperature evolutions in time at the exit of HL-LLW1 modules (C5 waste type) and HL-LLW2 modules (C6 waste type).

Table 1. *Standard deviations of weather fluctuations in temperature and relative humidity related to different periods of time*

Periods of time	Temperature (°C)	Moisture (%)
Hourly	7.6	20
Daily	7.1	15.5
Seasonal	6.8	9.73
Daily (seasonal adjusted)	3.7	11.2
Hourly (daily adjusted)	2.9	11.8

Characterization of the incoming air

The 7-year-long weather measurements on the Bure site have been used to define the average values and the range of weather fluctuations in air temperature and in relative humidity according to different periods of time (Table 1). Yearly averages of the temperature and the relative humidity are respectively 11.8 °C and 78%. The range of weather fluctuations is defined by the standard deviation from the average. The weather fluctuations in temperature and in relative humidity are assimilated to a normal law: the probability of a parameter being within the standard deviation is about 68%. The weather fluctuations have been analysed on different time-scales:

(1) the daily adjusted hourly variations (i.e. the daily rolling average is removed from hourly variations) are mostly due to the alternation of night and day (standard deviation in temperature of 2.9 °C);
(2) the seasonal-adjusted daily variations (i.e. the 2-month rolling average is removed from daily variations) can be interpreted as changes in the weather (standard deviation in temperature of 3.7 °C);
(3) the seasonal variations are significantly higher than the two previous variations (standard deviation in temperature of 6.8 °C).

Numerical methods

Heat and vapour fluxes in the air were evaluated on a geometric model of the air duct network composed of one-dimensional meshes connected to each other.

In the air duct, instantaneous fluxes were computed with a pure advective scheme. At the air duct intersection, the mixture of all incoming flows was considered as perfect. The time-dependent conditions at the inlet (weather fluctuations), on walls (temperature change with thermal conduction in wall; evaporation of *in situ* pore water) and in the air (heat released by radioactive wastes) were taken into account by solving a succession of steady states, which means that no genuine transient is solved. The retardation effects owing to vapour-and-heat storage in the air were neglected compared to the mean air residence time in the repository (few hours).

The thermal conduction inside walls was discretized with a linear finite element method on a two-dimensional axisymmetric mesh. The thermal transient in walls was computed with an implicit solver and a first-order scheme in time. The effect of nearby galleries (about 50 m between some of them) on wall temperature was neglected although the effect could be (slightly) sensitive after 50 years of exploitation according to a Fourier analysis. Both problems, the heat advection by the air and the thermal conduction in walls, were coupled via an iterative process (Picard algorithm) to force the overall heat balance.

Two types of transient were achieved: (a) the so-called secular transient describes the variation of the yearly average of ambient conditions by 1-year-long time-steps over 90 years, the yearly averages of temperature and the relative humidity being imposed at the inlet of the ventilation network; and (b) and the so-called seasonal transient describes the variations of ambient conditions by 4 h-long time-step over 2 years, the weather fluctuations being imposed at the inlet of the ventilation network. Among over things, the secular transient aimed at providing thermal initial conditions in the surrounding rock to the simulations of the seasonal transient.

The physical models and the algorithmic process coupling the air advection and the thermal conduction in walls were set up with the GIBIANE language. All of the modelling, meshing and problem solving have been achieved with the Cast3M software (Cast3M 2010).

Variation of ambient conditions throughout the repository at the end of the reversibility period

The results below describe the evolution of the temperature and the moisture in the repository ventilation network over 2 years after 90 years of operation (i.e. 12 years after the pre-closure ventilation had been generalized to the entire storage).

In the air supply shaft

The ventilation air enters the repository via a 10.5 m diameter shaft. Only 60% of the section is open to the ventilation. The air from the site surface undergoes a 60 mbar compression owing to the weight of the 550 m-high air column when arriving at the repository level through the air supply shaft. This quasi-isentropic compression results in a 5 °C increase in temperature and a significant decrease in humidity (−21% in yearly average). The convective heat exchanges with the shaft wall have no significant effect on the yearly average of temperatures and moistures at the repository level, although they slightly smooth out the weather fluctuations in temperature by (a) an 11% decrease in daily variations (i.e. standard deviation of seasonal-adjusted daily fluctuations), and (b) a 3% decrease in seasonal variations (i.e. standard deviation of seasonal fluctuations).

The water flux from surrounding rock (0.2 m^3 per year per shaft metre) has no significant effect on the air moisture in the shaft. The vapour density increases by 0.5% in winter, and by 0.2% in summer by entering the central zone. It should be noted that the evaporation fluxes in winter count for more than in summer, because the vapour density in the atmospheric air is often significantly lower in winter than in summer (i.e. the temperature-dependence of the vapour saturation pressure).

In the central zone of the repository

The central zone of the repository gives access to the IL-LLW and the HL-LLW storage zones. The fresh air flows through the central zone in full section via a 650 m-long gallery network, from the shaft bottom to the storage zones.

The convective heat exchanges with gallery walls mainly affect the temperature fluctuations in the air, given the high thermal capacity of wall materials. The wall thermal inertia effect on the air temperatures is particularly high on short-period fluctuations: about 95% of hourly fluctuations in temperature are absorbed, while 21% of daily fluctuations and only 6% of seasonal fluctuations are smoothed out by crossing the central zone. Consequently, the hourly fluctuations in temperature are neglected downstream of the central zone.

In the IL-LLW storage zones

The ventilation air enters the two IL-LLW1 and IL-LLW2 storage zones in the main gallery, and is distributed between all the storage modules opened to operation via secondary galleries. The distance covered to reach the remotest modules is

2.7 km through the IL-LLW1 zone, and 3.4 km through the IL-LLW2 zone.

After 90 years, the heat exchanges have no significant effect on the yearly average temperatures (see the solid curve in Fig. 4). The air temperature increases by 1.8 °C by flowing through (a) the main gallery (+0.7 °C), (b) the secondary galleries (+0.4 °C) and (c) the remotest module (+0.7 °C). Downstream of the modules, the hot air outgoing from cells containing exothermic wastes comes in addition to the air from other cells. Consequently, the yearly average of temperatures is higher in the air coming from the IL-LLW1 storage zone (21 °C) than from the IL-LLW2 storage zone (19 °C), where there is no exothermic waste (compare the temperatures downstream of the modules in Fig. 4a, b).

As through the central zone, the short-period fluctuations in temperature are smoothed out by the convective heat exchanges with the gallery walls. Indeed, the magnitude of daily fluctuations is reduced with the covered distance until it eventually falls to zero as crossing the storage module. The magnitude is reduced by 16% at the main gallery exit, 86% at the secondary gallery exit and 99% at the module exit. The seasonal fluctuations are less affected by crossing the IL-LLW zones. The fluctuation magnitude is reduced by 11% in the main gallery, 66% at the end of the secondary gallery network and 92% when exiting the module. The high attenuation of the temperature fluctuations through the storage module (see the curves getting closer through the module Fig. 4a, b) is induced by high heat exchanges with walls and concrete over-packs in the storage cell. Heat exchanges are promoted by a high heat-exchange perimeter (about 40 m), and a high air velocity (about 1.5 m s^{-1} during the pre-closure stage), because the ventilation cross-section is significantly smaller in a storage cell (about 2 m^2 between walls and over-packs) than in a secondary gallery (about 35 m^2).

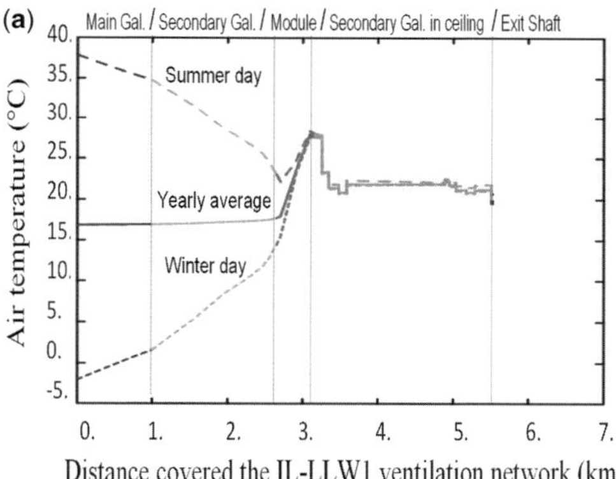

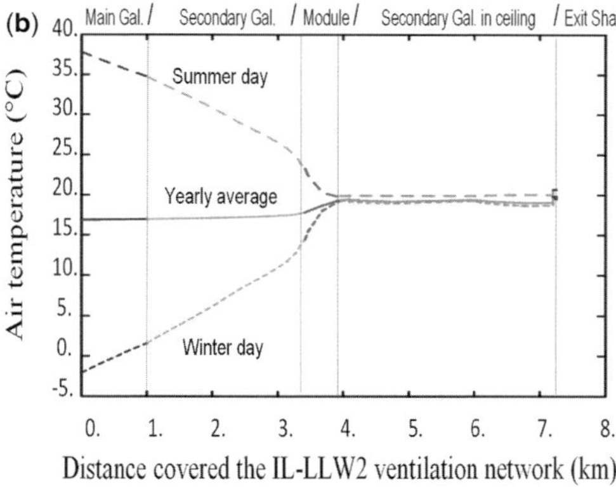

Fig. 4. Air temperature variations through the IL-LLW1 storage zone (**a**) and the IL-LLW2 storage zone (**b**) via the more distant modules; different weather conditions have been considered at the model inlet. The solid curves correspond to the yearly averaged conditions, the long-dotted-line curves correspond to a hot summer day conditions (30 °C) and the short-dotted-line curves correspond to a winter's day (−7 °C).

The yearly average of the air humidity decreases slightly, while the air temperature increases with the distance covered through the IL-LLW zones, because of the dependence of the relative humidity on air temperature. The evaporation of the *in situ* pore water has no sensitive effect on moisture (see Fig. 5b, the solid curve of yearly average of moisture, which is almost flat upstream and downstream of the module). It should be noted that the evaporation rate along the air network depends on the age of secondary-gallery segments which varies with the distance covered through the storage zone: from 90 years old (24 kg per year per gallery metre) to (a) 38 years old in the IL-LLW1 zone (33 kg per year per gallery metre), and (b) 18 years old in the IL-LLW1 zone (50 kg per year per gallery metre).

In winter, the low air temperature increases significantly in contact with hotter walls as the distance covered increases (see the tiny doted curves in Fig. 4a & b). The increase in air temperature results in a decrease in relative humidity because of the dependence on temperature of the saturation vapour pressure: in winter the air get drier by getting hotter, and the reverse occurs in summer. After 90 years, the yearly average of relative humidity is 58% at the entry of IL-LLW zones, 45% at the exit of the IL-LLW1 zone and 52% at the exit of the IL-LLW2 zone. The humidity variations with seasons are significant. At the exit of the storage zone, the humidity can drop to 15% during winter, and rise to 90% during summer.

The increase in vapour pressure in the atmosphere during May, June and July combined with the persistence of relatively low temperatures in the walls since the last winter results in condensation events over a few days per year. The most

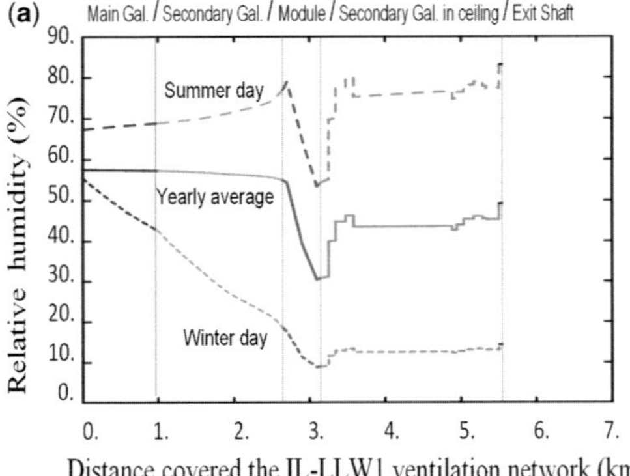

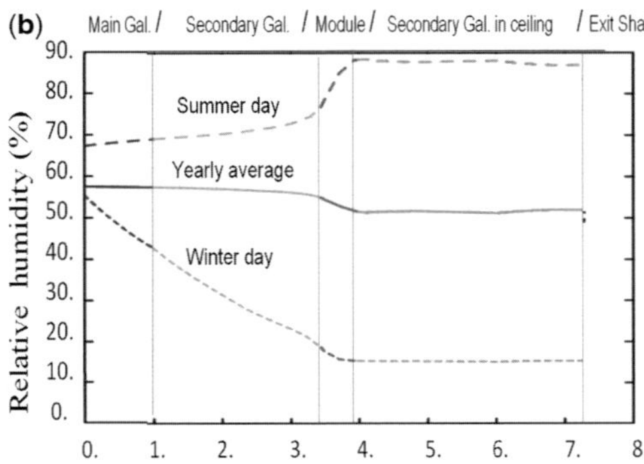

Fig. 5. Relative humidity variations through the IL-LLW1 storage zone (**a**) and the IL-LLW2 storage zone (**b**) via the more distant modules; different weather conditions have been considered at the model inlet. The solid curves correspond to the yearly averaged conditions (11.8 °C; 78% RH), the long-dotted- curves correspond to a wet summer day conditions (20 °C; 90% RH) and the short-dotted-line curves correspond to a winter's day (−7 °C; 80% RH).

distant modules are especially exposed (maximum of daily condensation rate from about 0.4–1. kg per gallery metre) as well as the most distant secondary galleries (maximum of daily condensation rate from about 1 to 10 kg m^{-1}). These condensation events results in an overall reduction of the vapour density in air throughout the IL-LLW (see Fig. 6b, the distance between green and red curves (a) above the red curve to appreciate the effect of the condensation events on the vapour density, and (b) under the red curve to appreciate the small effect of *in situ* pore water evaporation on the vapour density).

However, it should be noted that the assessment of magnitudes of moisture fluctuations in the ventilation and of wall-condensation rates is probably over-estimate, because the buffer effect of vapour exchanges with walls (i.e. the vapour exchanges between the ventilation air and the unsaturated porous media of walls) on the weather fluctuations in the air humidity is not modelled.

In the air exit shaft of the IL-LLW storage

The nuclear air going out of the IL-LLW storage zones is extracted from the repository through an 'exit shaft' separate from the HL-LLW air extraction disposal. The isentropic decrease in pressure from the base to the top of the shaft results in a −5 °C quasi-isentropic decrease of temperature. However, sometimes, mist formation in the air smooths out this temperature drop because of the latent heat released by vapour condensation. This reduction of the temperature drop reaches a maximum of 3 °C (see Fig. 6a, the thermal effect on the red curve), and occurs simultaneously (i.e. during late spring and early summer) with wall condensation events in the IL-LLW storage zones. In the exit shaft, the maximum of the mist-formation rate (up to 45 kg per metre of shaft per day) is higher than the wall-condensation rate in IL-LLW galleries, partly because the air flow rate is higher too: the overall ventilation air of the two storage zones is extracted through the same shaft.

In the HL-LLW storage zones

After 90 years, the HL-LLW1 and HL-LLW2 storage modules are in the pre-closure period for at least 42 and 22 years, respectively. At this time, the air enters the remotest HL-LLW1 module at 40 °C and exits at 60 °C. In the HL-LLW2 zone, the air temperatures are lower in the remotest module (34 and 45 °C, respectively) because the HL-LLW2 operation period starts 20 years later than the HL-LLW1 operation period.

The seasonal fluctuations of temperatures are significantly smoothed out on the way to the HL-LLW1 and HL-LLW2 storage modules (see Fig. 7, the large and tiny spotted curves upstream the storage module). They are respectively reduced by 99 and 98% at the entrance of the remotest module. The attenuation of temperature fluctuations is higher through HL-LLW storage zones than through the IL-LLW storage zones because (a) the distances covered by the ventilation air are longer (11–12 km instead of 5.5–7 km respectively) and (b) heat is exchanged between the fresh air and the nuclear air through the gallery partition: the nuclear air coming from the storage modules smooths out the temperature fluctuations of the incoming fresh air by increasing its temperature.

The heat collected in the storage modules by the ventilation has a significant effect on the humidity in the ventilation air downstream, but also upstream of the modules because of partitioned galleries. The

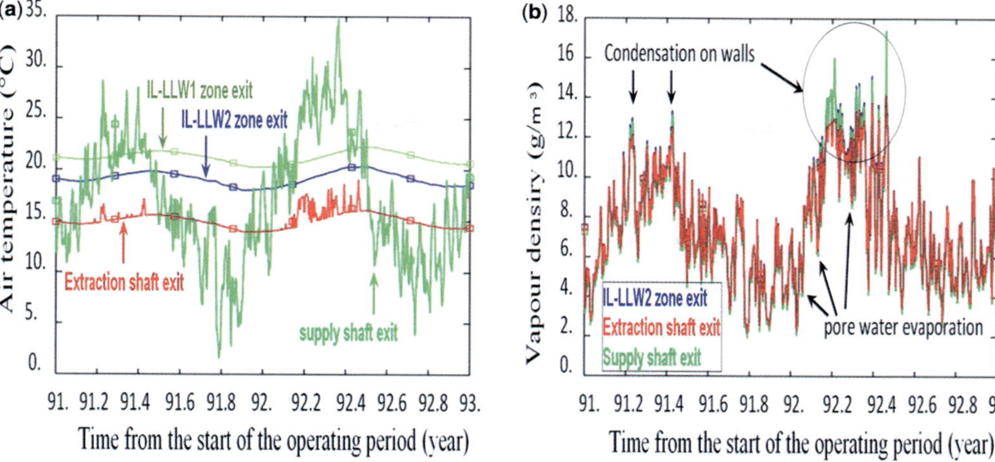

Fig. 6. Evolutions in time of (**a**) air temperature and (**b**) vapour density upstream of the IL-LLW storage zones (green), and downstream of the IL-LLW1 (olive green) and the IL-LLW2 storage zones (blue), and at the extraction shaft exit (red).

air is significantly dried by the overall temperature increase in the entire HL-LLW ventilation network: at the entrance of the remotest module, the yearly average of humidity drops to 16% in the HL-LLW1 storage zone and to 24% in the HL-LLW2 storage zone (see Fig. 8a, b, the solid curve over the module). The effect of the *in situ* pore water evaporation on the relative humidity is masked by the thermal effect. However, the evaporation of porous media contributes up to 30% to the overall vapour amount transported by the ventilation of the HL-LLW zones. This increase in vapour density is the consequence of (a) the long distance covered by the ventilation air, (b) the reduced ventilation rate during the pre-closure period (40 $m^3 s^{-1}$ for the two storage zones), which means a reduced capability to evacuate the collected vapour, and (c) the high evaporation rate inside HL-LLW storage modules – the *in situ* pore water evaporation is promoted by the increase in temperature of the rock around the storage cell, and subsequently around the module gallery, which induces (i) a water pressure build-up around the storage cell owing to the differential thermal expansion of liquid and solid phases in the host rock, (ii) an increase in hydraulic conductivity and in vapour diffusivity, and (iii) an increase in saturation vapour pressure.

Effect of the gradual extension of the repository

This analysis is based on results from the numerical simulation of the secular transient (see definition in the section 'Numerical methods'). The yearly average of weather conditions (11.8 °C; 78% relative humidity, RH) used at the model inlet, corresponds to constant air conditions of 16.8 °C and 57% RH at the repository level.

Operation of the IL-LLW storage zones

The 'active' operation period lasts 46 years in the IL-LLW1 zone, and 76 years in the IL-LLW2 zones. The two storage zones are assumed to extend gradually with a new module put in operation every 2 years. The ventilation rate of a module is (a) 10 $m^3 s^{-1}$ during the 'active' operation period and (b) 3 $m^3 s^{-1}$ in the pre-closure period.

In an IL-LLW1 storage module, the nuclear air gets hotter than the *in situ* temperature (22 °C) at the beginning of the operation period, because of the exothermic radioactive waste stored in the first two modules operated: the air temperature at the exit of the storage zone is about 24 °C after the first year of operation. In the outgoing air from a module, the temperature reaches a maximum of 36 °C about 10 years after the emplacement of exothermic wastes. During the following pre-closure period, the overall ventilation of the zone is reduced to 63 $m^3 s^{-1}$ and the yearly temperature at the exit of the storage zone decreases slowly by 1 °C and reaches 21.2 °C at 90 years (see Fig. 9, the IL-LLW1 curve).

In the ventilation air of the IL-LLW2 storage zone, the yearly temperature at the exit of the storage zone is 20 °C after the first year of operation, and then decreases slowly to 18.7 °C over

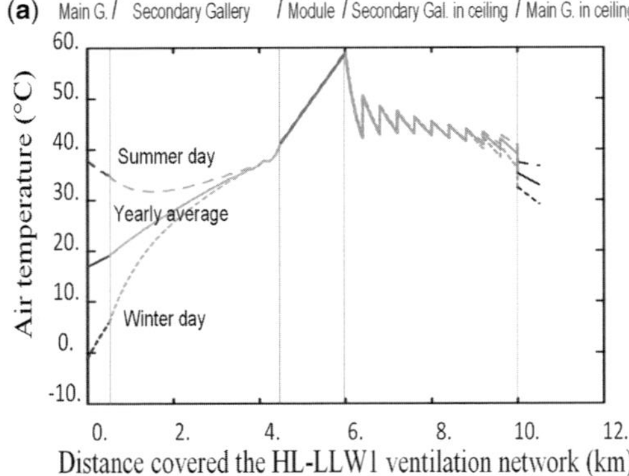

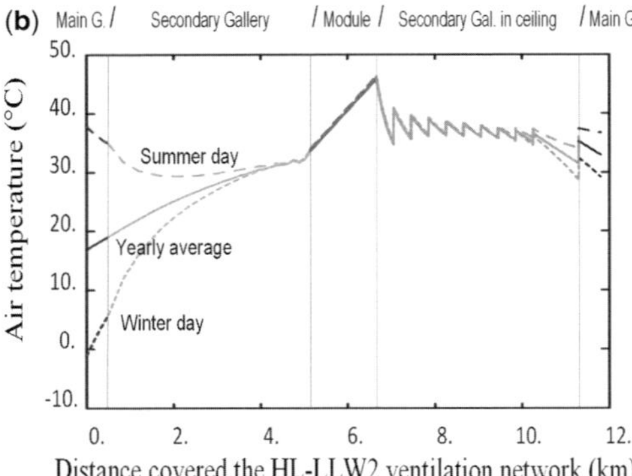

Fig. 7. Air temperature variations through the HL-LLW1 storage zone (**a**) and the HL-LLW2 storage zone (**b**) via the more distant modules; weather conditions have been considered at the model inlet. The solid curves correspond to the yearly averaged conditions, the long-dotted-line curves correspond to a hot summer day conditions (30 °C) and the short-dotted-line curves correspond to a winter's day (−7 °C).

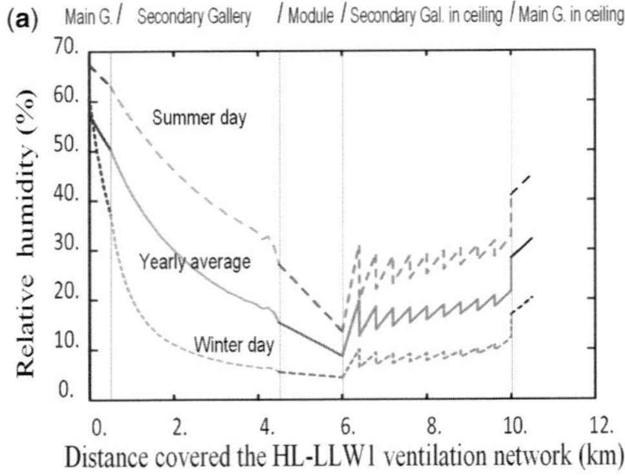

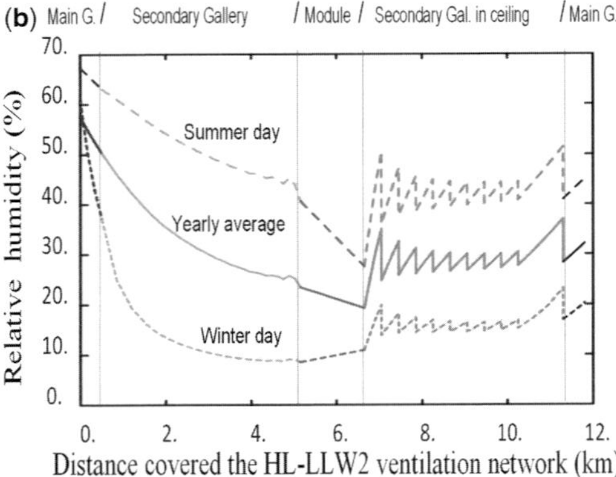

Fig. 8. Relative humidity variations through the HL-LLW1 storage zone (**a**) and HL-LLW2 via the more distant modules; different weather conditions have been considered at the model inlet. The solid curves correspond to the yearly averaged conditions (11.8 °C; 78% RH), the long-dotted-line curves correspond to a wet summer day conditions (20 °C; 90% RH) and the short-dotted-line curves correspond to a winter's day (−7 °C; 80% RH).

about 6 years, because the wall is gradually cooled by the ventilation. Afterwards, the air temperature at the exit of the storage zone remains around 19 °C, as the ventilation network becomes gradually longer with the steady extension of the storage zone (see Fig. 9, the IL-LLW2 curve). After 76 years, during the following pre-closure period, the overall ventilation of the zone is reduced to 108 $m^3\ s^{-1}$.

Operation of the HL-LLW storage zones

The construction of the HL-LLW storage zones starts 20 years after the IL-LLW storage zones, and the 'active' operation period lasts 44 years. During this period, only one storage zone gradually extends with a new module put in operation every 2 years. The construction of the HL-LLW2 zone is initiated after the last HL-LLW1 module is constructed. The ventilation rate of a module is (a) 30 $m^3\ s^{-1}$ during the 'active' operation period and (b) 2 $m^3\ s^{-1}$ during the pre- closure period.

During the 26 year-long operation period of the HL-LLW1 storage zone, the overall ventilation is high enough (104 $m^3\ s^{-1}$ maximum) to evacuate the heat released by radioactive waste that reaches the wall of module galleries. Downstream of the modules, the yearly average of the air temperature increases slightly from 20 to 24 °C thanks to the dilution of the hot air coming from pre-closure modules (2 $m^3\ s^{-1}$ per module) in the relatively cold air coming from modules still in 'active' operation (30 $m^3\ s^{-1}$ per module). During this period, the air temperatures reach from 30 to 60 °C in the modules already in the pre-closure period, and the

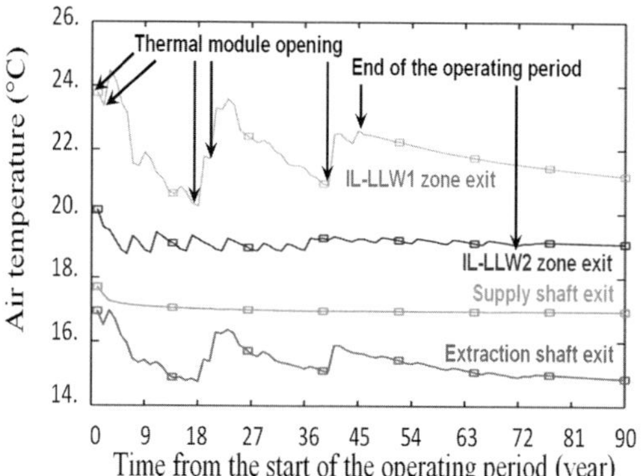

Fig. 9. Evolutions in time of the air temperature with time upstream of the IL-LLW zones (green in online version only; labelled 'supply shaft exit'), downstream of the IL-LLW1 (olive green in online version only; labelled 'IL-LLW1 zone exit') and IL-LLW2 zones (blue in online version only; labelled 'IL-LLW2 zone exit'), and at the shaft exit (red in online version only; labelled 'extraction shaft exit').

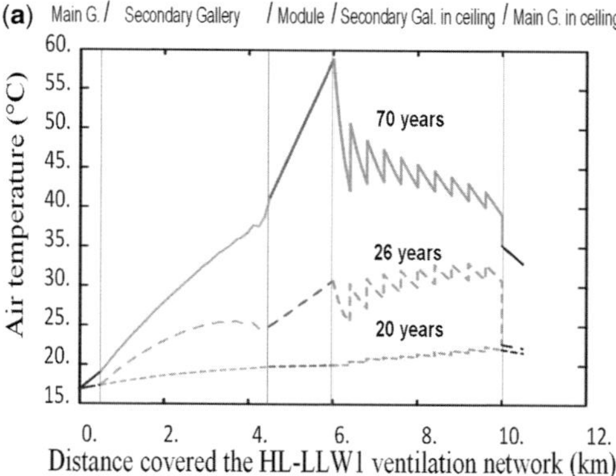

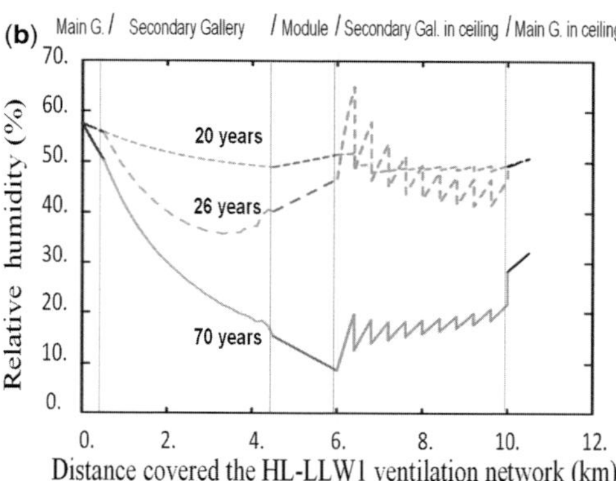

Fig. 10. Variations in the air temperature (**a**) and the relative humidity (**b**) through the HL-LLW1 storage zone via the more distant modules at various periods of operation; results from the simulation of the secular transient. The yearly averaged conditions (11.8 °C; 78% RH) have been taken at the model inlet.

relative humidity in the air slightly decreases from 57% at the entrance of storage zone to about 50% downstream the modules, because of the increase in air temperature (see Fig. 10a, b the variations in the air temperature and the relative humidity).

After 26 years of operation, the entire HL-LLW1 storage zone is in the pre-closure period and the overall ventilation rate is reduced to 20 $m^3 s^{-1}$. Consequently, the air temperature increases downstream and upstream of the modules, because of heat exchanges between nuclear air and fresh air through gallery partition. Furthermore, the low ventilation rate amplifies the effect of the *in situ* pore water evaporation on the increase in vapour density: the yearly average of vapour density rises from 8 to 13.5 g m^{-3} and the yearly average of relative humidity reaches a maximum of 65% at the 10th module exit (see Fig. 10b, the variations of the yearly average of relative humidity in the module). Table 2 gives the range of yearly average of the temperature in the fresh-air and the nuclear-air ventilation networks.

The operation in the HL-LLW2 storage zone starts 20 years after the HL-LLW1 storage zones. The evolutions of the air temperature and the humidity are typically the same. However, the air temperatures after 70 years of 'active' operation are lower (compare the air temperature ranges in both storage zone in Tables 2 and 3), because the waste packages are emplaced in the HL-LLW2 storage cells 20 years later than in the HL-LLW1 storage cells.

Some mist condensation events can occasionally occur in summer (maximum daily condensation rate estimated of 6 l m^{-1}) in the HL-LLW extraction shaft. However, such events vanish after 50 years of 'active' operation during the pre-closure period.

Sensitivity to physical hypotheses

Global warming

The change in weather conditions will be probably significant over the 100 year-long operation period of the repository because of global warming. The global warming scenario adopted in this test case assumes (a) a gradual rise in temperature of 3 °C over 90 years and (b) the same variations in relative humidity as those measured between 2002 and 2004, which means that the vapour density in the air will gradually get higher for 90 years. The calculation results are analysed at the end of the transient period.

Table 2. *Range of the yearly average of the air temperature and relative humidity in the HL-LLW1 zone after 20, 30 and 70 years of operation in the HL-LLW storage zones*

Time (years)	Upstream of the storage modules		Downstream of the storage modules	
	Air temperature (°C)	Moisture (%)	Air temperature (°C)	Moisture (%)
20	17–19	58–48	19–21	48–50
30	18–30	58–36	30–53	65–42
70	19–42	58–15	38–60	10–32

Gobal warming has no significant impact on temperature in the ventilation air of the HL-LLW storage zones. After 90 years, the average temperature in the outgoing air would increase by 0.2 °C compared with the reference results, which induces an increase in the relative humidity by 2–5%. In the IL-LLW storage zones, 3 °C global warming results in a 1.8 °C warming at the exit of IL-LLW zones, and a slight increase in frequency and intensity of condensation events in the exit shaft (mist formation) and on walls along the gallery network, probably because of higher vapour density at the repository entrance. Furthermore, there is no significant effect on the attenuation of weather fluctuations in temperature through the ventilation network.

Interstitial evaporation rate

The effect of weather fluctuations on pore-evaporation rate in the repository has not been modelled. In this test case, a 5-fold rise in pore-evaporation rate has been considered to cover this effect.

During the 'active' operation period, the ventilation is high enough to prevent a significant increase in vapour density despite the high pore-evaporation rate. In contrast, once the entire storage has switched to the pre-closure period, the high reduction of the ventilation rate induces in the ventilation a significant increase in vapour density because of the increase in air residence time in the storage. Consequently, wall condensation events tends to occur further in the repository, and to amplify in intensity and in frequency:

In both HL-LLW zones, chronic condensation of up to 9 kg per gallery metre per day has been assessed downstream of the storage modules, just after the generalization of the pre-closure ventilation over a storage zone. Then, the condensation rate gradually decreases to zero after about 20 years.

In the IL-LLW zones, condensation rate slightly increases in the exit shaft and in some storage modules. In addition, condensation grows downstream of the storage modules and reaches a maximum value of 2 kg m^{-1} per day instead of 0.4 kg m^{-1} per day with the reference assumptions.

Table 3. *Range of the yearly average of the air temperature in the HL-LLW2 zone after 40, 50 and 70 years of operation in the HL-LLW storage zones*

Time (years)	Air temperature (°C)	
	Fresh air	Nuclear air
40	17–22	22–40
50	18–31	28–45
70	19–34	32–51

Thermal properties

The heat exchanges between air and wall have a significant effect on the air temperature throughout the repository because of the high thermal capacity of concrete and rock. Thermal capacity and conductivity in walls globally contribute to cool the air in summer, to warm it in winter, and to smooth out the weather fluctuations in temperature through the ventilation network. However, a more or less 30% change of wall thermal properties has no great effect on the ambient conditions inside the repository: in yearly average, about ±0.8 °C at the end of the HL-LLW ventilation network, and about ±0.25 °C at the end of the IL-LLW ventilation network.

Sensitivity to the conceptual hypotheses

Ventilation reduction during the pre-closure period

One of the studied solutions to ensure that the disposal system remains easily reversible for a long time is to maintain ventilation in the access drifts of a module even after the waste disposal is complete. However, this has a cost directly linked to the ventilation rate. In this scenario, the ventilation rate of storage modules is reduced by 3 during the pre-closure period, which induces a 3-fold increase in the air residence time.

In the HL-LLW storage zones, the increase in the air residence time results in a significant increase in vapour density, particularly in the storage modules because of the ongoing evaporation in porous media. However, in contrast to the significant changes in vapour density, the ventilation reduction does not change the temperature in the outgoing air from storage modules, given the thermal equilibrium between air and wall at the end of the module galleries for low ventilation rate. Consequently, the ventilation removes 3-fold less heat from HL-LLW modules than with the reference assumptions, which results in (a) a decrease in air temperature by −4 to −6 °C and (b) a chronic condensation in ventilation network downstream of the modules. The yearly average of condensation rate rises up to 1.4 kg per gallery metre per day once the entire HL-LLW storage zone has switched to the pre-closure period, and then gradually decreases to zero 20 years later.

In the IL-LLW storage zones, the average amplitude of the temperature fluctuations is significantly reduced throughout the ventilation network, especially upstream of the storage modules. This reduction is about 30% at the end of the main galleries. At the end of the secondary galleries, the daily fluctuations disappear and the seasonal fluctuations are approximately 6 times lower than in the reference case. The condensation rates along the exit shaft and along the storage modules are divided by 3 as the ventilation transports 3 times less vapour from the atmosphere than in the reference case, and despite the ongoing evaporation in porous media.

Thermal insulation of the ventilation partition

In galleries of the HL-LLW zones, a 10-fold reduction of the thermal transmissivity of the ventilation partition (i.e. fresh air in full section and nuclear air in the ceiling) reduces by about 25% the warming of the incoming air by the hot outgoing air from storage modules.

Synthesis and conclusion

The ambient conditions in the nuclear waste ANDRA's repository have been described in terms of air temperature and relative humidity over the active operation and the pre-closure period. Temperature and humidity variations were assessed by performing numerical simulations of heat and vapour transfers throughout the ventilation network. The physical modelling covers: (a) air-to-wall heat exchanges owing to forced and free convection; (b) heat and vapour advection by the ventilation; (c) thermal storage and heat conduction in the concrete structure and in the host rock; (d) condensation fluxes on walls; (e) evaporation of *in situ* pore water through time functions; and (f) weather conditions at the inlet. During the operation period, the airflow rate is adapted to the extension of the repository by the construction of new storage modules every two years. The resulting analysis has highlighted the main physical mechanisms that determine the variations in time and in space of ambient conditions through the repository.

The ventilation air coming from the site surface arrives at the repository level about 5 °C hotter, because of the isentropic compression of the air by its own weight over the shaft height. The heat exchanges with the shaft wall have no significant effect on (short-term) air temperatures because the air residence time in the shaft is usually limited to few minutes at most.

In the repository, the surrounding rock mass buffers the weather fluctuations in temperature with the distance covered in the ventilation network. The hourly fluctuations in temperature vanish in the central zone on the way to the storage zones, while seasonal fluctuations are smoothed out further, as the air passes through the storage modules. The attenuation of temperature fluctuations induces significant changes in relative humidity through the ventilation network, because of the dependence in temperature of the vapour saturation pressure.

During spring and summer, hot and humid weather can induce condensation events in the IL-LLW storage, in and upstream of the more distant storage modules. The air entering the repository gets colder and consequently more humid, in contact with gallery walls that have been cooled during the previous winter. Condensation events are not the same from one year to another: frequency and intensity rise when a hot and stormy summer succeeds a severe winter.

In the HL-LLW zones, during the last decade of the reversibility period, the air becomes very dry, particularly in winter (down to 5% RH in very hot conditions – the air temperature reaches 60 °C in some storage modules). A significant part of the heat removed from the storage modules by the air renewal is recycled along partitioned galleries: incoming fresh air which flows in full section is warmed by outgoing hot air flowing in ceiling ducts through thermal conduction in the air ventilation partition.

The *in situ* pore water evaporation from the geological media has no significant effect on relative humidity during the operation stage. However, during the pre-closure period, the high evaporation rate in the HL-LLW storage modules combined with the ventilation reduction substantially increases the vapour concentration in the ventilation air.

It should be noted that the assessment of the humidity variations in the repository could be improved by coupling the vapour advection in the ventilation network with the hydraulic transient in gallery walls (Richards's equations) in order to take into account the buffer effect on the weather variations in humidity, which would probably reduce the magnitude and the frequency of condensation events, and increase the lowest values of humidity assessed.

Such results are valuable to ANDRA, among others elements, to evaluate if the coupling with other physical processes (i.e. the corrosion rate of metal in relation to water vapour in the air) is significant or not and to evaluate whether adaptation of the ventilation scheme is needed to change the ambient conditions in the repository in response to workers' safety and health requirements.

This work was funded by Andra and was based on data provided by Andra. The authors are grateful to P. Delage and an anonymous reviewer for their helpful and constructive remarks.

Reference

Cast3M 2010. http://www-cast3m.cea.fr/ [accessed 4th June 2014].

Gas injection test in the Callovo-Oxfordian claystone: data analysis and numerical modelling

RÉMI DE LA VAISSIÈRE[1]*, PIERRE GERARD[2,3], JEAN-POL RADU[2], ROBERT CHARLIER[2], FRÉDÉRIC COLLIN[2], SYLVIE GRANET[4], JEAN TALANDIER[1], MÉDÉRIC PIEDEVACHE[5] & BENJAMIN HELMLINGER[5]

[1]*Agence nationale pour la gestion des déchets radioactifs, RD960 55290 Bure, France*

[2]*Université de Liège, Département ArGEnCo, 1 chemin des Chevreuils – BAT 52, B4000 Liège, Belgium*

[3]*Département Batir, Université libre de Bruxelles, Belgium*

[4]*Laboratoire de Mécanique des Structures Industrielles Durables, UMR EDF/CNRS 2832, France*

[5]*Solexperts, France, Technopôle Nancy-Brabois 3, rue du Bois de la Champelle, 54500 Vandœuvre-lès-Nancy, France*

**Corresponding author (e-mail: remi.delavaissiere@andra.fr)*

Abstract: This paper describes a field-scale experiment on gas transport mechanisms performed at Andra's Underground Research Laboratory (URL) in a clay rock. The experimental layout consists of two parallel boreholes that are equipped with multiple packer completions isolating three intervals each, which have been continuously monitoring the pore pressure evolution of the clay rock. Nitrogen gas was injected in the middle test interval of one of the boreholes at increasing rates. The entire gas test comprised six periods of controlled gas injections, each followed by a shut-in pressure recovery phase. The experimental data are presented along with their interpretation by means of numerical modelling of two-phase flow of gas and water using different numerical codes and different geometrical approaches that include axisymmetric, half-space and full 3D models. An iterative modelling process was used to show step-by-step how an accurate description of each component of the experiment system produced a satisfactory reproduction of the experimental data and an improved understanding of the relevant phenomena. For instance, the initial volume of remaining water in the test interval, and the presence of a damaged zone around the boreholes, was important for the models to obtain good agreement with the field data.

Understanding the migration of gas produced by corrosion of metals, microbial degradation and radiolysis of water within a deep geological repository for radioactive waste is of great importance in the performance assessment and long-term evolution of the repository. If the rate of gas production exceeds the rate of diffusion of dissolved gas in the porewater of the host rock and the engineered barriers, a gas phase will form and accumulate until the associated pressure buildup is large enough to overcome the hydraulic pressure and the capillary resistance of the surrounding confining rock.

In an indurated claystone like the Callovo-Oxfordian or the Opalinus Clay, four gas migration mechanisms are usually referred to (Marschall *et al.* 2005): (a) migration by diffusion as dissolved gas in the porewater; (b) advective/dispersive visco-capillary gas flow in the partially gas-filled pore network of the clay; (c) gas flow along localized dilatant pathways which may or may not interact with the continuum stress field; and (d) gas fracturing of the rock at pressure levels above the minimum principal stress in the rock.

To investigate these mechanisms, the French national agency for the management of radioactive waste (Andra) has been conducting a field-scale experiment to examine the mechanisms controlling gas entry and gas migration in the Callovo-Oxfordian (COx) claystone, which is the proposed host rock for the French deep geological repository project. This experiment, called PGZ1, was designed to study the migration of nitrogen in the host rock using gas injection tests in packed off borehole sections. Similar tests were performed in the 1990s in the Opalinus Clay at the Mont-Terri URL. A review of the experimental data and analysis based on two-phase flow- and coupled

From: Norris, S., Bruno, J., Cathelineau, M., Delage, P., Fairhurst, C., Gaucher, E. C., Höhn, E. H., Kalinichev, A., Lalieux, P. & Sellin, P. (eds) 2014. *Clays in Natural and Engineered Barriers for Radioactive Waste Confinement*. Geological Society, London, Special Publications, **400**, 427–441.
First published online March 7, 2014, http://dx.doi.org/10.1144/SP400.10

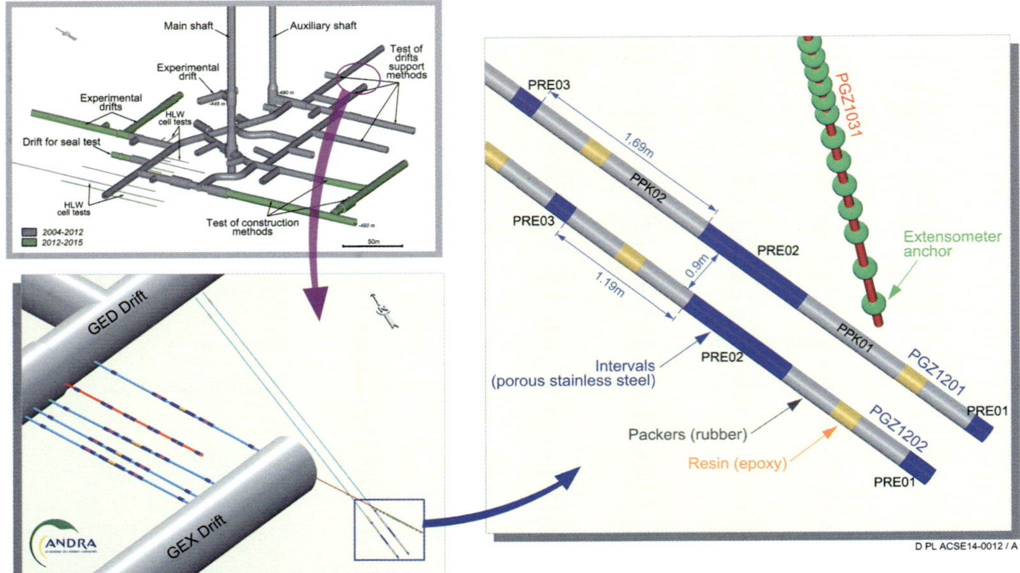

Fig. 1. General and detail layout of the large-scale gas injection experiment in the Meuse/Haute-Marne URL.

hydro-mechanical modelling is given in Croisé *et al.* (2006).

In this paper, a step-by-step presentation of the modelling is presented, highlighting how an accurate knowledge of the boundary conditions of the experiment and a close interaction between experimenters and modellers was necessary to obtain satisfactory numerical results. Three different numerical codes (LAGAMINE, Code_ASTER and TOUGH2/iTOUGH2) and different geometrical approaches (axisymmetric, half-space and full 3D models) were implemented.

Experimental setup and data

Geometry

The experimental layout consists of three boreholes (Fig. 1). Two parallel boreholes (named PGZ1201 and PGZ1202) were drilled with a diameter of 76 mm at an inclination of 35° from the GED drift (see Fig. 1). Both boreholes are oriented parallel to the direction of the *in situ* maximal stress and are equipped with a multiple packer system to monitor water/gas pressure in three isolated intervals, which are 1 m apart from each other along the borehole axis. The borehole PGZ1031 drilled with a diameter of 101 mm at an inclination of 48° from the GEX drift (see Fig. 1). PGZ1031 is equipped with a multiple magnetic extensometer probe to measure potential axial deformation during the gas injection tests. PGZ1031 intersects the vertical plane through PGZ1201 above the bottom packer of PGZ1201 at a distance of 2.2 m. All three boreholes were core drilled using air. The shortest (wall-to-wall) distances in 3D space between the six packed-off intervals and borehole PGZ1031 as indicated in Figure 1 are given in Table 1.

Test sequence

Hydraulic and gas injection tests were performed in interval 2 of borehole PGZ1201 (reference number in the Andra's data acquisition system: PGZ1201PRE02). The test sequence is summarized in Table 2. After the boreholes were drilled, the intervals were saturated with water of the same chemical composition as the formation porewater in July 2009. The GAS1 nitrogen gas injection phase started in February 2010 and finished in April 2011. The GAS1 phase consisted of six

Table 1. *Shortest wall-to-wall 3D distances between the measurement intervals and borehole PGZ1031*

	Interval 1 (PRE01)	Interval 2 (PRE02)	Interval 3 (PRE03)
PGZ1201	2.31 m	1.14 m	2.20 m
PGZ1202	2.74 m	1.90 m	2.15 m

Table 2. *Water and gas tests sequence*

Date	Elapsed time (days)	Phase	Duration (days)
2 July 2009	−25	Drilling PGZ1031	
27 July 2009	0	Drilling PGZ1201	
29 July 2009	2	Drilling PGZ1202	
28 August 2009	32	HYDRO1 – pulse withdrawal	
14 September 2009	49	HYDRO1 – constant pressure test (injection)	5
1 February 2010	189	Start GAS1: GRI1	11
12 February 2010	200	GRIS1	19
3 March 2010	219	GRI2	15
18 March 2010	234	GRIS2	20
7 April 2010	254	GRI3	7
14 April 2010	261	GRIS3	28
12 May 2010	289	GRI4	13
25 May 2010	302	GRIS4	29
24 June 2010	332	GRI5	11
5 July 2010	343	GRIS5	107
20 October 2010	450	GRI6	15
4 November 2010	465	GRIS6	146
30 March 2011	611	Stop GAS1 – start gas extraction GRE	
5 May 2011	647	Stop GRE + saturation + compressibility test	
9 June 2011	682	HYDRO2 – pulse withdrawal	
24 June 2011	697	HYDRO2 – constant pressure test (injection)	6

constant-rate gas injection steps (GRIx) of between 7 and 15 days' duration, each followed by a pressure recovery phase (GRISx) of between 19 and 146 days' duration. The pressure buildup during the last constant-rate injection step (GRI6) reached a maximum of 9.1 MPa, which is well below the minimum measured *in situ* principal stress component of about 12.3 MPa at the URL level (Senger *et al.* 2006). After the GAS1 phase, a gas extraction phase (GRE) was performed. On 5 May 2011, interval 2 was refilled with water and a second hydraulic test (HYDRO2) was performed in June 2011 with exactly the same test procedure as for HYDRO1.

Field data

Pressure and gas flowrate

Figure 2 shows the pressure and gas flow rate data monitored at boreholes PGZ1201 and PGZ1202 from July 2009 to September 2011. The axial deformation measurements at the fixed installed extensometer device in PGZ1031 are not presented in this paper, because they were not sensitive to the gas injection test. This shows that no significant mechanical deformations of the rock mass occurred at less than a 1.14 m distance from the injection borehole PGZ1201.

Hydraulic tests

Table 3 provides the analysis results of the HYDRO1 and HYDRO2 test sequence. Both hydraulic tests were interpreted using a composite radial flow model with an inner and outer zone (Baechler *et al.* 2011). The inner zone corresponds to the zone around the borehole, which was damaged during drilling, and the outer zone corresponds to the intact claystone. The extension of the inner zone was used as a fitting parameter in the inverse analysis of the hydraulic tests response. The results of the HYDRO1 analysis indicated the thickness of the borehole damaged zone to be a few millimetres. This value was conserved for the HYDRO2 test analysis. Damage zones are usually created around underground openings during the excavation process and also occur during borehole drilling (Bossart *et al.* 2002; Levasseur *et al.* 2010).

It should be noted that uncertainties in the estimated values for permeability and specific storage vary between a factor of 2 and 5. This indicates that no change in the hydraulic parameters occurred between HYDRO1 and HYDRO2.

Characteristics of the experimental data

Prior to the start of the GAS1 phase, the measured porewater pressures in boreholes PGZ1201 and

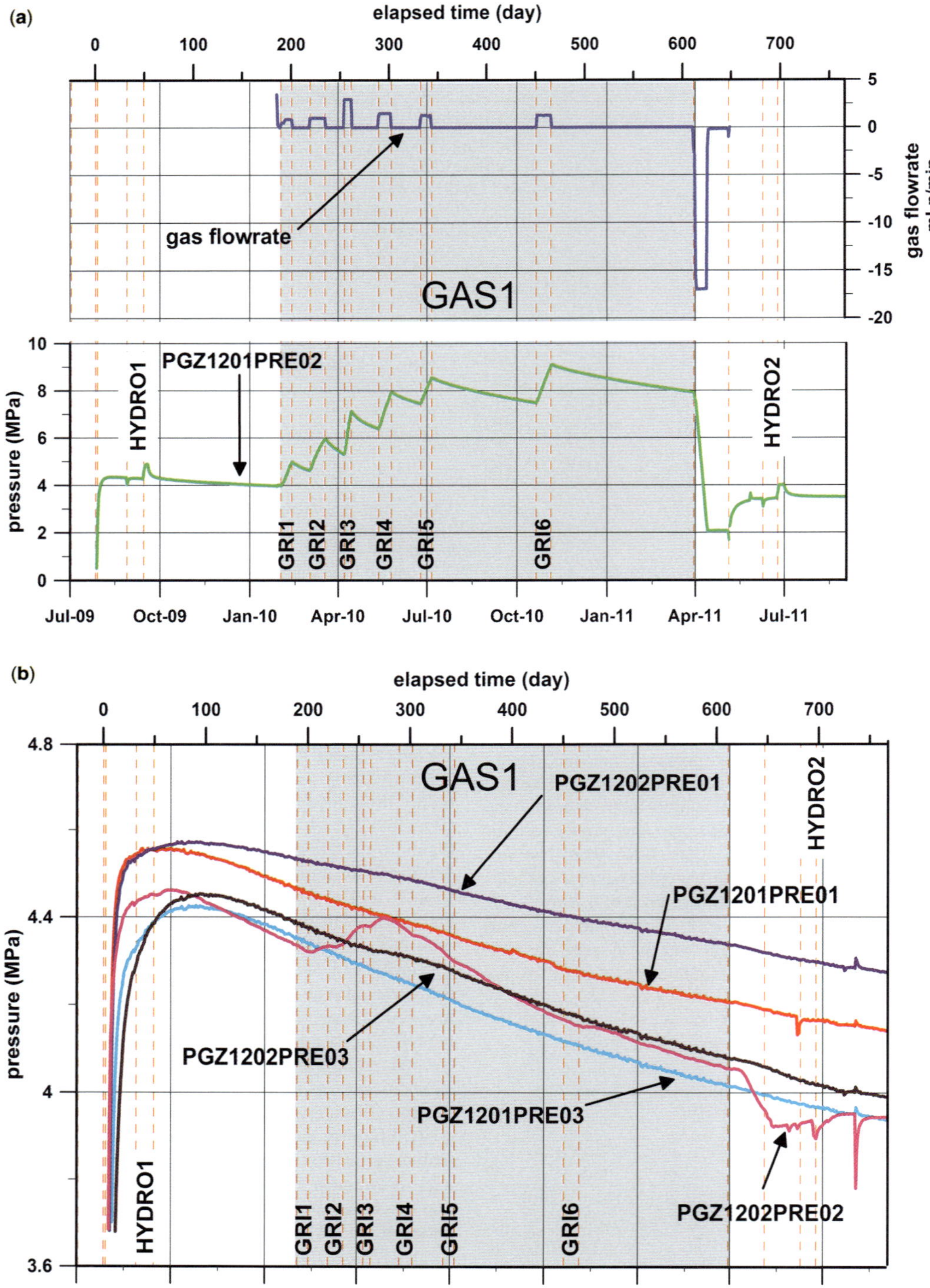

Fig. 2. Data from boreholes PGZ1201 and PGZ1202 between July 2009 and July 2011: (**a**) gas injection rate and gas or water pressure in the test interval; and (**b**) water pressures. Area shaded in grey: GAS1.

Table 3. *Results of hydraulic tests HYDRO1 and HYDRO2*

Parameters		HYDRO1 (September 2009)	HYDRO2 (June 2011)
Hydraulic conductivity (m s^{-1})	Inner zone	6.4×10^{-11}	2.7×10^{-11}
	Outer zone	2.6×10^{-13}	2.0×10^{-13}
Specific storage (m^{-1})	Inner zone	1×10^{-6}	1×10^{-6}
	Outer zone	3×10^{-6}	2.7×10^{-6}
Inner zone thickness (m)		0.004	0.004

PGZ1202 showed a gradual decline over time (Fig. 2) from October 2009 when the perturbation from the drilling of the borehole vanished. The pressure declines in the intervals 2 of PGZ1201 and PGZ1202, which are closest to PGZ1031, were steeper than the pressure declines in intervals 1 and 3. At the end of December 2009 the pressure decline was approximately −0.018 bar/day for the nearest interval 2 in PGZ1201, and −0.005 bar/day for the furthest interval 1 in PGZ1202.

During GAS1, a cross-hole hydraulic response was observed in interval 2 of borehole PGZ1202, located about 90 cm (wall to wall) from the gas injection interval, during the first three steps (GRI1 to GRI3). During the last steps (GRIS3 to GRIS6), the magnitude of the cross-hole response decreased while the injection pressure was the highest. In the observation intervals of borehole PGZ1201 (intervals 1 and 3), located below and above the test interval 2, no hydraulic interferences were observed during the entire GAS1 phase, indicating no hydraulic communication around the packers of the injection interval. A comparison of the six pressure recovery phases (Fig. 3) indicates a change in the characteristic behaviour during the course of the injection steps, with faster pressure declines occurring during GRIS1 to GRIS2 compared with those during the subsequent steps GRIS4 to GRIS6.

Preliminary modelling

A first set of modelling was performed to reproduce the measured data. One hydraulic simulation could reproduce the observed pore pressure decline observed for all the intervals before and after GAS1. In addition, two standard two-phase flow simulations and one coupled hydro-mechanical modelling were performed.

The geometry of the problem and a possible permeability anisotropy required a 3D modelling approach. However, in order to highlight the influence of each component of the system, a simpler 1D modelling approach was used.

Drainage effect

The observed pressure declines prior and after the GAS1 phase indicated that drainage affects the

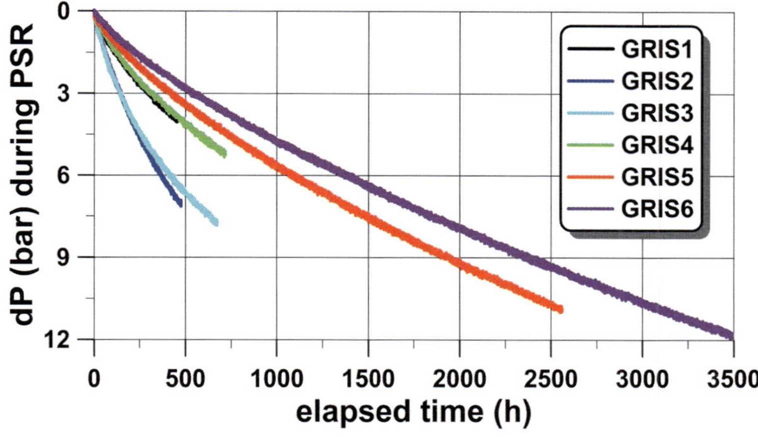

Fig. 3. Pressure data from interval 2 of PGZ1201 (PGZ1201PRE02) during the six pressure recovery steps (GRIS1 to GRIS6).

monitored pressure at the boreholes PGZ1201 and PGZ1202, even though the test interval locations are more than 20 m away from the GED drift. A numerical study of this effect was conducted with the TOUGH2 code (Pruess *et al.* 1999) using an axially symmetrical mesh of the borehole PGZ1031 and surrounding rock. The modelling confirmed the impact of drainage along the PGZ1031 borehole on the nearby PGZ1201 and PGZ1202 boreholes. This drainage was caused by the relatively high water permeability of the cement (*c.* 10^{-9} m s^{-1}) used to lock the extensometer device to the PGZ1031 borehole wall, thus, creating a hydraulic connection between the rock mass around PGZ1031 and the atmospheric pressure conditions in the GEX-drift. This hydraulic effect was taken into account in the numerical two-phase flow modelling of the GAS1 phase by correcting the experimental data to cancel the observed trend of decreasing pore pressure with time.

Two-phase flow modelling

A two-phase flow model was used to simulate transport of water and gas in partially saturated porous media. This model considered a liquid phase, composed of water and dissolved gas, and a gaseous phase, represented by an ideal mixture of dry gas and water vapour. The model accounts for advection in each phase described by the generalized Darcy's law, and for diffusion of the components within each phase (Fick's law). The constitutive relationships describing the capillary pressure between gas and water phases and the relative permeability of each phase as a function of the saturation are specified by the user's prescription.

Inverse modelling

A first modelling study was performed using the inverse code ITOUGH2 (Finsterle 2007) to estimate the hydraulic and two-phase flow parameters calibrated to the measured pressure response. For this, a fully three-dimensional mesh was constructed taking into account the drainage effect along borehole PGZ1031.

The two-phase flow parameters were based on the van Genuchten–Mualem model (Mualem, 1976; van Genuchten 1980) with parameter values of $P_r = 15$ MPa and $n = 1.49$ (Table 4), whereby the shape parameter from the retention curve was used for both water and gas relative permeability curves. An effective air-entry pressure was derived for the inner and outer zones based on the Horner extrapolation of the recovery responses (GRISx) by specifying the corresponding residual gas saturations in the capillary pressure curve (Marschall *et al.* 2005). In the model, an initial volume of the test interval of 810 cm^3 was used and was totally filled with gas.

The results of the inverse modelling reproduced well the measured GRI1 to GRI3 responses with permeabilities similar to those obtained from HYDRO1 test analysis, accounting for a perturbed inner zone and the outer zone corresponding to the intact claystone (see Table 3). To reproduce the pressure response for later gas injection sequences (GRI3S to GRI6S), a new calibration was necessary, suggesting a permeability decrease in the intact claystone. However, a mechanism that would explain such a change in permeability could not be identified. The inverse modelling was unable at this stage to give a consistent set of parameters to reproduce the entire sequence of the GAS1 phase.

Forward modelling

Forward modelling was performed using the LAGAMINE code (Charlier 1987; Collin *et al.* 2002), which is a numerical tool developed at the Université de Liège. For this simulation, hydraulic parameters for COx, given in Table 4, were taken from data reported by different investigation teams

Table 4. *Hydraulic and mechanical parameters for COx*

	Lagamine	Code_ASTER
Water permeability in saturated conditions (m^2)	4×10^{-20}	2.2×10^{-20}
Gas permeability in dried conditions (m^2)	4×10^{-18}	5.5×10^{-18}
Porosity (−)	0.18	0.18
Van Genuchten air entry pressure (MPa)	15	15
Van Genuchten parameter of retention curve (−)	1.49	1.49
Young's modulus (MPa)		4000
Poisson's ratio (−)		0.3
Biot coefficient (−)		0.6
Friction angle (deg)		15
Cohesion (MPa)		3

studying this rock (see Charlier *et al.* 2013 for a review of the retention characteristics of COx claystone).

The retention curve and the water relative permeability curve are given by the van Genuchten-Mualem's expression type. The gas relative permeability curve $k_{r,g}$ is a cubic function:

$$K_g = K_g^{dry} k_{r,g} = K_g^{dry}(1 - S_{r,w})^3 \quad (1)$$

with K_g the gas permeability and K_g^{dry} the gas permeability in dry conditions; $S_{r,w}$ the degree of water saturation.

An initial volume of the test interval equal to 1540 cm^3 was used and no damaged zone around the borehole was considered in this model. The test interval was assumed to be almost fully gas saturated at the start of gas injection; that is, all the water in the interval was displaced at the end of the water–gas exchange phase. In the model the initial degree of saturation $S_{r,w0}$ of this open void was assumed to be 0.05. The nitrogen injection was modelled through a specified gas flow rate that was prescribed at the test interval face, corresponding to the experimental gas injection rate (Fig. 2).

The simulated water and gas pressure evolutions in the test interval are shown together with the experimental pore pressures in Figure 4a. The test interval remains dry during the gas tests and the desaturation of claystone is very low (Fig. 4b). A comparison of the simulated results with the measured data shows large differences. As with inverse modelling above, the main difficulty is to reproduce the entire sequence with a unique set of parameters. This suggested that possible hydro-mechanical phenomena occurred, which could not be accounted for in the standard two-phase models.

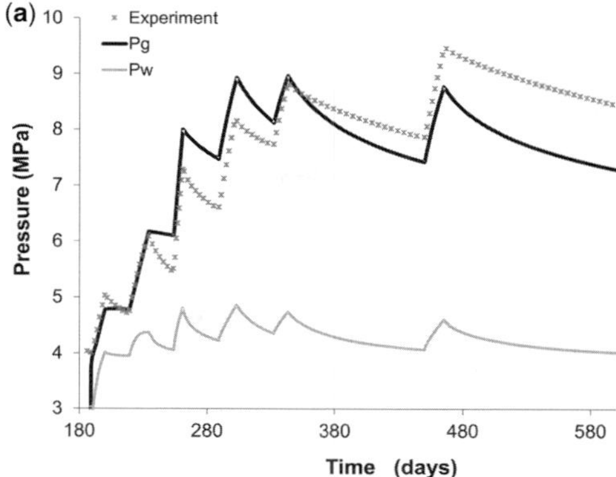

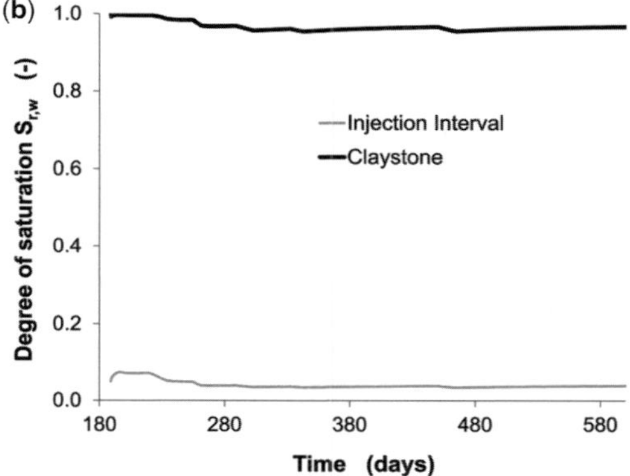

Fig. 4. Gas and water pressures evolution in the test interval: (**a**) comparison with pore pressures measurements; and (**b**) evolution of the degree of saturation in claystone and in the test interval for the first modelling (Lagamine).

Hydro-mechanical modelling

A hydro-hechanical model with the Code_ASTER (Chavant *et al.* 2007) was developed to evaluate the potential influence of coupled phenomena. Considering that anisotropy has a minor effect in this test, an axisymmetric model configuration was selected in a first approach. For the fully coupled hydromechanical model, only the test interval and surrounding claystone were considered.

The retention curve and the water relative permeability curve were represented by the van Genuchten–Mualem function, and the gas relative permeability curve was defined using a cubic function. The mechanical behaviour of the claystone was described by the classical elastoplastic Drucker–Prager law (see Granet & Meunier 2012 for details). This model assumes that intrinsic permeability is not affected by deformation, but conversely pore pressure affects deformation. Both mechanical and hydraulic parameters are given in Table 4.

Four simulation steps were performed: (a) drilling of the borehole during a 1 h period; (b) equilibration period for one day; (c) installation of packers (modelled by prescribing zero normal displacement around the borehole) with open test intervals (free surface), followed by an equilibration period for 189 days; and (d) gas injection. For this model a test interval volume of 1100 cm^3 was used, which was assumed to be filled with water at the start of the gas injection test.

The results of the geomechanical modelling show plastic deformation along a cross section perpendicular to the testing interval (Fig. 5): a small plastic area (2 cm) appears after borehole drilling and remains unchanged during gas injection. This indicates that the hydraulic to mechanical coupling

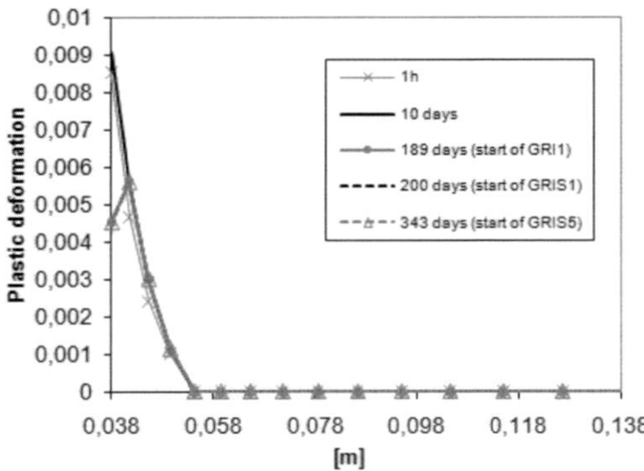

Fig. 5. Plastic deformation along a cross section perpendicular to borehole (Code_Aster).

is negligible. However, a borehole damaged zone has to be taken into account and is examined in more detail below. The simulated water pressures in the testing interval during GAS1 (Fig. 6) show a good fit with the observed pressures during GRI1 to GRI3. After GRI3, the differences increase as it becomes harder for the fluid to penetrate into the clay. During the gas injection phase we also observe that interval 2 is still fully water-saturated, which is contrary to the experimental observation (see the next section). For those reasons, the first simulation step of drilling of the borehole will not be considered in future modelling.

Partial conclusion

This preliminary modelling set was an important step in the understanding of the experimental data and it contributed to the global analysis presented below. Indeed, the standard two-phase flow and the hydro-mechanical modelling approaches failed

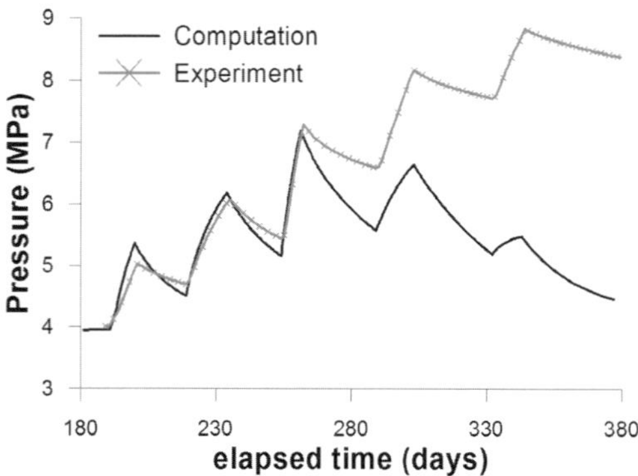

Fig. 6. Water pressure in the test interval for the HM computation during GAS1 (Code_Aster).

to reproduce the field data during GAS1. More importantly however, the preliminary modelling showed that it is more or less during the third injection step, GRI3, that something changed in the system. An in-depth analysis of the data was therefore performed to examine the role of each component of the system in order to improve the conceptual and numerical models.

Analysis of the pressure increase in the test interval using isothermal compression of gas in a closed volume

The injection pressure data indicated a change in behaviour during GAS1 that could not be reproduced by the preliminary numerical modelling. The pressure evolution during the gas-injection steps is highly controlled by the available compression volume. That is, accurate information on the volume of the injection interval 2 of PGZ1201 and the initial gas and water volumes in the test interval at the start of GAS1 is necessary for an accurate modelling. Although the theoretical volume of the test interval could be computed from the volumes of the lines and the porous filter screen, some uncertainty remained owing to the presence of breakouts and possible convergence at the borehole wall between drilling and the water–gas exchange phase. The estimated volume ranges between 804 and 1540 cm^3. The maximal volume corresponds to the case assuming no convergence of the rock mass and an annulus space between filter screen and borehole wall, whereas the minimal volume corresponds to the annulus space entirely closed as a result of the convergence of the clay around the filter screen of interval 2. In addition, the water in the test interval might have been only partially removed from the interval during the water–gas exchange phase because of the borehole inclination. This remaining water might be gradually displaced from the interval into the surrounding rock during the gas injection. The amount of water extracted from the test interval during the water–gas exchange prior to GAS1 was 810 cm^3, suggesting that a certain annular volume around the filter screen may have been present at the start of GAS1.

While the exact amount of remaining water in the test interval is unknown, the gas volume during the initial gas injection steps can be estimated by isothermal compression of the gas in a closed volume based on the ideal gas law. Figure 7 depicts the measured injection pressure increase for each injection step compared with the computed pressure increase based on isothermal compression for different gas volumes.

GRI1 to GRI3

The measured pressure at each step is located between the linear pressure curves calculated based on the minimum and maximum test-interval volumes of 804 and 1540 cm^3, respectively. The early part of GRI1 showed a best fit using a closed volume 1150 cm^3. After about 200 h the measured pressure increase deviated from the linear curve. This means that during these first 200 h the gas volume remained constant (compression only). The fitted gas compression volume increased to 1400 cm^3 at start of GRI2 and to 1550 cm^3 at start of GRI3. This indicated that an increase of 150 cm^3 in the gas volume occurred during GRI2 and GRIS2.

The best-fit volume for the first 50 h of GRI3 was almost identical to the maximum theoretical volume of the test interval. Thus, between the start of GRI1 and by the first 50 h of GRI3, all the remaining water was most likely expelled from the interval and gas started to migrate into the rock at the latest. At this time the interval gas pressure is of the order of 5.8 MPa, which corresponds to a gas entry pressure of approximately 2 MPa given that the water pressure is estimated at 3.8 MPa. This value is a maximum value as it depends on both the volume of remaining water in interval and the gas injection flow rate. For example, less remaining water would lead to earlier gas migration and, thus, a lower gas entry pressure. This maximum value for the gas entry pressure is relatively small and from Young–Laplace's law corresponds to an equivalent pore radius of 70 nm. This pore radius is high compared with the pore size distribution in the intact claystone (Andra 2005). Given the existence of a damaged zone around the borehole (see Table 3), this maximum gas entry pressure value clearly represents the upper bound of the gas entry pressure into the damaged zone around the borehole.

GRI4 to GRI6

For these last three gas injection steps the gas volumes estimated based on the isothermal compression were always greater than the maximum

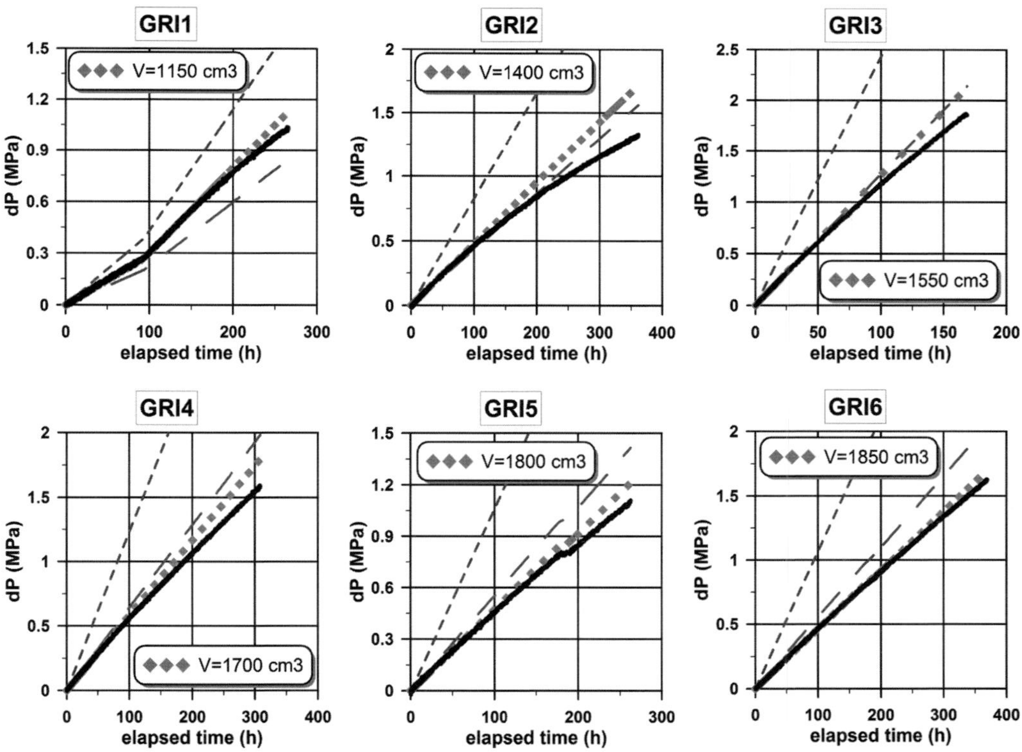

Fig. 7. Comparison of the pressure difference as measured (bold line) and that calculated at constant volume for the six gas injection steps GRI1 to GRI6. Large dashed line equals the minimum interval volume; the dotted line equals the maximum interval volume. The square dotted line is equal to the best fit volume.

theoretical volume of test interval. It is therefore more likely that, after GRI3, there was no remaining water in the test interval. The increase in gas volume at the start of these three steps reduced to 100 cm^3 between steps GRI4 and GRI5 and to 50 cm^3 between GRI5 and GRI6. Thus, an apparent higher resistance to gas migration into the rock was observed with time and increasing pressure. This observation on the GRI-phases is also consistent with the change in behaviour observed during the pressure recovery steps GRIS1 to GRIS6 (see Fig. 3).

New modelling activities

Based on the data analysis above, new two-phase flow modelling activities with LAGAMINE were performed, first with 1D modelling approach and in a section step with a 3D modelling approach.

1D modelling

The field data analysis highlighted the uncertainties on the volumes of the test interval and on the volume

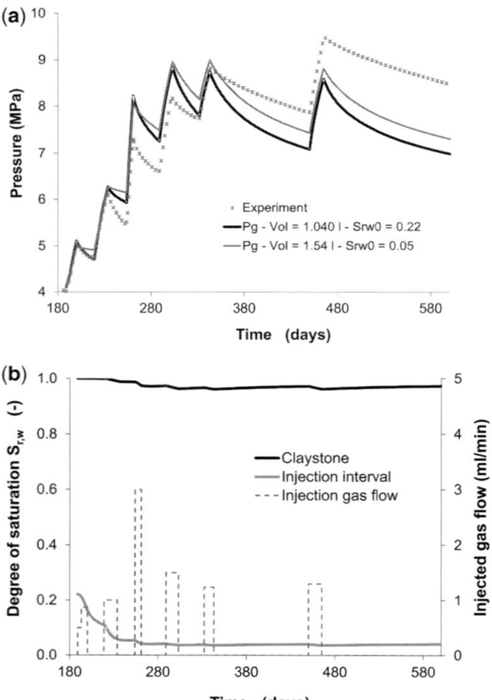

Fig. 8. Influence (**a**) of the interval volume and of the initial degree of saturation in the test interval and (**b**) evolution of the degree of saturation in claystone and in the test interval for the new 1D modelling (Lagamine).

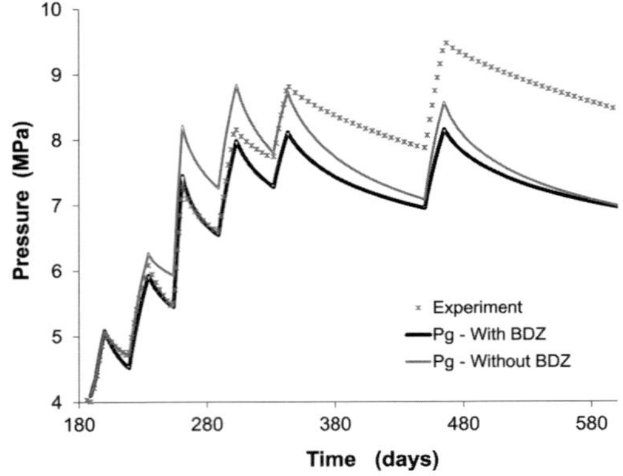

Fig. 9. Influence of the introduction of an excavated damaged zone on the time evolution of pore pressures in the test interval (Lagamine).

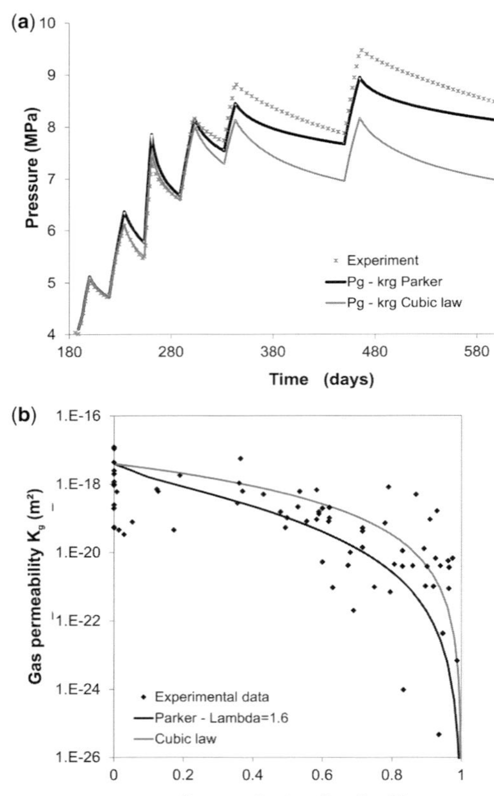

Fig. 10. (**a**) Influence of the gas relative permeability of the undisturbed claystone on the time evolution of pore pressure in the test interval. (**b**) Gas permeability against degree of saturation – comparisons with experimental data from Charlier *et al.* (2013) (Lagamine).

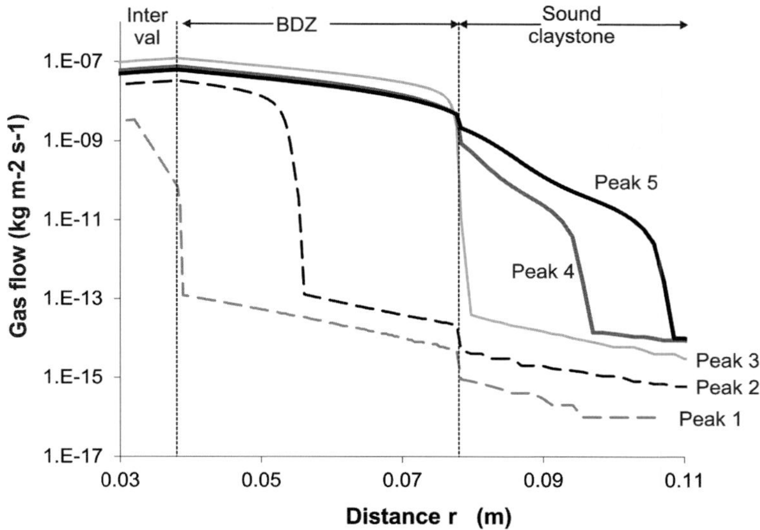

Fig. 11. Gas flows profiles at different gas injection peak – 1D modelling (Lagamine).

of the remaining water in the interval prior to the gas injection. Obtaining the best estimate of these two parameters was the first aim of the 1D modelling. Considering an interval with an initial volume of 1040 cm^3 and an initial degree of saturation of the air-filled void of 0.22 seemed to be satisfactory. The numerical results showed a significant improvement in the simulated pressures during GRI1 and GRI2 compared with the measured pressures in interval 2 of PGZ1201 (Fig. 8a). Note that these initial conditions do not correspond exactly to the ones deduced from the hydraulic data analysis. This could be explained by the gravitational effects where the remaining water accumulated at the bottom of the inclined test interval, which was not accounted for in the 1D model. Nevertheless the numerical results demonstrate that all the remaining water in the interval was expelled by the end of the third injection step (Fig. 8b), which is consistent with the volumetric data analysis.

To improve the numerical predictions of the later injection steps, a borehole damaged zone (BDZ) was introduced in the model. This zone corresponds to the inner zone identified during hydraulic tests (note that the BDZ thickness is 10 times higher than in Table 3), characterized by a higher permeability and a corresponding change in the retention properties compared with the intact host rock. For this damaged zone, a constant permeability was considered that was 500 times higher than that of the undisturbed claystone. The same factor was applied to the water and gas permeabilities $\left(K^{sat}_{w,BDZ} = 2 \times 10^{-17}\,m^2 - K^{dry}_{g,BDZ} = 2 \times 10^{-15}\,m^2\right)$. For the retention properties the parameter P_r of the van Genuchten's retention curve was decreased owing to the inferred microfracturing of the BDZs ($P_{r,BDZ} = 3$ MPa). The introduction of a damaged zone around the borehole significantly improved the simulated pressure response compared with the experimental observations through GRI4 (Fig. 9). The increase in the permeability and the decrease in the capillary-strength parameter P_r of the inner zone made it

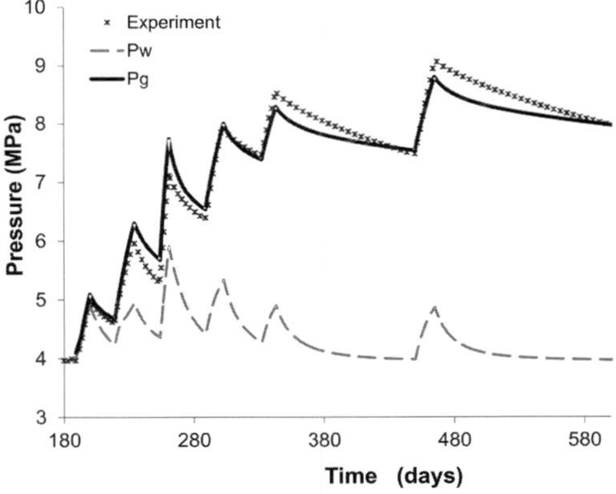

Fig. 12. Time evolution of pore pressures in the test interval – comparisons between experimental and numerical results (Lagamine).

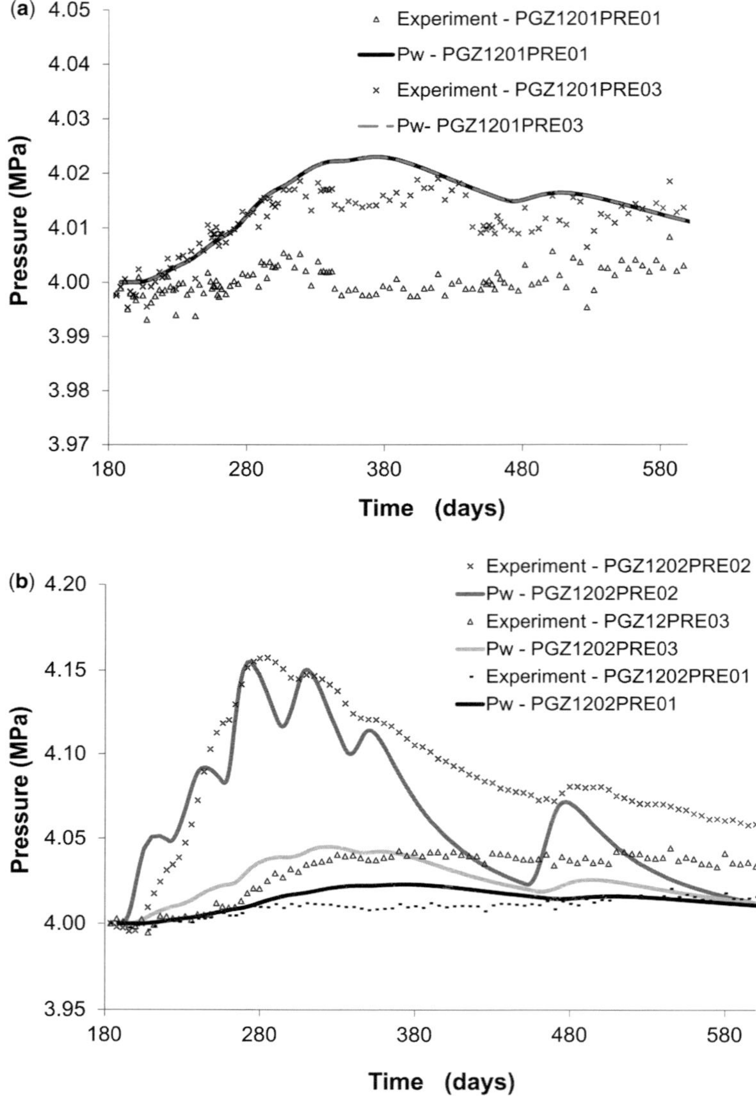

Fig. 13. Time evolution of pore pressures in (**a**) the two intervals of the injection borehole PGZ1201 and (**b**) in the three intervals of the measuring borehole PGZ1202 – comparisons between experimental and numerical results (Lagamine).

easier for the nitrogen to migrate into the rock, resulting in a decrease in the injection pressures during GRI3 and GRI4.

Despite the improvement of the model, the predictions were still unable to fit the experimental data for the entire GAS1 sequence. A modification of the COx claystone gas transport parameters was proposed to better reproduce the pore pressures during the last steps. The hydraulic characteristics of COx have been determined in numerous laboratory or field experimental studies (Charlier *et al.* 2013). Nevertheless, few experiments have investigated the gas transport properties of the COx near full saturation, owing to the experimental difficulties of accurately imposing a high degree of saturation and detecting and measuring very low gas permeabilities. Cubic law has typically been considered for the gas relative permeability function, even though we could not verify its relevance in the quasi-saturated domain. An alternative functional relationship was given by Parker (1987):

$$K_g = K_g^{\text{dry}} \times k_{r,g} = K_g^{\text{dry}} \times \sqrt{1 - S_{r,w}} \times (1 - S_{r,w}^{1/\lambda})^{2\lambda} \quad (2)$$

where λ is a fitting parameter of the model.

When the Parker relationship (with $\lambda = 1.6$) was used to define the gas relative permeability for the undisturbed rock and for the BDZ, a better agreement of the simulated pressures with the experimental pressures was obtained (Fig. 10a). The Parker function gave lower gas permeabilities than the cubic law (Fig. 10b), which resulted in higher injection pressure during the last two injection steps.

As confirmed by the gas-flow profiles obtained from the best 1D modelling in a domain where the gaseous transfers are predominant (Fig. 11), the first injection step (GRI1) tested only the behaviour of the test interval, while the responses of the second and the third injection steps (GRI2 and GRI3) were also influenced by the BDZ. It was only from the fourth injection step (GRI4) that nitrogen migrated into the undisturbed claystone as indicated by pore pressures measurements.

3D modelling

Based on the 1D modelling, a set of parameters characterizing the behaviour of the test interval, the BDZ and the undisturbed rock were derived. These parameters were then used in a 3D modelling of the entire GAS1 sequence, highlighting the role of the permeability anisotropy and the axial extent of the damaged zone along the borehole, that is, the role of preferential pathways along borehole wall.

Concerning the damaged zone, it is certainly not appropriate to keep the same properties along the entire length of the injection borehole, because of the installed packers separating the three intervals. Because the packer pressures were set to higher pressures than the gas injection pressures, two domains were defined for the BDZ: (a) along the open intervals; and (b) along the length of the packers. For both domains, the same radial extent as in the 1D modelling was assumed for the BDZ around the borehole, but with different hydraulic characteristics. Along the three intervals, the parameters from the 1D modelling were used in the 3D modelling. For the BDZ along the packers, we assumed that the relatively high packer pressures caused the properties to recover to those of the intact claystone. An anisotropic permeability ratio of 3 was introduced.

The results of the 3D modelling showed simulated pressures very close to the experimental data, as shown in Figure 12. In the two monitoring intervals (intervals 1 and 3 of PGZ1201) located on either side of the injection interval in PGZ1201, the simulated pressure evolution was similar to the measured pressures (Fig. 13a). In this zone, the rock mass remained saturated and the variations of the pore pressures were due to water overpressures induced by the nitrogen injection. In comparison, the measurements showed a different behaviour in the two intervals. This was due to the asymmetry of the drainage effect in PGZ1031 borehole (see Table 1), which was not explicitly simulated during the gas injection simulation. The order of magnitude of the pore pressure variations was very low in both intervals and the difference between experimental and numerical results is therefore not significant.

Three sensors were also installed along the parallel monitoring borehole (PGZ1202). There rock mass remained saturated and only water overpressures were predicted by the 3D model. The evolution of the pore pressures was well reproduced in the model, even though the magnitude in the pressure variation for each injection peak was slightly overestimated (Fig. 13b). The comparison between Figure 13a and b emphasizes the influence of permeability anisotropy, because pore pressure variations were higher in the monitoring intervals in PGZ1202 than in the monitoring intervals 1 and 3 in PGZ1201.

Conclusion

An *in situ* gas injection test was performed in the undisturbed COx claystone. Nitrogen gas was injected at different flow rates into a test interval of a small-diameter borehole, and the pressure buildup during gas injection and pressure recovery during shut-in were observed. 1D and 3D modelling of the GAS1 sequences was performed. The field data analysis and the modelling results showed the importance of understanding and accurately taking into account each component of the borehole/test interval system and their initial conditions (e.g. fluid saturation), the hydraulic and two-phase flow characteristics of the near borehole rock mass disturbed by drilling and those of the undisturbed rock mass. In particular, the way that water was removed from the test interval and displaced into the rock mass influenced strongly the analysis of the experimental observations. The numerical results were also strongly dependent on the definition of the gas relative permeability under near-saturated conditions of the rock mass.

Finally, the results demonstrated that a two-phase flow modelling based on an iterative approach was able to reproduce experimental observations from this large-scale experiment, as long as the injection flow rate and the gas pressures remained moderate. Taking into account the development of preferential gas pathways is certainly a crucial issue in the description of the laboratory experiment (Olivella & Alonso 2008), but seems to be negligible for this part of the test. More generally, the PGZ1 experiment has shown that gas would remain mainly confined in the BDZ. Even though gas migrates into the undisturbed claystone, the amount was low and was restricted to small distances from the test interval under the prescribed gas injection conditions.

In the context of long-term safety of the waste repository, a significant result was obtained with respect to the hydraulic permeability prior (HYDRO1) and after (HYDRO2) the gas injection test: the permeability of the COx clay rock was not affected by the long-lasting gas injection test, indicating that the COx rock was most likely not damaged at the pressure level of 9.1 MPa that was reached during the test.

A second gas injection test (GAS2) is currently being conducted to provide better insights into the potential phenomena associated with gas migration into the COx claystone, which might be of importance in the particular experimental setup. Future modelling work will also investigate the influence of the self-sealing of the COx claystone in the damaged zone on the transmissivity of the system, as well as the effects of the hydro-mechanical coupling on gas transport.

R. Senger, Intera, J.-M. Lavanchy and J. Croisé, AF-Consult, are gratefully acknowledged for their contribution to the analysis of the field data during the course of the experiment and for reviewing the English version. We also gratefully acknowledge the two anonymous reviewers for their improving comments. The research leading to these results has received funding from the European Atomic Energy Community's Seventh Framework Programme (FP7/2007–2011) under grant agreement no. 230357, the FORGE project.

References

ANDRA 2005. *Dossier 2005 Argile: Référentiel du site de Meuse/Haute-Marne, Tome 1: Histoire géologique et état actuel*. Andra, Bure, France.

Baechler, S., Lavanchy, J. M., Armand, G. & Cruchaudet, M. 2011. Characterisation of the hydraulic properties within the EDZ around drifts at level −490 m of the Meuse/Haute-Marne URL: a methodology for consistent interpretation of hydraulic tests. *Journal of Physics and Chemistry of the Earth*, **36**, 1922–1931, http://dx.doi.org/10.1016/j.pce.2011.10.005

Bossart, P., Meier, P. M., Moeri, A., Trick, T. & Mayor, J. C. 2002. Geological and hydraulic characterisation of the excavation disturbed zone in the Opalinus Clay of the Mont Terri Rock Laboratory. *Engineering Geology*, **66**, 19–38.

Charlier, R. 1987. *Approche unifiée de quelques problèmes non linéaires de mécanique des milieux continues par la méthode des éléments finis*. PhD thesis, Université de Liège, Belgium.

Charlier, R., Collin, F., Pardoen, B., Talandier, J., Radu, J.-P. & Gerard, P. 2013. An unsaturated hydro-mechanical modelling of two in situ experiments in Callovo-Oxfordian argillite. *Engineering Geology*, **165**, 46–63. http://dx.doi.org/10.1016/j.enggeo.2013.05.021

Chavant, C., Granet, S. & Fernandes, R. 2007. Thermo-hydro-mechanical numerical modelling: application to a geological nuclear waste disposal. *In*: *Conference on Computer Applications in Geotechnical Engineering*, 18–21 February 2007, Denver, CO. American Society of Civil Engineers. http://ascelibrary.org/doi/abs/10.1061/40901(220)12

Collin, F., Li, X. L., Radu, J.-P. & Charlier, R. 2002. Thermo-hydro-mechanical coupling in clay barriers. *Engineering Geology*, **64**, 179–193.

Croisé, J., Mayer, G., Marschall, P., Matray, J.-M., Tanaka, T. & Vogel, P. 2006. Gas threshold pressure test performed at the Mont Terri Rock Laboratory (Switzerland): Experimental data and data analysis. *Oil and Gas Science and Technology*, **61**, 631–645.

Finsterle, S. 2007. *iTOUGH2 User's Guide*. Report LBNL-40400 (updated reprint), Lawrence Berkeley National Laboratory, Berkeley, CA.

Granet, S. & Meunier, S. 2012. Numerical modelling on in situ experience of gas injection: PGZ1. *In*: *Actes du colloque national Transfert2012/workshop organised in the framework of the EU FP7 project FORGE*, Ecole Centrale de Lille/LML, 20–22 March 2012.

Levasseur, S., Charlier, R., Frieg, B. & Collin, F. 2010. Hydro-mechanical modelling of the excavation damaged zone around an underground excavation at Mont Terri Rock Laboratory. *International Journal of Rock Mechanics and Mining Sciences*, **47**, 414–425.

Marschall, P., Horseman, S. T. & Gimmi, T. 2005. Characterisation of gas transport properties of the opalinus clay, a potential host rock formation for radioactive waste disposal. *Oil and Gas Science and Technology*, **60**, 121–139.

Mualem, Y. 1976. A new model for predicting the hydraulic conductivity of unsaturated porous media. *Water Resources Research*, **12**, 513–522.

Olivella, S. & Alonso, E. E. 2008. Gas flow through clay barriers. *Géotechnique*, **58**, 157–168.

Parker, J. C., Lenhard, R. J. & Kuppusamy, T. 1987. A parametric model for constitutive properties governing multiphase flow in porous media. *Water Resources Research*, **23**, 618–624.

Pruess, K., Oldenburg, C. & Moridis, G. 1999. *TOUGH2 User's Guide, V2.0*. Lawrence Berkeley National Laboratory Report LBNL-43134, Berkeley, CA.

Senger, R. K., Enachescu, C., Doe, T., Distinguin, M., Delay, J. & Frieg, B. 2006. Design and analysis of a gas threshold pressure test in a low-permeability clay formation at Andra's Underground Research Laboratory, Bure (France). *In*: *Proceedings, TOUGH2 Symposium 2006*. Lawrence Berkeley National Laboratory, Berkeley, CA.

Van Genuchten, M. Th. 1980. A closed-form equation for predicting the hydraulic conductivity of unsaturated soils. *Soil Science Society of America Journal*, **44**, 892–898.

Full-scale 3D modelling of a nuclear waste repository in the Callovo-Oxfordian clay. Part 1: thermo-hydraulic two-phase transport of water and hydrogen

J. BROMMUNDT[1]*, TH. U. KAEMPFER[1], C. P. ENSSLE[1], G. MAYER[1] & J. WENDLING[2]

[1]*AF-Consult Switzerland Ltd, Groundwater Protection and Waste Disposal, Baden, Switzerland*

[2]*ANDRA, Chatenay-Malabry, France*

**Corresponding author (e-mail: juergen.brommundt@afconsult.com)*

Abstract: The development of deep geological repositories for nuclear waste requires a sound system understanding, to which performance analyses by means of numerical flow and transport simulations form an important contribution. In this context, we present modelling studies of the thermo-hydraulic two-phase transport of water and hydrogen in the planned French repository in this paper, part 1. Part 2 of this paper uses the same modelling concept extended to include radionuclide transport. The numerical TOUGH2-MP-model encompasses the host rock from cap to bed rock, including the repository in full 3D. We compare two repository layouts and assess their robustness through sensitivity simulations with varied key parameters and hydrogen generation regimes, and a coupling of hydrogen generation to water availability. The results show that a continuous gas phase functions as balancing system between connected waste areas. Its extension and continuity depend on the permeability and diffusion parameters, and the amount and temporal regime of hydrogen production. Naturally, the available storage volumes have an impact as well. The layout with larger storage volume behaves more benevolenty towards parameter variations. Overall, both layouts perform satisfactorily. Our methodology enables detailed studies of the global system behaviour and helps to compare different repository layouts.

All French high-level (HLW) and intermediate-level long-lived (ILW-LL) radioactive waste is planned to be disposed of in a single deep geological repository (Office parlementaire 2006). The French parliament installed Andra as national radioactive waste management agency under the authority of the Ministries for Energy, Research and Environment to implement this plan. The initial steps were the identification of a suitable host rock and a potential site for a repository. Since then, an extensive research programme has been launched by Andra to evaluate and demonstrate the feasibility of a safe repository implementation. The results of the on-going research have been compiled in two reports, the Dossier 2005 (Andra 2006) and the Jalon 2009 (Andra 2009*a*). These reports represent milestones and compile the relevant knowledge gathered as well as the identified open issues.

Andra identified the Callovo-Oxfordian (COx) argillites of the Meuse/Haute-Marne site in north-eastern France as a suitable host rock (Andra 2006). The repository is planned to be situated at around 500 m below the surface in the middle of the Callovo-Oxfordian clay. It is accessed through shafts and a ramp for the actual waste emplacement, and consists of a system of access, supply and workshop drifts. The waste is stored in several thousand waste emplacement cells in distinct sectors for HLW and ILW-LL. The final layout has not yet been selected; depending on the arrangement of the different structural components, the repository covers a horizontal area of 10–20 km^2 (Fig. 1).

The realization of the repository uses established underground construction and mining techniques. The repository will be excavated stepwise into the COx as the waste to be disposed of becomes ready for emplacement. Upon finalization of emplacement the drifts are backfilled and seals are erected at key positions. With the last waste emplaced the repository is closed and the ramps and shafts are backfilled and sealed. This constructional phase is expected to last roughly 100 years. After the sealing, the repository must enclose the waste safely for the long term, that is, up to one million years, without human intervention. A further requirement to the repository design is reversibility, meaning that for the first few centuries the waste must be retrievable with reasonable effort.

The total volume of the HLW for disposal predicted by the year 2030 totals to 5060 m^3, of

From: NORRIS, S., BRUNO, J., CATHELINEAU, M., DELAGE, P., FAIRHURST, C., GAUCHER, E. C., HÖHN, E. H., KALINICHEV, A., LALIEUX, P. & SELLIN, P. (eds) 2014. *Clays in Natural and Engineered Barriers for Radioactive Waste Confinement*. Geological Society, London, Special Publications, **400**, 443–467.
First published online May 7, 2014, http://dx.doi.org/10.1144/SP400.34

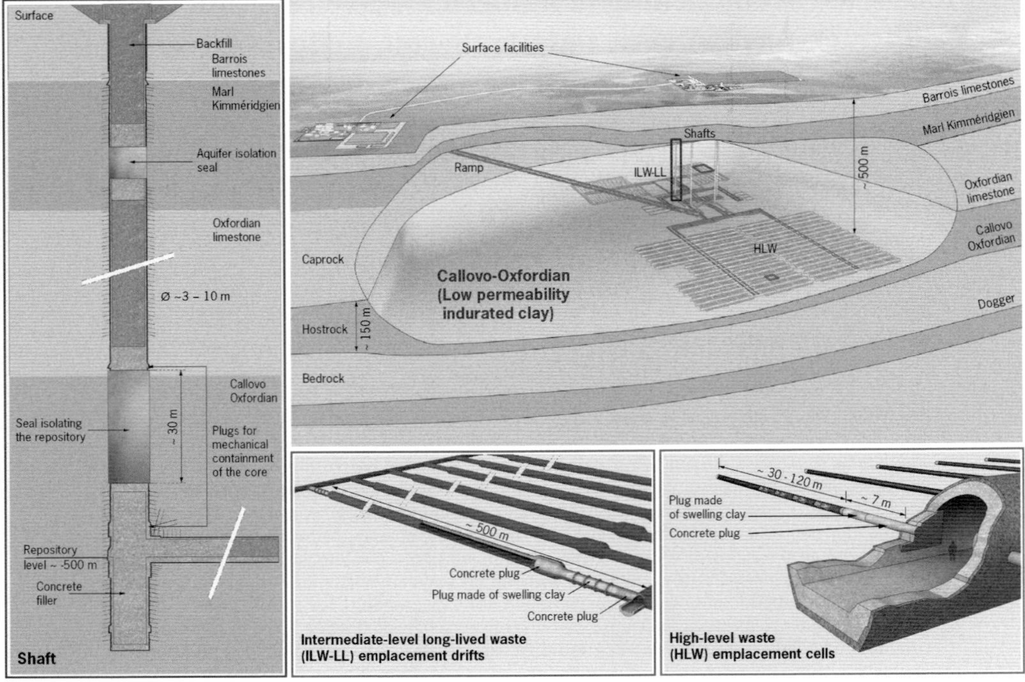

Fig. 1. Overview of repository and its main components (compiled from figures which are the property of Andra).

which 74 m^3 is spent nuclear fuel. The major part consists of a highly radioactive chemical solution produced during the recycling of nuclear fuel. For disposal this solution is calcined in the form of a powder and then incorporated into molten glass and poured into stainless steel containers. The continuing radioactive decay of their contents produces heat. Therefore, before emplacement the waste containers are allowed to cool in intermediate storage on the surface for some tens of years.

The total volume of ILW-LL for disposal predicted by the year 2030 totals to 51 009 m^3. This type of nuclear waste is mainly in solid state. For disposal this waste is compacted and stored in concrete and metal packages. Depending on the type of contents some of these packages release heat as well. Both HLW and ILW-LL waste eventually release radionuclides (Andra 2009*c*).

A major part of the feasibility evaluations for the repository is its long-term safety assessment. Obviously, the capability for the retention of contaminants, here first and foremost radionuclides, is the benchmark for the assessment of the repository. The host rock and to some extent the engineered barriers must maintain their retention capacity under the expected evolution of the system that is characterized by the pressure, saturation and temperature distributions as well as chemical conditions. The latter are not explicitly considered here since the focus of the present work is on two-phase transport of water and hydrogen.

For the site and potential repository layout in question, the processes having an essential influence on the mass fluxes and transport in and especially out of the repository are, besides diffusive processes, the gas pressure build-up owing to hydrogen generation by metal corrosion, the decay-heat generation with inherent volume expansion and pressure increase, and an increasing vertical gradient in hydraulic head causing an upward water flux through the repository that is also increasing over time (Andra 2006). The numerical modelling of these processes allows for a preliminary evaluation of the repository layout and is the basis for transport simulations considering radionuclides. Both are thus one of the bases for the long-term safety assessment.

A number of numerical modelling studies have been performed addressing these problems (Glascoe *et al.* 2002; Nagra 2008; Rutqvist *et al.* 2008; Poller *et al.* 2011). The challenge consists of the fully coupled description of these problems at the scale of the overall repository. With the complex processes, the huge number of constructional elements and the large range in scale – from the kilometre scale of the overall repository site to the decimetre scale of a single waste canister – an adequate and still manageable model representation poses a challenging task with respect to the setup and its numerical computation. Previous studies solved this problem by reducing the complexity using symmetries (Enssle *et al.* 2011, 2012; Poller *et al.* 2011) or reducing scale by modelling distinct parts of the repository individually (Poppei *et al.* 2003; Hoch & Wendling 2011). Other approaches include the sequential simulation on different scales, using a sequential coupling by transferring and incorporating simulation results from other scales (Andra 2007).

The present work concerns modelling studies of the transient non-isothermal transport of water, hydrogen, as well as a volatile and a highly soluble radionuclide on the scale of the entire repository and the host rock. We performed fully coupled thermo-hydraulic two-phase simulations of the operational and post-operational phase comparing two different repository layouts. The robustness of the system has been studied by sensitivity simulations with respect to physical key parameters and different hydrogen and gas generation regimes, including a sensitivity simulation on hydrogen generation that is coupled to water availability. Furthermore, fully coupled thermo-hydraulic single-phase simulations have been performed to assess the impact of the gas phase on radionuclide transport by comparison with the two-phase simulations.

The presentation of the simulation results of the studies is split into two parts. This paper contains part 1 and focuses on the thermo-hydraulic transients of hydrogen and water. Part 2 addresses the transport of two radionuclides in the repository and host rock (Enssle *et al.* 2014).

Approach and system description

The approach chosen for the evaluation of the non-isothermal two-phase flow and transport behaviour of a deep geological repository is the development, implementation and application of a physical model, also called the process model, and the numerical model of the entire repository including the host rock and the adjacent cap and bed rock. The development of the model requires the identification of the expected *phenomenology* and of the relevant *processes* and *time scales*. The research performed by Andra so far (Andra 2006, 2009*a*) provides a sound basis for this task. In the next step, a *physical model* is established considering the identified processes. From available data and repository layout plans the *geometrical model* is constructed. For the simulations a *numerical model* is set up which implements the physical model and applies it to the geometrical model considering the construction schedule, the temporal variation of the boundary conditions and the hydrogen and heat production by the waste.

Phenomenology

With the start of the construction of the repository, the natural conditions in the COx are disturbed. The repository is kept open during construction, waste emplacement and a period of potential retrievability, sealed afterwards and no human intervention is planned from then on. Important phenomenological aspects to be considered are (Andra 2006, 2009*a*):

(1) the creation of an excavation damaged zone (EDZ) around shafts and drifts;
(2) the depressurization and desaturation of the EDZ and the adjacent intact host rock as a result of the tunnel ventilation in the operational phase with a resaturation by pore water from the host rock and re-pressurization in the post-operational phase;
(3) the appearance, migration and disappearance of a gas phase owing to waste-type-specific release of gaseous substances, mainly hydrogen;
(4) the decay-heat-induced temperature increase leading to a pressure increase as a consequence of heat expansion of the fluids and the solids.

Processes

In the natural state, the relevant part of the system consists of a clay matrix fully saturated with formation water. During construction, air is introduced. In the post-operational phase, hydrogen is produced by various corrosion processes, waste degradation and radiolysis. In the long term and on the scale of the repository, only a minor amount of air is present in the system compared with the amount of hydrogen produced.

On the stratigraphic scale and for the natural conditions, water is vertically transported through the matrix of the COx by an upward pressure gradient from the bed rock to the cap rock. The comparatively good transmissive hydraulic conditions in the bed and cap rock can be considered as hydraulically fixed and unaffected by the processes inside the repository.

On the repository scale, the displacement of the fluid phases is driven by pressure gradients. Liquid and gas phases occur depending on the thermo-physical state including capillarity; their viscosity varies with the concentration of the mass constituents in the fluid and temperature; their macroscopic flux is influenced by the saturation-dependent tortuosity of the matrix. The mass constituents partition between the liquid and the gaseous phase, depending on their partial pressures. In addition to advection, the mass constituents are transported by diffusion, driven by concentration gradients.

On the micro-scale, all constituents are interacting with the matrix by, for example, binding effects on the pore space surface. The rock matrix reacts to pressure variations with elastic expansion and contraction of the pore space. At very high pressures the matrix might fracture, resulting in different flow processes.

All constituents as well as the matrix expand with warming and contract with cooling across the temperature range of interest. This affects the density and corresponding pressures. Differences in expansion and contraction between the matrix and the constituents result in fluid pressure variations.

In the context of this study, the listed processes are considered, albeit in different ways: the macro-scale processes are modelled explicitly while the micro-scale processes, that is, the matrix interactions, are parameterized, neglecting non-elastic transformation and fracturing completely.

Time scales

The objective is to model the thermo-hydraulic comportment inside the entire repository and the ambient host rock within a time-frame of 1 million years. This contrasts with the process time-scale that is on the order of years and determined from the spatial scale and the processes of interest as described above. This latter time-scale is relevant for the numerical time-stepping.

Model concept

Physical model

The processes listed in the section *Approach and system description – processes* can be formulated as generalized multiphase flow and transport problem through porous media (Zhang *et al.* 2008; Pruess *et al.* 1999). More specifically and in the terminology used in the numerical implementation to be adapted here, namely TOUGH2-MP, the problem consists in modelling:

- fluid flow through a porous medium in gaseous and liquid phase under pressure, viscous and gravity forces formulated using a multiphase extension of Darcy's law (Pruess *et al.* 1999, appendix A.5);
- interaction between the phases by means of functions for relative permeability and capillary pressure (various models, e.g. van Genuchten-Mualem, van Genuchten);
- dissolution of gases in the liquid phase described by Henry's law;
- pressure-dependent phase transition of water, respectively vapour;
- porosity changes owing to thermal expansion and compressibility;
- heat transport by conduction (dependent on saturation), convection in the liquid and gas phase and radiation (not considered here);
- heat storage dependent on saturation;
- temperature-dependent diffusion of all components in both phases described by Fick's law (Pruess *et al.* 1999);
- porosity and tortuosity dependency of effective diffusion (Pruess *et al.* 1999);
- components in the gas phase are treated as ideal gases except for water vapour;
- pressure- and temperature-dependent parameterization of the density, internal energy and viscosity of water and vapour.

For the mathematical formulation of above problem we refer to (Pruess *et al.* 1999, appendices A–E).

That such a generalized multiphase flow and transport formulation together with appropriate model parameters (see section on the numerical model below) is adequate for modelling the processes of interest in COx argillites follows from previous work performed by Andra. This work includes an extensive set of experimental on-site and laboratory tests as well as numerical studies and is documented, for example, in the Dossier 2005 (Andra 2006) and the Jalon 2009 (Andra 2009*a*).

Geometrical model

To date, the definite layout of the repository has not yet been chosen and different layout variants are currently under assessment by Andra. In the study presented here two preliminary layout variants are subject to the model simulations named LR 2009 and LV 2009. The first represents the reference, is rather elaborate and has been optimized in some iterations (Andra 2009*b*). The second, LV 2009, is rather compact and in a more preliminary state of design (Andra 2011).

Both repository layouts follow the same concept (Fig. 2): the main access ramps and shafts lead from the surface down to the repository level into a central zone where infrastructural components are located. Horizontally adjacent are the waste emplacement areas, with the zones named HLW, which are divided into subzones HLW-1 and HLW-2 for high-level wastes and subzone HLW-3 for spent fuel, and ILW-LL for intermediate-level long-lived wastes. All repository areas are developed by drifts for different functions such as access, transportation, ventilation, etc. The HLW canisters are

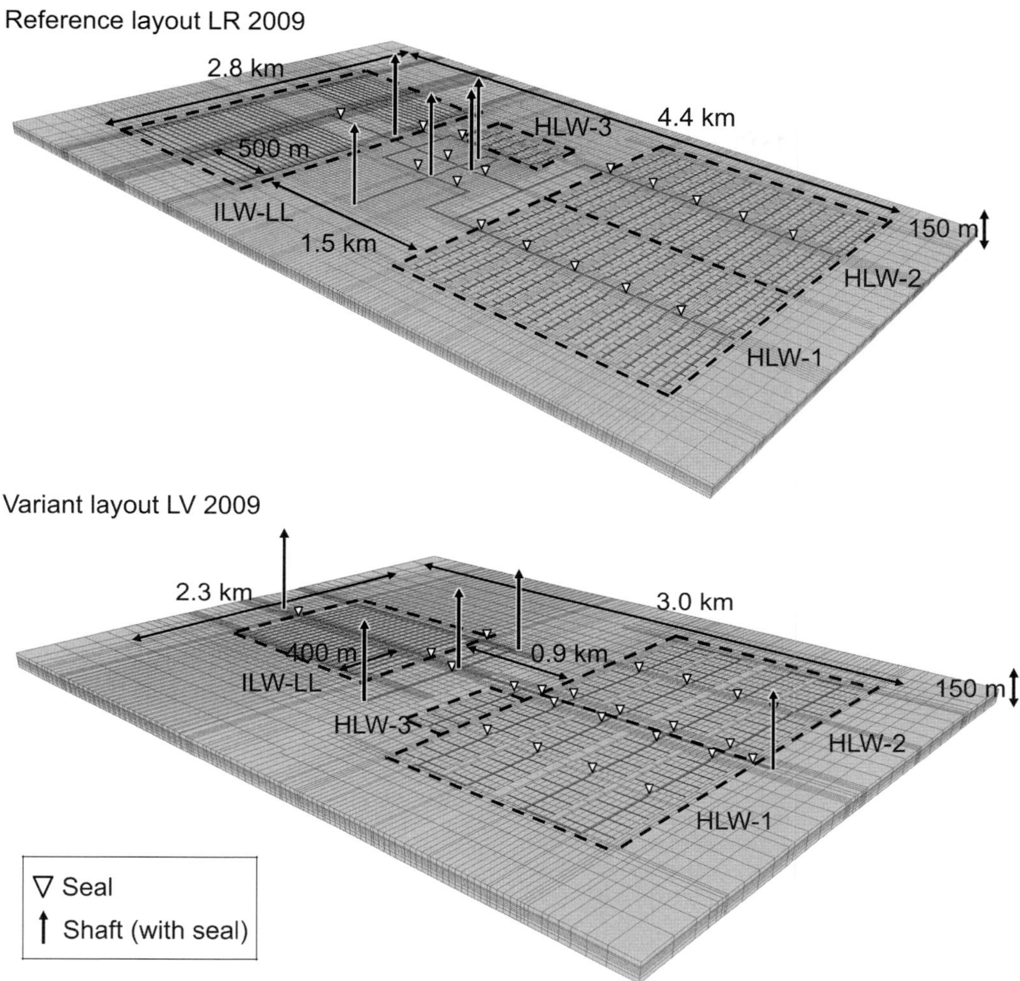

Fig. 2. Reference and variant layouts LR 2009 (top) and LV 2009 (bottom).

emplaced into several thousand emplacement cells of relatively small diameter (*c.* 0.7 m) and lengths between 30 and 120 m. The ILW-LL containers are stacked into emplacement drifts of around 10 m in diameter and 500 m in length. Each HL and IL-LL waste emplacement cell is sealed by a plug made from swelling clay (Fig. 1). When the repository is closed, all drifts and shafts are backfilled and further seals are located at various points in the drifts and shafts.

Table 1 provides a comparison of selected geometrical specifications of both repository layouts. The two layouts exhibit significant geometrical differences concerning the arrangement of the waste zones (especially the location of zone HLW-3, see Fig. 2), the amount of the stored HLW canisters and ILW-LL containers waste (+20% HLW, +80% ILW-LL in LR 2009), the lengths of the emplacement cells (*c.* factor 3 for the HLW emplacement cells in LV 2009), the location of the shafts and the total excavated volume.

Numerical model

Approach. The numerical model is based on the equation-of-state module EOS5 and EOS7R implemented in the numerical flow and transport simulator TOUGH2-MP (Pruess *et al.* 1999; Zhang *et al.* 2008). TOUGH2-MP EOS7R was modified in order to account for hydrogen instead of air as primary gaseous component.

Table 1. *Comparison of main geometrical specifications of the two repository layouts LR 2009 and LV 2009*

Item	Zone	Layout	
		LR 2009	LV 2009
Length HLW emplacement cell (type C5, C6)	HLW-1/2	35 m	120 m
Length HLW emplacement cell (type C0, CU3)	HLW-3	35 m	90 m
Length ILW-LL emplacement drift	ILW-LL	528 m	420 m
Total number of HLW emplacement cells	HLW-1/2/3	6030	1311
Total number of HLW canisters	HLW-1/2/3	*c.* 46 000	*c.* 38 000
Total number of ILW-LL emplacement drifts	ILW-LL	57	40
Total number of ILW-LL containers	ILW-LL	*c.* 15 000	*c.* 8000
Dimensions of the disposal area	–	2800×4400 m^2	2300×3000 m^2
Total length of drifts at repository level*	–	*c.* 90 km	*c.* 25 km
Total volume of drifts at repository level*	–	*c.* 3.9×10^6 m^3	*c.* 1.2×10^6 m^3
Shortest and longest distance from HLW emplacement cell to shaft end (along drifts)	–	*c.* 1.0/*c.* 4.5 km	*c.* 0.7/*c.* 2.0 km
Shortest and longest distance from ILW-LL waste container to shaft end (along drifts)	–	*c.* 0.5/*c.* 3.0 km	*c.* 0.4/*c.* 1.0 km

*Without ILW-LL emplacement drifts and HLW emplacement cells

Parameterization

Materials. All materials are represented by porous media. An overview of the reference material parameters for the most important materials is given in Table 2. These parameter values have been defined by Andra based on previous studies; see Dossier 2005 (Andra 2006), the Jalon 2009 (Andra 2009*a*) and references therein. Merged material properties for the EDZ, the HLW canisters and the ILW-LL containers are given in the last three columns. (Merged material properties were derived as a consequence of the topological mesh simplifications.) Relative permeabilities of liquid and gaseous phase k_{rl} and k_{rg} and capillary pressure p_c are represented following the modified van Genuchten model (Mualem 1976; van Genuchten 1980; Luckner *et al.* 1989) and are parameterized as follows:

$$k_{rl} = S_{ek}^{\eta} \cdot [1 - (1 - S_{ek}^{1/m})^m]^2$$

$$k_{rg} = (1 - S_{ek})^{\zeta}[1 - S_{ek}^{1/m}]^{2m}$$

$$p_c = \max\left(-\frac{1}{\alpha}\left[(S_{ek})^{-1/m} - 1\right]^{-1/n}, -p_{c,\max}\right)$$

with $S_{ek} = (S_l - S_{lrk})/(1 - S_{lrk} - S_{gr})$, $m = 1 - 1/n$, and parameters as described in Table 2.

Substance properties. Selected substance properties for water and hydrogen are given in Table 3.

Spatial discretization. The model domain is represented by a finite-volume mesh that contains all the relevant structures and components of the repository and the surrounding host rock. The host rock is represented in its full vertical depth between the cap and bed rock (*c.* 150 m), which represent the vertical bounds of the model. The repository level is located at the vertical midst of the host rock. The lateral bounds of the model domain are located at roughly 500 m away from the repository in order to minimize the influences of the lateral boundaries which are set as no-flow.

In view of the given large, and spatially heterogeneous repository domain and the physical complexity, for example, non-linear capillarity, and relative permeability, or sharp contrasts in permeability, simplifications and optimizations of the model mesh were inevitable. A set of measures including topological simplifications and a methodology we refer to as the *embedded meshing approach* was developed exploiting the meshing flexibilities inherent to the finite volume method in TOUGH2/-MP. (In the finite volume method in TOUGH2 space is discretized by grid blocks and connections between adjacent grid blocks. Normally, the arrangement and properties of the grid blocks and their connections are chosen in a geometrically realistic fashion; however, the user is free to modify the geometrical arrangement and properties such as, for example, to increase the volume of a certain grid block.)

Topological simplifications. (TS1) All tunnels and shafts including their concrete lining and EDZs are represented by square or rectangular instead of circular cross sections respecting the real cross-sectional area (Fig. 3). (The entailed increase in circumference and reduction of radial distance was less than 13 and 11%, respectively, and was neglected, yet this could be corrected for

Table 2. *Parameterization of materials, key parameters*

	Callovo-Oxfordian (COx)	Merged interior of HLW	EDZ (merged inner and outer EDZ)	Concrete (lining, seal)	Merged interior of ILW	Backfill	Bentonite (seal)
Porosity (–)	0.18	0.18	0.18	0.15	0.28	0.25	0.35
Permeability x, y (m^2)	4.80×10^{-20}	4.80×10^{-18}	9.59×10^{-17}*	9.59×10^{-18}	4.80×10^{-19}	9.59×10^{-17}	9.59×10^{-20}
Permeability z (m^2)	4.80×10^{-21}	4.80×10^{-18}	9.59×10^{-18}†	9.59×10^{-18}	4.80×10^{-19}	9.59×10^{-17}	9.59×10^{-20}
Specific storage (m^{-1})	2.00×10^{-6}	2.00×10^{-6}	2.00×10^{-6}	1.00×10^{-6}	3.03×10^{-6}	1.00×10^{-5}	5.00×10^{-6}
Pore compressibility (Pa^{-1})	6.96×10^{-10}	6.96×10^{-10}	6.96×10^{-10}	2.43×10^{-10}	6.66×10^{-10}	3.64×10^{-9}	1.02×10^{-9}
Pore expansivity (K^{-1})	2.00×10^{-5}	2.00×10^{-5}	2.00×10^{-5}	2.00×10^{-5}	2.00×10^{-5}	2.00×10^{-5}	2.00×10^{-5}
Relative permeability (modified van Genuchten-Mualem, see Mualem 1976; van Genuchten 1980; Luckner et al. *1989)*							
Residual liquid saturation S_{lrk}	0.01	0.01	0.01	0.01	0.01	0.01	0.4
Residual gas saturation S_{gr}	0.0	0.0	0.0	0.0	0.0	0.0	0.0
Parameter η	0.5	0.5	0.5	0.5	0.5	0.5	0.5
Parameter ζ	0.5	0.5	0.5	0.5	0.5	0.5	0.5
Capillary pressure model (modified van Genuchten, see van Genuchten 1980; Luckner et al. *1989)*							
Van Genuchten coefficient n	1.49	1.49	1.49	1.54	1.54	1.5	1.61
Pseudo gas entry pressure $1/\alpha$	1.47×10^{7}	1.47×10^{7}	1.96×10^{6}	1.96×10^{6}	1.96×10^{6}	1.96×10^{6}	1.77×10^{7}
maximum capillary pressure $p_{c,\max}$	1.00×10^{20}	1.00×10^{20}	1.00×10^{20}	1.00×10^{20}	1.00×10^{20}	1.00×10^{20}	1.00×10^{20}

*In direction of tunnel axis.

†In radial direction with respect to tunnel axis.

Table 3. *Selected substance properties*

Parameter	Water	Hydrogen
Molar mass (g mol^{-1})	18.0	2.0
Diffusion coefficient in bulk gas phase (m^2 s^{-1})	2.1×10^{-5}	9.5×10^{-5}
Diffusion coefficient in bulk liquid phase (m^2 s^{-1})	–	6.0×10^{-9}
Inverse Henry's constant (Pa^{-1})	–	1.379×10^{-10}

if sensitive.) Parallel tunnels are represented by a single tunnel with equivalent cross-sectional area.

(TS2) Inclined ramps are represented by horizontal drifts and vertical shafts with combined equivalent lengths.

(TS3) Meshing algorithms for regular grid block meshes such as MESHMAKER (Pruess *et al.* 1999) prolong the cell size of a grid block into the entire plane of the model mesh. Hence, small grid-block sizes – from the topological view-point only needed at certain locations – lead to a needless fine discretization at distant locations. In the study presented here such configurations result especially from the discretization of annulus shapes such as, for example, the EDZs. To avoid these, the concerned grid blocks are adapted to have a non-rectangular cuboidal shape removing the comparatively small corner grid blocks (Fig. 4). These new grid blocks do not strictly satisfy the Voronoi criterion that postulates orthogonality of the connection between the centres of adjacent grid blocks and their common interface. Further, the definition of the effect of gravity in these connections is ambiguous. The concerned connections occur, however, only at the annular space of the EDZs and enter therefore the computation of circular fluxes within these materials only. These fluxes are of negligible magnitude compared with the axial and radial fluxes where the mesh conforms as well to the Voronoi criterion as to gravity orientation; thus, the impact of the simplification on the simulation results is estimated to be minor.

(TS4) Certain small-scale components are merged to larger components with equivalent properties. This is, for example, applied to some components of the interior of a waste emplacement tunnel, for example, the ILW-LL primary waste packages and the waste containers, or the EDZs and the concrete lining of the tunnels and shafts. Incorporating these simplification measures the actual mesh generation of the entire model domain is outlined in the following section.

Embedded meshing approach. As a starting point the entire model domain, that is, the host rock including all waste emplacement zones, as well as the cap- and bed rocks, are discretized in a straightforward way with grid blocks of rectangular cuboidal shape using MESHMAKER (Pruess *et al.* 1999). This mesh is referred to as the *global mesh.*

Tunnels and shafts including their concrete lining and EDZs are embedded into the *global mesh* with grid blocks of non-rectangular cuboidal shape (cf. TS4) and rectangular cross-section (cf. TS1). This includes all ILW-LL emplacement tunnels.

The representation of the thousands of HLW emplacement cells is implemented by a technique of aggregation. (This approach partially adopts the previously developed approach of *Subdivision, Multiplication and Connection* (Poller *et al.* 2011).) This means that several adjacent identical emplacement cells, that is, same geometry and waste-type, are assembled into one aggregated emplacement cell with adapted geometry and properties. The adaptation is performed radially and the axial length kept constant. The aggregation factor f equals the number of assembled emplacement

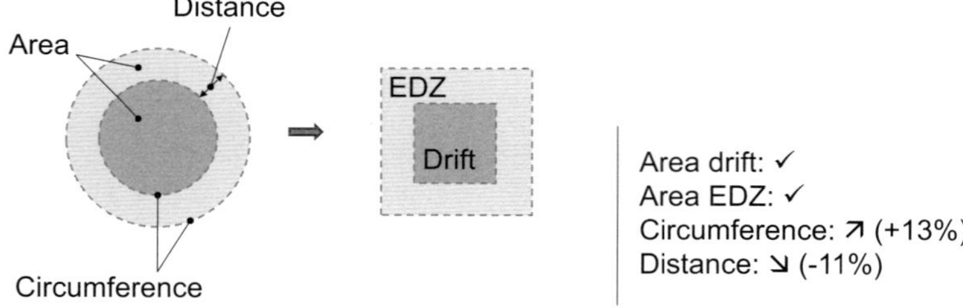

Fig. 3. Transition from circular to squared cross-sectional shape of tunnels and shafts.

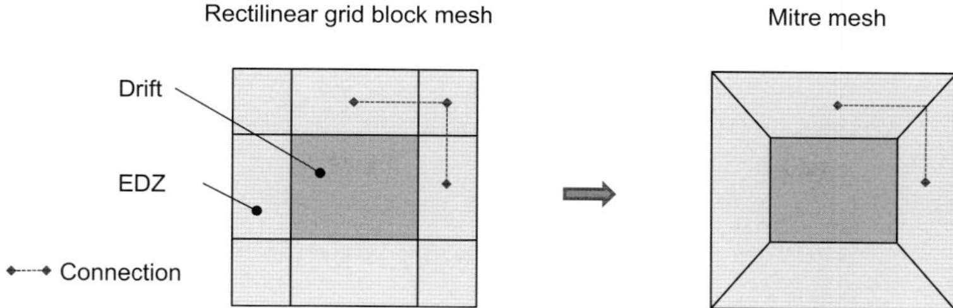

Fig. 4. Modification of rectilinear grid blocks. Left: drift cross-section discretized with nine rectangular grid blocks. Right: same drift configuration composed of five geometrically adapted grid blocks.

cells. (In the model mesh presented here, the aggregation factor was chosen to be around 10.)

The consequences for an aggregated HLW emplacement cell are as follows:

- Its extensive quantities are multiplied by *f*:
  - volume of grid blocks;
  - interface areas;
  - source terms.
- Its intensive quantities or properties remain unchanged, such as:
  - distances between grid blocks;
  - material properties;
  - state variables, that is, pressure, saturation and temperature.

Each aggregated emplacement cell is discretized with an axisymmetric submesh. This submesh contains the engineered components of the HLW emplacement cell, the EDZ, and the near-field host rock in a radius of *c*. 1.2 m (Fig. 5). The outer surrounding host rock is represented by brick-shaped grid blocks in the *global mesh* to which the submesh is connected. The embedding of each of the submeshes into the *global mesh* is done as follows:

(1) Each submesh is connected to the *global mesh* at its radial average location. The average location is the location of the emplacement cell situated in the centre of the set of emplacement cells assembled together (Fig. 5b, position '$(f+1)/2$'). The *global mesh* is constructed such that this location coincides with the centres of several specific grid blocks to which the corresponding submesh grid blocks are connected.

(2) The volumes of the *global mesh* grid blocks that are intersected by the submesh are adjusted by subtraction of the volume of the intersecting submesh (Fig. 5a).

(3) All grid blocks on the mantle of the submesh are connected to the corresponding (intersecting) grid blocks of the *global mesh*. The connection lengths are the distance between the centre of the mantle grid block and the mantle (*d1*) and the distance between the mantle and the centre of the corresponding grid block in the *global mesh*, that is, the mantle radius of the submesh (*d2*) (Fig. 5b).

Illustrations of the entire repository meshes for both layouts created with the *embedded meshing approach* are given in Figure 2. The complete meshes consist of some 250 000 grid blocks and around 700 000 connections for layout LR 2009 and *c*. 120 000 grid blocks and *c*. 370 000 connections for layout LV 2009. With 'conventional' mesh generation, that is, without topological simplifications, aggregation of HLW emplacement cells and their representation by axisymmetric grid blocks, the mesh size would reach some tens of millions of grid blocks and connections.

Discretization in time. Discretization in time is handled numerically in an adaptive way. However, several system changes have to be externally implemented.

The overall simulation was split into seven phases. The initial phase (#0) concerned the simulation of the natural conditions for the initialization of the model. Phases 1–5 concerned the operational phase with excavation, ventilation, waste emplacement and backfilling/sealing and a total duration of 100 years. Excavation and ventilation are considered in the sense that an EDZ appears around the excavated tunnels, tunnels are lined with concrete, and that air-filled void space is present for the period between excavation and backfilling; other mechanical phenomena, such as, for example, an explicit rock de- and re-compaction, are neglected as stated above. Figure 6 illustrates the conceptualized chronology of actions and conditions in the different areas of the repository during the operational phase. For simplification the occurrence of some sequentially occurring events such as the emplacement of waste-packages are temporarily condensed in the modelling. This means that for

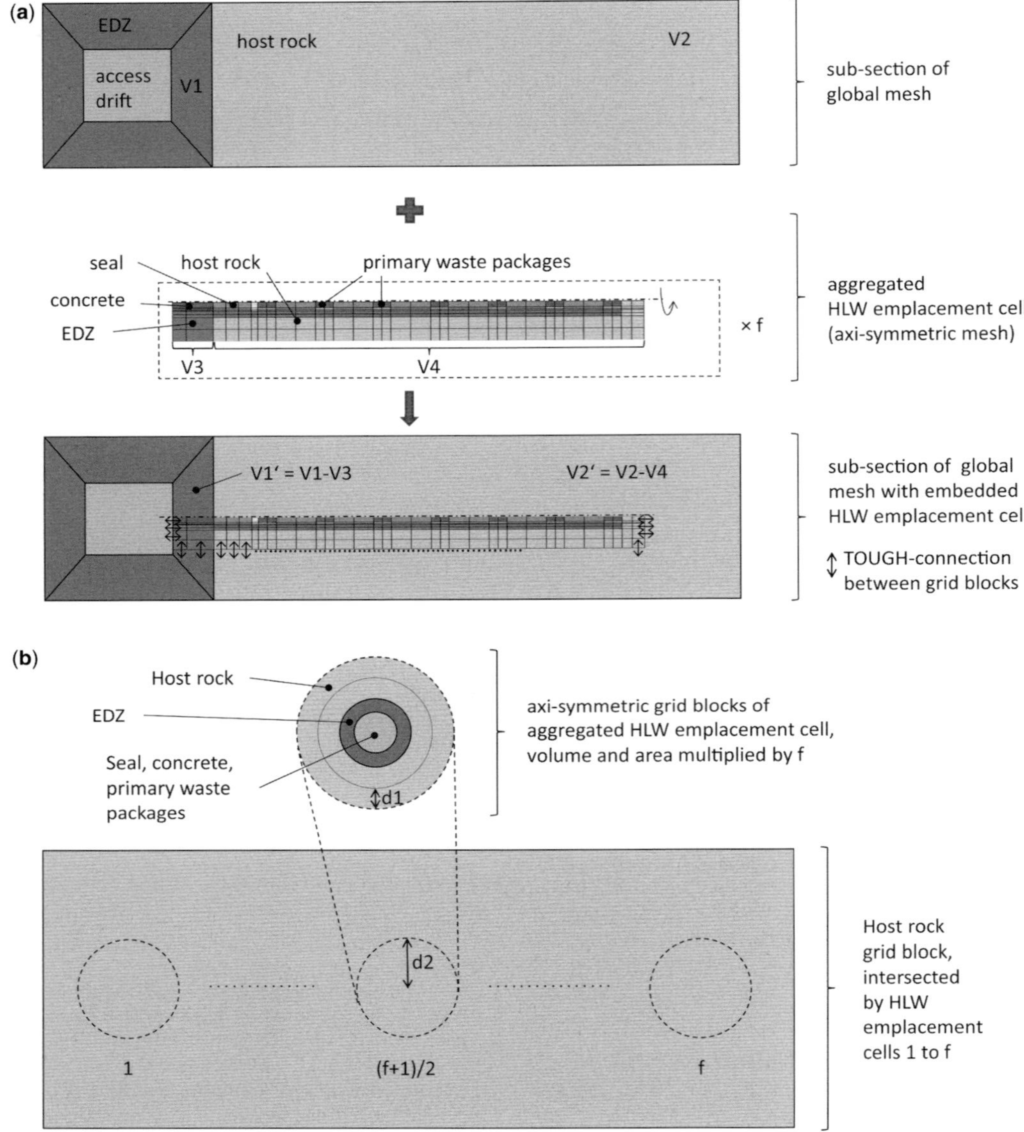

Fig. 5. Aggregation and embedding of HLW emplacement cells to global mesh. (**a**) Cross sectional view along an HLW emplacement cell. Top: grid blocks of the global mesh. Middle: grid blocks of the axisymmetric submesh. Bottom: grid blocks of the global mesh with embedded axisymmetric submesh. (**b**) Cross-sectional view along an access drift.

example all waste packages are emplaced at the same time. Phase 6 concerns the post-operational phase, that is, the time after the closure, between 100 and 1 million years.

Boundary and initial conditions

Conditions at the vertical and lateral model boundaries. At the bottom boundary of the model domain the conditions for pressure, liquid saturation, temperature and hydrogen concentration are set as constant over the entire simulation time with values of 5.140 MPa, 1, 24.35 °C and 0.0 kg m^{-3}. At the top boundary of the model domain the conditions for liquid saturation, heat, and hydrogen concentration are set as constant with values of 1, 20.65 °C and 0.0 kg m^{-3}. The pressure decreases

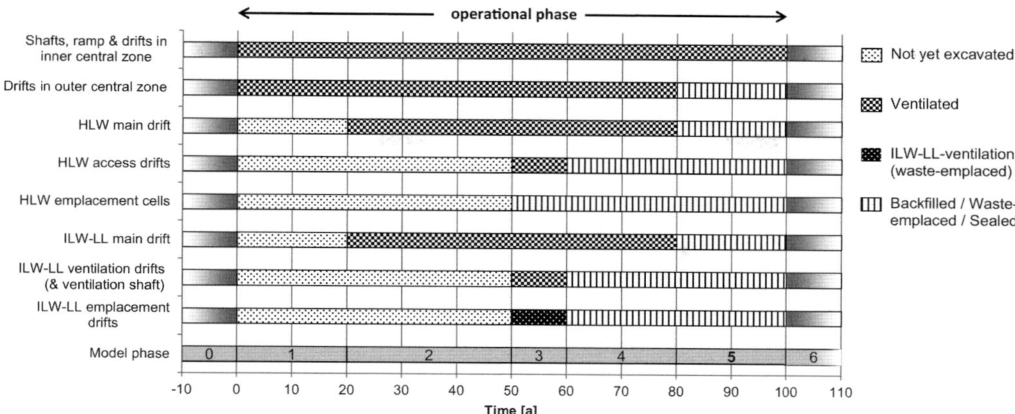

Fig. 6. Chronology of actions and conditions in different areas of the repository during operational phase and partitioning into simulation phases 0–6. (Note: 'Not yet excavated' = natural ambient conditions; 'Ventilated' = fixed boundary conditions with corresponding hygric conditions.)

linearly over time between 3.398 MPa at the start of the simulation and 3.108 MPa at the end. Hence the undisturbed hydraulic gradient evolves between 0.2 and 0.4 m m^{-1}; the thermal gradient is 0.025 K m^{-1}. The lateral boundaries are set as no-flow boundary conditions for mass and heat.

Initial conditions. The simulations start with today's natural conditions in the host rock. They are obtained from a steady-state run of the model under fully saturated conditions with the initial boundary conditions in which the entire model domain is parameterized as host rock. At the level of the disposal zones this leads to a fully saturated liquid pressure of 4.269 MPa, a temperature of 22.5 °C, and a hydrogen concentration of zero.

Ventilation conditions during operational phase. The conditions during ventilation are represented in the model as temporally fixed values for (equivalent) relative humidity, pressure and temperature with values of 50%, 0.101325 MPa, and 22.5 °C. Relative humidity is represented by the equivalent capillary pressure according to Kelvin's equation.

Source terms for mass and heat. The heat generation by radioactive decay is implemented with waste-specific heat source terms. Four different types are taken into account named *C5* for all of the HLW-1 canisters, *C6* for all of HLW-2 canisters, *C0* for some of the HLW-3 canisters and *CSD-C* for some of the ILW-LL containers. Some of the HLW-3 canisters and most of the ILW-LL containers do not emit heat. The waste type-specific heat generation rates are given in Figure 7 and already consider the cooling time before emplacement.

Hydrogen generation owing to the corrosion and degradation of the waste is implemented by time-dependent source terms (generation rates) for hydrogen. The hydrogen generation rates differ between the different waste types and are given in Figure 8. Hydrogen generation in the model starts after the waste is emplaced and sealed.

Results

Simulation cases

The above-described model has been run with several parameter variations, for two repository layouts, two different parameterizations for hydrogen production in zone ILW-LL and for simplified single-phase (SP) and fully coupled two-phase (2P) conditions (Table 4). The sensitivity simulations have been developed iteratively upon the results at hand. In total 11 simulation cases have been performed.

Notes:

- The waste specific and unique hydrogen generation in zone ILW-LL differ in total cumulative volume of released hydrogen. The scenario of unique hydrogen generation represents a conservative dimensioning and produces three times more hydrogen than the waste type-specific hydrogen generation scenario, which is more realistic according to the current state of knowledge.

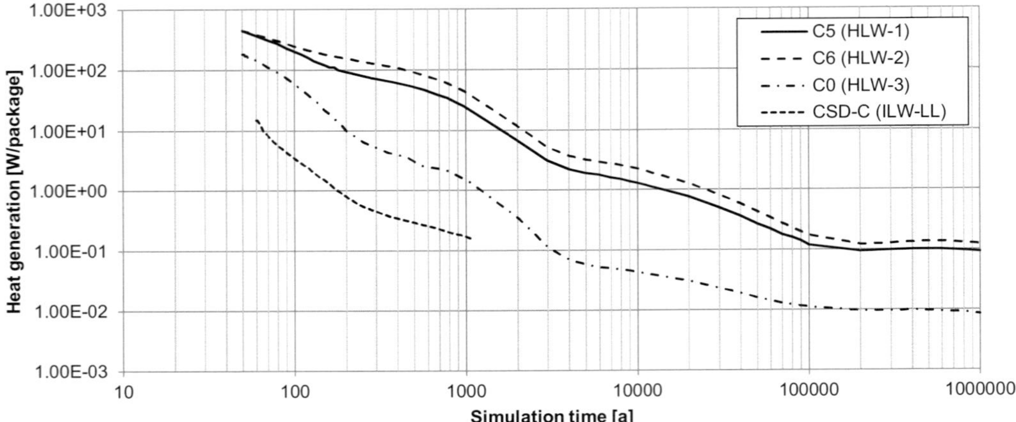

Fig. 7. Heat generation rates of the different waste types.

- The saturation-dependent hydrogen generation sensitivity run (S4 s) has been implemented using an approach where the potentially released hydrogen mass is determined from the total corrosion potential but released partially over time according to a water availability dependent corrosion rate, that is, if water is short corrosion reduces (Croisé *et al.* 2011).

For each simulation case, an extensive set of results has been obtained by post-processing the numerical output. A qualitative 3D image of the simulation results for the reference case Rs of both layouts LR 2009 and LV 2011 is given in Figure 9. More quantitative output that is used to discuss the results below includes the evolution of fluxes across repository zones (Fig. 10), performance indicators (Fig. 11) and mass balances per repository zone (Fig. 12).

Performance indicators

The focus of the analyses and comparisons of the results of the simulation cases is on the evolution of pressure, temperature and saturation conditions in the repository and the adjacent host rock. The magnitude of maxima and their time of occurrence are relevant for the first two parameters while for saturation the temporal evolution of the two-phase distribution, caused by desaturation owing to ventilation and gas generation and migration of gas and water, and resaturation times are relevant.

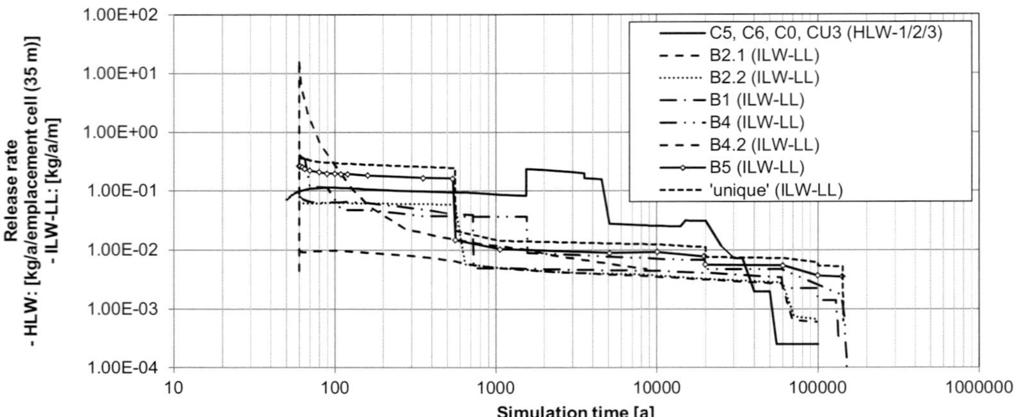

Fig. 8. Hydrogen (gas) generation rates of the different waste types.

Table 4. *Overview of simulations performed*

Name	Description	Hydrogen generation ILW-LL	Water and hydrogen		*Water, hydrogen & radionuclides**	
			LR 2009	LV 2009	*LR–2P*	*LR–SP*
Rs – reference	Reference parameterization	Waste type-specific	X	X	*X*	–
Ru – reference	Reference parameterization	Unique†	X	–	*X*	*X (R-SP)*
S1s – increased drift permeability	Permeability of backfill, concrete and EDZ (axial component) × 5 Diffusion coefficient of H_2 in gas and liquid phase/3	Waste type-specific	X	X	–	–
S1u – increased drift permeability	Permeability of backfill, concrete and EDZ (axial component) × 5 Diffusion coefficient of H_2 in gas and liquid phase/3	Unique†	X	–	*X*	*X (S1–SP)*
S2s – faster gas generation	Gas generation accelerated × 3	Waste type-specific	X	–	–	–
S2u – faster gas generation	Gas generation accelerated × 3	Unique†	X	–	*X*	–
S3s – increased drift permeability and faster gas generation	Permeability of backfill, concrete and EDZ (axial component) × 5 Diffusion coefficient of H_2 in gas and liquid phase/3 Gas generation accelerated × 3	Waste type-specific	X	–	*X*	–
S3u – increased drift permeability and faster gas generation	Permeability of backfill, concrete and EDZ (axial component) × 5 Diffusion coefficient of H_2 in gas and liquid phase/3 Gas generation accelerated × 3	Unique†	X	–	*X*	–
S4s – saturation-dependent gas generation	Gas as inventory with release dependent on water availability	Waste type-specific	X	–	–	–
S5u – increased diffusion of radionuclides	*Diffusion coefficient of ^{129}I in the liquid phase × 3* *Diffusion coefficient of ^{14}C in the gaseous phase × 3 (SP: in the liquid phase)*	*Unique†*	–	–	*X*	*X (S5-SP)*

*Radionuclide transfer simulations only with layout LR 2009.
†No hydrogen generation in the single-phase simulation cases.
Simulations formatted in *italics* are described in part 2 of this work (Enssle *et al.* 2014).

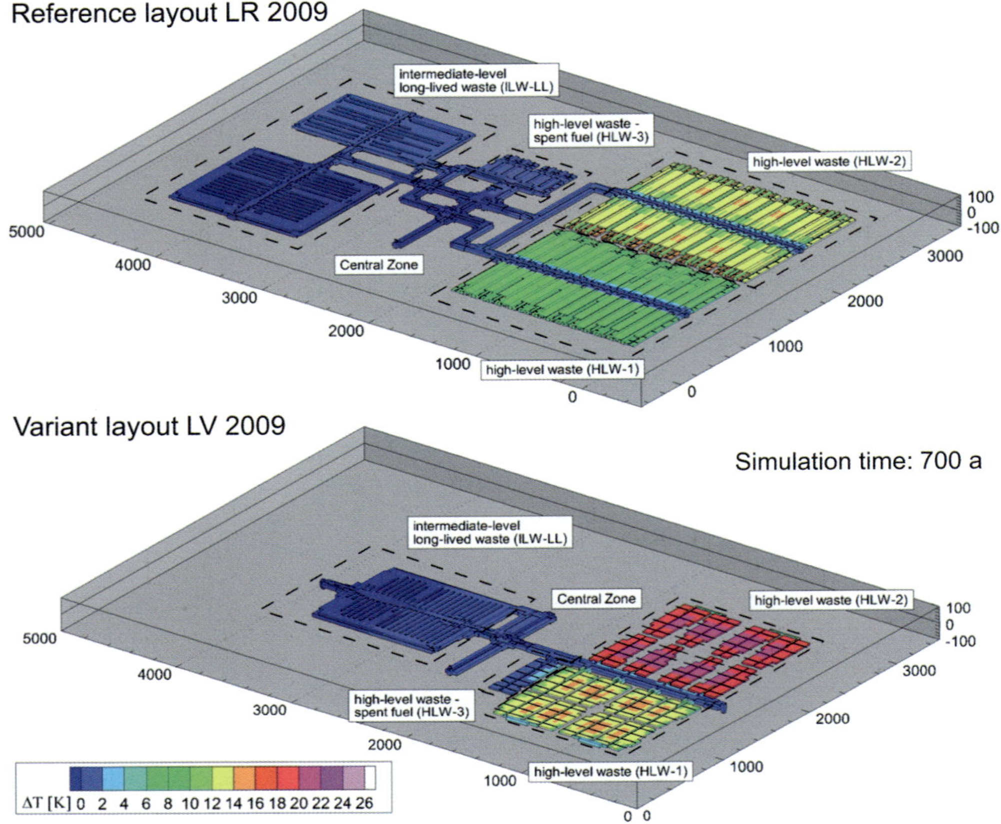

Fig. 9. Exemplary results for Rs at 700 years: temperature increase against the ambient temperature, projected on the surface of the desaturated parts of the repositories: LR 2009 (top), LV 2009 (bottom).

For the comparison of the different simulations, the following quantitative indicators per repository zone have been chosen:

- maximum temperature,
- volume of gas phase and resaturation time, that is, the time when the last grid block reaches full water saturation,
- maximum pressure in the gas phase.

Before presenting an analysis of the above performance indicators and how they are impacted by the different parameters, the reference case Rs of LR 2009 is analysed in detail to develop a systems understandings and then compared with the Rs of LV 2009.

Overall systems understanding of the reference case Rs: LR 2009

Ventilation of the shafts, ramps and access drifts in the initial 50 years of the simulation period leads to a desaturation of and inherent pressure decrease in the EDZs and the immediately adjacent COx. In the following 10 simulation years, the drifts and cells in zone ILW-LL are excavated; the waste is emplaced and ventilated, continuing the desaturation of the EDZs and the COx. At the beginning of this 10 year period, the waste in the HLW zones is modelled as instantaneously emplaced, immediately starting to produce heat and hydrogen (Fig. 11). All EDZs around the emplacement cells saturate quickly with water from the adjacent host rock. The starting heat production causes a significant increase of temperature and pressure, resulting in a first gas pressure peak in the heads of some emplacement cells owing to different expansivities of the water, gases and the matrix; at the same time, the hydrogen released by the waste in this zone during this period is not sufficient to create a continuous gas phase in the emplacement cells and dissolves completely. These processes occur very fast and very soon after the instantaneous insertion

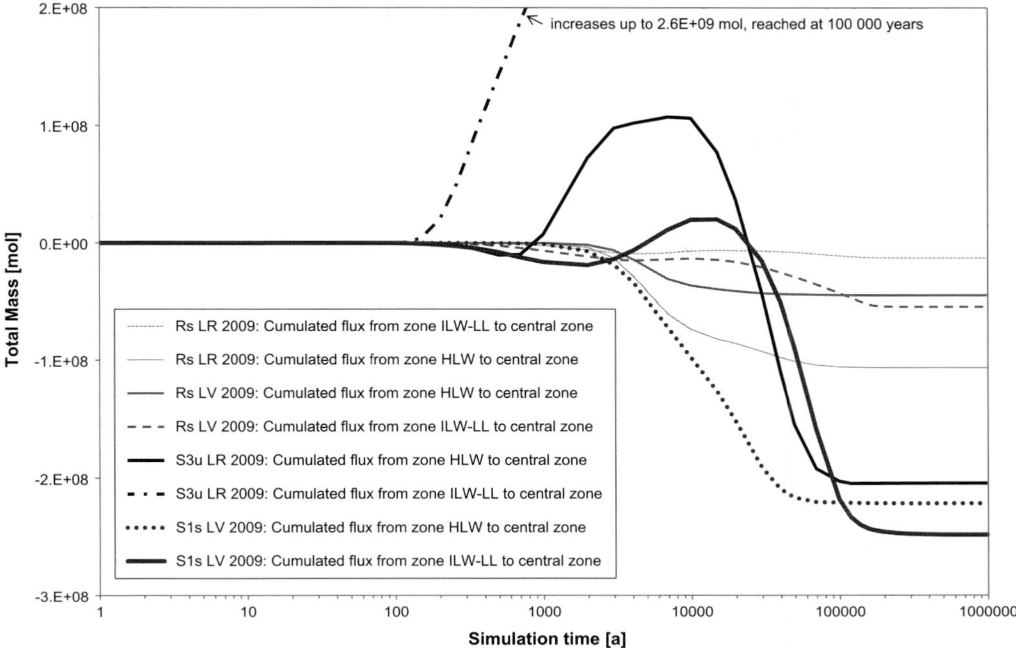

Fig. 10. Comparison of fluxes between zones for selected simulation cases.

of the HLW emplacement cells into the model. In reality, waste emplacement is not instantaneous, nor is the model in general designed to study the very early stage at fine detail, suggesting that these early peaks must be interpreted with care. Nevertheless, it is interesting to note that the CU3 waste in zone HLW-3 shows a different behaviour. This waste does not emit heat but only hydrogen. Therefore, there is no inherent pressure increase in these emplacement cells, enabling the almost immediate appearance of a gas phase.

In the following up to 80 years, all waste disposal zones are filled, sealed and backfilled. Desaturation of the host rock along the remaining open-ventilated main access drifts continues. In zone HLW, with the continuing heat production, the pressure increases further until enough hydrogen is generated to initiate gas phases in the EDZs of the emplacement drifts. The two-phase conditions lead to a reduction of the pressure, as the gas phase allows for a balancing of the pressure within the emplacement cells and the adjacent host rock. In zone ILW-LL, with the sealing of the emplacement drifts, hydrogen generation in this zone and heat emission of the CSD-C waste is modelled to start as well. Depending on the rate of the hydrogen generation, a further desaturation of the EDZ and the adjacent host rock is observable, especially in the vicinity of the B4.2 waste, which has a very high initial hydrogen generation rate (Fig. 8). The heat production of the CSD-C waste results in an increase in the temperature and an induced pressure increase in the emplacement drifts, delaying the creation of a gas phase compared with the B5 waste without heat generation. At the end of this period the EDZs of all the drifts are partially desaturated and the gas phases of the individual emplacement drifts are connected.

At 100 years the last main access drifts are backfilled and desaturation by ventilation stops there as well; all direct man-made perturbances of the system stop and the post-operational phase starts.

Until *c.* 1000 years, there are three typical types of behaviour distinguishable in different areas:

(1) Areas with hydrogen and heat producing waste – on the one hand, heat production and associated thermal expansion lead to a large pressure increase in saturated conditions, while in two-phase conditions this thermal effect is much smaller. On the other hand, hydrogen production and transport lead to the further appearance of two-phase conditions. Both effects interact particularly in the early stages when energy production is large: zones of high pressure build up around the emplacement cells and spread into their vicinity owing to heat transport. These zones are successively replaced by expanding zones of

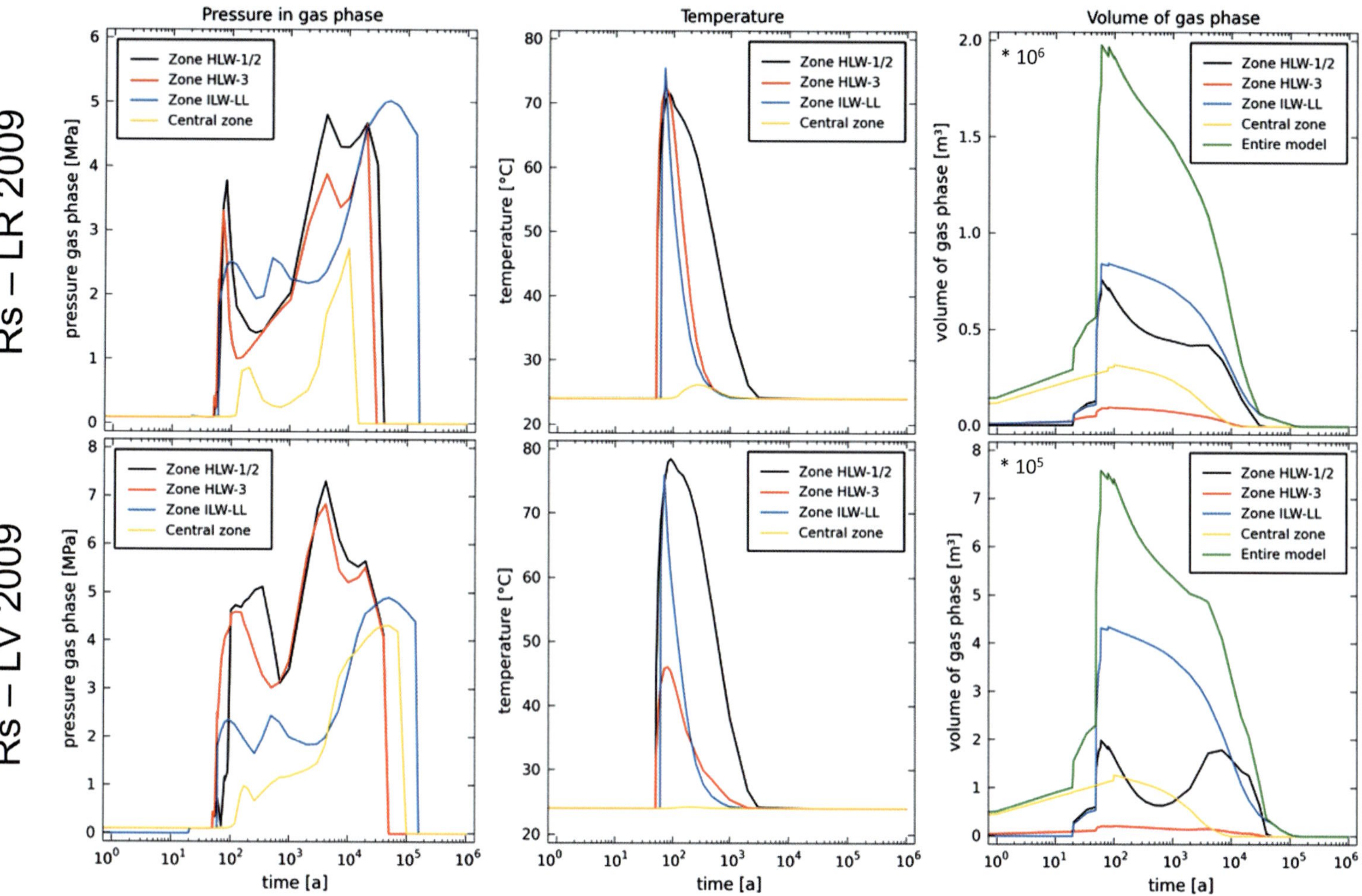

Fig. 11. Temporal evolution of performance indicators of selected simulation cases.

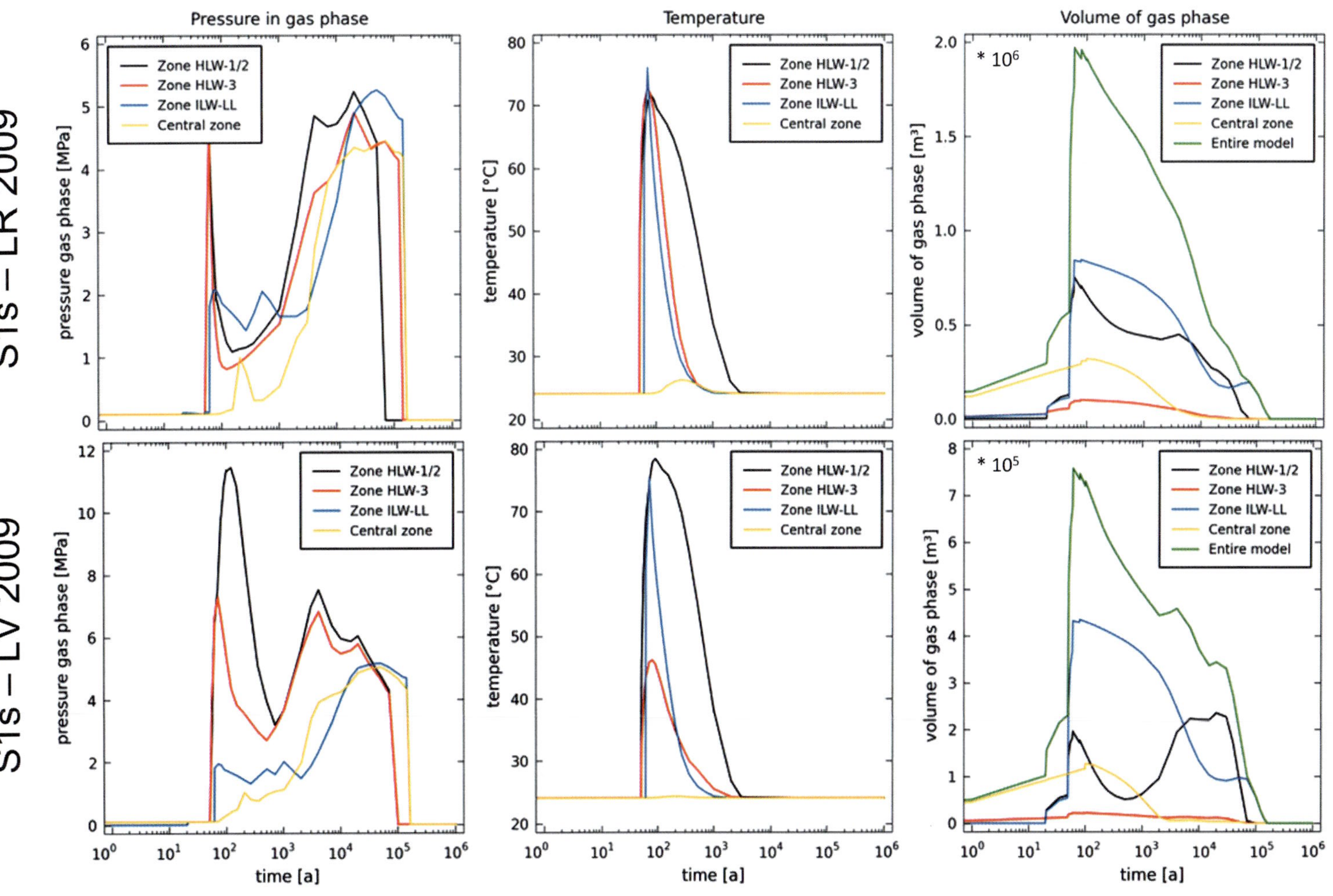

Fig. 11. *Continued*

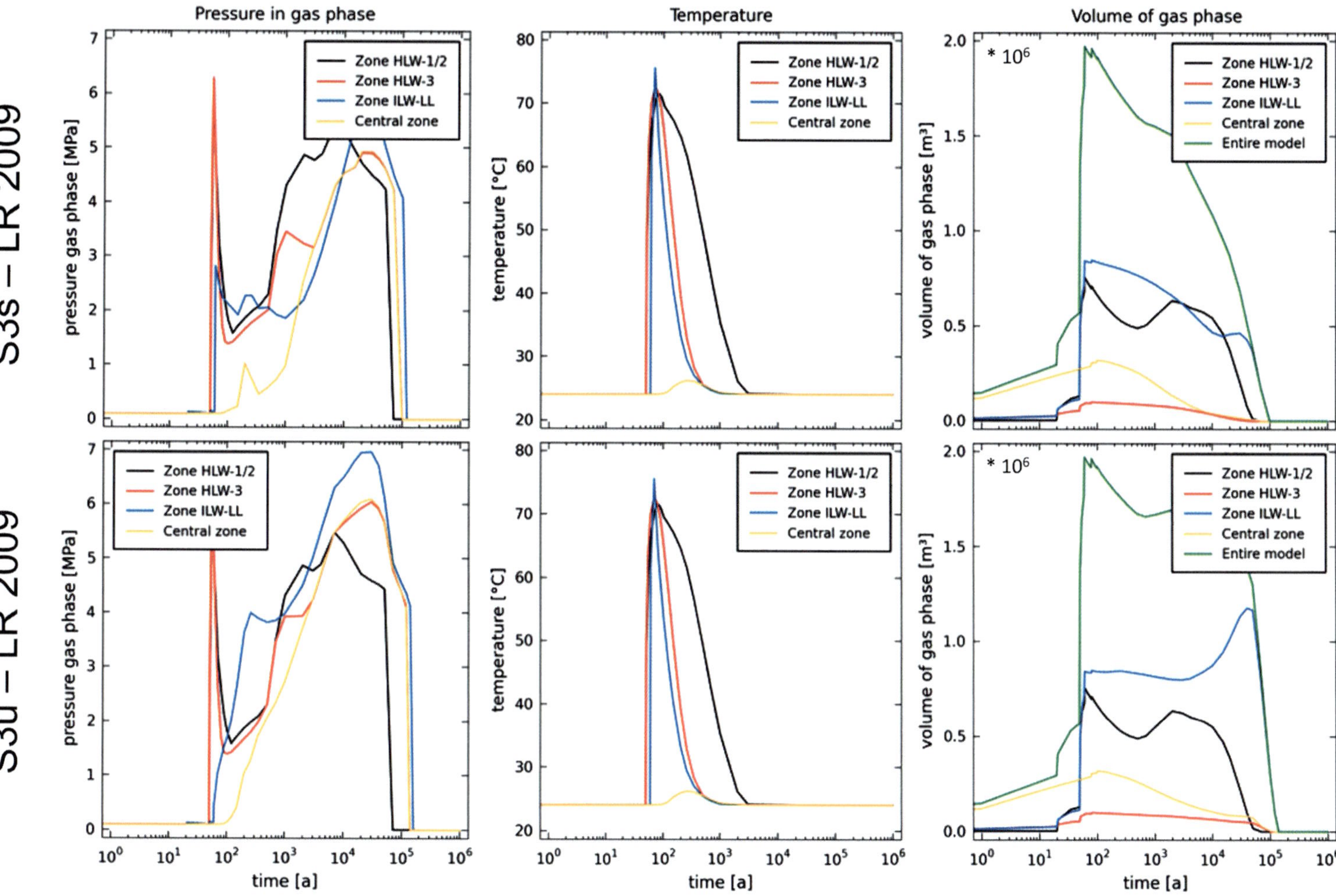

Fig. 11. *Continued*

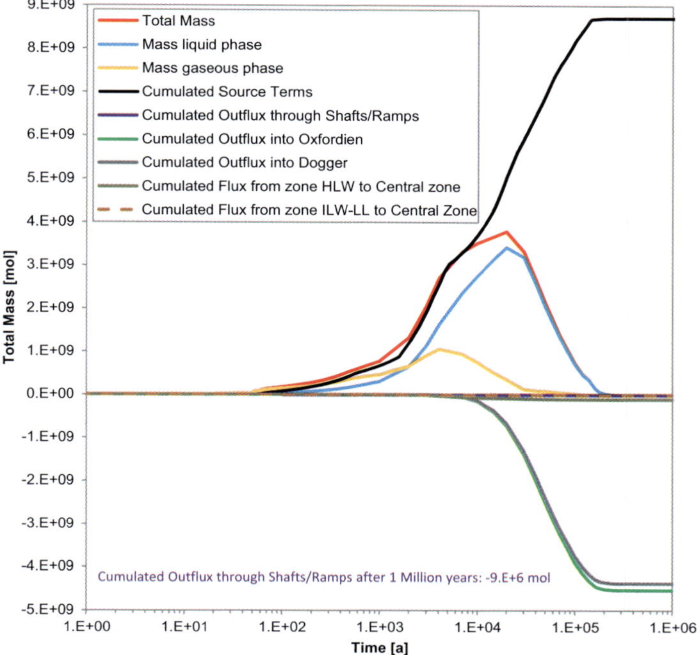

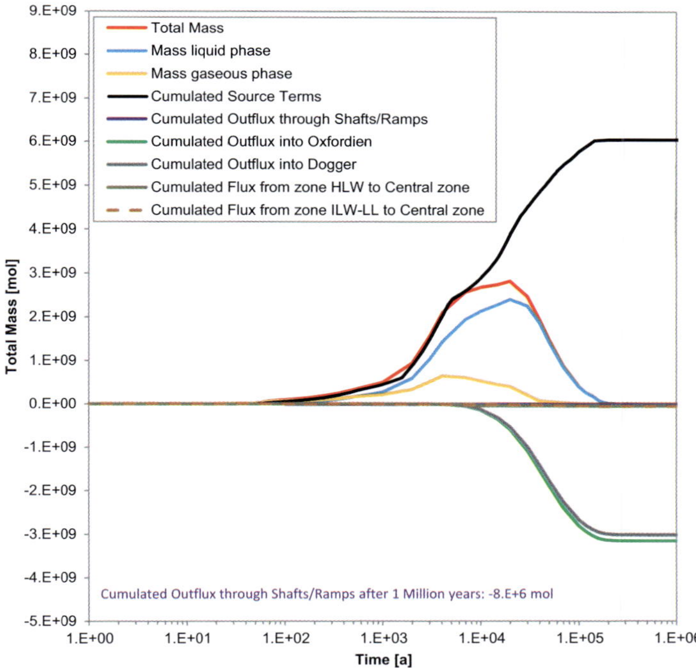

Fig. 12. Temporal evolution of hydrogen (gas) mass balances of selected simulation cases.

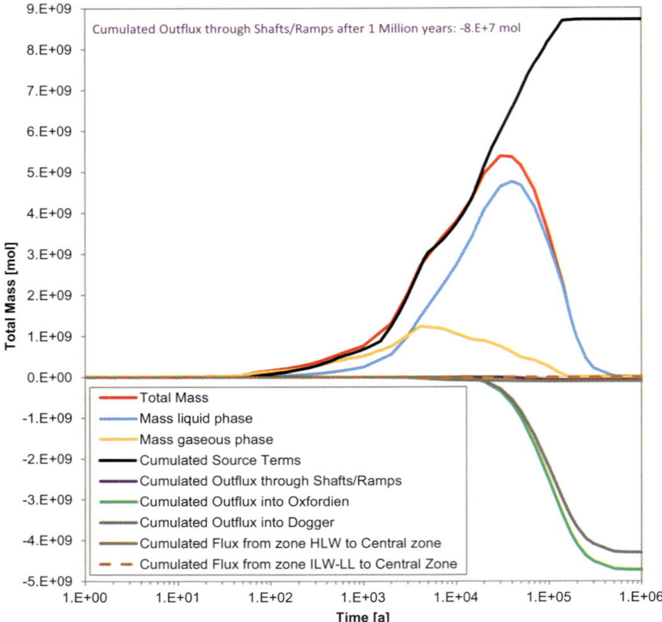

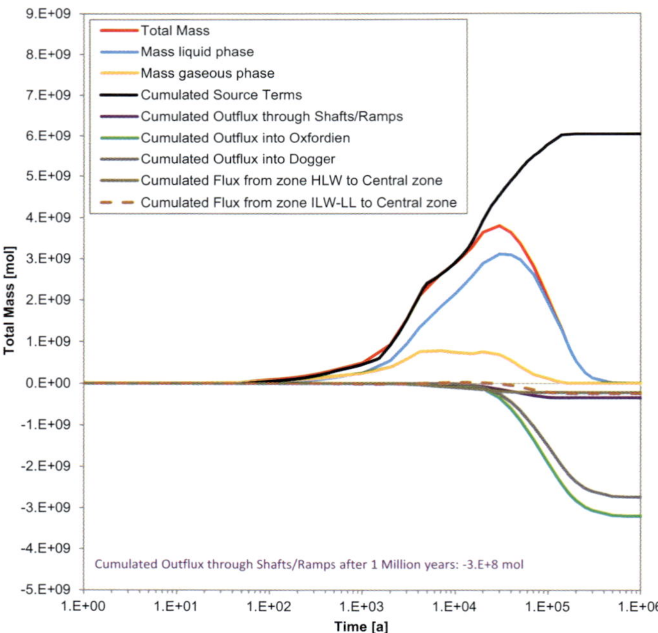

Fig. 12. *Continued*

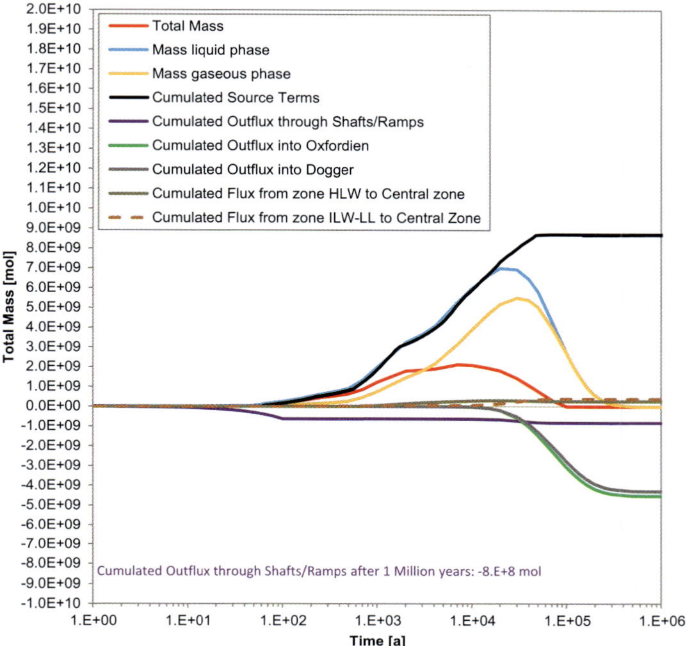

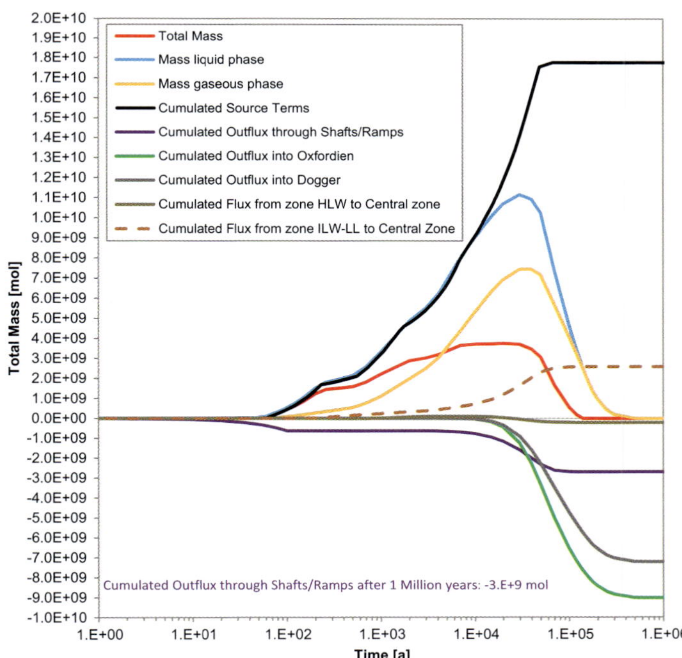

Fig. 12. *Continued*

two-phase conditions leading to a pressure reduction. After some 100 years the peak of heat production is passed (Fig. 7), resulting in a further reduction of pressure.

(2) Areas with hydrogen producing but non-exothermal waste – in areas where only hydrogen is produced, desaturation is accelerated compared with the areas where temperature increases as well – mainly because the pressure remains at a lower level. The extent of the desaturation into the adjacent host rock depends strongly on the individual hydrogen production rates. Pressure increase in these areas is exclusively attributed to the production of hydrogen exceeding the transport capacities into the vicinity. This hydrogen is, besides through diffusive processes, advectively transported in the gas phase, mainly along the backfilled drifts and their EDZs.

(3) Areas without heat or gas production – the drifts in the emplacement zones are kept at two-phase conditions by the hydrogen released from the emplacement cells through the different head seals and surrounding EDZs. They ensure a balancing of pressure and hydrogen between adjacent emplacement cells within the individual zones. In contrast, the access shaft in zone ILW-LL resaturates very early, starting with its closure after 80 years. The other shafts and the ramps in the central zone do not represent a diversion for hydrogen either, as they are resaturated laterally and from the top rather quickly. This implies that, already after 150 years, there is no continuous gas phase between their bottom and the overlying Oxfordian. Within the overall system, the central zone represents rather a storage volume for hydrogen than a pathway towards the surface.

In the following simulation phase up to *c.* 10 000 years the heat production reduces significantly; consequently the temperature at the repository level goes back to ambient conditions, that is, 22.5 °C. The saturation in the emplacement cells and adjacent areas is now dependent on the various hydrogen generation rates, which generally reduce over time. Pressure in the gas phase, which has generally dropped below the ambient pressure at the repository level during desaturation and the inherent two-phase conditions, generally increases with resaturation of the aforementioned areas from the host-rock. An exception occurs with the drastic increase in hydrogen production in zones HLW after 1550 years, which is attributed to the loss of impermeability of the metal tubing. The pressure in the gas phase reacts to this with a steep rise. In the period thereafter, pressure in this zone continues to be guided by the gas generation rate until *c.* 11 000 years when hydrogen generation drops.

In the subsequent simulation period, up to *c.* 100 000 years, the restoration of natural conditions at the repository level continues. The preceding drainage and desaturation of the host rock in the immediate vicinity of the repository during the constructional, operational and early post-operational phases leads to a depressurization of the large-scale vicinity of the repository, up to the Dogger and down to the Oxfordian. With the resaturation of the repository the natural pressure regime is restored. Furthermore, the remaining and the newly produced hydrogen dissolves and is transported mainly by diffusion in the liquid phase – which is by far the dominant transport process in the long run. This combination causes on the one hand a resaturation of parts where no more hydrogen is produced and on the other hand an increase in the pressure in the gas phase in the remaining two-phase areas as produced hydrogen must be stored in a smaller space. In many places the hydrogen production is no longer sufficient to maintain a gas phase under the predicted conditions and the hydrogen completely dissolves instantaneously as it is produced. This occurs everywhere except for those parts of the ILW-LL in the vicinity of the B1, B4 and B5 waste which still produce a large amount of hydrogen at these later times.

In the following period up to 1 000 000 years these gas phases disappear as well. After 200 000 years saturation has returned everywhere, whereas pressure perturbations cease after approximately 300 000 years.

Comparison with the reference case Rs: LV 2009

The simulations with the variant layout provided a similar picture with the following relevant deviations: (a) the early maximum pressure in the HLW zones is higher, which has to be attributed to the longer emplacement cells (Table 1) in this layout, resulting in a higher maximum temperature (Table 5); (b) pressure levels are generally higher, as the layout is more compact and thus the total pore volume available for a gas phase is half that of LR 2009 (Table 1), resulting in a smaller volume for the gas phase (Fig. 11) – at the same time the total hydrogen production is just two-thirds (Fig. 12), reducing, but not eliminating, the effect; and (c) the balancing of pressure occurs in LV 2009 also across zones with exchange of hydrogen between zones HLW and ILW-LL through the central zone (Fig. 10). This has to be attributed to the compact design of the layout (Fig. 2) and the very short distances between the zones (Table 1).

Table 5. *Results of performance indicators for simulations performed*

Name	Maximum temperature (°C)	Resaturation time (years)	Maximum gas pressure (MPa)
Rs – reference	LR 2009: 75.6 LV 2009: 78.5	LR 2009: 160 000 LV 2009: 160 000	LR 2009: 5.0 LV 2009: 7.3
Ru – reference	75.3	170 000	5.7
S1s – increased drift permeability	LR 2009: 76.0 LV 2009: 78.5	LR 2009: 160 000 LV 2009: 160 000	LR 2009: 5.3 LV 2009: 11.4
S1u – increased drift permeability	76.0	250 000	5.9
S2s – faster gas generation	76.0	70 000	7.8
S2u – faster gas generation	75.6	100 000	7.1
S3s – increased drift permeability and faster gas generation	75.6	120 000	6.3
S3u – increased drift permeability and faster gas generation	75.6	160 000	7.0
S4s – saturation-dependent gas generation	75.6	100 000	5.8

Summarizing, the overall impact of hydrogen and heat generation and the gas phase in the reference cases can be expressed as:

- Initially high pressures in the operational phase result from differences in temperature expansivity of the water and of the matrix. As soon as two-phase conditions are reached they are reduced. The gas phase is important for a reduction of pressure.
- A continuous gas phase is important for balancing of pressures between cells, drifts and zones.
- The by far dominating transport process of hydrogen from the repository is dissolution in the liquid phase and then diffusion into the cap and bed rock.
- Resaturation is delayed by the hydrogen production and follows its production regime to a certain extent.

Comparison of sensitivities

Maximum temperature. The maximum temperatures reached in all performed sensitivity simulations in the two layouts vary over a very small range. Variations result from the dependence of conductive heat transport from saturation, with a lower conductivity with increasing gas saturation. Advective transport of heated liquid or gas from the heat sources is negligible in the performed sensitivity simulations. Therefore, the characteristics do not differ, as can be seen in the comparison of the performance indicators shown in Table 5.

Resaturation time. The conservative unified hydrogen production rate applied in cases Ru, S1u, S2u, S3u, generally shows a prolonged resaturation time when compared with the corresponding cases with the waste type-specific hydrogen production. Considering that the amount of hydrogen production in zone ILW-LL differs by a factor of three, this is reasonable, and needs no further explanation.

The sensitivities S1 study the impact of a higher permeability of the backfill with a reduced diffusive transport in the host rock. The results show that this leads to a slower transport of hydrogen from the repository into the host rock, that is, a longer residence time in the repository (Fig. 12). At the same time the flux of hydrogen through the shafts increases. The fluxes between the zones increase as well (Fig. 10). The different resaturation behaviour is for all the cases (Rs–S1s, Ru–S1u for both layouts) apparent from the plots of the gas phase volume in Figure 11, even if for the cases with waste type-specific hydrogen production the overall resaturation times are similar, in fact equal at the time resolution computed here. The hypothetical acceleration of the gas generation studied in S2 results in higher gradients of hydrogen concentration, higher diffusive fluxes into the host rock, and therefore a faster resaturation.

The combination of a faster gas generation with higher backfill permeability and reduced diffusion coefficient of hydrogen in the host rock in simulation cases S3s and S3u results in a change in the transport behaviour of hydrogen. While the resaturation times stay between those of the reference and S1 cases, the fluxes and resulting mass allocation change significantly (Fig. 12). There is a significant increase of the flux of hydrogen through the shafts. In case S3u there is a flux of hydrogen from the zone ILW-LL into the zone HLW (Fig. 10), representing a significant change of the flux regime.

The parameterization of the hydrogen production in S4 s follows rather the fast gas generation scenario. Therefore, the resaturation time is in the same order of magnitude. This allows the conclusion to be drawn that, with the current model concept, there is no water shortage slowing down corrosion significantly.

Maximum pressure in gas phase. The discussion of the maximum pressure in the gas phase is constrained to the second (or third) peak (Fig. 11) as the first peak is affected by the constructional phase and its parameterization. In most cases this second (or third) peak is reached shortly before the gas phase collapses and resaturation occurs.

Generally, the cases with the unified hydrogen production in the ILW-LL zone show higher maximum pressures than those with the waste type-specific hydrogen production. This has to be attributed to the lower total volume of hydrogen and the different regime of its production.

With the effect of the slower hydrogen transport from the repository a higher maximum pressure is reached in all S1 sensitivity cases in both layouts as more gas (Fig. 12) is stored in approximately the same volume (Fig. 11). This effect is more pronounced in LV 2009 compared with LR 2009 owing to the compacter design.

In the S2 sensitivities the same effect is observable, here caused by the accelerated hydrogen production but with similar evolution of the gas phase volume, hence leading to even higher pressures in the gas phase.

In the S3 sensitivities the pressure increase is dampened compared with the S2 sensitivities by the enhanced permeability of the backfill, allowing for an easier transport of hydrogen along drifts and shafts.

Comparison of layouts

The maximum temperatures observed reflect the differences in spatial waste emplacement density and the different length of the HLW waste emplacement cells (Table 1), affecting the transport of heat from each emplacement cell. The different length is one of the reasons for the early peak of the maximum pressure in the gas phase, observable in LV 2009 (Fig. 11).

As said above, the role of the gas phase as a balancing system for pressure and hydrogen between zones is more pronounced in LV 2009 compared with LR 2009. This is clearly detectable in the results of sensitivities S1s. In both layouts resaturation is affected as described above, but the pressure increase is much more pronounced in LV 2009 owing to the more important role of the drifts as balancing systems, hence it is more sensitive towards the drift parameters.

Summary and conclusions

Within the presented study, two-phase flow modelling on the global scale of a repository for radioactive waste has been performed. Two repository layouts, a reference and a variant layout, have been compared.

The numerical model considers the operational and post-operational phase, where the former is approximated by a step-wise approach representing the individual construction stages. For example, drifts and shafts are dug or waste is emplaced instantaneously. This leads to discontinuities in the model and, in their vicinity, the model results are less representative with respect to a precise prediction of the system behaviour. On the large scale and in the post-operational phase – and this is the goal of the present studies – the simulations show consistent and plausible system behaviour and confidence in the simulation results is high. Furthermore, the model and its numerical implementation have been validated using convergence as well as mass and energy balance checks.

In particular, the comparison between the two repository layouts showed that the comparative analysis of different repository layouts is feasible using the present modelling approach. Also, the overall system comportment and parameter sensitivities with respect to key indicators can be assessed.

The computations have shown that the system – at least with respect to maximum pressures and hydrogen flow – is rather sensitive to the presence of a gas phase and its spatial distribution and connectivity. Moreover, hydrogen diffusion through the host rock has been identified to be a key process.

A connected continuous gas phase functions as a balancing system between connected areas. Its extension and continuity depend on the parameterization of the permeabilities and diffusion (S1s), the amount of hydrogen production (Rs, Ru) and the temporal regime of the hydrogen production (S2s/u, S3s/u). Naturally, the available storage volumes have an impact as well. These volumes differ significantly between the two compared layouts.

Considering this, the reference layout 2009 seems to be robust to the studied variations, in the sense that simulation results vary in their magnitude but not with respect to the overall qualitative picture. While this remark remains valid for the variant layout 2009, the results indicate that this layout reacts more strongly to variations.

This study has been performed on behalf of Andra, who provided the problem, the geometries of the repository layouts, and further input parameters. G. Pépin and J. Croisé supported the study with their know-how. K. Zhang performed test simulations on the Chinese supercomputer Tianhe-1A.

References

ANDRA 2006. *Dossier 2005 Argile – synthesis – evaluation of the feasibility of a geological repository in an argillaceous formation*. Agence nationale pour la gestion des déchets radioactifs (Andra) Report Series.

ANDRA 2007. *Etudes d'ingenierie relatives au stockage souterrain des dechets radioactifs HAVL – simulations thermiques – modelisation d'une zone de stockage HA*. Agence nationale pour la gestion des déchets radioactifs (Andra) technical report C.RP. 0ISL.07.009.

ANDRA 2009*a*. *Jalon 2009 Ha-Mavl referentiel du site Meuse/Haute-Marne, Presentation generale*. Agence nationale pour la gestion des déchets radioactifs (Andra) technical report C.RP.ADS.09.0007, 30 September.

ANDRA 2009*b*. *Architecture Ha-Mavl 2009 (G8-SB), Scénario de base, Revêtement galeries dimensionné sur 100 ans*. Agence nationale pour la gestion des déchets radioactifs (Andra) document C.PL.ASTE. 09.0552.A, 10 September.

ANDRA 2009*c*. *National Inventory of Radioactive Materials and Waste – synthesis report*. Agence nationale pour la gestion des déchets radioactifs (Andra) document DCAI-CO-09-0051.

ANDRA 2011. *Proposition d'architecture (SI) Janvier 2011*. Agence nationale pour la gestion des déchets radioactifs (Andra) document C.PL.ADBE.11.0020. A, 31 January.

Croisé, J., Mayer, G., Talandier, J. & Wendling, J. 2011. Impact of water consumption and saturation-dependent corrosion rate on hydrogen generation and migration from an intermediate-level radioactive waste repository. *Transport in Porous Media*, **90**, 59–75, http://dx.doi.org/10.1007/s11242-011-9803-0

Enssle, C. P., Croisé, J., Poller, A., Mayer, G. & Wendling, J. 2011. Full scale 3D-modelling of the coupled gas migration and heat dissipation in a planned repository for radioactive waste in the Callovo-Oxfordian clay. *Physics and Chemistry of the Earth*, **36**, 1754–1769, http://dx.doi.org/10.1016/j.pce.2011.07.033

Enssle, C. P., Brommundt, J., Kaempfer, Th. U., Mayer, G. & Wendling, J. 2012. 3D Modeling of the long-term behavior of a large geological repository for HLW and ILW-LL nuclear waste, considering heat, gas, and radionuclide release and transport – optimizations that allow for detailed large-scale modeling. *In*: *Proceedings of the TOUGH Symposium*, Berkeley, CA.

Enssle, C. P., Brommundt, J., Kaempfer, Th. U., Mayer, G. & Wendling, J. 2014. Full-scale 3D modelling of a nuclear waste repository in the Callovo-Oxfordian clay. Part 2: thermo-hydraulic two-phase transport of water, hydrogen, ^{14}C and ^{129}I. *In*: Norris, S., Bruno, J., Cathelineau, M., Delage, P., Fairhurst, C., Gaucher, E. C., Höhn, E. H., Kalinichev, A., Lalieux, P. & Sellin, P. (eds) *Clays in Natural and Engineered Barriers for Radioactive Waste Confinement*. Geological Society, London, Special Publications, **400**. First published online May 7, 2014, http://dx.doi.org/10.1144/SP400.35

Glascoe, L., Buscheck, T. A., Gansemer, J. & Sun, Y. 2002. *The Multi-scale Model Approach to Thermohydrology at Yucca Mountain*. UCRL-ID-149212. Lawrence Livermore National Laboratory, Livermore, CA, 27 March.

Hoch, A. & Wendling, J. 2011. Migration of gases around a cell containing high-activity vitrified wastes during the operational phase. *Physics and Chemistry of the Earth*, **36**, 1743–1753, http://dx.doi.org/10.1016/j.pce.2011.10.019

Luckner, L., van Genuchten, M. Th. & Nielsen, D. R. 1989. A consistent set of parametric models for the flow of immiscible fluids in the subsurface. *Water Resources Research*, **25**, 2187–2193.

Mualem, Y. 1976. A new model for predicting the hydraulic conductivity of unsaturated porous media. *Water Resources Research*, **12**, 513–522.

NAGRA 2008. *Effects of post-disposal gas generation in a repository for low- and intermediate-level waste sited in the Opalinus Clay of Northern Switzerland*. Technical Report 08-07. National Cooperative for the Disposal of Radioactive Waste (Nagra), Wettingen, Switzerland, http://www.nagra.ch/documents/database/dokumente/%24default/Default%20Folder/Publikationen/NTBs%202001-2010/e_ntb08-07.pdf

OFFICE PARLEMENTAIRE 2006. *Act 2006-739 of 28 June 2006: The 2006 Programme act on the sustainable management of radioactive materials and wastes*, http://annual-report2006.asn.fr/PDF/radioactive-waste-management-act-280606.pdf

Poller, A., Enssle, C. P., Mayer, G., Croisé, J. & Wending, J. 2011. Repository-scale modeling of the long-term hydraulic perturbation induced by gas and heat generation in a geological repository for high-level and intermediate-level radioactive waste: methodology and example of application. *Transport in Porous Media*, **90**, 77–94, http://dx.doi.org/10.1007/s11242-011-9725-x

Poppei, J., Mayer, G., Croisé, J. & Marschall, P. 2003. Modelling of resaturation, gas migration and thermal effect in a SF/ILW repository in low-permeability over-consolidated clay-shale. *In*: *Proceedings of the TOUGH Symposium*, Berkeley, CA.

Pruess, K., Oldenburg, C. & Moridis, G. 1999. *TOUGH2 User's Guide, Version 2.0*. Report LBNL-43134, Lawrence Berkeley National Laboratory, Berkeley, CA.

Rutqvist, J., Barr, D. et al. 2008. Results from an international simulation study on coupled thermal, hydrological, and mechanical (THM) processes near geological nuclear waste repositories. *Nuclear Technology*, **163**, 101–109.

van Genuchten, M. Th. 1980. A closed-form equation for predicting the hydraulic conductivity of unsaturated soils. *Soil Science Society of America Journal*, **44**, 892–898.

Zhang, K., Wu, Y.-S. & Pruess, K. 2008. *User's guide for TOUGH2-MP – a massively parallel version of the TOUGH2 code*. Report LBNL-315E, Lawrence Berkeley National Laboratory, Berkeley, CA.

Full-scale 3D modelling of a nuclear waste repository in the Callovo-Oxfordian clay. Part 2: thermo-hydraulic two-phase transport of water, hydrogen, ^{14}C and ^{129}I

C. P. ENSSLE[1], J. BROMMUNDT[1], TH. U. KAEMPFER[1]*, G. MAYER[1] & J. WENDLING[2]

[1]*AF-Consult Switzerland Ltd, Groundwater Protection and Waste Disposal, Baden, Switzerland*

[2]*ANDRA, Chatenay-Malabry, France*

**Corresponding author (e-mail: thomas.kaempfer@afconsult.com)*

Abstract: The development of deep geological repositories for nuclear waste requires a sound system understanding, to which performance analyses by means of numerical flow and transport simulations form an important contribution. In this context, we present modelling studies of the thermo-hydraulic two-phase transport of water and hydrogen in the planned French repository in part 1. This paper, part 2, uses the same modelling concept extended to include radionuclide transport. The numerical TOUGH2-MP-model encompasses the host rock from cap to bed rock including the repository in full 3D. We performed a series of single-phase and two-phase transport simulations of volatile radionuclides (^{14}C) and highly soluble ones (^{129}I). The sensitivity analyses concerned different conditions and scenarios, for example, early canister failure, varied hydrogen generation regimes and transport parameters. The results allow determination of the dominant radionuclide transport paths and transfer processes. They demonstrate that hydrogen generation inside the repository has a large impact on the transfer of ^{14}C, favouring its advective transfer in the gas phase along the backfilled tunnels, whereas it has only minor influence on the transfer of ^{129}I which is mainly transported by diffusion in the formation water of the host rock. The study demonstrates the feasibility of full-scale radionuclide transport simulations.

This article consists of part 2 of the combined paper 'Full-scale 3D modelling of a nuclear waste repository in the Callovo-Oxfordian clay' and presents results of numerical transport simulations concerning the release and migration of two radionuclide species in the planned French deep geological nuclear waste repository. The simulations take into account the transient thermo-hydraulic two-phase flow and solute transport inside the repository and the host rock influenced by its construction and the generation of hydrogen and heat by the wastes which is the focus of the simulations presented in part 1 of this combined paper (Brommundt *et al.* 2014). Owing to the large similarities concerning the model concepts and their implementation between the two studies, all common information is presented in part 1 and only briefly summarized here. However, all aspects related to the radionuclide transfer, as well as relevant differences between the two studies are presented in this paper, part 2. For the complete understanding of this article the reader is recommended to familiarize her-/himself with the contents of part 1 – most importantly sections 'Approach and system description' and 'Model concept'.

Approach and system description

In addition to the phenomena stated in part 1 – (a) the creation of an excavation damaged zone (EDZ) around shafts and drifts, (b) the depressurization and desaturation of the EDZ and the adjacent intact host rock as a result of the tunnel ventilation in the operational phase with a resaturation by pore water from the host rock and re-pressurization in the post-operational phase, (c) the appearance, migration, and disappearance of a gas phase owing to waste-type-specific release of gaseous substances, mainly hydrogen, (d) the decay-heat induced temperature increase leading to a pressure increase as a consequence of heat expansion of the fluids and the solids – here in part 2 we consider (e) the release of the radionuclide ^{14}C, in its volatile form of methane, and iodine ^{129}I owing to the degradation and corrosion of the waste packages and their migration by advection and diffusion. Their transport is influenced by decay and retardation owing to sorption processes.

The radioactive waste to be disposed of can be classified into two major groups. The first comprises high-level wastes (HLW), which mainly consist of

From: NORRIS, S., BRUNO, J., CATHELINEAU, M., DELAGE, P., FAIRHURST, C., GAUCHER, E. C., HÖHN, E. H., KALINICHEV, A., LALIEUX, P. & SELLIN, P. (eds) 2014. *Clays in Natural and Engineered Barriers for Radioactive Waste Confinement*. Geological Society, London, Special Publications, **400**, 469–481.
First published online May 7, 2014, http://dx.doi.org/10.1144/SP400.35

Table 1. *Substance properties of $^{14}CH_4$ and ^{129}I*

Parameter	^{14}C ($^{14}CH_4$)	^{129}I
Molar mass (g/mol)	18.0	129.0
Diffusion coefficient in bulk gas phase (m^2/s)	1.3×10^{-4}	–
Diffusion coefficient in bulk liquid phase (m^2/s)	4.9×10^{-11}	4.9×10^{-11}
Inverse Henry's constant (1/Pa)	3.0×10^{-10}	1.0×10^{50}
Half-life (years)	5.73×10^{3}	1.57×10^{7}
Adsorption coefficient K_d ($m^3\ kg^{-1}$) of Host rock (Callovo-Oxfordian clay)	1×10^{-4}	0.0
Adsorption coefficient K_d ($m^3\ kg^{-1}$) of Concrete	0.5	1×10^{-3}

highly radioactive calcined chemical solutions produced during the recycling of nuclear fuel and incorporated into molten glass and poured into stainless steel containers, and non-processed spent nuclear fuel (Andra 2009). The second group, the intermediate-level long-lived wastes (ILW-LL), comprises various mainly solid waste types that are compacted and stored in metal drums which are placed into concrete containers (Andra 2009).

The HLW and ILW-LL packages differ in terms of their temporal integrity with respect to the containment of the radionuclides. The ILW containers exhibit almost no barrier capacity and the release of radionuclides may begin immediately after their emplacement. In contrast, the HLW containers are designed to maintain their barrier capacity over several thousand years, yet an earlier canister failure cannot be excluded and needs to be taken into account in the safety assessment. In addition to the integrity aspect of the containment itself, the quantitative and transient regime of the radionuclide mobilization depends on further aspects such as the canister inventory, the canister degradation processes, the half-lives and the thermo-hydraulic and chemical conditions inside the container and in its vicinity, and on the transfer properties of the individual radionuclide constituents such as sorption and diffusion capacity to name just a few. One can distinguish between instantaneous point release, and long-term continuous, constant or varying release regimes.

The two radionuclides considered in this study, ^{14}C, as ^{14}C-labeled methane, and ^{129}I, exhibit significant differences with respect to the aforementioned release behaviour such that the release regime of ^{14}C is mostly continuous and lasts over periods of several 10 000 years; concerning ^{129}I, both continuous and point release regimes are likely to occur. The migration behaviour of the two radionuclides may differ strongly as the transfer of ^{14}C preferentially takes place within the gaseous phase (if present), whereas the ^{129}I exhibits a good solubility and hence migrates mainly in the formation water. Thus, the fate of the ^{14}C depends far more on the two-phase conditions than the ^{129}I.

Model concept

The physical model, the geometrical model concept, the numerical model and its parameterization – materials and substance properties, the spatial and temporal discretization, boundary and initial conditions, and the source terms for mass and heat – are fully adopted from part 1 (Brommundt *et al.* 2014). In the following only the adaptations related to the radionuclide transfer of ^{14}C and ^{129}I are outlined.

The considered radionuclides are representatives for (a) a highly volatile species with a relatively short half-life, little adsorption capacity onto the host rock and pronounced adsorption onto concrete and (b) a well soluble, poorly sorbing species with a very long half-life. Chemical reactions that might occur in the real system are neglected and both radionuclides are considered as non-reactive and treated as ideal gases. Both radionuclides partition between the aqueous and the gaseous phases according to Henry's law. They diffuse dependent on temperature and tortuosity in both phases

Table 2. *Cumulated ^{14}C and ^{129}I release in the different zones of the repository*

Radionuclide type	Zone HLW-1/2 (mol)	Zone HLW-3 (mol)	Zone ILW-LL (mol)	Total (mol)
^{14}C	1.8×10^{0}	2.7×10^{-3}	1.8×10^{2}	1.8×10^{2}
^{129}I	9.8×10^{2}	1.9×10^{2}	3.9×10^{2}	1.6×10^{3}

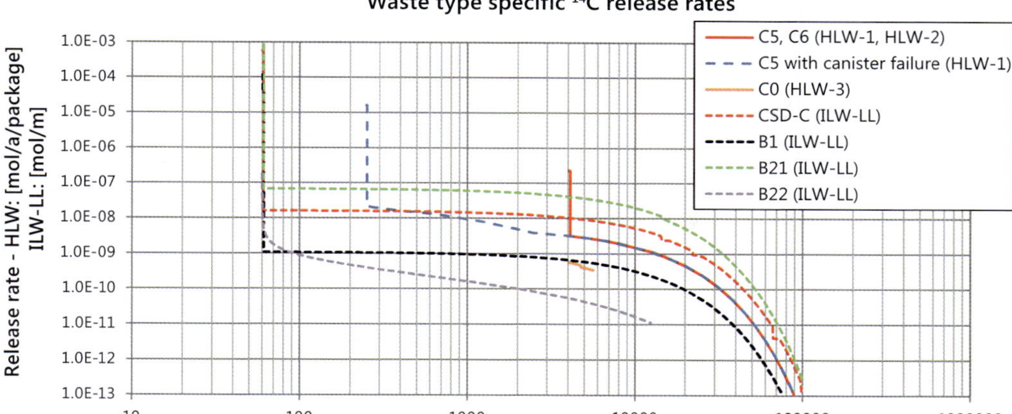

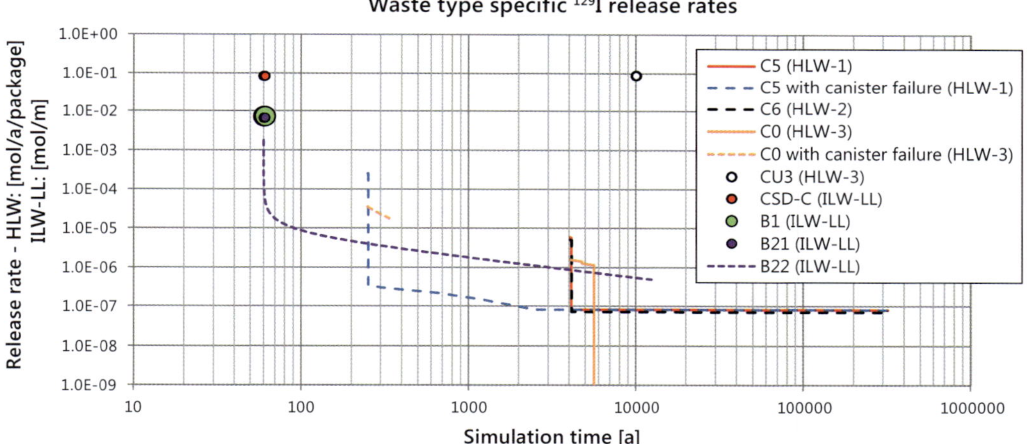

Fig. 1. Waste type-specific radionuclide release rates; top: ^{14}C, bottom: ^{129}I.

according to the Millington Quirk approach (Pruess *et al.* 1999). Adsorption of the radionuclides onto the solid grains is modelled as reversible instantaneous linear sorption. Decay products are not taken into account. The corresponding parameter values characterizing the species are specified in Table 1 and taken from Andra (2006).

For the model initialization, the radionuclide concentration fields are set to zero. At the model boundaries, Dirichlet conditions with zero radionuclide concentration are imposed. Furthermore, for the ventilation conditions during the operational phase, which are also implemented using Dirichlet conditions, concentrations are zero.

The release of the radionuclides is implemented by time- and waste type-dependent source-terms taking into account the emplacement year and the accurate location of each waste canister. The total amount of ^{14}C and ^{129}I released in the different zones over the whole simulation time of 1 million years is summarized in Table 2. The individual release rates per waste type and species are given in Figure 1, while the arrangement of the different waste types in the repository is visualized in Figure 2. The rates represent the effective release from the waste canister and not the total radionuclide inventory. The different waste type categories are named *C5* for wastes in zone HLW-1, *C6* for wastes in zone HLW-2, *C0* and *CU3* for wastes in zone HLW-3, and *CSD-C*, *B1*, *B2.1* and *B2.2* for wastes in zone ILW-LL. Early canister failure, that is, an earlier release of radionuclides, is considered for one HLW emplacement cell located in zone HLW-1 and one in zone HLW-3.

Table 3. *Overview of all simulation cases in both studies; simulation cases considering radionuclide transfer are indicated in the last two columns*

Name	Description	Hydrogen generation ILW-LL	*Water and hydrogen*		Water, hydrogen and radionuclides*	
			LR 2009	*LV 2009*	LR – 2P	LR – SP
Rs – reference	Reference parameterization	Waste type-specific	*X*	*X*	X	–
Ru – reference	Reference parameterization	Unique†	*X*	–	X	X (R-SP)
S1s – increased drift permeability	*Permeability of backfill, concrete and EDZ (axial component) × 5* *Diffusion coefficient of H_2 in gas and liquid phase/3*	*Waste type-specific*	*X*	*X*	–	–
S1u – increased drift permeability	Permeability of backfill, concrete and EDZ (axial component) × 5 Diffusion coefficient of H_2 in gas and liquid phase/3	Unique†	*X*	–	X	X (S1-SP)
S2s – faster gas generation	*Gas generation accelerated × 3*	*Waste type-specific*	*X*	–	–	–
S2u – faster gas generation	Gas generation accelerated × 3	Unique†	*X*	–	X	–
S3s – increased drift permeability and faster gas generation	Permeability of backfill, concrete and EDZ (axial component) × 5 Diffusion coefficient of H_2 in gas and liquid phase/3 Gas generation accelerated × 3	Waste type-specific	*X*	–	X	–
S3u – increased drift permeability and faster gas generation	Permeability of backfill, concrete and EDZ (axial component) × 5 Diffusion coefficient of H_2 in gas and liquid phase/3 Gas generation accelerated × 3	Unique†	*X*	–	X	–
S4s – saturation dependent gas generation	*Gas as inventory with release dependent on water availability*	*Waste type-specific*	*X*	–	–	–
S5u – increased diffusion of radionuclides	Diffusion coefficient of ^{129}I in the liquid phase × 3 Diffusion coefficient of ^{14}C in the gaseous phase × 3 (SP: in the liquid phase)	Unique†	–	–	X	X (S5-SP)

*Radionuclide transfer simulations only with layout LR2009.
†No hydrogen generation in the single-phase simulation cases.
Simulations formatted in *italics* are described in part 1 of this work (Brommundt *et al.* 2014).

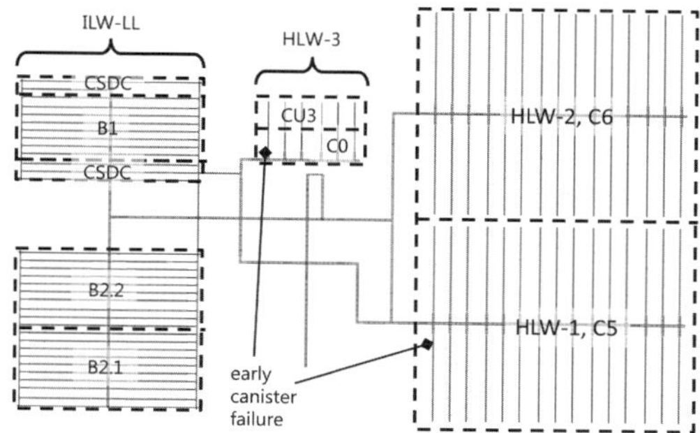

Fig. 2. Arrangement of the different waste types in the repository.

Figure 2 indicates their approximate locations. All parameter values are taken from Andra (2006).

Simulation results

Simulation cases and performance indicators

Table 3 gives an overview of all 21 simulation cases performed in both studies of which 10 cases consider the transfer of ^{14}C and ^{129}I. They are based on layout LR 2009 (see part 1, Brommundt *et al.* 2014). Three cases concern the radionuclide transfer under single-phase conditions and seven cases concern the radionuclide transfer under two-phase conditions. (In the single-phase simulations gas generation is not considered and the ventilation conditions during the operational phase are represented by fixed conditions of full saturation, atmospheric pressure and temperature.) The two-phase and single-phase radionuclide transfer reference cases, abbreviated as Ru and R–SP, use the reference parameterization as described in the section 'Numerical model' of part 1 of this article.

The sensitivity cases were chosen in order to study the repository's radionuclide transfer performance with respect to certain features, events and processes such as a lowered and more realistic gas generation, case Rs, an accelerated gas generation, case S2u, an increased permeability of the materials along the tunnels together with a reduced diffusion coefficient of hydrogen in the host rock, cases S1u and S1–SP, a combination of increased tunnel permeabilities and faster gas generation, cases S3u and S3s, and an increased diffusion capacity of the radionuclides, cases S5u and S5–SP.

For the individual and comparative performance assessment of the simulation results the following indicators addressing the radionuclide transfer behaviour were selected: (a) the time and duration of the radionuclides reaching the vertical model boundaries, that is, the cap rock, bed rock, and shaft ends; (b) the cumulative amount of radionuclides reaching the vertical model boundaries distinguishing between the transfer paths through the host rock and along the vertical shafts; (c) the time of occurrence and magnitude of maximum radionuclide flux to the vertical model boundaries; (d) the radionuclide transfer between the different zones of the repository; (e) the role of the different shafts with respect to their radionuclide transfer capacity; (f) the underlying radionuclide transfer processes; and (g) the contribution of each waste zone to the aforementioned balancing quantities. Figure 3 illustrates the assessed transfer paths and balancing zones for the analysis of the radionuclide transfer.

The repository performance with respect to the thermo-hydraulic comportment is the subject of

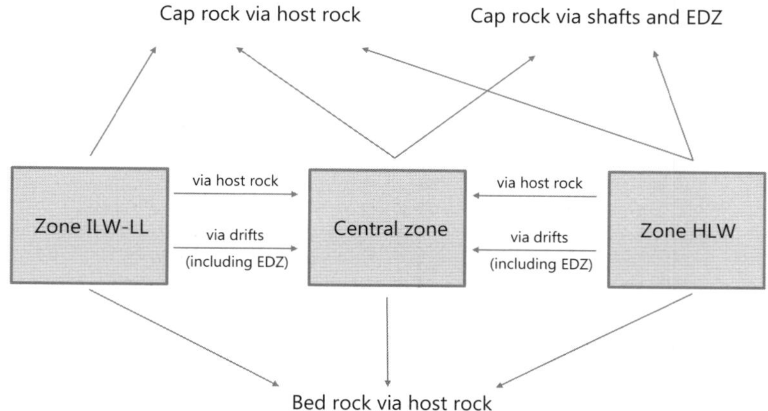

Fig. 3. Transfer paths for radionuclide flux balancing.

Table 4. *Two-phase reference case Ru –* ^{14}C *and* ^{129}I *transfer to the vertical model boundaries*

RN	Transfer path: host rock to …	Maximum flux rate (mol a^{-1})	Time of maximum flux rate (years)	Cumulated mass outflux after 1 Ma (mol)	Fraction of cumulated mass outflux along path and total cumulated outflux (−)	Fraction of cumulated mass outflux along path and total released mass (−)	Contribution of transfer processes (%)			
							Advection in		Diffusion in	
							liquid phase	gas phase	liquid phase	gas phase
^{14}C	Cap rock (without shafts)	2.4×10^{-10}	2.0×10^{4}	1.2×10^{-5}	0.08	6.5×10^{-8}	0.2	12.5	87.3	0
	Bed rock	9.6×10^{-11}	3.0×10^{4}	3.4×10^{-6}	0.02	1.9×10^{-8}	0	0	100	0
	Shaft outlets (sum of all shafts)	9.5×10^{-9}	2.0×10^{4}	1.5×10^{-4}	0.9	8.5×10^{-7}	0.1	98.3	1.6	0
	Sum	–	–	1.7×10^{-4}	1.0	9.3×10^{-7}	0.2	90.3	9.5	0
^{129}I	Cap rock (without shafts)	2.1×10^{-4}	1.0×10^{6}	89	0.49	0.057	1.3	0	98.7	0
	Bed rock	2.0×10^{-4}	1.0×10^{6}	92	0.51	0.059	0	0	100	0
	Shaft outlets (sum of all shafts)	5.4×10^{-11}	1.0×10^{6}	1.3×10^{-5}	$\ll 0.01$	8.2×10^{-9}	89.3	0	10.7	0
	Sum	–	–	181	1.0	0.12	0.6	0	99.4	0

part 1 of this article and we restrict the discussion to radionuclide transfer here.

Two-phase reference case Ru

The two-phase reference case in this part uses a unique gas generation rate for all ILW-LL wastes, which is conservatively over-estimated when compared with a waste type-specific gas generation rate (part 1 reference case Rs).

Radionuclide transfer to the vertical model boundaries. An overview of the radionuclide transfer to the vertical model boundaries, distinguishing between the transfer paths bed rock, cap rock (without shafts) and shafts (including their EDZs), is given in Table 4 for ^{14}C and ^{129}I. The evaluated quantities per transfer path are those described in the section above.

Concerning the transfer of ^{14}C, only around $10^{-4}\%$, that is, 1.7×10^{-4} mol, of the totally released ^{14}C into the model, 1.8×10^{2} mol, are transported to the vertical model boundaries. This rather small amount is attributed to the relatively short half-life when compared with the characteristic transport times of the species to the model boundaries. Around 90% of the ^{14}C transferred to the model boundaries reaches the cap rock via the shafts, mostly by advection in the gaseous phase which emerges in the waste zones and migrates preferentially along the comparatively highly permeable tunnel backfill and EDZ. Only 10% is transferred via the host rock to the cap and bed rock, mainly by diffusion in the liquid phase. The maximum flux rate to the cap rock of around 10^{-8} mol a^{-1} occurs at 20 000 years via the backfilled shafts; it coincides with the time of the maximum gas pressure and hence fastest gas velocities for the advective transfer of the highly volatile ^{14}C.

The transfer behaviour of ^{129}I to the vertical model boundaries shows a different picture. Owing to its strong solubility the released ^{129}I is mainly transferred to the vertical model boundaries via diffusion in the formation water of the host rock and fluxes along the drifts and shafts are negligible. During the simulation time of 1 million years around 12%, that is, 180 mol, of the totally released ^{129}I mass, 1.6×10^{3} mol, are transferred to the cap rock and bed rock. The significantly larger amount in comparison to the ^{14}C transfer is attributed to the much larger half-life of ^{129}I, which implies that soluble ^{129}I does not completely decay over the time it takes for it to diffuse through the liquid phase to the model boundaries. Consequently the maximum flux rates are significantly higher and occur at the end of the simulation time. The global flow balance into and out of the model after one million years is visualized in Figure 4.

Radionuclide transfer from the waste zones to the central zone. Key parameters for the radionuclide fluxes from the HL and IL-LL waste zones to the central zone are given in Table 5. Thus, the largest ^{14}C fluxes occur from the IL-LL waste zone to the central zone by advection in the gaseous phase, which is reasonable as 99% of the totally released ^{14}C originates from this zone and gas production is highest in this zone. Fluxes via the host rock only occur in desaturated locations, that is, in the EDZ and the host rock near the drifts. Inter-zonal

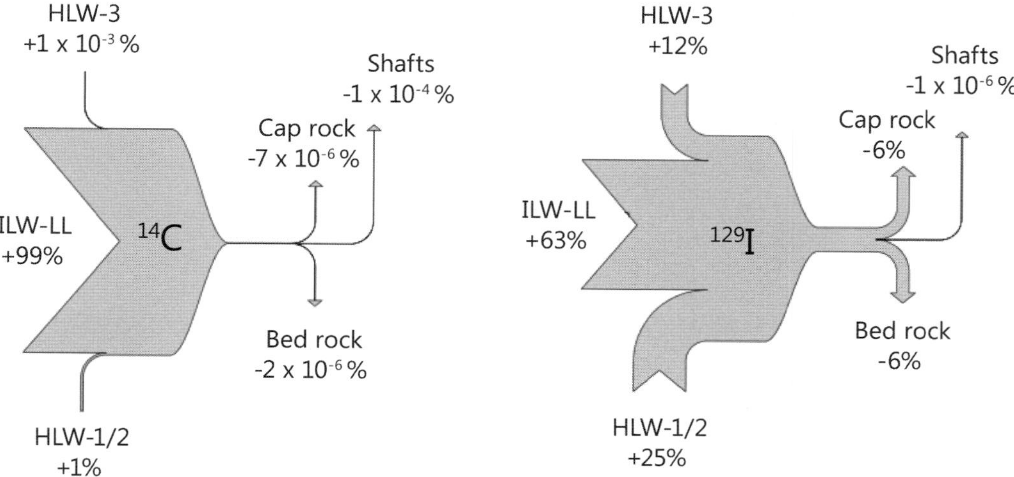

Fig. 4. Two-phase reference case Ru – flux chart of cumulative ^{14}C and ^{129}I model inflow and transfer to the vertical model boundaries after one million years.

Table 5. *Two-phase reference case Ru – inter-zonal radionuclide transfer distinguishing between fluxes via the drifts and fluxes via the host rock (maximum values in bold)*

Transfer path	RN	Cumulated mass after 1 Ma (mol)	Fraction of cumulated inter-zonal flux and total cumulated inter-zonal flux after 1 Ma (–)	Fraction of cumulated inter-zonal flux and total cumulated zonal RN release after 1 Ma (–)	Fraction of cumulated inter-zonal flux and total cumulated RN release after 1 Ma (–)	Dominating transfer process
HLW zone to central zone (via drifts)	^{14}C	1.3×10^{-3}	0.998	7.0×10^{-4}	5.0×10^{-4}	Advection in gas phase
	^{129}I	4.9×10^{-3}	0.004	$\ll 1 \times 10^{-5}$	4.0×10^{-3}	Advection in liquid phase
HLW zone to central zone (via host rock)	^{14}C	3.1×10^{-6}	0.002	$\ll 1 \times 10^{-5}$	$\ll 1 \times 10^{-5}$	Advection in gas phase in desaturated area around drifts
	$\mathbf{^{129}I}$	**1.1**	**0.996**	**0.001**	**0.995**	**Diffusion in liquid phase**
ILW-LL zone to central zone (via drifts)	$\mathbf{^{14}C}$	**2.3**	**0.985**	**0.012**	**0.987**	**Advection in gas phase**
	^{129}I	3.7×10^{-4}	0.61	$\ll 1 \times 10^{-5}$	3.0×10^{-4}	Advection in liquid phase
ILW-LL zone to central zone (via host rock)	^{14}C	3.1×10^{-2}	0.015	2.0×10^{-4}	0.013	Advection in gas phase in desaturated area around drifts
	^{129}I	2.4×10^{-4}	0.39	$\ll 1 \times 10^{-5}$	2.0×10^{-4}	Diffusion in liquid phase

diffusive fluxes of ^{14}C in the formation water are negligible. For the inter-zonal ^{129}I fluxes a contrary behaviour in terms of transfer path and transfer process occurs: the largest fluxes to the central zone, around 99.5%, originate from the HL waste zone and the dominating transfer process is diffusion in the formation water. Drift fluxes are of negligible order. The evolution of radionuclide

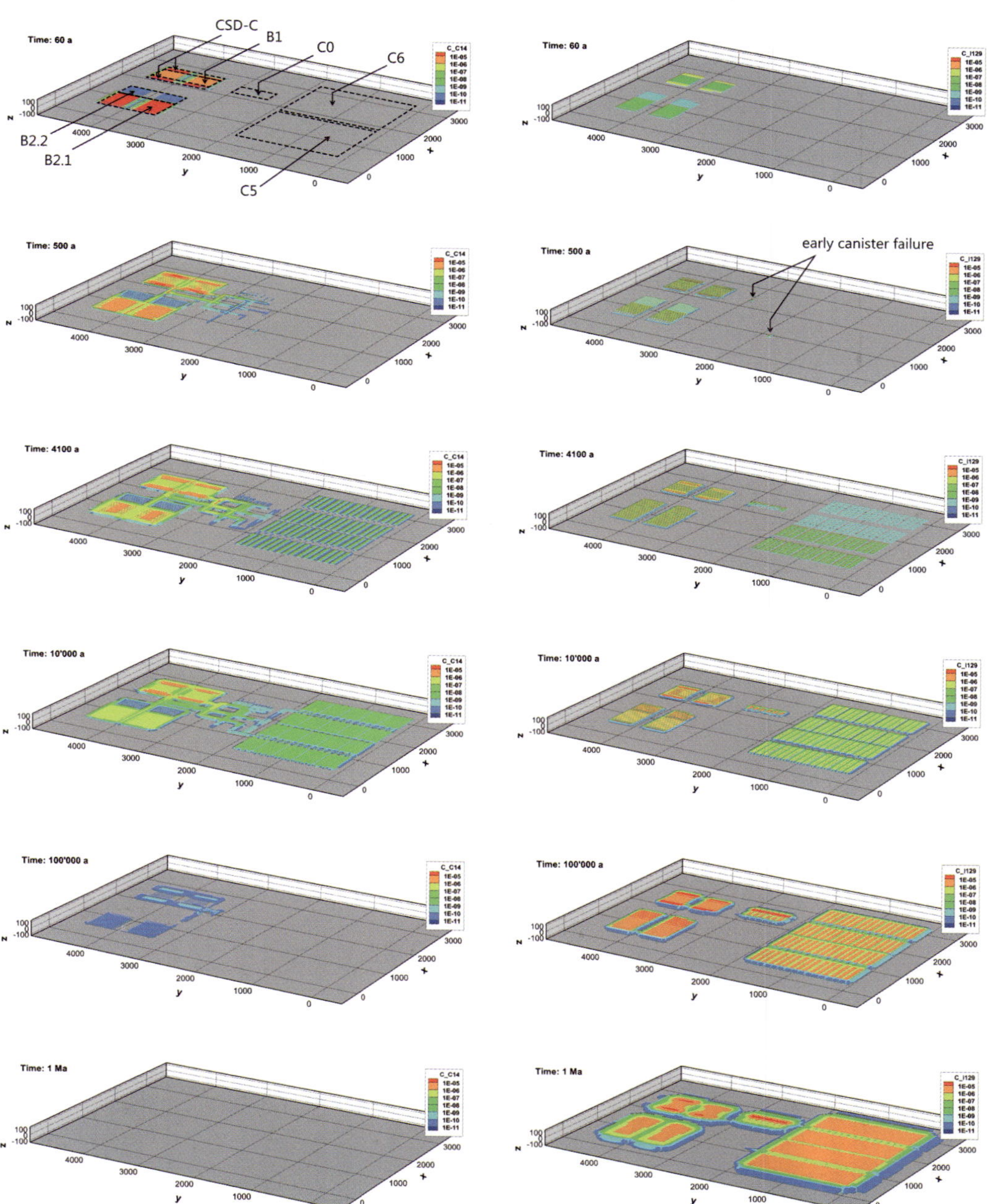

Fig. 5. Two-phase reference case Ru – distribution of radionuclide concentration (mol m^{-3}) for different simulation times; left column: ^{14}C, right column: ^{129}I; horizontal cut at emplacement level; concentrations below 1×10^{-11} mol m^{-3} are cut off.

concentrations of the two species inside the repository is illustrated in Figure 5.

Comparison of all simulation cases

Radionuclide transfer to the vertical model boundaries: ^{14}C. A comparison of the cumulated amounts of ^{14}C transported to the cap rock (either via the shafts or via the host rock) and to the bed rock between all simulation cases is given in Figure 6 (top) and compared there to the total amount released by the wastes (rightmost bar). The fractions of the underlying transport processes for each case are given in Figure 7 (top).

When gas generation is taken into account, ^{14}C is primarily transferred by advection in the gas phase – which evolves along the backfilled drifts and also enters the surrounding EDZs and undisturbed host rock – and reaches the cap rock (=top model boundary) via the shafts. The secondary transport path for ^{14}C is by diffusion and advection via the undisturbed host rock towards the cap rock and bed rock. In all two-phase simulation cases except case Rs (less gas generation) the transferred amount of ^{14}C via the host rock is by more than one order of magnitude lower than via the gaseous pathway along the drifts and shafts. When gas generation is not taken into account, only the second transport path is effective and much less ^{14}C leaves the model domain. The maximum value of around 0.5 mol occurs for two-phase sensitivity case S3u, which exhibits an increased permeability of the backfill and the EDZ, a lowered diffusion of hydrogen in the liquid phase, and faster gas generation. In

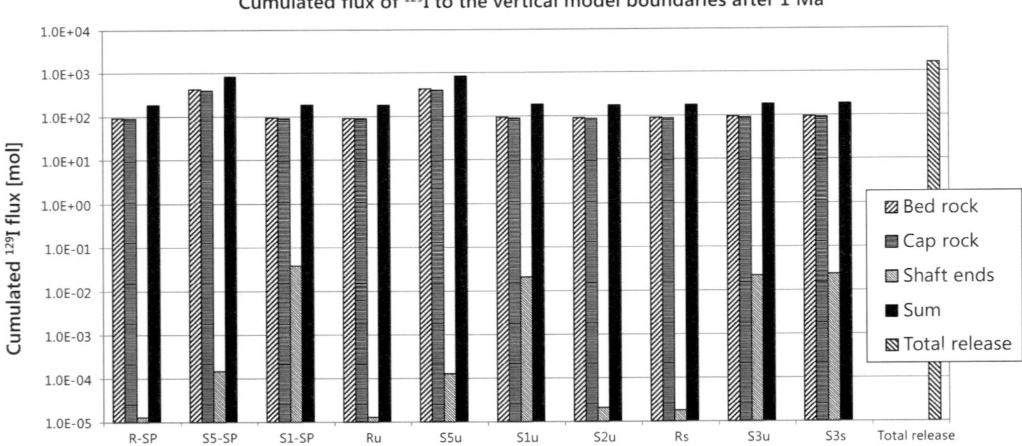

Fig. 6. Comparison of the cumulated radionuclide fluxes to the vertical model boundaries distinguishing between the transfer paths bed rock, cap rock (without shafts) and shafts (including their EDZs), top: ^{14}C, bottom: ^{129}I.

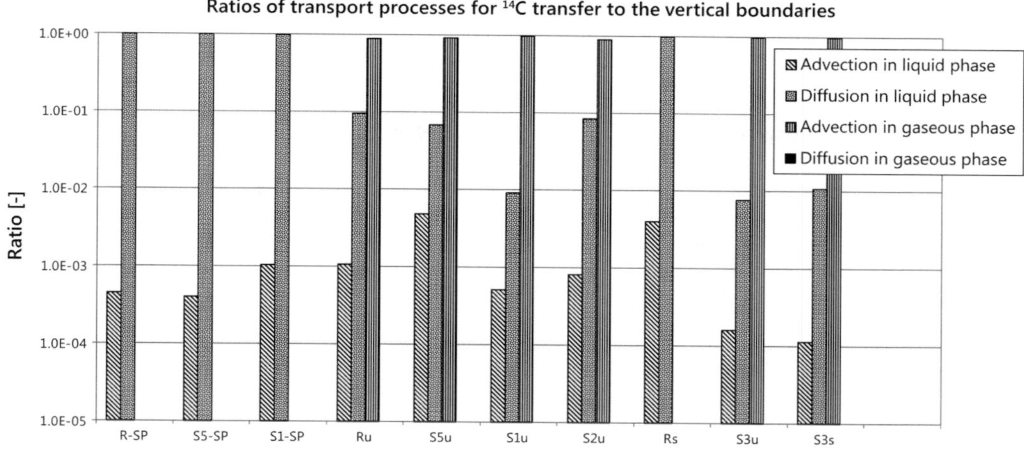

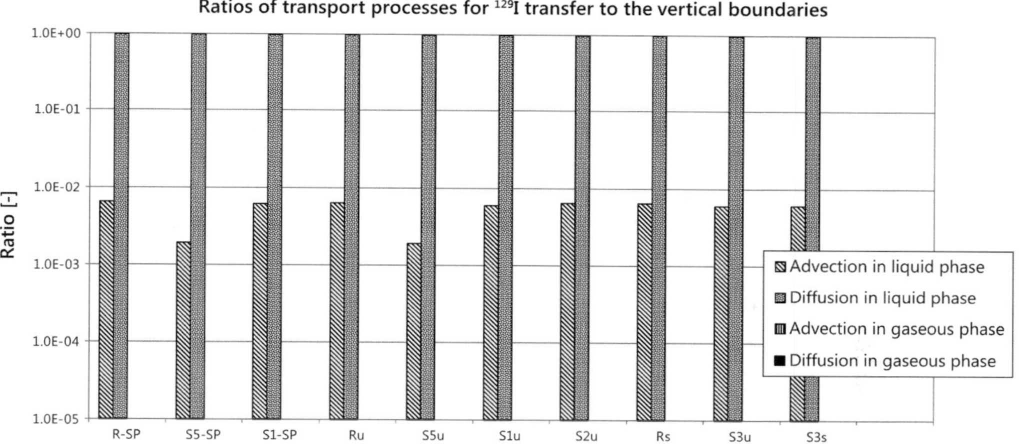

Fig. 7. Comparison of the proportions of the transport processes involved in the radionuclide transfer to the vertical model boundaries, top: ^{14}C, bottom: ^{129}I (see Fig. 6 for total fluxes).

the two-phase reference case Ru the corresponding value is around 1×10^{-4} mol; in the single-phase reference case R-SP it is around 5×10^{-9} mol. With respect to the totally released amount of ^{14}C, the transfer of ^{14}C to the vertical model boundaries is relatively small, with fractional values of 3×10^{-3} (S3u), 1×10^{-6} (Ru) and 3×10^{-11} (R-SP).

For the single-phase cases, the cumulative fluxes through the model boundaries are very small. That they are somewhat larger to the bed rock than to the cap rock is probably linked to a slightly coarser spatial discretization and thus larger numerical diffusion towards the bottom. Together with the short half-life of ^{14}C, an impact on cumulative fluxes can be observed, while it is negligible for ^{129}I.

In the two-phase cases the cumulative flux of ^{14}C to the cap rock is significantly larger than the flux to the bed rock (exception: case Rs). This behaviour can be explained by the fact that the flux to the cap rock is mainly by advection in a gas phase that evolves around the shafts and to a certain extent into their adjacent host rock. Only two-phase sensitivity case Rs (less gas generation) exhibits a similar behaviour to the single-phase cases, which is associated with the lack of a continuous gas phase at the shaft ends and in the host rock around the shafts.

The desaturation of the host rock by ventilation during the operational phase and gas generation in the two-phase cases leads also to a reduced vertical travel distance for ^{14}C by diffusion in the liquid phase to the cap and bed rock, which further increases the transfer of ^{14}C in comparison to the single-phase cases. That is, the first few metres of the vertical transfer are faster because of advection (and diffusion) in the gas phase.

The larger difference between the cumulative ^{14}C flux of the single-phase cases R-SP and S5-SP in comparison to the difference between two-phase cases Ru and S5u is attributed to the different parameterization of the ^{14}C diffusion. In the single-phase case the diffusion increase is effective for the liquid phase, whereas in the two-phase case the increase is effective for the gaseous phase only.

In conclusion, the cumulative flux of ^{14}C to the cap and bed rock is strongly underestimated when gas generation is not taken into account. However, even in the 'worst case' the amount of ^{14}C released totals to only 0.5 mol in 1 million years.

Radionuclide transfer to the vertical model boundaries: ^{129}I. The transfer of ^{129}I to the vertical model boundaries mainly takes place via diffusion in the liquid phase. The differences between the single-phase and two-phase cases are negligible. In all cases except sensitivity cases S5u and S5-SP around 200 mol of ^{129}I are transported to the vertical model boundaries during the simulation time of 1 million years. For the single-phase and two-phase sensitivity cases with ^{129}I diffusion increased by a factor of 3 (cases S5u and S5-SP) around 800 mol of ^{129}I are transferred to the vertical model boundaries (Fig. 6, bottom). The corresponding fractions of ^{129}I leaving the model domain with respect to the total ^{129}I release are 12% for all cases with the exception of cases S5u and S5-SP (increased radionuclide diffusion), where 52% are transferred. The fractions of the underlying transport processes for each case are given in Figure 7 (bottom).

Role of the different shafts. A comparison of the cumulated ^{14}C fluxes via the different shafts between the different simulation cases is given in Figure 8. When gas generation is considered, the transfer of ^{14}C to the cap rock is mainly taking place along and in the vicinity of the backfilled drifts and shafts. The largest fluxes occur via shaft B, which is situated within zone ILW-LL, in which around 99% of the total ^{14}C is released. All the other shafts and the ramp are situated in the central zone, that is, further away from zone ILW-LL, and exhibit smaller cumulated fluxes of ^{14}C to the cap rock. The dominance of shaft B is very significant in the cases with the reference values for permeability (cases Ru, S5u and S2u) – here the cumulated ^{14}C fluxes are at least 4 orders of magnitude higher than in the shafts situated in the central zone. With higher permeabilities of the materials along the drifts – cases S1u, S3u, S3s – the shafts in the central zone – A, C, D, E – become more important and their cumulated fluxes of ^{14}C are only around one order of magnitude lower in comparison to those of shaft B. The only exception is case Rs (waste type-specific and less gas generation in zone ILW-LL) with comparatively smaller ^{14}C fluxes via shaft B than via the shafts in the central zone.

The least cumulative fluxes occur via shaft A, which is obvious as this pathway exhibits the longest travel distance for the gas phase and is hence likely to resaturate before the arrival of a gas phase from the waste zones. Only in two-phase sensitivity case S3u, where gas is generated relatively fast and gas fluxes to the central zone are significantly increased in comparison to the other cases, does the ^{14}C flux via shaft A reach a similar order of

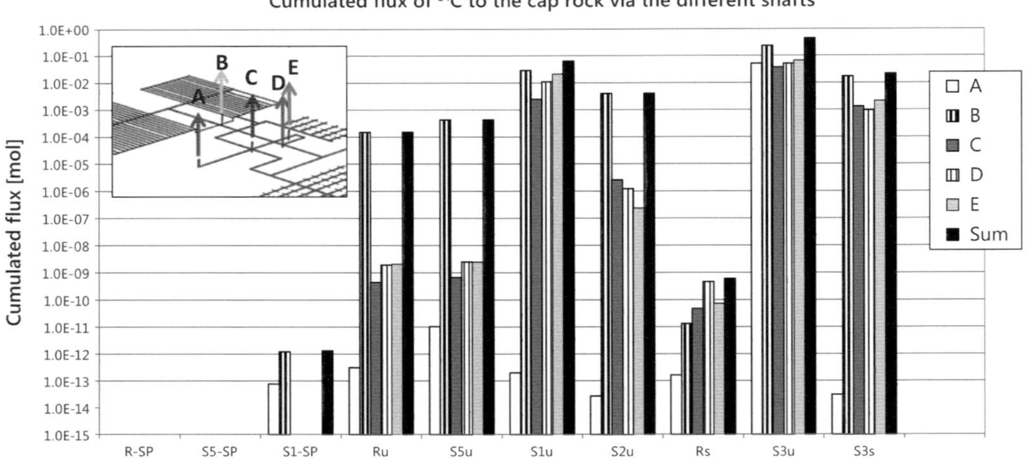

Fig. 8. Comparison of the cumulated ^{14}C fluxes at the different shaft ends.

magnitude in comparison to the other shafts. The ^{129}I fluxes through the shafts are of negligible magnitudes and hence not further discussed here.

Summary and conclusion

The single-phase and two-phase simulations of the study presented here produced a consistent dataset characterizing the repository performance with respect to water, hydrogen, ^{14}C and ^{129}I transport. The detailed analysis of the thermo-hydraulic behaviour of two potential repository layouts under various parameter sensitivity studies has been discussed in part 1 of the present work (Brommundt *et al.* 2014), while the behaviour with respect to radionuclide transport processes, focusing on one layout, has been the goal of the present part 2.

A strong dependency of the ^{14}C transfer on the considered processes and actions has been observed and must be expected for any highly volatile radionuclide. The most important aspect here is the consideration of gas generation, which leads to largely increased transfer rates of ^{14}C along the backfilled drifts and shafts. The ruling ^{14}C transport process is advection in the gas phase – hence, all factors influencing the gas phase with respect to its migration in space and time and its continuity are of great importance for the transport of ^{14}C. Favouring conditions for the transfer of ^{14}C to the cap rock are a faster gas generation (case S2u) and an increased permeability of the backfilled drifts, shafts and their EDZs (case S1u). The combination of these two leads to the largest transfer of ^{14}C to the cap rock (S3u). Moreover, this case showed significantly changed system behaviour with considerable gas and ^{14}C fluxes from the intermediate-level long-lived waste (ILW-LL) zone to the high-level waste (HLW) zone and a much better connection of the gas phase within the shafts in the central zone. In contrast, reduced gas generation rates (case Rs) showed a largely decreased ^{14}C transfer owing to a less distributed and less continuous gas phase.

Note, however, that in all the cases cumulative ^{14}C fluxes out of the model domain are very small and on the order of a maximum of 1/1000 of the amount released from the wastes. This is mostly due to the short half-life of ^{14}C with respect to the characteristic transport times.

The transfer of ^{129}I to the cap rock and bed rock is almost independent of the gas generation and non-sensitive to the performed parameter variations apart from that with increased ^{129}I diffusion in the liquid phase. The most important transfer process is vertical diffusion in the liquid phase to the cap rock and bed rock. Fluxes along the backfilled drifts and shafts are of comparatively small importance.

The performed sensitivity simulations identified important factors with respect to radionuclide transport such as, for example, the amount, distribution, and release regime of the hydrogen generation, the diffusivity in the liquid phase and the sorption capability (especially that of ^{14}C onto the concrete). The results also indicate that further factors are likely to considerably influence the radionuclide transfer such as the integrity of one or more main seals, the initial saturation of the backfill or the location of the drift in the intermediate-level long-lived waste zone.

The present study demonstrates the general feasibility of simulating the coupled thermo-hydraulic two-phase flow and radionuclide transfer processes on the scale of the entire repository and the embedding host rock between the cap rock (Oxfordian) and the bed rock (Dogger). The accomplishment of the simulations revealed that hardware resources are not necessarily the most important factor for successful model simulations. By choosing a skilful and suitable model conceptualization, including adequate meshing techniques and efficient data handling, the novel modelling concept of this study could be used and extended for comprehensive parameter studies.

This work was funded by Andra, and the modelling presented is based on data provided by Andra. Valuable input was provided by Guillaume Pépin and Jean Croisé. Thank you to Keni Zhang who kindly performed test calculations on the Chinese supercomputer Tianhe-1A.

References

ANDRA 2006. *Dossier 2005 Argile – synthesis – evaluation of the feasibility of a geological repository in an argillaceous formation.* Agence nationale pour la gestion des déchets radioactifs (Andra) Report Series.

ANDRA 2009*c*. *National inventory of radioactive materials and waste.* Synthesis Report, Agence nationale pour la gestion des déchets radioactifs (Andra), document DCAI-CO-09-0051.

Brommundt, J., Kaempfer, T. U., Enssle, C. P., Mayer, G. & Wendling, J. 2014. Full scale 3D modelling of a nuclear waste repository in the Callovo-Oxfordian clay – Part 1: thermo-hydraulic two-phase transport of water and hydrogen. *In*: Norris, S., Bruno, J., Cathelineau, M., Delage, P., Fairhurst, C., Gaucher, E. C., Höhn, E. H., Kalinichev, A., Lalieux, P. & Sellin, P. (eds) *Clays in Natural and Engineered Barriers for Radioactive Waste Confinement.* Geological Society, London, Special Publications, **400**. First published online May 7, 2014, http://dx.doi.org/10.1144/SP400.34

Pruess, K., Oldenburg, C. & Moridis, G. 1999. TOUGH2 User's Guide, Version 2.0. Report LBNL-43134, Lawrence Berkeley National Laboratory, Berkeley, CA.

Analysis of the long-term hydraulic-gas transient in the central zone of a deep clay repository

LUC-VINCENT BÉNET[1]*, CATALIN TULITA[1], ANTOINE PASTEAU[2] & JACQUES WENDLING[2]

[1]*SOCOTEC SA/Projets Industriels/AME, 1 Avenue du Parc, 78640 Montigny-le-Bretonneux, France*

[2]*Andra, 92298 Châtenay-Malabry Cedex, France*

**Corresponding author (e-mail: luc-vincent.benet@socotec.com)*

Abstract: In the Andra repository concept, a large amount of hydrogen gas will be generated through radiolysis of water and corrosion of metal parts from the infrastructure and the packages containing radioactive waste. The fate of the gas could have a significant impact on the rate of resaturation of the porous media, and needs to be accounted for in order to make accurate predictions of the long-term evolution of the repository. The central zone of the repository consists of a 4.5-km-long network of galleries which connect the storage zones with the surface through three shafts. These shafts, which will be backfilled and sealed after closure, will have a key role in gas and water exchange between the repository and the overlying aquifer. The aim of this work was to describe the evolution in time of both the water and gas flows in the central zone up to the end of the hydraulic-gas transient. The issue was addressed by means of numerical simulations. The simulations produced local predictions of flow rates and pressures, water saturation and amounts of dissolved gas. The results were analysed to determine how water and gas flows combine in the central zone, when the saturation of the seal will be complete, when the free gas will reach the overlying aquifer, and when the saturation of the central zone will be complete.

Since the end of the 1990s, the French agency for the management of nuclear wastes, Andra, has been studying the feasibility of locating a repository in the Callovo-Oxfordian argillites on the borders of the Meuse and Haute-Marne departments at a depth of about 500 m. Andra is concerned to understand the consequences of the disturbances to the hydrogeological system around the repository due to excavation, ventilation (heat and vapour exchange with air), and generation of both heat and gas. Gas generation is an important process because of (i) the large quantity generated (hydrogen will be produced mainly by the corrosion of steel in the packages and in reinforced concrete); (ii) the long duration (from the beginning of the operational phase up to several tens of thousands of years post-closure); (iii) the large area (all the regions of the repository, including the shafts, may be affected); and (iv) the effects on the water flow (the build-up of the gas pressure may delay the complete resaturation of the repository).

The repository has been designed with five storages zones dedicated to different types of radioactive wastes: three of them to High Level-Long Lived Wastes (HL-LLW) and two of them to Intermediate Level-Long Lived Wastes (IL-LLW) (Fig. 1). During the operational phase, access to the waste storage zones will be through a 4.5-km-long network of galleries, the so-called central zone. This network will be connected to the surface via three shafts.

During a century of operations, the evaporation of pore water due to ventilation dries the gallery walls; the concrete lining of the tunnels and the porous rocks become desaturated. Afterwards, the repository is completely closed; the galleries and shafts are backfilled and sealed. However, the backfilled central zone remains a potential pathway for gas and water to flow between the overlying Oxfordian aquifer and the storage zones. Because of their high permeability, porous media such as the backfill material and the Excavation Damaged Zones (EDZ) provide a potential preferential pathway for gas and water along the galleries and shafts, crossing the less permeable host rock.

When assessing the safety of the repository, the flow rate of water along the galleries and shafts will have a key role controlling the migration of dissolved radioactive species towards the overlying aquifers. In addition, advection with the gas flow and diffusion through the gaseous phase will have key roles in the migration of gaseous radioactive species (like those containing carbon-14).

The study focuses on the description of water and gas exchange through the shafts of the central zone between the repository and the overlying

From: Norris, S., Bruno, J., Cathelineau, M., Delage, P., Fairhurst, C., Gaucher, E. C., Höhn, E. H., Kalinichev, A., Lalieux, P. & Sellin, P. (eds) 2014. *Clays in Natural and Engineered Barriers for Radioactive Waste Confinement*. Geological Society, London, Special Publications, **400**, 483–496.
First published online April 9, 2014, http://dx.doi.org/10.1144/SP400.19

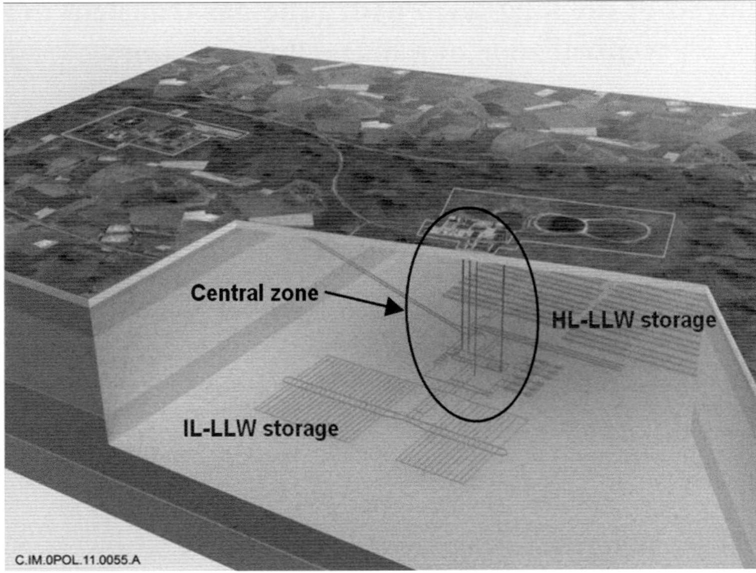

Fig. 1. Schematic view of the repository: architecture taken from Andra's 2009 design.

aquifer. Numerical simulations have been performed to determine how and when water from the overlying aquifer will enter the central zone, and how and when the gas will reach that aquifer from the central zone. Consequently, the shafts between the central zone in the host layer have been modelled with details including the concrete lining, the 45-m-long concrete plug at the bottom, and the 35-m-long bentonite plug above it. In contrast, the modelling of the gallery network has been simplified to focus on those parts of the network that are directly connected to the shafts, and the overlying aquifer has been partially modelled, because neither the details of the flow distribution through the gallery network nor the fate of the gas in the overlying aquifer are sought.

Geometric and hydrogeological modelling

The central zone is a complex network of galleries which cross each other about 30 times. This network of galleries is drastically simplified in the model geometry to focus on a fine description of the main hydraulic components between the Oxfordian layer and the repository. Only one third of the central zone is represented, using a simplified architecture of one shaft and four galleries which intersect each other. The length of the four galleries is chosen to be equal to one third of the cumulative length of all the galleries of the central zone. The geometry meshed can be further reduced by an eighth (see Fig. 2) with considering three of the four vertical planes of symmetry of the crossing galleries (i.e. the geometrical model includes an eighth of a shaft and half of a 377-m-long gallery).

In the lower part of the host rock, the EDZ around the gallery and the shaft is modelled by two different porous media: (i) the rupture zone EDZ-R, the thickness of which is half an excavation radius; and (ii) the micro-crack zone EDZ-E, the thickness of which is one third of an excavation radius.

In the upper part of the layer (which is 35 m thick) where the shaft seal is emplaced, the rock is less fragile due to higher concentration of carbonate. Consequently, the EDZ is modelled by a single micro-crack zone; its thickness is one tenth of an excavation radius.

The model geometry of both the underlying and overlying aquifers is simplified depending on their respective potential to exchange water or gas with the repository. On the one hand, the repository is separated from the underlying Dogger aquifer by a 75-m-thick argillite layer; on the other hand, the overlying Oxfordian aquifer is connected with the repository by the shafts which are sealed by a 35-m-high bentonite plug. Thus, the Dogger aquifer is modelled by fixing the gas and water pressures at the bottom boundary of the host rock, because no gas flow and very low exchanges of water and gas with the repository are expected.

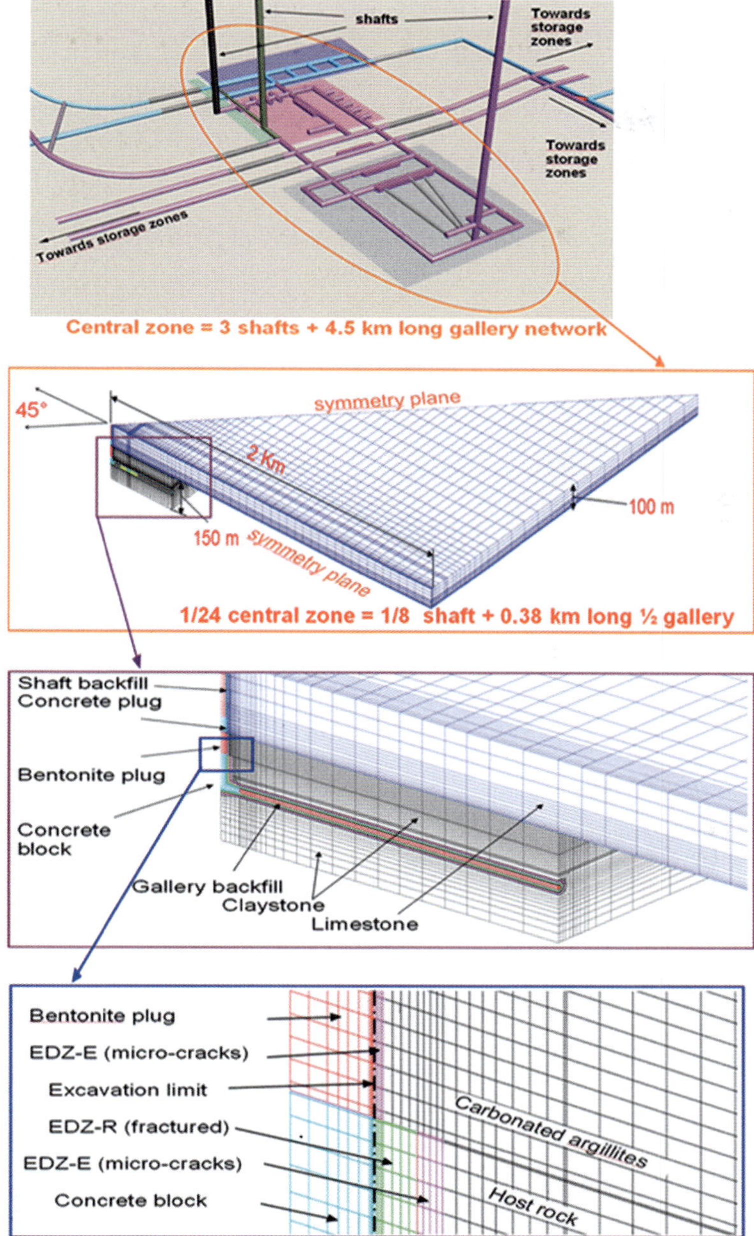

Fig. 2. Schema of the geometric modelling process from the central zone architecture to the meshing of the calculation domain taken from Andra's 2009 design.

In contrast to the Dogger aquifer, the Oxfordian aquifer is geometrically modelled to set hydraulic Dirichlet conditions on the upper boundaries far from the shaft seal, because significant gas and water exchanges between the aquifer and the repository are expected through the shafts. Consequently, the gas and water pressures are fixed on the edge of the Oxfordian mesh, on boundaries located 2 km away from the shafts, and a condition of no flux of gas or water is set on the upper boundary of the Oxfordian mesh (see sections 'During the Exploitation Period' and 'During the post-closure period' below for more details on boundary conditions) to take into account the overlying semi-permeable layer (Kimmeridgian), and only one third of the thickness of the aquifer is meshed to be consistent with the modelling of the third of the central zone.

It should be noted that the large horizontal dimension of the modelled Oxfordian aquifer model is useful to correctly simulate the effect of the Oxfordian-specific storability on the water flow entering the shaft. However, the calculations of gas fluxes through the aquifer are not expected to be correct, especially because the natural hydrogeological flow (i.e. the horizontal Darcy velocity can reach 5×10^{-2} m a^{-1} in some parts of the aquifer above the storage area) and its capability to dilute the gas dissolved in the water flowing through the Oxfordian aquifer have not been taken into account in the model. Therefore results about the gas fate through the Oxfordian aquifer are not discussed in this paper.

Finally, the mesh (about 41 100 elements) extends from the bottom of the host rock to 100 m above it (see Fig. 2 for details of the mesh).

Physical modelling

Basic definitions and assumptions

The heat released by the radioactive waste will not affect the temperature in the central zone because of the large distance from storages zones, which is about one kilometre in the 2009 design. Moreover, the geothermal gradient is neglected. Consequently, the physical properties are considered at the *in situ* temperature of 22 °C (i.e. at the repository level) in the entire model.

The physical model of the water and gas flows is based on generalized Darcy equations (for unsaturated flow). The values for the Darcy velocity for gas and water depend on the gradient of the dynamic gas pressure H_g and the dynamic water pressure H_w (or Darcy pressure) respectively.

The water pressure P_w is composed of two terms, the Darcy pressure H_w and the hydrostatic pressure P_{stat} due to the weight of the water. In the gas, dynamic and total pressures are considered equal (the weight of the gas is neglected). Note that before excavation (i.e. *in situ* conditions) the water pressure considered in the host rock at the repository level ($z = 0$ m) is 5 MPa, and the initial Darcy pressure is defined as a constant in the entire model (i.e. no water flux in the calculation domain):

$$P_w = H_w + P_{stat}$$

and within the calculation domain before excavation, $H_w^z = H_w^0 = P_w^0 = 5$ MPa.

The capillary pressure P_c, which is negative in unsaturated media and equal to zero in saturated media, is defined as the difference between the water pressure P_w and the gas pressure P_g:

$$P_c = P_w - P_g$$

Modelling of water behaviour in unsaturated media

The water retention curve depends on the capillary pressure. The water retention curve used has been proposed by van Genuchten (1980).

$$S_w = \left[1 + \left(\frac{-P_c}{P_r}\right)^n\right]^{-m}$$

where S_w is the water saturation, P_r is the reference pressure, n is the shape parameter and $m = 1 - 1/n$. The two parameters P_r and n have specific values related to the porous material. Note that the reference pressure of the van Genuchten model is of the same order of magnitude as the thickness of the more water-saturated part of the capillary fringe of a free aquifer, when expressed in terms of metres of water column. Thus, the water saturation is about 50% to 60% when considering the capillary pressure equal to minus the reference pressure for the porous materials used in this study (see the capillary parameters in Tables 1 & 2):

$$P_c = -P_r \Rightarrow 54.5\% \leq S_w \leq 58\%$$

The relative permeability of the water k_r^w depends on the water saturation. The formulation used has been established by combining the van Genuchten retention curve (van Genuchten 1980) with the generalized model by Mualem (1976).

$$k_r^w = \sqrt{S_w}\,[1 - (1 - S_w^{1/m})^m]^2$$

Furthermore, the water-flow model takes into account the compressibility of both the interstitial water and the solid matrix through the specific storativity parameter E, which depends on the porous material:

$$E = \omega \cdot \frac{d\rho_w}{\rho_w dP_w} + \rho_w \cdot \frac{d\omega}{\omega \cdot dP_w}$$

where ω is the porosity and ρ_w is the water density.

The water mass balance also takes account of vapour diffusive fluxes in the gas phase (Fick's law). The effective diffusivity in the gaseous fraction of a porous medium is related to the water

Table 1. *Reference values of physical property parameters in engineered materials*

Parameters	Symbols	Bentonite	Backfill	Concrete
Porosity	ω (%)	35	41.7	15
Specific storativity	E (m^{-1})	3.5×10^{-6}	10^{-5}	10^{-6}
Intrinsic permeability	k_i (m^2)	10^{-18}	6×10^{-17}	10^{-17}
Shape parameter (retention law)	n (−)	1.6	1.6	1.4
Reference pressure (retention law)	P_r (m)	1800	10	200
Initial water saturation	S_w (%)	70	34	50
Tortuosity	τ (−)	2	1	1

saturation and to the porous material according to the following formulation. For the vapour:

$$D^g_{\text{vap}} = (1 - S_w)\left(\frac{\omega}{\tau^2}\right)D^g_{\text{vap},0}$$

Where τ is the tortuosity of the porous medium, and $D^g_{\text{vap},0}$ is the diffusivity in open space (2×10^{-5} m^2 s^{-1}).

The vapour is supposed to be saturated, and its partial pressure depends on capillary pressure $P_c(S_w)$, according to the Kelvin equation:

$$P^g_{\text{vap}}(T, S_w) = \exp\left(\frac{M_w P_c(S_w)}{\rho_w RT}\right)P^{\text{sat}}_{\text{vap}}(T)$$

where M_w is the molar mass of water, R is the ideal gas constant, T is the temperature (in Kelvin in the formula), and $P^{\text{sat}}_{\text{vap}}$ is the vapour pressure at the state of saturation and without capillary forces.

Modelling of gas behaviour in unsaturated media

The gaseous phase is described as pure hydrogen gas with ideal gas behaviour. In fact, in the model, the vapour flux enters into the water mass balance, but is neglected in the gas balance on the basis that the vapour pressure can be neglected at the *in situ* temperature of 22 °C (i.e. under 26 mbar; Grigull 1982) and when it is compared to the atmospheric pressure during the exploitation phase, or to the high hydrogen pressure expected after the repository closure.

Furthermore, the air trapped in the porous media when the repository is closed is modelled as hydrogen at atmospheric conditions. Modelling of the residual air is therefore simplified because the build-up of hydrogen pressure in the storage zones is expected to reach several tenths of an atmosphere only after thousands of years.

As for the liquid phase, the relative permeability of the gas k^g_r is established by combining the van Genuchten retention curve (van Genuchten 1980) with the generalized model by Mualem (Mualem 1976):

$$k^g_r = \sqrt{1 - S_w}\left[1 - S_w^{1/m}\right]^{2m}$$

In unsaturated media, gaseous hydrogen is supposed to be in equilibrium with the concentration of hydrogen in the aqueous phase. For a non-condensable gas i, aqueous molar concentration n^w_i is related to partial pressure according to Henry's law:

$$n^w_i = P^g_i \cdot H_i$$

where H_i is the Henry parameter.

Table 2. *Reference values of physical property parameters in host rocks*

Parameters	Symbols	Clay	EDZ-R	EDZ-E
Porosity	ω (%)	18	18	18
Specific storativity	E (m^{-1})	2×10^{-6}	2×10^{-6}	2×10^{-6}
Horizontal intrinsic permeability	k^{xy}_i (m^2)	2.3×10^{-20}	5×10^{-16}	5×10^{-18}
Vertical intrinsic Permeability	k^z_i (m^2)	9.4×10^{-21}	5×10^{-16}	5×10^{-18}
Shape parameter (retention law)	n (−)	1.49	1.5	1.5
Reference pressure (retention law)	P_r (m)	1500	200	800
Initial water saturation	S_w (%)	100	100	100
Tortuosity	τ (−)	2	2	2
Thickness	e (m)	150	$0.5 \times R_{\text{excav}}$	$0.3 \times R_{\text{excav}}$; $0.1 \times R_{\text{excav}}$

The diffusion of dissolved hydrogen in water is modelled using Fick's law. The effective diffusivity of dissolved hydrogen in the porous media $D^w_{H_2}$ is related to the water saturation and the material properties:

$$D^w_{H_2} = S_w\left(\frac{\omega}{\tau^2}\right)D^w_{H_2,0}$$

where $D^w_{H_2,0}$ is the diffusivity of the dissolved hydrogen in free water (5.96×10^{-9} m^2 s^{-1}).

Material properties of porous media

The reference values of physical parameters are gathered in Tables 1 and 2 for engineered and natural materials respectively. The value of bentonite permeability of 10^{-18} m s^{-1} is specified by Andra.

It should be noted that the capillarity potential of the backfill material is relatively small (i.e. it has a small reference pressure). The backfill is considered to be mainly composed of clay rock waste from the excavation of the galleries. This material is characterized by a large porosity and very low water potential. The van Genuchten reference pressure value is considered to be 0.1 MPa. This means that if a gaseous phase appears in the backfill of the 280-m-high part of the shaft which will cross the Oxfordian aquifer, then the water and gas phases will be approximately stratified to within a 10-m-high capillary fringe.

Transient modelling

During the exploitation period

We have considered that shaft and gallery excavation, formation of the EDZ, and the building of infrastructure are all completed instantaneously at the initial calculation time. At closure of the repository, 100 years later, the emplacement of bentonite seals, supporting concrete blocks, and backfill in galleries and shafts is also considered to be instantaneous.

During the exploitation period, the evaporative effect of the ventilation air flow is modelled by fixing water and gas pressures within the mesh cells that will represent the future seals, concrete plugs and backfill to simulate the capillary pressure generated by evaporation on walls. The set capillary pressure corresponds to an assumed relative humidity (RH) of 80%, and is calculated using the Kelvin equation. Furthermore, in the overlying Oxfordian layer, a seepage boundary condition is implemented, recognizing that water will be supplied readily to the concrete wall of the shaft from the surrounding high permeability limestone. The seepage boundary condition consists of fixing the gas and water pressures at atmospheric pressure on the concrete wall, which corresponds to a water saturation condition (Fig. 3).

During the post-closure period

At the time of repository closure, the initial water saturation values in the bentonite plug, the concrete plugs and the backfill are set in the model. The gas and water pressures become free parameters. The flow of gas into the central zone is modelled by specifying a mass flux of hydrogen J_G on the backfill boundary at the end of the gallery. The no-water flux condition is maintained on this boundary (Fig. 4).

Regarding the hydrogen gas arriving from the storages zones, two scenarios have been considered:

(1) *Case R1*: a low rate of gas arriving from the storage zones is the more likely scenario. In the reference model for Andra's 2009 concept, the greatest part of the hydrogen generated is supposed to dissolve in the pore water along galleries and diffuse through the surrounding clay rock before reaching the central zone.

(2) *Case R2*: the scenario of a high rate of gas arriving from the storage zones assumes that the host rock has a low capacity to store dissolved gas, either because the storage zones will be more compact than in Andra's 2009 concept, or because the diffusivity and/or the solubility of hydrogen in the water of the clay rock are/is lower than the reference values considered.

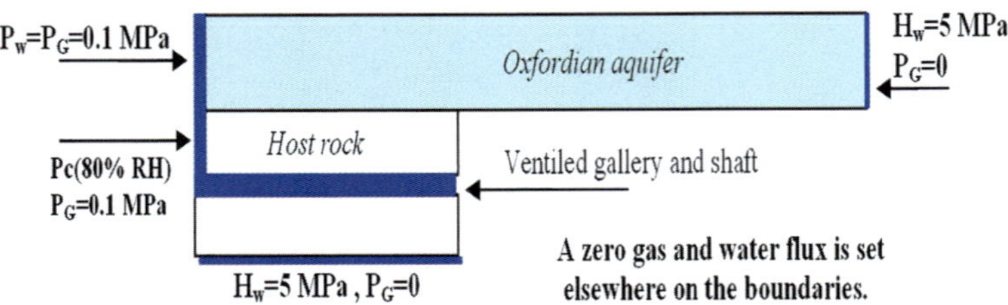

Fig. 3. Boundary conditions for the water and the gas during the exploitation period.

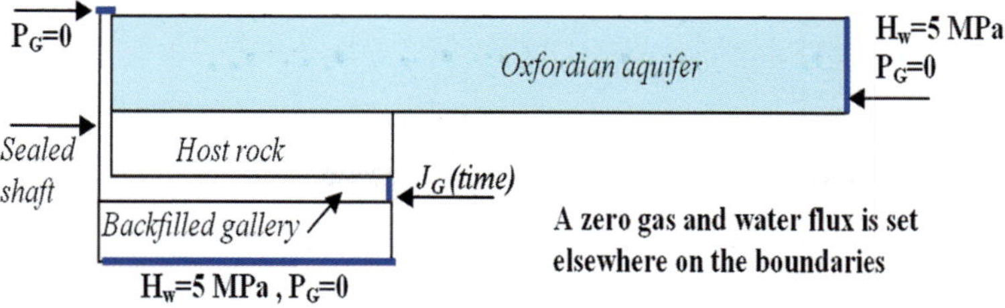

Fig. 4. Boundary conditions for the water and the gas during the post-closure period.

For both scenarios, the evolution in time of the incoming hydrogen flow is specified in Table 3. The incoming gas flow lasts 1400 years for the first scenario (case R1) and 150 000 years for the second one (case R2).

Numerical approach

The water and gas balance equations are discretized using a linear finite element method. The two resulting equations are solved to calculate water and gas pressures on nodes. The gas and water mass balances are coupled via the capillary pressure using an iterative process. The convergence of the algorithm is speeded up by using an Aitken's delta-squared process when it becomes linear (Boursier 2005).

The mesh work, physical model and solver algorithm have been set up with the GIBIANE language. The numerical problem has been solved with the Cast3M software (Cast3M).

Results and discussion

Low gas input scenario

In this first calculation case, named R1, we have considered the arrival of 1 kg a^{-1} of hydrogen over a period of 1400 years. Therefore the total quantity of hydrogen entering the central zone (14 000 m^3 STP) remains far smaller than the initial quantity of air trapped in the unsaturated media at the closure of the repository (about 130 000 m^3 STP).

In this scenario, most of the hydrogen generated in the repository is expected to dissolve and stay in the Callovo-Oxfordian argillites far away from the central zone, limiting the free gas escape from the repository through the central zone shafts. The gas phase remains trapped below the shaft seals, either in the engineered media (concrete, backfill) or in the host rock. The hydrogen is completely dissolved and the central zone fully resaturated after 4500 years.

Figure 5 shows the locations of seven observation points (from S1 to S7) and six surfaces (from F1 to F6) over which the calculated fluxes are integrated as black lines, the arrows showing the positive direction for each surface. The gas and water flow rate crossing the EDZ-E along the shaft or the gallery or towards the safe argillites are not shown because they are not significant.

After closure of the repository, water from the Oxfordian layer flows into the shaft backfill and the EDZ surrounding the shaft seal in the Callovo-Oxfordian argillites. It flows downwards into the EDZ at a high rate (110 m^3 a^{-1} just after closure) under the combined effects of gravity and the very high capillary pressures in the seal (see the curve F6 in Fig. 6). In contrast, the water from the gallery flows slightly upwards into the seal through the EDZ-R because of the high capillary potential of bentonite (see the reference pressure parameter of the retention law in Table 1), before reversing once the gravity overwhelms the draining effect of the capillary force due to the progressive resaturation of the seal (see the F3 curve in Fig. 6).

After 100 years, when resaturation of the bentonite plug is complete (see the curve S9 in Fig. 6), the water flow stabilizes at about 20 m^3 a^{-1} through the EDZ and 8 m^3 a^{-1} through the

Table 3. *Evolution in time of the incoming hydrogen flow (kg a^{-1}) from the storages zones into the central zone*

Period (years)	150–300	300–600	600–2000	2000–3000	3000–4000	4000–150 000
Case R1	0	0	1	0	0	0
Case R2	80	130	130	130	80	15

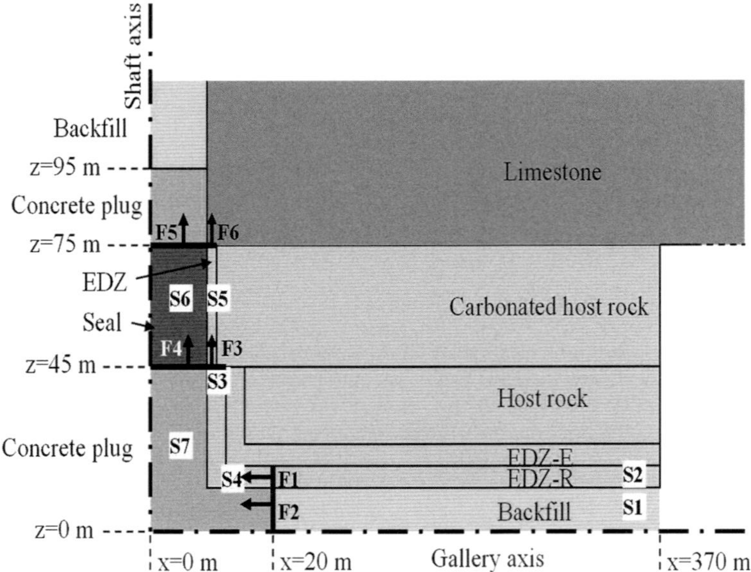

Fig. 5. Schema showing the observation points (S1–S7) and the surfaces (F1–F6) used to calculate the mean flow rates; the labels that identify the various surfaces and points are used in the subsequent plots.

bentonite (see the curves F6 and F5 in Fig. 6). This water invades the concrete block located at the bottom of the shaft, via both the top of the concrete block ($4\ m^3\ a^{-1}$) and the surrounding fractured EDZ ($24\ m^3\ a^{-1}$) (see the curves F4 and F3 in Fig. 6).

Then the amount of water entering the Callovo-Oxfordian argillites from the Oxfordian aquifer decreases smoothly, from $28\ m^3\ a^{-1}$ at 300 years to $23\ m^3\ a^{-1}$ at 4500 years, just before disappearing when complete resaturation occurs. At that time, the potential of Darcy pressure through the host rock between the Oxfordian aquifer and the gallery reaches 3.7 MPa.

It should be noted that water from the gallery backfill also contributes to resaturation of both the gallery EDZ and the bottom concrete block during the first 600 years, because of its very low capillary potential (see Table 1) compared to the other materials (see in Fig. 6 the water flowing from the gallery to the shaft on curves F1 and F2).

Up to 2700 years, the gas is pushed through the fractured EDZ towards the galleries by the water flowing from the Oxfordian aquifer. The gas is

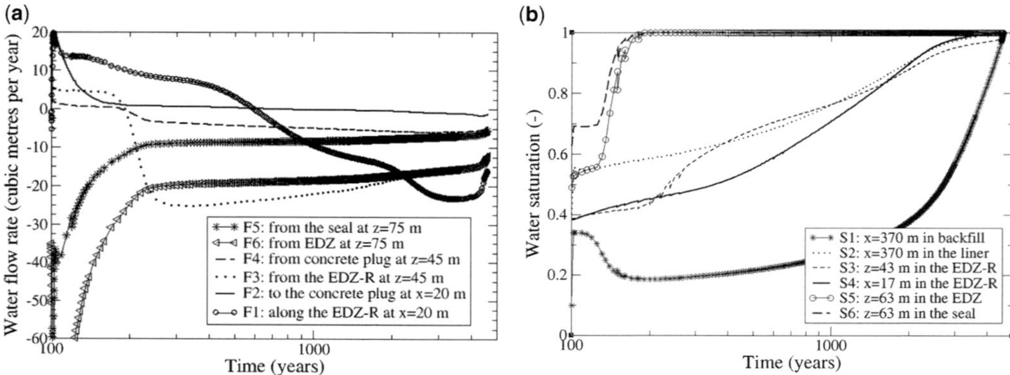

Fig. 6. Evolution in time of (**a**) the water flow per shaft and (**b**) the water saturation for the low incoming gas scenario (see Fig. 5 for more details on the labels in the legend).

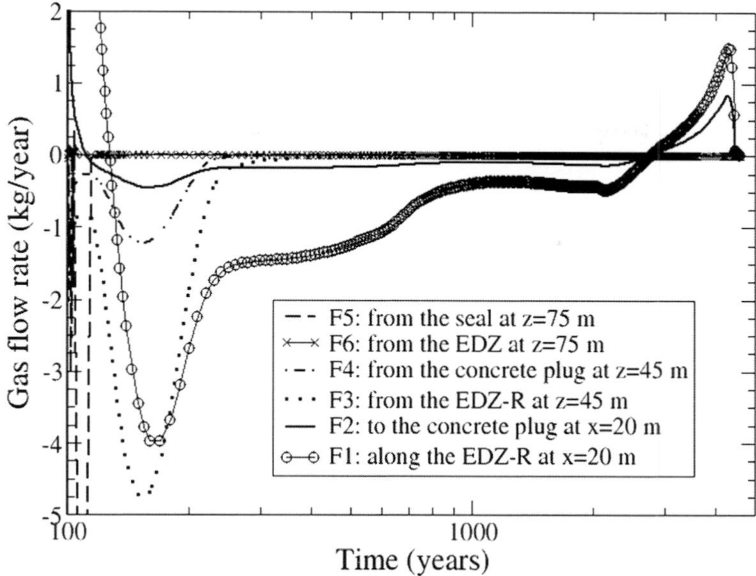

Fig. 7. Evolution in time of the gas flow per shaft for the low incoming gas scenario (see Fig. 5 for more details on the labels in the legend).

mainly confined to the backfill, as this porous medium offers a great void volume (porosity about 40%) that is easily accessible to the gas due to its weak capillary potential (see the negative gas fluxes during this period in Fig. 7).

After 2700 years, since all of the porous media except the backfill are saturated up to 95%, the rate of resaturation of the gallery backfill increases. Consequently, the gas is pushed backwards into the shaft via the EDZ-R due to the effect of hydrostatic pressure (see the positive gas fluxes on the curves F1 and F2 in Fig. 7), where it accumulates below the shaft seal.

Finally, as the resaturation goes on, the gas phase is confined to increasingly smaller volumes, giving rise to a gradual increase in gas pressure that increases dissolution of the hydrogen (according to Henry's law) until the gas phase vanishes. During this process, the incoming water flow from the Oxfordian layer remains quite constant. So, actually, the resaturation process is essentially dominated by gravitational forces along the shaft without significant impact of gas pressure build-up or capillary effects.

High gas input scenario

The water-gas transient. In contrast to the low incoming gas scenario, the evolution of the gas flow over time drastically depends on significant changes in gas input (see in Fig. 8 the curve of the gas input compared to the curve F1 showing the gas flowing out of the gallery through the EDZ-R). Thus, the gas input sharply begins at 150 years and generates a progressive increase in gas flow towards the shaft which lasts about 70 years. The following sharp increase of gas input which occurs at 300 years has a similar effect on the gas flow exiting the gallery. For the same reason, the two drops of gas input at 3000 and 4000 years induce drops in gas exchange between the gallery and the shaft (see the changes of both the gas input and gas flow (curve F1) in Fig. 8).

Moreover, the gas flow exiting the gallery diminishes from 350 to 1500 years, because the gas within the central zone is more and more confined and compressed by the water arriving from the Oxfordian aquifer. Then, it increases because the gas pressure build-up progressively stops the water arrival (see the water flow decrease after 1500 years in Fig. 8), and opens a pathway to the gas along the seal (see in Fig. 8 the evolution of gas flows through the EDZ-R from the gallery on curve F1, and in the shaft at 45 m height on curve F3).

Thus, during the first few thousand years, the gas flowing into the central zone from the storage zones remains confined within the galleries because of the complete saturation of both the bentonite seal of the shaft and the surrounding EDZ which occurs 100 years after closure. The resulting build-up of gas pressure lasts 3800 years. It induces: (i) a progressive drop of the water flow from the Oxfordian

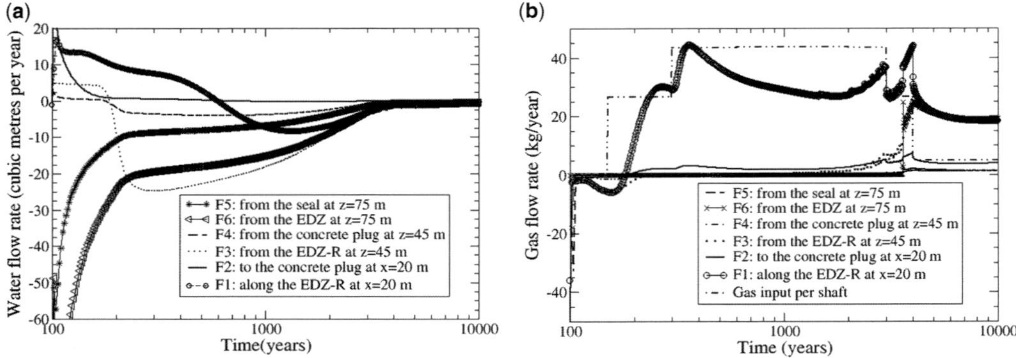

Fig. 8. Evolution in time of (**a**) the water flow and (**b**) gas flow per shaft for the high incoming gas scenario (see Fig. 5 for more details on the labels in the legend).

aquifer entering the Callovo-Oxfordian argillites over the period (it eventually drops below 1 $m^3 a^{-1}$ per shaft); and (ii) at the end of the period, the re-opening to gas flow of the EDZ surrounding the seal.

Indeed, this pathway out of the Callovo-Oxfordian argillites towards the Oxfordian aquifer is associated with a rapid increase in the gas flow along the gallery and the shaft EDZ. Subsequently, the porous media in and around the gallery and the shaft remain unsaturated for a long time, in contrast to the results obtained with the low incoming gas scenario.

The drop in the incoming Oxfordian water flow is due partly to the low effective permeability of the unsaturated porous media (especially the bentonite seal and the EDZ), and also to the high gas pressure in the central zone, which affects the water pressure through the capillary pressure.

It should be noted that there is a very low gas flow rate through the seal when desaturated compared to the flow bypassing the seal through the EDZ (the gas flow through the seal reaches about 1% of the gas flow through the EDZ; see the gas flows on curves F5 and F6 in Fig. 8). In contrast, the water flow rate through the seal reaches almost 50% of that of the EDZ when bypassing the seal (see the water flows on curves F5 and F6 in Fig. 8). This difference in the gas and water flow distributions between the seal and the EDZ is mainly due to the difference of capillary potential of both materials (see the reference pressures of the retention law in Tables 1 & 2). Actually, the capillary phenomenon does not affect the water flow bypassing the seal when saturated. In contrast, the desaturation of porous media especially reduces the effective permeability of the gas in the seal, because the capillary potential of bentonite is considered to be more than twice that of the EDZ (Fig. 9).

Thus, at 4000 years, the difference of Darcy pressures between the Oxfordian aquifer and the central zone drops to 0.68 MPa instead of 3.7 MPa in the low incoming gas case. At 150 000 years, the difference of Darcy pressures is reduced to 0.02 MPa. In fact, the water pressure increases in an asymptotic way towards its *in situ* values (hydrostatic conditions according to the study assumption) because the rate of gas arriving from the central zone remains constant from 4000 to 150 000 years and gives rise to a quasi-steady gas outflow. During this long time period, the maximum gas pressure increases slowly from 5.85 MPa at 4000 years to 6.05 MPa at 150 000 years, which represents an overpressure of 0.95 MPa relative to the *in situ* water pressure at the repository level.

When the gas input stops after 150 000 years, the Darcy pressure drops to 4.8 MPa in the central zone because free gas continues to dissolve, and the resaturation of the storage zone starts again. Water from the Oxfordian aquifer flows through the shaft and enters the central zone at a rate of 0.4 $m^3 a^{-1}$ per shaft. In this scenario, which assumes a large inflow of gas (about 2500 tonnes over 150 000 years), complete water saturation of the central zone occurs at around 210 000 years after repository closure, (i.e. about 60 000 years after the incoming gas flow stops; see Fig. 10).

The gas distribution. During the period when gas is flowing into the central zone, the mass of hydrogen in the gas phase remains steady while the dissolved mass increases. Dissolved gas enters the Dogger formation by diffusion after 20 000 years. At the end of the period, the fate of the 878 tonnes of cumulated incoming gas per shaft is distributed as follows: (i) 57% dissolved in the Callovo-Oxfordian argillites or in the Oxfordian aquifer; (ii) 28.2% dissolved in the Dogger aquifer; (iii) 11.4% in the gas

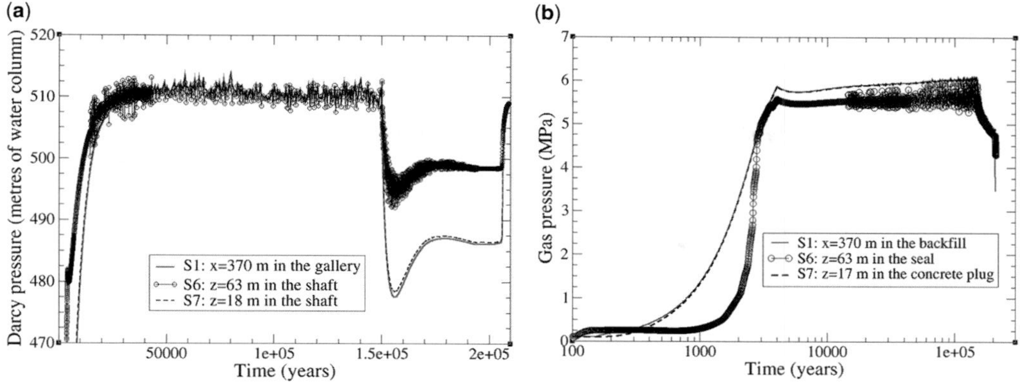

Fig. 9. Evolution in time of (**a**) the Darcy pressure and (**b**) the gas pressure along the gallery and the shaft for the high incoming gas scenario (see Fig. 5 for more details on the labels in the legend).

phase; and (iv) 3.4% diffused through the shaft beyond the upper boundaries of the calculation domain (see Fig. 11).

The gas phase is mainly present in the backfill because, in contrast to concrete and bentonite, the capillary forces cannot counter the build-up of gas pressure. Furthermore, the bentonite of the seal stays quasi-saturated due to high capillary forces, although the surrounding EDZ is unsaturated. At 150 000 years, the gallery backfill contains 80 tonnes of gaseous hydrogen, whereas the concrete block contains 0.9 tonnes and the bentonite 15 kg per shaft.

Numerical instabilities. After 20 000 years, the modelling of the gas phase in the Oxfordian layer encounters numerical difficulties. The dissolved gas diffusing upwards through the Oxfordian layer eventually turns into a gas phase, since the water pressure decreases with height like the hydrostatic pressure. However, it should be noted that, firstly, the mesh is too coarse to describe with accuracy

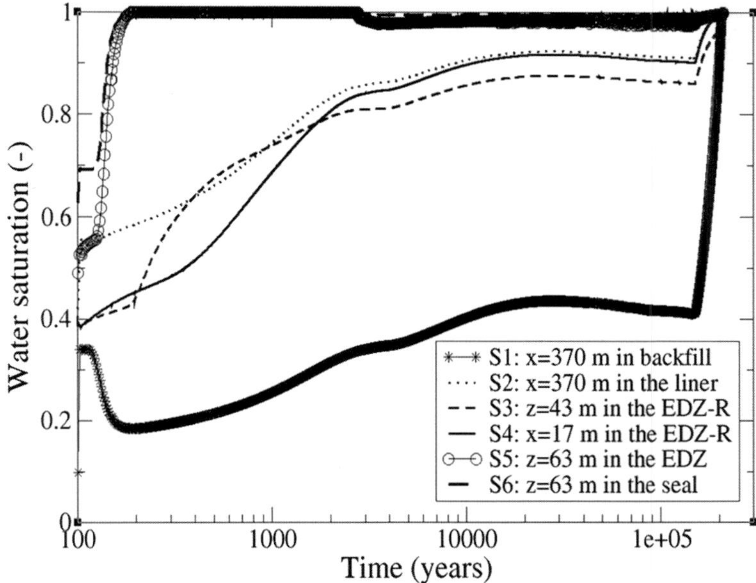

Fig. 10. Evolution in time of the water saturation along the gallery and the shaft for the high incoming gas scenario (see Fig. 5 for more details on the labels in the legend).

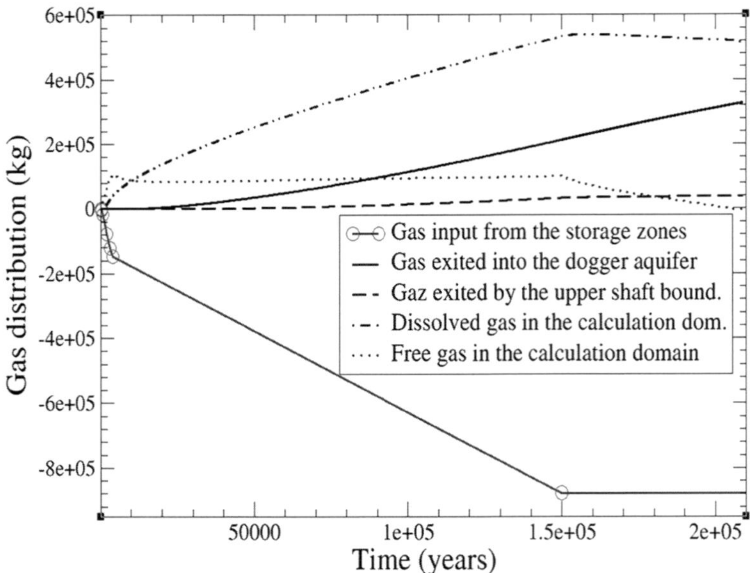

Fig. 11. Evolution in time of the hydrogen mass distribution between inlets and outlets, both in the gas phase and dissolved in the water, for the Callovo-Oxfordian argillites and the Oxfordian aquifer.

the gas phase behaviour in the upper part of the Oxfordian layer, and secondly, the time step is too large (50 or 100 years) to model the upward flow of the gas phase through the 80 m height of the shaft backfill, given the high permeability and the low capillarity of this material in the context of a two-phase flow driven by hydrostatic pressure.

These limitations of the numerical model generate changes in water saturation and gas pressure between calculation steps. However, these oscillations in some backfill cells located in the upper part of the Oxfordian layer have no effect in time on the mass balance and distribution (see Fig. 11). Also, they have no significant impact on gas pressures and water saturation inside the host layer (see Fig. 10).

Sensitivity study on physical parameters

A sensitivity study has been performed in order to assess the importance of various physical parameters (intrinsic permeability, diffusivity) or scenarios (backfill initial saturation, gas ingress). The main calculation cases are listed in Table 4.

The sensitivity cases are based on the high incoming gas scenario, for it is more discriminating as regards some important analysis criteria. In particular, the results of simulations are compared with the reference case using four analysis criteria listed in Table 5: (i) the water flow entering the Callovo-Oxfordian argillites through the EDZ surrounding the seal; (ii) the maximum gas pressure; (iii) the time for the gas phase to enter the Oxfordian layer along the shaft; and (iv) the time when the central zone reaches complete resaturation.

Water bypassing the seal. After closure of the repository, water from the Oxfordian layer flows into the shaft backfill and enters the Callovo-Oxfordian argillites via the seal and the EDZ surrounding the shaft seal. However, the main part of the flow is drained by the desaturated bentonite of the seal until it is fully resaturated, about 100 years after closure in both the reference cases. Only then does the water flow into the shaft and the gallery through the EDZ. The maximum water flow is taken at the bottom of the seal zone. It reaches 29 $m^3\ a^{-1}$ (25 $m^3\ a^{-1}$ through the EDZ and 4 $m^3\ a^{-1}$ through the seal) at 400 years. Both media have similar hydraulic conductivity (5×10^{-11} $m\ s^{-1}$ and 10^{-11} $m\ s^{-1}$ respectively). It should be noted that the contrast in distribution of water flows between both media at the bottom of the seal also depends on the underlying media: the very permeable EDZ-R (5×10^{-10} $m\ s^{-1}$) and the less permeable concrete block (10^{-11} $m\ s^{-1}$). The water flow bypassing the seal is increased in the following sensitive case:

- a tenfold more permeable EDZ (case K3: 180 $m^3\ a^{-1}$).

Table 4. *Sensitivity case definition*

Case	Parameters*	Materials	Units	Ref (R2)	Sensitivity
K1	Hydr. conductivity	Bentonite	(m s^{-1})	10^{-11}	10^{-12}
K2	Initial water sat.	Backfill	(−)	34%	70%
K3	Hydr. conductivity	EDZ-R	(m s^{-1})	5×10^{-9}	5×10^{-8}
	Hydr. conductivity	EDZ-E	(m s^{-1})	5×10^{-11}	5×10^{-10}
K4	Gas input doubled	–	–		
K5	Hydr. conductivity	EDZ-R	(m s^{-1})	5×10^{-9}	5×10^{-10}
	Hydr. conductivity	EDZ-E	(m s^{-1})	5×10^{-11}	5×10^{-12}
K7	H_2 diffusivity	Water	(m^2 s^-)	6×10^{-9}	2×10^{-9}

*Hydr. = Hydraulic; sat. = saturation.

It is decreased in the following sensitive cases:

- a tenfold less permeable EDZ (case K5: 14.5 m^3 a^{-1}; 45% through the seal and 55% through the EDZ);
- a tenfold less permeable seal (case K1: 15.1 m^3 a^{-1}; 8% through the seal and 92% through the EDZ).

Gas exit from the Callovo-Oxfordian argillites. The time period for the gas phase to bypass the shaft seal in the host layer and migrate upwards to the overlying aquifer is 3600 years in the reference case. Gas exit from the Callovo-Oxfordian argillites is speeded up in the following sensitive cases:

- twice the reference incoming gas flow (case K4: 2400 years);
- a tenfold more permeable EDZ (case K3: 3000 years); and
- twice the initial backfill saturation (K2: 3100 years).

It is slowed down in the following sensitive cases:

- a tenfold less permeable seal (case K1: 4000 years);
- a threefold lower H_2 diffusivity (case K7: 4000 years); and
- a tenfold less permeable EDZ (case K5: 4100 years).

Gas pressure maximum. The maximum gas pressure is gradually reached at the end of the period of incoming gas flow. This value changes by up to about 0.5 MPa depending on the sensitivity case. These pressure variations are significant when compared to the pressure difference between water and gas (capillary pressure), which is about 1.1 MPa in the reference case.

The most significant parameters are:

- The incoming gas flow from the storage zone: a doubled gas input leads to a +0.5 MPa rise in the maximum value of the gas pressure.
- The intrinsic permeability of the EDZ: a ten times higher permeability leads to a −0.6 MPa drop in the maximum value of the gas pressure.

Note that the initial backfill saturation also has a significant impact on the gas pressure by changing the free volume available to the gas inside the galleries. Thus an initial saturation of 70% instead of 35%, leads to a +0.4 MPa rise in the maximum gas pressure. Moreover, a tenfold less permeable EDZ (case K5) leads to a +0.15 MPa rise in the maximum gas pressure

Complete resaturation time. The time period to complete the water resaturation of the central zone after the incoming gas flow has stopped is 61 000 years in the reference case. This time period

Table 5. *Comparison of results from sensitivity and reference cases: more significant variations as regards four analysis criteria*

Criterion	Ref. values	Max. values		Min. values	
	(case R2)	Case	Value	Case	Value
Water flow bypassing the seal	29 m^3 a^{-1}	K3	180 m^3 a^{-1}	K5	14.5 m^3 a^{-1}
Gas pressure maximum	6.1 MPa	K4	6.6 MPa	K3	5.5 MPa
		K2	6.5 MPa		
Gas exit time	3.6 ky	K5	4.1 ky	K4	2.4 ky
Complete saturation time	0.21 My	K7	0.3 My	K3	0.175 My

changes from 25 000 years to 150 000 years depending on the sensitivity case.

The most significant parameters are:

- The diffusivity of dissolved H_2 gas: a threefold reduction leads to a longer time period of 150 000 years.
- The intrinsic permeability of the EDZ: a tenfold increase leads to a shorter time period of 25 000 years.

Conclusions

The numerical simulations performed in this study have allowed us to assess how the central zone, with its three shafts and its 4.5 km network of galleries, would contribute to the hydraulic-gas transient in the host rock around the repository. Our analysis has considered two scenarios for the incoming gas flow from the storage zones of the repository.

In the Callovo-Oxfordian argillites, the shaft seals are completely resaturated within a century after the closure of the repository, which prevents a free gas phase from exiting towards the overlying aquifer. Moreover, water from the Oxfordian aquifer bypasses the seal and flows down through the EDZ along the shaft (no healing of the EDZ has been considered) and into the backfilled central zone.

If the amount of hydrogen gas that enters the central zone remains significantly lower than the amount of air which was trapped when the repository was closed, then the central zone is completely flooded after 4500 years. So, without a significant input of gas, the water flow is mainly governed by gravitational forces.

However, with a large input of 2400 tonnes of hydrogen over a period lasting 150 000 years, the free gas phase bypasses the shaft seal, flows through the surrounding EDZ and starts to enter the Oxfordian aquifer at 3600 years. In the meantime, the water flowing from the Oxfordian aquifer to the repository is drastically reduced by the build-up of gas pressure. The amount of free gas in the model remains quite constant in time from 10 000 years until the time when the incoming gas flow stops. During this period, the mass of free gas represents 10% of the total mass of incoming gas, of which 99% is stored in the gallery backfill. So, the major part of the gas dissolves in the pore water of the host rock, as well as in the overlying and underlying aquifers. The central zone is completely resaturated after 210 000 years.

The most significant model parameters are those of the EDZ permeability (5×10^{-11} m s^{-1}) related to gas and water exchange between the central zone and the overlying aquifer. Indeed, the gas and water flows bypass the seal and flow through the more permeable EDZ which surrounds the seal, even though the intrinsic permeability of the bentonite is lower than 10^{-18} m^2 (the reference value). This parameter especially affects (i) the maximum gas pressure reached in the central zone after closure, (ii) the time period for the gas phase to bypass the shaft seal, and (iii) the time period for the central zone to be completely resaturated after the incoming gas flow from the storage zone has stopped. However, the possible self-healing of the EDZ which has not been modelled in this study, could give more importance to the hydraulic performance of the shaft seal because it would become the key pathway for water and gas exchange between the central zone and the overlying Oxfordian aquifer.

Finally, it should be noted that the effective diffusivity of the dissolved gas has a key role in determining the amount of dissolved gas staying in the central zone and indeed in the whole repository (i.e. in differentiating between the two incoming gas scenarios R1 and R2 considered in the study).

This work was funded by Andra. The authors are grateful to the anonymous reviewers for their helpful revision that significantly improved the final manuscript.

References

BOURSIER, I. 2005. *Deux approches de la simulation champ lointain d'un transport de solute en milieu poreux: modélisation par la technique d'homogénéisation de termes sources et méthode de décomposition de domaine adaptée à une equation de convection-diffusion*. PhD thesis, University of Lyon 1.

Cast3M website. http://www-cast3m.cea.fr/

GRIGULL, U. (ed.) 1982. *Properties of Water and Steam in SI-Units*. 3rd edn. Springer-Verlag, Berlin, Heidelberg, New York.

MUALEM, Y. 1976. A new model for predicting the hydraulic conductivity of unsaturated porous media. *Water Resources Research*, **12**, 513–522.

VAN GENUCHTEN, M. T. 1980. A closed-form equation for predicting the hydraulic conductivity of unsaturated soils. *Soil Science Society of America Journal*, **44**, 892–898.

Phenomena exposure from the large scale gas injection test (Lasgit) dataset using a bespoke data analysis toolkit

D. P. BENNETT[1]*, R. J. CUSS[2], P. J. VARDON[1,3], J. F. HARRINGTON[2] & H. R. THOMAS[1]

[1]*Geoenvironmental Research Centre, Cardiff School of Engineering, Cardiff University, Queen's Buildings, The Parade, Cardiff CF24 3AA, UK*

[2]*Transport Properties Research Laboratory, British Geological Survey, Keyworth, Nottingham NG12 5GG, UK*

[3]*Geo-Engineering Section, Department of Geoscience and Engineering, Delft University of Technology, PO Box 5048, 2600 GA Delft, The Netherlands*

**Corresponding author (e-mail: BennettDP@cardiff.ac.uk)*

Abstract: The Large Scale Gas Injection Test (Lasgit) is a field-scale experiment designed to study the impact of gas buildup and subsequent migration through an engineered barrier system. Lasgit has a substantial experimental dataset containing in excess of 21 million datum points. The dataset is anticipated to contain a wealth of information, ranging from long-term trends and system behaviours to small-scale or 'second-order' features. In order to interrogate the Lasgit dataset, a bespoke computational toolkit, designed to expose difficult to observe phenomena, has been developed and applied to the dataset. The preliminary application of the toolkit, presented here, has resulted in a large number of phenomena being indicated/quantified, including highlighting of second-order events (small gas flows, perturbations in stress/pore-water sensors, etc.) and quantification of temperature record frequency content. Localized system behaviour has been shown to occur along with systematic aberrant behaviours that remain unexplained.

The Large Scale Gas Injection Test (Lasgit) is a field-scale experiment operated by the British Geological Survey located at approximately 420 m depth in Svensk Kärnbränslehantering AB's (SKB's) Äspö Hard Rock Laboratory (HRL) in Sweden. Lasgit has been designed to study the impact of gas buildup and subsequent migration through the engineered barrier system (EBS) of the Swedish KBS-3 disposal concept for high-level radioactive waste (SKB 2006; Cuss *et al.* 2011). Specific experimental aims include improving understanding of the gas entry and flow mechanism into and through the EBS, and its effect on buffer saturation levels. Modifications to the KBS-3 concept for the Lasgit experiment include: sintered filters varying in size and location on the canister surface for the purpose of gas injection, simulating a canister defect and point source of gas; filter mats in and around the bentonite buffer material, used to accelerate the hydration of the system; a retained lid to simulate backfill of the repository; and a high number of sensors within the bentonite buffer material and at interfaces between the buffer and the canister/host rock (Cuss *et al.* 2010). Figure 1 depicts Lasgit's experimental arrangement, illustrating a canister emplaced in a deposition hole, the retaining structure simulating repository backfill, the hydration mats and the approximate distribution of sensors within the deposition hole.

Lasgit has been in continuous operation since February 2005, and is considered in the analysis presented here from 2007 to mid 2012. (This is due to the first two years of data possessing significant distortions and anomalies resulting from necessary operator interventions, such as the drilling and subsequent sealing of pressure relief holes; Harrington *et al.* 2008; Cuss *et al.* 2010.) Each of the *c.* 175 instruments have been recorded with a high frequency in comparison to the physical processes being measured. Since 2007, there exist approximately 120 000 records per sensor, averaging 2.5 logs/h. This has led to a substantial dataset containing in excess of 21 million datum points, approximately 18 million of which originate from sensors within and around the deposition hole, and the remainder being associated with the experimental control systems. The dataset is anticipated to contain a wealth of information, ranging from long-term trends and system behaviours to small-scale or 'second-order' features.

Instrumentation within the deposition hole consists of 32 total stress sensors (oriented either radially or axially), 26 pore-water pressure sensors and 56 temperature sensors. Additionally there are

From: NORRIS, S., BRUNO, J., CATHELINEAU, M., DELAGE, P., FAIRHURST, C., GAUCHER, E. C., HÖHN, E. H., KALINICHEV, A., LALIEUX, P. & SELLIN, P. (eds) 2014. *Clays in Natural and Engineered Barriers for Radioactive Waste Confinement*. Geological Society, London, Special Publications, **400**, 497–505.
First published online March 5, 2014, http://dx.doi.org/10.1144/SP400.5

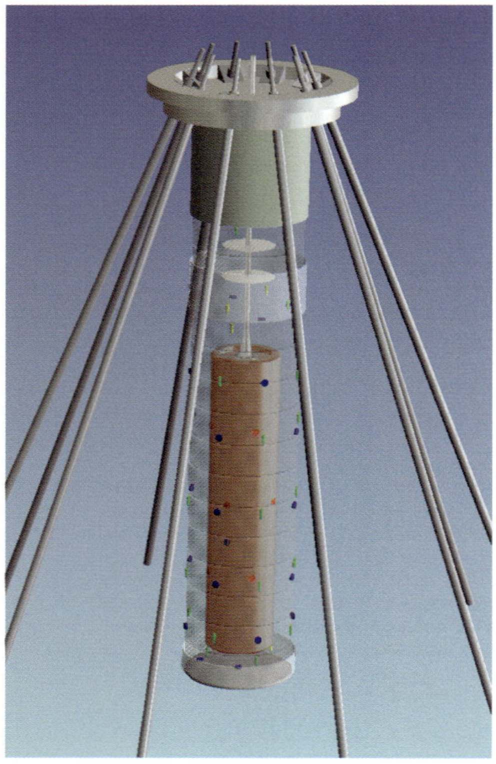

Fig. 1. Schematic depiction of Lasgit's experimental arrangement. A copper canister is shown in an emplacement borehole with hydration mats and retention simulating repository backfill. The approximate distribution of experimental instrumentation is shown by the coloured shapes (SKB 2008).

16 pore-water pressure transducers associated with the hydration mats and injection filters that act as sensors. Attached to the deposition hole retaining set-up there are 10 displacement sensors. Nine pore-water pressure sensors are located in boreholes adjacent to the deposition hole. The ambient air temperature of the facility at the experimental location is also monitored.

The experimental history consists of an initial period of natural and artificial hydration of the EBS lasting approximately 2 years, followed by the first gas injection test phase (during which major gas breakthrough did not occur). Hydration of the EBS was then continued for approximately one more year. Two more gas injection test phases, each lasting in excess of a year, were then performed back to back. Major gas breakthrough occurred during both of these gas test phases. Overviews of these major phases of the experimental history and the prominent events that occurred during each phase are reported in Harrington *et al.* (2008), Cuss *et al.* (2010) and Cuss *et al.* (2011). In addition to the small-scale and macro-scale phenomena associated with changes in flow regime that are considered in the analyses presented in these reports, a large number of second-order perturbations remain uninvestigated.

Systematic computational analysis of the Lasgit dataset implicitly reduces the subjectivity of any observations made. Additionally, computational analysis is required owing to the size of the dataset and the scale of the details of interest. Moreover, second-order phenomena can be exposed. However, owing to the length and complexity of the experiment, the Lasgit dataset is not typically suited to 'out of the box' computational analysis. Specifically, non-uniformity in the sampling rate (arising both intentionally and unavoidably in the case of the Lasgit dataset, see Fig. 2) is incompatible with any time series analysis algorithm or analysis technique that assumes a uniform sample rate. Such assumptions, while not ubiquitous, are typical in time series analysis (e.g. Box & Jenkins 1976). Applying such an algorithm to non-uniform data may result in a distortion of the physical meaning of the input. This is illustrated in Figure 3, where non-uniformly collected data are shown with a uniform spacing as significantly distorted. Solutions such as 'down-sampling' the dataset (i.e. using only the points coinciding with a lower common sample rate in the dataset) or interpolating between measured datum points to produce a uniform sample rate are possible. However, both potentially risk the loss of detail in the dataset (particularly at local maxima and minima), or, in the case of down-sampling, may not be possible if a common sample sub-rate is not present across the dataset.

Bespoke toolkit

In order to accommodate the non-uniform nature of the Lasgit dataset when performing an exploratory data analysis, a bespoke computational toolkit has been developed (Bennett *et al.* 2012). The toolkit implementation, in addition to avoiding the requirement for uniform input data, makes minimal assumptions and requires minimal knowledge about the nature of the input. This ensures that the applicability of the toolkit to long-term, large-scale experimental datasets is as broad as possible. Key toolkit capabilities (depicted in Fig. 4 and further elaborated in Bennett *et al.* 2012) include:

- non-parametric trend identification using singular spectrum analysis (e.g. Golyandina *et al.* 2001);
- statistical spike identification for isolated/individual aberrant/anomalous datum points;

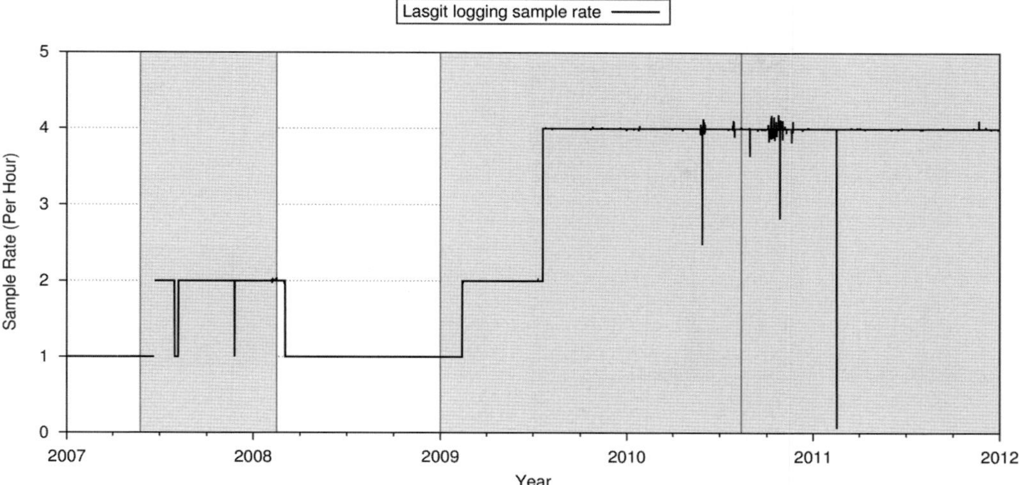

Fig. 2. Sample (logging) rate of Lasgit experiment varying with time. Grey-shaded regions correspond to gas injection test phases of the experimental history.

- event candidate identification, sensitive to second-order events;
- frequency domain analysis utilizing a discrete Fourier transform (DFT) (Bagchi & Mitra 1999) modified to accommodate non-uniform input data;
- smoothing and averaging functions that utilize moving windows defined as periods of time rather than a fixed number of points around a focus (time windowing);
- visualization technique for identifying where and when event candidates occur across multiple sensors/time series simultaneously.

The capabilities included in the toolkit were selected for their abilities to expose phenomena, that is, extract useful information from a series of measurements that would otherwise be obscured or occluded when observed. For example, trend identification and subsequent removal can expose

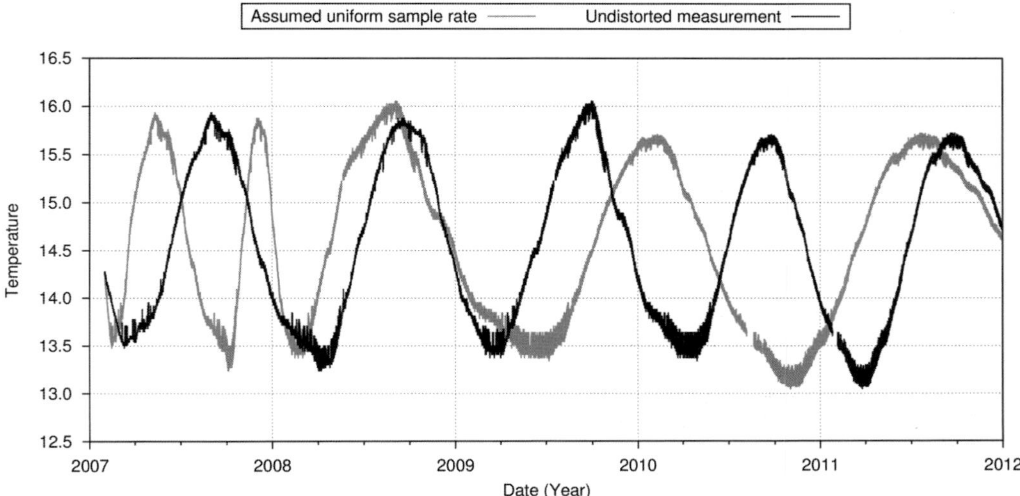

Fig. 3. Down-hole temperature record showing undistorted (black) temperature record and distorted (grey) temperature record resulting from assuming a uniform sample rate when analysing non-uniform data.

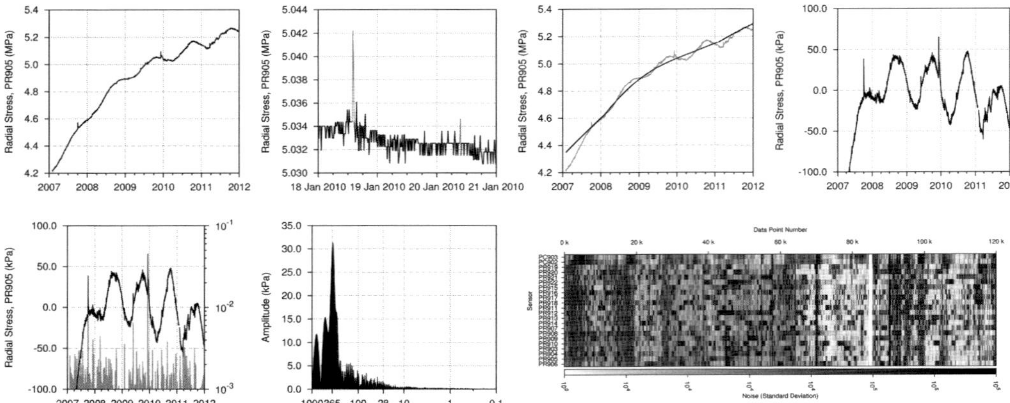

Fig. 4. Toolkit capabilities. Top row, from left to right: original signal; detected spikes in grey with remaining signal in black; singular spectrum analysis-derived trend in black over original signal in grey; and residual signal after trend removal. Bottom row, from left to right: second-order event impulses in grey with residual signal in black; frequency content (amplitude with wavelength) of residual; and intensity plot of second-order event impulses showing events occurring simultaneously across multiple sensors.

second-order events that may otherwise be subsumed within macro-scale system behaviours. This would also facilitate quantification of observable seasonal effects using frequency domain analysis; spike detection, smoothing and averaging functions allow noise to be reduced, improving the visibility and measurability of observable phenomenon; and visualization of simultaneous event candidates allows quick aggregation of the relevant information surrounding a second-order event in a large and complex system.

Implementation of each tool limits operator-defined parameters to statistical thresholds and time window length scales, and where appropriate generalizes the data analysis algorithms to accommodate non-uniform input. An example of such a generalization (the modification to the DFT) is shown in equations (1a) and (1b). The DFT is a common and well-understood method for accessing frequency domain information, and is simple to implement computationally. The modification described here pertains to the specification of frequency range inspected, with the term $2\pi kn/N$ being replaced by ωt_n, allowing arbitrary frequencies to be investigated and accurate time weighting of the input signal to be achieved. The modified non-uniform discrete Fourier transform (NDFT) no longer relies on the input time series possessing a uniform sample rate. The terms of equations (1a) and (1b) are defined in Table 1. Other toolkit generalizations are described in Bennett *et al.* (2012).

$$p(k) = \sum_{n=0}^{N-1} F_n \cdot e^{-i2\pi \frac{k}{N} n} \quad \text{(1a)}$$

Becomes:

$$p(\omega) = \sum_{n=0}^{N-1} F_n(t_n) \cdot e^{-i\omega t_n} \quad \text{(1b)}$$

Results

The procedural application of the toolkit to the Lasgit dataset consisted of: initial spike detection, utilizing a 48 h window and a spike threshold of three standard deviations from the mean, followed by exclusion of the detected spikes from subsequent analysis; trend identification and removal utilizing a 365 day singular spectrum analysis window to

Table 1. *Terms used in equations (1a) and (1b)*

Term	Definition
$P(k)$ and $P(\omega)$	Power (amplitude) of frequency in input time series at the frequency corresponding to k or ω
$F(n)$ and $F(t_n)$	Value of input time series at point n or time t_n
n	Index of record in input time series ($n = 0 \ldots N-1$)
N	Number of datum points in input time series
ω	Defined as $2\pi f$
k	DFT iterator to define frequency interrogation points ($k = 0 \ldots N-1$)
i	Square root of -1
f	Frequency at which to inspect time series (user defined)

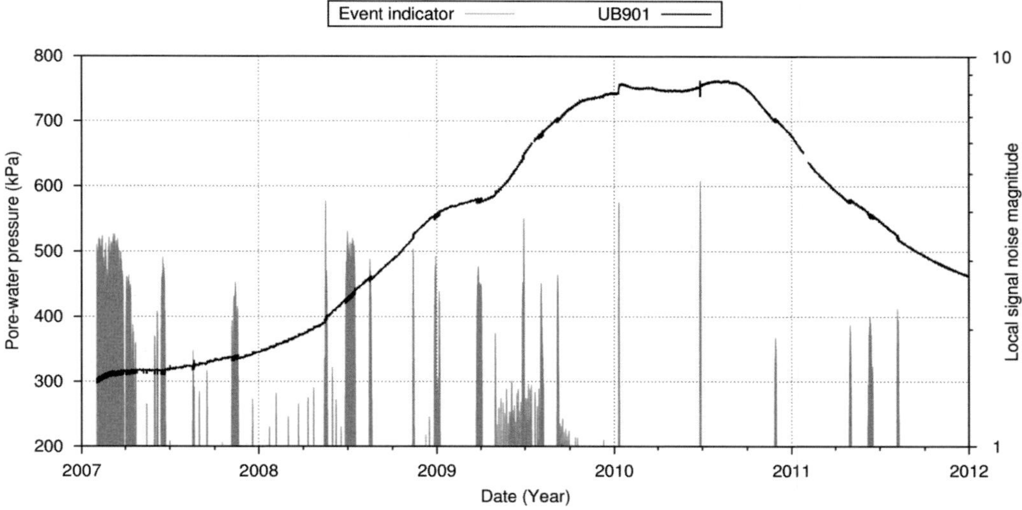

Fig. 5. Examples of second-order events in sensor UB901. Prominence of noise impulses indicates small-scale changes in sensor behaviour (i.e. second-order event candidates).

define trend scale; second-order event detection and frequency domain analysis on the residual; visualization of the second-order event information across sensor types to highlight any periods of synchronized behaviour; and a smoothing step to improve clarity of visualizations. A number of quantifications and observations were established as a result of the toolkit applications. A selection of salient points are presented here.

Approximately 130 000 datum points out of the *c.* 18 million originating from the deposition hole (0.7%) were identified as spikes, with a significant concentration (approximately 56% of all the spikes) occurring in the temperature sensor records (which account for 37% of the down-hole record). Additionally, temperature sensor spiking appears concentrated in the peaks and troughs of the records. Several hundred second-order event candidates were also highlighted, examples of which are shown in Figure 5 for sensor UB901. Strong seasonal (annual) cycles were found in every temperature sensor record, and in most stress and pore-water pressure sensors. The exceptions to this are those sensors containing large steps or discontinuities in their records, resulting from gas/water flow interception, or in some cases malfunction. Correlation between the seasonal effects (i.e. the change in cycle amplitude and time-offset) and depth were strong in the temperature sensors. While annual cycles are present in stress and pore-water pressure sensors, the strong correlation of amplitude and phase offset with depth is not present. Figure 6 shows the correlation between the seasonal effects and depth in the temperature sensors in the upper parts, and the lack of such a correlation in the stress and pore-water pressure sensors within the deposition hole in the lower parts.

Examples of specific details revealed by the toolkit are presented in Figures 7 & 8. Three pore-water pressure sensors have been identified as having anomalously large amplitude variations, UR912, UR914 and UR920. The locations of these sensors are presented in Figure 7 with respect to the fracture intersections with the emplacement borehole in order to consider the possible effect of the experimental boundary condition. Figure 8 shows a systematic aberration in a down-hole temperature sensor (UT901) exposed by (time window) smoothing of the signal. The aberration exposed is a sharp (but small) increase in the average temperature reported by the sensor followed by an equivalent decrease approximately 3 months later. This phenomenon occurs approximately 1.5 months after each annual minimum.

Second-order event candidates highlighted by the toolkit require detailed manual investigation on an individual basis. An initial survey of the highlighted events indicates that a mixture of sudden steps in stress, pore-water pressure, and gas injection pressure sensor records along with brief changes in sensor noise levels are represented. Manual investigation of the second-order event candidates highlighted by the toolkit application is ongoing, and beyond the scope of the initial results presentation.

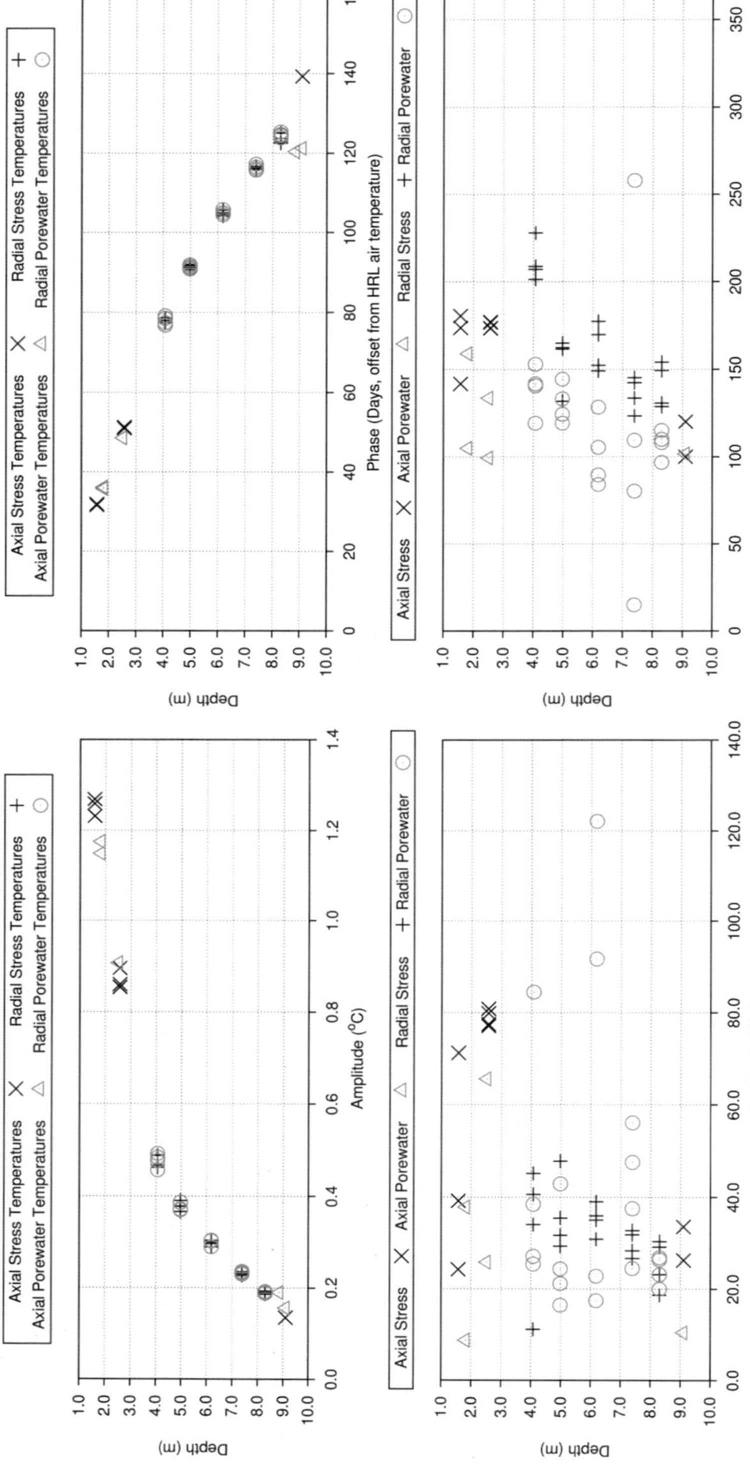

Fig. 6. Lasgit down-hole frequency analysis information. Top left: temperature sensor annual amplitudes with depth. Top right: temperature sensor annual cycle offsets (from HRL air temperature) with depth. Bottom left: stress and pore-water pressure sensor annual amplitudes with depth. Bottom right: stress and pore-water pressure sensors annual cycle offsets with depth.

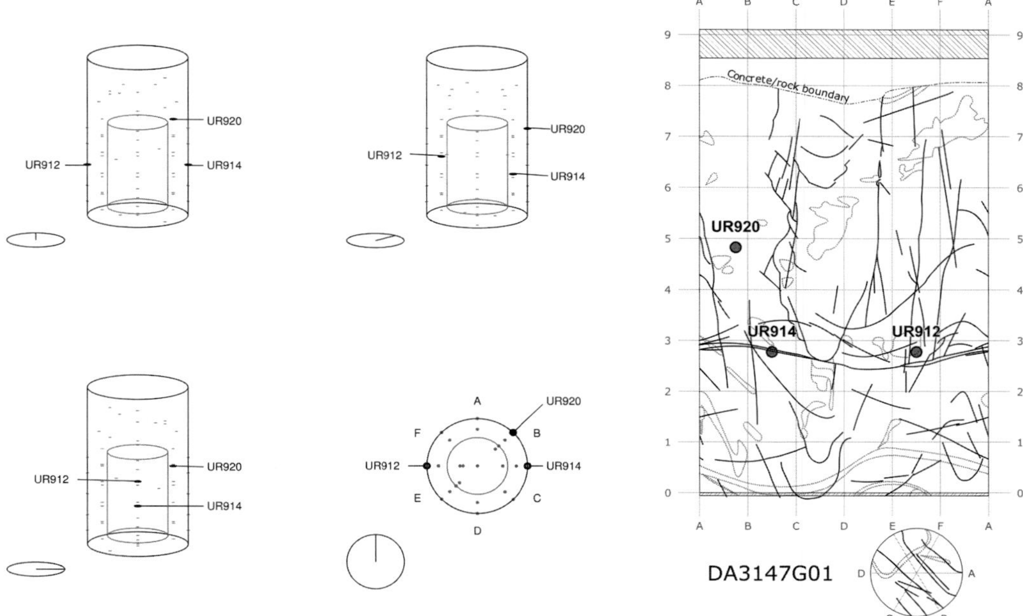

Fig. 7. High-amplitude pore-water pressure sensor locations spatially, highlighted in black (left section) and with respect to borehole fractures (right section). Fracture map reproduced from Hardenby & Lundin (2003).

Discussion

A possible statistical explanation for the concentration of spikes detected around the peaks and troughs in the temperature sensor records is the relatively low temperature variation recorded during those periods, leading to a local lowering of the threshold beyond which a point is considered a spike. The magnitudes of the spikes detected, however, suggest that this is not the case. Additionally,

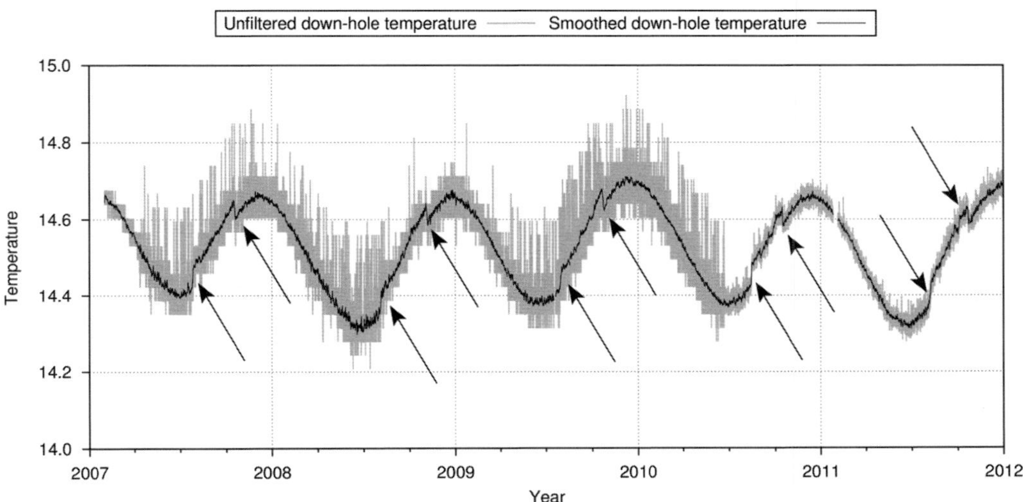

Fig. 8. Systematic aberration in down-hole temperature sensor record UT901. Original signal in grey, time windowed smoothed signal in black, showing an upward step followed by downward step each year indicated by arrows.

the 48 h window applied to the spike detection is too short for annual processes to exert large effects the variation within the window. This phenomenon of concentration of spikes is qualitatively similar, therefore, to other observable changes in the inherent 'noise' level in sensor records, as these changes are seemingly not linked to hydrological or geomechanical processes within the system. The cause therefore is unknown, and therefore warrants further exploration.

The strong correlation between depth and the temperature attributes (amplitude and phase, Fig. 6) implies that the driver for the down-hole temperature profile is the HRL air temperature (which is ventilated from the surface of the facility). The modified NDFT procedure, however, appears to be finely sensitive to experimental variations/distortions in the input data. The more complexly driven physical processes, such as those measured by the stress and pore-water pressure sensors, display detectable annual cycles (a prominent peak at a wavelength of approximately 365 days in almost every de-trended sensor record), but do not possess the tight correlation with depth that the more simply driven processes such as those measured by the temperature sensors do.

The three pore-water pressure sensors with high annual amplitudes (UR912, UR914 and UR920) cannot be correlated with the down-hole temperature profile, and are also not correlated spatially or obviously linked to boundary conditions. While two of these sensors are in close proximity to major fractures in the deposition hole rock wall (see Fig. 7), the upper sensor (UR920) does not appear to intersect such a feature. Additionally, sensors UR911 and UR913 (not shown on Fig. 7) lie on the same plane as UR912 and UR914 but do not have similar higher annual amplitudes. This does not exclude fracture influence on the relevant sensors as a possibility, however. It is conceivable that preferential flow of water to these locations occurs from a location that is more influenced by HRL air temperature, or that other material heterogeneities exist in the system.

The absence of distortions in the residual records (after trend removal) of the high-amplitude sensors indicates that they are driven by processes less complex than other nearby sensors; however, the processes influencing the different observed behaviours in these particular locations is not apparent from the boundary conditions of the experiment, such as the deposition hole fracture map. It is also possible that the sensors exhibiting high amplitudes are hydraulically connected by some other mechanism than the host rock fracture network.

The systematic aberration shown in the deepest temperature sensor (UT901, shown in Fig. 8), in the absence of other phenomena that coincide or synchronize with the steps, suggests a sensor or logging equipment error. The step changes (in the order of 0.05 °C) are subsumed almost entirely by the sensor noise for the majority of the record. As seasonal temperature changes of approximately ± 1 °C appear to be driving stress and pore-water pressure changes in the order of tens of kilopascal, the errors introduced by malfunctions of this magnitude are likely to be insignificant.

Conclusions

A bespoke analytical toolkit has been developed (Bennett *et al.* 2012) and applied to the Lasgit dataset in order to investigate 'second-order' phenomenon present in the dataset. The developed toolkit has been shown to be accommodating of non-uniform time series input and sensitive to second-order events, exposing phenomena that would otherwise be occluded during an exploratory data analysis. The toolkit components implemented have been chosen for broad applicability to a variety of physical processes and inputs.

A large number of second-order event candidates have been identified by the application of the toolkit, providing guidance towards perturbations in the dataset of potential interest for further investigation. They include changes in gas flow, stress and pore-water pressure, and changes in system noise. This ability to locate small areas of potential interest is of major use given the amount of data and the scale of the perturbations in comparison to the overall sensor magnitudes. The identification of these event candidates is also not dependent on the separability of the seasonal components from the non-seasonal components in the dataset, nor on the macroscale behaviours of the experiment.

Otherwise unobservable phenomenon, for example, small systematic behaviours and localized differences in behaviours, have also been exposed through application of the toolkit. In the cases where there is an inability to firmly identify any obscured phenomena that synchronizes with observable phenomena, there is an improved confidence in the conclusion of the absence of such additional phenomena. The small recurring steps in UT901's sensor record are not measurable or observable elsewhere in the dataset, suggesting that they are an isolated but unexplained phenomenon.

Expected system behaviours, such as a highly ordered and depth correlated down-hole temperature profile, have been quantitatively confirmed. However, the stress and pore-water pressure sensors, while possessing similar quantifiable seasonal behaviours, lack this highly correlated profile. The additional non-seasonal aspects of the stress and pore-water pressure sensor records in comparison to the temperature sensor records appear to prevent the measurement of this correlation with depth, should it exist underlying the distortions, while not preventing the measurement of the presence of the seasonal components. The development of the toolkit has therefore enabled the quantification of observable phenomena, and improved the efficiency of the manual investigation of small-scale or 'second-order' analysis of a large scale, non-uniform dataset.

The research leading to these results has received funding from the European Atomic Energy Community's Seventh Framework Programme (FP7/2007–2011) under grant agreement no. 230357, the FORGE project.

References

BAGCHI, S. & MITRA, S. K. 1999. *The Nonuniform Discrete Fourier Transform and its Applications in Signal Processing*. Kluwer Academic, Dordrecht.

BENNETT, D. P., CUSS, R. J., VARDON, P. J., HARRINGTON, J. F., PHILP, R. N. & THOMAS, H. R. 2012. Data analysis toolkit for long-term, large-scale experiments. *Mineralogical Magazine*, **76**, 3355–3364.

BOX, G. E. P. & JENKINS, G. M. 1976. *Time Series Analysis: Forecasting and Control*, revised edn. Holden-Day, San Francisco, CA.

CUSS, R. J., HARRINGTON, J. F. & NOY, D. J. 2010. *Large Scale Gas Injection Test (Lasgit) Performed at the Äspö Hard Rock Laboratory: Summary Report 2008*. Technical Report **TR-10-38**, Svensk Kärnbränslehantering AB.

CUSS, R. J., HARRINGTON, J. F., NOY, D. J., WIKMAN, A. & SELIN, P. 2011. Large scale gas injection test (Lasgit): results from two gas injection tests. *Physics and Chemistry of the Earth*, **36**, 1729–1742.

GOLYANDINA, N., NEKRUTKIN, V. & ZHIGLJAVSKY, A. 2001. *Analysis of Time Series Structure: SSA and Related Techniques*. Chapman & Hall/CRC, London.

HARDENBY, C. & LUNDIN, J. 2003. *Äspö Hard Rock Laboratory – TBM Assembly Hall: Geological Mapping of the Assembly Hall and Deposition Hole*. International Progress Report **IPR-03-28**. Svensk Kärnbränslehantering AB.

HARRINGTON, J. F., BIRCHALL, D. J., NOY, D. J. & CUSS, R. J. 2008. *Large Scale Gas Injection Test (Lasgit) Performed at the Äspö Hard Rock Laboratory: Summary Report 2007*. Technical Report **CR/07/211**, British Geological Survey.

SKB 2006 *Long-term Safety for KBS-3 Repositories at Forsmark and Laxemar – a First Evaluation: Main Report of the SR-Can project*. Technical Report **TR-06-09**, Svensk Kärnbränslehantering AB.

SKB 2008. *Äspö Hard Rock Laboratory*. Annual Report 2007. Technical Report TR-08-10, Svensk Kärnbränslehantering AB.

Experimental observations of mechanical dilation at the onset of gas flow in Callovo-Oxfordian claystone

ROBERT CUSS[1]*, JON HARRINGTON[1], RICHARD GIOT[2] & CHRISTOPHE AUVRAY[2]

[1]*Transport Properties Research Laboratory, British Geological Survey, Keyworth, Nottingham, NG12 5GG, UK*

[2]*LAEGO-ENSG-Université de Lorraine, Rue du Doyen Marcel Roubault, B.P. 40, F-54501 Vandœuvre-Lès-Nancy, France*

**Corresponding author (e-mail: rjcu@bgs.ac.uk)*

Abstract: Understanding the mechanisms controlling the advective movement of gas and its potential impact on a geological disposal facility (GDF) for radioactive waste is important to performance assessment. In a clay-based GDF, four primary phenomenological models can be defined to describe gas flow: (i) diffusion and/or solution within interstitial water; (ii) visco-capillary (or two-phase) flow in the original porosity of the fabric; (iii) flow along localized dilatant pathways (micro-fissuring); and (iv) gas fracturing of the rock. To investigate which mechanism(s) control the movement of gas, two independent experimental studies on Callovo-Oxfordian claystone (COx) have been undertaken at the British Geological Survey (BGS) and LAEGO–ENSG Nancy (LAEGO).

The study conducted at BGS used a triaxial apparatus specifically designed to resolve very small volumetric (axial and radial) strains potentially associated with the onset of gas flow. The LAEGO study utilized a triaxial setup with axial and radial strains measured by strain gauges glued to the sample. Both studies were conducted on COx at *in situ* stresses representative of the Bure Underground Research Laboratory (URL), with flux and pressure of gas and water carefully monitored throughout long-duration experiments.

A four-stage model has been postulated to explain the experimental results. Stage 1: gas enters at the gas entry pressure. Gas propagation is along dilatant pathways that exploit the pore network of the material. Around each pathway the fabric compresses, which may lead to localized movement of water away from the pathways. Stage 2: the dendritic flow path network has reached the mid-plane of the sample, resulting in acceleration of the observed radial strain. During this stage, outflow from the sample also develops. Stage 3: gas has reached the backpressure end of the sample with end-to-end movement of gas. Dilation continues, indicating that gas pathway numbers have increased. Stage 4: gas-fracturing occurs with a significant tensile fracture forming, resulting in failure of the sample.

Both studies clearly showed that as gas started to move through the COx, the sample underwent mechanical dilation (i.e. an increase in sample volume). Under *in situ* conditions, the onset of dilation (micro-fissuring) is a necessary precursor for the advective movement of gas.

In a geological disposal facility (GDF) for radioactive waste, corrosion of ferrous materials under anoxic conditions, combined with the radioactive decay of the waste and the radiolysis of water, will lead to the formation of hydrogen. A full understanding of the migration behaviour of this gas is of fundamental importance to the development of a GDF. If the rate of gas production exceeds the rate of gas diffusion within the pores of the barrier or host rock, a discrete gas phase will form (Wikramaratna *et al.* 1993; Ortiz *et al.* 2002; Weetjens & Sillen 2006). If this occurs, capillary restrictions (Aziz & Settari 1979) on the movement of gas will result in the build-up of pressure to a critical value when pressure becomes sufficiently large for it to enter the surrounding material and move through advective processes.

To accurately predict the movement of gas through argillaceous materials (Neretnieks 1984; Kreis 1991; Askarieh *et al.* 2000; Ekeroth *et al.* 2006; Smart *et al.* 2006), it is first necessary to define the correct conceptual model that best represents the empirical data. In a clay-based GDF, four primary phenomenological models can be defined to describe gas flow, as proposed by Marschall *et al.* (2005; Fig. 1): (i) gas movement by solution and/or diffusion, governed by Henry's and Fick's laws respectively, within interstitial fluids along prevailing hydraulic gradients; (ii) gas flow in the original porosity of the fabric, governed by a generalized form of Darcy's Law, commonly referred to as visco-capillary (or two-phase) flow; (iii) gas flow along localized dilatant pathways (micro-fissuring),

From: Norris, S., Bruno, J., Cathelineau, M., Delage, P., Fairhurst, C., Gaucher, E. C., Höhn, E. H., Kalinichev, A., Lalieux, P. & Sellin, P. (eds) 2014. *Clays in Natural and Engineered Barriers for Radioactive Waste Confinement*. Geological Society, London, Special Publications, **400**, 507–519.
First published online May 8, 2014, http://dx.doi.org/10.1144/SP400.26

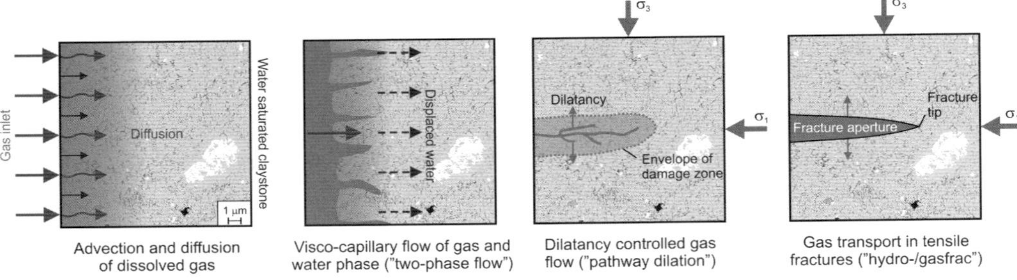

Fig. 1. Description of the four main processes of gas movement in clays (after Marschall *et al.* 2005).

which may or may not interact with the continuum stress field, the permeability of which is dependent on an interplay between local gas pressure and the effective stress state; and (iv) gas flow along macro fractures similar in form to those observed in hydrofracture activities during reservoir stimulation, where fracture initiation occurs when the gas pressure exceeds the sum of the minor principle stress and tensile strength. For engineering problems such as radioactive waste disposal, underground gas storage and carbon dioxide sequestration, most interest is focused on understanding the processes and mechanisms governing the advective transport of gas (mechanisms ii–iv).

There is now a growing body of evidence (Horseman *et al.* 1996, 2004; Harrington & Horseman 1999; Angeli *et al.* 2009; Harrington *et al.* 2009, 2012) that in the case of plastic clays and in particular bentonite, classic concepts of porous-medium two-phase flow are inappropriate. Flow through dilatant pathways has been shown in a number of argillaceous materials (Horseman & Harrington 1994; Harrington & Horseman 1997, 1999; Ortiz *et al.* 1996, 2002; Gallé & Tanai 1998; Horseman *et al.* 1999; Autio *et al.* 2006; Angeli *et al.* 2009). The pathways are pressure-induced and their aperture is a function of their internal gas pressure and structural constraints within the clay. However, the exact mechanisms controlling gas entry, flow and pathway sealing in clay-rich media are not fully understood and the 'memory' of such pathways could impair barrier performance.

To investigate which mechanism(s) control the advective movement of gas, two independent experimental studies on Callovo-Oxfordian claystone (COx) have been undertaken at the British Geological Survey (BGS, UK) and LAEGO–ENSG–Université de Lorraine (Nancy, France).

Experimental apparatus at BGS

The bespoke stress-path permeameter (SPP, Fig. 2) was used to investigate water and gas (helium) flow in COx from the Bure Underground Research Laboratory (URL) in the eastern part of the Paris Basin under *in situ* conditions (see Table 1 for test material parameters and experimental boundary conditions). See Cuss & Harrington (2010, 2011); Cuss *et al.* (2012) for a full description of the test apparatus and experimental history.

The triaxial SPP testing rig (Fig. 2) was designed to observe sample volume changes during flow experiments conducted along an evolving stress path; please note that the results presented here were for a static triaxial boundary condition. The apparatus had a thick-walled pressure vessel that imposed a confining pressure (12.5 MPa) through the compression of glycerol by an ISCO syringe pump. The test sample (Fig. 2c) was cylindrical and of 56 mm diameter and 82 mm length. The sample was jacketed in a Hoek sleeve, which had been thinned to reduce compressibility and had three brass plates cemented to the outside of the jacket. Three pressure-balanced devices (dash-pots) allowed direct measurement of the displacement of these plates to give radial displacement. The sample was axially loaded by an Enerpac hydraulic ram driven by an ISCO syringe pump, giving a stress of 13 MPa in the axial direction. Axial displacement was directly recorded by a Mitutoyo digital micrometer. The pore-pressure system had two ISCO syringe pumps; one acted as an injection pump and the other maintained constant back-pressure (4.5 MPa). A test was conducted using stages of constant pressure or constant-flow pressure ramps in the injection system in order to initiate gas or water flow. The experimental rig was completed by a number of pressure sensors, thermocouples and a digital acquisition system, which logged data every 2 min.

One experimental uncertainty in transport testing is the short-circuiting of the flow system along the jacket of the test sample. The addition of a 6-mm-wide, 2-mm-deep, porous stainless-steel annular filter along the outer edge of each platen (Fig. 2b) allowed porewater pressure to be monitored and discounted unwanted sidewall flow. These two

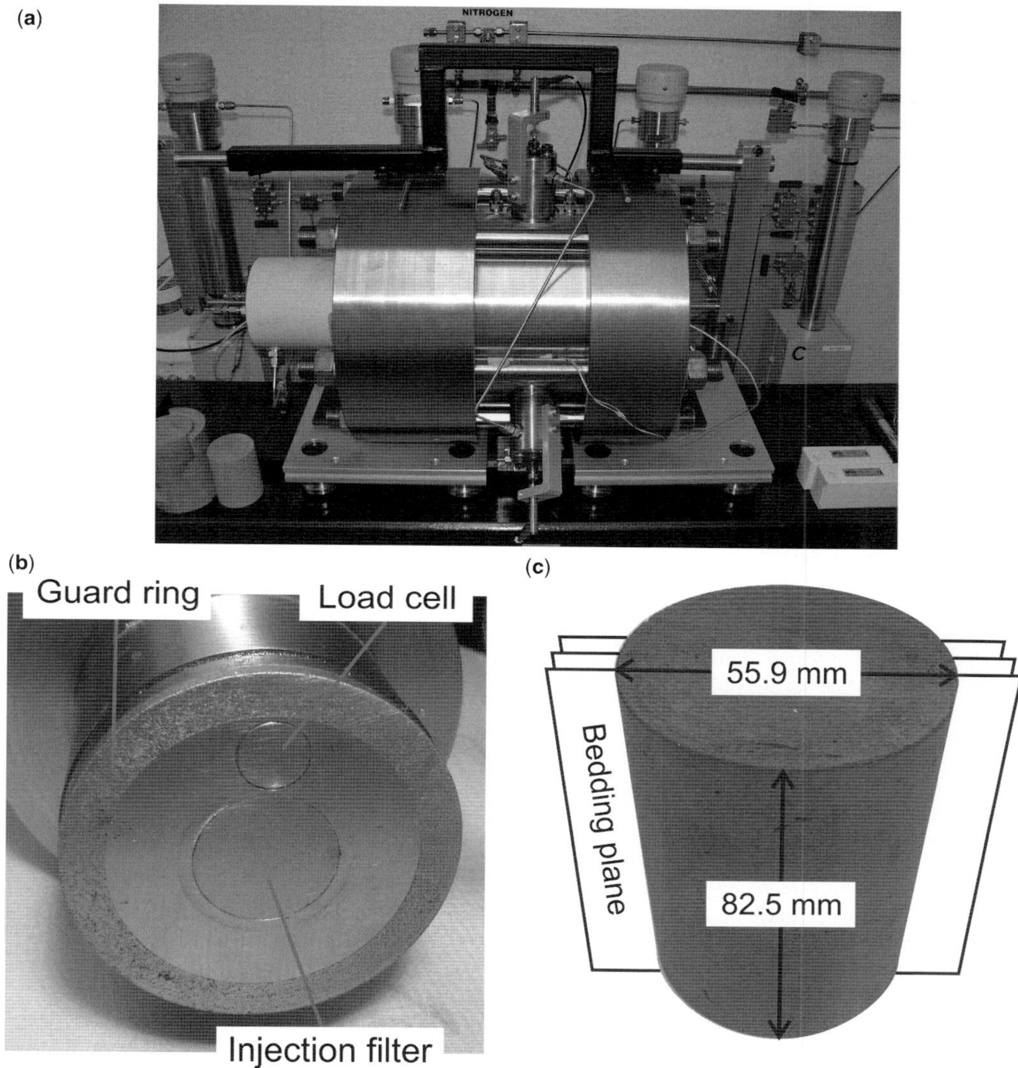

Fig. 2. Triaxial test setup: (**a**) triaxial SPP apparatus; (**b**) end platen showing the injection filter and guard ring arrangement; (**c**) sample of COx prior to testing.

guard rings (Harrington *et al.* 2003) were each connected to a pressure transducer and the complete guard-ring system (filter, pipework and pressure sensor) was saturated with water and flushed in order to eliminate gas from the system. The inlet/outlet filter was comprised of a porous disc 20 mm in diameter and 2 mm depth. The control board of the apparatus allowed the guard rings to be either connected to the injection system to assist in hydration, or isolated to give an independent measure of pore pressure. As well as being able to eliminate sidewall flow as a possible transport mechanism, the guard rings meant pore pressure was measured at four different points on the test sample (injection pressure, injection guard-ring pressure, backpressure, back guard-ring pressure), therefore providing data on hydraulic anisotropy within the sample.

The SPP was used to conduct experiments on COx under *in situ* stress conditions. Wenk *et al.* (2008) reports clay (25–55 wt%), 23–44% carbonates and 20–31% silt (essentially quartz + feldspar). Clay minerals are reported to include illite and illite-smectite with subordinate kaolinite

Table 1. *Description of pre-test COx material properties and experimental boundary conditions*

		Triaxial geometry BGS	Triaxial geometry LAEGO	Units
Sample properties	Sample reference	SPP_COx-2	EST28906-6	
	Location	Bure URL, France	Bure URL, France	
	Borehole/drill core	OHZ1201/EST30341	EST28906	
	Core direction	Parallel to bedding	Normal to bedding	
	Average length	82.45 ± 0.03	76.69	mm
	Average diameter/width	55.85 ± 0.04	37.89	mm
	Volume	2.020×10^{-4}	8.647×10^{-5}	m^3
	Average weight	495.02	203.22	g
	Density	2.451	2.35	g cc^{-1}
	Grain density	2.7	2.71	g cc^{-1}
	Moisture weight	28.7	11.95	g
	Moisture content	6.2	6.3	%
	Dry weight	466	191.27	g
	Dry density	2.31	2.21	g cc^{-1}
	Void ratio	0.174		
	Porosity	14.8	18.5	%
	Degree of saturation	96	85	%
Experimental boundary conditions	Confining pressure	12.5	12	MPa
	Axial load	13	12	MPa
	Pore pressure	4.5–10.5	4/6	MPa
	Back pressure	4.5	3.5	MPa
	Pore fluid chemistry	Chemically balanced pore fluid†		

*An assumed specific gravity for the mineral phases of 2.70 Mg m^{-3} (Zhang *et al.* 2007) was used in these calculations.
†227 mg l^{-1} Ca^{2+}, 125 mg l^{-1} Mg^{2+}, 1012 mg l^{-1} Na^{+}, 35.7 mg l^{-1} K^{+}, 1266 mg l^{-1} SO_4^{2-}, 4.59 mg l^{-1} Si, 9.83 mg l^{-1} SiO_2, 13.5 mg l^{-1} Sr, 423 mg l^{-1} total S, 0.941 mg l^{-1} total Fe.

and chlorite. Upon receipt of the preserved T1-cell core barrels at BGS, the material was catalogued and stored under refrigerated conditions of 4 °C to minimize biological and chemical degradation. The test sample was prepared by dry machine-lathing parallel to bedding, and the ends were ground flat and parallel giving an 82.45 mm length sample. The starting water saturation was 96%, with the early stages of testing designed to raise this to full saturation. Table 1 summarizes the geotechnical properties of the starting material; on completion of the full test programme the sample was seen to have saturation close to 100%.

Test SPP_COx-2 lasted a total of 566 days, with the results from the first 320 days up to the point of attainment of near-steady-state gas flow presented in this paper (see Table 2 for a summary of test boundary conditions). At the start of testing the sample was loaded to the *in situ* conditions at Bure with a confining pressure of 12.5 MPa, axial stress of 13 MPa and a pore pressure of 4.5 MPa. The initial stage of testing was designed to resaturate the sample fully and lasted 47 days. The sample exhibited a negative volumetric strain (i.e. volume increase) due to swelling. A constant head hydraulic test was subsequently performed in order to both remove any residual gas from the sample and to define the baseline hydraulic properties (permeability and specific storage) of the sample; this stage of testing lasted 79 days. This was followed by a gas injection experiment.

Experimental apparatus at LAEGO–ENSG–Université de Lorraine (Nancy)

A laboratory test was developed at LAEGO–ENSG–Université de Lorraine (Nancy) in order to give rise to dilatancy-controlled gas flow (Fig. 1, mode iii) and to gas transport in tensile fractures (Fig. 1, mode iv). The apparatus (Fig. 3) allowed gas flow to be monitored in samples submitted to a stress state close to the natural stress state under different gas pressures. The occurrence of either/both transport modes (iii, iv) induces an increase of gas and water flow as well as dilatant sample strain. The measurement of flow and strain should make it possible to distinguish between either transport mode.

The experimental apparatus consisted of a triaxial load cell, equipped with two fluid circuits, one for water and one for gas. Pressure sensors were installed on both circuits to measure the evolution of gas and water pressures at different places throughout the entire test history.

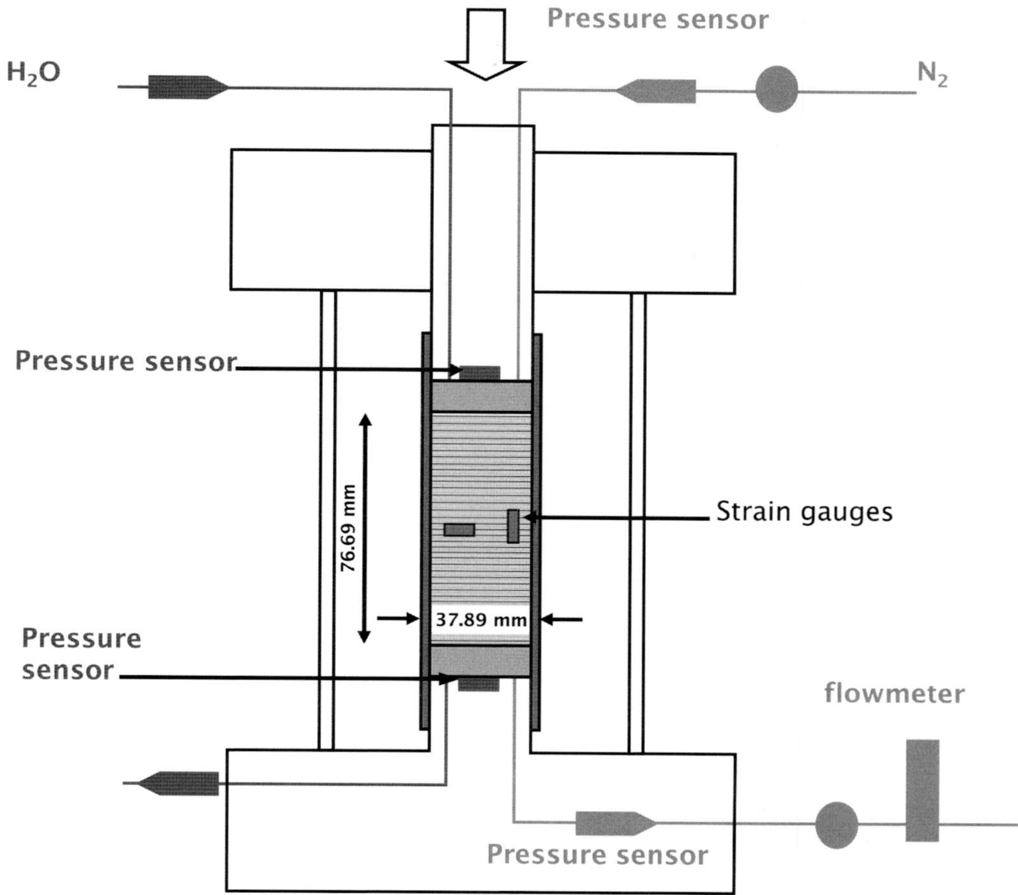

Fig. 3. Schematic of the experimental setup at LAEGO.

The presence of both gas and water required specific flow meters, with respective precisions of 10^{-3} g h^{-1} and 10^{-4} l min^{-1}. Strain gauges were glued directly on the sample to continuously measure axial and lateral deformations of the sample. Hydrostatic (for confining pressure) and pore (during saturation phase) pressures were produced through ISCO syringe pumps. Gas pressure was produced by a third ISCO syringe pump coupled with an oil-to-gas exchanger located between the pump and the cell. The oil-to-gas exchanger comprised a reservoir consisting of two compartments separated by a hermetic membrane; pressure was generated in the compartment containing oil and transmitted, through the membrane, to the compartment containing the gas.

The test consisted of three steps:

(1) Saturation of the sample. A sample was saturated under a hydrostatic stress state of 12 MPa with a difference between injected upstream and downstream pressures of 0.5 MPa. This step could typically last two to three months, depending on the sample. Axial and transversal strains were measured with strains gauges.

(2) Increase of gas pressure. An initial gas pressure of 4 MPa was applied upstream to the sample. Strains and outflowing gas and water flux were measured; pressure sensors were located along the system to check for constant gas pressure. After stabilization of strain and flux, a second gas pressure of 6 MPa was applied, followed by a third gas pressure level of 8 MPa.

(3) A third step can be considered if both the metrology and the sample allow it: a deviatoric stress can be applied to induce rupture (or the maximal stress of 20 MPa of the experimental apparatus).

Table 2. *Summary of experimental history conducted at BGS showing stage number, description of stage, axial stress and confining stress*

Stage number	Description	Axial stress (MPa)	Confining stress (MPa)	Injection pore pressure (MPa)	Test time at start (days)	Length of stage (days)
1	Saturation and swelling	12.5	11.5	4.5	0.1	21.7
2	Equilibration	13.0	12.5	4.5	21.8	26.1
3a	Hydraulic testing	13.0	12.5	8.5	47.9	50.1
3b	Hydraulic testing	13.0	12.5	4.5	98.0	29
4a	Gas injection ramp (constant flow)	13.0	12.5	4.5–9.5	127.0	35.3
4b	Gas injection at constant pressure	13.0	12.5	9.5	162.3	25.5
4c	Gas injection ramp (constant flow)	13.0	12.5	9.5–10	187.8	22.2
4d	Gas injection at constant pressure	13.0	12.5	10	210.0	26.1
4e	Gas injection ramp (constant flow)	13.0	12.5	10–10.5	236.1	18.9
4f	Gas injection at constant pressure	13.0	12.5	10.5	255.0	320.0

The axis of the sample was oriented normal to the bedding plane, with a diameter of 37.89 mm and a height of 76.69 mm. The carbonate content was measured as 18%, initial saturation was 85%, with the first stage of the test designed to achieve full saturation. The saturation at the end of the test could not be determined as the sample was contaminated with confining fluid while the test was decommissioned. Table 1 summarizes the geotechnical properties of the starting material.

Test results at BGS

Gas testing began on day 131, with a constant-flow pressure ramp that raised the injection gas pressure from 4.5 to 9.5 MPa over a 35-day period (see Fig. 4a). Gas pressure was held constant for 25 days and no indication of flow was seen. A second pressure ramp with a duration of 22 days was initiated on day 187, raising the gas injection pressure to 10 MPa. The pressure was held constant for 26 days and again no evidence of gas flow was seen. A final pressure ramp was initiated on day 236, raising the gas injection pressure to 10.5 MPa over a 19-day period.

Gas entry was subsequently identified as first occurring on day 220, with an anomalous reading in the load cell at day 221.875 signifying that gas had propagated to this sink. No further gas propagation seems to have occurred during this constant-pressure stage.

By day 241 it was apparent that gas had started flowing, so it took a total of 119 days to carefully raise the injection pressure from *in situ* levels to that necessary to initiate gas flow.

A second response was seen in the load cell reading at day 237.32. This suggests that the first gas propogation event pathway(s) closed. A day after the load cell reading response the water pressure in the guard ring started to increase (see Fig. 5a); a day after this, the first indication of deformation at the mid-plane was seen (Fig. 5b). The following day, the water pressure in the backpressure guard ring started to increase. These changes are interpreted as being a hydromechanical response to the onset of gas migration and the displacement of water from the injection guard ring due to the geometry of the test apparatus.

Gas migration was seen to result in an accelerated dilation at the mid-plane at around day 255 (see Figs 4b & 5b) that cannot be described by the mechanical response of a change in effective stress alone. The sample behaviour suggests that gas had migrated along dilational pathways. Gas reached the backpressure guard ring by around day 273. Marked deformation occurred following the onset of gas flow with a total of 48 μm of length dilation occurring and an average radial dilation of *c.* 16 μm. This represents a total volumetric strain of 0.18%, or a 360 μl change in sample volume. During the remainder of the gas tests, the radial strain trace exhibited the same functional form as that for the outflow of gas from the sample, providing conclusive proof that gas flow was accompanied by dilation of the COx.

Dilatancy observed at LAEGO–ENSG–Université de Lorraine (Nancy)

The sample of COx was fully saturated under an isotropic stress of 12 MPa, close to the *in situ* state of stress for Bure, for four months. Axial and radial strains were measured using strain gauges glued to the sample, and gas and water flux and pressure were monitored. An initial gas pressure of 4 MPa was applied to the sample for 24 days (Fig. 6a). The pressure downstream of the sample initially decreased and then stabilized after ten days, with

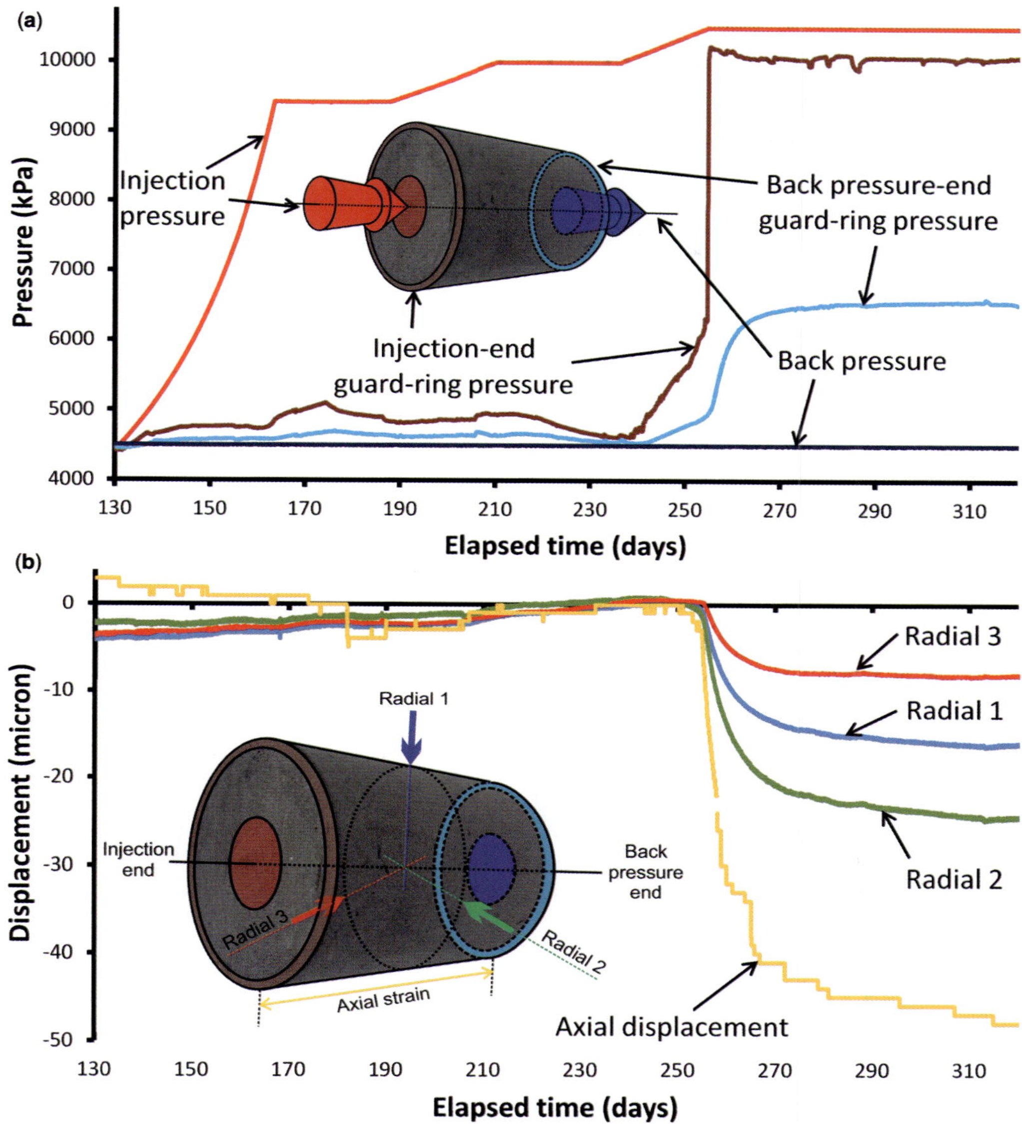

Fig. 4. Experimental results from the BGS test: (**a**) pressure recorded at four points within the experimental setup; (**b**) deformation recorded at four locations on the sample, clearly showing dilation.

the axial strain (Fig. 6b) stabilizing four days later. On day 25, the gas pressure was increased to 6 MPa, at which point the downstream pressure suddenly increased to 4.2 MPa. Thereafter, back-pressure decreased steadily without showing any conspicuous sign of stabilizing. In parallel, the axial and lateral strains increased rapidly and appeared to stabilize by day 38. This was followed by an acceleration of axial strain up to day 42. Alongside these accelerating strains, upstream and downstream gas pressures dropped suddenly, which seems to indicate that the sample underwent micro-cracking. These observations show that the onset of gas migration is described by model (iii), corresponding to gas micro-fissuring of the saturated sample. The maximum pressure that was reached seems lower than that conventionally measured, and the micro-fissuring phase corresponds to an

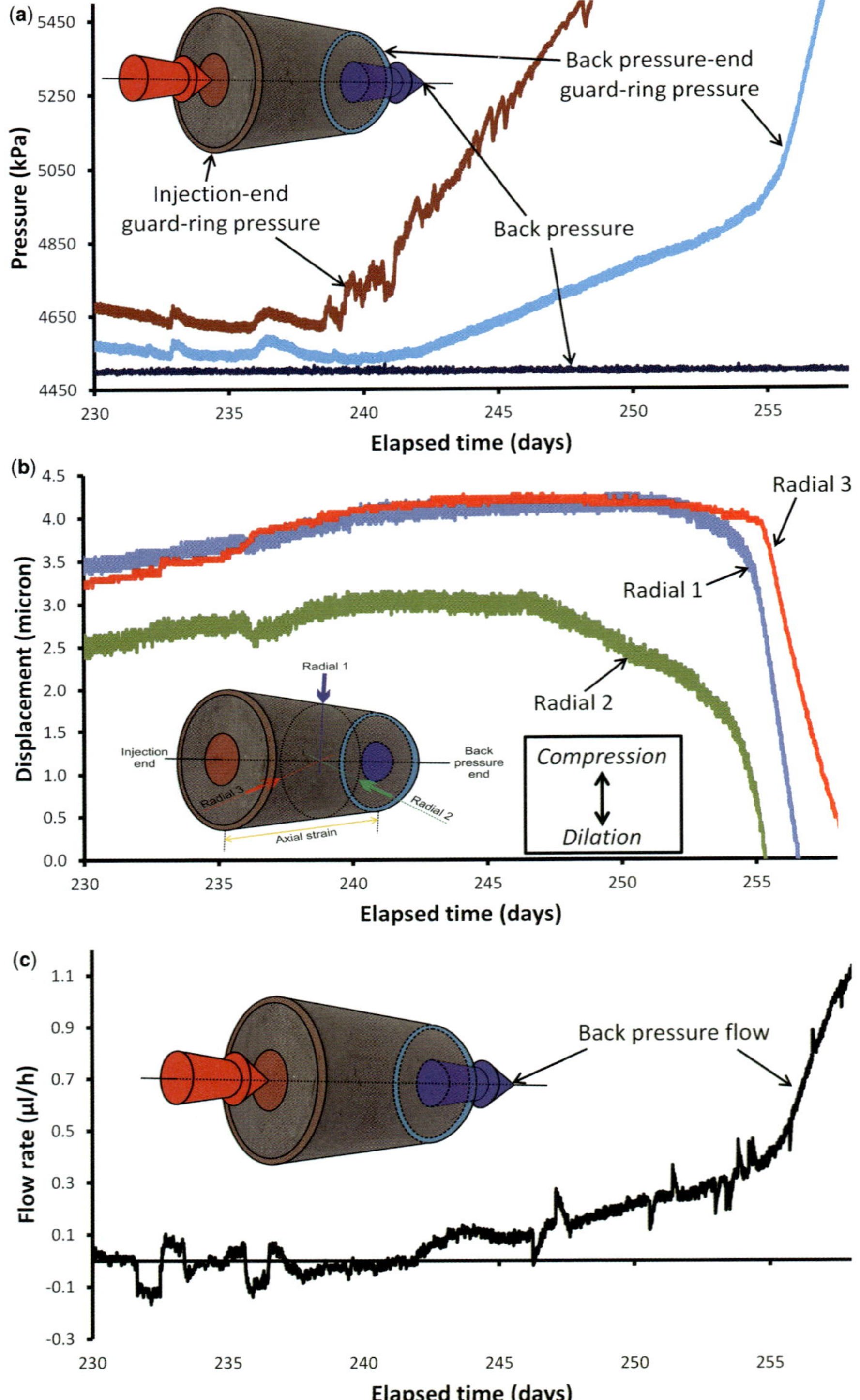

Fig. 5. Detail of data at the onset of gas flow: (**a**) pressure response; (**b**) deformation; (**c**) outflow from the sample.

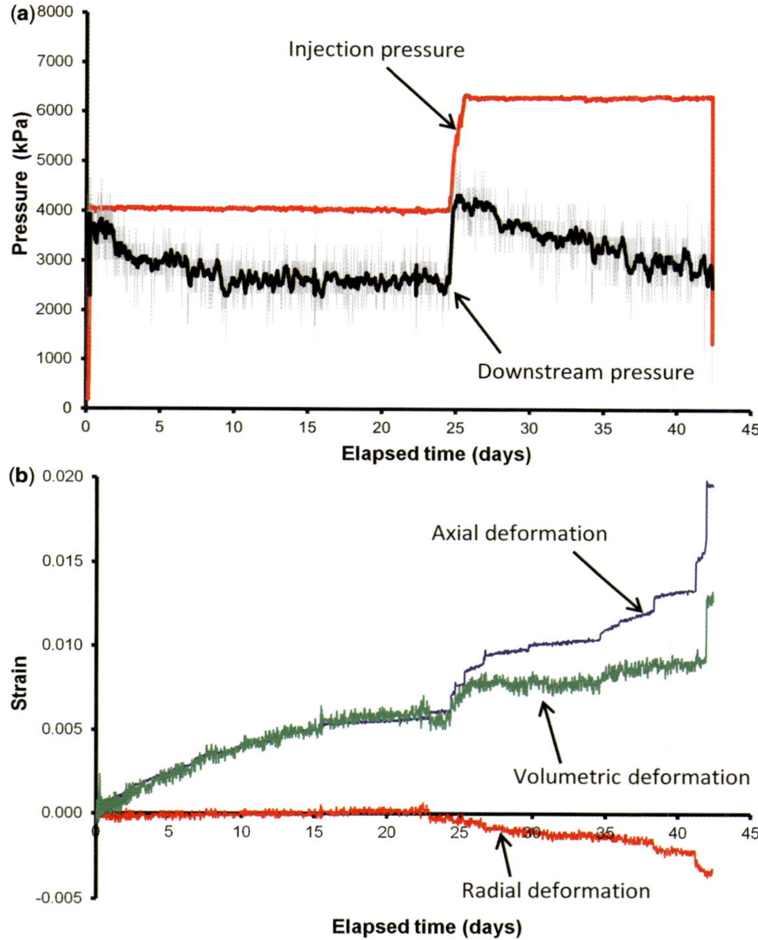

Fig. 6. Experimental results from the LAEGO test: (**a**) gas pressure recorded upstream and downstream of the sample; (**b**) deformation of the sample, clearly showing dilation. Positive strains are extensional/dilational.

acceleration of the axial and lateral strains of the sample until failure.

Discussion

Two independent studies have been conducted specifically to determine the mode of gas propagation in COx. Both studies used a bespoke apparatus designed to measure subtle volumetric strains observed during gas flow experimentation. In the case of BGS, three pressure 'dash-pots' directly measured the radial strain at three points at the mid-plane of a 56-mm-diameter sample. Axial strain was additionally measured by a digital micrometer. The approach at LAEGO used strain gauges cemented directly to the sample surface, giving axial and lateral strains. In both studies, strains of much less than 1% could be resolved with a high degree of confidence. Both experimental systems were seen to perform effectively and showed marked deformation at the point of gas flow. Both studies clearly show dilational volumetric deformation at the onset of gas flow.

Figure 7 shows the interpretation of the onset of gas migration in COx. A three-stage interpretation has been postulated, which fits all the observations of the BGS and LAEGO, and also previously published results (Horseman *et al.* 1996, 2004; Harrington & Horseman 1999; Angeli *et al.* 2009; Harrington *et al.* 2009, 2012).

Stage 1 (Fig. 7a). The determination of the point of gas entry is a difficult task and one that has eluded many research teams. If gas entry occurs during a

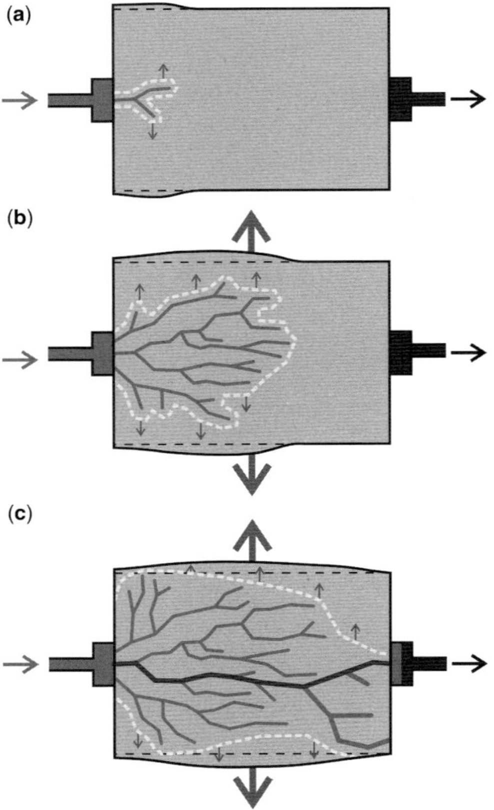

Fig. 7. Interpretation of the onset of gas flow in COx. (**a**) Onset of gas flow when gas started to enter the sample at the injection port. Each gas pathway causes compaction of COx around the pathway, causing bulk sample dilation. (**b**) Dendritic pattern of gas pathways formed, slowly propagating through the sample. Significant dilation was observed at the mid-plane as the pathways propagate past this location. (**c**) Gas migration had reached the backpressure end of the sample as seen in a rise of outflow. As more pathways were opened the sample continued to dilate.

stage of constant injection pressure ramp, it can be inferred by comparing the expected gas pressure from Boyle's Law with the measured pressure. (Boyle's Law states $P_1V_1 = P_2V_2$, where P is pressure and V is volume, so $P_{\text{predicted}} = P_1 \cdot V_1/V$.) At the onset of gas entry, a deviation between the two will occur. However, initial gas entry may only involve small volumes of gas, so the measured pressure may be undiscernibly close to the predicted pressure. If gas entry occurs during a constant pressure stage, the identification of gas flow means that the pressure is greater than the gas entry criteria. Because of the experimental geometry used at the BGS, secondary evidence of gas entry was identified, whereas LAEGO could not identify the pressure at which gas entry occurred.

At about day 220 in the BGS test, gas started to enter the sample when the injection pressure was 10 MPa. Gas migrated as far as the load cell (see Fig. 2b) located on the injection-end platen, where it was identified as an anomalous reading. It is probable that gas migrated along a single bedding layer of the COx as the test sample was oriented parallel to bedding. Gas propagation was along dilatant pathways that exploited the pore network of the material; the pathways can be visualized as 'ruptures' in the pore network. Around each pathway the fabric compressed, which may lead to localized movement of water away from the pathways.

Stage 1a. This stage is derived purely from the geometry of the BGS apparatus. However, it offers insight into gas flow. It is likely that a dendritic pattern of flow pathways was formed, similar in form to that proposed by Hildenbrand *et al.* (2002). At some time, gas migration reached the corner of the sample and interacted with the water-saturated guard rings. As gas entered this sink, water was displaced into the sample until the guard ring was saturated with gas. This resulted in sample swelling and was identified by a slow increase in injection guard-ring pressure. The start of the stage occurred at day 237.32 with a second anomalous reading in the injection-end load cell signifying the movement of gas. Pressure in the guard ring started to increase at day 238.55. A day later, at day 239.8, the first indication of deformation was recorded at the mid-plane array of radial strain measurement devices. This initial strain may have been a consequence of water expulsion from the guard ring and small amounts of sample swelling.

Stage 2 (Fig. 7b). The dendritic network of dilatant pathways propagated further into the sample and reached the mid-plane. In the BGS test this resulted in an acceleration of dilatant deformation around day 255, some 32 days after the first evidence of gas migration. Differences were seen in the timing of the accelerated deformation (see Fig. 5b): radial 2 (day 254), radial 1 (day 254.7), radial 3 (day 255.13), then axial deformation (day 265.03). If all of the observed radial deformation was a result of water being displaced from the guard ring (stage 1b) into the sample, it would be expected that axial deformation would have reacted first. Stage 2 also showed the onset of outflow from the sample, which may have derived from slug flow as gas migrating to the injection-end guard ring resulted in water being expelled into the sample and increased swelling. If all deformation was explained as derived from swelling this would not result in outflow from the sample. Therefore, even though the BGS experimental geometry was likely to result in swelling, the majority of observed deformation was caused by the dilatant propagation of gas along pathways and cannot be attributed to swelling alone.

Stage 3 (Fig. 7c). Gas had propagated through the sample as far as the backpressure end. The onset of gas reaching the backpressure guard ring (BGS test) at day 273 indicated that gas was now moving throughout the entire length of the sample, although a true steady-state flow had not been achieved, as seen by a mass imbalance between gas entering and exiting the sample. During this stage, the sample continued to slowly dilate, which may indicate that the number of gas pathways was continuing to increase. In the LAEGO test, the increase in upstream gas pressure was immediately followed by an increase in downstream gas pressure, indicating a connection between the ends of the sample. The increase of axial strains seems to show a dilatancy-controlled gas flow, but the low lateral strains suggest that this flow happened in micro-fissures rather than in a major conductive feature. At day 25, with the increase of upstream gas pressure from 4 to 6 MPa, the evolution of downstream gas pressure still suggests a connection between both ends of the sample. At this gas pressure level, the lateral strains began to increase, which suggest, as seen in the BGS test, the initiation in the sample of a major conductive feature (Fig. 7c).

Stage 4. In the LAEGO test, at day 42 the sudden drop in upstream and downstream gas pressures, as well as the sudden increase in axial and lateral strains, seem to show that the major conductive feature resulted in fracturing of the sample, accompanied by gas transport in this tensile fracture (Fig. 1, model iv). The BGS test showed no evidence of gas fracturing. This may indicate that the BGS test was held at a pressure marginally above the gas entry pressure, whereas the second pressure step (6 MPa) in the LAEGO test was greater than the gas entry pressure.

In the BGS test, the acceleration of strain at the mid-plane occurred approximately at day 255, whereas pressure at the backpressure guard ring initiated at day 240.5 (nearly 15 days earlier). This suggests that the backpressure response is a hydromechanical response, whereas the accelerated dilatation is one of gas movement. The pressure increase in the injection pressure guard ring is unstable and this is indicative of gas propagation, whereas the pressure increase at the backpressure guard ring is gradual and suggests that gas had not reached the backpressure end of the sample and is the result of the hydromechanical coupling of the sample. In the LAEGO test, dilatancy-controlled gas flow and gas transport in tensile fractures are barely dissociable, and only the mechanical response of the sample, measured with the strain gauges, allows an understanding of the response of the sample to gas injection. Both research teams have therefore clearly demonstrated that the hydromechanical coupling associated with gas movement and the mechanical response of the sample are essential information for fully understanding gas transport mechanisms. The mechanical deformation can be measured either by direct measurement, as in the BGS test using pressure-balanced dash-pots, or by strain gauges directly located on the sample, as in the LAEGO test.

Both experimental studies have shown a strong hydromechanical coupling, which could lead to the question of whether the observed deformation is derived purely as a mechanical response to a change in porewater pressure within the sample. Water is a compressible medium with a bulk modulus of 2.2 GPa. In the BGS test the starting moisture content of the sample was calculated to be 28.7 g (i.e. 28.7 ml), which, given a change in porewater pressure from 4.5 to 10.5 MPa gas pressure (i.e. a change in pore pressure of 6 MPa) would result in 78 μl change in bulk volume of the sample. This is considerably lower than the recorded total sample deformation of 360 μl. In addition, large pressure differentials were seen throughout the test sample, as seen in Figure 4a. Accordingly, the 'bulk sample' does not experience the full 6 MPa change in porewater pressure. A mean pore pressure of 7.9 MPa is observed from the four pressures recorded, giving rise to a change in average pore pressure of 3.4 MPa. This would result in an elastic deformation of 40 μl. Therefore, the full deformation seen by the sample cannot be explained by elastic deformation of the sample water alone. The gas-fracturing of the LAEGO test sample also clearly demonstrates that the strain recorded was not purely derived from elastic deformation of the porewater.

It should be noted that the two independent studies reported here yield significantly different gas entry pressures. The BGS study reports an excess gas entry pressure of 10.5 MPa (backpressure of 4.5 MPa), whereas LAEGO reports 6 MPa (backpressure of *c.* 3 MPa). Similar discrepancies in gas entry pressure for COx have been reported by Davy *et al.* (2012) and have been attributed to micro- or macro-cracking of COx, due to variations in confining pressure, initial sample state or sample preparation method. The sample used by BGS was prepared by machine lathing, whereas the LAEGO sample was prepared by diamond coring. The orientations of the tests samples were also different. For the BGS test the sample was orientated with the long axis parallel to the bedding direction, whereas the LAEGO study was performed perpendicular to bedding. Despite the difference in core orientation, gas injection testing reported by Harrington *et al.* (2012) also gave a high gas entry pressure similar to that reported here for a sample of COx orientated perpendicular to bedding. Therefore bedding direction does not have a major role in determining gas entry pressures and does not influence the underlying physics governing gas flow.

The current studies have attempted to determine the gas entry pressure, the pressure at which gas can be defined as entering the test sample. The BGS study used a series of pressure ramps and constant pressure stages to observe gas behaviour, whereas the LAEGO study used two constant pressure steps. In the LAEGO test, the two-step approach means that the recorded gas entry and breakthrough pressures are the same. In the BGS test, the gas entry pressure was seen to be 10 MPa, and breakthrough was not seen until a pressure of 10.5 MPa. The gas pressure ramp to increase pressure from 10 to 10.5 MPa was initiated prior to the identification of gas entry and it is uncertain whether gas breakthrough would have occurred at 10 MPa. It can be noted clearly that little difference is observed between gas entry and breakthrough pressures.

Conclusions

Two independent studies have been conducted on COx at the BGS and LAEGO. Both studies were conducted at *in situ* stresses representative of the Bure URL and examined the onset of gas flow through COx. The study conducted at BGS used a bespoke triaxial apparatus specifically designed to be able to resolve very small volumetric (axial and radial) strains potentially associated with the onset of gas flow. The LAEGO study utilized a triaxial setup with axial and radial strains measured by strain gauges glued to the sample. Both studies carefully monitored gas and water flux and pressures throughout long-duration experiments.

A four-stage model has been postulated to explain the experimental results. In stage 1, gas is seen to enter the sample at the gas entry pressure, probably in a dendritic pattern of flow paths. Gas propagation is along dilatant pathways that exploit the pore network of the material, and the pathways can be visualized as 'ruptures' in the pore network. Around each pathway the fabric compresses, which may lead to localized movement of water away from the pathways. In stage 2 the dendritic flow path network has reached the mid-plane of the sample, resulting in an acceleration of radial strain. During this stage, outflow from the sample also develops. In stage 3, gas has reached the back-pressure end of the sample, and free movement of gas occurs from sample end to end. The sample continues to dilate, which may indicate that the number of gas pathways is increasing. Both experimental studies show subtle lateral strain, suggesting that gas flow occurs along micro-fissures rather than in major conductive features (i.e. tensile fractures). Stage 4 represents the gas-fracturing of COx and was only seen in the test conducted by LAEGO. A significant tensile fracture forms, resulting in failure of the sample.

The deformation observed showed considerable hydromechanical coupling. Simple elastic deformation of the porewater could not fully describe the deformation seen by the sample, and the gas-fracturing of the LAEGO experiment clearly demonstrated that gas flow occurred by dilatant pathway formation. The gas entry pressures measured by the two experimental groups provide significantly different gas entry pressures (BGS, 10.5 MPa; LAEGO 6 MPa). This may be explained by differences in sample preparation technique.

Funding for the study was provided by Agence Nationale pour la Gestion des Déchets Radioactifs, Andra (within the auspices of the 'Transfert de Gaz' initiative), the European Atomic Energy Community's Seventh Framework Programme (FP7/2007–2011) under grant agreement no. 230357 (the FORGE project), and the BGS through its well-founded laboratory programme and the Geosphere Containment project (part of the BGS core strategic programme). This paper is published with the permission of the Executive Director, British Geological Survey (NERC).

References

Angeli, M., Soldal, M., Skurtveit, E. & Aker, E. 2009. Experimental percolation of supercritical CO_2 through a caprock. *Energy Procedia*, **1**, 3351–3358.

Askarieh, M. M., Chambers, A. V., Dabniel, F. B. D., Fitzgerald, P. L., Holtom, G. J., Pilkington, N. J. & Rees, J. H. 2000. The chemical and microbial degradation of cellulose in the near field of a repository for radioactive wastes. *Waste Management*, **20**, 93–106.

Autio, J., Gribi, P., Johnson, L. & Marschall, P. 2006. Effect of excavation damage zone on gas migration in a KBS-3H type repository at Olkiluotu. *Physics and Chemistry of the Earth*, **31**, 649–653.

Aziz, K. & Settari, A. 1979. *Petroleum Reservoir Simulation*. Applied Science, London.

Cuss, R. J. & Harrington, J. F. 2010. *Effect of Stress Field and Mechanical Deformation on Permeability and Fracture Self-Sealing: Progress Report on the Stress Path Permeameter Experiment Conducted on Callovo-Oxfordian Claystone*. British Geological Survey Commissioned Report **CR/10/151**.

Cuss, R. J. & Harrington, J. F. 2011. *Update on Dilatancy Associated with Onset of Gas Flow in Callovo-Oxfordian Claystone; Progress Report on Test SPP_COx-2*. British Geological Survey Commissioned Report **CR/11/110**.

Cuss, R. J., Harrington, J. F. & Noy, D. J. 2012. *Final Report of FORGE WP4.1.1: The Stress-path Permeameter Experiment Conducted on Callovo-Oxfordian*

Claystone. British Geological Survey Commissioned Report **CR/12/140**.

Davy, C. A., M'Jahad, S., Skoczylas, F., Talandier, J. & Ghayaza, M. 2012. Evidence of discontinuous and continuous gas migration through undisturbed and self-sealed COx claystone. Oral Presentation O/04/2, *Clays in Natural and Engineered Barriers for Radioactive Waste Confinement*, 22–25 October 2012, Montpellier, France.

Ekeroth, E., Roth, O. & Jonsson, M. 2006. The relative impact of radiolysis products in radiation-induced oxidative dissolution of UO_2. *Journal of Nuclear Materials*, **355**, 38–46.

Gallé, C. & Tanai, K. 1998. Evaluation of gas transport properties of backfill materials for waste disposal: H_2 migration experiments in compacted Fo-Ca clay. *Clays and Clay Minerals*, **46**, 498–508.

Harrington, J. F. & Horseman, S. T. 1997. Projects on the effects of gas in underground storage facilities for radioactive waste (Pegasus project): Gas migration in clay. *Proceedings of a Progress Meeting held in Mol, Belgium*, 28–29 May 1997, European Science and Technology Series (1998) EUR 18167 EN, 153–173.

Harrington, J. F. & Horseman, S. T. 1999. Gas transport properties of clays and mudrocks. *In*: Aplin, A. C., Fleet, A. J. & Macquaker, J. H. S. (eds) *Muds and Mudstones: Physical and Fluid Flow Properties*. Geological Society, London, Special Publications, **158**, 107–124.

Harrington, J. F., Horseman, S. T. & Noy, D. J. 2003. *Measurements of Water and Gas Flow in Opalinus Clay using a Novel Guard-Ring Methodology*. British Geological Survey Technical Report **CR/03/32**.

Harrington, J. F., Noy, D. J., Horseman, S. T., Birchall, J. D. & Chadwick, R. A. 2009. Laboratory study of gas and water flow in the Nordland Shale, Sleipner, North Sea. *In*: Grobe, M., Pashin, J. C. & Dodge, R. L. (eds) *Carbon Dioxide Sequestration in Geological Media—State of the Science*. AAPG, Tulsa, Oklahoma, Studies in Geology, **59**, 521–543.

Harrington, J. F., de la Vaissière, R., Noy, D. J., Cuss, R. J. & Talandier, J. 2012. Gas flow in Callovo-Oxfordian Clay (COx): Results from laboratory and field-scale measurements. *Mineralogical Magazine*, **76**, 3303–3318.

Hildenbrand, A., Schlömer, S. & Kroos, B. M. 2002. Gas breakthrough experiments on fine-grained sedimentary rocks. *Geofluids*, **2**, 3–23.

Horseman, S. T. & Harrington, J. F. 1994. *Migration of Repository Gases in an Overconsolidated Clay*. British Geological Survey Technical Report **WE/94/7**.

Horseman, S. T., Harrington, J. F. & Sellin, P. 1996. Gas migration in Mx80 buffer bentonite. *In*: *Symposium on the Scientific Basis for Nuclear Waste Management XX (Boston)*. Materials Research Society, **465**, 1003–1010. http://dx.doi.org/10.1557/PROC-465-1003

Horseman, S. T., Harrington, J. F. & Sellin, P. 1999. Gas migration in clay barriers. *Engineering Geology*, **54**, 139–149.

Horseman, S. T., Harrington, J. F. & Sellin, P. 2004. Water and gas flow in Mx80 bentonite buffer clay. *In*: *Symposium on the Scientific Basis for Nuclear Waste Management XXVII (Kalmar)*. Materials Research Society, **807**, 715–720. http://dx.doi.org/10.1557/PROC-807-715

Kreis, P. 1991. *Hydrogen Evolution from Corrosion of Iron and Steel in Low/Intermediate Level Waste Repositories*. Nagra Technical Report **NTB 91-21**.

Marschall, P., Horseman, S. & Gimmi, T. 2005. Characterisation of gas transport properties of the Opalinus Clay, a potential host rock formation for radioactive waste disposal. *Oil & Gas Science and Technology – Revue de l'Institute Frances Petrole*, **60**, 121–139.

Neretnieks, I. 1984. Impact of alpha-radiolysis on the release of radionuclides from spent fuel in a geologic repository. *Proceedings of the Materials Research Society Symposium*, Boston, MA, 14 November 1983, **26**, 1009–1022. http://dx.doi.org/10.1557/PROC-26-1009

Ortiz, L., Volcharet, G. *et al.* 1996. MEGAS Modelling and Experiments on Gas Migration in Repository Host-rocks. Nuclear Science and Technology Series, EUR 16746 EN, Luxembourg, 127–147.

Ortiz, L., Volckaert, G. & Mallants, D. 2002. Gas generation and migration in Boom Clay, a potential host rock formation for nuclear waste storage. *Engineering Geology*, **64**, 287–296.

Smart, N. R., Carlson, L., Hunter, F. M. I., Karnland, O., Pritchard, A. M., Rance, A. P. & Werme, L. O. 2006. *Interactions Between Iron Corrosion Products and Bentonite*. Serco Assurance Report **SA/EIG/12156/C001**.

Weetjens, E. & Sillen, X. 2006. Gas generation and migration in the near field of a supercontainer-based disposal system for vitrified high-level radioactive waste. *In*: *Proceedings of the 11th International High-Level Radioactive Waste Management Conference (IHLRWM)*, Las Vegas, Nevada, 30 April–4 May 2006. American Nuclear Society (ANS), 1–8.

Wenk, H.-R., Voltolini, M., Mazurek, M., Van Loon, L. R. & Vinsot, A. 2008. *Preferred orientations and anisotropy in shales: Callovo-Oxfordian Shale (France) and Opalinus Clay (Switzerland)*. Clays and Clay Minerals, **56**, 285–306.

Wikramaratna, R. S., Goodfield, M., Rodwell, W. R., Nash, P. J. & Agg, P. J. 1993. *A Preliminary Assessment of Gas Migration from the Copper/Steel Canister*. SKB Technical report **TR93-31**.

Zhang, C.-L., Rothfuchs, T., Su, K. & Hoteit, N. 2007. Experimental study of the thermo-hydro-mechanical behaviour of indurated clays. *Physics and Chemistry of the Earth*, **32**, 957–965.

Laboratory gas injection tests of compacted bentonite buffer material for TRU waste disposal

KAZUTO NAMIKI[1], HIDEKAZU ASANO[1]*, SHINICHI TAKAHASHI[2], TOMOYUKI SHIMURA[2] & KEN HIROTA[3]

[1]*Radioactive Waste Management Funding and Research Center (RWMC), 1-15-7, Tsukishima, Chuo-ku, Tokyo, Japan*

[2]*Obayashi Corporation, 2-15-2, Konan, Minato-ku, Tokyo, Japan*

[3]*Toyo Engineering Corporation, 2-8-1 Akanehama, Narashino-city, Chiba, Japan*

**Corresponding author (e-mail: asano@rwmc.or.jp)*

Abstract: The Radioactive Waste Management Funding and Research Center (RWMC) is leading a research programme to evaluate the gas transport mechanisms through a TRU (TRans-Uranium) waste disposal facility in Japan and acquire information on gas migration properties. In this paper we describe a series of laboratory gas injection tests using the bentonite adopted for use in Japanese TRU disposal, as well as an attempted visualization of the gas migration path inside the bentonite used as a buffer. By building a conceptual model from the results of these tests, the characteristics of gas migration through to breakthrough for bentonite can now be better understood in the context of Japanese TRU disposal. Thanks to the outcome of this research project, advanced knowledge may be applied to the conceptualization of TRU waste disposal.

TRU waste is generated during the operation and decommissioning of reprocessing plants and Mixed Oxide (MOX) fuel fabrication plants (NUMO 2008). In the context of TRU waste disposal it is important to assess the behaviour of gas generated inside the waste, and to evaluate gas transport mechanisms through a TRU waste disposal facility it is essential to understand the gas migration phenomena observed in previous research. Previous large-scale gas migration tests have mainly been performed for the purpose of understanding relevant phenomena under realistic site conditions.

The acquisition and expansion of fundamental data regarding the bentonite buffer material (Kunigel V1, with an effective clay density of 1.36 Mg m^{-3}) adopted for use in Japanese TRU disposal is also important in order to achieve a realistic understanding of the effects of gas migration (Fig. 1).

In research into gas migration with bentonite as a buffer material, the results of gas injection tests using MX80 (Horseman *et al.* 1999) reported the existence of a threshold value in breakthrough pressure and the formation of a migration path that does not depend on two-phase flow. Graham *et al.* (2002) established the relationship between breakthrough pressure and degree of saturation from the results of injection tests using bentonite from Canada. They also showed that breakthrough pressure rises rapidly near saturation. In Japan, Tanai & Yamamoto (2003) performed injection tests with Kunigel V1 and attempted visualization of the sample after breakthrough by using Computerized Axial Tomography (CAT) radiography. However, there are relatively few examples of previous gas injection tests in Japan. The Radioactive Waste Management Funding and Research Center (RWMC) is therefore carrying out a series of laboratory gas injection tests with a view to acquiring data on the gas migration properties of relevant materials under assumed disposal conditions in Japan.

Experimental concept

In this study we designed experiments to study three phenomena: (re-)saturation, gas migration and gas breakthrough (sudden flow increase) in a gas migration scenario (Fig. 2). Two sizes of bentonite columns were taken, with heights of 50 mm and 25 mm, respectively. Both types of column had a diameter of 60 mm and a dry density of 1.36 Mg m^{-3}. The bentonite (Kunigel V1) had a 46–49 wt% smectite content of Na type. Figure 3 shows an overview of the experimental system, which consisted of an operation panel, a data logger, a water tank, a nitrogen gas cylinder and a column for storing the bentonite. Figure 4 presents a detailed sketch of the structure inside a column. The test apparatus consisted of a lower loading platform, a bentonite column mould and an upper loading platform. The bentonite columns were saturated with water and then gas was injected at the lower ends of the columns. A load cell was installed in the lower loading platform, and the

From: NORRIS, S., BRUNO, J., CATHELINEAU, M., DELAGE, P., FAIRHURST, C., GAUCHER, E. C., HÖHN, E. H., KALINICHEV, A., LALIEUX, P. & SELLIN, P. (eds) 2014. *Clays in Natural and Engineered Barriers for Radioactive Waste Confinement*. Geological Society, London, Special Publications, **400**, 521–529.
First published online May 12, 2014, http://dx.doi.org/10.1144/SP400.27

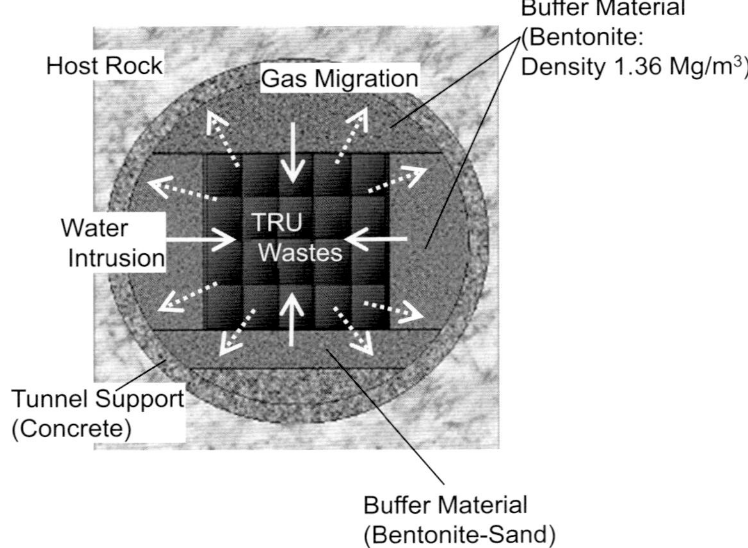

Fig. 1. Gas migration considered in the context of TRU waste disposal in Japan (adapted from JAEA *et al.* 2007).

swelling pressure of the sample was measured. Bentonite columns with an initial water saturation of 90% were used in order to reduce the time needed for saturation. Table 1 presents the test specifications and conditions, and Figure 5 shows the procedure used in the gas injection experiment, which consisted of a production phase (adjustment of density and degree of saturation with compacting bentonite powder after adding water) of the buffer model, a saturation phase and a gas injection phase. The saturation phase was maintained until water injected (with constant pressure, example shown in Fig. 6) at the bottom of the column was discharged from the top surface of the column. The amount of water outflow and the load at the bottom were measured at this time. In the gas injection phase, a stepwise pressurization (0.1 MPa/2 days) approach was adopted. We carried out about ten trials using the abovementioned test system. In general, each test was conducted until breakthrough occurred (with a sudden increase of displacement), and the displacement volumes of water and gas from the top surface of the column were measured at this time (separation of nitrogen gas and water was performed using a burette). In addition, a visualization investigation

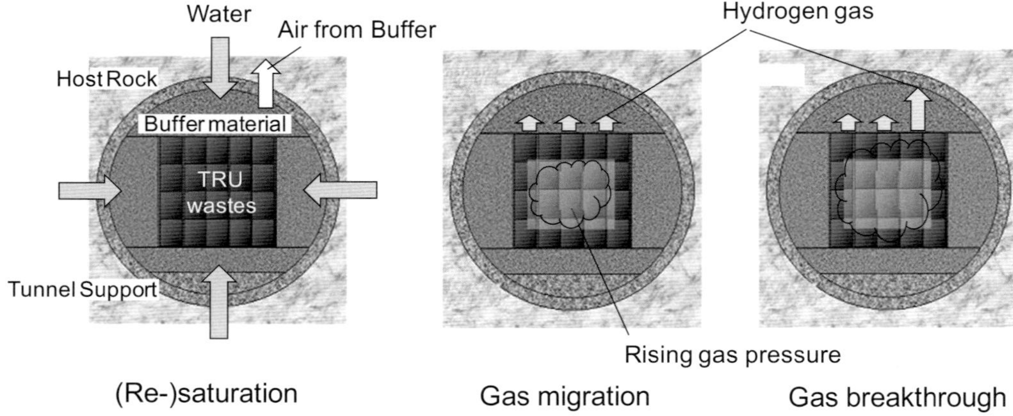

Fig. 2. Scenario used in the experiments.

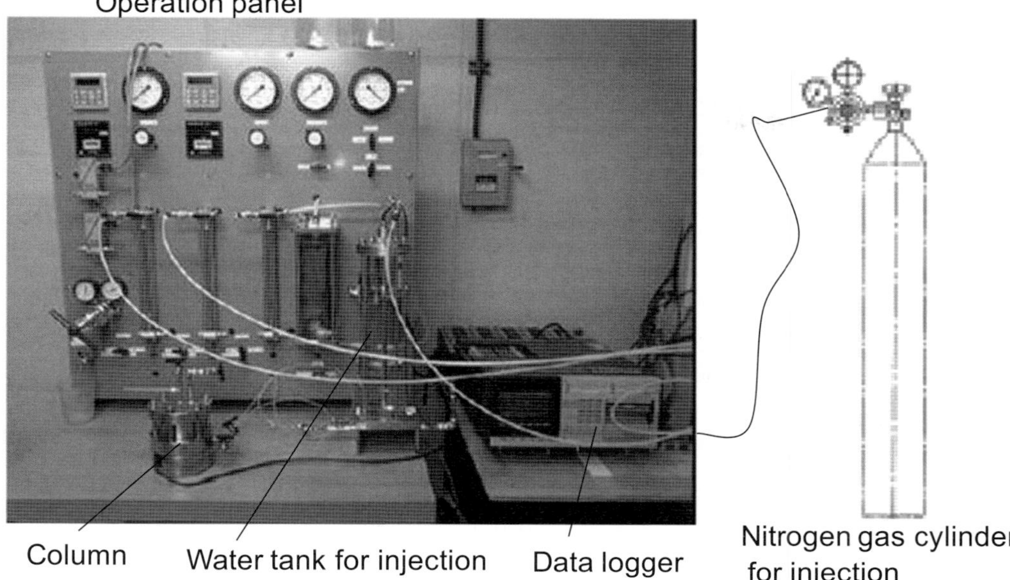

Fig. 3. Overview of the experimental system.

was conducted to assess the situation of the breakthrough course after breakthrough.

Test results

Gas permeability in saturated bentonite

For samples of both sizes (heights of 25 mm and 50 mm) we carried out ten tests and confirmed the reproducibility of the observed phenomena regarding both the water-saturation phase and the gas-migration phase. During the water-saturation phase of all tests, drainage from the top surface of a sample could be checked by injecting a quantity far exceeding the calculated amount of water required for saturation. Figures 6–8 show the experimental results of a typical gas injection test (column height, 50 mm; pressure increment, 0.1 MPa over 2 days). Figure 6 shows the total amount of water outflow from the top surface and changes in the swelling pressure as estimated according to the load at the bottom in the saturation phase. Water outflow from the top surface was checked by injecting an amount of water exceeding the amount at saturation by *c.* 5%, as calculated from the water budget and evaporation from the column. Figure 7 shows changes in the amount of water outflow from the top surface v. gas permeability during the gas injection phase. Water outflow from the outer section of the top surface started to increase rapidly after the gas injection pressure reached

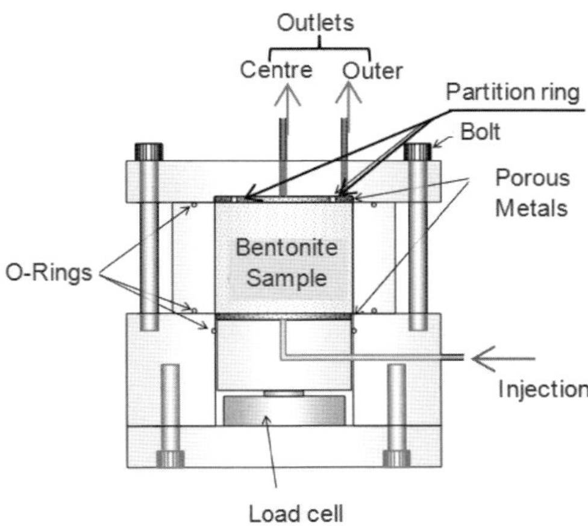

Fig. 4. Column details.

Table 1. *Test specifications*

Specifications	
Material	Kunigel V1
Dry density	1.36 Mg m^{-3}
Saturation ratio	90%
Diameter	60 mm
Height	25 mm, 50 mm

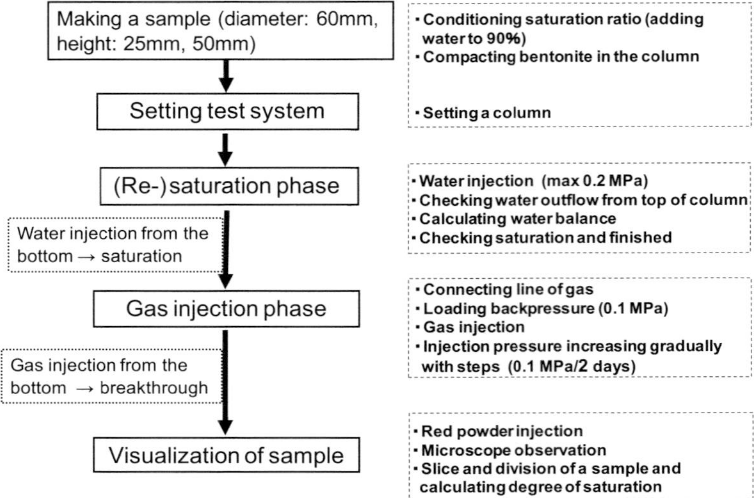

Fig. 5. Flow chart of the experiment procedure.

1.7 MPa, and gas breakthrough then occurred at an injection pressure of 1.8 MPa. Figure 8 shows changes in the amount of water and gas outflow from the top surface of the column at the time of breakthrough.

Once gas breakthrough occurred, the amount of gas outflow from the outer section increased uniformly until injection ended. During this time, water outflow decreased immediately after the start of gas breakthrough and then began to increase again until the end of the test.

Following breakthrough on the 34th day there was little variation in the amount of water outflow with respect to the process of increasing gas outflow, but there was a noticeable variation in gas outflow. In the central section of the bentonite column there was no increase in either gas outflow or water outflow, although both values increased in the outer section.

Visualization of the gas flow paths

Additional tests were performed in order to understand the existence of the gas flow paths that led to a breakthrough.

Red powder injection. After breakthrough, gas injection continued for 24 h and the gas pathways

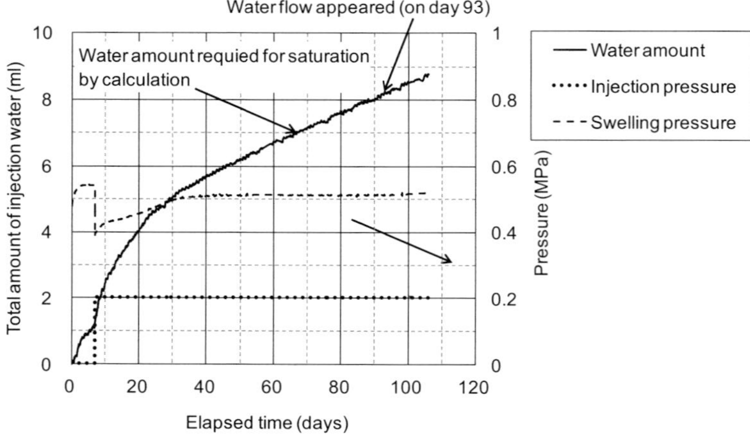

Fig. 6. Gas and water outflow from bentonite in water saturation phase.

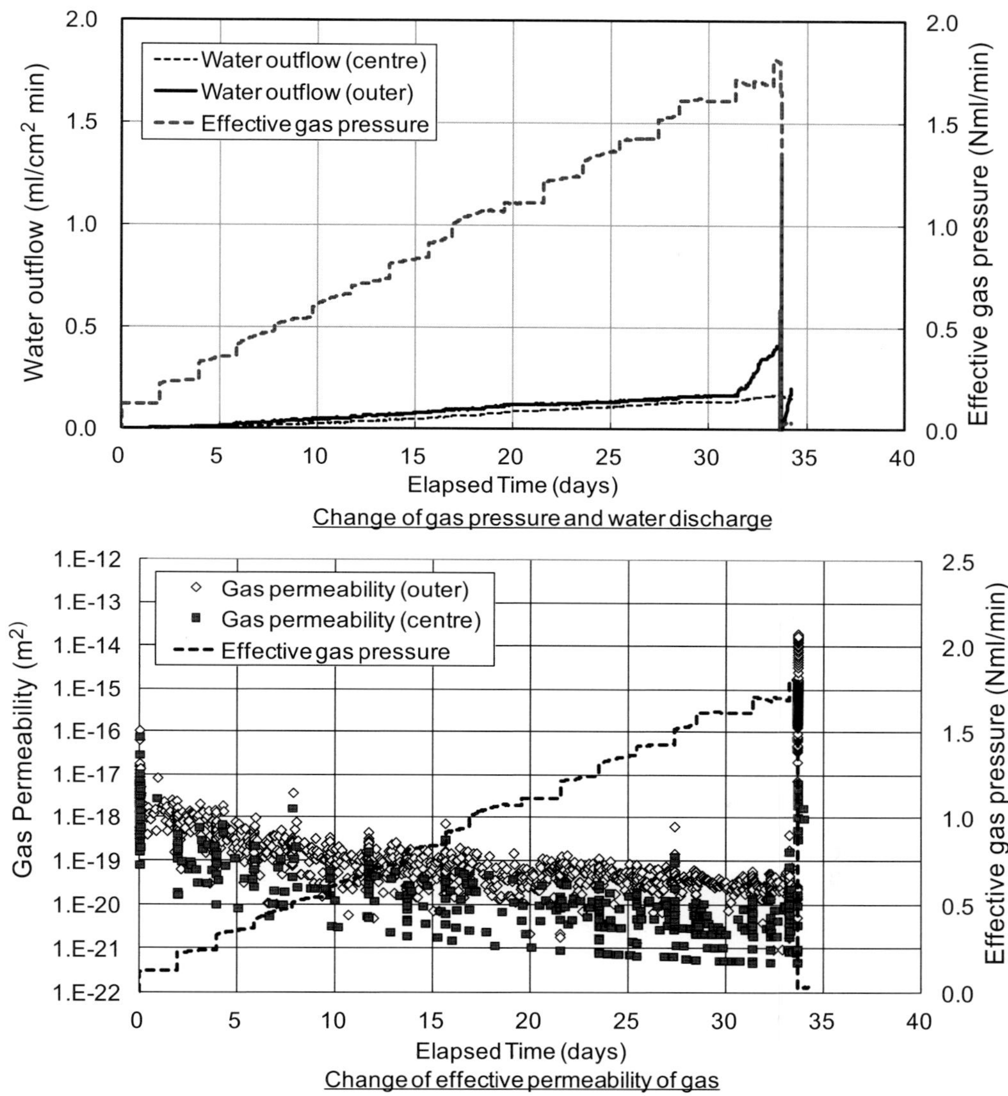

Fig. 7. Outflow and permeability of gas and water in the gas injection phase.

were allowed to dry. A red powder (synthetic pigment for food with a particle diameter of ≤ 10 μm) was then injected into the bentonite from the bottom using compressed air to visualize the gas flow pathways and develop an understanding of gas breakthrough. Figure 9 shows the bentonite dismantled in slices to illustrate the nature of the gas flow paths. Although many gas flow paths can be identified at the section near the bottom of the column, the paths gradually concentrate in one location at the periphery as the gas moves to the upper part. This result aligns with dilatant pathway formation, as observed in work undertaken previously, for example, by Horseman *et al.* (1999).

Microscope observations. Microscope observations at the bottom of the bentonite column and water content measurement of the bentonite were also carried out after gas breakthrough. Figure 10 shows the procedure for sample observation using a microscope, as well as a representative sample of the photographs taken. After the test (column height, 25 mm; pressure increment, 0.1 MPa over 2 days), the bentonite sample that had a

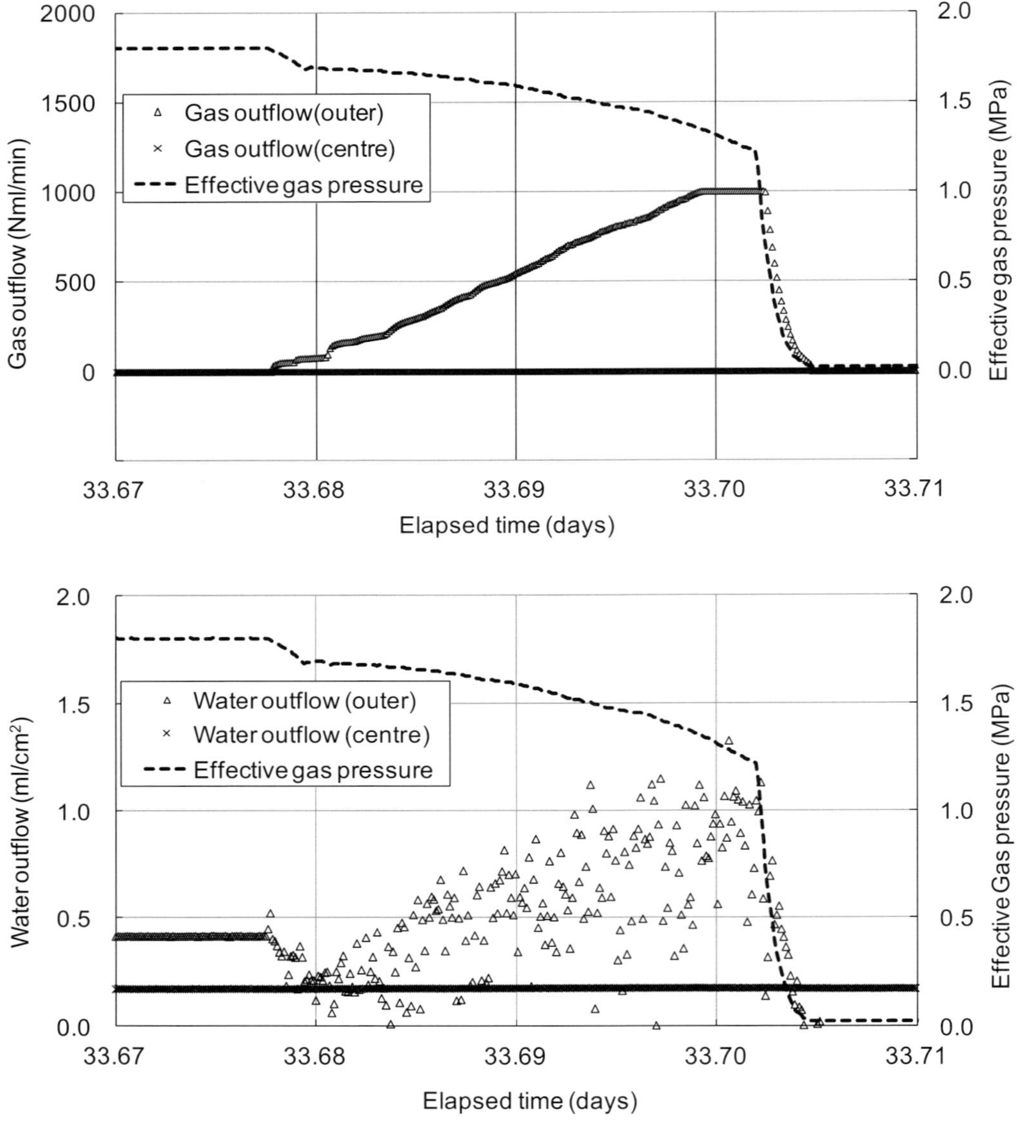

Fig. 8. Outflow and permeability of gas and water at breakthrough in the gas injection phase.

breakthrough at the end of the experiment was removed from the column, and the bottom of the sample was observed with a microscope. When observed at $\times 20$ magnification, a short crack was seen with shallow depth and little continuity, as well as a crack with great depth in which continuity could be recognized. The continuous deep crack was examined further by increasing the magnification to $\times 150$ and observing the inside of the crack. The depth of the continuous crack was *c.* 2–3 mm, and the crack differed from the one extending to the upper part as shown by the results of the red powder injection. This phenomenon was interpreted to indicate that, the crack in the bentonite sample had partially resealed, because an injection such as that with the red powder had not been performed before the microscope observation was carried out. We also tried to observe the interior of the sample, but were unable to do so because the sample could not be cut in such a way that the internal structures, including cracks, would remain.

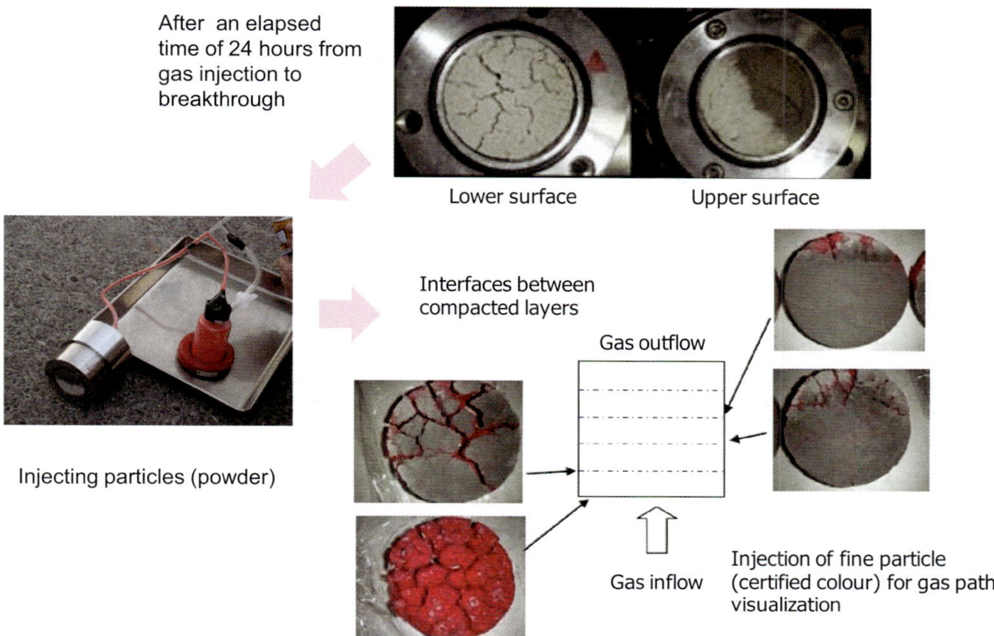

Fig. 9. Visualization of the gas flow paths by injecting red powder after breakthrough.

Evaluation of degree of saturation inside a sample. The bentonite sample that had not been subject to red powder injection was divided into sections at intervals of 5 mm and cut into a central portion (A) and four peripheral portions (B–E) after microscope observation. The mass of each fragment and

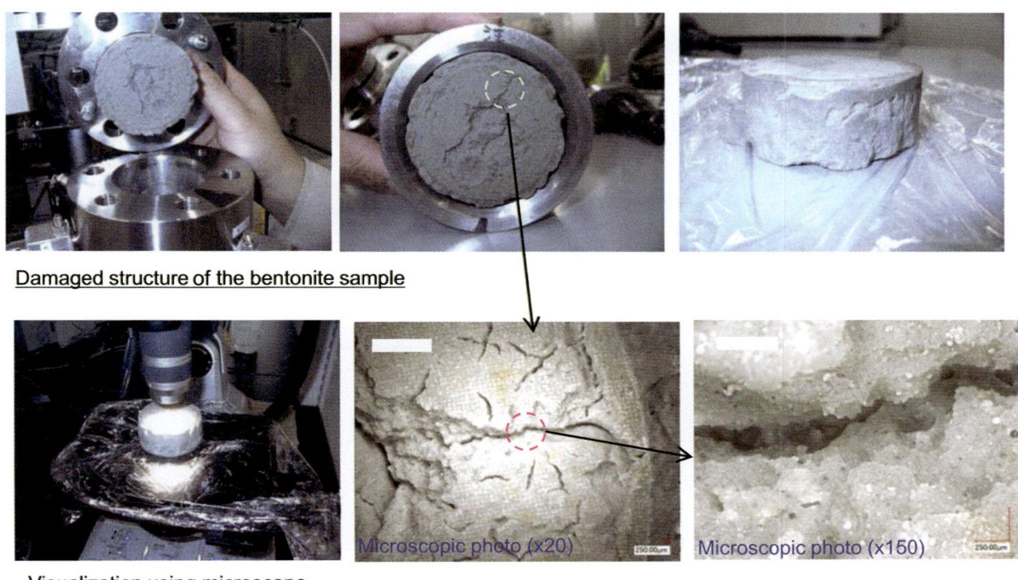

Fig. 10. Observation of the breakthrough path using a microscope after the gas injection test.

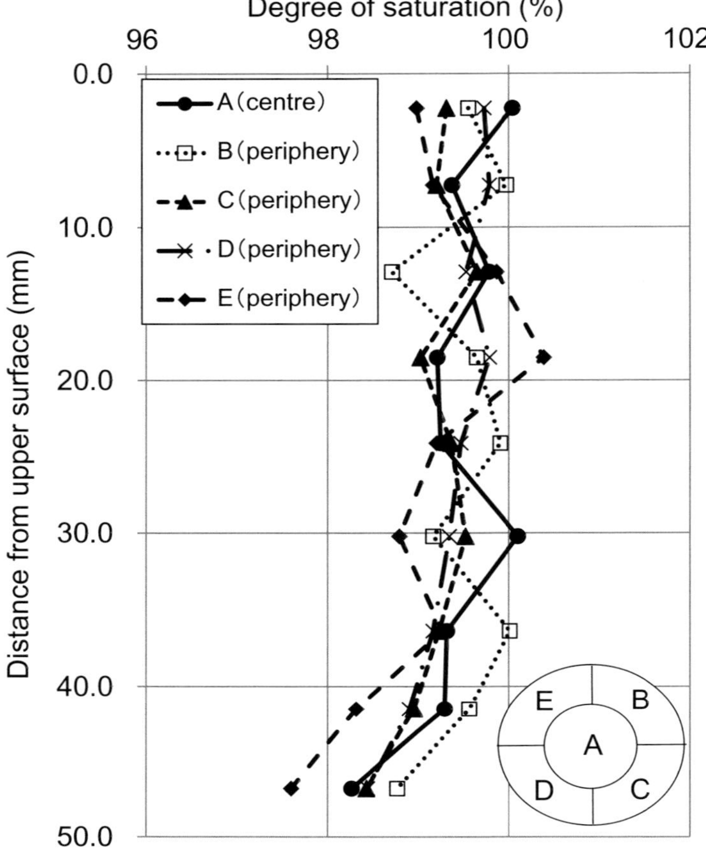

Fig. 11. Measurement of saturation degree inside the bentonite after breakthrough.

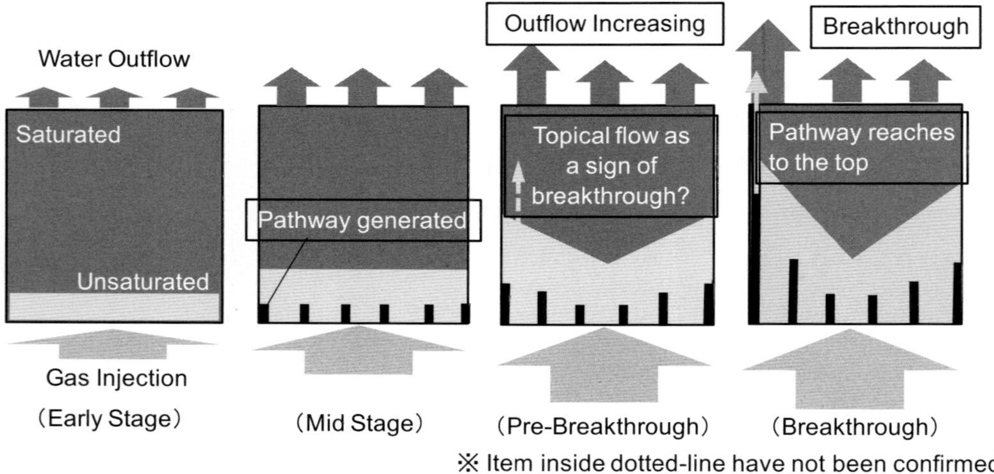

Fig. 12. The concept of gas migration derived from this empirical study.

its degree of saturation were then measured and plotted on a profile graph (Fig. 11). The inside of the test piece after breakthrough showed a high degree of saturation in excess of 97% in all portions. A comparison of the central and peripheral portions in the section that produced the breakthrough revealed higher saturation at the centre, with a difference of *c.* 1%.

A comparison of values in the peripheral portions revealed that portion E had lower saturation in an area extending *c.* 25 mm from the bottom, but the other profile sections were low in the upper range. We believe that although the gas had uniform upward penetration inside the material, after migrating within the material for part of its length breakthrough resulted at a certain distance from the bottom after the gas rose near a wall surface that formed an interface with the material near the top surface.

Discussion

A conceptual model of gas migration was derived based on the series of experimental results shown in this paper. A conceptual model is shown in Figure 12. The model consists of the following four stages:

(1) *Early stage.* In early stages of gas injection, an unsaturated zone occurs near the bottom and extends upwards, in a generally homogeneous pattern.
(2) *Mid stage.* In the gas migration process, pathways generate and extend upwards from the bottom of sample, and water outflow will occur homogeneously.
(3) *Pre-breakthrough.* When the gas pressure reaches a value corresponding to the breakthrough pressure, drainage in the periphery will increase compared with the centre. There is a possibility of an increase in topical flow as a sign of breakthrough.
(4) *Breakthrough.* Breakthrough occurs as the water outflow at the periphery increases further, generating a partial breakthrough at the top of the sample.

Although gas outflow associated with two-phase flow has been observed in the pre-breakthrough stage in a similar experiment (Tanaka *et al.* 2007), this gas outflow could not be verified in our series of experiments. This is because small gas outflows cannot be measured by the measurement system, so there is a possibility that a small gas outflow has occurred together with the outflow of water.

Conclusion

Laboratory injection tests were carried out to investigate gas migration through the candidate buffer (Kunigel V1 bentonite) in the Japanese TRU waste disposal concept. The observations are indicative of the migration of gas and water, resulting from gas breakthrough. A visualization technique was used to trace gas migration in a sample that had experienced breakthrough. The test results have provided significant basic data to support the understanding of the mechanisms of gas migration through the bentonite buffer material used for Engineering Barrier System (EBS) design and a performance assessment of TRU waste disposal.

Although future studies based on the results of sufficient laboratory tests will also be required to instil reliability into the evaluation of gas behaviour, the results of this paper will contribute to progress in the development of Japanese TRU waste disposal.

This study includes part of the results of the project 'Development of the technique for the evaluation of long-term performance of EBS, FY2010-11' under a grant from the Agency of Natural Resources and Energy, the Ministry of Economy Trade and Industry of Japan.

References

Graham, J., Halayko, K. G., Hume, H., Kirkham, T., Malcolm, G. & Oscarson, D. 2002. A capillary-adjective model for gas break-through in clays. *Engineering Geology*, **64**, 273–286.

Horseman, S. T., Harrington, J. F. & Sellin, P. 1999. Gas migration in clay barriers. *Engineering Geology*, **54**, 139–149.

JAPAN ATOMIC ENERGY AGENCY (JAEA) AND THE FEDERATION OF ELECTRIC POWER COMPANIES OF JAPAN 2007. *Second Progress Report on Research and Development for TRU Waste Disposal in Japan.* Repository Design, Safety Assessment and Means of Implementation in the Generic Phase.

NUCLEAR WASTE MANAGEMENT ORGANIZATION OF JAPAN (NUMO) 2008. *Geological Disposal of TRU Waste.*

Tanai, K. & Yamamoto, M. 2003. *Experimental and Modeling Studies on Gas Migration in Kunigel V1 Bentonite.* JNC TN8400 **2003-024**, Japan Nuclear Cycle Development Institute.

Tanaka, Y., Hironaga, M. & Kudo, K. 2007. *Gas Migration Mechanism of Saturated Dense Bentonite and its Modelin.* (CRIEPI Report **N7005** in Japanese).

Characterization of gas flow through low-permeability claystone: laboratory experiments and two-phase flow analyses

RAINER SENGER[1]*, ENRIQUE ROMERO[2], ALESSIO FERRARI[3] & PAUL MARSCHALL[4]

[1]*INTERA Incorporated, Swiss Branch, Ennetbaden, Switzerland*

[2]*Department of Geotechnical Engineering and Geosciences, Universitat Politècnica de Catalunya (UPC), Barcelona, Spain*

[3]*École Polytechnique Fédérale de Lausanne (EPFL), Laboratory for Soil Mechanics, Switzerland*

[4]*Nagra, Lausanne, Wettingen, Switzerland*

**Corresponding author (e-mail: rsenger@intera.com)*

Abstract: For the characterization of gas migration through a low-permeability clay host rock for deep underground repositories, a comprehensive understanding of the relevant phenomena of gas and fluid flow through low-permeability clay is required. The National Cooperative for the Disposal of Radioactive Waste (Nagra) in Switzerland has developed a comprehensive programme to characterize gas flow in low-permeability Opalinus Clay through laboratory tests and detailed numerical analyses for developing appropriate constitutive models.

Laboratory tests were performed on cores by two different laboratories, the Laboratory for Soil Mechanics at EPFL and the Department of Geotechnical Engineering and Geosciences at UPC. Loading tests were performed by both laboratories to study rock compressibility at different stress levels and water permeability dependence on void ratio. The water retention behaviour demonstrated by EPFL and UPC produced comparable results. Water permeability tests and fast controlled-volume air injection experiments were performed in a triaxial cell under isotropic stress conditions on two samples with flow parallel and normal to the bedding planes. A confining stress of 15 MPa was applied during gas testing, corresponding to a lithostatic pressure at a depth of *c.* 600 m below ground.

For detailed analyses, the two-phase flow code TOUGH2 (Pruess *et al.* 1999) was used. This considers fluid flow in both liquid and gas phases under the influence of pressure, viscous and gravity forces, according to Darcy's law. The standard analyses could not reproduce the measured pressure responses well, and the calibrated hydraulic and two-phase parameters were not consistent with the preceding water test and laboratory analyses. Implementing the non-linear behaviour in terms of the observed relationship between changes in void ratio and associated changes in permeability under different stress conditions significantly improved the simulated results, resulting in a conceptual model that well reproduced the observed injection pressure and outflow responses for both tests, parallel and normal to bedding, using a consistent parameter set.

The characterization of gas migration through a low-permeability clay host rock for repositories is important, because significant amounts of waste-generated gas (produced mainly by the anaerobic corrosion of metals and the degradation of organic materials) are expected to migrate from low- and intermediate-level waste (L/ILW) and high-level waste (HLW) repositories into the surrounding host rock (Nagra 2004, 2008). To assess the long-term safety of a repository, it is necessary to have a comprehensive understanding of the relevant phenomena of gas and fluid flow through low-permeability clay. The assessment of gas migration from the repository is carried out using large-scale numerical models, which incorporate the two-phase flow and associated constitutive models needed to properly represent the relevant processes. The National Cooperative for the Disposal of Radioactive Waste (Nagra) in Switzerland has proposed the Opalinus Clay (OPA) as one of the host rocks chosen for Stage 1 of the Sectoral Plan process. Nagra has therefore developed a comprehensive programme to characterize gas flow in OPA with laboratory tests in order to determine the relevant hydraulic, geomechanical and two-phase properties, and to develop appropriate constitutive models through numerical analyses of the laboratory tests (Nagra 2009).

Laboratory tests on OPA cores from boreholes in the Mont Terri Underground Rock Laboratory

From: Norris, S., Bruno, J., Cathelineau, M., Delage, P., Fairhurst, C., Gaucher, E. C., Höhn, E. H., Kalinichev, A., Lalieux, P. & Sellin, P. (eds) 2014. *Clays in Natural and Engineered Barriers for Radioactive Waste Confinement*. Geological Society, London, Special Publications, **400**, 531–543.
First published online April 9, 2014, http://dx.doi.org/10.1144/SP400.15

(URL) in Mont Terri (NW Switzerland) were performed by two different laboratories. The Laboratory for Soil Mechanics at EPFL performed laboratory tests on OPA cores from the BHG-D1 borehole, which have been described in detail in Ferrari & Laloui (2012) and are summarized in the following. The Department of Geotechnical Engineering and Geosciences at UPC analysed core samples from borehole BHA-8/1 at the MI niche and from borehole BDR 1_06 14 at the DR site in Mont Terri. These have been described in Romero *et al.* (2012) and are also discussed in the following.

Laboratory tests

Whereas EPFL focused on water retention behaviour and geomechanical tests, UPC performed specific water-permeability and air-injection tests to determine single-phase liquid and two-phase properties. Oedometer and isotropic compression tests were performed by the laboratories to study rock compressibility at different stress levels as well as the dependence of water permeability on void ratio.

Water retention behaviour

The procedure for determining water retention curves for the BHG-D1 core samples is described in detail in Ferrari & Laloui (2012) and is summarized in the following. The initial suction of the samples was measured by a WP4c dew-point psychrometer. The volume of the core sample (measured using the kerdane method; Laloui *et al.* 2012) and the initial water content, void ratio and degree of saturation were also determined (Ferrari & Laloui 2012).

To determine the water retention curve (i.e. suction as a function of water content or degree of saturation), core samples were prepared with specified water contents in order to have at least seven series of experimental water content values for both drying and wetting paths. For each water-content point, the suction and volume of the sample were measured, and from these were calculated the void ratio and degree of saturation (Ferrari *et al.* 2012). In addition, mercury intrusion porosimetry (MIP) was used to determine the retention behaviour for comparison. The results are shown in terms of degree of saturation as a function of suction (Fig. 1). A separate laboratory analysis of the retention behaviour of a core sample from Mont Terri (core BHA-8/1; MI-niche) was performed by the Department of Geotechnical Engineering and Geosciences at UPC using a WP4 dew-point psychrometer and MIP measurements, which showed similar water retention results (Fig. 1), providing further support for the laboratory analyses and confirming the earlier derived parameters for the van Genuchten (1980) model.

Compression tests

Oedometer tests and isotropic compression tests were performed to study rock compressibility at different stress levels as well as the dependence of water permeability on void ratio, as described in detail in Romero *et al.* (2012) and in Ferrari & Laloui (2012). Samples normal and parallel to bedding were taken from BHA-8/1 and BDR 1_06 14, respectively, and their axial strain responses as a function of mean total stress are shown in Figure 2. The tests demonstrate axial strains for different mean total stresses at different pore pressures. The initial conditions of the two tested samples are summarized in Table 1.

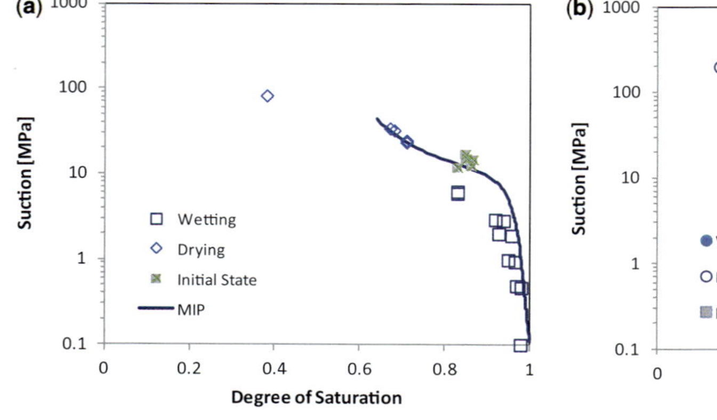

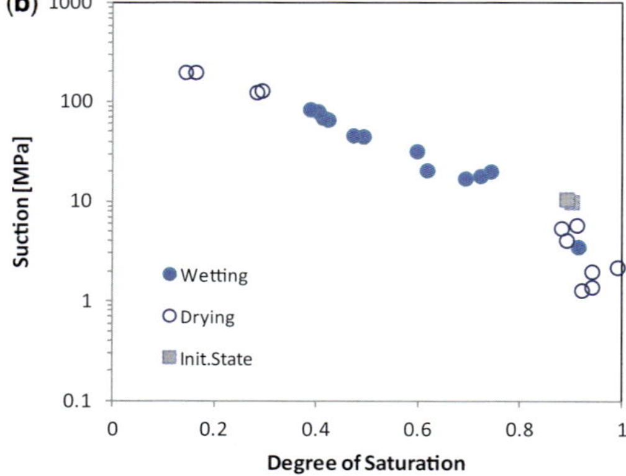

Fig. 1. Total suction v. degree of saturation: (**a**) sample BHA-8/1 (UPC), (**b**) core sample BHG-D1 (EPFL).

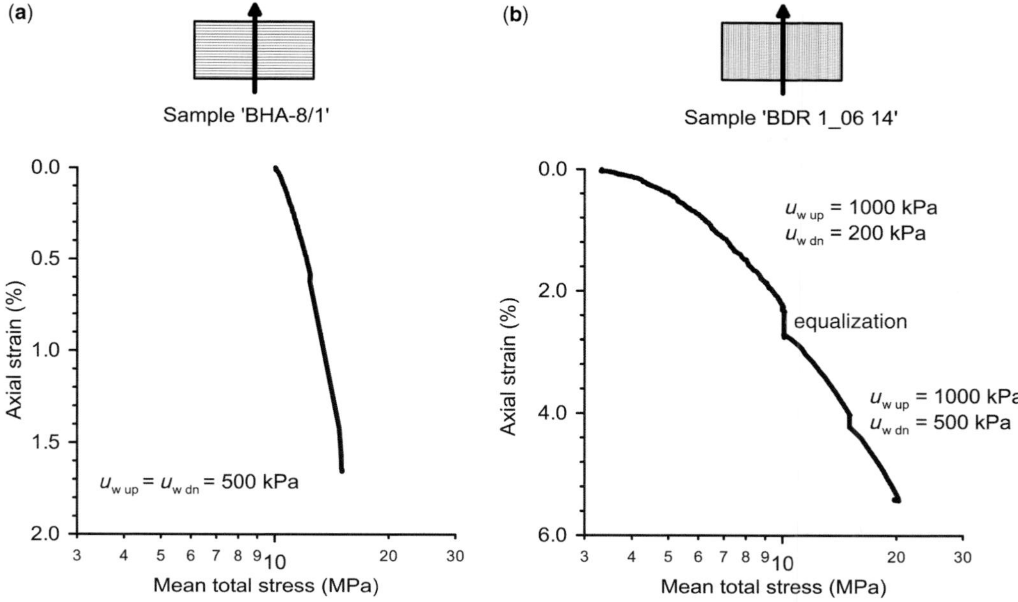

Fig. 2. Axial strain during isotropic compression tests for samples with flow (**a**) normal and (**b**) parallel to bedding at the given upstream ($u_{w\ up}$) and downstream ($u_{w\ dn}$) water pressures.

The compression index κ of the OPA samples can be estimated from the compression tests according to

$$\kappa = -\sigma' \cdot \frac{de}{d\sigma'} = \sigma' \cdot (1+e) \cdot \frac{d\varepsilon_v}{d\sigma'} \quad (1)$$

where σ' is the effective stress, e is the void ratio, and ε_v is the volumetric strain.

The effective stress is related to the total stress and pore pressure by

$$\sigma' = \sigma - \zeta P \quad (2)$$

where σ is the mean total stress and ζ is the Biot coefficient. For $\zeta = 1$ the equation represents the standard effective stress law (e.g. Azizi 2000), and

Table 1. *Initial conditions for samples BHA-8/1 and BDR 1_06 14*

Property	Value
Density	2.34–2.38 Mg m^{-3}*, 2.31 Mg m^{-3}†
Density of solids	2.70 Mg m^{-3}*
Dry density	2.20–2.23 Mg m^{-3}*, 2.16 Mg m^{-3}†
Void ratio	0.21–0.24*, 0.25†
Porosity	0.17–0.20*, 0.20†
Water content	6.6–6.9%*, 6.9–7.7%†
Degree of saturation	77–88%*, 79%†
Total suction	15 MPa*
Dominant pore mode (MIP‡)	23 nm*
Air-entry value (MIP‡)	13 MPa*
Liquid limit	38 ± 5%*
Plastic limit	23 ± 2%*

*Samples with flow normal to bedding (BHA-8/1).
†Samples with flow parallel to bedding (BDR 1_06 14).
‡Mercury intrusion porosimetry.

the corresponding effective stresses and associated strain behaviour provide a bounding value of the compression index.

From the oedometer tests, permeability k can be calculated from the loading steps according to

$$k = \frac{c_V \gamma}{E_{\text{oed}}} \quad (3)$$

where c_v is the coefficient of consolidation, γ is the specific weight of water, and E_{oed} is the oedometric modulus. These tests can then be used to derive relationships between the void ratio and permeability at different effective or total mean stresses, as described in Ferrari & Laloui (2012).

Water injection tests

A typical test for determining hydraulic conductivity prescribes a pressure gradient across the sample and measures the water flow rate. For fully water-saturated samples, water tests parallel to the bedding (sample BDR 1_06 14) and normal to the bedding (BHA-8/1) were performed under constant hydraulic gradients subject to different confining stresses. The test configuration, shown schematically in Figure 3, is also used for the fast air-injection tests, described in the following.

With this test configuration, the upstream and downstream volume changes and axial deformation were monitored. The prescribed gradient conditions were maintained for about two to three weeks to achieve approximately stationary flow

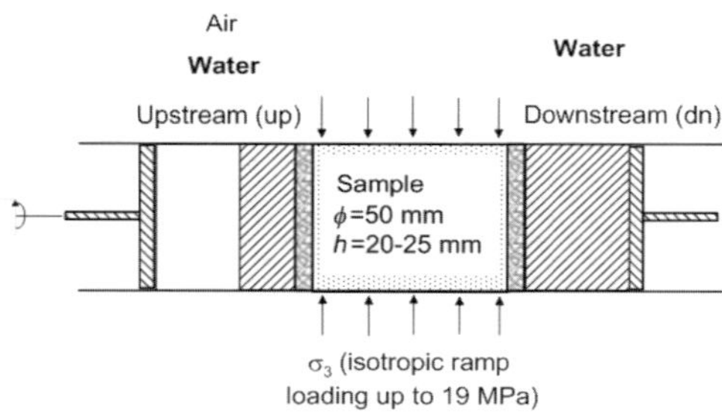

A Water permeability (starting from saturated state): P_{up}=const.$_1$ (water); P_{dn}=const.$_2$ (water) dV/dt (upstream and downstream volumes)

B Pressure pulse / decay air test: Sudden P_{up} (air); then V_{up}=const.$_1$ (air) P_{dn}=const.$_2$ (water) (initial condition) time evolution of P_{up} and V_{dn}

Fig. 3. Schematic test configuration for controlled-gradient water permeability tests and fast air-injection tests followed by recovery period at constant volume.

conditions, from which the hydraulic conductivities were then determined. For tests under different confining stress conditions, the measured axial deformation was used to estimate the change in void ratio. The results of the water tests are summarized in Figure 4 in terms of the derived relationship between void ratio and stress and the corresponding permeability as a function of void ratio.

Air-injection tests

The focus of this paper is on the air-injection tests, which were performed on two core samples with flow parallel and normal to the bedding. Figures 5 and 6 show the time evolution of air injection pressure during fast controlled volumetric-rate tests (100 ml/min) at 15 MPa isotropic confining stress on samples parallel (BDR 1_06 14) and normal (BHA-8/1) to the bedding, respectively. The injection pressure increased up to *c.* 12 and 13 MPa (depending on the orientation), followed by a shut-in and recovery period. For the flow parallel to the bedding, outflow response was observed immediately after shut-in, corresponding to a sudden drop in the injection pressure, followed by a subsequent gradual decline. The pressure in the fixed-volume outflow chamber rapidly increased until reaching 2 MPa, at which point a constant pressure was ensured by a release valve.

For flow perpendicular to the bedding, the injection pressure increased to 12 MPa and remained relatively flat after shut-in. The outflow response was significantly delayed compared to the case with flow parallel to the bedding. Only after apparent gas breakthrough did the injection pressure show a steep decline. The pressure in the fixed-volume downstream chamber increased until reaching 2 MPa, when a constant pressure was prescribed by a release valve. This test indicated gas migration into the sample for a certain amount of time prior to gas outflow (i.e. gas breakthrough). In other words, continuous gas flow occurs at an injection pressure of 12 MPa, which is significantly below the vG-P_0 value of 18 MPa (van Genuchten's capillary strength parameter P_0; see also equation (5)). As shown in the following, the vG-P_0 parameter does not necessarily represent the air-entry value for gas flow into the sample, but has sometimes been interpreted as representing the capillary pressure at which a continuous gas path is established.

Examination of the measured axial deformation revealed different deformation regimes. For the test parallel to the bedding (Fig. 5), the initial pressure increase shows axial deformation to negative values, indicating expansion. During the early period after shut-in, the pressure decreased slightly, but the axial strain continued to increase. Afterwards, as the pressure continued to decrease, the

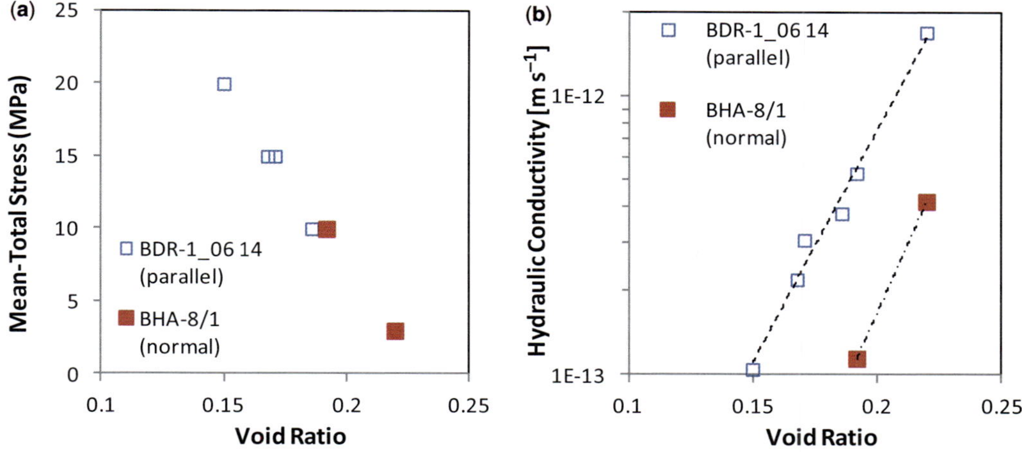

Fig. 4. (**a**) Compressibility on isotropic loading of samples with flow parallel and normal to bedding. (**b**) Results of the water permeability tests on samples parallel and normal to bedding in terms of void ratio and hydraulic conductivity.

axial deformation reversed, indicating compression as the effective stress increased.

For the gas test with flow normal to the bedding (Fig. 5), the axial deformation indicates a similar pattern, although no measurements were available during the early injection period. During the early period after shut-in, very little change in pressure occurred, whereas the axial deformation became negative, indicating expansion. As the test progressed, the axial deformation reversed as the pressure decline steepened, indicating compression of the sample. The axial deformation, reflecting the changes in void ratio, indicates that, during the injection period, expansion and a corresponding increase in void ratio occurred associated with gas migration into the pore space of the core sample and effective stress decrease due to pore pressure increase. This expansion continued beyond the shut-in as the gas pressure front propagated into the sample, causing the fluid pressure to increase

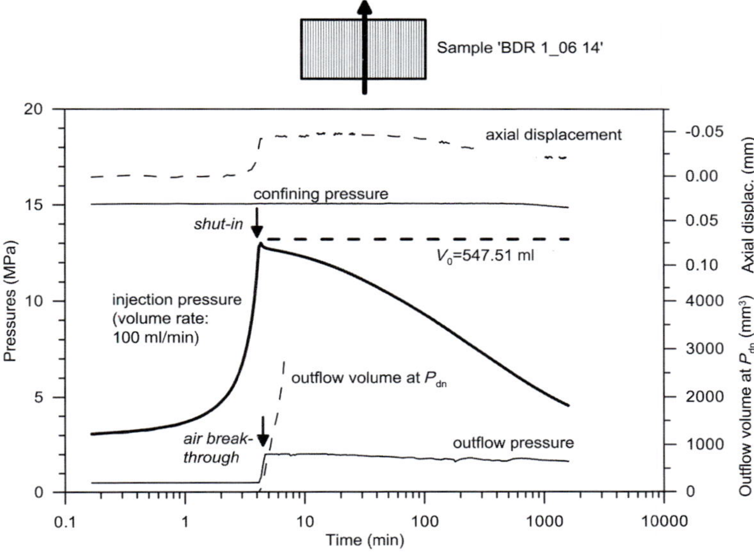

Fig. 5. Measured pressures at the injection and outflow sides together with outflow volume and axial displacements for the test with flow parallel to bedding.

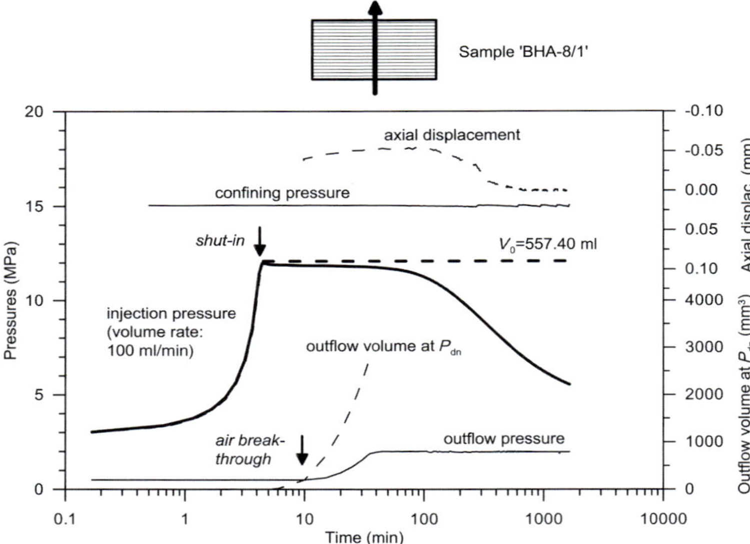

Fig. 6. Measured pressures at the injection and outflow sides together with outflow volume and axial displacements for the test with flow normal to bedding.

and the effective stress to decrease. Some time later the pressure in the outflow chamber started to increase, the injection pressure declined and the effective stress increased, which induced compression. This assumes air pressure acting on all pore space and that pore deformation is not localized.

The magnitude of the expansion after shut-in differs significantly for the two tests. For the gas test with flow normal to the bedding (Fig. 6), the outflow response is significantly later than for the gas test with flow parallel to the bedding (Fig. 5). Even with the lower intrinsic permeability, the observed pressure decline indicates that a larger volume of the gas injected from the upstream chamber has to be stored in the sample compared to that for the gas test with flow parallel to the bedding. This corresponds to the differences in the measured axial strain and corresponding expansion during the early part of the test (Figs 5 & 6). Even though radial deformation could not be measured, the axial deformation can be assumed to represent the relative change in void ratio. This change in void ratio is defined by the pore compressibility of the material derived from the loading tests (Fig. 2) and by the pressure change during the tests (Figs 5 & 6).

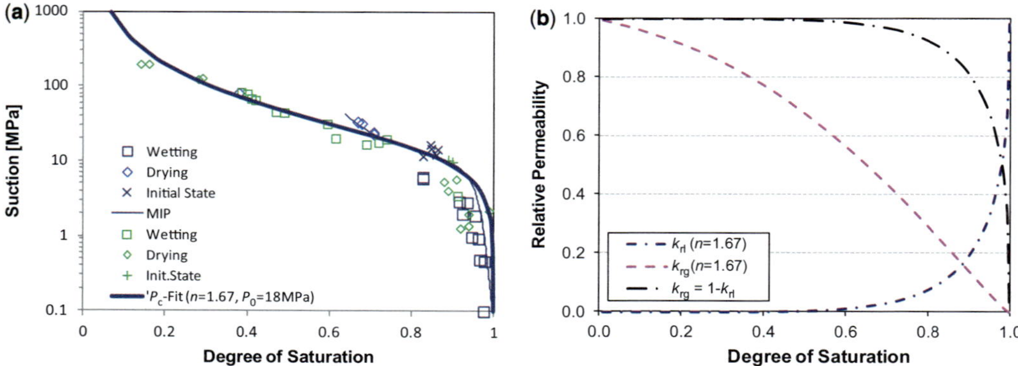

Fig. 7. (**a**) Fitted water retention curve using the van Genuchten model. (**b**) Corresponding relative permeability curves using the same shape parameter n.

Numerical analyses

To investigate the measured responses in terms of injection and outflow pressures, numerical models were developed using the geometry of the rock core and the boundary conditions on the injection and outflow sides.

For the detailed analyses, the two-phase flow code TOUGH2 (Pruess *et al.* 1999) was used, which takes into account fluid flow in both liquid and gas phases occurring under pressure, viscous and gravity forces according to Darcy's law. The corresponding inverse code, ITOUGH2 (Finsterle 1999), was used to estimate two-phase and hydraulic parameters.

In a first step, the air-injection tests were simulated using the measured results for permeability, porosity and pore compressibility, as well as the two-phase parameters derived from the water retention behaviour. The resulting simulations could not reproduce the observed pressure response for both air-injection tests (parallel and normal to the bedding). In a second step, inverse modelling was used to estimate the relevant parameters in order to better reproduce the measurements. Finally, a revised approach was developed to better account for the phenomena from the preceding test results.

Standard two-phase flow analysis

The initial analysis assumed standard two-phase flow behaviour and constant intrinsic permeability. For inverse modelling the measured pressures in the inflow and outflow chambers were used for

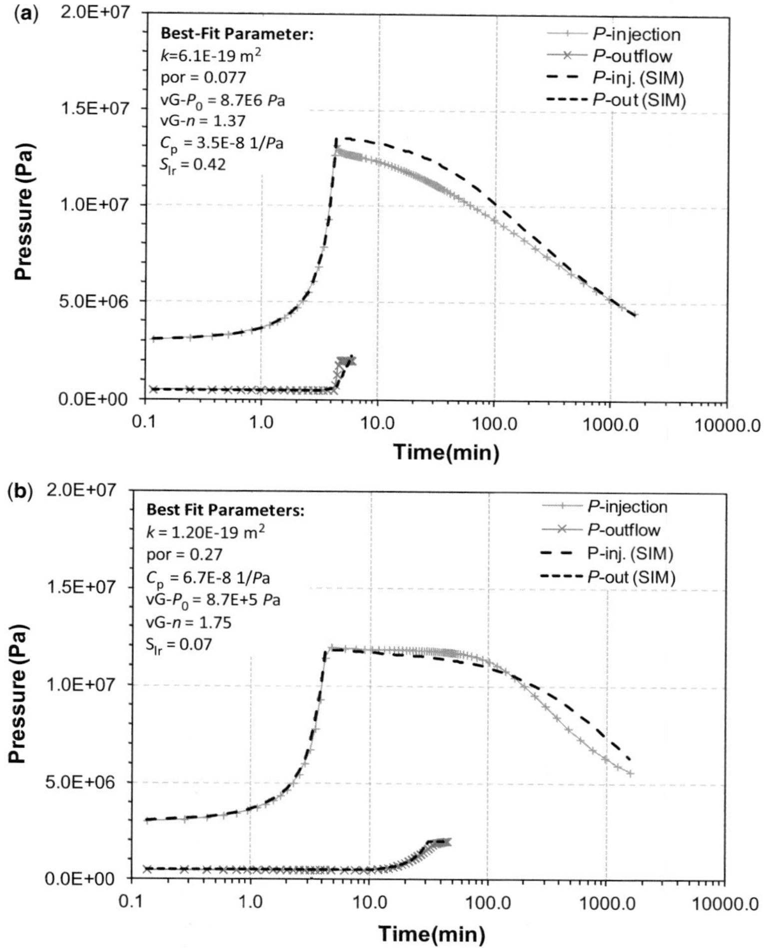

Fig. 8. Simulated (dashed lines) and measured injection and outflow pressure responses with the best-fit parameters for flow (**a**) parallel and (**b**) normal to bedding.

optimization, and only outflow pressures up to 2 MPa were considered. The parameters to be estimated included the hydraulic parameters (permeability k, porosity ϕ, pore compressibility C_p), as well as two-phase parameters of the van Genuchten (1980) model (capillary strength parameter P_0, shape parameter n, residual water saturation for relative permeability S_{lr}). The shape parameter of the relative permeability curves used the same value fitted to the retention curve, based on the relationship between capillarity and relative permeability (Mualem 1976), which is described by the following equations for k_r and P_c, as a function of saturation (Fig. 7):

$$k_{r,l} = S_{el}^{1/2} \cdot \left[1 - \left(1 - S_{el}^{1/m}\right)^m\right]^2; \quad S_{el} = \frac{S_l - S_{lr}}{1 - S_{lr}}$$

$$k_{r,g} = (1 - S_{eg})^{1/3} \cdot \left(1 - S_{eg}^{1/m}\right)^{2m}; \quad S_{eg} = \frac{1}{1 - S_{gr}}$$

$$P_c = P_0 \cdot \left(S_e^{-1/m} - 1\right)^{1-m}; \quad S_e = \frac{S_l - S_{lr}}{1 - S_{lr}} \qquad (4)$$

The fit of the retention data in Figure 1 compares well with the data from earlier studies, shown in Figure 7. The van Genuchten model implies that the capillary pressure approaches zero as the saturation is one.

The residual water saturation (S_{lr}), used as a fitting parameter for the relative permeability curves, would shift the curves to the right (Fig. 7) and represents the immobile water that cannot be displaced by the injected gas. The intersection of the gas and water relative permeability curves indicates significant phase interference (i.e. the sum of k_{rg} and k_{rl} is significantly less than 1). Alternatively, if the two phases use separate paths due to heterogeneous pore structures (i.e. gas tends to migrate through the larger pores), the gas relative permeability is typically described by $k_{rg} = 1 - k_{rl}$ (Grant 1977), as shown in Figure 7. Using the same shape parameter for the capillary pressure and relative permeability functions reduces the number of fitting parameters and improves the identifiability of the inverse solution and reduces the uncertainty of the parameter estimates.

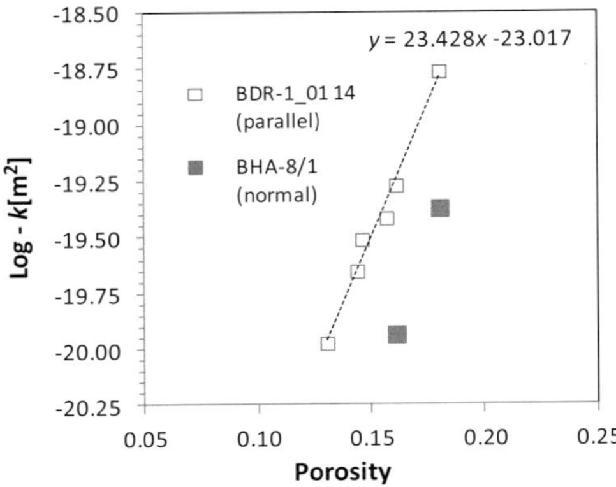

Fig. 9. Derived linear relation between log k and porosity based on data presented in Figure 4. The fitted linear relation is associated with BDR 1_0614 (parallel).

The results of the inverse modelling of both air-injection tests are shown in Figure 8. The results indicate that the outflow pressure response is reproduced well, whereas the injection pressure response is not. Moreover, the best estimates of hydraulic conductivity were not consistent with those from the preceding water tests and the two-phase parameter did not compare well with the fitted parameters from the water retention data (Fig. 1).

For the test with flow parallel to the bedding, the sudden drop in pressure at the peak pressure is probably associated with a compliance effect and is not accounted for in the simulation. However, the shape of the subsequent recovery shows a slower early recovery, whereas after *c.* 100 min

Table 2. *Model input parameters*

	Test (parallel)	Test (normal)	Comments
Permeability k(m^2)	1.7×10^{-19} (1.4×10^{-19})	4.3×10^{-20}	Water test (analysis with TOUGH)
Porosity (−)	0.20	0.18	Measured (initial)
Pore compressibility C_p (1/Pa)	2.0×10^{-8}	1.5×10^{-8}	Estimated from compression tests
van Genuchten P_0 (Pa)	18.0×10^6	18.0×10^6	Retention curve fit
van Genuchten $n - P_c$	1.67	1.67	Retention curve fit
van Genuchten $n - k_r$	1.67	1.67	Assumed same n for k_r curves
Residual water saturation S_{lr}	0.01	0.01	Assumed
Residual gas saturation S_{gr}	0	0	Assumed
Initial saturation S_l	1	1	Assumed fully water-saturated conditions

the simulated pressure decline is steeper than the measured pressures. The parameter estimates from the inverse modelling indicate a very low value for ϕ and a high value for S_{lr}, as well as a higher intrinsic permeability compared to that derived from the water test.

The inverse simulation of the test with flow normal to the bedding also reproduced the pressure response well in the outflow chamber, but did not reproduce the injection pressure response well (Fig. 8). The estimated parameters also indicate a higher permeability compared to that derived from the water test and a relatively low value for the vG-P_0 parameter. In this case, the simulated pressure after shut-in shows a distinct decline at early times, followed by a delayed pressure decline after *c.* 100 min. That is, at early times the effective gas permeability is too high, and at late times the effective permeability is not high enough to reproduce the steeper pressure decline. This suggests that some non-linear behaviour is not accounted for in the model.

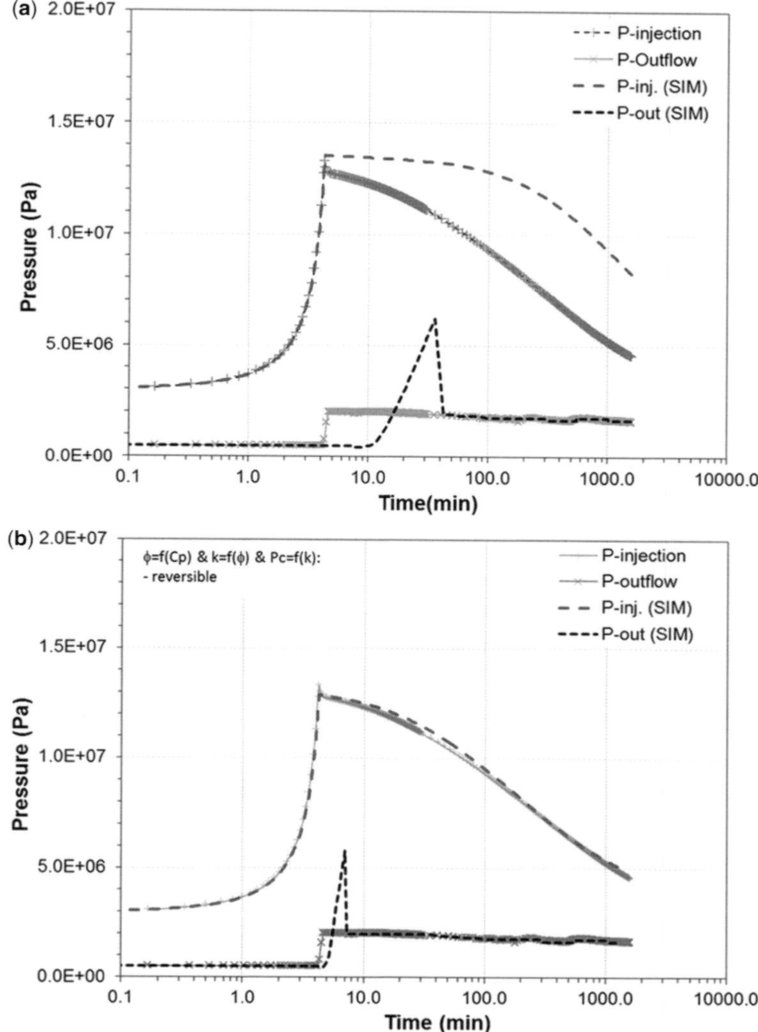

Fig. 10. Simulated (dashed lines) and measured pressures for the air-injection test with flow parallel to bedding for (**a**) the case without coupling and (**b**) the case with coupling and adjustment of the initial volume of the injection chamber. The simulated peaks in the outflow chamber are an artefact of the modelling, where the prescribed pressure was specified at a certain time instead of at the specific air release valve pressure.

Revised analysis

The non-linearity is indicated in the relations between void ratio and stresses and the corresponding change in permeability (Fig. 4). Whereas the changes in porosity are accounted for in the standard TOUGH2 code through the pore compressibility, the potential change in permeability is not. Assuming largely linear deformation associated with the gas injection test, the inferred change in void ratio can be accounted for by the pore compressibility. The internal coupling of the change in void ratio or porosity with permeability accounts for the geomechanical deformation using the standard two-phase flow code.

The pore compressibility C_p is related to the compression index, void ratio and effective stress by

$$C_p = \frac{\kappa}{e \cdot \sigma'} \tag{5}$$

The relations in Figure 4 can be converted to a linear relation between log k and ϕ, as shown in Figure 9.

With the confining stress kept constant at 15 MPa during the air-injection tests, the variation

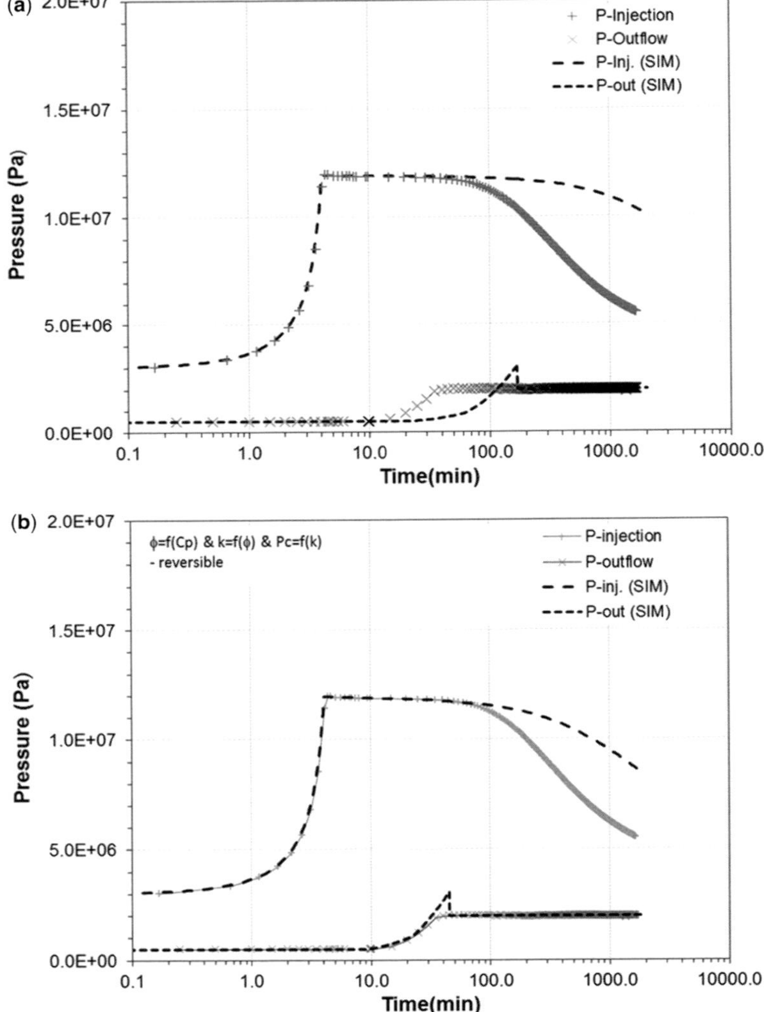

Fig. 11. Simulated (dashed lines) and measured pressures for the air-injection test with flow normal to bedding for (**a**) the case without coupling and (**b**) the case with coupling.

in effective stress can be related to the variation in pressures (equation (2)). In TOUGH2, the effect of compressibility is accounted for by the change in porosity ($d\phi$) in response to a change in fluid pressure (dP) as

$$d\phi = \phi C_p dP \quad (6)$$

The corresponding change in permeability is then given by the slope of the curves in Figure 9 as

$$d(\log k) = 23.43 d\phi \quad (7)$$

Even though only two data points are available for the case with flow normal to the bedding, the overall trend between flow parallel and normal is similar and the same relationship was used in the modelling in the following.

For two-phase flow, the change in porosity/permeability also affects the capillary pressure, which can be accounted for by the Leverett function (Leverett 1941), given by

$$P_C = P_{C0} \frac{1}{\sqrt{k/k_0}} \quad (8)$$

where P_{C0} and k_0 are the reference values for capillary pressure and permeability, respectively.

The revised analysis used a modified version of the standard two-phase flow code, with internal coupling of changes in porosity with changes in permeability and capillary pressures using the explicit functional relationships described by equations (7) and (8). For the revised analysis, only forward simulations were performed, using the two-phase parameters fitted to the measured retention curve data and the estimated permeability from the water test. A summary of the input parameter for the simulations of the two tests parallel (Fig. 5) and normal (Fig. 6) to the bedding is given in Table 2.

The results of two simulations for the air-pulse test parallel to the bedding are shown in Figure 10. A reference simulation was performed with constant permeability (i.e. uncoupled) and a simulation incorporating the coupling between porosity, permeability and capillary pressure, as described above. The reference simulation indicates that the simulated injection pressure remains relatively high after the shut-in. At the same time, the simulated pressure response in the outflow chamber is significantly delayed. Also note that the pressure peak at the start of the shut-in could not be reproduced. An adjustment was made in the volume of the injection chamber in order to limit the pressure build-up during injection, resulting in a shift of the shut-in pressure curve to slightly lower pressures. The simulation incorporating the coupling between the porosity change arising from the increased pore pressure and the corresponding permeability change shows a much better fit of both the injection pressure response and the outflow pressure response. As described above, the inverse simulations could reproduce the general response, but only with parameters that were significantly different from those derived from the retention curve data and water permeability tests.

The results of the two simulations for the air-injection test normal to bedding are shown in

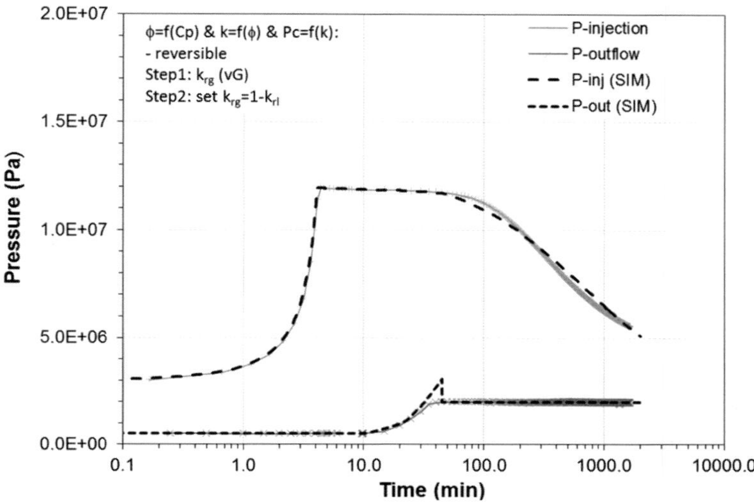

Fig. 12. Simulated (dashed lines) and measured pressures for the air-injection test with flow normal to bedding, assuming a preferential gas path after air breakthrough (i.e. assume $k_{rg} = 1 - k_{rl}$).

Figure 11. Similar to the case above, a reference simulation was performed with constant permeability (i.e. uncoupled) and a simulation incorporating the coupling between porosity, permeability and capillary pressure (as described above). The reference simulation indicates that the simulated injection pressure remains relatively high after the shut-in and does not reproduce the pressure recovery at late times. Also, the simulated pressure response in the outflow chamber is significantly delayed compared to the measured response.

The simulation incorporating the coupling between the porosity change arising from the pore pressure increase and the corresponding permeability change shows a much better fit of the outflow pressure response. The injection pressure response improved somewhat, but the magnitude of the recovery at late times was not reproduced.

As indicated, the outflow response for the test with flow normal to the bedding is significantly later than for the test with flow parallel to the bedding (Fig. 11). That is, gas continues to migrate into the expanding pores prior to the breakthrough response. One can assume that preferential gas pathways are established, resulting in higher gas mobility and less phase interference. This can be represented through a Grant model for the gas relative permeability in the numerical model. The results are shown in Figure 12, indicating a significantly faster recovery of the injection pressure, which corresponds well with the measured response.

Conclusions

Detailed analyses of laboratory experiments on Opalinus Clay cores from boreholes BHA-8/1and BDR-1 at Mont Terri, Switzerland, were used to develop a conceptual and numerical model for simulating two-phase flow of gas through a low-permeability clay formation. Relevant information from the laboratory experiments in terms of the stress dependence of void ratios and associated changes in permeability were taken into consideration. These phenomena were implemented in the numerical model by relating the stress dependency through the corresponding effect of pore compressibility on porosity/permeability as a function of pore pressure through the effective stress concept at fixed total stress. For the two-phase flow properties, the laboratory experiments on water retention behaviour were used, where changes in porosity/permeability were accounted for by scaling the capillary strength parameter P_0 with permeability using the Leverett function.

The numerical simulation of the fast controlled-volume air-injection experiments produced a significant improvement in terms of reproducing the observed injection pressure and outflow pressure responses for both the test with flow parallel to the bedding and the test with flow normal to the bedding. The experiments with flow normal to the bedding indicated a significant delay in the outflow response after full injection of air volume from the injection chamber compared to the experiment with flow parallel to the bedding, indicating a relatively rapid outflow pressure response. The injection pressure response suggests that preferential gas paths were developed as gas was migrating into the expanding pores of the samples at increasing injection pressures, resulting in higher gas mobility and a corresponding rapid pressure recovery following gas breakthrough. This delay in the outflow pressure response is also observed in core samples from a deep borehole in the Opalinus Clay on which similar laboratory experiments have been conducted (Romero *et al.* 2012).

This study has been performed under contract from the National Cooperative for the Disposal of Radioactive Waste (NAGRA) Switzerland. The comments of the reviewers were very helpful in improving the paper and are greatly appreciated.

References

Azizi, F. 2000. *Applied Analyses in Geotechnics*. Spon Press, London.

Ferrari, A. & Laloui, L. 2012. Advances in the testing of the hydro-mechanical behaviour of shales. *In*: Laloui, L. & Ferrari, A. (eds) *Multiphysical Testing of Soils and Shales*. Springer, Berlin, 57–68.

Ferrari, A., Witteveen, P. & Laloui, L. 2012. *Material Properties and Geomechanical Tests on BHG-D1 Cores: Mont Terri HG-D Experiment, Phase 15: Mont Terri HG-D Experiment, Phase 15*. Mont Terri Technical Report **TN 2010-52**.

Finsterle, S. 1999. *ITOUGH2 User's Guide*. Lawrence Berkeley National Laboratory **LBNL-40040**.

Grant, M. A. 1977. *Permeability Reduction Factors at Wairakei*. AICHEASME Heat Transfer Conference, Salt Lake City, Utah, August 15–17, 1977, paper 77-HT-52.

Laloui, L., Ferrari, A. & Salager, S. 2012. *Testing the Thermo-hydro-mechanical Behaviour of a shale*. 3rd EAGE Shale Workshop, Barcelona, Spain, 23–25 January 2012, paper no. D05.

Leverett, M. C. 1941. Capillary behaviour in porous solids. *Transactions of the AIME*, **142**, 159–172.

Mualem, Y. A. 1976. New model for predicting the hydraulic conductivity of unsaturated porous media. *Water Resource Research*, **12**, 513–522.

NAGRA 2004. *Effects of Post-disposal Gas Generation in a Repository for Spent Fuel, High-level Waste and Long-lived Intermediate-level Waste Sited in Opalinus Clay*. Nagra Technical Report **04-06**.

NAGRA 2008. *Effects of Post-disposal Gas Generation in a Repository for Low- and Intermediate-level Waste*

Sited in the Opalinus Clay of Northern Switzerland. Nagra Technical Report **08-07**.

NAGRA 2009. *The Nagra Research, Development and Demonstration (RD&D) Plan for the Disposal of Radioactive Waste in Switzerland.* Nagra Technical Report **09-06**.

PRUESS, K., OLDENBURG, C. & MORIDIS, G. 1999. *TOUGH2 User's Guide, Version 2.0.* Lawrence Berkeley National Laboratory **LBNL-43134**.

ROMERO, E., SENGER, R. & MARSCHALL, P. 2012. *Air Injection Laboratory Experiments on Opalinus Clay. Experimental techniques.* 3rd EAGE Shale Workshop, Barcelona, Spain, 23–25 January 2012, Results and Analyses.

VAN GENUCHTEN, M. Th. A. 1980. Closed-form equation for predicting the hydraulic conductivity of unsaturated soils. *Soil Science Society*, **44**, 892–898.

Extended two-phase flow model with mechanical capability to simulate gas migration in bentonite

Y. TAWARA[1]*, A. HAZART[1], K. MORI[1], K. TADA[1], T. SHIMURA[2], S. SATO[2], S. YAMAMOTO[2], H. ASANO[3] & K. NAMIKI[3]

[1]*Geosphere Environmental Technology Corp., NCO Kanda Awajicho-Build. 3F, 2-1 Kanda Awajicho, Chiyoda-ku, Tokyo 101-0063, Japan*

[2]*Obayashi Corp., Shinagawa Intercity Tower B, 2-15-2, Konan, Minato-ku, Tokyo, 108-8502, Japan*

[3]*Radioactive Waste Management Funding and Research Center (RWMC), Pacific Marks Tsukishima, 8th Floor, 1-15-7, Tsukishima, Chuo-ku, Tokyo, 104-0052, Japan*

**Corresponding author (e-mail: tawara@getc.co.jp)*

Abstract: Long-term gas migration through clays cannot be simulated by conventional two-phase flow models alone owing to the presence of material deformation. In this article, an extended two-phase flow model that incorporates mechanical effects is proposed. The model allows the formation of preferential pathway and considers the relation between pore moisture and pore deformation. It was carried out with the intention of avoiding the complexity of a fully coupled thermal, hydraulic and mechanical modelling. In the new model, porosity, permeability, swelling pressure and pathways formation threshold depend on the water saturation. The model is validated on different gas injection experiments with controlled flow rate and controlled pressure. Some experiments are well known in the literature; some are new. In each case, an inverse approach is used to identify the model parameters. The results confirm that, depending on the type of bentonite (MX80, Avonlea, KunigelV1), modelling the gas migration could require the existence of a pressure-induced saturation-depending preferential pathway. In laboratory-scale experiments, the model leads to an accurate evaluation of the long-term gas migration trends, including not just the gas migration stage but also the water re-saturation level. In a field-scale experiment, the behaviour of the model in a realistic context is revealed.

In the concept of deep geological disposal, the waste is enclosed by a geological medium at a depth of approximately 500 m. The aim is to isolate the waste from the environment and human activity during the necessary period. Because of the very long half-life of radionuclides present in some wastes, the duration of storage should be granted for a period that usually exceeds 100 000 years. In most disposal systems, an engineered barrier system (EBS) must be present to isolate effectively the radioactive substances and chemical elements from the environment. This is the case in Japan, where the host rock does not form a sufficient protection. The study presented here focuses on clay-based engineered barrier systems, also called buffers. In such systems, multiple barriers that depend on the type of waste are built primarily to prevent water circulation, since water can degrade the waste packages and transport the radioactivity in the host rock. As certain types of waste will produce heat in the early period of storage, the behaviour of the buffer when the temperature increases and remains high must be considered. Among the different gases that can be generated in a repository, some can have a direct radiological impact. Other non-radioactive gases can have an indirect impact by favouring the release of radioactive gases from the repository (Rodwell *et al.* 1999). This is the case for hydrogen (H_2), methane (CH_4) and carbon dioxide (CO_2), which are likely to be generated by radiolysis, degradation of organic material and corrosion.

As a result, the gas generation could definitively damage the EBS, affect the volume of expelled water and potentially generate transient pathways for relatively rapid gas migration into the host rock. This is especially true when the buffer is made from low-permeability material. Designing a buffer that ensures a safe storage in the long term requires an accurate understanding of the gas migration process. Several factors make the efficient modelling of those processes a difficult task.

Modelling issues

International research on gas migration in deep geological disposal is less mature than the research

From: Norris, S., Bruno, J., Cathelineau, M., Delage, P., Fairhurst, C., Gaucher, E. C., Höhn, E. H., Kalinichev, A., Lalieux, P. & Sellin, P. (eds) 2014. *Clays in Natural and Engineered Barriers for Radioactive Waste Confinement*. Geological Society, London, Special Publications, **400**, 545–562.
First published online March 11, 2014, http://dx.doi.org/10.1144/SP400.7

on water flow. Nevertheless, the importance of considering a coupled water–gas system is well known. While the rate of gas generation could be affected by the availability of water, conversely gas generation could potentially reduce the flow into the repository. Although conventional two-phase flow models are commonly stated to be adequate for simulating gas transport through cementitious barriers and crystalline rock, their validity in clay barriers (engineered or geological) is questionable (Rodwell *et al.* 2003). Indeed, the need to consider the pressures over which gas migration would occur is well documented (JAEA 2007; Alonso *et al.* 1999). In Alonso *et al.* (1999), the authors highlight the limitations of current two-phase flow codes in modelling highly expansive clays where coupled hydromechanical phenomena between different layers within the material are likely to occur. The problem involves different modes of gas migration (dissolution, diffusion, two-phase flow, preferential flow, micro fissuring, macro fracturing, etc.); some of these modes are complex. Moreover, re-saturation and gas migration are supposed to take place simultaneously, leading to gas migration in partially saturated material.

Bentonite, although having favourable properties for designing waste disposal buffers, is a complex material that is not yet perfectly understood. Favourable properties are low hydraulic conductivity, high capillary pressure at partially saturated condition, swelling behaviour in contact with water, low diffusion coefficient and high sorption capacity. Complex properties are the swelling and shrinkage of the pores that can create fissures and fractures.

The computational cost of a fully coupled THM (thermo-hydro-mechanical) model would be too expensive to be performed in a reasonable amount of time. The approach followed here is to develop the new model by incorporating mechanical phenomena into an existing TH coupled code, here the mature multiphase multicomponent simulator GETFLOWS (Tosaka *et al.* 2000, 2010). The idea behind our approach is to keep the model simple by incorporating only the selected mechanical phenomena. The extended model requires only slight modification of GETFLOWS.

Mechanical effects

Following several studies (e.g. Graham *et al.* 2002; Komine & Ogata 2004), we assume that the gas migration in clays is highly dependent on two principal mechanical effects. First, at near fully saturated initial condition, the gas begins to flow once the pressure of the gas inflow has reached a given value, called the breakthrough pressure. It is observed that the gas pressure reaches a peak at the time of breakthrough and then decreases. Based on this observation, Horseman *et al.* (1999) conclude that the gas essentially flows through preferential paths that appear in the clay. Moreover, Graham *et al.* (2002) analyse the influence of the initial water content on the pressure breakthrough. They conclude that, if the initial water saturation is high, some preferential gas pathway may appear that should be modelled to reproduce the experimental results. Note that the breakthrough pressure depends not only on the material properties or the initial water saturation but also on other factors like the boundary conditions, the length of the specimen and its shape.

Second, the montmorillonite that mostly composes bentonite material swells when the hydration starts. During this process, water is trapped in montmorillonite interlayers, resulting in modification of the micropore/macropore distribution (Komine & Ogata 2004). Although the overall porosity might remain constant depending on the boundary conditions, the micropore size is supposed to increase and the macropores size to decrease, as described in Figure 1. In dry conditions, owing to the presence of macropores, any gas released flows smoothly into the bentonite (Graham *et al.* 2002). As the bentonite becomes saturated, the montmorillonite starts to swell, resulting in the reduction of the macro void space. This behaviour is assumed to affect the properties of the bentonite in the unsaturated condition, particularly in increasing the swelling pressure.

Since these mechanical phenomena have been identified, some authors have proposed incorporating them into existing coupling models. Regarding the pathway dilation, several modifications of the numerical model TOUGH2 (Pruess *et al.* 1999) have been proposed to provide pathways for gas flow. In Calder *et al.* (2006), the authors modify the permeability and the capillary pressure to include a dependency with the pressure. The modified permeability is increasing linearly with the pressure between two pressure thresholds, and the capillary pressure is decreasing with the permeability (corresponding of a reduction of the air entry pressure). Directionality of dilation is also considered by modifying the permeability in vertical and horizontal direction only. The model is validated in a gas migration model of a full-scale test of gas transport in bentonite. In Navarro (2009), the model TOUGH2-PD is proposed to model pathway dilation, which does not affect the water flow and the gas flow equally. They modify the generalized Darcy equation of the gas phase by adding a term describing the gas flow in a fracture network, and they incorporate the dependence of the permeability and porosity with the gas pressure after

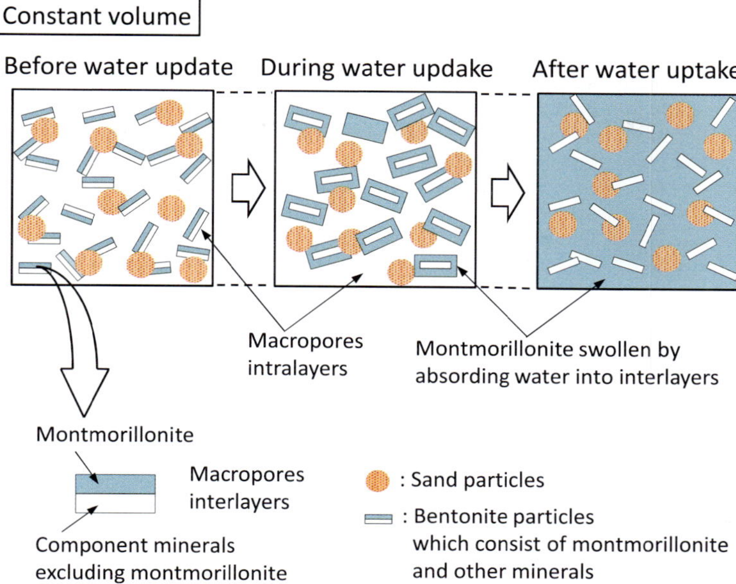

Fig. 1. Micropores/macropores deformation at constant volume (after Komine & Ogata 2004).

a threshold is reached (respectively with an exponential and linear relationship). The modified TOUGH2 model is tested on a hypothetical disposal of radioactive wastes in clay stone. In Finsterle (2009), a gas pressure threshold that depends on the depth and the local stress field is used. The permeability in the vertical direction varies exponentially with the pressure after the pressure overpasses the threshold and the capillary pressure depends on the modified permeability. With a different approach, Olivella & Alonso (2008) propose an extension that allows the development of preferential pathways. They simulate a gas migration test in a specimen of Opalinus clay shale (laboratory scale) and gas flow in a sand–bentonite material (field scale). Recently, Gerard *et al.* (2012) applied a fracture model based on Olivella & Alonso (2008) where a zone prone to the development of preferential pathway is preliminarily defined in the specimen. To validate their model, the authors consider a long-term gas injection test on Callove–Oxfordian argillite (laboratory scale). Regarding the pore deformation process, Jobman (1998) incorporates an increasing swelling pressure as a function of the water content into a two-phase flow model. Alonso *et al.* (1999) propose a THM coupling code that includes the modelling of the macropores invasion by the expansion of micropores.

The model proposed in this paper is an extension of a conventional two-phase flow model that incorporates the two mechanical effects described above (preferential pathways and micro/macro pores deformation). Firstly, the pathway dilation aspect is introduced to allow the formation of preferential pathways in the bentonite. Unlike in Olivella & Alonso (2008) and Gerard *et al.* (2012), it does not require the prior knowledge of the location or the diameter of the potential fracture in the buffer. The model can be seen as an adaptation to the variable threshold of the model proposed in Mori *et al.* (2006), where a fixed threshold for the formation of pathways is used. Secondly, the micropore swelling aspect is introduced to incorporate the micro/macropore deformation effect. It resolves some inconsistencies observed when the pathway dilation model is applied under unsaturated conditions. Note that the common assumption of constant volume boundary condition is considered in the model. It is usually justified by the constraints enforced by the rock surrounding the disposal. It is therefore assumed that the global swelling necessary induced by the formation of pathways is limited and can be neglected. The mathematical formulation of the conventional two-phase flow mode, the pathway dilation model and the micropore swelling model are described in the next section.

Mathematical model

Conventional two-phase flow

The conventional two-phase flow model used in this article is GETFLOWS (Tosaka *et al.* 2000, 2010) (free version for non-commercial use available). It supports multiphase multicomponent flow, gas dissolution to water, solute transport, heat transport and fully coupled surface–subsurface fluid flow. In this modelling study, the fluid movement in the engineer barrier system was simulated by using water–gas two-phase flow equations based on the generalized Darcy's law:

$$\nabla\left(\rho_w \frac{Kk_{rw}}{\mu_w}\nabla\Psi_w\right) - \rho_w q_w = \frac{\partial}{\partial t}(\rho_w \phi S_w) \quad (1)$$

$$\nabla\left(\rho_g \frac{Kk_{rg}}{\mu_g}\nabla\Psi_g\right) - \rho_g q_g = \frac{\partial}{\partial t}(\rho_g \phi S_g) \quad (2)$$

where K is the intrinsic permeability (m^2), k_{rp} is the relative permeability of phase p ($p = w$ for water, $p = g$ for gas), Ψ_p is the hydraulic potential of p phase (Pa), ϕ is the effective porosity, ρ_p is the density of phase p (kg m^{-3}), q_p is the sink/source rate of phase p (m^3 m^{-3} s^{-1}), S_p is the saturation of phase p (no dimension), μ_p is the viscosity of phase p (Pa s) and ∇ is the Hamilton operator. Note that the sink/source rate is defined as the volumetric production/injection rate of phase p measured at standard conditions in a unit volume of porous media.

The model incorporates also the dissolved gas, given by the following equation:

$$\nabla\left(R_w\rho_w \frac{Kk_{rw}}{\mu_w}\nabla\Psi_w\right) + \nabla(D_w \nabla\rho_w R_w)$$
$$- R_w\rho_w q_w + f_{g-w} = \frac{\partial}{\partial t}(R_w\rho_w\phi S_w) \quad (3)$$

with R_w the concentration of dissolved gas in water phase (m^3 m^{-3}), D_w the hydrodynamic dispersion coefficient (m^2 s^{-1}) and f_{g-w} the rate of mass transfer between gas and water phase. The average mass transfer between phases was described by the penetration theory of Higbie, assuming a one-dimensional transient diffusion in a semi-infinite media.

The thermal convection-diffusion equation for liquid and solid are given respectively by:

$$\nabla\left(\rho_w \frac{Kk_{rw}H_w}{\mu_w}\nabla\Psi_w\right) + \nabla\left(\rho_g \frac{Kk_{rg}H_g}{\mu_g}\nabla\Psi_g\right)$$
$$+ \nabla\lambda_f(\nabla T_f) + E_{f\leftrightarrow s} - \rho_{wS}q_{wS}H_w - \rho_{gS}q_{gS}H_g$$
$$= \frac{\partial}{\partial t}(\rho_{wS}\phi S_w U_w + \rho_{gS}\phi S_g U_g) \quad (4)$$

$$\nabla\lambda_s(\nabla T_s) - E_{f\leftrightarrow s} = \frac{\partial}{\partial t}[\rho_s(1-\phi)U_s] \quad (5)$$

where $E_{f\leftrightarrow s}$ is the heat exchange rate between fluid and solid (J s^{-1}), H_p is the enthalpy of phase p, λ_f and λ_s are the thermal conductivity of fluid and solid phase, T_f and T_s are the temperature of fluid and solid phase, respectively, and U_p is the internal energy of phase p. These equations express the material balance of the water phase, gas phase, thermal condition of the fluid phase and thermal condition of the solid phase. They are solved simultaneously with a fully implicit integrated finite difference scheme. Note that, in the conventional two-phase flow model, the intrinsic permeability is a constant, noted K_0 in this article.

Pathway dilation

Owing to the gas generated inside of the repository, it is assumed that pore deformation can occur locally in the bentonite, leading to an increase in the intrinsic permeability. The physical assumption is that a network of connected fractures is created in the bentonite when the gas pressure increases. To avoid the difficulty of modelling such a network, the choice is made to allow the formation of a pathway and to let the model adjusting its location automatically. The model proposed to implement this concept is a heuristic hydromechanical coupling process of gas pressure, porosity and permeability. A parametric function relating the pressure applied on pores and the porosity is proposed. This formulation assumes a linear relationship between porosity and pressure, with a very small positive slope (small elastic deformation) for pressure up to a preset threshold pressure. Over the threshold, the change rate is assumed to become significantly larger (yield point). Defining the threshold pressure as the sum of the swelling and water pressures, the equation of the porosity as a function of the pore pressure value takes the following form:

When $P_g \le P_w + P_s$ then

$$\phi = \phi_0(1 + c_r(P_g - P_0)) \quad (6)$$

$$K = K' \quad (7)$$

Once $P_g > P_w + P$ then

$$\phi = \phi_0(1 + c_r(P_s - P_0) + ac_r(P_g - P_s)) \quad (8)$$

$$K = bK'\left(\frac{\phi}{\phi_0}\right)^c \quad (9)$$

where P_g is the gas pressure, P_s is the swelling pressure, P_0 is the atmospheric pressure, ϕ is the porosity, K is the intrinsic permeability, ϕ_0 is the porosity of bentonite when no constrain/load is considered ($P_g = P_0$), and c_r is the rock compressibility. Note that K' is given by equation (15) if the micro-pores swelling aspect is considered or by $K' = K_0$ with K_0 the conventional intrinsic permeability

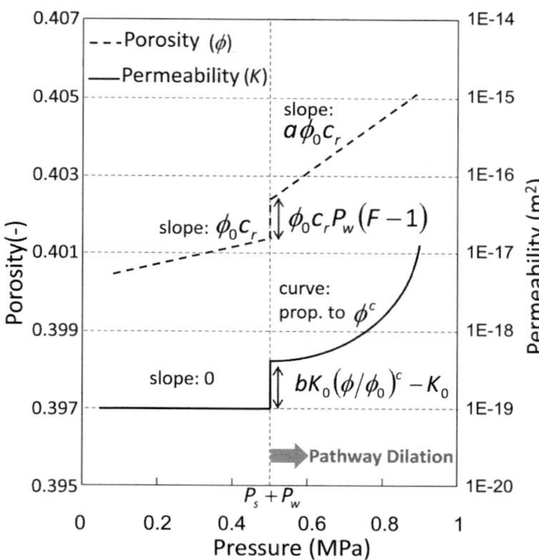

Fig. 2. Example of curves obtained with the pathway dilation model (equations (6)–(9)).

if only the pathway dilation aspect is used. The parameters a, b, c and P_s are constants that have to be calibrated. An illustrative representation of equations (6)–(9) is given in Figure 2.

As shown in equation (9), the variation of the permeability with the pressure is expressed via the relation with the porosity, more specifically with the porosity ratio (porosity/initial porosity). For a, b, c and P_s, plausible values are first selected from experience and then adjusted by inverse modelling. Note that the swelling pressure P_s is the most sensitive parameter, as it controls the threshold of the pathway dilation formation. In the current implementation of the model, once the gas pressure has overpassed the threshold, the second set of equations is applied (equations (8) and (9)) even if the gas pressure decreases below the threshold. This corresponds to the hypothesis that the preferential pathways remain open. Also, even if no local or directional constraints are considered explicitly to ensure a localization of gas flow, it is expected that the model will create local heterogeneity at the grid scale that is sufficient to represent the fracture network. Indeed, the gas pressure and the water pressure are potentially different in each grid element, resulting in strong potential heterogeneity in the localization of the pathways.

Several others approaches have been proposed to implement the pathway dilation effect into an existing two-phase flow model (e.g. Calder *et al.* 2006; Finsterle 2009; Navarro 2009). A large group of models, including the model proposed here, are based on the same principle: adding the variability of the permeability, the porosity and the capillary pressure with the gas pressure. The main difference between the models resides in the assumed relationship (linear, exponential, etc.) and the expression of the gas pressure threshold. Here, a linear law for the porosity and a power law for the permeability were considered and with a threshold that depends linearly with the swelling pressure. These assumptions were based on existing studies and experimental results where there was observed a strong influence of the swelling pressure of the bentonite on the gas flow rate and expelled water.

Micropore swelling

The modification of the pore distribution in montmorillonite that follows the hydration process is schematized in Figure 3. Although the variations of the global volume are negligible, constrained by the boundary conditions, the distribution of the pores in the porous part of bentonite is modified by the entry of water. The interlayer space expands resulting in an increase of the porosity ϕ_s and a decrease of the macro porosity ϕ_n following the relation ($\phi_n = \phi - \phi_s$) (Komine & Ogata 2004). While water might be able to move in the micropores, it is assumed that the quantity would be negligible. Therefore, in our model, the fluids (water and gas) can only move in the macropores (the non-swelling pores). This assumption leads to the interpretation that, if no water is expelled after a gas injection starts (e.g. Horseman *et al.* 1999), then the macropores are non-existent and all the water is trapped in the micropores. As a result, the residual water saturation corresponds to the water trapped in the interlayers and the permeability depends on the macropore distribution. We suppose here that this effect is responsible for the variation in permeability measured in bentonite material when in the unsaturated condition or fully saturated condition (e.g. in JAEA 2007). In dry conditions, the permeability $K = K_{dry}$ is higher than at fully saturated conditions. The swelling pressure P_s and the residual water saturation are equal to zero and the interlayer porosity is equal to the fixed dry porosity. As the water infiltrates the bentonite, the swelling pressure increases and so does the microporosity. Owing to the increasing quantity of water trapped in the interlayer, the residual water saturation S_{wr} increases.

One purpose of the model is to allow the reproduction of experimental results with a unique set of parameters for the re-saturation and gas migration stages. To do so, the model is developed to add some degrees of freedom to the two-phase flow curves. The mathematical model that formalizes the mechanism of micropore deformation in the conventional two-phase flow code consists of defining the expression of the swelling pressure,

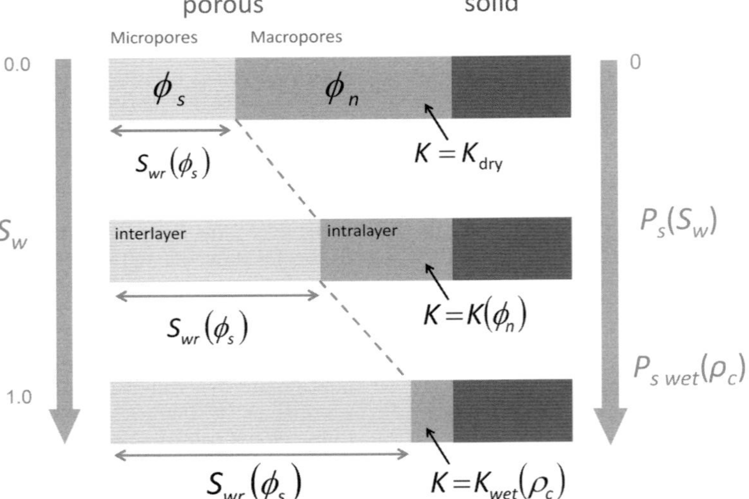

Fig. 3. Schematic view of the micropore swelling and the macropore shrinkage model.

the porosity, the residual water saturation and the permeability. The variation of the swelling pressure with the water saturation is expressed as:

$$P_s = P_{s\,wet} S_w^{\alpha} \tag{10}$$

where P_{ssat} is the swelling pressure at fully saturated condition and α is a parameter to be adjusted, as established by experimentation (JAEA 2007). It extends the linear relation used in Jobman (1998) to a non-linear relation that better fits the measured swelling pressure values. The relationship between porosity and swelling pressure is given by the following equations:

$$\phi_s = (\phi_{s\,wet} - \phi_{s\,dry})\left(\frac{P_s}{P_{s\,wet}}\right)^{\beta} + \phi_{s\,dry} \tag{11}$$

$$\phi_n = \phi - \phi_s. \tag{12}$$

Equation (11) was proposed by Jobman (equation (1) of Jobman 1998) but is expressed here as a function of the difference between the microporosity in saturated and dry conditions ($\phi_{s\,wet} - \phi_{s\,dry}$). The parameter β is the adjusting parameter. Equation (12) expresses the relationship between porosity, microporosity and macro porosity. Our model also considers the variation of the residual water saturation with the water saturation, given by the following system of equations:

$$S_{wr} = S_w^{\gamma}\left(\frac{\phi_s}{\phi}\right) \quad \text{if } S_w^{\gamma}\left(\frac{\phi_s}{\phi}\right) \leq S_w \tag{13}$$

$$S_{wr} = S_w \quad \text{if } S_w^{\gamma}\left(\frac{\phi_s}{\phi}\right) \geq S_w. \tag{14}$$

Equations (13) and (14) are original. They are based on the fact that, in the conceptual model, the residual water saturation S_{wr} depends on the micropore size and the water saturation. Note that equation (14) is used only to ensure that S_{wr} is always inferior or equal to S_w. Finally, the variation of the permeability is given by:

$$K' = \left(\frac{\phi_n}{\phi_{n\,wet}}\right)^{\lambda} K_{wet} \tag{15}$$

This equation was proposed in Jobman (1998) to express the relationship between the intrinsic permeability and the porosity (equation (2) of Jobman 1998).

As the water saturation is computed in each grid element of the discretized domain, it allows a strong heterogeneity in the swelling parameters. Contrary to the conventional two-phase flow model where the permeability is kept constant, here the permeability is decreasing with the water saturation. The variable residual water saturation of equation (13) simulates the increase of the water trapped into the bentonite interlayer, in the micropores, and the resulting shrinkage of the macropores where the transport of water and gas is assumed to occur. By allowing S_{wr} to vary, the model adds a degree of freedom to the two-phase flow curves (capillary pressure and relative permeability) via the effective saturation $S_{we} = (S_w - S_{wr})/(1 - S_{wr} - S_{gr})$. With this model, the shapes of two-phase flow curves are updated depending on the amount of water saturation. Therefore, we expect to obtain a unified model of the two-phase flow curves for the imbibition and drainage stages.

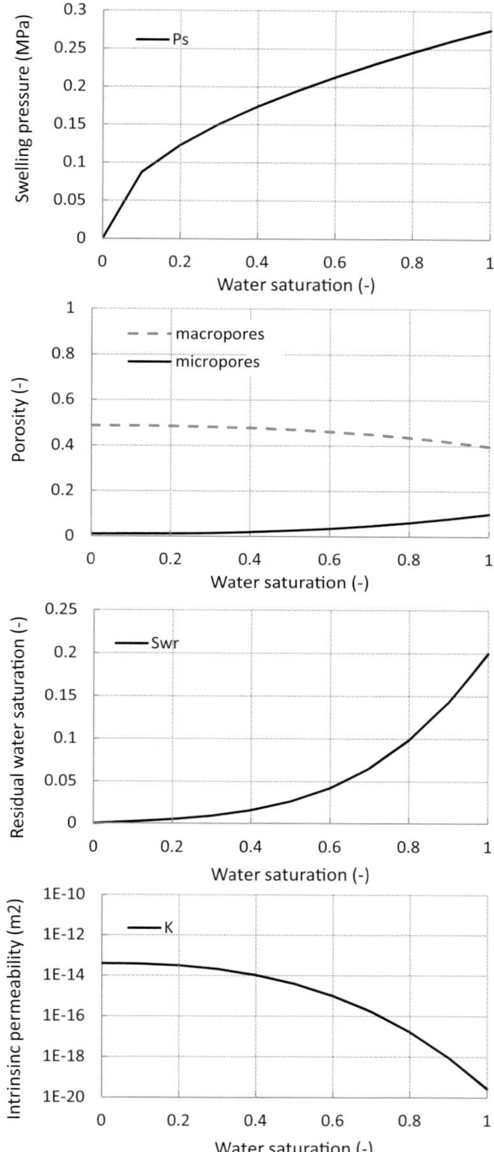

Fig. 4. Example of water-content dependency of the hydrologic parameters expressed by equations (10)–(15) with the parameters given in Table 4.

A representation of equations (10)–(15) is given in Figure 4 with the parameter values of Table 4. It shows the increasing swelling pressure, microporosity and residual saturation with the water saturation. In these equations, the parameters α, β, γ and λ have to be calibrated, either manually or with an inverse method.

Inverse approach for model calibration

Even if the number of parameters of the model was kept relatively small, fewer than 10, the calibration by trial-and-error simulations would require a large number of trials to obtain an accurate estimation. An alternative method is to use an inverse approach to estimate automatically the parameter values that minimize the discrepancy between simulations and observations. Here, a determinist optimization method was chosen for its relative simplicity and its good behaviour on weakly ill-conditioned problems (Nocedal & Wright 2006). In such a method, an optimal set of parameters that minimize a given objective function is obtained by an optimization algorithm. Here, the objective function is the classical weighted least-square function. It quantifies the discrepancy between the model output and the observations:

$$J(X) = \sum_{i=1}^{m} a_i(y_i - h_i(X))^2 \tag{16}$$

where J is the objective function, $X = [x_1, \ldots, x_p]^t$ is the vector of parameters to be adjusted, $Y = [y_1, \ldots, y_m]^t$ is the vector of observations of size m, $H(X) = [h_1(X), \ldots, h_m(X)]^t$ is the vector of model output and the weights a_i are fixed to adjust the confidence on the observations. Note that, owing to the highly non-linear relationship between the parameters to estimate X and the observations Y, the problem is a non-linear inverse problem. The minimization was conducted with a modified Gauss–Newton algorithm with second derivative approximation that allowed a relatively fast convergence. As the algorithm implements a local search, there is no guarantee of obtaining the global minimizer of the objective function. It is therefore essential to provide good initial values for the parameters. In addition, constraints on the domain of existence of the parameters were considered in order to restrict the domain to explore and ensure that plausible values of the parameters are obtained.

Regarding the implementation, the program UCODE2005 (Poeter *et al.* 2005) was selected because of its robustness and flexibility. The parameters to calibrate include the parameters of the retention curve model (three parameters), the relative permeability models (three parameters) and the pathway dilation model (three parameters). The parameters of the micropore deformation model (four parameters) were calibrated manually by trial-and-error or estimated from preliminary simulations. The most sensitive parameters were $P_{s\ wet}$, which controls the threshold of the dilation, b, which controls the magnitude of the dilation, and γ, which influences the residual water saturation. Initial values were based on previous experiments, expert knowledge and trial-and-error. Regarding the permeability and the swelling pressure at saturated conditions (K_{wet}, $P_{s\ wet}$), initial values were based on experimental results from the Japan Atomic Energy Agency (JNC 1999; JAEA 2007):

$$K_{wet\ init} = \exp(-2.1232\rho_c^2 + 1.11447\rho_c - 42) \quad (17)$$

$$P_{s\ wet\ init} = \exp(3.8497\rho_c^2 - 7.3332\rho_c + 2.0856) \quad (18)$$

where ρ_c is the dry density of the bentonite. These expressions were established after testing a large set of bentonite specimens of different dry density. Examples of initial values, minimum and maximum values used in the following section are given in Table 1. In each application, a few executions of the inverse method with different initial values were required to obtain a good match with the measured data.

Applications

Four applications of the proposed model are presented. First, following a literature survey on available experimental data related to gas migration in compacted bentonite (see e.g. Olivella & Alonso 2008), the flow rate controlled-gas injection experiment of Horseman *et al.* (1999) and the pressure-controlled-gas injection test on a wide range of clay density and water content values of Halayko (1998) (see also Graham *et al.* 2002) were selected. This allows the validation of the model on the gas migration stage, under fully saturated and various unsaturated conditions, respectively (RWMC 2009; see also Namiki *et al.* this volume, in press). Then, regarding the re-saturation stage, owing to the lack of available experimental data in the literature, an experiment was conducted at the RWMC (Radioactive Waste Management Funding and Research Center, Japan) that allowed the complete validation of the model. Finally, in order to show the behaviour of the model on a field-scale case, a realistic deep geological waste disposal was simulated. As experimental data are not available for this case, it is proposed for illustrative purposes only. In all cases, the two-phase flow model extended to consider pathway dilation and micropore deformation was used.

Laboratory-scale experiments of gas migration (Horseman)

The flow rate controlled gas injection experiments described in Horseman *et al.* (1999) were selected to validate the pathway dilation model. In these experiments, the clay specimen is confined by an isotropic stress that remains constant during the gas injection. Helium gas, as a safe replacement for hydrogen, is repeatedly injected into a fully saturated bentonite specimen (MX80, dry density equals 1.6 mg m^{-3}). A scheme of the experiment is represented in Figure 5. The upper volume is used to control the flow rate of the gas injection into the specimen. Note that, because the gas is injected in fully saturated bentonite only, the micropore swelling effect does not arise. Some preliminary results on these experiments have been presented in Tawara *et al.* (2010).

Figure 6a, b shows the results obtained by a conventional two-phase flow model and the extended two-phase flow model, respectively. The corresponding identified parameters are given in Tables 2 and 3. In these tables, c_r is the rock compressibility, K_w and K_g are the absolute permeability of water phase and gas phase, respectively, and S_{wr} and S_{gr} are the residual saturation of water phase and gas phase, respectively. The models of retention curve we used are the Narasimhan model for the capillary pressure and the vanGenuchten model for the relative permeability (water phase and gas phase). As a reminder, the Narasimhan model of capillary pressure is given by the formula:

$$P_c = P_{aep} + P_{Nar}\left(\frac{1 - S_{wr}}{S_w - S_{wr}}\right)^{\frac{1}{n}} \quad (19)$$

with P_{aep} the air entry pressure, S_w the water saturation, S_{wr} the residual water saturation, and P_{Nar}

Table 1. *Example of values used for the calibration of the model (RWMC experiment)*

Parameter	Unit	Minimum value	Maximum value	Initial value
K_{wet}	m^2	0.10	10.0	1.00
S_{wr}	(−)	0.00	0.90	0.00
$n(K_{wr})$	(−)	1.20	20.0	2.50
$n(K_{gr})$	(−)	1.20	20.0	19.2
$n(P_c)$	(−)	1.50	10.0	6.2
P_{vG}	MPa	1.0	20.00	1.43
a	(−)	1.00	2.00	1.00
c	(−)	0.10	5.00	0.4129

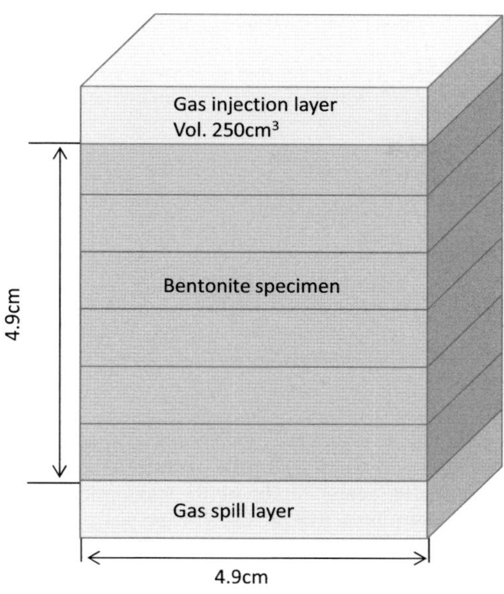

Fig. 5. Schematic representation of the Horseman *et al.* (1999) experiments.

and n two parameters to be calibrated. The vanGenuchten model of relative permeability for each phase is given by:

$$k_{wr} = \sqrt{S_{we}}\,(1 - (1 - S_{we}^{1/m})^m)^2 \qquad (20)$$

$$k_{gr} = \sqrt{S_{ge}}\,(1 - (1 - S_{ge}^{1/m})^m)^2 \qquad (21)$$

where S_{we} and S_{ge} are the effective saturation of water phase and gas phase respectively, n is a parameter to be calibrated and $m = 1 - 1/n$. Note that $S_{we} = (S_w - S_{wr})/(1 - S_{wr} - S_{gr})$ and $S_{ge} = (S_g - S_{gr})/(1 - S_{wr} - S_{gr})$.

Figure 6a, b shows the flow rate of expelled gas and gas pressure. The peaks are only well reproduced when the pathway dilation model is considered, which confirms the necessity to incorporate the effect of preferential pathways. It should be highlighted that, for both conventional and extended models, the identified values of the residual water saturation (0.9944 and 0.9842, respectively) and the residual gas (0.0002073 and 0.002835, respectively) are extreme. As a result, the models can be applied only on experiments with fully saturated initial conditions that match this experiment (the bentonite remains almost fully saturated and no water is expelled). This excludes application on re-saturation experiments. Simulation with the thermo-hydro-mechanical code CODE_BRIGHT (DIT-UPC 2000) was also conducted and the results are shown in Figure 6c. It can be seen that,

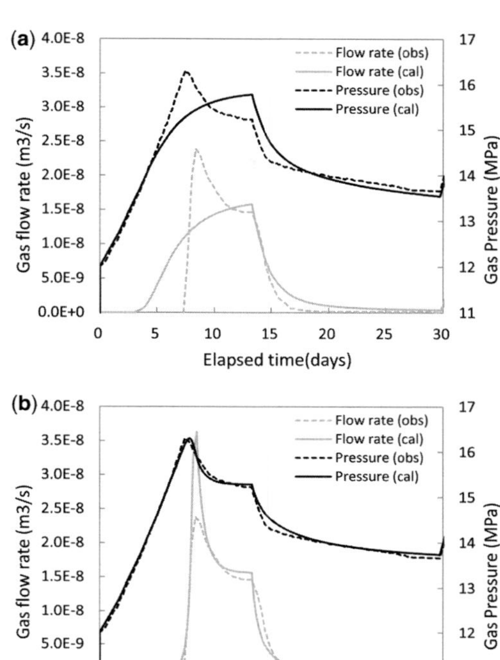

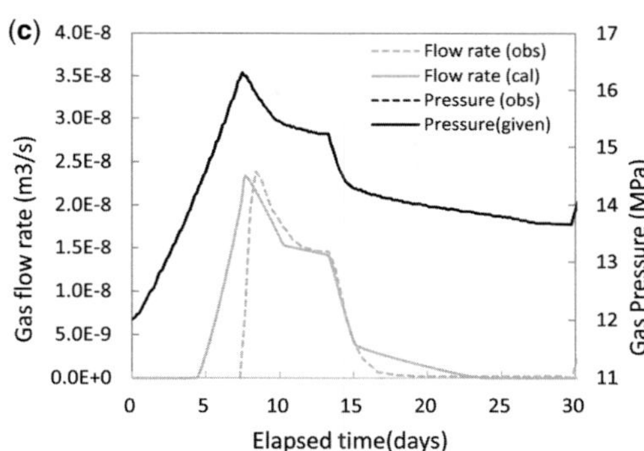

Fig. 6. Experimental results (dashed line) and calculation results (solid line) of gas flow rate (grey) and gas pressure (black) (Horseman *et al.* 1999). (**a**) Conventional two-phase flow model (GETFLOWS); (**b**) two-phase flow model with pathway dilation (GETFLOWS) and (**c**) CODE_BRIGHT.

if the peak of gas flow rate is present, a significant delay is noticed for its apparition. Therefore, this seems to show that mechanical modelling that does not consider coupled phenomena between micro and macro structures is not sufficient to model preferential pathways, as noted in Alonso *et al.* (1999) about the Barcelona Basic Model. Note that, in this last case, we assume that the gas

Table 2. *Conventional two-phase flow parameters to reproduce the Horseman* et al. *(1999) experiment*

Model	Parameter	Unit	Value	Calibration
Two-phase flow	c_r	1/MPa	1.754×10^{-3}	No
	K (K_0)	m^2	2.427×10^{-20}	Auto
	S_{wr}	(−)	0.9944	Auto
	S_{gr}	(−)	0.0002073	Auto
	$n(K_{wr})$	(−)	2.879	Auto
	$n(K_{gr})$	(−)	1.892	Auto
	$n(P_c)$	(−)	0.9770	Auto
	P_{Nar}	MPa	0.0009807	Auto
	P_{aep}	MPa	11.0	No

pressure is given; it is therefore impossible to conclude on the capabilities of CODE_BRIGHT to reproduce such an experiment with accuracy.

Laboratory-scale experiments of gas migration (Halayko)

When the initial condition of the bentonite specimen is not fully saturated, the existence of a saturation threshold over which gas breakthrough occurred has been observed (Graham *et al.* 2002). Among the experimental results presented in the literature, the set of experiments no. 10 of Halayko (1998) was selected to validate our model. These tests are interesting because they use specimens made from 100% bentonite (important for the consistency with the other experiments), with a dry filter (to avoid an initial uncontrolled re-saturation), without a saturation stage (usually not documented enough to set up a model) and with published measures of gas pressure (to allow the validation of the model).

In these tests, the specimen is made from Avonlea bentonite. The inlet pressure is increased by 0.2 MPa every 5 min. As the gas outlet is isolated, any increase in pressure of the gas collection tank is an indication of gas breakthrough. Bentonite specimens with different dry densities and different initial conditions of water saturation are set up, leading to 29 experiments. After removing the inconsistent experiments (experiments where the water saturation before the gas injection was lower than after the gas injection), 13 tests remained that were reproduced by the extended two-phase flow model. Only the pathway dilation model was considered (expected to be effective when the specimen is almost fully saturated). In unsaturated conditions, the permeability was adjusted empirically. The same as for the Horseman experiments, the retention curve and relative permeability curves were given by the Narasimhan model and the van-Genuchten model, respectively.

Figure 7 shows the measured gas breakthrough pressure and the computed ones as a function of

Table 3. *Extended two-phase flow with pathway dilation parameters to reproduce the Horseman* et al. *(1999) experiment*

Model	Parameter	Unit	Value	Calibration
Two-phase flow	c_r	1/MPa	3.405×10^{-3}	No
	K (K_0)	m^2	0.8939×10^{20}	Auto
	S_{wr}	(−)	0.9842	Auto
	S_{gr}	(−)	0.002835	Auto
	$n(K_{wr})$	(−)	1.113	Auto
	n (K_{gr})	(−)	5.136	Auto
	$n(P_c)$	(−)	0.5414	Auto
	P_{Nar}	MPa	0.001037	Auto
	P_{aep}	MPa	11.0	No
Pathway dilation	a	(−)	1.467	Auto
	b	(−)	56.23	Manual
	c	(−)	0.9867	Auto
	P_s	MPa	14.8	No

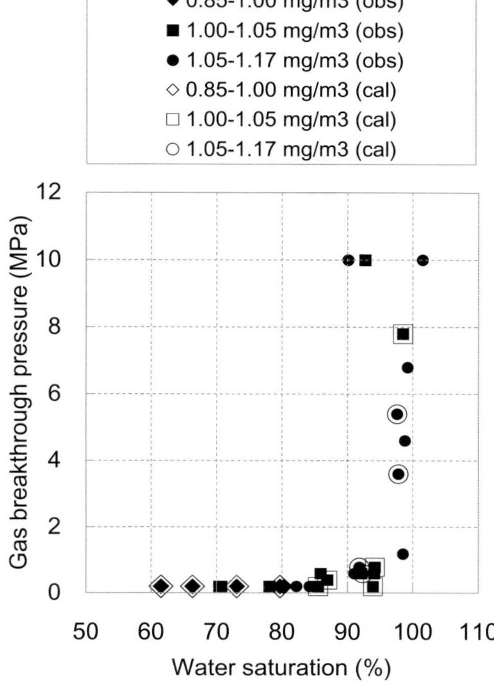

Fig. 7. Gas pressure breakthrough as a function of the initial saturation of the bentonite specimen for specimens with dry density from 0.85 to 1.17 mg m^{-3}. Experimental results from Halayko (1998) (solid markers) and simulated results with the proposed model (contour markers). It confirms the existence of the pathways when the initial saturation of the bentonite specimen is over 90%.

the initial water saturation. In the figure, the results are classified into three groups depending on the dry density of the bentonite, which directly influences the swelling pressure value. This gives a clue as to the formation of preferential pathways into the bentonite at high-saturation and high-pressure injection. It can be seen that the gas breakthrough values are well estimated by the model.

Laboratory-scale experiment of re-saturation and gas migration

In order to validate the full model (with pathway dilation model and micropores deformation model), a laboratory-scale experiment was established in collaboration with RWMC. In this case, both the re-saturation stage and the gas migration stage were considered. Figure 8 shows the schematic representation of the experiment. The specimen consists of 100% bentonite (KunigelV1) with a

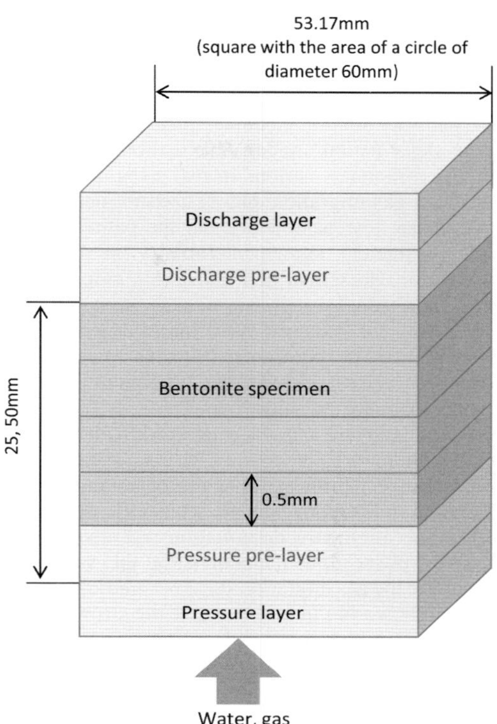

Fig. 8. Schematic representation of the RWMC experiment.

dry density of 1.36 mg m^{-3}. Constant boundary pressure was applied on the lateral sides of the specimen, and atmospheric pressure was imposed at the top and bottom. The global volume was kept constant. Water and gas were injected from the bottom layer and expelled water flow and gas flow are measured at the upper part. As the side interfaces were set up to be impermeable, all the injected water and gas could be trapped in the bentonite or expelled at the discharge layer. The expression for the retention curves and the capillary pressure are given respectively by the vanGenuchten models of equations (20) and (21) and by the vanGenuchten model:

$$P_c = P_{vG}(S_{we}^{-1/m} - 1)^{1/n} \tag{22}$$

with S_{we} the effective saturation given by $S_{we} = (S_w - S_{wr})/(1 - S_{wr} - S_{gr})$, n and P_{vG} two parameters to be calibrated and $m = 1 - 1/n$. Table 4 lists the model parameters including the parameters calibrated with the inverse approach.

In the re-saturation stage, water injection is controlled by a one-step function for the water pressure, as shown in Figure 9 (grey dash line). Then, in the gas migration stage, gas is injected following to a multistep function for the gas pressure, as shown in Figure 10 (grey dashed line).

Two models are used to reproduce the experimental data: the pathway dilation model only and the pathway dilation model coupled with the micropore deformation model. The calibration of the model parameters is done in two steps. In the first step, the gas migration data is used to identify the two-phase flow parameters and the pathway dilation parameters, shown in Table 4 (in the same way as in Horseman and Halayko's cases). Note that the residual water saturation S_{wr} remains variable; only the value at fully saturated conditions is identified. In the second step, the re-saturation data are used to identify the micropore swelling model parameters, shown in Table 4. Note that $\phi_{s\ dry}$ is fixed and $\phi_{s\ wet}$ is calculated from equation (13) with $S_w = 1$.

Accurate reproduction of the observations was obtained with both models (Fig. 9). However, when the micropore deformation was considered, the two-phase flow parameters of the saturation stage and the gas migration stage became consistent. This is shown in Figure 11, where the arrow next to the capillary pressure and relative permeability curves indicates the corresponding stage (left to right, saturation or imbibition; right to left, gas migration or drainage). Such consistency is important to ensure the robustness of the model against any initial condition of water saturation. In contrast, the results of the conventional model are consistent only for dry or fully saturated initial conditions. However, the curve of the relative permeability of gas phase still shows a strong hysteresis. This could be a sign that a phenomenon not considered in the model is interfering with the gas migration.

Figure 8 shows the expelled water volume and the gas flow rate along with the injected gas pressure (right axis). There is good reproduction of the expelled water volume and gas flow rate during the gas migration stage. Note that the expelled water volume evolves smoothly, which means that no breakthrough seems to occur in the bentonite until 28 days. This is certainly due to the type of bentonite used: contrary to the experiment of Horseman *et al.* (1999) where no water was expelled (water was trapped into the micropores, the macropores being almost nonexistent), here the macropores space still exists and allows the water to be expelled from the beginning of the experiment. Then, the gas flow rate suddenly increases around 28 days, indicating the formation of preferential pathways in the macropores of the bentonite. These results validate the applicability of the proposed model on the hydration and gas migration stage, when breakthroughs induced by the gas pressure are formed. Moreover, consistent two-phase flow parameters for imbibition and drainage are used.

Field-scale simulation of re-saturation and gas migration

The objective of this section is to reveal the behaviour of the proposed model on a real field-scale experiment, namely a re-saturation and gas migration test in a geological transuranic (TRU)

Table 4. *Extended two-phase flow with pathway dilation and micropores parameters to reproduce the RWMC experiment*

Model	Parameter	Unit	Value	Calibration
Two-phase flow	c_r	1/MPa	1.0×10^{-3}	No
	K_{wet}	m^2	2.68×10^{-20}	Auto
	$S_{wr}(wet)$	(−)	0.200	Auto
	S_{gr}	(−)	0	No
	$n\ (K_{wr})$	(−)	2.884	Auto
	$n\ (K_{gr})$	(−)	2.751	Auto
	$n\ (P_c)$	(−)	3.709	Auto
	P_{vG}	MPa	0.333	Auto
Pathway dilation	a	(−)	1.00	Auto
	b	(−)	6.058×10^4	Manual
	c	(−)	0.413	Auto
Micropore swelling	$P_{s\ wet}$	MPa	0.275	Manual
	$\phi_{s\ dry}$	(−)	0.01	No
	$\phi_{s\ wet}$	(−)	0.09926	No
	α	(−)	0.5	Manual
	β	(−)	5	Manual
	γ	(−)	1	Manual
	λ	(−)	70	Manual

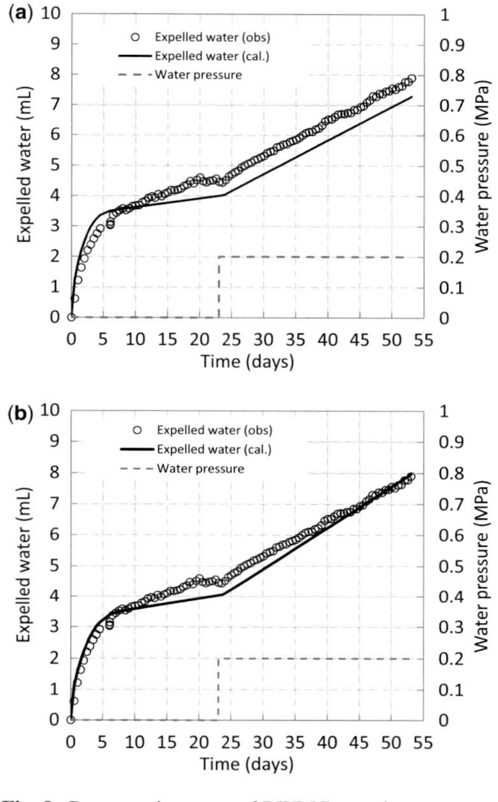

Fig. 9. Re-saturation stage of RWMC experiment. Observed expelled water (circle), computed expelled water (line) and input water pressure (grey dashed line) are represented. Similar results were obtained without micropore deformation aspect (**a**) and with micropore deformation aspect (**b**) but only one set of parameters of the two-phase flow curves is needed in (b) for the imbibition and drainage stages (except for the relative permeability of gas).

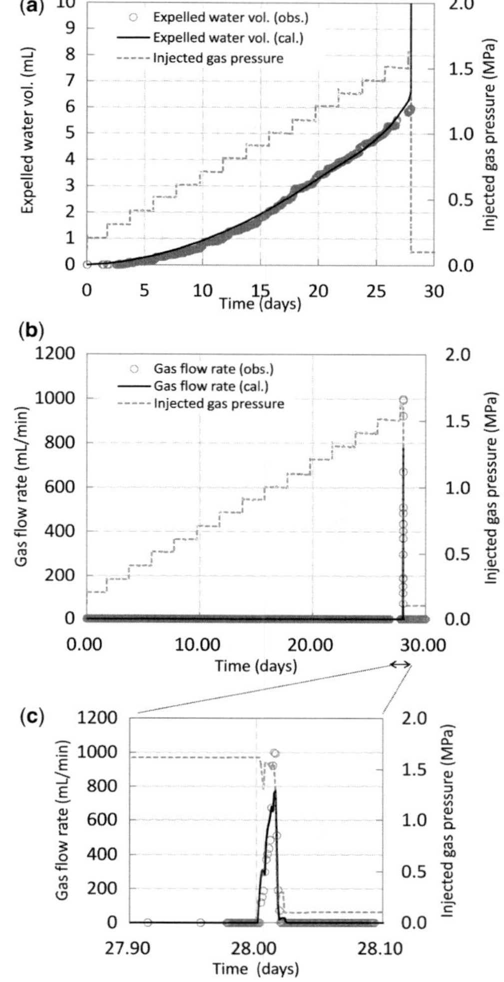

Fig. 10. Gas migration stage of RWMC experiment. Expelled water (**a**) and gas flow rate (**b**, **c**) are shown. The multistep function of injected gas pressure is represented as a grey dashed line. Water started to be expelled from the beginning of the experiment. Gas flow rate increased suddenly after about 28 days.

waste disposal (transuranic refers to artificial atoms that are heavier than uranium). The designed numerical model consists of a two-dimensional column of 1000 m height, as represented in Figure 12. Thanks to the symmetry of the disposal, only one-half of the disposal has to be modelled. The structure includes the TRU waste package (located at a depth of 500 m), the bentonite buffer, the excavation damage zone and the host rock. On the column of soil, 'no flow' boundary constraints are applied on both sides; constant pressure and constant temperature are set up on the bottom and top boundaries. A geothermal gradient of 0.03°C is applied. Under initial conditions, the host rock is assumed to be fully saturated, and the other materials are unsaturated. It is assumed that the re-saturation stage and the gas migration stage start simultaneously at the beginning of the simulation. Owing to the large amount of time required by the computation, a Monte-Carlo sensitivity analysis was preferred over the inverse approach for the identification of the model parameters.

Additional to the primary variables (water saturation, gas pressure and temperature), the model allows the visualization of a second set of variables (swelling pressure, porosity, residual saturation and intrinsic permeability). Figure 13 shows the values of the variables at four time steps (10, 50,

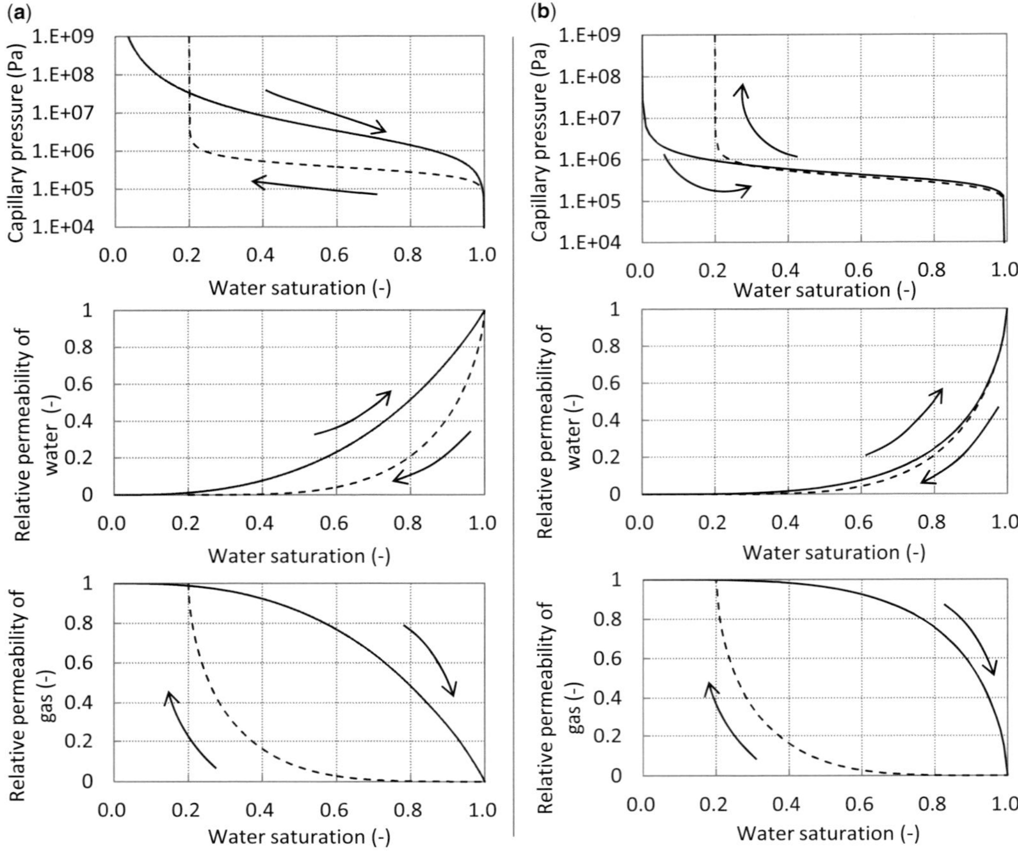

Fig. 11. Representation of the two-phase flow parameters curves obtained (**a**) without and (**b**) with micropore deformation model. Line from top to bottom: capillary pressure (vanGenuchten model), relative permeability of the water phase (vanGenuchten model) and relative permeability of gas phase (Grant model). Solid line, imbibition; dashed line, drainage.

100 and 500 years) represented by colour maps. In this figure, we can notice the gradual re-saturation during the period when the gas pressure is still low (up to about 70 years). Then, the gas pressure increases gradually and the water starts to be expelled from the engineered barrier. Owing to the thermal gradient, the temperature quickly increases from zero to 10 years (not visible) before decreasing gradually owing to the diffusion phenomenon. Following the water saturation fluctuations, the swelling pressure P_s and the residual water saturation increase until 100 years and then slowly decrease until 500 years; the intrinsic permeability K and the macroporosity ϕ_n evolve in the same manner but inversely (decreasing and then slight increasing after 100 years). Note that, if isothermal conditions were considered, the residual saturation S_{wr} would remain at a high value. The diminution is interpreted as the evaporation of the water present in both micropores and macropores. These behaviours match the expectations and show the influence of the pore deformation effect on the gas migration.

It is notable that the model allows the visualization of the pathways formed in the bentonite during the simulation by representing the map of the gas pressure that overpasses the gas breakthrough threshold (last column of Fig. 13). It is shown that the formation of fractures starts at the outer interface and then propagates into the buffer with lateral structures until reaching a homogenous state.

Owing to a lack of space, only the application of the full model (with pathway dilation and micropores swelling) has been presented. To test the influence of each aspect of the model, the model

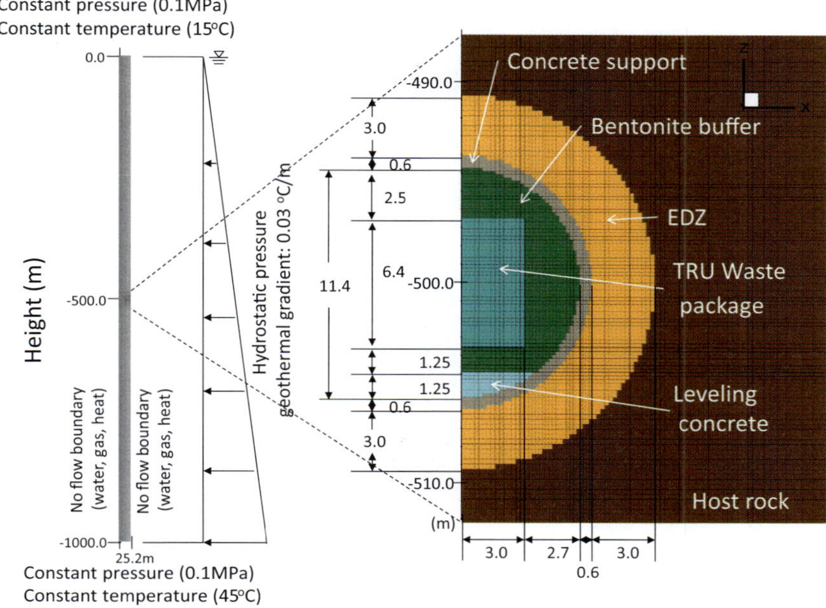

Fig. 12. Geometry and mesh of the field-scale case.

was applied also with the pathway dilation effect only. It was found that the mechanical effect of swelling/shrinkage-induced deformation of macropores affected the performance measure in term of maximum gas pressure and expelled water volume, and should be considered in the gas migration analysis.

Discussion

The proposed model was applied successfully in gas migration experiments with the pathway dilation part (Horseman *et al.* 1999) and on experiments including re-saturation and gas migration with pathway dilation and micropores swelling parts (Halayko 1998, RWMC experiments). Despite this extensive study, it is notable that the validation tests were focused on the reproduction of the observations. Even if the number of adjusting parameters is kept small, the possibility of overfitting exists. A quantitative study of the predictive capability of the model would be relevant to evaluate this possibility. Also, the heterogeneity was introduced on the grid scale. As the volumes of the grid element are usually large compared with the dimensions of the modelled phenomena, this implies significant simplification of the phenomena.

Note that the concept of preferential pathway is implemented in such a way as to make it possible to visualize the pathways created by the model to reproduce the observations. Our study does not provide any clue on the real localization of the gas flow, particularly because the pathways are created at a scale that is disconnected from the scale of the physical phenomena. Nevertheless, we consider that the representation of these pathways is relevant to evaluating the behaviour of the model (cf. last column of Fig. 13).

The results obtained with three different types of bentonite (MX80, Avonlea and KunigelV1) confirm the importance of considering the existence of preferential pathways in the simulation of the gas migration. They also underline the influence of the bentonite characteristics on the breakthrough pressure: the swelling pressure of these clays varies by several orders of magnitude and the gas pressure required to create the breakthrough varies also by several orders of magnitude.

One of the purposes of the model was to handle the two-phase flow parameters in a unified manner in order to reproduce the experiments with a unique set of parameters for the drainage and the imbibition curves. It was found that the pathway dilation effect, which is effective at highly saturated conditions, was not sufficient to obtain a model with a unique set of parameters for both stages. To do so, the concept of micropore/macropore redistribution that leads to the micropores swelling model is successful in unifying the capillary pressure and

Fig. 13. Computed results of the field-scale case (see text for comments).

the relative permeability of water. It simulates the increase in water trapped in the bentonite interlayer, in the micropores, and the resulting shrinkage of the macropores where the transport of water and gas is assumed to occur. Two sets of parameters for the relative permeability of gas were still necessary to reproduce the observations (for imbibition and drainage). While this could be explained by a strong hysteresis, we consider that a model that can unify the imbibition and drainage curves would be more accurate to simulate the gas migration in buffer with variable condition of water saturation, where re-saturation and gas migration may happen simultaneously.

Finally, the persistence of a preferential pathway after their formation is a point that should be considered with caution, as some experiments have shown the instability of the pathways (e.g. Horseman 1999; Graham *et al.* 2012). Here, it is simply considered that the pathway remains open once it is formed (the porosity and the permeability continue to follow the constitutive laws of the pathway dilation model). This hypothesis might be too conservative as it leads to the wide- open pathway shown in the field-scale case (last column of Fig. 13).

Conclusion

An extended two-phase flow model that considers mechanical effects has been proposed for the purpose of simulating the gas migration in deep radioactive waste disposal. To avoid the computational cost of a fully mechanical coupling, the model extends a conventional two-phase flow model by incorporating two specific effects: the preferential pathway formation and the micropores/macropores redistribution. It is notable that the proposed model requires only slight modifications of the conventional two-phase flow model. One advantage of the model over some existing approaches is that no prior knowledge of the location and the diameter of the potential fractures is necessary. Also, thank to the modelling of the micropore swelling effect, the model can handle the two-phase flow parameters in a unified manner using the same set of parameters for the imbibition and the drainage stage. The micropore swelling effect can explain why the European bentonite (MX80) and the Japanese bentonite (KunigelV1) behave differently regarding the expelled water flow induced by gas injection. It should be noticed that the model assumes significant simplifications of the physical phenomena and therefore it should be considered with caution.

The model was applied on three laboratory experiments and one field test. In the laboratory experiments, it confirmed that the measured test data could be successfully reproduced by the new modelling approach using a unique set of parameters for the retention curve and the permeability of the water phase. The permeability of gas phase still required two set of parameters, resulting in a strong hysteresis between the re-saturation and gas migration stages. In the field-scale experiment, the results obtained by our model were consistent with the expectations regarding the re-saturation and gas migration stages. In a comparison of our model with and without the micropore swelling effect (not shown owing to a lack of space), it was found that the mechanical effect of swelling/shrinkage deformation of macropores affected the performance measure and should be considered in the gas migration analysis. Future work will focus on the question of the hysteresis of the gas phase permeability and the phenomenon of closing pathways after a first gas injection has led to the preferential path formation.

This study includes a part of the result of ‘Development of the technique for the evaluation of long-term performance of EBS, FY2011’ under a grant from the Agency of Natural Resources and Energy, the Ministry of Economy Trade and Industry of Japan. The authors would like to thank the two anonymous reviewers for their especially careful review and their constructive comments.

References

Alonso, E., Vaunat, J. & Gens, A. 1999. Modelling the mechanical behaviour of expansive clays. *Engineering Geology*, **54**, 173–183.

Calder, N., Avis, J., Senger, R. & Leung, H. 2006. Modifying TOUGH2 to support modeling of gas transport through saturated compacted bentonite as part of the large-scale gas injection test (LASGIT) in Sweden. *In*: *Proceedings of TOUGH Symposium 2006*, Lawrence Berkeley National Laboratory, Berkley, CA, 15–17 May 2006.

DIT-UPC. 2000. *CODE_BRIGHT, A 3-D Program for Thermo-Hydro-Mechanical analysis in Geological Media. User's Guide*. Centro Internacional de Metodos Numericos en Ingenieria, Barcelona.

Finsterle, S. 2009. *iTOUGH2-IFC: An Integrated Flow Code in Support of Nagra's Probabilistic Safety Assessment*. User's Guide and Model Description. Technical Report LBNL-1441E, Lawrence Berkeley National Laboratory, Berkley, CA.

Gerard, P., Harrington, J., Charlier, R. & Collin, F. 2012. Hydro-mechanical modelling of the development of preferential gas pathways in claystone. *In*: *2nd European Conference on Unsaturated Soils*, 20–22 June 2012, Napoli, Italy, 175–180.

Graham, C. C., Harrington, J. F., Cuss, R. J., Sellin, P. & Evans, N. 2012. Gas migration experiments in bentonite: implications for numerical modeling. *Mineralogical Magazine*, **76**, 3279–3292.

Graham, J., Halayko, K. G., Hume, H., Kirkham, T., Gray, M. & Oscarson, D. 2002. A capillarity-advective model for gas break-through in clays. *Engineering Geology*, **64**, 273–286.

Halayko, K. S. G. 1998. *Gas flow in compacted clays*. MSc thesis, University of Manitoba, Winnipeg MB.

Horseman, S. T., Harrington, J. J. & Sellin, P. 1999. Gas migration in clay barriers. *Engineering Geology*, **54**, 139–149.

JAEA (JAPAN ATOMIC ENERGY AGENCY) 2007. *Second Progress Report on Research and Development for TRU Waste Disposal in Japan*. JAEA Report JAEA-Review 2007-010.

JNC (JAPAN NUCLEAR CYCLE) 1999. *Swelling properties of bentonite buffer*. JNC TN8400 99-038 (in Japanese).

Jobman, M. 1998. Modification and application of the TOUGH2 code for modeling of water flow through swelling unsaturated sealing constructions. *In*: *Proceedings TOUGH Workshop '98*, 4–6 May 1998, Berkeley, CA.

Komine, H. & Ogata, N. 2004. Predicting swelling characteristics of bentonites. *Journal of Geotechnical and Geoenvironmental Engineering*, **130**, 818–829.

Mori, K., Tada, K. *et al.* 2006. A 2-PHASE, 3-d flow modelling for the gas migration test. *In*: *International High-Level Radioactive Waste Management Conference*, Las Vegas, NV, 30 April to 4 May 2006.

Namiki, K., Asano, H., Takahashi, S., Shimura, T. & Hirota, K. in press. Laboratory gas injection tests of compacted bentonite buffer material for TRU waste disposal. *In*: Norris, S., Bruno, J., Cathelineau, M., Delage, P., Fairhurst, C., Gaucher, E. C., Höhn, E. H., Kalinichev, A., Lalieux, P. & Sellin, P. (eds) *Clays in Natural and Engineered Barriers for Radioactive Waste Confinement*. Geological Society, London, Special Publications, **400**, http://dx.doi.org/10.1144/SP400.27

Navarro, M. 2009. *Simulating the migration of repositorygases through argillaceous rock by implementing the mechanism of pathway dilation into the code TOUGH2 (TOUGH2-PD)*. PAMINA project, public milestone 3.2.14. Gesellschaft für Anlagen- und Reaktorsicherheit (GRS) mbH, Cologne.

Nocedal, J. & Wright, S. 2006. *Numerical Optimization*, 2nd edn. Springer Series in Operations Research and Financial Engineering, **XXII**. Springer, Berlin, 664.

Olivella, S. & Alonso, E. E. 2008. Gas flow through clay barriers. *Geotechnique*, **58**, 157–176.

Poeter, E., Hill, M., Banta, E., Mehl, S. & Christensen, S. 2005. *UCODE_2005 and Six Other Computer Codes for Universal Sensitivity Analysis, Calibration, and Uncertainty Evaluation*. US Geological Survey, Reston, VA.

Pruess, K., Oldenburg, C. & Moridis, D. 1999. *TOUGH2 User's Guide. Version 2.0, LBNL-43134*. Lawrence Berkeley National Laboratory, Berkeley, CA.

RWMC 2009. *Radioactive Waste Management Funding and Research Center*. Project report for FY2008 of Development of the technique for the evaluation of long term performance of EBS for TRU-waste disposal (in Japanese).

Rodwell, W.R., Harris, A.W. *et al.* 1999. *Gas migration and two-phase flow through engineered and geological barriers for a deep repository for radioactive waste: a joint EC/NEA Status Report*. Nuclear Energy Agency and European Commission Nuclear Science and Technology 88 Report. EUR 19122 EN ISBN 92-828-8132-6. European Commission, Luxembourg.

Rodwell, W. R., Norris, S. *et al.* 2003. *A Thematic Network, on Gas Issues in Safety Assessment of Deep Repositories for Radioactive Waste (GASNET)*. European Commission Report, EUR 20620 ISBN92-894-6401-1. European Commission, Luxembourg.

Tawara, Y., Mori, K., Tada, K., Shimura, T., Sato, S., Yamamoto, S. & Hayashi, H. 2010. A validation study for the gas migration modelling of the compacted bentonite using existing experiment data. *In*: *4th International Meeting of Clays in Natural & Engineered Barriers for Radioactive Waste Confinement*, Nantes.

Tosaka, H., Itho, K. & Furuno, T. 2000. Fully coupled formation of surface flow with 2-phase subsurface flow for hydorological simulation. *Hydrological Process*, **14**, 449–464.

Tosaka, H., Mori, K., Tada, K., Tawara, Y. & Yamashita, K. 2010. A general-purpose terrestrial fluids/heat flow simulator for watershed system management. *In*: *IAHR International Groundwater Symposium*.

In situ diffusion test of hydrogen gas in the Opalinus Clay

A. VINSOT[1]*, C. A. J. APPELO[2], M. LUNDY[1], S. WECHNER[3], Y. LETTRY[4], C. LEROUGE[5], A. M. FERNÁNDEZ[6], M. LABAT[7], C. TOURNASSAT[5], P. DE CANNIERE[8], B. SCHWYN[9], J. MCKELVIE[10], S. DEWONCK[1], P. BOSSART[11] & J. DELAY[1]

[1]*Andra, CMHM, F-55290 Bure, France*

[2]*NL-1071 MB Amsterdam, The Netherlands*

[3]*Hydroisotop GmbH, D-85301 Schweitenkirchen, Germany*

[4]*Solexperts AG, CH-8617 Mönchaltorf, Switzerland*

[5]*BRGM, F-45060 Orléans, France*

[6]*Ciemat, S-28040 Madrid, Spain*

[7]*IRD, AMU-MIO, F-13288 Marseille, France*

[8]*FANC, B-1000 Brussels, Belgium*

[9]*Nagra, CH-5430 Wettingen, Switzerland*

[10]*NWMO, Toronto, ON, M4T 2S3, Canada*

[11]*Swisstopo, CH-2882 St-Ursanne, Switzerland*

**Corresponding author (e-mail: agnes.vinsot@andra.fr)*

Abstract: Hydrogen gas was injected, together with helium and neon, into a borehole in the low-diffusivity Opalinus Clay rock. The hydrogen partial pressure was at most 60 mbar. A water production flow rate from the surrounding rock of *c.* 15 ml/day had been obtained previously, indicating that the test interval wall was presumably saturated with water. Helium and neon concentrations decreased as expected while taking into account dissolution and diffusion processes in the porewater. In contrast, the disappearance rate of hydrogen observed (2×10^{-4} to 3×10^{-4} mol/day/m^2) was *c.* 20 times larger than the calculated rate considering only dissolution and diffusion. The same rate was observed following a new hydrogen injection and over a six-month semi-continuous injection phase. Simultaneously, sulphate and iron concentrations decreased in the water, whereas sulphide became detectable. These evolutions may be due to biotic processes involving hydrogen oxidation, sulphate reduction and Fe(III) reduction.

Several national nuclear waste management programmes are considering using steel components in the design of their deep geological repositories for high-level and long-lived radioactive waste. After closure of the underground disposal facility, oxygen will be consumed and anoxic corrosion of steel is expected to produce hydrogen gas. Studies of hydrogen interactions and transfer in argillaceous rocks aim to evaluate what will happen to this hydrogen and how the hydrogen pressure will evolve with time in the underground repositories. Among these studies is the 'Hydrogen Transfer' (HT) experiment, which was installed in 2009 in the Mont Terri Rock Laboratory to determine *in situ* the effective diffusion coefficient of hydrogen in Opalinus Clay. A second objective of this experiment was to evaluate whether a hydrogen reaction with the argillaceous rock can be detected *in situ* at low temperature (15–16 °C) and whether microorganisms play a role in this potential consumption.

The experimental concept of the HT experiment is based on gas circulation within a borehole (Vinsot *et al.* 2008*a*, *b*). After an initial phase during which the rock natural gas and porewater production and composition at the test location were followed up, the first hydrogen injection was performed in June 2011.

This paper describes the design of this *in situ* test and presents the experimental results regarding the evolution of hydrogen concentration in the circulating gas. It also reports the observed seepage

From: NORRIS, S., BRUNO, J., CATHELINEAU, M., DELAGE, P., FAIRHURST, C., GAUCHER, E. C., HÖHN, E. H., KALINICHEV, A., LALIEUX, P. & SELLIN, P. (eds) 2014. *Clays in Natural and Engineered Barriers for Radioactive Waste Confinement*. Geological Society, London, Special Publications, **400**, 563–578.
First published online April 2, 2014, http://dx.doi.org/10.1144/SP400.12

water composition evolution. Based on these experimental results, the processes involved in the observed evolution of hydrogen concentration are discussed.

Geological context and rock characteristics

The Mont Terri Rock Laboratory is located in a tunnel in the Jura Mountains, in northwestern Switzerland at a depth of *c.* 300 m below ground level. It is a 'methodological laboratory' (cf. Delay *et al.* 2014) in the Opalinus Clay, which is a Jurassic-age well-consolidated claystone with a hydraulic conductivity below 3×10^{-12} m s^{-1} (Thury & Bossart 1999).

Based on the study of five core samples, the total dry weight percentage of clay minerals (illite, kaolinite, chlorite and illite–smectite mixed layers) varies between 56 and 61 wt%, quartz varies between 12 and 19 wt%, and calcite varies between 10 and 18 wt% at the test location (Fernández *et al.* 2009; Lerouge *et al.* 2010). Other compounds include feldspars (2–8 wt%), dolomite (2–3 wt%), siderite (0.2–1 wt%), pyrite (0.5–1.4 wt%), organic matter (*c.* 1 wt% organic C) and TiO_2 (0.5 wt%). These values correspond to the ranges described by Pearson *et al.* (2003) over the Opalinus Clay at Mont Terri.

From thin section observations, the rock displays a fine grain fabric with heterogeneities. The clay matrix is formed of particles of less than 1 μm with detrital components (biotite, muscovite and chlorite) of 10–50 μm. It contains automorphic crystals of calcite, dolomite and siderite as well as framboidal pyrites. The heterogeneities are due to the occurrence of fossil debris ranging in size from 50 μm to a few millimetres. The fossil debris includes particles of organic matter, carbonate test and shell fragments, and phosphate elements. Despite a careful search, celestite was not found in the core samples. The $\delta^{13}C$ and $\delta^{18}O$ values measured on calcites vary from −0.8 to −2.0‰ PDB (Pee Dee Belemnite) and from +23.8 to +25.3‰ SMOW (standard mean ocean water), respectively (Lerouge *et al.* 2010).

On the studied core samples, the water content lies between 7.0% and 7.7% by mass (with respect to the dry rock) and the total porosity lies between 17.2% and 18.6% by volume (Lerouge *et al.* 2010). Based on leaching and squeezing tests, the chloride accessible porosity was evaluated to be *c.* 57% of the total porosity (Fernández *et al.* 2009).

Measurements of exchangeable cations (Na^+, K^+, Mg^{2+}, Ca^{2+}, Sr^{2+}) were made on four core samples (Lerouge *et al.* 2010). The sum of exchangeable cations obtained lies between 14.1 and 15.6 meq/100 g, with site occupancies of *c.* 43.5% for Na^+, 6.5% for K^+, 20% for Mg^{2+} and 30% for Ca^{2+}. The exchangeable Sr^{2+} value was 0.002 meq/100 g.

Experimental methods

Experimental principles

The experimental setup consists of a 15-m-long and 76-mm-diameter inclined ascending borehole, of which the last 5 m constitute the test interval. The borehole is perpendicular to the bedding. The theoretical rock surface area in the test interval is 1.2 m^2. Precautions were taken during the coring and equipment fitting phases to minimize the introduction of microorganisms and to avoid the introduction of organic matter into the borehole. The last 6 m of the borehole were cored with argon as a drilling fluid in order to protect the rock from any contact with atmospheric oxygen (oxidation) and in order to measure the concentration of dissolved nitrogen gas naturally present in the clay rock. Two gas flow lines link the test interval to a gas circulation module, which is located in the drift and allows monitoring of the gas composition. In addition, the gas circulation module makes it possible to inject pure hydrogen at a controlled flow rate.

The gas pressure in the test interval has been set and maintained from the beginning at a value between 1.3 and 2.5 bar. This pressure is much lower than the pore pressure in the surrounding rock. Under the effect of the hydraulic pressure difference between the borehole and the surrounding rock, and in spite of the low hydraulic conductivity of the rock, water flows into the borehole interval, where it accumulates and is then pumped out through a water-sampling line. In the drift, a water-sampling module connected to this line makes it possible to extract the water in order to maintain a constant water column height in the test interval and to monitor the water composition.

Experimental setup

The experimental setup for this test consisted of equipment installed in the ascending borehole BHT-1, a gas circulation module and a water-sampling module (Lettry & Fierz 2009) (Fig. 1). These are described in the following.

Borehole equipment. The borehole equipment consists of a multi-packer completion including a 5-m-long test interval for gas circulation and porewater collection at the far end of the borehole and a 50-cm-long observation interval for measuring pore pressure below the test interval. The observation interval is separated from the test

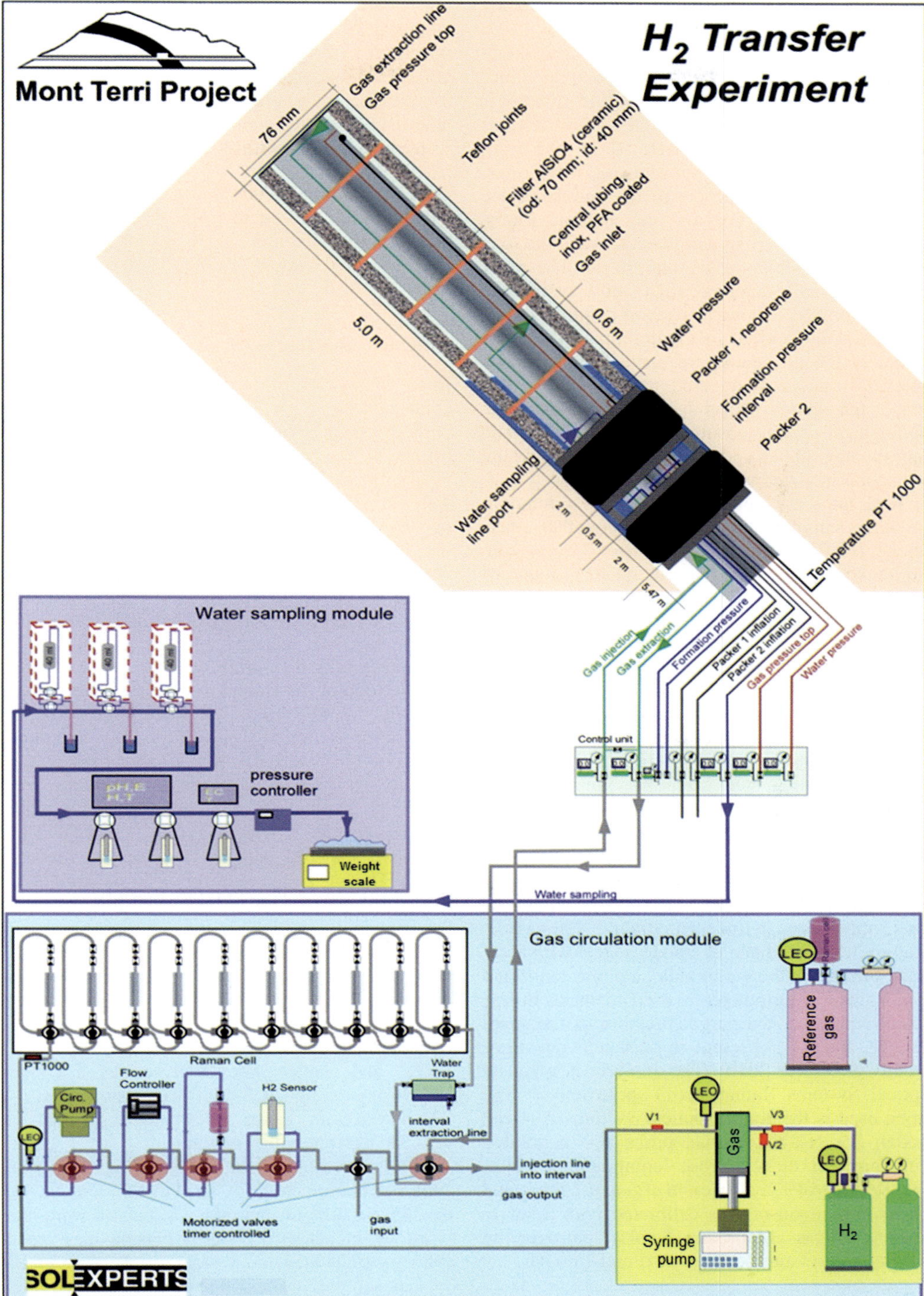

Fig. 1. Experimental layout of the borehole gas injection equipment, the gas circulation system and the water-sampling device.

interval by a 2-m-long packer filled with water (Fig. 1).

The test interval comprises a perfluoroalkoxy (PFA)-coated, stainless steel inner tube (outer diameter OD = 40 mm) fitted with several, 15-mm thick, 42% porosity, $AlSiO_4$ ceramic filter ring sections (OD = 70 mm, inner diameter ID = 40.5 mm). Teflon spacers are fitted between the filter rings. These spacers form barriers in the annular space between the inner tube and the ceramic filters, forcing the gas to circulate in contact with the rock and not just within the annular space. The inner tube and ceramic filter occupy *c.* 60% of the test interval volume, leaving a space of *c.* 9 l for fluids.

The completion has two hydraulic packers. The rubber of the first packer in contact with the test interval is neoprene-covered natural rubber, and that of the second packer is natural rubber alone. Neoprene is used to ensure better chemical inertia than natural rubber and so as not to influence the composition of the water.

Five lines are connected to the test interval: the water-sampling line made of poly-ether-ether-ketone (PEEK) (ID = 1.57 mm) set 12 cm from the packer, the two gas circulation lines made of stainless steel (ID = 2.4 mm) set 72 and 416 cm from the packer, respectively, and three pressure control lines. In the drift, each line is connected to a Keller LEO 3 pressure gauge in a control unit.

Gas circulation module. The gas circulation module includes a KNF circulation pump, a Bronkhorst gas-flow controller (range 0–100 ml min^{-1}), 10 Swagelok stainless-steel sampling cylinders (75 or 150 ml) that can be disconnected to perform gas sample analyses, a HY-OPTIMA 740 by H2scan H_2-specific solid detector probe, and a Teledyne ISCO D-500 gas injection pump. The gas volume in the gas circulation module is *c.* 0.9 l when the ten 75-ml-volume cylinders are online. The gas circulation lines are made of stainless steel with Swagelok fittings. In the gas module, a probe connected to a Raman spectrometer is used to make in-line measurements of the partial pressure of the gases (H_2, N_2 and CH_4) present in sufficient quantities (Lundy & Vinsot 2010). The analyser is a Kaiser Optical Systems Raman Rxn3 spectrometer. The probe used is the Kaiser Optical Systems AirHead model. For H_2, the Raman probe was calibrated with three mixtures of gases comprising 1 vol%, 3 vol% or 5 vol% hydrogen in argon at 1.5 bar and the hydrogen sensor was calibrated with a single mixture of gas comprising 5 vol% hydrogen by varying its pressure between 1.05 and 1.80 bar.

Water-sampling module. The water-sampling module includes an El-Press Bronkhorst water-pressure control device, which regulates the water flow rate, three water-sampling cylinders and a Sartorius weight scale. The water from the borehole passes through the sampling vials before collecting in a 600 ml Tedlar bag standing on the scale. The water-sampling cylinders are made of polytetrafluoroethylene (PTFE)-coated stainless steel or of PEEK. The water lines and the fittings are made of PEEK.

Data acquisition system. All the sensors (pressure, gas flow rate, hydrogen, scale, etc.) are connected to a central database, which acquires and stores the measured values every 5–20 min (Tabani *et al.* 2010).

Chemical analyses

In the laboratory, the water sample cylinders were connected to a special cell to measure the pH without contact with the ambient air. A few millilitres of the water were immediately used for alkalinity measurement by titration. Another 2 ml were immediately transferred into one arm of a two-armed glass vessel, the second arm containing H_3PO_4 (85%). The vessel was then flushed with pure nitrogen and closed. Inclining the vessel caused the H_3PO_4 to mix with the sample, and as a result the CO_2 from all dissolved inorganic carbon compounds was released into the gas phase. The gas phase was then transferred to the evacuated sample loop of the isotope ratio mass spectrometer with dual inlet system for measurement (IRMS, MAT-250; resolution (5% valley), 200; abundance sensitivity, 1.3×10^{-6} for 44/45; high-voltage stability, 1×10^{-5}). The measured results were corrected to carbon standards NBS18–Calcite, NBS19–Calcite, NBS23–$SrCO_3$, IMEP8–CO_2, CO_2–lab standard and Hydroisotop lab standard (DIC). The one sigma error for clean standard material is $\pm 0.2‰$. Results were related to Vienna Pee Dee Belemnite (VPDB) in the delta notation.

Cations Li^+, Na^+, K^+, Mg^{2+}, Ca^{2+} and Sr^{2+} were analysed by a Dionex ion chromatograph using an IonPac CG 12 A, 4×50 mm guard column and a CS 12 A, 4×250 mm analytical column. Anions Cl^-, NO_3^-, SO_4^{2-}, Br^-, I^-, PO_4^{3-}, $S_2O_3^{2-}$ and acetate were analysed by a Dionex ion chromatograph ICS 1500 using an IonPac AG 22, 4×50 mm guard column and an AS 22, 4×250 mm analytical column.

NH_4^+ and total sulphide (H_2S, S^{2-}, HS^-) were analysed photometrically. Cr, Fe, Mn, Co, Ni, Cu, Zn, As, Se, Rb, Zr, Nb, Mo, Cs and Ba were analysed by inductively coupled plasma mass spectrometry (ICP-MS).

Gas samples were collected in 25 ml stainless-steel cells. The gas composition was analysed using a Shimadzu GC17A gas chromatograph (GC) equipped with two capillary columns (column 1, plot fused silica, Molsieve 5A, 50 × 0.53 mm, film thickness df = 50 μm, Varian; column 2, plot fused silica, CP Poraplot Q-HT, 25 × 0.53 mm, df = 10 μm, Varian) and two detectors (detector 1, micro-volume thermal conductivity detector (micro-TCD), VICI Instruments; detector 2, flame ionization detector (FID), Shimadzu). Sample attachment was carried out with a special quick connection (Swagelok), allowing direct connection of the sample cells to the evacuated inlet system of the gas chromatograph (eight-port dual external sample injector, Valco Europe) without contact with the outside atmospheric gases.

The carbon-13 content of the alkanes was analysed by gas chromatography isotope ratio mass spectrometry (GC-IRMS) (MAT-250, Varian MAT; resolution (5% valley) 200; abundance sensitivity, 1.3×10^{-6} for 44/45; high-voltage stability, 1×10^{-5}). The alkanes were oxidized completely to CO_2 in a combustion interface and then measured in the isotope ratio mass spectrometer. Measured results were corrected to carbon standards NBS18–Calcite, IAEA CO–1, NBS23–$SrCO_3$, CO_2–lab standard 1 and CO_2–lab standard 2.

Chronology of the test

Initial phase. The borehole equipment was installed just after drilling. The test interval was then filled with pure argon at a pressure of 2.3 bar and the circulation of this gas was started. The composition of this circulating gas was monitored over almost two years. It was observed that its composition evolved: concentrations of nitrogen, methane and other light alkane up to C6 (hexane) increased with time; helium and CO_2 were also detected. The flow of nitrogen in the test interval was *c.* 0.15 mmol/day over the first year (half of the test). During the same period, the flow of methane was close to 0.019 mmol/day and the flow of ethane was *c.* 0.0016 mmol/day. All of these gases were originally dissolved in the rock porewater. Indeed, their occurrence in Opalinus Clay porewater has been described (Pearson *et al.* 2003; Vinsot *et al.* 2008*a*; Cailteau *et al.* 2011). The surface area of the rock supplying these gases in the test interval was *c.* 1.2 m^2. The ratios between nitrogen, ethane and propane, respectively, and methane deduced from the measured volume fractions were 7.745 for N_2/CH_4, 0.085 for C_2H_6/CH_4 and 0.058 for C_3H_8/CH_4 (Vinsot 2012).

During the year preceding the first injection of hydrogen, the water flow rate from the surrounding rock was between 10 and 20 ml/day into the borehole and the water composition was determined. As the rock surrounding the test interval was homogeneous and not fractured, this flow rate indicated that the wall of the test interval was saturated with water.

Hydrogen injection phase. The first hydrogen injection was performed by replacing the previous circulating gas with a mixture of gases containing 5 vol% H_2, 5 vol% He, 5 vol% Ne and 85 vol% Ar at a total pressure close to 1.5 bar. Because the initial gas was not completely eliminated after the gas replacement operation, the largest hydrogen partial pressure value was close to 0.06 bar. Helium and neon served as reference non-reactive gases because changes in their content should only depend on dissolution and diffusion processes in the rock porewater. As a consequence, they can help to calibrate the transport part in a reactive transport model.

Following this first injection in the test interval, the hydrogen concentration dropped below the detection limit in 65 days. In November 2011, pure hydrogen was added to the circulating gas to obtain, once more, a hydrogen partial pressure of 0.06 bar in the test interval. After this second injection, the added hydrogen disappeared once again in 65 days.

In February 2012, a semi-continuous hydrogen injection phase was launched. This consisted of injecting pure hydrogen regularly to maintain a partial pressure close to 0.06 bar in the test interval.

Hydrogen tightness test. Before the first hydrogen injection, on-site tightness tests were conducted with the gas module alone (i.e. without connection to the borehole) with a mixture of gases comprising 2.2 vol% hydrogen in argon at a total pressure of 1.1 bar over 200 days. The outcome of these tightness tests was an average total gas leak rate of 0.5 mbar/day, corresponding to 0.5 ml/day at standard ambient temperature and pressure (SATP). Hydrogen contributed to 9% of the gas loss. This rate was four times higher than the hydrogen content of the gas, showing that hydrogen leaked more easily than argon. The pure hydrogen leak rate was therefore estimated to be 0.04 mbar/day or 1.6×10^{-6} mol/day.

Experimental results

Gas composition

Between June 2011 and September 2012, 17 gas sampling cylinders were disconnected from the gas module for laboratory analyses (Table 1). New gas sampling cylinders were added three times over this time period – 28 June 2011, 26 August 2011 and 9 February 2012. The added gas sampling cylinders contained pure argon at a pressure between 10 and 20 mbar.

Table 1. *Gas analyses*

Lab no.	–	–	227432	226613	226614	227311	227312	227313	227431	229318	229319
Sampling date	–	–	Gas mixture	10 June 2011	13 June 2011	16 June 2011	21 June 2011	28 June 2011	5 July 2011	15 July 2011	29 July 2011
Sampling time	–	(days)	1	4	7	12	19	26	36	50	76
Total pressure	–	(bar)	1.52	1.52	1.51	1.51	1.51	1.46	1.46	1.45	1.44
Hydrogen	H_2	(vol%)	5.05	3.95	3.85	3.55	3.35	3.18	2.65	2.08	0.52
Helium	He	(vol%)	4.77	4.05	4.01	4	3.9	3.9	3.75	3.65	3.8
Neon	Ne	(vol%)	4.98	3.95	4	4.03	4	3.9	3.75	3.9	3.85
Argon	Ar	(vol%)	85.2	85.7	86.8	87	87.4	87.7	88.5	89	90.1
Oxygen	O_2	(vol%)	< 0.05	< 0.05	< 0.05	< 0.05	< 0.05	< 0.05	< 0.05	< 0.05	< 0.05
Nitrogen	N_2	(vol%)	< 0.05	2.05	1.24	1.2	1.11	1.09	1.09	1.51	1.87
Carbon dioxide	CO_2	(vol%)	< 0.03	0.15	0.03	0.05	0.03	0.03	0.03	0.01	< 0.01
Methane	CH_4	(vpm)	< 1	790	733	895	994	1120	1220	1300	1680
Ethane	C_2H_6	(vpm)	< 1	126	120	149	163	183	209	220	262
Propane	C_3H_8	(vpm)	< 1	126	122	148	161	181	216	216	250
Carbon-13-CH_4	$\delta^{13}C$-CH_4	VPDB‰	–	− 36.9	− 22.7	− 22.7	− 26	− 37.7	− 38.6	− 38.1	− 38.4
Deuterium-H_2	δ^2H-H_2	VSMOW‰	− 850 ± 15	− 864	− 854	− 862	n.a.	n.a.	− 854	n.a.	n.a.

Table 1. (*Continued*)

Lab no.	–	–	231003	231715	233563	233936	233937	234665	236349	240299
Sampling date	–	–	3 November 2011	23 November 2011	9 February 2012	22 February 2012	28 February 2012	24 March 2012	14 May 2012	5 September 2012
Sampling time	–	(days)	147	167	245	258	264	289	340	454
Total pressure	–	(bar)	1.40	1.44	1.40	1.34	1.36	1.36	1.35	1.35
Hydrogen	H_2	(vol%)	< 0.05	2.9	< 0.1	< 0.05	4.9	3.83	2.96	2.86
Helium	He	(vol%)	3.1	2.97	3.05	3.1	2.87	3.1	3.05	2.92
Neon	Ne	(vol%)	3.79	3.7	3.65	3.65	3.5	3.55	3.5	3.35
Argon	Ar	(vol%)	88.8	86	87	86.6	82.3	82.6	82.5	77.4
Oxygen	O_2	(vol%)	< 0.05	< 0.05	< 0.1	< 0.1	< 0.1	< 0.1	< 0.1	< 0.02
Nitrogen	N_2	(vol%)	3.68	3.9	5.5	5.8	5.6	6.1	7.12	12.2
Carbon dioxide	CO_2	(vol%)	0.08	0.012	0.04	0.02	0.02	0.02	0.02	0.09
Methane	CH_4	(vpm)	3200	3600	6000	6700	6400	6700	7900	10300
Ethane	C_2H_6	(vpm)	436	445	550	570	560	595	685	850
Propane	C_3H_8	(vpm)	393	415	452	470	460	475	540	585
Carbon-13-CH_4	$\delta^{13}C$-CH_4	VPDB‰	− 33.1	− 38.5	− 40.3	− 39.8	− 38.4	− 38.7	− 38.8	− 47.9
Deuterium-H_2	δ^2H-H_2	VSMOW‰	n.a.	n.a.	n.a.	n.a.	− 846	− 826	− 835	− 807

The initial time for the calculation of the sampling time corresponded to the date of the first hydrogen injection; vpm, volumic part per million; n.a., not available.

The three methods used for hydrogen monitoring (H_2-specific sensor, Raman spectrometry and gas chromatography analyses performed on gas samples) gave similar results, except when rapid fluctuations affected the gas chromatography and Raman analyses more than the H_2-specific sensor (Fig. 2). Regarding the first hydrogen injection, the concentration of hydrogen, helium and neon at the time of injection was 4.0 ± 0.2vol%; this value was lower than the concentration in the injected gas mixture because the injected gas mixture did not entirely replace the existing gas in the borehole (Vinsot 2012).

Following this injection, the hydrogen vanished completely in 65 days, whereas helium and neon losses were less than 12% of the initial content over the same time period and less than 28% over 454 days (Fig. 3). When pure hydrogen was added to the circulating gas in November 2011, the 0.017 moles of added hydrogen disappeared at the same rate. For both injections, the change in the volume concentration of hydrogen was virtually linear with

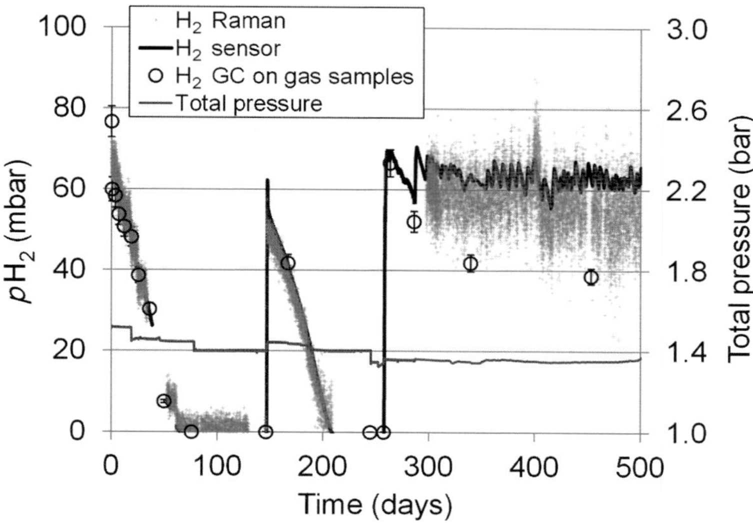

Fig. 2. Hydrogen partial pressure and total pressure measured in the gas.

a slope close to 0.06 vol% per day, corresponding to an average hydrogen loss of *c.* 2.7×10^{-4} mol/day.

Over the semi-continuous hydrogen injection phase started 22 February 2012, the hydrogen partial pressure varied from 40 to 80 mbar. From the beginning of this phase (day 258) until day 454 (date of the last water sampling, 5 September 2012), the quantity of hydrogen added was 0.085 moles (Fig. 4) and the average hydrogen loss rate was 3.3×10^{-4} mol/day.

The hydrogen loss rates observed are two orders of magnitude higher than that of the hydrogen leak rate estimated from the on-site tightness tests. Another tightness test was performed in June 2012. It consisted in bypassing the borehole for a period of 28 h: the hydrogen content value obtained with the specific sensor remained constant over this time period. This result confirmed that the hydrogen loss observed was not due to a leak from the gas module circuit.

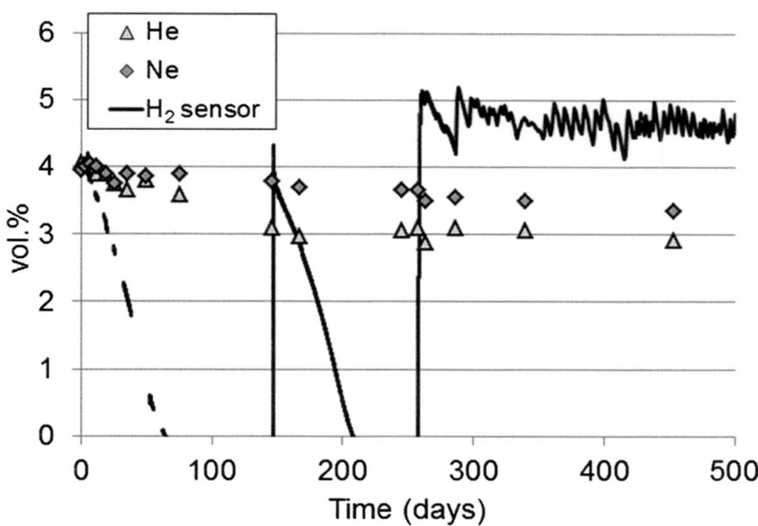

Fig. 3. Volume fraction of helium, neon and hydrogen measured in the gas.

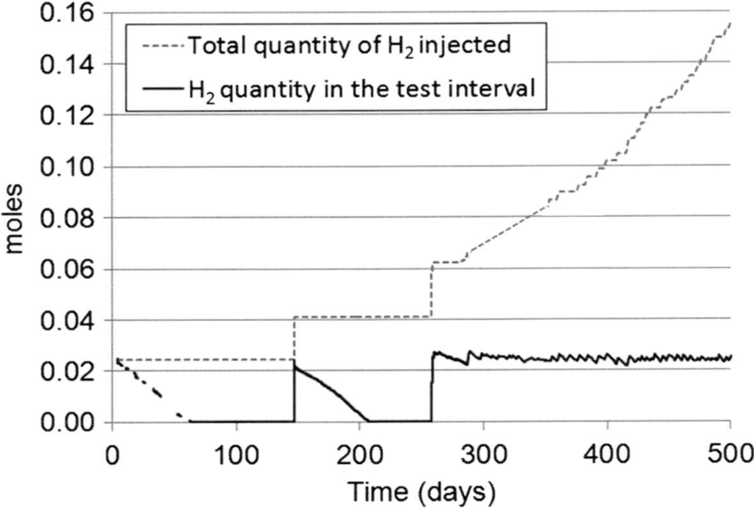

Fig. 4. Evolution of the quantity of hydrogen existing in the test interval and added since the beginning of the test.

Nitrogen and alkane concentrations evolved similarly before and after the first hydrogen injection (Table 1 and Vinsot 2012). N_2/CH_4 and C_2H_6/CH_4 ratios calculated during the hydrogen phase (i.e. after the first hydrogen injection) were equal or slightly bigger than those obtained during the initial phase (i.e. before the first hydrogen injection). During the initial phase, the $\delta^{13}C$ values of methane varied by 3–10‰VPDB when the gas circuit was open for gas extraction or gas injection (Vinsot 2012). Other than during these events, the $\delta^{13}C$ values of methane remained in the range −40.3 to −36.3‰VPDB over the whole test and did not significantly differ between the initial phase and the hydrogen phase.

The CO_2 concentrations measured in the gas were between 0.003 and 0.14 vol%, corresponding to CO_2 partial pressures ranging between 0.08 and 3.4 mbar (log pCO_2 of −4.1 to −2.5). The CO_2 partial pressure was much lower than that calculated from equilibrium with the seepage water using pH, alkalinity and total inorganic carbon (log pCO_2 around −2, cf. Table 2). The measurements of the gases have been checked and are not at issue. The highest values were associated with the events of gas extraction or injection. The processes involved are not yet understood. The measured $\delta^{13}C$ values of CO_2, between −10.1 and −7.5‰VPDB, and the $\delta^{18}O$ values of CO_2, between 30.1 and 34.7‰VSMOW, fit the ranges already described for the Opalinus Clay at Mont Terri (Pearson *et al.* 2003; Girard *et al.* 2005).

The H_2S concentrations measured in the gas were always below the detection limit value (1 vol%).

Water composition

Table 2 presents the results of analyses of eight samples of BHT-1 borehole water. The first sample was taken about 14 months after the start of the experiment by disconnecting an inline vial into which the borehole water had been flowing for almost two months. The second sample was taken after about 26 months, just before the first hydrogen injection. The third sample was taken a little less than one month after the first hydrogen injection. The fourth water sample and the next four water samples were taken before and during the semi-continuous hydrogen injection phase, respectively.

Overall, the samples have similar concentrations of the major species (Fig. 5 and Table 2). The pH measured in the laboratory immediately after opening of the vials is between 6.9 and 7.3 (Fig. 6). The BHT-1 water fits the chloride profile of a NW–SE cross-section of the Mont Terri laboratory derived from a summary of geochemical data in Pearson *et al.* (2003). Based on the Br^-/Cl^- and SO_4^{2-}/Cl^- ratio values, the BHT-1 borehole water corresponds to dilute seawater, like all the water sampled from the Opalinus Clay at Mont Terri (Pearson *et al.* 2003, 2011).

After the first hydrogen injection, the concentrations of sulphate and strontium have decreased a little, together with total iron (Fig. 7). Sulphide was detected only in the last two samples (Table 2).

The strontium isotope ratios of the BHT-1 borehole water were between 0.707651 ± 0.000013 and 0.707770 ± 0.000012 and fall within the range for Mont Terri water, which extends from 0.707651 to

Table 2. *Water analyses and parameters calculated from water composition*

Lab no.	Unit	215145	226615	227433	233659	233935	235660	236066	241154
Sampling date	–	15 June 2010	8 June 2011	5 July 2011	9 February 2012	28 February 2012	21 March 2012	26 April 2012	5 September 2012
Sampling time	(days)	−359	−1	26	245	264	286	322	454
Electrical conductivity (25 °C) Lab	(μS cm^{-1})	30 000	29 800	29 800	29 600	29 600	29 700	29 800	29 400
pH value (20 °C) Lab	–	7.1	7.4	7.2	7.0	7.0	7.0	7.0	6.9
Alkalinity pH4.3	(meq l^{-1})	2.33	2.09	2.15	2.06	1.78	2.00	2.53	2.08
Alkalinity pH3.3	(meq l^{-1})	2.95	2.75	2.8	2.75	2.55	3.33	2.75	2.85
TIC (measured)	(mmol l^{-1})	2.42	2.38	2.45	2.1	1.9	2.7	3.1	2.5
Na	(mol l^{-1})	2.42E-01	2.62E-01	2.67E-01	2.63E-01	2.66E-01	2.53E-01	2.53E-01	2.55E-01
K	(mol l^{-1})	2.14E-03	1.54E-03	1.52E-03	1.60E-03	1.59E-03	1.38E-03	1.39E-03	1.47E-03
Ca	(mol l^{-1})	1.63E-02	1.65E-02	1.65E-02	1.77E-02	1.77E-02	1.62E-02	1.62E-02	1.63E-02
Mg	(mol l^{-1})	1.97E-02	1.87E-02	1.89E-02	1.95E-02	1.93E-02	1.83E-02	1.85E-02	1.86E-02
NH_4	(mol l^{-1})	6.04E-04	6.22E-04	5.83E-04	5.75E-04	5.91E-04	6.16E-04	4.80E-04	5.24E-04
Cl	(mol l^{-1})	2.80E-01	2.93E-01	3.05E-01	3.05E-01	3.08E-01	2.90E-01	2.90E-01	2.95E-01
SO_4	(mol l^{-1})	1.56E-02	1.70E-02	1.70E-02	1.54E-02	1.55E-02	1.51E-02	1.51E-02	1.44E-02
Br	(mol l^{-1})	4.64E-04	4.39E-04	4.34E-04	4.49E-04	4.74E-04	4.27E-04	4.55E-04	4.48E-04
I	(mol l^{-1})	2.17E-05	1.13E-05	1.37E-05	8.84E-06	8.84E-06	6.42E-06	8.03E-06	6.42E-06
S (−II)	(mol l^{-1})	<1.5E-05	<3.0E-06	<1.5E-05	<6.0E-06	<1.5E-05	<1.5E-05	1.11E-04	1.05E-04
Ba	(mol l^{-1})	4.90E-07	3.71E-07	4.09E-07	3.34E-07	3.34E-07	5.94E-07	5.19E-07	4.45E-07
Fe	(mol l^{-1})	1.97E-05	7.85E-05	1.55E-04	2.28E-06	3.84E-06	1.64E-06	2.74E-06	2.92E-06
Li	(mol l^{-1})	6.75E-05	6.76E-05	6.76E-05	6.76E-05	6.91E-05	5.87E-05	6.17E-05	6.17E-05
Mn	(mol l^{-1})	3.34E-06	2.88E-06	4.09E-06	3.71E-06	2.14E-06	3.43E-06	3.34E-06	2.50E-06
Sr	(mol l^{-1})	4.66E-04	5.56E-04	5.58E-04	4.87E-04	4.98E-04	4.01E-04	3.93E-04	3.69E-04
B	(mol l^{-1})	2.26E-04	2.97E-04	3.02E-04	2.12E-04	2.22E-04	3.35E-04	3.44E-04	2.88E-04
Se	(mol l^{-1})	1.70E-06	<7.0E-08	<7.0E-07	2.78E-06	2.45E-06	<7.0E-08	4.39E-07	9.68E-08
Si	(mol l^{-1})	1.94E-04	5.99E-04	7.26E-04	4.90E-04	5.63E-04	5.99E-04	6.89E-04	6.71E-04
DOC	(mol C l^{-1})	6.50E-04	2.42E-03	2.39E-03	2.13E-04	1.28E-03	1.31E-03	1.46E-03	1.07E-03

(*Continued*)

Table 2. *Continued*

Lab no.	Unit	215145	226615	227433	233659	233935	235660	236066	241154
$\delta^{18}O-H_2O$	(‰ VSMOW)	−8.3	−8.53	−8.4	−8.71	−8.68	−8.61	−8.66	−8.46
$\delta^{2}H-H_2O$	(‰ VSMOW)	−51	−53.9	−53	−54	−53.4	−53.4	−53.6	−53.2
$\delta^{13}C-TIC$	(‰ VPDB)	−6.99	−13	−17.6	−15.1	−12.2	−12.6	−14	−17.7
Ionic strength	(M)	3.31E-01	3.46E-01	3.54E-01	3.54E-01	3.56E-01	3.38E-01	3.38E-01	3.41E-01
Charge balance	(%)	0.66	0.8	−0.12	0.26	0.24	0.04	0.16	0.11
Calcite SI	–	−0.06	0.1	0	−0.17	−0.2	−0.11	−0.20	−0.28
Celestite SI	–	−0.05	0.04	0.04	−0.06	−0.05	−0.14	−0.15	−0.19
log pCO_2 (meas. alk)	log(bar)	−2.10	−2.34	−2.23	−2.03	−2.07	−1.95	−2.04	−1.92
log pCO_2 (meas. TIC)	log(bar)	−2.23	−2.42	−2.32	−2.21	−2.25	−2.09	−2.03	−2.04
Dolomite SI	–	−0.01	0.27	0.08	−0.26	−0.34	−0.14	−0.32	−0.48
Strontianite SI	–	−0.78	−0.55	−0.65	−0.9	−0.93	−0.89	−0.99	−1.1
Gypsum SI	–	−0.49	−0.47	−0.47	−0.49	−0.48	−0.51	−0.51	−0.53
Barite SI	–	0.46	0.35	0.39	0.27	0.27	0.52	0.47	0.38
Rhodochrosite SI	–	−1.35	−1.38	−1.27	−1.39	−1.64	−1.33	−1.4	−1.59
Siderite SI	–	−0.73	0.00	0.21	−1.80	−1.61	−1.86	−1.82	−1.86
Br/Cl	(mol/mol)	5.57E-02	5.79E-02	5.57E-02	5.05E-02	5.04E-02	5.19E-02	5.19E-02	4.90E-02
SO_4/Cl	(mol/mol)	1.66E-03	1.50E-03	1.42E-03	1.47E-03	1.54E-03	1.47E-03	1.57E-03	1.52E-03

Presented ionic strength, charge balance, saturations indices and CO_2 partial pressures have been calculated at 15 °C with PHREEQC (Parkhurst & Appelo 1999) using the Thermochimie-V8 database (Duro *et al.* 2012).

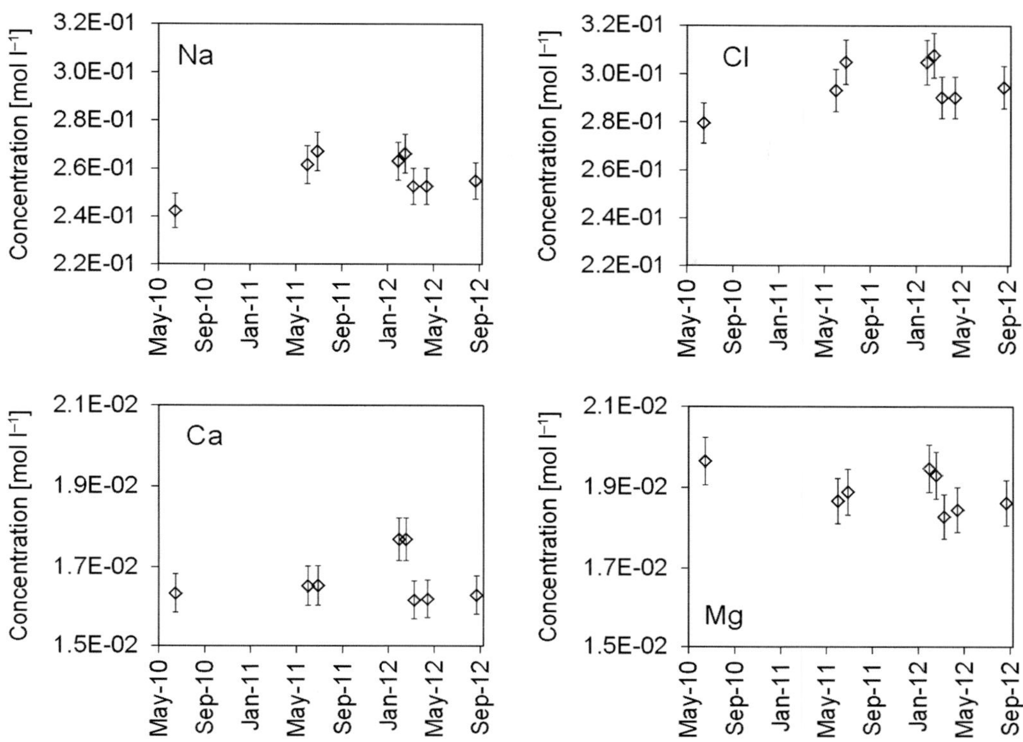

Fig. 5. Evolution of four dissolved species concentrations measured in the borehole water. Error bars represent the uncertainty in the measured values.

0.707774 (Pearson *et al.* 2003). The ratios do not show any specific tendency in their evolution.

The measured $\delta^{13}C$ values of total inorganic carbon (TIC) were between −17 and −7‰ VPDB. The lowest values likely indicate a contribution of oxidized organic matter to the TIC (Girard *et al.* 2005).

Speciation calculations were performed with PHREEQC (Parkhurst & Appelo 1999) and the Thermochimie-V8 database (Duro *et al.* 2012).

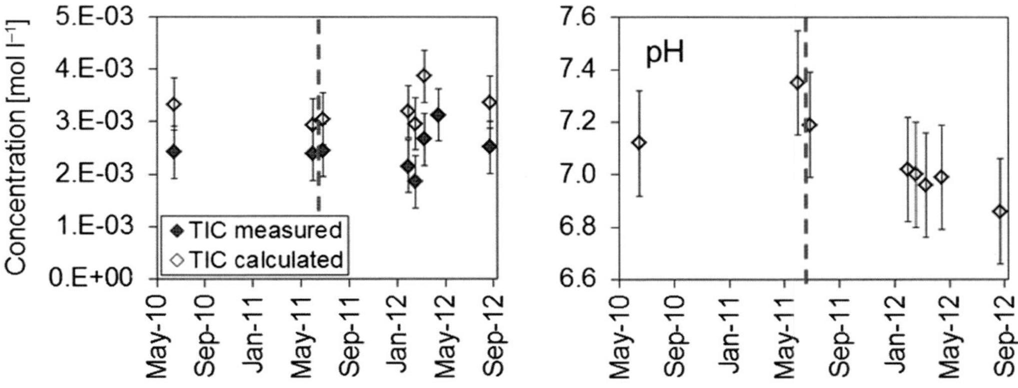

Fig. 6. Evolution of the TIC content and pH in the borehole water. Error bars represent the uncertainty in the values.

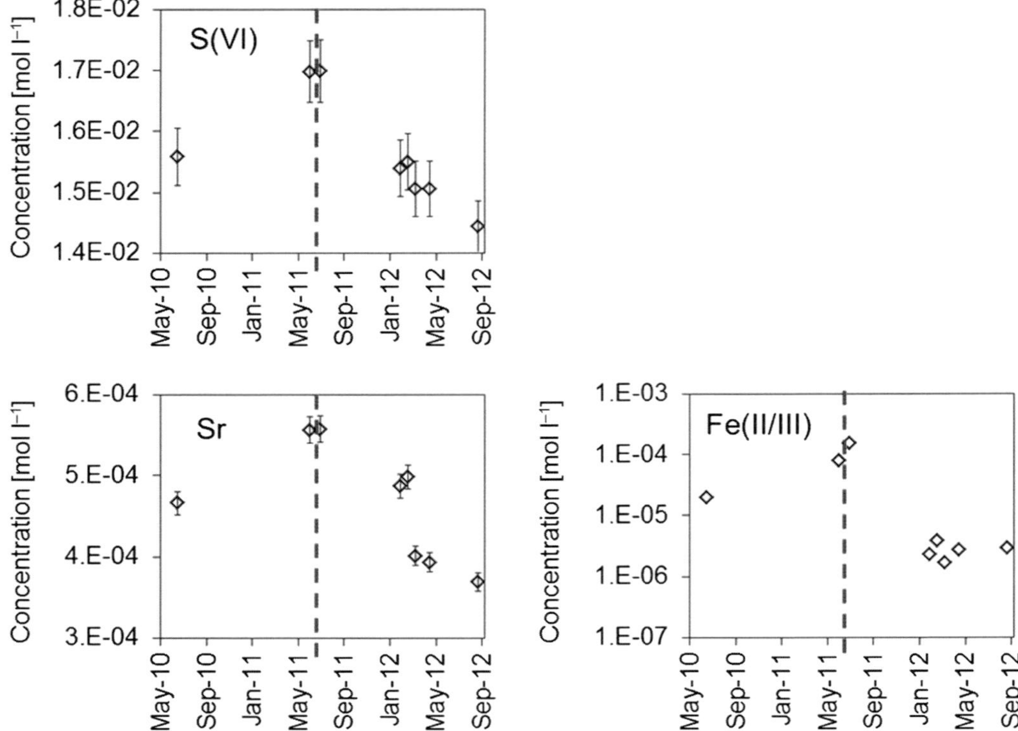

Fig. 7. Evolution of the concentrations of sulphate, strontium and iron measured in the borehole water. Error bars represent the uncertainty in the measured values.

Results are presented in Table 2. The measured alkalinity was used for the calculations. The calculated log pCO_2 values are between -2.4 and -1.7. When available, the measured TIC was used to calculate log pCO_2. The difference between the log pCO_2 values based on measured alkalinity and on measured TIC is smaller than 0.2 (Fig. 8). This shows a rather good internal consistency among the measured pH, TIC and alkalinity values. All the analyses are charge balanced to within 2%. Before hydrogen injection, the sampled water was close to equilibrium with calcite, dolomite and celestite, with almost all the absolute values of the saturation indices with respect to these minerals below 0.1. After the first hydrogen injection, these saturation indices decreased (Fig. 8), but the values obtained remained within the range of uncertainty. Over the experiment, the siderite saturation index varied between -1.9 and 0.2.

The minor species fit with the previous observations summarized by Pearson *et al.* (2003, 2011).

The composition of the BHT-1 seepage water gives complementary insights into the Opalinus Clay porewater, and seems to be in agreement with the current understanding (Pearson *et al.* 2011).

Microbial analyses

Six water samples were taken in sterile glass or plastic tubes between June 2011 and April 2012 to carry out preliminary investigations (using cultural methods) on the microbial population. The results of these investigations show that acetoclastic organisms, sulphate-reducing organisms and thiosulphate-reducing organisms developed over the sampling time period. No methanogens or hydrogenotrophics were found, in spite of a specific search for these organisms.

Discussion

Gases

The evolution of H_2, He and Ne concentrations in the gas was calculated taking into account dissolution and diffusion (Appelo & Vinsot 2012). The modelling was carried out with PHREEQC in a one-dimensional radial configuration, using the multi-component diffusion option (Appelo & Wersin 2007). Diffusion over the water/gas interface at the borehole perimeter was modelled with a

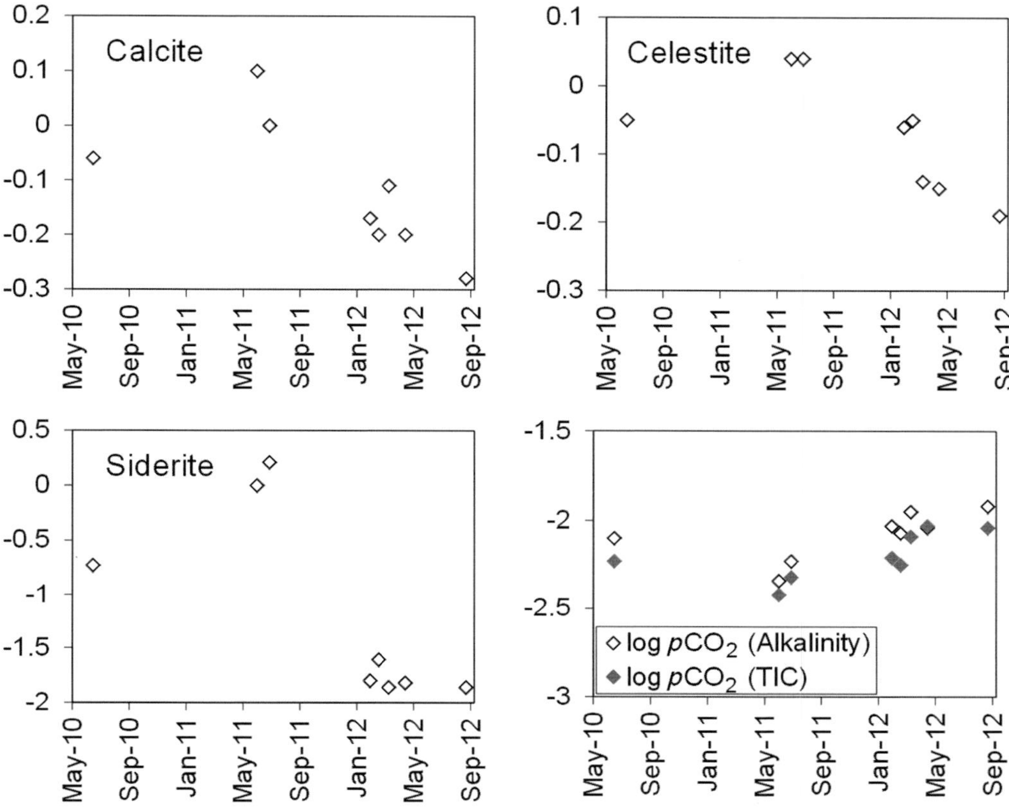

Fig. 8. Saturation indices with respect to calcite, celestite and siderite and log pCO_2 calculated for the borehole water.

two-film model (Liss & Slater 1974). The water-film thickness was assumed to be 0.25 mm (in agreement with calculations for other systems, cf. Appelo & Postma 2005). The Henry constants K_H (atm l mol^{-1}) and diffusion coefficients in water D_w (Jähne *et al.* 1987) are listed in Table 3, together with the effective diffusion coefficients $D_e = \varepsilon D_w/\theta^2$ at 15 °C. Porosity $\varepsilon = 0.16$ and tortuosity factor $\theta^2 = \varepsilon^{-1.1} = 7.5$ were the same for the three dissolved gases. The calculated evolutions were in good agreement with the measured data for neon and slightly above the measured data for helium (Fig. 9). BRGM (2012) tested the influence of the tortuosity value on the calculated evolution of the three gases, but for all the tested values, the decrease in gas concentration rates were still an order of magnitude greater than those observed for hydrogen. In contrast, the measured hydrogen concentration decreased very quickly compared to the model. From these results it appears that the

Table 3. *Solubilities and tracer diffusion coefficients for the gases in the model*

	$-\log K_H$	D_w (10^{-9} m^2 s^{-1})	D_e (10^{-11} m^2 s^{-1})
H_2	3.10	5.13	8.12
He	3.41	7.29	11.53
Ne	3.35	4.04	6.39

The solubilities and diffusion coefficients are for 25 °C; solubilities are corrected to 15 °C using Van't Hoff's equation with a polynomial; diffusion coefficients are corrected by accounting for the viscosity change of water with temperature.

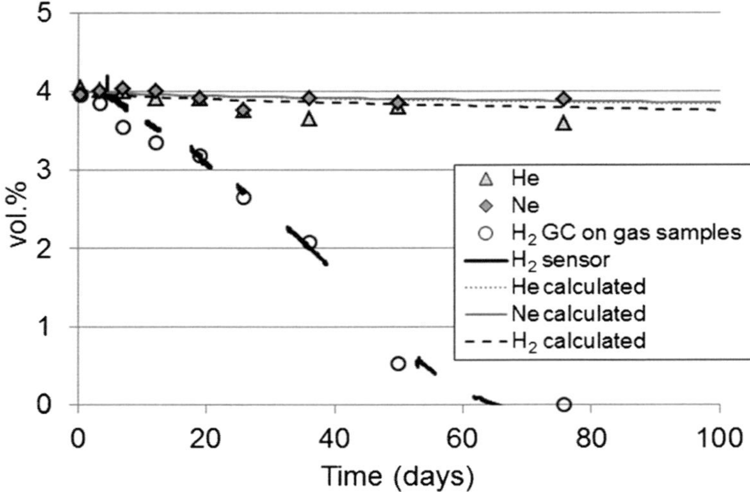

Fig. 9. Observed and calculated evolution of hydrogen, helium and neon over 100 days after the first hydrogen injection.

helium and neon concentration evolutions can be reproduced with a model taking into account dissolution and diffusion in the porewater. The difference observed between the evolution of the helium concentration and the evolution of the neon concentration could be due to a small leak of helium. However, the difference observed between the change in concentration in hydrogen and the change in concentration of helium is evidence of behaviour particular to hydrogen.

The total quantity of hydrogen, summed up to day 454, was *c.* 126 mmoles. The hydrogen quantity consumed due to dissolution and diffusion over the same time period was *c.* 6 mmoles. As a consequence, the loss of *c.* 120 mmoles of hydrogen may be due to another process.

Water

The concentrations of three dissolved species (sulphate, iron and strontium) evolved slightly after hydrogen injection. Furthermore, sulphide was detected in the last two water samples.

The measured concentrations of sulphate, iron and sulphide in the borehole water, together with the water production flow rate, were used to calculate the total quantities that have been lost (sulphate and iron) or gained (sulphide) in the water since the first hydrogen injection. Quantities obtained were *c.* −23 mmoles for sulphate, −1 mmole for iron and +1 mmole for sulphide. As a consequence, the average rate of sulphate concentration decrease was *c.* 5×10^{-5} mol/day.

Processes

Biotic and abiotic processes involving hydrogen consumption in argillaceous rocks have been intensively investigated (e.g. Truche *et al.* 2009, 2010; Libert *et al.* 2011; Stucki 2011; Didier *et al.* 2012).

Reaction (1), which consumes hydrogen, sulphate and Fe(II) and produces sulphide, could explain the observed phenomena:

$$4H_2 + SO_4^{2-} + Fe^{2+} = FeS + 4H_2O \quad (1)$$

Pyrrhotite (FeS) precipitation is thermodynamically possible in the borehole water as the calculated saturation indices (on the two last water samples containing detectable sulphide) were higher than 3.5. Furthermore, this reaction is in a rather good agreement with the observed consumption of *c.* 120 mmoles of hydrogen and 23 mmoles of sulphate, with the mass balance of hydrogen and sulphate having a ratio of 4/1.

The hypothesis that this reaction could describe the observed processes implies that more than 20 mmoles Fe(II) have become available. The iron source is unknown. The ionic exchanger could contribute a part to the Fe(II) needed for the reaction. However, a part of the Fe(II) could also come from Fe(III) reduction.

In any case, these reactions should be controlled by kinetics, as it was shown that the observed data cannot be reproduced with a local equilibrium hypothesis in a reactive transport model (BRGM 2012).

Moreover, these reactions should involve microorganisms, because it was shown that abiotic reactions did not occur in the experimental conditions (Truche *et al.* 2009, 2010; Didier *et al.* 2012). Hydrogenotrophs have not been identified yet, but this result, based on cultural methods, cannot be considered as a proof of their absence. Iron-reducing bacteria also need to be sought.

This hypothesis corresponds to a preliminary and simplified conclusion. It has to be checked quantitatively, taking into account diffusion. It is also necessary to look for other potential processes.

On the other hand, it can be observed that methane production associated with carbonate reduction by hydrogen did not seem to occur. The N_2/CH_4 and C_2H_6/CH_4 ratios were superior or equal before and after the first H_2 injection, and the $\delta^{13}C$ values of methane remained constant before and after the first H_2 injection.

Conclusion

It has been possible to inject hydrogen into a borehole in the Mont Terri Rock Laboratory, controlling precisely the quantity of hydrogen added. Helium and neon were also injected into the borehole. The evolution of the concentration of these two gases was reproduced with a diffusion model in a one-dimensional radial configuration, considering dissolution and diffusion in the porewater. The effective diffusion coefficient was 11.53×10^{-11} m^2 s^{-1} for helium and 6.39×10^{-11} m^2 s^{-1} for neon. These values are in good agreement with the known transfer properties of the rock.

Hydrogen disappeared much faster than helium. Simultaneously, the evolution of the borehole water composition gave indications that hydrogen could have been consumed by reactions involving sulphate and iron reduction.

These preliminary conclusions need to be confirmed and quantified, with the help of reactive transport modelling. Methods from molecular biology should be used to search for hydrogenotrophic and iron-reducing organisms.

We appreciate fruitful discussions with D. Traber, U. Mäder, E. Valcke, N. Waber, S. Daumas and R. Bernier-Latmani during the Mont Terri geochemical meetings. Many thanks also go to G. Lorenz, T. Fierz, T. Theurillat, P. Tabani, A. Mangeot, P. Delage, A. Garnier, A. Bagnoud and C. Nussbaum for their continuous support of the project. We acknowledge the two anonymous reviewers for their helpful comments.

References

Appelo, C. A. J. & Postma, D. 2005. *Geochemistry, Groundwater and Pollution*. 2nd edn. Balkema, Leiden.

Appelo, C. A. J. & Vinsot, A. 2012. *HT (Hydrogen Transfer) Experiment: Building a Model for Calculating the Diffusion of H_2, He, Ne and Ar in a Borehole in Opalinus Clay*. Mont Terri Project Technical Note **TN2009-40**.

Appelo, C. A. J. & Wersin, P. 2007. Multicomponent diffusion modeling in clay systems with application to the diffusion of tritium, iodide, and sodium in Opalinus clay. *Environmental Science and Technology*, **41**, 5002–5007.

BRGM 2012. *Reactive Modeling Transport of Hydrogen (HT Experiment)*. Mont Terri Project Technical Note **TN2011-31**.

Cailteau, C., Pironon, J., de Donato, Ph., Vinsot, A., Fierz, T., Garnier, C. & Barres, O. 2011. FT-IR metrology aspects for on-line monitoring of CO_2 and CH_4 in underground laboratory conditions. *Analytical Methods*, **3**, 877–887.

Delay, J., Bossart, P. et al. 2014. Three decades of Underground Research Laboratories. What have we learned? *In*: Norris, S., Bruno, J. et al.. (eds) *Clays in Natural and Engineered Barriers for Radioactive Waste Confinement*. Geological Society, London, Special Publications, **400**. First published online March 5, 2014, http://dx.doi.org/10.1144/SP400.1.

Didier, M., Leone, L., Greneche, J.-M., Giffaut, E. & Charlet, L. 2012. Adsorption of hydrogen gas and redox processes in clays. *Environmental Science and Technology*, **46**, 3574–3579.

Duro, L., Grivé, M. & Giffaut, E. 2012. ThermoChimie, the ANDRA Thermodynamic Database. *MRS Proceedings*, 1475, imrc11-1475-nw35-o71, http://dx.doi.org/10.1557/opl.2012.637.

Fernández, A. M., Melón, A. M. et al. 2009. *HT Experiment. Physical, Geochemical and Mineralogical Characterization of the Opalinus Clay at Mont Terri from Core Samples of the Borehole BHT-1 Located in the Gallery 08*. Mont Terri Project Technical Note **TN2009-46**.

Girard, J.-P., Flehoc, C. & Gaucher, E. 2005. Stable isotope composition of CO_2 outgassed from cores of argillites: a simple method to constrain $\delta^{18}O$ of porewater and $\delta^{13}C$ of dissolved carbon in mudrocks. *Applied Geochemistry*, **20**, 713–725.

Jähne, B., Heinz, G. & Dietrich, W. 1987. Measurement of the diffusion coefficients of sparingly soluble gases in water. *Journal of Geophysical Research*, **92**, 10767–10776.

Lerouge, C., Gailhanou-Vigier, H. et al. 2010. *HT (Hydrogen Transfer) Experiment: Mineralogy and Geochemistry of Cores of the BHT-1 Borehole*. Mont Terri Project Technical Note **TN2009-45**.

Lettry, Y. & Fierz, T. 2009. *HT (Hydrogen Transfer) Experiment. Description of Surface and Downhole Equipment, Site Instrumentation*. Installation Report. Mont Terri Project Technical Note **TN2009-20**.

Libert, M., Bildstein, O., Esnault, L., Jullien, M. & Sellier, R. 2011. Molecular hydrogen: An abundant energy source for bacterial activity in nuclear waste repositories. *Physics and Chemistry of the Earth A/B/C*, **36**, 1616–1623.

Liss, P. S. & Slater, P. G. 1974. Flux of gases across the air–sea interface. *Nature*, **247**, 181–184.

LUNDY, M. & VINSOT, A. 2010. Implementation of Raman and mass spectrometry for on line measurement of gas composition in boreholes. *In*: ANDRA, (ed.) *Clays in Natural & Engineered Barriers for Radioactive Waste Confinement, 4th International Meeting*, 29 March–1 April 2010, Nantes, France, Abstracts 535–536.

PARKHURST, D. L. & APPELO, C. A. J. 1999. *User's Guide to PHREEQC (version 2) – A Computer Program for Speciation, Batch Reaction, One-Dimensional Transport and Inverse Geochemical Calculations. USGS Water Resource Invest* Report **99/4259**.

PEARSON, F. J., TOURNASSAT, C. & GAUCHER, E. C. 2011. Biogeochemical processes in a clay formation *in situ* experiment: Part E – equilibrium controls on chemistry of pore water from the Opalinus Clay, Mont Terri Underground Research Laboratory, Switzerland. *Applied Geochemistry*, **26**, 990–1008.

PEARSON, J., ARCOS, D. ET AL. 2003. *Mont Terri Project, Geochemistry of Water in the Opalinus Clay Formation at the Mont Terri Rock Laboratory*. Reports of the Swiss Federal Office for Water and Geology (FOWG), Bern, Geology Series, No. 5.

TABANI, P., HERMAND, G., DELAY, J. & MANGEOT, A. 2010. Geoscientific data acquisition and management system (SAGD) of the Andra Meuse/Haute-Marne research center. *In*: ANDRA, (ed.) *Clays in Natural and Engineered Barriers for Radioactive Waste Confinement, 4th International Meeting*, 29 March–1 April 2010, Nantes, France, Abstracts, 253–254.

THURY, M. & BOSSART, P. 1999. Mont Terri Rock Laboratory. Results of the hydrogeological, geochemical and geotechnical experiments performed in 1996 and 1997. SNHGS Geological Report **23**.

TRUCHE, L., BERGER, G., DESTRIGNEVILLE, C., PAGES, A., GUILLAUME, D., GIFFAUT, E. & JACQUOT, E. 2009. Experimental reduction of aqueous sulphate by hydrogen under hydrothermal conditions: Implication for the nuclear waste storage. *Geochimica Cosmochimica Acta*, **73**, 4824–4835.

TRUCHE, L., BERGER, G., DESTRIGNEVILLE, C., GUILLAUME, D. & GIFFAUT, E. 2010. Kinetics of pyrite to pyrrhotite reduction by hydrogen in calcite buffered solutions between 90 and 180 °C: Implications for nuclear waste disposal. *Geochimica Cosmochimica Acta*, **74**, 2894–2914.

STUCKI, J. W. 2011. A review of the effects of iron redox cycles on smectite properties. *Comptes Rendus Geoscience*, **343**, 199–209.

VINSOT, A. 2012. *Hydrogen Transfer (HT) Experiment, Progress Report*. Mont Terri Project Technical Note **TN 2012-77**.

VINSOT, A., APPELO, C. A. J. ET AL. 2008*a*. CO_2 data on gas and pore water sampled *in situ* in the Opalinus Clay at the Mont Terri rock laboratory. *Physics and Chemistry of the Earth*, **33**, S54–S60.

VINSOT, A., METTLER, S. & WECHNER, S. 2008*b*. *In-situ* characterization of the Callovo–Oxfordian pore water composition. *Physics and Chemistry of the Earth*, **33S1**, S75–S86.

Experimental study on diffusion of tritiated water and anions under variable water-saturation and clay mineral content: comparison with the Callovo-Oxfordian claystones

S. SAVOYE[1]*, C. IMBERT[2], A. FAYETTE[1] & D. COELHO[3]

[1]*CEA, DEN, DPC, Laboratory of Radionuclides Migration Measurements and Modeling, F-91191 Gif-sur-Yvette, France*

[2]*CEA, DEN, DPC, Laboratory of Concrete and Clay Behaviour Studies, F-91191 Gif-sur-Yvette, France*

[3]*Andra, DRD, 1/7 rue Jean-Monnet, Châtenay-Malabry 92298, France*

**Corresponding author (e-mail: sebastien.savoye@cea.fr)*

Abstract: The diffusion of tritiated water and anionic species was studied through unsaturated compacted materials with variable clay content. These materials were designed to be analogous to the Callovo-Oxfordian claystone, a potential host-rock for disposal of high-level radioactive wastes. The diffusion parameters under such conditions were determined using modified through-diffusion cells in which the suction is generated by the process of osmosis. This device leads to saturation degree values ranging from about 70 to 100%. The results show that the diffusion through the unsaturated samples is clearly slower than that in the fully saturated samples, with steady-state fluxes decreasing by a factor up to 10 for tritiated water and up to 16 for anionic species. This tritiated water diffusivity (D_e/D_0) decrease is similar to that obtained in the Callovo-Oxfordian claystone samples, whereas it is lower than that obtained in claystone samples for anionic species. This difference can be accounted for by the distinct hydro-mechanic behaviour of these materials, the pore-volume of which depends on their rigidity and clay content when hydrating. Finally, these diffusive behaviours have been modelled by means of an excluded-volume expansion of Archie's second law, taking into account a critical water saturation below which no tracer can diffuse.

Diffusive transport in rocks plays an important role in a wide variety of processes of environmental concern, such as the spread of hazardous wastes in the ground (Hamamoto *et al.* 2009), oil recovery (Cui *et al.* 2009), CO_2 geological sequestration (e.g. Berthe *et al.* 2011) and the containment of nuclear wastes (e.g. Van Loon *et al.* 2004; Savoye *et al.* 2012*a–c*). For example, the safety of geological disposal of high-level radioactive waste relies on the very impervious properties of the engineered components (packed clay barrier) and the natural system (e.g. argillaceous rocks in France or in Switzerland) so that the slow process of diffusion is assumed to be dominant with respect to the advective one (Croisé *et al.* 2004). Therefore, considerable effort has been devoted to estimating diffusion coefficients for solutes transported through all these types of natural materials.

Moreover, there are many situations where the soils or the rocks can be unsaturated, leading to a decrease in the diffusive rate of solute species. For example, many landfills are constructed in arid or semiarid environments, where the soil can be unsaturated at great depths (Fityus *et al.* 1999). In this case, the values of the diffusion coefficient for solutes in unsaturated soils can be up to 10 or 20 times lower than the corresponding values in saturated soils (Shackelford 1991; Hamamoto *et al.* 2009). The presence of a radioactive waste repository in the consolidated argillaceous formations is also known to induce a hydraulic disequilibrium in the host rocks, and thus, their partial dehydration, since the underground drifts and shafts will be ventilated during their construction and the operation phase (Andra 2005; Jougnot *et al.* 2010).

However, since determining the solute diffusion coefficients through water-saturated rocks is already a challenging task, especially through these low-permeability clay rocks (e.g. Van Loon *et al.* 2004; Savoye *et al.* 2011), few diffusion experiments carried out under partially saturated conditions are reported in literature. Most of the diffusion testing is based on a transient approach, such as the decreasing source concentration method (Badv & Faridfard 2004) or the half-cell method (Schaefer *et al.* 1995; Hamamoto *et al.* 2009), known to suffer from a partial contact between the two half-cells (Shackelford 1991). The suction potential is generally

From: Norris, S., Bruno, J., Cathelineau, M., Delage, P., Fairhurst, C., Gaucher, E. C., Höhn, E. H., Kalinichev, A., Lalieux, P. & Sellin, P. (eds) 2014. *Clays in Natural and Engineered Barriers for Radioactive Waste Confinement*. Geological Society, London, Special Publications, **400**, 579–588.
First published online March 5, 2014, http://dx.doi.org/10.1144/SP400.9

applied either by the hanging water suction method or by pressure plate apparatus. While the former method cannot bring samples to suction values higher than 30 kPa (e.g. no significant dehydration in clayrocks), the latter may work well in the wet range (suction values up to 500 kPa), but beyond that, water potential values will typically not be accurate (Campbell 1988). Nevertheless, diffusion testing in partially saturated materials containing clay minerals requires suction methods capable of applying higher suction values, since the clay minerals are known to strongly retain water. Delage *et al.* (1998) developed and extended the osmotic technique to higher suctions (up to 10 MPa) for studying the hydro-mechanical behaviour of highly compacted unsaturated clays used in engineered clay barriers for nuclear waste disposal. This approach was taken up by Savoye *et al.* (2010), who modified their through-diffusion cells so that the diffusion of uncharged tritiated water and anionic iodide was investigated through consolidated clay rock samples with saturation degrees ranging from 81 to 100%. The rock samples originated from the Callovo-Oxfordian (COx) argillaceous formation, studied in France by the National Radioactive Waste Management Agency (Andra) as a potential host rock for nuclear waste disposal. In a previous study, we showed that the diffusion through unsaturated samples was clearly slower than through fully saturated samples, with steady-state fluxes decreasing by a factor up to 7 for tritiated water and up to 50 for iodide (Savoye *et al.* 2010). The diffusion cells were also adapted so that the diffusive behaviour of a sorbing cation, that is caesium, was studied in COx samples by acquiring both the caesium concentration in the source reservoir and the post-mortem caesium rock profiles (Savoye *et al.* 2012*b*). We showed that the effective diffusion coefficient (D_e) for caesium decreased by a factor up to 60 from the fully saturated sample to the more dehydrated one.

Such sharp drops in the diffusion coefficients when dehydrating even weakly the COx clay rock samples suggest that consolidated samples would be strongly sensitive to dehydration, probably because of their rigid structure. Kristensen *et al.* (2010) already pointed out that the gas-phase diffusivity in structured and intact soils differs from that of structureless repacked soils. Therefore, the aim of this paper is to address the following issue: how could the diffusive properties of repacked materials evolve with a mineralogical composition close to the COx claystones, when dehydrating them under the same suction range? For that purpose, the diffusion of uncharged tritiated water and iodide was investigated through two types of compacted mineral mixtures (mainly illite/calcite/quartz) under the same suction range as previously applied by Savoye *et al.* (2010) with the osmotic method (i.e. 0, 1.9, 6.3 and 9 MPa).

Materials and methods

Sample origin and sample preparation

Two mixtures were considered with variable clay mineral content for estimating the impact of these minerals with high water retention capacities on the diffusion when dehydrating. The choice of the mixture composition was based on that of the COx claystone core investigated by Savoye *et al.* (2010) and located at about 484.5 m below ground level. At this level, Gaucher *et al.* (2004) determined the following mineralogical composition: 35–54 wt% of clay minerals, 20–31 wt% of carbonates, 22–33 wt% of tectosilicates including quartz and less than 5 wt% of accessory minerals. Therefore, we used for the mixtures a very pure silica sand from Fontainebleau (France) and natural Illite du Puy (IdP) (Gabis 1958), containing 70 wt% of illite, 20 wt% of calcite, 5 wt% of kaolinite and 5 wt% of quartz (Poinssot *et al.* 1999). The two chosen IdP–sand mixtures with relative proportions of 80:20 wt% and 60:40 wt% led to the following compositions: clay minerals (60 or 45 wt%), carbonates (16 or 12 wt%) and sand (24 or 43 wt%), respectively.

Reagents of highest purity were obtained from Fluka (Buchs, Switzerland) or Merck (Dietikon, Switzerland). Deionized water (Milli-Q® water) was used throughout for the preparation of solutions. The silica sand was at first sieved (100 μm mesh), then washed with 0.1 mol l^{-1} HNO_3 and afterwards with deionized water (18.2 MΩ cm^{-1}). The silica sand was equilibrated three times in series with 0.1 mol l^{-1} KCl, then three times with 10^{-2} mol l^{-1} KCl. The IdP was primarily washed with ultrapure deionized water, then sieved (50 μm mesh), and equilibrated with 0.1 mol l^{-1} KCl and, afterwards with 10^{-2} mol l^{-1} KCl. After a last centrifugation for removing the supernatants, the sand and the IdP were stored at 30.0 ± 0.2 °C in desiccators containing an NaCl-oversaturated solution (suction = 39 MPa), until suction equilibrium (indicated by mass stabilization), as made by Savoye *et al.* (2010, 2012*b*) for COx samples. Then, the two IdP–sand mixtures were prepared by weighting the corresponding amounts of sand and IdP, by homogenizing the mixtures for 2 days with a shaking mixer, and by storing them in desiccators containing an NaCl-oversaturated solution for four days. This initial dehydration of the samples at a suction level lower than those imposed by the osmotic method prevents any shrinkage of samples (only hydration pathway). Mixtures were directly compacted into the stainless-steel diffusion cells.

Re-saturation procedure

The suction is generated by the osmosis process between the pore-water (present in the pores of the sample) and a highly concentrated solution with large-sized molecules of polyethylene glycol (PEG; for more details see Savoye *et al.* 2010). The sample is separated from the PEG-solution by a semipermeable membrane (which is permeable to all except PEG). The exclusion of the PEG from the sample results in a chemical-potential imbalance between the water in the clay sample and the water in the reservoir chambers. This osmotic suction has the effect of keeping the sample unsaturated. Moreover, the value of the imposed suction and thus the saturation state of the sample depend on the PEG concentration in solution (Delage *et al.* 1998). The different saturation states were reached with chemical solutions prepared with increasing PEG concentrations (0, 0.42, 0.76, and 0.95 g per g of solution), leading to increasing suction values (0, 1.9, 6.3 and 9 MPa, respectively) (Savoye *et al.* 2010, 2012*b*).

Samples were gradually re-saturated with solutions prepared with ultrapure deionized water (18.2 $M\Omega\ cm^{-1}$), PEG 6000 (Merck, Germany), KCl (10^{-2} mol l^{-1}) and $Na_2S_2O_3$ (5×10^{-4} mol l^{-1}). The concentration of KCl was chosen for having solutions with ionic strength close to that of the COx pore-water (Vinsot *et al.* 2007) and thiosulphate was added to ensure that iodide used as tracer remains in the same redox state (Savoye *et al.* 2012*c*; Wittebroodt *et al.* 2012). The experimental set-up comprises a diffusion cell, as described above, a 16-channel peristaltic pump (IPS, Ismatec, Idex Corporation, USA) and 100 cm^3 reservoirs. During the hydric equilibrium phase, both sides of the samples were in contact with the same synthetic water and PEG solution, using one 100 cm^3 reservoir. One month was shown to be sufficient to achieve the hydric equilibrium for the more impervious COx rock sample (Savoye *et al.* 2010), so the same duration was chosen for the IdP–sand mixtures. Two series (80:20 and 60:40) of four compacted samples each were prepared. The main characteristics of these samples (thickness, bulk dry and grain densities) are reported in Table 1.

Experimental method for petrophysical characterizations

Measurement of both bulk and grain densities allows the determination of the total porosity, which is the sum of the opened and closed pore volumes. Bulk densities (ρ) were determined using a hydrostatic weighting in petroleum. After crushing and drying, powder samples were used for the grain density (ρ_s) measurements with a helium pycnometer (Accupyc II 1340, Micromeritics Instrument Norcross, GA, USA). The calculation of the volumetric moisture content, θ, the total porosity, ϕ, and the saturation degree, S_w, is described in Savoye *et al.* (2010).

Diffusion experiments

After one month of saturation treatment, the eight cells were connected to two distinct reservoirs. The upstream reservoir was filled with 100 cm^3 of a fresh solution labelled with tritiated water (HTO) and a 20 cm^3 downstream reservoir filled with a fresh solution without tracer was connected to the cell. During the through-diffusion experiment, careful attention was taken to keep a nearly constant radiotracer concentration in the upstream reservoir (more than 95% of its initial value, adjusted with radiotracers) and low concentration in the downstream reservoir (less than 3% of the upstream reservoir concentration, renewed if higher). At the end of the HTO through-diffusion experiment, out-diffusion was performed by replacing the solutions

Table 1. *Characteristics of the samples used for the diffusion experiments and overview of the tracer used and their activity*

Sample	[PEG], $g\ g^{-1}$ of solution	Suction MPa	Thickness (mm)	Bulk dry density $\rho_{d,b}$ ($kg\ m^{-3}$)	Grain density ρ_s ($kg\ m^{-3}$)	Diffusion experiments activity ($kBq\ kg^{-1}$ of solution)		
						HTO	$^{36}Cl^-$	$^{125}I^-$
80:20	0	0	10.557	1765	2,731	6340 ± 400	5860 ± 350	1995 ± 150
80:20	0.42	1.9	10.539	1775		7415 ± 500		1990 ± 150
80:20	0.76	6.9	10.319	1808		7415 ± 500		1989 ± 150
80:20	0.95	9	10.117	1840		7138 ± 500		1994 ± 150
60:40	0	0	10.865	1807	2,698	6110 ± 400	5580 ± 350	1999 ± 150
60:40	0.42	1.9	10.728	1827		7363 ± 500		1995 ± 150
60:40	0.76	6.9	10.904	1802		7779 ± 500		1990 ± 150
60:40	0.95	9	10.686	1832		7151 ± 500		1985 ± 150

in both reservoirs with fresh synthetic water without tracer for up to 3 weeks. Through-diffusion with $^{125}I^-$ could start as soon as the HTO out-diffusion flux was negligible, that is, for a residual HTO activity close to the detection limit (0.5 Bq). In addition to the radio-tracer, $^{125}I^-$, stable iodide at a concentration of 10^{-3} mol l^{-1} was added in the upstream reservoir in order to limit the influence of iodine uptake, as previously observed on claystones by Wittebroodt *et al.* (2012) and Savoye *et al.* (2012*c*) for lower concentrations ($<10^{-4}$ mol l^{-1}). $^{125}I^-$ was preferred to $^{36}Cl^-$ because of the lack of waste management means for organic solutions containing long-lived radionuclides, such as $^{36}Cl^-$. Nevertheless, through-diffusion experiments were performed with HTO, $^{36}Cl^-$ and $^{125}I^-$ on the two cells without PEG in order to compare the diffusive behaviours of iodide and chloride. HTO and $^{36}Cl^-$ activities were measured by liquid scintillation (Tricarb 2500, Canberra-Packard, USA), while $^{125}I^-$ was measured by gamma spectrometry (Packard 1480 Wizard, USA). An overview of the diffusion experiments is given in Table 1 with the tracers used and their activity. The interpretation of the experimental data was carried out by means of fully analytical solutions of the 1D Fick's second law (See Savoye *et al.* 2010 and 2012*c*, for details).

Results

Petrophysical data

A general trend can be highlighted from Figure 1, as a function of the PEG concentration, when (a) gathering the samples having undergone the two highest suction values (i.e. 0.76 g and 0.95 g of PEG per g of solution) and (b) without taking into account the occurrence of possible error bar overlapping: the smaller the applied suction is (i.e. the less concentrated the solution in PEG), the higher the volumetric water content of the samples and the higher the saturation degree (from about 71 to 79%, for solutions with PEG). It is noteworthy that, in addition to the water content, the 80:20 sample series displays a progressive increase in its total porosity when decreasing PEG concentration, leading to a change in the saturation degree values. This suggests a possible reorganization of the porosity related to the higher amount of clay minerals, when hydrating from the initial hydric state imposed by the saline method (suction *c.* 39 MPa) to the different suction levels. The impossibility of using filter plates in the set-up leads indeed to the existence, at the sample–cell interface, of some dead-volumes capable of being slightly filled by the materials when hydrating, so that the total porosity can increase.

Through-diffusion results

Figure 2a–d shows the flux and the cumulative tracer activity for the eight through-diffusion experiments performed with HTO and $^{125}I^-$. Increasing the suction tends to decrease both the instantaneous flux and the cumulative mass relative to the fully saturated samples. For example, the steady-state fluxes are decreased by factors of approximately 10 and 16 for HTO and $^{125}I^-$, respectively, independently of the mixtures considered. An overview of the estimated diffusive parameter values for HTO, $^{36}Cl^-$ and $^{125}I^-$ is given in Table 2. In the fully saturated samples, $^{125}I^-$ and $^{36}Cl^-$ show similar

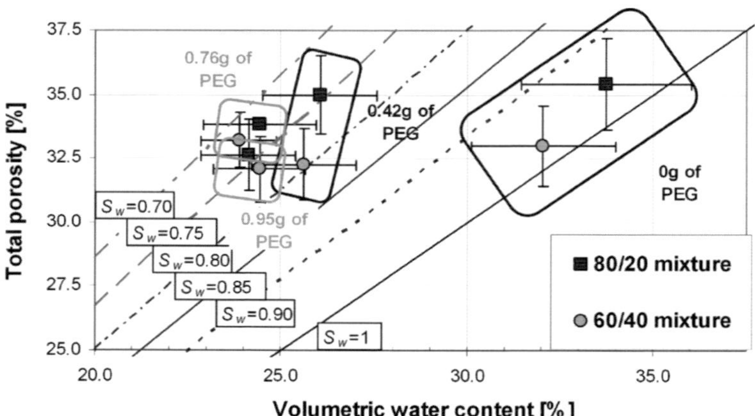

Fig. 1. Values of the total porosity as a function of the volumetric water content determined on compacted samples having undergone variable suction by means of the osmotic method. The higher the value of PEG concentration is, the higher the suction. Error bars are calculated from the Gauss propagation equation.

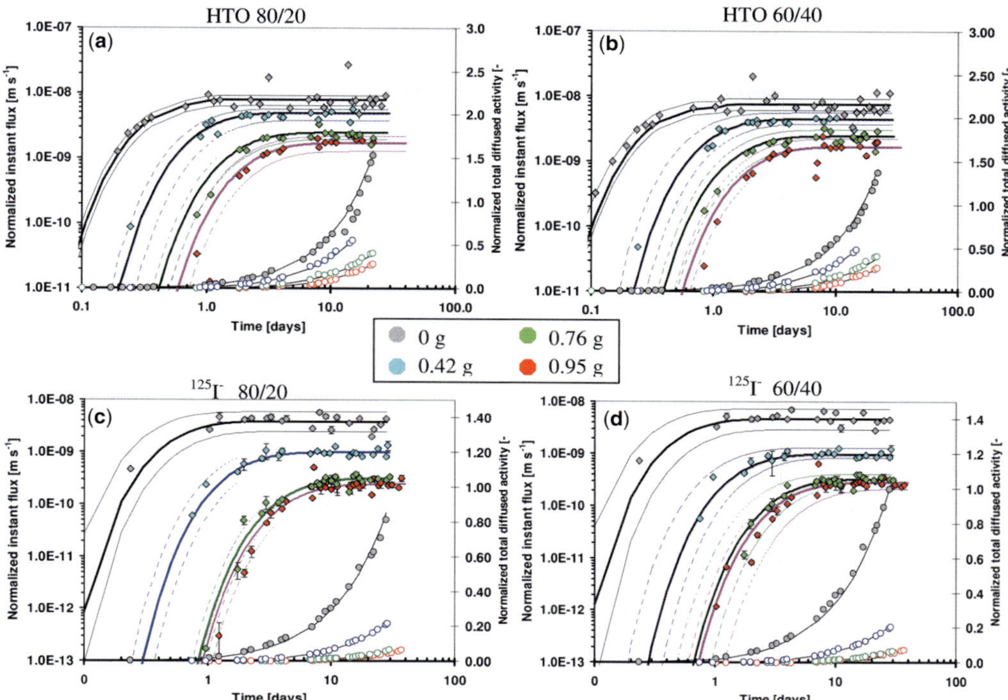

Fig. 2. Total tracer activity and normalized instant fluxes at different PEG concentration for HTO (**a**, **b**) and for $^{125}I^-$ (**c**, **d**). The solid and dashed curves for the fluxes were calculated using the analytical solutions with the parameters specified in Table 2, solid curves from the mean values and dashed curves from the minimum and maximum values of the diffusive parameters. Normalized flux is the ratio of instantaneous flux (in Bq m^{-2} s^{-1}) over the activity in the upstream reservoir (in Bq m^{-3}). The normalized total diffused tracer amount is the ratio of total diffused tracer activity (in Bq) over the activity in the upstream reservoir (in Bq m^{-3}) and the volume of the sample (in m^3).

diffusive parameters (Table 2). This confirms that iodide at such a high concentration (10^{-3} mol l^{-1}) behaves as a conservative tracer, as previously shown by Savoye *et al.* (2010, 2012*c*) and Wittebroodt *et al.* (2012). Moreover, at full saturation, a slight effect of the clay mineral content on the diffusive parameters for all the tracers can be observed, when focusing only on the mean diffusive

Table 2. *Values of effective diffusion coefficients (D_e) and porosities (θ_{acc}) for the diffusion of HTO, $^{36}Cl^-$, and $^{125}I^-$ for various imposed suctions*

Sample	[PEG], g g^{-1} of solution	D_e (HTO) (m^2 s^{-1}) $\times 10^{10}$	θ'_{diff} (HTO) (%)	$1/G$ (−) $\times 10^2$	Anion tracer	D_e (m^2 s^{-1}) $\times 10^{11}$	θ'_{diff} (%)	$1/G$ (−) $\times 10^2$
80:20	0	1.65 (1.30–1.90)	32 (26–35)	25.7	$^{36}Cl^-$	4.0 (2.0–6.0)	12 (10–14)	18.9
					$^{125}I^-$ *	4.0 (2.5–6.0)	13 (10–14)	17.2
80:20	0.42	0.50 (0.40–0.60)	25 (24–26)	10.0	$^{125}I^-$ *	1.0 (0.8–1.2)	11 (10–12)	5.4
80:20	0.76	0.25 (0.20–0.30)	25 (23–27)	5.0	$^{125}I^-$ *	0.33 (0.26–0.40)	9 (8–10)	2.0
80:20	0.95	0.17 (0.13–0.22)	24 (21–27)	3.6	$^{125}I^-$ *	0.25 (0.20–0.34)	7.5 (7–10)	1.9
60:40	0	1.55 (1.25–1.90)	29 (27–35)	26.6	$^{36}Cl^-$	5.0 (3.5.0–6.5)	13 (11–15)	21.8
					$^{125}I^-$ *	5.0 (3–7.5)	15 (13–17)	18.7
60:40	0.42	0.50 (0.40–0.60)	25 (24–26)	9.5	$^{125}I^-$ *	1.1 (0.9–1.4)	10 (9–11)	6.2
60:40	0.76	0.27 (0.20–0.30)	22 (18–25)	6.0	$^{125}I^-$ *	0.35 (0.30–0.45)	7 (7–8)	2.8
60:40	0.95	0.18 (0.14–0.24)	21 (18–25)	4.3	$^{125}I^-$ *	0.31 (0.23–0.38)	7 (7–8)	2.5

*In addition to I-125, stable iodide (10^{-3} M) was added in the upstream reservoir.
The values in brackets indicate the uncertainty ranges.

parameter values, that is, without taking into account the uncertainty ranges. The higher the clay mineral content in the samples, the higher the diffusive parameters for HTO and the lower the diffusive parameters for the anionic species, that is, $^{125}I^-$ and $^{36}Cl^-$. The increase in clay minerals means, on the one hand, a higher total porosity accessible to HTO, and, on the other hand, a higher extent of the anionic exclusion for $^{125}I^-$ and $^{36}Cl^-$, as already mentioned by Iida *et al.* (2011) for the diffusion of selenium under variable bentonite content.

Discussion

A comparison of the diffusive behaviour of the two IdP–sand mixtures with the COx samples having undergone the same suction range is given in Figure 3a, b with the evolution of the diffusivity (D_e/D_0) for HTO and $^{125}I^-$ (respectively), as a function of the volumetric water content (105 °C), θ. Remember that D_0 is the free-solution (aqueous) diffusion coefficient ($m^2\ s^{-1}$). Given the associated uncertainties, the extent of diffusivity decrease for HTO is similar for all the types of materials, with values ranging from 7 to 10, while for $^{125}I^-$, the decrease is more pronounced for the COx samples (up to 50) than for the mixture samples (up to 16).

Several authors have developed empirical models to predict the diffusion coefficient from physical characteristics of natural media. A widely used model presented as Archie's second law (Archie 1942) links the diffusivity (D_e/D_0) to the volumetric water content (θ) and the total porosity (ϕ), as

$$\frac{D_e}{D_0} = \left(\frac{\theta}{\phi}\right)^n \phi^m, \tag{1}$$

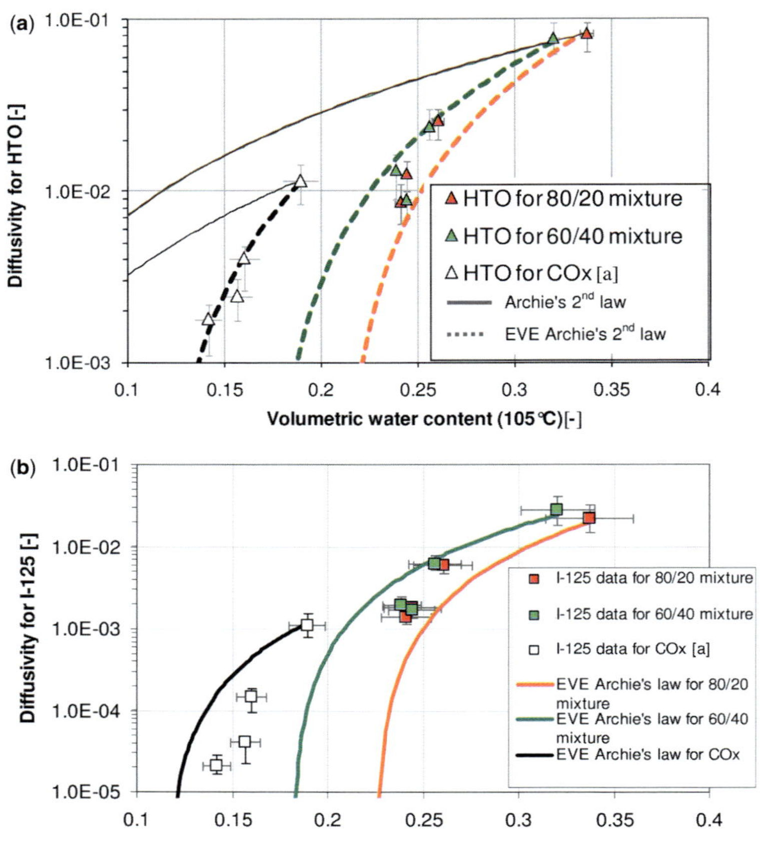

Fig. 3. HTO (**a**) and $^{125}I^-$ (**b**) diffusivity (D_e/D_0) as a function of the volumetric water content. [a] Data from Savoye *et al.* (2010).

where n is the saturation index, generally assumed to be 2.0 (Revil & Jougnot 2008), and m is the cementation exponent. As shown in Figure 3a, this law tends to underestimate the decrease in the diffusivity for tritiated water. Many authors (Martys 1999; Moldrup *et al.* 2007; Hamamoto *et al.* 2010) have suggested that there is a critical water saturation at which the water phase is no longer connected at the scale of a representative volume in the rock samples. When the water saturation reaches this critical value, the tracer diffusive pathway goes to infinity and its transport through the pore-water is no longer possible. This phenomenon has been taken into account by introducing a percolation threshold in Archie's second law, as proposed by Hamamoto *et al.* (2010). Equation (1) can be modified to form an excluded-volume expansion of Archie's second law,

$$\frac{D_e^t}{D_0^t} = \left(\frac{\theta - \phi_{th}^t}{\phi - \phi_{th}^t}\right)^n (\phi - \phi_{th}^{*t})^m, \qquad (2)$$

where ϕ_{th}^{*t} is the inactive liquid phase for the given transport process of the tracer t at water saturation $(-)$, as suggested by Balberg (1986) and ϕ_{th}^t is the percolation threshold for the tracer t at water-unsaturated condition $(-)$. ϕ_{th}^* is justified for anion species by the anionic exclusion phenomenon, leading to the existence of a porosity type from which the anions are excluded, owing to the presence of negative charges at the surface of clay minerals. For a water tracer like HTO, this parameter is necessarily null since HTO is known to diffuse into the whole porosity under full-saturation conditions (Savoye *et al.* 2011). Hence, we assumed in a first approach, (a) for $^{125}I^-$, that ϕ_{th}^{*I-125} is equal to ϕ_{th}^{I-125}, and (b) for HTO, that ϕ_{th}^{*HTO} is equal to zero and ϕ_{th}^{HTO} is equal to ϕ_{th}^{I-125}. Note that ϕ_{th}^{*I-125} was calculated from the values of the porosity accessible to the diffusion of $^{125}I^-$ and the values of the volumetric water content, both being determined under fully saturated conditions (Table 2). The value of n was kept equal to 2.0, as suggested by Revil & Jougnot (2008) and Hamamoto *et al.* (2010), and m was estimated from the experimental data obtained at full saturation.

The curves calculated from the equation (2) using the parameters n, m, ϕ_{th}^* and ϕ, given in Table 3, are reported in Figure 3a and b for HTO and $^{125}I^-$, respectively. The experimental data are roughly reproduced by the excluded-volume expansion Archie's second law, independently of the tracer considered, except to some extent for some of the data. This is mainly the case for $^{125}I^-$ data, with the COx samples and the 60:40 mixture (Fig. 3b). The m values, especially for $^{125}I^-$, are quite consistent with what is expected for these different types of materials, that is $2 < m < 3$ for consolidated argillaceous rocks and $m < 2.2$ for unconsolidated materials (Hu & Wang 2003; Descostes *et al.* 2008; Jougnot *et al.* 2010; Peng *et al.* 2012). Finally, the model proposed by Hamamoto *et al.* (2010) seems to be relatively robust, compared with the basic Archie's second law, for simulating diffusive transport phenomenon both in consolidated and unconsolidated media, even though some deviations occur for the more consolidated media.

Table 3. *Parameters used for the excluded-volume expansion of Archie's second law given in equation (2)*

Tracer	Sample	ϕ_{th}^* (−)	ϕ (−)	n	m
HTO	80:20 Mixture	0.224	0.354	2	2.2
	60:40 Mixture	0.180	0.329	2	2.2
	COx sample	0.114*	0.189*	2	2.7
$^{125}I^-$	80:20 Mixture	0.224	0.354	2	1.8
	60:40 Mixture	0.180	0.329	2	1.9
	COx sample	0.114*	0.189*	2	2.6

*Data from Savoye *et al.* (2010).

A further investigation of the evolution of the diffusivity with saturation can be achieved by considering the expression of the effective diffusion coefficient, D_e, given by Grathwohl (1998):

$$D_e = \frac{1}{G} D_0 \theta_{diff}^t, \qquad (3)$$

where G is the geometrical factor $(-)$. This means that the diffusivity (D_e/D_0) is related to the ratio of θ_{diff}^t over the geometrical factor, G, linked to the geometrical structure of the diffusion pathway (*i.e.* tortuosity and constrictivity). The relative contribution of θ_{diff}^t to the decrease in D_e/D_0 can be evidenced by reporting the evolution of θ_{diff}^t as a function of θ (105 °C) (Fig. 4). Even though the θ_{diff}^{HTO} values for the COx samples were lower than those for the two mixtures, they evolved similarly by decreasing linearly with θ when dehydrating. Conversely, for the θ_{diff}^{I-125} values, a tendency can be drawn from the 80:20 mixture to the COx samples. The θ_{diff}^{I-125} values for the 80:20 mixture decrease linearly with θ, meaning that the extent of anion exclusion remains roughly constant ($\theta_{diff}^{I-125}/\theta \approx 1/3$) when dehydrating. Conversely, the θ_{diff}^{I-125} values for the COx samples and to a lesser extent, the 60:40 mixture deviate from the constant ratio $\theta_{diff}^{I-125}/\theta$ by sharply decreasing. This indicates that some pores, into which iodide could normally diffuse under fully saturated conditions, become isolated from the diffusive pathway when dehydrating.

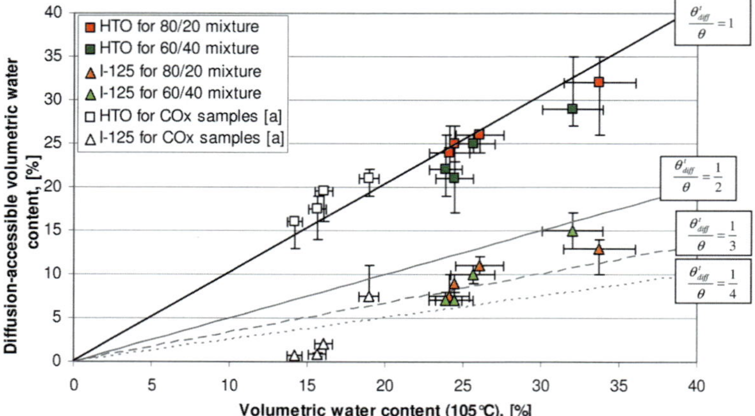

Fig. 4. HTO and $^{125}I^-$ accessible volumetric water content as a function of the volumetric water content for the two mixtures and the COx samples under the same range of suction. [a] Data from Savoye *et al.* (2010).

Figure 5 shows the evolution of $1/G$ as a function of θ^t_{diff}, each parameter having been normalized by the values they took at full saturation for each material. Independently of the tracer (HTO and $^{125}I^-$) and the type of studied material (80:20 and 60:40 mixtures, and COx), all the reverses of normalized geometrical factor evolved roughly to the same extent (decrease of up to 80–90%), as well as all the normalized θ^{HTO}_{diff}(decrease of up to 75%). Conversely, as already mentioned above, the normalized θ^{I-125}_{diff} clearly exhibits a distinct behaviour, with a weak decrease of less than 40% for the 80:20 mixture, a more important one for the 60:40 mixture (up to 50%) and a stronger one for the COx sample (up to 90%). The origin of this difference (that would mainly be responsible for the difference in the extent of the diffusivity decrease for $^{125}I^-$) can be inferred to the distinct hydro-mechanic behaviour of the compared materials. On the one hand, the 80:20 mixture is an unconsolidated material in which the clay minerals are present in the larger amount, so that, when dehydrating or hydrating, their shrinkage or their swelling, even to a low extent, could induce a more

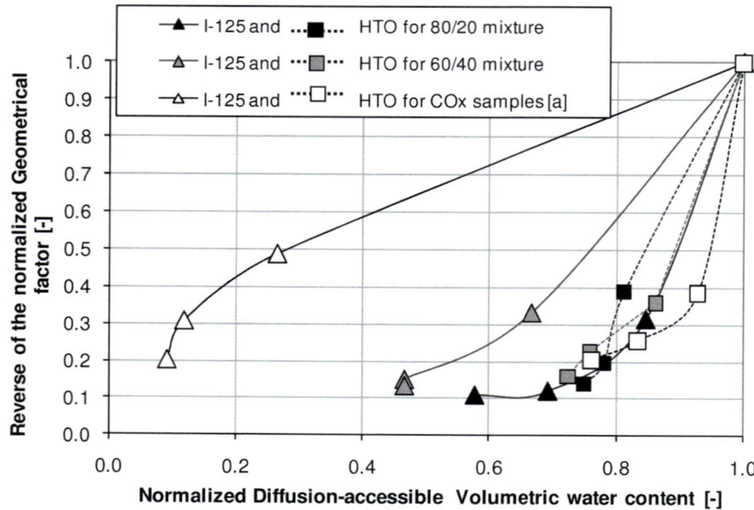

Fig. 5. Reverse of the normalized geometrical factor as a function of the normalized diffusive-accessible volumetric water content. [a] Data from Savoye *et al.* (2010).

significant change in the pore volume, and therefore, facilitate the continuity of the liquid phase for diffusion. Recall that the petrophysical characterization showed that the total porosity of the 80:20 mixture increased when hydrating. In the other unconsolidated material, that is the 60:40 mixture, a weaker reorganization of the pore network can be expected owing to the smaller amount of clay minerals. On the other hand, the COx medium, owing to its consolidated structure, should show less deformations of its skeleton. Therefore, the water film continuity would be more easily broken into the large emptied pores, reaching more rapidly, even under the same applied suction, a 'percolation threshold', from which the diffusion would not occur anymore.

Conclusion

In this work, the through-diffusion cells modified by Savoye *et al.* (2010), in which the suction is generated by the process of osmosis, has been successfully used to measure the diffusion of uncharged tritiated water and iodide through unsaturated compacted materials with variable clay content. Their mineralogical compositions have been chosen close to that of the Callovo-Oxfordian claystone in order to compare the diffusive behaviours of unconsolidated and consolidated media when dehydrating. At full saturation, the resultant diffusive parameters are controlled by the mineralogy. The higher the clay content, the higher the effective diffusion coefficient (D_e) and the porosity for tritium and the lower D_e and the porosity for iodide, owing to the anionic exclusion phenomenon. Moreover, the diffusion through the unsaturated samples is clearly slower than that in the fully saturated samples, with steady-state fluxes decreasing by a factor up to 10 for tritium and up to 16 for the anionic species.

Comparison with the diffusive data obtained under the same suction conditions in the Callovo-Oxfordian claystones indicates a similar extent of the diffusivity (D_e/D_0) decrease for tritium, and a sharper D_e/D_0 decrease of the claystones for anionic species when dehydrating. The similarity observed for tritium suggests that, for tracers diffusing into the whole porosity, the analogy between the COx and the two Illite du Puy–sand mixtures still remains valid, since this type of tracer diffuses mainly into the clay matrix porosity, almost unaffected by dehydration. Conversely, the difference observed for the anionic species, that is $^{125}I^-$, has been accounted for by the distinct hydro-mechanic behaviour of the studied materials. Depending on the rigidity and the clay content of these media, their pore volume evolved more or less when hydrating, facilitating or not the continuity of the liquid phase for diffusion. This would especially affect the anionic species constrained to diffuse into the largest pores, primarily impacted by dehydration. Finally, such diffusive behaviours have been modelled by means of Archie's second law modified by Hamamoto *et al.* (2010), by taking into account a critical water saturation below which no tracer can percolate.

This work received financial support from Andra and CEA. The authors are grateful to T. Martin for petrophysical measurements and mercury injection porosimetry, and S. Bassot for the determination of the grain density by He pycnometry. We would also like to thank the editor and reviewers for their insightful comments.

References

ANDRA 2005. *Dossier 2005 argile – Tome – Evolution phénoménologique du stockage géologique*. Rapport Andra no. C.RP.ADS.04.0025, France, http://www.andra.fr/download/site-principal/document/editions/182.pdf

Archie, G. E. 1942. The electrical resistivity log as an aid in determining some reservoir characteristics. *Transport AIME*, **146**, 54–62.

Badv, K. & Faridfard, M. R. 2004. Laboratory determination of water retention and diffusion coefficient in unsaturated sand. *Water Air Soil Pollution*, **161**, 25–38.

Balberg, I. 1986. Excluded-volume explanation of Archie's law. *Physical Review B*, **33**, 3618–3620, http://dx.doi.org/10.1103/PhysRevB.33.3618

Berthe, G., Savoye, S., Wittebroodt, C. & Michelot, J.-L. 2011. Changes in containment properties of claystone caprocks induced by dissolved CO_2 seepage. *Energy Procedia*, **4**, 5314–5319, http://dx.doi.org/10.1016/j.egypro.2011.02.512.

Campbell, G. S. 1988. Soil water potential measurement: an overview. *Irrigation Science*, **9**, 265–273.

Croisé, J., Schlickenrieder, L., Marschall, P., Boisson, J.-Y., Vogel, P. & Yamamoto, S. 2004. Hydrogeological investigations in a low permeability claystone formation: the Mont Terri Rock Laboratory. *Physics and Chemistry of the Earth*, **29**, 3–15, http://dx.doi.org/10.1016/j.pce.2003.11.008

Cui, X., Bustin, A. & Bustin, R. 2009. Measurements of gas permeability and diffusivity of tight reservoir rocks: Different approaches and their applications. *Geofluids*, **9**, 208–223.

Delage, P., Howat, M. D. & Cui, Y. J. 1998. The relationship between suction and swelling properties in a heavily compacted unsaturated clay. *Engineering Geology*, **50**, 31–48.

Descostes, M., Blin, V. *et al.* 2008. Diffusion of anionic species in Callovo-Oxfordian argillites and Oxfordian limestones (Meuse/Haute-Marne, France). *Applied Geochemistry*, **23**, 655–677, http://dx.doi.org/10.1016/j.apgeochem.2007.11.003

Fityus, S. G., Smith, D. W. & Booker, J. R. 1999. Contaminant transport through an unsaturated soil liner beneath a landfill. *Canadian Geotechnical Journal*, **36**, 330–354.

GABIS, V. 1958. Etude préliminaire des argiles oligocènes du Puy-en-Velay (Haute-Loire). *Bulletin de la Société Française de Minéralogie et Cristallographie*, **81**, 183–185.

GAUCHER, E., ROBELIN, C. ET AL. 2004. ANDRA underground research laboratory: interpretation of the mineralogical and geochemical data acquired in the Callovian–Oxfordian formation by investigative drilling. *Journal of Physics and Chemistry of the Earth*, **29**, 55–77.

GRATHWOHL, P. 1998. *Diffusion in Porous Media*. Springer, Berlin.

HAMAMOTO, S., PERERA, M. S. A., RESURRECCION, A., KAWAMOTO, K., HASEGAWAD, S., KOMATSU, T. & MOLDRUP, P. 2009. The solute diffusion coefficient in variably compacted, unsaturated volcanic ash soils. *Vadoze Zone Journal*, **8**, 942–952.

HAMAMOTO, S., MOLDRUP, P., KAWAMOTO, K. & KOMATSU, T. 2010. Excluded-volume expansion of Archie's law for gas and solute diffusivities and electrical and thermal conductivities in variably saturated porous media. *Water Resources Research*, **46**, W06514, http://dx.doi.org/10.1029/2009WR008424

HU, Q. & WANG, J. S. Y. 2003. Aqueous-phase diffusion in unsaturated geologic media: a review. *Critical Reviews in Environmental Science and Technology*, **33**, 275–297.

IIDA, Y., YAMAGUCHI, T. & TANAKA, T. 2011. Experimental and modelling study on diffusion of selenium under variable bentonite content and porewater salinity. *Journal of Nuclear Science and Technology*, **48**, 1–14.

JOUGNOT, D., REVIL, A., LU, N. & WAYLLACE, A. 2010. Transport properties of the Callovo-Oxfordian rock under partially saturated conditions. *Water Resources Research*, **46**, W08514, http://dx.doi.org/10.1029/2009WR008552

KRISTENSEN, A. H., THORBJORN, A., JENSEN, M. P., PEDERSEN, M. & MOLDRUP, P. 2010. Gas-phase diffusivity and tortuosity of structured soils. *Journal of Contaminant Hydrology*, **115**, 26–33.

MARTYS, N. S. 1999. Diffusion in partially-saturated porous materials. *Materials and Structures*, **32**, 555–562.

MOLDRUP, P., OLESEN, T., BLENDSTRUP, H., KOMATSU, T., DE JONGE, L. W. & ROLSTON, D. E. 2007. D.E. predictive-descriptive models for gas and solute diffusion coefficients in variably saturated porous media coupled to pore-size distribution: IV. Solute diffusivity and the liquid phase impedance factor. *Soil Science*, **172**, 471–750.

PENG, S., HU, Q. & HAMAMOTO, S. 2012. Diffusivity of rocks: Gas diffusion measurement and correlation to porosity and pore size distribution. *Water Resources Research*, **48**, W02507, http://dx.doi.org/10.1029/2011WR011098

POINSSOT, C., BAEYENS, B. & BRADBURY, M. H. 1999. Experimental and modelling studies of caesium sorption on illite. *Geochimica Cosmochimica Acta*, **63**, 3217–3227.

REVIL, A. & JOUGNOT, D. 2008. Diffusion of ions in unsaturated porous materials. *Journal of Colloid and Interface Science*, **319**, 226–235.

SAVOYE, S., PAGE, J., PUENTE, C., IMBERT, C. & COELHO, D. 2010. A new experimental approach for studying diffusion through an intact and unsaturated medium: a case study with Callovo-Oxfordian argillite. *Environmental Science and Technology*, **44**, 3698–3704.

SAVOYE, S., GOUTELARD, F. ET AL. 2011. Effect of temperature on the containment properties of argillaceous rocks: the case study of Callovo–Oxfordian claystones. *Journal of Contaminant Hydrology*, **125**, 102–112.

SAVOYE, S., MICHELOT, J. L., MATRAY, J. M., WITTEBROODT, C. & MIFSUD, A. 2012*a*. A laboratory experiment for determining both the hydraulic and diffusive properties and the initial pore-water composition of an argillaceous rock sample: a test with the Opalinus clay (Mont Terri, Switzerland). *Journal of Contaminant Hydrology*, **127**, 47–57.

SAVOYE, S., BEAUCAIRE, C., FAYETTE, A., HERBETTE, M. & COELHO, D. 2012*b*. Mobility of caesium through the Callovo-Oxfordian claystones under partially saturated conditions. *Environmental Science and Technology*, **46**, 2633–2641.

SAVOYE, S., FRASCA, B., GRENUT, B. & FAYETTE, A. 2012*c*. How mobile is iodide in the Callovo–Oxfordian claystones under experimental conditions close to the in situ ones? *Journal of Contaminant Hydrology*, **142–143**, 82–92, http://dx.doi.org/org/10.1016/j.jconhyd.2012.10.003

SCHAEFER, C. E., ARANDS, R. R., VAN DER SLOOT, H. A. & KOSSON, D. S. 1995. Prediction and experimental validation of liquid-phase diffusion resistance in unsaturated soils. *Journal of Contaminant Hydrology*, **20**, 145–166.

SHACKELFORD, C. 1991. Laboratory diffusion testing for waste disposal. *Journal of Contaminant Hydrology*, **7**, 177–217.

VAN LOON, L. R., SOLER, J. M., MÜLLER, W. & BRADBURY, M. H. 2004. Anisotropic diffusion in layered argillaceous rock: a case study with Opalinus Clay. *Environmental Science and Technology*, **38**, 5721–5728.

VINSOT, A., METTLER, S. & WECHNER, S. 2007. In situ characterization of the Callovo-Oxfordian pore water composition. *Journal of Physics and Chemistry of the Earth*, **33**, S75–S86.

WITTEBROODT, C., SAVOYE, S., FRASCA, B., GOUZE, P. & MICHELOT, J.-L. 2012. Diffusion of HTO, $^{36}Cl^-$ and $^{125}I^-$ in Upper Toarcian argillite samples from Tournemire: effects of initial iodide concentration and ionic strength. *Applied Geochemistry*, **27**, 1432–1441, http://dx.doi.org/10.1016/j.apgeochem.2011.12.017

Long-term impact of temperature on the hydraulic permeability of bentonite

J. F. HARRINGTON[1]*, G. VOLCKAERT[2] & D. J. NOY[1]

[1]*Transport Properties Research Laboratory, British Geological Survey, Keyworth, Nottingham NG12 5GG, UK*

[2]*Studiecentrum voor Kernenergie (SCK-CEN), Boeretang 200, 2400 Mol, Belgium*

**Corresponding author (e-mail: jfha@bgs.ac.uk)*

Abstract: In a Swedish repository for the disposal of heat-emitting waste, the long-term thermal stability of the bentonite engineered barrier forms a key component of the safety case. Central to such consideration is the evolution of hydraulic permeability and a potential degradation of hydraulic properties, in response to prolonged thermal exposure of the clay. To address this issue, a detailed programme of laboratory-based experiments has been undertaken at both the British Geological Survey and Studiecentrum voor Kernenergie/Centre d'Etude de L'Energie Nucleaire, in order to examine the hydraulic behaviour of bentonite that had previously been exposed to elevated temperatures. Hydraulic properties were calculated from both steady-state pressure gradients and from analysis of the pressure transients. Inspection of the data found no significant difference in hydraulic behaviour between the virgin material and clay samples taken from the Canister Retrieval Test. Based on these observations, the authors find no evidence for an adverse increase in hydraulic conductivity of bentonite as a result of prolonged thermal exposure to temperatures of 80 °C.

The thermal alteration of clay minerals resulting in the transition of smectite to illite is a well-established phenomenon and has been empirically observed by numerous researchers in the past (e.g. Velde *et al.* 1986; Bruce *et al.* 1987; Freed & Peacor 1989; Velde & Vasseur 1992; Lanson *et al.* 2009; Mosser-Ruck *et al.* 2010). A review of thermal stability of bentonite is given by Meunier *et al.* (1998). If such a reaction were to occur in a repository for heat-emitting waste, this thermally driven alteration and restructuring of the clay lattice to a lower swelling form of clay mineral would have a profound impact on the hydromechanical and geochemical properties of the resultant material. There is no single thermal activation energy for such reactions to occur, as these processes are highly dependent on the geochemistry of the system (e.g. Mosser-Ruck *et al.* 2010). However, recent work by Pusch *et al.* (2010) suggests that early-onset hydraulic degradation (i.e. an increase in hydraulic permeability) can occur in compact bentonite subject to prolonged temperatures of only 80 °C. While the experimental detail underpinning these observations is unclear, the authors report a two order of magnitude increase in permeability for two clay samples taken from the Canister Retrieval Test (CRT) test performed at the Äspö underground research laboratory. If correct, such an increase in permeability would represent a major alteration of the smectite component in the bentonite and would compromise repository safety performance.

However, this work is in direct contradiction to earlier results presented by Pusch (1985), who, after examining bentonite samples from the buffer mass test (BMT) performed at Stripa in the early 1980s, reported no significant changes in permeability and concluded that 'the influence of heating was insignificant and that the important physical properties, swelling and permeability, were not altered'. While temperatures in the BMT were higher than those reported in the CRT (i.e. 125 v. 80 °C, respectively), the test was of much shorter duration (i.e. 1 year as opposed to 5 years for the CRT). Villar & Gómez-Espina (2009) performed a series of hydraulic tests on Febex bentonite compacted to a range of dry densities from 1.5 to 1.7 $m^3\ mg^{-1}$. While no consistent results were obtained, permeability was seen to increase with temperature. However, while the increase in permeability could not solely be explained by changes in viscosity, the magnitude of permeability increase was an order of magnitude less than that report by Pusch & Weston (2012), and no consistent behaviour was observed.

To investigate this issue, an independent programme of laboratory-based experiments was undertaken in parallel at the British Geological

From: Norris, S., Bruno, J., Cathelineau, M., Delage, P., Fairhurst, C., Gaucher, E. C., Höhn, E. H., Kalinichev, A., Lalieux, P. & Sellin, P. (eds) 2014. *Clays in Natural and Engineered Barriers for Radioactive Waste Confinement*. Geological Society, London, Special Publications, **400**, 589–601.
First published online May 12, 2014, http://dx.doi.org/10.1144/SP400.31

Survey (BGS) and Studiecentrum voor Kernenergie (SCK-CEN), to examine the hydraulic behaviour of bentonite exposed to elevated temperatures. To prevent bias and to ensure impartiality, the studies were conducted in isolation. Complementary approaches were adopted by each institute (described below) to measure hydraulic permeability under a range of test conditions. In an attempt to replicate the experiments of Pusch *et al.* (2010), preserved test material was taken from a number of locations within the same CRT experiment. In addition, a control sample of 'virgin' bentonite, supplied by Clay Technology AB, was examined to define the hydraulic properties of unaltered material without thermal loading.

To minimize perturbation of the clay during testing, the BGS opted to perform a series of long-term controlled flow-rate experiments in which the magnitude of the hydraulic gradient imposed across each sample is a dependent variable, simply related to the volumetric flow rate. In these tests, head gradient is allowed to slowly evolve in response to the imposed flow rate. As such, if hydraulic permeability had increased owing to thermal degradation of the buffer, then the differential pressure for a given flow rate would scale accordingly, resulting in proportionately smaller differential pressures. Ascending and descending flow cycles were imposed across each sample to examine any underlying hysteresis while the specimen was subject to a fixed backpressure.

At the same time, a complimentary programme of constant pressure tests was performed at SCK-CEN. In these experiments flux in and out of the sample was carefully controlled by two high-precision syringe pumps, in order to accurately measure hydraulic behaviour under steady-state conditions. As part of the test programme experiments were performed for different durations to examine the impact of calculating permeability under non-equilibrium conditions during the transient phase of testing. An attempt was also made to measure the hydraulic conductivity using a much smaller gradient, that is, a 1 m water column to investigate what effect, if any, this may have on permeability.

Data reduction

Data were transferred to a spread sheet for processing and plotting. Hydraulic transients were very well-defined and largely free of experimental noise.

The one-dimensional equation of flow is

$$S_s \frac{\partial h}{\partial t} = K \frac{\partial^2 h}{\partial x^2} \tag{1}$$

where S_s (m^{-1}) is the specific storage, K (m s^{-1}) is the hydraulic conductivity, h (m) is the hydraulic head and x (m) is distance in the flow direction. This equation must be solved subject to the boundary conditions:

$$q = \frac{Q}{A_s} = -K \frac{\partial h}{\partial x}\bigg|_{x=0} \tag{2}$$

and

$$h = 0 \quad \text{at } x = L_s \tag{3}$$

where q (m s^{-1}) is the Darcy velocity, Q (m^3 s^{-1}) is the volumetric flow rate, A_s (m^2) and L_s (m) are the cross-sectional area and length of the specimen, respectively. Hydraulic head is related to the pore pressure, P (Pa), by

$$h = \frac{P}{\rho_w g} \tag{4}$$

where ρ_w (kg m^{-3}) is the density of water and g (=9.81 m s^{-2}) is the acceleration due to gravity.

The head at $x = 0$ and flow at $x = L_s$ as functions of time were obtained by numerically inverting the Laplace transform solution given in Appendix 1. Five parameters are required to define the solution. Three are experimentally determined: Q, A_s and L_s. The remaining two are the material properties that the test is designed to determine (i.e. K and S_s). In order to estimate the values of these parameters, a general nonlinear least squares fitting routine was used to minimize the differences between the calculated curves and the measured head data.

Hydraulic conductivity was also calculated from the head gradient at steady state, enabling the two values to be compared. Permeability, k (m^2),

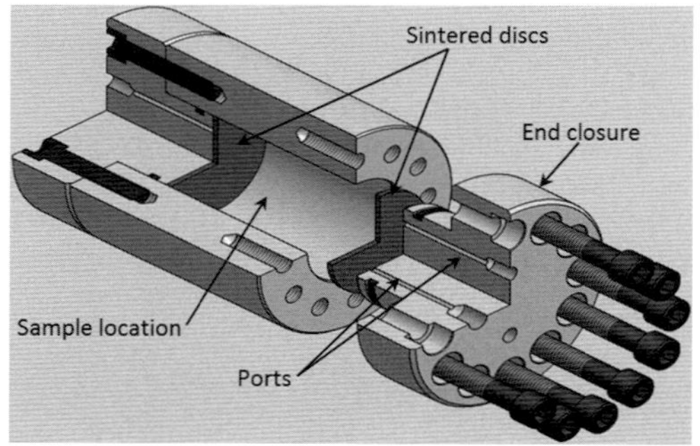

Fig. 1. Cut-through isometric diagram of the test vessel showing the location of the injection and backpressure ports, position of sintered filters and location of cap screws. One end-closure is shown removed from of the bore of the vessel to illustrate individual system components.

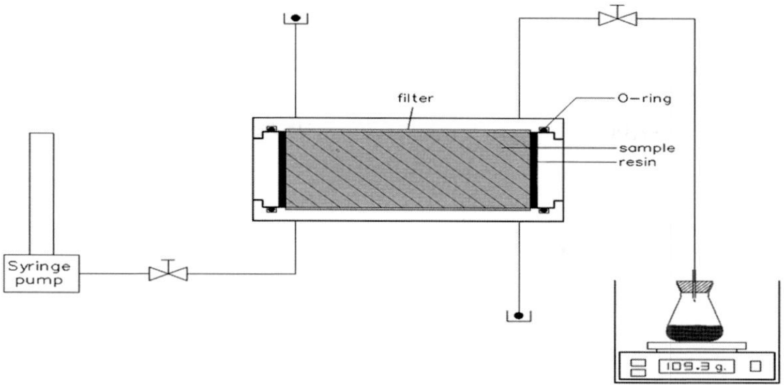

Fig. 2. Schematic of the high pressure 'constant head' experimental set-up with syringe pump.

was calculated from hydraulic conductivity using the relationship

$$k = \frac{K\eta_w}{\rho_w g} \tag{5}$$

where η_w is the viscosity of water (=0.001002 Pa s).

Experimental system

Independent laboratory studies were undertaken at the BGS and SCK-CEN on supplied samples of compact bentonite. Each laboratory performed a series of flow measurements using custom-designed apparatus imposing a constant volume boundary condition. These systems are described in detail below.

Constant volume apparatus (BGS)

Figure 1 shows a schematic of the constant volume permeameter designed and commissioned specifically for this experimental study. In this geometry the specimen is volumetrically constrained, preventing dilation of the clay in any direction. The apparatus consists of four main components: (a) a thick-walled dual-closure stainless steel pressure vessel; (b) a fluid injection system; (c) a backpressure system; and (d) a data acquisition system operating within a LabView™ environment. Pressure in the injection and backpressure circuits is continuously monitored through independent pressure transducers providing a check on system pressures. Testing was performed in an air-conditioned laboratory at a nominal air temperature of around 20 ± 0.5 °C.

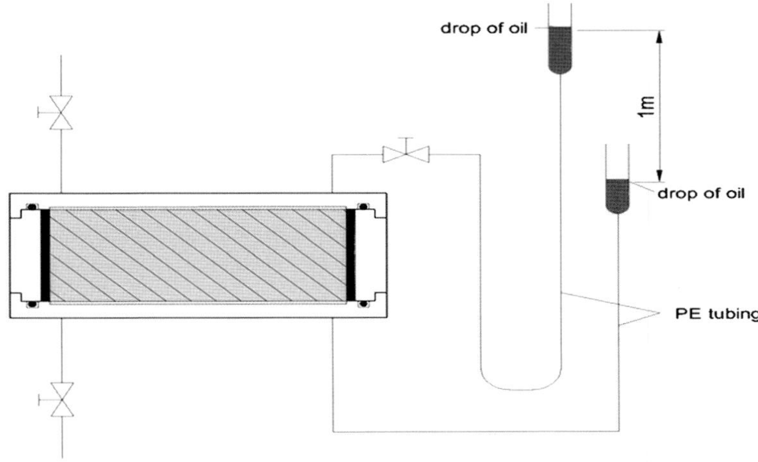

Fig. 3. Schematic of the low pressure 'constant head' experimental set-up using water columns to set the gradient.

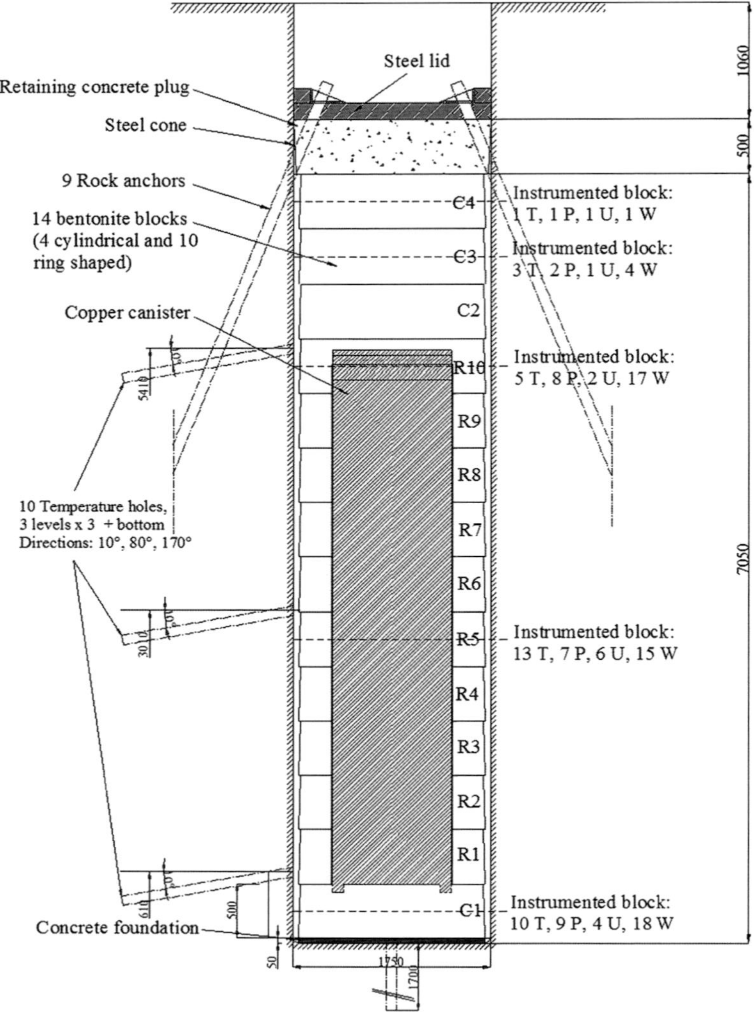

Fig. 4. Schematic layout of the Canister Retrieval Test, CRT (Eng 2008). Temperature (T), pressure (P), pore pressure (U) and relative humidity (W) sensors were located in five of the blocks. Outputs from these devices were used to track the state of hydration within the clay. Samples for this study came from rings R7 and R8. Copyright SKB.

The pressure vessel is comprised of a dual-closure tubular vessel manufactured from 316 stainless steel rated to 70 MPa. Each end-closure is secured by 12 high-tensile steel cap screws that can also be used to apply a small pre-stress to the specimen if required. The 60 mm internal bore of the pressure vessel is honed to give a polished surface. Each end-closure has a dedicated port through which the pressure in each of the filters can be independently monitored.

Volumetric flow rates are controlled or monitored using a pair of high-precision ISCO-100, Series DM, syringe pumps operated from a single digital control unit. The position of each pump piston is determined by a digital encoder with each step equivalent to a change in volume of 4.8 nL, yielding a flow accuracy of 0.5% of the set-point. Movement of the pump piston is controlled by a micro-processor, which continuously monitors and adjusts the rate of rotation of the encoded disc using a DC-motor connected to the piston assembly via a geared worm drive. This allows each pump to operate in either constant pressure or constant flow modes. A programme written in LabVIEW™ elicits data from the pump at pre-set time intervals of 2 min.

Fig. 5. Photograph of a preserved sample taken from block R8, direction 300° at a distance of 540–620 mm from the centre (CRT R8:300:540). The location of the orange (online only) pin indicates the surface closest to the canister. Photograph courtesy of Clay Technology AB. Copyright SKB.

Both ISCO syringe pumps and associated pressure transducers were calibrated to a known laboratory standards, with increments and decrements in pressure to quantify hysteresis. Using a spreadsheet, least-squares regression fits were calculated and the parameters used to correct the raw data.

Constant volume apparatus (SCK-CEN)

Two experimental geometries were designed by SCK-CEN to define the hydraulic behaviour of the bentonite. Both systems were configured to perform constant head hydraulic conductivity tests in which the compact clay samples were confined within a constant volume cell, with flow in and out measured as a function of time. In this way, it is possible to specify the hydraulic gradient imposed across the sample while continuously monitoring in- and outflow until equal. Analysis of the transient phase provides an indication of the initial degree of saturation. By measuring flux in and out of the core, it is possible to identify any leaks that might be present. The minimal hydraulic gradient at which it is possible to perform the experiment is defined by the minimal in- and outflow that can be precisely measured. An ISCO-100, Series DM, syringe pump was used to impose the inlet pressure and measure inflow, while at the outlet, no excess backpressure was imposed (i.e. P_{out} = atmospheric pressure) as a high-precision balance used to measure outflow (Fig. 2). Both were calibrated using a high-precision balance to a known laboratory standard.

To minimize evaporation at the outlet, stainless steel tubing was used to connect the sample holder to the collection bottle, the latter closed with a septum. This allowed evaporation rates through the septum to be determined, which were found to range from 3 to 10 μl/week. It was therefore decided to perform all measurements at an inlet pressure of around 1 MPa. Since the samples have a diameter of 50 mm and length of 50 mm (and are fully confined in the test cell), this resulted in a gradient of around 2000 m m^{-1} applied to the sample. In a similar manner to BGS, test samples were machine turned on a lathe to achieve the required dimensions.

A second apparatus was constructed to try and measure the hydraulic conductivity at a much lower gradient, that is, a gradient of 20 instead of 2000. In this test system, a water column of 1.5 m in height was connected to the inlet of the sample holder and a second column of 0.5 m height connected at the outlet (Fig. 3). A drop of silicon oil was placed on the top of each column to avoid evaporation. A pipet graduated in 1 μl intervals was then connected to both columns to provide an accurate measure of flow. Unlike the previous high-pressure apparatus (Fig. 2), connections from the sample holder to the water column were made with standard laboratory polyethylene tubing (Fig. 3).

Table 1. *Shows the basic physical properties of all test samples prior to testing assuming a specific gravity for the mineral solids of 2.77 mg m^{-3}*

Sample number	Sampling interval (code)	Water content (wt%)	Bulk density (mg m^{-3})	Dry density (mg m^{-3})	Void ratio	Porosity	Degree of saturation (%)
CRT-1	R8:300:540	24.8	2.00	1.61	0.736	0.421	0.95 [–]
CRT-2	R8:300:850	23.5	2.05	1.66	0.67	0.400	0.98 [1.0]
CRT-3	Virgin material	27.1	1.98	1.56	0.772	0.436	0.97 [1.0]
CRT-4	R7–225-665	23.0	2.08	1.69	0.64	0.389	0.92 [1.0]
CRT-5	R7-225-765	22.2	2.09	1.71	0.621	0.383	0.90 [1.0]
CRT-6	R7-225-790	23.7	2.07	1.67	0.65	0.396	0.93 [1.0]

Inspection of the table indicates all samples started with a slight degree of desaturation ranging from 95 to 98% for ring 8 and from 90 to 93% for ring 7 (the latter measurements are based on off-cut material from the neighbouring core).

Table 2. *Flow in and out, injection and backpressure and head gradient for each test stage of sample CRT-1 at steady-state*

Sample	Step no. and stage	Flow in (μl h^{-1})	Flow out (μl h^{-1})	Injection pressure (MPa)	Backpressure (MPa)
CRT-1	1 (EQ)	–	–	1.0	1.00
	2 (CFR)	1.6	1.5	2.71	1.00
	3 (CFR)	2.6	2.4	3.67	1.00
	4 (CFR)	3.0	2.7	4.11	1.00
	5 (CFR)	4.6	4.3	5.83	1.00
	6 (CFR)	2.6	2.3	3.88	1.00
	7 (CFR)	1.6	1.4	2.79	1.00
CRT-2	1 (EQ)	–	–	1.00	1.00
	2 (CFR)	1.6	1.5	2.99	1.00
	3 (CFR)	2.6	2.5	4.41	1.00
	4 (CFR)	1.6	1.5	3.09	1.00
CRT-3	1 (EQ)	–	–	1.00	1.00
	2 (CFR)	1.6	1.5	2.84	1.00
	3 (CFR)	2.6	2.5	4.12	1.00
	4 (CFR)	3.0	3.0	4.67	1.00
	5 (CFR)	2.6	2.5	4.18	1.00
	6 (CFR)	1.6	1.5	2.93	1.00
	7 (CFR)	3.0	2.7	4.66	1.00
	8 (CFR)	1.6	1.5	2.94	1.00

Inspection of the data indicates a good mass balance between flow in and out with difference less than or equal to 0.3 μl h^{-1}.

Test material

BGS

Testing was performed on a section of bentonite retrieved from ring 8 of the CRT (Fig. 4). Block samples of bentonite were taken in a tangential pattern from the clay–canister interface outwards. From this material BGS selected two samples, one close to the canister surface (Fig. 5) and the other from the opposite side of the ring next to the rock wall. A third sample of ‘virgin’ clay freshly manufactured by Clay Technology AB was selected as

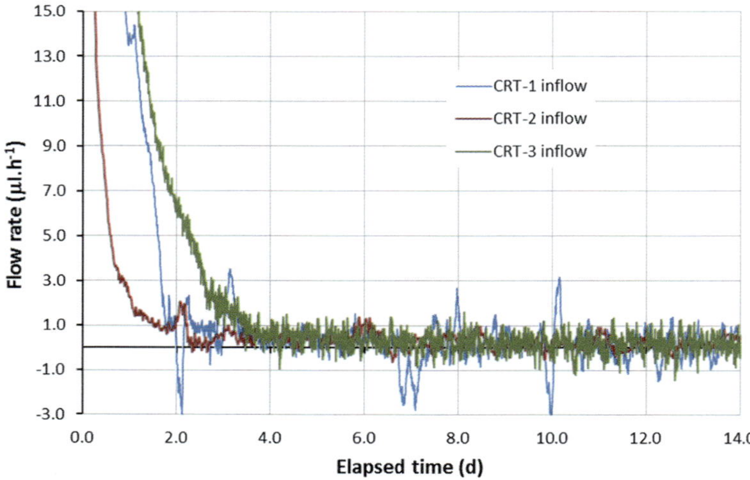

Fig. 6. Flow into specimens CRT-1 to -3 during the initial EQ stages (Table 2). Positive flow represent uptake of distilled water by the specimens. Inspection of the data indicates the bulk of the inflow occurs within the first 5 days which then tails off to near zero flow from 12 days onwards.

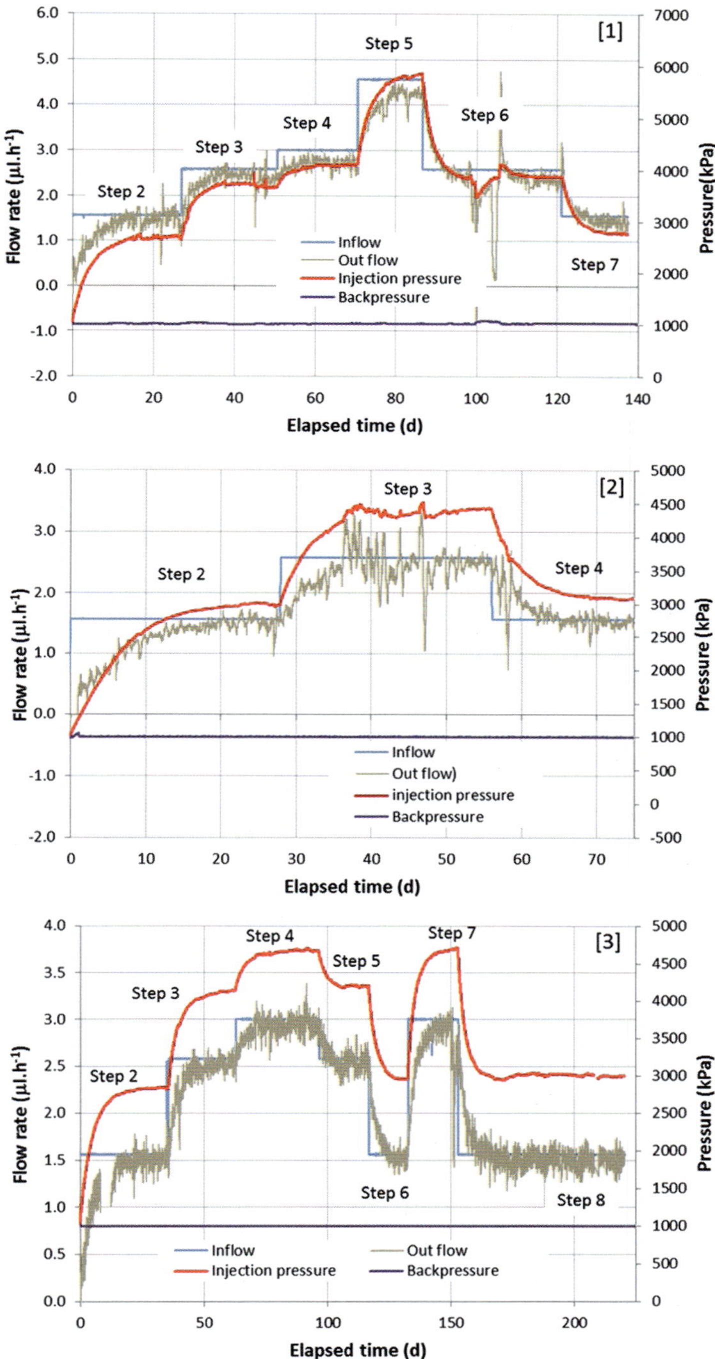

Fig. 7. Flow and pressure data for tests CRT-1, -2 and -3 respectively. Inspection of the data shows well-defined transients leading to steady-state conditions. Discrete spikes in data in [1] relate to small temperature fluctuations within the laboratory. Problems with the air conditioning system between 100 and 107 days resulted in a considerable noise within the data. However, projection of the flux and pressure asymptotes prior to this event suggests that it had no long term deleterious effect on the data.

Table 3. *Hydraulic properties based on steady-state and transient analysis of the data, showing hydraulic permeability and specific storage values for each test stage*

Sample no.	Stage no.	Hydraulic permeability k ($m^2 \times 10^{21}$)		Modelled permeability k ($m^2 \times 10^{21}$)	Modelled specific storage S_s ($m^{-1} \times 10^5$)
		Flow in	Flow out		
CRT-1	Step 2	4.9	4.6	4.7	1.9
	Step 3	5.2	4.7	4.8	1.2
	Step 4	5.2	4.6	5.9	2.8
	Step 5	5.0	4.7	4.6	1.4
	Step 6	4.8	4.3	5.2	1.2
	Step 7	4.7	4.2	4.5	1.4
CRT-2	Step 2	4.7	4.0	4.6	1.6
	Step 3	4.5	3.9	4.3	1.1
	Step 4	4.4	3.8	4.5	1.1
CRT-3	Step 2	5.0	4.9	4.9	1.7
	Step 3	4.9	4.8	4.7	1.3
	Step 4	4.8	4.8	4.8	1.2
	Step 5	4.8	4.7	4.8	0.9
	Step 6	4.8	4.6	4.8	0.9
	Step 7	4.8	4.4	4.8	0.9
	Step 8	4.6	4.4	4.8	0.9

a control. Cylindrical test specimens with an external diameter of 60 mm were manufactured by a combination of hand-trimming using a tubular former with a sharpened leading edge (Horseman & Harrington 1994) and machine lathing (Harrington *et al.* 2012). Sample lengths were 53.9, 60.0 and 59.7 mm, respectively, for samples CRT-1, CRT-2 and CRT-3.

Table 1 shows the basic physical properties of the test specimens. Porosity and degree of saturation are based on an average grain density of 2.77 mg m^{-3}. Post-test measurements of saturation indicate the samples were fully hydrated. This is confirmed by measurements of inflow during the equilibrium (EQ) stages of each test.

SCK-CEN

SCK-CEN received samples from ring 7 of the CRT at 225°. Samples were taken at 665, 765 and 790 mm radial distance. At each of these distances samples were taken, from which cylindrical cores with a diameter of 38 mm were prepared using a lathe. Samples with a length of 20 mm were prepared to measure bulk density and water content and the derived properties as given in Table 1. Samples 50 mm long were prepared for the hydraulic conductivity measurements.

Results

Controlled flow rate experiments (BGS)

A series of well-constrained hydraulic measurements were undertaken by BGS on three specimens of compact bentonite. Each test comprised a series of stages (Table 2) designed to initially hydrate, equilibrate and then determine baseline hydraulic behaviour. During the initial EQ stage, the specimen is exposed to distilled water on both faces of the core, with the same pressure at either end of the sample. A constant flow rate (CFR) stage is used to evaluate permeability and involves pumping distilled water through the filter disc at low volumetric flow rates, with the same solution as the backpressuing fluid at the downstream end. Specific storage is defined by numerical modelling of the flow transients.

During each equilibrium stage the sample rapidly hydrates, exhibiting a well-defined transient leading to a near-zero flow condition after around 13 days (Fig. 6). By the end of each EQ stage, volumetric flow into the clay had reduced to an average flux of less than 0.1 $\mu l\ h^{-1}$. Post-test measurements of saturation indicate that all test samples were fully hydrated.

Table 2 shows steady-state values for flow in and out of the specimens as well as injection and backpressure pressures for each stage of hydraulic testing (Fig. 7). The small discrepancy between fluxes relates to minor leakage ($\leq 0.3\ \mu l\ h^{-1}$) from one of the test systems. To accommodate this, hydraulic properties for all specimens have been calculated for both inflow and outflow data (Table 3). Transient analysis of the pressure data has been undertaken in a similar manner, yielding a second estimate for conductivity/permeability and an indication for the specific storage of the samples (Table 3, Fig. 8; Appendix 1). Examination of the conductivity data shows little variation in

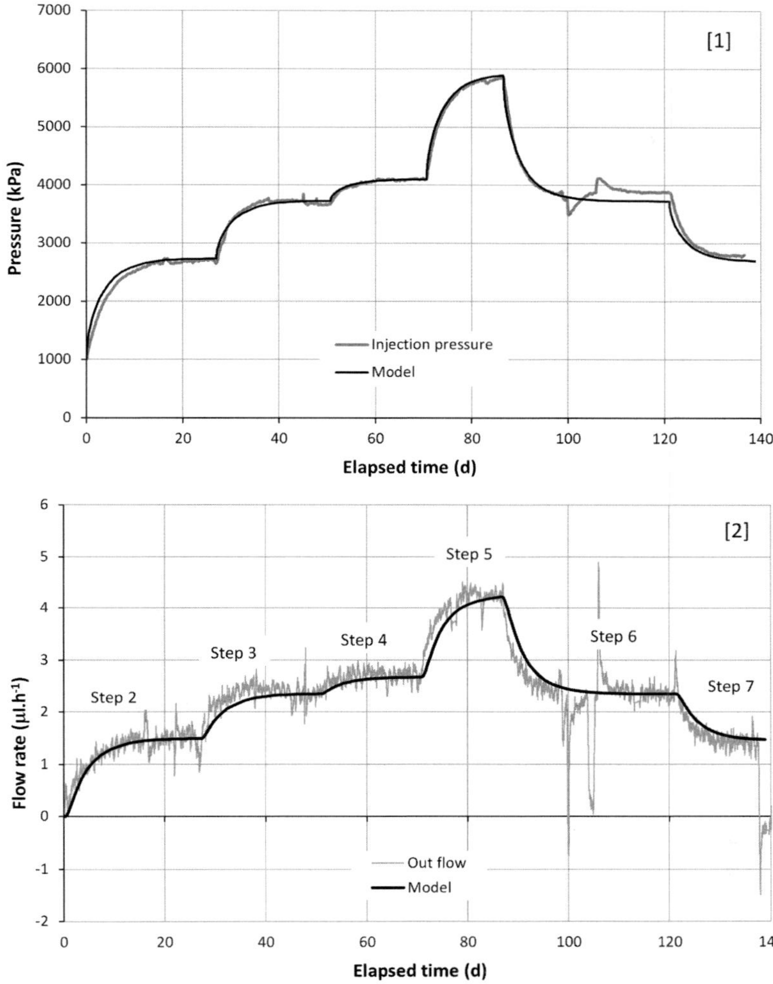

Fig. 8. Typical fits for hydraulic transient data (for CRT-1) based on backpressure flow rates. Plot [1] yields a good fit between predicted and measured pressures. In plot [2] there is a fairly good correlation between predicted and measured flux though some of the detail of the transients are less well-represented by the model. To improve model fits, data from test stages were individually fitted using an automated least-squares procedure optimized against the injection pressure (presented in Table 3).

values between the data processing methods, indicative of the quality of the raw data and the good mass balance obtained during hydraulic testing.

Based on analysis of the complete dataset (including analysis of the transient data), for fluxes in the range 1.6–4.6 μl h^{-1}, the mean hydraulic permeability is 4.7×10^{-21} m^2 (standard deviation = 0.4×10^{-21} m^2). The mean specific storage (S_s) for the same range of flux is 1.5×10^{-5} m^{-1} (standard deviation = 0.5×10^{-5} m^{-1}). These small standard deviations clearly demonstrate very little variation in permeability across the range of flow rates imposed. Inspection of the data provides no evidence for the alteration of bentonite following sustained thermal exposure of up to 5 years in duration. This observation is supported by Dueck *et al.* (2010) who also noted similar results for tests on CRT material.

In these tests, head gradient was allowed to slowly evolve during each test stage as a dependent variable, simply related to the volumetric flow rate. As such, if hydraulic permeability had increased owing to thermal degradation of the buffer (Pusch *et al.* 2010), then the differential pressure for a given flow rate would scale accordingly, resulting in proportionately smaller differential pressures.

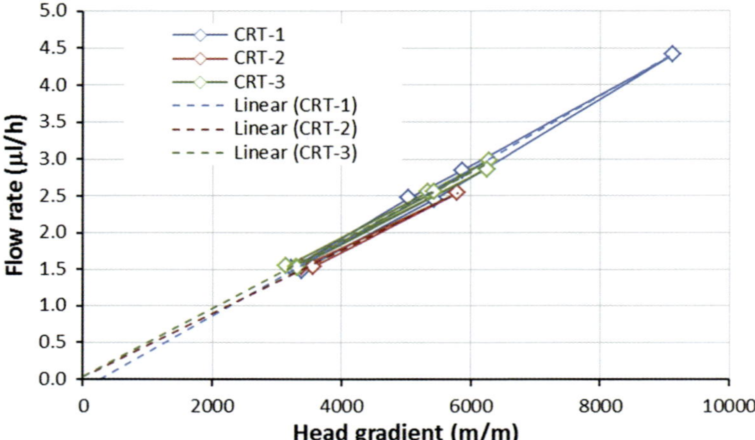

Fig. 9. Flow in and out of the specimens plotted against head gradient. A small amount of hysteresis in flow between advancing and decreasing cycles is observed in the data. Extrapolation of the results towards the *y*-axis indicates no significant threshold (i.e. non-Darcian behaviour) to flow.

A cross plot of flow rate against head gradient (Fig. 9) clearly shows that large head gradients are a necessary requirement for the movement of very small volumetric flow rates across the sample. The data exhibits only minor hysteresis between increasing and decreasing flow cycles. While controlled flow rates were varied from 1.6 to 4.6 μl h^{-1}, values for head gradient (a dependent variable in this study) ranged from 3150 m m^{-1} (at the lowest flow) to 9131 m m^{-1} (at the highest). Linear regression of the data in Figure 9 yields threshold values close to zero, indicating the bentonite exhibits little if any significant threshold (i.e. non-Darcian behaviour) to hydraulic flow. This observation is supported by recent work by Villar & Gómez-Espina (2009) who undertook a series of hydraulic and hydromechanical tests for a range of clay densities at varying temperature. Their results, for clays with similar dry density, yielded only one test where flow was not detected. This observation, performed at a low hydraulic gradient, may reflect the resolution of the measurement system rather than a true threshold to flow.

Constant head: controlled flow rate experiments (SCK-CEN)

Flow data for the sample CRT-6 (R7-225-790) at 1 MPa inlet pressure are shown in Figure 10. The cumulative inflow during the first 25 days was significantly higher than afterwards. It was also significantly higher than the outflow even when taking into account the evaporation at the outlet. This clearly shows that the sample was not fully saturated at the onset of testing and that, when subject to relatively modest injection pressures of 1.0 MPa, can take 90 days or more to reach full saturation. The initial degree of saturation of the samples was found to be between 90 and 93%. This may relate to desiccation during sample preparation (drying during adjustment on the lathe) or reflect the state of saturation of the material as shipped. However, what is clear is that, owing to the resaturation of the samples in the first month, the derivation of hydraulic conductivity during this time would result in an overestimation of the fully saturated hydraulic conductivity/permeability values (Table 4). As

Table 4. *Hydraulic properties derived from constant head experiments on samples CRT-4 to 6*

Sample	Sampling interval	Conductivity (m s^{-1})	Estimated error	Permeability (m^2)
CRT-4	R7-225-665 1	1.3×10^{-13}	0.8×10^{-14}	1.3×10^{-20}
CRT-5	R7-225-765 1	1.1×10^{-13}	0.8×10^{-14}	1.1×10^{-20}
CRT-4	R7-225-665 5	4.1×10^{-14}	0.8×10^{-14}	4.2×10^{-21}
CRT-6	R7-225-790 5	4.3×10^{-14}	0.8×10^{-14}	4.4×10^{-21}

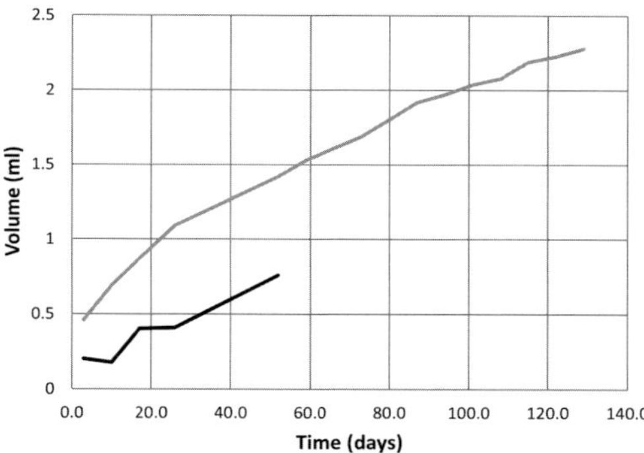

Fig. 10. Flow into sample CRT-6 at *c.* 1 MPa inlet pressure as function of time after initial resaturation with 1 m of water column. Grey and black lines represent cumulative in- and outflow respectively.

long as the sample is not fully saturated, the hydraulic gradient is mainly determined by the suction potential in the unsaturated bentonite sample.

Low-gradient constant head tests (SCK-CEN)

An attempt was also made to measure hydraulic conductivity at a much lower gradient, that is, 20 instead of 2000. However after 30 days it was found that the rate of evaporation through the PE tubing was around 0.2 ml/week. This value was checked by isolating the sample from the tubework and measuring the loss of fluid from the apparatus. This yielded a similar value or 0.2 ml/week, indicating that little if any flow into the clay could be detected following the initial resaturation of the clay close to the injection filter.

Importantly, if this value were to be used as the inflow parameter in the calculation the hydraulic conductivity, it would result in a gross overestimation of conductivity, yielding a value around 10^{-19} m^2. While this value would be in line with that report by Pusch *et al.* (2010), it cannot be reconciled with the data from these studies on samples of saturated compact bentonite. It should be noted that performing experiments at very low heads requires the inclusion of many precautions to avoid evaporation and temperature fluctuations. It also requires a great deal of time to get in a fully saturated state and equilibrium flow (i.e. in- and outflow equal).

Conclusions

It has been suggested that bentonite that has been subjected to sustained thermal exposure will exhibit a significant and permanent increase in hydraulic permeability. To examine this issue, BGS and SCK-CEN undertook a series of independent, well-constrained hydraulic measurements on a number of specimens of compact bentonite taken from the CRT performed at the Äspö Hard Rock Laboratory. To prevent bias and to ensure impartiality, the studies were conducted in isolation. Complementary approaches were adopted by each institute to measure the hydraulic properties under a range of test conditions in order to best examine potential changes in permeability.

Specific attention was paid to the choice of test technique to minimize the head gradient applied to the specimen during each test stage. In total, 19 constant flow rate tests were performed with flow rates ranging from 1.6 to 4.6 μl h^{-1}. The duration of each test stage ranged from 14 to 64 days, with the data exhibiting well-defined transient and steady-state behaviour.

Hydraulic permeability based on both steady-state pressure and transient analysis yielded a mean value of 4.7×10^{-21} m^2 ($\pm 0.4 \times 10^{-21}$ m^2) for fluxes in the range of 1.6–4.6 μl h^{-1} and a mean specific storage coefficient of flux of 1.5×10^{-5} m^{-1} ($\pm 0.5 \times 10^{-5}$ m^{-1}). Inspection of the data clearly demonstrates no significant difference in hydraulic behaviour between the virgin material and the clay from the CRT test.

At the same time, a complimentary programme of constant pressure gradient (*c.* 1 MPa over 5 cm) tests were performed at SCK-CEN. In these experiments, saturated hydraulic permeability ranged from 4.2 and 4.3×10^{-21} m^2 with an estimated error of $\pm 0.9 \times 10^{-21}$ m^2 owing to temperature fluctuations within the laboratory. These values are in close agreement with those derived from the BGS study.

Problems associated with low hydraulic gradient testing (owing to the prolonged restoration times and evaporation through the PE tubing) can lead to a gross overestimation of conductivity. This transient value, around 10^{-19} m^2, does not reflect the actual hydraulic properties of the bentonite when fully saturated. Under the latter conditions the permeability is clay is around 4.4×10^{-21} m^2.

In contrast to values reported by Pusch *et al.* (2010), the data from this study is in close agreement with observations for unaltered saturated bentonite. It is important to note that, if the permeability of the bentonite had indeed increased by two orders of magnitude following thermal exposure, then the application of these small volumetric fluxes would have resulted in the generation of head gradients ranging from 8 to 23 m m^{-1}.

Based on these observations, the authors find no evidence for an adverse increase in hydraulic conductivity of bentonite, similar to that proposed by

Pusch *et al.* (2010), as a result of thermal exposure of the clay to temperatures of 80 °C for up to 5 years.

Funding for this study was provided by the Swedish Nuclear Fuel and Waste Management Co. (SKB). This paper is published with the permission of the Director, British Geological Survey (NERC).

Appendix 1

Consider a specimen of length L_s and cross-sectional area A_s with hydraulic head initially everywhere at zero. A fluid flow of Q is initiated at $t = 0$ into the specimen at the end $x = 0$ and the response of the hydraulic head at $x = 0$ is sought as a function of time.

The equation of one-dimensional flow is given by

$$S_s \frac{\partial h}{\partial t} = K \frac{\partial^2 h}{\partial x^2} \tag{A1}$$

where S_s is the specific storage, K is the hydraulic conductivity and h is the hydraulic head. This equation must be solved subject to the boundary conditions

$$q = \frac{Q}{A_s} = -K \frac{\partial h}{\partial x}\bigg|_{x=0} \tag{A2}$$

and

$$h = 0 \quad \text{at } x = L_s \tag{A3}$$

To obtain the solution to equation (A1), we take its Laplace transform,

$$pS_s\bar{h} = K \frac{\partial^2 \bar{h}}{\partial x^2} \tag{A4}$$

where p is the transform parameter and $\bar{h}$ is the Laplace Transform of the head. The solution to this may be written as

$$\bar{h}(x) = Ae^{(\lambda x)} + Be^{(-\lambda x)} \tag{A5}$$

where A and B are constants to be determined from the boundary conditions and

$$\lambda = \sqrt{\frac{pS_s}{K}} \tag{A6}$$

From the boundary condition in equation (A3), we have

$$Ae^{(\lambda L)} + Be^{(-\lambda L)} = 0 \tag{A7}$$

Taking the Laplace transform of equation (A2), we have

$$A\lambda - B\lambda = -\left(\frac{q}{K}\right)\frac{1}{p} \tag{A8}$$

Substituting using equation (A7) and re-arranging, we have

$$A = -\left(\frac{q}{K\lambda}\right)\frac{1}{p}\frac{e^{(-\lambda L)}}{(e^{(\lambda L)} + e^{(-\lambda L)})} \tag{A9}$$

and

$$B = \left(\frac{q}{K\lambda}\right)\frac{1}{p}\frac{e^{(\lambda L)}}{(e^{(\lambda L)} + e^{(-\lambda L)})} \tag{A10}$$

Thus we may write the Laplace transform of the head at $x = 0$ as

$$\bar{h}(x = 0) = \frac{q}{Kp\lambda} \tanh(\lambda L_s) \tag{A11}$$

Similarly, we may write the Laplace Transform of the flow at $x = L_s$ as

$$\bar{q}(x = L_s) = -K\frac{\partial \bar{h}}{\partial x}\bigg|_{x=L_s} = \frac{q}{p\cosh(\lambda L_s)} \tag{A12}$$

The head at $x = 0$ and flow at $x = L_s$ as functions of time are obtained by numerically inverting the Laplace transform solutions given in equations (A11) and (A12) using the method of Talbot (1979). Five parameters are required to define the solution. Three are experimentally determined: Q, A_s and L_s. The remaining two are the material properties that the test is designed to determine (i.e. K and S_s). In order to estimate the values of these parameters, a general nonlinear least squares fitting routine was used to minimize the differences between the calculated curves and the measured head data.

References

Bruce, S. M., Rosenberg, P. E. & Kittrick, J. A. 1987. The stability of illite/smectite during diagenesis: an experimental study. *Geochimica et Cosmochimica Acta*, **51**, 2103–2115.

Dueck, A., Johannesson, L.-E., Kristensson, O. & Olsson, S. 2010. *Canister Retrieval Test at the Äspö Hard Rock Laboratory*. CRT project. Report on hydro-mechanical and chemical-mineralogical properties of the bentonite. SKB report, Stockholm.

Eng, A. 2008. *Äspö Hard Rock Laboratory. Canister Retrieval Test. Retrieval phase*. Project report. SKB IPR-08-13, Svensk Kärnbränslehantering AB.

Freed, R. L. & Peacor, D. R. 1989. Variability in temperature of the smectite/illite reaction in gulf Coast sediments. *Caly Minreals*, **24**, 171–180.

Harrington, J. F., Vaissiere, De La, Noy, D. J., Cuss, R. J. & Talandier, J. 2012. Gas flow in Callovo-Oxfordian Clay (COx): results from laboratory and field-scale. *Mineralogical Magazine*, **76**, 3303–3318.

Horseman, S. T. & Harrington, J. F. 1994. *Migration of Repository Gases in an Overconsolidated Clay*. British Geological Survey, Technical report WE/94/7.

Lanson, B., Sakharov, B. A., Claret, F. & Drits, V. A. 2009. Diagenetic smectite-to-illite transition in clay-rich sediments: a reappraisal od X-ray diffraction results using the multi-specimen method. *American Journal of Science*, **309**, 476–516.

MEUNIER, A., VELDE, B. & DRIFFAULT, L. 1998. The reactivity of bentonites: a review. An application to clay barrier stability for nucleasr waste storage. *Clay Minerals*, **33**, 9187–196.

MOSSER-RUCK, R., CATHELINEASU, M., GUILLAUME, D., CHARPENTIER, D., ROUSSET, D., BARRES, O. & MICHAU, N. 2010. Effects of temperature, pH and iron/caly and liquid/clay ratios on experimental conversion of dioctahedral smectite to berthierine, chlorite, vermiculite, or saponite. *Clays and Clay Minerals*, **58**, 280–291.

PUSCH, R. 1985. *Final report of the buffer mass test – volume III: chemical and physical stability of the buffer materials*. Report **85-14**. Svensk Kärbränslehantering AB, Stockholm, Sweden.

PUSCH, R. & WESTON, R. 2012. Superior techniques for disposal of highly radioactive waste (HLW). *Environmental Earth Science*, http://dx.doi.org/10.1007/s12665-012-1545-y

PUSCH, R., KASBOHM, J. & THAO, H. T. M. 2010. Chemical stability of montmorillonite buffer clay under repository-like conditions – a synthesis of relavent experimental data. *Applied Clay Science*, **47**, 113–119.

TALBOT, A. 1979. The accurate numerical inversion of Laplace transforms. *Journal of the Institute of Mathematics and its Applications*, **23**, 97.

VILLAR, M. V. & GÓMEZ-ESPINA, R. 2009. Report on thermo-hydro-mechanical laboratory teststs performed by CIEMAT on Febex bentonite 2004–2008. Departamento de Medio Ambiente, report no. **1178**. Conditions – a synthesis of relavent experimental data. *Applied Clay Science*, **47**, 113–119.

VELDE, B. & VASSEUR, G. 1992. Estimation of the diagenetic smectite to illite transformation in time-temperature space. *American Mineralogist*, **77**, 967–976.

VELDE, B., SUZUKI, T. & NICOT, E. 1986. Pressure–tempertaure–composition of illite/smectite mixed-layer minerals: Niger Delta mudstones and other examples. *Clays and Clay Minerals*, **34**, 435–441.

The uncertainties associated with the application of through-diffusion, the steady-state method: a case study of strontium diffusion

J. GONDOLLI* & P. VEČERNÍK

Fuel Cycle Chemistry and Waste Management Division, ÚJV Řež, a. s., Hlavní 130, Řež, 250 68 Husinec, Czech Republic

**Corresponding author (e-mail: Jenny.Gondolli@ujv.cz)*

Abstract: In this study, the uncertainties associated with through-diffusion, the steady-state method used for determination of the pore diffusion coefficient (D_p), were analysed and evaluated. The diffusion of strontium through the sample of compacted bentonite was studied to evaluate various sources of uncertainties and compare their contribution with uncertainty of D_p. Different diffusion experiment arrangements were used to test the effect of solution accessibility to the filter membranes. Three main general factors influencing the uncertainty of D_p at given conditions were identified: (a) the compacted sample preparation and its physical properties; (b) the properties of the diffusion cell; and (c) the strontium concentration analyses. Some of the identified sources can be treated as fixed with limited extent (e.g. the deformation of components of the diffusion cell), but some sources must be treated as variable with wide extent depending on the experimental conditions (e.g. the physical properties of the compacted bentonite sample – moisture, density, porosity). The variable uncertainty sources have to be taken into account, especially in repeatability and reproducibility tests. It was concluded that each of the diffusion cells with compacted bentonite sample represents a unique diffusion system. It is preferable to present the results of diffusion experiments as a range of values rather than as one average value with an uncertainty.

The Czech concept of a high-level radioactive waste repository considers the use of local resources for both waste packaging and isolation barriers (buffer and backfill). One of these barrier materials is factory-processed local bentonite, whose physical, chemical and retardation parameters should be determined. Many of these parameters will be used for the safety assessment and in models describing the repository's long-term behaviour. To date, much experimental work has been performed, using natural or factory-processed local bentonite from Rokle deposit as a reference bentonite for the Czech concept (e.g. Vejsada *et al.* 2005; Vopálka *et al.* 2006; Vokál *et al.* 2010). However, there is a lack of reliable data, especially diffusion data, for assessment of selected bentonite. The main problem lies in the large variability of bentonite (from the Rokle deposit) composition and in different approaches to the realization of diffusion experiments.

Diffusion is considered as the main mechanism of contaminant transport in materials with low hydraulic conductivity, such as clay materials. The mechanism and magnitude of diffusive transport in bentonite is of major concern for the safety assessment of waste repositories, especially for a high-level radioactive waste. To date, much theoretical and experimental work has been performed in this field and the results generally show dependencies on various parameters (concentration of diffusing element, density of clay material, chemical form of the element; e.g. Kozaki *et al.* 2005; Birgersson & Karnland 2009; Jakob *et al.* 2009). In order to evaluate the clay barrier function, the diffusion properties have to be characterized. The overview of different laboratory diffusion testing techniques is presented in Shackelford (1991) for waste disposal applications. In a number of other publications (e.g. Boving & Gratwohl 2001; García-Gutiérrez *et al.* 2001, 2006; Appelo *et al.* 2010; Glaus *et al.* 2011), practical extensions to bentonites and other clay materials (especially argillaceous rocks) are described. Most of through-diffusion experiments are performed in an arrangement with solution circulation along the filters used to separate compacted clay samples from solutions (e.g. Eriksen & Jansson 1996; Van Loon *et al.* 2003).

This study aims to answer some methodological questions that are important for the performance and evaluation of diffusion experiments. A new diffusion cell was developed to minimize the effect of the filters and to provide the equipment with identical parameters for repeatability and reproducibility tests in long-term experiments. The through-diffusion steady-state method used for the pore diffusion coefficient D_p calculation was used with the evaluation described in Shackelford (1991) to provide the data for later analysis and verification of diffusion processes in the selected compacted bentonite. The purpose of the study was to show the effect

From: NORRIS, S., BRUNO, J., CATHELINEAU, M., DELAGE, P., FAIRHURST, C., GAUCHER, E. C., HÖHN, E. H., KALINICHEV, A., LALIEUX, P. & SELLIN, P. (eds) 2014. *Clays in Natural and Engineered Barriers for Radioactive Waste Confinement*. Geological Society, London, Special Publications, **400**, 603–612.
First published online March 5, 2014, http://dx.doi.org/10.1144/SP400.3

of different uncertainty sources that comes into account in diffusion coefficient determination using the selected method. All experiments and determinations were focused on supporting uncertainty analysis by experimental data.

Materials and methods

Bentonite sample

The source bentonite comes from the Rokle deposit (Cenozoic neovolcanic area, NW Bohemia) and is processed in the Obrnice factory by the mining company Keramost a.s., Czech Republic. The company produces various types of bentonite products (including modified bentonites) according the requests of customers (industry, civil engineering, ecology). For all of the performed experiments, Bentonite 75 (B75) was used. This bentonite belongs to the natural bentonite product group, with no additives or modifications applied (Keramost 2012). More than 100 kg from one production lot was obtained (as a fine-grained final product in 40 kg paper bags) and various analyses of this material were performed (e.g. chemical analysis, mineral composition analysis, cation exchange capacity, natural ion exchange complex composition). The results are summarized in Tables 1–3.

Diffusion cell

The cell consists of a sample holder, two end components and two identical reservoirs for liquid phase (Fig. 1). The sample holder and end components are made from stainless steel with parts made from a carbon composite (inner part of the sample holder and filter support and saturation part in each end component). The clay sample is separated from the solution using two thin stainless steel filter membranes with well-defined thickness and pore size (produced by Euro SITEX s.r.o., Czech Republic). The inner diameter of the sample holder is 3×10^{-2} (1×10^{-5}) m and the length is 5×10^{-3} (1×10^{-5}) m. (The standard uncertainty of a value x is given in parentheses as $u(x)$ in the following text.) The filter membranes are made from high-resistance CrNi(Mo) steel, the membrane in contact with the clay sample has a thickness of $9.0(5) \times 10^{-5}$ m and a pore size of $9–10 \times 10^{-6}$ m, and the membrane in contact with the saturation part of the cell has a thickness of $6.2(1) \times 10^{-4}$ m and a pore size of $1.15–1.25 \times 10^{-4}$ m. Each reservoir for the liquid phase is made from transparent polycarbonate glass and its nominal volume is 0.1925(4)l. Complete diffusion cell is designed to avoid the contact of the solution and a clay sample with metal parts of the cell, except the filter membranes.

Table 1. *Chemical composition of B75 (content in weight-per cent)*

	Content (wt%)
SiO_2	51.91
Al_2O_3	15.52
TiO_2	2.28
Fe_2O_3	8.89
FeO	2.95
MnO	0.11
MgO	2.22
CaO	4.60
Na_2O	1.21
K_2O	1.27
P_2O_5	0.40
CO_2	5.15
Loss of ignition	10.65
Natural amount of water	6.94 ± 0.43

Table 2. *Mineral composition of B75 based on chemical composition and X-ray diffraction analysis (content in weight-per cent)*

Minerals	Content (wt%)
Montmorillonite	75.5
Quartz	8.1
Illite	3.9
Kaolinite	3.1
Calcite	3.1
Anatase	2.6
Siderite	1.8
Cristobalite	1.4
Ankerite	0.5

Sample preparation

The required amount of B75 for the compacted dry density 1.6 g cm^{-3} was weighed on a Precisa 240A (PAG Oerlikon, Switzerland) balance and

Table 3. *Cation exchange capacity (CEC) of B75 and natural Rokle bentonite and ion exchange complex composition (meq/100 g) determined using Cu-trien method (Meier & Kahr 1999)*

	B75	Natural Rokle
CEC_{total}	56.8 ± 1.0	69.0 ± 0.8
Na	36.9 ± 0.4	0.6 ± 0.1
K	3.6 ± 0.1	1.6 ± 0.1
Ca	2.0 ± 0.7	54.7 ± 1.4
Mg	26.4 ± 0.7	18.8 ± 0.7

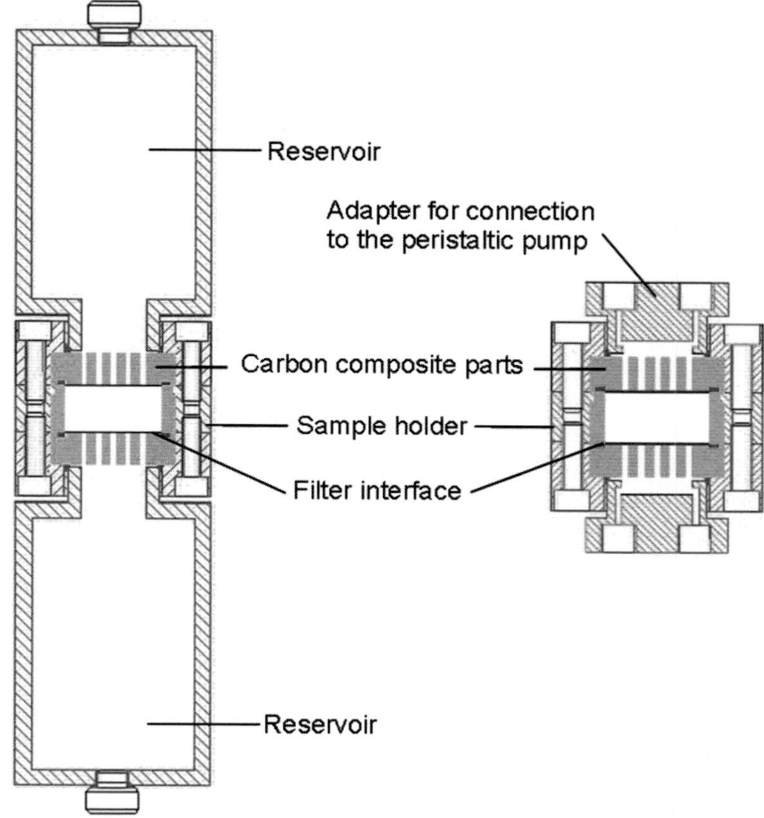

Fig. 1. Schematic drawing of diffusion cell. The standard arrangement with reservoirs is displayed on the left; the arrangement with adapters for connection to the peristaltic pump is displayed on the right.

compacted directly to the sample holder using a PC-controlled hydraulic press MEGA 11-300 DM1S (Form + Test, Seidner & Co. GmbH, Germany). Corrections to the natural amount of water present in the bentonite and to the loss of the material during the compaction procedure were applied. The exact weight of the compacted sample was determined after the compaction and checked after the diffusion experiment. Then the final compacted dry density was calculated for later use in diffusion evaluation.

All compacted samples were saturated by background electrolyte using a gas–liquid pressure multiplier at three saturation pressure steps 1, 5 and 8 MPa. The main aim of the selected procedure was full water saturation of the sample and significant removal of the activation reagent residues from the bentonite. After the saturation, both source and target reservoirs were connected to the diffusion cell and filled by the background electrolyte to reach pressure equilibrium.

Liquid phase

As a liquid phase, synthetic granitic groundwater with the composition given in Table 4 was used as

Table 4. *Composition of the synthetic granitic groundwater*

Species	Concentration (mol l^{-1})
Na^+	4.63×10^{-4}
K^+	4.60×10^{-5}
Ca^{2+}	6.74×10^{-4}
Mg^{2+}	2.64×10^{-4}
Cl^-	1.20×10^{-3}
SO_4^{2-}	2.88×10^{-4}
NO_3^-	1.02×10^{-4}
HCO_3^-	4.98×10^{-4}
F^-	1.05×10^{-5}
Ionic strength	3.61×10^{-3}
pH	7.8

a background electrolyte in all diffusion experiments. For the source solution, the appropriate amount of $Sr(NO_3)_2$ was added to obtain 0.01 (6.3×10^{-6}) mol l^{-1} Sr^{2+}. For the target solution, pure synthetic granitic groundwater was used (with background Sr^{2+} concentration 2.1(3) $\times 10^{-7}$ mol l^{-1}).

Diffusion experiment procedure

The basic steps – sample preparation and saturation (described above) – were identical for all samples. The effect of liquid phase movement in reservoirs and accessibility of the solution to the sample interface (filter membrane) was tested in three separate experiments: without movement (sample set A, sample B1), with movement (using a laboratory shaker, sample set C) and with liquid-phase continuous circulation (using special adapters and peristaltic pump instead of reservoirs, sample set D). In the first and second experiment, source and target reservoirs were filled with appropriate solutions and connected to the diffusion cells. In the third experiment, the source side and target side adapters were connected to their reservoirs via the peristaltic pump. The design of the connection is identical to the design described in Van Loon *et al.* (2003) except for the volume of the reservoirs. To maintain a constant concentration gradient across the sample and to maintain steady-state conditions, both source and target reservoirs content in the first and second experiment were replaced by fresh solutions in every sampling step. In the third experiment, a large volume source reservoir (5 l) was used where the possible decrease in strontium concentration was assumed to be insignificant and a small volume target reservoir (total volume 0.109(1) l) with its content replaced by fresh solution in every sampling step. During the sampling approximately 25 ml of aqueous phase was collected from every target reservoir for the subsequent total strontium content analysis (inductively coupled plasma mass spectrometry (ICP-MS) and flame atomic absorption spectroscopy (FAAS)). In addition, every newly prepared source stock solution was sampled to check the strontium concentration. All diffusion experiments were performed under normal laboratory conditions (temperature range 22–25 °C).

Uncertainty analysis

For the uncertainty analysis, the experimental arrangement, conditions and evaluation were selected with the aim of obtaining all required parameters experimentally with appropriate uncertainties. The arrangement of the through-diffusion method with an evaluation from the steady state as described in Shackelford (1991) was chosen. The analysis is divided into three steps. The first step is to find out possible uncertainty sources for the required parameters of given mathematical evaluation of the diffusion process at steady state. The second step is to identify experimental uncertainties connected with experimental arrangement for exactly one diffusion cell. The third step is to test repeatability and reproducibility at given conditions and to find out complex uncertainty sources. All uncertainty analyses were performed according Ellison & Williams (2012).

In the first step, parameters required in equation (1) (Shackelford 1991) are analysed and their uncertainties described.

$$D_{\mathrm{p}} = -\frac{L}{n \times A \times \Delta c} \times \frac{\Delta m}{\Delta t} \tag{1}$$

where L is the length, A is the cross-sectional area and n is the porosity of the sample, Δc is the concentration difference and Δm is the change in the mass of the chemical specie in an increment of time Δt.

Individual groups of components of uncertainty expressed as standard uncertainties ($u(x_{\mathrm{i}})$) are calculated to the combined standard uncertainty ($u_c(y)$). The general relationship between the combined standard uncertainty $u_c(y)$ of a value y and the uncertainty of the independent parameters x_1, x_2, … x_n on which it depends is

$$u_c(y(x_1, x_2, \ldots)) = \sqrt{\sum_{i=1,n} c_i^2 u(x_{\mathrm{i}})^2} = \sqrt{\sum_{i=1,n} u(y, x_{\mathrm{i}})^2} \tag{2}$$

where $y(x_1, x_2, \ldots)$ is a function of several parameters $x_1, x_2, \ldots$, c_i is a sensitivity coefficient evaluated as $c_i = \partial y/\partial x_{\mathrm{i}}$, the partial differential of y with respect to x_{i} and $u(y, x_{\mathrm{i}})$ denotes the uncertainty in y arising from the uncertainty in x_{i}.

Length L (m). The length of the sample is identical to the length of the inner composite ring in the sample holder. The combined uncertainty is given by the standard uncertainty of the length measurement $u(L)$ and the uncertainty caused by one-directional deformations (in the z-axis, where x and y are axes in the plane of the sample) of the sample and diffusion cell's components (support and saturation part in each end piece, both filter membranes in each end piece) – deformation length expressed as $u(d)$. The $u(d)$ takes positive values only and has a maximum value given by the construction design of the cell. The maximum value of the $u(d)$ was measured using a length difference sensor of the hydraulic press up to $P_{max} =$ 10 MPa. Only the deformation of the filter membrane set was observed; a maximum deformation of 1.2×10^{-4} m was achieved at 7.9 MPa. All composite and stainless steel parts of the diffusion cell can be considered as not deformable at given pressure conditions (up to 10 MPa). The final $u(d)$ is dependent on the swelling pressure of bentonite sample at given density and can be determined from the continuous deformation curve obtained during the deformation measurement. In all cases, the rising deformation length leads to an increase in the length of the sample and the sample density decrease at full water saturation.

Porosity n (−). The total porosity of the sample is calculated from the following parameters: weight of the sample at natural moisture content m, natural moisture content w, volume of the sample V and mineralogical density of the solid particles ρ_m. All parameters except the mineralogical density have their own standard uncertainty values $u(m)$, $u(w)$ and $u(V)$ determined directly by measurements. Only the standard uncertainty of mineralogical density of the solid particles was estimated because the density was calculated based on the equation presented in Jaksa & Kaggwa (1992). This calculation requires as input parameters the densities of the individual mineral phases and their amount in the sample. Although the densities of the individual mineral phases are known with sufficient accuracy (except the clay minerals from the smectite group), their amount in the sample is determined with high uncertainty because of the method limits (X-ray diffraction with phase calculation based on chemical analysis). In this case, the calculated mineralogical density of the solid particles is 2772(30) kg m^{-3} for the B75 sample with the composition listed in the Table 2, which is consistent with the published recommended values for bentonites (e.g. SKB 2010). Note that the natural moisture content was found to be variable depending on actual air moisture content and amount of material from which a sample is taken. This variability was found to be less than 1% of the natural moisture content of the sample. The uncertainty of the natural moisture content reflects this fact.

Cross-sectional area A (m^2). The cross-sectional area contains only the standard uncertainty of the sample diameter that is determined directly by the measurement. The sample holder and a sample inside it are not deformable in the x- and y-axis at given conditions.

Concentration difference Δc (mol m^{-3}). According to Ficks's first law for one-dimensional diffusive transport, $J_D = -D \times gradC$ or for porous material $J_D = -D_p \times n \times gradC$, where J_D is the diffusive flux, D_p is the pore diffusion coefficient, n is the porosity and $gradC$ is the concentration gradient. Then $gradC$ can be approximated by $\Delta c/L$ at steady state, where Δc is the concentration difference at the boundary with length L. It is assumed that both filter membranes at each end of the sample are fully saturated with appropriate solution (source side with strontium concentration C_S and target side with strontium concentration C_T) and the concentration gradient in the filters is negligible. Then Δc can be expressed as $\Delta c = C_T - C_S \cong -C_S$ for $C_S \gg C_T$. In this case, at given conditions, the initial difference of C_S and C_T after the reservoir content change is four orders of magnitude. Assuming that at steady state the decrease in strontium amount in the source reservoir is equal to the increase in strontium amount in the target reservoir, Δc was set as a constant value equal to C_S. The following uncertainties were included: uncertainty of background strontium concentration and the combined uncertainty of total strontium concentration in the source solution.

Change in the mass of the chemical specie Δm (mol). The change in the mass of the chemical specie is calculated based on a measurement of its concentration in the sample taken from the target reservoir and contains standard uncertainty of the concentration measurement $u(c)$ and target reservoir volume $u(V_2)$. In this case, the following simplifications were introduced:

(1) According to the chemical speciation modelling at given conditions, only strontium as Sr^{2+} was calculated, although in the sample total strontium was measured. The presence of other strontium species in the solution represents the uncertainty in its total amount, the variable, which could be quantified using the speciation analysis of each sample solution. Based on speciation modelling under given conditions, Sr^{2+} was identified as the dominant specie; minor species identified were $Sr(NO_3)^+$ in the source reservoir and $SrSO_4(aq)$ in both reservoirs, but their content in the samples was not determined.

(2) The correction to the background Sr^{2+} concentration was taken into account, but according the results obtained, it could be omitted because of its two orders of magnitude lower value compared with measured concentrations at steady state.

(3) The uncertainty of the target reservoir volume was found to be an operator-dependent variable. Thus, the combined uncertainty (for one reservoir plus variability in a group of

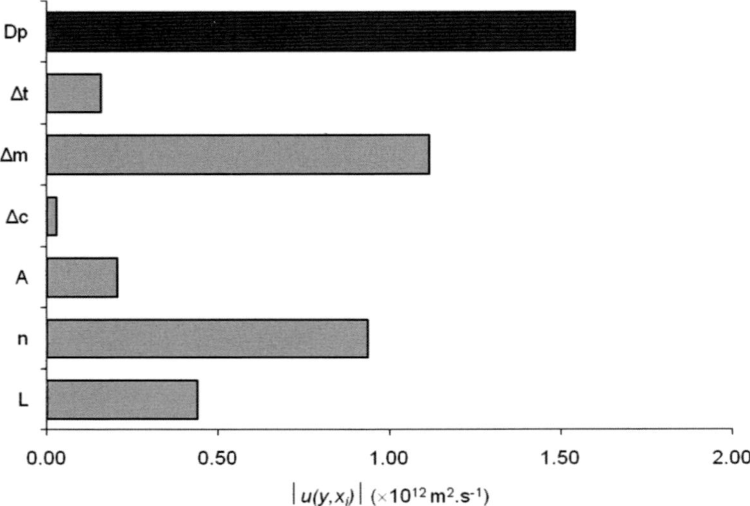

Fig. 2. Uncertainty contributions in D_p calculation. Example for A1 sample, values multiplied by 10^{12}.

reservoirs in the case of replicated cells) was extended for a contribution of uncertainty caused by different operators performing the samplings.

Increment of time Δt (s). The increment of time for which the change in the mass of the chemical specie in the target reservoir is measured contains the uncertainty of the sampling time record $u(t)$ only. This uncertainty was found to be operator-dependent and has a maximum value given by the sampling requirements. According to more than 100 samplings performed, the uncertainty of the recorded time was estimated to 600 s for each sampling point no matter which operator performed the sampling.

Results and discussion

Bentonite sample

The bentonite sample supplied by the producer was declared to be a natural bentonite. Performed analyses and comparison with truly natural bentonite samples from the Rokle deposit (see e.g. Vejsada *et al.* 2005; Vejsada 2006) confirmed its partial activation or contamination with activation reagent. According to the results (Table 1), sodium carbonate was one of the possible reagents. It was not possible to decide whether it is an activation or a contamination. The cation exchange capacity results suggested partial activation (see Table 3), but the pH measurement of a suspension (pH in a range 9.3–9.6) and total alkalinity determination suggested contamination with activation reagent. In addition, the chemical analysis of water leachate, especially the significantly higher amount of sodium and carbonates in solution in comparison to the natural Rokle bentonite leachate, supported the assumption, that the B75 sample was partially activated and partially contaminated with a activation reagent.

Experimental uncertainties for exactly one diffusion cell

The calculation of combined standard uncertainty of D_p for exactly one diffusion cell was performed in this step of the analysis and is illustrated in Figure 2 and summarized in Table 5. As can be seen, generally the highest contribution came from the change in the mass of the chemical specie Δm. Two reasons were identified: the uncertainty of strontium determination using the FAAS or ICP-MS technique and the operator effect in sampling and reservoir solution change (expressed as uncertainty of total reservoir volume). The second highest contribution to the combined standard uncertainty came from the porosity n parameter. This parameter contains various uncertainty sources and some of them cannot be determined with sufficient certainty (e.g. mineralogical density of bentonite). The third highest contribution came from the length L parameter. The main reason for this consists of the deformations of both filter membranes caused by the swelling of compacted bentonite sample at full water saturation. This is one of the variable uncertainties that depends on experimental

Table 5. *Contributions of uncertainties $|u(y, x_i)|$ in D_p calculation for compacted bentonite samples*

Sample	$u(L)$ (m)	$u(n)$ (−)	$u(A)$ (m^2)	$u(\Delta c)$ (mol m^{-3})	$u(\Delta m)$ (mol)	$u(\Delta t)$ (s)	$u_c(D_p)$ ($m^2\ s^{-1}$)
A1	0.44	0.93	0.21	0.03	1.12	0.16	1.54
A2	0.49	1.04	0.23	0.03	1.24	0.17	1.71
B1	0.61	1.21	0.28	0.04	1.55	0.14	2.08
C1	0.57	1.17	0.27	0.04	1.45	0.13	1.97
C2	0.59	1.23	0.28	0.04	1.50	0.14	2.05
D1	0.52	1.08	0.23	0.03	0.67	0.11	1.40
D2	0.44	0.92	0.20	0.03	0.55	0.10	1.18

Values multiplied by 10^{12}. A–D indicate different experiments.

conditions (e.g. resulting swelling pressure at given bentonite density and water saturation state). The remaining contributions (from the A, Δc and Δt) were significantly lower because of the nature of these parameters.

Repeatability and reproducibility, complex uncertainties

As can be seen from the Table 5, the general trend of contributions to diffusion coefficient uncertainty described in previous paragraph was identical for studied samples A–C, but differed in Δm uncertainty for sample D. In this case, a different FAAS instrument with better parameters was used for strontium analyses, which resulted in significantly lower contribution of Δm uncertainty. All samples in sets A, C and D were duplicated to test the repeatability and one sample (B1) was used to test the reproducibility (same method, different time and operator). Figure 3 illustrates the differences among samples presented as flux J through the compacted bentonite sample. From the repeatability point of view, experimental data in the sets A (A1, A2) and C (C1, C2) were in good agreement, especially in

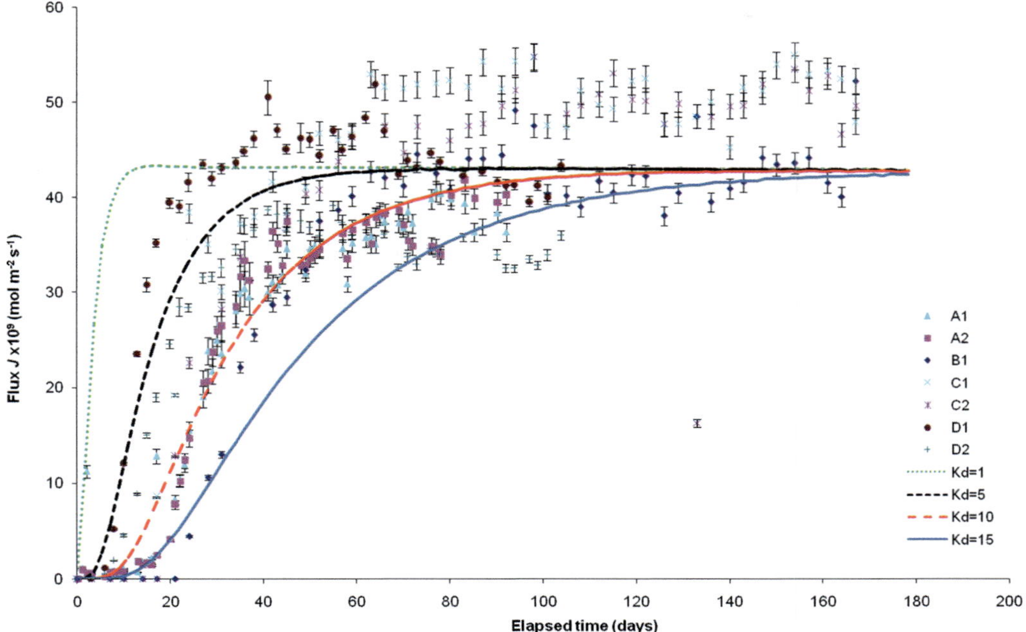

Fig. 3. Comparison of all samples presented as a strontium flux J (mol m^{-2} s^{-1}). Error bars represent combined standard uncertainty of the flux. Lines represent model curves of strontium flux calculated for four different values of distribution coefficients K_d (dm^3 kg^{-1}) in the model.

Table 6. *Initial dry density, related porosity and calculated D_p of compacted bentonite samples*

Sample	Dry density ρ_d (g cm^{-3})	$u(\rho_d)$ (g cm^{-3})	Porosity n (–)	$u(n)$ (–)	D_p (m^2 s^{-1})	$u(D_p)$ (m^2 s^{-1})
A1	1.601	0.019	0.423	0.009	4.65×10^{-11}	1.66×10^{-12}
A2	1.601	0.019	0.423	0.009	4.52×10^{-11}	1.61×10^{-12}
B1	1.556	0.018	0.439	0.009	4.90×10^{-11}	1.70×10^{-12}
C1	1.578	0.019	0.431	0.009	5.95×10^{-11}	2.08×10^{-12}
C2	1.583	0.019	0.429	0.009	5.96×10^{-11}	2.09×10^{-12}
D1	1.600	0.020	0.423	0.009	4.76×10^{-11}	1.18×10^{-12}
D2	1.600	0.019	0.423	0.009	4.05×10^{-11}	9.71×10^{-13}

A–D indicate different experiments.

the steady state. Experimental data in the set D were different for each sample (D1, D2), which was probably caused by the experimental device used. It was confirmed that, at very low flow velocities, each head of the peristaltic pump provided different flow that was not constant in the long-duration experiments. From the reproducibility point of view, experimental data for the sample B1 were in good agreement with the data for the set A, although minor differences existed in the steady state. This can be attributed to the effect of the operator performing samplings and measurements and the effect of the measuring equipment.

The effect of liquid phase movement in reservoirs and accessibility of the solution to the sample interface (filter membrane) was tested in this step. Many of diffusion experiments described in the literature are performed in the arrangement with a solution circulation along the filter interface (the arrangement described, for example, in Van Loon *et al.* 2003). According to the design of the diffusion cell used in this study, we compared this arrangement with the arrangement where the solutions in the reservoirs are in movement for the duration of the experiment and to the arrangement where the solutions in the reservoirs are not moving. To provide the movement, the diffusion cells with both reservoirs not filled to their maximum volume were placed in the laboratory shaker.

As can be seen from Figure 3, significant differences among samples were present, particularly in the transient state (up to approximately 60 days). Four model curves were calculated with four different distribution coefficients K_d used in the calculation (all other parameters were held constant). The fit to the experimental data showed that in the transient state the parameter K_d (or retardation factor) may differ by a factor *c.* 2–3 among samples. With the transition to the steady-state, differences among samples still existed, especially the higher flux for samples C1 and C2. This is attributed to the effect of the liquid phase accessibility to the sample interface, but this cannot be confirmed with respect to the data obtained for samples D1 and D2.

No significant difference between diffusion cell with and without circulation as described in Suzuki *et al.* (2003) was observed, which might indicate a negligible effect of sample interface (the liquid phase accessibility to the sample interface, respectively) on diffusion process. On the other hand, the highest flux was observed for the samples C1 and C2 where the content of reservoirs was in motion. These samples also have lower initial dry density and related porosity in comparison to samples from the sets A and D that can be even lower after the sample volume expansion owing to deformations of the filter interfaces. From this point of view, the combined effect of liquid phase accessibility to the sample interface and the lower density of the sample to the diffusion process may result in the observed flux.

One of the possible explanations for differences among samples in the observed flux (Fig. 3) at steady state and the calculated D_p (Table 6) found in this study consists of the different properties of each sample during the diffusion process, especially the true density under given conditions. All samples were prepared carefully to obtain a homogeneous sample pool with respect to the dry density, but differences among samples still existed (initial dry density for all samples is in range 1.556–1.601 g cm^{-3}; Table 6) and these differences may have been later enhanced by the bentonite swelling and filter interface deformations. The second possible explanation is the effect of the filter interface where the membrane with a pore size of *c.* 10 μm may be clogged with bentonite particles causing alteration of the concentration gradients in the filter interface. The third possible explanation consists of the parameters of the filter interface (structure and thickness) combined with the effect of moving liquid phase along the filter membrane that may result in bentonite colloid release to the liquid phase through the membrane and change the properties of the filter with time.

Conclusion

Experiments confirmed that the selected method of diffusion coefficient determination and evaluation contains various sources of uncertainty and brings relevant uncertainty to the calculated diffusion coefficient. The insight into its complex structure allows the quantification of an experimental source and sources that are associated with arrangement of the experiments and repeatability and reproducibility.

According to the results it can be concluded that a single diffusion cell acts as a unique system with a determined parameter set valid only for this cell under given experimental conditions. It was found that the preparation of two or more identical samples is problematic; for example, the density of the sample may change depending on actual swelling pressure and associated deformation of the filter interfaces. The bentonite heterogeneity is generally an important source of uncertainty and at the laboratory scale the mineral heterogeneity becomes essential because the compacted samples are prepared in separate steps. Moreover, it is more important for reproducibility tests and interlaboratory comparison where experiments should be made on different samples prepared from the same lot of material but under different conditions. In this study, the uncertainty arising from the bentonite heterogeneity was minimized by the use of an industrial product, but in general, this uncertainty has to be taken into account. On the other hand, the use of an industrial product presents a risk of the presence of production artefacts (e.g. the presence of activation reagent in studied bentonite) that may significantly influence diffusion experiments.

The design and properties of the diffusion cell were identified as an important uncertainty source, especially the filter membranes that affect the length of the compacted sample because of pressure deformations. In addition, the variability of the volume of the reservoirs affects the calculations of the specie's mass change. The main problem consists of the fact that these uncertainties are unique for each diffusion cell and, of course, for each experiment (it is especiallyimportant for reproducibility tests).

Because the diffusion experiments last months to years (according to the arrangements and requirements), the operator effect and measuring equipment effect at the appropriate time scale have to be taken into account. The results confirmed that both effects are identifiable in obtained data.

This study has shown that the selected experimental procedure brings various sources of uncertainties and a great part of overall uncertainty of the diffusion coefficient comes from this source. Many sources of uncertainties in a diffusion methodology are dependent on experimental conditions, and their proper identification and quantification can help in understanding their contribution to the final calculations.

This work was supported by the Ministry of Trade and Industry of the Czech Republic under contract FR-TI1/362. The authors are grateful to K. Videnská (Institute of Chemical Technology, Prague) for strontium speciation modelling. The comments of two anonymous reviewers are very much appreciated.

References

Appelo, C. A. J., Van Loon, L. R. & Wersin, P. 2010. Multicomponent diffusion of a suite of tracers (HTO, Cl, Br, I, Na, Sr, Cs) in a single sample of Opalinus Clay. *Geochimica et Cosmochimica Acta*, **74**, 1201–1219.

Birgersson, M. & Karnland, O. 2009. Ion equilibrium between montmorillonite interlayer space and an external solution – consequences for diffusional transport. *Geochimica et Cosmochimica Acta*, **73**, 1908–1923.

Boving, T. B. & Grathwohl, P. 2001. Tracer diffusion coefficients in sedimentary rocks: correlation to porosity and hydraulic conductivity. *Journal of Contaminant Hydrology*, **53**, 85–100.

Ellison, S. L. R. & Williams, A. (eds) 2012. *Eurachem/CITAC Guide: Quantifying Uncertainty in Analytical Measurement* 3rd edn. http://www.eurachem.org/index.php/publications/guides/quam

Eriksen, T. E. & Jansson, M. 1996. *Diffusion of I^-, Cs^+ and Sr^{2+} in Compacted Bentonite – Anion Exclusion and Surface Diffusion*. SKB TR-96-16, Svensk Kärnbränslehantering AB, Stockholm.

García-Gutiérrez, M., Cormenzana, J. L., Missana, T., Mingarro, M. & Molinero, J. 2001. Overview of laboratory methods employed for obtaining diffusion coefficients in FEBEX compacted bentonite. *Journal of Iberian Geology*, **32**, 37–53.

García-Gutiérrez, M., Cormenzana, J. L., Missana, T., Mingarro, M. & Martín, P. L. 2006. Large-scale laboratory diffusion experiments in clay rocks. *Journal of Physics and Chemistry of the Earth*, **31**, 523–530.

Glaus, M. A., Frick, S., Rossé, R. & Van Loon, L. R. 2011. Consistent interpretation of the results of through-, out-diffusion and tracer profile analysis for trace anion diffusion in compacted montmorillonite. *Journal of Contaminant Hydrology*, **123**, 1–10.

Jakob, A., Pfingsten, W. & Van Loon, L. R. 2009. Effects of sorption competition on caesium diffusion through compacted argillaceous rock. *Geochimica et Cosmochimica Acta*, **73**, 2441–2456.

Jaksa, M. B. & Kaggwa, W. S. 1992. Degree of saturation of the Keswick clay within the Adelaide city area above the general groundwater table. *Proceedings of 6th Australia New Zealand Conference on Geomechanics*, Christchurch, 336–341.

Keramost, 2012. Technical data sheet, 4 June 2012, http://www.keramost.cz/dokumenty/tds-bentonite-nonactivated.pdf

KOZAKI, T., FUJISHIMA, A., SAITO, N., SATO, S. & OHASHI, H. 2005. Effects of dry density and exchangeable cations on the diffusion process of sodium ions in compacted montmorillonite. *Engineering Geology*, **81**, 246–254.

MEIER, L. P. & KAHR, G. 1999. Determination of the exchange capacity (CEC) of clay minerals using the complexes of copper (II) ion with triethylenetetramine and tetraethylenepentamine. *Clays and Clay Minerals*, **47**, 386–388.

SHACKELFORD, C. D. 1991. Laboratory diffusion testing for waste disposal – a review. *Journal of Contaminant Hydrology*, **7**, 177–217.

SKB. 2010. *Data Report for the Safety Assessment SR-Site*, SKB TR-10-52. Svensk Kärnbränslehantering AB, Stockholm.

SUZUKI, S., SATO, H. & TACHI, Y. 2003. A technical problem in the through-diffusion experiments for compacted bentonite. *Journal of Nuclear Science and Technology*, **40**, 698–701.

VAN LOON, L. R., SOLER, J. M. & BRADBURY, M. H. 2003. Diffusion of HTO, $^{36}Cl^-$ and $^{125}I^-$ in Opalinus Clay samples from Mont Terri: effect of confining pressure. *Journal of Contaminant Hydrology*, **61**, 73–83.

VEJSADA, J. 2006. The uncertainties associated with the application of batch technique for distribution coefficients determination – a case study of cesium adsorption on four different bentonites. *Applied Radiation and Isotopes*, **64**, 1538–1548.

VEJSADA, J., HRADIL, D., ŘANDA, Z., JELÍNEK, E. & ŠTULÍK, K. 2005. Sorption of cesium on Czech smectite-rich clays – a comparative study. *Applied Clay Science*, **30**, 53–66.

VOKÁL, A., VEČERNÍK, P. & VOPÁLKA, D. 2010. An approach for acquiring data for description of diffusion in safety assessment of radioactive waste repositories. *Journal of Radioanalytical and Nuclear Chemistry*, **286**, 751–757.

VOPÁLKA, D., FILIPSKÁ, H. & VOKÁL, A. 2006. Some methodological modifications of determination of diffusion coefficients in compacted bentonite. *Materials Research Society Symposium Proceedings*, **932**, http://dx.doi.org/10.1557/PROC-932-11.1

Model validation based on *in situ* radionuclide migration tests in Boom Clay: status of a large-scale migration experiment, 24 years after injection

E. WEETJENS*, N. MAES & L. VAN RAVESTYN

SCK•CEN, Belgian Nuclear Research Centre, Institute for Environment, Health and Safety, Boeretang 200, B-2400 Mol, Belgium

**Corresponding author (e-mail: eef.weetjens@sckcen.be)*

Abstract: Demonstration of the long-term safety of a nuclear waste repository relies on earth science models, integrated in a performance assessment model chain. These models are subject to quality assurance procedures and principles of which model validation, qualification and verification are essential elements. However, in the context of performance assessment, model validation is often limited owing to extreme timescales and the use of natural barriers that can never be entirely characterized. Nevertheless, it is often possible to demonstrate that the models are valid or qualified to describe the processes at hand. In case of geological disposal in Boom Clay, the host formation is the dominant barrier for radionuclide migration and releases to the biosphere. Therefore, it has to be demonstrated that migration of solutes through Boom Clay at relevant scale is adequately understood. Large-scale and long-term *in situ* migration experiments, such as the CP1 experiment, form a cornerstone in this confidence-building process. In this experiment, accurate predictions of the tracer's breakthrough curves up to 3 m from the source have been obtained using the conventional advection–dispersion–reaction equation to describe solute transport and parameters obtained from (small-scale) migration experiments in the laboratory.

Any country deploying nuclear power reactors for electricity production has to deal with high- and intermediate-level radioactive waste. In Belgium, the Boom Clay has been studied since 1974 as candidate formation to host a geological radwaste repository. In view of meeting the principal safety objectives of isolation and confinement, the plastic Boom Clay shows excellent properties: pronounced self-sealing capacity, limited water flow and a substantial sorption capacity. The low hydraulic conductivity and low hydraulic gradient over the formation make molecular diffusion the dominant solute transport mechanism. Together with other retarding processes like sorption and (co)precipitation, radionuclide migration through the clay barrier is extremely slow so that a large fraction of the initial radioactivity present in the disposed waste has decayed before reaching the biosphere.

Safety (SA) and performance assessment (PA) calculations respectively aim to quantitatively demonstrate the long-term radiological safety of the disposal system, and to evaluate the long-term barrier performance. They generally consist of a chain of model calculations assessing the radionuclides release, their migration through engineered barriers and host formation, the dilution in surrounding aquifer layers, transfers between biosphere components and eventually health effects for a reference person. For the Belgian reference disposal solution, these types of calculations have repeatedly shown that the Boom Clay is the dominant barrier in providing long-term safety (ONDRAF/NIRAS 2001; Marivoet *et al.* 2002).

Since the demonstration of safety relies on earth science models that overspan extreme timescales, it is important to demonstrate the validity of those models so that stakeholders can have confidence in the safety assessment outcome. The models are therefore subject to quality assurance procedures of which model validation, qualification and verification are essential elements (ONDRAF/NIRAS 2008). Validation and verification of groundwater flow and transport models, whenever they are used in important decision-making processes, are not only a matter of good practice, but in the context of radioactive waste disposal are also a requirement of the IAEA (2009) and the Belgian nuclear safety agency FANC. However, it is acknowledged that, in the context of SA/PA, model validation is often limited because of two reasons: (a) the extreme timescale covered by the assessment; and (b) the use of natural barriers, for which complete characterization is impossible (NEA 1991, 1999). Nowadays, validation is considered a confidence-building process, aimed at demonstrating that the model is consistent with the scientific understanding and that it adequately represents the considered phenomena and interactions relevant to the assessment case.

From: Norris, S., Bruno, J., Cathelineau, M., Delage, P., Fairhurst, C., Gaucher, E. C., Höhn, E. H., Kalinichev, A., Lalieux, P. & Sellin, P. (eds) 2014. *Clays in Natural and Engineered Barriers for Radioactive Waste Confinement*. Geological Society, London, Special Publications, **400**, 613–623.
First published online May 15, 2014, http://dx.doi.org/10.1144/SP400.39

In order to build stakeholder confidence, underground research laboratories, or URLs, play an important role. These facilities are essential to provide scientific and technical information and practical experience that are needed for the design and construction of disposal facilities and, importantly, for the development of the safety case that must be presented at various stages of repository development (NEA 2001). Between 1980 and 1983, SCK•CEN built an underground research facility, HADES, at a depth of 223 m in the Boom Clay layer underneath the nuclear site in Mol (Belgium), in order to assess the feasibility of repository construction in this plastic clay formation, and to perform *in situ* testing at relevant depth. This paper focuses on validation and verification of solute transport models implemented in SA based on one of these large-scale *in situ* migration experiments with tritiated water (HTO) as quasi-conservative tracer: the so-called CP1 test. This test started in 1988 and is still on-going today (2014), making it the longest running radionuclide migration experiment in a clay environment worldwide. At present well-developed breakthrough curves have been obtained up to a distance of 3 m from the injection filter. This paper demonstrates that blind predictions based on parameters obtained from small-scale samples correspond very well with the measured tracer concentrations and that including a better description of the hydrological evolution around the HADES facility even further increases the goodness-of-fit.

Validation: a concise literature overview

During recent decades, groundwater models have become more and more complex and have been extensively used in important decision-making processes. Model validation has therefore become an important point of discussion and even controversy within the groundwater modelling and geoscientific community. Numerous publications can be found and in 1992 a two-part special issue of *Advances in Water Resources* (vol. 15, nos 1 and 3) was entirely dedicated to the subject. In a prominent article in this special issue, Konikow and Bredehoeft stated that 'groundwater models cannot be validated' (Konikow & Bredehoeft 1992; Bredehoeft & Konikow 1993). They argue that validation *in stricto sensu* is impossible, because of the non-uniqueness of the solution (or underdetermination of the problem). This view is shared by Oreskes *et al.* (1994): 'Validation of numerical models of natural systems is impossible. This is because natural systems are never closed and because model results are always non-unique.' This non-uniqueness usually refers to situations where there are insufficient data to constrain the computations to a unique model and multiple conceptual models are plausible. They discuss the problem from hydrogeological point of view (which often comprises boundary value problems), where it is often impossible to discriminate between two models (e.g. Theis solution for confined aquifers v. Hantush 'modified leaky aquifer' solution) based on a limited dataset. Although different models may reproduce the data equally well (history matching), longer-term predictions may substantially diverge, because of uncertainty on future stresses on the system (boundary conditions). Their theory is rooted in the philosophical view of Karl Popper, who states that scientists cannot validate a hypothesis, only invalidate it. Each time an observation agrees with the proposed model or theory, it survives and confidence in its validity is increased, but if ever a new observation is found to disagree, the model needs to be modified. Often, however, practice teaches that models or theories are seldomly abandoned, rather slightly modified or refined as process understanding grows.

The statement that validation is impossible has been discussed and disputed many times especially during the 1990s, among others by De Marsily *et al.* (1992). According to Leijnse & Hassanizadeh (1994), the confusion following the publication of Konikow and Bredehoeft was simply a matter of semantics. They argue that the words 'validation' and 'model' do not have a clear and unambiguous meaning and they therefore make a distinction between weak and strong definitions: the weak definition of a model refers only to the conceptual model or the description of the main processes that characterize a particular system, while the strong definition includes also the values of the parameters appearing in the equations as well as the initial and boundary conditions. Note that the terms 'weak' and 'strong' by no means entail a quality judgement. Similarly, validation in the weak sense refers to the validity of the conceptual part of the model, while validation in the strong sense refers to the validity of the model of a given system as whole, including all parameter values (Leijnse & Hassanizadeh 1994). As such, they agree that groundwater models cannot be validated in the strong sense. According to Nordstrom (2012), part of the controversy could also be due to misunderstanding of the term 'prediction', and the need to distinguish phenomenological or logical from chronological or temporal prediction, a remark made from and especially relevant to the field of geochemistry modelling.

The introduction of definitions may have helped in structuring the discussion, but it did not resolve the controversy completely since the definitions themselves are often perceived as unsatisfactory. In the field of radioactive waste disposal development, several international/EC projects (INTRACOIN, HYDROCOIN, INTRAVAL and GEOVAL) were devoted to model validation in the early 1990s. Especially in the INTRAVAL project, progress was made towards defining what validation means, but the participants did not reach a consensus (Pescatore 1995). Since then numerous definitions of 'model', 'validation' and associated terminology have been reported in safety assessment related literature. A fairly comprehensive overview of these running definitions can be found in the appendix of the review paper by Nordstrom (2012). As an example, the definitions of model validation and verification from the IAEA document 'Safety assessment for facilities and activities' (IAEA 2009) are given below:

- Model validation is the process of determining whether a mathematical model is an adequate representation of the real system being modelled, by comparing the predictions of the model with observations of the real system.
- Verification is the process of determining that a computational model correctly implements the intended conceptual model or mathematical model.

Pescatore (1995) proposed to categorize the nature of these definitions into three classes:

(1) the purist view, aiming at predicting the physical world as faithfully as possible;
(2) the operational view, suggesting that validation is accomplished only when results of blind prediction models can accurately describe experimental results;
(3) validation as a confidence-building process, aiming at obtaining reasonable assurance that a repository will perform as intended over the assessment timeframe.

The latter is a rather broad interpretation invoking, besides a careful model and parameter selection, a whole range of quality assurance (QA) principles such as peer-review, knowledge management, traceability and application of best available practices. Also key is that validation and confidence-building are considered processes, which should occur iteratively and progressively (NEA 2012).

Many of the above-mentioned definitions use terms suc as 'reasonable assurance' or 'adequate representation' which inevitably entail a subjective judgement. When is a model sufficiently validated? Safety assessments of nuclear waste repositories often emphasize the value of *robustness* as an objective of the safety concept and repository design. Robustness means that the performance of the barriers is rather insensitive to uncertainties in future evolution. The 'confidence building' interpretation of validation is therefore consistent with the aim of safety assessment, which is not to predict the future, but rather to provide broad (e.g. through multiple lines of reasoning) and sound support to a conclusion that possible impacts will not exceed certain acceptable limits (NEA 1999). McCombie *et al.* (1991) states that, for robust systems and models, validation requirements may relax. 'A model can be considered robust if we have confidence that the results are either correct (i.e. sufficiently realistic) or that they overpredict detriment (i.e. err on the side of conservatism). Robust models with no excessive demands on validation result from a combination of a simple, well understood system with large reserves of performance and a conservative approach for the choice of methods and data.'

Nowadays, there seems to be consensus on the limitations of conventional validation in the fields of geosciences, environmental engineering and disposal of nuclear and other hazardous waste, and it is sometimes recommended to abandon or avoid the word altogether (e.g. NEA 1991, 1999; NEA 2012; Nordstrom 2012). Nordstrom concludes that the emphasis on validation was 'a carryover from engineering practice into environmental investigations in which complexity and the nature of science prevents useful application of model validation'. As alternative, the concept of *model qualification* has been introduced (NEA 2012) with the more humble objective of demonstrating that the (conceptual) model is consistent with the scientific understanding within the assessment basis, and that it adequately represents the considered phenomena and interactions relevant to the assessment case. Indeed, it is the quality of the conceptual model that determines the usefulness and relevance of any mathematical modelling.

In the Belgian safety assessment methodology (ONDRAF/NIRAS 2009), both the terms 'qualification' and 'validation' are defined and the schematic illustration given in Figure 1 is adopted to explain the relation between the different QA and model concepts. Depending on the temporal and spatial scales concerned, validation can be understood in the classical scientific sense or must be understood as the process of building confidence that a computer model adequately represents a real system for a specific purpose. In the case study elaborated in this paper verification activities were also performed by benchmarking numerical models with (semi-)analytical models, being a standard QA activity. Nevertheless, the emphasis will be on model qualification, or specifically, the demonstration that advective–dispersive transport in Boom Clay, being the dominant barrier of the multicomponent disposal system, is well understood. Limits to this validation exercise will be indicated, where appropriate.

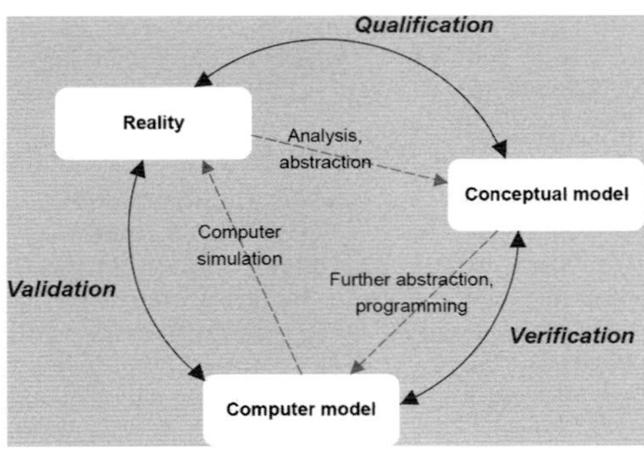

Fig. 1. Schematic illustration of the concepts of qualification, verification and validation (ONDRAF/NIRAS 2009).

The CP1 test setup

Geological setting

The Boom Clay in NE Belgium has been intensively studied since 1974 because of its favourable properties to host a geological repository for high-level and intermediate-level radioactive waste. Boom Clay provides both a physical (limited water flow) and a chemical (retention) barrier for radionuclide transport. The Boom Clay layer is a Cenozoic formation belonging to the Rupel group, which was sedimented during the Lower Oligocene, Rupelian Stage, between 28.4 and 33.9 myr ago. It consists of an alternation of silty clay and clayey silt, with a high content of pyrite and glauconite in the silty layers. The Boom Clay has a typical banded nature, with variations in silt and clay content, carbonates (enriched in so-called septaria-bearing layers) and organic matter (Vandenberghe 1978; Vandenberghe & Laga 1986; Van Keer & De Craen 2001). The Belgian radioactive waste management organization ONDRAF/NIRAS currently considers a reference site located in the Mol–Dessel nuclear zone, where the Boom Clay is approximately 100 m thick and can be subdivided into four members (see Fig. 2):

- the Belsele Waas member (*c.* 15 m thick)–the lower part of the Boom Clay, which is characterized by two thick bands of silts and by the absence of organic matter horizons;
- the Terhagen member (*c.* 16 m thick)–the middle part of the Boom Clay that contains less silt but is characterized by two black bands rich in organic matter;
- the Putte member (*c.* 45 m thick)–the upper part of the Boom Clay, characterized by a systematic presence of black bands rich in organic matter and silty horizons;
- the Boeretang member (transition zone on Fig. 2; *c.* 25 m thick)–located at the top of the formation, and consisting of alternating layers of silt and clay.

The Boom Clay mineralogy consists of 30–60% clay minerals (dominant clay minerals are illite, smectite and illite/smectite mixed layers), with an important fraction of pyrite (which controls redox conditions) and carbonates (which control pH). The Boom Clay pore water at Mol is mainly a 0.014 mol l^{-1} $NaHCO_3$ type of water with a considerably high content in humics. Because of the high smectite content and high porosity, the Boom Clay is plastic, which favours fast self-sealing behaviour. Figure 2 also shows the homogeneity (by means of gamma logging) and the low variability of the Boom Clay's vertical hydraulic conductivity and migration parameters for HTO and I^-, except for the lower part of the Belsele–Waas member. The modelled migration characteristic is the product of porosity and retardation factor (ηR), which can be assumed equal to the porosity for both tracers, since both are considered non-retarded species in Boom Clay. The Boom Clay thickness considered of good quality as safety barrier corresponds therefore to maximum 90 m in the safety assessment.

The CP1 experiment

The CP1 experiment, which stands for 'Concrete Plug 1' experiment, was installed in 1986 at the end of the URL part of the HADES facility where a concrete plug was installed (see Fig. 4). One multiple-screen tube with diameter 4.6 cm, penetrates 10 m into the Boom Clay and contains nine filters with a pitch of 1 m. In 1988, 1.25 GBq of HTO was injected in the central filter (see Fig. 3) and pore water samples were taken from source and neighbouring filters at regular time intervals. The objective of the CP1 experiment was to validate model and parameter values for transport of radionuclides in Boom Clay, and more specifically to investigate whether transport parameter values determined in laboratory experiments are also valid for larger temporal and spatial scales. The experiment was part of the test cases for the international INTRAVAL phase 2 validation project (Larsson *et al.* 1997) and is presently still running (2014).

Tritiated water is water in which a hydrogen atom is replaced by a tritium atom, which has a radioactive decay half-life of 12.3 years. Because of its neutral charge, it is not subject to electrostatic interactions, and therefore redistributes itself over the entire diffusion accessible pore space. Also, its conservative behaviour allows for a better

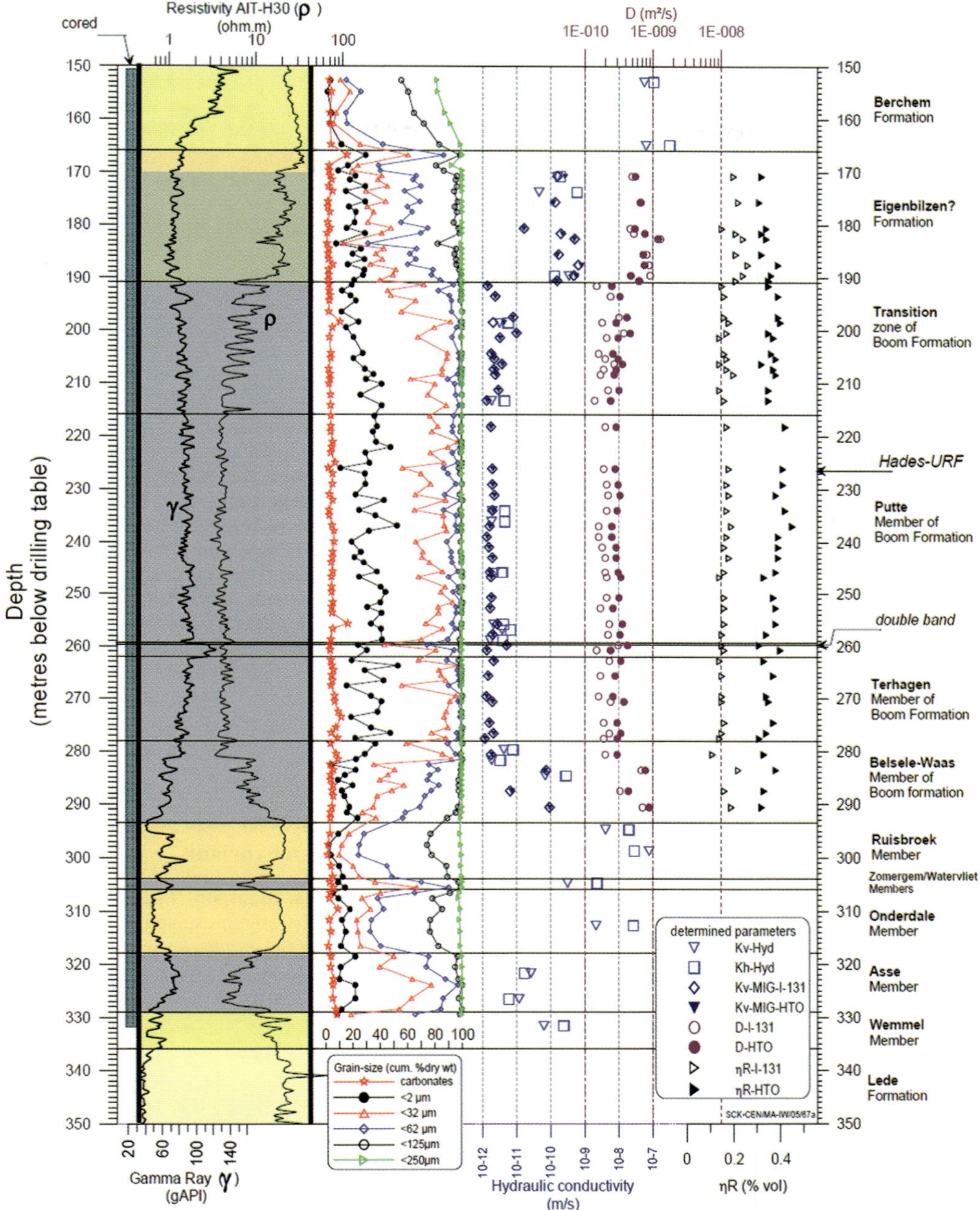

Fig. 2. Vertical profile of hydraulic parameters in the Boom Clay and part of overlaying/underlying layers for the Mol-1 drilling (May 1997): resistivity log, hydraulic conductivity, dispersion coefficient and product of porosity and retardation factor ηR (after Aertsens *et al.* 2005).

understanding of migration mechanisms of other radionuclides in Boom Clay. Therefore, HTO is ideally suited for studying the transport characteristics inherent to the clay formation itself, and it serves as a reference molecule for migration parameters of a whole range of solutes (other

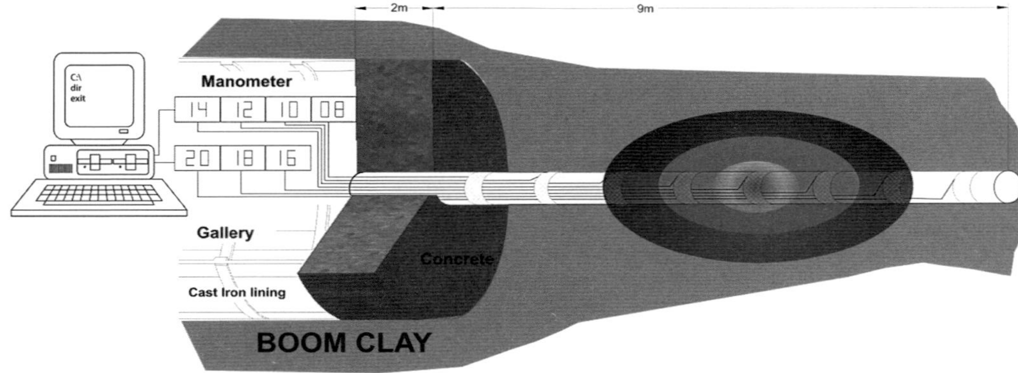

Fig. 3. Setup of the horizontal piezometer nest for the *in situ* HTO injection experiment CP1.

neutral but also cationic species; Bruggeman *et al.* 2013).

In the mid-1990s, several similar large-scale *in situ* migration experiments were initiated (see overview in Aertsens 2012, or Weetjens *et al.* 2011). One especially worth mentioning here is the 'Tribicarb-3D' experiment, dedicated to assessing anisotropy of radionuclide migration. The locations of these experiments are shown in Figure 4.

The CP1 experiment, now on-going for 24 years, gives well-developed breakthrough curves up to a distance of 3 m from the injection filter. Intermediate results with measurements up to 1 and 2 m from the source filter were previously communicated in Put *et al.* (1993) and De Cannière *et al.* (2007) respectively. The concentration measurements were compared with blind predictions using the MICOF code, which solves the advection–dispersion transport equation analytically in 3D (Aertsens 2012). More recently, results were updated to include the tracer breakthrough at three metres distance from the injection filter. The model prediction is based on HTO migration properties obtained from centimetre-scale cores. Since both parameter values and measured datapoints have evolved, as well as the modelling capabilities of flow and transport in clay, including 3D, complex geometries and coupled problems, it was considered worthwile to perform a thorough update and show the sensitivity to some conceptual model variants.

Update of parameter values and considered conceptual model variants

For most solutes, transport through (saturated) Boom Clay can be well represented by the following

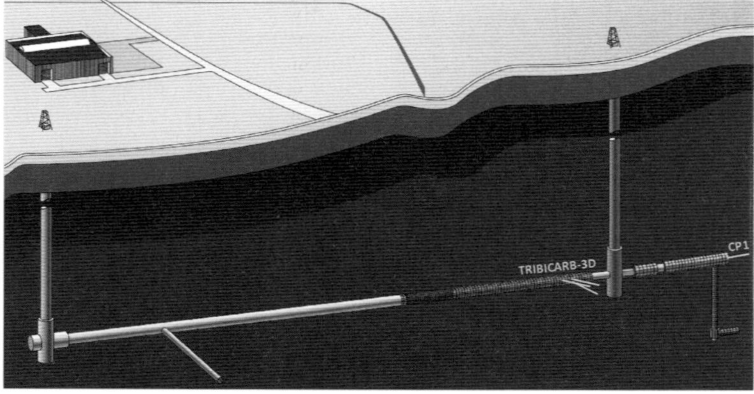

Fig. 4. Location of the *in situ* migration experiments CP1 and Tribicarb-3D at the HADES URL.

conservation equation for advective–dispersive-reactive transport (modified from Fetter 1999):

$$(\eta + \rho_b K_d)\frac{\partial c}{\partial t} - \nabla \cdot [\eta(D + D_p)\nabla c + u_{Darcy}c]$$
$$= -\lambda(\eta + \rho_b K_d)c + r + S \quad (1)$$

where c is the concentration of a given solute in the pore water, η is the porosity, ρ_b is the bulk density, K_d is the distribution coefficient, D is the hydrodynamic dispersion tensor, D_p is the pore diffusion coefficient tensor, u_{Darcy} is the Darcy velocity, λ is the decay rate, r is the reaction rate (precipitation/dissolution) and S is the solute source. Since HTO does not react with either sorb or clay minerals, eqution (1) can be simplified. On the other hand, although the permeability of the Boom Clay is very low, advection cannot be neglected in the case of the CP1 experiment owing to drainage towards the open gallery infrastructure. The values for the required parameters of equation (1) are obtained from laboratory migration experiments on Boom Clay cores of a few centimetres length (Henrion *et al.* 1991; Put *et al.* 1991; De Preter *et al.* 1992; Put 1992). A comparison between the values used in De Cannière *et al.* (2007) and the most recent parameter dataset for migration of HTO in Boom Clay (Aertsens *et al.* 2005; Bruggeman *et al.* 2013) is given in Table 1. For the ratio between horizontal and vertical pore diffusion coefficients, a value of 2 is considered as best estimate, although the anisotropy ratio is preferably expressed as a range between 2 and 3, depending on the degree of compaction (Bruggeman *et al.* 2013).

As a base case, we assume tracer transport by isotropic diffusion and a fixed Darcy velocity towards the gallery of 6.0×10^{-11} m s^{-1}. This value was obtained by multiplying the hydraulic gradient i measured between two CP1 filters ($i = 18.9$ m/m) with a hydraulic conductivity of 3.2×10^{-12} m s^{-1}. In variant 1, anisotropy of diffusion is introduced. Variant 2 introduces Darcy velocities as function of distance to the gallery instead of a fixed value input. This function is obtained by detailed modelling of the hydrological evolution around the gallery, which gives the evolution of pore water pressures and corresponding Darcy velocities. Both a 2D axisymmetric model (isotropic) and a 3D anisotropic model were implemented where the URL was represented as an open cylinder ($r = 1.75$ m) at atmospheric pressure in a certain Boom Clay volume with following flow properties: vertical hydraulic conductivity $K_v = 2.1 \times 10^{-12}$ m s^{-1}, horizontal hydraulic conductivity $K_h = 4.5 \times 10^{-12}$ m s^{-1} and specific storage factor 10^{-5} m^{-1}. These values are assumed best estimate values for the Boom Clay from a whole range of tests at various scales (Wemaere *et al.* 2008; Yu *et al.* 2011). Figure 5 shows that the difference between the obtained pseudo-steady-state flow profiles of the 2D and 3D modelling cases is marginal. Nevertheless, it is clear that the pressure gradients and Darcy velocities are higher than the fixed value input closer to the gallery and smaller at larger distance from the gallery, which might have a (small) impact on the predictions of tracer breakthroughs. The models were implemented in COMSOL multiphysics, Earth Science Module (2008).

Results

Figure 6 compares measured (diamonds) and predicted (lines) HTO breakthrough curves with updated migration parameters for the base case, considering isotropic diffusion and a fixed Darcy velocity. Although diffusion is the dominant transport mechanism, it can be observed that the measured concentrations in the direction of the URL are slightly higher because of the drainage towards the gallery, which is most pronounced for the furthest filters (at 3 m from the injection filter). An excellent agreement between measured and fitted concentrations is obtained as is evidenced by the very high R^2-value as indicator of the goodness-of-fit. Introduction of diffusion anisotropy changes slightly the concentration gradients, resulting in an improved prediction for the source concentration filter (variant 1 in Fig. 7). Introduction of the flow velocity profile as obtained in Figure 5 (variant 2 in Fig. 7) has a very small effect particularly noticeable on the furthest filter concentrations.

Table 1. *HTO migration parameters used in predictive modelling of the CP1 experiment*

			De Cannière *et al.* (2007)	Bruggeman *et al.* (2013)
Boom Clay porosity	η	–	0.35	0.37
Dispersion length	α	m	2.00×10^{-3}	4.00×10^{-3}
Pore diffusion coefficient, horizontal	$D_{p,h}$	m^2 s^{-1}	4.10×10^{-10}	4.60×10^{-10}
Pore diffusion coefficient, vertical	$D_{p,v}$	m^2 s^{-1}	2.05×10^{-10}	2.30×10^{-10}

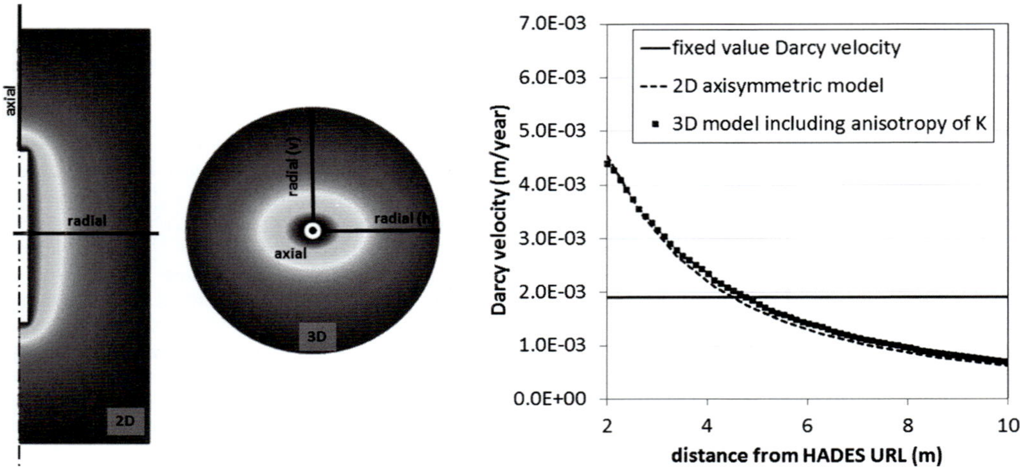

Fig. 5. Modelled pseudo steady-state Darcy velocity as function of distance to the URL gallery for a 2D (axisymmetric, isotropic) and 3D (anisotropic) model, and comparison with previously applied fixed value. The HTO tracer injection occurred at 6 m distance from the gallery.

Combination of both anisotropy and the flow velocity profile results in slight overestimations of the concentrations at 2 and 3 m distance, resulting in a slightly decreased goodness-of-fit ($R^2 = 0.9762$, results not shown).

In general, the results are hardly sensitive to the considered refinements in the conceptual model. The robust base model, governed by the horizontal diffusion coefficient, is considered sufficient to describe the tracer transport in this system. Although based on log-transformed concentrations, the goodness-of-fit of the source concentrations still dictates the R^2-value, while the interest is evidently in the accuracy of the predictions the furthest from the source. Therefore, these R^2-values are not an ideal comparator.

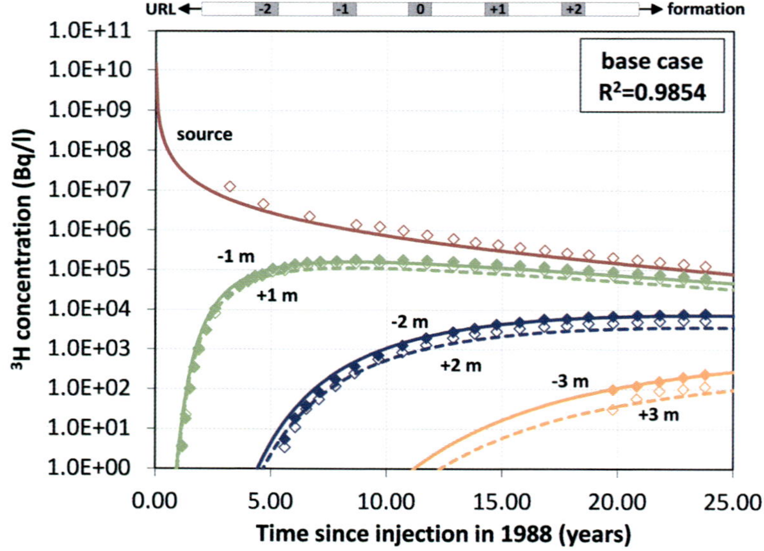

Fig. 6. Model predictions (lines) v. measured HTO or tritium concentrations (diamonds) for the CP1 *in situ* experiment. Base case results implementing isotropic diffusion and a fixed Darcy velocity.

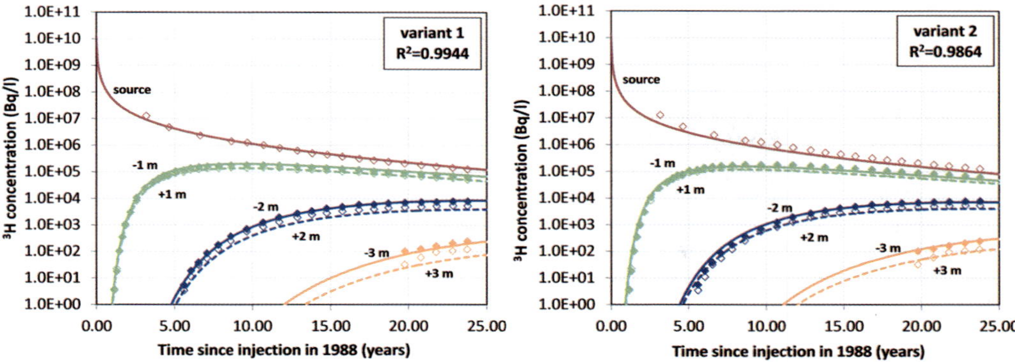

Fig. 7. Variant 1 (left) implementing anisotropic diffusion; variant 2 (right) implementing the Darcy profile of Figure 5.

Discussion and safety case relevance

The results in Figure 6 show clearly that the HTO diffusion coefficient in the direction of interest (*in casu* horizontal) is the dominant parameter governing HTO migration, and that diffusion parameters obtained from centimetre-scale laboratory tests are well representative for a metre-scale (i.e. approaching the real-scale) problem. In order to grasp how sensitive the results are to the diffusion parameters, the model was coupled to the MATLAB optimization toolbox to determine the D_p and D_v variables by least squares curve fitting. The obtained 95% confidence intervals are very narrow: for D_p between 4.19×10^{-10} and 4.25×10^{-10} $m^2 s^{-1}$, and for D_v between 2.31×10^{-10} and 2.69×10^{-10} $m^2 s^{-1}$. The latter results are not used to adjust the best estimate HTO migration parameters, in order to not undermine the spirit of the validation test as it is conceived.

The excellent agreement between model and measurements in the CP1 test demonstrates validation, within limits, of solute transport in Boom Clay at the level of the conceptual model ('weak' validation) and parameters ('strong' validation). As an initial value problem, the model's predictive capability does not depend on boundary condition settings, as long as they are not reached. It is reassuring that parameters obtained from migration experiments on small-scale Boom Clay cores can be applied for metre-scale solute transport problems. In safety assessment, however, the direction of interest is vertical, since releases to the aquifer systems surrounding Boom Clay are to be quantified. Therefore, complementary large-scale migration experiments have been performed in the vertical direction (see overview of *in situ* tests with different geometries and types of tracers in Aertsens (2012) and Aertsens *et al.* (2013)). The impact of spatial variability on radionuclide transport, which is very low for Boom Clay, as can be observed from Figure 2, has been quantified by Huysmans & Dassargues (2006) to be limited to 25% at PA scale.

Contrary to HTO, most radionuclides present in the radioactive waste do not migrate conservatively through Boom Clay, but they are strongly retarded by various types of sorption and other interaction processes on clay minerals or organic matter. Confidence in the migration models for these retarded nuclides is obtained through application of multiple lines of evidence, where the conceptual transport model culminates from integration of batch tests, multiscale migration tests and more sophisticated reactive transport modelling (e.g. multicomponent diffusion modelling). Together, these elements are believed to provide reasonable assurance that the transport models through the Boom Clay, being the dominant safety barrier, are qualified for purpose.

Conclusion

The demonstration of the long-term safety of a nuclear waste repository relies on earth science models, integrated in a total repository system safety assessment. For stakeholder confidence, it is important to demonstrate the validity of those models and the associated parameters. Although the extreme timescales and use of natural materials in the field of radioactive waste management are acknowledged to put limits on conventional validation, the very good agreement between predictive model results and measurements in the large-scale *in situ* migration experiment CP1 contributes significantly to the confidence in Boom Clay transport modelling at a relevant scale. The results further show that a simplified base model with only diffusion and isotropic parameters in the direction of interest is sufficient to give a good correspondence with the observed breakthrough curves. Including more detail in the conceptual model results only in a marginal improvement. Large-scale *in situ* migration experiments, such as the CP1 test, can be considered an important cornerstone of the confidence-building process, and together with application of other confidence-building principles, such as multiple lines of reasoning, they should contribute to a set of convincing arguments supporting the argument that the long-term radiological safety of a geological repository in Boom Clay can be guaranteed.

This work is performed in close cooperation with, and with the financial support of ONDRAF/NIRAS, the Belgian Agency for Radioactive Waste and Fissile Materials, as part of the programme on geological disposal of high-level/long-lived radioactive waste that is carried out by ONDRAF/NIRAS.

References

AERTSENS, M. 2012. *Overview of migration experiments in the HADES Underground Research Facility*. External report of the Belgian Nuclear Research Centre, Mol, Belgium, SCK•CEN-ER-164.

AERTSENS, M., DIERCKX, A. ET AL. 2005. *Determination of the hydraulic conductivity, the product ηR of the porosity η and the retardation factor R, and the apparent diffusion coefficient Dp on Boom Clay cores from the Mol-1 drilling*. Restricted Report of the Belgian Nuclear Research Centre, Mol, Belgium, SCK•CEN-R-3503.

AERTSENS, M., MAES, N., VAN RAVESTYN, L. & BRASSINES, S. 2013. Overview of radionuclide migration experiments in the HADES Underground Research Facility at Mol (Belgium). *Clay Minerals*, **48**, 153–166.

BREDEHOEFT, J. D. & KONIKOW, L. F. 1993. Groundwater models: validate or invalidate. *Ground Water*, **31**, 178–179.

BRUGGEMAN, C., MAES, N., AERTSENS, M. & DE CANNIÈRE, P. 2013. *Tritiated water retention and migration behaviour in Boom Clay*. External report of the Belgian Nuclear Research Centre, Mol, SCK•CEN-ER-248.

COMSOL 2008. *COMSOL Multiphysics 3.5a, Earth Science Module*. COMSOL AB, Stockholm.

DE CANNIÈRE, P., VOLCKAERT, G., ORTIZ, L., PUT, M., SNEYERS, A. & NEERDAEL, B. 2007. Transport of solutes and gas in soft clay: experience from the HADES URL. *In*: *Mont Terri Project: Proceedings of the 10 Year Anniversary Workshop* Federal Office of Topography Swisstopo, 61–64.

DE MARSILY, G., COMBES, P. & GOBLET, P. 1992. Comment on 'Ground-water models cannot be validated' by L.F. Konikow and J.D. Bredehoeft. *Advances in Water Resources*, **15**, 367–369.

DE PRETER, P., PUT, M., DE CANNIÈRE, P. & MOORS, H. 1992. *Migration of radionuclides in Boom Clay. State-of-the-Art report June 1992*. NIROND report, Brussels, NIROND **92-07**.

FETTER, C. W. 1999. *Contaminant Hydrogeology*. Prentice Hall, Englewood Cliffs, NJ.

HENRION, P., PUT, M. & VAN GOMPEL, M. 1991. The influence of compaction on the diffusion of non-sorbed species in Boom Clay. *Radioactive Waste Management and the Nuclear Fuel Cycle*, **16**, 1–14.

HUYSMANS, M. & DASSARGUES, A. 2006. Stochastic analysis of the effect of spatial variability of diffusion parameters on radionuclide transport in a low permeability clay layer. *Hydrogeology*, **14**, 1094–1106.

IAEA 2009. *Safety Assessment for Facilities and Activities*. International Atomic Energy Agency, Vienna, IAEA Safety Standard Series No. GSR Part 4.

KONIKOW, L. F. & BREDEHOEFT, J. D. 1992. Groundwater models cannot be validated. *Advances in Water Resources*, **15**, 75–83.

LARSSON, A., PERS, K., SKAGIUS, K. & DVERSTORP, B. 1997. *The International INTRAVAL Project to Study Validation of Geosphere Transport Models for Performance Assessment of Nuclear Waste Disposal*. OECD Nuclear Energy Agency, Paris, France.

LEIJNSE, A. & HASSANIZADEH, S. 1994. Short communication: model definition and model validation. *Advances in Water Resources*, **17**, 197–200.

MARIVOET, J., SILLEN, X., MALLANTS, D. & DE PRETER, P. 2002. Performance assessment of geological disposal of high-level radioactive waste in a plastic clay formation. *Material Research Society Symposium Proceedings*, **713**, 189–200.

MCCOMBIE, C., MCKINLEY, I. G. & ZUIDEMA, P. 1991. Sufficient validation: the value of robustness in performance assessment and system design. In: *Validation of Geosphere Flow and Transport Models (GbEOVAL), Proceedings of a NEA/SKI Symposium*, Stockholm, 14–17 May 1990. OECD Nuclear Energy Agency, Paris.

NEA 1991. *Disposal of Radioactive Waste: Review of Safety Assessment methods*. OECD Nuclear Energy Agency, Paris.

NEA 1999. *Confidence in the Long-term Safety of Deep Geological Repositories: Its Development and Communication.* OECD Nuclear Energy Agency, Paris.

NEA 2001. *The Role of Underground Laboratories in Nuclear Waste Disposal Programmes.* OECD Nuclear Energy Agency, Paris.

NEA 2012. *Methods for Safety Assessment for Geological Disposal Facilities for Radioactive Waste. Outcomes of the NEA MeSA Initiative*. OECD Nuclear Energy Agency, Paris.

NORDSTROM, D. K. 2012. Models, validation, and applied geochemistry: issues in science, communication, and philosophy. *Applied Geochemistry*, **27**, 1899–1919.

ONDRAF/NIRAS 2001. *SAFIR-2: Second safety assessment and interim report.* NIROND report, Brussels, NIROND **2001-06E**.

ONDRAF/NIRAS 2008. *Simulation tools used in long-term radiological safety assessments. NIRAS milestone MP5-01 of the Project Near Surface Disposal of Category A Waste at Dessel.* NIROND report, Brussels, NIROND-TR **2008-11E**.

ONDRAF/NIRAS 2009. The *long-term safety assessment methodology for the geological disposal of radioactive waste, second full* draft. NIROND report, Brussels, NIROND-TR **2009–14E**.

Oreskes, N., Schrader–Frechette, K. & Belitz, K. 1994. Verification, validation and confirmation of numerical models in the earth sciences. *Science*, **263**, 641–646.

Pescatore, C. 1995. Validation: the eluding definition. *Radioactive Waste Management and Environmental Restoration*, **20**, 13–22.

Put, M. 1992. *Three Dimensional in situ migration experiment in the Boom Clay formation at the Mol Site in Belgium. INTRAVAL Phase 2 test case description and data*. Extract from NIROND report, Brussels, NIROND **92-07**.

Put, M., Monsecour, M., Fonteyne, F. & Yoshida, H. 1991. Estimation of the migration parameters for the Boom Clay formation by percolation experiments on undisturbed clay cores. *In*: *Material Research Society Symposium Proceedings*, **212**, 823–829.

Put, M., De Cannière, P., Moors, H. & Fonteyne, A. 1993. Validation of performance assessment model by large scale *in situ* migration experiments. *In*: *Proceedings of an International Symposium on Geologic Disposal of Spent Fuel, High Level and Alpha Bearing Wastes*, Organized by IAEA/EC/OECD NEA, Antwerp, 19–23 October 1992. IAEA, Vienna.

Vandenberghe, N. 1978. *Sedimentology of the Boom Clay (Rupelian) in Belgium*. Verhandeling Koninklijke Academie voor Wetenschappen, Letteren en Schone Kunsten van België, Klasse Wetenschappen XL.

Vandenberghe, N. & Laga, P. 1986. The septaria of the Boom Clay (Rupelian) in its type area in Belgium. *Aardkundige Mededelingen*, **3**, 229–238.

Van Keer, I. & De Craen, M. 2001. *Sedimentology and diagenetic evolution of the Boom Clay: State of the art*. Restricted report of the Belgian Nuclear Research Centre, Mol. SCK•CEN-R-3483.

Weetjens, E., Govaerts, J. & Aertsens, M. 2011. *Model and parameter validation based on in situ experiments in Boom Clay*. External report of the Belgian Nuclear Research Centre, Mol, SCK•CEN-ER-171.

Wemaere, I., Marivoet, J. & Labat, S. 2008. Hydraulic conductivity variability of the Boom Clay in north-east Belgium based on four core drilled boreholes. *Physics and Chemistry of the Earth*, **33**, S24–S36.

Yu, L., Gedeon, M., Wemaere, I., Marivoet, J. & De Craen, M. 2011. *Boom Clay hydraulic conductivity*. A synthesis of 30 years of Research. External report of the Belgian Nuclear Research Centre, Mol, SCK•CEN-ER-122.

Index

Page numbers in *italic* denote figures. Page numbers in **bold** denote tables.